Lattice-ordered Rings and Modules

Lattice-ordered Rings and Modules

Stuart A. Steinberg
Toledo, OH, USA

Springer

Stuart A. Steinberg
Department of Mathematics
University of Toledo
Toledo, OH 43601
USA
stuart.steinberg@utoledo.edu

ISBN 978-1-4899-8297-1 ISBN 978-1-4419-1723-2 (eBook)
DOI 10.1007/978-1-4419-1721-8
Springer New York Dordrecht Heidelberg London

Mathematics Subject Classification (2010): 06F25, 13J25, 16W60, 16W80, 06F15, 12J15, 13J05, 13J30, 12D15

Printed on acid-free paper

Springer is part of Springer Science+Business Media (www.springer.com)

To Diane
Stephen, David, and Julia

Preface

A lattice-ordered ring is a ring that is also a lattice in which each additive translation is order preserving and the product of two positive elements is positive. Many ring constructions produce a ring that can be lattice-ordered in more than one way. This text is an account of the algebraic aspects of the theories of lattice-ordered rings and of those lattice-ordered modules which can be embedded in a product of totally ordered modules—the f-modules. It is written at a level which is suitable for a second-year graduate student in mathematics, and it can serve either as a text for a course in lattice-ordered rings or as a monograph for a researcher who wishes to learn about the subject; there are over 800 exercises of various degrees of difficulty which appear at the ends of the sections. Included in the text is all of the relevant background information that is needed in order to to make the theories that are developed and the results that are presented comprehensible to readers with various backgrounds.

In order to make this book as self-contained as possible it was necessary to include a large amount of background material. Thus, in the first chapter we have constructed the Dedekind and MacNeille completions of a partially ordered set (poset) and developed enough of universal algebra so that we can present Birkhoff's characterization of a variety and so that we can also verify the existence of free objects in a variety of algebras. Much of the material on lattice-ordered groups (ℓ-groups) in the second chapter appears in those books devoted to the subject. What is new in this book is the emphasis on ℓ-groups with operators. This allows for the common development of basic results about ℓ-groups, f-rings, and f-modules. The Amitsur-Kurosh theory of radicals is developed for the class of ℓ-rings in the second section of Chapter 2 . Still more background material is given in the first two sections of Chapter 4 where the injective hull of a module, the Utumi maximal right quotient ring, and the ring of quotients and the module of quotients with respect to a hereditary torsion theory are constructed and studied for a ring which is not necessarily unital. Also, the Artin–Schrier theory of totally ordered fields is given in the first section of Chapter 5, and enough of the theory of valuations on a field is presented in the second section so that a complete proof of the Hahn embedding theorem for a well-conditioned commutative lattice-ordered domain can be given.

Chapters 3, 4, 5, and 6 constitute the heart of the book. While not every known result on the topics included is presented, enough is presented so as to make the text, by which I mean the exercises also, reasonably complete. The first section of Chapter 3 develops the basic theory of ℓ-rings including the fact that canonically ordered matrix rings have no unital f-modules. Section 4 shows that the fundamental process of embedding an f-algebra in a unital f-algebra is more complicated than the analogous embedding for algebras and cannot always be carried out. The fifth section shows how to construct power series type examples of ℓ-rings and ℓ-modules using a poset which is a partial semigroup and which is rooted in the sense that the set of upper bounds of each element is a chain. The basic structure of f-rings is given in the third section of Chapter 3 and some of the richer structure of archimedean f-rings is given in the sixth section. In the last two sections the structure of ℓ-rings in other varieties is examined. The seventh section studies those ℓ-rings that have squares positive and gives conditions on a partially ordered generalized semigroup for the lexicographically ordered semigroup ring to have this property. The last section considers those ℓ-rings which satisfy polynomial constraints more general than that of squares being positive. One effect of these constraints is to coalesce the set of nilpotent elements into a subring or an ideal and to force an ℓ-semiprime ring to lack nilpotent elements. Also included in this section is a proof of the commutativity of an archimedean almost f-ring.

Chapter 4 concentrates on the category of f-modules. The most conclusive results occur for a semiprime f-ring whose maximal right quotient ring is an f-ring extension and whose Boolean algebra of annihilators is atomic. In the third section necessary and sufficient conditions for the module (or ring) of quotients to be an f-module (or an f-ring) extension are given, and the structure of right self-injective f-rings is given. The unique totally ordered right self-injective ring that does not have an identity element is exhibited. The module and order theoretic properties that determine when a nonsingular f-module is relatively injective are given in the fourth section—there are no injectives in this category of f-modules. A useful representation of the free nonsingular f-module is given in the last section and the size of a disjoint set in a free f-module is determined.

In a totally ordered field the set of values—those convex subgroups which are maximal with respect to not containing a given nonzero element—becomes a totally ordered group under the operation induced on it by multiplication in the field. A proof of the Hahn Embedding Theorem for totally ordered fields is given in the second section of Chapter 5; namely, a totally ordered field is embedded in a power series field where the exponents belong to this value group of the field and the coefficients are real numbers. Also, a totally ordered division ring is embedded in a totally ordered division algebra over the reals. In the third section of Chapter 5 the Hahn Embedding Theorem is given for a lattice-ordered commutative domain which satisfies a finiteness condition, and another embedding theorem for a suitably conditioned ℓ-field into a formal power series crossed product ℓ-ring is given. Also, the theory of archimedean ℓ-fields is presented and lattice orders other than the usual total order are constructed for the field of real numbers.

Chapter 6 begins with a characterization of the canonically ordered real semigroup ℓ-algebra over a locally finite left cancellative semigroup, and semigroup ℓ-rings in which squares are positive are studied in more detail. In the second section it is shown that in an ℓ-algebra in which the nonzero f-elements are not zero-divisors each algebraic f-element is central. A complete description is also given of those rings which have the property that each partial order is contained in a total order. In the third section more commutativity theorems are presented. It is shown that a totally ordered domain which is co-ℓ-simple and which has a positive semidefinite form with a nontrivial solution must be commutative. A similar conclusion giving the centrality of f-elements appears for an ℓ-ring in which the commutators are suitably bounded. A proof of Artin's solution to Hilbert's 17th problem which is mainly dependent on the variety of f-rings generated by the real numbers is also included in this section. Lattice orders on the $n \times n$ matrix algebra over a totally ordered field are considered in the last section. When the field is archimedean or when $n = 2$ all of the lattice orders are described, and it is shown that in these cases the usual order is the only lattice order in which the identity element is positive.

A few words about the method of referencing are in order. An exercise is referenced by its number alone when the exercise occurs in the section in which it is referenced, and, otherwise, it is referenced by its number preceded by the chapter and section numbers in which it occurs. A reference to a theorem or a numbered line uses all three of its numbers.

Paul Taylor's package was used in preparing the diagrams in the text. I wish to thank Joanne Guttman and Shirley Michel for their splendid job of typing and preparation of the manuscript. I would also like to thank my colleague Charles Odenthal for making the diagrams fit.

Toledo, Ohio *Stuart Steinberg*
August 17, 2009

Contents

List of Symbols

$\lvert X\rvert$, card (X)	cardinality of X
$\subseteq, \supseteq$	inclusion
$\subset, \supset$	proper inclusion
$\dot\cup$	disjoint union
$\mathscr{P}(X)$	power set of X; $\mathscr{P}$-radical of the ℓ-ring X
ΠX_i	cartesian product of $\{X_i :\in I\}$
$(x_i)_{i\in I}$	an element of ΠX_i
X^I (X^n)	cartesian product of I (n) copies of X
X^*	set of nonzero elements in the monoid or ring X; annihilator of X in a reduced ring
Pog (Poag)	category of (abelian) po-groups
Log (Loag)	category of (abelian) ℓ-groups
$\mathscr{M}_R$ (u-$\mathscr{M}_R$)	category of (unital) right R-modules
po-(f-, ℓ-)$\mathscr{M}_R$	category of right po-modules (f-modules, ℓ-modules) over R
ns$f$-$\mathscr{M}_R$	category of nonsingular right f-modules over R
$\mathscr{U}(R)$	group of units in the monoid R
$GL(n,K)$	group of units in the $n \times n$ matrix ring over K
ord(T)	ordinal number of the well-ordered set T
$W(\alpha)$	initial segment of the ordinals determined by α
$\aleph_\alpha$	infinite cardinal number
ω_α	first ordinal whose cardinality is $\aleph_\alpha$
$U(X), U_P(X)$	set of upper bounds of X (in P)
$L(X), L_P(X)$	set of lower bounds of X (in P)
$L_s(X)$ $(U_s(X))$	set of strict lower (upper) bounds of X
lub$X$, lub$_P X$	least upper bound of X (in P)
sup$X$, sup$_P X$	least upper bound of X (in P)
$\bigvee_{i\in I} x_i$	least upper bound of $\{x_i : i \in I\}$
$a \vee b$	least upper bound of $\{a, b\}$
glb$X$, glb$_P X$	greatest lower bound of X (in P)
inf$X$, inf$_P X$	greatest lower bound of X (in P)
$\bigwedge_{i\in I} x_i$	greatest lower bound of $\{x_i : i \in I\}$

$a \wedge b$	greatest lower bound of $\{a,b\}$
$a\|\|b$	a is incomparable to b
$[a,b]$	closed interval; commutator
$X \leq Y$	$x \leq y$ for each $x \in X$ and each $y \in Y$
$P \underset{\leftarrow}{\times} Q$	antilexicograhically ordered product
$P \underset{\rightarrow}{\times} Q$	lexicographically ordered product
$M(P)$	MacNeille completion of the poset P
$D(P)$	Dedekind completion of the poset P; subring generated by the d-elements on an ℓ-module P or in an ℓ-ring P
$\underrightarrow{\lim}$	direct limit
$\underleftarrow{\lim}$	inverse limit
$\mathscr{C}[A,B]$, $[A,B]$	morphisms from A to B in the category $\mathscr{C}$
Ω	operator domain
$\Omega(n)$	set of n-ary operators
$\omega(a_1,\ldots,a_n)$	evaluation of the n-ary operation ω at $a_1,\ldots,a_n$
ω_A	a constant in the Ω-algebra A
$S_\Omega(X)$	subalgebra of an Ω-algebra generated by X
$\mathscr{A}(\Omega)$	the class (category) of Ω-algebras
$W(\Omega,X)$	the Ω-row algebra on X
$F_\Omega(X), F(X)$	Ω-word algebra on X, free Ω-algebra on X
$\mathscr{V}_\Omega(S,X), S^*$	variety of Ω-algebras satisfying the identities in S on the alphabet X
$\mathscr{V}(\mathscr{C})$	variety of Ω-algebras generated by the class $\mathscr{C}$
$\mathscr{V}_C(R)$ ($\mathscr{V}(R)$)	variety of C-ℓ-algebras (ℓ-rings) generated by R
$\mathscr{C}^*$	set of identities satisfied by each Ω-algebra in the class $\mathscr{C}$
G^+	set of positive elements in the po-group G
$\boxplus_{i\in I}G_i$	direct sum of the groups, rings, or modules G_i
$\oplus_{i\in I}G_i$	direct sum of the po-groups, po-rings, or po-modules G_i
$\mathrm{Aut}(A)$	automorphism group of A
$\mathrm{End}(G)^+$	set of po-homomorphisms of the po-group G
x^+ (x^-)	positive (negative) part of x
$\|x\|$	absolute value of x
$d(M)$	divisible hull of the abelian group M; set of d-elements on the module M
$d(P,S)$	divisible closure of the partial order P in S
$d_r(R)$	set of right d-elements in the ring R
$V(\Gamma,G_\gamma)$	Hahn product of the po-groups G_γ indexed by the poset Γ
$\Sigma(\Gamma,G_\gamma)$	subgroup of elements in the Hahn product whose support is finite
$W(\Gamma,G_\gamma)$	subgroup of elements in the Hahn product whose support is a W-set
D_α	domain of the function α; right inner derivation determined by the ring element α
supp α	support of the function α

$S(\alpha)$	closure of the support of α
A^- (A°)	closure (interior) of A in a topological space
$\mathbb{R}^-$	$\mathbb{R}\cup\{\infty,-\infty\}$
$N_\varepsilon(x)$	ε-neighborhood of x
$\lim x_n$	limit of a convergent net
$\mathscr{N}(x)$	neighborhood system of x
$C(X)$ $(C^*(X))$	set of continuous (bounded) real-valued functions on the topological space X
$E(X)$	set of continuous extended real-valued functions on X
$D(X)$	set of elements in $E(X)$ which are real-valued on a dense subset of X
$C(X)$ $(C^G(X))$	convex ℓ-subgroup (of G) generated by X
$C_\Omega(X)$ $(C_\Omega^G(X))$	convex ℓ-Ω-subgroup (of G) generated by X
$[X]$	ℓ-subgroup generated by X
$R(1)$	$C_C^R(1)$ for the ℓ-subalgebra R of a unital C-ℓ-algebra
$M(1)$	set of elements in $R(1)$ which are infinitely smaller than 1 with respect to C
$\mathscr{C}(G)$ $(\mathscr{C}_\Omega(G))$	lattice of convex ℓ-subgroups (convex ℓ-Ω-subgroups) of the $(\Omega$-$)\ell$-group G
$A^\perp$ $(A^{\perp_R})$	polar of A (in R)
$\mathscr{B}(G)$	Boolean algebra of polars of the ℓ-group G
G^e	$c\ell$-essential closure of the archimedean ℓ-group G
lex B	lexicographic extension of the convex ℓ-subgroup B
A_Ω	largest convex ℓ-Ω-subgroup contained in A
trunk (P)	trunk of the poset or ℓ-group P
$\Gamma_\Omega(a)$ $(\Gamma_\Omega(a,G))$	set of Ω-values of a (in G)
$\Gamma(a)$ $(\Gamma(a,G))$	set of values of a (in G)
$\Gamma(G)$ $(\Gamma_\Omega(G))$	$(\Omega$-$)$value set of the ℓ-group $(\Omega$-f-group$)$ G
Spec(R)	set of prime ideals of the ring or distributive lattice R
R_n	$n\times n$ matrix ring over the ring R
$R[x_1,\ldots,x_n]$	polynomial ring over R in the noncommuting or commuting indeterminates $x_1,\ldots,x_n$
$D[x;\sigma,\delta]$	left skew polynomial ring
$D[\delta,\sigma;x]$	right skew polynomial ring
$R[X]_0$	polynomials over R in the set of indeterminates X with zero constant term
$P(a_1,\ldots,a_n)$	set of polynomials which are positive when evaluated at $(a_1,\ldots,a_n)$
$P^*(a_1,\ldots,a_n)$	set of nonconstant polynomials in $P(a_1,\ldots,a_n)$
$p'(x)$	derivative of a polynomial
$f(M)$ $(F(M)$	set (subring) of f-elements on the ℓ-module or ℓ-ring M
$f_r(R)$ $(F_r(R))$	set (subring) of right f-elements in the ℓ-ring R
$\bar{f}(M)$ $(\bar{F}(M))$	set (subring) of elements multiplication by which are f-maps on the ℓ-module or ℓ-ring M
$f_r(\Delta)$ $(f_\ell(\Delta))$	set of elements in the pops Δ right (left) translation by which

	preserves incomparability with respect to any given element
$\langle X\rangle$ ($\langle X\rangle_r$)	(right) ℓ-ideal generated by X
XA	subgroup generated by all xa with x in X and a in A
$A^{[n]}$	ℓ-ideal generated by A^n
$r(X;R)$ ($r(X)$)	right annihilator of X in R
$\ell(A;M)$ ($\ell(A)$)	left annihilator of A in M
$\mathrm{Ann}(R)$	Boolean algebra of annihilator ideals of a semiprime ring
$r_\ell(X;R)$ ($r_\ell(X)$)	right ℓ-annihilator of X in R
$\ell_\ell(A;M)$ ($\ell_\ell(A)$)	left ℓ-annihilator of A in M
$i(V)$	elements in an F-vector lattice V which are F-infinitely smaller than each F-strong order unit
$\mathscr{S}\mathscr{P}$	class of $\mathscr{P}$-semisimple ℓ-rings
$\mathscr{U}\mathscr{N}$	upper radical determined by the class of ℓ-rings $\mathscr{N}$
$\mathscr{L}\mathscr{A}$	lower radical determined by the class of ℓ-rings $\mathscr{A}$
ℓ-β	lower ℓ-nil radical
ℓ-Nil	upper ℓ-nil radical
ℓ-N_g	generalized ℓ-nil radical
$\mathscr{R}(R)$	intersection of the modular maximal right ℓ-ideals of the ℓ-ring R
$\mathscr{O}(R)$	intersection of the right ℓ-primitive ℓ-ideals of the ℓ-ring R
$\mathscr{J}$ ($\mathscr{J}_{left}$, $\mathscr{J}^+$)	(left, positive) Johnson radical of the ℓ-ring R
$N(R)$ ($N_n(R)$)	set of nilpotent elements (of index at most n) in the ring R
$M(R)$	set of elements in the ℓ-ring R whose absolute values are nilpotent
$\mathscr{S}(\mathscr{V})$	class of ℓ-rings which, modulo their lower ℓ-nil radical, belong to the variety $\mathscr{V}$
$\mathscr{S}(f)$	$\mathscr{S}(\mathscr{V})$ for $\mathscr{V}$ the variety of f-rings
$\mathscr{S}_{\mathscr{P}}(\mathscr{V})$	class of ℓ-rings which, modulo their $\mathscr{P}$-radical, belong to the variety $\mathscr{V}$
R_u	unital cover of the f-ring R
$R_{C\text{-}u}$	C-unital cover of the f-algebra R
$x \ll_A y$ ($x \ll y$)	x is infinitely smaller than y with respect to A ($\mathbb{Z}$)
$\Sigma(A*\Delta)$	crossed product
$V(A*\Delta)$	formal power series crossed product
$N_n(\Delta)$	set of elements α in the pops Δ for which $n\alpha$ is not defined
$\mathscr{E}(M)$	set of essential submodules of M
$t(M)$	torsion submodule of M
$(Y:X)$	set of ring elements r with $Xr \subseteq Y$
D_s	$(R:s)$
$E(M)$	injective hull of the module M
$\mathbb{D}(R)$	set of dense right ideals of the ring R
$Z(M);Z(R)$	singular submodule of the module M, center of the ring R
$Z_r(R)$	right singular ideal of the ring R
$c\ell(N)$	closure of the submodule N
$Q(R)$ ($Q_r(R)$)	maximal right quotient ring of the ring R

$Q_2(R)$	maximal two-sided quotient ring of R
$Q(P)$	localization of the positive cone P of a field obtained by inverting the elements of P
$\leq_q$	partial order determined by $Q(P)$
$\leq_u$	usual total order of $\mathbb{R}$
$(\mathscr{T},\mathscr{F})$	a torsion theory
$t_{\mathscr{T}}(M)$	sum of the $\mathscr{T}$-submodules of M
$\mathscr{T}_t$ $(\mathscr{F}_t)$	torsion class (torsion-free class) determined by the left exact radical t
$\mathbb{F}$ $(\mathbb{F}_t)$	right topology (determined by the left exact radical t)
$t_{\mathbb{F}}$	radical determined by the topology $\mathbb{F}$
$\mathbb{G}$	Goldie topology
$M_{\mathbb{F}}$, $Q(M)$, (M_{Σ})	module of quotients of the module M (with respect to the multiplicatively closed set Σ)
$Q_c(R)$ (R_{Σ})	classical right quotient ring of R (with respect to the multiplicatively closed set Σ)
$c\ell_{\mathbb{F}}(A)$	$\mathbb{F}$-closure of the submodule A
$\mathscr{C}_{\mathbb{F}}(M)$	lattice of those submodules of M for which the quotient is $\mathbb{F}$-torsion-free
$\mathscr{C}(M)$ $(\mathscr{C}_r(R))$	$\mathscr{C}_{\mathbb{G}}(M)$ $(\mathscr{C}(R_R))$
$M \otimes_R N$	tensor product of the modules M and N
$M \otimes_R^{po} N$	po-tensor product of the po-modules M and N
$M \otimes_R^{\ell} N$	ℓ-tensor product of the po-modules M and N
$M \otimes_R^{\ell\ell} N$	ℓ-tensor product of the ℓ-modules M and N
F_M	free representable f-module over the po-module M in a category $\mathscr{C}$
F_n	free nonsingular f-module of rank n
Γ_D (U_D)	value group (group of units) of the valued division ring D
R_D (J_D)	(maximal ideal of the) valuation ring of the valued division ring D
$\hat{F}$	Cauchy completion of the totally ordered division ring F
G_a	the largest totally ordered subgroup of the archimedean ℓ-group G which contains $a > 0$
$C^n(S,K^{+*})$	group of positive n-cochains of the semigroup S with coefficients in the totally ordered field K
$Z^2(S,K^{+*})$	group of positive 2-cocycles of S over K
$B^2(S,K^{+*})$	group of positive coboundaries of S over K
$H^2(S,K^{+*})$	second cohomology group of S over K
f^*	homomorphism on cohomology groups induced by the semigroup homomorphism f
Δ_0, $\Delta_{\Gamma_1,\Gamma_2}$	new pops obtained by modifying the partial orders of the pops Δ
P_0 (P_{Γ_1,Γ_2})	Hahn ordering of the generalized semigroup ring $A[\Delta_0]$ $(A[\Delta_{\Gamma_1\Gamma_2}])$
$P_{n,\alpha,\beta}$ $(P_{n,\alpha})$	partial orders of a generalized semigroup ring
tr a (det a)	trace (determinant) of the matrix a
a^t	transpose of the matrix a
$a^{(j)}$ $(a_{(j)})$	jth column (row) of the matrix a

$d(\delta_1,\ldots,\delta_n)$	diagonal matrix with diagonal entries $\delta_1,\ldots,\delta_n$
$P(a)$	the partial order $(K^+)_n a^t$ of the matrix ring K_n over the po-ring K
$\mathrm{Irr}(a,K)$	irreducible polynomial of the algebraic element a over the field K

Chapter 1
Partially Ordered Sets and Lattices

In this chapter we present some basic facts about partially ordered sets and lattices which are fundamental for our study of lattice-ordered groups, rings, and modules. The material presented includes Zorn's Lemma and some of its equivalences in Section 1.1, standard characterizations of distributive lattices and Boolean algebras in Section 1.2, and the construction of the MacNeille and Dedekind completions of a partially ordered set in Section 1.3. We also introduce some of the basic language of category theory and present enough of the subject of universal algebra so that we can establish the existence of free algebras in varieties.

The symbols " $\subseteq$ " and " $\supseteq$ " will be used for set inclusion and " $\subset$ " and " $\supset$ " will be used for proper inclusion. The letters $\mathbb{N}$, $\mathbb{Z}$, $\mathbb{Q}$, $\mathbb{R}$, and $\mathbb{C}$ denote the sets of natural numbers (excluding 0), integers, rational numbers, real numbers, and complex numbers, respectively. The symbols "$|X|$" or "card (X)" denote the cardinal number of the set X.

1.1 Partially Ordered Sets

A *relation from the set X to the set Y* is a subset α of the Cartesian product $X \times Y$. We will usually denote the fact that $(x,y) \in \alpha$ by writing $x\alpha y$. If $X = Y$, then α is called a *relation on X*. A relation $\leq$ on the set P is called a *partial order* of P if it is *reflexive, antisymmetric*, and *transitive*; that is, for all $x,y,z \in P$,

($P1$) $x \leq x$;

($P2$) if $x \leq y$ and $y \leq x$, then $x = y$;

($P3$) if $x \leq y$ and $y \leq z$, then $x \leq z$.

The pair $(P, \leq)$ is called a *poset* (*partially ordered set*), and it will usually be denoted by P alone. Frequently, $a \leq b$ will also be written as $b \geq a$. Also, $a < b$ (respectively,

S.A. Steinberg, *Lattice-ordered Rings and Modules*,
DOI 10.1007/978-1-4419-1721-8_1, © Springer Science + Business Media, LLC 2010

$b > a$) means that $a \leq b$ (respectively, $b \geq a$) and $a \neq b$. A poset P is called a *totally ordered set* or a *simply ordered set* or a *chain* if any two of its elements are comparable : $x \leq y$ or $y \leq x$ for all $x, y \in P$. In this case $\leq$ is called a *total order* of P. If two elements a and b of a poset are incomparable we will write $a||b$. Clearly, each subset X of a poset P is itself a poset with the partial order $\leq \cap (X \times X)$. An *upper bound* of the subset X of P is an element $u \in P$ such that $x \leq u$ for each x in X. This will sometimes be written as $X \leq u$; more generally, for subsets X and Y of P, $X \leq Y$ means that $x \leq y$ for each $x \in X$ and for each $y \in Y$. P itself can have at most one upper bound, and an upper bound for P, if it exists, is called the *greatest element* of P. Analogous definitions can be given for a *lower bound* of X and for the *least element* of P. The least (respectively, greatest) element of P will sometimes be denoted by 0 (respectively, 1). Note that each element in P is an upper bound and also a lower bound of the empty set. Let $U(X) = U_P(X)$ (respectively, $L(X) = L_P(X)$) denote the set of upper bounds (respectively, lower bounds) of X. If $U(X)$ has a least element s, then s is called the *least upper bound* of X or the *supremum* of X and is denoted by $s = \text{lub}X = \text{lub}_P X$ or $s = \sup X = \sup_P X$. Also, if the subset X is indexed by the set I, $X = \{x_i : i \in I\}$, then s will frequently be written as

$$s = \bigvee_{i \in I} x_i,$$

or if $I = \{1, 2, \ldots, n\}$ is finite or $X = \{a, b\}$, then

$$s = x_1 \vee x_2 \vee \cdots \vee x_n \quad or \quad s = a \vee b.$$

Analogously, the greatest element ℓ in $L(X)$, if it exists, is the *greatest lower bound* or *infimum* of X and is denoted by

$$\ell = \text{glb}X = \text{glb}_P X = \inf X = \inf_P X = \bigwedge_{i \in I} x_i,$$

or

$$\ell = x_1 \wedge x_2 \wedge \cdots \wedge x_n \quad or \quad \ell = a \wedge b$$

if $I = \{1, 2, \ldots, n\}$, or $X = \{a, b\}$. Note that each of the equations

$$\inf X \wedge \inf Y = \inf(X \cup Y)\ , \ \sup X \vee \sup Y = \sup(X \cup Y) \qquad (1.1.1)$$

holds exactly when, in the first equation, for example, $\inf X$ and $\inf Y$ both exist and either side exists.

A *minimal* (respectively, *maximal*) *element* in P is an element $p \in P$ which exceeds (respectively, is exceeded by) no other element of P : $q \leq p$ (respectively, $p \leq q$) implies that $p = q$. If each nonempty subset of P has a minimal (respectively, maximal) element, then P is said to satisfy the *minimum* (respectively, *maximum*) *condition*. A *well-ordered set* is a totally ordered set which satisfies the minimum condition. We will find it essential to be able to determine that certain posets that arise do indeed have maximal elements. The most important tool to be used in this regard is the well-known set-theoretic axiom known as Zorn's Lemma. In the first theorem we state this axiom together with some of the better known and most useful statements to which it is equivalent. A few definitions are needed first. The *power set* of the set X is the collection $\mathscr{P}(X)$ of all subsets of X. It becomes a poset when it is supplied with the inclusion relation $\subseteq$. A *choice function* for X is a function $c : \mathscr{P}(X) \setminus \{\phi\} \longrightarrow X$ with the property that $c(A) \in A$ for each nonempty subset A of X. The *cartesian product* of the indexed family of sets $\{X_i : i \in I\}$ is the set

$$\prod_{i \in I} X_i = \left\{ I \xrightarrow{f} \bigcup_{i \in I} X_i : \forall i \in I, f(i) \in X_i \right\}.$$

As is customary, a function $f \in \Pi X_i$ will frequently be denoted by indexing its range as the I-*tuple* $(x_i)_{i \in I}$ where $f(i) = x_i$. If each $X_i = X$ then ΠX_i will be denoted by X^I. In particular, if $n \in \mathbb{N}$ then X^n will denote the Cartesian product of n copies of X. The *projections* $\pi_i : \Pi X_i \longrightarrow X_i$ are given by $\pi_i(f) = f(i)$ or $\pi_i((x_i)_i) = x_i$. A *maximal chain* in a poset P is a totally ordered subset of P that is not a proper subset of any totally ordered subset of P.

Theorem 1.1.1. *The following statements are equivalent.*

(a) (Axiom of Choice) Each nonempty set has a choice function.
(b) The cartesian product of a nonempty family of nonempty sets is nonempty.
(c) (Zorn's Lemma, first form) Each chain in a poset is contained in a maximal chain.
(d) (Zorn's Lemma, second form) If each chain in a nonempty poset has an upper bound, then the poset has a maximal element.
(e) (Zorn's Lemma, third form) If each nonempty chain in a nonempty poset has a least upper bound, then the poset has a maximal element.
(f) Each set can be well-ordered.

Proof. To see the equivalence of (a) and (b) let $\{X_i : i \in I\}$ be a nonempty family of nonempty sets, and let $X = \cup X_i$ be the union of this family. If c is a choice function for X, then the composite of the indexing function $I \longrightarrow \{X_i : i \in I\}$ and the restriction of c to $\{X_i : i \in I\}$ is an element of the cartesian product of the family. Conversely, let X be a nonempty set and index the collection of its nonempty subsets by the collection itself: $P(X) \setminus \{\phi\} = \{A_A : A_A = A\}$. Then each element in the cartesian product of this family is a choice function for X.

(a) implies (c). This implication is a consequence of the following theorem.

Let P be a nonempty poset in which each nonempty chain has a least upper bound. Suppose that $f : P \longrightarrow P$ is a function such that $f(x) \geq x$ for each x in P, and if $x \leq y \leq f(x)$ then $y = x$ or $y = f(x)$. Then $f(x) = x$ for some x in P.

To prove this let us fix an element x_0 in P. A subset Q of P is called a *B*-*set* if it has the following three properties:

(i) x_0 is the least element of Q;
(ii) $f(Q) \subseteq Q$;
(iii) if S is a nonempty chain in Q then $\sup_P S \in Q$.

For example, $\{x \in P : x \geq x_0\}$ is a B-set. Let B_0 be the intersection of all of the B-sets in P. Clearly, B_0 is the smallest B-set in P. Our goal is to show that B_0 is a chain. Toward this end let x be an element of B_0 which is comparable to every other element of B_0, and let $B_1 = B_1(x) = \{y \in B_0 : y \leq x \text{ or } f(x) \leq y\}$. We claim that $B_1 = B_0$; to show this it suffices to verify that B_1 is a B-set. Note that $x, f(x)$ and x_0 are all elements of B_1, and, in particular, B_1 satisfies (i). Suppose that $y \in B_1$. If $f(x) \leq y$, then $f(x) \leq y \leq f(y)$ shows that $f(y) \in B_1$. On the other hand, suppose that $y \leq x$. Since x is comparable to $f(y)$ we either have $y \leq x \leq f(y)$, in which case $y = x$ or $x = f(y)$, and in both of these cases $f(y) \in B_1$, or $f(y) \leq x$ and again $f(y) \in B_1$. Thus B_1 satisfies (ii). As for (iii) let S be a chain in B_1. If $s \geq f(x)$ for some s in S, then $\sup_P S \geq s \geq f(x)$ and $\sup_P S \in B_1$. Otherwise, x is an upper bound of S, $\sup_P S \leq x$, and $\sup_P S \in B_1$.

We show next that the set B_2 of all those elements of B_0 which are comparable to each element of B_0 is a B-set. Certainly $x_0 \in B_2$; and if $x \in B_2$ then, for any $y \in B_0$, $y \leq x \leq f(x)$ or $f(x) \leq y$ since $B_0 = B_1(x)$. In either case $f(x)$ is comparable to y, whence $f(x) \in B_2$. Finally, let S be a chain in B_2 and take $x \in B_0$. If $s \geq x$ for some s in S, then $\sup_P S \geq x$, and otherwise, x is an upper bound of S and $\sup_P S \leq x$. In either case $\sup_P S$ is comparable to x and hence B_2 satisfies (iii). Thus, $B_0 = B_2$ and B_0 is a chain. Now, $s = \sup_P B_0 \in B_0$ and hence $s \geq f(s) \geq s$; so $f(s) = s$. This proves the theorem.

Now let Q be a chain in the poset P and let $\mathscr{T}$ be the collection of all those chains in P which contain Q. $\mathscr{T}$ is a subposet of $\mathscr{P}(P)$. If $A \in \mathscr{T}$ is not a maximal chain let $A^* = \{x \in P \backslash A : A \cup \{x\} \text{ is a chain }\}$, and let c be a choice function for P. Define the function $f : \mathscr{T} \longrightarrow \mathscr{T}$ by

$$f(A) = \begin{cases} A & \text{if } A \text{ is a maximal chain,} \\ A \cup \{c(A^*)\} & \text{if } A \text{ is not maximal.} \end{cases}$$

Then $\mathscr{T}$ and f satisfy the hypotheses of the theorem. So f has a fixed point A, and A is a maximal chain that contains Q.

(c) implies (d). Suppose that P is a nonempty poset in which each nonempty chain has an upper bound. If Q is a maximal chain in P and $x \in U_P(Q)$, then x is a maximal element in P.

(d) implies (e). This is trivial.

(e) implies (f). The subset S of the well-ordered set T is called an *initial segment* of T if there is an element t in T such that $S = \{s \in T : s < t\}$. Now let X be a set and let

$$\mathscr{T} = \{(Y, \leq) : Y \subseteq X \text{ and } \leq \text{ is a well-ordering of } Y\}.$$

We make $\mathscr{T}$ into a poset by defining the relation $\underline{\sigma}$ on $\mathscr{T}$ by $(Y, \leq)\underline{\sigma}(Z, \leq')$ if $(Y, \leq)$ is a subposet of $(Z, \leq')$ and either $Y = Z$ or Y is an initial segment of Z. Using the facts that a well-ordered set T is not an initial segment of itself and that an initial segment of an initial segment of T is an initial segment of T, it is easily verified that $\underline{\sigma}$ is a partial order of $\mathscr{T}$. Suppose that $\{(Y_i, \leq_i) : i \in I\}$ is a chain in $\mathscr{T}$. Let $Y = \cup Y_i$ and let $\leq = \cup \leq_i$. Then $(Y, \leq)$ is a totally ordered set and each $(Y_i, \leq_i)$ is a subposet of $(Y, \leq)$. We claim that $(Y, \leq)$ is the least upper bound of the chain

$\{(Y_i, \leq_i)\}$ in $\mathscr{T}$. First, we will check that $\leq$ well-orders Y. Suppose that $A \subseteq Y$ and $A \cap Y_i \neq \phi$. Let m be the minimal element of $A \cap Y_i$. If $a \in A$ then $a, m \in Y_j$ for some $j \in I$. If $a \in Y_i$ then $m \leq a$; if $a \notin Y_i$ then $Y_i = \{y \in Y_j : y < x\}$ for some x in Y_j, and hence $m < x \leq a$. Thus, m is the least element of A; so $(Y, \leq) \in \mathscr{T}$.

Next, we check that $(Y, \leq)$ is an upper bound of $\{(Y_i, \leq_i) : i \in I\}$ in $\mathscr{T}$. To see this first note that if $a \in Y$ and $b \in Y_i$ with $a < b$, then $a \in Y_i$. For, a is in some Y_j, and either $a \in Y_j \subseteq Y_i$, or $Y_i = \{c \in Y_j : c < d\}$ for some d in Y_j, in which case, again, $a \in Y_i$. Now, if some $Y_i \subset Y$ let x be the least element in $Y \setminus Y_i$. Then it easily follows that $Y_i = \{y \in Y : y < x\}$.

Finally, $(Y, \leq)$ is the least upper bound of $\{(Y_i, \leq_i)\}$. For, suppose that $(Y_i, \leq_i)$ $\underline{\sigma}$ $(Z, \leq')$ for each $i \in I$. If $a \in Z, b \in Y$ and $a <' b$, then $a \in Y$ since $b \in Y_i$ for some i and Y_i is an initial segment of Z, or $Y_i = Z$. So, again, if $Y \subset Z$ then $Y = \{y \in Z : y <' z\}$ where z is the least element of $Z \setminus Y$.

Now, by (e) $\mathscr{T}$ has a maximal element $(Y_0, \leq)$. If $Y_0 \subset X$ take $x \in X \setminus Y_0$ and partial order $Y = Y_0 \cup \{x\}$ with $\leq' = \leq \cup \{(y,x) : y \in Y\}$. Then $(Y, \leq') \in \mathscr{T}$ and $(Y_0, \leq) \underline{\sigma} (Y, \leq')$ which contradicts the maximality of $(Y_0, \leq)$. So $Y_0 = X$ and X can be well-ordered.

(f) implies (a). Let X be a nonempty set and let $\leq$ be a well- ordering of X. Then the function which picks out the least element of each nonempty subset of X is a choice function for X. □

A poset P is said to satisfy the *ascending* (respectively, *descending*) *chain condition,* or to have *a.c.c.* (respectively, *d.c.c.*), or to be *noetherian* (respectively, *artinian*) if for each increasing (respectively, decreasing) sequence $a_1 \leq a_2 \leq \cdots$ (respectively, $a_1 \geq a_2 \geq \cdots$) in P there is an integer n such that $a_m = a_n$ whenever $m \geq n$. One consequence of the Axiom of Choice is that this condition is equivalent to the maximum (respectively, minimum) condition.

The following important combinatorial result is useful in the proof of one interesting characterization of a noetherian poset that has no infinite trivially ordered subsets; see Exercise 11. This type of poset will come up in our study of power series rings in Section 3.5.

For the set X let 2-X denote the collection of all those subsets of X whose cardinality is 2.

Theorem 1.1.2. *Suppose that* 2-$\mathbb{N} = A_1 \cup \cdots \cup A_p$ *is a partition of* 2-$\mathbb{N}$. *Then there is an infinite subset* X *of* $\mathbb{N}$ *and an integer* i *with* $1 \leq i \leq p$ *such that* 2-$X \subseteq A_i$.

Proof. We will recursively define distinct elements $x_1, \ldots, x_n$ of $\mathbb{N}$, infinite subsets $Y_1 \supseteq Y_2 \supseteq \cdots \supseteq Y_n$ of $\mathbb{N}$ and integers $i_1, \ldots, i_n$ in $\{1, \ldots, p\}$ with the following properties:

(i) $1 \leq j < n \Rightarrow x_{j+1} \in Y_j$;
(ii) $1 \leq j \leq n \Rightarrow \{x, x_j\} \in A_{i_j}$ for each $x \in Y_n$.

Let $x_1 = 1$ and choose $i_1 \in \{1, \ldots, p\}$ such that $Y_1 = \{x \in \mathbb{N} : \{x_1, x\} \in A_{i_1}\}$ is infinite. Now suppose that we have constructed $x_1, \ldots, x_n, Y_1, \ldots, Y_n$ and $i_1, \ldots, i_n$ with the stated properties. Let x_{n+1} be the least element of Y_n. Then there is a choice of

$i_{n+1} \in \{1,\dots,p\}$ such that $Y_{n+1} = \{x \in Y_n : \{x, x_{n+1}\} \in A_{i_{n+1}}\}$ is infinite, and hence $x_1,\dots,x_{n+1}, Y_1,\dots,Y_{n+1}$, and $i_1,\dots,i_{n+1}$ satisfy (i) and (ii). For some $i_0 \in \{1,\dots,p\}$ the set $\{n \in \mathbb{N} : i_n = i_0\}$ is infinite. Thus, $X = \{x_n : i_n = i_0\}$ is infinite and 2-$X \subseteq A_{i_0}$. For if x_n, x_m are two elements of X with $n \neq m$, then $m \geq n+1$ or $n \geq m+1$. Hence, either $x_m \in Y_{m-1} \subseteq Y_n$ and $\{x_m, x_n\} \in A_{i_n} = A_{i_0}$ or $\{x_n, x_m\} \in A_{i_m} = A_{i_0}$. □

We next present a few examples of posets. If P is any set, then $(P, =)$ is a poset which is said to be *trivially ordered*. The set of real numbers $\mathbb{R}$ with its usual ordering is a totally ordered set. So, of course, are its subsets $\mathbb{N}$, $\mathbb{Z}$ and $\mathbb{Q}$. The *direct product* of the family of posets $\{(P_i, \leq_i)\}_{i \in I}$ is the poset $(P = \Pi P_i, \leq)$ where $\leq$ is the coordinatewise partial order of $P : (x_i) \leq (y_i)$ if and only if $x_i \leq_i y_i$ for each $i \in I$. If I is a poset with the maximum condition then another reasonable partial order $\leq$ can be given to the cartesian product P by defining $(x_i) \leq (y_i)$ if $x_j \leq y_j$ for each maximal element j in the set $\{i \in I : x_i \neq y_i\}$. This poset is called the *ordinal product* of the family $\{P_i : i \in I\}$ (see Exercise 1). The ordinal product of two posets P and Q will be denoted by $P \overleftarrow{\times} Q$. It has the *antilexicographical* order: $(p_1, q_1) \leq (p_2, q_2)$ if $q_1 < q_2$, or $q_1 = q_2$ and $p_1 \leq p_2$. The *lexicographical* order is denoted by $P \overrightarrow{\times} Q$ and is defined by : $(p_1, q_1) \leq (p_2, q_2)$ if $p_1 < p_2$, or $p_1 = p_2$ and $q_1 \leq q_2$. Either of these orderings of $P \times Q$ has a convenient picture. For example, for $P \overleftarrow{\times} Q$ the partial order in the PQ-plane is upward along vertical lines and from left to right; that is, if a and b are two points in this plane, then $a < b$ if either a and b lie on the same vertical line and b is higher than a, or if the vertical line through b is to the right of the vertical line through a.

Again, if I is a poset, then the *ordinal sum* of the family of posets $\{P_i : i \in I\}$ is the disjoint union $P = \dot{\cup} P_i$ of the P_i supplied with the following partial order : if $x, y \in P$ and $x \in P_i$ and $y \in P_j$, then $x \leq y$ if $i = j$ and $x \leq y$ in P_i, or if $i < j$. If I is trivially ordered, then the ordinal sum is called the *cardinal sum* .

The subset S of the poset P is called *cofinal* (respectively, *coinitial*) in P if for every $p \in P$ there is some $s \in S$ with $s \geq p$ (respectively, $s \leq p$). If S is both coinitial and cofinal in P it is called *coterminal* in P. These concepts are used in Exercise 8 and also arise in later sections.

It will occasionally be useful to use the language of category theory, and so we present the basic definitions here. A *category* $\mathscr{C}$ consists of a pair of classes, $\mathscr{O}b(\mathscr{C})$, the *objects* of the category, and $\mathscr{M}or(\mathscr{C})$, the *morphisms* of the category, which satisfy certain conditions that we now elucidate. With each pair of objects A and B in $\mathscr{C}$ there is associated a set $\mathscr{C}[A,B]$ of morphisms. If $f \in \mathscr{C}[A,B]$ then f is called a *morphism from A to B* and we write $f : A \longrightarrow B$. Moreover, for each triple of objects A, B, C in $\mathscr{C}$ there is a function, called *composition*,

$$\mathscr{C}[B,C] \times \mathscr{C}[A,B] \longrightarrow \mathscr{C}[A,C], \ \ (f,g) \longmapsto fg,$$

such that for all objects A, B, C, D in $\mathscr{C}$:

(i) If $(A,B) \neq (C,D)$, then $\mathscr{C}[A,B] \cap \mathscr{C}[C,D] = \phi$;

(ii) If $f \in \mathscr{C}[A,B]$, $g \in \mathscr{C}[B,C]$ and $h \in \mathscr{C}[C,D]$, then $h(gf) = (hg)f$;

(iii) There is a morphism $1_A \in \mathscr{C}[A,A]$ such that if $f \in \mathscr{C}[A,B]$ and $g \in \mathscr{C}[C,A]$, then $f1_A = f$ and $1_A g = g$.

The category $\mathscr{D}$ is a *subcategory* of the category $\mathscr{C}$ if $\mathscr{O}b(\mathscr{D}) \subseteq \mathscr{O}b(\mathscr{C})$ and for all A, $B \in \mathscr{O}b(\mathscr{D})$, $\mathscr{D}[A,B] \subseteq \mathscr{C}[A,B]$, composition in $\mathscr{D}$ is the same as in $\mathscr{C}$, and $1_A \in \mathscr{D}[A,A]$. If $\mathscr{D}[A,B] = \mathscr{C}[A,B]$ for all objects A, $B \in \mathscr{D}$, then $\mathscr{D}$ is called a *full* subcategory of $\mathscr{C}$. For example, the category of groups is not a full subcategory of the category of sets and has the category of abelian groups as a full subcategory.

The morphism 1_A, which is necessarily unique, is called the *identity morphism* for A. A morphism $f \in \mathscr{C}[A,B]$ is an *isomorphism* if there is a morphism $g \in \mathscr{C}[B,A]$ such that $gf = 1_A$ and $fg = 1_B$. As is usual we denote g by f^{-1} and call f^{-1} the *inverse* of f. The morphism f is called *monic* or is a *monomorphism* if $g = h$ whenever g and h are morphisms with $fg = fh$; dually, f is *epic* or is an *epimorphism* if $g = h$ whenever $gf = hf$. We will frequently call a function in a particular category an epimorphism when we mean that it has the possibly stronger property of being a surjection; that is, it is an epimorphism in the category of sets.

A *covariant functor* from the category $\mathscr{C}$ to the category $\mathscr{D}$ is a pair of functions $\mathscr{O}b(\mathscr{C}) \longrightarrow \mathscr{O}b(\mathscr{D})$ and $\mathscr{M}or(\mathscr{C}) \longrightarrow \mathscr{M}or(\mathscr{D})$, each of which is denoted by F, such that for objects A and B in $\mathscr{C}$

$$F : \mathscr{C}[A,B] \longrightarrow \mathscr{D}[F(A),F(B)],$$

$F(fg) = F(f)F(g)$ provided that the composite fg is defined, and $F(1_A) = 1_{F(A)}$. A *contravariant functor* from $\mathscr{C}$ to $\mathscr{D}$ is a function F such that

$$F : \mathscr{C}[A,B] \longrightarrow \mathscr{D}[F(B),F(A)],$$

$F(fg) = F(g)F(f)$, and $F(1_A) = 1_{F(A)}$. We will denote the *identity functor* on $\mathscr{C}$ by $1_{\mathscr{C}}$. If $F : \mathscr{C} \longrightarrow \mathscr{D}$ and $G : \mathscr{D} \longrightarrow \mathscr{E}$ are functors, then it should be clear what is meant by the *composite* functor $GF : \mathscr{C} \longrightarrow \mathscr{E}$.

If $F, G : \mathscr{C} \longrightarrow \mathscr{D}$ are two covariant functors, then a *natural transformation* from F to G is a function $\alpha : \mathscr{O}b(\mathscr{C}) \longrightarrow \mathscr{M}or(\mathscr{D})$ which satisfies the conditions that for each object $A \in \mathscr{C}$, $\alpha(A) \in \mathscr{D}[F(A),G(A)]$, and for each $f \in \mathscr{C}[A,B]$ the diagram

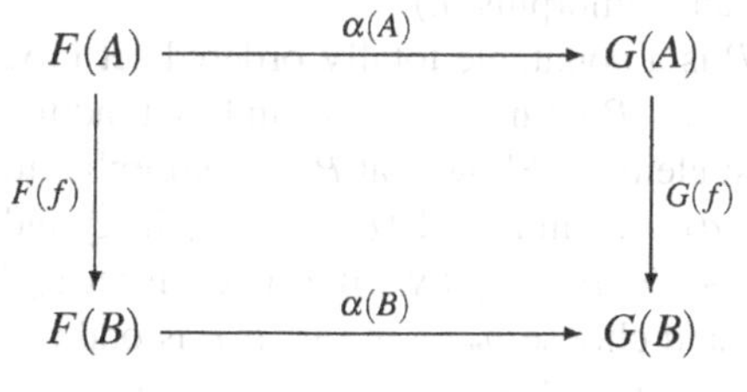

is *commutative*; that is, $G(f)\alpha(A) = \alpha(B)F(f)$. If each $\alpha(A)$ is an isomorphism, then α is called a *natural equivalence*. If F and G are both contravariant, then, of course, the above diagram has to be replaced by

(a) P has a.c.c. and each trivially ordered subset of P is finite.
(b) Each sequence in P has a decreasing subsequence.
(c) P has a.c.c. on ideals. (An *ideal* of P is a subset I such that if $x \le y$ with $x \in P$ and $y \in I$, then $x \in I$.)
(d) For each subset X of P there is a finite subset Y of X with $I(X) = I(Y)$ where $I(X)$ is the ideal of P generated by X.
(e) Each ideal of P is finitely generated.
(f) The set of maximal elements of each nonempty subset of P is a nonempty finite set.

11. Use Theorem 1.1.2 to prove that the following statement is equivalent to (a) of Exercise 10. If $(x_n)_{n\in\mathbb{N}}$ is a sequence in P, then $x_n \ge x_m$ for some indices $n < m$.

1.2 Lattices

A *lattice* is a poset L in which $a \wedge b$ and $a \vee b$ exist for any two elements a and b in L. In conjunction with the equations (1.1.1) a simple use of mathematical induction shows that L is a lattice exactly when each nonempty finite subset of L has a sup and an inf. A subset M of the poset L is a *sublattice* provided that $\inf_L\{a,b\}$ and $\sup_L\{a,b\}$ exist and are in M whenever $a,b \in M$. The next theorem shows that a lattice may be viewed as an abstract algebra with two operations that satisfies six identities (see Section 1.4).

Theorem 1.2.1. *For all elements a,b,c in a lattice L,*

(L1) $a \wedge (b \wedge c) = (a \wedge b) \wedge c\ ,\ a \vee (b \vee c) = (a \vee b) \vee c$ *(associative law);*

(L2) $a \wedge b = b \wedge a\ ,\ a \vee b = b \vee a$ *(commutative law);*

(L3) $a \wedge (a \vee b) = a = a \vee (a \wedge b)$ *(absorption law).*

Conversely, if L is a set with two binary operations $\wedge$ and $\vee$ which satisfy (L1), (L2), and (L3), then the relation $\le$ defined on L by $a \le b$ if and only if $a = a \wedge b$ is a partial order on L; moreover L is a lattice with $\sup\{a,b\} = a \vee b$ *and* $\inf\{a,b\} = a \wedge b$.

Proof. In a lattice L ($L1$) is a special case of equations (1.1.1), and ($L2$) and ($L3$) are obviously true. Conversely, suppose that $(L,\wedge)$ and $(L,\vee)$ are commutative semigroups whose operations are entwined by the absorption laws ($L3$). Consider the relation $\le$ defined on L by $a \le b \Leftrightarrow a = a \wedge b$. To check that $\le$ is antisymmetric, let $a \le b$ and $b \le a$. Then $a = a \wedge b = b \wedge a = b$. For transitivity, if $a \le b$ and $b \le c$, then, $a = a \wedge b = a \wedge b \wedge c = a \wedge c$, so $a \le c$. Finally, notice that reflexivity of $\le$ is equivalent to the condition that each element of the semigroup $(L,\wedge)$ is idempotent. But $a \wedge a = a \wedge\ [a \vee (a \wedge a)] = a$ by two uses of ($L3$); so $a \le a$. Let $x = a \wedge b$; we claim that $x = \inf\{a,b\}$. Since $x \wedge a = a \wedge b \wedge a = a \wedge b = x$, $x \le a$, and, similarly,

$x \le b$. If $y \le \{a,b\}$, then $y = y \wedge a = y \wedge b$; so $y \wedge x = y \wedge a \wedge b = y \wedge b = y$, and hence $y \le x$. Now, by this same argument the relation $\le'$ given by $a \le' b \Leftrightarrow b = a \vee b$ is a partial order of L and $a \vee b = \sup\{a,b\}$ in $(L, \le')$. Hence, to finish the proof it suffices to show that $a \le b \Leftrightarrow b = a \vee b$. But if $a \le b$, then $a \vee b = (a \wedge b) \vee b = b$, and if $b = a \vee b$, then $a \wedge b = a \wedge (a \vee b) = a$, so $a \le b$. □

Each of the axioms $(L1)$, $(L2)$, and $(L3)$ is recovered when $\wedge$ and $\vee$ are interchanged. This implies the *principle of duality*, namely, if this interchange is performed in any theorem about lattices, together with the interchange of $\le$ and $\ge$, another (not necessarily different) theorem is obtained.

The lattices that occur in this book will usually have the property that each of the operations $\wedge$ and $\vee$ is distributive over the other, and we now briefly examine these kinds of lattices. First note that in any lattice L the following dual distributive inequalities are valid: For all $a,b,c \in L$,

$$(D1) \quad a \wedge (b \vee c) \ge (a \wedge b) \vee (a \wedge c),$$

$$(D2) \quad a \vee (b \wedge c) \le (a \vee b) \wedge (a \vee c).$$

If equality holds in $(D1)$ (or, as we will see, in $(D2)$) the lattice is called *distributive*. More generally, each of these inequalities implies the self-dual implication

$$(M) \quad a \ge b \Rightarrow a \wedge (b \vee c) \ge b \vee (a \wedge c) \qquad \forall a,b,c \in L.$$

If the inequality is replaced by equality in the conclusion of the implication in (M) the lattice is called *modular*. Clearly, each distributive lattice is modular. Now, we have the following characterization of a distributive lattice.

Theorem 1.2.2. *The following conditions are equivalent in a lattice L.*

$$(D) \quad a \wedge (b \vee c) = (a \wedge b) \vee (a \wedge c) \quad \forall a,b,c \in L$$

$$(D') \quad a \vee (b \wedge c) = (a \vee b) \wedge (a \vee c) \quad \forall a,b,c \in L$$

$$(D'') \quad a \wedge (b \vee c) \le b \vee (a \wedge c) \qquad \forall a,b,c \in L$$

Proof. Since (D) is the dual of (D') and since (D'') is self-dual it suffices to verify the equivalence of (D) and (D''). Since $a \wedge b \le b$ it is clear that (D) implies (D''). Assume that (D'') holds; to verify (D), by $(D1)$, it suffices to show that

$$a \wedge (b \vee c) \le (a \wedge b) \vee (a \wedge c).$$

But,

$$a \wedge (b \vee c) = a \wedge a \wedge (b \vee c) \le a \wedge [b \vee (a \wedge c)] = a \wedge [(a \wedge c) \vee b] \le (a \wedge c) \vee (a \wedge b)$$

by two uses of (D''). □

In a distributive lattice the following generalized distributive identities hold for any finite sets I and J:

$$\bigwedge_{i\in I}\bigvee_{j\in J} a_{ij} = \bigvee_{f\in J^I}\bigwedge_{i\in I} a_{if(i)}\,, \quad \bigvee_{i\in I}\bigwedge_{j\in J} a_{ij} = \bigwedge_{f\in J^I}\bigvee_{i\in I} a_{if(i)} \tag{1.2.1}$$

The proof is left to the reader (see Exercise 2).

Let L be a lattice with a least element 0 and a greatest element 1. A *complement* of the element $a \in L$ is an element $b \in L$ such that $a \wedge b = 0$ and $a \vee b = 1$. In a distributive lattice complements are unique. For if b and c are both complements of a, then

$$b = b \wedge 1 = b \wedge (a \vee c) = (b \wedge a) \vee (b \wedge c) = 0 \vee (b \wedge c) = b \wedge c.$$

So $b \le c$, $c \le b$ and hence $b = c$. If each element in a lattice L has a complement, then L is said to be *complemented.* A *Boolean algebra* is defined to be a complemented distributive lattice. So in a Boolean algebra complements are unique, and if a' denotes the complement of a, then $a = a''$. The lattice of all subsets of a given set (ordered by inclusion) is a Boolean algebra. The following characterization of a Boolean algebra is quite useful.

Theorem 1.2.3. *Let B be a Boolean algebra with least element 0, and denote the complement of an element a in B by a'. Then for all $a, b \in B$,*

$$(B) \quad a \wedge b' = 0 \Leftrightarrow a \wedge b = a.$$

Conversely, let B be a set which contains a distinguished element 0 and which has a binary operation $\wedge$ and a unary operation $'$. Suppose that $(B, \wedge)$ is a commutative semigroup in which each element is idempotent and which satisfies (B). Then B is a Boolean algebra with $\inf\{a,b\} = a \wedge b$, *0 is the least element of B, and a' is the complement of a in B.*

Proof. We first show that a Boolean algebra B satisfies (B). Suppose that $a \wedge b' = 0$. Then

$$a = a \wedge 1 = a \wedge (b \vee b') = (a \wedge b) \vee (a \wedge b') = (a \wedge b) \vee 0 = a \wedge b;$$

and if $a \le b$, then $0 \le a \wedge b' \le b \wedge b' = 0$. So $a \wedge b' = 0$ and (B) holds. For the converse, first note that the proof of Theorem 1.2.1 yields that the relation defined by $a \le b$ if and only if $a = a \wedge b$ is a partial order of B in which $a \wedge b = \inf\{a,b\}$ for all $a, b \in B$. Thus (B) translates to

$$(B') \quad a \le b \Leftrightarrow a \wedge b' = 0 \quad \forall a, b \in B.$$

Next, we verify that for all $a, b \in B$,

$$a = a'' \tag{1.2.2}$$

and

$$a \le b \Leftrightarrow b' \le a'. \tag{1.2.3}$$

Since $a' \le a'$, $a'' \wedge a' = 0$ and we have that $a'' \le a$, by (B'); also, 0 is the least element of B since $0 = a \wedge a' \le a$. If a is replaced by a' in the inequality $a'' \le a$ we get that $a''' \le a'$; hence, $0 \le a''' \wedge a \le a' \wedge a = 0$, $a''' \wedge a = 0$, $a \le a''$, and, consequently, $a = a''$. If $a \le b$, then $0 = a \wedge b' = a'' \wedge b'$ gives that $b' \le a'$ and this proves (1.2.3). Now define a binary operation on B by, for all $a, b \in B$,

$$a \vee b = (a' \wedge b')'.$$

Then $x \ge a, b \Leftrightarrow x' \le a', b' \Leftrightarrow x' \le a' \wedge b' \Leftrightarrow x \ge a \vee b$; so $a \vee b = \sup\{a, b\}$, and hence B is a complemented lattice. However,

$$\begin{aligned} a \wedge (b \vee c) \le x \ &\Leftrightarrow a \wedge (b \vee c) \wedge x' = 0 \\ &\Leftrightarrow a \wedge x' \le (b \vee c)' = b' \wedge c' \\ &\Leftrightarrow a \wedge x' \le b', c' \\ &\Leftrightarrow a \wedge x' \wedge b = 0 = a \wedge x' \wedge c \\ &\Leftrightarrow a \wedge b, a \wedge c \le x \\ &\Leftrightarrow (a \wedge b) \vee (a \wedge c) \le x, \end{aligned}$$

and so B is a distributive lattice. □

If P and Q are posets a function $f : P \longrightarrow Q$ is *isotone* or *order preserving* (respectively, *antitone* or *order reversing*) if $x \le y$ implies that $f(x) \le f(y)$(respectively, $f(x) \ge f(y)$) for all $x, y \in P$. If, also, $f(x) \le f(y)$ implies that $x \le y$, then f is called an *embedding* or a *monomorphism* . This terminology does not agree with the categorical terminology. A morphism in the category of posets and isotone maps is monic if and only if it is one-to-one, and it need not be a monomorphism as just defined. Nevertheless, in all subcategories of the category of posets with which we will be concerned, our use of the word *monomorphism* (or even *monic*) will entail this stronger meaning. If f is an onto monomorphism it is called an *isomorphism* ; P and Q are then said to be *order isomorphic* . As usual, an isomorphism between P and itself is an *automorphism* . A function $f : P \longrightarrow Q$ is called *complete* if it preserves all infs and sups that exist in P; that is, if whenever X is a subset of P such that either $p = \sup_P X$ or $q = \inf_P X$ exists, then $f(p) = \sup_Q f(X)$ or $f(q) = \inf_Q f(X)$. A *lattice homomorphism* is a function $f : P \longrightarrow Q$ between lattices P and Q such that $f(x \vee y) = f(x) \vee f(y)$ and $f(x \wedge y) = f(x) \wedge f(y)$ for all x and y in P. Each lattice homomorphism is isotone and an isomorphism between posets is complete, but a poset monomorphism between lattices need not be a lattice homomorphism. However, a lattice homomorphism that is one-to-one is a monomorphism and a bijection f between posets is an isomorphism if and only if f and f^{-1} are both isotone.

Exercises.

1. Find two distinct lattice polynomials $p(x,y,z)$ and $q(x,y,z)$ such that a lattice L is modular iff it satisfies the identity $p(x,y,z) = q(x,y,z)$.
2. (a) Prove the generalized distributive equations (1.2.1).
   (b) Show that in a lattice L the equation
$$\bigwedge_{i\in I}\bigvee_{j\in J} a_{ij} = \bigvee_{f\in J^I}\bigwedge_{i\in I} a_{if(i)}$$
   holds whenever both sides exist if and only if the equation
$$\bigwedge_{i\in I}\bigvee_{j\in J_i} a_{ij} = \bigvee_{f\in \Pi J_i}\bigwedge_{i\in I} a_{if(i)}$$
   holds whenever both sides exist, and dually (for arbitrary sets I, J, and J_i).
   (c) A lattice L is *infinitely distributive* if whenever $\vee x_i$ exists, then, for each y in L, $\vee(y\wedge x_i)$ exists and $y\wedge(\vee x_i) = \vee(y\wedge x_i)$, and the dual also holds. Show that each Boolean algebra is infinitely distributive. (If $u \geq y\wedge x_i$, then $x_i \leq u\vee y'$.)
3. Prove the equivalence of the following statements in the lattice L.
   (a) L is modular.
   (b) If two of the three elements a,b,c are comparable, then $a\wedge(b\vee c) = (a\wedge b)\vee(a\wedge c)$.
   (c) If two of the three elements a,b,c are comparable, then $a\vee(b\wedge c) = (a\vee b)\wedge(a\vee c)$.
4. The element a in the poset P is said to have a *successor* if the set $\{p\in P : a < p\}$ has a least element, and this least element is called the successor of a; it is denoted by a_+. P is a *poset with successors* if each of its elements has a successor. The *predecessor* of a, denoted by a_-, and a *poset with predecessors* are defined dually.
   (a) If P has predecessors (respectively, successors) and Q is any poset, then the ordinal product $P \overleftarrow{\times} Q$ has predecessors (respectively, successors).
   (b) The ordinal sum of a family of posets, each of which has successors (respectively, predecessors) also has successors (respectively, predecessors).
   (c) If P has predecessors and successors, then $a = a_{+-} = a_{-+}$ for each a in P.
   (d) If P is a lattice with successors (respectively, predecessors), then P is totally ordered.
5. Let P and Q be nonempty posets. Show that the ordinal product $P \overleftarrow{\times} Q$ is a lattice if and only if the following conditions are satisfied by P and Q.
   (a) Q is a lattice;
   (b) if Q is not totally ordered, then $0, 1 \in P$;

(c) $\forall x,y \in P$, $x \vee y$ exists if and only if x and y have a common upper bound, and dually;
(d) if $x \vee y$ does not exist for some $x,y \in P$, then $0 \in P$ and Q is a poset with successors, and dually.

6. Let $f : L \longrightarrow M$ be a function between lattices.
   a. If L is a chain show that f is a lattice homomorphism iff it is isotone.
   b. Find an example where f is a monomorphism of posets but it is not a lattice homomorphism.
   c. Suppose that f is a bijection. Show that f is a lattice homomorphism iff f^{-1} is a lattice homomorphism iff f and f^{-1} are both isotone.
7. Let R be a ring with no nonzero nilpotent ideals (equivalently, no nonzero nilpotent one-sided ideals). For each ideal A of R let $A' = \{x \in R : xA = 0\} = \{x \in R : Ax = 0\}$ be its annihilator, and let $\mathrm{Ann}(R)$ be the set of annihilator ideals of R. Show that $(\mathrm{Ann}(R), \subseteq, ')$ is a complete Boolean algebra (see Section 1.3).
8. A *Boolean ring* is a ring B such that $x^2 = x$ for each x in B.

   (a) Show that each Boolean algebra B becomes a unital Boolean ring when it is given the operations: $ab = a \wedge b$ and $a + b = (a' \wedge b) \vee (a \wedge b')$.
   (b) Show that each unital Boolean ring becomes a Boolean algebra when it is given the operations: $a \wedge b = ab$, $a' = 1 - a$ and $a \vee b = a + b - ab$.
   (c) Let B^- denote the Boolean algebra (respectively, unital Boolean ring) associated with the unital Boolean ring (respectively, Boolean algebra) B. Show that $B^{--} = B$.
   (d) Let B be the set of central idempotents of the unital ring R. Show that B becomes a Boolean algebra when it is given the operations defined in (b). What is addition in B^-? If $e, f \in B$ and $n \in \mathbb{N}$ when is $e + f - nef \in B$?
   (e) A *generalized Boolean algebra* is a distributive lattice L with least element 0 which is *relatively complemented*; this means that complements exist in each closed interval $[a,b] = \{x \in L : a \leq x \leq b\}$. Formulate and prove (a), (b), (c), and (d) for a generalized Boolean algebra. Show that such an algebra is infinitely distributive (Exercise 2).

9. An *ideal* of a lattice L is a nonempty subset I such that if $a,b \in I$ and $c \in L$ with $c \leq a$, then $a \vee b \in I$ and $c \in I$. I is a *prime* ideal of L if I is a proper ideal and whenever $a,b \in L$ with $a \wedge b \in I$, then $a \in I$ or $b \in I$. Let I be a subset of the generalized Boolean algebra B. Show that I is an ideal of B if and only if I is an ideal of the ring B (see Exercise 8). Moreover, the following are equivalent for the proper ideal I.

   (a) I is a prime ideal.
   (b) I is a maximal ideal.
   (c) If $x \in B$ and y is its complement in some interval $[a,b]$, then $x \in I$ or $y \in I$.
   (d) card $(B/I) = 2$.

10. (a) If I is an ideal of the lattice L and $a \in L$ then $\{x \in L : x \leq y \vee a \text{ for some } y \in I\}$ is the ideal of L generated by I and a (that is, it is the intersection of all the ideals of L which contain $I \cup \{a\}$).
(b) If L is distributive and I is an ideal maximal with respect to not containing some element $a \in L$, then I is prime (see Exercise 9).
(c) Each proper ideal in a distributive lattice is the intersecion of prime ideals.
(d) Let Spec(L) be the set of prime ideals of the distributive lattice L. Then the mapping $f : L \longrightarrow \mathscr{P}(\text{Spec(L)})$ given by $f(a) = \{I \in \text{Spec (L)} : a \notin I\}$ is a lattice monomorphism.

11. Let $P = \Pi P_i$ be the product of the family of posets $\{P_i\}_{i \in I}$, and let $a = (a_i) \in P$. Let $\pi_i : P \longrightarrow P_i$ be the projection onto P_i and let $\alpha_i : P_i \longrightarrow P$ be defined by $\pi_i \alpha_i = 1_{P_i}$ and $\pi_j \alpha_i(P_i) = a_j$ if $j \neq i$.

(a) Show that α_i and π_i are complete.
(b) Let $P_a = \{p \in P : \pi_i(p) = a_i \text{ except for a finite subset of I}\}$. Show that P_a is closed under any finite sups or infs that exist in P, and $\pi_i : P_a \longrightarrow P_i$ is complete.

12. In a lattice L the *medians* of three elements x, y, z are defined by $(x, y, z) = (x \wedge y) \vee (y \wedge z) \vee (z \wedge x)$ and $[x, y, z] = (x \vee y) \wedge (y \vee z) \wedge (z \vee x)$.

(a) Both medians are symmetric in x, y, and z.
(b) If $x \leq y \leq z$ then $(x, y, z) = [x, y, z] = y$.
(c) If $z \geq x, y$ then $(x, y, z) = [x, y, z] = x \vee y$, and dually.
(d) L is distributive if and only if it satisfies the identity $(x, y, z) = [x, y, z]$. (First show that L is modular and then compute $x \wedge (x, y, z)$.)
(e) Let $f : L \longrightarrow M$ be an isotone map between lattices. The following statements are equivalent.
 (i) f is a lattice homomorphism.
 (ii) $f((x, y, z)) = (f(x), f(y), f(z))$ $\forall x, y, z \in L$.
 (iii) $f([x, y, z]) = [f(x), f(y), f(z)]$ $\forall x, y, z \in L$.

1.3 Completion

If each subset of the poset P has both an inf and a sup in P, then P is said to be *complete*. Clearly, a complete poset is a lattice. Complete lattices are of course densely packed, at least intuitively, and arise naturally as subsets of power sets; for example, the lattice of subgroups of a group. The real numbers are essentially complete and, in fact, become complete when a largest and a smallest element are adjoined. In this short section we construct the completion of a poset and show that some posets also have a smaller completion in which, like the real numbers, every bounded subset has an inf and a sup. This type of lattice arises forcefully in Section 2.3 where archimedean lattice-ordered groups are studied. It is easy to see that in order for P to be complete it suffices that each of its subsets has an inf in P. In particular, any subset of a complete lattice which contains the inf of each of its subsets is itself a complete lattice, though not necessarily a sublattice.

A *closure operator* on a poset P is an isotone function $c : P \longrightarrow P$ such that for all $x \in P$,

(C1) $x \leq c(x)$;

(C2) $c(c(x)) = c(x)$.

An element $x \in P$ is called *closed* if it is a fixed point of $c : c(x) = x$. If X is a set of closed elements and $y = \inf X$, then y is closed. For, $y \leq c(y) \leq c(x) = x$, for each x in X, and hence $y = c(y)$. This proves

Theorem 1.3.1. *Let c be a closure operator on a complete lattice L. The subset C of all closed elements in L is itself a complete lattice, and the inf operator in C is the restriction to C of the inf operator in L.* □

This result will now be applied to a closure operator on the power set $\mathscr{P}(P)$ of the poset P which is, of course, a complete Boolean algebra. In fact, the construction to be given is just the familiar cut construction of the reals from the rationals, applied to an arbitrary poset. First note that the upper bound and lower bound functions on $\mathscr{P}(P)$ are inclusion reversing : if $X \subseteq Y \subseteq P$, then $L_P(Y) \subseteq L_P(X)$ and $U_P(Y) \subseteq U_P(X)$. A subset S of the poset T is said to be *dense* (or *order dense*) in T if, for each t in T, $t = \sup_T L_S(t)$ and $t = \inf_T U_S(t)$, where $L_S(t) = L_T(t) \cap S$ and $U_S(t) = U_T(t) \cap S$. It is easy to see that S is dense in T exactly when each element of T is the sup and the inf of some subsets of S. A *completion of the poset P* is a pair (C, φ) where C is a complete lattice and $\varphi : P \longrightarrow C$ is an embedding of P onto a dense subset of C.

Theorem 1.3.2. *(a) Each poset P has a completion.*

(b) If (C, φ) is a completion of P, then φ is a complete monomorphism.

(c) If (C, φ) and (C_1, φ_1) are two completions of P, then there is a unique isomorphism $\psi : C \longrightarrow C_1$ such that the following diagram is commutative.

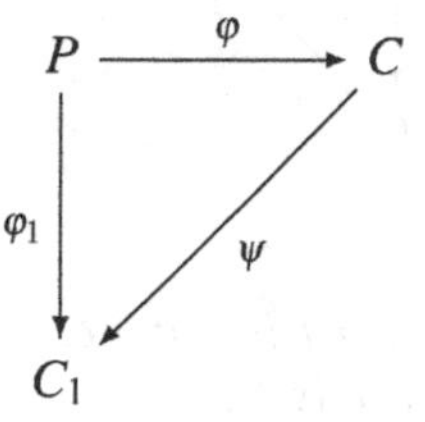

Proof. Consider the operator $LU : \mathscr{P}(P) \longrightarrow \mathscr{P}(P)$, $LU(X) = L(U(X))$. Since L and U each reverses inclusion LU is isotone, and clearly $X \subseteq LU(X)$; so LU satisfies $(C1)$. Thus, to show that LU is a closure operator it suffices to show that $LULU(X) = LU(X)$. But this will follow once the equation $ULU(X) = U(X)$ is verified. However, $X \subseteq LUX$ and $Y \subseteq ULY$ yields that $ULUX \subseteq UX$, and $UX \subseteq UL(UX)$. Let

$$M(P) = \{X \subseteq P : LU(X) = X\}.$$

According to Theorem 1.3.1 $M(P)$ is a complete lattice in which $\inf\{X_i : i \in I\} = \cap\{X_i : i \in I\}$. Define $\varphi : P \longrightarrow M(P)$ by $\varphi(p) = L(p)$. Since $LUL = L$, $\varphi(p) \in M(P)$, and since $p \leq q \Leftrightarrow L(p) \subseteq L(q)$, φ is an embedding. That $\varphi(P)$ is dense in $M(P)$ follows from the fact that if $LU(X) = X$, then

$$\cup\{L(x) : x \in X\} = X = \cap\{L(p) : p \in U(X)\}. \tag{1.3.1}$$

The first equality arises from: $a \leq x \in X \Rightarrow a \in LU(a) \subseteq LU(x) \subseteq LU(X) = X$; and the second from: $a \leq p$ for each $p \in U(X) \Rightarrow a \in LU(X) = X$. This completes the proof of the first statement. For the second statement it suffices to show that if P is a dense subset of the poset C and if $X \subseteq P$ has the inf (respectively, sup) a in P, then a is the inf (respectively, sup) of X in C. We will only do the inf case, since the sup case will follow dually. Suppose, then, that $c \in C$ and $c \leq X$. Then $L_P(c) \subseteq L_P(X)$. So if $p \in L_P(c)$, then $p \leq a$; that is, $L_P(c) \leq a$. But $c = \sup_C L_P(c)$, so $c \leq a$.

The last statement is a consequence of the next theorem. □

The completion $M(P)$ constructed above, together with the canonical embedding $\varphi = \varphi_P : P \longrightarrow M(P)$, is called the *MacNeille completion* of P. The poset P is usually identified with its canonical image in $M(P)$.

Theorem 1.3.3. *Let $\varphi : P \longrightarrow M(P)$ be the MacNeille completion of the poset P, and let $f : P \longrightarrow C$ be a function into the complete lattice C.*

(a) There is an isotone function $f_* : M(P) \longrightarrow C$ such that the diagram below

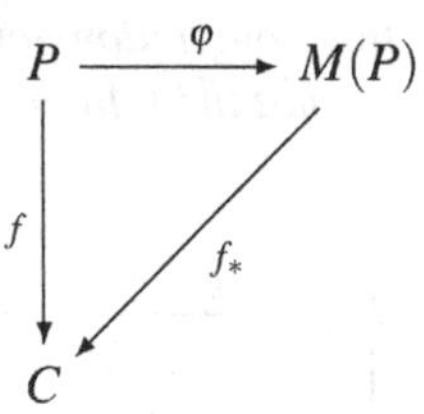

is commutative exactly when f is isotone.

(b) If f is an embedding then so is f_*.

(c) If f is an embedding onto a dense subset of C, then f_* is an isomorphism and it is the unique embedding that makes the diagram commutative.

Proof. (a) Let $X \in M(P)$ and define $f_*(X) = \sup_C f(X)$. Then for $X \subseteq Y \in M(P)$ we have $f_*(X) \leq f_*(Y)$, and hence f_* is isotone. If f is isotone and $p \in P$, then $f(p)$ is the largest element in $f(L(p))$; consequently, $f_*(\varphi(p)) = \sup_C f(L(p)) = f(p)$.

(b) Suppose that $X, Y \in M(P)$ with $f_*(X) \leq f_*(Y)$. We claim that $U_P(Y) \subseteq U_P(X)$, and hence that $X = LU(X) \subseteq LU(Y) = Y$. For if $q \in X$ and $p \in U_P(Y)$, then, since $L(q) \subseteq X$ (by (1.3.1)), we have $f(q) = f_*(L(q)) \leq f_*(X) \leq f_*(Y) \leq f(p)$. Thus, $q \leq p$ since f is an embedding.

(c) Suppose now that f is an embedding and $f(P)$ is dense in C. It remains to show that f_* is onto. Let $c \in C$, and let $X = \{p \in P : f(p) \leq c\}$ and $Y = \{p \in P : c \leq f(p)\}$. Then $c = \sup_C f(X) = \inf_C f(Y)$. But $X = LU(X) \in M(P)$; for if $p \in LU(X)$, then $p \leq Y$ since $Y \subseteq U(X)$. So $f(p) \leq f(Y)$, $f(p) \leq c$ and hence $p \in X$. Thus, $f_*(X) = c$ and f_* is an isomorphism. If $g : M(P) \longrightarrow C$ is an embedding with $g \circ \varphi = f$, then for $X \in M(P)$ we have

$$g(X) = g\left(\bigcup_{p \in X} L(p)\right) = \bigvee_{p \in X} g(L(p)) = \sup_C f(X) = f_*(X)$$

since $g(M(P))$ is dense in C and hence g preserves sups by Theorem 1.3.2. □

For some purposes a completion of P that is slightly smaller than $M(P)$ is more useful than $M(P)$ (for example, when P is a group). A subset X of P is said to be *bounded above* (respectively, *below*) if it has an upper (respectively, lower) bound, and it is called *bounded* if it is bounded both above and below. The poset P is *conditionally complete* if each nonempty bounded subset of P has an inf and a sup in P; P is *directed* (respectively, *directed up, directed down*) if each two element subset of P is bounded (respectively, bounded above, bounded below).

Theorem 1.3.4. *Let P be a directed poset, and let $D(P) = P \cup (M(P) \setminus \{0,1\})$.*

(a) $D(P)$ is a conditionally complete lattice.
(b) P is dense in $D(P)$.
(c) If P has neither a first nor a last element, then Theorem 1.3.3 holds if $M(P)$ is replaced by $D(P)$, provided C is a conditionally complete lattice with no first or last element.

Proof. If $a, b \in D(P)$, then the closed interval $[a,b] = \{x \in M(P) : a \leq x \leq b\}$ is contained in $D(P)$. So if X is a nonempty bounded subset of $D(P)$, say $a \leq X \leq b$, then $\inf_{M(P)} X, \sup_{M(P)} X \in [a,b] \subseteq D(P)$, and hence $D(P)$ is conditionally complete. Now, if $a \in D(P)$, then there exists $p \in P$ with $p \leq a$. For, $a = \sup_{M(P)} L_P(a)$, and if $L_P(a) = \phi$, then $0 = a \in D(P)$; consequently, $0 \in L_P(a)$ which is impossible. Similarly, $a \leq q$ for some $q \in P$. So if $a, b \in D(P)$ then we can find $p, p_1, p_2, q, q_1, q_2 \in P$ with $p_1 \leq a \leq q_1$, $p_2 \leq b \leq q_2$, $p \leq p_1, p_2$ and $q_1, q_2 \leq q$. So $\{a,b\}$ is bounded, and hence $D(P)$ is a conditionally complete sublattice of $M(P)$. If S is a dense subset of the poset T and $S \subseteq W \subseteq T$, then S is dense in W. So P is dense in $D(P)$. Suppose that neither P nor C has a 0 or 1, C is a conditionally complete lattice and $f : P \longrightarrow C$ is isotone. Since it is easily verified that $M(C) = C \cup \{0,1\}$ we have the following diagram

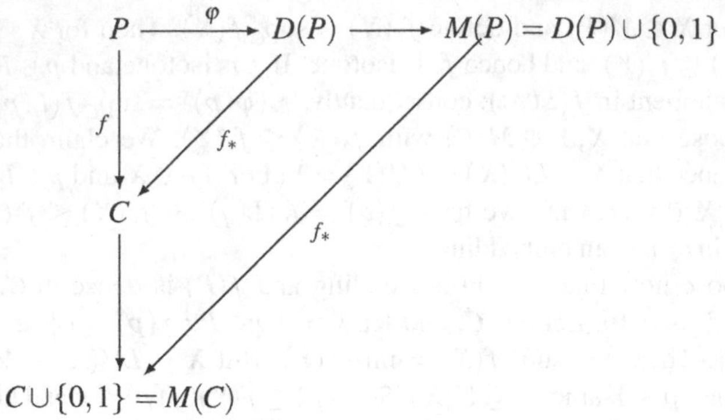

where f_* is defined by $f_*(X) = \sup_{M(C)} f(X)$ for $X \in M(P)$ just as in the previous theorem. If $X \in D(P)$, then $U(X) \neq \phi$; so $X \leq p$ for some $p \in P$. But then $0 < f(X) \leq f(p)$ and $f_*(X) \in C$. Then the restriction of f_* to $D(P)$ has its image in C, is an embedding if f is an embedding, and is an isomorphism if also $f(P)$ is dense in C, since then $f(P)$ is dense in $M(C)$. □

The completion $D(P)$ is called the *Dedekind completion* of P.

Exercises.

1. (See Exercise 1.2.5.) Let P and Q be nonempty posets.
   (a) Find an example where $P \underset{\leftarrow}{\times} Q$ is a lattice, Q is totally ordered, but P is not a lattice.
   (b) If P and Q are lattices and $0, 1 \in P$ show that $P \underset{\leftarrow}{\times} Q$ is a lattice.
   (c) If Q has maximal as well as minimal elements, show that $P \underset{\leftarrow}{\times} Q$ is a lattice if and only if P and Q are lattices, and $0, 1 \in P$ in case Q is not a chain.
   (d) If P and Q are complete lattices, show that $P \underset{\leftarrow}{\times} Q$ is a complete lattice.
   (e) Show that there is no Q for which $P \underset{\leftarrow}{\times} Q$ is a lattice for every P.
   (f) Find Q for which $P \underset{\leftarrow}{\times} Q$ is a lattice for every P that contains 0 and 1.
   (g) Show that there is a poset P with more than one element for which $P \underset{\leftarrow}{\times} Q$ is a lattice for every lattice Q.
2. Prove that the MacNeille completion of a totally ordered set is totally ordered.
3. Find the MacNeille completion of the poset P in each of the following cases: P is trivially ordered, or, P is $\mathbb{Z}$, $\mathbb{Q}$, $\mathbb{R}$ or $\mathbb{Q} \underset{\leftarrow}{\times} \mathbb{Q}$. Find the Dedekind completion in all of these cases. In the last case identify the completion inside of $C \underset{\leftarrow}{\times} C$ where C is the MacNeille completion of $\mathbb{Q}$.
4. Prove that the following statements are equivalent for the lattice L.
   (a) L is conditionally complete.
   (b) Each nonempty bounded subset of L has an inf in L.
   (c) Each nonempty subset of L that is bounded below has an inf in L.

(d) The dual of (b).
(e) The dual of (c).

5. (a) Prove that the MacNeille completion of a Boolean algebra B is a Boolean algebra and B is a subalgebra of $M(B)$.
(b) Show that (a) holds for a generalized Boolean algebra (Exercise 1.2.8). (A is a subalgebra of the generalized Boolean algebra B if $0 \in A$, A is a sublattice of B, and A is a generalized Boolean algebra; equivalently, A is a subring of the ring B.)
6. An *atom* in a poset P with 0 is an element $x \in P$ with $0 < x$ and such that $0 < y \leq x$ implies that $y = x$. P is called *atomic* if for each $0 < z \in P$ there is an atom x with $x \leq z$. Show that the Boolean algebra B is isomorphic to the power set of some set X if and only if B is complete and atomic. (*Hint*: Let X be the set of atoms of B.)
7. Let X be a topological space. A subset U of X is called *clopen* if it is both open and closed. Show that the set of all of the clopen subsets of X is a Boolean subalgebra of the power set of X.
8. Let X be a topological space. Denote the interior, complement, and closure of the subset A of X by $A^\circ, A', $ and A^-, respectively. A is called *regular* if $A = A^{-\circ}$. Let $A^* = A^{-\prime}$. Show that

(a) $A^\circ = A^{\prime-\prime}$;
(b) A is regular iff $A = A^{**}$;
(c) the set $\mathscr{B}$ of all of the regular open subsets of X is a complete Boolean algebra with the operations: for $A, B \in \mathscr{B}$, A^*, as defined above, is the complement of A in $\mathscr{B}$, $A \wedge B = A \cap B$ and $A \vee B = (A \cup B)^{**}$.

1.4 Universal Algebra

Our main concern in this book is with rings and groups that are also lattices. These structures are given as sets upon which certain operations are defined. Since the basic consequences of these operations are assumed to be familiar to the reader it is not particularly useful to consider these structures as special cases of a more general abstract structure. However, we will have some interest in free objects in varieties, and the existence of these free objects in general may not be so familiar. Consequently, we will develop enough of the basic ingredients of universal algebra so as to make these concepts meaningful. Specifically, we will construct free algebras of arbitrary rank in any nontrivial variety and give an internal characterization of a variety as a class of algebras closed under subalgebras, homomorphic images, and products.

An *operator domain* is a set Ω together with a function $a : \Omega \longrightarrow \mathbb{N}_0 = \mathbb{N} \cup \{0\}$; the elements of Ω are called *operators*, and if $\omega \in \Omega$ with $a(\omega) = n$, then we will say that the *arity* of ω is n or that ω is an *n-ary operator*. The set of all n-ary operators in Ω will be denoted by $\Omega(n)$. An *Ω-algebra* is a set A together with a sequence of functions

$$\Omega(n) \longrightarrow A^{(A^n)}.$$

When no confusion is likely an Ω-algebra A will be referred to as just an algebra. Each $\omega \in \Omega(n)$ determines an n-ary operation on A, and the value of this operation at $(a_1,\dots,a_n) \in A^n$ will be denoted by $\omega(a_1,\dots,a_n)$. If ω is a 0-ary operator, then by definition $A^0 = A^\emptyset = \{\emptyset\}$, and ω determines some distinguished element or *constant* ω_A of A. For this reason a 0-ary operator is frequently called a *constant operator*. For example, the identity element in a group is the result of a constant operator.

A subset B of the algebra A is a *subalgebra* of A if $\omega(b_1,\dots,b_n) \in B$ for each $n \in \mathbb{N}_0$, for each $\omega \in \Omega(n)$, and for any $b_1,\dots,b_n \in B$. Each subalgebra B contains the set of constants $\{\omega_A : \omega \in \Omega(0)\}$. An algebra is said to be *trivial* if it has just one element. If X is a subset of A there is a smallest subalgebra of A that contains X, namely the intersection of all of those subalgebras that contain X. This subalgebra, which we denote by $S_\Omega(X)$, is called the *subalgebra generated by* X. A description of the elements of $S_\Omega(X)$ is given inductively as follows. Let

$$\begin{cases} X_0 &= X \cup \{\omega_A : \omega \in \Omega(0)\} \text{ and} \\ X_{k+1} &= X_k \cup \{a \in A : a = \omega(y_1,\dots,y_n) \\ & \qquad \text{where } n \in \mathbb{N}, \omega \in \Omega \text{ and } (y_1,\dots,y_n) \in X_k^n\}. \end{cases} \tag{1.4.1}$$

Then (see Exercise 1)

$$S_\Omega(X) = \bigcup_{k=0}^{\infty} X_k. \tag{1.4.2}$$

If A and B are Ω-algebras the function $f : A \longrightarrow B$ is called an *Ω-homomorphism* (or just a *homomorphism* or *morphism*) if $f(\omega(a_1,\dots,a_n)) = \omega(f(a_1),\dots,f(a_n))$ for each $\omega \in \Omega$ and each n-tuple $(a_1,\dots,a_n) \in A^n$, where $n \in \mathbb{N}_0$. The familiar terminology associated with homomorphisms will be used. So, an injective (respectively, surjective) morphism is a *monomorphism* (respectively, an *epimorphism*), and a bijective morphism is an *isomorphism*. An *endomorphism* is a morphism from an algebra to itself, and an endomorphism that is an isomorphism is an *automorphism*. If there is an epimorphism from A to B, then B is called a *homomorphic image* of A. The class of all Ω-algebras $\mathscr{A}(\Omega)$ together with all homomorphisms between Ω-algebras is a category which will also be denoted by $\mathscr{A}(\Omega)$. Likewise, we will not distinguish between a subclass $\mathscr{C}$ of $\mathscr{A}(\Omega)$ and the subcategory of $\mathscr{A}(\Omega)$ that it determines.

The cartesian product ΠA_i of a family of Ω-algebras $\{A_i\}_{i \in I}$ becomes an Ω-algebra in the following natural way. If $a_1,\dots,a_n \in \Pi A_i$ and $\omega \in \Omega(n)$, then $\omega(a_1,\dots,a_n)$ is defined by $\pi_i(\omega(a_1,\dots,a_n)) = \omega(\pi_i(a_1),\dots,\pi_i(a_n))$, for each $i \in I$, where $\pi_i : \Pi A_i \longrightarrow A_i$ is the ith projection. This Ω-algebra is called the *direct product* of the family $\{A_i\}_{i \in I}$. Of course, this is the only Ω-algebra structure on the cartesian product for which each projection is an Ω-homomorphism. The direct product ΠA_i is a categorical product in the category of Ω-algebras (see Exercise 3).

An equivalence relation $\sim$ on a set A induces an equivalence relation $\sim_n$ on A^n which is defined by $(a_1,\ldots,a_n) \sim_n (b_1,\ldots,b_n)$ if $a_i \sim b_i$ for each $i = 1,\ldots,n$. The equivalence relation $\sim$ on the Ω-algebra A is called a *congruence* on A if for each $n \in \mathbb{N}$ and each $\omega \in \Omega(n)$, $\omega(a_1,\ldots,a_n) \sim \omega(b_1,\ldots,b_n)$ whenever $(a_1,\ldots,a_n) \sim_n (b_1,\ldots,b_n)$. Since the intersection of a set of congruences on A is a congruence, the collection of congruences on A is a complete lattice with least element the *diagonal* $\Delta = \{(a,a) : a \in A\}$ and greatest element A^2. The diagonal is called the *trivial congruence* on A and any congruence other than A^2 is called *proper*. If $\sim$ is an equivalence relation on A and $a \in A$, then $\bar{a}$, or $[a]$, will denote the equivalence class of A determined by a. The set of all of the equivalence classes will be denoted by $A/\sim$ and is called the *quotient* of A determined by $\sim$, and the function $\eta : A \longrightarrow A/\sim$ which is given by $\eta(a) = \bar{a}$ is the *natural mapping* associated with $\sim$. If $f : A \longrightarrow B$ is any function, then the *kernel* of f (abbreviated $\ker f$) is the equivalence relation on A determined by the partition $\{f^{-1}(b) : b \in B\}$ of A. If $\sim$ is a congruence relation on an algebra A, then the quotient $A/\sim$ becomes an Ω-algebra in a natural way and is called the *quotient algebra* of A determined by $\sim$. The Ω-algebra structure of $A/\sim$ is given in the following theorem whose proof is left to the reader.

Theorem 1.4.1. *Let $\sim$ be a congruence relation on the Ω-algebra A. For $\omega \in \Omega(n)$ and $a_1,\ldots,a_n \in A$ define $\omega(\bar{a}_1,\ldots,\bar{a}_n)$ by $\omega(\bar{a}_1,\ldots,\bar{a}_n) = [\omega(a_1,\ldots,a_n)]$.*

(a) The quotient $A/\sim$ is an Ω-algebra, and $\eta : A \longrightarrow A/\sim$ is a morphism of Ω-algebras.
(b) If $f : A \longrightarrow B$ is a morphism of Ω-algebras, then $\ker f$ is a congruence on A, $f(A)$ is a subalgebra of B, and $f = j\eta$ where $j : A/\ker f \longrightarrow B$ is the morphism $j(\bar{a}) = f(a)$; j induces an isomorphism between the quotient algebra $A/\ker f$ and the image of f.
(c) If $f : A \longrightarrow B$ is a morphism, then there is a (unique) morphism $g : A/\sim \longrightarrow B$ with $g\eta = f$ if and only if $\sim \subseteq \ker f$. □

If $\mathscr{C}$ is a class of Ω-algebras and F is an algebra in $\mathscr{C}$, then F is called a *free $\mathscr{C}$-algebra on the set* X if there is an injective function $i : X \longrightarrow F$ (which will usually be an inclusion function) such that the image of X generates F and each function $f : X \longrightarrow A$ into an Ω-algebra A in $\mathscr{C}$ can be extended to a morphism $\bar{f} : F \longrightarrow A$ in the sense that $f = \bar{f}i$. Since $i(X)$ generates F there is at most one such extension (Exercise 4). In order to prove the existence of free algebras, with some degree of generality, we will first construct free algebras in the class $\mathscr{A}(\Omega)$ of all Ω -algebras.

Let X be a set which we will assume to be disjoint from the operator domain Ω; otherwise, take the disjoint union of X and Ω. The *Ω-row algebra on X* is defined to be the set

$$W(\Omega,X) = \bigcup_{m=1}^{\infty} (\Omega \cup X)^m$$

together with the Ω-algebra structure defined by juxtaposition: if $a_1,\ldots,a_n \in W(\Omega,X)$ and $\omega \in \Omega(n)$, then

$$\omega(a_1,\ldots,a_n) = (\omega,a_1,\ldots,a_n).$$

More explicitly, if $a_i = (a_{i1},\ldots,a_{im_i})$ for $1 \le i \le n$, where $a_{ij} \in \Omega \cup X$, then

$$\omega(a_1,\ldots,a_n) = (\omega,a_{11},\ldots,a_{1m_1},a_{21},\ldots,a_{nm_n}).$$

An element of the algebra $W(\Omega,X)$ will be called an *Ω-row in X*. The subalgebra of $W(\Omega,X)$ generated by X, which is denoted by $F_\Omega(X)$ or $F(X)$, is called the *Ω-word algebra on X*, and its elements will be called *Ω-words in the alphabet X*. According to (1.4.2) the elements of $F_\Omega(X)$ are either in $X_0 = X \cup \Omega(0)$ or are in X_k for some $k \ge 1$. The elements in $X_1 \setminus X_0$ are of the form $\omega(a_1,\ldots,a_n) = (\omega,a_1,\ldots,a_n)$ where $n \in \mathbb{N}$, $\omega \in \Omega(n)$ and each a_i is either an element of X or is a constant operator.

As expected, the Ω-word algebra $F_\Omega(X)$ is a free algebra in $\mathscr{A}(\Omega)$. In order to verify this and to also show the uniqueness of the spelling of an Ω-word we will introduce two integer-valued functions on $W(\Omega,X)$. These functions will make the required bookkeeping easier.

An Ω-row $w = (y_1,\ldots,y_m) \in (\Omega \cup X)^m$ is said to have *length* $m : \ell(w) = m$; and its *valence* $v(w)$ is defined by

$$v(w) = \sum_{i=1}^{m} v(y_i),$$

where $v(y) = 1$ if $y \in X$ and $v(y) = 1-n$ if $y \in \Omega(n)$. A *segment* of $w = (y_1,\ldots,y_m)$ is an Ω-row of the form $u = (y_i,y_{i+1},\ldots,y_j)$ where $1 \le i \le j \le m$. The segment u is an *initial segment* of w if $i = 1$ and is a *tail* of w if $j = m$. If $w = (u_1,\ldots,u_k)$, where each u_ℓ is a segment of w, then the k Ω-rows $u_1,\ldots,u_k$ form a *partition* of w.

Theorem 1.4.2. *Let $w \in W(\Omega,X)$.*

(a) The necessary and sufficient conditions for w to have a partition into r Ω-words, with $r \ge 1$, is that $v(w) = r$ and each tail of w has positive valence.
(b) $w \in F_\Omega(X)$ if and only if $v(w) = 1$ and each tail of w has positive valence.
(c) Suppose that $w \in F_\Omega(X)$. Then $\ell(w) = 1$ if and only if $w \in X \cup \Omega(0)$; if $\ell(w) \ge 2$, then w has a unique partition of the form $w = (\omega,w_1,\ldots,w_n)$ where $n \ge 1$, $\omega \in \Omega(n)$ and each $w_i \in F_\Omega(X)$.

Proof. If $\ell(w) = 1$, then, according to (1.4.1) and (1.4.2), $w \in F_\Omega(X)$ if and only if $w \in X \cup \Omega(0)$. Consequently, the theorem holds for words of length 1, and, except for the hidden beginning of an inductive argument, we will assume that $\ell(w) \ge 2$. Note also that (b) is a special case of (a).

(a) We use induction on $\ell(w)$. Suppose that $w = (w_1,\ldots,w_r)$ where each w_i is an Ω-word. By (1.4.1) and (1.4.2) $w_i \in X \cup \Omega(0)$ and $v(w_i) = 1$, or $w_i = (\omega,a_1,\ldots,a_n)$ where $n \ge 1$, $\omega \in \Omega(n)$ and $a_j \in F_\Omega(X)$. Then $v(a_j) = 1$ and each tail of a_j has positive valence for $1 \le j \le n$ since $\ell(a_j) < \ell(w)$, and hence $v(w_i) = 1-n+n = 1$ and $v(w) = r$. Also, if u is a proper tail of w, then $u = (t,w_{i+1},\ldots,w_r)$ where either t is a proper tail of w_i, or $t = w_i$ and $i \ge 2$. In the second case $v(u) = r-i+1 \ge 1$. In the first case $t = (s, a_{j+1}, \ldots, a_n)$ where s is a tail of a_j and $v(t) = v(s) + n - j > 0$, and hence $v(u) = v(t) + v(w_{i+1}) + \cdots + v(w_r) > 0$. Conversely, suppose that w is an Ω-row with $v(w) = r \geq 1$ and each tail of w has positive valence. Then $w = (y, w_1)$, where $y \in X \cup \Omega$ and $v(w_1) = s \geq 1$. Since $\ell(w_1) < \ell(w)$, $w_1 = (u_1, \ldots, u_s)$ where $u_j \in F_\Omega(X)$. If $y \in X \cup \Omega(0)$, then $w = (y, u_1, \ldots, u_s)$ is a partition of w into $1+s$ Ω-words and $1 + s = v(y) + v(w_1) = r$. If $y \in \Omega(n)$ with $n \geq 1$, then $r = 1 - n + s$ gives that $n = s - (r-1) \leq s$. So $w = (y, u_1, \ldots, u_n, u_{n+1}, \ldots, u_s) = (a, u_{n+1}, \ldots, u_s)$ is a partition of w into $s - n + 1 = r$ words where $a = (y, u_1, \ldots, u_n) \in F_\Omega(X)$.

(c) By (1.4.2) $w = (\omega, w_1, \ldots, w_n)$ where $\omega \in \Omega(n)$, $n \geq 1$ and $w_i \in F_\Omega(X)$. To prove the uniqueness of this partition write $w = (\omega, y_1, \ldots, y_m)$ where $y_j \in X \cup \Omega$. We will first show that if $w' = (y_i, \ldots, y_j)$ is a proper segment of w that is itself a word, and y_i occurs in w_p, then w' is a segment of w_p. Otherwise, $w' = (z, w_{p+1}, \ldots, w_{\ell-1}, u)$ where z is a tail of w_p, u is an initial segment of w_ℓ, and $\ell \geq p + 1 \geq 2$. From (b) we have that both u and z have strictly positive valence since each is the tail of an Ω-word. This gives the contradiction

$$1 = v(w') = v(z) + \ell - 1 - p + v(u) \geq 2.$$

Now if $w = (\omega, w_1, \ldots, w_n) = (\bar{\omega}, \bar{w}_1, \ldots, \bar{w}_m)$ with $\bar{w}_i \in F_\Omega(X)$, then clearly $\omega = \bar{\omega}$ and $n = m$, and by the previous discussion, first, $w_1 = \bar{w}_1$, and then $w_2 = \bar{w}_2, \ldots, w_n = \bar{w}_n$. □

We can now prove

Theorem 1.4.3. *For each set X the Ω-word algebra $F_\Omega(X)$ is a free $\mathscr{A}(\Omega)$-algebra on X.*

Proof. Let $f : X \longrightarrow A$ be a function into the Ω-algebra A. Since, by (1.4.2),

$$F_\Omega(X) = \bigcup_{k=0}^{\infty} X_k$$

we can construct the morphism $\bar{f} : F_\Omega(X) \longrightarrow A$ which extends f by defining it inductively on each X_k, as follows. If $\omega \in \Omega(0)$ let $\bar{f}(\omega) = \omega_A$. Now let w be an Ω-word with $\ell(w) \geq 2$, and let $k \geq 1$ be minimal with $w \in X_k$. Then by (c) of Theorem 1.4.2 w can be written uniquely as $w = \omega(w_1, \ldots, w_n)$ where $n \geq 1$, $\omega \in \Omega(n)$, and $w_i \in X_{k-1}$. Assuming that $\bar{f}(w_i)$ is defined for $1 \leq i \leq n$, let $\bar{f}(w) = \omega(\bar{f}(w_1), \ldots, \bar{f}(w_n))$. This defines $\bar{f}$ on $F_\Omega(X)$, and clearly $\bar{f}$ is a morphism of Ω-algebras. □

Suppose that $w = (y_1, \ldots, y_m)$ is an Ω-word where $y_i \in \Omega \cup X$. If $\{y_1, \ldots, y_m\} \cap X \subseteq \{x_1, \ldots, x_n\}$ we will write $w = w(x_1, \ldots, x_n)$ to indicate that $x_1, \ldots, x_n$ are the only possible elements of X that occur in w. Now suppose that $a_1, \ldots, a_n$ are elements of an Ω-algebra A and $f : X \longrightarrow A$ is a function with $f(x_i) = a_i$ for $1 \leq i \leq n$. We will denote by $w(a_1, \ldots, a_n)$ the image of $w(x_1, \ldots, x_n)$ under the unique morphism $F_\Omega(X) \longrightarrow A$ induced by f. Thus, $w(a_1, \ldots, a_n)$ is the element of A obtained from $w(x_1, \ldots, x_n)$ by substituting a_i for x_i.

The algebras that we are interested in (rings, groups, and lattices) are all algebras that satisfy certain identities. We will soon make this statement more precise, but we first introduce a related useful concept. A *Galois connection* between two partially ordered classes P and Q is a pair of antitone mappings ${}^* : P \longrightarrow Q$ and ${}^* : Q \longrightarrow P$ such that $x \leq x^{**}$ for each x in $P \cup Q$. The mappings $P \longrightarrow P$ and $Q \longrightarrow Q$, given by $x \mapsto x^{**}$, which are induced by the Galois connection, are closure operators. For, $x^* \leq (x^*)^{**}$, and $x \leq x^{**}$ implies that $x^{***} \leq x^*$; so $x^* = x^{***}$ and hence $x^{**} = x^{****}$ for each x in $P \cup Q$. The elements of P or of Q that are closed are those elements that are in the images of *, and the mappings ${}^* : P^* \longrightarrow Q^*$ and ${}^* : Q^* \longrightarrow P^*$ are inverse antitone bijections between the classes of closed elements in Q and P. We will now examine a particular Galois connection. An element (w_1, w_2) of the cartesian product $F_\Omega(X)^2$ is called an *identity* (or a *law*) in the alphabet X. The identity (w_1, w_2) will frequently be written as $w_1 = w_2$. The algebra A is said to *satisfy the identity* $w_1 = w_2$ if $\varphi(w_1) = \varphi(w_2)$ for each morphism $\varphi : F_\Omega(X) \longrightarrow A$. If $w_i = w_i(x_1, \ldots, x_n)$ for $i = 1, 2$ this just means that $w_1(a_1, \ldots, a_n) = w_2(a_1, \ldots, a_n)$ for all $a_1, \ldots, a_n \in A$. For example, a semigroup is abelian if it satisfies the identity $x_1x_2 = x_2x_1$. Now if S is any set of identities in X let $S^* = \mathscr{V}_\Omega(S, X)$ be the class of all Ω-algebras which satisfy each identity in S; $\mathscr{V}_\Omega(S, X)$ is called the *variety* determined by S. If $\mathscr{C}$ is any class of Ω-algebras let $\mathscr{C}^*$ denote the set of all identities in $F_\Omega(X)^2$ which are satisfied by each algebra in $\mathscr{C}$. It is easy to see that these two mappings form a Galois connection between the power set P of $F_\Omega(X)^2$ and the power class Q of $\mathscr{A}(\Omega)$.

Let us note that the varieties $\mathscr{V}_\Omega(S, Y)$ that are obtained using various alphabets Y can all be obtained from a fixed infinite alphabet (Exercise 10). One important consequence of this observation and the previous remarks about Galois connections is that the class of varieties in $\mathscr{A}(\Omega)$ is a set. Another consequence is the following. If $\mathscr{C}$ is any class of Ω-algebras the variety generated by $\mathscr{C}$, $\mathscr{V}(\mathscr{C})$, is, by definition, the intersection of all those varieties of Ω-algebras that contain $\mathscr{C}$. If X is an infinite set and * is the Galois connection relative to the alphabet X, then it is easily seen that $\mathscr{V}(\mathscr{C}) = \mathscr{C}^{**}$.

A class $\mathscr{C}$ of Ω-algebras is called *hereditary* if each subalgebra of an algebra A in $\mathscr{C}$ is also in $\mathscr{C}$; it is called *homomorphically closed* if for each epimorphism $A \longrightarrow B$, if $A \in \mathscr{C}$, then $B \in \mathscr{C}$ (*isomorphically closed* is defined analogously). The class $\mathscr{C}$ is *productive* if the direct product ΠA_i is in $\mathscr{C}$ for each subset $\{A_i\}_{i \in I}$ of $\mathscr{C}$. Each variety of Ω-algebras has these three properties (Exercise 13) and we will soon see that these properties actually characterize varieties. But first we will establish the existence of free algebras in varieties.

An Ω-algebra A is said to be a *subdirect product* of the family of Ω-algebras $\{A_i\}_{i \in I}$ if there is an injective morphism $f : A \longrightarrow \Pi A_i$ such that, for each $i \in I$, $\pi_i f : A \longrightarrow A_i$ is an epimorphism. A congruence $\sim$ on an Ω-algebra A is called *fully invariant* if $a \sim b$ implies that $\varphi(a) \sim \varphi(b)$ for each endomorphism φ of A.

Theorem 1.4.4. *Let $\mathscr{C}$ be a class of Ω-algebras, and let $S = \mathscr{C}^* \subseteq F_\Omega(X)^2$ be the set of those identities in the alphabet X which are satisfied by all of the algebras in the class $\mathscr{C}$.*

(a) S is a fully invariant congruence on $F_\Omega(X)$.

(b) Let $i : X \longrightarrow F_\Omega(X)/S$ be the restriction to X of the natural map $F_\Omega(X) \longrightarrow F_\Omega(X)/S$, and denote $F_\Omega(X)/S$ by $F_\Omega(\mathscr{C},X)$. If $A \in \mathscr{C}$ and if $f : X \longrightarrow A$ is any function, then there is a unique morphism $g : F_\Omega(\mathscr{C},X) \longrightarrow A$ such that the following diagram is commutative:

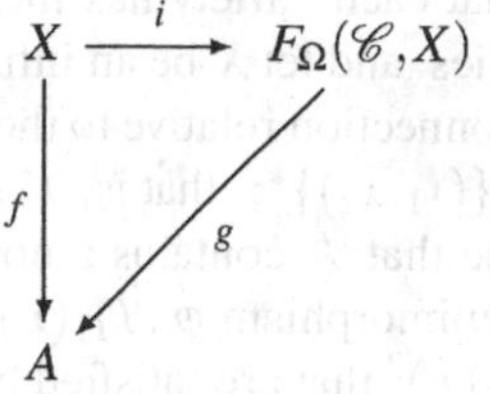

In particular, each algebra A in $\mathscr{C}$ is a homomorphic image of $F_\Omega(\mathscr{C},X)$ for some X.

(c) If $\mathscr{C}$ is hereditary and is closed under the formation of subdirect products, then $F_\Omega(\mathscr{C},X) \in \mathscr{C}$, and if, also, some nontrivial algebra is in $\mathscr{C}$, then $F_\Omega(\mathscr{C},X)$ is a free $\mathscr{C}$-algebra on X.

Proof. (a) Let $\{\sim_j\}_{j\in J}$ be the set of those congruences on $F_\Omega(X)$ such that $F_\Omega(X)/\sim_j$ is isomorphic to a subalgebra of an algebra in $\mathscr{C}$. Then

$$S = \bigcap_{j\in J} \sim_j . \tag{1.4.3}$$

For if $(w_1, w_2) \in S$ and if $\eta_j : F_\Omega(X) \longrightarrow F_\Omega(X)/\sim_j$ is the natural map, then $\eta_j(w_1) = \eta_j(w_2)$; that is, $w_1 \sim_j w_2$. On the other hand, if $w_1 \sim_j w_2$ for each $j \in J$ and if $\varphi : F_\Omega(X) \longrightarrow A$ is a morphism into an algebra A in $\mathscr{C}$, then $\ker \varphi = \sim_j$ for some $j \in J$ (Theorem 1.4.1); so $\varphi(w_1) = \varphi(w_2)$ and $(w_1, w_2) \in S$. Thus, S is a congruence on $F_\Omega(X)$. Furthermore, if $(w_1, w_2) \in S$ and if ψ is any endomorphism of $F_\Omega(X)$, then for any morphism $\varphi : F_\Omega(X) \longrightarrow A$ into an algebra A in $\mathscr{C}$ we have $\varphi\psi(w_1) = \varphi\psi(w_2)$. Hence $(\psi(w_1), \psi(w_2)) \in S$ and thus S is fully invariant.

(b) Let $\bar{f} : F_\Omega(X) \longrightarrow A$ be the morphism that extends the function $f : X \longrightarrow A$. Then $\bar{f}(w_1) = \bar{f}(w_2)$ for each $(w_1, w_2) \in S$. So $\bar{f}$ induces a morphism $g : F_\Omega(X)/S \longrightarrow A$ (Theorem 1.4.1), and $gi(x) = \bar{f}(x) = f(x)$. The uniqueness of g follows from the fact that X generates $F_\Omega(X)$. The last statement follows by taking $f : X \longrightarrow A$ to be the inclusion map where X is a generating set for A.

(c) By (1.4.3) and Exercise 11, $F_\Omega(X)/S$ is a subdirect product of algebras in $\mathscr{C}$ and hence is in $\mathscr{C}$. Also, if $A \in \mathscr{C}$ is nontrivial and if $x \neq y$ in X, then $x = y$ is not an identity in A; so $i(x) \neq i(y)$ and i is injective. □

The converse of part (a) of Theorem 1.4.4 is also true; that is, a fully invariant congruence of $F_\Omega(X)$ is closed in the Galois correspondence (Exercise 15). Note also that if $\mathscr{C}$ is hereditary, then it is closed under subdirect products if and only if it is productive and isomorphically closed.

We now have the promised internal characterization of varieties.

Theorem 1.4.5. *A class $\mathscr{V}$ of Ω-algebras is a variety if and only if it is hereditary, homomorphically closed and productive.*

Proof. We have already noted that each variety has these properties. Suppose that $\mathscr{V}$ has these three closure properties, and let X be an infinite set. We claim that $\mathscr{V} = \mathscr{V}^{**}$ where * denotes the Galois connection relative to the alphabet X. If each algebra in $\mathscr{V}$ is trivial, then, in fact, $\mathscr{V} = \{(x_1,x_2)\}^*$; that is, $\mathscr{V}$ is the variety determined by the identity $x_1 = x_2$. Now, assume that $\mathscr{V}$ contains a nontrivial algebra and let $A \in \mathscr{V}^{**}$. We can find a set Y and an epimorphism $\varphi : F_\Omega(Y) \longrightarrow A$. Let $S = S(\mathscr{V},Y)$ be the set of all those identities in $F_\Omega(Y)^2$ that are satisfied by each algebra in $\mathscr{V}$. By the previous theorem $F_\Omega(\mathscr{V},Y) = F_\Omega(Y)/S$ is in $\mathscr{V}$. If $(w_1(y_1,\dots,y_n), w_2(y_1,\dots,y_n)) \in S$, then, for all $a_1,\dots,a_n \in A$, $w_1(a_1,\dots,a_n) = w_2(a_1,\dots,a_n)$ since $(w_1(x_1,\dots,x_n), w_2(x_2,\dots x_n)) \in \mathscr{V}^*$. So φ induces an epimorphism $\psi : F_\Omega(\mathscr{V},Y) \longrightarrow A$, and hence A is in $\mathscr{V}$. □

Exercises.

1. Verify equation (1.4.2).
2. (a) Show that the function $f : A \longrightarrow B$ between Ω-algebras is a homomorphism if and only if f is a subalgebra of the direct product $A \times B$.
   (b) Show that an equivalence relation $\sim$ on the Ω-algebra A is a congruence if and only if $\sim$ is a subalgebra of A^2.
3. Show that the direct product of a family of Ω-algebras is a product in the category $\mathscr{A}(\Omega)$.
4. Show that two homomorphisms $f,g : A \longrightarrow B$ between the Ω-algebras A and B are identical if and only if they agree on a generating set for A.
5. Let $\sim$ be a congruence on A. Show that there is a lattice isomorphism between the lattice of congruences on $A/\sim$ and the sublattice of the congruence lattice of A that consists of those congruences on A which contain $\sim$.
6. Let $\sim$ be a congruence relation on A and let A_1 be a subalgebra of A.
   (a) $\sim_1 = \sim \cap A_1^2$ is a congruence on A_1.
   (b) $A_2 = \{b \in A : \exists a \in A_1 \text{ with } \bar{a} = \bar{b}\}$ is a subalgebra of A.
   (c) $A_1/\sim_1$ is isomorphic to $A_2/\sim_2$.
7. If $\sim \subseteq \approx$ are congruences on A, then the function $f : A/\sim \longrightarrow A/\approx$ given by $\bar{a} \mapsto \bar{a}$ is an epimorphism. If ker f is denoted by $\approx/\sim$, then $(A/\sim)/(\approx/\sim)$ is isomorphic to $A/\approx$.
8. Let G be a group. If $\sim$ is a congruence on G, then $\bar{1}$ is a normal subgroup, and, conversely, each normal subgroup determines a congruence on G.

9. (a) Let $\sim$ be the relation defined on the Ω-algebra A by $a \sim b$ if $a = b$ or $a = \omega(a_1,\ldots,a_n)$ and $b = \omega_1(b_1,\ldots,b_m)$ for some ω, $\omega_1 \in \Omega$ and some $a_i, b_j \in A$. Show that $\sim$ is a congruence on A.
   (b) Prove that $F_\Omega(X)$ is isomorphic to $F_\Omega(Y)$ if and only if card $(X) =$ card (Y).
10. Let $\mathscr{V} = \mathscr{V}(S,Y)$ be the variety determined by the set of words $S \subseteq F_\Omega(Y)^2$ in the alphabet Y. If X is infinite show that there is a subset $T \subseteq F_\Omega(X)^2$ such that $\mathscr{V} = \mathscr{V}(T,X)$.
11. (a) If $\sim$ is the intersection of a family of congruences $\{\sim_i\}_{i\in I}$ of the Ω-algebra A, then $A/\sim$ is a subdirect product of the set of Ω-algebras $\{A/\sim_i\}_{i\in I}$. Conversely, if A is a subdirect product of $\{A_i\}_{i\in I}$, then there is a family of congruences $\{\sim_i\}_{i\in I}$ on A such that each A_i is isomorphic to $A/\sim_i$ and $\cap \sim_i = \Delta$ (the diagonal of A).
   (b) Show that each free $\mathscr{V}$-algebra in the variety $\mathscr{V}$ is a subdirect product of finitely generated free $\mathscr{V}$-algebras.
12. The Ω-algebra A is called *subdirectly irreducible* if whenever A is a subdirect product of a family $\{A_i\}_{i\in I}$ there exists some $j \in I$ such that the morphism $A \longrightarrow \Pi A_i \xrightarrow{\pi_j} A_j$ is an isomorphism.
   (a) Show that A is subdirectly irreducible if and only if the poset of nontrivial congruences on A has a least element.
   (b) Show that each Ω-algebra A is a subdirect product of a family of subdirectly irreducible Ω-algebras. (If $a \neq b$ in A let $\sim$ be a congruence on A that is maximal with respect to $(a,b) \notin \sim$).
13. Show that each variety in $\mathscr{A}(\Omega)$ is hereditary, homomorphically closed and productive.
14. Show that each free $\mathscr{C}$-algebra F is *projective*; that is, if $A, B \in \mathscr{C}$, where $\mathscr{C}$ is a class of Ω-algebras, and $h : A \longrightarrow B$ is an epimorphism, then for each morphism $f : F \longrightarrow B$ there is a morphism $g : F \longrightarrow A$ with $hg = f$:

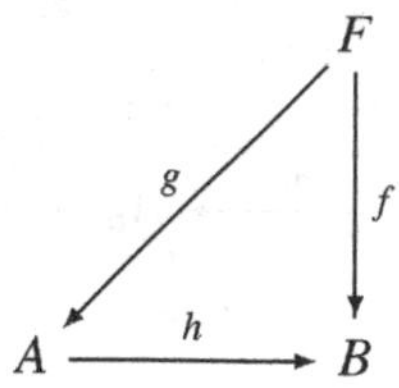

15. Show that each fully invariant congruence S on $F_\Omega(X)$ is the set of identities of some variety of Ω-algebras; that is, $S = S^{**}$. (First use Exercise 14 to verify that $F_\Omega(X)/S \in S^*$.)

16. Let $\mathscr{V}$ be the variety of Ω-algebras generated by the class $\mathscr{C} \subseteq \mathscr{A}(\Omega)$. Show that the algebra A is in $\mathscr{V}$ iff A is a homomorphic image of a subalgebra of a direct product of a family of algebras in $\mathscr{C}$.

17. Show that an Ω-algebra A is in a variety $\mathscr{V}$ iff each finitely generated subalgebra of A is in $\mathscr{V}$.

18. An algebra A is called a *generic* algebra in the variety $\mathscr{V}$ if $\mathscr{V}$ is the variety generated by $\{A\}$. Show that each variety has generic algebras. Specifically, a free $\mathscr{V}$-algebra on an infinite alphabet is generic in the nontrivial variety $\mathscr{V}$.

19. (a) Let $\mathscr{V}$ be the variety generated by the class $\mathscr{C} \subseteq \mathscr{A}(\Omega)$, and let $F \in \mathscr{V}$ be generated by X. Then F is $\mathscr{V}$-free on X if and only if each function $\varphi : X \longrightarrow A$ from X into an algebra A in $\mathscr{C}$ can be extended to a morphism $\psi : F \longrightarrow A$. (Use Exercise 16.)

    (b) Suppose that the Ω-algebra A is generated by X. Show that A is $\mathscr{V}$-free on X, for some variety $\mathscr{V}$, if and only if each function $\varphi : X \longrightarrow A$ can be extended to an endomorphism $\psi : A \longrightarrow A$.

20. Let $\mathscr{V}$ be the variety generated by the Ω-algebra A, let X be a set, let $P = A^{(A^X)}$ be the direct product of A^X copies of A, and let $F = S_\Omega(\{\pi_x\}_{x \in X})$ be the subalgebra of P generated by the projections $\pi_x : A^X \longrightarrow A$. Show that F is $\mathscr{V}$-free on X (or $\{\pi_x\}_{x \in X}$). (Use Exercise 19(a).)

21. Let $\{A_\alpha\}_{\alpha \in \Lambda}$ be a family of Ω-algebras in the variety $\mathscr{V}$. For each $\alpha \in \Lambda$ let $F_\alpha = F(\mathscr{V}, X_\alpha)$ be a $\mathscr{V}$-free algebra on X_α and let $p_\alpha : F_\alpha \longrightarrow A_\alpha$ be an epimorphism. Let X be the disjoint union of the X_α and let $i_\alpha : F_\alpha \longrightarrow F$ be the monomorphism induced by the injection $X_\alpha \longrightarrow X$ where $F = F(\mathscr{V}, X)$ is $\mathscr{V}$-free on X. Let $\sim$ be the congruence on F generated by

$$\bigcup_\alpha (i_\alpha \times i_\alpha)(\ker p_\alpha),$$

let $A = F/\!\sim$ and let $\kappa_\alpha : A_\alpha \longrightarrow A$ be the unique morphism that makes the diagram

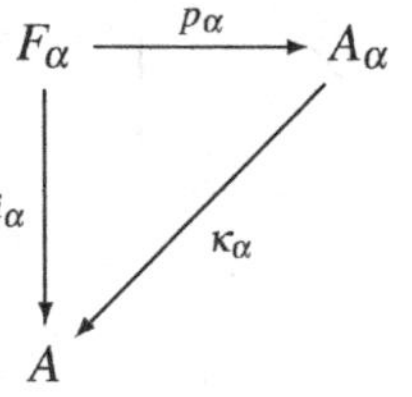

commutative, where $\eta : F \longrightarrow A$ is the natural map.

(a) Show that $(A, \{\kappa_\alpha\}_{\alpha \in \Lambda})$ is a coproduct of $\{A_\alpha\}$ in $\mathscr{V}$.

(b) The coproduct A of $\{A_\alpha\}$ is called a *free product* if the canonical morphisms κ_α are monomorphisms. Show that a coproduct is a free product if for each $\alpha \neq \beta$ there is a morphism $A_\alpha \longrightarrow A_\beta$.
(c) If each A_α contains a trivial subalgebra, then the coproduct is a free product, and, also, $\kappa_\alpha(A_\alpha) \cap \kappa_\beta(A_\beta)$ is contained in a trivial subalgebra of A, for each $\alpha \neq \beta$.
(d) Suppose that M_α is the smallest subalgebra of A_α and that M_α and M_β are isomorphic for each $\alpha, \beta \in \Lambda$. If, for each α, there is a morphism $A_\alpha \longrightarrow M_\alpha$ whose restriction to M_α is the identity, then the coproduct A of the $\{A_\alpha\}$ is a free product. Show, also, that $\kappa_\alpha(A_\alpha) \cap \kappa_\beta(A_\beta)$ is the smallest subalgebra of A, for each $\alpha \neq \beta$.

22. Show that each nontrivial variety contains a minimal nontrivial variety. (If S is a fully invariant congruence on $F_\Omega(X)$ and $x \neq y$ in X consider a fully invariant congruence that contains S and is maximal with respect to excluding (x,y).)

23. Let $(\{A_i\}_{i\in I}, \{\psi_{ij}\}_{i\leq j})$ be a direct system of Ω-algebras and suppose that I is directed up (see Exercise 1.1.7). Let $\sim$ be the relation on the disjoint union $\dot\cup A_i$ defined by $a_i \sim a_j$, for $a_i \in A_i$ and $a_j \in A_j$, if $\psi_{ik}(a_i) = \psi_{jk}(a_j)$ for some $k \geq i, j$.

(a) Show that $\sim$ is an equivalence relation on $\dot\cup A_i$ and that $\sim$ induces a congruence relation on each A_i.
(b) Let $A = (\dot\cup A_i)/\sim$ be the quotient determined by $\sim$. For $\omega \in \Omega(n)$ and $[a_{i_1}], \ldots, [a_{i_n}] \in A$, where $a_{i_\nu} \in A_{i_\nu}$ for $1 \leq \nu \leq n$, define $\omega([a_{i_1}], \ldots, [a_{i_n}]) = [\omega(\psi_{i_1 i}(a_{i_1}), \ldots \psi_{i_n i}(a_{i_n}))]$, where $i_\nu \leq i$ for each $\nu = 1, \ldots, n$. Let $\alpha_i : A_i \longrightarrow A$ be given by $\alpha_i(a_i) = [a_i]$. Then $\varinjlim A_i = (A, \{\alpha_i\}_{i\in I})$.
(c) If each A_i is in the variety $\mathscr{V}$ show that $\varinjlim A_i \in \mathscr{V}$.

24. Let $R \subseteq X \times Y$ be a relation from the set X to the set Y. If $A \subseteq X$ and $B \subseteq Y$ let $A^* = \{y \in Y : A \times \{y\} \subseteq R\}$ and $B^* = \{x \in X : \{x\} \times B \subseteq R\}$. Show that these mappings give a Galois connection between the power sets of X and Y.

Notes. Good references for the material in the first three sections are the notes by Weinberg [WE6] and the books by Birkhoff [BIR3], Kelley [KEL], Lambek [LA], and Rosenstein [ROS]. The results on universal algebra given in the fourth section appear in the book by Cohn [C1] and our presentation is based on that of his.

Chapter 2
Lattice-ordered Groups

In this chapter we present the most basic parts of the theory of lattice-ordered groups. Though our main concern will eventually be with abelian groups it is not appreciably harder to develop this material within the class of all groups. Moreover, the additional generality allows us to digress somewhat (if it is possible to digress before one begins) and to present some of the classical theorems in the subject. The fundamental interactions between the lattice structure and the group structure of a lattice-ordered group are dealt with first. Included are the characterization of the latticc structure in terms of the subsemigroup of positive elements as well as the elementary identities which result from the two structures. We then examine the morphisms in the category of lattice-ordered groups and the various kinds of subobjects. What is significant here is that the lattice of kernels in a lattice-ordered group is a distributive lattice and so is the corresponding lattice of subobjects that arises by dropping the normality requirement. These latter subobjects are precisely those for which the corresponding partition of cosets of the group is a lattice homomorphic image of the underlying lattice of the lattice-ordered group.

Those lattice-ordered groups that lack bounded subgroups are those that admit a completion, and they will be represented as extended real-valued continuous functions on a compact topological space. This representation bears fruit not only in the study of these kinds of lattice-ordered groups but also in the study of those lattice-ordered fields of which they are the underlying group.

In order to represent other lattice-ordered groups as lattice-ordered groups of functions it is essential to find totally ordered sets that arise naturally from the group. This is accomplished by taking the partition of cosets determined by a not necessarily normal kernel that is maximal with respect to excluding a given element. These totally ordered sets can be joined together and used to represent a lattice-ordered group as a subobject of the lattice-ordered group of all automorphisms of a totally ordered set. They will also be used to represent an abelian lattice-ordered group as a subobject of an antilexicographically ordered lattice-ordered group of real-valued functions.

S.A. Steinberg, *Lattice-ordered Rings and Modules*,
DOI 10.1007/978-1-4419-1721-8_2, © Springer Science + Business Media, LLC 2010

2.1 Basic Identities and Examples

Numerous identities and inequalities arise from the coexistence of the group and lattice structures. The two that will become most prominent are the unique decomposition of an element as the difference of two disjoint centralizing elements and a modified triangle inequality that can be replaced by the usual triangle inequality exactly when the group is abelian. Before this is established we will see that the lattice makes the group torsion-free and the group makes the lattice distributive. Let G be a group which is also a partially ordered set. We will denote the group operation in G additively even if G is not assumed to be abelian. G is called a *po-group* if translations in G are order preserving:

$$x \leq y \Rightarrow a+x+b \leq a+y+b \quad \forall x,\ y,\ a,\ b \in G. \tag{2.1.1}$$

If only the right (respectively, left) translations are isotone, then G is a *right* (respectively, *left*) *po-group*. In a right po-group each right translation is, in fact, an automorphism of the underlying poset since the inverse of a right translation is a right translation. One consequence of this fact is that in a po-group G, if $\bigvee x_i$ (or $\bigwedge x_i$) exists and each x_i commutes with x, then $\bigvee x_i$ (or $\bigwedge x_i$) commutes with x – that is, each centralizer in G contains the sup or inf of any of its subsets that exist in G. For, $x+(\bigvee x_i) = \bigvee(x+x_i) = \bigvee(x_i+x) = (\bigvee x_i)+x$, and dually. Also, in a po-group G, inversion is an anti-automorphism of the underlying poset since

$$x \leq y \Rightarrow -y = -y+x-x \leq -y+y-x = -x.$$

The po-group G is called a *totally ordered group*, a *lattice-ordered group* (ℓ-group) or a *directed group*, accordingly, as its underlying poset is totally ordered, a lattice, or is directed. The *positive cone* of G is

$$G^+ = \{g \in G \,:\, 0 \leq g\}$$

and the elements of G^+ are called *positive*. As the following theorem demonstrates the partial orders of the group G that make it into a po-group are in one-to-one correspondence with the positive cones of G. For this reason we will frequently refer to G^+ as a partial order (or a total order or a lattice-order) of G.

Theorem 2.1.1. *If G is a po-group with positive cone $P = G^+$, then*

(a) $P+P \subseteq P$ (P is a subsemigroup of G);
(b) $-g+P+g \subseteq P\ \forall g \in G$ (P is normal in G);
(c) $P \cap -P = 0$;
(d) $x \leq y \Leftrightarrow y-x \in P\ \ \forall x,y \in G$.

Conversely, if G is a group and P is a normal subsemigroup of G that satisfies (c), then the relation defined in (d) is a partial order of G which makes G into a po-group with positive cone P.

Proof. That (a) is true follows from the more general observation that if $a \le b$ and $c \le d$ in the po-group G, then $a+c \le b+d$; for, $a+c \le b+c \le b+d$. As for (b), if $0 \le x$ then $0 \le -g+x+g$. If $x \in P \cap -P$, then $0 \le x$ and $x = -y$ for some $y \ge 0$. So $x \le 0$ and hence $x = 0$. This proves (c), and (d) is a consequence of the fact that right translation by either x or $-x$ is order preserving. Conversely, given a normal subsemigroup P that satisfies (c), the relation $\le$ defined in (d) is reflexive since $0 \in P$. Also, if $x \le y$ and $y \le x$, then $y-x,\ x-y \in P$; so $x-y=0$ by (c). If $x \le y$ and $y \le z$, then $z-x = (z-y)+(y-x) \in P$ by (a); so $x \le z$ and $\le$ is a partial order of G. Finally, (2.1.1) is satisfied since, by (b), $(a+y+b)-(a+x+b) = a+(y-x)-a \in P$, if $y-x \in P$; and $G^+ = P$ by (d). □

Since two elements in the po-group G are comparable precisely when their difference is comparable to 0, G is totally ordered if and only if $G = G^+ \cup -G^+$. Also, since inversion is an order anti-automorphism, G is directed exactly when it is directed down (or up). Moreover, for $a, b \in G$

$$-(a \wedge b) = -a \vee -b = [0 \vee (-b+a)] - a$$

provided any one of the three expressions exists. So G is an ℓ-group if and only if $x \vee 0$ exists for each x in G. This proves the more transparent parts of

Theorem 2.1.2. *Let G be a po-group.*

(a) *G is totally ordered iff $G = G^+ \cup -G^+$.*
(b) *G is directed iff G is directed up (down).*
(c) *G is directed iff G^+ generates G. Moreover, if G^+ generates G, then $G = G^+ - G^+ = \{a-b : a, b \in G^+\}$.*
(d) *G is an ℓ-group iff $x \vee 0$ (respectively, $x \wedge 0$) exists for each $x \in G$.*
(e) *G is an ℓ-group iff G^+ is a lattice and generates G.*

Proof. By the previous remarks we only have to prove (c) and (e). If G is directed let $g \in G$ and take $x \in G$ with $x \le 0,\ g$. Then $g = g - x - (-x) \in G^+ - G^+$. Conversely, if G^+ generates G, then for $a, b \in G$ there exists $g_1, \ldots, g_n \in G^+$ such that $a - b = g_1 \pm g_2 \pm \cdots \pm g_n \le g_1 + g_2 + \cdots + g_n = x$, and then $a-b,\ 0 \le x$. So, $a,\ b \le x+b$. Certainly G^+ is a sublattice of G and G^+ generates G (by (c)) provided that G is an ℓ-group. On the other hand, suppose that G^+ is a lattice and generates G. If $a, b \in G^+$ and $g = \sup_{G^+}\{a, b\}$, then $g = \sup_G\{a, b\}$. For if $h \ge a, b$, then $h \in G^+$; so $h \ge g$. But then if $x \in G$, $x = a - b$ with $a, b \in G^+$, and $(a \vee b) - b = (a-b) \vee 0 = x \vee 0$. □

The proof of (e) in the previous theorem shows that G is an ℓ-group provided that it is directed and each pair of elements in G^+ has a sup in G^+. The dual is also true (Exercise 9).

Let us give some examples of po-groups and ℓ-groups.

(i) Each group is a po-group with positive cone $P = \{0\}$.

(ii) The group and poset direct product $G = \prod G_i$ of a family of po-groups $\{G_i : i \in I\}$ is a po-group that is called the *direct product* of the family. Recall that the group operation in G is $(x_i) + (y_i) = (x_i + y_i)$ and the partial order is given by $(x_i) \le$

(y_i) if $x_i \leq y_i$ for each $i \in I$. G is an ℓ-group if each G_i is an ℓ-group. Also, the group direct sum of the G_i, denoted by $\boxplus_{i\in I} G_i$, is a po-subgroup of G which is an ℓ-subgroup if each G_i is an ℓ-group; it is called the *direct sum* of the family of po-groups $\{G_i : i \in I\}$. It will be denoted by

$$\bigoplus_{i\in I} G_i$$

or by $G_1 \oplus \cdots \oplus G_n$ if $I = \{1,\ldots,n\}$. Of course, the I-tuple $(x_i) \in \bigoplus G_i$ iff $\{i \in I : x_i \neq 0\}$ is finite. If each $G_i = H$ then the direct sum will be denoted by $H^{(I)}$.

(iii) Let A and B be nonzero po-groups and let $\varphi : B \longrightarrow \mathrm{Aut}(A)$ be a group homomorphism into the group of automorphisms of the po-group A. Then the ordinal semidirect product $A \overleftarrow{\times}_\varphi B$ is a po-group. Here, $(a,b) \leq (c,d)$ if $b < d$, or $b = d$ and $a \leq c$; and the group operation is $(a,b)+(c,d) = (a+{}^b c, b+d)$ where ${}^b c = \varphi(b)(c)$. $A \overleftarrow{\times}_\varphi B$ is an ℓ-group exactly when B is totally ordered and A is an ℓ-group (Exercise 10). If φ is the trivial homomorphism, then we get the ordinary ordinal product $A \overleftarrow{\times} B$.

(iv) If $\{G_i : i \in I\}$ is a family of totally ordered groups and I is a totally ordered set which satisfies the maximum condition, then the ordinal product $V(I, G_i)$ of the G_i is a totally ordered group which is called a *Hahn product*. Here, the underlying group of V is the direct product of the G_i, and $v \in V^+$ provided that $v(i) > 0$ if i is maximal with $v(i) \neq 0$. In particular, each torsion-free abelian group can be made into a totally ordered group. For each such group is embedded in a rational vector space (Exercise 2) and hence in the direct product $\mathbb{Q}^I$ for some set I (Exercise 1). According to Theorem 1.1.1 I can be totally ordered in such a way that it has the maximum condition; hence $\mathbb{Q}^I = V(I, \mathbb{Q})$ is a totally ordered group.

(v) If P is a poset then its automorphism group $\mathrm{Aut}(P)$ becomes a po-group when it is given the pointwise partial order : $f \leq g$ if and only if $f(p) \leq g(p)$ for each p in P; so it is a subposet of the direct product P^P. If P is totally ordered, then $\mathrm{Aut}(P)$ is an ℓ-group (Exercise 5).

(vi) Let G be a topological ℓ-group. This means that G is a topological space and an ℓ-group for which the inversion function $- : G \longrightarrow G$ is continuous, and also the functions $+, \vee, \wedge : G \times G \longrightarrow G$ are continuous if $G \times G$ is given the product topology. If X is a topological space, then $C(X,G) = \{X \xrightarrow{f} G : f \text{ is continuous}\}$ is an ℓ-subgroup of the direct product G^X. In particular, if G is the additive group of $\mathbb{R}$ we get that $C(X)$, the set of all continuous real-valued functions on X, is an ℓ-group.

(vii) Let G be the group of real 3×3 matrices of the form

$$A = \begin{pmatrix} 1 & a & c \\ 0 & 1 & b \\ 0 & 0 & 1 \end{pmatrix}.$$

Then G is a totally ordered group if A is defined to be positive if $a > 0$, or $a = 0$ and $b > 0$, or $a = b = 0$ and $c \geq 0$. This also works for $n \times n$ matrices.

(viii) Various other examples can be found in analysis. For instance, the set of measurable functions on a measure space is an ℓ-group.

Two of the most important consequences of the interaction of the lattice and group structures in an ℓ-group are that the underlying group is torsion-free and the underlying lattice is distributive. Moreover, there are several basic identities and inequalities that hold in any ℓ-group. We present these facts next but first we need to give some definitions. The po-group G is said to be *semiclosed* if $nx \geq 0$ implies that $x \geq 0$ for any x in G and any integer $n \geq 1$. Recall from Exercise 1.2.2 that the lattice L is infinitely distributive if whenever $\{x_i\}$ is a subset of L for which $\bigvee x_i$ exists, then, for each y in L, $\bigvee(y \wedge x_i)$ exists and $y \wedge \bigvee x_i = \bigvee(y \wedge x_i)$, and the dual also holds.

Theorem 2.1.3. *Let G be an ℓ-group.*

(a) G is infinitely distributive and hence is distributive.
(b) G is semiclosed and hence it has no nonzero elements of finite order.

Proof. (a) Suppose that $x = \bigvee x_i$ exists in G. Then, for $y \in G$,

$$y \wedge x \leq [(x - x_i) + y] \wedge x = (x - x_i) + (y \wedge x_i).$$

So

$$0 \leq (y \wedge x) - (y \wedge x_i) \leq x - x_i.$$

But

$$0 = x - x = x + \bigwedge -x_i = \bigwedge (x - x_i);$$

so

$$\bigwedge [(y \wedge x) - (y \wedge x_i)] = 0,$$
$$(y \wedge x) + \bigwedge -(y \wedge x_i) = 0,$$

and $y \wedge x = \bigvee(y \wedge x_i)$. The dual follows by inversion.

(b) To see that G is semiclosed we show by induction that

$$n(x \wedge 0) = nx \wedge (n-1)x \wedge \cdots \wedge x \wedge 0$$

for any $x \in G$ and each integer $n \geq 1$. If this equation holds for some n, then

$$\begin{aligned}(n+1)(x \wedge 0) &= x \wedge 0 + [nx \wedge \cdots \wedge x \wedge 0] \\ &= [x + (nx \wedge \cdots \wedge x \wedge 0)] \wedge nx \wedge \cdots \wedge x \wedge 0 \\ &= (n+1)x \wedge \cdots \wedge 2x \wedge x \wedge nx \wedge \cdots \wedge x \wedge 0 \\ &= (n+1)x \wedge nx \wedge \cdots \wedge x \wedge 0.\end{aligned}$$

So if $kx \geq 0$ for some k, then $k(x \wedge 0) = (k-1)(x \wedge 0)$; and hence $x \wedge 0 = 0$. □

Let x be an element of the ℓ-group G. The *positive part, negative part,* and *absolute value* of x are defined by

$$x^+ = x \vee 0,$$

$$x^- = (-x) \vee 0,$$

and

$$|x| = x \vee (-x),$$

respectively. Since each centralizer in an ℓ-group is a sublattice any pair of elements in $\{x, x^+, x^-, |x|\}$ commute. If $x, y \in G$, then x and y are said to be *disjoint* if

$$|x| \wedge |y| = 0.$$

The subset X of G is called *disjoint* if any two distinct elements in X are disjoint.

Theorem 2.1.4. *Let G be an ℓ-group. The following hold for all x, y, $z \in G$.*

(a) $x - (x \wedge y) + y = x \vee y$.
(b) $(x - x \wedge y) \wedge (y - x \wedge y) = 0 = (-(x \wedge y) + x) \wedge (-(x \wedge y) + y)$.
(c) $|x| \leq y \Leftrightarrow -y \leq x \leq y$.
(d) $|x| = |-x| \geq 0$.
(e) $|x - y| = (x \vee y) - (x \wedge y)$.
(f) $x = x^+ - x^-$.
(g) $|x| = x^+ + x^-$.
(h) $|x| = 0 \Leftrightarrow x = 0$.
(i) $x^+ \wedge x^- = 0$; consequently, disjoint elements commute.
(j) The following are equivalent:

$$(*) \qquad x \wedge y = 0;$$
$$(**) \qquad x + y = x \vee y;$$
$$(***) \qquad x = (x - y)^+ \text{ and } y = (x - y)^-.$$

(k) (Riesz decomposition property) If x, y, $z \in G^+$ and $x \leq y + z$, then $x = y_1 + z_1$ where $0 \leq y_1 \leq y$ and $0 \leq z_1 \leq z$.
(l) If x, y, $z \in G^+$, then $x \wedge (y + z) \leq (x \wedge y) + (x \wedge z)$.
(m) $x \wedge y = x \wedge z = 0 \Rightarrow x \wedge (y + z) = 0$.
(n) $n(x \vee y) = nx \vee ny$ and $n(x \wedge y) = nx \wedge ny$ if $n \in \mathbb{N}$ and x and y commute.
(o) $|x + y| \leq (|x| + |y| + |x|) \wedge (|y| + |x| + |y|)$.
(p) G is abelian if and only if $|x + y| \leq |x| + |y|$ for all x and y in G.

Proof. (a) $x - (x \wedge y) + y = x + (-x \vee -y) + y = y \vee x$.
(b) $(x - x \wedge y) \wedge (y - x \wedge y) = x \wedge y - x \wedge y = 0$.
(c) $|x| = x \vee (-x) \leq y \Leftrightarrow x, -x \leq y \Leftrightarrow -y \leq x \leq y$.
(d) By (c), $-|x| \leq x \leq |x|$; so $2|x| \geq 0$ and hence $|x| \geq 0$ since G is semiclosed.
(e) $(x \vee y) - (x \wedge y) = [x + (-x \vee -y)] \vee [y + (-x \vee -y)] = 0 \vee (x - y) \vee (y - x) \vee 0 = |x - y|$.
(f) $x + x^- = x + (-x \vee 0) = 0 \vee x = x^+$.
(g) Put $y = 0$ in (e).
(h) $0 \leq x^+$, $x^- \leq |x|$ by (g). So if $|x| = 0$, then $x = 0$ by (f).
(i) $x^+ \wedge x^- = (x \vee 0) \wedge (-x \vee 0) = (x \wedge -x) \vee 0 = -|x| \vee 0 = 0$ by (d) and the

fact that G is a distributive lattice. If $|x| \wedge |y| = 0$, then any two distinct elements in $\{x^+, x^-, y^+, y^-\}$ are disjoint and hence commute by (a). So x and y commute by (f).

(j) The equivalence of $(*)$ and $(**)$ is an easy consequence of (a), and that $(***)$ implies $(*)$ follows from (i). Also, $(*)$ implies $(***)$ since $x = x + 0 = x + (-x \vee -y) = (x-y)^+$ and $y = (y-x)^+ = (x-y)^-$.

(k) Let $y_1 = x \wedge y$ and $z_1 = -y_1 + x$. Then $0 \le y_1 \le y$, $z_1 = (-x \vee -y) + x = 0 \vee (-y + x) \le z$, $z_1 \ge 0$, and $x = y_1 + z_1$.

(l) By (k) $x \wedge (y+z) = y_1 + z_1$ with $0 \le y_1 \le y$ and $0 \le z_1 \le z$; but also, y_1, $z_1 \le y_1 + z_1 \le x$; so $x \wedge (y+z) \le (x \wedge y) + (x \wedge z)$.

(m) This is an immediate consequence of (l).

(n) First, let $y = 0$. Since $nx^+ \wedge nx^- = 0$ by (i) and (m), $(nx)^+ = (nx^+ - nx^-)^+ = nx^+$ by (f) and (j). In general, $n(x \vee y) = n[(x-y)^+ + y] = (nx - ny)^+ + ny = nx \vee ny$. That $n(x \wedge y) = nx \wedge ny$ follows by the inversion anti-automorphism (or use (a)).

(o) and (p). $-|x| - |y| - |x| \le -|x| - |y| \le x + y \le |x| + |y| \le |x| + |y| + |x|$. So, by (c), $|x+y| \le |x| + |y| + |x|$ and $|x+y| \le |x| + |y|$ if x and y commute. If the triangle inequality holds and $x, y \in G^+$, then $x + y = |x+y| = |-y-x| \le |-y| + |-x| = y + x$. Interchanging x and y we get that $x + y \ge y + x$; so $x + y = y + x$, G^+ is abelian, and hence so is G since it is generated by G^+. □

The statements in the preceding theorem will frequently be used without an explicit reference.

Exercises.

1. Use Zorn's Lemma to show that each vector space has a basis.

2. The group D is said to be *divisible* if $nD = D$ for each $n \in \mathbb{N}$ where $nD = \{nd : d \in D\}$. A subgroup G of the group D is called *essential* in D if $G \cap A \neq 0$ for each nonzero subgroup A of D. If G is an essential subgroup of the divisible abelian group D, then D is called a *divisible hull* of G.

   (a) Show that a torsion-free abelian group is divisible if and only if it is a vector space over the rationals.
   (b) Show that any two divisible hulls of G are isomorphic.
   (c) Show that each torsion-free abelian group G has a divisible hull $d(G)$. (*Hint*: Imitate the construction of the rationals from the integers.)

3. Let $d(G)$ be the divisible hull of the torsion-free abelian po-group G (Exercise 2), and let
$$P = \{x \in d(G) : \exists n \in \mathbb{N} \text{ with } nx \in G^+\}.$$
   (a) Show that P is a positive cone for $d(G)$ and $(d(G), P)$ is semiclosed.
   (b) Show that G is semiclosed iff $G^+ = P \cap G$.
   (c) If G is an ℓ-group (or is directed or is totally ordered), show that $d(G)$ is an ℓ-group (or is directed or is totally ordered) and G is a sublattice of $d(G)$.

4. Suppose that each element g of the po-group G has a decomposition $g = x - y$ where $x \wedge y = 0$. Show that G is an ℓ-group.

5. If T is a totally ordered set show that the group Aut(T) of all order automorphisms of T is a sublattice of the direct product T^T and hence is an ℓ-group. Show that each disjoint set of positive elements in Aut(T) has a least upper bound in Aut(T). An ℓ-group with this property is called *laterally complete.*

6. Let G be a group. Show that there is a bijection between the set of partial orders of G that make G into a right po-group and the set of subsemigroups P of G with the property that $P \cap -P = 0$. In this correspondence $(G, \leq)$ is a totally ordered right po-group iff $P \cup -P = G$, and then P is called a *right order* of G and G is called a *right O-group.*

7. In a torsion-free abelian group show that each partial order is contained in a total order.

8. Prove that an abelian po-group G is semiclosed if and only if G^+ is the intersection of total orders of G.

9. Let G be a po-group.

   (a) Suppose that $a, b, c, d, e \in G^+$ and $c = \inf_{G^+}\{a, b\}$ and $e = \inf_{G^+}\{a+d,\ b+d\}$. Show that $e = c + d$.
   (b) Show that G is an ℓ-group if and only if G is directed and any two elements of G^+ have an inf in G^+.

10. Prove that $A \underset{\leftarrow}{\times_\varphi} B$ is an ℓ-group iff B is a totally ordered group and A is an ℓ-group.

11. Let $G = (\mathbb{Z} \oplus \mathbb{Z}) \underset{\leftarrow}{\times_\varphi} \mathbb{Z}$ where $\varphi : \mathbb{Z} \longrightarrow \mathrm{Aut}(\mathbb{Z} \oplus \mathbb{Z})$ is given by $\varphi(1) = \tau$ where $\tau(x, y) = (y, x)$. Then G is an ℓ-group but the subgroup H of G generated by $a = (1, -1, 0)$ and $b = (0, 0, 1)$ cannot be made into an ℓ-group. (*Hint*: $-b + a + b = -a$; show that a^+ cannot exist in H.)

12. Let R be a commutative integral domain with identity element, and let F be its field of quotients. If $a, b \in F$, then a divides b relative to R (notation : $a \,|\, b$) if $b = ar$ for some $r \in R$. Let F^* denote the multiplicative group of nonzero elements of F, and let U be the group of invertible elements in R. Define a relation $\leq$ in the quotient group F^*/U by $aU \leq bU$ if $a \,|\, b$. Verify each of the following.

    (a) F^*/U is a directed po-group. (It is called the *group of divisibility* of R.)
    (b) F^*/U is an ℓ-group iff any two elements in R have a greatest common divisor. (R is called a *GCD domain.*)
    (c) F^*/U is totally ordered iff R is a *valuation domain*—that is, the lattice of ideals of R is a chain.

13. Prove that the following hold in any ℓ-group G.

    (a) If $\bigvee_i x_i$ and $\bigvee_j y_j$ exist, then $(\bigvee_i x_i) + (\bigvee_j y_j) = \bigvee_{i,j}(x_i + y_j)$, and dually.
    (b) $a \geq 0$ iff $2a \vee a \geq 0$.

(c) $x \wedge y = 0$ iff $x \in G^+$ and $x + y = |x - y|$. (See Exercise 11 for an example where this equation holds with $0 \not< x < y$.)
(d) If x and y commute or if G can be embedded into a product of totally ordered groups, and $x + y = |x - y|$, then $x \wedge y = 0$.
(e) $|(x \wedge y) - (x \wedge z)| \leq |y - z|$.
(f) $x^+ \wedge y^+ \leq (x + y)^+$.
(g) $|(x \vee z) - (y \vee z)| + |(x \wedge z) - (y \wedge z)| = |x - y|$. (*Hint*: Let $s = |x - y|$ and $t = x \wedge y$ and, using (e) of Theorem 2.1.4, express $x \vee y$ and each of the absolute values in terms of s, t and z.)
(h) $|x^+ - y^+| \leq |x - y|$.
(i) $|x^- - y^-| \leq |x - y|$ if x and y commute. Give an example of an ℓ-group in which this inequality fails.
(j) $(x + y)^+ = x^+ - (x^- \wedge y^+) - (x^+ \wedge y^-) + y^+$ (*Hint*: $(x + y)^+ = (x \vee -y) + y$; compute $(x \vee -y)^+$ and $(x \vee -y)^-$ in terms of x^+, x^-, y^+ and y^-.) Also, $(u + v)^+ = u^+ + v^+$ if and only if $(u + v)^+ \geq u \vee v$.
(k) If $A = \{a_1, \ldots, a_n\}$ is a finite subset of G^+, then $\inf A = 0$ if and only if, for each $0 < x \in G$, there exists $y \in G$ and $a_i \in A$ such that $0 < y \leq x$ and $y \wedge a_i = 0$.

14. This exercise shows that each abelian ℓ-group is a group of divisibility (see Exercise 12). If F is a ring and G is a semigroup the *semigroup ring of G over F*, denoted by $F[G]$, is defined as the ring with G as an F-basis and with multiplication induced by the semigroup operation in G. More specifically, the elements of $F[G]$ are functions from G into F which are written as

$$\alpha = \sum_{g \in G} a_g x^g$$

where $a_g \in F$ and $a_g = 0$ except for finitely many $g \in G$. If $\beta = \sum b_g x^g \in F[G]$, then $\alpha + \beta = \sum (a_g + b_g) x^g$ and

$$\alpha\beta = \sum_g \left(\sum_{h+k=g} a_h b_k \right) x^g.$$

The *support of* α is defined to be supp $\alpha = \{g \in G : a_g \neq 0\}$.

(a) If F is a domain (that is, $ab \neq 0$ if $a \neq 0$ and $b \neq 0$) and G is a right O-group (Exercise 6), then $F[G]$ is a domain.
(b) Let F be a domain and let G be an ℓ-group. (G is a right O-group by Exercise 2.4.2(c).) Define the function $v : F[G] \longrightarrow G \cup \{\infty\}$ by $v(\alpha) = \inf(\text{supp } \alpha)$. Show that $v(\alpha + \beta) \geq v(\alpha) \wedge v(\beta)$ and $v(\alpha\beta) = v(\alpha) + v(\beta)$. (*Hint*: For the latter, reduce to the case that $v(\alpha) = v(\beta) = 0$ and use Exercise 13(k) three times. Use it twice to show that if $0 < x \in G$, then there exists $0 < z \leq x$ such that $A = z^\perp \cap \text{supp } \alpha \neq \phi$ and $B = z^\perp \cap \text{supp } \beta \neq \phi$, where $z^\perp = \{g \in G : |g| \wedge z = 0\}$. Let $\alpha^* \in F[G]$ have supp $\alpha^* = A$ and agree with α on A

and similarly for β^*. Then $\alpha\beta = \alpha^*\beta^* + \gamma$ with supp $\alpha^*\beta^* \cap$ supp $\gamma = \phi$. Use it again to show that $v(\alpha\beta) = 0$.)

(c) Let Q be the field of quotients of $F[G]$ where F is a field and G is an abelian ℓ-group. Show that there is a unique function $v : Q \longrightarrow G \cup \{\infty\}$ that extends v on $F[G]$ and has the two properties in (b).

(d) Let $R = \{\gamma \in Q : v(\gamma) \geq 0\}$. Show that R is a subring of Q, and Q is the field of quotients of R.

(e) Show that the group of divisibility of R is isomorphic to G.

(f) Show that each finitely generated ideal of R is principal; that is, R is a *Bezout domain*.

15. Let x and y be elements of the po-group G.

(a) If $x + y \leq y + x$, then $nx + ny \leq n(x+y) \quad \forall n \in \mathbb{N}$.

(b) If $y + x \leq x + y$, then $ny + nx \leq n(x+y) \quad \forall n \in \mathbb{N}$.

16. Show that a group homomorphism $\varphi : H \longrightarrow D$ between two po-groups is complete if and only if $0 = \wedge \varphi(a_i)$ in D whenever $0 = \wedge a_i$ in H.

17. The po-group G is called *n-semiclosed* $(n \in \mathbb{N})$ if $nx \geq 0$ implies that $x \geq 0$.

(a) If G is abelian and n-semiclosed then it satisfies the condition:

$$(*)\ \text{If } z \in G,\ X \subseteq G \text{ and } z \geq nX,\ \text{ then } z \geq x_1 + \cdots + x_n \ \text{ for all } x_1, \ldots, x_n \in X.$$

(b) If G satisfies $(*)$ then multiplication by n, $G \xrightarrow{n\cdot} G$, is a complete map.

(c) If X is an upward directed subset of G and $a = \sup X$, then $na = \sup nX$.

(d) If G is abelian and semiclosed and $d(G)$ is the divisible hull of G, then the inclusion map $G \longrightarrow d(G)$ is complete.

18. The po-group G is called a *Riesz* group if it has the following property: For all subsets $X = \{x_1, x_2\}$ and $Y = \{y_1, y_2\}$ of G

$$(*) \qquad X \leq Y \Rightarrow \exists z \in G \text{ with } X \leq z \leq Y.$$

Show that the following are equivalent for the po-group G.

(a) G is a Riesz group.

(b) (*) holds if $0 \in X$.

(c) (*) holds for all finite nonempty subsets X and Y of G.

(d) The set $U(X)$ of upper bounds of a finite nonempty subset X of G is downward directed.

(e) If X and Y are finite and nonempty, then $U(X) + U(Y) = U(X+Y)$.

(f) For all x, $y \in G^+$, $[0,x] + [0,y] = [0, x+y]$.

(g) If $x, y_1, \ldots, y_m \in G^+$ with $x \leq y_1 + \cdots + y_m$, then there exist $x_1, \ldots, x_m \in G^+$ with $x_j \leq y_j$ for each j, and $x = x_1 + \cdots + x_m$.

If G is abelian show that the following may be added to the previous list.

(h) If $x_1,x_2,y_1,y_2 \in G^+$ with $x_1+x_2 = y_1+y_2$, then there is a 2×2 matrix (x_{ij}) with entries in G^+ such that x_i is the sum of the entries in the ith row of (x_{ij}) and y_j is the jth column sum of (x_{ij}).

(i) If $x_1,\dots,x_n,y_1,\dots,y_m \in G^+$ with $x_1+\cdots+x_n = y_1+\cdots+y_m$, then there is an $n\times m$ matrix (x_{ij}) with entries in G^+ such that x_i (respectively, y_j) is the ith row sum (respectively, jth column sum) of (x_{ij}).

(Show the equivalence of (a) with each of (b), (c), (d), and (e), the implications (a) $\Rightarrow$ (f) $\Rightarrow$ (g) $\Rightarrow$ (b), and the equivalence of (f) with (h).)

2.2 Subobjects and Homomorphisms

In this section we examine homomorphisms between two ℓ-groups as well as those subobjects of an ℓ-group that arise from various order-theoretic conditions. It is shown that those subgroups of an ℓ-group modulo which the lattice structure is sustained constitute a complete and distributive sublattice of the subgroup lattice which satisfies one of the infinitely distributive equations but not the other. Among these subgroups are the polars, which arise as the closed elements of the Galois connection associated with disjointness. The lattice of polars is a complete Boolean algebra, and in the exercises an investigation is undertaken of when the Boolean algebra of polars of a subobject is naturally isomorphic to that of the full group. In subsequent sections we will see that this occurs under different guises when we are able to form a "completion."

Let C be a subgroup of the po-group G, and let G/C denote the set of all of the left cosets of C in G. The relation defined on G/C by

$$x+C \le y+C \text{ if } x \le y+c \text{ for some } c \in C \tag{2.2.1}$$

is independent of the coset representatives; for, if $x \le y+c$, $x = x_1+d$ and $y = y_1+e$ where $c,\ d,\ e \in C$, then $x_1+d \le y_1+e+c$ yields that $x_1 \le y_1+e+c-d$. This relation is reflexive and transitive but not necessarily antisymmetric. We are, of course, interested in the situation when it is a partial order of G/C. The subset X of the poset P is called *convex* if $x \le p \le y$ with $x,y \in X$ implies that $p \in X$. It is easily seen that a subgroup C of the po-group G is convex precisely when it contains the closed interval $[0,c]$ whenever $c \in C$. The subgroup C of a po-group is called an *ℓ-subgroup* if it is also a sublattice. Since $x \vee y = (x-y)^+ + y$ and $x \wedge y = -(-x \vee -y)$, C is an ℓ-subgroup exactly when $x \in C$ implies that $x^+ \in C$ (whenever x^+ exists in the larger group).

Theorem 2.2.1. *Let C be a subgroup of the po-group G.*

(a) The relation defined by (2.2.1) is a partial order of G/C if and only if C is convex. If C is convex then the natural map $G \longrightarrow G/C$ is isotone, and the map $G \longrightarrow \mathrm{Aut}(G/C)$ induced by the left translations in G is a po-group homomorphism into the po-group of automorphisms of the poset G/C.

(b) Suppose that G is an ℓ-group and that C is a convex subgroup of G. Then the following statements are equivalent.

(i) G/C is a (distributive) lattice and the natural map $G \longrightarrow G/C$ is a lattice homomorphism.

(ii) C is an ℓ-subgroup of G.

(c) If C is normal and convex, then G/C is a po-group.

(d) If G is an ℓ-group and C is a normal convex ℓ-subgroup, then G/C is an ℓ-group.

Proof. (a) If C is convex and $x+C \leq y+C$ and $y+C \leq x+C$, then $x \leq y+c$ and $y \leq x+d$ for some $c,\ d \in C$. But then $-d \leq -y+x \leq c$, so $-y+x \in C$ and G/C is a poset. Conversely, suppose that (2.2.1) defines a partial order of G/C. If $c \leq x \leq d$ with $c,\ d \in C$, then $C = c+C \leq x+C \leq d+C = C$; so $x \in C$ and C is convex. The last statement is easily verified.

(b) Let C be a convex ℓ-subgroup of the ℓ-group G. If $z+C \geq x+C,\ y+C$, then $z+c \geq x$ and $z+d \geq y$ for some $c,d \in C$. So $z+(c \vee d) \geq x \vee y$, and hence $z+C \geq x \vee y + C$. Thus $x \vee y + C = \sup_{G/C}\{x+C,\ y+C\}$, and dually. Conversely, let $G \longrightarrow G/C$ be a lattice homomorphism. Then if $x \in C$, $C = (x+C) \vee C = x^+ + C$; so $x^+ \in C$ and C is an ℓ-subgroup.

(c) If C is a normal convex subgroup of G, then (a) implies that each translation in G/C is isotone, so G/C is a po-group.

(d) This is a consequence of (b) and (c). □

If G and H are ℓ-groups (respectively, po-groups) then a group homomorphism $f\colon G \longrightarrow H$ which is also a lattice homomorphism (respectively, isotone) is called an *ℓ-homomorphism* (respectively, a po-*homomorphism*). So a group homomorphism f is a po-homomorphism exactly when $f(G^+) \subseteq H^+$. If the context is clear an ℓ-homomorphism or po-homomorphism will just be called a homomorphism. If there is an ℓ-homomorphism (respectively, a po-homomorphism) from G to H which is an order isomorphism we say that G and H are *isomorphic* and we write $G \cong H$. It is easily verified that the kernel of an ℓ-homomorphism (respectively, a po-homomorphism) is a convex ℓ-subgroup (respectively, convex subgroup) of its domain. The usual isomorphism theorems hold in the variety of ℓ-groups, but before we state them we give some useful criteria for a group homomorphism to be an ℓ-homomorphism.

For the rest of this chapter, unless stated otherwise, all groups will be ℓ-groups.

Theorem 2.2.2. *Let $f\colon G \longrightarrow H$ be a group homomorphism between the ℓ-groups G and H. Then the following statements are equivalent.*

(a) f is an ℓ-homomorphism.
(b) $f(x^+) = f(x)^+$ for each x in G.
(c) $x \wedge y = 0 \Rightarrow f(x) \wedge f(y) = 0$ for all x,y in G.
(d) $f(|x|) = |f(x)|$ for each x in G.

Proof. It is clear that (a) implies (c), and (a) also implies (d) since $f(|x|) = f(x \vee -x) = f(x) \vee -f(x) = |f(x)|$.

(c) implies (b). Since $f(x^+) \wedge f(x^-) = 0$, $f(x)^+ = [f(x^+) - f(x^-)]^+ = f(x^+)$ by (j) of Theorem 2.1.4.

(b) implies (a). $f(x \vee y) = f((x-y)^+ + y) = f((x-y)^+) + f(y) = [f(x) - f(y)]^+ + f(y) = f(x) \vee f(y)$, and $f(x \wedge y) = -f(-x \vee -y) = -(f(-x) \vee f(-y)) = f(x) \wedge f(y)$.

(d) implies (b). Since $2f(x^+) = f(2x^+) = f(x + |x|) = f(x) + |f(x)| = 2f(x)^+$, and $f(x^+)$ and $f(x)^+$ commute, and H is torsion-free, we have $f(x^+) = f(x)^+$. □

Theorem 2.2.3. *Let N be a normal convex ℓ-subgroup of the ℓ-group G.*

(a) If $f\colon G \longrightarrow H$ is an ℓ-homomorphism with kernel N, then $f(G)$ is an ℓ-subgroup of H and $G/N \cong f(G)$.
(b) The mapping $A \longmapsto A/N$ is a lattice isomorphism between the lattice of convex subgroups (respectively, ℓ-subgroups) of G that contain N and the lattice of convex subgroups (respectively, ℓ-subgroups) of G/N.
(c) If A is an ℓ-subgroup of G, then $A+N$ is an ℓ-subgroup of G and $(A+N)/N \cong A/A \cap N$.
(d) If K is a normal convex ℓ-subgroup of G with $N \subseteq K$, then $(G/N)/(K/N) \cong G/K$.

Proof. We will leave it to the reader to check that the lattice isomorphism in (b) between the subgroup lattices restricts to an isomorphism on the indicated sublattices. Each of the isomorphisms in (a), (c), and (d) is the well-known group isomorphism, so it suffices to verify that each is an ℓ-isomorphism. For (d) this follows from (a) (for example). As for (a), if f_* denotes this isomorphism, then $f_*((x+N)^+) = f_*(x^+ + N) = f(x^+) = f(x)^+ = f_*(x+N)^+$; so f_* is an ℓ-isomorphism by the previous theorem. Finally, (c) follows from (a) provided that $A+N$ is an ℓ-subgroup. But, in fact, if A is just a sublattice of G, then $A+N$ is also a sublattice. For if $a, b \in A$ and $n, m \in N$, then $(a+n) \vee (b+m) + N = (a+N) \vee (b+N) = a \vee b + N$ in the lattice G/N; so $(a+n) \vee (b+m) \in A+N$, and dually. □

In the next two results we give some fundamental properties of the subobjects of an ℓ-group. If X is a subset of the ℓ-group G, then $C(X) = C^G(X)$ will denote the convex ℓ-subgroup of G generated by X, and $[X]$ will denote the ℓ-subgroup generated by X. The *polar* of X is defined by

$$X^{\perp} = \{a \in G\colon |a| \wedge |x| = 0 \ \forall x \in X\}.$$

Note that $X \subseteq X^{\perp\perp}$, ${}^{\perp}$ reverses inclusion and $X^{\perp} = X^{\perp\perp\perp}$. Let $\mathscr{C}(G)$ denote the set of all convex ℓ-subgroups of G, and let $\mathscr{B}(G)$ denote the set of polars of G. Since $\mathscr{C}(G)$ is closed under intersections it is a complete lattice.

Theorem 2.2.4. *Let G be an ℓ-group.*

(a) A subgroup C of G is a convex ℓ-subgroup if and only if $|x| \leq |c|$ with $x \in G$ and $c \in C$ implies that $x \in C$.

(b) If $X \subseteq G$, then

$$C(X) = \{g \in G : |g| \leq |x_1| + \cdots + |x_n| \text{ for some } x_1, \ldots, x_n \in X\}.$$

(c) The subgroup of G generated by a family of convex ℓ-subgroups is a convex ℓ-subgroup, and its positive cone is the subsemigroup of G^+ generated by the corresponding family of positive cones.

(d) If X, $Y \subseteq G$, and D, $E \in \mathscr{C}(G)$ and $a, b \in G$, then

(i) $C(X) \vee C(Y) = C(X \cup Y) = C(\{|x| \vee |y| : x \in X,\ y \in Y\})$;

(ii) $C(X) \cap C(Y) = C(\{|x| \wedge |y| : x \in X,\ y \in Y\})$;

(iii) $C(D,a) \cap C(D,b) = D \vee C(|a| \wedge |b|)$;

(iv) $D \cap E = 0$ iff $D \subseteq E^\perp$.

(e) If $X \subseteq G$, then $X^\perp \in \mathscr{C}(G)$.

(f) The sublattice of G generated by a subgroup S is an ℓ-subgroup and, in fact,

$$[S] = \left\{ \bigwedge_i \bigvee_j s_{ij} \ :\ s_{ij} \in S,\ 1 \leq i \leq n,\ 1 \leq j \leq m \right\}$$

$$= \left\{ \bigvee_i \bigwedge_j s_{ij} \ :\ s_{ij} \in S,\ 1 \leq i \leq n,\ 1 \leq j \leq m \right\}.$$

Proof. (a) If C is a convex ℓ-subgroup of G and $|x| \leq |c|$ with $c \in C$, then $0 \leq x^+, x^- \leq |x| \leq |c|$ gives that $x = x^+ - x^- \in C$. Conversely, if the subgroup C has this property, then it is certainly convex, and it is also an ℓ-subgroup since $0 \leq c^+ \leq |c|$.

(b) Since any convex ℓ-subgroup which contains X must contain g if $|g| \leq |x_1| + \cdots + |x_n|$ with $x_i \in X$, it suffices to show that the set C of all such g is a convex ℓ-subgroup. But if also $|h| \leq |y_1| + \cdots + |y_m|$ with $y_j \in X$, then by (o) of Theorem 2.1.4

$$|g - h| \leq |g| + |h| + |g| \leq |x_1| + \cdots + |x_n| + |y_1| + \cdots + |y_m| + |x_1| + \cdots + |x_n|.$$

So C is a subgroup which is a convex ℓ-subgroup, by (a).

(c) Let $\{C_i : i \in I\}$ be a family of convex ℓ-subgroups of G, and let C be the subgroup of G generated by the C_i. If $x = c_1 + \cdots + c_n \in C$ where each c_j is in some C_i, then $|x| \leq d_1 + \cdots + d_m$ with each d_j in some C_i^+, by Theorem 2.1.4 (o). So if $|y| \leq |x|$, then $y^+ = e_1 + \cdots + e_m$ and $y^- = f_1 + \cdots + f_m$ where $0 \leq e_j,\ f_j \leq d_j$, by Theorem 2.1.4 (k). Hence $y = y^+ - y^- \in C$ and C is a convex ℓ-subgroup by (a). The last statement follows by specializing to $x = |x| = y$.

(d) The equalities in (i) are clear and, since $|x| \wedge |y| \in C(X) \cap C(Y)$ if $x \in X$ and $y \in Y$, $C(\{|x| \wedge |y| : x \in X, y \in Y\}) \subseteq C(X) \cap C(Y)$. But if $g \in C(X) \cap C(Y)$, then

$$|g| \leq (|x_1| + \cdots + |x_n|) \wedge (|y_1| + \cdots + |y_m|) \leq |x_1| \wedge |y_1| + \cdots + |x_n| \wedge |y_m|$$

for some $x_i \in X$, $y_j \in Y$. This proves (ii) which, together with (i), readily gives (iii) and (iv).

(e) If $a, b \in X^\perp$ and $x \in X$, then by (o) and (m) of Theorem 2.1.4, $0 \le |a-b| \wedge |x| \le (|a|+|b|+|a|) \wedge |x| = 0$. So $a-b \in X^\perp$, and hence $X^\perp$ is a convex ℓ-subgroup by (a).

(f) It clearly suffices to show that the set

$$H = \left\{ \bigwedge_I \bigvee_J s_{ij} : s_{ij} \in S, I \text{ and } J \text{ finite} \right\}$$

is an ℓ-subgroup of G, and this follows readily from the general distributivity equations (1.2.1) which hold in any distributive lattice. For,

$$-\left(\bigwedge_I \bigvee_J s_{ij}\right) = \bigvee_I \bigwedge_J (-s_{ij}) = \bigwedge_{J^I} \bigvee_I (-s_{if(i)})$$

and

$$\left(\bigwedge_I \bigvee_J s_{ij}\right) + \left(\bigwedge_K \bigvee_L t_{kl}\right) = \bigwedge_I \bigvee_J \bigwedge_K \bigvee_L (s_{ij} + t_{kl}) = \bigwedge_I \left(\bigwedge_{K^J} \bigvee_J\right) \bigvee_L (s_{ij} + t_{f(j)l})$$

$$= \bigwedge_{I \times K^J} \bigvee_{J \times L} (s_{ij} + t_{f(j)l}).$$

Thus, H is a subgroup; it is an ℓ-subgroup since

$$\left(\bigwedge_I \bigvee_J s_{ij}\right)^+ = \bigwedge_I \bigvee_J (s_{ij} \vee 0) \in H.$$

□

Theorem 2.2.5. *Let G be an ℓ-group.*

(a) $\mathscr{C}(G)$ is a complete distributive sublattice of the lattice of subgroups of G, and, in fact, it satisfies the infinite distributive law

$$A \cap \left(\bigvee_i B_i\right) = \bigvee_i (A \cap B_i).$$

(b) The mapping $C \mapsto C^+$ is a lattice isomorphism between $\mathscr{C}(G)$ and the lattice of convex ℓ-subsemigroups of G^+ which contain 0.

(c) $\mathscr{B}(G)$ is a complete Boolean algebra.

Proof. (a) That $\mathscr{C}(G)$ is a sublattice of the subgroup lattice of G follows from part (c) of the previous theorem. If $0 \le x \in A \cap (\vee B_i)$, then $x = b_1 + \cdots + b_n$ with each b_j in some B_i^+ (again, by (c) of Theorem 2.2.4). But then each $b_j \in A$ since $0 \le b_j \le x$; so $x \in \bigvee (A \cap B_i)$. The other inclusion is obvious.

(b) Since C^+ generates C it suffices to verify that the mapping is onto. So, let

S be a convex ℓ-subsemigroup of G^+ with $0 \in S$, and let C be the subgroup that it generates. We claim that $C = \{a - b : a, b \in S\} \in \mathscr{C}(G)$ and $C^+ = S$. Let $a, b, c, d \in S$. Then $0 \le -(b \wedge c) + b = b_1 \le b$ and $0 \le -(b \wedge c) + c = c_1 \le c$; so b_1 and $c_1 \in S$. Also, b_1 and c_1 are disjoint and so commute. But then $c + b_1 - c_1 = c - c_1 + b_1 = b \wedge c + b_1 = b$. Thus $(a - b) + (c - d) = a + c_1 - b_1 - d = (a + c_1) - (d + b_1)$. So $\{a - b : a, b \in S\}$ is a subgroup which must be C. If $|x| \le |a - b| \le a + b + a$, where $a, b \in S$, then x^+ and $x^- \in S$ and so $x = x^+ - x^- \in C$. So C is a convex ℓ-subgroup. If $a, b \in S$, then $0 \le a \vee b - b = (a - b)^+ \le a \vee b$; so $(a - b)^+ \in S$ and $C^+ = \{x^+ : x \in C\} = S$.

(c) $\mathscr{B}(G)$ is a complete lattice since

$$\bigcap_i X_i^\perp = \left(\bigcup_i X_i\right)^\perp.$$

To show that $\mathscr{B}(G)$ is a Boolean algebra it suffices, by Theorem 1.2.3, to verify that for polars A and B, $A \subseteq B$ if and only if $A \cap B^\perp = 0$. But this follows from Theorem 2.2.4(d)(iv). □

An example of an ℓ-group G in which the other infinite distributive equation fails is obtained by letting G be the ℓ-group of all real-valued continuous functions on the closed interval $[0, 1]$. If B_a is the set of those functions in G which vanish at $a \in [0, 1]$, then B_a is a maximal ideal of the ring G and

$$B_0 \vee \left(\bigwedge_{a \neq 0} B_a\right) = B_0 \subset G = \bigwedge_{a \neq 0} (B_0 \vee B_a).$$

As we indicate below, the fact that $\mathscr{C}(G)$ is half infinitely distributive has an effect on the direct sum decompositions of G. The ℓ-group G is the *direct sum* of its convex ℓ-subgroups $\{C_i : i \in I\}$ if the map $(c_i) \mapsto \Sigma c_i$ is an isomorphism of the ℓ-group $\bigoplus C_i$ onto G. In this case we write

$$G = \bigoplus_{i \in I} C_i.$$

The conditions for this to hold are the familiar group theoretic ones: G must be generated by the C_i, each C_i must be normal in G, and, for each i,

$$C_i \cap \left(\bigvee_{j \neq i} C_j\right) = 0.$$

Under these condition the mapping is certainly a group isomorphism; but it is also an ℓ-isomorphism. For if $c_1 + \cdots + c_n \ge 0$ where $c_k \in C_{i_k}$ and $i_1, \cdots, i_n$ are distinct indices, then $-c_k \le c_1 + \cdots + c_{k-1} + c_{k+1} + \cdots + c_n = d_k$; so $c_k^- \le d_k^+$ and

$$c_k^- \in C_{i_k} \cap \left(\bigvee_{j \neq i_k} C_j \right) = 0.$$

Thus, each $c_k \geq 0$ and the mapping is an order isomorphism.

These conditions can be weakened slightly.

Theorem 2.2.6. *Let G be an ℓ-group.*

(a) Suppose that $\{C_i : i \in I\}$ is a family of convex ℓ-subgroups of G that generates G. Then the following statements are equivalent.

(i) $G = \bigoplus C_i$.

(ii) $C_i \cap C_j = 0$ if $i \neq j$.

(iii) If $x_1 + \cdots + x_n \geq 0$ where $x_j \in C_{i_j}$ and $i_1, \ldots, i_n$ are distinct indices, then each $x_j \geq 0$.

(b) If $C_i, D_j \in \mathscr{C}(G)$ and

$$G = \bigoplus_{i \in I} C_i = \bigoplus_{j \in J} D_j,$$

then

$$G = \bigoplus_{i,j} C_i \cap D_j.$$

(c) If $A, B \in \mathscr{C}(G)$ and $G = A \oplus B$, then $B = A^\perp$.

Proof. (a) Certainly, (i) implies (iii) and (iii) implies (ii). If (ii) holds, then, as we have seen, the elements in C_i are disjoint from those in C_j and hence commute with those in C_j. So each C_i is normal and

$$C_i \cap \bigvee_{j \neq i} C_j = \bigvee_{j \neq i} (C_i \cap C_j) = 0;$$

thus (ii) implies (i).

As for (b),

$$C_i = C_i \cap G = \bigvee_j (C_i \cap D_j) = \bigoplus_j (C_i \cap D_j);$$

so

$$G = \bigoplus_i \bigoplus_j (C_i \cap D_j).$$

For the proof of (c) just note that $A^\perp = A^\perp \cap (A \oplus B) = A^\perp \cap B$ yields that $A^\perp \subseteq B \subseteq A^\perp$. □

Given two decompositions $G = \oplus C_i = \oplus D_j$ of G, $\{D_j\}$ is a *refinement* of $\{C_i\}$ if for each j there is an i with $D_j \subseteq C_i$. An easy induction using (b) of the previous theorem gives that any finite number of decompositions of G have a common refinement.

Recall that an ℓ-group G is a subdirect product of the family $\{G_i : i \in I\}$ of ℓ-groups if there is a monomorphism $f : G \longrightarrow \Pi G_i$ such that each composite $\pi_i f$ is an epimorphism, and G is subdirectly irreducible if, in any such representation of G there is an index i such that $\pi_i f$ is an isomorphism. Now, a function $f: G \longrightarrow \Pi G_i$ on an ℓ-group G is uniquely determined by the family $\{\pi_i f : i \in I\}$, and f is an ℓ-homomorphism if and only if each $\pi_i f$ is an ℓ-homomorphism. In this case $\ker f = \cap_i \ker(\pi_i f)$ and $G/\ker(\pi_i f) \cong$ image of $\pi_i f$. Consequently, each family $\{N_i : i \in I\}$ of normal convex ℓ-subgroups of the ℓ-group G determines a homomorphism $G \longrightarrow \Pi G/N_i$ with kernel $N = \cap_{i\in I} N_i$, and G/N is a subdirect product of the family $\{G/N_i : i \in I\}$; and all subdirect product representations of G/N essentially arise in this way. Clearly, a nonzero ℓ-group G is subdirectly irreducible if and only if it has a smallest nonzero normal convex ℓ-subgroup. As a specific instance of Birkhoff's theorem for abstract algebras (see Exercise 1.4.12) we have

Theorem 2.2.7. *Each ℓ-group is a subdirect product of a family of subdirectly irreducible ℓ-groups.*

Proof. Let G be an ℓ-group. If $0 \neq a \in G$ let N_a be a normal convex ℓ-group of G which is maximal with respect to excluding a. The existence of N_a is given by Zorn's Lemma. Since each normal convex ℓ-subgroup of G that properly contains N_a must contain a, G/N_a is subdirectly irreducible. But

$$\bigcap_{a\neq 0} N_a = 0,$$

so G is isomorphic to a subdirect product of the G/N_a. □

Exercises.

1. The following statements are equivalent for the ℓ-group G.
   (a) G is totally ordered.
   (b) Each subset of G is a sublattice.
   (c) Each convex subset is a sublattice.
   (d) Each convex subgroup is a sublattice.

   (The first three are equivalent in any poset G.)
2. In an ℓ-group G a minimal element in $G^+ \setminus \{0\}$ is called an *atom*. Prove:
   (a) The subgroup of G generated by the atoms is a normal, abelian, convex l-subgroup.
   (b) The following statements are equivalent:
      (i) G is generated by its atoms.
      (ii) G^+ has the minimum condition.
      (iii) G is isomorphic to a direct sum of copies of $\mathbb{Z}$.
   (c) If R is a commutative unital domain with quotient field F, then R is a unique factorization domain if and only if its group of divisibility F^*/U is an ℓ-group which is generated by its atoms (see Exercise 2.1.12).

3. The following statements are equivalent for the subgroup C of the ℓ-group G.

   (a) C is a convex ℓ-subgroup.
   (b) C is a convex directed subgroup.
   (c) If $a, b \in G$ and $a \wedge b = 0$, then $a \wedge (b+c) \in C$ for each $c \in C$.
   (d) If $a, b \in G$ and $a \wedge b \in C$, then $a \wedge (b+c) \in C$ for each $c \in C$.

4. Let A and B be ℓ-groups and let $\mathrm{Aut}(A)$(respectively, ℓ-$\mathrm{Aut}(A)$) be the group of automorphisms of the group (respectively, ℓ-group) A. Suppose that $\varphi : B \longrightarrow \mathrm{Aut}(A)$ is a group homomorphism and let $G = A \times_\varphi B$ be the semidirect product of A by B supplied with the partial order of the direct product: $(a,b) \leq (c,d)$ if $a \leq c$ and $b \leq d$. Show that the following statements are equivalent.

   (a) G is an ℓ-group.
   (b) $\varphi(B) = 1$.
   (c) φ is isotone and its image is contained in ℓ-$\mathrm{Aut}(A)$, where the partial order of ℓ-$\mathrm{Aut}(A)$ is coordinatewise: $f \leq g$ if $f(x) \leq g(x)$ for each x in A.

5. Let $C \in \mathscr{C}(G)$. Prove:

   (a) If C is finitely generated as a convex ℓ-subgroup, then C has a single generator.
   (b) If C is finitely generated as an ℓ-subgroup, then it need not have a single generator.

6. (a) Show that each convex ℓ-subgroup of a divisible ℓ-group is divisible (see Exercises 2.1.2 and 2.1.3).
   (b) If $d(G)$ is the divisible hull of the abelian ℓ-group G, show that the map $C \mapsto d(C)$ is a lattice isomorphism from $\mathscr{C}(G)$ onto $\mathscr{C}(d(G))$.
   (c) Show that the isomorphism in (b) restricts to an isomorphism between the Boolean algebras $\mathscr{B}(G)$ and $\mathscr{B}(d(G))$.

7. Let $\overline{P}$ be the MacNeille completion of the poset P. Show that there is an embedding of po-groups $\varphi : \mathrm{Aut}(P) \longrightarrow \mathrm{Aut}(\overline{P})$ that preserves any infs or sups that exist. In particular, if P is totally ordered, then φ is an embedding of ℓ-groups (Exercise 2.1.5).

8. Let X be a poset, H a po-group and P the set of isotone maps from X into H. Then the subset P of the po-group H^X (the direct product of X copies of H) has the following properties.

   (a) $P + P \subseteq P$.
   (b) $P \cap -P = \{f \in H^X : x \leq y \text{ implies } f(x) = f(y)\}$.
   (c) If H is abelian and X is a directed po-group and $\mathrm{Hom}_{\mathbb{Z}}(X,H)$ is the group of homomorphisms from X to H, then, $P \cap \mathrm{Hom}_{\mathbb{Z}}(X,H)$, the set of po-group homomorphisms from X to H, is a positive cone for the group $\mathrm{Hom}_{\mathbb{Z}}(X,H)$.

9. The category whose objects are po-groups (respectively, abelian po-groups) and whose morphisms are po-group homomorphisms will be called Pog (respectively, Poag), while the category whose objects are ℓ-groups (respectively,

abelian ℓ-groups) and whose morphisms are ℓ-homomorphisms will be called Log (respectively, Loag).

(a) The direct product is a product in each of the categories Pog, Poag, Log, and Loag.
(b) Let $(\{G_i\}_{i\in I}, \{\psi_{ij}\}_{i\geq j})$ be an inverse system in any one of these four categories, and let $G = \{(g_i) \in \Pi G_i : \psi_{ij}(g_i) = g_j \text{ if } i \geq j\}$. Then $\varprojlim G_i = (G, \{p_i\})$ where p_i is the restriction of the projection π_i (see Exercise 1.1.6).

10. (a) The direct sum is a free product in Poag (see Exercise 1.4.21).
(b) If $G_i \neq 0$ for two indices, then $\oplus G_i$ is not a coproduct of $\{G_i\}_{i\in I}$ in Loag. (If $G_i \neq 0$, then G_i has a nonzero totally ordered homomorphic image by Theorem 2.4.4.)
(c) Show that the direct limit construction that is given in Exercise 1.4.23, which is valid in the varieties Log and Loag, also gives the direct limit in the categories Pog and Poag.

11. Let Sgp (respectively, $\mathscr{L}$) denote the category whose objects are semigroups (respectively, lattices) and whose morphisms are semigroup (respectively, lattice) homomorphisms. Suppose that $G, H \in$ Pog with G directed.

(a) Show that the restriction map $\text{Pog}[G,H] \longrightarrow \text{Sgp}[(G^+,+),(H^+,+)]$ is a bijection.
(b) If $G, H \in$ Log, then the image of Log$[G,H]$ under this bijection is Sgp $[(G^+,+),(H^+,+)] \cap \mathscr{L}[G^+,H^+] = \text{Sgp}[(G^+,+),(H^+,+)] \cap \text{Sgp}[(G^+,\vee), (H^+,\vee)] = \text{Sgp}[(G^+,+),(H^+,+)] \cap \text{Sgp}[(G^+,\wedge),(H^+,\wedge)]$.

12. Let N be a normal subgroup of the group G, and suppose that N and G/N are po-groups. Let $P = N^+ \cup \{x \in G : 0 \neq x + N \in (G/N)^+\}$. Then P is a positive cone for G iff N^+ is normal in G. Assume that P is a positive cone.

(a) The subgroup (N, N^+) is a convex subgroup of (G,P) and $(G/N)^+ = P/N$.
(b) If $g \in P \setminus N$, then $g > N$.
(c) The group (G,P) is totally ordered iff (N,N^+) and $(G/N,(G/N)^+)$ are totally ordered.
(d) If $N \neq 0$, then (G,P) is an ℓ-group iff (N,N^+) is an ℓ-group and $(G/N, (G/N)^+)$ is totally ordered.
(e) If (G,P) is an ℓ-group, then the inclusion map $N \longrightarrow G$ is complete.

13. (a) Let X be a subset of the ℓ-group G. Then $X^\perp = \langle X\rangle^\perp = [X]^\perp = C(X)^\perp$, where $\langle X\rangle$ is the subgroup generated by X.
(b) If $\{B_i : i \in I\} \subseteq \mathscr{B}(G)$, then $\bigvee_i B_i = (\bigcup_i B_i)^{\perp\perp}$.
(c) If $B \in \mathscr{B}(G)$, then $B = \bigvee_{b\in B} b^{\perp\perp}$ (where $b^\perp = \{b\}^\perp$).
(d) For each $n \in \mathbb{N}$ and $a \in G$, $a^\perp = |a|^\perp = (na)^\perp$.
(e) The function $a \mapsto a^{\perp\perp}$ is a lattice homomorphism from G^+ to $\mathscr{B}(G)$ which preserves all sups that exist in G^+.
(f) If $a, b \in G^+$, then $(a+b)^{\perp\perp} = (a \vee b)^{\perp\perp}$.

14. Let H be a subset of the ℓ-group D which contains 0. If $X \subseteq D$, then $X^{\perp_D} = X^\perp$ and $X^{\perp_H} = X^\perp \cap H$. Let $\mathscr{B}(H) = \{A \subseteq H : A = A^{\perp_H \perp_H}\}$. Let $\mathscr{C}(H)$ denote the collection of all those convex subsets of H which contain 0, and define the functions $\varphi : \mathscr{B}(H) \longrightarrow \mathscr{B}(D)$ and $\psi : \mathscr{B}(D) \longrightarrow \mathscr{C}(H)$ by $\varphi(A) = A^{\perp_H \perp_D}$ and $\psi(B) = B \cap H$. $\mathscr{B}(H)$ and $\mathscr{B}(D)$ are called *canonically isomorphic* if φ and ψ are inverse isomorphisms between $\mathscr{B}(H)$ and $\mathscr{B}(D)$; that is, if $\psi(\mathscr{B}(D) \subseteq \mathscr{B}(H)$ and $\varphi\psi = 1_{\mathscr{B}(D)}$ and $\psi\varphi = 1_{\mathscr{B}(H)}$. Verify each of the following.
    (a) $\mathscr{B}(H)$ is a complete lattice. If $|x| \wedge |y| \in H$ whenever $x, y \in H$, then $\mathscr{B}(H)$ is a Boolean algebra.
    (b) $\psi\varphi = 1_{\mathscr{B}(H)}$.
    (c) $\psi(\mathscr{B}(D)) \subseteq \mathscr{B}(H)$ iff $\psi(\mathscr{B}(D)) = \mathscr{B}(H)$.
    (d) If $X \subseteq D$, then $X^{\perp_D \perp_H} \subseteq X^{\perp_H \perp_H}$ and $X^{\perp_D \perp_D} \subseteq X^{\perp_H \perp_D}$.
    (e) The following statements are equivalent.
        (i) $\mathscr{B}(H)$ and $\mathscr{B}(D)$ are canonically isomorphic.
        (ii) φ is surjective.
        (iii) $\psi(\mathscr{B}(D)) \subseteq \mathscr{B}(H)$ and ψ is injective.
        (iv) If $X \subseteq D$, then $X^{\perp_H \perp_D} = X^{\perp_D \perp_D}$
    (f) If $\mathscr{B}(H)$ and $\mathscr{B}(D)$ are canonically isomorphic, then $X^{\perp_D \perp_H} = X^{\perp_H \perp_H}$ for each subset X of D. An example which shows that the converse fails is given by the direct sum $D = H \oplus K$ where H and K are nonzero ℓ-groups. In this example $\psi(\mathscr{B}(D)) \subset \mathscr{B}(H)$.
    (g) Suppose that Ω is a set of operators on D^+; so, each $w \in \Omega$ induces a function $w : D^+ \longrightarrow D^+$. Assume:
        (i) each $w \in \Omega$ is isotone;
        (ii) $a \wedge b = 0 \Longrightarrow aw \wedge b = 0$, for all $a, b \in D$ and every $w \in \Omega$;
        (iii) if $0 < a \in D$, then $0 < aw \in H$ for some $w \in \Omega$.
        Then $\mathscr{B}(H)$ and $\mathscr{B}(D)$ are canonically isomorphic. (Show that (iv) of (e) holds.)
    (h) H is called a *cℓ-essential subset* of D if $H^+ \cap C \neq 0$ for each nonzero convex ℓ-subgroup C of D. If H is a cℓ-essential subset of D, then $\mathscr{B}(H)$ and $\mathscr{B}(D)$ are canonically isomorphic. (Let $\Omega = \{n(\cdot) \wedge d : n \in \mathbb{N}, d \in D^+\}$, where $a(n(\cdot) \wedge d) = na \wedge d$ and use (g).)

15. A monomorphism $\varphi : H \longrightarrow D$ of ℓ-groups is called *cℓ-essential* if $\varphi(H)$ is a cℓ-essential ℓ-subgroup of D (Exercise 14). Show that if φ is a cℓ-essential monomorphism, then φ is complete if either D is abelian or, for each $0 < d \in D$, there exists $h \in H$ with $0 < \varphi(h) \leq d$ (use Exercises 2.1.16 and 2.1.17).

16. Show that the ℓ-subgroup G of the ℓ-group H is cℓ-essential in H if and only if, for each convex subgroup C of H, $G \cap C \neq 0$ or C is trivially ordered.

17. Let G be a po-group, let S be a convex subset of G^+ with $0 \in S$ and let H be the subgroup of G that is generated by S.
    (a) The following conditions are equivalent.
        (i) S is the positive cone for some po-subgroup of G.

(ii) $S+S \subseteq S$.
(iii) S is directed up, and if $a \in S$ then $2a \in S$.
(iv) $H^+ = S$.
(v) $H = \{a-b : a,b \in S\}$.

(b) If the conditions in (a) hold or if G has the Riesz decomposition property (that is, G satisfies the condition in Theorem 2.1.4(k)), then H is convex.
(c) If H is convex, then H is closed under any finite sups and infs that exist in the po-group G.
(d) If G is an ℓ-group, then H is a convex ℓ-subgroup of G, and the conditions in (a) are equivalent to
(vi) $H = \{a \in G : |a| \in S\}$.

18. Let G and A be ℓ-subgroups of the ℓ-group H, and suppose that $G+A$ is a subgroup of H.

(a) Show that $(g+a)^+ \in G+A$ whenever $g \in G$, $a \in A$ and one of them is comparable to 0 iff $g \wedge a \in G+A$ for every $g \in G^+$ and $a \in A^+$. (Assuming the meet condition show that $(g+a) \wedge 0 \in G+A$ by considering the elements $g^- + (g+a) \wedge 0$ if $a \geq 0$ and $(g+a) \wedge 0 - a + g^-$ if $a < 0$.)
(b) If A is totally ordered show that $G+A$ is an ℓ-subgroup of H iff $g \wedge a \in G+A$ for every pair $(g,a) \in G^+ \times A^+$.
(c) If $g \wedge a \in G$ for every $(g,a) \in G^+ \times A^+$ show that G is convex in $G+A^+$.
(d) If A is totally ordered show that G is a convex ℓ-subgroup of the ℓ-group $G+A$ iff $g \wedge a \in G$ for every pair of elements $(g,a) \in G^+ \times A^+$.

19. Let C be a convex directed normal subgroup of the Riesz group G (see Exercise 2.1.18). Show that if $S = \{x_1+C, x_2+C, \ldots\}$ is a countable subset of G/C, then there exists a subset $T = \{t_1, t_2, \ldots\}$ of G such that $t_n + C = x_n + C$ for each n and $T \longrightarrow S$ is an order isomorphism. (Given $t_1, \ldots, t_{n-1}$ let $X = \{t_i : t_i + C < x_n + C\}$ and $Y = \{t_i : x_n + C < t_i + C\}$. Take $x, y \in G$ with $X < x$, $y < Y$ and $x + C = x_n + C = y + C$. Then $X \cup \{x - c_2\} \leq t_n \leq Y \cup \{x + c_1\}$ where $y = x + c_1 - c_2$ with $c_1, c_2 \in C^+$.)

20. Let $f : L \longrightarrow M$ be a lattice homomorphism from the lattice L onto the lattice M. Suppose $S = \{f(x_1), f(x_2), \ldots\}$ is a countable subset of M. Show that there is a subset $T = \{t_1, t_2, \ldots\}$ of L with $f(t_n) = f(x_n)$ for each n and $f : T \longrightarrow S$ is an order isomorphism. (In the previous exercise let $x = \text{lub } X$, $y = \text{glb } Y$ and $t_n = (x \vee x_n) \wedge y$.)

2.3 Archimedean ℓ-groups

In this section we are concerned with those ℓ-groups which, like the additive group of the reals, have the property that they do not have any nonzero bounded subgroups. As we will see, each such ℓ-group is abelian, and its completion is also an ℓ-group. Also, a representation theorem for these ℓ-groups as extended real-valued continuous functions on a topological space will be given. The target of this representation is rather complete and has the property that it is a summand of each ℓ-group in the current category in which it is embedded as a convex ℓ-subgroup. This latter property is also investigated here and other ℓ-groups which have it are identified. In preparation for the representation theorem we will establish various topological results. One of these is the duality between the category of Boolean algebras and the category of those compact Hausdorff spaces in which each open subset is a union of sets which are both open and closed. The topological space on which the representing functions are defined is obtained from the Boolean algebra of polars through this duality.

A po-group G is called *integrally closed* if, for all $a, b \in G$, $\mathbb{N}a \leq b$ implies that $a \leq 0$; and it is called *archimedean* if $\mathbb{Z}a \leq b$ implies that $a = 0$. An ℓ-group G is *complete* if its underlying lattice is conditionally complete. Recall that this means that each nonempty bounded subset of G has an inf and a sup. It is σ*-complete* if each nonempty bounded countable subset has an inf and a sup. While this terminology is different than that for posets no confusion should arise since the only complete (in the sense of posets) po-group is 0. The basic connections between these concepts are given in

Theorem 2.3.1. *(a) An integrally closed po-group is archimedean.*

(b) An archimedean po-group need not be integrally closed.

(c) An ℓ-group is archimedean iff it is integrally closed.

(d) Each σ-complete ℓ-group is integrally closed.

Proof. Certainly, (a) is clear, and an example of an archimedean po-group that is not integrally closed is given in Exercise 5. Suppose that the ℓ-group G is archimedean and $na \leq b$ for each $n \in \mathbb{N}$. Then $na^+ \leq b^+$ for each $n \in \mathbb{Z}$; so $a^+ = 0$ and $a \leq 0$. If G is σ-complete and $na \leq b$ for each $n \in \mathbb{N}$, then $c = \sup \mathbb{N}a \in G$. Since $c - a \geq na$ for each $n \in \mathbb{N}$, $c - a \geq c$ and $a \leq 0$. So G is integrally closed. □

Let G be a directed po-group, and let $D(G)$ be its Dedekind completion. Recall from Section 1.3 that

$$D(G) = \{X \subseteq G : X = LUX,\ \ \phi \subset X \subset G\}$$

$$= \{LUX : X \subseteq G,\ X \neq \phi \text{ and } UX \neq \phi\}.$$

We wish to extend the group operation from G to $D(G)$. It will be convenient to denote LUX by X^*, for $X \subseteq G$. Since $*$ is a closure operator on the power set of G

$$X^* = \cap\{A \subseteq G : X \subseteq A \text{ and } A = A^*\}. \tag{2.3.1}$$

A *partially ordered semigroup* is a semigroup S which is also a poset in which each left and each right translation is order preserving:

$$\forall x, y, z \in S\ \ x \leq y \Rightarrow\ z + x \leq z + y \text{ and } x + z \leq y + z.$$

A *monoid* is a semigroup with identity.

If X and Y are subsets of the group G, then, as usual, $X+Y=\{x+y : x\in X \text{ and } y\in Y\}$ and $X-Y=\{x-y : x\in X \text{ and } y\in Y\}$.

Theorem 2.3.2. *Let G be a directed po-group. If $X,Y\in D(G)$ define $X\oplus Y=(X+Y)^*$. Then $D(G)$ is a partially ordered monoid with zero element 0^*, and G is a po-subgroup of the po-group of units of $D(G)$.*

Proof. It suffices to verify the following four statements, for $X,Y,Z\in D(G)$ and $g,h\in G$, since, clearly, $X\oplus Y\in D(G)$:
(a) $[(X+Y)^*+Z]^*=(X+Y+Z)^*=[X+(Y+Z)^*]^*$;
(b) $[X+L(0)]^*=X=[L(0)+X]^*$;
(c) $X\subseteq Y\Rightarrow (X+Z)^*\subseteq (Y+Z)^*$ and $(Z+X)^*\subseteq (Z+Y)^*$;
(d) $L(g)+L(h)=L(g+h)$.
To verify the first equation in (a) it suffices, by equation (2.3.1), to check that if $A=A^*$, then $(X+Y)^*+Z\subseteq A$ if and only if $X+Y+Z\subseteq A$. "Only if" is a consequence of the inclusion $X+Y\subseteq (X+Y)^*$; as for "if", for each $z\in Z$, $X+Y\subseteq A-z=LUA-z=LU(A-z)$ implies that $(X+Y)^*\subseteq (A-z)^*=A-z$. So $(X+Y)^*+Z\subseteq A$. The second equation in (a) follows in a similar way. For (b), if $x\in X$ and $y\in L(0)$, then $x+y\leq x$; so $x+y\in L(x)\subseteq X$ and $X+L(0)\subseteq X\subseteq X+L(0)$. Similarly, $X=L(0)+X$. Also, (c) is obviously true; and as for (d), $L(g)+L(h)\subseteq L(g+h)=L(g)+h\subseteq L(g)+L(h)$. □

In the next result we determine when $D(G)$ is an ℓ-group.

Theorem 2.3.3. *A directed po-group can be embedded into a complete ℓ-group if and only if it is integrally closed. In particular, each archimedean ℓ-group can be embedded into a complete ℓ-group.*

Proof. By Theorem 2.3.1(d) each subgroup of a complete ℓ-group is integrally closed. Suppose that G is integrally closed. We claim that $D(G)$ is a group and hence it is a complete ℓ-group. It suffices to show that if $X\in D(G)$, then $Y=-U(X)=L(-X)$ is a right inverse for X in $(D(G),\oplus)$; and for this we need that $LU(X+L(-X))=L(0)$. If $x\in X$ and $y\leq -X$ then $x+y\leq 0$; that is, $LU(X+L(-X))\subseteq L(0)$. For the reverse inclusion we merely need to verify that $0\in LU[X+L(-X)]$, or, if $a\in U[X+L(-X)]$, then $0\leq a$. We will show below, by induction on n, that $na\in U[X+L(-X)]$ for each $n\in\mathbb{N}$. Assuming this, then $na\geq x+y$ for $x\in X$ and $y\in L(-X)$. So $n(-a)\leq -(x+y)$, $-a\leq 0$ and $a\geq 0$. Suppose then, by induction, that $ka\in U[X+L(-X)]$. If $x\in X$ and $y\leq -X$, then $ka\geq X+y$, $ka-y\in U(X)$, $y-ka\in L(-X)$, and $a\geq x+y-ka$. So $(k+1)a\geq x+y$ and the induction is complete. □

A *completion* of the ℓ-group G is a pair (H,φ) where H is a complete ℓ-group and $\varphi : G\longrightarrow H$ is a monomorphism whose image is dense in H. Each archimedean ℓ-group has a unique completion.

Theorem 2.3.4. *An ℓ-group has a completion if and only if it is archimedean. If (A,α) and (B,β) are two completions of the ℓ-group G, then there is a unique isomorphism $\rho : A\longrightarrow B$ such that the diagram*

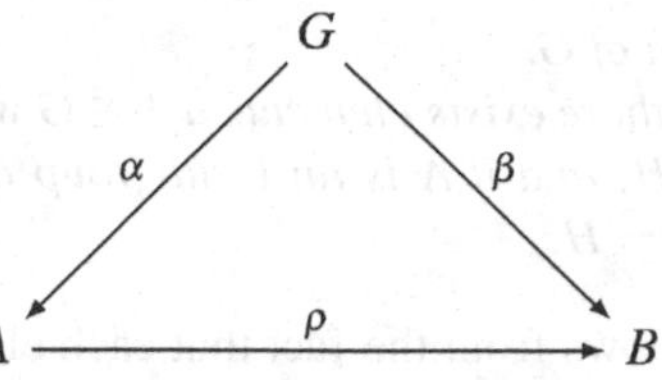

is commutative.

Proof. The first statement is a consequence of Theorems 2.3.1 and 2.3.3. According to Theorem 1.3.4 there is a unique lattice isomorphism ρ that makes the diagram commutative. If $a \in A$ and $g \in G$, then $\alpha(g) \leq a$ iff $\beta(g) \leq \rho(a)$; so if $a, b \in A$, then

$$\rho(a+b) = \rho\left(\bigvee_{\alpha(g)\leq a} \alpha(g) + \bigvee_{\alpha(h)\leq b} \alpha(h)\right) = \rho\left(\bigvee_{\substack{\alpha(g)\leq a \\ \alpha(h)\leq b}} \alpha(g+h)\right)$$

$$= \bigvee_{\substack{\beta(g)\leq\rho(a) \\ \beta(h)\leq\rho(b)}} (\beta(g)+\beta(h)) = \bigvee_{\beta(g)\leq\rho(a)} \beta(g) + \bigvee_{\beta(h)\leq\rho(b)} \beta(h)$$

$$= \rho(a)+\rho(b).$$

Thus, ρ is an isomorphism of ℓ-groups. □

We give next a useful characterization of the completion that does not explicitly mention sups or infs; but first we present the following preliminary result. The subset G of the ℓ-group H is called *left dense in* H if $G^+\backslash\{0\}$ is coinitial in $H^+\backslash\{0\}$, and G is called *right dense in* H if G^+ is cofinal in H^+. If G is a convex subset of H^+ and $0 \in G$, then G is left dense in $G^{\perp\perp}$, as is easily verified, and each abelian ℓ-group is right dense in its divisible hull.

Theorem 2.3.5. *Let G be a subset of the archimedean ℓ-group H. Assume that $g \in G$ implies that $\mathbb{N}g^+ \subseteq G$. Then the following statements are equivalent.*

(a) *G is left dense in H.*
(b) *If $0 < h \in H$, then $h = \text{lub}_H L_G(h)$.*

Proof. If (a) holds and $h > 0$, then each upper bound in H of $L_G(h)$ is strictly positive. Suppose that $h \neq \text{lub}_H L_G(h)$. Then there is an $x \in H$ with $x \geq L_G(h)$ and $h \nleq x$. So $L_G(h) \leq h \wedge x = y < h$. Let $g \in G$ with $0 < g \leq h - y$. Then $g \in L_G(h)$, and, inductively, if $ng \in L_G(h)$, then $(1+n)g \leq g + y \leq h$. Since H is archimedean this gives that $g \leq 0$ which contradicts that $g > 0$. Conversely, if (b) holds, then for each $0 < h \in H$ there exists $g \in L_G(h)$ with $g \nleq 0$. So $h \geq g^+ > 0$. □

Theorem 2.3.6. *Let G be an ℓ-subgroup of the complete ℓ-group H. Then the following statements are equivalent.*

(a) H is the completion of G.
(b) If $0 < h \in H$, then there exists elements $a, b \in G$ with $0 < a \leq h \leq b$.
(c) G is left dense in H, and if A is an ℓ-subgroup of H that is complete and contains G, then $A = H$.

Proof. (a) $\Rightarrow$ (b). This follows from the fact that each element of H is the sup and the inf of subsets of G.

(b) $\Rightarrow$ (c). Let A be an ℓ-subgroup of H which contains G and which is complete. To show that $A = H$ it suffices to show that A is a convex subgroup of H. Suppose that $0 < h < x$ with $h \in H$ and $x \in A$. Since A is complete $\text{lub}_A L_G(h)$ exists. But A is a $c\ell$-essential ℓ-subgroup of H, so $\text{lub}_A L_G(h) = \text{lub}_H L_G(h) = h$ by Exercise 2.2.15 and Theorem 2.3.5; thus, $h \in A$.

(c) $\Rightarrow$ (a). Since $C^H(G)$ is complete, $H = C^H(G)$. If $h \in H$, then $h \leq g$ for some $g \in G$, and hence $g - h = \text{lub}_H X$ for some subset X of G, by Theorem 2.3.5. Thus, $h = \inf_H(-X + g)$, and, if h is replaced by $-h$, we get that $h = \text{lub}_H Y$ for some subset Y of G. □

Suppose that G is a left dense ℓ-subgroup of the complete ℓ-group H. An immediate consequence of the previous result is that if A is a convex ℓ-subgroup of G, then the completion of A is the convex ℓ-subgroup of H generated by A.

Our next goal is to show that an archimedean ℓ-group is abelian. We will accomplish this by showing that a complete ℓ-group is abelian. Another proof is given in Exercise 2.4.14. We first need two results, each of which is interesting in its own right. The subset T of the poset P is called *completely closed* in P provided that T contains the least upper bound or greatest lower bound of any of its nonempty subsets, if either exists in P.

Theorem 2.3.7. *Consider the following conditions on the convex ℓ-subgroup A of the ℓ-group G:*

(a) A is a summand of G.
(b) A is a polar.
(c) A is completely closed in G.

Then (a) implies (b), (b) implies (c), and, if G is complete, (c) implies (a).

Proof. That (a) implies (b) is given in Theorem 2.2.6 (c). Suppose that $A = A^{\perp\perp}$, $\{a_i\} \subseteq A$ and $a = \bigvee a_i \in G$. According to Theorem 2.1.3 G is infinitely distributive; so if $0 \leq b \in A^{\perp}$, then $b \wedge a^+ = b \wedge \bigvee a_i^+ = \bigvee(b \wedge a_i^+) = 0$, and $b \wedge a^- = b \wedge \bigwedge a_i^- = \bigwedge(b \wedge a_i^-) = 0$ since $a^- = (\bigwedge -a_i) \vee 0 = \bigwedge a_i^-$. Thus, $b \wedge |a| = b \wedge (a^+ + a^-) = 0$ and $a \in A^{\perp\perp} = A$; so A is completely closed. Assume that G is complete and that A is completely closed. For any $g \in G^+$, $g \geq \{a \wedge g : a \in A^+\}$, so

$$g \geq \bigvee_{a \in A^+} (a \wedge g) = b \in A^+.$$

Let $c = g - b$. Then for $a \in A^+$, $0 \leq a \wedge c = a \wedge (g - b) = (a + b) \wedge g - b \leq 0$ since $a + b \in A^+$. Thus $c \in A^{\perp}$ and $G = A \oplus A^{\perp}$. □

For elements a and b of the group G recall that the b-conjugate of a is denoted by $a^b = -b + a + b$, and denote the *commutator* of a and b by $[a,b] = -a - b + a + b$. Note that $[a,b] = -[b,a]$.

Theorem 2.3.8. *The following identities hold in the group G, for each $n \geq 1$.*

(a) $[a,nb] = \sum_{i=0}^{n-1} [a,b]^{ib}$.

(b) $[-nb,a] = \sum_{i=n}^{1} [a,b]^{-ib}$.

Proof. We use induction on n to prove (a), the case $n = 1$ being trivial. Suppose that $[a,(n-1)b] = \sum_{i=0}^{n-2} [a,b]^{ib}$.

Then,

$$\begin{aligned}
[a,nb] &= -a - nb + a + nb \\
&= -a - b + a + b - b - a - (n-1)b + a + (n-1)b + b \\
&= [a,b] - b + [a,(n-1)b] + b \\
&= [a,b] + \left(\sum_{i=0}^{n-2} [a,b]^{ib} \right)^b \\
&= \sum_{i=0}^{n-1} [a,b]^{ib}.
\end{aligned}$$

As for (b),

$$\begin{aligned}
[-nb,a] &= -[a,-nb] = -\sum_{i=0}^{n-1} [a,-b]^{-ib} \\
&= \sum_{i=n-1}^{0} -[a,-b]^{-ib} = \sum_{i=n-1}^{0} [-b,a]^{-ib} \\
&= \sum_{i=n-1}^{0} ib + b - a - b + a - ib \\
&= \sum_{i=n-1}^{0} (i+1)b - a - b + a + b - (i+1)b \\
&= \sum_{i=n}^{1} [a,b]^{-ib}.
\end{aligned}$$

□

We are now able to prove

Theorem 2.3.9. *A directed integrally closed po-group is abelian.*

Proof. By Theorem 2.3.3 we may assume that the group is a complete ℓ-group G. If $x \in G$, then $G = (x^+)^\perp \oplus (x^+)^{\perp\perp}$ by Theorem 2.3.7; and the components of x in this decomposition (namely, $-x^-$ and x^+) are comparable to 0. By the remarks after Theorem 2.2.6 this means that for each finite subset X of G there is a decomposition $G = C_1 \oplus \cdots \oplus C_n$ of G for which the components of each element of X are comparable to 0. Now, let $a, b \in G^+$. By taking the aforementioned decomposition we may assume that $b - a$ and $[a,b]$ are comparable to 0. Since the case $a \leq b$ and $0 \leq [a,b]$ is identical to the case $b \leq a$ and $[a,b] \leq 0$, and the case $a \leq b$ and $[a,b] \leq 0$ is identical to the case $b \leq a$ and $0 \leq [a,b]$, the four cases reduce to two. By taking a further refinement we may assume that $[[a,b],b]$ is also comparable to 0. Thus, there are four possibilities:

(i) $a \leq b,\ 0 \leq [a,b],\ 0 \leq [[a,b],b]$;
(ii) $a \leq b,\ 0 \leq [a,b],\ [[a,b],b] \leq 0$;
(iii) $a \leq b,\ [a,b] \leq 0,\ 0 \leq [[a,b],b]$;
(iv) $a \leq b,\ [a,b] \leq 0,\ [[a,b],b] \leq 0$.

We will show that $[a,b] = 0$ in each of the first two cases and leave the second two for the reader. In case (i) we have that $0 \leq -[a,b] - b + [a,b] + b$, or $[a,b] \leq [a,b]^b$. So $[a,b] \leq [a,b]^{ib}$ for each $i \geq 0$, and hence $n[a,b] \leq [a,nb]$ for each $n \geq 1$ by (a) of Theorem 2.3.8. But $[a,nb] = -a - nb + a + nb = -a + a^{nb} \leq -a + b$ since $a^{nb} \leq b^{nb} = b$. So $n[a,b] \leq -a + b$ and $[a,b] = 0$. For (ii), we have that $-[a,b] - b + [a,b] + b \leq 0$; so $[a,b] \leq [a,b]^{-ib}$ for $i \geq 1$, and, by (b) of Theorem 2.3.8, $n[a,b] \leq [-nb,a] = nb - a - nb + a \leq a$, for each $n \geq 1$, since $-a^{-nb} \leq 0$. Thus, $[a,b] = 0$. □

An ℓ-group is called *ℓ-simple* if it has exactly two normal convex ℓ-subgroups. The next theorem establishes the fact that the only abelian ℓ-simple ℓ-groups are those that can be embedded in the additive group of the reals.

Theorem 2.3.10. *(Hölder) The following statements are equivalent for the ℓ-group G.*

(a) The only convex ℓ-subgroups of G are 0 and G.
(b) G is totally ordered and archimedean.
(c) G can be embedded in $\mathbb{R}$.

Proof. (a) implies (b). If $a \wedge b = 0$, then $C(a) \cap C(b) = 0$. Hence $a = 0$ or $b = 0$ and G is totally ordered. If $\mathbb{Z}a \leq b$ and $a \neq 0$, then since $b \in C(a)$, $\mathbb{Z}a \leq ka$ for some $k \in \mathbb{Z}$; but then $\mathbb{Z}a \leq 0$ which is impossible. Thus $a = 0$ and G is also archimedean.

(b) implies (c). By Theorem 2.3.9 G is abelian. By passing to the divisible hull $d(G)$ of G we may assume that G is divisible (Exercise 2.1.3) since $d(G)$ is also archimedean. So G is a vector space over $\mathbb{Q}$ and $\mathbb{Q}^+G^+ \subseteq G^+$. Suppose that $0 \neq A$ is a subspace of G and $\alpha : A \longrightarrow \mathbb{R}$ is an ℓ-group embedding. If A is a proper subspace take $x \in G^+ \backslash A$ and let $A_- = \{a \in A : a < x\}$ and $A_+ = \{a \in A : x < a\}$. Then $\alpha(A_-) < \alpha(A_+)$ so $\alpha(A_-) \leq r \leq \alpha(A_+)$ for some $r \in \mathbb{R}$. Now we can extend α to the subspace $A + \mathbb{Q}x$ by defining $\beta : A + \mathbb{Q}x \longrightarrow \mathbb{R}$ by $\beta(a + px) = \alpha(a) + pr$. This is clearly a

$\mathbb{Q}$-linear transformation, and we check that it is isotone. Suppose that $a+px \geq 0$. If $p>0$, then $x \geq -\frac{1}{p}a$ and $r \geq -\frac{1}{p}\alpha(a)$; that is $\beta(a+px)=\alpha(a)+pr \geq 0$. If $p<0$, then $-\frac{1}{p}a \geq x$; so $-\frac{1}{p}\alpha(a) \geq r$ and $\beta(a+px)=\alpha(a)+pr \geq 0$. If $p=0$, then $\beta(a)=\alpha(a) \geq 0$. Thus, β is isotone, and, since $A+\mathbb{Q}x$ does not have a nontrivial convex subgroup, β is an embedding. Now, consider the set $\{(A,\alpha) : A$ is a subspace of G and $\alpha : A \longrightarrow \mathbb{R}$ is an embedding$\}$, supplied with the partial order given by : $(A,\alpha) \leq (B,\beta)$ if $A \subseteq B$ and $\beta(a)=\alpha(a)$ for each a in A. By Zorn's Lemma this poset has a maximal element (A,α), and the preceding argument implies that $A=G$. □

We next wish to extend Theorem 2.3.10 to archimedean ℓ-groups. More specifically, we wish to represent the elements in such an ℓ-group by extended real-valued continuous functions on a topological space. The topological space will arise from the Boolean algebra of polars of the ℓ-group. We turn now to some topological considerations and to the construction of this space.

If A is a subset of the topological space X, then the closure of A in X will be denoted by A^-, and its interior will be denoted by A°. The subset A is *dense* in X if $A^- = X$. This is easily seen to be equivalent to the condition that $A \cap U \neq \emptyset$ for each nonempty open subset U of X. The subset B of X is *nowhere dense* in X if each nonempty open set U in X contains a nonempty open subset V such that $V \cap B = \emptyset$. Clearly, if B is nowhere dense, then so is B^-. Also, it is straightforward to verify that B^- is nowhere dense exactly when its complement $B^{-\prime}$ is dense, or, equivalently, $B^{-\circ}$ is empty. One example of a nowhere dense set is obtained by taking the *boundary* $V^- \cap V'^-$ of an open set V, as can readily be verified. A subset of X is said to be of the *first category* if it is a union of a countable number of nowhere dense subsets of X. Recall that X is *compact* if each of its open covers has a finite subcover, and is *Hausdorff* if distinct elements of X have disjoint neighborhoods. For later use we record the following result; the one after it will be put to immediate use.

Theorem 2.3.11. *(Baire category theorem.) If X is a compact Hausdorff space, then each subset of X of the first category has an empty interior. Equivalently, a countable intersection of dense open sets is dense.*

Proof. We will first show that if $x \in U$ is open, then there exists V open with $x \in V$ and $V^- \subseteq U$. For each $y \in U'$ there are disjoint open sets V_y and W_y with $x \in V_y$ and $y \in W_y$. Since U' is compact a finite number of the W_y, say, $W_1,\ldots,W_n$, cover U'. Let $V_1,\ldots,V_n$ be the corresponding V_y's. If

$$W=\bigcup_{i=1}^{n} W_i \text{ and } V=\bigcap_{i=1}^{n} V_i,$$

then $x \in V$, $U' \subseteq W$ and $V \cap W = \emptyset$. So $V^- \cap W = \emptyset$ and $x \in V \subseteq V^- \subseteq U$. Now, suppose that

$$B=\bigcup_{n=1}^{\infty} B_n$$

where each B_n is nowhere dense. To show that $B^\circ = \emptyset$ it suffices to show that if U is a nonempty open subset of X, then $U \backslash B \neq \emptyset$. Take V_1 open and nonempty with $V_1^- \subseteq U$, and let $U_1 \subseteq V_1$ be a nonempty open set with $U_1 \cap B_1 = \emptyset$. Inductively, given the decreasing chain

$$U \supseteq U_1^- \supseteq U_1 \supseteq U_2^- \supseteq \cdots \supseteq U_{n-1}^- \supseteq U_{n-1} \supseteq U_n^- \supseteq U_n$$

of nonempty open subsets $U_1, \ldots, U_n$ with $U_k \cap B_k = \emptyset$ for $1 \leq k \leq n$, let U_{n+1} and V_{n+1} be nonempty open subsets with $V_{n+1}^- \subseteq U_n$, $U_{n+1} \subseteq V_{n+1}$ and $U_{n+1} \cap B_{n+1} = \emptyset$. Then $C = \cap U_n = \cap U_n^-$ is not empty since X is compact, and $C \cap B = \emptyset$ since $C \cap B_n = \emptyset$ for each n. The verification of the last statement is left to the reader. □

A subset of X which is both open and closed will be called *clopen*. The collection of all clopen subsets of X is a Boolean subalgebra of the Boolean algebra of all subsets of X; it is called the *dual algebra of* X. Recall that a collection $\mathscr{B}$ of open sets is a *base* for the topology of X if each open set in X is a union of some of the sets in $\mathscr{B}$, and $\mathscr{B}$ is a *subbase* if the collection of all intersections of finitely many members of $\mathscr{B}$ is a base. A compact Hausdorff space whose clopen sets form a base is called a *Boolean space*. If F is a closed subset of the Boolean space X and U is open in F, then there is a family $\{P_i\}$ of clopen sets in X such that $U = F \cap (\cup P_i) = \cup (P_i \cap F)$. Thus, F is also a Boolean space. Moreover, if U is clopen in F, then a finite number of the P_i, say $P_1, \ldots, P_n$, cover U. But then $U = P \cap F$ where $P = P_1 \cup \cdots \cup P_n$; that is, each clopen set in F is the intersection with F of a clopen set in X.

Theorem 2.3.12. *Let X be a compact Hausdorff space. Suppose that $\mathscr{A}$ is a subalgebra of the dual algebra $\mathscr{B}$ of X, and $\mathscr{A}$ has the property that for each pair of distinct elements x, y in X there is a P in $\mathscr{A}$ such that $x \in P$ and $y \in P'$. Then X is a Boolean space and $\mathscr{A} = \mathscr{B}$.*

Proof. We first show that if x is not in the closed set F, then there exists a clopen set $P \in \mathscr{A}$ with $F \subseteq P$ and $x \notin P$. For each $y \in F$ there is a set $P_y \in \mathscr{A}$ with $y \in P_y$ and $x \in P_y'$. Since F is compact a finite number of the P_y, say, $P_1, \ldots, P_n$, cover F. Let $P = P_1 \cup \cdots \cup P_n$. Then $F \subseteq P \in \mathscr{A}$ and $x \in P'$. Thus, each open set is a union of sets from $\mathscr{A}$, and hence X is a Boolean space. Also, each clopen set is a finite union of sets from $\mathscr{A}$, and hence is in $\mathscr{A}$; so $\mathscr{A} = \mathscr{B}$. □

A *Boolean homomorphism* between two Boolean algebras is a lattice homomorphism that preserves complements (equivalently, it is an identity preserving ring homomorphism between the corresponding Boolean rings (Exericse 1.2.8)). The set $\mathbf{2} = \{0, 1\}$ is *the* totally ordered Boolean algebra; we will also consider $\mathbf{2}$ as a topological space, giving it the discrete topology. Given the Boolean algebra B, the set $X = \text{Bool}[B, \mathbf{2}]$ of all $\mathbf{2}$-valued homomorphisms on B is a subspace of the product space $\mathbf{2}^B$. X is called the *dual space of* B.

Theorem 2.3.13. *The dual space $X = \text{Bool}[B, \mathbf{2}]$ of the Boolean algebra B is a Boolean space. Moreover, if $\mathscr{B}$ is the dual algebra of X, then the function $\alpha : B \longrightarrow \mathscr{B}$ given by*

$$\alpha(b) = \{f \in X : f(b) = 1\}$$

is an algebra isomorphism.

Proof. By Tychonoff's theorem the product space $\mathbf{2}^B$ is compact, and it is also Hausdorff. Recall that a base for $\mathbf{2}^B$ consists of all those sets of the following form. Given a finite subset $\{b_1, \ldots, b_n\}$ of B and elements $\varepsilon_1, \ldots, \varepsilon_n \in \mathbf{2}$, let

$$V(b_1, \ldots, b_n;\ \varepsilon_1, \ldots, \varepsilon_n) = \{f \in \mathbf{2}^B : f(b_i) = \varepsilon_i \text{ for } i = 1, \ldots, n\}.$$

Since $V(b;\varepsilon)' = V(b;\varepsilon')$ and since

$$V(b_1, \ldots, b_n; \varepsilon_1, \ldots, \varepsilon_n) = \bigcap_{i=1}^{n} V(b_i; \varepsilon_i),$$

each basic open set is clopen. So $\mathbf{2}^B$ is a Boolean space. We claim that X is a closed subset of $\mathbf{2}^B$. Toward this end let $b, c \in B$, and let $Y_{b,c} = \{f \in \mathbf{2}^B : f(b \vee c) = f(b) \vee f(c)\}$. If $\pi_b : \mathbf{2}^B \longrightarrow \mathbf{2}$ denotes the projection (or evaluation) at b, given by $\pi_b(f) = f(b)$, then π_b is continuous. Moreover, the function $\pi_b \vee \pi_c : \mathbf{2}^B \longrightarrow \mathbf{2}$ given by $(\pi_b \vee \pi_c)(f) = \pi_b(f) \vee \pi_c(f)$ is continuous since it is the composite of the continuous functions

$$\mathbf{2}^B \xrightarrow{\Delta} \mathbf{2}^B \times \mathbf{2}^B \xrightarrow{\pi_b \times \pi_c} \mathbf{2} \times \mathbf{2} \xrightarrow{\vee} \mathbf{2},$$

where Δ is the diagonal map given by $\Delta(f) = (f, f)$. Since $Y_{b,c} = \{f \in \mathbf{2}^B : (\pi_b \vee \pi_c)(f) = \pi_{b \vee c}(f)\}$ is the set at which two continuous functions into a Hausdorff space agree, $Y_{b,c}$ is closed. Hence so is the intersection $Y = \{f \in \mathbf{2}^B : \forall b, c \in B, f(b) \vee f(c) = f(b \vee c)\}$ of all of the $Y_{b,c}$. Similarly, $Z_b = \{f \in \mathbf{2}^B : \pi_{b'}(f) = \pi_b(f)' = ({}' \circ \pi_b)(f)\}$ is closed, and so is the intersection $Z = \{f \in \mathbf{2}^B : \forall b \in B,\ f(b') = f(b)'\}$ of all of the Z_b. Also, $X = Y \cap Z$ is closed.

We now show that α is an isomorphism. First, $\alpha(b) = \pi_b^{-1}(1) \cap X \in \mathscr{B}$. Next, $\alpha(b \vee c) = \{f \in X : f(b) \vee f(c) = 1\} = \alpha(b) \cup \alpha(c)$, and $\alpha(b') = \{f \in X : f(b)' = 1\} = \alpha(b)'$. So α is a homomorphism. If M is a maximal ideal of B that excludes b , then B/M is a two element Boolean algebra (Exercises 1.2.8 and 1.2.9), and hence M gives rise to a homomorphism $f \in X$ with $f(b) = 1$; that is, if $b \neq 0$ then $\alpha(b) \neq \emptyset$. So α is a monomorphism and $\alpha(B)$ is a subalgebra of $\mathscr{B}$. If f and g are distinct elements of X, then, for some $b \in B$, $f(b) = 1$ and $g(b) = 0$. So $f \in \alpha(b)$ and $g \notin \alpha(b)$; thus $\alpha(B) = \mathscr{B}$ by Theorem 2.3.12. □

A $\mathbf{2}$-valued homomorphism of the Boolean algebra B is completely determined by its kernel which is a maximal ideal of B. Thus, the dual space $X = \text{Bool}[B, \mathbf{2}]$ may be replaced by the set Spec(B) which consists of all of the maximal ideals of B, and we will make this replacement whenever it is convenient to do so. The basic clopen sets of Spec(B) are of the form $V(b) = \{M \in \text{Spec}(B) : b \notin M\}$, where $b \in B$, and, according to Theorem 2.3.13, V is an isomorphism between B and the algebra of clopen sets in Spec(B). This topology on Spec(B) is called the *Zariski topology* or the *hull-kernel topology* of Spec(B).

The previous theorem asserts that each Boolean algebra is isomorphic to its second dual. It is also true that each Boolean space is isomorphic to its second dual.

Theorem 2.3.14. *Let X be a Boolean space, let A be the dual algebra of X, and let $Y =$ Bool$[A,\mathbf{2}]$ be the dual space of A. Then the function $\beta : X \longrightarrow Y$ given by $\beta(x)(P) = 1$ if $x \in P$ and $\beta(x)(P) = 0$ if $x \notin P$ is a homeomorphism.*

Proof. If P and Q are clopen sets in X, then it is clear that $\beta(x)(P') = [\beta(x)(P)]'$; also, $\beta(x)(P \cup Q) = \beta(x)(P) \vee \beta(x)(Q)$ since $\beta(x)(P) \vee \beta(x)(Q) = 1$ exactly when $x \in P \cup Q$. Thus $\beta(x) \in Y$ for each x in X. To see that β is continuous let U be a clopen set in Y. Then, by Theorem 2.3.13, $U = \alpha(P) = \{f \in Y : f(P) = 1\}$ for some (unique) clopen set P in X. But then $\beta^{-1}(U) = \{x \in X : \beta(x)(P) = 1\} = P$. Since the clopen sets in Y form a base β is continuous. This also shows that β is onto. For if $U \neq \emptyset$, then $\beta(X) \cap U \neq \emptyset$; so $\beta(X)$ is dense and closed in $Y (\beta(X)$ is compact), and hence $\beta(X) = Y$. If $x \neq y$ in X, then there is a clopen set P in X with $x \in P$ and $y \in P'$. So $\beta(x)(P) \neq \beta(y)(P)$, $\beta(x) \neq \beta(y)$, and β is one-to-one. Since β takes closed sets to closed sets it is a homeomorphism. □

We now wish to examine the duality that connects homomorphisms between Boolean algebras with continuous functions between Boolean spaces. This is most conveniently expressed using the language of category theory. A contravariant (respectively, covariant) functor $F : \mathscr{C} \longrightarrow \mathscr{D}$ is called a *duality* (respectively, an *equivalence*) if there is a contravariant (respectively, covariant) functor $G : \mathscr{D} \longrightarrow \mathscr{C}$ and two natural equivalences $\alpha : 1_{\mathscr{C}} \longrightarrow GF$ and $\beta : 1_{\mathscr{D}} \longrightarrow FG$. Of course, G is also a duality (respectively, an equivalence), and we say that the categories $\mathscr{C}$ and $\mathscr{D}$ are *dual* (respectively, *equivalent*). The basic fact that we need is

Theorem 2.3.15. *If $F : \mathscr{C} \longrightarrow \mathscr{D}$ is a duality, then for all objects A and B in $\mathscr{C}$ the function $F : \mathscr{C}[A,B] \longrightarrow \mathscr{D}[FB,FA]$ is a bijection.*

Proof. Let $G : \mathscr{D} \longrightarrow \mathscr{C}$, $\alpha : 1_{\mathscr{C}} \longrightarrow GF$ and $\beta : 1_{\mathscr{D}} \longrightarrow FG$ be a functor and natural equivalences associated with F. Define the function $H : \mathscr{D}[FB,FA] \longrightarrow \mathscr{C}[A,B]$ to be the composite $H(g) = \alpha_B^{-1} G(g) \alpha_A : A \xrightarrow{\alpha_A} GFA \xrightarrow{G(g)} GFB \xrightarrow{\alpha_B^{-1}} B$. If $f \in \mathscr{C}[A,B]$ then the commutativity of the diagram

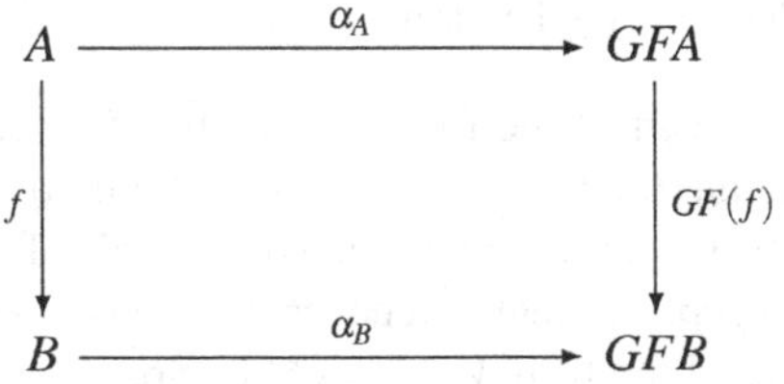

gives that $HF(f) = f$; so F is one-to-one and H is onto. Similarly, G is one-to-one, and therefore H is a bijection. Thus H is the inverse of F and F is a bijection. □

Now let Bool denote the category of Boolean algebras and Boolean algebra homomorphisms, and let $\mathscr{J}$ denote the category of Boolean spaces and continuous functions. Then Theorems 2.3.13 and 2.3.14 give a duality between Bool and $\mathscr{J}$. More precisely, we have

Theorem 2.3.16. *Let* Bool *and* $\mathscr{J}$ *be the categories of Boolean algebras and Boolean spaces, respectively. Then* $F : \text{Bool} \longrightarrow \mathscr{J}$ *and* $G : \mathscr{J} \longrightarrow \text{Bool}$ *are paired dualities where* $F(B)$ *is the dual space of* B *and* $G(X)$ *is the dual algebra of* X.

Proof. Let $\alpha : 1_{\text{Bool}} \longrightarrow GF$ and $\beta : 1_{\mathscr{J}} \longrightarrow FG$ be the functions given in Theorems 2.3.13 and 2.3.14. To complete the definition of F we must define the morphism

$$F(f) : F(B) = \text{Bool}[B, \mathbf{2}] \longrightarrow \text{Bool}[A, \mathbf{2}] = F(A)$$

in $\mathscr{J}$ for each morphism $f : A \longrightarrow B$ in Bool. Let $F(f)$ be right multiplication by $f : (F(f))(g) = gf$. The function $F(f)$ is continuous because if U is clopen in $F(A)$, then, according to Theorem 2.3.13, $U = \alpha(A)(a) = \{f \in \text{Bool}[A, \mathbf{2}] : f(a) = 1\}$ for some a in A. So $(F(f))^{-1}(U) = \{g \in \text{Bool}[B, \mathbf{2}] : (gf)(a) = 1\} = \alpha(B)(f(a))$ is clopen in $F(B)$.

To complete the definition of G we must define the morphism $G(f) : G(Y) \longrightarrow G(X)$ in Bool for any morphism $f : X \longrightarrow Y$ in $\mathscr{J}$. Let $G(f)$ be the inverse image map induced by f restricted to the dual algebra of Y : $(G(f))(Q) = f^{-1}(Q)$.

To complete the proof we merely must check that α and β are natural transformations since Theorems 2.3.13 and 2.3.14 will then give that they are natural equivalences. Let $f \in \text{Bool}[A, B]$. We must verify the commutativity of the diagram

$$\begin{array}{ccc} A & \xrightarrow{\alpha(A)} & GF(A) \\ \Big\downarrow{\scriptstyle f} & & \Big\downarrow{\scriptstyle GF(f)} \\ B & \xrightarrow{\alpha(B)} & GF(B) \end{array} .$$

But $(GF(f))(\alpha(A)(a)) = F(f)^{-1}(\alpha(A)(a)) = \alpha(B)(f(a))$, as we have seen in the first paragraph above. Now, let $g \in \mathscr{J}[X, Y]$, and consider the diagram

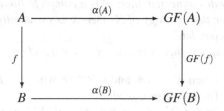

$$\begin{array}{ccc} X & \xrightarrow{\beta(X)} & FG(X) \\ \big\downarrow g & & \big\downarrow FG(g) \\ Y & \xrightarrow{\beta(Y)} & FG(Y) \end{array} \quad .$$

We have $FG(g)(\beta(X)(x) = \beta(X)(x)G(g) : G(Y) \longrightarrow G(X) \longrightarrow \mathbf{2}$; so if P is a clopen set in Y, then $[FG(g)(\beta(X)(x))](P) = \beta(X)(x)(g^{-1}(P)) = 1$ if and only if $g(x) \in P$. But $[\beta(Y)(g(x))](P) = 1$ if and only if $g(x) \in P$. So this diagram is also commutative. □

The Boolean algebra of polars of an ℓ-group is complete, and this property is reflected in its dual space and is an essential ingredient for the representation of an archimedean ℓ-group. We will now consider the topological anologue of completeness. A topological space is called *extremally disconnected* if the closure of each of its open sets is open. A *Stone space* is an extremally disconnected Boolean space.

Theorem 2.3.17. *Let $\mathscr{B}$ be the dual algebra of the topological space X.*

(a) If $\{P_i : i \in I\}$ is a subset of $\mathscr{B}$ and $W = \cup\{P_i : i \in I\}$, then W^- is open if and only if $W^- = \bigvee\{P_i : i \in I\}$ in $\mathscr{B}$.
(b) Suppose that each nonempty open subset of X contains a nonempty element of $\mathscr{B}$, and let P_i and W be as in (a). If $P = \bigvee\{P_i : i \in I\}$ exists in $\mathscr{B}$, then $P = W^-$.
(c) If X is extremally disconnected, then $\mathscr{B}$ is a complete Boolean algebra.
(d) If $\mathscr{B}$ is a base for the topology of X, then X is extremally disconnected if and only if $\mathscr{B}$ is complete.
(e) If X is a Boolean space, then X is a Stone space if and only if $\mathscr{B}$ is complete.

Proof. (a) If W^- is open then $W^- \in \mathscr{B}$, and if $P \in \mathscr{B}$ with $P_i \subseteq P$ for each i, then $W^- \subseteq P$. So $W^- = \bigvee P_i$. The converse is trivial.

(b) Clearly $W^- \subseteq P$; and if Q is clopen with $Q \subseteq P \setminus W^-$, then $P \setminus Q$ is clopen and $P_i \subseteq P \setminus Q$ for each $i \in I$. Thus $P \subseteq P \setminus Q$, $Q = \phi$, and hence $P = W^-$.

Now (c) follows from (a), (d) is a consequence of (c) and (b), and (e) is a special case of (d). □

We now need to investigate some fundamental properties of an extremely disconnected space. In particular, we wish to establish the fact that continuous functions defined on subspaces can sometimes be lifted to the entire space.

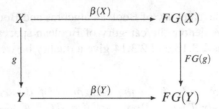

As is usual, we will denote the ℓ-group of all real-valued continuous functions defined on the topological space X by $C(X)$. $C(X)$ is an ℓ-subgroup of the product $\mathbb{R}^X$; that is, the partial order in $C(X)$ is pointwise: $f \leq g$ if $f(x) \leq g(x)$ for each x in X. The ℓ-subgroup of $C(X)$ which consists of the bounded functions will be denoted by $C^*(X)$. A subset A of X is said to be *C-embedded* (respectively, *C^*-embedded*)

if each $f \in C(A)$ (respectively, $f \in C^*(A)$) can be extended to an element of $C(X)$. Note that if $f \in C^*(A)$ is bounded by $r \in \mathbb{R}^+$ (that is $-r \leq f \leq r$) and has the extension $g \in C(X)$, then it has the extension $(-r \vee g) \wedge r)$ in $C^*(X)$ which also is bounded by r. Two subsets A and B of X are said to be *completely separated* in X if there is a function f in $C(X)$ such that $f(A) = 0$ and $f(B) = 1$. By replacing f by $0 \vee (f \wedge 1)$ we may assume that $0 \leq f \leq 1$. Also, any numbers $r < s$ may replace 0 and 1 in this definition.

Theorem 2.3.18. *Let Y be a subset of the topological space X. Then Y is C^*-embedded in X if and only if each pair of sets that is completely separated in Y is also completely separated in X.*

Proof. Suppose that Y is C^*-embedded in X and that A and B are completely separated in Y. If $f \in C^*(Y)$ with $f(A) = 0$ and $f(B) = 1$ and $g \in C(X)$ is an extension of f, then certainly $g(A) = 0$ and $g(B) = 1$.

Conversely, suppose that the property of being completely separated lifts from Y to X, and that $f \in C^*(Y)$. Then $|f| \leq m$ for some $m \in \mathbb{N}$. Define the sequence (r_n) of real numbers by $r_n = \frac{m}{2}\left(\frac{2}{3}\right)^n$. We claim that there are sequences of functions (f_n) in $C^*(Y)$ and (g_n) in $C^*(X)$ with $f_1 = f$, $|f_n| \leq 3r_n$, $|g_n| \leq r_n$ and $f_{n+1} = f_n - g_n$ on Y. Assume that $f_1, \ldots, f_n$ and $g_1, \ldots, g_{n-1}$ have been constructed. Let $A_n = f_n^{-1}((-\infty, -r_n])$ and $B_n = f_n^{-1}([r_n, \infty))$. Then A_n and B_n are completely separated in Y by $(-r_n \vee f_n) \wedge r_n$; so there exists $g_n \in C^*(X)$ such that $|g_n| \leq r_n$, $g_n(A_n) = -r_n$ and $g_n(B_n) = r_n$. Now let $f_{n+1} = f_n - g_n \in C^*(Y)$. On A_n we have that $-3r_n \leq f_n \leq -r_n = g_n$; on B_n we have that $g_n = r_n \leq f_n \leq 3r_n$; and on $Y \backslash (A_n \cup B_n)$ we have that $-r_n \leq f_n, g_n \leq r_n$. Thus, $|f_{n+1}| \leq 2r_n = 3r_{n+1}$. Define the function g on X by

$$g(x) = \sum_{n=1}^{\infty} g_n(x).$$

Since the series converges uniformly, $g \in C(X)$. Also, on Y, $g_1 + \cdots + g_n = (f_1 - f_2) + \cdots + (f_n - f_{n+1}) = f_1 - f_{n+1}$. So g is an extension of f since the sequence $(f_n(y))$ converges to 0 for each y in Y. □

We now return to extremally disconnected spaces.

Theorem 2.3.19. *The following statements are equivalent for the topological space X.*

(a) X is extremally disconnected.
(b) The interior of each closed subset of X is closed.
(c) If U and V are open then $(U \cap V)^- = U^- \cap V^-$.
(d) Any two disjoint open sets have disjoint closures.
(e) Each open subset of X is C^*-embedded in X.
(f) If U is open, then each pair of completely separated sets in U is completely separated in X.
(g) Any two disjoint open sets are completely separated in X.

Proof. Recall that $A^\circ = A'^{-\prime}$ for each subset A of X. If U is open and $F = U'$, then $F^\circ = F'^{-\prime} = U^{-\prime}$ is closed if and only if U^- is open. Thus (a) and (b) are equivalent; also, the equivalence of (e) and (f) is a consequence of Theorem 2.3.18.

(a)$\Rightarrow$ (c). In any space X, if P is clopen and $A \subseteq X$, then $(P \cap A)^- = P \cap A^-$. For, $(P \cap A)^- \subseteq P \cap A^-$ since $P \cap A^-$ is closed and contains $P \cap A$. On the other hand, if $x \in P \cap A^-$ and W is open with $x \in W$, then $W \cap P \cap A \neq \phi$; so $x \in (P \cap A)^-$. Since, clearly, $(U \cap V)^- \subseteq U^- \cap V^-$, we only need to establish the reverse inclusion or, equivalently, that $P = (U \cap V)^{-\prime} \subseteq (U^- \cap V^-)'$. But $P \cap U \cap V = \phi$ yields that $P \cap U \cap V^- = \phi$, since $P \cap U$ is open, and consequently that $P \cap (U \cap V^-)^- = \phi$. However, $P \cap (U \cap V^-)^- = P \cap U^- \cap V^-$ since V^- is clopen.

(c) $\Rightarrow$ (d). This is trivial.

(d) $\Rightarrow$ (a). If U is open, then $U \cap U^{-\prime} = \phi$ gives that $U^- \cap U^{-\prime -} = \phi$. So $U^{-\prime -} \subseteq U^{-\prime}$ and $U^{-\prime}$ is closed; that is, U^- is open.

(a) $\Rightarrow$ (f). Suppose that $A, B \subseteq U$ and $f \in C^*(U)$ with $f(A) = 0$ and $f(B) = 1$. Then there exist disjoint open sets V and W in U which contain A and B, respectively. By (d), $V^- \cap W^- = \phi$ and V^- and W^- are both clopen. Hence, the characteristic function of W^- is in $C(X)$ and separates A and B.

(f) $\Rightarrow$ (g). If U and V are disjoint open sets, then U and V are completely separated in $U \cup V$, and hence are completely separated in X.

(g) $\Rightarrow$ (d). This is obvious. □

From the definition of an extremally disconnected space it is easily seen that each open subset of such a space is also extremally disconnected. This is also true for a dense subset. Before we verify this we state some definitions. Let $\mathbb{R}^- = \mathbb{R} \cup \{\infty, -\infty\} = [-\infty, \infty]$ denote the two-point compactification of the reals $\mathbb{R}$. The set of all extended real-valued continuous functions on the space X will be denoted by

$$E(X) = \{f : X \longrightarrow \mathbb{R}^- : f \text{ is continuous}\}.$$

Clearly, $E(X)$ is a poset with respect to the coordinatewise partial order: $f \leq g$ if $f(x) \leq g(x)$ for each x in X, and, in fact, $E(X)$ is a sublattice of the product $(\mathbb{R}^-)^X$ (Exercise 1 (c)). Note that $f \leq g$ provided that $f(x) \leq g(x)$ for each x in some dense subset of X (Exercise 1 (d)). Also, let

$$D(X) = \{f \in E(X) : f^{-1}(\mathbb{R}) \text{ is dense in } X\}$$

be the set of those continuous functions which are real-valued on a dense (and open) subset of X. The subset $D(X)$ is a sublattice of $E(X)$ since the intersection of two dense open subsets is dense. A subset A of X is said to be *E-embedded* (respectively, *D-embedded*) in X if each $f \in E(A)$ (respectively, $f \in D(A)$) is the restriction of some $g \in E(X)$ (respectively, $g \in D(X)$).

Theorem 2.3.20. *Let A be a dense subset of the extremally disconnected space X. Then:*

(a) A is extremally disconnected; and

(b) A is C^*-embedded, E-embedded, and D-embedded in X.

Proof. If U and V are disjoint open sets in A, then $U = U_1 \cap A$ and $V = V_1 \cap A$ where U_1 and V_1 are open in X. But then $U_1 \cap V_1 \cap A = \phi$ and hence $U_1 \cap V_1 = \phi$. By (d) of Theorem 2.3.19 $U_1^- \cap V_1^- = \phi$ and, consequently, $U_A^- \cap V_A^- = U^- \cap A \cap V^- \cap A \subseteq U_1^- \cap V_1^- = \phi$, where U_A^- is the closure of U in A. Thus, A also satisfies condition (d) of Theorem 2.3.19, and hence it is extremally disconnected. This argument, together with Theorem 2.3.18 and (g) of Theorem 2.3.19, also gives that A is C^*-embedded in X. Let $f \in E(A)$ and let $\varphi : \mathbb{R}^- \longrightarrow [0,1]$ be a homeomorphism. Then φf has an extension $g \in C^*(X)$ such that $0 \leq g \leq 1$; so $\varphi^{-1} g \in E(X)$ is then an extension of f. If, in fact, $f \in D(A)$, then also $g \in D(X)$, since $f^{-1}(\mathbb{R}) \subseteq g^{-1}(\mathbb{R})$ and density is a transitive property. □

In order to examine the structure of $D(X)$ we first prove

Theorem 2.3.21. *Suppose that X is a set and $\{U^t : t \in \mathbb{R}^-\}$ is a collection of subsets of X such that $U^t \subseteq U^s$ whenever $t > s$. Define the function $f : X \longrightarrow \mathbb{R}^-$ by*

$$f(x) = \vee\{t \in \mathbb{R}^- : x \in U^t\}.$$

Then for each $s \in \mathbb{R}^-$,

(a) $f^{-1}((s,\infty]) = \cup\{U^t : t > s\}$;
(b) $f^{-1}([s,\infty]) = \cap\{U^t : t < s\}$.

Proof. (a) If $f(x) > s$, then there exists $t > s$ with $x \in U^t$. Conversely, if $x \in U^t$ with $t > s$, then $f(x) \geq t > s$.

(b) If $f(x) \geq s$ and $t < s$, then there is an $r > t$ with $x \in U^r$. So $x \in U^r \subseteq U^t$. Conversely, if $x \in \cap\{U^t : t < s\}$, then $f(x) \geq t$ for each $t < s$; so $f(x) \geq s$. □

Theorem 2.3.22. *Let X be an extremally disconnected space. Then $E(X)$ is a complete lattice, and $D(X)$ is an ℓ-group and is a conditionally complete sublattice of $E(X)$. Also, $D(X)$ is a commutative ring in which the product of two positive elements is positive.*

Proof. We have already noted that $E(X)$ is a sublattice of $(\mathbb{R}^-)^X$. Given a family $\{f_i : i \in I\} \subseteq E(X)$ and $r \in \mathbb{R}^-$ define

$$U^r = \left[\bigcup_{i \in I}\{x : f_i(x) > r\}\right]^-,$$

and

$$f(x) = \vee\{r : x \in U^r\}.$$

If $i \in I$ and $f_i(x) > r$, then $x \in U^r$ and $f(x) \geq r$. So $f(x) \geq f_i(x)$ for each $i \in I$ and each x in X. If $s < t$ and some $f_i(x) > t$, then $f_i(x) > s$; hence $U^t \subseteq U^s$. Now, each U^r is clopen; so for each $s \in \mathbb{R}^-$, $f^{-1}((s,\infty])$ is open and $f^{-1}([s,\infty])$ is closed, by the previous theorem. Thus, $f \in E(X)$. Suppose that $g \in E(X)$ and $g \geq f_i$ for each $i \in I$. If $x \in U^r$ and $V = g^{-1}([-\infty, r))$, then V is open and

$$V \cap \bigcup_i \{y : f_i(y) > r\} = \phi$$

since $g \geq f_i$ for each i. Thus, $V \cap U^r = \phi$, $g(x) \geq r$ and hence $g(x) \geq f(x)$. So $f = \sup_{E(X)}\{f_i : i \in I\}$ and $E(X)$ is complete.

Since $f^{-1}(\mathbb{R}) \cap g^{-1}(\mathbb{R}) \subseteq (f \vee g)^{-1}(\mathbb{R}) \cap (f \wedge g)^{-1}(\mathbb{R})$ for all $f, g \in E(X)$, $D(X)$ is a sublattice of $E(X)$. If $\{f_i\} \subseteq D(X)$, $g \in D(X)$ and $f_i \leq g$ for each i, then $f_i \leq f = \vee f_i \leq g$. But then $f_i^{-1}(\mathbb{R}) \cap g^{-1}(\mathbb{R}) \subseteq f^{-1}(\mathbb{R})$; so $f \in D(X)$, and hence $D(X)$ is conditionally complete (Exercise 1.3.4). If $f, g \in D(X)$ then the function h defined by $h(x) = f(x) + g(x)$ for $x \in A = f^{-1}(\mathbb{R}) \cap g^{-1}(\mathbb{R})$ is in $D(A)$, and it has a unique extension, also called h, in $D(X)$. This defines addition in $D(X)$ as $f + g = h$. If $f \leq g$ and $k \in D(X)$ then $f(x) + k(x) \leq g(x) + k(x)$ for each x in some dense subset of X. So $f + k \leq g + k$ and $D(X)$ is an ℓ-group. The product fg is defined similarly and makes $D(X)$ into a commutative ring in which the product of any two positive elements is positive. □

Note that if $f, g \in D(X)$, then $(f + g)(x) = f(x) + g(x)$ and $(fg)(x) = f(x)g(x)$ whenever the right sides of these equations are defined (Exercise 7(a)).

Let $X = \mathrm{Spec}(\mathscr{B}(H))$ be the Stone space of the Boolean algebra $\mathscr{B}(H)$ of polars of the ℓ-group H. Recall that X is the set of maximal ideals of $\mathscr{B}(H)$ supplied with the Zariski topology. According to Theorem 2.3.13 and the remarks after it each clopen set in X is of the form

$$V(A) = \{m \in X : A \notin m\} \tag{2.3.2}$$

for some unique polar $A \in \mathscr{B}(H)$, and V is an isomorphism between $\mathscr{B}(H)$ and the dual algebra of X. If $a \in H$ we write $V(a) = V(a^{\perp\perp})$. As a consequence of Exercise 2.2.13 we have that the mapping

$$V : H^+ \longrightarrow \text{dual algebra of } X \tag{2.3.3}$$

is a lattice homomorphism which preserves all sups that exist in H^+. Moreover, $V(a) = V(|a|) = V(na)$ if $a \in H$ and $0 \neq n \in \mathbb{Z}$, and $V(a + b) = V(a \vee b)$ if $a, b \in H^+$. Also, by Theorem 2.3.17, since

$$A = \bigvee_{a \in A} a^{\perp\perp},$$

$$V(A) = \bigvee_{a \in A} V(a) = \left[\bigcup_{a \in A} V(a)\right]^-. \tag{2.3.4}$$

Recall from Exercises 2.2.14 and 2.2.15 that an ℓ-subgroup H of an ℓ-group D is a $c\ell$-essential ℓ-subgroup if $H \cap C \neq 0$ for each nonzero convex ℓ-subgroup C of D, and a monomorphism of ℓ-groups is a $c\ell$-essential monomorphism if its image is a $c\ell$-essential ℓ-subgroup.

Theorem 2.3.23. *Let H be an archimedean ℓ-group, and let $X = Spec(\mathscr{B}(H))$ be the Stone space of the Boolean algebra $\mathscr{B}(H)$ of polars of H. Then there is a (complete) $c\ell$-essential monomorphism from H into $D(X)$.*

Proof. No harm is done if H is replaced by its divisible hull $d(H)$ since $d(H)$ is certainly archimedean and $\mathscr{B}(H) \cong \mathscr{B}(d(H))$ (Exercise 2.2.6). Thus, we will assume that H is a vector space over the rationals. Also, since $a^{\perp} = (pa)^{\perp}$ for each $0 \neq p \in \mathbb{Q}, V(a) = V(pa)$. Let $\{e_i : i \in I\}$ be a maximal disjoint set of nonzero elements in H^+. For $a \in H^+$ and $q \in \mathbb{Q}$ let

$$W(q,a) = \bigcup_{i \in I} V((qe_i - a)^+),$$

and define the function $f_a : X \longrightarrow \mathbb{R}^-$ by, if $m \in X$,

$$f_a(m) = \wedge\{q \in \mathbb{Q} : m \in W(q,a)^-\}.$$

Since $(qe_i - a)^+ = 0$ if $q \leq 0$, we have that $0 < \{q \in \mathbb{Q} : m \in W(q,a)^-\}$; so $f_a(m) \geq 0$, and in the definition of $f_a(m)$ we may replace "$q \in \mathbb{Q}$" by either "$q \in \mathbb{Q}^+$" or "$0 < q \in \mathbb{Q}$." Note that W is isotone in its first variable and antitone in its second variable (as a mapping into the open subsets of X). Also note that since $\{e_i\}$ is a maximal disjoint set in H, $W(q,0)$ is dense in X for each $q > 0$, and hence $f_0 = 0$.

Now, if $p > f_a(m)$, then $m \in W(p,a)^-$ since there is a $q \in \mathbb{Q}$ with $f_a(m) \leq q < p$ such that $m \in W(q,a)^- \subseteq W(p,a)^-$, while if $p < f_a(m)$, then $m \notin W(p,a)^-$. The proof will proceed by a sequence of steps.

(i) f_a is continuous. Let $r \in \mathbb{R}^-$. Then

$$\{m \in X : f_a(m) < r\} = \bigcup_{p<r} W(p,a)^-$$

is open since X is extremally disconnected, and

$$\{m \in X : f_a(m) \leq r\} = \bigcap_{r<p} W(p,a)^-$$

is closed. Thus $f_a \in E(X)$.

(ii) $f_a \in D(X)$. We will first verify that for $a, e \in H^+$

$$e^{\perp\perp} = \bigvee_{p>0} ((pe - a)^+)^{\perp\perp}$$

where the sup taken here is in $\mathscr{B}(H)$. Since $(pe - a)^+ \leq pe$, we have that $((pe - a)^+)^{\perp\perp} \subseteq e^{\perp\perp}$ and hence

$$\bigvee_{p>0} ((pe - a)^+)^{\perp\perp} \subseteq e^{\perp\perp}.$$

For the other inclusion it suffices to verify that

$$\bigcap_{p>0}((pe-a)^+)^\perp \subseteq e^\perp.$$

But if $b\wedge(pe-a)^+=0$, then $p(b\wedge e)-a=(pb-a)\wedge(pe-a)\le pb\wedge(pe-a)^+=0$; so $p(b\wedge e)\le a$, and if this holds for each $p>0$, then $b\wedge e=0$ since H is archimedean. In terms of X we have, by Theorem 2.3.17, that

$$V(e)=\left[\bigcup_{p>0}V((pe-a)^+)\right]^-$$

since V is an isomorphism of Boolean algebras.

To see that $f_a\in D(X)$ suppose that $f_a^{-1}(\mathbb{R})\cap V(C)=\phi$ for some basic clopen set $V(C)$. Then $f_a^{-1}(\mathbb{R})\cap V(b)=\phi$ for each $b\in C$ and hence $f_a(V(b))=\infty$. Thus $V(b)\cap W(p,a)=\phi$ for each $0<p\in\mathbb{Q}$. Consequently, $V(b)\cap V((pe_i-a)^+)=\phi$ for each $i\in I$ and each $p>0$; so $b\wedge e_i=0$ for each i, and $b=0$ since $\{e_i\}$ is a maximal disjoint subset. Thus $C=0$ and $V(C)=\phi$.

(iii) $f_{a+b}=f_a+f_b$. We will show first that if $m\in X$, then $f_{a+b}(m)\le f_a(m)+f_b(m)$. Suppose, then, that $f_a(m)<p$ and $f_b(m)<q$ with $p,q\in\mathbb{Q}$. Then $m\in W(p,a)^-\cap W(q,b)^-=[W(p,a)\cap W(q,b)]^-$ by Theorem 2.3.19. Since $(pe_i-a)^+\wedge(qe_j-b)^+=0$ if $i\neq j$, and $u^+\wedge v^+\le(u+v)^+$ (Exercise 2.1.13 (f)),

$$W(p,a)\cap W(q,b)=\bigcup_{i,j}[V((pe_i-a)^+)\cap V((qe_j-b)^+)]$$

$$=\bigcup_i V((pe_i-a)^+\wedge(qe_i-b)^+)\subseteq\bigcup_i V(((p+q)e_i-(a+b))^+).$$

So $m\in W(p+q,a+b)^-$ and $f_{a+b}(m)\le p+q$. Thus

$$f_a(m)+f_b(m)=\wedge\{p:f_a(m)<p\}+\wedge\{q:f_b(m)<q\}$$

$$=\wedge\{p+q:f_a(m)<p \text{ and } f_b(m)<q\}$$

$$\ge f_{a+b}(m).$$

For the reverse inequality let $p,q\in\mathbb{Q}^+$ with $p<f_a(m)$ and $q<f_b(m)$. Then $m\notin W(p,a)^-\cup W(q,b)^-=[W(p,a)\cup W(q,b)]^-$. Now,

$$W(p,a)\cup W(q,b)=\bigcup_i V((pe_i-a)^+)\cup V((qe_i-b)^+)$$

$$=\bigcup_i V((pe_i-a)^+ + (qe_i-b)^+),$$

and $V([(p+q)e_i-(a+b)]^+)\subseteq V((pe_i-a)^+ + (qe_i-b)^+)$, since $(u+v)^+\le u^++v^+$. So $m\notin W(p+q,a+b)^-$ and $f_{a+b}(m)\ge p+q$. Thus,

$$f_a(m)+f_b(m) = \vee\{p\in\mathbb{Q}^+ : p<f_a(m)\}+\vee\{q\in\mathbb{Q}^+ : q<f_b(m)\}$$

$$= \vee\{p+q : p<f_a(m),\, q<f_b(m)\}$$

$$\leq f_{a+b}(m),$$

and hence $f_{a+b}=f_a+f_b$.

(iv) $f_a(m)=0$ if $m\notin V(a)$. If $0<p\in\mathbb{Q}$, then $V(e_i)=V(pe_i-a+a)\subseteq V((pe_i-a)^+ +a)=V((pe_i-a)^+)\cup V(a)$. Since $\cup V(e_i)$ is dense in X, $X=V(a)\cup W(p,a)^-$. So if $m\notin V(a)$, then $m\in W(p,a)^-$ and $f_a(m)\leq p$ for each $p>0$; thus $f_a(m)=0$.

(v) $f_{a\vee b}=f_a\vee f_b$ and $f_{a\wedge b}=f_a\wedge f_b$. If $a\wedge b=0$, then $V(a)\cap V(b)=\phi$, and hence by (iv), $f_a\wedge f_b=0=f_{a\wedge b}$. In general, by (iii), $f_a\wedge f_b=(f_{a-a\wedge b}\wedge f_{b-a\wedge b})+f_{a\wedge b}=f_{a\wedge b}$; and $f_{a\vee b}+f_{a\wedge b}=f_a+f_b=f_a\vee f_b+f_a\wedge f_b$. So $f_{a\vee b}=f_a\vee f_b$.

(vi) For each i, f_{e_i} is the characteristic function of $V(e_i)$. If $q\in\mathbb{Q}^+$, then $V((qe_i-e_i)^+)=V((q-1)^+e_i)$ and $V((qe_j-e_i)^+)=V(qe_j)$ if $j\neq i$. So

$$W(q,e_i)=\bigcup_{j\neq i}V(qe_j) \text{ if } 0\leq q\leq 1,$$

and $W(q,e_i)^-=X$ if $q>1$. Let $m\in V(e_i)$. Then $m\in W(q,e_i)^-$ if and only if $1<q$; so $f_{e_i}(m)=1$, and we are done by (iv).

Define thc function $\varphi : H\longrightarrow D(X)$ by $\varphi(a)=f_a=f_{a^+}-f_{a^-}$.

(vii) φ is an ℓ-homomorphism. Since $f_{a^+}=(f_a)^+$, by (v), we only have to check that $f_{a+b}=f_a+f_b$. But $(a+b)^+ +a^- +b^- =(a+b)^- +a^+ +b^+$ gives that $f_{(a+b)^+}+f_{a^-}+f_{b^-}=f_{(a+b)^-}+f_{a^+}+f_{b^+}$, by (iii), and hence $f_{a+b}=f_a+f_b$. Alternatively, we may use Exercise 2.2.11.

(viii) φ is a monomorphism. Assume that $\varphi(a)=0$ and that $a\geq 0$. Let $0<p\in\mathbb{Q}$ and $i\in I$, and suppose that $m\in V([e_i-p(a\wedge e_i)]^-)$. Since $V([e_i-p(a\wedge e_i)]^-)\subseteq V(a\wedge e_i)\subseteq V(e_i)$,

$$1=f_{e_i}(m)=f_{e_i-p(a\wedge e_i)}(m)=-f_{[e_i-p(a\wedge e_i)]^-}(m)\leq 0.$$

Thus, $V([e_i-p(a\wedge e_i)]^-)=\phi$ and $p(a\wedge e_i)\leq e_i$ for each $p\in\mathbb{Q}^+$. So $a\wedge e_i=0$ for each i and hence $a=0$.

(ix) φ is a complete monomorphism. This follows from (x) and Exercise 2.2.15. However, we give another proof and show, additionally, that infs and sups are computed pointwise outside of a set of the first category. By Exercise 2.1.16 it suffices to show that if

$$\bigwedge_\alpha a_\alpha=0,$$

then

$$\bigwedge_\alpha f_{a_\alpha}=0.$$

If $0 < p \in \mathbb{Q}$, then

$$pe_i = pe_i + \bigvee_\alpha -a_\alpha = \bigvee_\alpha (pe_i - a_\alpha)^+.$$

So

$$V(e_i) = \left[\bigcup_\alpha V((pe_i - a_\alpha)^+)\right]^-,$$

and hence

$$X = \left[\bigcup_i V(e_i)\right]^- = \left[\bigcup_i \left[\bigcup_\alpha V((pe_i - a_\alpha)^+)\right]^-\right]^-$$
$$= \left[\bigcup_\alpha \bigcup_i V((pe_i - a_\alpha)^+)\right]^-$$
$$= \left[\bigcup_\alpha W(p, a_\alpha)^-\right]^-.$$

Thus,

$$A_p = \left[\bigcup_\alpha W(p, a_\alpha)^-\right]'$$

is nowhere dense in X. Suppose that $m \in X$ and

$$0 < \bigwedge_\alpha f_{a_\alpha}(m).$$

Then for each p with

$$0 < p < \bigwedge_\alpha f_{a_\alpha}(m)$$

we have that $m \in A_p$. Let A be the union of those A_p for which $p \in \mathbb{Q}$ satisfies the previous inequality for some $m \in X$. Then A is a set of the first category, and if $m \notin A$, then

$$\bigwedge_\alpha f_{a_\alpha}(m) = 0.$$

Now, suppose that $f \in D(X)$ and that $f \leq f_{a_\alpha}$ for each α. We claim that $f \leq 0$. If not, there exists an open set $U \neq \phi$ such that $f(m) > 0$ for each $m \in U$. But then $U \subseteq A^\circ$, yet $A^\circ = \phi$ by Theorem 2.3.11.

(x) φ is a $c\ell$-essential monomorphism of H into $D(X)$. Suppose that $0 < f \in D(X)$. Then there exists a nonempty clopen set U in X and an $n \in \mathbb{N}$ such that $f(m) \geq \frac{1}{n}$ for each m in U. Hence, for some $0 < b \in H$, $f(m) \geq \frac{1}{n}$ for each $m \in V(b) \subseteq U$. For some $i \in I$, $a = b \wedge e_i > 0$, and then $f_a \leq nf$ by (iv) and (vi). □

If f is a positive unit in the ring $D(X)$, then the representation φ of H given in Theorem 2.3.23 followed by multiplication by f is another similar representation of H since f^{-1} is also positive. The following result shows that these are essentially the only such representations of H, and it also shows that any Stone space used to represent H is uniquely determined by H.

The *support* of a function $f \in D(X)$ is defined to be the clopen subset $S(f) = \{x \in X : f(x) \neq 0\}^-$ of X. According to Exercise 15, $V(a) = S(\varphi(a))$ for each $a \in H$.

Theorem 2.3.24. *Let H be an archimedean ℓ-group, let Y be a Stone space, and let $\varphi_1 : H \longrightarrow D(Y)$ be a $c\ell$-essential monomorphism. If $\varphi : H \longrightarrow D(X)$ is the representation given in Theorem 2.3.23, then there is a homeomorphism $\tau : X \longrightarrow Y$ and a positive unit f in $D(Y)$ such that for each $a \in H$ and each $m \in X$,*

$$\varphi_1(a)(\tau(m)) = f(\tau(m))\varphi(a)(m), \tag{2.3.5}$$

whenever the right side of this equation is defined. That is, the diagram

$$\begin{array}{ccc} H & \xrightarrow{\quad \varphi \quad} & D(X) \\ \Big\downarrow{\scriptstyle \varphi_1} & & \Big\downarrow{\scriptstyle \mu_{\tau^*(f)}} \\ D(Y) & \xrightarrow{\quad \tau^* \quad} & D(X) \end{array}$$

is commutative where τ^* is the isomorphism that is induced by τ ($\tau^*(g) = g \circ \tau$) and $\mu_{\tau^*(f)}$ is multiplication by the unit $\tau^*(f)$.

Proof. Again, we will assume that H is divisible. Let $G(X)$ be the dual algebra of X. As we have previously seen in Theorem 2.3.13 and the remarks following it, and in (2.3.2) and (2.3.4), the mapping $V : \mathscr{B}(H) \longrightarrow G(X)$ given by

$$V(A) = \bigvee_{a \in A} V(a) = \left[\bigcup_{a \in A} S(\varphi(a))\right]^-$$

is an isomorphism of Boolean algebras. By Exercise 16(e) the function $\psi_1 : \mathscr{B}(H) \longrightarrow G(Y)$ defined by

$$\psi_1(A) = \left[\bigcup_{a \in A} S(\varphi_1(a))\right]^-$$

is also an isomorphism. Let $\delta = V\psi_1^{-1} : G(Y) \longrightarrow G(X)$. By Theorems 2.3.15 and 2.3.16 there is a unique homeomorphism $\tau : X \longrightarrow Y$ such that $G(\tau) = \delta$; that is, $\delta(P) = \tau^{-1}(P)$ for each clopen set P in Y. So $\tau^{-1}\psi_1(A) = V\psi_1^{-1}\psi_1(A) = V(A)$, and hence $\tau V(A) = \psi_1(A)$ for each polar $A \in \mathscr{B}(H)$. In particular, since $V(a^{\perp\perp}) =$

$S(\varphi(a))$ and $\psi_1(a^{\perp\perp}) = S(\varphi_1(a))$ (Exercise 16), for each $a \in H$, we have that $\tau(S(\varphi(a))) = S(\varphi_1(a))$.

Since $\varphi_1(H)$ is a $c\ell$-essential ℓ-subgroup of $D(Y)$, $\{\varphi_1(e_i) : i \in I\}$ is a maximal disjoint subset of $D(Y)$. Thus, by Exercises 12 and 13, $\{S(\varphi_1(e_i))\}_{i\in I}$ is a family of pairwise disjoint clopen sets in Y whose union Z is dense in Y.

Define the function f on Z by stipulating that, for each $i \in I$, f agrees with $\varphi_1(e_i)$ on $S(\varphi_1(e_i))$. This defines f as a continuous extended real-valued function on the dense subset Z of Y, and since each $\varphi_1(e_i) \in D(Y), f \in D(Z)$. Since Z is D-embedded in Y (by Theorem 2.3.20) f has a unique extension to an element in $D(Y)$ which is also called f. Since $\{y \in Y : f(y) > 0\}$ contains the union $\cup\{y \in Y : \varphi_1(e_i)(y) > 0\}$ and the latter is dense in Y, f is a positive unit of $D(Y)$ (Exercise 13).

To verify (2.3.5) we may assume that $a \in H^+$, and we may restrict m to the dense subset

$$T = (f\tau)^{-1}(\mathbb{R}) \cap \varphi(a)^{-1}(\mathbb{R}) \cap \left[\bigcup_i S(\varphi(e_i))\right]$$

of X. Suppose, then, that $m \in S(\varphi(e_j))$ for some j, and that $f(\tau(m))$ and $\varphi(a)(m)$ are both real. Let $p \in \mathbb{Q}$ with $p > \varphi(a)(m)$. Since $\varphi(e_j)$ is the characteristic function of $S(\varphi(e_j))$, $\varphi(pe_j - a)(m) > 0$. Thus, $m \in S(\varphi(pe_j - a)^+)$ and $\tau(m) \in S(\varphi_1(pe_j - a)^+) \subseteq S(\varphi_1(e_j))$. Since f agrees with $\varphi_1(e_j)$ on $S(\varphi_1(e_j))$, $pf(\tau(m)) \geq \varphi_1(a)(\tau(m))$. So

$$\begin{aligned} f(\tau(m))\varphi(a)(m) &= f(\tau(m))(\wedge\{p \in \mathbb{Q} : p > \varphi(a)(m)\}) \\ &= \wedge\{f(\tau(m))p : p > \varphi(a)(m)\} \\ &\geq \varphi_1(a)(\tau(m)). \end{aligned}$$

Similarly, if $p < \varphi(a)(m)$, then $pf(\tau(m)) \leq \varphi_1(a)(\tau(m))$, and hence $f(\tau(m))\varphi(a)(m) \leq \varphi_1(a)\tau(m)$. So $\varphi_1(a)(\tau(m)) = f(\tau(m))\varphi(a)(m)$. □

Recall that an ℓ-group is laterally complete if each disjoint subset of its positive cone has a sup in the ℓ-group. As an application of Theorem 2.3.23 we have

Theorem 2.3.25. *Let G be an ℓ-group.*

(a) G is divisible and complete if and only if there is a Stone space X such that G is isomorphic to a convex ℓ-subgroup of $D(X)$.
(b) G is divisible, complete and laterally complete if and only if there is a Stone space X such that G is isomorphic to $D(X)$.

Proof. Suppose that G is divisible and let $\varphi : G \longrightarrow D(X)$ be the representation given in Theorem 2.3.23. For simplicity we will identify G with its image $\varphi(G)$. If H is the convex ℓ-subgroup of $D(X)$ that is generated by G, then, since G is a $c\ell$-essential divisible ℓ-subgroup of H, H is the completion of G, by Theorem 2.3.6. Thus, if G is complete, then $G = H$. Suppose that G is also laterally complete. If $\{e_i\}$ is the maximal disjoint subset of G^+ used in the proof of Theorem 2.3.23, then e_i is the characteristic function of $S(e_i)$ and

$$1 = \bigvee_i e_i.$$

Since G is laterally complete, and sups in G are also sups in $D(X)$, $1 \in G$. Thus, $C(X) \subseteq G$. Suppose that $1 \leq f \in D(X)$. Let $\{g_\lambda\}$ be a disjoint subset of G^+ maximal with respect to the property that for each λ, $g_\lambda = f$ on $S(g_\lambda)$. Then

$$g = \bigvee_\lambda g_\lambda \in G$$

and g agrees with f on

$$U = \bigcup_\lambda S(g_\lambda).$$

If U is not dense in X then there exists a nonempty clopen subset $V \subseteq f^{-1}(\mathbb{R})$ with $U \cap V = \phi$. Let $h \in C(X)$ be defined by $h = f$ on V and $h = 0$ on V'. Then $0 < h \in G$ and $g_\lambda \wedge h = 0$ for each λ. Thus, by the maximality of the family $\{g_\lambda\}$, U is dense and $f = g \in G$. Now, if $f \in D(X)^+$, then $f \vee 1 \in G$, and also $f \in G$; so $G = D(X)$. Both converses have already been mentioned (see Exercise 13). □

If G is a nonzero ℓ-group, then the lexicographic extension $G \overleftarrow{\times} \mathbb{Z}$ is a proper $c\ell$-essential extension of G. The next result, which is basically a rephrasing of Theorem 2.3.25(b), characterizes those ℓ-groups in the category of archimedean ℓ-groups which do not have any proper $c\ell$-essential extensions in this category. An archimedean ℓ-group G is *cℓ-essentially closed* if each $c\ell$-essential monomorphism from G into an archimedean ℓ-group is an isomorphism. A *cℓ-essential closure of G* is an archimedean ℓ-group that is $c\ell$-essentially closed and is a $c\ell$-essential extension of G.

Theorem 2.3.26. *The following statements are equivalent for the archimedean ℓ-group G.*

(a) G is cℓ-essentially closed.
(b) G is divisible, complete, and laterally complete.
(c) The representation $\varphi : G \longrightarrow D(X)$ of Theorem 2.3.23 is an isomorphism.
(d) There is a Stone space Y such that $G \cong D(Y)$.

Proof. We only need to verify the implication $(d) \Rightarrow (a)$. Suppose that Y is a Stone space and $\alpha : D(Y) \longrightarrow H$ is a $c\ell$-essential monomorphism into the archimedean ℓ-group H. Let X be the Stone space of $\mathscr{B}(H)$. By Theorem 2.3.23 there is a $c\ell$-essential monomorphism $\varphi : H \longrightarrow D(X)$ with $\varphi(\alpha(1)) = 1$ since $\{\alpha(1)\}$ is a maximal disjoint subset of H (see Exercise 17). By the proof of Theorem 2.3.25 $\varphi\alpha(D(Y)) = D(X)$. Thus, $\varphi\alpha$ is an isomorphism and hence so are φ and α. □

An immediate consequence of Theorems 2.3.23 and 2.3.26 is that each archimedean ℓ-group has a $c\ell$-essential closure which is unique up to an isomorphism, by Theorem 2.3.24. A little more is true. The $c\ell$-essential closure of G will henceforth be denoted by G^e.

Theorem 2.3.27. *For $i = 1,2$ let $\varphi_i : G \longrightarrow D_i$ be a $c\ell$-essential monomorphism of the ℓ-group G into the $c\ell$-essentially closed archimedean ℓ-group D_i. Then there is a unique isomorphism $\psi : D_1 \longrightarrow D_2$ such that $\varphi_2 = \psi\varphi_1$.*

Proof. If X_G is the Stone space of $\mathscr{B}(G)$, then from Exercise 23 we obtain the commutative diagram

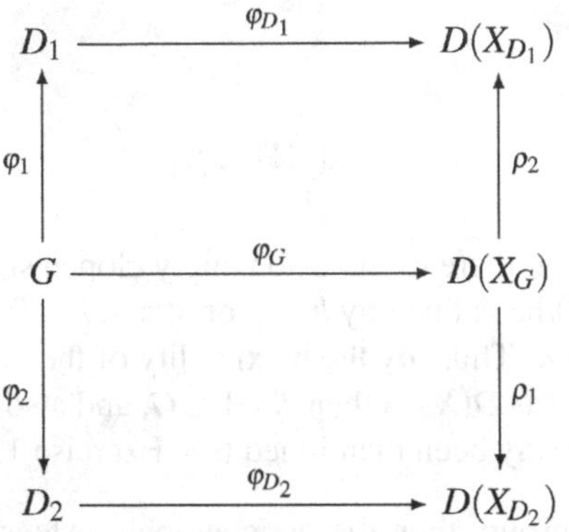

where ρ_i and φ_{D_i} are isomorphisms. Thus, $\psi = \varphi_{D_2}^{-1}\rho_1\rho_2^{-1}\varphi_{D_1}$ is the desired isomorphism. The uniqueness of ψ is a consequence of the fact that, by Theorem 2.3.5, each element of D_1^+ is the sup of elements in the divisible closure of $\varphi_1(G)$; also see Exercise 30. □

Using the analogy of injective modules in a module category one would expect a $c\ell$-essentially closed archimedean ℓ-group to be a summand of every archimedean ℓ-group in which it is embedded as a convex ℓ-subgroup. This turns out to be true, but this property is not a characterization of $c\ell$-essentially closed ℓ-groups. An archimedean ℓ-group is said to have the *splitting property* if it is a summand of each archimedean ℓ-group in which it is a convex ℓ-subgroup.

Theorem 2.3.28. *The following statements are equivalent for the archimedean ℓ-group G.*

(a) G has the splitting property.
(b) If H is a $c\ell$-essential archimedean extension of G and G is convex in H, then $G = H$.
(c) If G is a convex ℓ-subgroup of the ℓ-subgroup H of G^e, then $G = H$.
(d) If G is a convex ℓ-subgroup of the archimedean ℓ-group H, then $H = G^\perp \oplus G^{\perp\perp}$.
(e) For each $0 < a \in G^e \backslash D(d(G))$ there are elements $0 < g \in G$ and $n \in \mathbb{N}$ such that $g \wedge na \notin G$.

Proof. The implications (a) $\Rightarrow$ (b) $\Rightarrow$ (c) $\Rightarrow$ (b) and (a) $\Rightarrow$ (d) are obvious.

(b) $\Rightarrow$ (a). Suppose that $G \in \mathscr{C}(H)$. Since G is $c\ell$-essential in $G^{\perp\perp}$ we have that G is a polar of H. Since $G \longrightarrow H/G^{\perp}$ is a $c\ell$-essential monomorphism, $H = G \oplus G^{\perp}$.

(d) $\Rightarrow$ (b). Suppose that $G \in \mathscr{C}(H)$ and G is a $c\ell$-essential ℓ-subgroup of H. Then $G^{\perp_H} = 0$. If $0 < h \in H \backslash G$ let $K = G_1 \oplus G_2 + \mathbb{Z}(h,h) \subseteq H \oplus H$, where $G_1 = G_2 = G$. Since $G_1 \oplus G_2$ is a convex ℓ-subgroup of $H \oplus H$ and $\mathbb{Z}(h,h)$ is an ℓ-subgroup, K is an ℓ-subgroup of $H \oplus H$. We claim that $G_1^{\perp_K} = G_2$. For if $0 \leq (g,a) + n(h,h) \in G_1^{\perp_K}$, then $(g+nh, a+nh) \wedge (b,0) = 0$ for each $b \in G^+$. So $nh = -g \in G$ and hence $h \in G$ unless $n = 0$, in which case $g = 0$, also. Similarly, $G_2^{\perp_K} = G_1$ and $G_1^{\perp_K} \oplus G_1^{\perp_K \perp_K} = G_2 \oplus G_1 \subset K$. Thus, $H = G$.

(e) $\Rightarrow$ (c). Suppose that $G \subseteq H \subseteq G^e, G$ is convex in H and $0 < a \in H \backslash G$. Note that $C^{G^e}((d(G)) = D(d(G))$ by Theorem 2.3.6. If $a \in C^{G^e}(d(G))$, then $0 < a < g/n \leq g$ for some $g \in G$ and $n \in \mathbb{N}$. But then we have the contradiction that $a \in G$. Thus, $a \notin D(d(G))$, but $g \wedge na \in G$ for each $g \in G^+$ and each $n \in \mathbb{N}$.

(c) $\Rightarrow$ (e). If $g \wedge na \in G$ for each $g \in G^+$ and $n \in \mathbb{N}$, then $H = G + \mathbb{Z}a$ is an ℓ-subgroup of G^e and G is convex in H by Exercise 2.2.19. □

Our next goal is to show that each laterally complete archimedean ℓ-group has the splitting property. The first essential ingredient of the proof is the following technical result.

Theorem 2.3.29. *Let G be an abelian ℓ-group and let $x, y \in G^+$. For each $n \in \mathbb{Z}^+$ let*

$$w_n = ((n+2)x - y)^+ \wedge (y - nx)^+.$$

Suppose u is an upper bound for the disjoint set $\{(2n+1)w_{2n} : n \in \mathbb{Z}^+\}$ and v is an upper bound for the disjoint set $\{(2n+2)w_{2n+1} : n \in \mathbb{Z}^+\}$. Then $2y$ is an upper bound for the subgroup generated by $x \wedge (y - y \wedge (u+v))$.

Proof. Note that if $m \geq n+2$, then $w_n \wedge w_m = 0$ since

$$0 \leq w_n \wedge w_m \leq ((n+2)x - y)^+ \wedge (mx - y)^- \leq (mx - y)^+ \wedge (mx - y)^- = 0.$$

We will show by induction on $n \in \mathbb{Z}^+$ that

$$\sum_{k=0}^{n} (k+1) w_k \geq [(n+1)((n+2)x - y)^+] \wedge y \geq (nx - y)^+ \wedge y.$$

Of course, it is the first inequality that needs to be established since the second is obvious. For $n = 0$ this inequality is $w_0 \geq w_0$. The inductive step requires that we verify the inequality

$$([(n+1)((n+2)x - y)^+] \wedge y) + (n+2) w_{n+1} \geq [(n+2)((n+3)x - y)^+] \wedge y.$$

But,

$$[(n+2)((n+3)x - y)^+] \wedge y =$$

$$[(n+2)((n+3)x - y)^+] \wedge [((n+1)((n+2)x - y)) + ((n+2)(y - (n+1)x))] \wedge y \leq$$

$$
\begin{aligned}
&[((n+2)((n+3)x-y)^+ \wedge (n+1)((n+2)x-y)^+)\\
&+((n+2)((n+3)x-y)^+ \wedge ((n+2)(y-(n+1)x)^+)] \wedge y \leq\\
&[((n+1)((n+2)x-y)^+) + (n+2)(((n+3)x-y)^+ \wedge (y-(n+1)x)^+)] \wedge y \leq\\
&[((n+1)((n+2)x-y)^+) + (n+2)w_{n+1}] \wedge y \leq\\
&[(n+1)((n+2)x-y)^+] \wedge y + (n+2)w_{n+1}.
\end{aligned}
$$

Note that for each $n \in \mathbb{Z}^+$

$$\sum_{k=0}^{n}(k+1)w_k = \bigvee_{k \text{ even}}(k+1)w_k + \bigvee_{k \text{ odd}}(k+1)w_k \leq u+v.$$

Hence,

$$
\begin{aligned}
x \wedge [y-(y\wedge(u+v))] &\leq x \wedge \left[y - \left(y \wedge \sum_{k=0}^{n}(k+1\right)w_k\right]\\
&\leq x \wedge [y-(y\wedge(nx-y)^+)]\\
&= x \wedge (2y-nx)^+\\
&\leq (2y-nx)^+ \wedge [((n+1)x-2y) \vee x]\\
&= [2y+(-2y \vee -nx)] \wedge [(n+1)x+(-2y \vee -nx)]\\
&= (2y \wedge (n+1)x) - (2y \wedge nx).
\end{aligned}
$$

Thus, for $k \in \mathbb{N}$,

$$
\begin{aligned}
k(x \wedge [y-(y\wedge(u+v))]) &\leq \sum_{n=0}^{k-1}[(2y\wedge(n+1)x)-(2y\wedge nx)]\\
&= (2y \wedge kx) - (2y \wedge 0) \leq 2y.
\end{aligned}
$$

□

An ℓ-group is called *projectable* if each principal polar is a direct summand. The previous result gives the following.

Theorem 2.3.30. *Let G be an archimedean ℓ-group. Suppose that each positive disjoint subset of every principal bipolar $x^{\perp\perp}$ is bounded in $x^{\perp\perp}$. Then G is projectable. In particular, a laterally complete archimedean ℓ-group is projectable.*

Proof. Continuing with the notation of Theorem 2.3.29 we have $y - y \wedge (u+v) \in x^{\perp}$. Since $w_n \leq (n+2)x$, the elements u and v may be chosen in $x^{\perp\perp}$. So $y = (y - y \wedge$

$(u+v))+y\wedge(u+v)\in x^{\perp}+x^{\perp\perp}$. The second statement follows immediately from the fact that each polar is a completely closed convex ℓ-subgroup. □

The second essential ingredient needed to show that a laterally complete archimedean ℓ-group has the splitting property is the following.

Theorem 2.3.31. *Let G be an archimedean ℓ-group, and let H be the ℓ-subgroup of G^e generated by the least upper bounds of disjoint subsets of G^+. Then $C^{G^e}(H)=G^e$.*

Proof. Let X be the Stone space of the Boolean algebra of polars of G. From Theorem 2.3.23 and its proof we may assume $G\subseteq D(X)=G^e$, $\{e_\lambda:\lambda\in\Lambda\}$ is a maximal disjoint subset of G^+, e_λ is the characteristic function of the set X_λ, and $\{X_\lambda:\lambda\in\Lambda\}$ is a family of disjoint clopen sets whose union is dense in X. Let $0<h\in D(X)$. For each $n\in\mathbb{Z}^+$ and each $\lambda\in\Lambda$, let $U_n=h^{-1}((n-1,n))^-$, $V_n=h^{-1}(n)^{\circ-}$, $U_{n,\lambda}=U_n\cap X_\lambda$, and $V_{n,\lambda}=V_n\cap X_\lambda$. Now, $\{U_n,V_n:n\in\mathbb{Z}^+\}$ is a family of disjoint clopen sets, $h^{-1}(\mathbb{R})\cap(\bigcup_\lambda X_\lambda)$ is a dense open subset of X, and

$$h^{-1}(\mathbb{R})\cap\left(\bigcup_\lambda X_\lambda\right)=\bigcup_{(n,\lambda)\in\mathbb{Z}^+\times\Lambda}(U_{n,\lambda}\cup V_{n,\lambda}) \tag{2.3.6}$$

For $(n,\lambda)\in\mathbb{Z}^+\times\Lambda$ let $\{g_\alpha:\alpha\in\Gamma_{n,\lambda}\}$ and $\{h_\beta:\beta\in\Sigma_{n,\lambda}\}$ be maximal disjoint subsets of G^+ with $S(g_\alpha)\subseteq U_{n,\lambda}$ and $S(h_\beta)\subseteq V_{n,\lambda}$, respectively. For $n,m\in\mathbb{Z}^+$, $\lambda\in\Lambda$, $\alpha\in\Gamma_{n,\lambda}$ and $\beta\in\Sigma_{n,\lambda}$ define

$$w_{n,\lambda,\alpha,m}=((m+2)g_\alpha-ne_\lambda)^+\wedge(ne_\lambda-mg_\alpha)^+,$$

$$v_{n,\lambda,\beta,m}=((m+2)h_\beta-ne_\lambda)^+\wedge(ne_\lambda-mh_\beta)^+.$$

Each of the sets

$$A=\{(2m+1)w_{n,\lambda,\alpha,2m}:(\lambda,\alpha,n,m)\in\Lambda\times\Gamma_{n,\lambda}\times\mathbb{Z}^+\times\mathbb{Z}^+\}$$

$$\cup\{(2m+1)v_{n,\lambda,\beta,2m}:(\lambda,\beta,n,m)\in\Lambda\times\Sigma_{n,\lambda}\times\mathbb{Z}^+\times\mathbb{Z}^+\}$$

and

$$B=\{(2m+2)w_{n,\lambda,\alpha,2m+1}:(\lambda,\alpha,n,m)\in\Lambda\times\Gamma_{n,\lambda}\times\mathbb{Z}^+\times\mathbb{Z}^+\}$$

$$\cup\{(2m+2)v_{n,\lambda,\beta,2m+1}:(\lambda,\alpha,n,m)\in\Lambda\times\Sigma_{n,\lambda}\times\mathbb{Z}^+\times\mathbb{Z}^+\}$$ is

a disjoint subset of G^+.

Let $u=\mathrm{lub}_{G^e}A$ and $v=\mathrm{lub}_{G^e}B$. We will show $u+v\geq h$, and to do so it suffices to verify that $(u+v)(x)\geq h(x)$ for each x in the dense subset given in (2.3.6). Now, there exists $(n,\lambda)\in\mathbb{Z}^+\times\Lambda$ with $x\in U_{n,\lambda}\cup V_{n,\lambda}$; so $h(x)\leq n$. By Theorem 2.3.29, for each $\alpha\in\Gamma_{n,\lambda}$ and each $\beta\in\Sigma_{n,\lambda}$, we have

$$ne_\lambda-(ne_\lambda\wedge(u+v))\in g_\alpha^{\perp_H}\cap h_\beta^{\perp_H}.$$

Since

$$\bigcup_{\alpha\in\Gamma_{n,\alpha}} S(g_\alpha)\cup\bigcup_{\beta\in\Sigma_{n,\beta}} S(h_\beta)$$

is dense in $U_{n,\lambda}\cup V_{n,\lambda}$, for each $y\in U_{n,\lambda}\cup V_{n,\lambda}, ne_\lambda(y)=ne_\lambda(y)\wedge(u+v)(y)$; that is, $n\le(u+v)(y)$. In particular, $(u+v)(x)\ge n\ge h(x)$. Thus, $h\in C^{G^e}(H)$. □

Theorem 2.3.32. *Let H be an archimedean ℓ-group that is a $c\ell$-essential extension of the laterally complete ℓ-group G. Then H is projectable and it has the splitting property.*

Proof. Since $G\subseteq H\subseteq G^e=H^e$ and $C^{H^e}(H)=H^e$ by the previous result, H has the splitting property by Exercise 25(a). Let $0<x,y\in H$ and take $g\in G$ with $g>y$. Suppose $\{x_\lambda:\lambda\in\Lambda\}$ is a maximal disjoint subset of $(x^{\perp_H\perp_G})^+$, and let $z=\vee_\lambda x_\lambda\in G$. Then $\{x_\lambda\}$ is a maximal disjoint subset of $x^{\perp_H\perp_H}$ and $x^{\perp_H}=z^{\perp_H}$. For, if $0<h\in x^{\perp_H\perp_H}$ and $0<a\le nh$ with $a\in G$, then $a\in x^{\perp_H\perp_G}$ and $a\wedge x_\lambda>0$ for some $\lambda\in\Lambda$; so $h\wedge x_\lambda>0$. Also, if $0<h\in x^{\perp_H}$, then $h\wedge x_\lambda=0$ for each $\lambda\in\Lambda$; so $h\wedge z=0$. Conversely, if $h\wedge z=0$, then $h\wedge x\in x^{\perp_H\perp_H}$ and $h\wedge x\wedge x_\lambda=0$ for each $\lambda\in\Lambda$; so $h\wedge x=0$. By Theorem 2.3.30 G is projectable and so there exist $g_1\in z^{\perp_G}$ and $g_2\in z^{\perp_G\perp_G}$ with $g=g_1+g_2$. But then

$$y=y\wedge g=(y\wedge g_1)+(y\wedge g_2)\in z^{\perp_H}+z^{\perp_H\perp_H}=x^{\perp_H}+x^{\perp_H\perp_H}$$

since $g_2\in z^{\perp_G\perp_G}=z^{\perp_H\perp_G}\subseteq z^{\perp_H\perp_H}$ by Exercise 2.2.14. Thus, $H=x^{\perp_H}\oplus x^{\perp_H\perp_H}$ and H is projectable. □

Exercises.

1. Let T be a totally ordered set which is topologized by the *interval topology*; that is, the topology has as a subbase all intervals of the form $\{x\in T:x<a\}$ or $\{x\in T:x>a\}$, where $a\in T$. Hence the collection of finite intersections of these intervals is a base for the interval topology.
   (a) If S is an order dense subset of T, then S is a topologically dense subset of T and the interval topology of S coincides with the subspace topology of S.
   (b) If for each pair of elements $a,b\in T$ with $a<b$, there exists $c\in T$ with $a<c<b$, then each topologically dense subset of T is order dense in T.
   (c) If $T\times T$ has the product topology, then the inf and sup functions inf, sup : $T\times T\longrightarrow T$ are both continuous.
   (d) Let X be a topological space and let f,g be two continuous functions from X into T. If $f(x)\le g(x)$ for each x in a dense subset of X show that $f(x)\le g(x)$ for each x in X.
2. Show that a subgroup of $\mathbb{R}$ is either cyclic or is topologically dense in $\mathbb{R}$, and a subring of $\mathbb{R}$ is either contained in $\mathbb{Z}$ or is dense.

3. Prove that a nonzero totally ordered group is complete if and only if it is isomorphic to $\mathbb{Z}$ or to $\mathbb{R}$.

4. This proof of Hölder's theorem does not use Theorem 2.3.3 or Theorem 2.3.9. Let e be a strictly positive element of the totally ordered archimedean group G. If $a \in G$ let $\ell(a) = \{\frac{m}{n} \in \mathbb{Q} : me \leq na\}$ and $u(a) = \{\frac{m}{n} \in \mathbb{Q} : me \geq na\}$ $(n \in \mathbb{N})$. Define $\varphi : G \longrightarrow \mathbb{R}$ by $\varphi(a) = \mathrm{lub}_{\mathbb{R}}\ell(a)$. Then φ is an embedding. (*Hint*: Use Exercise 2.1.15 to show that $\ell(a)+\ell(b) \subseteq \ell(a+b)$ and $u(a)+u(b) \subseteq u(a+b)$.)

5. (a) Let the group $G = \mathbb{Z}\times\mathbb{Z}$ have the positive cone $G^+ = \{(a,b) : a > 0 \text{ and } b > 0\} \cup \{(0,0)\}$. Then G is a directed archimedean po-group that is not integrally closed.
   (b) If G is a directed integrally closed po-group show that $(\mathrm{End}_{\mathbb{Z}}(G), \mathrm{End}_{\mathbb{Z}}(G)^+)$ is integrally closed where $\mathrm{End}_{\mathbb{Z}}(G)^+$ is the set of isotone endomorphisms of G.

6. Show that an ℓ-group can be embedded in a product $\mathbb{R}^I$, for some set I, if and only if it has a collection of normal maximal convex ℓ-subgroups whose intersection is 0.

7. Let A be a dense subset of the topological space Y, and suppose that $f,g,h \in E(Y)$.
   (a) If $h(y) = f(y)+g(y)$ (respectively, $h(y) = f(y)g(y)$) whenever $y \in A$ show that $h(y) = f(y)+g(y)$ (respectively, $h(y) = f(y)g(y)$) whenever $f(y)+g(y)$ (respectively, $f(y)g(y)$) is defined. ($a+b$ is defined unless $\{a,b\} = \{\infty,-\infty\}$ and ab is defined unless $\{a,b\} = \{0,\infty\}$ or $\{a,b\} = \{0,-\infty\}$.)
   (b) If $f,g,h \in D(Y)$ and $h(y) = f(y)+g(y)$ for each $y \in A$, then we write $h = f+g$. This defines a partial addition in $D(Y)$. A nonempty sublattice of $D(Y)$ that is closed under addition and negation is called an ℓ-subgroup of $D(Y)$. Show that each element of $D(Y)$ is contained in an ℓ-subgroup of $D(Y)$. In fact, it is contained in a real ℓ-subspace of $D(Y)$ that contains $C(Y)$. (If $f \in D(Y)$ and $g \in C(Y)$ consider f^+, f^- and $f \vee g$.)

8. Let $I = [0,1]$ be the closed unit interval in $\mathbb{R}$, and let $G = \{f \in D(I) : f(x) = g(x) + \sum_{i=1}^{n} \frac{b_i}{(x-a_i)^2}$, where $a_i \in I$, $b_i \in \mathbb{R}$ and $g \in C(I)\}$. Then:
   (a) G is an archimedean ℓ-group. (First, assume $g = 0$ and construct f^+ by considering the partition of I determined by the zeros of f; show that $f^+(x) = \sum_{b_i>0} \frac{b_i}{(x-a_i)^2} + h(x)$ where h is continuous.)
   (b) Each maximal convex ℓ-subgroup of G contains $C(I)$. (Show that if the maximal convex ℓ-subgroup M of G does not contain $C(I)$, then its intersection with $C(I)$ is a maximal ring ideal of $C(I)$.)
   (c) G cannot be embedded in a product of copies of $\mathbb{R}$.

9. Let F be the field of quotients of the commutative unital domain R. The element $q \in F$ is *almost integral over* R if there exists a nonzero element r in R such that $rq^n \in R$ for each $n \in \mathbb{N}$. R is called *completely integrally closed* if each element

of F that is almost integral over R is in R. Show that R is completely integrally closed if and only if its group of divisibility is integrally closed (Exercise 2.1.12).

10. Show that if I is an infinite set, then $\mathbb{Z}^I/\mathbb{Z}^{(I)}$ is not archimedean.

11. The Hausdorff space X is *completely regular* if for each closed set F and each $x \notin F$ there exists $f \in C(X)$ such that $f(x) = 1$ and $f(F) = 0$. Show that the following statements are equivalent for the completely regular space X.

    (a) X is extremally disconnected.
    (b) $D(X)$ is a conditionally complete lattice.
    (c) $C(X)$ is a complete ℓ-group.

    (For $(c) \Rightarrow (a)$, if U is an open subset of X, use the set $\{f \in C(X) : f \leq 1$ and $f(U') = 0\}$ to construct a function which separates U and $U^{-\prime}$ and then use (g) of Theorem 2.3.19.)

12. Let X be an extremally disconnected space and let $f, g \in D(X)$.

    (a) $S(|f| \wedge |g|) = S(f) \cap S(g)$.
    (b) $S(|f| + |g|) = S(|f| \vee |g|) = S(f) \cup S(g)$.
    (c) $|f| \wedge |g| = 0$ iff $S(f) \cap S(g) = \phi$.
    (d) $f^{\perp\perp} = \{h \in D(X) : S(h) \subseteq S(f)\}$.

13. Let Y be an extremally disconnected space.

    (a) Show that each nonempty disjoint subset of $D(Y)$ has its sup and inf in $D(Y)$; hence, $D(Y)$ is laterally complete.
    (b) A disjoint subset $\{f_i : i \in I\}$ of $D(Y)$ is a maximal disjoint subset if and only if $\cup S(f_i)$ is dense in Y.
    (c) An element $f \in D(Y)$ is a unit iff $S(f) = Y$ iff $f^{\perp} = 0$.
    (d) There is a Stone space X such that $D(Y)$ is isomorphic to $D(X)$.

14. Let $\{G_i : i \in I\}$ be a collection of nonzero ℓ-groups.

    (a) ΠG_i is laterally complete if and only if each G_i is laterally complete.
    (b) $\oplus G_i$ is laterally complete if and only if I is finite and each G_i is laterally complete.
    (c) If I is infinite then the divisible hull of $\mathbb{Z}^I$ is not laterally complete.

15. Let X be the Stone space of the Boolean algebra of polars of the archimedean ℓ-group H. Show that $V(a) = S(f_a)$ for each $a \in H$. (Use step (iv) in the proof of Theorem 2.3.23 and Exercise 12(c).)

16. Let $G(Y)$ be the dual algebra of the extremally disconnected space Y, let H be a $c\ell$-essential ℓ-subgroup of $D(Y)$, and let $\mathscr{B}(D(Y))$ (respectively, $\mathscr{B}(H)$) be the Boolean algebra of polars of $D(Y)$ (respectively, H).

    (a) If $f, g \in D(Y)$, then $fg = 0$ if and only if $|f| \wedge |g| = 0$.

(b) A subset of $D(Y)$ is a polar if and only if it is an annihilator ideal of $D(Y)$ (see Exercise 1.2.7).
(c) $\mathscr{B}(D(Y))$ is isomorphic to the Boolean algebra of idempotents of $D(Y)$ (see Exercise 1.2.11).
(d) The mapping $\rho : \mathscr{B}(D(Y)) \longrightarrow G(Y)$ given by

$$\rho(B) = \left[\bigcup_{f \in B} S(f)\right]^{-}$$

is an isomorphism of Boolean algebras. (Use (c) and Exercise 12).
(e) If $\psi_1 : \mathscr{B}(H) \longrightarrow G(Y)$ is the isomorphism

$$\psi_1(A) = \left[\bigcup_{a \in A^{\perp_H \perp_D}} S(a)\right]^{-}$$

given by (d) and Exercise 2.2.14, then $\psi_1(a^{\perp_H \perp_H}) = S(a)$ for each $a \in H$ and

$$\psi_1(A) = \left[\bigcup_{a \in A} S(a)\right]^{-}$$

for each $A \in \mathscr{B}(H)$.
(f) Suppose that Y is a Stone space and X_D is the Stone space of $\mathscr{B}(D(Y))$. Give an explicit homeomorphism from Y onto X_D.

17. A *weak order unit* in the ℓ-group H is an element $e \in H^+$ with $e^\perp = 0$. Let H be archimedean and let $\varphi : H \longrightarrow D(X)$ be the embedding given in Theorem 2.3.23.

(a) H has a weak order unit if and only if $\varphi(H)$ contains a unit of $D(X)$.
(b) If H has a weak order unit e, then there is an embedding φ with $\varphi(e) = 1$.

18. (a) Let G be an ℓ-group and let $U = \mathscr{U}(D(G))$ be the group of units of the po-monoid $D(G)$. Show that G is an ℓ-subgroup of U.
(b) If G is the ℓ-group $\mathbb{Q} \overleftarrow{\times} \mathbb{Q}$ determine $\mathscr{U}(D(G))$.

19. Let H be an archimedean ℓ-group, and let $\varphi : H \longrightarrow D(X)$ be the embedding given in Theorem 2.3.23. Suppose that F is a subring of $\mathbb{R}$ and H is an F-module such that $F^+H^+ \subseteq H^+$ and $rx = 0$ implies that $r = 0$ or $x = 0$ $(r \in F,\ x \in H)$. Show that φ is an F-module homomorphism.

20. Suppose that H is an archimedean ℓ-group and, for $i = 1, 2, Y_i$ is a Stone space and $\varphi_i : H \longrightarrow D(Y_i)$ is a $c\ell$-essential monomorphism. Show that there is a homeomorphism $\tau : Y_1 \longrightarrow Y_2$ and a positive unit $f \in D(Y_2)$ such that, for all $a \in H$ and every $y_1 \in Y_1$, $\varphi_1(a)(y_1) = \varphi_2(a)(\tau y_1)f(\tau(y_1))$ whenever the right side is defined. That is, $\varphi_1 = \tau^* \circ \ell_f \circ \varphi_2$ where $\tau^* : D(Y_2) \longrightarrow D(Y_1)$ is the isomorphism induced by $\tau, \tau^*(g) = g \circ \tau$, and $\ell_f : D(Y_2) \longrightarrow D(Y_2)$ is

the isomorphism induced by $f, \ell_f(g) = fg$. Show that f and τ are unique. (If $\varphi_1 = \sigma^* \circ \ell_g \circ \varphi_2$ use Theorem 2.3.16 to show that $\tau^* \circ \ell_f = \sigma^* \circ \ell_g$.)

21. The following statements are equivalent for the convex ℓ-subgroup C of the ℓ-group G.

    (a) C is completely closed in G.
    (b) If $\{c_i\}_{i \in I} \subseteq C$ and $c = \vee c_i$ exists in G, then $c \in C$.
    (c) If $\{c_i\}_{i \in I} \subseteq C$ and $c = \wedge c_i$ exists in G, then $c \in C$.
    (d) If $\{c_i\}_{i \in I} \subseteq C^+$ and $c = \vee c_i$ exists in G, then $c \in C$.
    (e) The natural map $G \longrightarrow G/C$ is complete.

22. (a) Show that if A is a polar in the archimedean ℓ-group G, then G/A is archimedean. (Give an elementary proof.)
    (b) Show that if A is a completely closed convex ℓ-subgroup of the archimedean ℓ-group G, then A is a polar. (Use Theorem 2.3.5.)

23. Let $\alpha : G \longrightarrow D$ be a monomorphism of the ℓ-group G into the archimedean ℓ-group D, and suppose that $\mathscr{B}(\alpha(G))$ and $\mathscr{B}(D)$ are canonically isomorphic (see Exercise 2.2.14).

    (a) Show that α preserves maximal disjoint subsets. (Here, D need not be archimedean.)
    (b) Let $\{e_i : i \in I\}$ be a maximal disjoint set in G, and let X_G and X_D be the Stone spaces of $\mathscr{B}(G)$ and $\mathscr{B}(D)$, respectively. Show that the following diagram commutes.

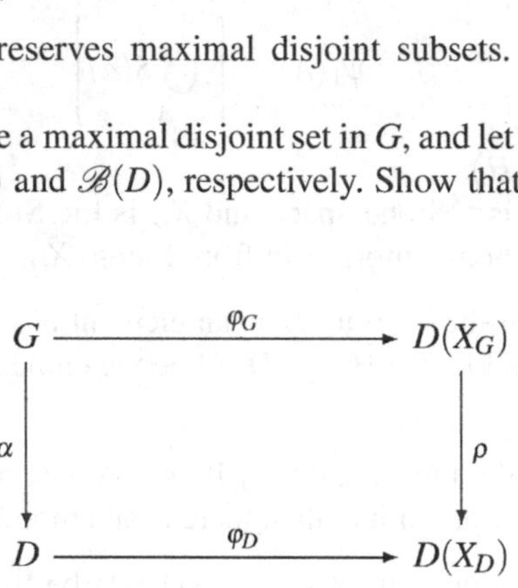

    Here, φ_G and φ_D are the embeddings given in Theorem 2.3.23 that are based on $\{e_i\}$ and $\{\alpha(e_i)\}$, respectively, and ρ is the isomorphism induced by the canonical isomorphism $\mathscr{B}(G) \longrightarrow \mathscr{B}(D)$; that is, ρ is induced by the homeomorphism $X_D \longrightarrow X_G$ given by $m \longmapsto \{\alpha^{-1}(B \cap \alpha(G)) : B \in m\}$. In particular, if G and $\alpha(G)$ are identified and X_G and X_D are identified, then φ_D is an extension of φ_G.
    (c) Show that α is a $c\ell$-essential monomorphism.

24. Let H be an ℓ-subgroup of the archimedean ℓ-group D, and assume that $B \cap H \neq 0$ for each nonzero polar B in D. Show that D is a $c\ell$-essential extension of H. (Assume that $D = D(Y)$ for some Stone space Y. Use Exercise 16(c) to verify (iv) of Exercise 2.2.14 (e), and then apply Exercise 23(c).)

25. Let G be an archimedean ℓ-group.

 (a) If $G^e = C^{G^e}(G)$, show that G has the splitting property.
 (b) Suppose that G is an ℓ-subgroup of $\prod_i T_i = T$ which contains $\oplus_i T_i$, where $\{T_i : i \in I\}$ is a collection of nonzero subgroups of $\mathbb{R}$. If $T = C^T(G)$, show that G has the splitting property. In particular, if some $T_i \subset \mathbb{R}$, then T has the splitting property, but $T \subset T^e = \mathbb{R}^I$.

26. (a) Show that an ℓ-group G is complete iff G^+ is conditionally complete (see Exercise 1.3.4).
 (b) If $G = \varprojlim G_i$ is an inverse limit in the category of ℓ-groups and each G_i is complete show that G is complete (see Exercise 2.2.9).
 (c) Suppose that $G = \varinjlim G_i$ is a direct limit in the category of ℓ-groups where the index set I is directed up. Assume that each map $\psi_{ij} : G_i \longrightarrow G_j$ is a monomorphism and its image is convex. If each G_i is complete show that G is complete (see Exercise 1.4.22).
 (d) If $\{G_i : i \in I\}$ is a family of archimedean ℓ-groups show that $D(\oplus G_i) = \oplus D(G_i)$ and $D(\prod G_i) = \prod D(G_i)$.

27. (a) Suppose that $\{A_i : i \in I\}$ is a family of nonzero ℓ-subgroups of $\mathbb{R}$, and G is an ℓ-subgroup of $\prod A_i$ that contains $\oplus A_i$ (that is, G is archimedean and has a basis - see Exercise 2.5.27). Show that $D(G)$ is the convex ℓ-subgroup of $\prod D(A_i)$ generated by G (see Exercise 2).
 (b) Let A be the ℓ-group of all convergent real-valued sequences and let B be its ℓ-subgroup of eventually constant sequences. Find $D(A)$ and $D(B)$.
 (c) Show that an embedding of ℓ-groups $\alpha : G \longrightarrow \mathbb{R}^I$ can be extended to an embedding $\beta : D(G) \longrightarrow \mathbb{R}^I$. (Assume that, for each $i \in I, A_i = \ker \alpha_i \subset G$ where $\alpha_i = \pi_i \alpha$ and $\pi_i : \mathbb{R}^I \longrightarrow \mathbb{R}$ is the ith projection. For each $i \in I$, let $g_i \in G^+ \backslash A_i$, and let B_i be a convex ℓ-subgroup of $D(G)$ that contains A_i and is maximal with respect to excluding g_i. Then $B_i \cap G = A_i$ and B_i is a maximal convex ℓ-subgroup of $D(G)$. Use Theorem 3.3.1 (a).)

28. (a) If H is archimedean show that its ℓ-subgroup G is dense in H iff it is left and right dense in H.
 (b) Suppose that $G \subseteq D$ are ℓ-subgroups of $H = \mathbb{R}^I$. Show that the following statements are equivalent: (i) D is a completion of G. (ii) D is maximal among those ℓ-subgroups of $C^H(G)$ in which G is left dense. (iii) $D = C^T(G)$ where T is an ℓ-subgroup of H maximal with respect to containing G as a left dense ℓ-subgroup.

29. Let G be an archimedean ℓ-group.

 (a) If $a \in G$ show that $C^G(a)$ is a subdirect product of a family of subgroups of $\mathbb{R}$ (Use Exercises 6 and 2.5.32 (a)).
 (b) If $0 \leq a \leq b$ show that $C^G(a)$ has a unique completion in $D(C^G(b))$; let $\psi_{ab} : D(C^G(a)) \longrightarrow D(C^G(b))$ be the unique embedding.
 (c) Show that $(\{D(C^G(a)) : a \in G^+\}, \{\psi_{ab} : a \leq b\})$ is a direct system of ℓ-groups and $D(G) = \varinjlim D(C^G(a))$ (see Exercise 26(c)).

30. Let G be a left dense convex ℓ-subgroup of $D(X)$, where X is a Stone space.
    (a) Show that each element of $D(X)^+$ is the sup of a disjoint subset of $C(X)^+$. (Use the argument in the proof of Theorem 2.3.25.)
    (b) Show that 1 is the sup of a disjoint set of idempotents from G. (Each nonempty clopen subset of X contains a nonempty clopen subset whose characteristic function is in G.)
    (c) Show that each element in $D(X)^+$ is the sup of a disjoint subset of G^+.
    (d) If H is a $c\ell$-essential ℓ-subgroup of $D(X)$ show that each element of $D(X)^+$ is the sup of a disjoint subset of $D(d(H))$.

2.4 Prime Subgroups, Representability, and Operator Sets

The lattices in which it is easiest to compute are the totally ordered sets. Thus, to study an ℓ-group it would be useful to have enough totally ordered sets (or groups) arising from the ℓ-group to represent it. We show here that this is the case, and verify that the class of ℓ-groups which can be embedded into a product of totally ordered groups is a variety determined by a two variable identity. With an eye toward rings and modules we introduce ℓ-groups with operators and several results are expressed in this format. The basic theory concerning the totally ordered convex subgroups of an ℓ-group is presented in the exercises and it will be used in subsequent sections. A convex ℓ-subgroup C of the ℓ-group G is called a *prime subgroup* if whenever $a,b \in G$ with $a \wedge b \in C$, then $a \in C$ or $b \in C$.

Theorem 2.4.1. *The following statements are equivalent for the convex ℓ-subgroup C of G.*

(a) C is a prime subgroup.
(b) If $a,b \in G$ with $a \wedge b = 0$, then $a \in C$ or $b \in C$.
(c) The lattice of left cosets G/C is totally ordered.
(d) The lattice of convex ℓ-subgroups of G that contain C is totally ordered.

Proof. Obviously, (a) implies (b). To see that (b) implies (a) and (c), let $a,b \in G$; then $[-(a\wedge b)+a] \wedge [-(a\wedge b)+b] = 0$ implies that $-(a\wedge b)+a \in C$ or $-(a\wedge b)+b \in C$. So, if $a\wedge b \in C$, then $a \in C$ or $b \in C$ and hence C is prime. Also, $a+C = a\wedge b+C = (a+C)\wedge(b+C)$ or $b+C = a\wedge b+C = (a+C)\wedge(b+C)$; thus G/C is totally ordered.

(c) implies (d). Let A and B be convex ℓ-subgroups of G that contain C. If $a \in A^+\backslash B$ and $b \in B^+$, then, since G/C is totally ordered, there exists $c \in C^+$ such that $0 \le a \le b+c \in B$ or $0 \le b \le a+c \in A$. Thus $b \in A$, and hence $B \subseteq A$.

(d) implies (b). Let $a\wedge b = 0$ and let A (respectively, B) be the convex ℓ-subgroup generated by a (respectively, b) and C. Then $A \subseteq B$ or $B \subseteq A$; but $A \cap B = C$ by Theorem 2.2.4(d). So either $a \in C$ or $b \in C$. □

Each subgroup C that contains a prime subgroup is an ℓ-subgroup, and hence is itself prime if it is convex. For, if $x \in C$, then, since $x^+ \wedge x^- = 0$, $x^+ = x+x^- \in C$.

Also, it is easy to see that the intersection of any chain of prime subgroups is prime. In particular, if P is a prime subgroup and $\{P\}$ is enlarged to a maximal chain $\{P_i : i \in I\}$ of prime subgroups, then $\cap P_i$ is a minimal prime contained in P. We will give a useful characterization of minimal primes which is modelled after the theory of prime ideals in a commutative ring. But first we construct some primes and show that the primes are plentiful.

A subset S of G is *meet closed* if $a, b \in S$ implies that $a \wedge b \in S$. Note that a convex ℓ-subgroup P is prime iff $G^+ \setminus P$ (or $G \setminus P$) is meet closed.

Theorem 2.4.2. *Let G be an ℓ-group.*

(a) If S is a meet closed subset of G^+ and P is a convex ℓ-subgroup maximal with respect to $P \cap S = \phi$, then P is prime.
(b) If P is a minimal prime, then $G^+ \setminus P$ is a meet closed subset of G^+ maximal with respect to excluding 0.
(c) Each convex ℓ-subgroup is the intersection of prime subgroups.

Proof. (a) Suppose that $a \wedge b = 0$ and let A (respectively, B) be the convex ℓ-subgroup generated by P and a (respectively, P and b). Then $A \cap B = P$ by Theorem 2.2.4 (d). If $a, b \notin P$, then there exist $s \in A \cap S$ and $t \in B \cap S$; so $s \wedge t \in P \cap S$, which is impossible.

(b) Suppose that T is meet closed, $0 \notin T$ and $G^+ \setminus P \subseteq T \subseteq G^+$. Let Q be a convex ℓ-subgroup that is maximal with respect to $Q \cap T = \phi$. Then Q is prime, and $Q^+ \subseteq G^+ \setminus T \subseteq P^+$; so $Q = P$ by the minimality of P, and $Q^+ = G^+ \setminus T = P^+$. Thus $T = G^+ \setminus P$.

(c) Let C be a convex ℓ-subgroup of G, and for each $a \in G^+ \setminus C$ let P_a be a convex ℓ-subgroup that contains C and is maximal with respect to excluding a. Then $C = \cap\{P_a : a \in G^+ \setminus C\}$; for if $0 \leq b \in (\cap P_a) \setminus C$, then $b \in P_b$. Also, each P_a is prime by (a). □

A convex ℓ-subgroup that is maximal with respect to excluding some element is called a *regular* subgroup. The proof of (c) in the previous theorem shows that each $C \in \mathscr{C}(G)$ is the intersection of regular subgroups. It is easy to see that a convex ℓ-subgroup C is regular if and only if there is a least element D in the poset of all convex ℓ-subgroups that properly contain C. We then say that D *covers* C. We will have more to say about these covering pairs later.

Theorem 2.4.3. *A prime subgroup P of the nonzero ℓ-group G is a minimal prime if and only if, for each $x \in P$, $x^\perp \nsubseteq P$.*

Proof. Suppose that P has this property and that Q is a convex ℓ-subgroup properly contained in P. If $x \in P \setminus Q$ and $y \in x^\perp \setminus P$, then $|x| \wedge |y| = 0$ but neither x nor y is in Q. So Q is not prime and P is a minimal prime. For the converse, suppose that P is a minimal prime and let $S = G^+ \setminus P$. If $0 < x \in P$ let $T = S \cup \{x \wedge s : s \in S\}$. Then T is meet closed and contains S properly since $x \wedge s \in P$. By Theorem 2.4.2, part (b), $0 \in T$; that is, $x \wedge s = 0$ for some $s \notin P$. □

An ℓ-group is called *representable* if it can be embedded into a product of a family of totally ordered groups. This condition is clearly equivalent to the requirement that the ℓ-group is a subdirect product of a family of totally ordered groups. Since 0 is an intersection of primes (Theorem 2.4.2 (c)) each abelian ℓ-group is representable. The next result shows that an ℓ-group is representable if and only if it satisfies certain identities. Recall that if a and g are elements of a group then the conjugate of *a determined by g* is denoted by $a^g = -g+a+g$. Note that in an ℓ-group conjugation by g is an ℓ-automorphism.

Theorem 2.4.4. *The following statements are equivalent for the ℓ-group G.*

(a) G is representable.
(b) Each subdirectly irreducible homomorphic image of G is totally ordered.
(c) Each minimal prime subgroup is normal.
(d) The largest normal subgroup of G contained in a prime subgroup is itself prime.
(e) Each polar is normal.
(f) If $a \wedge b = 0$ then $a^y \wedge b = 0$ for each y in G.
(g) $(x^+)^y \wedge x^- = 0$ for all $x, y \in G$.
(h) $2(x \wedge y) = 2x \wedge 2y$ (equivalently, $2(x \vee y) = 2x \vee 2y$) for all $x, y \in G$.
(i) $n(x \wedge y) = nx \wedge ny$ (equivalently, $n(x \vee y) = nx \vee ny$) for all $x, y \in G$ and for all $n \in \mathbb{N}$.
(j) If a and x are elements of G and $a \wedge a^x = 0$, then $a = 0$.

Proof. Since (i) holds in any totally ordered group it holds in any product of totally ordered groups. So (a) implies (i) and the latter implies (h).

(h) implies (g). $2x \wedge 2y = (x \wedge y) + (x \wedge y) = 2x \wedge 2y \wedge (x+y) \wedge (y+x)$ yields that $2x \wedge 2y \leq x+y$. Replacing x by $y+x$ gives that $(y+x+y+x) \wedge (y+y) \leq y+x+y$, or $x \wedge (-y-x+y) \leq 0$. So $0 = [x \wedge (-y-x+y)]^+ = x^+ \wedge [(-x)^y]^+$ and $(x^+)^{-y} \wedge x^- = 0$.

(g) implies (f). This follows since $a = x^+$ and $b = x^-$ for $x = a-b$.

(f) implies (e). According to (f) each polar of the form $b^\perp$ is normal; but, clearly, each polar is an intersection of such polars.

(e) implies (d). Let

$$K = \bigcap_{g \in G} P^g$$

be the largest normal subgroup of G that is contained in the prime subgroup P. If $a \wedge b = 0$ and $b \notin K$, then $b^h \notin P$ for some h. But $a^g \wedge b^h = 0$ for each $g \in G$, so each $a^g \in P$; that is, $a \in K$.

(d) implies (c). This is obvious.

(c) implies (a). Since the intersection of all of the primes is 0 and since each prime contains a minimal prime, G is a subdirect product of the family of totally ordered groups $\{G/P : P \text{ a minimal prime}\}$.

(b) implies (a). This follows from Theorem 2.2.7.

(a) implies (b). Let $\overline{G}$ be a subdirectly irreducible homomorphic image of G. Because of the equivalence of (a) and (g) $\overline{G}$ is representable and hence is totally ordered.

(a) implies (j). Since (j) holds in any totally ordered group, and hence in any product of totally ordered groups, it hold in a representable ℓ-group.

(j) implies (f). If $a \wedge b = 0$, then, for each $y \in G, (a^y \wedge b) \wedge (a^y \wedge b)^{-y} = 0$. So $a^y \wedge b = 0$. □

As a consequence of the previous theorem we see that the class of representable ℓ-groups is a variety of ℓ-groups.

We next consider operators on an ℓ-group. A function $\omega : G \longrightarrow G$ on the ℓ-group G is called a *d-map* (respectively, an *f-map*) if $x \wedge y = 0$ implies that $\omega x \wedge \omega y = 0$ (respectively, $\omega x \wedge y = 0$). If ω is a group homomorphism as well as an f-map it is called an *f-homomorphism*. The basic properties of these maps are given in the following result.

Theorem 2.4.5. *Let ω and λ be functions from the ℓ-group G to itself.*

(a) If ω is an f-map, then it is a d-map.
(b) If ω is a d-map, then $\omega 0 = 0$ and $\omega G^+ \subseteq G^+$. If G is totally ordered, then any function ω with these two properties is an f-map.
(c) If ω and λ are f-maps then so is $\omega + \lambda$, where $(\omega+\lambda)(x) = \omega(x) + \lambda(x)$.
(d) ω is an f-map iff $\omega G^+ \subseteq G^+$ and $\omega + 1$ is a d-map.
(e) If ω and λ are d-maps (respectively, f-maps, f-homomorphisms), then so is the composite $\omega\lambda$.
(f) If ω and λ are f-homomorphisms and $|x| \wedge |y| = 0$, then $(\omega+\lambda)(x+y) = (\omega+\lambda)x + (\omega+\lambda)y$.
(g) If ω is a d-map (respectively, an f-map) and $\lambda(x) \le \omega(x)$ for each $x \in G^+$, then λ is a d-map (respectively, an f-map) iff $\lambda G^+ \subseteq G^+$.

Proof. (a) is obvious, (b) follows from $\omega 0 = \omega 0 \wedge \omega 0 = 0$ and $\omega x \wedge 0 = 0$ if $x \in G^+$, and (c) follows from the fact that if $x \wedge y = 0$, then $(\omega + \lambda)x \wedge y = (\omega x + \lambda x) \wedge y = 0$. If $\omega G^+ \subseteq G^+$ and $\omega + 1$ is a d-map, then $x \wedge y = 0$ implies that $0 \le \omega x \wedge y \le (\omega x + x) \wedge (\omega y + y) = (\omega + 1)x \wedge (\omega + 1)y = 0$. This, together with (c), proves (d). (e) is obvious, and as for (f), if $|x| \wedge |y| = 0$, then $0 = \lambda|x| \wedge \omega|y| = |\lambda x| \wedge |\omega y|$ if ω and λ are f-homomorphisms (Theorem 2.2.2); hence λx and ωy commute and $(\omega+\lambda)(x+y) = \omega x + \omega y + \lambda x + \lambda y = (\omega+\lambda)x + (\omega+\lambda)y$. For (g), suppose that $\lambda(x) \le \omega(x)$ for each $x \in G^+$, $\lambda G^+ \subseteq G^+$ and ω is a d-map. If $x \wedge y = 0$, then since $0 \le \lambda x \wedge \lambda y \le \omega x \wedge \omega y = 0$, λ is a d-map. Similarly, if ω is an f-map then so is λ. □

Let Ω be a set and let G be an ℓ-group. Suppose that $\varphi : \Omega \longrightarrow \mathrm{Set}[G,G]$ is a function from Ω to the set of set maps on G. If $\omega \in \Omega$ and $g \in G$, then, as usual, we write $\omega g = \varphi(\omega)g$. G is called an *Ω-group* if $\omega 0 = 0$ and $\omega G^+ \subseteq G^+$ for each $\omega \in \Omega$. G is called a *weak Ω-d-group* if the image of φ is contained in the set of d-maps of G, and G is an *Ω-d-group* if the image of φ is contained in the set of ℓ-endomorphisms of G. Similarly, G is called a *weak Ω-f-group* or an *Ω-f-group* depending on whether the image of φ is contained in the set of f-maps or the set of f-endomorphisms of G. An *Ω-subgroup* of G is a subgroup A with $\omega A^+ \subseteq A^+$ for each $\omega \in \Omega$. In this generality an Ω-subgroup A may not be an Ω-group (in the group sense) since we may not have $\omega A \subseteq A$ for $\omega \in \Omega$. But if ω induces a group homomorphism and if A is a directed subgroup, then $\omega A \subseteq A$. If G and H are Ω-groups, then an ℓ-homomorphism $f : G \longrightarrow H$ is an *Ω-homomorphism* if $f(\omega x) = \omega f(x)$ for each $\omega \in \Omega$ and each $x \in G$. It is easily seen that if A is a normal convex ℓ-Ω-subgroup of the Ω-d-group G, then G/A is an Ω-d-group and the natural map $G \longrightarrow G/A$ is an Ω-homomorphism. As a consequence of Theorem 2.1.4(m) each ℓ-group is a weak $\mathbb{N}$-f-group. Also, if Ω is the group of inner automorphisms of G then, by Theorem 2.4.4, G is representable if and only if G is an Ω-f-group. The direct product $\prod G_i$ of a family of Ω-groups is an Ω-group with Ω-action given by $\omega((x_i)) = (\omega x_i)$. In general, an Ω-group is said to be *representable* if it can be embedded into a product of a family of totally ordered Ω-groups. The next two results are companions to Theorem 2.4.4.

Theorem 2.4.6. *The following statements are equivalent for the Ω-group G.*

(a) G is a weak Ω-f-group.
(b) Each minimal prime subgroup is an Ω-subgroup.
(c) Each polar is an Ω-subgroup.

Proof. (a) $\Rightarrow$ (b). Let P be a minimal prime subgroup, and let $x \in P^+$, $0 \leq y \in x^\perp \setminus P$ (Theorem 2.4.3) and $\omega \in \Omega$. Then $\omega x \wedge y = 0$ yields that $\omega x \in P^+$.

(b) $\Rightarrow$ (c). This is a conseqeuence of the fact that each polar A is an intersection of a family of minimal primes. For, if K is the intersection of all those minimal primes that do not contain $A^\perp$ and L is the intersection of the remaining minimal primes, then $K \cap L = 0$. So $A \subseteq K \subseteq L^\perp$ and $A^\perp \subseteq L$, and, therefore, $A \subseteq K \subseteq L^\perp \subseteq A^{\perp\perp} = A$.

(c) $\Rightarrow$ (a). If $\omega \in \Omega$ and $x \wedge y = 0$, then $\omega x \in (y^\perp)^+$; so $\omega x \wedge y = 0$. □

Theorem 2.4.7. *Let G be an Ω-group.*

(a) If G is a representable Ω-group, then G is a weak Ω-f-group.
(b) If G is a representable ℓ-group and is also an Ω-f-group, then G is a representable Ω-f-group.
(c) G is a representable Ω-f-group if and only if each Ω-subdirectly irreducible homomorphic image of G is a totally ordered Ω-f-group.

Proof. Since any product of a family of totally ordered Ω-groups is a weak Ω-f-group, by (b) of Theorem 2.4.5, each representable Ω-group is a weak Ω-f-group. Conversely, suppose that G is representable and is an Ω-f-group, and let P be a minimal prime subgroup of G. Then P is a normal Ω-subgroup, and G/P is a totally ordered Ω-group. Thus, G is a subdirect product of the family of totally ordered Ω-f-groups $\{G/P : P \text{ a minimal prime}\}$, and hence it is a representable Ω-f-group. This proves (a) and (b) and, consequently, each homomorphic image H of a representable Ω-f-group is a representable Ω-f-group. So if H is subdirectly irreducible it must be totally ordered. The other implication in (c) follows from Exercise 1.4.12 (or, more explicitly, the proof of Theorem 2.2.7 works if we assume that each $\omega \in \Omega$ induces a group endomorphism of G). □

We also give the following condition for an Ω-group to be a weak Ω-f-group. If $x \in G$, then $\Omega x = \{\omega x : \omega \in \Omega\}$.

Theorem 2.4.8. *Suppose that G is an ℓ-group and $\Omega \longrightarrow G^G$ is a function such that:*

(i) Each $\omega \in \Omega$ induces an isotone function on G and $\omega 0 \geq 0$.
(ii) If $\omega, \lambda \in \Omega$, then there is a d-map α with $\omega\lambda x, \omega x \leq \alpha x$ for each x in G^+.

Then, for any $x \in G$ and any $\lambda \in \Omega$, $\Omega(\lambda x^+ \wedge x^-) = 0$. Consequently, G is a weak Ω-f-group if, for each $0 < x \in G$, $\omega x \neq 0$ for some $\omega \in \Omega$.

Proof. Suppose that $a \wedge b = 0$, λ, $\omega \in \Omega$, and let α be a d-map which satisfies the second condition. Then

$$0 \leq \omega(\lambda a \wedge b) \leq \omega\lambda a \wedge \omega b \leq \alpha a \wedge \alpha b = 0,$$

and hence $\omega(\lambda a \wedge b) = 0$. □

Exercises.

1. Prove that each ℓ-group G can be embedded into an ℓ-group Aut(T) (see Exercise 2.1.5) for some totally ordered set T. (*Hint*: If $\{P_i : i \in I\}$ is a family of prime subgroups of G with $\cap P_i = 0$ totally order I and give $\cup(G/P_i)$ the ordinal sum order.)

2. Let $(T, \leq)$ be a totally ordered set and let Aut(T) be the ℓ-group of all automorphisms of T. If $\underline{\alpha}$ is a well-ordering of T let

$$P_{\underline{\alpha}} = \{f \in \text{Aut}(T) : f(t_0) > t_0 \text{ where } t_0 \text{ is the first element in } (\{t \in T : f(t) \neq t\}, \underline{\alpha})\}.$$

Prove each of the following statements.

(a) $P_{\underline{\alpha}}$ is a right order of Aut(T).
(b) $\text{Aut}(T)^+ = \cap P_{\underline{\alpha}}$ where $\underline{\alpha}$ runs over all of the well-orderings of T.
(c) The positive cone of each ℓ-group is an intersection of right orders.

3. Show that the positive cone of a representable ℓ-group is an intersection of total orders. (Hint: Reduce to the case that the ℓ-group is a product of totally ordered groups and use Hahn products.)

4. Let $C, C_1, \ldots, C_n \in \mathscr{C}(G)$. Show that the following statements are equivalent.

(a) C is prime.
(b) C is finitely meet irreducible in $\mathscr{C}(G)$: $C = C_1 \cap \cdots \cap C_n \Rightarrow C = C_i$ for some i.
(c) If $C_1 \cap \cdots \cap C_n \subseteq C$ then, for some i, $C_i \subseteq C$.

5. Show that $C \in \mathscr{C}(G)$ is regular iff it is meet irreducible in $\mathscr{C}(C)$:

$$C = \bigcap_{i \in I} C_i$$

with $C_i \in \mathscr{C}(G) \Rightarrow C = C_i$ for some i.

6. Prove that the following statements are equivalent for an ℓ-group G.

   (a) G is representable.
   (b) If P is a prime subgroup, then P is comparable to each of its conjugates $P^g = -g + P + g$.
   (c) If C is a regular subgroup, then C is comparable to each of its conjugates.

7. Show that the following ℓ-groups are not representable.

   (a) The ℓ-group in Exercise 2.1.11.
   (b) $\mathrm{Aut}(\mathbb{R})$.

8. Show that if $\leq$ is a partial order of the free group F and $(F, \leq)$ is an ℓ-group, then $\leq$ is a total order. (*Hint*: This follows from the two facts: (i) F is an indecomposable group (easy), and (ii) each subgroup of F is free (hard).) It is known that a free group can be made into a totally ordered group (see Theorem 3.5.10).

9. If P is a minimal prime subgroup of the ℓ-group G, show that $P = \cup\{a^\perp : a \in G^+ \backslash P\}$. (So each minimal prime is a (directed) union of polars, and in the proof of Theorem 2.4.6 it was seen that each polar is an intersection of minimal primes.)

10. Let A and B be totally ordered convex subgroups of the ℓ-group G. Then A contains B, or is contained in B, or $A \cap B = 0$.

11. Let $0 \neq a \in G$, where G is an ℓ-group. If $a > 0$ and $[0, a] = \{x \in G : 0 \leq x \leq a\}$ is totally ordered, then a is called a *basic element* of G.

    (a) If $|a|$ is basic, then $a \in G^+ \cup -G^+$.
    (b) The following statements are equivalent.
        (i) $|a|$ is basic.
        (ii) $C(a)$ is totally ordered.
        (iii) $a^{\perp\perp}$ is totally ordered.
        (iv) $b^\perp = a^\perp$ for each $0 \neq b \in a^{\perp\perp}$.
    (c) If a is basic, then $a^{\perp\perp}$ is maximal among those convex subgroups of G that are totally ordered.
    (d) Show that the subgroup of G generated by its set of basic elements is a convex ℓ-subgroup and is a direct sum of totally ordered groups.

12. Show that the following statements are equivalent for the element $0 < a$ in the ℓ-group G.

    (a) a is basic.
    (b) $a^\perp$ is a prime subgroup.
    (c) $a^\perp$ is a minimal prime subgroup.
    (d) $a^{\perp\perp}$ is totally ordered.
    (e) $a^{\perp\perp}$ is a minimal polar (minimal among the nonzero polars).
    (f) $a^\perp$ is a maximal polar (maximal among the proper polars).

13. A subset S of basic elements in the ℓ-group G is called a *basis of G* if S is a maximal disjoint subset of G.
    (a) Show that the following statements are equivalent.
        (i) G has a basis.
        (ii) If $0 < g \in G$, then there is a basic element b with $b \leq g$.
        (iii) The Boolean algebra $\mathscr{B}(G)$ of polars of G is atomic (see Exercise 1.3.6).
    (b) Show that any two bases of G have the same cardinality.

14. Here is a proof of the fact that an archimedean ℓ-group is abelian that doesn't use the Dedekind completion. Let G be an ℓ-group.
    (a) If $a,x,y \in G^+$, then $a^x \wedge a^y \geq [-(x+y)+a]^+$.
    (b) If $b,c \in G$ and $b \wedge b^c = 0$, then $\forall n \in \mathbb{N}$, $nb \leq |c|$. (Apply (a) with $a = nb, x = c^-$ and $y = c^+$.)
    (c) If G is totally ordered and $a,b \in G^+$, then $n[a,b] \leq a \vee b$ for each $n \in \mathbb{N}$. (*Hint*: If A is the convex subgroup maximal with respect to excluding $[a,b]$ and A^* is its cover, then $A \lhd A^*$ and A^*/A is abelian (Exercise 2.3.4).
    (d) If G is archimedean, then G is abelian.

15. (a) The following statements are equivalent for the ℓ-group G.
        (i) Each homomorphic image of G is archimedean (G is then called *hyper-archimedean*).
        (ii) Each proper prime subgroup of G is a maximal prime.
        (iii) Each proper prime subgroup is a minimal prime.
        (iv) For each $x \in G$, $G = C(x) \oplus x^\perp$.
        (v) If $x,y \in G^+$, then for some $n \in \mathbb{N}$, $[y-(nx \wedge y)] \wedge x = 0$.
        (vi) There is an embedding $\varphi : G \longrightarrow \mathbb{R}^I$ with the property : For each x,y in G^+ there exists $n \in \mathbb{N}$ such that if $\varphi(x)(i) > 0$, then $n\varphi(x)(i) > \varphi(y)(i)$.
        (vii) If $x,y \in G^+$, then there is an $n \in \mathbb{N}$ such that $y \wedge nx = y \wedge (n+1)x$. (First, give a cyclic proof of (i) through (v).)
    (b) For any set I, $G = \{f \in \mathbb{R}^I : f(I) \text{ is finite}\}$ is a hyper-archimedean ℓ-group.

16. Let G be a representable ℓ-group. Suppose that $a,b \in G$ and n and m are nonzero integers such that na and mb commute. Show that a and b commute (use Theorem 2.3.8).

17. Let G be an ℓ-group and let $H = \{g \in G : x \wedge y = 0 \Rightarrow x^g \wedge y = 0\}$. Then H is a normal ℓ-subgroup of G which contains the center of G. Also, G is representable iff $G = H$.

18. Give an example of an abelian, Ω-simple (and hence Ω-subdirectly irreducible) weak Ω-f-group that is not totally ordered. (Ω-simple means there is no proper nonzero normal convex ℓ-Ω-subgroup.)

19. Let $f(M)$ be the set of f-maps of the ℓ-group M, and let $fh(M)$ be the set of f-endomorphisms of M.

(a) Show that there are surjections $f(\oplus_\lambda M_\lambda) \longrightarrow \Pi f(M_\lambda)$ and $f(\Pi_\lambda M_\lambda) \longrightarrow \Pi_\lambda f(M_\lambda)$ that preserve addition.
(b) Show that there are bijections $fh(\oplus M_\lambda) \longrightarrow \Pi fh(M_\lambda)$ and $fh(\Pi M_\lambda) \longrightarrow \Pi fh(M_\lambda)$.
(c) Show that the bijections in (b) preserve addition if each M_λ is abelian.

20. Suppose that α is an f-map of the ℓ-group G and, for all $x \in G^+, \alpha x = 0$ iff $x = 0$. Show that if β is a d-map of G that is comparable to α ($\beta \leq \alpha$ if $\beta x \leq \alpha x$ for each $x \in G^+$), then β is an f-map.

21. Suppose that ${}_\Omega G$ and G_Λ are f-groups. Show that $|\omega g|, |g\lambda| \leq |\omega g + g\lambda|$ for any $\omega \in \Omega, g \in G$ and $\lambda \in \Lambda$; and $|\omega g| + |g\lambda| = |\omega g + g\lambda|$ iff ωg^- and $|g|\lambda$ commute iff ωg^- and $g^-\lambda$ commute.

22. Let G and K be Ω-d-groups and suppose $f : S \longrightarrow K$ is an Ω-group homomorphism where S is an Ω-subgroup of G. Show that there is an (unique) Ω-ℓ-group homomorphism $g : [S] \longrightarrow K$ which extends f if and only if $\bigvee_I \bigwedge_J f(s_{ij}) = 0$ in K whenever $\bigvee_I \bigwedge_J s_{ij} = 0$ in $[S]$. (Use Theorem 2.2.4.)

23. Let G be an Ω-ℓ-subgroup of the Ω-f-group H and let P be a prime Ω-subgroup of G. Suppose Q is a convex Ω-ℓ-subgroup of H which is maximal with respect to $Q \cap G = P$. Show that Q is a prime Ω-subgroup of H.

24. Suppose G is a representable Ω-f-group and let $C_\Omega = \{g \in G :$ the convex ℓ-Ω-subgroup generated by g is bounded$\}$; see Section 2.5. Verify each of the following:

(a) C_Ω is a convex ℓ-Ω-subgroup of G.
(b) If α is an Ω-ℓ-endomorphism of G, then $\alpha(C_\Omega) \subseteq C_\Omega$.
(c) If $-g + \omega x + g = \omega(-g + x + g)$ for all $g, x \in G$ and $\omega \in \Omega$, then C_Ω is a normal subgroup.
(d) If for all $\omega \in \Omega$ and $g \in G$ there is an integer $n \in \mathbb{N}$ with $\omega g \leq ng$, then C_Ω contains the derived subgroup G'. (Use the argument in the verification of (i)–(iv) in the proof of Theorem 2.3.9, or use Exercise 14(c), and the equation $[-b, a] = [a, b]^{-b}$.)
(e) If G is finitely generated as an Ω-group or, more generally, as a convex ℓ-Ω-subgroup, then $C_\Omega \subset G$.
(f) Give an example of a countable abelian ℓ-group with $C_\Omega = G$.

2.5 Values

Throughout this section we assume that G is an Ω-f-group. This includes the case that G is just an ℓ-group if we set $\Omega = \{1\} \subseteq \mathbb{N}$. Most of the results of Sections 2.2 and 2.4 hold for an Ω-f-group. In particular, the collection $\mathscr{C}_\Omega(G)$ of all convex ℓ-Ω-subgroups of G is a complete sublattice of $\mathscr{C}(G)$, and Theorem 2.2.4 holds for Ω-subgroups, with some modifications. If $\omega_1, \ldots, \omega_n \in \Omega$ and $x \in G$ we will abbreviate $\omega_1 \cdots \omega_n x$ by Wx (n could be 0, in which case $Wx = x$) and we will abuse notation by writing $W = \omega_1 \cdots \omega_n \in \Omega^\infty$. The first thing to note is that the convex ℓ-Ω-subgroup generated by $X \subseteq G$ is given by

$$C_\Omega(X) = \{g \in G : |g| \leq W_1|x_1| + \cdots + W_n|x_n|, \text{ where } x_i \in X \text{ and } W_i \in \Omega^\infty\}.$$

The other changes in Theorem 2.2.4 are then:

$$C_\Omega(X) \cap C_\Omega(Y) = C_\Omega(\{W|x| \wedge V|y| : x \in X, y \in Y, \text{ and } W, V \in \Omega^\infty\}); \tag{2.5.1}$$

$$C_\Omega(D,a) \cap C_\Omega(D,b) = D \vee C_\Omega(\{W|a| \wedge V|b| : W, V \in \Omega^\infty\}) \quad \text{if } D \in C_\Omega(G) \text{ and } a, b \in G; \tag{2.5.2}$$

$$\text{if } S \text{ is a subgroup with } \Omega S \subseteq S, \text{ then } \Omega[S] \subseteq [S]. \tag{2.5.3}$$

Also, Theorems 2.2.5 and 2.4.2 hold for convex ℓ-Ω-subgroups.

Those convex ℓ-Ω-subgroups with a successor in $\mathscr{C}_\Omega(G)$ (the values) are studied for their elucidation of the structure of the Ω-f-group. A tight relation between the values in $\mathscr{C}_\Omega(G)$ and the values in $\mathscr{C}(G)$ is established. By imposing a finite condition on the values of an element a sharp local structure theorem is obtained which survives in an infinite form when the condition is relaxed. A related relation between two convex ℓ-Ω-subgroups is introduced and is used in an exercise to establish how a finitely conditioned Ω-f-group can be put together by means of totally ordered groups. Also, this relation is used to describe when some convex ℓ-Ω-subgroups are summands, and this description will be used heavily in later sections. The set of values will be used in the next section to get a deep embedding theorem.

If $a \in G$ and P is a convex ℓ-Ω-subgroup maximal with respect to not containing a, then P is called an *Ω-value* of a and also an *Ω-regular* Ω-subgroup. By Theorem 2.4.2 and the remarks following it each Ω-regular Ω-subgroup is prime, and $P \in \mathscr{C}_\Omega(G)$ is Ω-regular if and only if it is covered by some $M \in \mathscr{C}_\Omega(G)$. Also, if M covers P, then P is an Ω-value of a precisely when $a \in M \setminus P$. Let $\Gamma_\Omega(a) = \Gamma_\Omega(a, G)$ denote the trivially ordered subset of $\mathscr{C}_\Omega(G)$ which consists of the Ω-values of a. If $\Omega = \{1\}$, then an Ω-value will be called a *value* of a and $\Gamma_\Omega(a)$ will be denoted by $\Gamma(a)$.

A poset P is called *rooted* if for each p in P the set $U(p) = \{x \in P : x \geq p\}$ is a chain; equivalently, if no two incomparable elements of P have a common lower bound. Clearly, the collection of all of the prime Ω-subgroups of the Ω-f-group G is a rooted subset of $\mathscr{C}_\Omega(G)$, and so is its subset $\Gamma_\Omega(G)$ which consists of all of the Ω-regular Ω-subgroups of G. If $\Omega = \{1\}$, then $\Gamma_\Omega(G)$ will be denoted by $\Gamma(G)$. The rooted poset $\Gamma_\Omega(G) = \Gamma_\Omega$ (respectively, $\Gamma(G) = \Gamma$) is called the *Ω-value set* (respectively, *value set*) of G. We have seen in Theorem 2.4.2 that each convex ℓ-Ω-subgroup is an intersection of Ω-regular Ω-subgroups. As another indication of the importance of Ω-values we have

Theorem 2.5.1. *(a)* $\Gamma_\Omega(a) = \Gamma_\Omega(|a|) = \Gamma_\Omega(a^+) \cup \Gamma_\Omega(a^-)$ *and* $\Gamma_\Omega(a^+) \cap \Gamma_\Omega(a^-) = \phi$.

(b) $a \geq 0$ *iff* $a + P \geq P$ *in* G/P *for each* $P \in \Gamma_\Omega(a)$.

(c) $|a| \wedge |b| = 0$ *iff* $\Gamma_\Omega(a) \cup \Gamma_\Omega(b)$ *is a disjoint union and is trivially ordered.*

Proof. That $\Gamma_\Omega(a) = \Gamma_\Omega(|a|)$ is clear. Suppose that $P \subset M$ is a covering pair in $\mathscr{C}_\Omega(G)$. Since P is prime, $a^+ \in P$ or $a^- \in P$. Consequently, $a = a^+ - a^- \in M \setminus P$ iff $a^- \in M \setminus P$ or $a^+ \in M \setminus P$; thus, (a) is proven. Suppose that $a + P > P$ for each $P \in \Gamma_\Omega(a)$. If $a^- > 0$ and $P \in \Gamma_\Omega(a^-) \subseteq \Gamma_\Omega(a)$, then $a^+ \in P$, and $P = a^+ + P = (a + P) \vee P = a + P > P$. Thus, $a^- = 0$ and $a \geq 0$. This proves (b) since the other implication is trivial. Suppose that $\Gamma_\Omega(a) \cup \Gamma_\Omega(b)$ is a disjoint union and is trivially ordered. If $|a| \wedge |b| > 0$ let P be one of its Ω-values. Then $|a|$ (respectively, $|b|$) has an Ω-value Q_1 (respectively, Q_2) containing P. But then Q_1 and Q_2 are comparable which is impossible. Hence $|a| \wedge |b| = 0$. This finishes the proof of (c) since the converse follows from (a). □

We next wish to compare the Ω-values of $a \in G$ with its Ω-values in a convex ℓ-Ω-subgroup. Toward this end we have the basic

Theorem 2.5.2. *Let* $A \in \mathscr{C}_\Omega(G)$ *and let* $\mathscr{P}_\Omega(G)$ *be the poset consisting of the prime* Ω*-subgroups of G. Then we have the isotone mappings*

$$\mathscr{P}_\Omega(G) \longrightarrow \mathscr{P}_\Omega(A), \quad P \longmapsto P \cap A = P^c$$

$$\mathscr{P}_\Omega(A) \longrightarrow \mathscr{P}_\Omega(G), \quad Q \longmapsto Q^e = \{x \in G : C_\Omega(x) \cap A \subseteq Q\}.$$

Furthermore, these mappings are inverse order isomorphisms when restricted to the following subsets:

$$\{P \in \mathscr{P}_\Omega(G) : A \not\subseteq P\} \longleftrightarrow \mathscr{P}_\Omega(A) \setminus \{A\}.$$

Proof. If $P \in \mathscr{P}_\Omega(G)$, then it is clear that $P^c \in \mathscr{P}_\Omega(A)$ and also that $A \not\subseteq P$ iff $P^c \neq A$. If $Q \in \mathscr{P}_\Omega(A)$, then Q^e is a convex ℓ-Ω-subgroup of G. For, if $x, y \in Q^e$ and $z \in G$ with $|z| \leq |x|$, and $\omega \in \Omega$, then $C_\Omega(\omega|x|) \cap A$, $C_\Omega(z) \cap A$, $C_\Omega(x - y) \cap A \subseteq C_\Omega(x, y) \cap A = [C_\Omega(x) \vee C_\Omega(y)] \cap A = [C_\Omega(x) \cap A] \vee [C_\Omega(y) \cap A] \subseteq Q$. Also, Q^e is prime since if $x \wedge y = 0$ in G, then $C_\Omega(x) \cap C_\Omega(y) \cap A = 0$, and hence $C_\Omega(x) \cap A \subseteq Q$ or $C_\Omega(y) \cap A \subseteq Q$ by Exercise 2.4.4. To see that the restrictions are inverse isomorphisms first note that if Q is a proper prime Ω-subgroup of A, then $A \not\subseteq Q^e$ and

$$Q^{ec} = Q^e \cap A = \{x \in A : C_\Omega(x) \cap A \subseteq Q\} = Q.$$

Secondly, if P is a prime Ω-subgroup that does not contain A, then

$$\begin{aligned} P^{ce} &= \{x \in G : C_\Omega(x) \cap A \subseteq P \cap A\} \\ &= \{x \in G : C_\Omega(x) \cap A \subseteq P\} \\ &= \{x \in G : C_\Omega(x) \subseteq P\} = P. \end{aligned}$$

□

As a consequence of this result we have

Theorem 2.5.3. *If $a \in A \in \mathscr{C}_\Omega(G)$, then the mapping*

$$\Gamma_\Omega(a,G) \longrightarrow \Gamma_\Omega(a,A), \quad P \longmapsto P \cap A$$

is a bijection, and its globalization

$$\{P \in \Gamma_\Omega(G) : A \nsubseteq P\} \longrightarrow \Gamma_\Omega(A), \quad P \longmapsto P \cap A$$

is an isomorphism.

Proof. If $P \in \Gamma_\Omega(a,G)$, then $a \notin P \cap A$. Suppose that $A \supseteq B \supset P \cap A$ with $B \in C_\Omega(A)$. If $A \supset B$, then $B^e \supset P$ and $a \in B^{ec} = B$ by the previous theorem. Thus $P \cap A$ is an Ω-value of a in A. Now let $Q \in \Gamma_\Omega(a,A)$. Then $a \notin Q^e$, and if C is a convex ℓ-Ω-subgroup of G which properly contains Q^e, then $C \cap A \supset Q^e \cap A = Q$, again by the previous theorem. So $a \in C$ and Q^e is an Ω-value of a in G.

For the second part take $P \in \Gamma_\Omega(G)$ with its cover K, and suppose that $A \nsubseteq P$. Then $P^c = P \cap A \subseteq K \cap A = K^c \subseteq A$. By Theorem 2.5.2, if $A \nsubseteq K$ then P^c is covered by K^c. But also, if $A \subseteq K$ then P^c is covered by A; for if $P^c \subset Q^c \subset A$ then $P \subset Q$, and hence $A \subseteq K \subseteq Q$ and $Q^c = A$. In both cases $P^c \in \Gamma_\Omega(A)$. On the other hand, take $Q \in \Gamma_\Omega(A)$. If Q is covered by A and $a \in A \setminus Q$, then Q^e is an Ω-value of a. For if $Q^e \subset C$ and $a \notin C$, then $Q \subset C^c \subset A$ gives a contradiction. The other possibility is that Q is covered by $K \subset A$, but then Q^e is covered by K^e. □

We consider next the relationship between the regular subgroups and the Ω-regular Ω-subgroups of G.

Theorem 2.5.4. *For $P \in \mathscr{C}(G)$ let P_Ω be the largest convex ℓ-Ω-subgroup of G which is contained in P.*

(a) P is prime if and only if P_Ω is prime.
(b) For each $a \in G$ the mapping

$$\Gamma(a) \longrightarrow \Gamma_\Omega(a) \ , \ P \longmapsto P_\Omega$$

is a bijection.
(c) There is a natural embedding of posets $\Gamma_\Omega \longrightarrow \Gamma$.

Proof. Suppose that P is prime and that $a \wedge b = 0$. Then $C_\Omega(a) \cap C_\Omega(b) = 0$; so $C_\Omega(a) \subseteq P$ or $C_\Omega(b) \subseteq P$. Thus $a \in P_\Omega$ or $b \in P_\Omega$, and P_Ω is prime. Conversely, if P_Ω is prime then so is P since $P_\Omega \subseteq P$. This proves (a). Now suppose that P is a value of a, and let C be a convex ℓ-Ω-subgroup which properly contains P_Ω. Then P and C are comparable but C is not contained in P. Hence $P \subset C$ and $a \in C$. This proves that P_Ω is an Ω-value of a. If $P_\Omega = Q_\Omega$ where Q is also a value of a, then P and Q are comparable and hence $P = Q$. So the mapping $P \mapsto P_\Omega$ is one-to-one. It is also onto; for, suppose that Q is an Ω-value of a. Then there is a unique value P

of a which contains Q. If C is any convex ℓ-Ω-subgroup of G which is contained in P, then $Q \subseteq Q \vee C \subseteq P$ and $a \notin Q \vee C$. Hence $C \subseteq Q$ and $Q = P_\Omega$. This proves (b).

In order to prove (c) note that, by (b), the mapping $K \longmapsto K_\Omega$, when restricted to Γ, maps Γ onto Γ_Ω, and it is order preserving. But then each right inverse of this function is an embedding of Γ_Ω into Γ, as the following result shows.

> Let X and Y be rooted subsets of some poset, and let $\rho : X \longrightarrow Y$ be an order preserving onto mapping such that $\rho(x) \leq x$ for each x in X. Then each of the right inverses of ρ is an embedding of Y into X.

For if $\rho(x_1) < \rho(x_2)$, then, since $\rho(x_i) \leq x_i$, x_1 and x_2 are comparable. But then $x_1 < x_2$ since ρ is isotone. Thus, each right inverse φ of ρ is isotone. Also, if $\varphi(y_1) < \varphi(y_2)$, then $y_1 = \rho\varphi(y_1) < \rho\varphi(y_2) = y_2$; so φ is an embedding. □

Since each $C \in \mathscr{C}_\Omega(G)$ is the intersection of a set of Ω-values it is clear that $\Gamma(G) = \Gamma_\Omega(G)$ if and only if $\mathscr{C}(G) = \mathscr{C}_\Omega(G)$. In fact, these equalities hold provided the embedding given in Theorem 2.5.4(c) is an isomorphism.

Theorem 2.5.5. *Let G be an Ω-f-group. If K is an Ω-value (or a value) in G let K^* denote its cover in the lattice $\mathscr{C}_\Omega(G)$ (or $\mathscr{C}(G)$). Then the following statements are equivalent.*

(a) The mapping $\Gamma(G) \longrightarrow \Gamma_\Omega(G)$ given by $K \longmapsto K_\Omega$ is an isomorphism.
(b) $K^* \setminus K = (K_\Omega)^* \setminus K_\Omega$ for each K in $\Gamma(G)$.
(c) $\Gamma(G) = \Gamma_\Omega(G)$.
(d) $(K_\Omega)^* = (K^*)_\Omega$ for each K in $\Gamma(G)$.

Proof. (a) ⇒ (b). If $K \in \Gamma(G)$, then $K^* \setminus K \subseteq (K_\Omega)^* \setminus K_\Omega$ by Theorem 2.5.4. Let $x \in (K_\Omega)^* \setminus K_\Omega$. Then x has a value L which contains K_Ω. Since $L_\Omega \supseteq K_\Omega$ and $x \notin L_\Omega$, necessarily $L_\Omega = K_\Omega$. Thus $L = K$ and $x \in K^* \setminus K$.

(b) ⇒ (c). Suppose that $K \in \Gamma$ and $x \in K \setminus K_\Omega$. Then x has an Ω-value L that contains K_Ω. Since L and K are comparable $L \subset K$, and hence $L = K_\Omega$. So $x \in (K_\Omega)^* \setminus K_\Omega = K^* \setminus K$. This contradiction gives that $K = K_\Omega$.

(d) ⇒ (b). We need to verify that $(K_\Omega)^* \setminus K_\Omega \subseteq K^* \setminus K$. If $x \in (K_\Omega)^* \setminus K_\Omega$ then $x \in (K_\Omega)^* = (K^*)_\Omega \subseteq K^*$. According to Theorem 2.5.4, x has a value L with $L_\Omega = K_\Omega$. Then L and K are comparable, and if $K \subseteq L$ then $x \in K^* \setminus K$. Suppose that $L \subset K$. Then $L^* \subseteq K$ and hence $(L_\Omega)^* = (L^*)_\Omega \subseteq K_\Omega = L_\Omega$, which is nonsense.

Since (c) obviously implies (a) and (d) the proof is complete. □

Let $A \subseteq B$ be convex ℓ-Ω-subgroups of G. B is called a *lexicographic extension* of A, and we write B = lex A, if A is a prime Ω-subgroup of B and $b > A$ for each $b \in B^+ \setminus A$. Note that if $A \neq 0$, then this latter condition implies that A is prime in B; in fact, it implies that both elements of each pair of nonzero disjoint elements of B are in A. Note also that if $A \subseteq B \subseteq G$ and if G = lex B and B = lex A, then G = lex A; the converse is a consequence of Theorem 2.5.6. A totally ordered group is a lexicographic extension of each of its convex subgroups. Also, if $G = A \overleftarrow{\times}_\varphi B$ is the ordinal semidirect product of the ℓ-group A by the totally ordered group B, then G = lex A.

A maximal chain in a rooted poset P is called a *root*, and the intersection of all of its roots is defined to be the *trunk* of P. This intersection will be denoted by trunk (P). Clearly, if $p \in P$, then $p \in \text{trunk}(P)$ if and only if p is comparable to each element of P. The *Ω-trunk* of the Ω-f-group G is defined to be the intersection of all of the Ω-regular Ω-subgroups in the trunk of its Ω-value set:

$$\text{trunk}_\Omega(G) = \cap\{A : A \in \text{trunk}(\Gamma_\Omega(G))\}.$$

If G is just an ℓ-group (that is, $\Omega = \{1\}$), then $\text{trunk}_\Omega(G)$ will be denoted by trunk(G). Since $\text{trunk}_\Omega(G)$ is the intersection of a chain of prime Ω-subgroups it is a prime Ω-subgroup. The relationship between the Ω-trunk of G and those convex ℓ-Ω-subgroups of G of which G is a lexicographic extension is very tight.

Theorem 2.5.6. *The following statements are equivalent for $C \in \mathscr{C}_\Omega(G)$.*

(a) $G = lex\, C$.
(b) $G \backslash C \subseteq G^+ \cup -G^+$.
(c) If $a \in G \backslash C$, then $a + C \subseteq G^+$ or $a + C \subseteq -G^+$.
(d) C is the intersection of some subset of the trunk of $\Gamma_\Omega(G)$.
(e) $C \supseteq trunk_\Omega(G)$.

Proof. (a) $\Rightarrow$ (b). Let $a \in G \backslash C$. Then, either $a^+ \in G \backslash C$ and $a^- \in C$, or $a^- \in G \backslash C$ and $a^+ \in C$. Thus, $a^+ > a^-$ or $a^- > a^+$, and hence $a \in G^+ \cup -G^+$.

(b) $\Rightarrow$ (c). This follows from the fact that G/C is a poset, and hence the coset $a + C$ cannot contain both positive and negative elements.

(c) $\Rightarrow$ (a). If $C = 0$ then G is totally ordered; and if $C \neq 0$ and $a \in G^+ \backslash C$, then $a > C$.

(a) $\Rightarrow$ (d). By Theorem 2.4.2(c) for Ω-f-groups, $C = \cap A_j$ for some chain $\{A_j\}_{j \in J}$ in Γ_Ω. We claim that each $A_j \in$ trunk(Γ_Ω). For if $B \in \Gamma_\Omega$, then either $B \subseteq C$, or there is an element $b \in B^+ \backslash C$. In the latter case $b > C$ and so $C \subseteq B$. In either case the set $\{A_j, B\}_{j \in J}$ is a chain and hence $\{A_j\}_{j \in J} \subseteq$ trunk(Γ_Ω).

(d) $\Rightarrow$ (e). This is obvious.

(e) $\Rightarrow$ (b). We may assume that $C = \text{trunk}_\Omega(G)$. First note that $\text{trunk}_\Omega(G)$ is comparable to each $A \in \Gamma_\Omega$. For, either $A \in$ trunk(Γ_Ω), in which case $A \supseteq \text{trunk}_\Omega(G)$, or $A \subset B$ for each $B \in$ trunk(Γ_Ω) and $A \subseteq \text{trunk}_\Omega(G)$. Now, suppose that $a \in G \setminus \text{trunk}_\Omega(G)$, and let A be an Ω-value of a which contains $\text{trunk}_\Omega(G)$. Then $A \in \Gamma_\Omega(a^+)$ or $A \in \Gamma_\Omega(a^-)$ by Theorem 2.5.1. If $A \in \Gamma_\Omega(a^+)$ and $B \in \Gamma_\Omega(a^-)$, then $B \subseteq \text{trunk}_\Omega(G)$ or $\text{trunk}_\Omega(G) \subseteq B$. In either case A and B are comparable which is impossible by Theorem 2.5.1. Thus $a^- = 0$ and $a \in G^+$. Similarly, if $A \in \Gamma_\Omega(a^-)$, then $a \in -G^+$. □

The next result gives, among other things, another description of $\text{trunk}_\Omega(G)$.

Theorem 2.5.7. *(a) If $G = lex\, A$, then $\Gamma_\Omega(a, A) = \Gamma_\Omega(a, G)$ for each $a \in A$. So $\Gamma_\Omega(A) \subseteq \Gamma_\Omega(G)$.*
(b) The trunk of the poset $\Gamma_\Omega(trunk_\Omega(G))$ is empty; that is, $trunk_\Omega(trunk_\Omega(G))$ $= trunk_\Omega(G)$.

(c) If N is the subgroup of G generated by $\{a \in G : a^{\perp} \neq 0\}$, then N = trunk$_{\Omega}(G)$.
(d) trunk$_{\Omega}(G)$ = trunk(G) is a normal subgroup of G which is independent of the operator set Ω.

Proof. (a) Let $a \in A$. Since A is comparable to each $P \in \Gamma_{\Omega}$ (Theorem 2.5.6), if $P \in \Gamma_{\Omega}(a,G)$, then $P \subset A$. Thus $\Gamma_{\Omega}(a,A) = \Gamma_{\Omega}(a,G)$ by Theorem 2.5.3.

(b) Let A = trunk$_{\Omega}(G)$ and B = trunk$_{\Omega}(A)$. By the previous theorem, G = lex A, A = lex B, G = lex B, and $A \subseteq B$. So $A = B$ and the trunk of $\Gamma_{\Omega}(G)$ is empty.

(c) N is a convex ℓ-Ω-subgroup of G. For if $|y| \leq |x|$ with $x \in N$, then $|y| \leq |a_1| + \cdots + |a_n|$ with $a_i^{\perp} \neq 0$. So $y^+ = b_1 + \cdots + b_n$ and $y^- = c_1 + \cdots + c_n$ with $0 \leq b_i, c_i \leq |a_i|$. Thus $b_i, c_i \in N$ and $y = y^+ - y^- \in N$. If $\omega \in \Omega$ and $y \in N^+$, then $y = b_1 + \cdots + b_n$ with $b_i \in N^+$ and $b_i^{\perp} \neq 0$. Then $\omega y = \omega b_1 + \cdots + \omega b_n$ and $(\omega b_i)^{\perp} \neq 0$. Thus $\omega y \in N$. It is clear that $N \subseteq$ trunk$_{\Omega}(G)$ since each of its generators lies in trunk$_{\Omega}(G)$. Now, G = lex N; for if $a \in G \backslash N$, then either $a^+ = 0$ or $a^- = 0$. Thus, $G \backslash N \subseteq G^+ \cup -G^+$, G = lex N, and $N \supseteq$ trunk$_{\Omega}(G)$.

(d) trunk$_{\Omega}(G) = N =$ trunk(G) is normal since N is generated by a normal subset (or, since trunk $\Gamma(G)$ is invariant under each ℓ-automorphism of G). □

The element $a \in G$ is called *Ω-special* if it has a unique Ω-value. As a consequence of Theorem 2.5.4, a is Ω-special if and only if it is $\{1\}$-special in G. Consequently, an Ω-special element will frequently just be called a *special* element. If G = lex A and $a \in G \backslash A$, then a is special. For a has a value N such that $N \supseteq$ trunk(G). But any value of a is comparable to trunk(G) and hence is comparable to N; so N is the only value of a. These remarks are summarized in

Theorem 2.5.8. *The following statements are equivalent for $a \in G$.*

(a) a is Ω-special in G.
(b) a is Ω-special in $C_{\Omega}(a)$.
(c) If $A \in \mathscr{C}_{\Omega}(G)$ and $a \in A$, then a is Ω-special in A.
(d) a is special in G.
(e) $C_{\Omega}(a)$ = lex B for some proper convex ℓ-Ω-subgroup of $C_{\Omega}(a)$.
(f) $C_{\Omega}(a)$ = lex B for some proper convex ℓ-subgroup of $C_{\Omega}(a)$.

Proof. The preceding remarks show that (a) and (d) are equivalent and that (e) implies (b). Conversely, suppose that B is the unique Ω-value of a in $\mathscr{C}_{\Omega}(a)$. Then B is the unique maximal convex ℓ-Ω-subgroup of $C_{\Omega}(a)$. So $B \in$ trunk $\Gamma_{\Omega}(C_{\Omega}(a))$, and hence $C_{\Omega}(a)$ = lex B by Theorem 2.5.6. The equivalence of (a), (b) and (c) is given by Theorem 2.5.3, and the equivalence of (e) and (f) is given by Theorem 2.5.6. □

In order to further examine special elements we need a technical result which is of interest in its own right and which will also be useful later.

Theorem 2.5.9. *Suppose that $B \in \mathscr{C}_{\Omega}(G)$, A is a proper convex ℓ-Ω-subgroup of B and B = lex A. Then*

$$(B \oplus B^{\perp})^+ = \{x \in G^+ : x \not> B\}.$$

Consequently, if B is unbounded then it is a summand of G.

Proof. Let $x \in G^+$ with $x \not> B$. Then $x \not> b$ for some $b \in B^+ \setminus A$. To see that $x \in (B \oplus B^\perp)^+$ let $y = -(x \wedge 2b) + x$ and $c = -(x \wedge 2b) + 2b$, and note that $y \wedge c = 0$. Then $x = (x \wedge 2b) + y$ and $x \wedge 2b \in B$; so it suffices to show that $y \in B^\perp$. First note that $c \in B^+ \setminus A$; for if $c \in A$, then $x + c \geq x \wedge 2b + c = 2b > b + c$ yields the contradiction $x > b$. Now, if $d \in B^+$ then $d \wedge y \wedge c = 0$ gives that $d \wedge y \in A$. So $c > d \wedge y$ and hence $d \wedge y = 0$. The other inclusion is obvious. □

Theorem 2.5.10. *Suppose that a is a special element of G, and let M be its value in G and N its value in C(a). For $A \in \mathscr{C}(G)$ let A_Ω be the largest convex ℓ-Ω-subgroup of G that is contained in A. Then:*

(a) $N_\Omega = M_\Omega \cap C_\Omega(a)$ is the Ω-value of a in $C_\Omega(a)$, and $C_\Omega(a)$ = lex N_Ω.
(b) $M_\Omega = N_\Omega \oplus a^\perp$.
(c) $N = M \cap C_\Omega(a)$ is the value of a in $C_\Omega(a)$.
(d) $M/N \cong M_\Omega/N_\Omega$ as ℓ-groups.
(e) $C_\Omega(a)$ = lex $C(a)$.
(f) If C(a) is not totally ordered, then C(a) contains a nonzero convex ℓ-Ω-subgroup of G.

Proof. By Theorems 2.5.3 and 2.5.4, $A = M_\Omega \cap C_\Omega(a)$ is the unique maximal convex ℓ-Ω-subgroup of $C_\Omega(a)$. By Theorems 2.5.6 and 2.5.8, $C_\Omega(a)$ = lex A, and since $C_\Omega(a)^\perp = a^\perp$,

$$(C_\Omega(a) \oplus a^\perp)^+ = \{x \in G^+ : x \not> C_\Omega(a)\} \supseteq M_\Omega^+$$

by Theorem 2.5.9. Thus,

$$M_\Omega = [M_\Omega \cap C_\Omega(a)] \oplus a^\perp \qquad (2.5.4)$$

since $a^\perp \subseteq M_\Omega$. In particular, using $\Omega = \{1\}$ and taking $C_\Omega(a)$ for G, we get from (2.5.4) that $M \cap C_\Omega(a) = N \oplus (a^\perp \cap C_\Omega(a)) = N$. But $(X \cap Y)_\Omega = X_\Omega \cap Y_\Omega$ for $X, Y \in \mathscr{C}(G)$, so $N_\Omega = M_\Omega \cap C_\Omega(a) = A$. This proves (a), (b) and (c). Now, (d) follows from (b), and (e) and (f) are consequences of (a) since $C(a) \supseteq N \supseteq N_\Omega$. □

We consider next those elements in the Ω-f-group G that have only a finite number of values, and we give the important local structure theorem for such elements. If $a \in G$ has only a finite number of Ω-values, then a is called *finite valued*. If $M \in \Gamma_\Omega(G)$ (respectively, $M \in \Gamma(G)$) is the unique Ω-value (respectively, value) of some special element, then M is also called *Ω-special* (respectively, special). Clearly, if $M \in \mathscr{C}(G)$ is a special subgroup, then M_Ω is an Ω-special Ω-subgroup; but the converse is not true.

Theorem 2.5.11. *The following statements are equivalent for $0 \neq a \in G$.*

(a) a is finite valued.
(b) Each Ω-value of a is Ω-special.
(c) $a = a_1 + \cdots + a_n$ where each a_i is special and $|a_i| \wedge |a_j| = 0$ if $i \neq j$.

Moreover, the decomposition of a as a sum of pairwise disjoint special elements is unique and gives the unique indecomposable decomposition of $C_\Omega(a) = C_\Omega(a_1) \oplus \cdots \oplus C_\Omega(a_n)$.

Proof. We first note that if $M \in \Gamma_\Omega(a)$, then M is Ω-special if and only if $M \cap C_\Omega(a)$ is Ω-special in $C_\Omega(a)$. For, if M is the unique Ω-value of $b \in G$, then, since $\Gamma_\Omega(x) \cap \Gamma_\Omega(y) \subseteq \Gamma_\Omega(|x| \wedge |y|)$, we have that M is an Ω-value of $|a| \wedge |b|$. But each Ω-value of $|a| \wedge |b|$ is contained in an Ω-value of b; so $|a| \wedge |b|$ is special with M as its unique Ω-value. Thus, by Theorem 2.5.3, $M \cap C_\Omega(a)$ is Ω-special in $C_\Omega(a)$. Conversely, if $M \cap C_\Omega(a)$ is the only Ω-value of b in $C_\Omega(a)$, where $0 \le b \le |a|$, then M is the only Ω-value of b. For, if $M \subset L$ where $L \in \mathscr{C}_\Omega(G)$, then $a \in L$, and hence $b \in L$ (or we can use Theorem 2.5.3). So we may assume that $G = C_\Omega(a)$.

(a) $\Rightarrow$ (b). Suppose that $A_1, \ldots, A_n$ are all of the maximal convex ℓ-Ω-subgroups of G and assume that $n > 1$. For each $i = 2, \ldots, n$ choose $0 < b_i \in A_1 \setminus A_i$ and $0 < c_i \in A_i \setminus A_1$. Then for

$$b = \bigvee_{i \ge 2} b_i \in A_1 \setminus \bigcup_{i \ge 2} A_i \text{ and } c = \bigwedge_{i \ge 2} c_i \in \left(\bigcap_{i \ge 2} A_i \right) \setminus A_1,$$

we have that $d = b - b \wedge c$ and $e = c - b \wedge c$ are strictly positive disjoint elements. Furthermore, $A_2, \ldots, A_n$ are Ω-values of d and A_1 is an Ω-value of e. Let N be another Ω-value of e. Then $N \subseteq A_i$ for some $i \ge 2$; but $d \wedge e = 0$ gives the contradiction that $d \in N$ or $e \in N$. Thus, A_1 is Ω-special and similarly, each A_i is Ω-special.

(b) $\Rightarrow$ (c). Let $\Gamma_\Omega(a) = \{A_i : i \in I\}$ be the set of maximal convex ℓ-Ω-subgroups of G, and suppose that A_i is the only Ω-value of b_i. Then $\{b_i : i \in I\}$ is a pairwise disjoint subset of G (Theorem 2.5.1), and hence

$$\bigoplus_{i \in I} C_\Omega(b_i) \subseteq G.$$

Since this containment cannot be proper $G = \oplus_{i \in I} C_\Omega(b_i)$ and I is finite; say, $I = \{1, 2, \ldots, n\}$. Then $a = a_1 + \cdots + a_n$ with $a_i \in C_\Omega(b_i) = C_\Omega(a_i)$, and each a_i is special by Theorem 2.5.8.

(c) $\Rightarrow$ (a). This is a consequence of Theorem 2.5.1.

The uniqueness statements follow from the fact that $\mathscr{C}_\Omega(G)$ is a distributive lattice (Theorems 2.2.5 and 2.2.6). □

The Ω-f-group G is called *finite valued* if each of its elements has only a finite number of values. According to the previous theorem G is finite valued precisely when each Ω-regular Ω-subgroup is Ω-special. It is even possible to determine whether or not G is finite valued by examining its lattice of convex ℓ-Ω-subgroups, as we will now show.

A complete lattice L is called *completely distributive* if for every doubly indexed subset $\{a_{ij} : i \in I,\ j \in J\}$ the following equation and its dual hold in L:

$$\bigwedge_{i\in I}\bigvee_{j\in J} a_{ij} = \bigvee_{f\in J^I}\bigwedge_{i\in I} a_{if(i)}.$$

The power set of any set is completely distributive, as is easily checked. A *dual ideal* of a poset P is a subset D of P such that if $d < p$ with $d \in D$, then $p \in D$. The set P' of all of the dual ideals of P is a complete sublattice of the power set of P and hence is completely distributive. Each element A in $\mathscr{C}_\Omega(G)$ is the intersection of a dual ideal of Γ_Ω, by Theorem 2.4.2(c). If each element in $\mathscr{C}_\Omega(G)$ is the intersection of a unique dual ideal of Γ_Ω, then $\mathscr{C}_\Omega(G)$ is said to be *freely generated* by Γ_Ω.

Theorem 2.5.12. *The following statements are equivalent for G.*

(a) *$\mathscr{C}_\Omega(G)$ is freely generated by $\Gamma_\Omega(G)$.*
(b) *$\mathscr{C}_\Omega(G)$ is completely distributive.*
(c) *$\mathscr{C}(G)$ is completely distributive.*
(d) *$B \vee (\bigwedge_{i\in I} A_i) = \bigwedge_{i\in I}(B \vee A_i)$ for each subset $\{A_i : i \in I\} \cup \{B\}$ of $\Gamma_\Omega(G)$.*
(e) *G is finite valued.*

Proof. If $A \in \mathscr{C}_\Omega(G)$ let $A' = \{N \in \Gamma_\Omega : A \subseteq N\}$. Then A' is a dual ideal of Γ_Ω, and under the assumption of (a), the mapping $A \longmapsto A'$ is a lattice anti-isomorphism of $\mathscr{C}_\Omega(G)$ onto the completely distributive lattice Γ'_Ω. Thus, (a) implies (b), and clearly (b) implies (d). Suppose that (d) holds, and let $\{B_j : j \in J\}$ and $\{A_i : i \in I\}$ be dual ideals of Γ_Ω with $\cap B_j = \cap A_i$. Then for each $j \in J$,

$$B_j = B_j \vee \left(\bigcap_k B_k\right) = B_j \vee \left(\bigcap_i A_i\right) = \bigcap_i (B_j \vee A_i).$$

So $B_j \supseteq A_i$ for some $i \in I$, by Exercise 2.4.5 for $\mathscr{C}_\Omega(G)$, and hence $B_j \in \{A_i : i \in I\}$. By symmetry $\{B_j\} = \{A_i\}$, and thus (d) implies (a).

We will show that (a) is equivalent to the condition that each Ω-regular Ω-subgroup is Ω-special. It will then follow from Theorem 2.5.11 that (a) is equivalent to (e). Toward this end, if $M \in \Gamma_\Omega(G)$ let $\delta(M) = \{N \in \Gamma_\Omega(G) : N \nsubseteq M\}$, and let $M^* = \cap\{N : N \in \delta(M)\}$. Then $\delta(M)$ is a dual ideal of $\Gamma_\Omega(G)$, and it can readily be seen that

$$M^* = \{a \in G : \text{ each } \Omega\text{-value of } a \text{ is contained in } M\}. \tag{2.5.5}$$

Now suppose that M is not Ω-special. Then $M^* \subseteq M$. For if $a \in M^*$ and A is an Ω-value of a, then A is properly contained in M, and hence $a \in M$. If $\rho(M) = \delta(M) \cup M'$, then $\rho(M)$ is a dual ideal distinct from $\delta(M)$, yet

$$\cap_{N\in\delta(M)} N = M^* = \cap_{N\in\rho(M)} N.$$

Thus, $\mathscr{C}_\Omega(G)$ is not freely generated by $\Gamma_\Omega(G)$. Conversely, if $\mathscr{C}_\Omega(G)$ is not freely generated by $\Gamma_\Omega(G)$, then there is a dual ideal δ of Γ_Ω and $M \in \Gamma_\Omega \setminus \delta$ such that

$$\cap_{N\in\delta} N \subseteq M.$$

But then $\delta \subseteq \delta(M)$ and hence $M^* \subseteq M$. If M is Ω-special and $a \in G$ has M as its unique Ω-value, then $a \in M^* \subseteq M$ by (2.5.5). This contradiction shows that M is not

Ω-special. To complete the proof it is only necessary to note that the equivalence of (b) with (e) also gives the equivalence of (c) with (e). □

An Ω-f-group is finite valued if and only if each positive element is the sup of a finite set of positive disjoint special elements. A description of the infinite analogue of this property is given next. A subset Δ of $\Gamma_\Omega(G)$ is called *plenary* if $\Gamma_\Omega(g) \cap \Delta \neq \emptyset$ for each $0 \neq g \in G$, and whenever $g \notin N \in \Delta$, then there exists $M \in \Delta \cap \Gamma_\Omega(g)$ with $M \supseteq N$. Another description of a plenary subset is given in Exercise 33.

Theorem 2.5.13. *Suppose Δ is a plenary subset of $\Gamma_\Omega(G)$. If $g \in G$ and $\Gamma_\Omega(g) \cap \Delta$ is finite, then $\Gamma_\Omega(g) \subseteq \Delta$.*

Proof. By Exercise 34 $\Delta^c = \{N \cap C_\Omega(g) : N \in \Delta \text{ and } g \notin N\}$ is a plenary subset of $\Gamma_\Omega(C_\Omega(g))$, and by Theorem 2.5.3 we may assume $G = C_\Omega(g)$. Let $\Gamma_\Omega(g) \cap \Delta = \{P_1, \ldots, P_n\}$. Suppose $P \in \Gamma_\Omega(g) \backslash \Delta$. For $i = 1, \ldots, n$ take $a_i \in P^+ \backslash P_i$ and $b_i \in P_i^+ \backslash P$, and let $a = a_1 \vee \cdots \vee a_n$ and $b = b_1 \wedge \cdots \wedge b_n$. Then $a \in P \backslash P_i$ and $b \in P_i \backslash P$ for each i; so $0 < c = a - a \wedge b$, $d = b - a \wedge b$ and $c \wedge d = 0$. Let $Q \in \Delta \cap \Gamma_\Omega(d)$. Since $g \notin Q$ we have $Q \subseteq P_i$ for some i; but now we have the contradiction $a = c + a \wedge b \in Q + P_i \subseteq P_i$. □

An Ω-ℓ-subgroup H of G is *laterally completely closed* if for each disjoint subset $\{h_i : i \in I\}$ of H^+, $\bigvee_i h_i \in H$ whenever $\bigvee_i h_i \in G$. Each Ω-special Ω-value is laterally completely closed; see Exercise 36. In one case these are the only Ω-values with this property. The element $0 < g \in G$ is *special valued* if $g = \bigvee_{i \in I} g_i$ for some disjoint set of positive special elements $\{g_i : i \in I\}$, and G is *special valued* if each $0 < g \in G$ has this property. According to Exercise 37 any such representation of g is unique.

Theorem 2.5.14. *Let Δ be the set of Ω-special Ω-values of G. If G is special valued, then Δ is a plenary subset of $\Gamma_\Omega(G)$. Conversely, if Δ is plenary and each $N \in \Delta$ is normal in its cover N^*, then G is special valued. Moreover, if $0 < g \in G$, then $\Gamma_\Omega(g) \cap \Delta = \bigcup_{i \in I} \Gamma_\Omega(g_i)$ where $\{g_i : i \in I\}$ is the disjoint set of special elements with $g = \bigvee_{i \in I} g_i$.*

Proof. Assume first that G is special valued. Suppose $0 < g = \bigvee_i g_i$ where $\{g_i : i \in I\}$ is a disjoint set of special elements in G. Then $\bigcup_i \Gamma_\Omega(g_i) \subseteq \Gamma_\Omega(g)$ by Exercise 37, and if $g \notin N \in \Delta$, then $g_i \notin N$ for some $i \in I$ by Exercise 36. Consequently the Ω-value N_i of g_i contains N and is an Ω-value of g; so Δ is plenary. Assume now that Δ is plenary and N is normal in N^* for each $N \in \Delta$. Suppose $0 < g \in G$, $\Gamma_\Omega(g) \cap \Delta = \{N_i : i \in I\}$, and $0 < x_i \in G$ is special with Ω-value N_i. Now, $N_i^* = C_\Omega(x_i) + N_i$ and $g = y_i + a_i < x_i + y_i + a_i$ with $y_i \in C_\Omega(x_i)^+$ and $a_i \in N_i^+$. Since $C_\Omega(x_i) = C_\Omega(x_i + y_i)$, $z_i = x_i + y_i$ is special with N_i as its Ω-value. Also, $z_i + N_i > g + N_i$ since $g < z_i + a_i$ and $-g + z_i = -a_i - y_i + x_i + y_i \notin N_i$. Let $g_i = g \wedge z_i$. We claim that g_i is special with N_i as its Ω-value. Clearly, $g_i \in N_i^* \backslash N_i$ and by Theorem 2.5.13 it suffices to show that N_i is the only Ω-special Ω-value of g_i. Suppose $N \in \Delta$. If $N \subseteq N_i$, then $N \notin \Gamma_\Omega(g_i)$ unless $N = N_i$. If $N \not\subseteq N_i$, then $z_i \in N$ and hence $g_i \in N$. Thus, $\Gamma_\Omega(g_i) = \{N_i\}$. By (c) of Theorem 2.5.1, $g_i \wedge g_j = 0$ if $i \neq j$. To see that $g = \bigvee_{i \in I} g_i$ take $h \in G$ with $h \geq g_i$ for each $i \in I$, and let $N \in \Gamma_\Omega(h - g) \cap \Delta$. If $g \in N$, then $h - g + N = h + N > N$. If $g \notin$

N, then $N \subseteq N_i$ for some $i \in I$. Since $-g+N_i > -z_i+N_i$ we have $-g+N > -z_i+N$, and hence $-g_i+N = -g \vee -z_i+N = -g+N$; so $h-g+N = h-g_i+N \geq N$. Since the proof of (b) of Theorem 2.5.1 shows, in fact, that (b) holds when the Ω-values come from a given plenary subset of $\Gamma_\Omega(G)$, we have $h \geq g$. □

Exercises.

Unless stated otherwise G is an Ω-f-group in the following exercises.

1. Let G be a weak Ω-f-group and let

$$\Omega_f = \{\omega \in \Omega : \omega \text{ induces an } f\text{-endomorphism of } G\}.$$

G is called a *pseudo Ω-f-group* if it satisfies the condition

$$(*) \qquad \forall \omega \in \Omega,\ \forall x \in G^+,\ \exists \lambda \in \Omega_f,\ \exists n \in \mathbb{N} \text{ such that } \omega x \leq n\lambda x.$$

(a) Show that each of the theorems in this section holds for a pseudo Ω-f-group.
(b) Give an example of a pseudo Ω-f-group that is not an Ω-f-group.

2. Suppose that $P \in \Gamma(G)$ is covered by P^*, $Q \in \Gamma_\Omega(G)$ is covered by Q^* and $\alpha : G \longrightarrow G$ is an Ω-f-endomorphism of G.

(a) Show that either $(P^*)_\Omega = P_\Omega$ or $(P^*)_\Omega$ is the cover of P_Ω in $\mathscr{C}_\Omega(G)$.
(b) Show that $\alpha Q^* \not\subseteq Q^*$ iff $\alpha(Q^* \backslash Q) \cap Q^* = \phi$. In particular, P^* is an Ω-subgroup of G iff $\Omega a \subseteq P^*$ for some $a \in P^* \backslash P$.
(c) Suppose that Q is normal in Q^*. Show that if $\alpha Q^* \subseteq Q^*$, then $\alpha Q \subseteq Q$. (If N is a minimal prime subgroup of G contained in Q, then α induces an isotone map on G/N; and $Q^*/N \longrightarrow Q^*/Q$ is also isotone.)
(d) If G is normal-valued, show that either P and P^* are Ω-subgroups, or for some $\omega \in \Omega, \omega(P^* \backslash P) \cap P^* = \phi$ (see Exercise 18).

3. Suppose that x is a special element of G with Ω-value P and that $\{Q \in \Gamma_\Omega(G) : Q \subseteq P\}$ is a chain. Show that x is basic (Exercise 2.4.11).

4. Let H be an ℓ-Ω-subgroup of G.

(a) If $a \in H$ is special in G, then a is special in H.
(b) If $0 \neq a \in G$, then a is special in some ℓ-subgroup of G iff $a \in G^+ \cup -G^+$.
(c) If $a \in H$, then $|\Gamma_\Omega(a,H)| \leq |\Gamma_\Omega(a,G)|$.

5. If P is a rooted poset, then trunk(P) is a dual ideal of P.

6. Show that $\text{trunk}(G) = 0$ iff G is totally ordered.

7. Suppose that $A \in \mathscr{C}_\Omega(G)$.

(a) If $A \supseteq \text{trunk}(G)$, then each convex ℓ-Ω-subgroup of G is comparable to A.

(b) If A is comparable to each convex ℓ-Ω-subgroup of G, then $A = 0$ or $A \supseteq \mathrm{trunk}(G)$.

8. Verify that each of the following subsets of G is equal to trunk (G):

   (a) the subgroup generated by $\{a \in G^+ : a^\perp \neq 0\}$;
   (b) $\mathrm{lub}_{\mathscr{C}(G)}\{P : P$ is a minimal prime subgroup of $G\}$;
   (c) the subgroup generated by $\{a \in G : a||0\}$;
   (d) $\{a \in G :$ for some $n \geq 0$ there exist $a_1, \ldots, a_n \in G$ with $a||a_1||\cdots||a_n||0\}$; show it is a subgroup.

9. (a) $G \setminus \mathrm{trunk}(G) = \{a \in G : a$ is special and $a^\perp = 0\}$.
   (b) If a is special in G then $a^{\perp\perp}$ has a proper trunk.

10. Let $\mathscr{P}_\Omega(G)$ be the rooted poset consisting of all of the prime Ω-subgroups of G, and let $\mathscr{P}_0$ be the set of minimal prime subgroups. Then |set of roots of $\Gamma_\Omega| =$ |set of roots of $\mathscr{P}_\Omega(G)| = |\mathscr{P}_0| =$ |set of roots of Γ|.

11. (a) $\Gamma_\Omega(G)$ has precisely n roots ($n \in \mathbb{N}$) iff G has a subset of n pairwise disjoint nonzero elements but no such subset with $n+1$ elements. In this case G is called *finitely rooted*.
    (b) If $\Gamma_\Omega(G)$ has at most n roots and H is an ℓ-Ω-subgroup of G, then $\Gamma_\Omega(H)$ has at most n roots.
    (c) If G is finitely rooted, then G is a direct sum of a finite number of indecomposable Ω-f-groups.

12. Let N be a normal convex ℓ-Ω-subgroup of G and set $\bar{G} = G/N$. For an element $a \in G$ denote its image in $\bar{G}$ by $\bar{a}$.

    (a) $|\Gamma(\bar{a}, \bar{G})| \leq |\Gamma(a, G)|$.
    (b) If G is finite valued so is $\bar{G}$.
    (c) If G is finitely rooted so is $\bar{G}$.

13. If $G = \bigoplus_{i \in I} G_i$ is a direct sum, then $\Gamma_\Omega(G) = \bigcup_{i \in I} \Gamma_\Omega(G_i)$ is a cardinal sum.

14. Let P be a rooted poset. Define a relation $\sim$ on P by: $p \sim q$ if p and q have a common upper bound. Show that $\sim$ is an equivalence relation on P and that P is the cardinal sum of its equivalence classes.

15. Suppose that G is finite valued.

    (a) If $\Gamma_\Omega(G) = \Gamma_1 \cup \cdots \cup \Gamma_k$ is a cardinal sum show that $G = G_1 \oplus \cdots \oplus G_k$ with $\Gamma_\Omega(G_i) \cong \Gamma_i$.
    (b) If G is indecomposable and finitely rooted show that G has a proper trunk (use Exercise 14).

16. Let $\mathscr{A}$ be a class of totally ordered Ω-f-groups which contains 0, and let $\mathscr{L} = \mathscr{L}(\mathscr{A})$ be the smallest class of Ω-f-groups which contains $\mathscr{A}$ and which is closed under isomophism, finite direct sums, and lexicographic extensions. So if $A, B \in \mathscr{L}$ and G is an Ω-f-group, then

(i) $G \cong A \Rightarrow G \in \mathscr{L}$;

(ii) $A \oplus B \in \mathscr{L}$;

(iii) $G = \text{lex}\, A$, $A \lhd G$ and $G/A \cong B \Rightarrow G \in \mathscr{L}$.

(a) Show that if $G \in \mathscr{L}(\mathscr{A})$, then G can be obtained from $\mathscr{A}$ in a finite number of steps by using (i), (ii) or (iii).
(b) If G is a finitely rooted Ω-f-group, show that there is a finite set $\{T_1, \ldots, T_k\}$ of totally ordered Ω-f-groups such that $G \in \mathscr{L}(\{T_1, \ldots, T_k\})$ (use Exercise 15).

17. Let $G = A \overleftarrow{\times}_{\varphi} B$ be an ordinal semidirect product of the ℓ-group A by the totally ordered group B. Suppose that A and B are Ω-groups, and let Ω operate on G coordinatewise.

(a) Show that G is an Ω-group if and only if the following condition is satisfied:

$$\text{If } \omega \in \Omega \text{ and } 0 < b \in B \text{ with } \omega b = 0, \text{ then } \omega A \subseteq A^+.$$

In (b), (c) and (d) it is assumed that G is an Ω-group.

(b) If A is an Ω-f-group, then so is G.
(c) G is an Ω-d-group (respectively, an Ω-f-group) iff A and B are Ω-d-groups (respectively, Ω-f-groups) and $\omega({}^{b}a) = {}^{\omega b}\omega a$ for each $\omega \in \Omega$ and each $(a,b) \in G$.
(d) If G is an Ω-f-group, then $\mathscr{C}_{\Omega}(G) \cong \mathscr{C}_{\Omega}(A) \overleftarrow{\times} \mathscr{C}_{\Omega}(B)$ and $\Gamma_{\Omega}(G) \cong \Gamma_{\Omega}(A) \overleftarrow{\times} \Gamma_{\Omega}(B)$.

18. The *normalizer* of the subset X of G is $N_G(X) = \{g \in G : X^g = -g + X + g = X\}$. Using the notation of Theorem 2.5.2 with $\Omega = \{1\}$, let $P \in \mathscr{P}(G)$ with $A \nsubseteq P$. Then

(a) $N_G(P) \cap A = N_A(P \cap A)$.
(b) $A \subseteq N_G(P)$ iff $P \cap A \lhd A$.
(c) If $a \in A$ and $P \in \Gamma(a, G)$ is covered by $P^* \in \mathscr{C}(G)$, then the following statements are equivalent.
   (i) $P \lhd P^*$.
   (ii) $P \cap A \lhd (P \cap A)^*$ where $(P \cap A)^*$ is the cover of $P \cap A$ in A.
   (iii) $P \cap C(a) \lhd C(a)$.

In this case P is called a *normal value* of a. If each $P \in \Gamma(a, G)$ is normal in its cover P^*, then a is said to be *normal valued*; and if each $a \in G$ is normal valued, then G is called a *normal valued* ℓ-group.

(d) If a is finite valued, then a is normal valued.
(e) If G is representable, then G is normal valued.

19. (a) Suppose that G is representable or is finite valued. If $\Gamma(G)$ satisfies the ascending chain condition show that $\mathscr{C}(G) = \mathscr{C}_\Omega(G)$. (Use (e) of Theorem 2.5.10 to first show that $C(a) = C_\Omega(a)$ if a is special.)
   (b) Give an example of a totally ordered abelian Ω-f-group G for which $\Gamma(G)$ satisfies the descending chain condition but in which $\mathscr{C}_\Omega(G) \subset \mathscr{C}(G)$.

20. If $G = \text{lex}\, A$ where $A \in \mathscr{C}(G)$ and A is a group summand of G with a complement B, then $G = A \overleftarrow{\times} B$.

21. This exercise gives an example of a totally ordered abelian group G with a convex subgroup A that is not a group summand of G. In fact, $G = \text{lex}\, A$ does not imply that $G \cong A \overleftarrow{\times} B$. Let $\{p_n : n \in \mathbb{N}\}$ be the increasing sequence of primes, and let G be the subgroup of $\mathbb{Q} \overleftarrow{\times} \mathbb{Q}$ generated by the set $\{(n/p_n,\, 1/p_n) : n \in \mathbb{N}\}$. Let $A = \mathbb{Z} \times \{0\} \subseteq \mathbb{Q} \overleftarrow{\times} \mathbb{Q}$.
   (a) $G \cap (\mathbb{Q} \times \{0\}) = A$.
   (b) A is the only nonzero proper convex subgroup of G.
   (c) A is not a group summand of G.

22. Show that G is isomorphic to a direct sum of a family of totally ordered Ω-f-groups iff G is representable, finite valued and each special element is basic (use Exercises 2.4.10, 2.4.11, and 2.4.12).

23. Let $A_1, A_2 \in \mathscr{C}_\Omega(G)$ and suppose that $A_i = \text{lex}\, B_i$ for some $B_i \subset A_i$. Show that A_1 and A_2 are comparable, or $A_1 \cap A_2 = 0$. Consequently, $\{A \in \mathscr{C}_\Omega(G) : A$ has a proper trunk$\}$ is a rooted subset of $\mathscr{C}_\Omega(G)$.

24. Show that G is isomorphic to a direct product of a family of totally ordered Ω-f-groups if and only if G is laterally complete, has a basis (Exercise 2.4.13) and is projectable (that is, each polar $a^{\perp\perp}$ is a direct summand of G).

25. A representation

$$\varphi : G \longrightarrow \prod_{i \in I} G_i$$

of G as a subdirect product of the family $\{G_i : i \in I\}$ of Ω-f-groups is called *irredundant* if $\varphi(G) \cap G_i \neq 0$ for each $i \in I$. (Equivalently, each induced map

$$\varphi_i : G \longrightarrow \prod_{j \neq i} G_j$$

has a nonzero kernel.)
   (a) Show that G is an irredundant subdirect product of a family of totally ordered Ω-f-groups if and only if G is representable and $\mathscr{B}(G)$ is atomic (see Exercise 2.4.13).
   (b) Suppose that $\varphi : G \longrightarrow \prod G_i$ is an irredundant subdirect product representation of G where each G_i is totally ordered. Show that there is a bijection between I and the set of maximal polars of G such that for each maximal

polar A_i, $i \in I$, $G_i \cong G/A_i$. Furthermore, the diagram

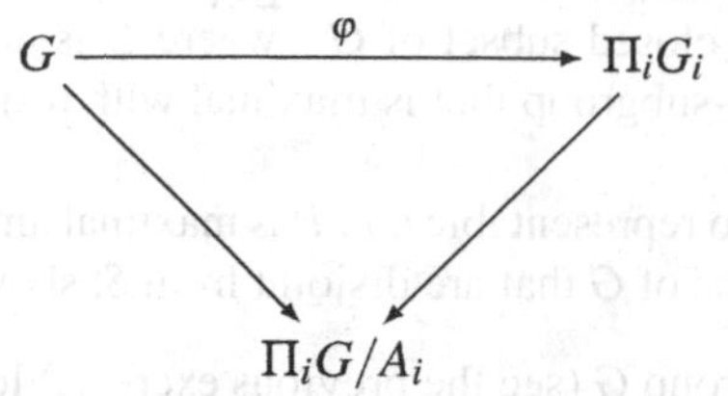

commutes (see Exercise 2.4.12). Show that $\mathscr{B}(G) \cong \mathscr{B}(\Pi_i G_i) \cong \mathscr{P}(I)$ (see Exercise 2.2.14(e)).

26. (a) Show that the following statements are equivalent for G.
    (i) G contains a maximal disjoint subset $\{a_i : i \in I\}$ such that each a_i is special and $a_i^{\perp\perp}$ is unbounded in G.
    (ii) There is a family $\{G_i : i \in I\}$ of Ω-f-groups such that each G_i has a proper trunk, and there is an embedding of G into the direct product ΠG_i such that the image of G contains the direct sum $\oplus G_i$. (For (i) $\Rightarrow$ (ii) use Theorem 2.5.9 and Exercise 9.)
    (b) If G is special valued and each special element is basic show that G has the properties given in (a).
    (c) Give an example of a special valued ℓ-group with a basis which does not satisfy the conditions in (a).
27. Show that the following statements are equivalent for G.
    (a) G is archimedean and has a basis.
    (b) There is a family $\{G_i : i \in I\}$ where each G_i is an Ω-subgroup of the reals, and there is an embedding of G into the direct product ΠG_i such that the image of G contains the direct sum $\oplus G_i$ (see Exercises 26, 2.4.12, and 2.4.13).
28. If $G =$ lex A, show that A is completely closed in G and the inclusion map $A \longrightarrow G$ is complete (see Exercise 2.3.21).
29. Let $A, B \in \mathscr{C}_\Omega(G)$ with $A =$ lex B and $B \subset A$. Show that $G =$ lex A iff $A^\perp = 0$.
30. Let G be a po-Ω-Λ-ℓ-group; that is, G has two operator sets Ω and Λ with associated functions $\Omega \longrightarrow \mathrm{End}(G)^+$ and $\Lambda \longrightarrow \mathrm{End}(G)^+$ such that $(\omega x)\lambda = \omega(x\lambda)$ for all $\omega \in \Omega$, $\lambda \in \Lambda$, and $x \in G$.
    (a) If X is a subset of G give a description of the elements in $C_{\Omega\text{-}\Lambda}(X)$, the convex ℓ-Ω-Λ-subgroup of G that is generated by X, in terms of the elements in X, Ω and Λ.
    (b) Describe the elements in $C^{(G)}_{\Omega\text{-}\Lambda}(X)$, the normal convex ℓ-Ω-Λ-subgroup that is generated by X.

(c) Suppose that G is an Ω-Λ-f-group and $a \wedge b = 0$. If D is a convex ℓ-Ω-Λ-subgroup, show that $C_{\Omega\text{-}\Lambda}(D,a) \cap C_{\Omega\text{-}\Lambda}(D,b) = D$. If G is also representable and D is normal, show that $C^{(G)}_{\Omega\text{-}\Lambda}(D,a) \cap C^{(G)}_{\Omega\text{-}\Lambda}(D,b) = D$.
(d) Let S be a meet closed subset of G^+ where G is an Ω-Λ-f-group. If P is a convex ℓ-Ω-Λ-subgroup that is maximal with respect to $P \cap S = \phi$, show that P is prime.
(e) In (d), if G is also representable and P is maximal among the normal convex ℓ-Ω-Λ-subgroups of G that are disjoint from S, show that P is prime.

31. For the po-Ω-Λ-ℓ-group G (see the previous exercise) let

$$\rho(G) = \{x \in G : \text{for each sequence } (\omega_n) \text{ in } \Omega \text{ there exists an } n \in \mathbb{N} \text{ such that } \omega_n \omega_{n-1} \cdots \omega_1 |x| = 0\}.$$

(a) Show that $\rho(G)$ is a normal convex ℓ-Ω-Λ-subgroup of G.
(b) Show that $\rho(G/\rho(G)) = 0$; that is, $G/\rho(G)$ is *ρ-semisimple*.
(c) If N is a normal convex ℓ-Ω-Λ-subgroup of G, show that $\rho(G) + N/N \subseteq \rho(G/N)$.
(d) Show that $\rho(G) = 0$ iff $r(\Omega;G) = 0$ where $r(\Omega;G) = \{x \in G; \omega x = 0$ for each $\omega \in \Omega\}$.
(e) Assume that G is a representable Ω-Λ-f-group. Let $\{P_i : i \in I\}$ be the collection of normal prime Ω-Λ-subgroups of G, and let $\rho(G/P_i) = K_i/P_i$. Show that $\rho(G) = \cap_i K_i$. (If $x \notin \rho(G)$ because of the sequence (ω_n), let S be the meet closed subset of G^+generated by the elements $\omega_n \cdots \omega_1 |x|$ and use Exercise 30.)
(f) If G is a representable Ω-Λ-f-group, show that G is ρ-semisimple if and only if G is a subdirect product of totally ordered ρ-semisimple Ω-Λ-f-groups.

32. Let G be an Ω-f-group and for $x \in G$ let $\Omega x = \{\omega x : \omega \in \Omega\}$. G is called *Ω-archimedean* if for all $x, y \in G, \Omega x \leq y$ implies that $x \leq 0$.

(a) If G is Ω-archimedean and $x \in M$, show that $x^\perp = \cap\{N : N \in \Gamma_\Omega(x)\}$. (If $y \notin x^\perp, 0 < z = x \wedge y$ and $\omega \in \Omega$ with $\omega z \nleq x$, consider the elements $g = \omega z - x \wedge \omega z$ and $h = x - x \wedge \omega z$ and take $P \in \Gamma_\Omega(g)$.)
(b) Conversely, suppose that $x, y \in G$ with $x^\perp = \cap\{N : N \in \Gamma_\Omega(x)\}$, and N is normal in its cover N^* for each $N \in \Gamma_\Omega(x)$. Show that if $C_\Omega(y) \leq x$, then $y = 0$.
(c) Suppose that $x^\perp = \cap\{N : N \in \Gamma_\Omega(x)\}$ for each $x \in G$. Show that if $x, y \in G$, then $x^\perp = \cap\{N : N \in \Gamma_\Omega(x)$ and $y \in N^*\}$. (Assume that $x, y \in G^+$ and $x \leq y$. If $0 < z \notin x^\perp$, then $x \wedge z \notin y^\perp$.)

33. For $\Delta \subseteq \Gamma_\Omega(G)$ and $\Lambda \subseteq \Gamma(G)$ let $\Delta' = \{A \in \Gamma(G) : A_\Omega \in \Delta\}$ and $\Lambda_\Omega = \{A_\Omega : A \in \Lambda\}$.

(a) Show that Δ is a plenary subset of $\Gamma_\Omega(G)$ iff Δ is a dual ideal of $\Gamma_\Omega(G)$ and $\cap_{N \in \Delta} N = 0$.
(b) Verify that $(\Delta')_\Omega = \Delta$.

(c) Show that Δ is a plenary subset of $\Gamma_\Omega(G)$ iff Δ' is a plenary subset of $\Gamma(G)$.
(d) If Λ is a plenary subset of $\Gamma_\Omega(G)$ show that Λ_Ω is a plenary subset of $\Gamma_\Omega(G)$.
(e) Give an example to show that Λ_Ω can be a plenary subset of $\Gamma_\Omega(G)$ without Λ being a plenary subset of $\Gamma(G)$.

34. Suppose Δ is a plenary subset of $\Gamma_\Omega(G)$ and $A \in \mathscr{C}_\Omega(G)$. Show that $\Delta^c = \{N \cap A : N \in \Delta \text{ and } A \nsubseteq N\} = \{N \cap A : N \in \Delta \cap \Gamma_\Omega(a) \text{ for some } a \in A\}$ is a plenary subset of $\Gamma_\Omega(A)$.

35. Suppose $g = \bigvee_{i \in I} a_i$ where $\{a_i : i \in I\}$ is a disjoint subset of G^+.

(a) Show that for each $j \in I$, $g = a_j + \bigvee_{i \neq j} a_i = \bigvee_{i \neq j} a_i + a_j$.
(b) Show that $ng = \bigvee_{i \in I} na_i$ for every $n \in \mathbb{N}$.

36. (a) If $N \in \Gamma_\Omega(G)$ is Ω-special show that N is laterally completely closed. (Suppose $\Gamma_\Omega(h) = \{N\}$, $\{g_i : i \in I\} \subseteq N$ is a disjoint set of positive elements with $g = \bigvee_i g_i \notin N$. If $x = g \wedge h$ show that $\Gamma_\Omega(x) = \{N\}$, and use Theorems 2.6.8 and 2.6.10 together with Exercise 35 (b) to show that $x \geq 2x$).
(b) If g is a special valued element of G and $N \in \Gamma_\Omega(g)$ is not Ω-special show that N is not laterally completely closed.

37. The element $a \in G$ is called *indecomposable* if $|a| = b + c$ with $b \wedge c = 0$ implies $b = 0$ or $c = 0$. Equivalently, $C_\Omega(a)$ is an indecomposable Ω-f-group.

(a) Show that each special element of G is indecomposable.
(b) Suppose $g = \bigvee_i a_i = \bigvee_j b_j$ where $\{a_i : i \in I\}$ and $\{b_j : j \in J\}$ are each disjoint sets of positive indecomposable elements. Show that $\{a_i : i \in I\} = \{b_j : j \in J\}$ and $\bigcup_i \Gamma_\Omega(a_i) \subseteq \Gamma_\Omega(g)$. (Use Exercise 35.)

2.6 Hahn Products and the Embedding Theorem

Representing an abstract algebra as an algebra of functions is always satisfying. Here, a lexicographically inspired partial order is constructed on a subgroup of a group direct product of a family of po-groups indexed by a poset, and the conditions for the resulting po-group to be an ℓ-group are determined. The value set of an abelian ℓ-group together with this construction is used to represent the ℓ-group as an ℓ-group of real-valued functions.

Let G be a po-group and let End$(G)^+$ denote the set of po-group endomorphisms of G. If Ω is a set which operates on G, then G is called an *Ω-po-group* if the associated function maps Ω to End$(G)^+$. Clearly, each Ω-d-group is an Ω-po-group and each po-group is an ϕ-po-group as well as an $\{1\}$-po-group, but a pseudo Ω-f-group as defined in Exercise 2.5.1 need not be an Ω-po-group. An element $\omega \in \Omega$ is *strict on* G if $\omega x > 0$ for each $0 < x \in G$. If each $\omega \in \Omega$ is strict on G, then G is called a *strict Ω-po-group*.

Suppose that Γ is a poset and for each $\gamma \in \Gamma$ let G_γ be an Ω-po-group. Note that the group direct product $\prod G_\gamma$ admits Ω as an operator set : $(\omega v)(\gamma) = \omega v(\gamma)$

for $v \in \prod G_\gamma$, $\omega \in \Omega$ and $\gamma \in \Gamma$. Whenever we refer to a subgroup of the direct product $\prod G_\gamma$ as an Ω-group we mean, as usual, that it is an Ω-subgroup of the product with this Ω-action. We intend to generalize the Hahn product construction given in Example (iv) of Section 2.1. Specifically, for each element $v \in \prod G_\gamma$ the *support* of v (abbreviated supp v) is defined by

$$\text{supp } v = \{\gamma \in \Gamma : v(\gamma) \neq 0\},$$

and the maximal elements in supp v will be denoted by maxsupp v. Let

$$V = V(\Gamma, G_\gamma) = \{v \in \prod_\gamma G_\gamma : \text{ supp } v \text{ has the maximum condition}\},$$

and set

$$V^+ = \{v \in V : v(\gamma) > 0 \text{ for each } \gamma \in \text{maxsupp } v\}.$$

If each $G_\gamma = G$ we will denote $V(\Gamma, G_\gamma)$ by $V(\Gamma, G)$. For each $\gamma \in \Gamma$ and $x_\gamma \in G_\gamma$ the natural group embedding $G_\gamma \longrightarrow V$ will be indicated by $x_\gamma \longmapsto \overline{x}_\gamma$.

Theorem 2.6.1. *Let G_γ be an Ω-po-group for each element γ in the poset Γ. Then $V(\Gamma, G_\gamma)$ is an Ω-subgroup of $\prod G_\gamma$, and V^+ is a positive cone for $V(\Gamma, G_\gamma)$. Moreover, V is an Ω-po-group if and only if whenever $\omega \in \Omega$ is not strict on some G_γ, then $\omega G_\delta = 0$ for each $\delta < \gamma$.*

Proof. V is an Ω-subgroup of the product $\prod G_\gamma$ since if $u, v \in V$, and $\omega \in \Omega$, then supp $\omega v \cup$ supp $(u - v) \subseteq$ supp $u \cup$ supp v, and the union of two subsets of Γ, each of which satisfies the maximum condition, also satisfies the maximum condition. To see that $V^+ + V^+ \subseteq V^+$ let $u, v \in V^+$. We claim that

$$\gamma \in \text{ maxsupp } (u+v) \Rightarrow \gamma \in \text{ maxsupp } u \text{ or } u(\gamma) = 0. \tag{2.6.1}$$

For if $u(\gamma) \neq 0$, take $\delta \in$ maxsupp u with $\delta \geq \gamma$. If $\delta > \gamma$, then $v(\delta) = -u(\delta) < 0$; so there exists $\alpha > \delta$ with $\alpha \in$ maxsupp v. But then $0 = (u+v)(\alpha) = v(\alpha) > 0$. Thus, $\gamma = \delta \in$ maxsupp u. Similarly, either $\gamma \in$ maxsupp v or $v(\gamma) = 0$. So, in each of the three possible cases, $(u+v)(\gamma) > 0$ and hence $u + v \in V^+$. Clearly, $V^+ \cap -V^+ = 0$ and V^+ is normal in V since supp $(-u+v+u) =$ supp v.

Suppose that V is an Ω-po-group. Let $\delta < \gamma$ in Γ and let ω be an element of Ω which is not strict on G_γ. Take $0 < x_\gamma \in G_\gamma$ with $\omega x_\gamma = 0$, and let $x_\delta \in G_\delta$. Then $\overline{x}_\gamma + \overline{x}_\delta \in V^+$, so $\omega x_\delta \geq 0$. Thus $\omega G_\delta = 0$. Conversely, suppose that this condition is satisfied, and let $\omega \in \Omega$. We need to verify that the homomorphism of V induced by ω is isotone. Suppose that $v \in V^+$ and let $\delta \in$ maxsupp ωv. Then there exists $\gamma \in$ maxsupp v with $\gamma \geq \delta$. If $\gamma > \delta$, then $\omega v(\gamma) = 0$ and hence $\omega G_\delta = 0$ since ω is not strict on G_γ. But this contradicts that $\delta \in$ maxsupp ωv; so $\gamma = \delta$ and hence $\omega v(\delta) > 0$. Thus $\omega v \in V^+$. □

An Ω-po-group of the form $(V(\Gamma, G_\gamma), V^+)$ will be called a *Hahn product*. We next determine when a Hahn product is an ℓ-group as well as when it is an Ω-*d*-group or an Ω-*f*-group. The conditions on the G_γ for V to be an ℓ-group arise from the lexicographic conditions in Exercise 2.1.10. For convenience we will always assume that each $G_\gamma \neq 0$ in the Hahn product construction. Note that the map $G_\gamma \longrightarrow V$, $x_\gamma \longmapsto \overline{x}_\gamma$, is an embedding of po-groups.

Theorem 2.6.2. *The Hahn product $V(\Gamma, G_\gamma)$ is an ℓ-group if and only if the following conditions are satisfied.*

(a) Each G_γ is an ℓ-group.
(b) G_γ is totally ordered if γ is not a minimal element of Γ.
(c) Γ is a rooted poset.

Moreover, if $V(\Gamma, G_\gamma)$ is an ℓ-group and an Ω-po-group, then it is an Ω-d-group (respectively, an Ω-f-group) if and only if each G_γ is an Ω-d-group (respectively, an Ω-f-group).

Proof. Suppose that V is an ℓ-group. We will first verify (a). Let $a_\gamma \in G_\gamma$ and assume that a_γ is not comparable to 0. Then $\overline{a}_\gamma^+ > 0, \overline{a}_\gamma$. We claim that if $\alpha \in$ maxsupp $\overline{a}_\gamma^+$, then $\gamma \not< \alpha$. For, if $\gamma < \alpha$, then, since supp $\overline{a}_\gamma^+ \setminus \{\gamma\} =$ supp $(\overline{a}_\gamma + \overline{a}_\gamma^+) \setminus \{\gamma\}$, $\alpha \in$ maxsupp $(\overline{a}_\gamma + \overline{a}_\gamma^+)$, and, in fact, maxsupp $\overline{a}_\gamma^+ =$ maxsupp $(\overline{a}_\gamma + \overline{a}_\gamma^+)$; so if $\beta \in$ maxsupp $\overline{a}_\gamma^+$, then $(\overline{a}_\gamma + \overline{a}_\gamma^+)(\beta) = \overline{a}_\gamma^+(\beta) > 0$. Thus, $\overline{a}_\gamma + \overline{a}_\gamma^+ > 0$, $\overline{a}_\gamma^- = 0$ and $\overline{a}_\gamma \geq 0$, contrary to our assumption. Now, $\gamma \in$ supp $\overline{a}_\gamma^+$ (and, hence, $\gamma \in$ maxsupp $\overline{a}_\gamma^+$). For, otherwise, there exists $\alpha \in$ maxsupp $(\overline{a}_\gamma^+ - \overline{a}_\gamma)$ with $\alpha \geq \gamma$, and, in fact, $\alpha > \gamma$ since $-a_\gamma \not> 0$. But since $\alpha \in$ maxsupp $\overline{a}_\gamma^+$ we have a contradiction. So $\overline{a}_\gamma^+(\gamma) > 0$, $\gamma \in$ maxsupp $(\overline{a}_\gamma^+ - \overline{a}_\gamma)$, and $\overline{a}_\gamma^+(\gamma) > a_\gamma$. We now claim that $\overline{a}_\gamma^+(\gamma) = a_\gamma^+$. For, suppose that $b_\gamma \in G_\gamma$ with $b_\gamma > 0, a_\gamma$ and $b_\gamma \neq \overline{a}_\gamma^+(\gamma)$. Then $\overline{b}_\gamma \geq \overline{a}_\gamma^+$ and there exists $\alpha \in$ maxsupp $(\overline{b}_\gamma - \overline{a}_\gamma^+)$ with $\alpha \geq \gamma$. If $\alpha > \gamma$, then $\alpha \in$ maxsupp $\overline{a}_\gamma^+$ gives a contradiction. So $\alpha = \gamma$ and $b_\gamma > \overline{a}_\gamma^+(\gamma)$. Thus G_γ is an ℓ-group and $a_\gamma^+ = \overline{a}_\gamma^+(\gamma)$.

In order to prove (b) we first note that the map $a_\gamma \longmapsto \overline{a}_\gamma$ is an ℓ-group embedding of G_γ into V. For, if $\overline{a_\gamma^+} > \overline{a}_\gamma^+$ and $\beta \in$ maxsupp $(\overline{a_\gamma^+} - \overline{a}_\gamma^+)$, then, since $\overline{a_\gamma^+}(\gamma) = a_\gamma^+$, $\gamma \neq \beta \in$ maxsupp $\overline{a}_\gamma^+$ and $\overline{a}_\gamma^+(\beta) < 0$. This contradiction gives that $\overline{a_\gamma^+} = \overline{a}_\gamma^+$. Now, if $\delta < \gamma$ and if $a_\gamma \in G_\gamma$ is not comparable to 0, then $0, \overline{a}_\gamma < \overline{a}_\delta + \overline{a}_\gamma^+$ for each $a_\delta \in G_\delta$. But then $\overline{a}_\gamma^+ \leq \overline{a}_\delta + \overline{a}_\gamma^+$ and hence $G_\delta = 0$. Thus, a_γ is comparable to 0 and G_γ is totally ordered.

Now suppose that α and β are incomparable elements of Γ with a common lower bound γ. Let $0 < a_\alpha \in G_\alpha$, $0 < a_\beta \in G_\beta$ and $v \in V^+$. We claim that $v \leq \overline{a}_\alpha, \overline{a}_\beta$ if and only if maxsupp $v < \alpha, \beta$. For, suppose that $v \leq \overline{a}_\alpha, \overline{a}_\beta$ and $\delta \in$ maxsupp v with $\delta \not\leq \alpha$. Then $\delta \in$ maxsupp $(\overline{a}_\alpha - v)$ and $v(\delta) < 0$, a contradiction. Thus maxsupp $v \leq \alpha, \beta$, and hence maxsupp $v < \alpha, \beta$. The converse is obvious. But now if $0 \leq v \leq \overline{a}_\alpha$, $\overline{a}_\beta$ and $0 < a_\gamma \in G_\gamma$, then $v < v + \overline{a}_\gamma < \overline{a}_\alpha, \overline{a}_\beta$. Thus $\overline{a}_\alpha \wedge \overline{a}_\beta$ cannot exist. This proves that Γ is rooted.

Conversely, assume that conditions (a), (b), and (c) are satisfied, and first consider the case in which Γ is totally ordered. If Γ does not have a least element, then $V(\Gamma, G_\gamma)$ is totally ordered, and if Γ does have a least element α, then $V(\Gamma, G_\gamma) = G_\alpha \overleftarrow{\times} V(\Gamma \setminus \{\alpha\}, G_\gamma)$ is an ordinal product and hence is an ℓ-group by Exercise 2.1.10. Next, suppose that Γ is an arbitrary rooted poset and let M be a root of Γ.

Then $V_M = \{v \in V : \text{supp } v \subseteq M\}$ is a subgroup of V, and V_M is an ℓ-group since $V_M \cong V(M, G_\gamma)$. The projection $V \longrightarrow V_M$ which sends v to v_M, where v_M is given by $v_M(\gamma) = v(\gamma)$ if $\gamma \in M$ and $v_M(\gamma) = 0$ if $\gamma \notin M$, is a group homomorphism. Moreover, since Γ is rooted if and only if each root is a dual ideal, maxsupp $v_M \subseteq$ maxsupp v and

$$\text{maxsupp } v = \bigcup_M \text{maxsupp } v_M;$$

so $v \geq 0$ if and only if $v_M \geq 0$ for each root M. Now, for $v \in V$ define w by $w(\gamma) = v_M^+(\gamma)$ if γ is in the root M of Γ. To see that w is well-defined, suppose that γ is in the distinct roots M and N of Γ. Then γ is not a minimal element of Γ. Since supp $v_P^+ \subseteq$ supp $v_P \subseteq$ supp v for any root P, if $v(\gamma) = 0$, then $v_M^+(\gamma) = v_N^+(\gamma) = 0$. On the other hand, if $v(\gamma) \neq 0$ and $\beta \geq \gamma$ with $\beta \in$ maxsupp v, then $\beta \in M \cap N$, and $v_M^+(\gamma) = v(\gamma) = v_N^+(\gamma)$ or $v_M^+(\gamma) = 0 = v_N^+(\gamma)$ depending on whether $v(\beta) > 0$ or $v(\beta) < 0$. Since supp $w \subseteq \cup_M$ supp $v_M^+ \subseteq$ supp v we have that $w \in V$. Finally, if $u \in V$ with $u \geq 0, v$, then $u_M \geq v_M^+ = w_M \geq 0, v_M$ for each root M; so $u \geq w$ and $w = v^+$.

Now suppose that $V(\Gamma, G_\gamma)$ is an ℓ-group and an Ω-po-group. Then for any root M of Γ, V_M is an ℓ-subgroup of V since supp $v^+ \subseteq$ supp v; and, by Theorem 2.6.1, V_M is an Ω-po-group. Also, each projection $V \longrightarrow V_M$ is an ℓ-homomorphism since, using the notation of the previous paragraph, $(v^+)_M = w_M = (v_M)^+$, as well as an Ω-homomorphism. Consequently, V is isomorphic to a subdirect product of the family $\{V_M : M \text{ is a root of } \Gamma\}$ as an Ω-po-group and as an ℓ-group. So, to prove the last statement of the theorem, we may assume that Γ is totally ordered. If V is an Ω-d-group and $\gamma \in \Gamma$, then, since the embedding $G_\gamma \longrightarrow V$, $a_\gamma \mapsto \overline{a_\gamma}$ is one of ℓ-groups and Ω-po-groups, G_γ is an Ω-d-group (of course, G_γ is a totally ordered Ω-f-group if γ is not minimal). Conversely, suppose that G_γ is an Ω-d-group where γ is the minimal element of Γ. If $u \wedge v = 0$ in V and $0 < u, v$, then $\{\gamma\} =$ maxsupp $u =$ maxsupp v (since $V = G_\gamma \overleftarrow{\times} V(\Gamma \setminus \{\gamma\}, G_\alpha)$). So $u = \overline{a}_\gamma$ and $v = \overline{b}_\gamma$ for some $a_\gamma, b_\gamma \in G_\gamma$, and hence $\omega u \wedge \omega v = 0$ for each $\omega \in \Omega$. The same argument works for the Ω-f-group case. □

One reason for the importance of Hahn products is that each abelian ℓ-group can be embedded in $V(\Gamma, \mathbb{R})$ for an appropriate rooted poset Γ. Before we prove this we first need a definition and an interesting fact about vector spaces.

Let $\mathscr{L}_\Omega(G)$ denote the lattice of Ω-subgroups of the Ω-group G, and let $\mathscr{S}$ be a nonempty subset of $\mathscr{L}_\Omega(G)$. A *partial complementation for G with respect to $\mathscr{S}$* is an antitone function $f : \mathscr{S} \longrightarrow \mathscr{L}_\Omega(G)$ such that, for each X in $\mathscr{S}$, G is the direct sum of its Ω-subgroups X and $f(X)$. If $\mathscr{S} = \mathscr{L}_\Omega(G)$, then f is called a *complementation for G*.

Theorem 2.6.3. *Let V be a left vector space over a division ring D. Suppose that $f : \mathscr{S} \longrightarrow \mathscr{L}_D(H)$ is a partial complementation for the subspace H of V. Let $\overline{\mathscr{S}} = \{X \in \mathscr{L}_D(V) : X \cap H \in \mathscr{S}\}$. Then there is a partial complementation $\overline{f} : \overline{\mathscr{S}} \longrightarrow \mathscr{L}_D(V)$ for V such that $f(X \cap H) \subseteq \overline{f}(X)$ for each $X \in \overline{\mathscr{S}}$. In particular, V has a complementation.*

Proof. The last statement follows by setting $H = 0$. Let $\overline{\mathscr{S}}_W = \overline{\mathscr{S}} \cap \mathscr{L}_D(W)$ for the subspace W of V, and set

$$\mathscr{P} = \{(W,\alpha) : H \subseteq W,\ \overline{\mathscr{S}}_W \xrightarrow{\alpha} \mathscr{L}_D(W) \text{ is a partial complementation for } W, \text{ and } f(X \cap H) \subseteq \alpha(X) \text{ for each } X \in \overline{\mathscr{S}}_W\}.$$

The relation defined on $\mathscr{P}$ by $(W,\alpha) \leq (U,\beta)$ if $W \subseteq U$ and $\alpha(X \cap W) \subseteq \beta(X)$ for each $X \in \overline{\mathscr{S}}_U$ is easily seen to be a partial order of $\mathscr{P}$. If $\{(W_i,\alpha_i)\}$ is a chain in $\mathscr{P}$ let $W = \bigcup_i W_i$, and define $\alpha : \overline{\mathscr{S}}_W \longrightarrow \mathscr{L}_D(W)$ by

$$\alpha(X) = \bigcup_i \alpha_i(X \cap W_i).$$

Then $(W,\alpha) \in \mathscr{P}$. For, α is antitone since each α_i is antitone. If $X \in \overline{\mathscr{S}}_W$, then

$$X \cap \alpha(X) = \bigcup_{i,j} (X \cap W_i) \cap \alpha_j(X \cap W_j),$$

and, for each i and j, either $W_i \subseteq W_j$ and $X \cap W_i \cap \alpha_j(X \cap W_j) \subseteq X \cap W_j \cap \alpha_j(X \cap W_j) = 0$, or $W_j \subseteq W_i$ and $X \cap W_i \cap \alpha_j(X \cap W_j) \subseteq X \cap W_i \cap \alpha_i(X \cap W_i) = 0$. So $X \cap \alpha(X) = 0$ and, also,

$$W = \bigcup_i [(X \cap W_i) + \alpha_i(X \cap W_i)] \subseteq \bigcup_i (X \cap W_i) + \bigcup_i \alpha_i(X \cap W_i) = X + \alpha(X).$$

Clearly, (W,α) is an upper bound for the set $\{(W_i,\alpha_i)\}$. By Zorn's Lemma $\mathscr{P}$ has a maximal element (A,β). If $A \subset V$ and $v \in V \setminus A$, let $U = A + Dv$, and define $\gamma : \overline{\mathscr{S}}_U \longrightarrow \mathscr{L}_D(U)$ by $\gamma(X) = \beta(X) + Dv$ if $X \subseteq A$ and $\gamma(X) = \beta(X \cap A)$ if $X \nsubseteq A$. If $X \subseteq A$, then $X + \gamma(X) = X + \beta(X) + Dv = A + Dv = U$ and $X \cap \gamma(X) = X \cap [\beta(X) + Dv] = 0$. On the other hand, if $X \nsubseteq A$, then $U = A + X = X \cap A + \beta(X \cap A) + X = X + \gamma(X)$ and $X \cap \gamma(X) = X \cap A \cap \beta(X \cap A) = 0$. So in both cases U is the direct sum of X and $\gamma(X)$. It is easy to see that γ is antitone and that $(A,\beta) < (U,\gamma)$. This contradiction gives that $A = V$ and that β is a partial complementation for V with respect to $\mathscr{S}$. □

Let $\Gamma_\Omega = \{G_\gamma\}$ be the rooted poset of Ω-values of the Ω-f-group G. We will find it useful to identify γ with G_γ, that is, to index Γ_Ω with itself. If G^γ denotes the cover of G_γ in $\mathscr{C}_\Omega(G)$ and $\alpha, \beta \in \Gamma_\Omega$, then we have that $\alpha \leq \beta$ (respectively, $\alpha < \beta$) precisely when $G_\alpha \subseteq G_\beta$ (respectively, $G^\alpha \subseteq G_\beta$).

We present next the promised embedding theorem.

Theorem 2.6.4. *Let G be an abelian Ω-f-group with Ω-value set Γ_Ω, and let $\mathscr{S}$ be the set of all those convex ℓ-Ω-subgroups of G which are unions of countable chains from $\{G^\gamma : \gamma \in \Gamma_\Omega\}$. Suppose that G has a partial complementation with respect to $\mathscr{S}$ and that $V(\Gamma_\Omega, G^\gamma/G_\gamma)$ is an Ω-po-group. Then there is an Ω-f-group embedding $\varphi : G \longrightarrow V(\Gamma_\Omega, G^\gamma/G_\gamma)$ such that, for each $g \in G, \Gamma_\Omega(g) =$ maxsupp $\varphi(g)$.*

Proof. Let $f : \mathscr{S} \longrightarrow \mathscr{L}_\Omega(G)$ be a partial complementation for G. For each $\gamma \in \Gamma_\Omega$ we have the Ω-group direct sum $G = G^\gamma + f(G^\gamma)$. Let $\varphi_\gamma : G \longrightarrow G^\gamma/G_\gamma$ be the composite $G \longrightarrow G^\gamma \longrightarrow G^\gamma/G_\gamma$ of the projection onto G^γ followed by the natural homomorphism onto G^γ/G_γ. Explicitly, if $g \in G$, then $g = g_\gamma + a_\gamma$ with $g_\gamma \in G^\gamma$ and $a_\gamma \in f(G^\gamma)$, and $\varphi_\gamma(g) = g_\gamma + G_\gamma$; so $\varphi_\gamma(g) = g + G_\gamma$ if $g \in G^\gamma$, and ker $\varphi_\gamma = G_\gamma + f(G^\gamma)$. Now, the family $\{\varphi_\gamma : \gamma \in \Gamma_\Omega\}$ determines a unique Ω-group monomorphism

$$\varphi : G \longrightarrow \prod_{\gamma\in\Gamma_\Omega} G^\gamma/G_\gamma$$

given by $\varphi(g)(\gamma) = \varphi_\gamma(g) = g_\gamma + G_\gamma$. Note that supp $\varphi(g) = \{\gamma \in \Gamma_\Omega : g_\gamma \notin G_\gamma\} = \{\gamma \in \Gamma_\Omega : g \notin G_\gamma + f(G^\gamma)\}$. We claim that $\varphi(G) \subseteq V(\Gamma_\Omega, G^\gamma/G_\gamma)$. For, suppose that $\gamma_1 \leq \gamma_2 \leq \cdots$ is an increasing sequence in supp $\varphi(g)$. If

$$A = \bigcup_k G^{\gamma_k},$$

then $G = A + f(A)$, $g = a + b$ with $a \in A$ and $b \in f(A)$, and $a \in G^{\gamma_n}$ for some γ_n. But then $\gamma_n = \gamma_{n+1} = \cdots$; if not, $\gamma_n < \gamma_m$ for some m and consequently $G^{\gamma_n} \subseteq G_{\gamma_m}$. So $a \in G_{\gamma_m}$; but also $b \in f(A) \subseteq f(G^{\gamma_m})$, and hence $g \in G_{\gamma_m} + f(G^{\gamma_m}) = \ker \varphi_{\gamma_m}$. This contradicts the assumption that $\gamma_m \in$ supp $\varphi(g)$.

To see that $\Gamma_\Omega(g) =$ maxsupp $\varphi(g)$, let $\gamma \in \Gamma_\Omega(g)$. Then $\varphi(g)(\gamma) = \varphi_\gamma(g) = g + G_\gamma$, so $\gamma \in$ supp $\varphi(g)$. Also, if $\delta > \gamma$, then $g \in G^\gamma \subseteq G_\delta$ and $\varphi(g)(\delta) = \varphi_\delta(g) = 0$. Thus $\gamma \in$ maxsupp $\varphi(g)$. Conversely, if $\gamma \in$ maxsupp $\varphi(g)$, then $g \notin G_\gamma$. If also $g \notin G^\gamma$, then g has an Ω-value G_δ which contains G^γ. But then $\delta > \gamma$ and $\delta \in$ maxsupp $\varphi(g)$, a contradiction. It is now easy to see that φ is an ℓ-homomorphism. For if $g \wedge h = 0$, then $\Gamma_\Omega(g) \cap \Gamma_\Omega(h) = \phi$ and $\Gamma_\Omega(g) \cup \Gamma_\Omega(h)$ is trivially ordered (Theorem 2.5.1), and therefore no element in maxsupp $\varphi(g)$ is comparable to any element in maxsupp $\varphi(h)$. However, this last statement is equivalent to $\varphi(g) \wedge \varphi(h) = 0$ (Exercise 1). □

The most important case of this theorem is given in

Theorem 2.6.5. *Let G be an abelian ℓ-group with value set Γ. Then there is an embedding $\varphi : G \longrightarrow V(\Gamma, \mathbb{R})$ such that, for each $g \in G$, $\Gamma(g) =$ maxsupp $\varphi(g)$.*

Proof. The divisible closure $d(G)$ of G is a rational vector space and so Theroem 2.6.3 guarantees that it has a complementation. Also, the mapping $G_\gamma \longrightarrow d(G_\gamma)$ is an isomorphism between $\Gamma(G)$ and $\Gamma(d(G))$ and $d(G^\gamma/G_\gamma) \cong d(G^\gamma)/d(G_\gamma)$ (Exercise 2.2.6); so $\Gamma_G(g) = \Gamma_{d(G)}(g)$. By Hölder's theorem (Thereom 2.3.10) $d(G^\gamma/G_\gamma)$ is isomorphic to a $\mathbb{Q}$-subspace of $\mathbb{R}$. Thus, the embedding $\varphi : d(G) \longrightarrow V(\Gamma, d(G^\gamma/G_\gamma))$ given by Theorem 2.6.4, when restricted to G, embeds G into $V(\Gamma, d(G^\gamma/G_\gamma))$ and hence into $V(\Gamma, \mathbb{R})$. □

An example of a totally ordered abelian group G with a finite value set Γ which cannot be embedded in $V(\Gamma, G^\gamma/G_\gamma)$ is given in Exercise 8.

A *totally ordered division ring* is a division ring D whose additive group is a totally ordered group and whose positive cone is closed under multiplication. A left vector space E over D which is also an ℓ-group is called a *vector lattice over D* if $D^+E^+ \subseteq E^+$. If $0 < d \in D$, then $d^{-1} = dd^{-2} \in D^+$. Consequently, multiplication by d is an order automorphism of E, and hence it is an f-map by Theorem 2.4.8 (or Theorem 2.4.5). Thus, each vector lattice E over D is a D^+-f-group, and by Theorems 2.6.2, 2.6.3, and 2.6.4, E can be embedded in the vector lattice $V(\Gamma_D(E), E^\gamma/E_\gamma)$. If E is finite valued, then it is possible to arrange this embedding so that its range contains each E^γ/E_γ.

Theorem 2.6.6. *Let E be a finite valued vector lattice over the totally ordered division ring D. Then there is an embedding $\varphi : E \longrightarrow V(\Gamma_D(E), E^\gamma/E_\gamma)$ such that $\varphi(E) \supseteq \Sigma(\Gamma_D(E), E^\gamma/E_\gamma) = \{v \in V(\Gamma_D(E), E^\gamma/E_\gamma) : \text{supp } v \text{ is finite}\}$.*

Proof. For each $\gamma \in \Gamma_D(E)$ take $0 < a_\gamma \in E$ with E_γ as its only D-value (Theorem 2.5.11). Then $C_D(a_\gamma) = \text{lex } (E_\gamma \cap C_D(a_\gamma))$ and $E_\gamma \cap C_D(a_\gamma)$ is the maximal convex ℓ-subspace of $C_D(a_\gamma)$ (Theorems 2.5.3 and 2.5.8). Let A_γ be a subspace of $C_D(a_\gamma)$ such that $C_D(a_\gamma) = (E_\gamma \cap C_D(a_\gamma)) + A_\gamma$ is a D-direct sum. Then A_γ is an ℓ-simple (necessarily, totally ordered) vector lattice. Now, each nonzero element of A_γ has $E_\gamma \cap C_D(a_\gamma)$ as its only D-value in $C_D(a_\gamma)$, and hence it has E_γ as its only D-value in E, by Theorem 2.5.3. Since $E^\gamma = E_\gamma + C_D(a_\gamma) = E_\gamma + A_\gamma$ and $E_\gamma \cap A_\gamma = E_\gamma \cap C_D(a_\gamma) \cap A_\gamma = 0$, E^γ is the D-direct sum of E_γ and A_γ. Also, $A = \sum_\gamma A_\gamma$ is a D-direct sum. For suppose that $b_1 + \cdots + b_n = 0$ with $0 \neq b_i \in A_{\gamma_i}$, where $\gamma_1, \ldots, \gamma_n$ are distinct and γ_1 is a maximal element of $\{\gamma_1, \ldots, \gamma_n\}$; then $b_i \in E_{\gamma_1}$ for each $i \geq 2$. But now we have the contradiction that $b_1 \in E_{\gamma_1}$. We next show that if $C \in \mathscr{C}_D(E)$, then $C \cap A = \Sigma\{A_\gamma : A_\gamma \subseteq C\}$. If $0 \neq x \in C \cap A$, then $x = x_1 + \cdots + x_n$ where $0 \neq x_i \in A_{\gamma_i}$ and $\gamma_1, \ldots, \gamma_n$ are distinct. By induction on n, if some $A_{\gamma_i} \subseteq C$, then each $A_{\gamma_i} \subseteq C$. So, assume that $A_{\gamma_i} \nsubseteq C$ for each i. Since A_{γ_i} is ℓ-simple, $A_{\gamma_i} \cap C = 0$. Thus $x_i \notin C$ and $C \subseteq E_{\gamma_i}$; so, $x \in E_{\gamma_i}$ for each i. If γ_j is a maximal element of $\{\gamma_1, \ldots, \gamma_n\}$, then $x_i \in E_{\gamma_j}$ for each $i \neq j$. But this gives the contradiction that $x_j = x - (x_1 + \cdots + x_{j-1} + x_{j+1} + \cdots + x_n) \in E_{\gamma_j}$.

Now, let $\mathscr{S} = \{C \cap A : C \in \mathscr{C}_D(E)\}$, and define $f : \mathscr{S} \longrightarrow \mathscr{L}_D(A)$ by

$$f(C \cap A) = \Sigma\{A_\gamma : A_\gamma \nsubseteq C\} = \Sigma\{A_\gamma : A_\gamma \cap C = 0\}.$$

Then, by the previous paragraph, $A = (C \cap A) + f(C \cap A)$ is a D-direct sum, and f is a partial complementation for A with respect to $\mathscr{S}$. By Theorem 2.6.3 there is a partial complementation $\bar{f} : \mathscr{C}_D(E) \longrightarrow \mathscr{L}_D(E)$ of E with respect to $\mathscr{C}_D(E)$ such that $f(C \cap A) \subseteq \bar{f}(C)$ for each $C \in \mathscr{C}_D(E)$. Let $\varphi : E \longrightarrow V(\Gamma_D(E), E^\gamma/E_\gamma)$ be the corresponding embedding given by Theorem 2.6.4. So if $0 \neq b + E_\gamma \in E^\gamma/E_\gamma$, then $\varphi(b)(\gamma) = b + E_\gamma$. Since $E^\gamma = E_\gamma + A_\gamma$ we can assume that $b \in A_\gamma$, and we claim that supp $\varphi(b) = \{\gamma\}$; that is $\varphi(b) = \overline{b + E_\gamma}$. For, take $\alpha \in \Gamma_D$ with $\alpha \neq \gamma$, and write $b = x + y$ with $x \in E^\alpha$ and $y \in \bar{f}(E^\alpha)$. Then $\varphi(b)(\alpha) = x + E_\alpha$. If $\alpha \nless \gamma$, then $b \in E_\alpha$ since E_γ is the only D-value of b, and $b = x$; so $\varphi(b)(\alpha) = 0$. If $\alpha < \gamma$, then $E^\alpha \subseteq E_\gamma$; so $A_\gamma \nsubseteq E^\alpha$ and $A_\gamma \subseteq f(E^\alpha \cap A) \subseteq \bar{f}(E^\alpha)$. Thus $x = 0$ and, again, $\varphi(b)(\alpha) = 0$. □

Exercises.

1. Let $V = V(\Gamma, G_\gamma)$ be an ℓ-group. Recall that $\alpha || \beta$ means that the elements α and β of Γ are not comparable.
   (a) For $u, v \in V^+$, then $u \wedge v = 0$ if and only if whenever $\alpha \in$ supp u and $\beta \in$ supp v are comparable, then $\alpha = \beta$ is a minimal element of Γ and $u(\alpha) \wedge v(\alpha) = 0$.
   (b) If $u, v \in V$ give a description of the functions $u^+, u \vee v$, and $u \wedge v$, and show that supp $(u \vee v) \cup$ supp $(u \wedge v) \subseteq$ supp $u \cup$ supp v.
   (c) If $u \in V$, then supp $|u| =$ supp u.
   (d) If $u \in V$, then $|u|(\beta) = |u(\beta)|$ for each $\beta \in \Gamma$ if and only if whenever $\rho, \delta \in$ supp u and $u(\rho)$ and $u(\delta)$ have opposite sign, then $\rho || \delta$.
   (e) Suppose that each G_γ is totally ordered and let $u, v \in V^+$. Show that the following are equivalent statements.
      (i) $u \wedge v = 0$.
      (ii) If $\alpha \in$ supp u and $\beta \in$ supp v, then $\alpha || \beta$.
      (iii) If $\alpha \in$ maxsupp u and $\beta \in$ maxsupp v, then $\alpha || \beta$.
2. (a) Let $V = V(\Gamma, G_\gamma)$ be a Hahn product of po-groups and let $\Sigma = \Sigma(\Gamma, G_\gamma) = \{v \in V :$ supp v is finite$\}$ and $W = W(\Gamma, G_\gamma) = \{v \in V :$ supp v is a union of a finite number of chains $\}$. Show that V is an ℓ-group iff W is an ℓ-group, iff Σ is an ℓ-group.
   (b) If each G_γ is an Ω-f-group and V is an Ω-f-group show that Σ, W and $T(\Gamma, G_\gamma) = \{v \in V :$ maxsupp v is finite$\}$ are Ω-f-subgroups of V. More generally, for each infinite cardinal $\aleph, V_\aleph = \{v \in V :$ card (supp v) $< \aleph\}$ and $T_\aleph = \{v \in V :$ card (maxsupp v) $< \aleph\}$ are Ω-f-subgroups of V.
3. Let G be an abelian Ω-f-group and suppose that $V(\Gamma_\Omega(G), G^\gamma/G_\gamma)$ is an Ω-f-group. An Ω-group homomorphism $\varphi : G \longrightarrow V(\Gamma_\Omega(G), G^\gamma/G_\gamma)$ is *value-preserving* or an Ω-v-homomorphism if for each $g \in G, \Gamma_\Omega(g) =$ maxsupp $\varphi(g)$, and $\varphi(g)(\gamma) = g + G_\gamma$ if $\gamma \in \Gamma_\Omega(g)$.
   (a) Each Ω-v-homomorphism is an embedding of Ω-f-groups.
   (b) If there is an Ω-v-homomorphism whose image contains $\Sigma = \Sigma(\Gamma_\Omega(G), G^\gamma/G_\gamma)$, then G is finite valued.
   (c) Suppose that φ is an Ω-v-homomorphism and $\varphi(G) \supseteq \Sigma$. If $g \in G$ show that the set of maximal elements in supp $\varphi(g) \setminus$ maxsupp $\varphi(g)$ is finite.
   (d) Suppose that G is a vector lattice over the totally ordered division ring D and that $\Gamma_D(G)$ satisfies the minimum condition. Then there is a D-v-homomorphism $\varphi : G \longrightarrow V(\Gamma_D(G), G^\gamma/G_\gamma)$ with $\varphi(G) = \Sigma(\Gamma_D(G), G^\gamma/G_\gamma)$ if and only if G is finite valued.
4. Suppose that $V(\Gamma, G_\gamma)$ is an ℓ-group. Show that if α is a minimal element of Γ, then G_α is completely closed in $V(\Gamma, G_\gamma)$.
5. If $V(\Gamma, G_\gamma)$ is an ℓ-group show that $V(\Gamma, G_\gamma)$ is laterally complete iff each G_γ is laterally complete.

6. Let $V(\Gamma, G_\gamma)$ be an ℓ-group. Suppose that $\{\Gamma_i\}_{i\in I}$ is the set of equivalence classes determined by the equivalence relation that is given in Exercise 2.5.14. Show that

$$V(\Gamma, G_\gamma) \cong \prod_i V(\Gamma_i, G_\gamma)$$

and that each $V(\Gamma_i, G_\gamma)$ is an indecomposable ℓ-group.

7. Suppose that each G_γ is a nonzero subgroup of $\mathbb{R}$ and that $V = V(\Gamma, G_\gamma)$ is an ℓ-group. For $\gamma \in \Gamma$ let $V_\gamma = \{v \in V : v(\alpha) = 0 \text{ if } \alpha \geq \gamma\}$ and let $V^\gamma = \{v \in V : v(\alpha) = 0 \text{ if } \alpha > \gamma\}$.
   (a) Show that V^γ covers V_γ in $\mathscr{C}(V)$ and that V_γ is special.
   (b) Show that Γ can be embedded in $\Gamma(V)$.
   (c) If Γ is finite, show that the embedding in (b) is an isomorphism.
   (d) Find an example where this embedding is not onto.
   (e) Suppose Γ is a chain and for $\Delta \subseteq \Gamma$ let $V(\Delta, G_\gamma) = \{v \in V : \text{supp } v \subseteq \Delta\}$. Show that the mapping $\Delta \mapsto V(\Delta, G_\gamma)$ is an isomorphism between the posets of ideals of Γ and convex subgroups of V.

8. Let G be the ℓ-group of Exercise 2.5.21. Show that G cannot be embedded in $V(\Gamma(G), G^\gamma/G_\gamma)$.

9. A right module M over the ring R is called *semisimple* if it is the sum of simple submodules. (M is simple if $M \neq 0$ and 0 and M are the only submodules of M).
   (a) If $0 \neq M = xR + \mathbb{Z}x$ is cyclic show that M has a maximal submodule.
   (b) Suppose that

$$M = \sum_{i\in I} N_i$$

   where each N_i is a simple submodule, and let A be a submodule of M. Show that I has a subset J such that

$$M = A \boxplus \boxplus_{i\in J} N_i.$$

   (c) Show that the following statements are equivalent.
      (i) M is semisimple.
      (ii) M is the direct sum of a family of simple submodules.
      (iii) Each submodule of M is a direct summand of M.
      (iv) M has a complementation.
      (v) If N is a submodule of M and $f : \mathscr{S} \longrightarrow \mathscr{L}(N)$ is a partial complementation for N, then there is a partial complementation $\overline{f} : \overline{\mathscr{S}} \longrightarrow \mathscr{L}(M)$ of M such that $f(A \cap N) \subseteq \overline{f}(A)$ for each $A \in \overline{\mathscr{S}} = \{A \in \mathscr{L}(M) : A \cap N \in \mathscr{S}\}$.
   (d) Let L be a lattice with 0 and 1. A partial complementation for $k \in L$ is an antitone function $f : S \longrightarrow [0,k]$ such that $f(x)$ is a complement of x in $[0,k]$, for each $x \in S$, where $\phi \neq S \subseteq [0,k] = \{x \in L : x \leq k\}$. If $k = 1$ and $S = L$ then f is a complementation of L. A lifting of f is a partial complementation

$\bar{f}:\bar{S}\longrightarrow L$ where $\bar{S}=\{x\in L: x\wedge k\in S\}$ and $f(x\wedge k)\leq\bar{f}(x)$ for each $x\in\bar{S}$. Suppose that L is modular and

(i) each chain in L has a sup;
(ii) $a\wedge(\vee_i c_i)=\vee_i(a\wedge c_i)$ for each $a\in L$ and each chain $\{c_i\}$;
(iii) if $x\wedge b=p\wedge(x\vee b)=0$, then $x\wedge(b\vee p)=0$ where $x,b,p\in L$ and p is an atom;
(iv) 1 is the join of atoms.

Show that each partial complementation of each element of L has a lifting, and hence L has a complementation.

10. This exercise gives a proof of the embedding theorem that does not use a complementation. Let G be an abelian ℓ-group and assume that G^γ/G_γ is divisible for each $\gamma\in\Gamma=\Gamma(G)$. Let $p_\gamma: G^\gamma\longrightarrow G^\gamma/G_\gamma$ be the natural map, and let H be a subgroup of G. Let $\varphi: H\longrightarrow V=V(\Gamma,G^\gamma/G_\gamma)$ be a group homomorphism with the property that if $g\in H\cap G^\gamma$, then $\varphi(g)(\gamma)=p_\gamma(g)$. A homomorphism φ with this property is called a *p-mapping*.
    (a) Each p-mapping φ is monic, and if $a,b\in H$ with $a\wedge b=0$, then $\varphi(a)\wedge\varphi(b)=0$.
    (b) If $a\in H$, then $\Gamma(a)=$ maxsupp $\varphi(a)$.
    (c) If $x\in G\setminus H$ then φ can be extended to a p-mapping $\psi: H+\mathbb{Z}x\longrightarrow V$. (Define ψ by $\psi(h+nx)=\varphi(h)+nv$ where $v\in V$ is defined by $v(\gamma)=0$ if $(H+G^\gamma)\cap\mathbb{Z}x=0$ and $v(\gamma)=\frac{1}{n}[\varphi(h)(\gamma)+p_\gamma(a)]$ if $0\neq nx=h+a\in(H+G^\gamma)\cap\mathbb{Z}x$, with $h\in H$ and $a\in G^\gamma$.)
    (d) There exists a p-mapping from G into V.

11. Let $\Gamma=P\overleftarrow{\times}Q$ be the ordinal product of the nonempty posets P and Q. Show that Γ is rooted iff P and Q are rooted and P is totally ordered or Q is trivially ordered.

12. Suppose $V(\Gamma,G_\gamma)$ is an ℓ-group and G is an ℓ-subgroup of $V(\Gamma,G_\gamma)$ which contains $\Sigma(\Gamma,G_\gamma)$. For $X\subseteq\Gamma$ let ${}'X=\{\gamma\in\Gamma:\gamma$ is comparable to some element of $X\}$. Show that the following statements are equivalent for $v\in G$.
    (a) v is basic in G.
    (b) v is basic in $V(\Gamma,G_\gamma)$.
    (c) ${}'(\text{supp } v)$ is totally ordered and G_γ is totally ordered if $\gamma\in{}'(\text{supp } v)$.
    (d) There is a unique root M of Γ with supp $v\subseteq M$, the set of lower bounds of maxsupp v is totally ordered, and G_γ is totally ordered for $\gamma\in M$.

Notes. The abstract theory of ℓ-groups began with Birkhoff's paper [BIR1]. Much of the basic theory presented here appears in the notes by Weinberg [WE6] and Conrad [CON10]. Other references are the books by Bigard, Keimel, and Wolfenstein [BKW], Darnel [D], Glass [GL1] and [GL2], Anderson and Feil [AF], Birkhoff [BIR3], and Fuchs [F]. The exposition given in Exercise 2.1.14 of Jaffard's Theorem [JA] that each abelian ℓ-group is a group of divisibility follows Gilmer [GIL, p. 215].

Good references for the toplogical results presented in section seven are the books by Kelley [KEL] and Gillman and Jerison [GJ]. The characterization of the completion given in Theorem 2.3.6 is due to Conrad and McAlister [CM]. The proof that an archimedean ℓ-group is abelian given in Theorem 2.3.10 is due to Birkhoff [BIR3, p. 317]; the proof outlined in Exercise 2.4.14 is due to Wolfenstein; see p. 1.24 of Conrad's notes [CON10]. A good source for the Stone Representation Theorem, that is, the duality between the categories of Boolean algebras and Boolean spaces, is Halmos' book [HA]. The representation theorem for an archimedean ℓ-group given in Theorems 2.3.23, 2.3.24, and 2.3.25 is due to Bernau [BER1] while the concepts of the *cl*-essential closure and the splitting property as well as Theorem 2.3.26 is due to Conrad [CON7]. Theorems 2.3.28 through 2.3.32 come from Anderson, Conrad, and Kenny [ACK], while the proof of Theorem 2.3.29 as well as the special case of Theorem 2.3.32 when H is projectable is due to Bernau [BER3]. Exercise 2.3.8 is attributed to Kaplansky in Conrad, Harvey and Holland [CHH, p. 166], Exercise 2.3.21 is due to Weinberg [WE1], Exercises 2.3.22, 2.3.25, and 2.3.30 come from Conrad [CON8], and Exercises 2.3.26–2.3.29 are due to Conrad and McAlister [CM].

The use of operator sets in Section 2.4 originated in Steinberg [ST1] and allows for a uniform treatment of basic properties of ℓ-groups, ℓ-modules and ℓ-rings. Most of the ℓ-group results in Sections 2.4 and 2.5 are due to Conrad and appear in [CON10]. Theorem 2.4.3 for real vector lattices is in Johnson and Kist [JK]. Attributions for the various parts of Theorem 2.4.4 are given in Conrad [CON10, p. 1.21], Exercises 2.4.1 and 2.4.2 are due to Holland [HO], and Exercises 2.4.10 through 2.4.13 come from Conrad [CON6]. Hyper-archimedean ℓ-groups appear in Bigard [BI1, Chapter 6] and Exercise 2.4.15 appears in Conrad [CON10, p. 2.17]. Most of the material on values in section 2.5 occurs in Conrad [CON10] and originated in Conrad's papers [CON4], [CON6], and [CON7]; the translation to Ω-values and the relations between values and Ω-values comes from Steinberg [ST1] and [ST21]. What we have called the trunk of an ℓ-group is called the lex-kernel in Conrad [CON10, p. 2.25]; we have followed Weinberg's terminology [WE6, p. II–71]. Exercises 2.5.21 and 2.5.26 come from Conrad [CON10], p. 4.9 and p. 2.35, respectively; Exercise 2.5.31 comes from Steinberg [ST23] and is a generalization of the corresponding result for f-rings given by Pierce [PI1], and Exercise 2.5.32 is a result of Bigard [BI, Theorem 5.1] and Bigard and Keimel [BK, Lemma 4], in a more general setting. The Hahn embedding theorem for totally ordered abelian groups originated in Hahn [H], and for abelian ℓ-groups it is due to Conrad, Harvey, and Holland [CHH]. Theorem 2.5.3 about complementations for vector spaces is due to Banaschewski [BA] who has also noted the characterization of semisimple modules using complementations that is given in Exercise 2.6.9.

Chapter 3
Lattice-ordered Rings

Lattice-ordered rings occur as polynomial rings, power series rings, and semigroup rings, as do the perhaps more familiar totally ordered rings, but they also occur as matrix rings and endomorphism rings. We are concerned with the theory and structure of lattice-ordered rings and lattice-ordered modules and consequently a diverse number of topics appears. After initially supplying examples and identifying interesting classes of these objects we present the theory of radicals in the variety of lattice-ordered rings. Additional examples are provided when generalized power series rings are studied.

As in ring theory radicals are connected with the structure of a lattice-ordered ring by means of the intent to factor out a bad radical and end up with a good quotient. The most useful radical comes from the class of nilpotent lattice-ordered rings. For this radical the good quotient has no nilpotent kernels and in the right situation it even lacks positive nilpotent elements. By concentrating on the variety of f-rings, which is the variety generated by the class of totally ordered rings, we are able to obtain more fruitful structural results. For example, in this variety the building blocks for the ℓ-ring analogue of the Jacobson semisimple rings are the ℓ-simple totally ordered domains with an identity element. This is also true for some varieties larger than the f-ring variety, but no such definitive identification is likely in general.

Because of the complications inherent in the interaction between the lattice structure with the ring multiplication the question of unitability, that is, of when an identity element can be adjoined to a lattice-ordered ring so that membership in a variety is preserved, is quite daunting. For the variety of f-rings and, more generally for f-algebras, answers to this question are provided. The natural questions concerning the uniqueness of this adjunction as well as when the f-algebra is convex in its enlargement are also considered.

As might be expected, archimedean f-rings are extremely well behaved. In particular, the previous embedding theorem for an archimedean ℓ-group is shown to be an embedding theorem for an archimedean f-ring modulo its nilpotent radical. Thus, an archimedean f-ring with zero nilpotent radical is a ring of almost real-valued continuous functions on a compact Hausdorff space. This representation will allow us to obtain definitive answers to several interesting questions, examples of

S.A. Steinberg, *Lattice-ordered Rings and Modules*,
DOI 10.1007/978-1-4419-1721-8_3, © Springer Science + Business Media, LLC 2010

which are: (i) Can an archimedean ℓ-group be the additive ℓ-group of distinct f-rings with identity elements? (ii) Under what circumstances is an archimedean ℓ-group the additive ℓ-group of an f-ring with identity?

If we make the reasonable assumption that the variety of f-rings is, in many ways, the most desirable variety of f-rings, then it becomes intriguing to investigate which structural properties are preserved in larger varieties; for example, the variety in which squares are positive or a variety determined by some other polynomial constraint. This is done in the latter part of this chapter and in the exercises.

3.1 Basics, Examples, and Nonexamples

Each ring will be associative but need not have an identity element. A ring which has an identity will be called *unital*, as will a module M if $x1 = x$ for each x in M.

A *partially ordered ring* (*po-ring*) is a ring R whose additive group is a po-group and in which the product of positive elements is positive; that is, if $a \geq 0$ and $b \geq 0$ then $ab \geq 0$. If $(R, +, \leq)$ is directed, or is an ℓ-group or is totally ordered, then R is called a *directed po-ring*, or a *lattice-ordered ring* (*ℓ-ring*) or a *totally ordered ring*, respectively. If R is unital and $1 > 0$ then R will be called *po-unital* (or *ℓ-unital* if R is an ℓ-ring). In addition to collecting some basic facts about ℓ-rings and lattice-ordered modules we present some examples and identify several types of ℓ-rings which will be studied in later sections. Also, a useful f-subring of an ℓ-ring is identified.

The ring analogue of Theorem 2.1.1, whose proof is left to the reader, states that there is a bijection between the set of partial orders of the ring R which make it into a po-ring and the set of those subsets P of R with the properties:

$$P + P \subseteq P \quad (P \text{ is a subsemigroup of } (R, +));$$
$$PP \subseteq P \quad (P \text{ is a subsemigroup of } (R, \cdot));$$
$$P \cap -P = 0.$$

A subset of a ring R that satisfies the first two of these three properties is called a *subsemiring* of R. As is the case for groups, we will frequently identify the partial order of a po-ring with its positive cone.

Each ring with its trivial partial order is a po-ring, and each abelian po-group with trivial multiplication is a po-ring; equally obvious is the fact that each subring of the reals with the usual order is a totally ordered ring. Most ring theoretic constructions will produce a po-ring if they start with ordered objects. So, for example, the direct product or the direct sum of a family of po-rings becomes a po-ring if either is ordered coordinatewise, and each will be directed or be an ℓ-ring if each member of the family is directed or is an ℓ-ring. Similarly, R_n, the $n \times n$ matrix ring over the po-ring R, together with the canonical positive cone $(R^+)_n$, is a po-ring which is directed or is an ℓ-ring if R has either of these properties. An identical statement can be made for the polynomial ring $R[x]$ with positive cone $R^+[x]$. Whenever a matrix ring is considered as a po-ring it will be assumed that it has its canonical partial order, unless indicated otherwise. The po-ring (respectively, ℓ-ring) R is called a *po-domain* (respectively, an *ℓ-domain*) if $a > 0$ and $b > 0$ imply that $ab > 0$. Recall that the ring R is a *domain* if $a \neq 0$ and $b \neq 0$ imply that $ab \neq 0$. When R is a po-domain other partial orders on $R[x]$ are possible. In particular, $R[x]$ can be ordered *lexicographically* or *antilexicographically*; let us write a nonzero polynomial $p(x) \in R[x]$ with increasing exponents:

$$p(x) = a_k x^k + a_{k+1} x^{k+1} + \cdots + a_n x^n, \qquad 0 \leq k \leq n,\ a_k, a_n \neq 0.$$

In the antilexicographic ordering $p(x) > 0$ if and only if $a_k > 0$, whereas in the lexicographic ordering $p(x) > 0$ if and only if $a_n > 0$. In both cases $R[x]$ is a po-domain and is directed if R is directed. If R is not trivially ordered, then $R[x]$ is always directed in the lexicographic ordering since $p(x), 0 < p(x) + ax^{n+1}$ if $0 < a \in R$. Of course, if R is a totally ordered domain, then so is $R[x]$ in either of these two orderings. If A and B are po-rings then the ring direct product supplied with the lexicographic ordering, $R = A \underleftarrow{\times} B$, is a po-group as we have already seen; but it is not always a po-ring. The conditions for it to be a po-ring are given in Exercise 1. Here we just note that if $A^2 = 0$, then $A \underleftarrow{\times} B$ is a po-ring; and it is an ℓ-ring exactly when A is an ℓ-group and B is totally ordered (see Exercise 2.1.10).

The material in Section 2.2 on subobjects and factor groups of a po-group or an ℓ-group and on morphisms between two ordered groups carries over to po-rings and ℓ-rings in the standard way. In particular, if I is a convex ideal of the po-ring R, then R/I is a po-ring with positive cone $(R/I)^+ = \{a + I : a \in R^+\}$. An *$\ell$-ideal* of an ℓ-ring R is a convex ℓ-subgroup I of R that is also an ideal of R; if I is only a right (respectively, left) ideal then it is called a *right* (respectively, *left*) *ℓ-ideal* of R. If X is a subset of R, then $\langle X \rangle_r$, $\langle X \rangle_\ell$, and $\langle X \rangle$ will denote the right, left or two-sided ℓ-ideal of R generated by X, respectively. The ℓ-ideal I is called *irreducible* if R/I is totally ordered; that is, if I is an ideal and a prime subgroup of R. A *homomorphism* from one ℓ-ring (respectively, po-ring) to another is a ring homomorphism that is also a lattice homomorphism (respectively, isotone); sometimes this will be called an *ℓ-homomorphism*. Conditions for a ring homomorphism to be an ℓ-homomorphism are given in Theorem 2.2.2. The class of ℓ-rings is a variety within the class of all Ω-algebras, where $\Omega = \{+, \cdot, \vee, \wedge, -, 0\}$, since the condition that the product of positive elements is positive can be expressed as the identity $(x^+ y^+)^- = 0$. Thus, all of the properties attached to a variety hold in the class of ℓ-rings. We also note that the kernel of a morphism in the category of ℓ-rings is an ℓ-ideal, and each ℓ-ideal is the kernel of a morphism. It is also clear that the set of ℓ-ideals of the ℓ-ring R is a complete sublattice of the lattice of convex ℓ-subgroups of R. If this sublattice has exactly two elements and $R^2 \neq 0$, then R is called *ℓ-simple*.

If R is a po-ring and M is a po-group that is also a right R-module, then M is called a (*right*) *po-module over R* if $xa \in M^+$ whenever $x \in M^+$ and $a \in R^+$; a *left po-R-module* is defined analogously. Note that the module M_R is a po-module if and only if it is an R^+-po-group (see Section 2.6). The concept of a po-module is equivalent to that of a *po-representation* of R. If M is considered as a right $\mathrm{End}_{\mathbb{Z}}(M)$-module and $\varphi : R \longrightarrow \mathrm{End}_{\mathbb{Z}}(M)$ is the representation corresponding to the R-module structure of M, then the po-group M is a po-R-module if and only if φ is relation preserving; that is, $\varphi(R^+)$ is contained in the set of po-group endomorphisms of M (see Exercise 2.2.8). If M is an ℓ-group as well as a po-R-module it is called an *ℓ-module over R*. Some immediate examples of po-modules include each right ideal I of the po-ring R as well as the factor module R/I if I is also convex; similar ℓ-examples arise using the appropriate ℓ-subobjects of the ℓ-ring R. Without further elaboration we note that the remarks that were made in the previous paragraph about the morphisms and their kernels in the category of ℓ-rings also hold for the category of ℓ-modules over the po-ring R. An *ℓ-submodule* of an ℓ-module is, of course, a submodule that is also a sublattice. Note that if R is not directed then an R^+-ℓ-subgroup of M need not be an ℓ-submodule of M_R, though it will be an ℓ-submodule of M_S where $S = R^+ - R^+$ is the subring of R generated by R^+.

Let M_R be an ℓ-module over the po-ring R. The element $a \in R^+$ is called a *distributive element* or a *d-element on M* if

$$\forall x, y \in M, \ \ x \wedge y = 0 \text{ implies } xa \wedge ya = 0;$$

a is called an *f-element on M* if

$$\forall x, y \in M, \ \ x \wedge y = 0 \text{ implies } xa \wedge y = 0.$$

Each f-element on M is a d-element on M. Analogously, if M_R is a po-module over the ℓ-ring R, then $x \in M^+$ is called a *d-element on R* if

$$\forall a, b \in R, \ \ a \wedge b = 0 \text{ implies } xa \wedge xb = 0.$$

The ℓ-module M is called a *d-module* (respectively, an *f-module*) if each element of R^+ is a d-element (respectively, an f-element) on M. So, a d-module is just an R-module which is an R^+-d-group, and an f-module is an R-module which is an R^+-f-group. If each positive element of the po-module M is a d-element on R, then M is called *strong*. For example, each po-module over a totally ordered ring is strong. In particular, each abelian ℓ-group is a unital strong f-module over $\mathbb{Z}$. Note also that submodules and products of strong po-modules are strong. Moreover, if N is a convex directed submodule of the strong po-module M, then M/N is strong since disjoint elements in M map to disjoint elements in M/N.

We will denote the set of d-elements on M by $d(M_R) = d(M)$, the set of f-elements on M by $f(M_R) = f(M)$, and the additive subgroups that they generate by $D(M_R) = D(M)$ and $F(M) = F(M_R)$, respectively.

If R is an ℓ-ring, then an f-element or a d-element on the right ℓ-module R_R is called a *right f-element* or a *right d-element* of R, respectively. If the ℓ-module R_R is an f-module or a d-module then R is called a *right f-ring* or a *right d-ring*, respectively. Similar definitions apply on the left. An *f-ring* is an ℓ-ring that is both a left f-ring and a right f-ring, and a d-ring is an ℓ-ring that is both a left d-ring and a right d-ring. Also, an *f-element* or a *d-element* of R is an element that is an f-element or a d-element on both sides.

If M is a right R-module and $X \subseteq M$ and $A \subseteq R$, then

$$XA = \left\{ \sum_{i=1}^{n} m_i x_i a_i : x_i \in X, a_i \in A, m_i \in \mathbb{Z} \right\}$$

will denote the subgroup of M generated by $\{xa : x \in X, a \in A\}$. Occasionally, XA will denote the additive subsemigroup generated by this set but the context will always make this clear. The *right annihilator of X in R* and the *left annihilator of A in M* are defined by

$$r(X) = r(X;R) = \{a \in R : Xa = 0\} = \{a \in R : xa = 0 \ \forall x \in X\}$$

and

$$\ell(A) = \ell(A;M) = \{x \in M : xA = 0\} = \{x \in M : xa = 0 \ \forall a \in A\}.$$

In general, $r(X)$ is a right ideal of R which is an ideal if X is a submodule, and $\ell(A)$ is a subgroup of M which is a submodule if A is a left ideal. If M is an ℓ-module over the ℓ-ring R, then the *ℓ-annihilators* are given by

$$r_\ell(X) = r_\ell(X;R) = \{a \in R : |x||a| = 0 \ \forall x \in X\}$$

and

$$\ell_\ell(A) = \ell_\ell(A;M) = \{x \in M : |x||a| = 0 \ \forall a \in A\}.$$

The verification that $r_\ell(X)$ is a right ℓ-ideal of R that is contained in $r(X)$ and $\ell_\ell(A)$ is a convex ℓ-subgroup of M that is contained in $\ell(A)$, as well as other facts about annihilators, is left as an exercise (Exercises 16 and 17). We do note here that, for any po-module M_R, each annihilator of a positive subset is convex. If X consists of a single element x, then we write $r(x)$ for $r(\{x\})$, and, similarly, for the other annihilators. The element a in the ring R is called *right* (respectively, *left*) *regular* if $r(a) = r(a;R) = 0$ (respectively, $\ell(a) = 0$). If $r(a) = \ell(a) = 0$, then a is called *regular*. If $r(M) = 0$, then M is a *faithful* R-module, and if $r_\ell(M) = 0$, then M is called *ℓ-faithful*.

The theory of Ω-f-groups developed in the last three sections applies directly to the categories of R-f-modules and of f-rings. Though it is not necessary to restate this theory in the present context in order to use it, we will, nevertheless, restate and slightly embellish Theorems 2.4.7 and 2.4.8.

If R and S are rings and $_SM$ and M_R are modules, then $_SM_R$ is called a *(left S-right R-) bimodule* if $s(xr) = (sx)r$ for each $x \in M, s \in S$ and $r \in R$. If $_SM$ and M_R are f-modules (respectively, ℓ-modules, po-modules), then M is called an *f-bimodule* (respectively, *ℓ-bimodule*, *po-bimodule*).

Theorem 3.1.1. *Let M_R be an ℓ-module over the po-ring R, let $f(M)$ be the set of f-elements on M, and let $F(M)$ be the additive subgroup of R that is generated by $f(M)$.*

(a) $f(M)$ is a convex subsemiring of R^+ that is closed under any finite sups and infs that exist in R.

(b) $F(M)$ is a convex directed subring of R that is closed under any finite sups and infs that exist in R.

(c) $F(M)^+ = f(M)$ *and* $F(M)$ *is the largest directed subring of* R *over which* M *is an* f*-module.*

(d) If R *is an* ℓ*-ring, then* $F(M)$ *is a convex* ℓ*-subring of* R*, and*

$$F(M) = \{a \in R : |a| \in f(M)\}.$$

(e) If M *is a* d*-module and* $N = \ell(R^+;M)$*, then* M/N *is an* f*-module over the directed po-ring* $S = R^+ - R^+$*.*

(f) If R *is an* ℓ*-ring, then, for each* $x \in M$ *and each* $a \in R$, $|xa| \leq |x||a|$*.*

(g) If R *is an* ℓ*-ring, then* M *satisfies the identities* $|xa| = |x||a|$ *if and only if it is a strong* d*-module.*

(h) If R *is an* ℓ*-ring, then* $F(R) = F(R_R) \cap F({}_RR)$ *is an* f*-ring, and it is the largest directed subring of* R *over which* R *is an* f*-bimodule.*

Proof. Since $0 \leq \wedge a_i, \vee a_i \leq \Sigma a_i$ if $\{a_1,\ldots,a_n\} \subseteq R^+$ and either $\vee a_i$ or $\wedge a_i$ exists, (a) is a consequence of Theorem 2.4.5. Also, (b), (c), and (d) are consequences of Exercise 2.2.18, (e) follows from Theorem 2.4.8 since N is a convex ℓ-S-submodule of M by Exercise 17(d), and (h) is a specialization of (c) and (d). If R is an ℓ-ring, then

$$\begin{aligned} |xa| &= |x^+a^+ + x^-a^- - x^+a^- - x^-a^+| \\ &\leq x^+a^+ + x^-a^- + x^+a^- + x^-a^+ \\ &= |x||a|; \end{aligned}$$

so (f) is proven. As for (g), if M satisfies the identity $|xa| = |x||a|$, then for any $x \in M^+$ and any $a \in R^+$ the group homomorphisms $x\cdot : R \longrightarrow M$ and $\cdot a : M \longrightarrow M$ induced by x and a are both lattice homomorphisms, by Theorem 2.2.2; that is, M is a strong d-module. Conversely, if these maps are lattice homomorphisms, then, for all $x \in M$ and all $a \in R$, xa^+ and xa^- are disjoint, since

$$|xa^+| \wedge |xa^-| = |x|a^+ \wedge |x|a^- = |x|(a^+ \wedge a^-) = 0.$$

So

$$\begin{aligned} |x||a| &= |x|a^+ + |x|a^- = |xa^+| + |xa^-| \\ &= |xa^+ - xa^-| = |xa|. \end{aligned}$$

□

From Theorems 2.4.6 and 2.4.7 we immediately get

Theorem 3.1.2. *Let* M_R *be an* ℓ*-module over the po-ring* R*, and let* $S = R^+ - R^+$ *be the largest directed subring of* R*. The following statements are equivalent.*

(a) M_R *is an* f*-module.*

(b) Each minimal prime subgroup of M *is an* S*-submodule.*

(c) Each polar of M *is an* S*-submodule.*

(d) The ℓ-module M_S can be embedded into a direct product of a family of totally ordered S-modules.

(e) Each subdirectly irreducible homomorphic image of the ℓ-module M_S is totally ordered. □

The class of ℓ-modules over the po-ring R is a variety since, just as for an ℓ-ring, the identities $(x^+a)^- = 0$, for $x \in M$ and $a \in R^+$, express the condition $M^+R^+ \subseteq M^+$. The subclasses of d-modules and f-modules are also varieties which are determined by the respective identities $|xa| = |x|a$ and $x^+a \wedge x^- = 0$, where $x \in M$ and $a \in R^+$; also, the identities $|x^+a| = x^+|a|$, where $x \in M$ and $a \in R$, determine the variety of strong ℓ-modules over the ℓ-ring R.

The preceding two results can be applied to an ℓ-ring R since R is an R-R-ℓ-bimodule, and R is also an ℓ-module over the directed po-ring S that is generated by the left and right multiplication maps in $\mathrm{End}_{\mathbb{Z}}(R)$. In particular, a d-ring, being just an ℓ-ring that is a strong right d-module over itself, is characterized by the identity $|xy| = |x||y|$, and a unital d-ring is an f-ring. An f-ring is characterized as an ℓ-ring that can be embedded in a product of totally ordered rings since R is an f-ring if and only if R is an f-module over S and the S-submodules are just the ideals of R; that is, the class of f-rings is the variety of ℓ-rings generated by the class of totally ordered rings. Another identity that describes a d-ring is

$$(x^+y^+ \wedge x^-y^+) \vee (y^+x^+ \wedge y^+x^-) = 0,$$

and an identity that describes an f-ring is

$$(x^+y^+ \wedge x^-) \vee (y^+x^+ \wedge x^-) = 0.$$

A po-ring (respectively, an ℓ-ring) is said to have squares positive or to be an *sp-ring* or an *sp-po-ring* (respectively, an *sp-ℓ-ring*) if it satisfies the inequality $x^2 \geq 0$, or, equivalently if it satisfies the identity $(x^2)^- = 0$. An ℓ-ring is called an *almost f-ring* if it satisfies the identity $x^+x^- = 0$; this is equivalent to the condition: $a \wedge b = 0$ implies $ab = 0$. Since both of these identities hold in a totally ordered ring they hold in any f-ring. Alternatively, in an f-ring, if $a \wedge b = 0$, then $ab \wedge b = 0$ and $ab = ab \wedge ab = 0$; and in an almost f-ring $x^2 = (x^+ - x^-)^2 = (x^+)^2 + (x^-)^2 \geq 0$. So an almost f-ring is an sp-ℓ-ring.

An example of a right f-ring which is a d-ring but is not an f-ring is given by the ℓ-subring $R = \begin{pmatrix} \mathbb{Q} & 0 \\ \mathbb{Q} & 0 \end{pmatrix}$ of the canonically ordered matrix ring $\mathbb{Q}_2$. An example of a unital right f-ring (which, as we will see in Theorem 3.8.10, is necessarily an almost f-ring) that is not an f-ring is given in Exercise 3.8.13; and an example of a unital sp-ℓ-ring that is not an almost f-ring is given in Exercise 3.7.9; also see Exercise 3.8.7.

We exhibit next two diametrically opposite types of po-rings: one for which each unital ℓ-module is an f-module and one which has no nontrivial f-modules. Interestingly, totally ordered division rings are of the first type whereas their lattice-ordered matrix rings are of the second type, even though the two unital module categories they determine are indistinguishable. Recall that a subset X of the poset P is cofinal in P if for each p in P there is some x in X with $x \geq p$. We will denote the group of units of the unital ring or monoid R by $\mathscr{U}(R)$. If R is a directed po-ring and R^+ is a monoid, then R is po-unital.

Theorem 3.1.3. *Let M_R be a unital po-module over the po-ring R and let u be a unit of R.*

(a) If $u \in \mathscr{U}(R^+)$, then multiplication by u is an automorphism of the po-group M.
(b) Suppose that R is an ℓ-ring. Then $u \in \mathscr{U}(R^+)$ if and only if u is a right (left) d-element of R.
(c) Suppose that $1 \in R^+$ and $\mathscr{U}(R^+)$ is cofinal in R^+. If M is an ℓ-module, then it is an f-module.
(d) If R is a totally ordered division ring and M is an ℓ-module, then M is an f-module.

Proof. (a) is obvious and gives one implication of (b). For the other implication, if u is a right d-element then $1^- u = u^- = 0$ gives that $1 > 0$. Again, $(u^{-1})^- u = 1^- = 0$ yields that $u^{-1} \in R^+$. If $u, u^{-1} \in R^+$ then, according to (a), u is a d-element on the ℓ-module M; hence (c) follows from Theorem 2.4.8. Clearly, (d) is a special case of (c). □

The class of po-rings determined by the condition in (c) of the preceding result is homomorphically closed and productive and includes the f-rings $C(X)$ where X is a topological space (Exercise 11). A unital f-module over a division ring will be called a *vector lattice*.

If I is a set, then a set $\{e_{ij} : i, j \in I\}$ of nonzero elements of the ring R is called a set of *matrix units of degree card* (I) if $e_{ij}e_{k\ell} = \delta_{jk}e_{i\ell}$, for all $i, j, k, \ell \in I$, where δ_{jk} is the Kronecker delta : $\delta_{jk} = 1$ if $j = k$ and $\delta_{jk} = 0$ if $j \neq k$.

Theorem 3.1.4. *Let $\{e_{ij} : i, j \in I\} \subseteq R^+$ be a set of matrix units and let M be a po-module over R.*

(a) If $x \in M$, then for each $i \in I$ the set $\{xe_{ij} : j \in I\}$ is trivially ordered.
(b) If M is a d-module over R and card $(I) \geq 2$, then $Me_{ij} = 0$ for all i, j.
(c) If $R = T_n$ is the canonically ordered matrix ring of degree $n \geq 2$ over the po-unital po-ring T, then R has no nontrivial d-modules.

Proof. (a) Let $x_{ij} = xe_{ij}$. If some $x_{ij} \neq 0$, then $x_{ik} \neq 0$ for each k since $x_{ik}e_{kj} = x_{ij}$; also, $x_{ij} \neq x_{ik}$ if $j \neq k$ since $0 = x_{ij}e_{kj}$, whereas $x_{ik}e_{kj} = x_{ij}$. If $x_{ij} > x_{ik}$, then

$$x_{im} = x_{ij}e_{jm} \geq x_{ik}e_{jm} = 0$$

for each $m \in I$. But then

$$0 = x_{ij}e_{km} \geq x_{ik}e_{km} = x_{im} > 0.$$

(b) Let $S = \Sigma \mathbb{Z} e_{ij}$ be the subgroup of R generated by the e_{ij}. Then S is a subring of R with positive cone $S^+ = S \cap R^+ = \Sigma \mathbb{Z}^+ e_{ij}$. For if $s = \Sigma n_{ij} e_{ij} \in S \cap R^+$, then $0 \leq e_{kk} s e_{mm} = n_{km} e_{km}$ for each $k, m \in I$. But then $n_{km} \in \mathbb{Z}^+$. Let $N = \ell(S; M)$. Since S is directed $N = \ell(S^+; M)$, and since M is a d-module $\bar{M} = M/N$ is an f-module over S by (e) of Theorem 3.1.1. Since $\bar{M}$ is a subdirect product of totally ordered S-modules we may assume that $\bar{M}$ is totally ordered. But then $\bar{x} e_{ij} = 0$ for each $x \in M$ and each $i, j \in I$, by (a). Thus $xS = xS^2 = 0$ and $M = N$. This proves (b) and also (c). □

The ring R is called a *C-algebra* over the unital commutative ring C if R is a unital C-module and $\alpha(xy) = (\alpha x)y = x(\alpha y)$ for all $x, y \in R$ and $\alpha \in C$. If R is a C-algebra and a po-module over C, then R is called a *po-algebra over C*. If the po-algebra R is an ℓ-ring, then it is called a *C-ℓ-algebra* provided that R is an f-module over C. Unless stated otherwise, when considering po-algebras over C it will be assumed that $1 \in C^+$. Each ℓ-ring is an ℓ-algebra over $\mathbb{Z}$, and each ℓ-ring that is a po-algebra over a totally ordered field C is an ℓ-algebra. A module M over the C-algebra R is called an *R-algebra module* if it is also a unital C-module such that $(x\alpha)r = x(r\alpha) = (xr)\alpha$ for all $x \in M$, $\alpha \in C$ and $r \in R$. An *R-algebra po-module* over R is defined to be an R-algebra module that is a po-module over R and over C, and an *R-algebra ℓ-module* is an R-algebra po-module that is an f-module over C.

More generally, suppose that R and T are rings and R is a T-T bimodule. Then R is called a *T-ring* if multiplication in $R \cup T$ is associative; if all the actions are unital this just means that there is a ring homomorphism $T \longrightarrow R$ which preserves the identity. If T and R are po-rings, then R is a *po-T-ring* if it is a T-ring and $R^+ \cup T^+$ is closed under multiplication and R is an *ℓ-T-ring* if R is also an ℓ-ring.

A po-module M_R is said to have a *canonical basis* if M is isomorphic to a direct sum of copies of $R_R : M \cong R^{(I)}$. So,

$$M = \bigoplus_{i \in I} e_i R, \; M^+ = \sum_{i \in I} e_i R^+ \text{ and } e_i a \geq 0 \text{ iff } a \in R^+.$$

If R is not unital the e_i may not be in M but can be assumed to lie in a po-module extension of M (Exercise 2). Of course, M is an ℓ-module exactly when R is an ℓ-ring. Suppose that R is a T-ring over the po-unital ring T, R is a unital T-bimodule and ${}_T R$ is a po-module with a canonical basis $\{e_i\}_{i \in I}$. Then R is a po-ring if and only if $a e_i b e_j \in R^+$ for any indexes i and j and all $a, b \in T^+$. This condition can be expressed by asserting that the *structural constants* lie in T^+:

$$e_i b = \sum_k a_{ik} e_k, \; e_i e_j = \sum_k a_{ijk} e_k \text{ where } b, a_{ik}, a_{ijk} \in T^+.$$

By taking a subring of T the assumption that T is unital may be dropped. For example, R could be a matrix ring or a semigroup ring over T.

Exercises.

1. Let A and B be po-rings and let $A \overleftarrow{\times} B$ be the ring direct product supplied with the lexicographic partial order.

   (a) Show that $A \overleftarrow{\times} B$ is a po-ring if and only if
      (i) $B^+ = 0$, or
      (ii) $A^2 = 0$, or
      (iii) $B^+ \neq 0$, B is a po-domain and $A^+ \subseteq \ell(A) \cap r(A)$.
   (b) Let B be a nonzero totally ordered domain and suppose that $A^+ = 0$. Show that $(A \overleftarrow{\times} B)^+$ is a maximal partial order of the ring $A \times B$ if and only if $\ell(A) \cap r(A) \subseteq t(A)$, where $t(A)$ is the torsion ideal of A.

2. Let R be a C-algebra and let R_1 be the C-algebra obtained by freely adjoining C to R. So $R_1 = R \times C$ as C-modules and the multiplication in R_1 is given by $(r,\alpha)(s,\beta) = (r\beta + s\alpha + rs, \alpha\beta)$ for $r,s \in R$ and $\alpha,\beta \in C$.

   (a) Show that R_1 is a unital C-algebra and R is an ideal of R_1.
   (b) If M is an R-algebra module, then the scalar multiplication $x(r,\alpha) = xr + x\alpha$ makes M into a unital R_1-algebra module, and the R_1-submodules of M are precisely the R-algebra submodules.
   (c) If R is a po-algebra, then so is R_1 if it is ordered coordinatewise. If R is an ℓ-algebra, then R_1 is an ℓ-algebra if and only if C is an f-ring.
   (d) If $S = R + C \cdot 1$ is a unital C-algebra (po-algebra) which contains R, show that there is a unique algebra (po-algebra) homomorphism from R_1 onto S.
   (e) If M is an R-algebra po-module or an ℓ-module, then M is a po-module or an ℓ-module over R_1. Moreover, the following statements are equivalent.
      (i) M is an f-module over R.
      (ii) M is an f-module over R_1.
      (iii) M is a d-module over R_1.
   (f) Show that R_1 is a po-domain if and only if R and C are po-domains and R_C is strict.

3. (a) The following statements are equivalent for the ℓ-module M_R.
      (i) M is an f-module.
      (ii) $x^\perp \subseteq (xa)^\perp$ if $x \in M$ and $a \in S = R^+ - R^+$.
      (iii) $X^\perp \subseteq (XA)^\perp$ if $X \subseteq M$ and $A \subseteq S$.
   (b) The following statements are equivalent for the ℓ-ring R.
      (i) R is an f-ring.
      (ii) $(ab)^{\perp\perp} \subseteq a^{\perp\perp} \cap b^{\perp\perp}$ for all $a,b \in R$ (or R^+).
      (iii) $(AB)^{\perp\perp} \subseteq A^{\perp\perp} \cap B^{\perp\perp}$ for all subsets A and B of R (or R^+).

4. In each of the cases (a), (b), and (c) prove that the statements are equivalent for the ℓ-module M_R.

   (a) (i) M is a d-module.
      (ii) $|x| \wedge |y| = 0 \Longrightarrow |xa| \wedge |ya| = 0$ for $x,y \in M$ and $a \in S = R^+ - R^+$.
      (iii) $x^\perp a \subseteq (xa)^\perp$ if $x \in M$ and $a \in S$.
      (iv) $X^\perp a \subseteq (Xa)^\perp$ if $X \subseteq M$ and $a \in S$.

(b) First, formulate the analogues of the last three statements in (a) if the first is replaced by "M_R is a strong po-module over the ℓ-ring R."

(c) (i) M_R is a strong d-module over the ℓ-ring R.

(ii) $(xa)^+ = x^+a^+ + x^-a^-$ if $x \in M$ and $a \in R$.

(iii) $(xa)^- = x^+a^- + x^-a^+$ if $x \in M$ and $a \in R$.

5. Show that an ℓ-ring is an almost f-ring if and only if it satisfies the identity $x^2 = |x|^2$.

6. Let M_R be an ℓ-module.

(a) Show that $d(M)$ is a convex multiplicatively closed subset of R, and $D(M)$ is a directed subring of R.

(b) The following statements are equivalent (see Exercise 2.2.17).

(i) $d(M)$ is additively closed.

(ii) $d(M)$ is directed up.

(iii) $D(M)^+ = d(M)$.

(iv) $D(M) = \{a - b : a, b \in d(M)\}$.

(v) M is a d-module over $D(M)$.

(c) If the conditions in (b) hold, then

(i) $D(M)$ is convex ($D(M)$ is also convex if R has the Riesz decomposition property; see Theorem 2.1.4(k));

(ii) $d(M)$ and $D(M)$ are closed under any finite sups and infs that exist in the po-ring R;

(iii) $D(M)$ is the largest directed subring of R over which M is a d-module.

(d) $d(M) = f(M)$ if and only if $D(M) = F(M)$.

(e) If $\ell(d(M);M)^+ = 0$, then each of the conditions in (b) is equivalent to $d(M) = f(M)$.

(f) If R is an ℓ-ring, then $D(M)$ is a convex ℓ-subring of R, and the conditions in (b) are equivalent to $D(M) = \{a \in R : |a| \in d(M)\}$.

(g) Let N be a $D(M)$-submodule of M. Show that the sublattice of M generated by N is a $D(M)$-ℓ-submodule of M (use Theorem 2.2.4(f)).

(h) Let ${}_RM_R$ be an ℓ-bimodule and put $d({}_RM_R) = d(M_R) \cap d({}_RM)$, $\bar{D}({}_RM_R) = D(M_R) \cap D({}_RM)$, and let $D({}_RM_R)$ be the additive subgroup of R generated by $d({}_RM_R)$. If R has the Riesz decomposition property, show that $D({}_RM_R) = \bar{D}({}_RM_R)$.

(i) Let ${}_RM_R$ be an ℓ-bimodule. Show that M is a $D({}_RM_R)$-d-bimodule iff M is a $\bar{D}({}_RM_R)$-d-bimoldule, iff $D(M)^+ = d(M)$, iff $\bar{D}(M)^+ = d(M)$.

(j) If ${}_RM_R$ is an ℓ-bimodule and $A \subseteq R$ the *centralizer of A in M* is $C_M(A) = \{x \in M : \forall a \in A,\ xa = ax\}$. If $A \subseteq d({}_RM_R)$, show that $C_M(A)$ is an ℓ-subgroup of M. Also, show that $C_M(D({}_RM_R))$ is an ℓ-subgroup.

(k) If R is an ℓ-ring we will denote $d({}_RR_R)$ and $D({}_RR_R)$ by $d(R)$ and $D(R)$, respectively. Show that the center of $D(R)$ is an ℓ-subring of R.

7. Let M be an ℓ-module over the ℓ-ring R, let $d_M(R)$ be the set of those elements in M that are d-elements on R, and let $D_M(R)$ be the subgroup of M generated by $d_M(R)$.

(a) State and prove the analogues of (a), (b) and (f) in Exercise 6.
(b) Show that $D_M(R)$ is a right ℓ-module over $D({}_RR)$.
(c) If $D_M(R)^+ = d_M(R)$ and $D({}_RR)^+ = d({}_RR)$, then $D_M(R)$ is a strong d-module over $D({}_RR)$.

8. (a) Let $\varphi : R \longrightarrow S$ be a po-ring homomorphism and let M_S be a po-module. Then φ induces a po-R-module structure on M given by $xr = x\varphi(r)$.
(b) Let M_R be an ℓ-module and let $\bar{f}(M) = \{a \in R : x \wedge y = 0 \Longrightarrow xa \wedge y = 0,\ \forall x, y \in M\}$. Use (a) to show that $\bar{f}(M)$ is a convex subsemiring of R and that the subgroup $\bar{F}(M)$ of $(R,+)$ generated by $\bar{f}(M)$ is a convex subring of R. Also, verify that $\bar{F}(M)^+ = f(M)$ (and hence $\bar{F}(M) = F(M)$) if and only if $\bar{F}(M)$ is directed.
(c) If M is a strong ℓ-module show that $a \in \bar{f}(M)$ if and only if $a^+ \in f(M)$ and $a^- \in r(M)$; hence $\bar{F}(M) = F(M)$.

9. Show that each of the following conditions implies that the po-ring S is directed.

(a) S is an sp-ring and for each $x \in S$ there is an $e \in S$ such that $x \leq ex + xe$ or $x \leq -(ex + xe)$.
(b) S is an sp-subring of an ℓ-ring and for each $x \in S$ there is an $e \in S$ with $x \leq |ex + xe|$.
(c) S is a unital sp-ring which is 2-semiclosed (see Exercise 2.1.16).
(d) S is a po-subring of an almost f-ring, and for each $x \in S$ there is an $e \in S$ with $|x| \leq |x|e$ or $|x| \leq e|x|$.

10. Let u be a unit in the po-ring R. Show that the following statements are equivalent.

(a) $u \in \mathscr{U}(R^+)$.
(b) $u \in R^+$ and multiplication by u is a po-group automorphism of each unital po-module over R.
(c) $u \in R^+$ and left multiplication by u is an automorphism of the right po-module R_R.
(d) $1 \in R^+$ and $uR^+ = R^+$.

If R is an ℓ-ring the following statements may be added to this list.

(e) There is a unital ℓ-module M_R such that u is a d-element on M, and M contains an element x which is a d-element on R and $r(x) = 0$.
(f) u is a d-element on each unital ℓ-module.

11. Let R be a unital po-ring with the property that for each $a \in R$ there is an $n \in \mathbb{N}$ with $a^n \in R^+$. Show that $\mathscr{U}(R^+)$ is cofinal in R^+ if R satisfies any one of the following conditions.

(a) $\mathscr{U}(R)$ is cofinal in R.
(b) R has *bounded inversion*; that is, if $a \geq 1$ then $a \in \mathscr{U}(R)$. (The rings $C(X)$, where X is a topological space, and $D(X)$, where X is extremely disconnected, both have bounded inversion.)

(c) R is *local*; that is, R/J is a division ring where J is the Jacobson radical of R, or, equivalently, the set of nonunits is an ideal.

12. Let D be an ℓ-unital ℓ-ring.
    (a) $\mathscr{U}(D) \cap F(D) = \mathscr{U}(F(D))$.
    (b) If R is a local sp-subring with the same identity as D, then $R \subseteq F(D)$ (use Exercises 9, 10, and 11).
    (c) If D is local, then $F(D)$ is the largest local sp-subring of D and also the largest directed local subring in which some power of each unit is positive.
    (d) If D is an ℓ-domain, then $F(D)$ is a totally ordered domain even when D is not unital.
    (e) If D is a division ring, then $F(D)$ is a totally ordered subdivision ring which contains each totally ordered subdivision ring of D. Also,
    $$F(D) = \{d \in D : (1+|d|)^{-1} \in D^+\}.$$
    (f) Give an example of po-rings R and D as in (b) except that the identity of R is not the identity of D and R is not contained in $F(D)$.

13. Let F be the ℓ-field given by $F = \mathbb{Q}(\sqrt{2})$ and $F^+ = \mathbb{Q}^+(1+\sqrt{2}) + \mathbb{Q}^+(2+\sqrt{2})$. Then $1 \notin F^+$. If $R = \mathbb{Z}[\sqrt{2}] = \mathbb{Z} + \mathbb{Z}\sqrt{2}$ is the ring of integers in F, then R is an ℓ-subring of F and $\mathscr{U}(R)^+$ is cofinal in R.

14. Let T be an ℓ-ring and let R be one of the following canonically ordered ℓ-rings (that is, each has the obvious canonical basis).

$T_{|I|} =$ the ring of $I \times I$ column finite matrices over T (the elements of $T_{|I|}$ are matrices $A = (a_{ij})_{(i,j) \in I \times I}$ such that each column of A has only a finite number of nonzero entries);

$u\Delta(T_{|I|}) =$ the subring of $T_{|I|}$ of upper triangular matrices (here, the set I comes with a total order);

$T[x;\sigma] =$ the skew polynomial ring with coefficients on the left; here $\sigma : T \longrightarrow T$ is an ℓ-ring endomorphism and the multiplication is given by

$$\left(\sum_i a_i x^i\right)\left(\sum_j b_j x^j\right) = \sum_k \left(\sum_{i+j=k} a_i \sigma^i(b_j)\right) x^k;$$

$T[[x;\sigma]] =$ the skew formal power series ring with multiplication given above.

In each case determine $f({}_RR)$, $F({}_RR)$, $f(R_R)$, $F(R_R)$, $F(R)$, $d({}_RR)$, $D({}_RR)$, $d(R_R)$, $D(R_R)$, and $D(R) = D({}_RR) \cap D(R_R)$. For the matrix rings show that $f({}_RR) = f({}_RT^{(I)})$, etc. Some answers are given below.

(a) If $R = T_n$, then

$$d(_RR) = \left\{ (a_{ij}) \in d(_TT)_n : \text{ for each } i \text{ and } j, a_{ij}T \subseteq \left(\sum_{k \neq j} a_{ik}T \right)^{\perp} \right\},$$

$D(_RR) = D(_TT)_n$, and $F(_RR)$ consists of all those matrices with diagonal entries from $F(_TT)$ and off diagonal entries from $\ell_\ell(T)$.

(b) If $R = T[x]$, then $D(_RR) = D(_TT)[x]$ and $F(_RR) = F(_TT) + x\ell_\ell(T)[x]$.

(c) If $R = T[[x]]$, then $F(_RR) = F(_TT) + x\ell_\ell(T)[[x]]$; and if T is finitely rooted then $D(_RR) = \ell_\ell(T)[[x]] + D(_TT)[x]$.

(d) If $R = T[x;\sigma]$ then $D(R_R) = S[x;\sigma]$ where S is a σ-invariant convex ℓ-subring of $D(T_T)$.

15. (a) Show that for each module M_R the annihilator maps $r(\ ) : \mathscr{P}(M) \longrightarrow \mathscr{P}(R)$ and $\ell(\ ) : \mathscr{P}(R) \longrightarrow \mathscr{P}(M)$ constitute the Galois connection between the power sets of M and R that is determined by the relation $xa = 0$ if $(x,a) \in M \times R$ (see Exercise 1.4.24).

(b) Similarly, if M is an ℓ-module over the ℓ-ring R, then $r_\ell(\ )$ and $\ell_\ell(\ )$ constitute the Galois connection that is determined by the relation $|x||a| = 0$ if $(x,a) \in M \times R$.

16. Let M be an ℓ-module over the ℓ-ring R, and let $X \subseteq M$.

(a) $r_\ell(X) = r_\ell(X^p) \cap r_\ell(X^-) = r_\ell(|X|)$ is a right ℓ-ideal of R contained in $r(X)$, where $X^p = \{x^+ : x \in X\}$, $X^- = \{x^- : x \in X\}$ and $|X| = \{|x| : x \in X\}$.

(b) $r_\ell(X)$ is the largest directed subgroup (right ideal) of R that is contained in $r(X)$ iff $r(X)^+ \subseteq r(|X|)$, iff $r_\ell(X) = r(X)^+ - r(X)^+$.

(c) Each of the following conditions implies that $r_\ell(X)$ is the largest convex ℓ-subgroup (right ℓ-ideal) of R that is contained in the convex right ideal $r(X)$.

  (i) X is contained in the subgroup of M that is generated by X^+.

  (ii) $r(X)^+ \subseteq d(M)$.

  (iii) If $x \in X$, then $X \cap \{x^+, x^-, |x|\} \neq \phi$.

(d) Formulate analogous statements for $\ell(A;M)$ and $\ell_\ell(A;M)$ if $A \subseteq R$.

17. Suppose that M is a d-module over the directed ring R and $X \subseteq M$.

(a) $r(X) = r(X^p) \cap r(X^-) = r(|X|)$ (see Exercise 16).

(b) If R is an ℓ-ring and $|X| \subseteq d_M(R)$ (Exercise 7), then $r(X) = r_\ell(X)$.

(c) Suppose that R is an ℓ-ring and $x \in M^+$. Then $x \in d_M(R)$ and xR is totally ordered if and only if $r(x)$ is a prime subgroup of R. (Here, M is just an ℓ-module over R.)

(d) If $A \subseteq R$, show that $\ell(A)$ is an ℓ-subgroup of M.

18. (a) If the ℓ-module M is a subdirect product of the family of ℓ-modules $\{M_i\}_{i \in I}$, then $f(M) = \cap f(M_i)$.

(b) If the ℓ-ring R is a subdirect product of the family of ℓ-rings $\{R_i\}_{i\in I}$, then $F(R_R) \subseteq \Pi F(R_{iR_i})$ and equality holds if R is the full product.

19. If R is a right d-ring with a left identity e, then $|e|$ is a left identity. If R is a right f-ring then $e^\perp = 0$.

20. Let M be a po-module over the po-ring R, and let e be an idempotent of R with $M^+e \subseteq M^+$. If M is an ℓ-module, let $\bar{d}(M) = \{a \in R : a$ induces an ℓ-endomorphism of $M\}$.

    (a) Show that $eRe \cap R^+ = eR^+e$ if $eR^+ + R^+e \subseteq R^+$.
    (b) If $K = \ell(e;M)$, then K is a convex subgroup of M and $M = Me \boxplus K$ as groups.
    (c) If M_R is an ℓ-module and $e \in \bar{d}(M)$, then Me and K are ℓ-subgroups of M. Moreover, M_{eRe} is a d-module if and only if $(Me)_{eRe}$ is a d-module.
    (d) Suppose that R is unital and M is an ℓ-module and $1 \in \bar{d}(M)$. Then M is a d-module if and only if $M1$ is a d-module.
    (e) If M is an ℓ-module, and $M =$ lex K, then M_{eRe} is an f-module.
    (f) If M is totally ordered, then $M = K \underset{\leftarrow}{\times} Me$.

21. Let M be an n-dimensional vector lattice over $\mathbb{R}$.

    (a) M is the direct sum of a finite number of indecomposable vector lattices.
    (b) If M is indecomposable, then $M \cong N \underset{\leftarrow}{\times} \mathbb{R}$ for an $(n-1)$-dimensional vector lattice N (use Theorems 2.5.8 and 2.5.11 and Exercise 2.5.20).
    (c) If M is totally ordered, then $M \cong \mathbb{R}\underset{\longleftarrow}{\times\cdots\times}\mathbb{R}$.
    (d) If M is archimedean, then $M \cong \mathbb{R}^n$.
    (e) Suppose M is a subspace (not a po-subspace) of a Euclidean space $\mathbb{R}^m$. Show that M is totally ordered and M^+ is closed in $\mathbb{R}^m$ iff $n = 1$.

22. Let $E = \mathrm{End}_{\mathbb{R}}(M)$ be the po-algebra of $\mathbb{R}$-linear transformations of the finite dimensional real vector lattice M; so $E^+ = \{f \in E : f(M^+) \subseteq M^+\}$. Show that the following statements are equivalent.

    (a) M is archimedean.
    (b) E is an ℓ-ring (ℓ-algebra).
    (c) E is directed.

23. This exercise determines all two-dimensional algebras over $\mathbb{R}$ (there are eight), and the next exercise determines all such ℓ-algebras (up to an isomorphism). Let R be a two-dimensional real algebra and let N be the largest nilpotent ideal of R. N is an algebra ideal.

    (a) If $N = R$, then either $R^2 = 0$ (equivalently, R satisfies $x^2 = 0$), or $R^3 = 0$ but $R^2 \neq 0$ (equivalently, R has a basis $\{a, a^2\}$ with $a^3 = 0$).
    (b) If $N = 0$, then $R \cong \mathbb{C}$ (complex numbers) or $R \cong \mathbb{R} \times \mathbb{R}$.
    (c) If $0 \subset N \subset R$, then $N^2 = 0$ and R has a nonzero idempotent e. (If $a^2 - a \in N$, then $a^2 = a^3(2-a)$, $a^2(2-a)^2 = e$ is idempotent and $e - a \in N$.)

(i) If $e = 1$, then $R \cong \mathbb{R}_0 \dot{\times} \mathbb{R}$ where $\mathbb{R}_0$ is the additive group of the reals with $\mathbb{R}_0^2 = 0$ and $\mathbb{R}_0 \dot{\times} \mathbb{R}$ is the algebra obtained by freely adjoining $\mathbb{R}$ to $\mathbb{R}_0$ (Exercise 2).

(ii) If $e \neq 1$, then R is isomorphic to one of the algebras $\mathbb{R} \times \mathbb{R}_0$, $\begin{pmatrix} \mathbb{R} & \mathbb{R} \\ 0 & 0 \end{pmatrix}$ or $\begin{pmatrix} \mathbb{R} & 0 \\ \mathbb{R} & 0 \end{pmatrix}$.

24. Let R be a two-dimensional real ℓ-algebra. As a vector space assume that $R = \mathbb{R}^2$ with basis e_1, e_2, where $e_1 = (1,0)$ and $e_2 = (0,1)$. As a vector lattice, by Exercise 21, $R = \mathbb{R}u \oplus \mathbb{R}v$ (the archimedean case) or $R = \mathbb{R}u \underset{\rightarrow}{\times} \mathbb{R}v$ (the totally ordered case) for some vectors u and v in R. Thus, for some angles α and β the positive cone can be described in one of the following ways.

If R is archimedean: $R^+ = \{w \in R : \alpha \leq \arg w \leq \beta$, where $0 < \beta - \alpha < \pi\}$;
if R is non-archimedean: $R^+ = \{w \in R : \alpha < \arg w \leq \alpha + \pi\}$ or $R^+ = \{w \in R : \alpha \leq \arg w < \alpha + \pi\}$.

Use the description of the algebras given in Exercise 23, with e_1 and e_2 as the standard basis in each case, to show that the following additional conditions must be satisfied by the positive cone. In the cases (c), (d), and (e) the algebra isomorphism $e_1 \mapsto e_1, e_2 \mapsto -e_2$ may be used to reduce to the given conditions since the ℓ-algebras that are obtained otherwise are isomorphic to those described here.

(a) $R^3 = 0, R^2 \neq 0 : e_1^2 = e_2$, $e_1 e_2 = e_2 e_1 = e_2^2 = 0$, $e_2 \in R^+$ and $\mathbb{R}e_2$ is an ℓ-ideal.
archimedean: $\alpha = \pi/2$ or $\beta = \pi/2$.
non-archimedean: $\alpha = -\pi/2$ and $-\pi/2 < \arg w \leq \pi/2$.

(b) $R = \mathbb{C} : R$ is not an ℓ-algebra.
$R = \mathbb{R} \times \mathbb{R} : e_i e_j = \delta_{ij} e_i$.
archimedean: either $\alpha = 0$ and $\beta = \pi/2$, or $-\pi/4 \leq \alpha \leq 0$ and $\tan^{-1}(\tan^2 \alpha) \leq \beta \leq \pi/4$.
non-archimedean: there are no non-archimedean ℓ-orderings.

(c) $R = \mathbb{R} \dot{\times} \mathbb{R}_0 : e_1 = 1$, $e_2^2 = 0$, $\mathbb{R}e_2$ is an ℓ-ideal.
archimedean: $0 \leq \alpha < \beta \leq \pi/2$.
non-archimedean: same as (a).

(d) $R = \mathbb{R} \times \mathbb{R}_0 : e_i e_j = \delta_{i1} \delta_{j1} e_1$.
archimedean: $e_1 \in R^+$; $-\pi/2 \leq \alpha$ and $\beta \leq \pi/2$.
non-archimedean: same as (a).

(e) $R = \begin{pmatrix} \mathbb{R} & \mathbb{R} \\ 0 & 0 \end{pmatrix} : e_1^2 = e_1$, $e_1 e_2 = e_2$, $e_2^2 = e_2 e_1 = 0$.
archimedean: $-\pi/2 \leq \alpha < \beta \leq \pi/2$.
non-archimedean: same as (a).

These conditions also apply to the opposite algebra of R (the "transpose" of R).

25. Let M_R be a unital po-module with a canonical basis. If R is a right f-ring show that $\mathrm{End}_R(M)$ is an ℓ-subring of $\mathrm{End}_{\mathbb{Z}}(M)$ (that is, $\mathrm{End}_R(M)$ is a sublattice of the po-group $\mathrm{End}_{\mathbb{Z}}(M)$).

26. Let R be an ℓ-ring, T a po-ring, ${}_TM_R$ an ℓ-bimodule, and let $\varphi : R \longrightarrow \mathrm{End}_T(M) = H$ be the corresponding representation. M is called a *(right) $\ell\ell$-module* if $R\varphi$ is a lattice in the partial order it inherits from H and φ is an ℓ-homomorphism (this condition is independent of T). If, additionally, $R\varphi$ is a sublattice of H, then M is called an *$\ell\ell$-T-module*.

    (a) The following statements are equivalent.
        (i) ${}_TM_R$ is an $\ell\ell$-module.
        (ii) If $a \wedge b = 0$ in R, then $a\varphi \wedge b\varphi = 0$ in the po-subring $R\varphi$ of $\mathrm{End}_T(M)$.
        (iii) $\{c \in R : \forall x \in M,\ \forall a \in R,\ x^+c \leq x^+a^+ \wedge x^+a^-\} \subseteq \{c \in R : M^+c \subseteq -M^+\}$.
        (iv) $\{a \in R : M^+a \subseteq M^+\} = \{a \in R : a^- \in r(M)\}$.
        (v) $r(M)$ is an ℓ-ideal of R and $\{a \in R : M^+a \subseteq M^+\} \subseteq \{a \in R : a^- \in r(M)\}$.
    (b) Give an example of a totally ordered faithful ℓ-bimodule ${}_TM_R$ which is not an $\ell\ell$-module.
    (c) If $M^+ = T^+d_M(R) + D_M(R)^+$ (see Exercise 7) and N is a convex ℓ-submodule of ${}_TM_R$, then M/N is an $\ell\ell$-T-module.
    (d) Conversely, if $1 \in T^+$ and M is an $\ell\ell$-T-module with a canonical T-basis $\{e_i\}_{i\in I}$, then $\{e_i\}_{i\in I} \subseteq d_M(R)$. (*Hint*: If $a \wedge b = 0$ in R define $f \in H$ by $e_i f = e_i a \wedge e_i b$.)
    (e) If $M = D_M(R)$ and N is a convex ℓ-submodule of ${}_TM_R$, then N and M/N are $\ell\ell$-$\mathbb{Z}$-modules (and hence $\ell\ell$-T-modules).

27. Let F be a commutative unital totally ordered domain and let $\delta \in F$ with $0 < \delta < 1$. Let R be the following ℓ-algebra over F : $R = F \times F$ as an algebra and it has the canonical basis $1, u = (\delta+1, -1)$ over F (see Exercise 24(b)). Let C be a commutative totally ordered domain which contains F.

    (a) R is an ℓ-domain, and each ℓ-ideal of R is of the form $J1 + Ju$ for some ℓ-ideal J of F.
    (b) If ${}_CM_R$ is a unital ℓ-bimodule with $\alpha x = x\alpha$ for $x \in M$ and $\alpha \in F$ and ${}_CM$ has a canonical basis, then M is not a right $\ell\ell$-C-module. (*Hint*: It suffices to show that there is no row-finite matrix $A = (c_{ij})$ (of any size) with entries in C^+ such that $A^2 = \delta A + (1+\delta)I$ and $c_{ii} = 0$ for each i. If A is such a matrix, then (i) $\Sigma_k c_{ik}c_{kj} = \delta c_{ij}$ if $i \neq j$, and (ii) $\Sigma_k c_{ik}c_{ki} = 1+\delta$ for each i. (I) Use (i) to show that $c_{ij}c_{ji} \leq \delta^2$ and hence $(c_{ij}c_{ji})^2 \leq c_{ij}c_{ji}$ for any i and j; for, $c_{ij}c_{jk}c_{ji}c_{ik} \leq \delta^2 c_{ik}c_{jk}$, and if $0 < c_{ij}c_{ji}$, then there is a $k \neq i, j$ with $c_{ik}c_{kj} > 0$ and hence $c_{jk} > 0$. (II) Now, if $i \neq k$ then $\delta^2 c_{ik}c_{ki} = (\Sigma_j c_{ij}c_{jk})(\Sigma_m c_{km}c_{mi}) \geq \Sigma_j (c_{ij}c_{ji})(c_{jk}c_{kj})$; sum over all $k \neq i$ to get a contradiction.)

28. An ℓ-T-ring R is called (*right*) ℓ-*regular* if R_R is an $\ell\ell$-module (see Exercise 26); additionally, R is called T-*regular* if ${}_TR_R$ is an $\ell\ell$-T-module.

   (a) If $1 \in R$, then R is ℓ-regular iff $1 \in R^+$.
   (b) If $R^+ = T^+d({}_RR) + D({}_RR)^+$, then R is T-regular and so is R/A where A is an ℓ-ideal that is a left T-submodule.
   (c) If $R = D({}_RR)$ and A is a convex ℓ-subring of R, then A is $\mathbb{Z}$-regular.
   (d) If $1 \in T^+$ and ${}_TR$ has a canonical basis $\{e_i\}$, then R is T-regular iff $\{e_i\} \subseteq d({}_RR)$.
   (e) Let T be an ℓ-ring for which $T^+ = T^+d({}_TT) + D({}_TT)^+$, and let R be one of the canonically ordered ℓ-rings $T[x;\sigma]$, T_n or the semigroup ring $T[G]$ where G is a left cancellative semigroup (see Exercise 14). Then R is T-regular.
   (f) The direct sum of a family of T-regular ℓ-rings is T-regular.
   (g) Let $R = F[G]$ be the canonically ordered group algebra of the group G over the totally ordered field F, and let Z be the center of R. Let C_x denote the conjugacy class of $x \in G$, $\triangle = \{x \in G : \text{card}(C_x) < \infty\}$, and e_x the sum of the elements in C_x, for $x \in \triangle$. Then the distinct e_x form a canonical basis for Z. Moreover, $e_xe_y = \Sigma_z n_{xyz}e_z$ where $n_{xyz} = |\{(x',y') \in C_x \times C_y : x'y' = z\}|$. The following statements are equivalent for the finite group G.
      (i) Z is F-regular.
      (ii) $\forall x,y \in G,\ e_xe_y \in \mathbb{N}e_{xy}$.
      (iii) G is abelian.
   Thus, an ℓ-subalgebra of an F-regular ℓ-algebra need not be F-regular.

29. Show that the following statements are equivalent for the f-module M_R over the ℓ-ring R.

   (a) M is strong.
   (b) If N is a prime submodule of M, then $r(M/N)$ is an irreducible ℓ-ideal of R.
   (c) If N is a prime submodule of M, then $r_\ell(M/N)$ is an irreducible ℓ-ideal. (Here, of course, $r(M/N) = r_\ell(M/N)$.)

30. Let M be an abelian ℓ-group, $E = \text{End}_{\mathbb{Z}}(M)$ and $F = F({}_EM)$. Show that the following statements are equivalent. Moreover, if these conditions hold, then F is an ℓ-subring of E.

   (a) F is an ℓ-ring and M is a strong f-module over F.
   (b) If $\alpha,\beta \in F^+$ and $a,b \in M^+$, then $\alpha(a+b) \vee \beta(a+b) = (\alpha a \vee \beta a) + (\alpha b \vee \beta b)$.
   (c) F is an f-ring and M is a strong f-module over F.

31. (a) Let S be the C-subalgebra of the C-algebra R that is generated by the Boolean algebra B of central idempotents of R. Show that each element of S is a C-linear combination of a finite orthogonal subset of B. (Reduce to the case that B is finite and use Exercise 1.2.6.)
   (b) If each element of B is an f-element of the ℓ-ring R, show that S is an ℓ-subring of R.

32. Let F be a nonzero subring of $\mathbb{R}$ and let G and H be F-semiclosed po-modules over F with G directed and H archimedean. So, if $0 < r \in F$ and $x \in G \cup H$ with $rx \geq 0$, then $x \geq 0$. If $\alpha : G \longrightarrow H$ is an isotone group homomorphism show that α is an F-homomorphism. (Show that it may be assumed that F contains the rationals. For $0 < r \in F$ take $p, q \in \mathbb{Q}$ with $0 < p \leq r \leq q$.)

Notes. The algebraic study of ℓ-rings as well as of the subvarieties of f-rings, d-rings and sp-ℓ-rings was initiated by Birkhoff and Pierce [BP]. Two other fundamental papers on f-rings are those by Johnson [JOH1] and Henriksen and Isbell [HI]. The study of f-modules has its origins in Bigard [BI1] and Steinberg [ST1] though vector lattices and totally ordered modules appeared earlier. Most of the results in this section either appear in or are similar to those that do appear in these papers. Exercise 21(e) comes from Ma and Wojciechowski [MW3], Exercise 30 comes from Bigard and Keimel [BK], and Exercise 32 comes from Conrad and Diem [CDI].

3.2 Radical Theory

In this section we present the general theory of radicals for ℓ-rings. Simultaneously, by dropping all reference to the partial order, the basic ingredients of the theory of radicals for rings are also obtained. Specific radicals are also considered; in particular, the nil radicals and Johnson's radical, the ℓ-ring analogue of the Jacobson radical, are investigated. Our main concern with radicals is in their application to the structure of ℓ-rings. However, some aspects of the theory will be presented for their inherent interest.

Recall from Section 1.4 that a class $\mathscr{C}$ of ℓ-rings is homomorphically closed if each homomorphic image of an ℓ-ring in $\mathscr{C}$ is also in $\mathscr{C}$, and $\mathscr{C}$ is productive if it is closed under the formation of direct products. Also, $\mathscr{C}$ is hereditary if each ℓ-subring of each ℓ-ring in $\mathscr{C}$ is also in $\mathscr{C}$. The class $\mathscr{C}$ will be called *i-hereditary* (respectively, *left i-hereditary* or *right i-hereditary*) provided that each ℓ-ideal (respectively, left ℓ-ideal or right ℓ-ideal) of each member of $\mathscr{C}$ is also in $\mathscr{C}$. It should be noted that the general theory, as it is developed here, is valid in any i-hereditary homomorphically closed class of ℓ-rings $\mathscr{C}$. However, since all of the radicals that are considered can be constructed in the class of all ℓ-rings we will assume that $\mathscr{C}$ is the class of all ℓ-rings. Each subclass of $\mathscr{C}$ that is considered will always be a nonempty subclass.

The class $\mathscr{P}$ is called *extensionally closed* if whenever A is an ℓ-ideal of R with A and R/A in $\mathscr{P}$, then $R \in \mathscr{P}$. An ℓ-ring in $\mathscr{P}$ will be called a *$\mathscr{P}$-ℓ-ring*. If A is a $\mathscr{P}$-ℓ-ring which is an ℓ-ideal of R it will be called a *$\mathscr{P}$-ℓ-ideal of R*. An ℓ-ring is called *$\mathscr{P}$-semisimple* if it has no nonzero $\mathscr{P}$-ℓ-ideals.

A class $\mathscr{P}$ is called a *radical class* or just a *radical* if it satisfies the following two conditions.

$$\mathscr{P} \text{ is homomorphically closed.} \tag{3.2.1}$$

$$\text{If } R \notin \mathscr{P} \text{ then } R \text{ has a nonzero } \mathscr{P}\text{-semisimple homomorphic image.} \tag{3.2.2}$$

Some useful alternative descriptions of a radical class are given in

Theorem 3.2.1. *The following statements are equivalent for the class of ℓ-rings $\mathscr{P}$.*

(a) $\mathscr{P}$ is a radical class.
(b) $\mathscr{P}$ is homomorphically closed, and each ℓ-ring R has a $\mathscr{P}$-ℓ-ideal P such that R/P is $\mathscr{P}$-semisimple.
(c) $R \in \mathscr{P}$ iff each of its nonzero homomorphic images has a nonzero $\mathscr{P}$-ℓ-ideal.
(d) $\mathscr{P}$ is homomorphically closed, extensionally closed, and the union of each chain of $\mathscr{P}$-ℓ-ideals in any ℓ-ring is a $\mathscr{P}$-ℓ-ideal.

Proof. (a) $\Rightarrow$ (b). Let $\{A_i\}_{i\in I}$ be the set of all of the $\mathscr{P}$-ℓ-ideals of the ℓ-ring R, and let $P = \Sigma_i A_i$. If $P \notin \mathscr{P}$, then P has a nonzero homomorphic image $\varphi(P)$ which is $\mathscr{P}$-semisimple. But then $\varphi(A_i) = 0$ for each $i \in I$, and hence $\varphi(P) = 0$. Thus, $P \in \mathscr{P}$. Let A/P be a $\mathscr{P}$-ℓ-ideal of R/P. If $A \notin \mathscr{P}$, then A has a proper ℓ-ideal K such that A/K is $\mathscr{P}$-semisimple. Since, for each $i \in I$, $A_i + K/K$ is a $\mathscr{P}$-ℓ-ideal of A/K, $A_i \subseteq K$ and therefore $P \subseteq K$. Thus, $A/K \cong (A/P)/(K/P) \in \mathscr{P}$, and $A = K$, which is contrary to K being a proper ℓ-ideal of A. So $A \in \mathscr{P}$, $A = P$, and R/P is $\mathscr{P}$-semisimple.

(b) $\Rightarrow$ (c). This is obvious.

(c) $\Rightarrow$ (d). Since a homomorphic image of a homomorphic image of R is a homomorphic image of R it is clear that $\mathscr{P}$ is homomorphically closed. Suppose that A is an ℓ-ideal of R and A and R/A are both in $\mathscr{P}$. If $0 \neq R/K$ is $\mathscr{P}$-semisimple, then $A \subseteq K$ and hence $R/K \cong (R/A)/(K/A) \in \mathscr{P}$. Thus R has no nonzero homomorphic image which is $\mathscr{P}$-semisimple, and hence $R \in \mathscr{P}$. This shows that $\mathscr{P}$ is closed under extensions. Let $\{A_i\}_{i\in I}$ be a chain of $\mathscr{P}$-ℓ-ideals of the ℓ-ring R, and put $A = \cup A_i$. The argument in the first paragraph of this proof shows that $A = \Sigma A_i$ has no nonzero $\mathscr{P}$-semisimple homomorphic images; so A is a $\mathscr{P}$-ℓ-ideal of R.

(d) $\Rightarrow$ (a). Suppose that $R \notin \mathscr{P}$. By Zorn's Lemma R has a maximal $\mathscr{P}$-ℓ-ideal P. If B/P is a $\mathscr{P}$-ℓ-ideal of R/P, then $B \in \mathscr{P}$ since $\mathscr{P}$ is extensionally closed, and hence $B = P$ by maximality of P. So R/P is a nonzero $\mathscr{P}$-semisimple homomorphic image of R, and $\mathscr{P}$ is a radical. □

If $\mathscr{P}$ is a radical, then the ℓ-ideal of the ℓ-ring R given in (b) of the preceding theorem is called the $\mathscr{P}$-*radical* of R and will be denoted by $\mathscr{P}(R)$. The ℓ-ideal $\mathscr{P}(R)$ is characterized among the ℓ-ideals of R by any one of the following conditions.

$$\mathscr{P}(R) \text{ is the largest } \mathscr{P}\text{-}\ell\text{-ideal of } R. \tag{3.2.3}$$

$$\mathscr{P}(R) \in \mathscr{P} \text{ and } R/\mathscr{P}(R) \text{ is } \mathscr{P}\text{-semisimple.} \tag{3.2.4}$$

$$\mathscr{P}(R) \text{ is the smallest } \ell\text{-ideal of } R \text{ such that } R/\mathscr{P}(R) \text{ is } \mathscr{P}\text{-semisimple.} \quad (3.2.5)$$

Instead of starting with a radical class we could equally well start with the associated semisimple class. Before we make this statement more precise we will consider both classes simultaneously. A *radical pair* is a pair of classes $(\mathscr{P}, \mathscr{M})$ such that

$$R \in \mathscr{P} \text{ iff } R \text{ has no nonzero homomorphic image in } \mathscr{M}. \quad (3.2.6)$$

$$R \in \mathscr{M} \text{ iff } R \text{ has no nonzero } \mathscr{P}\text{-}\ell\text{-ideal.} \quad (3.2.7)$$

$\mathscr{P}$ is called the *lower class* and $\mathscr{M}$ is the *upper class* of the radical pair $(\mathscr{P}, \mathscr{M})$. The following theorem is an immediate consequence of the definitions and Theorem 3.2.1.

Theorem 3.2.2. *$(\mathscr{P}, \mathscr{M})$ is a radical pair if and only if $\mathscr{P}$ is a radical class and $\mathscr{M}$ is the associated class of $\mathscr{P}$-semisimple ℓ-rings. Moreover, if $\mathscr{M}$ is the class of semisimple ℓ-rings for each of the radical classes $\mathscr{P}_1$ and $\mathscr{P}_2$, then $\mathscr{P}_1 = \mathscr{P}_2$.* □

Now, a *semisimple class* is a class $\mathscr{M}$ that satisfies the following two conditions.

$$\text{If } R \in \mathscr{M}, \text{ then each nonzero } \ell\text{-ideal of } R \text{ has a nonzero homomorphic image in } \mathscr{M}. \quad (3.2.8)$$

$$\text{If } R \notin \mathscr{M}, \text{ then } R \text{ has a nonzero } \ell\text{-ideal which has no nonzero homomorphic image in } \mathscr{M}. \quad (3.2.9)$$

Equivalently, and dual to (c) of Theorem 3.2.1, a semisimple class is easily seen to be characterized by the following condition.

$$R \in \mathscr{M} \text{ iff each nonzero } \ell\text{-ideal of } R \text{ has a nonzero homomorphic image in } \mathscr{M}. \quad (3.2.10)$$

For any class $\mathscr{P}$ we will denote the class of $\mathscr{P}$-semisimple ℓ-rings by $\mathscr{S}\mathscr{P}$. We have

Theorem 3.2.3. *(a) If $\mathscr{P}$ is a radical, then $\mathscr{S}\mathscr{P}$ is a semisimple class.*
(b) If $\mathscr{M}$ is a semisimple class, then the class $\mathscr{R}\mathscr{M}$ of all those ℓ-rings which have no nonzero homomorphic images in $\mathscr{M}$ is a radical class, and $\mathscr{S}\mathscr{R}\mathscr{M} = \mathscr{M}$.

Proof. From the definition of a radical class and from (3.2.10) it is immediate that $\mathscr{S}\mathscr{P}$ is a semisimple class if $\mathscr{P}$ is a radical class. Let $\mathscr{M}$ be a semisimple class and put $\mathscr{P}=\mathscr{R}\mathscr{M}$. Then $(\mathscr{P}, \mathscr{M})$ is a radical pair. For, certainly (3.2.6) holds, and (3.2.7) is just a rewording of (3.2.10). So (b) is a consequence of Theorem 3.2.2. □

In each of the next two results we will construct a radical from a given class of ℓ-rings. The first construction will initially produce a semisimple class whereas the second will produce a radical class.

Theorem 3.2.4. *(The upper radical construction) Let $\mathcal{N}$ be a class of ℓ-rings which satisfies (3.2.8), and let $\mathcal{M}$ be the class defined by*

$$\mathcal{M} = \{R: \text{ each nonzero } \ell\text{-ideal of } R \text{ has a nonzero homomorphic image in } \mathcal{N}\}.$$

(a) $\mathcal{M}$ is the smallest semisimple class that contains $\mathcal{N}$.
(b) The radical class determined by $\mathcal{M}$ is given by $\mathcal{U}\mathcal{N} = \{R : R$ has no nonzero homomorphic image in $\mathcal{N}\}$.
(c) $\mathcal{U}\mathcal{N}$ is the largest radical class for which each ℓ-ring in $\mathcal{N}$ is semisimple.

Proof. (a) Clearly $\mathcal{N} \subseteq \mathcal{M}$, and therefore $\mathcal{M}$ satisfies (3.2.8). If $R \notin \mathcal{M}$, then R has a nonzero ℓ-ideal A which has no nonzero homomorphic image in $\mathcal{N}$. If φA is a nonzero homomorphic image in $\mathcal{M}$, then φA, and hence A also, has a nonzero homomorphic image in $\mathcal{N}$. Thus, φA doesn't exist and $\mathcal{M}$ satisfies (3.2.9). If $\mathcal{M}_1$ is any semisimple class containing $\mathcal{N}$, then $\mathcal{M} \subseteq \mathcal{M}_1$ since $\mathcal{M}_1$ satisfies (3.2.10).

(b) It suffices to note that if R has no nonzero homomorphic image in $\mathcal{N}$, then it has no nonzero homomorphic image in $\mathcal{M}$.

(c) Suppose that $\mathcal{P}$ is a radical class and $\mathcal{N} \subseteq \mathcal{S}\mathcal{P}$. Then $\mathcal{M} \subseteq \mathcal{S}\mathcal{P}$ by (a), and hence $\mathcal{P} \subseteq \mathcal{U}\mathcal{N}$. □

The radical $\mathcal{U}\mathcal{N}$ constructed in the previous theorem is called the *upper radical* determined by $\mathcal{N}$.

For the second construction of a radical let $\mathcal{A}$ be any class of ℓ-rings and let $\mathcal{P}_0$ be the homomorphic closure of $\mathcal{A}$. So $R \in \mathcal{P}_0$ if and only if R is a homomorphic image of an ℓ-ring in $\mathcal{A}$. Suppose that μ is an ordinal and that the class $\mathcal{P}_\nu$ has been defined for each ordinal $\nu < \mu$. Then $\mathcal{P}_\mu$ is defined by

$$\begin{aligned}\mathcal{P}_\mu = \{R : &\text{ if } \varphi(R) \text{ is a nonzero homomorphic image of } R, \text{ then} \\ &\exists\, \nu < \mu \text{ such that } \varphi(R) \text{ has a nonzero } \mathcal{P}_\nu\text{-}\ell\text{-ideal}\}.\end{aligned}$$

Clearly, each class $\mathcal{P}_\mu$ is homomorphically closed, and $\mathcal{P}_\nu \subseteq \mathcal{P}_\mu$ if $\nu \leq \mu$. Let $\mathcal{P}$ be the union of all of the classes $\mathcal{P}_\nu$.

Theorem 3.2.5. *(The lower radical construction) The class $\mathcal{P}$ constructed above is the smallest radical class containing $\mathcal{A}$.*

Proof. Suppose that $\mathcal{Q}$ is a radical class and $\mathcal{A} \subseteq \mathcal{Q}$. Then $\mathcal{P}_o \subseteq \mathcal{Q}$ by (3.2.1). Assume, by induction, that $\mu > 0$ and $\mathcal{P}_\nu \subseteq \mathcal{Q}$ for each $\nu < \mu$. Take $R \in \mathcal{P}_\mu$ and let $\varphi(R)$ be one of its nonzero homomorphic images. Then $\varphi(R)$ has a nonzero $\mathcal{P}_\nu$-ℓ-ideal for some $\nu < \mu$, and hence $\varphi(R)$ has a nonzero $\mathcal{Q}$-ℓ-ideal. So $R \in \mathcal{Q}$ by (3.2.2), $\mathcal{P}_\mu \subseteq \mathcal{Q}$, and thus $\mathcal{Q}$ contains $\mathcal{P}$. To see that $\mathcal{P}$ is a radical we will check that $\mathcal{P}$ satisfies condition (c) of Theorem 3.2.1. We have already noted that each class $\mathcal{P}_\mu$ is homomorphically closed, and hence so is $\mathcal{P}$. Conversely, suppose that each nonzero homomorphic image of the ℓ-ring R has a nonzero $\mathscr{P}$-ℓ-ideal. Then for each proper ℓ-ideal A of R there is an ordinal ν_A and an ℓ-ideal A_{ν_A} of R which properly contains A such that A_{ν_A}/A is a $\mathscr{P}_{\nu_A}$-ℓ-ideal of R/A. Let μ be an ordinal with $\nu_A < \mu$ for each A. If $\varphi(R)$ is a nonzero homomorphic image of R, then $\varphi(R) \cong R/A$ for some ℓ-ideal A. Since $\mathscr{P}_{\nu_A}$ is isomorphically closed $\varphi(R)$ has a nonzero $\mathscr{P}_{\nu_A}$-ℓ-ideal. Thus $R \in \mathscr{P}_\mu \subseteq \mathscr{P}$ and $\mathscr{P}$ is a radical. □

The radical $\mathscr{P}$ constructed in Theorem 3.2.5 is called the *lower radical* determined by $\mathscr{A}$ and will be denoted by $\mathscr{L}\mathscr{A}$.

If R is an ℓ-algebra over a totally ordered field it would be nice to know that each radical ℓ-ideal of R is a subalgebra. In order to establish a more general fact we first need a definition.

Let $H = \mathrm{End}_{\mathbb{Z}}(R)$ be the endomorphism ring of the additive group of R. The *centroid* of R is the subring C of H that consists of all those endomorphisms $\alpha \in H$ such that $\alpha(xy) = (\alpha x)y = x(\alpha y)$, for all $x, y \in R$. If E is the subring of H that is generated by the set of left and right multiplication maps: $a \mapsto xa$, $a \mapsto ya$, where $x, y \in R$, then C is the centralizer of E in H. Note that if R is a module over the commutative ring D, then R is a D-algebra if and only if the image of the corresponding representation $D \longrightarrow H$ is a subring of C.

Theorem 3.2.6. *Suppose that α is an ℓ-endomorphism in the centroid of R and αI is convex for each ℓ-ideal I of R. If $\mathscr{P}$ is any radical, then $\alpha\mathscr{P}(R) \subseteq \mathscr{P}(R)$. In particular, if R is an ℓ-algebra over a directed po-field, then $\mathscr{P}(R)$ is a subalgebra.*

Proof. By our assumptions about $\alpha J = \mathscr{P}(R) + \alpha\mathscr{P}(R)$ is an ℓ-ideal of R. Since $J^2 \subseteq \mathscr{P}(R)$ the map $\varphi : \mathscr{P}(R) \mapsto J/\mathscr{P}(R)$ given by $\varphi(x) = \alpha x + \mathscr{P}(R)$ is an epimorphism of ℓ-rings. But then J is a $\mathscr{P}$-ℓ-ideal of R since $\mathscr{P}$ is extensionally closed; so $\alpha\mathscr{P}(R) \subseteq \mathscr{P}(R)$. If R is an ℓ-algebra over the po-field F and $0 < f \in F$, then multiplication by f satisfies the conditions on α. So $f\mathscr{P}(R) \subseteq \mathscr{P}(R)$ and $\mathscr{P}(R)$ is a subalgebra if F is directed. □

The lower ℓ-nil radical

If $A_1, \ldots, A_n$ are subsets of the ℓ-ring R, then $A_1 \cdots A_n$ denotes the additive subgroup of R generated by the set of products $\{a_1 \cdots a_n : a_i \in A_i\}$. This notation agrees with that of the product XA given in the previous section. As usual, $C(A_1 \cdots A_n)$ denotes the convex ℓ-subgroup generated by $A_1 \cdots A_n$ (or by the set of products):

$$C(A_1 \cdots A_n) = \{x \in R : |x| \leq \sum_{i_1,\ldots,i_n} |a_{i_1} \cdots a_{i_n}|, \text{ where } a_{i_j} \in A_j\}. \tag{3.2.11}$$

If each A_i is a directed subgroup of R, then

$$\begin{aligned} C(A_1 \cdots A_n) &= \{x \in R : |x| \leq a_1 \cdots a_n \text{ where each } a_j \in A_j\} \\ &= C(C(A_1) \cdots C(A_n)). \end{aligned} \tag{3.2.12}$$

If, additionally, A_1 is a left ideal, then $A_1 \cdots A_n$ is a left ideal, and $C(A_1 \cdots A_n)$ is a left ℓ-ideal; similarly, $A_1 \cdots A_n$ and $C(A_1 \cdots A_n)$ are right ideals if A_n is a right ideal.

If $A_i = A$ for $i = 1, \ldots, n$, then $A_1 \cdots A_n = A^n$, and $C(A^n)$ will be denoted by $A^{[n]}$; that is, $A^{[n]}$ is the convex ℓ-subgroup generated by A^n. A is said to be *nilpotent of index n* if $A^n = 0$, or, equivalently, $A^{[n]} = 0$, but $A^k \neq 0$ for $k < n$. If A is a directed subgroup, then

$$A^{[n]} = \{x \in R : |x| \leq a^n \text{ for some } a \in A^+\}, \tag{3.2.13}$$

$$C(A^{[n]}A^{[m]}) = A^{[n+m]}, \tag{3.2.14}$$

and

$$(A^n)^{[m]} = (A^{[n]})^{[m]} = A^{[nm]}. \tag{3.2.15}$$

In particular, an ℓ-ring R is nilpotent if and only if it satisfies an identity $x^n = 0$ for some n. If a ring R is nilpotent of index at most 2 it will be called a *zero ring*. If $R = R^{[2]}$ then R is called *ℓ-idempotent*. If R is ℓ-idempotent, then $R = R^{[n]}$ for each $n \in \mathbb{N}$. For, if $R = R^{[k]}$, then $R^{[k+1]} = C(R^{[k]}R) = C(R^2) = R^{[2]}$, by (3.2.14).

Since one important use of radical theory is to get rid of a "bad" property – the $\mathscr{P}$-radical of an ℓ-ring – by factoring $\mathscr{P}(R)$ out of R so that $R/\mathscr{P}(R)$ is now a "good" ℓ-ring, it is natural to first consider the lower radical determined by the class of zero ℓ-rings. We will call this radical the *lower ℓ-nil radical* and we denote it by ℓ-β : ℓ-$\beta = \mathscr{L}$ (zero ℓ-rings). Before we examine this radical in detail we will first show that its construction stops rather quickly.

Theorem 3.2.7. *Let A be an ℓ-ideal of the ℓ-ring R and let B be an ℓ-ideal of A. If D is the ℓ-ideal of R generated by B, then $D^{[3]} \subseteq B$. Consequently, if A/B does not contain any nonzero nilpotent ℓ-ideals, then B is an ℓ-ideal of R.*

Proof. $D = B + C(RB) + C(BR) + C(RBR) \subseteq A$ and $D^{[3]} = C(D^3)$. But

$$D^3 \subseteq A(B + C(RB) + C(BR) + C(RBR))A \subseteq B.$$

The last statement is obvious. □

Theorem 3.2.8. *Let $\mathscr{A}$ be a homomorphically closed and i-hereditary class of ℓ-rings which contains each zero ℓ-ring. Then $\mathscr{L}\mathscr{A} = \mathscr{P}_1$. In particular,*

$$\ell\text{-}\beta = \{R : \textit{each nonzero homomorphic image of } R \textit{ has a nonzero } \ell\textit{-ideal whose square is zero}\}.$$

Proof. By Exercise 5 it suffices to show that $\mathscr{P}_1 = \mathscr{P}_2$. If $0 \neq R \in \mathscr{P}_2$, then R has a nonzero ℓ-ideal A with $A \in \mathscr{P}_1$, and A has a nonzero ℓ-ideal B with $B \in \mathscr{P}_0 = \mathscr{A}$. If D is the ℓ-ideal of R generated by B, then $D^{[3]} \subseteq B$; hence, $D^{[3]}$ is a $\mathscr{P}_0$-ℓ-ideal of R since $\mathscr{A}$ is i-hereditary. If $D^{[3]} = 0$, then $(D^{[2]})^2 = 0$ and $D^{[2]} \in \mathscr{P}_0$. So either $D^{[3]}$, $D^{[2]}$, or D is a nonzero $\mathscr{P}_0$-ℓ-ideal of R. Since $\mathscr{P}_2$ is homomorphically closed each nonzero homomorphic image of R has a nonzero $\mathscr{P}_0$-ℓ-ideal; thus, $R \in \mathscr{P}_1$. □

Before we identify the ℓ-β-semisimple ℓ-rings we collect some information about the nilpotent ℓ-ideals of an ℓ-ring.

Theorem 3.2.9. *Let R be an ℓ-ring.*

(a) *The ℓ-ideal of R generated by a nilpotent one-sided ℓ-ideal is nilpotent.*
(b) *The sum of a finite number of nilpotent one-sided ℓ-ideals is nilpotent.*
(c) *If ℓ-$N(R)$ is the sum of all of the nilpotent ℓ-ideals of R, then*

$$\begin{aligned} \ell\text{-}N(R) &= \{a \in R : |a|R \text{ is nilpotent}\} \\ &= \{a \in R : (|a|x)^n = 0 \text{ for some } n \in \mathbb{N} \text{ and all } x \in R\}. \end{aligned}$$

Consequently, ℓ-$N(R)$ contains each nilpotent one-sided or two-sided ideal that has positive generators.

Proof. If I is a nilpotent ℓ-subring and J is a nilpotent ℓ-ideal of R, then $I+J$ is a nilpotent ℓ-subring since $I+J/J \cong I/I \cap J$ is nilpotent. In particular, if I is a nilpotent right ℓ-ideal let

$$\begin{aligned} J = C(RI) &= \{x \in R : |x| \leq \Sigma_i r_i a_i \text{ for some } r_i \in R \text{ and } a_i \in I\} \\ &= \{x \in R : |x| \leq ra \text{ for some } r \in R \text{ and } a \in I\}. \end{aligned}$$

Since $(RI)^n = R(IR)^{n-1}I \subseteq RI^n, RI$ is a nilpotent ideal, and hence J is a nilpotent ℓ-ideal. Thus, $I+J$, the ℓ-ideal generated by I, is nilpotent. This proves (a), and (b) follows from (a) by induction since the sum of two nilpotent ℓ-ideals has been shown to be nilpotent. If $|a|R$ is nilpotent, then so is the right ℓ-ideal $J = C(|a|R) = \{x \in R : |x| \leq |a|r \text{ for some } r \subset R\}$ that is generated by $|a|R$. The right ℓ-ideal generated by a is $S = J + C(a)$. Since $S^2 \subseteq J$, S is nilpotent and hence $a \in$ ℓ-$N(R)$. This proves (c). □

In the presence of chain conditions the ℓ-ideal ℓ-$N(R)$ of Theorem 3.2.9 is nilpotent. In fact, more is true. The subset X of R is *nil* if each of its elements is nilpotent; that is, if $a \in X$, then $a^n = 0$ for some $n \in \mathbb{N}$.

Theorem 3.2.10. *If the ℓ-ring R satisfies the ascending chain condition on ℓ-ideals, then R has a largest nilpotent ℓ-ideal. If R satisfies the descending chain condition on ℓ-ideals, then each nil ℓ-ideal is nilpotent.*

Proof. If R has a maximal nilpotent ℓ-ideal N, then, by (b) of Theorem 3.2.9, N is the largest nilpotent ℓ-ideal of R. This proves the first statement. Suppose that R has d.c.c. on ℓ-ideals, and let A be a nil ℓ-ideal of R. Since $A \supseteq A^{[2]} \supseteq \cdots$ is a descending chain of ℓ-ideals, for some integer $n, A^{[n]} = A^{[n+1]} = \cdots$. If $B = A^{[n]} \neq 0$, then B is ℓ-idempotent since $B^{[2]} = C(A^{[n]}A^{[n]}) = A^{[2n]} = B$ by (3.2.14); so B contains a nonzero ℓ-ideal D of R which is minimal with respect to $BDB \neq 0$. If $0 < d \in D$ with $BdB \neq 0$, then $BC(BdB)B \neq 0$ since $B = B^{[2]}$; so $D = C(BdB)$ by the minimality of D. But then $d \leq bdb$ with $b \in B^+$ and $b^m = 0$ for some m; so $d \leq bdb \leq b^2db^2 \leq \cdots \leq b^m db^m = 0$. Thus, $B = 0$ and A is nilpotent. □

An ℓ-ring is called *ℓ-semiprime* if it contains no nonzero nilpotent ℓ-ideals.

Theorem 3.2.11. *The following statements are equivalent for an ℓ-ring R.*

(a) R is ℓ-semiprime.
(b) ℓ-$\beta(R) = 0$; that is, R is ℓ-β-semisimple.
(c) If A is an ℓ-ideal of R and $A^2 = 0$, then $A = 0$.
(d) If A is a right (respectively, left) ℓ-ideal of R and $A^2 = 0$, then $A = 0$.
(e) If $a \in R^+$ and $aRa = 0$ (equivalently, $aR^+a = 0$), then $a = 0$.

Proof. (a)$\Rightarrow$ (b). If ℓ-$\beta(R) \neq 0$, then ℓ-$\beta(R)$ has a nonzero ℓ-ideal B with $B^2 = 0$, by Theorem 3.2.8. By Theorem 3.2.7, then, $D^6 = 0$ where D is the ℓ-ideal of R generated by B. Thus, ℓ-$\beta(R) = 0$.

(b)$\Rightarrow$ (c). This is immediate since R contains no nonzero zero ℓ-ideals.

(c) $\Rightarrow$ (d). If A is any right ideal, then $(A+RA)^n \subseteq A^n + RA^n$ for each $n \in \mathbb{N}$. So if A is a right ℓ-ideal with $A^2 = 0$, then $A + C(RA)$ is an ℓ-ideal whose square is zero; thus $A = 0$.

(d) $\Rightarrow$ (e). If $aRa = 0$ with $a \in R^+$ then $(aR)^{[2]} = 0$; so $aR = 0$ and $a \in A = \ell_\ell(R)$. Since $A^2 = 0$, $A = 0$ and hence $a = 0$.

(e) $\Rightarrow$ (a). If A is a nilpotent ℓ-ideal of index $n \geq 2$ then $B = A^{[n-1]}$ is a nilpotent ℓ-ideal of index 2 by (3.2.15). But then $B = 0$ since $bRb = 0$ for each $b \in B^+$. Thus, R is ℓ-semiprime. □

The ℓ-ideal P of the ℓ-ring R is an *ℓ-prime ℓ-ideal* if whenever I and J are ℓ-ideals of R with $IJ \subseteq P$, then $I \subseteq P$ or $J \subseteq P$. If 0 is an ℓ-prime ℓ-ideal of R, then R is called an *ℓ-prime ℓ-ring*. Clearly, P is an ℓ-prime ℓ-ideal of R if and only if the factor ring R/P is an ℓ-prime ℓ-ring. Each ℓ-prime ℓ-ideal contains a minimal ℓ-prime ℓ-ideal (Exercise 14).

Theorem 3.2.12. *The following statements are equivalent for the ℓ-ring R.*

(a) R is ℓ-prime.
(b) If I and J are right (respectively, left) ℓ-ideals of R and $IJ = 0$, then $I = 0$ or $J = 0$.
(c) If I is a nonzero right (respectively, left) ℓ-ideal of R, then $r_\ell(I) = 0$ (respectively, $\ell_\ell(I) = 0$).
(d) If $a, b \in R^+$ and $aRb = 0$ (equivalently, $aR^+b = 0$), then $a = 0$ or $b = 0$.

Proof. (a) $\Rightarrow$ (b). If $I_1 = I + C(RI)$ and $J_1 = J + C(RJ)$, then I_1 and J_1 are ℓ-ideals and $I_1J_1 = 0$; so $I = 0$ or $J = 0$.

(b) $\Rightarrow$ (c). This is clear since $J = r_\ell(I)$ is an ℓ-ideal of R and $IJ = 0$.

(c) $\Rightarrow$ (d). Since $aRb = 0$, $C(aR)b = 0$, and hence $aR = 0$ or $b = 0$. If $aR = 0$, then $a \in \ell_\ell(R) = 0$.

(d) $\Rightarrow$ (a). If I and J are ℓ-ideals with $IJ = 0$ and if $0 \neq a \in I^+$, then, for each $b \in J^+$, $aRb \subseteq IJ$; so $aRb = 0$, $b = 0$ and hence $J = 0$. □

A commutative ℓ-ring is ℓ-prime if and only if it is an ℓ-domain. Since a commutative ℓ-domain need not be a domain, an ℓ-prime ℓ-ring need not be a prime ring. An example of a noncommutative prime ℓ-ring is the ring of linear transformations of a left vector space over a totally ordered division ring (see Exercise 2.4.25).

We will now investigate the connection between ℓ-semiprime ℓ-rings and ℓ-prime ℓ-rings, and in the process we will get a description of the elements in ℓ-$\beta(R)$.

Let R be an ℓ-ring. A nonempty subset S of R^+ is called an *M^+-system* if for every pair $a, b \in S$ there exists an $x \in R^+$ with $axb \in S$. A sequence (a_n) in R^+ is called an *M^+-sequence* if for each $n \in \mathbb{N}$ there is an $x_n \in R^+$ such that $a_{n+1} = a_n x_n a_n$. The next result connects M^+-systems with ℓ-prime ℓ-ideals and the one after it connects both of these with ℓ-$\beta(R)$.

Theorem 3.2.13. *Let R be an ℓ-ring.*

(a) Each M^+-sequence in R is an M^+-system.
(b) If $a \in S$ and S is an M^+-system, then S contains an M^+-sequence that begins with a.
(c) A proper ℓ-ideal P of R is an ℓ-prime ℓ-ideal iff $R^+ \backslash P$ is an M^+-system.
(d) If S is an M^+-system of R and P is an ℓ-ideal of R that is maximal with respect to $P \cap S = \phi$, then P is ℓ-prime.
(e) If $a \in F(R)$ and $a^2 = 0$, then $aRa = 0$.

Proof. The proofs of the first three statements are left for the exercises (Exercise 13). To see that P is ℓ-prime in (d), suppose that I and J are ℓ-ideals of R that properly contain P and $IJ \subseteq P$. Then we can find elements $s \in I \cap S$, $t \in J \cap S$ and $x \in R^+$ with $sxt \in S$. But $sxt \in IJ \subseteq P$ gives a contradiction. For (e), suppose first that a is an f-element and $z \in R^+$. Then

$$aza = aza \wedge aza = (az - za)^+ a \wedge a(az - za)^- = 0.$$

In general, $|aza| \leq |a||z||a| = 0$; so $aza = 0$. □

Theorem 3.2.14. *Let R be an ℓ-ring and let A and B be the subsets of R given by*

$$A = \cap\{P : P \text{ is an } \ell\text{-prime } \ell\text{-ideal of } R\}, \tag{3.2.16}$$

$$B = \{a \in R : \text{ each } M^+\text{-system that contains } |a| \text{ also contains } 0\}. \tag{3.2.17}$$

Then ℓ-$\beta(R) = A = B$ and ℓ-$\beta(R)$ is a nil ℓ-ideal of R.

Proof. Suppose that $a \in A$ and S is an M^+-system that contains $|a|$. If $0 \notin S$, then, by Zorn's Lemma and by (d) of Theorem 3.2.13, R has an ℓ-prime ℓ-ideal P that is disjoint from S; but then $|a| \in P \cap S$. So S must contain 0 and hence $A \subseteq B$. On the other hand, if $a \notin A$ and P is an ℓ-prime ℓ-ideal of R with $a \notin P$, then $R^+ \backslash P$ is an M^+-system that contains $|a|$ but not 0. So $a \notin B$, $B \subseteq A$, and $A = B$. Since an ℓ-prime ℓ-ring is ℓ-semiprime, ℓ-$\beta(R)$ is contained in each ℓ-prime ℓ-ideal of R, by Theorem 3.2.11; hence ℓ-$\beta(R) \subseteq A$. Since $A = B$ each M^+-system in A contains 0. Let $0 < a_1 \in A^+$. If $a_1 A a_1 \neq 0$ let $0 < a_2 \in a_1 A a_1$. Given the elements $a_1, \ldots, a_n$ in $A^+ \backslash \{0\}$ with $a_{k+1} \in a_k A a_k$, either $a_n A a_n = 0$ or there exists $0 < a_{n+1} \in a_n A a_n$. Since this construction must stop, $bAb = 0$ for some $b \in A^+ \backslash \{0\}$ and A is not ℓ-semiprime. If $\bar{A}$ is a nonzero homomorphic image of A, then $\bar{A}$ does not have a proper ℓ-prime ℓ-ideal since A has no such ℓ-ideal. Hence, as we have just seen, $\bar{A}$ has a nonzero nilpotent ℓ-ideal. Thus, A is an ℓ-β-ℓ-ideal of R by Theorem 3.2.8; so $A \subseteq \ell$-$\beta(R)$.

Finally, if $a \in \ell\text{-}\beta(R)$, then $\{|a|^n : n \in \mathbb{N}\}$ is an M^+-system that contains $|a|$; so $|a|^n = 0$ for some n and $a^n = 0$ since $|a^n| \leq |a|^n$. □

It is now easy to get some interesting properties of $\ell\text{-}\beta$.

Theorem 3.2.15. *(a) $\ell\text{-}\beta$ is hereditary.*

(b) For each ℓ-ring $R, \ell\text{-}\beta(R)$ contains every one-sided $\ell\text{-}\beta\text{-}\ell$-ideal of R.

(c) Let β be the lower radical of rings *determined by the class of all zero rings. If R is an ℓ-ring, then $\ell\text{-}\beta(F(R)) \subseteq \ell\text{-}\beta(R) \subseteq \beta(R)$.*

Proof. (a) By Theorem 3.2.14 each M^+-system of a subring of an $\ell\text{-}\beta\text{-}\ell$-ring contains 0, and hence each ℓ-subring of an $\ell\text{-}\beta\text{-}\ell$-ring is an $\ell\text{-}\beta\text{-}\ell$-ring.

(b) Suppose that A is a left $\ell\text{-}\beta\text{-}\ell$-ideal of R and $A \not\subseteq \ell\text{-}\beta(R)$. By passing to $R/\ell\text{-}\beta(R)$ we may assume that R is ℓ-semiprime and $A \neq 0$. Then $A^2 \neq 0$ so $Aa_1 \neq 0$ for some a_1 in A^+. But then $C(Aa_1)^2 \neq 0$ and hence $uv \neq 0$ for some $0 \leq u \leq ba_1$ and $0 \leq v \leq ca_1$ with $b, c \in A^+$. Consequently $0 < ba_1ca_1$ and $Aa_2 \neq 0$ with $a_2 = a_1ca_1$. Continuing, we get an M^+-sequence (a_n) in A which excludes 0. This is impossible by Theorem 3.2.14.

(c) Let $a \in \ell\text{-}\beta(R)$ and let (a_n) be an M-sequence in R beginning with a; so $a_1 = a$ and there is a sequence (x_n) in R with $a_{n+1} = a_nx_na_n$ for each $n \in \mathbb{N}$. Let $b_1 = |a_1|$ and $b_{n+1} = b_n|x_n|b_n$. Then (b_n) is an M^+-sequence in R and $|a_n| \leq b_n$ for each n. By Theorem 3.2.14 $b_n = 0$ for some n; so $a_n = 0$ and hence $a \in \beta(R)$ by the ring analogue of Exercise 15. To see that $\ell\text{-}\beta(F(R)) \subseteq \ell\text{-}\beta(R)$ take $a \in F(R)$ with $0 = a^n = |a|^n$. If $n = 2$, then $|a|R|a| = 0$ by Theorem 3.2.13(e) and $a \in \ell\text{-}N(R) \subseteq \ell\text{-}\beta(R)$ by Theorem 3.2.9. Since $(a^2)^{n-1} = 0$, $a^2 \in \ell\text{-}\beta(R)$ by induction on n. Thus, again, $|a|R|a| \subseteq \ell\text{-}\beta(R)$ and $a \in \ell\text{-}\beta(R)$. □

In contrast to the situation for f-elements there are nilpotent d-elements that are not in $\ell\text{-}\beta(R)$. For an example just take a canonically ordered matrix ring over an f-ring.

Special Radicals

Since the class of ℓ-prime ℓ-rings is i-hereditary, it follows from Theorem 3.2.14 that $\ell\text{-}\beta$ is the upper radical determined by the class of ℓ-prime ℓ-rings. We wish to now consider upper radicals determined by other classes of ℓ-prime ℓ-rings. We first note the following.

Theorem 3.2.16. *Let $\mathscr{P}$ be a radical and let A be an ℓ-ideal of the ℓ-ring R.*

(a) $\mathscr{P}$ is i-hereditary if and only if for each R and A, $\mathscr{P}(R) \cap A \subseteq \mathscr{P}(A)$.
(b) $\mathscr{S}\mathscr{P}$ is i-hereditary if and only if for each R and A, $\mathscr{P}(A) \subseteq \mathscr{P}(R) \cap A$.
(c) If for each R and A, $\mathscr{P}(A)$ is an ℓ-ideal of R, then $\mathscr{S}\mathscr{P}$ is i-hereditary.
(d) If $\ell\text{-}\beta \subseteq \mathscr{P}$, then $\mathscr{P}(A)$ is an ℓ-ideal of R; so $\mathscr{S}\mathscr{P}$ is i-hereditary.

Proof. If $\mathscr{P}$ is i-hereditary, then $\mathscr{P}(R) \cap A$ is a $\mathscr{P}$-ℓ-ideal of A, so $\mathscr{P}(R) \cap A \subseteq \mathscr{P}(A)$. For the converse in (a), if $R = \mathscr{P}(R)$, then $A = \mathscr{P}(R) \cap A \subseteq \mathscr{P}(A)$, and hence $A = \mathscr{P}(A)$. If $\mathscr{S}\mathscr{P}$ is i-hereditary, then $(A + \mathscr{P}(R))/\mathscr{P}(R)$ is $\mathscr{P}$-semisimple and contains the $\mathscr{P}$-ℓ-ideal $(\mathscr{P}(A) + \mathscr{P}(R))/\mathscr{P}(R)$; so $\mathscr{P}(A) \subseteq \mathscr{P}(R)$. For the converse in (b), if $\mathscr{P}(R) = 0$, then certainly $\mathscr{P}(A) = 0$. Now, (c) is obvious, and (d) is a consequence of Theorem 3.2.7 since $A/\mathscr{P}(A)$ is ℓ-semiprime. □

Each semisimple class of rings is i-hereditary (see Exercise 24). It is not known if this is true for ℓ-rings. If there is a semisimple class $\mathscr{S}\mathscr{P}$ of ℓ-rings which is not i-hereditary, then, by (d) of the previous theorem, $\mathscr{S}\mathscr{P}$ must contain a nonzero zero ℓ-ring. However, if $\mathscr{P}$ is the upper radical determined by the class of all zero ℓ-rings, then $\mathscr{S}\mathscr{P}$ is i-hereditary (see Exercise 25).

A radical which is i-hereditary and which contains the class of zero ℓ-rings is called *supernilpotent*. We will now construct supernilpotent radicals. If A is an ℓ-ideal of the ℓ-ring R let $A^* = \ell_\ell(A;R) \cap r_\ell(A;R)$; so

$$A^* = \{x \in R : |x|A = 0 = A|x|\}. \tag{3.2.18}$$

Note that $\ell_\ell(A;R) \subseteq r_\ell(A;R)$ if $\ell_\ell(A;A) = 0$ and equality holds if also $r_\ell(A;A) = 0$ An i-hereditary class $\mathscr{M}$ of ℓ-prime ℓ-rings is called a *special class* if it satisfies the condition:

$$\text{if } A \in \mathscr{M} \text{ and } A \text{ is an } \ell\text{-ideal of } R, \text{ then } R/A^* \in \mathscr{M}. \tag{3.2.19}$$

Since a special class satisfies (3.2.8) it determines an upper radical which is called a *special radical.*

Theorem 3.2.17. *Let $\mathscr{P}_\mathscr{M}$ be the special radical determined by the special class $\mathscr{M}$. Then $\mathscr{P}_\mathscr{M}$ is supernilpotent. For each ℓ-ring R*

$$\mathscr{P}_\mathscr{M}(R) = \cap\{A : A \text{ is an } \ell\text{-ideal of } R \text{ and } R/A \in \mathscr{M}\}. \tag{3.2.20}$$

Thus, $R \in \mathscr{S}\mathscr{P}_\mathscr{M}$ iff R is a subdirect product of a subset of $\mathscr{M}$.

Proof. Since $\mathscr{M} \subseteq \mathscr{S}\ell$-$\beta$, $\mathscr{S}\mathscr{P}_\mathscr{M} \subseteq \mathscr{S}\ell$-$\beta$ and ℓ-$\beta \subseteq \mathscr{P}_\mathscr{M}$, by Theorem 3.2.4. To see that $\mathscr{P}_\mathscr{M}$ is i-hereditary take $R \in \mathscr{P}_\mathscr{M}$ and let A be an ℓ-ideal of R. Suppose that B is an ℓ-ideal of A and $A/B \in \mathscr{M}$. Then B is an ℓ-ideal of R (Theorem 3.2.7) and $R/B/(A/B)^* \in \mathscr{M}$. But R has no nonzero homomorphic images in $\mathscr{M}$; so $(A/B)^* = R/B$. Thus, $AR \subseteq B$, $A^2 \subseteq B$ and therefore $A = B$. So A has no nonzero homomorphic images in $\mathscr{M}$ and $A \in \mathscr{P}_\mathscr{M}$.

Let C be the right-side of (3.2.20). If $R/A \in \mathscr{M}$, then $R/A \in \mathscr{S}\mathscr{P}_\mathscr{M}$ and $\mathscr{P}_\mathscr{M}(R) \subseteq A$; so $\mathscr{P}_\mathscr{M}(R) \subseteq C$. If C is a $\mathscr{P}_\mathscr{M}$-ℓ-ring, then $C \subseteq \mathscr{P}_\mathscr{M}(R)$ and hence $C = \mathscr{P}_\mathscr{M}(R)$. If C is not a $\mathscr{P}_\mathscr{M}$-ℓ-ring, then C contains a proper ℓ-ideal D with $C/D \in \mathscr{M}$. Then D is an ℓ-ideal of R and $(R/D)/(C/D)^* \in \mathscr{M}$ by (3.2.19). Now, $(C/D)^* = B/D$ where $B = \{x \in R : |x|C \subseteq D \text{ and } C|x| \subseteq D\}$. So $R/B \in \mathscr{M}$ and hence $C \subseteq B$. But then we have the contradiction that $(C/D)^2 = 0$.

The last statement is a consequence of (3.2.20). □

The class of all ℓ-prime ℓ-rings is special (Exercise 18); so ℓ-β is the smallest special radical.

The Upper ℓ-nil Radical

Let ℓ-Nil be the class of nil ℓ-rings. It is easily seen that ℓ-Nil satisfies the conditions in (d) of Theorem 3.2.1, and hence it is a radical that is called the *upper ℓ-nil radical*. A ring is *locally nilpotent* if each of its finitely generated subrings is nilpotent, and an ℓ-ring is *ℓ-locally nilpotent* if each of its ℓ-subrings which is finitely generated (as an ℓ-ring) is nilpotent. The radical class of nil rings properly contains the radical class of locally nilpotent rings. However, as we will now show, these radicals coincide for ℓ-rings.

Theorem 3.2.18. *Suppose that the subset $X = \{x_1, \ldots, x_n\}$ generates the ℓ-ring R. Then the following statements are equivalent.*

(a) R is nilpotent.

(b) $\bigvee_{i=1}^{n} |x_i|$ is nilpotent.

(c) There is an integer N such that $|x_{i_1}| \cdots |x_{i_N}| = 0$ for all $x_{i_1}, \ldots, x_{i_N} \in X$.

Proof. Clearly, (a) implies (b), and if $x^N = 0$ where $x = \vee |x_i|$, then $0 \leq |x_{i_1}| \cdots |x_{i_N}| \leq x^N = 0$; so (b) implies (c).

(c) implies (a). Let $\mathbb{Z}[Y]$ denote the subring of R generated by the subset Y, and let $L(Y)$ denote the sublattice generated by Y. Now put $R_1 = \mathbb{Z}[X]$, $L_1 = L(R_1)$ and, inductively, let $R_{k+1} = \mathbb{Z}[L_k]$ and $L_{k+1} = L(R_{k+1})$. Then R is the union of the ascending chain of subrings $R_1 \subseteq R_2 \subseteq \cdots$. We claim that $|t_1| \cdots |t_N| = 0$ if $t_1, \ldots, t_N \in R_k$, and hence $R^N = 0$. By assumption, any monomial in $|x_1|, \ldots, |x_n|$ of length N is 0; so $|t_1| \cdots |t_N| = 0$ if $t_1, \ldots, t_N \in R_1$. Assume the claim is true for R_k and let $t_1, \ldots, t_N \in L_k = L(R_k)$. Then, by (f) of Theorem 2.2.4, for some finite sets I and J,

$$t_\nu = \bigwedge_{i \in I} \bigvee_{j \in J} a_{ij}^{(\nu)}$$

where $a_{ij}^{(\nu)} \in R_k$. Now, for each ν,

$$|t_\nu| \leq \sum_{i,j} |a_{ij}^{(\nu)}|$$

since this inequality holds in each of the totally ordered homomorphic images of the additive group of R, and R is an f-module over $\mathbb{Z}$. But then $|t_1| \cdots |t_N| \leq \Sigma_m |b_{m1}| \cdots |b_{mN}|$ where each $b_{m\nu} \in R_k$; so $|t_1| \cdots |t_N| = 0$. Now, if $p_1, \ldots, p_N \in R_{k+1}$, then $|p_1| \cdots |p_N|$ is dominated by a sum of terms of the form $|t_1| \cdots |t_s|$ where $s \geq N$ and each $t_m \in L_k$; so $|p_1| \cdots |p_N| = 0$. □

An immediate consequence of Theorem 3.2.18 is

Theorem 3.2.19. *The following statements are equivalent for the ℓ-ring R.*

(a) *R is nil.*
(b) *R is ℓ-locally nilpotent.*
(c) *R is locally nilpotent.* □

In spite of Theorem 3.2.19 it is possible to have ℓ-$\beta(R) = \ell$-Nil$(R) \subset$ Nil$(R) = \beta(R)$ for an ℓ-ring R; see Exercise 30.

We will show next that the upper ℓ-nil radical has some of the good properties of the lower ℓ-nil radical.

Theorem 3.2.20. *(a) The upper ℓ-nil radical is a special radical.*
(b) If R is an ℓ-ring, then ℓ-Nil(R) contains each one-sided nil ℓ-ideal of R.

Proof. (a) Let $\mathscr{M}$ be the class of ℓ-prime ℓ-nil-semisimple ℓ-rings. By Exercise 18 and (d) of Theorem 3.2.16 $\mathscr{M}$ is i-hereditary, and hence $\mathscr{M}$ is a special class provided it satisfies (3.2.19). Suppose that A is an ℓ-ideal of R and $A \in \mathscr{M}$. If B/A^* is a nil ℓ-ideal of R/A^* and $b \in A \cap B$, then, for some $n \in \mathbb{N}, b^n \in A \cap A^* = 0$. Thus, $A \cap B$ is nil, $A \cap B = 0$, and $B = A^*$. Consequently, $R/A^* \in \mathscr{M}$ by Exercise 18, and $\mathscr{M}$ does satisfy (3.2.19).

To show that ℓ-Nil is the upper radical determined by $\mathscr{M}$ it suffices to verify that if R is an ℓ-ring and

$$B = \bigcap_i \{B_i : B_i \text{ is an } \ell\text{-ideal of } R \text{ and } R/B_i \in \mathscr{M}\},$$

then ℓ-Nil$(R) = B$. Certainly ℓ-Nil$(R) \subseteq B$. If B is not nil, then B has a positive element b that is not nilpotent and ℓ-Nil$(R) \cap \{b^n\} = \phi$. Let P be an ℓ-ideal of R that contains ℓ-Nil(R) and is maximal with respect to $P \cap \{b^n\} = \phi$. By (d) of Theorem 3.2.13 P is an ℓ-prime ℓ-ideal of R. If A/P is a nonzero ℓ-ideal of R/P, then $b^n \in A$ for some n, and hence A/P is not nil. So $R/P \in \mathscr{M}, P = B_i$ for some i, and $b \in P$. This contradiction proves that B is nil, and so $B = \ell$-Nil(R).

(b) Let A be a nil right ℓ-ideal of R. If $x \in A + C(RA)$, the ℓ-ideal generated by A, then $|x| \le a + ra = (1+r)a$ for some $a \in A^+$ and $r \in R^+$. Since $((1+r)a)^{n+1} = (1+r)(a(1+r))^n a$ for each n and $a(1+r)$ is nilpotent, $(1+r)a$ is nilpotent. So $A + C(RA)$ is a nil ℓ-ideal and $A \subseteq \ell$-Nil (R). □

It is an open question in the theory of rings (the Koethe conjecture) whether or not a one-sided nil ideal generates a two-sided nil ideal.

The Generalized ℓ-Nil Radical

Let $\mathscr{M}$ be the class of ℓ-domains. Since $\mathscr{M}$ is hereditary it determines an upper radical ℓ-N_g which is called the *generalized ℓ-nil radical.* Since $\mathscr{M} \subseteq \mathscr{S}\ell$-Nil, necessarily ℓ-Nil $\subseteq \ell$-N_g, and the latter inclusion is proper because, for example, the matrix ring $\mathbb{R}_n$ is an ℓ-N_g-ℓ-ring which is ℓ-Nil-semisimple.

A ring is *reduced* if it has no nonzero nilpotent elements; similarly, a po-ring is *po-reduced* if it has no nonzero positive nilpotent elements, and a po-reduced ℓ-ring is called *ℓ-reduced.* Note that if R is po-reduced and if $a, b, x \in R^+$ with $ab = 0$, then $axb = 0$. For, $(ba)^2 = 0$, $ba = 0$, and hence $(axb)^2 = 0$ and $axb = 0$. In particular, an ℓ-reduced ℓ-prime ℓ-ring is an ℓ-domain.

Theorem 3.2.21. *(a) The class of ℓ-domains is special, and hence the generalized ℓ-nil radical is a special radical.*
(b) The following statements are equivalent for the ℓ-ring R.
(i) R is ℓ-reduced.
(ii) $\ell\text{-}N_g(R) = 0$.
(iii) R is a subdirect product of ℓ-domains.

Proof. (a) Let A be an ℓ-domain and suppose that A is an ℓ-ideal of R. If $x \in R^+$, then $xA = 0$ iff $AxA = 0$ iff $Ax = 0$. Now, if $a, b \in R^+$ with $abA = 0$, then either $Aa = 0$ or $bA = 0$. Hence, $a \in A^*$ or $b \in A^*$ and R/A^* is an ℓ-domain.

(b) The equivalence of (ii) and (iii) follows from (a) and Theorem 3.2.17. Clearly, (iii) implies (i); so it remains to verify that (iii) is a consequence of (i). Suppose that R is ℓ-reduced. Let P be a minimal ℓ-prime ℓ-ideal of R and let $S = \{a_1 \cdots a_n : a_i \in R^+ \backslash P\}$ be the multiplicative subsemigroup of R^+ that is generated by the M^+-system $R^+ \backslash P$. If $R^+ \backslash P = S$, then R/P is an ℓ-domain. If $R^+ \backslash P \subset S$, then, since $R^+ \backslash P$ is an M^+-system which is maximal with respect to excluding 0 (Exercise 14), there exist $a_1, \ldots, a_n \in R^+ \backslash P$ with $a_1 \cdots a_n = 0$. Now, we can find $x_1, \ldots, x_{n-1} \in R^+$ such that $a_1 x_1 a_2 \in R^+ \backslash P, \ldots, a_1 x_1 a_2 x_2 a_3 \cdots a_{n-1} x_{n-1} a_n \in R^+ \backslash P$. But then we have the contradiction that $0 = a_1 x_1 a_2 \cdots a_{n-1} x_{n-1} a_n \in R^+ \backslash P$. Since the intersection of all of the minimal ℓ-prime ℓ-ideals of R is 0 (Exercise 15) R is a subdirect product of ℓ-domains. □

An ℓ-ideal (respectively, ideal) P of R is *completely ℓ-prime* (respectively, *completely prime*) if R/P is an ℓ-domain (respectively, a domain). We have just seen that a minimal ℓ-prime ℓ-ideal in an ℓ-reduced ℓ-ring is completely ℓ-prime and, of course, the analogous ring theoretic statement is also true. A characterization of these ℓ-ideals (ideals) is given next (see Theorem 2.4.3).

Theorem 3.2.22. *An ℓ-prime ℓ-ideal P of the nonzero ℓ-reduced ℓ-ring R is a minimal ℓ-prime ℓ-ideal if and only if $a^* (= \ell_\ell(a)) \nsubseteq P$ for each $a \in P^+$.*

Proof. Suppose that P is a minimal ℓ-prime ℓ-ideal and let $a \in P^+$. Let S be the multiplicatively closed subset of R^+ given by

$$S = R^+ \backslash P \cup \{a_1 a a_2 \cdots a_n a a_{n+1} : a_i \in R^+ \backslash P \text{ and } n \geq 1\}.$$

Then $0 \in S$, by Exercise 14(c), since $R^+ \backslash P \subset S$; so $0 = a_1 a a_2 \cdots a_n a a_{n+1} = a_1 \cdots a_{n+1} a^n = a_1 \cdots a_{n+1} a$ for some $a_1, \cdots, a_{n+1} \in R^+ \backslash P$, since R is embeddable in a product of ℓ-domains (or use Exercise 26). Thus $a_1 \cdots a_{n+1} \in \ell_\ell(a) \backslash P$. Conversely, suppose that the condition holds and let Q be a minimal ℓ-prime ℓ-ideal that is contained in P. If $Q \subset P$ and $a \in P^+ \backslash Q$, then $ba = 0$ for some $0 < b \notin P$. But Q is a completely ℓ-prime ℓ-ideal; so $b \in Q \subseteq P$. Consequently, $P = Q$ is minimal. □

We immediately use this result to show

Theorem 3.2.23. *The following statements are equivalent for the reduced ℓ-ring R.*

(a) The set of minimal prime ideals of R is equal to the set of minimal ℓ-prime ℓ-ideals of R.

(b) *Each minimal ℓ-prime ℓ-ideal is completely prime.*
(c) *Each minimal ℓ-prime ℓ-ideal contains a prime ideal, and $ab = 0$ if and only if $|a||b| = 0$.*

Proof. (a) $\Rightarrow$ (b). This is a consequence of the remark preceding Theorem 3.2.22 that a minimal prime ideal in a reduced ring is completely prime.

(b) $\Rightarrow$ (c). If $\{P_i\}_{i\in I}$ is the set of minimal ℓ-prime ℓ-ideals, then R is a subdirect product of the family $\{R/P_i\}_{i\in I}$ and each R/P_i is a domain. So if $ab = 0$, then $|a||b| = 0$; and, of course, the converse always holds since $|ab| \leq |a||b|$.

(c) $\Rightarrow$ (a). Let Q be a minimal prime ideal and let $|x| \leq |b|$ with $b \in Q$. By the ring analogue of Theorem 3.2.22 there is an element $a \in R\backslash Q$ with $ab = 0$, and therefore $|a||x| = 0$. So $ax = 0$ and hence $x \in Q$ since Q is completely prime. Now, let P be a minimal ℓ-prime ℓ-ideal. Then P contains a minimal prime ideal Q which is an ℓ-prime ℓ-ideal. So $P = Q$ is a minimal prime. Also, each minimal prime ideal Q contains a minimal ℓ-prime ℓ-ideal P which is a minimal prime; so $Q = P$. □

It is interesting and worthwhile to determine when an ℓ-prime ℓ-ideal is completely ℓ-prime.

Theorem 3.2.24. *Let P be an ℓ-ideal of the ℓ-ring R. Then P is completely ℓ-prime if and only if it satisfies the following conditions.*

(a) *P is ℓ-prime.*
(b) *If $a \wedge b = 0$ and $ab \in P$, then $ba \in P$.*
(c) *If $a, b \in R^+$ with $a^2 + b^2 \in P$, then $ab \in P$.*

Moreover, P is completely prime if and only if it satisfies (a), (b), (c) and

(d) *If $ab \in P$ then $|a||b| \in P$.*

Proof. Suppose that the first three conditions are satisfied. By passing to R/P we may assume that $P = 0$ and that R is ℓ-prime. Since an ℓ-reduced ℓ-prime ℓ-ring is an ℓ-domain it suffices to show that R does not have a nonzero positive element whose square is 0. Suppose, then, that $a \in R^+$ with $a^2 = 0$. We will show that $aR^+a = 0$ and hence $a = 0$. Let $z \in R^+$. If $az \leq za$ or $za \leq az$, then $aza = 0$. Otherwise, $(za - az)^- > 0$ and $(za - az)^+ > 0$. Now, $0 \leq (za - az)^+(za - az)^- \leq zaaz = 0$; so, by (b), $(za - az)^-(za - az)^+ = 0$. If $y \in R^+$, then $[(za - az)^+ y(za - az)^-]^2 = 0$; so, by (c), $a(za - az)^+ y(za - az)^- = 0$. Thus, $a(za - az)^+ = 0$ and similarly, $a(za - az)^- = 0$. So $0 = a[(za - az)^+ - (za - az)^-] = a(za - az) = aza$.

If the fourth condition also holds then clearly R is a domain. The converses are trivial. □

The Johnson Radical

The Johnson radical of an ℓ-ring is the ℓ-ring analogue of the Jacobson radical of a ring. The basic ingredients for a structure theory using the Johnson radical are the (right) ℓ-primitive ℓ-rings. However, except in special cases, for example, the variety of f-rings, these building blocks are not very well understood.

The right ideal I of the ring R is *modular* if there is an element $e \in R$, called *a left identity modulo I*, such that $ex - x \in I$ for each $x \in R$. If I is modular and I is contained in the right ideal J, then J is also modular with e as a left identity modulo J. Equally clear is that $I = R$ exactly when $e \in I$. If e is a left identity modulo the proper right ℓ-ideal I of the ℓ-ring R and J is a right ℓ-ideal of R that contains I and is maximal with respect to not containing e, then J is a maximal modular right ℓ-ideal. Equivalently, J is a modular maximal right ℓ-ideal.

An element a in the ring R is *right quasi-regular* (abbreviated *right QR*) if $a \in (1-a)R$. The element a in the ℓ-ring R is *right ℓ-quasi-regular in R* (abbreviated *right ℓ-QR*) if a is in the right ℓ-ideal generated by $(1-a)R$. So a is right ℓ-*QR* in R if and only if there are elements $x_1, \ldots, x_n, y_1, \ldots, y_m, z$ in R with

$$|a| \leq \sum_{i=1}^{n} |x_i - ax_i| + \sum_{j=1}^{m} |y_j - ay_j| z.$$

Since

$$|x - ax| = |(x^+ - ax^+) - (x^- - ax^-)| \leq |x^+ - ax^+| + |x^- - ax^-|$$

we may take x_i, y_j, z in R^+. Each right *QR* element is right ℓ-*QR*; in particular, if a is nilpotent of index n, then a is right ℓ-*QR* since $a = (1-a)(a + a^2 + \cdots + a^{n-1})$.

The *circle operation* is defined in R by $a \circ b = a + b - ab$. In any unital ring containing R we have that $1 - (a \circ b) = (1-a)(1-b)$. This implies that $(R, \circ)$ is a monoid with identity element 0 which is isomorphic to the multiplicative submonoid $1 - R$ of the ring. It also implies the equivalence of the following statements: (i) a is right *QR*; (ii) a has a right inverse in $(R, \circ)$; (iii) $1 - a$ has a right inverse in $1 - R$.

The element a in the ℓ-ring R is called *n-right ℓ-QR* if there are elements $x_1, \ldots, x_n$ in R such that $|a| \leq |x_1 - ax_1| + \cdots + |x_n - ax_n|$. If $a \leq 0$, then a is 1-right ℓ-*QR* since $|a| = -a \leq (-a) - a(-a)$. This inequality is equivalent to $0 \leq a \circ (-a)$. Some relations between the element a being right ℓ-*QR* and the comparability to 0 of elements in $a \circ R$ are given in Exercises 3.3.13.

The right ℓ-ideal A is a *right ℓ-QR right ℓ-ideal of R* if each of its elements is right ℓ-*QR* in R. If each element of R is right ℓ-*QR* then R is called a *right ℓ-QR ℓ-ring*.

An ℓ-ideal P of R is a *right ℓ-primitive ℓ-ideal of R* if there is a modular maximal right ℓ-ideal I of R with $P = r_\ell(R/I)$. Since I is modular it is easy to see that P is the largest ℓ-ideal of R that is contained in I. The ℓ-ring R is called (*right*) *ℓ-primitive* if 0 is a right ℓ-primitive ℓ-ideal of R. Clearly, R/P is an ℓ-primitve ℓ-ring if and only if P is a right ℓ-primitive ℓ-ideal of R.

Let

$$\mathscr{R}(R) = \cap \{I : I \text{ is a modular maximal right } \ell\text{-ideal of } R\}, \tag{3.2.21}$$

$$\mathscr{O}(R) = \cap \{P : P \text{ is a right } \ell\text{-primitive } \ell\text{-ideal of } R\}. \tag{3.2.22}$$

Theorem 3.2.25. *(a) For each ℓ-ring R, $\mathscr{O}(R) \subseteq \mathscr{R}(R)$ and the inclusion could be proper.*

(b) $\mathscr{R}(R)$ is the largest right ℓ-QR right ℓ-ideal of R, and $\mathscr{O}(R)$ is the largest right ℓ-QR ℓ-ideal of R.
(c) $\mathscr{R}(R/\mathscr{O}(R)) = \mathscr{R}(R)/\mathscr{O}(R)$ and $\mathscr{O}(R/\mathscr{O}(R)) = 0$.

Proof. The inclusion in (a) and the equations in (c) are obvious. An example where $\mathscr{O}(R) \subset \mathscr{R}(R)$ is given in Exercise 28(b). As for (b), suppose that $a \in \mathscr{R}(R)$. If the right ℓ-ideal generated by $(1-a)R$ is proper, then there is a maximal modular right ℓ-ideal K of R modulo which a is a left identity. Since $a \in K$ we get the contradiction that $I = R$. So a is right ℓ-*QR* and $\mathscr{R}(R)$ is a right ℓ-*QR* right ℓ-ideal of R. Suppose that J is a right ℓ-*QR* right ℓ-ideal of R, and I is a modular maximal right ℓ-ideal of R with e as a left identity modulo I. If $J \not\subseteq I$, then $R = I + J$ and $e = a + b$ with $a \in I$ and $b \in J$. Then

$$
\begin{aligned}
|b| &\leq \sum_i |x_i - bx_i| + \sum_j |y_j - by_j|r \\
&= \sum_i |(x_i - ex_i) + ax_i| + \sum_j |(y_j - ey_j) + ay_j|r
\end{aligned}
$$

and $\{x_i - ex_i, ax_i, y_j - ey_j, ay_j\} \subseteq I$. So $b \in I$ and hence $e \in I$. This contradiction gives that $J \subseteq I$; so $J \subseteq \mathscr{R}(R)$ and $\mathscr{R}(R)$ is the largest right ℓ-*QR* right ℓ-ideal of R. If A is a right ℓ-*QR* ℓ-ideal of R then $A \subseteq \mathscr{R}(R)$, and hence A is contained in each right ℓ-primitive ℓ-ideal; so $A \subseteq \mathscr{O}(R)$ and $\mathscr{O}(R)$ is the largest right ℓ-*QR* ℓ-ideal of R. □

It is now easy to see that the class $\mathscr{J}$, which is defined as the class of all of the right ℓ-*QR* ℓ-rings, is a radical class which contains ℓ-Nil. For it is clearly homomorphically closed, and it satisfies (3.2.2) since $R/\mathscr{O}(R)$ is $\mathscr{J}$-semisimple by Theorem 3.2.25. Note that $\mathscr{J}(R)$ is the largest ℓ-ideal of R which is a right ℓ-*QR* ℓ-ring, and $\mathscr{J}(R) \subseteq \mathscr{O}(R)$. However, since $\mathscr{O}(R)$ need not be an ℓ-*QR* ℓ-ring it is possible to have $\mathscr{J}(R) \subset \mathscr{O}(R)$ (see Exercise 28(a)). But, of course, if one of $\mathscr{J}(R), \mathscr{O}(R)$ or $\mathscr{R}(R)$ is all of R so are the other two.

$\mathscr{J}$ is the upper radical determined by the class of right ℓ-primitve ℓ-rings. To see this first note that if A is a nonzero ℓ-ideal of a right ℓ-primitive ℓ-ring R, then A is not a right ℓ-*QR* ℓ-ring, (otherwise, $A \subseteq \mathscr{O}(R) = 0$); so $\mathscr{J}(A) \subset A$ and A has right ℓ-primitive ℓ-ideals. Hence, the class of right ℓ-primitive ℓ-rings satisfies (3.2.8) and determines an upper radical $\mathscr{P}$. According to Theorem 3.2.4, $R \in \mathscr{P}$ if and only if R has no right ℓ-primitive homomorphic image, and this condition is equivalent to having $\mathscr{O}(R) = R$ and hence $\mathscr{J}(R) = R$, as we have previously noted. We summarize this discussion in

Theorem 3.2.26. *The class $\mathscr{J}$ of right ℓ-QR ℓ-rings is the upper radical determined by the class of right ℓ-primitive ℓ-rings.* □

$\mathscr{J}$ is not *i*-hereditary (Exercise 30), and so it is neither special nor supernilpotent. However, we do have

Theorem 3.2.27. *The class of right ℓ-primitive ℓ-rings is an i-hereditary class of ℓ-prime ℓ-rings.*

Proof. Suppose that R is a right ℓ-primitive ℓ-ring, and let I be a modular maximal right ℓ-ideal of R with $r_\ell(R/I) = 0$. Let A be a nonzero ℓ-ideal of R. Then $R = I + A$. If B is an ℓ-ideal of R with $AB = 0$, then $RB = IB \subseteq I$ and $B \subseteq r_\ell(R/I) = 0$. So R is ℓ-prime.

To see that A is also a right ℓ-primitive ℓ-ring we will first show that $A \cap I$ is a modular maximal right ℓ-ideal of A. If $e \in R$ is a left identity modulo I, then $e = i + f$ where $i \in I$ and $f \in A \backslash I$. So for any $a \in A, fa - a = (e-i)a - a = (ea-a) - ia \in A \cap I$, and $A \cap I$ is a proper modular right ℓ-ideal of A. To see that $A \cap I$ is maximal let $a \in A^+ \backslash A \cap I$. Then $aA \not\subseteq I$ since $R/I = C_R(a+I)$ is an ℓ-faithful ℓ-simple ℓ-module. So $R = I + C(aA)$ and $A = A \cap I + C(aA)$. In particular, if J is a right ℓ-ideal of A which properly contains $A \cap I$ and $a \in J^+ \backslash A \cap I$, then $A = A \cap I + C(aA) \subseteq J$; so $J = A$.

To finish the proof it suffices to show that $A \cap I$ contains no nonzero ℓ-ideals of A. Let B be an ℓ-ideal of A with $B \subseteq A \cap I$, and let D be the ℓ-ideal of R generated by B. Then $D^3 \subseteq B \subseteq I$ (Theorem 3.2.7); so $D^3 = 0$, $D = 0$ since R is ℓ-prime, and hence $B = 0$. □

General conditions on a radical $\mathscr{P}$ are given in Exercise 21 that ensure that the equation $\mathscr{P}(R_n) = \mathscr{P}(R)_n$ holds for each ℓ-ring R and each $n \in \mathbb{N}$. The radical $\mathscr{J}$ does not satisfy these conditions. Nevertheless, the equation $\mathscr{J}(R_n) = \mathscr{J}(R)_n$ is still valid.

If I is a right ideal of R and $1 \le i \le n$, then $I(i) = e_{11}R_n \oplus \cdots \oplus e_{ii}I_n \oplus \cdots \oplus e_{nn}R_n$ will denote the right ideal of R_n which consists of all those matrices whose ith row has entries from I. Here, the e_{ij} are the usual matrix units and are not elements of R_n if R is not unital.

Theorem 3.2.28. *(a) Let J be a right ℓ-ideal of the matrix ℓ-ring R_n. Then J is a maximal right ℓ-ideal of R_n and $R_n^2 \not\subseteq J$ if and only if there is a maximal right ℓ-ideal I of R with $R^2 \not\subseteq I$ and an integer i, $1 \le i \le n$, such that $J = I(i)$.*
(b) Moreover, J is modular if and only if I is modular.
(c) The mapping $P \mapsto P_n$ is a bijection between the set of right ℓ-primitive ℓ-ideals of R and the set of right ℓ-primitive ℓ-ideals of R_n.

Proof. (a) Let J be a maximal right ℓ-ideal of R_n with $R_n^2 \not\subseteq J$. Then $J = (J \cap e_{11}R_n) \oplus \cdots \oplus (J \cap e_{nn}R_n)$ and $J \cap e_{ii}R_n \subset e_{ii}R_n$ for some i. Suppose that $0 \le v \in e_{ii}R_n \backslash J$. Then $C(vR_n) \not\subseteq J$ since $\ell_\ell(R_n; R_n/J) = 0$; so $R_n = J + C(vR_n)$. If $x \in R^+$, then, for every p and q, $xe_{pq} = (y_{k\ell}) + (z_{k\ell})$ where $(y_{k\ell}) \in J^+$ and $(z_{k\ell}) = \Sigma_\ell z_{i\ell}e_{i\ell} \in C(vR_n)^+ \subseteq e_{ii}R_n$. If $p \ne i$, then $x = y_{pq}$ and $0 \le xe_{pq} \le (y_{k\ell})$; so $xe_{pq} \in J$ and $e_{pp}R_n \subseteq J$. Let

$$I = \left\{ a \in R : \sum_{j=1}^{n} ae_{ij} \in J \right\}.$$

Then I is a right ℓ-ideal of R and we will show that $J \cap e_{ii}R_n = e_{ii}I_n$. If $0 \le \Sigma_\ell x_\ell e_{i\ell} \in e_{ii}I_n$, then $x = \vee_\ell x_\ell \in I$ and $\Sigma_\ell x_\ell e_{i\ell} \le \Sigma_\ell xe_{i\ell} \in J$; so $\Sigma_\ell x_\ell e_{i\ell} \in J \cap e_{ii}R_n$. Now, let $0 \le \Sigma_\ell x_\ell e_{i\ell} \in J \cap e_{ii}R_n$. Then, for any $(y_{pq}) \in R_n$, and each j,

$$\left(\sum_\ell x_j e_{i\ell}\right)\left(\sum y_{pq}e_{pq}\right) = \sum_\ell x_j y_\ell e_{i\ell} = \left(\sum_\ell x_\ell e_{i\ell}\right)\left(\sum_\ell y_\ell e_{j\ell}\right) \in J,$$

where $y_\ell = \Sigma_p\, y_{p\ell}$ is the ℓth column sum of (y_{pq}). So $\Sigma_\ell\, x_j e_{i\ell} \in J, x_j \in I$, and $\Sigma_\ell\, x_\ell e_{i\ell} \in e_{ii}I_n$. Thus, $J = I(i)$. If I_1 is a right ℓ-ideal of R with $I \subseteq I_1$, then $J \subseteq I_1(i)$ and hence $I_1(i) = J$ or R_n; so $I_1 = I$ or R and I is maximal. Also, $R^2 \not\subseteq I$ since $(R^2)_n = R_n^2 \not\subseteq J = I(i)$.

Conversely, let $J = I(i)$ where I is a maximal right ℓ-ideal of R with $R^2 \not\subseteq I$. If $0 \leq u = \Sigma_\ell\, a_\ell e_{i\ell} \in e_{ii}R_n \backslash J$, then $a_j \in R^+ \backslash I$ for some j, and, putting $a = a_j$ and $v = ae_{ij}$,

$$vR_n = \left\{\sum_\ell ax_\ell e_{i\ell} : x_\ell \in R\right\} \not\subseteq J$$

since $aR \not\subseteq I$. Since $0 \leq v \leq u$, $uR_n \not\subseteq J$. If $b_1, \ldots, b_n$ are any elements of R^+, then $0 \leq b_j \leq c + ar$ for some $c \in I^+$ and $r \in R^+$ since $R = I + C(aR)$. But then

$$0 \leq \sum_\ell b_\ell e_{i\ell} \leq \sum_\ell ce_{i\ell} + \sum_\ell are_{i\ell} \in J + C(vR_n).$$

Thus, $R_n = J + C(uR_n)$ and J is a maximal right ℓ-ideal of R_n.

(b) If $e \in R$ is a left identity modulo I, then $ee_{ii} \in R_n$ is a left identity modulo $J = I(i)$ since the ith row of $ee_{ii}(x_{k\ell}) - (x_{k\ell})$ is $(ex_{i1} - x_{i1} \cdots ex_{in} - x_{in})$. On the other hand if $f = (f_{k\ell})$ is a left identity modulo $J = I(i)$, then the ith diagonal element of $f(xe_{ii}) - xe_{ii}$, which is in I, is $f_{ii}x - x$. So f_{ii} is a left identity modulo I.

(c) This follows because each right ℓ-primitive ℓ-ideal is an ℓ-prime ℓ-ideal (Theorem 3.2.27) and Q is an ℓ-prime ℓ-ideal of R_n exactly when $Q = P_n$ for some ℓ-prime ℓ-ideal P of R (see Exercise 22(d)). Hence, P is the largest ℓ-ideal of R that is contained in the modular maximal right ℓ-ideal I of R if and only if P_n is the largest ℓ-ideal of R_n that is contained in the modular maximal right ℓ-ideal $J = I(i)$ of R_n. □

This result gives

Theorem 3.2.29. *For each ℓ-ring R, $\mathscr{R}(R_n) = \mathscr{R}(R)_n$, $\mathscr{O}(R_n) = \mathscr{O}(R)_n$ and $\mathscr{J}(R_n) = \mathscr{J}(R)_n$.*

Proof. Let I be a variable for the maximal modular right ℓ-ideals of R, and let P be a variable for the right ℓ-primitive ℓ-ideals of R. Then from (3.2.21), (3.2.22), and the previous result,

$$\mathscr{R}(R_n) = \bigcap_I \bigcap_{i=1}^n I(i) = \bigcap_I I_n = \left(\bigcap_I I\right)_n = \mathscr{R}(R)_n$$

and

$$\mathscr{O}(R_n) = \bigcap_P P_n = \left(\bigcap_P P\right)_n = \mathscr{O}(R)_n.$$

Since $R_n/\mathscr{J}(R_n)$ is ℓ-semiprime $\mathscr{J}(R_n) = T_n$ for some ℓ-ideal T of R (Exercise 22(d)). Since T_n has no proper modular right ℓ-ideals, neither does T by the previous result. Hence T is a $\mathscr{J}$-ℓ-ideal of R and $\mathscr{J}(R_n) = T_n \subseteq \mathscr{J}(R)_n$. For the same reasons $\mathscr{J}(R)_n$ is a $\mathscr{J}$-ℓ-ideal of R_n; so $\mathscr{J}(R)_n \subseteq \mathscr{J}(R_n)$ and we have the equality $\mathscr{J}(R_n) = \mathscr{J}(R)_n$. □

One can, of course, construct $\mathscr{J}_{\text{left}}$, the *left Johnson radical*, using left ℓ-quasi-regularity, maximal modular left ℓ-ideals, and left ℓ-primitivity. In general, the radicals $\mathscr{J}$ and $\mathscr{J}_{\text{left}}$ are distinct (see Exercise 29).

Exercises.

$\mathscr{P}$ denotes a radical class of ℓ-rings.

1. Verify the characterizations of $\mathscr{P}(R)$ given in (3.2.3), (3.2.4), and (3.2.5).
2. (a) If $\varphi(R)$ is a homomorphic image of R, then $\varphi(\mathscr{P}(R)) \subseteq \mathscr{P}(\varphi(R))$. Give an example where the inclusion is proper.
   (b) If A is an ℓ-ideal of R with $A \subseteq \mathscr{P}(R)$, then $\mathscr{P}(R/A) = \mathscr{P}(R)/A$.
3. (a) If $\{A_i\}_i$ is a family of $\mathscr{P}$-ℓ-ideals of R, then ΣA_i is a $\mathscr{P}$-ℓ-ideal of R.
   (b) $\mathscr{P}(\oplus_i R_i) = \oplus_i \mathscr{P}(R_i)$ for any family $\{R_i\}$ of ℓ-rings.
   (c) $\mathscr{P}(\Pi_i R_i) \subseteq \Pi_i \mathscr{P}(R_i)$ and the inclusion could be proper.
4. The class $\mathscr{S}\mathscr{P}$ is extensionally closed and is also closed with respect to the formation of subdirect products.
5. In the lower radical construction, if $\mathscr{P}_\nu = \mathscr{P}_{\nu+1}$, then $\mathscr{P} = \mathscr{P}_\nu$.
6. Let $\mathbb{Z}_0$ be the additive group of the integers with zero multiplication, and let $\mathscr{N}$ be the class of nilpotent ℓ-rings. Then ℓ-$\beta = \mathscr{L}(\mathbb{Z}_0) = \mathscr{L}(\mathscr{N})$.
7. Let ℓ-$N_1(R) = \ell$-$N(R)$ be the ℓ-ideal of Theorem 3.2.9. For a limit ordinal ν let ℓ-$N_\nu(R) = \bigcup_{\mu<\nu} \ell$-$N_\mu(R)$, and, for each ordinal ν, let ℓ-$N_{\nu+1}(R)/\ell$-$N_\nu(R) = \ell$-$N(R/\ell$-$N_\nu(R))$. If ℓ-$N_\nu(R) = \ell$-$N_{\nu+1}(R)$, show that ℓ-$\beta(R) = \ell$-$N_\nu(R)$. If A is an ℓ-ideal of R show that, for each ordinal ν, ℓ-$N_\nu(A) = \ell$-$N_\nu(R) \cap A$. Show that ℓ-$N(R_n) = [\ell$-$N(R)]_n$.
8. Let F be a totally ordered field and let $S = F_2$. Let R be the ℓ-F-algebra of all $\mathbb{N} \times \mathbb{N}$ strictly upper triangular row finite matrices over S with only finitely many entries not in Fe_{12} ($e_{12} = \left(\begin{smallmatrix} 0 & 1 \\ 0 & 0 \end{smallmatrix}\right)$). Show that ℓ-$N(R)$ consists of all those matrices in R which have only a finite number of nonzero entries, and $R^2 \subseteq \ell$-$N(R)$. Hence $R = \ell$-$\beta(R) = \ell$-$N_2(R) \supset \ell$-$N_1(R)$ (see Exercise 7).
9. Let C be the centroid of the ring R; see the discussion prior to Theorem 3.2.6.

   (a) If $1 \in R$, then $C \cong Z(R)$ ($Z(R)$ is the center of R).
   (b) If $\alpha, \beta \in C$, then $(\alpha\beta - \beta\alpha)(R^2) = 0$.
   (c) If $R^2 = R$, or $\ell(R) = 0$, or $r(R) = 0$, then C is commutative.

(d) If $\ell(R) = r(R) = 0$ and $\alpha \in \mathrm{End}_{\mathbb{Z}}(R)$ with $\alpha(xzy) = x(\alpha z)y$ for all x, y, z in R, then $\alpha \in C$.
(e) If R is semiprime (respectively, prime), then C is reduced (respectively, a domain).
(f) If R is simple, then C is a field.
(g) If R is an ℓ-semiprime (respectively, ℓ-prime) ℓ-ring, then C is po-reduced (respectively, a po-domain).

10. Suppose that $\ell\text{-}\beta \subseteq \mathscr{P}$ and R is an ℓ-ring.

(a) If C is the centroid of R and $\alpha \in C^+$, then $\alpha\mathscr{P}(R) \subseteq \mathscr{P}(R)$.
(b) If R is a po-algebra over D and D is directed, then $\mathscr{P}(R)$ is a subalgebra.

11. If $\ell\text{-}\beta \subseteq \mathscr{P}$ and A is an ideal and a sublattice of the ℓ-ring R, then $\mathscr{P}(A) \subseteq \mathscr{P}(R)$.

12. Let the polynomial rings $F = \mathbb{Z}[x]$ and $R = F[y]$ each be totally ordered with the antilexicographic order. Let P be the ideal of R generated by $x^2 + y^2$. Show that P is a prime ideal of R but the convex ideal $C(P)$ that is generated by P is not prime.

13. Prove (a), (b), and (c) of Theorem 3.2.13.

14. (a) The intersection of a chain of ℓ-prime ℓ-ideals in an ℓ-ring R is an ℓ-prime ℓ-ideal.
(b) Each ℓ-prime ℓ-ideal of R contains a minimal ℓ-prime ℓ-ideal.
(c) The ℓ-ideal P is a minimal ℓ-prime ℓ-ideal of R if and only if $R^+ \backslash P$ is an M^+-system that is maximal with respect to excluding 0.

15. Show that $\ell\text{-}\beta(R) = \cap\{P : P \text{ is a minimal } \ell\text{-prime } \ell\text{-ideal of } R\} = \{a \in R : \text{each } M^+\text{-sequence in } R \text{ that begins with } |a| \text{ contains } 0\}$.

16. (a) Show that the ℓ-ring R is ℓ-semiprime if and only if it has no nonzero directed nilpotent ideals (or no nonzero directed nilpotent right ideals).
(b) Show that R is ℓ-prime if and only if $IJ \neq 0$ for any nonzero directed ideals I and J.

17. An ℓ-ideal A of the ℓ-ring R is an *ℓ-essential ℓ-ideal* of R if $A \cap B \neq 0$ for each nonzero ℓ-ideal B of R.

(a) If $A \cap A^* = 0$, then A is ℓ-essential in R/A^*.
(b) Suppose that $A \neq 0$. Show that R is ℓ-prime if and only if A is an ℓ-prime ℓ-ring and A is ℓ-essential in R.
(c) If $\mathscr{P}$ is a special radical and $\mathscr{P}(A) = 0$, show that $\mathscr{P}(R/A^*) = 0$.

18. Show that the class of all ℓ-prime ℓ-rings is a special class.

19. Let $\mathscr{A}$ be the class of all archimedean ℓ-rings. Show that $\mathscr{A}$ satisfies the condition in (3.2.19). Consequently, if $\mathscr{M}$ is a special class of ℓ-rings, then so is $\mathscr{M} \cap \mathscr{A}$.

20. For the ℓ-ring R let $N = \{a \in R : a \text{ is nilpotent}\}$ and let $M = \{a \in R : |a| \text{ is nilpotent}\}$.

    (a) Show that $\ell\text{-}\beta(R) = N_g(R)$ iff $\ell\text{-}\beta(R) = N$.
    (b) Show that $\ell\text{-}\beta(R) = \ell\text{-}N_g(R)$ iff $\ell\text{-}\beta(R) = M$.
    (c) Suppose that $\ell\text{-}\beta(R) = N_g(R)$. In Theorem 3.2.23 replace the last phrase in (c) by "ab is nilpotent if and only if $|a||b|$ is nilpotent," and prove the theorem for this R.

21. A class of ℓ-rings is *matrix-closed* if whenever R is in the class so is the canonically ordered $n \times n$ matrix ring R_n, for each $n \in \mathbb{N}$.

    (a) Show that $\mathscr{P}$ and $\mathscr{S}\mathscr{P}$ are both matrix-closed if and only if $\mathscr{P}(R_n) = \mathscr{P}(R)_n$ for each ℓ-ring R and each $n \in \mathbb{N}$.
    (b) Suppose that $\mathscr{P}$ is left i-hereditary, $\ell\text{-}\beta \subseteq \mathscr{P}$, and $\mathscr{P}(R)$ contains each left $\mathscr{P}$-ℓ-ideal of R (for any R). Then $\mathscr{P}(R_n) = \mathscr{P}(R)_n$. (*Hint*: If $R \in \mathscr{P}$, show that each of the columns $R_n e_{ii}$ of R_n is in $\mathscr{P}$. On the other hand, if A is a nonzero $\mathscr{P}$-ℓ-ideal of R_n, then $B = C(R_n A R_n) \in \mathscr{P}$, and $B = D_n$ where $D = C(\Sigma RaR)$ and a varies over the entries of the matrices in A; by considering the first column Be_{11} of B, show that $D \in \mathscr{P}$.)
    (c) The upper and lower ℓ-nil radicals satisfy the conditions in (b).
    (d) $\mathscr{S}\ell\text{-}N_g$ is not matrix-closed. Show that $\ell\text{-}N_g$ fails to satisfy just one of the conditions in (b).

22. (a) For each ℓ-ring R and each $n \in \mathbb{N}$ the mapping $I \mapsto I_n$ is a lattice monomorphism between the lattice of ℓ-ideals of R and the lattice of ℓ-ideals of the matrix ring R_n.
    (b) If $a \in C(RaR)$ for each $a \in R^+$, show that this mapping is an isomorphism.
    (c) Let J be an ℓ-ideal of R_n and let I be the additive subgroup of R that is generated by all of the entries of the matrices in J^+. Show that I is an ℓ-ideal of $R, J \subseteq I_n$, and $I_n^3 \subseteq J$.
    (d) Show that the mapping in (a) restricts to a bijection between the set of ℓ-prime (respectively, ℓ-semiprime) ℓ-ideals of R and the set of ℓ-prime (respectively, ℓ-semiprime) ℓ-ideals of R_n. (Here, A is an ℓ-semiprime ℓ-ideal of R if R/A is an ℓ-semiprime ℓ-ring.)

23. Let R_Γ be the ℓ-ring of $\Gamma \times \Gamma$ row finite matrices over the ℓ-ring R. If I is a right ℓ-ideal of R and $\lambda_0 \in \Gamma$, let $I(\lambda_0) = \{(a_{\lambda\mu}) \in R_\Gamma : a_{\lambda_0\mu} \in I, \forall \mu \in \Gamma\}$. Prove each of the following.

    (a) $I(\lambda_0)$ is a right ℓ-ideal of R.
    (b) $I(\lambda_0)$ is modular if and only if I is modular.
    (c) $R_\Gamma^2 \nsubseteq I(\lambda_0)$ if and only if $R^2 \nsubseteq I$.
    (d) A right ℓ-ideal J of R_Γ is maximal, has $R_\Gamma^2 \nsubseteq J$, and $\oplus_\lambda e_{\lambda\lambda} R_\Gamma \nsubseteq J$ if and only if $J = I(\lambda_0)$ for some $\lambda_0 \in \Gamma$ and some maximal right ℓ-ideal I of R with $R^2 \nsubseteq I$.
    (e) If P is a right ℓ-primitive ℓ-ideal of R, then P_Γ is a right ℓ-primitive ℓ-ideal of R_Γ, but not every right ℓ-primitive ℓ-ideal of R_Γ is of this form. (If F is

a totally ordered field consider the ℓ-ideal Q of $F_{\mathbb{N}}$ consisting of all of the matrices of finite rank : $(a_{ij}) \in Q$ iff $\exists n \in \mathbb{N}$ such that $\forall i, \forall m \geq n, a_{im} = 0$.)

24. Let $\mathscr{R}$ be a radical class of rings. If A is an ideal of the ring R show that $\mathscr{R}(A)$ is an ideal of R; hence $\mathscr{S}\mathscr{R}$ is i-hereditary. (If $x \in R$ show that the map $\mathscr{R}(A) \longrightarrow (x\mathscr{R}(A) + \mathscr{R}(A))/\mathscr{R}(A)$ given by left multiplication by x is a homomorphism).

25. Let $\mathscr{Z}$ be the class of zero ℓ-rings and let $\mathscr{N}$ be the class of nilpotent ℓ-rings. Show that $\mathscr{U}\mathscr{Z} = \mathscr{U}\mathscr{N} = \{R : R = R^{[2]}\}$, that $\mathscr{S}\mathscr{U}\mathscr{Z} = \{R : R$ has no nonzero ℓ-idempotent ℓ-ideals$\}$, and that $\mathscr{S}\mathscr{U}\mathscr{Z}$ is i-hereditary.

26. Let G be a multiplicative semigroup with 0 which has no nonzero nilpotent elements. Suppose that $a_1, \ldots, a_n \in G$ and $b_i = a_{\tau(i)}$ for $i = 1, \ldots, n$, where τ is a permutation. If $a_1 \cdots a_n = 0$ show that $b_1 \cdots b_n = 0$. (Show that $ab = 0 \Rightarrow ba = bxa = 0$, and $abc = 0 \Rightarrow bac = 0$.)

27. (a) Let e be a positive nonzero idempotent of R, and suppose that e is a left d-element of R. Then e is not right ℓ-QR. However, e may be right ℓ-QR even if $R = D(R)$. (Try 2×2 matrices.)
    (b) If a^n is right ℓ-QR for some $n \in \mathbb{N}$, then a is right ℓ-QR.

28. (a) Let the ℓ-ring R be given by $R = \mathbb{Q}a \oplus \mathbb{Q}1$ as an ℓ-group with $a^2 = a$. Show that $\mathscr{J}(R) \subset \mathscr{O}(R)$.
    (b) Let $F = \mathbb{R}[y]$ be the lexicographically ordered polynomial ring over the reals, and let $\alpha^{(k)}$ be the kth derivative of $\alpha \in F$. Let $R = F[x;\,']$ be the Orc extension determined by $'$. So the elements of R are polynomials $\Sigma_i x^i \alpha_i$, addition is as usual and multiplication is given by $\alpha x = x\alpha + \alpha'$ for $\alpha \in F$; more generally,

$$\alpha x^n = \sum_{k=0}^{n} \binom{n}{k} x^{n-k} \alpha^{(k)}.$$

    Then R is an ℓ-ring if its positive cone is given by $\Sigma_i x^i F^+$. Show that R is ℓ-simple (in fact, R is simple), $\mathscr{O}(R) = 0$ and $\mathscr{R}(R) = xR$.

29. Let $R = \mathbb{Q}e \oplus \mathbb{Q}a$ be the ℓ-algebra with the canonical basis $\{e, a\}$ and with multiplication given by $e^2 = e$, $ea = e$, $ae = a$, $a^2 = a$. Show that R is an ℓ-simple right f-ring which is left ℓ-primitive and right ℓ-QR.

30. Let $R = \mathbb{Q}a \oplus \mathbb{Q}b$ be the ℓ-algebra with the canonical basis $\{a, b\}$ and with multiplication given by $a^2 = ab = ba = b^2 = a$. Show that R is a $\mathscr{J}$-ℓ-ring with a nonzero $\mathscr{J}$-semisimple ℓ-ideal.

31. Let $\mathscr{J}^+$ be the class given by : $R \in \mathscr{J}^+$ if and only if each element of R^+ is right ℓ-QR. Show that $\mathscr{J}^+$ is a radical class and $\mathscr{J} \subset \mathscr{J}^+$. (Call a right ℓ-ideal I of R *po-modular* if R has a positive left identity modulo I, and call it *right po-ℓ-QR* if each element of I^+ is right ℓ-QR in R. Then $\mathscr{J}(R) \subseteq \mathscr{J}^+(R)$, $\mathscr{O}(R) \subseteq \mathscr{O}^+(R)$, and $\mathscr{R}(R) \subseteq \mathscr{R}^+(R)$, where $\mathscr{O}^+$ and $\mathscr{R}^+$ have the obvious meaning.) Show that $\mathscr{J}^+(R_n) = (\mathscr{J}^+(R))_n$ for each ℓ-ring R and each $n \in \mathbb{N}$. If each maximal modular right ℓ-ideal of R is po-modular,

show that $\mathscr{J}(R_n) = \mathscr{J}^+(R_n)$. Show that R has this property if $\bar{R} = D(\bar{R}_{\bar{R}})$ where $\bar{R} = R/\mathscr{O}(R)$.

32. (a) Show that a minimal ℓ-ideal in an ℓ-ring is either ℓ-simple or is a zero ℓ-ring.
(b) The minimal ℓ-ideal in a subdirectly irreducible ℓ-ring is called its *heart*. Let $\mathscr{N}$ be an isomorphically closed class of ℓ-simple ℓ-rings and let $\mathscr{M} = \{R : R$ is subdirectly irreducible and its heart is in $\mathscr{N}\}$. Show that $\mathscr{M}$ is a special class.
(c) Let $\mathscr{M}$ be the class of all subdirectly irreducible ℓ-rings with an ℓ-idempotent heart; so $\mathscr{N}$ is the class of all ℓ-simple ℓ-rings. The upper radical determined by $\mathscr{M}$ is called the *antisimple* radical and will be denoted by ℓ-$\mathscr{A}$. Show that $R \in \ell$-$\mathscr{A}$ iff R has no ℓ-prime subdirectly irreducible homomorphic image.
(d) Prove that ℓ-Nil $\subseteq \ell$-$\mathscr{A}$.
(e) Assume that for each $a \in R$ the chain $\langle a\rangle \supseteq \langle a^2\rangle \supseteq \cdots$ is finite. Show that ℓ-Nil(R) $= \ell$-$\mathscr{A}(R)$. (If $\langle a^n\rangle = \langle a^{n+1}\rangle$ and C is an ℓ-ideal of $\langle a^n\rangle$ which is maximal with respect to not containing a^n, show that C is a maximal ℓ-ideal of $\langle a^n\rangle$.)
(f) If R has d.c.c. on ℓ-ideals, show that ℓ-$\mathscr{A}(R)$ is nilpotent.

33. Show that the following statements are equivalent for an ℓ-ring R. The class of ℓ-rings that satisfy these conditions will be denoted by ℓ-$\mathscr{I}$.
(a) Each subdirectly irreducible homomorphic image of R is ℓ-prime.
(b) Each nonzero homomorphic image of R is a subdirect product of a family of subdirectly irreducible ℓ-prime ℓ-rings.
(c) If A is an ℓ-ideal of R, then $A = \cap\{P : P$ is an ℓ-ideal of $R, A \subseteq P$ and R/P is subdirectly irreducible with an ℓ-idempotent heart$\}$.
(d) Each ℓ-ideal of R is ℓ-idempotent.
(e) Each principal ℓ-ideal of R is ℓ-idempotent.
(f) Each homomorphic image of R is ℓ-semiprime.

34. (a) Show that the class ℓ-$\mathscr{I}$ that is defined in the previous exercise is an i-hereditary radical.
(b) Prove that a subdirectly irreducible ℓ-ring with a nilpotent heart is ℓ-$\mathscr{I}$-semisimple.
(c) Let A be an ℓ-ideal of R. Show that $A \subseteq \ell$-$\mathscr{I}(R)$ iff each (principal) ℓ-ideal of R that is contained in A is ℓ-idempotent.
(d) Show that ℓ-$\mathscr{I}(R) = \cap\{M : R/M$ is subdirectly irreducible with a nilpotent heart$\}$. (If $a \notin \ell$-$\mathscr{I}(R)$, let $b \in \langle a\rangle$ with $\langle b\rangle \neq \langle b\rangle^{[2]}$, and let M be an ℓ-ideal of R containing $\langle b\rangle^{[2]}$ which is maximal with respect to not containing b.)

35. (a) Show that ℓ-$\mathscr{I} \subseteq \mathscr{S}\ell$-$\mathscr{A}$, ℓ-$\mathscr{A} \subseteq \mathscr{S}\ell$-$\mathscr{I}$ and ℓ-$\mathscr{I} \cap \ell$-$\mathscr{A} = 0$.
(b) Show that ℓ-$\mathscr{I}$ is the largest radical $\mathscr{P}$ with $\mathscr{P} \cap \ell$-$\mathscr{A} = 0$.
(c) Show that ℓ-$\mathscr{A}$ is the largest radical $\mathscr{P}$ with $\mathscr{P} \cap \ell$-$\mathscr{I} = 0$.

36. Let R be an ℓ-semiprime ℓ-ring. Show that $\ell(F(R);R) + r(F(R);R) \subseteq F^{\perp}$.

37. Let V be a vector lattice over the totally ordered field F. An element $u \in V^+$ is an *F-strong order unit of* V if it generates V as a convex ℓ-subspace: $V = C_F(u)$. The *i-radical* of V is defined by

$$i(V) = \{x \in V : F|x| \leq u \text{ for every } F\text{-strong order unit } u \text{ of } V\}.$$

Verify each of the following.

(a) $i(V) = 0$ iff V has an F-strong order unit and is F-archimedean (that is, V has no nonzero bounded subspaces).
(b) If u is an F-strong order unit of V, then $i(V) = \{x \in V : F|x| \leq u\}$ is a proper subset of V.
(c) $i(V)$ is a convex ℓ-subspace of V and $i(V/i(V)) = 0$.
(d) No element of $i(V)$ is an F-strong order unit of V.
(e) If R is an ℓ-algebra over F, then $i(R)$ is an ℓ-ideal of R.
(f) If R is an ℓ-algebra over F and $I = i(R)$ has an F-strong order unit, then $i(I^{[n]}) \supseteq I^{[n+1]}$.
(g) If R is an F-ℓ-algebra and $i(R)$ is finite dimensional over F, then $i(R)$ is nilpotent.

Notes. The general theory of radicals in rings began with the papers by Amitsur [AM1], [AM2], [AM3], and the paper by Kurosh [KU]; Amitsur considers radicals in the more general setting of a complete lattice. The books by Divinsky [DIV], Gardner and Wiegandt [GW] and Szász [SZ] are good references. The translation of the general theory to ℓ-radicals for ℓ-rings occurs in the papers by Shatalova [SH1] and [SH2] and later, independently, by Steinberg [ST7]. Earlier, Birkhoff and Pierce [BP] considered some aspects of the lower ℓ-nil radical and Johnson [JOH1] defined and studied the analogue for f-rings of the Jacobson radical of a ring. Other papers that are concerned with specific ℓ-radicals and the structure of ℓ-rings are Pierce [PI1], Diem [DI], Keimel [KE4], Ma [M2], Shyr [SHY], Shyr and Viswanathan [SV], and Steinberg [ST23]. The fact that an ℓ-ring is nil if and only if it is locally nilpotent (Theorem 3.2.19), which is not true for rings, is due to Shatalova [SH2], and the characterization of a completely ℓ-prime ℓ-ideal given in Theorem 3.2.24 was given by Diem [DI] for an *sp*-ℓ-ring, and it was noted by Steinberg [ST8] that it holds in any ℓ-ring. Exercises 32 through 35 come from Shatalova [SH1] and Exercise 37 is due to Birkhoff and Pierce [BP].

3.3 f-Rings

Because of the relative ease with which one can compute in a totally ordered ring the structure of an f-ring is easier to determine than the structure of an arbitrary ℓ-ring. Moreover, the answers to many questions about ℓ-rings are frequently more readily available in the variety of f-rings. Here, we present the basic theory of f-rings. We will see that the set of nilpotent elements is an ideal and the ℓ-primitive f-rings are all unital and ℓ-simple. We start with a characterization of subrings of the reals $\mathbb{R}$ that is analogous to the characterization of its additive subgroups given in Theorem 2.3.10.

If A and B are subgroups of the additive group of a ring R let

$$I(A,B) = \{r \in R : rA \subseteq B\}. \tag{3.3.1}$$

Then $I(A,B)$ is a subgroup of R, $I(A) = I(A,A)$ is a subring of R, and $I(A,B)$ is a left-$I(B)$-right-$I(A)$-bimodule.

Recall from Exercise 2.2.8 that $\mathrm{Hom}_{\mathbb{Z}}(A,B)^+$, the set of isotone group homomorphisms from the directed abelian po-group A to the abelian po-group B, is a partial order of the group $\mathrm{Hom}_{\mathbb{Z}}(A,B)$. Let $F(A,B)$ be the subgroup of $\mathrm{Hom}_{\mathbb{Z}}(A,B)$ that is generated by $\mathrm{Hom}_{\mathbb{Z}}(A,B)^+$, and put $F(A) = F(A,A)$ and $F(B) = F(B,B)$.

Theorem 3.3.1. *Let A and B be nonzero subgroups of $\mathbb{R}$.*

(a) If $f : A \longrightarrow B$ is an isotone group homomorphism, then there is a unique element $t \in \mathbb{R}^+$ such that $f(a) = ta$ for each $a \in A$.
(b) The map $\psi_A : F(A) \longrightarrow I(A)$ that arises from (a) is an isomorphism of totally ordered rings, and the map $\psi_{A,B} : {}_{F(B)}F(A,B)_{F(A)} \longrightarrow {}_{I(B)}I(A,B)_{I(A)}$ is an isomorphism of totally ordered bimodules.

Proof. (a) If $\ker f = A$ let $t = 0$; otherwise, $\ker f = 0$. Let $0 < a_1, a_2 \in A$. If $f(a_1)/f(a_2) < a_1/a_2$, then $f(a_1)/f(a_2) < \frac{m}{n} < a_1/a_2$ for some $m/n \in \mathbb{Q}$ with $n > 0$. Then $f(na_1) < f(ma_2)$, but $ma_2 < na_1$. Similarly, $f(a_1)/f(a_2)) > a_1/a_2$ is impossible. Thus, $f(a_1)/f(a_2) = a_1/a_2$ and $f(a_1)/a_1 = f(a_2)/a_2 = t$; hence, $f(a) = ta$ for each $a \in A$. The uniqueness of t is clear.

(b) If $f, g \in \mathrm{Hom}_{\mathbb{Z}}(A,B)^+$ and $f(a) = sa$, $g(a) = ta$ and $s \geq t$, then $f(a) \geq g(a)$ for each $a \in A^+$. Thus, $f \geq g$, and $F(A,B) = \{f - g : f, g \in \mathrm{Hom}_{\mathbb{Z}}(A,B)^+\}$ is a totally ordered group. The map $\psi_{A,B} : F(A,B) \longrightarrow I(A,B)$ is given by $\psi_{A,B}(h) = u$, where $h(a) = ua$ for each $a \in A$. The rest of the proof is left to the reader. □

Theorem 3.3.2. *Let R be a totally ordered archimedean ℓ-ring. Then, either $R^2 = 0$ and $(R,+)$ is isomorphic to a subgroup of $\mathbb{R}$, or there is a unique embedding of R into $\mathbb{R}$.*

Proof. By Theorem 2.3.10 the additive group of R can be embedded in $\mathbb{R}$. Let $f : R \longrightarrow A = f(R) \subseteq \mathbb{R}$ be an isomorphism of ℓ-groups. If R is not a domain, then $ab = 0$ for some $0 < a, b \in R$. If $c, d \in R^+$, then $c \leq na$ and $d \leq mb$ for some $n, m \in \mathbb{N}$. So $cd = 0$ and hence $R^2 = 0$. Suppose that R is a domain. Let E be the subring of $\mathrm{End}_{\mathbb{Z}}(R)$ consisting of the left multiplication maps. Then the composite $\psi_A \varphi \rho$,

$$R \xrightarrow{\rho} E \xrightarrow{\varphi} F(A) \xrightarrow{\psi_A} I(A),$$

is an embedding of R into $\mathbb{R}$. Here, $\rho(x)$ is left multiplication by $x, \varphi(\rho(x)) = f\rho(x)f^{-1}$, and ψ_A is the isomorphism given in Theorem 3.3.1. So $(\psi_A \varphi \rho)(x) = t_x$ where $t_x f(u) = f(xu)$ for each $u \in R$. To show the uniqueness of this embedding it suffices to show that if $\gamma : A \longrightarrow B$ is an isomorphism between subrings of $\mathbb{R}$, then $\gamma = 1$. By (a) of Theorem 3.3.1 γ is multiplication by some $t \in \mathbb{R}^+$. If $0 \neq a \in A$, then $ta^2 = \gamma(a^2) = t^2a^2$ gives that $t = 1$. □

The theory of radicals in the variety of f-rings is much nicer than it is in the variety of all ℓ-rings. In order to investigate the nil radical of an f-ring we first note the following.

Theorem 3.3.3. *Let a and b be elements of the po-ring R with $ab \leq ba$ and let $n \in \mathbb{N}$.*

(a) If $aR^+ + R^+a \subseteq R^+$, then $a^nb \leq ba^n$.
(b) If $bR^+ + R^+b \subseteq R^+$, then $ab^n \leq b^na$.
(c) If $aR^+ + bR^+ + R^+a + R^+b \subseteq R^+$, then $a^nb^n \leq (ab)^n \leq (ba)^n \leq b^na^n$.

Proof. The conclusions all follow by induction. For (a), if $a^nb \leq ba^n$, then $a^{n+1}b \leq aba^n \leq ba^{n+1}$, and similarly for (b). For (c), if $a^nb^n \leq (ab)^n \leq (ba)^n \leq b^na^n$, then $a^{n+1}b^{n+1} = a(a^nb^n)b \leq a(ba)^nb = (ab)^{n+1} \leq (ba)^{n+1} = b(ab)^na \leq b(ba)^na \leq b^{n+1}a^{n+1}$. □

For the ring R let

$$N_n = \{a \in R : a^n = 0\}. \tag{3.3.2}$$

If $\mathscr{V}$ is a variety of ℓ-rings, then $\mathscr{S}(\mathscr{V})$ will denote the class of all ℓ-rings R for which $R/\ell\text{-}\beta(R)$ is in $\mathscr{V}$. We will refer to those ℓ-rings in $\mathscr{S}(\mathscr{V})$ as *extended $\mathscr{V}$-ℓ-rings*. The class of extended f-rings will be denoted by $\mathscr{S}(f)$. Among the ℓ-rings in $\mathscr{S}(f)$ are the triangular matrix rings over f-rings, as well as each almost f-ring (Theorem 3.2.24) and each unital one-sided f-ring (Theorem 3.8.10).

Theorem 3.3.4. *If R is an f-ring, then N_n is a nilpotent ℓ-ideal of R of index at most n and*

$$\ell\text{-}\beta(R) = \bigcup_{n=1}^{\infty} N_n = \sum_{n=1}^{\infty} N_n.$$

Consequently, if R is an extended f-ring, then $\ell\text{-}\beta(R) = \beta(R) = \ell\text{-}N_g(R) = N_g(R)$.

Proof. First, assume that R is totally ordered. If $a, b \in N_n$ and $|a| \leq |b|$, then

$$|(a-b)^n| = |a-b|^n \leq (|a| + |b|)^n \leq 2^n|b|^n = 0;$$

so $a - b \in N_n$. Also, if $x \in R$, then $|a||x| = |ax|$ and $|x||a| = |xa|$ are comparable; so ax and $xa \in N_n$ by Theorem 3.3.3(c). Now let R be an f-ring. If $a, b \in N_n$ and $x \in R$, then $(\overline{a} - \overline{b})^n = (\overline{ax})^n = (\overline{xa})^n = 0$ in any totally ordered homomorphic image $\overline{R}$ of R. Thus, $a - b, ax$ and xa are in N_n since R is a subdirect product of totally ordered rings. Also, if $|a| \leq |b|$ with $b \in N_n$, then $|a^n| \leq |b|^n = |b^n| = 0$; so $a \in N_n$. Clearly, $N_n^n = 0$ (see (3.2.13)). For the second statement, since $R/\ell\text{-}\beta(R)$ is a reduced f-ring, $\ell\text{-}N_g(R) \subseteq \ell\text{-}\beta(R), \beta(R) \subseteq \ell\text{-}\beta(R)$, and $N_g(R) \subseteq \ell\text{-}\beta(R)$. Thus, the four radicals are equal since the reverse inclusions hold in any ℓ-ring. □

Note that the second statement in Theorem 3.3.4 applies to any d-ring since $\ell_\ell(R) = r_\ell(R) = 0$ in an ℓ-semiprime ℓ-ring, and a d-ring in which these annihilators vanish is an f-ring. It also applies to an ℓ-ring R for which R/A is an f-ring for some ℓ-ideal A of R that is contained in ℓ-$\beta(R)$. However, we may still have that ℓ-$N(R) \subset \ell$-$\beta(R)$ in an ℓ-ring for which R/ℓ-$N(R)$ is an f-ring (see Exercise 3.2.8).

Theorem 3.3.4 gives some information about the ℓ-prime ℓ-ideals in an f-ring.

Theorem 3.3.5. *Let R be an extended f-ring.*

(a) Each ℓ-prime ℓ-ideal of R is irreducible and is a completely prime ideal.

(b) The set of minimal prime ideals of R is equal to the set of minimal ℓ-prime ℓ-ideals of R.

(c) The following statements are equivalent.

(i) R is ℓ-semiprime.

(ii) For $a, b \in R$, $ab = 0$ iff $|a| \wedge |b| = 0$.

(iii) $X^\perp = \ell(X)$ for each subset X of R.

(iv) The set of minimal prime ideals of R is equal to the set of minimal prime subgroups of R.

Proof. (a) is a consequence of Theorem 3.2.24. However, we will give a slightly more direct proof. We may assume that R is an ℓ-prime f-ring. If $a \wedge b = 0$, then $\langle a \rangle \cap \langle b \rangle = 0$ (see 2.5.2), and hence $\langle a \rangle \langle b \rangle = 0$. Thus $a = 0$ or $b = 0$ and R is totally ordered. If $ab = 0$ and $|a| \leq |b|$, then $a^2 = |a|^2 \leq |a||b| = 0$. But R is reduced by the previous result; so R is a domain.

(b) This follows from (a) and Theorems 3.3.4 and 3.2.23 since each prime ideal and each ℓ-prime ℓ-ideal contains ℓ-$\beta(R)$.

(c) If R is ℓ-semiprime and $ab = 0$, then $(|a| \wedge |b|)^2 \leq |ab| = 0$. Since R is reduced, $|a| \wedge |b| = 0$. On the other hand, if (ii) holds and $a^2 = 0$, then $|a| = |a| \wedge |a| = 0$; so R is reduced. Thus, (i) and (ii) are equivalent, and clearly (ii) and (iii) are equivalent. To see that (i) implies (iv) let R be a semiprime f-ring and let P be a minimal prime subgroup of R. Then P is an ℓ-ideal by Theorem 3.1.2. If $a^2 \in P$, then $a^2 \wedge b = 0$ for some $b \notin P$ by Theorem 2.4.3. But then $(|a| \wedge b)^3 \leq a^2 b = 0$; so $|a| \wedge b = 0$, $a \in P$, and P is a prime ℓ-ideal of R. If Q is a prime ℓ-ideal with $Q \subseteq P$, then Q is a prime subgroup by (a). So $Q = P$ and P is a minimal prime ideal by (b). On the other hand, each minimal prime ideal P contains a minimal prime subgroup which necessarily must be equal to P. Thus, (i) implies (iv). Now, if (iv) holds, then it follows from (b) and Exercise 3.2.15 (or Theorem 2.4.2) that ℓ-$\beta(R) = 0$. This completes the proof. □

We turn now to the Johnson radical, and we first examine the maximal modular right ℓ-ideals and the right ℓ-primitive ℓ-ideals in an f-ring.

Theorem 3.3.6. *Let R be an f-ring.*

(a) If I is a maximal modular right ℓ-ideal of R, then I is a maximal ℓ-ideal.

(b) The ℓ-ideal P of R is right ℓ-primitive if and only if R/P is an ℓ-simple unital totally ordered domain.

Proof. Let P be the largest ℓ-ideal of R that is contained in the maximal modular right ℓ-ideal I. Since P is an ℓ-prime ℓ-ideal (Theorem 3.2.27), R/P is a totally ordered domain (Theorem 3.3.5). By passing to R/P we may assume that $P = 0$. Let $a \in R^+ \backslash I$. If $aI \not\subseteq I$, then $I \subset C(aI) = R$. If e is a left identity modulo I, then $ae \leq ai$ for some $i \in I$; so $e \leq i$ and $e \in I$. Thus, $aI \subseteq I$, $I = P$ is an ℓ-ideal, and R/P is unital and ℓ-simple. This proves (a) and one implication of (b). Conversely, any ℓ-simple unital ℓ-ring is certainly right ℓ-primitive. □

We can now see that the Johnson radical is a good radical in the variety of f-rings.

Theorem 3.3.7. *The Johnson radical $\mathscr{J}$ is a special radical in $\mathscr{S}(f)$, the class of extended f-rings. If $R \in \mathscr{S}(f)$, then, for each $n \in \mathbb{N}$, $\mathscr{J}(R_n) = \mathscr{O}(R_n) = \mathscr{R}(R_n) = \mathscr{J}_{\text{left}}(R_n)$ is the largest right (respectively, left) ℓ-QR right (respectively, left) ℓ-ideal of R_n.*

Proof. According to Exercise 7 the class $\mathscr{S}(f)$ is hereditary and homomorphically closed. Let $\mathscr{M}$ be the class of ℓ-simple unital totally ordered domains. If $A \in \mathscr{M}, R \in \mathscr{S}(f)$, and A is an ℓ-ideal of R, then R/A^* is an ℓ-prime ℓ-ring (Exercise 3.2.18). So ℓ-$\beta(R) \subseteq A^*$, R/A^* is an f-ring, and hence R/A^* is a totally ordered domain. But then the embedding $A \longrightarrow R/A^*$ is an isomorphism. Thus, $\mathscr{M}$ satisfies the condition (3.2.19), and so it is a special class in $\mathscr{S}(f)$. Since $\mathscr{J}$ is the upper radical determined by $\mathscr{M}$ (Theorems 3.3.6 and 3.2.26), $\mathscr{J}$ is a special radical in $\mathscr{S}(f)$. Also, by Theorems 3.3.6 and 3.2.17, $\mathscr{J}(R) = \mathscr{O}(R) = \mathscr{R}(R)$ and $\mathscr{J}(R) = \mathscr{J}_{\text{left}}(R)$; and the same equations hold for R_n, by Theorem 3.2.29. □

One consequence of Theorem 3.3.6 is that a unital ℓ-simple f-ring does not have any nonzero proper one-sided ℓ-ideals. In fact, this is true for any ℓ-simple f-ring, and this leads to the general result that the sets of maximal ℓ-ideals and maximal one-sided ℓ-ideals coincide in any f-ring; see Exercise 25. In order to see this in the ℓ-simple case, as well as for other purposes, it is convenient to give the following definition. The element e in the po-ring R is a *left superunit* if $ex \geq x$ for each $x \in R^+$. A *right superunit* is defined analogously, and a *superunit* is an element that is both a left and a right superunit. The ℓ-ring R is called *superunital* if it has a superunit. As we will now show, each ℓ-simple f-ring is superunital.

Theorem 3.3.8. *The following statements are equivalent for the f-algebra R over the directed po-ring C.*

(a) R is ℓ-simple.
(b) For all $0 \neq c, d \in R$ there exists $x \in R$ with $|d| < |cx|$.
(c) For all $0 \neq c, d \in R$ there exists $x \in R$ with $|d| < |xcx|$.
(d) For all $0 \neq c, d \in R$ there exists $x \in R$ with $|d| < |xc|$.
(e) $R^2 \neq 0$ and R has no nonzero proper one-sided ℓ-ideals.
(f) R is subdirectly irreducible with an ℓ-idempotent heart.

Proof. We first show that (c) implies (b). Under the assumption of (c) R must be totally ordered since if c and d are disjoint so are xcx and d. Also, R is a domain since it is, in fact, an ℓ-simple f-ring. If $|ab| \leq |a| \wedge |b|$ holds identically in R, then we get the contradiction $|c| < |xcx| \leq |x| \wedge |c| \leq |c|$. We claim that, in any totally ordered domain, if $a, b \in R^+$ and $ab > a$ or $ba > a$, then b is a superunit. For, if $ab > a$, then $abr > ar$ and $br > r$ for each $r > 0$. So $b^2 > b$ and hence $rb > r$ for each $r > 0$. Let e be any superunit of R. If $ab > e$, then ba is a superunit since $(ba)^2 = baba > bea \geq ba$. Now, $e < x|c|x$ implies that $|c|x^2$ is a superunit; so $|d| < 2|c|x^2|d| = |c|y$. Since (a) implies (c) and also (b) implies (a), by symmetry we now have that the first five statements are equivalent. To finish the proof we will show that a subdirectly irreducible f-algebra R with an ℓ-idempotent heart H is ℓ-simple. It is easy to see that R is a subdirectly irreducible ℓ-ring with heart H. Recall from Exercise 3.2.32 that H is the minimal ℓ-ideal of R. Now, R is a totally ordered domain, H is ℓ-simple, and, as we have just seen, H contains an element e with $e^2 > e > 0$, and e is a superunit of R. Thus, $R = H$. □

An example of a "minimal" totally ordered domain that has a right ℓ-ideal that is not an ℓ-ideal is given in Exercise 3.4.35 and Theorem 3.4.17. An ℓ-unital ℓ-ring that is a simple domain can have nonzero proper right ℓ-ideals; see Exercise 3.2.28(b). However, in a class of ℓ-rings that contains all ℓ-unital ℓ-rings that are either ℓ-reduced or satisfy an identity of the form $p(x) \geq 0$ with $0 \neq p(x) \in \mathbb{Z}^+[x]$, the number of maximal ℓ-ideals is the same as the number of maximal right ℓ-ideals; see Exercise 26 and Theorem 3.8.4.

The f-ring analogue of the Wedderburn-Artin theorem is presented next.

Theorem 3.3.9. *Let R be an extended f-ring with d.c.c. on ℓ-ideals. Then ℓ-$\beta(R)$ is nilpotent and R/ℓ-$\beta(R)$ is a direct sum of ℓ-simple totally ordered domains.*

Proof. We have already seen that ℓ-$\beta(R)$ is nilpotent (Theorem 3.2.10) and we may assume that R is an ℓ-semiprime f-ring. By Theorem 3.3.8 each ℓ-prime ℓ-ideal of R is a maximal ℓ-ideal. If $P_1, P_2, \ldots, P_n, \ldots$ are countably many distinct ℓ-prime ℓ-ideals, then the descending chain $P_1 \supseteq P_1 \cap P_2 \supseteq \cdots$ gives that $P_1 \cap \cdots \cap P_n = P_1 \cap \cdots \cap P_n \cap P_{n+1}$ for some n. But then we have the contradiction that $P_i \subseteq P_{n+1}$ for some $i \leq n$. Suppose that $P_1, \ldots, P_n$ are all of the proper ℓ-prime ℓ-ideals of R. Then, for each i,

$$P_i + \bigcap_{j \neq i} P_j = \bigcap_{j \neq i} (P_i + P_j) = R,$$

and hence the canonical embedding $R \longrightarrow R/P_1 \oplus \cdots \oplus R/P_n$ is onto. For, if $a_1, \ldots, a_n \in R$, then $a_i = b_i + c_i$ with $b_i \in P_i$ and $c_i \in P_j$ for each $j \neq i$. Let $x = c_1 + \cdots + c_n$. Then $x + P_i = c_i + P_i = a_i + P_i$ for each i. □

Note that the ring of even integers is an example of an ℓ-simple f-ring with $\mathscr{J}(R)$ not nilpotent. However, if R is an extended f-ring with d.c.c. on ℓ-ideals and a one-sided identity element, then $\mathscr{J}(R)$ is nilpotent. For, $\mathscr{J}(R/\ell\text{-}\beta(R)) = 0$ by Theorem 3.3.9 and therefore $\mathscr{J}(R) \subseteq \ell$-$\beta(R)$. Analogous results for ℓ-rings more general than f-rings are given in Exercises 27 and 28; also, see Exercises 29, 30, and 3.6.26.

Exercises.

1. This is a strict version of Theorem 3.3.3.

   (a) Let G and H be po-groups and let $\varphi : G \longrightarrow H$ be a group homomorphism. Show that φ is isotone and $\ker\varphi$ is trivially ordered iff $\varphi(x) < \varphi(y)$ whenever $x < y$.
   (b) Suppose that R is a po-ring, $a, b \in R$, $aR^+ + R^+a \subseteq R^+$ and $ab < ba$. If $r(a)^+ = 0$ or $\ell(a)^+ = 0$, show that $a^n b < ba^n$ if $n \geq 1$.
   (c) If $ab < ba$, $bR^+ + R^+b \subseteq R^+$, and $r(b)^+ = 0$ or $\ell(b)^+ = 0$, show that $ab^n < b^n a$ if $n \geq 1$.
   (d) Suppose that $ab < ba$ and $aR^+ + R^+a + bR^+ + R^+b \subseteq R^+$. If $n \geq 2$ and $r(a)^+ = \ell(b)^+ = 0$, show that $a^n b^n < (ab)^n$; if $r(b)^+ = \ell(a)^+ = 0$, show that $(ba)^n < b^n a^n$.

2. Let e be a left superunit of the po-ring R.

   (a) If $f \geq e$, show that f is a left superunit.
   (b) If R is directed, prove that R has a positive left superunit.
   (c) If R is an ℓ-ring, show that $e^- \in \ell_\ell(D(R_R))$.
   (d) If R is an ℓ-ring and $e \in F(R_R)$, show that $e \geq 0$.
   (e) Find an example of a left f-ring that is also a d-ring which has a nonpositive left identity element.
   (f) Show that a directed po-ring has a superunit if and only if it has a left superunit and a right superunit.
   (g) Suppose that R is directed and assume that for each $x \in R^+$ there is an element $f \in R$ with $fx \geq x$. If X is a finite subset of R show that there is an element f in R^+ with $fx \geq x$ for every $x \in X$.

3. Let d be an element of the f-ring R and let $2 \leq n \in \mathbb{N}$.

   (a) If $d^n \geq d$, show that $d^2 \geq d$.
   (b) If $d^n \leq d$, show that $d^3 \leq d$; and $d^2 \leq d$ iff $d \geq 0$.
   (c) If $d^n = d$, show that $|d| = d^2$ and $d^3 = d$.

4. Suppose that R is a po-ring and $0 < y \in R$. Assume that, for any $a \in R$, if $ay > 0$ (respectively, $ya > 0$), then $a > 0$. If $xy > y$ (respectively, $yx > y$), show that $(2x)^2 > 2x$. If $F[x]$ is the canonically ordered polynomial ring over the po-domain F, find elements $y \in F[x]$ such that $ya > 0$ iff $a > 0$.

5. Let a, b, and e be elements of the po-ring R. Suppose that $\ell(a)^+ = r(a)^+ = 0$ and $aR^+ + R^+a \subseteq R^+$. If each of the products ab and ba is comparable to e and $ae = ea = a$, show that the order relation between ab and e is the same as that between ba and e.

6. Let R be a po-reduced po-ring. Show that R is an f-ring iff each $x \in R$ can be written as $x = x_1 - x_2$ with $x_i \in R^+$ and $x_1 x_2 = x_2 x_1 = 0$. (*Hint*: If $a, b \in R^+$, show that $ab = 0$ iff $a \wedge b = 0$.)

7. Let $\mathscr{V}$ be a variety of ℓ-rings, let $\mathscr{P}$ be a radical, and let $\mathscr{S}_{\mathscr{P}}(\mathscr{V})$ be the class of all ℓ-rings R such that $R/\mathscr{P}(R) \in \mathscr{V}$.

   (a) Show that $\mathscr{S}_{\mathscr{P}}(\mathscr{V})$ is homomorphically closed.
   (b) If $\mathscr{P}$ is hereditary, i-hereditary, or left i-hereditary, show that $\mathscr{S}_{\mathscr{P}}(\mathscr{V})$ has the corresponding property.
   (c) If $\mathscr{P}$ is a special radical determined by the special class $\mathscr{M}$, show that $R \in \mathscr{S}_{\mathscr{P}}(\mathscr{V})$ iff $R/P \in \mathscr{V}$ whenever P is an ℓ-ideal of R with $R/P \in \mathscr{M}$.

8. (a) If the ℓ-ring R has an ℓ-faithful ℓ-simple ℓ-module M_R (that is, $r_\ell(M) = 0$ and 0 and M are the only convex ℓ-submodules), show that R is an ℓ-prime ℓ-ring.
   (b) Prove that an ℓ-ring R is a right ℓ-primitive f-ring iff it has an ℓ-faithful ℓ-simple right f-module M which is cyclic ($M = m_0R$) with a generator m_0 that is a d-element on R and such that $m_0 = m_0e$ with $|e| \geq r_\ell(m_0)$. (Show that $r_\ell(x) = r_\ell(m_0)$ for each $0 < x \in M$).
   (c) Prove that an extended f-ring is a right ℓ-primitive f-ring iff it has an ℓ-faithful ℓ-simple ℓ-module which is a cyclic module with a positive generator.

9. (a) Let $\{A_i : i \in I\}$ be a family of independent right ℓ-ideals of the ℓ-ring R. If each A_i is a right ℓ-QR ℓ-ring, show that $A = \oplus_i A_i \subseteq R$ is a right ℓ-QR ℓ-ring.
   (b) If R is an extended f-ring and I is a right ℓ-QR right ℓ-ideal of R_n, show that I is a right ℓ-QR right ℓ-ring. (Use (a) to reduce to the case that $n = 1$.)
   (c) If $R/\mathscr{J}(R)$ is an f-ring, show that $\mathscr{J}(R_n)$ is the largest right ℓ-ideal of R_n with the property that each of its elements is right or left ℓ-QR in R_n. (Use Theorem 3.2.28(c) to show that a right ℓ-ideal with this property is contained in each right ℓ-primitive ℓ-ideal of R_n.)

10. If A is a right ℓ-ideal of the extended f-ring R, show that $\mathscr{J}(A) = A \cap \mathscr{J}(R)$.

11. Let a be an element of the ℓ-ring R and let $\overline{R} = R/B$ where B is a right ℓ-QR ℓ-ideal of R.

   (a) Show that a is right ℓ-QR in R iff $\overline{a}$ is right ℓ-QR in $\overline{R}$.
   (b) Assume that $\overline{R} = D(R_{\overline{R}})$. If $|a|$ is right ℓ-QR, show that a is right ℓ-QR.
   (c) If $\overline{R}$ is a right f-ring, show that a is right ℓ-QR iff a^+ is right ℓ-QR.
   (d) For each element x in $\{a, -a, a^+, a^-, |a|\}$ find an ℓ-ring in which x is right ℓ-QR and the other four elements are not right ℓ-QR. What if R is required to be an f-ring, or to be totally ordered?

12. Let R be a po-ring. Recall that if $x, y \in R$, then $x \circ y = x + y - xy$.

   (a) Show that if $x, y, z \in R$, then $x \circ z \leq y \circ z$ iff $(y - x)(1 - z) \in R^+$.
   (b) Let $a \in R$ and define the maps $f, g : R \longrightarrow R$ by $f(x) = ax$ and $g(x) = a \circ x$. If $f(x \vee y) = f(x) \vee f(y)$ whenever $x \vee y \in R$, show that $g^{-1}(-R^+) = \{x \in R : a \circ x \leq 0\}$ and $g^{-1}(R^+) = \{x \in R : a \circ x \geq 0\}$ are sublattices of R. If $a \in R^+$, show that $g^{-1}(-R^+)$ is closed under addition. If $a > 1$ (respectively, $a < 1$) in some unital po-ring extension of R, show that $g^{-1}(-R^+)$ is a dual ideal

(respectively, an ideal) and $g^{-1}(R^+)$ is an ideal (respectively, a dual ideal) of the po-set R.

(c) If $x,y \in R$ and $x \circ y \leq p$ for some positive nilpotent element p, show that $x \circ z \leq 0$ for some z in R.

13. Let a be an element of the ℓ-ring R and let $\overline{R} = R/\ell\text{-Nil}(R)$.

(a) Show that a is 1-right ℓ-*QR* iff there exists $y \in R$ with $(a \circ y) \wedge (a \circ -y) \leq 0 \leq (a \circ y) \vee (a \circ -y)$.

(b) Suppose that R is an ℓ-subring of a unital right d-ring. Show that a is right ℓ-*QR* in R iff a is 1-right ℓ-*QR* in R.

(c) Assume that $\overline{R}$ is an f-ring. Show that a is right ℓ-*QR* iff there exist elements $p,y \in R$, with p nilpotent, such that $[(a^+ \circ y) \wedge (a^+ \circ -y)] \circ p \leq 0$. ($\overline{R}$ satisfies the hypothesis of (b) by Theorem 3.4.3.)

(d) If there are elements x,y in R with $a \circ x \leq 0 \leq a \circ y$, show that a is 2-right ℓ-*QR*.

(e) If $a^2 \geq 0$ and there exists $x \in R$ with $a \circ x \leq 0$, show that a is 2-right ℓ-*QR*.

(f) Show that R satisfies $(*)$ iff $\overline{R}$ satisfies $(*)$.

$(*)$ a is right ℓ-*QR* iff $a \circ x \leq 0$ for some x in R.

Show that the class of ℓ-rings that satisfy $(*)$ is finitely productive, is productive for *sp*-ℓ-rings, and contains $\varinjlim R_i$ if each R_i satisfies $(*)$ and the index set I is directed up (see Exercise 1.4.23).

(g) Show that R satisfies $(*)$ provided that $\overline{R}$ is either a unital f-ring or is totally ordered. If R is a unital almost f-ring, show that each right ℓ-*QR* element of R is 2-right ℓ-*QR*.

(h) If $\overline{R}$ is an f-ring, show that a^2 is right ℓ-*QR* iff $a^2 \circ x \leq 0$ for some x in R (see (c)).

(i) Suppose that R is an ℓ-subring of a unital ℓ-ring and a is comparable to 1. If a is right and left ℓ-*QR*, show that there is an element x in R such that $a \circ x \leq 0$ and $x \circ a \leq 0$.

(j) Let R be an f-ring. Show that a is right ℓ-*QR* iff there are elements $y_1,\ldots,y_n$ in an f-ring extension of R such that $|y_1|,\ldots,|y_n| \in R$ and $a \circ (y_1 + \cdots + y_n) \leq 0$.

(k) Find an example of an element $a \in \mathscr{J}(R)$ such that for any $r \in R$, $a \circ r || 0$. (Try a polynomial ring.)

14. Let R be an ℓ-subalgebra of the ℓ-algebra V, and let P be an ℓ-ideal of R that is an ideal of V. Show that $C(P)$ is an ℓ-ideal of V, and if $Q = \{v \in V : |v| \wedge |r| \in C(P)$ for each $r \in R\}$, then Q is a convex ℓ-subgroup of V, $P = C(P) \cap R = Q \cap R$, and $(R/P)^{\perp_{V/Q}} = 0$.

15. Suppose that ${}_TR_R$ is an ℓ-bimodule where R is an ℓ-semiprime f-ring and T is a po-ring. Show that ${}_TR$ is an f-module.

16. Let R be an f-ring with $\ell(R) = 0$, and let K be the centroid of R. Show that K is an f-ring and R is a strong f-algebra over K. (See Exercises 2.2.11 and 3.2.9.)

17. Let C be a commutative domain with field of quotients Q, and let ${}_CM$ be a C-module. The *torsion submodule* of M is defined by $t(M) = \{x \in M : \alpha x = 0$ for some $0 \neq \alpha \in C\}$, and M is *torsion-free* if $t(M) = 0$. The *closure* of the submodule N of M is defined by $clN/N = t(M/N)$, and N is *closed* if $clN = N$.

 (a) Show that $M/t(M)$ is torsion-free and clN is closed.
 (b) Define a *module of quotients* $Q(M)$ of M by imitating the construction of Q from C. So the elements of $Q(M)$ are equivalence classes $[(x,\alpha)] = \alpha^{-1}x$ with $x \in M$ and $0 \neq \alpha \in C$. Show that $Q(M)$ is a vector space over Q.
 (c) Show that the mapping $M \longrightarrow Q(M)$ given by $x \mapsto \alpha^{-1}(\alpha x)$ is a C-homomorphism whose kernel is $t(M)$.
 (d) Show that $Q(M)$ is contained in each Q-vector space that contains $M/t(M)$.
 (e) Show that $Q(M) \cong M \otimes_C Q \cong M/t(M) \otimes_C Q$.
 (f) If R is an algebra over C, show that $t(R)$ is an ideal of R and $Q(R)$ is an algebra over Q.

18. We continue with the notation from the previous exercise, but now assume that C is totally ordered, Q is the totally ordered field extension of $C(\alpha^{-1}\beta \in Q^+$ iff $\alpha\beta \geq 0)$ and M is a d-module over C.

 (a) Show that the closure of a convex submodule or of an ℓ-submodule is convex or is an ℓ-submodule, respectively. In particular, $t(M)$ is a convex ℓ-submodule.
 (b) Prove the equivalence of the following statements.
  (i) M is torsion-free.
  (ii) Each polar of M is a closed submodule.
  (iii) Each minimal prime subgroup is a closed submodule.
  (iv) M is a subdirect product of totally ordered torsion-free modules.
 (c) Show that $Q(M)$ can be made into a vector lattice over Q such that $M/t(M)$ is an f-C-submodule of $Q(M)$ in one and only one way.
 (d) If R is an f-algebra over C, show that $Q(R)$ is an f-algebra over Q.

19. This exercise is an application of Exercise 2.5.31. Let R be a directed po-ring, C a commutative totally ordered domain, and ${}_RM_C$ an f-bimodule.

 a. Show that $r(R;M) = 0$ and M_C is torsion-free iff M is a subdirect product of totally ordered C-torsion-free f-bimodules $\{M_\lambda\}$ with $r(R;M_\lambda) = 0$. (*Hint*: Modify the definition of S in the proof of Exercise 2.5.31.)
 b. Suppose that R is a torsion-free f-algebra over C. Show that $r(R) = 0$ (respectively, $\ell(R) = 0$) iff R is a subdirect product of totally ordered torsion-free C-algebras R_λ with $r(R_\lambda) = 0$ (respectively, $\ell(R_\lambda) = 0$).
 c. Let R be as in (b). Show that $\ell(R) \cap r(R) = 0$ iff R is a subdirect product of totally ordered torsion-free C-algebras R_λ with $\ell(R_\lambda) = 0$ or $r(R_\lambda) = 0$.

d. Let R be a torsion-free ℓ-algebra over C which satisfies the identity $y^+x^+y^+ \wedge x^- = 0$. Show that $\ell(R) = r(R) = 0$ iff R is a subdirect product of totally ordered torsion-free C-algebras R_λ with $\ell(R_\lambda) = r(R_\lambda) = 0$.

20. As in Exercise 17, a module M_R over the domain R is torsion-free if $0 \neq xr$ whenever $0 \neq x \in M$ and $0 \neq r \in R$. Show that the following statements are equivalent for the totally ordered domain R.

    (a) R is ℓ-simple.
    (b) If M_R is a d-module, then $M/\ell(R;M)$ is a torsion-free f-module.
    (c) If N is a convex ℓ-submodule of the torsion-free f-module M, then M/N is torsion-free.

21. Prove that an ultraproduct of ℓ-primitive f-rings is ℓ-primitive.

22. Show that a totally ordered domain can be embedded in an ℓ-primitive f-ring.

23. Suppose that R is an ℓ-ring such that $R/\ell\text{-Nil}(R)$ is an f-ring; see Exercise 7. Recall from Exercise 3.2.32 that $\ell\text{-}\mathscr{A}$ denotes the antisimple radical.

    (a) Show that $\ell\text{-}\mathscr{A}(R_n) \subseteq \mathscr{J}(R_n)$ for each $n \in \mathbb{N}$.
    (b) If R is unital show that equality holds in (a).
    (c) If R has d.c.c. on principal ℓ-ideals show that $\ell\text{-}\mathscr{A}(R_n) = \ell\text{-}\beta(R_n)$ is a nilpotent ℓ-ideal.

24. Show that the ℓ-ring R with center $Z(R)$ is a commutative semiprime f-ring iff $F(R) \cap Z(R)$ is semiprime and each closed interval of R^+ with distinct end points contains an element of $F(R) \cap Z(R)$.

25. (a) If R is an f-ring, show that A is a maximal one-sided ℓ-ideal of R iff A is a maximal ℓ-ideal.
    (b) If $R/\ell\text{-Nil}(R)$ is an f-ring, show that each maximal ℓ-ideal of R is a maximal one-sided ℓ-ideal.
    (c) If $R/\mathscr{O}(R) \cap \mathscr{O}_{\text{left}}(R)$ is an f-ring, show that a maximal modular one-sided ℓ-ideal of R is a maximal ℓ-ideal.

26. Let P be an ℓ-prime ℓ-ideal of the ℓ-ring R in which $\ell\text{-}\beta(R) = \ell\text{-}N_g(R)$ and put $\overline{R} = R/P$. Suppose S is an ℓ-subring of R and $\ell(\overline{F(S)}; \overline{S}) \cap r(\overline{F(S)}; \overline{S}) = 0$.

    (a) Show that $\overline{S}$ is an almost f-ring iff $\overline{S}$ is totally ordered. (If $a \wedge b = 0$ in S and $x, y \in \underline{f(S)}$, then $[\overline{(a \wedge x)R(b \wedge y)}]^2 = 0$.)
    (b) If $\ell(\overline{F(R)}; F(\overline{R})) \cap r(\overline{F(R)}; F(\overline{R})) = 0$ show that $F(\overline{R})$ is a totally ordered domain (use Theorem 3.2.13(e)).
    (c) Suppose $F(R) \nsubseteq P$. Show that $F(\overline{R})$ is totally ordered iff $\ell(\overline{F(R)}; F(\overline{R})) = 0$, iff $r(\overline{F(R)}; F(\overline{R})) = 0$, iff $\ell(\overline{F(R)}; F(\overline{R})) \cap r(\overline{F(R)}; F(\overline{R})) = 0$.

27. Let R be an ℓ-ring in which $\ell\text{-}\beta(R) = \ell\text{-}N_g(R)$ and which has a left f-superunit.

    (a) Show that there is a bijection between the set of maximal ℓ-ideals of R and the set of maximal ℓ-ideals of $F(R)$. (If P is a maximal ℓ-ideal of R let P_f be the maximal ℓ-ideal of $F(R)$ which contains $P \cap F(R)$; see Exercise 26. If Q is a maximal ℓ-ideal of $F(R)$ apply (d) of Theorem 3.2.13 to $F(R)^+ \backslash Q$.)

(b) Show that each ℓ-prime ℓ-ideal of R is contained in a unique maximal ℓ-ideal and a unique maximal right ℓ-ideal of R.

(c) Show that each maximal right ℓ-ideal of R contains a maximal ℓ-ideal of R and there is a bijection between the sets of maximal right ℓ-ideals and maximal ℓ-ideals of R.

28. Let R be an ℓ-reduced ℓ-ring whose Boolean algebra $\ell\text{-Ann}(R)$ of ℓ-annihilator ℓ-ideals is atomic.
   (a) Show that R is an irredundant subdirect product of ℓ-domains (See Exercises 1.3.6 and 4.1.33 and Theorem 4.1.14.)
   (b) Suppose $\ell\text{-Ann}(R)$ has a.c.c. and $F(R)$ contains a left superunit of R. Show that $R/\mathscr{O}(R)$ is isomorphic to a direct product of ℓ-simple right ℓ-primitive ℓ-rings.

29. (a) Let $\mathscr{S}$ be a collection of convex ℓ-subgroups of the ℓ-group G with the property that any two distinct maximal elements of $\mathscr{S}$ generate G. Suppose $0 \neq S$ is a totally ordered convex subgroup of G which is not contained in any maximal element of $\mathscr{S}$. Show that $\mathscr{S}$ has at most one maximal element.
   (b) Suppose R is an ℓ-ring which has a left identity element that is basic. Show that R is ℓ-simple iff it is right ℓ-primitive.
   (c) Suppose that $F(R)$ contains a left identity element of the ℓ-ring R and ℓ-$\beta(R) = \ell\text{-}N_g(R)$. Show that the ℓ-ideal P of R is right ℓ-primitive iff it is a maximal ℓ-ideal. (Use Exercise 26.)
   (d) If the ℓ-ring R in (c) has d.c.c. on ℓ-ideals show that $R/\mathscr{O}(R)$ is isomorphic to the direct product of ℓ-simple ℓ-rings.

30. Let R be an ℓ-ring in which $f(R)^*$ is not empty and consists of regular elements of R. Show that R is ℓ-simple iff (i) R has d.c.c. on ℓ-ideals or R is subdirectly irreducible; (ii) the ℓ-ideal of R generated by F is R; (iii) $F^{\perp}$ does not contain a nonzero ℓ-ideal of R; (iv) for each proper minimal ℓ-ideal A of R, and $a \in A \cap f(R)$ and $0 \leq b \in A \cap F^{\perp}$, there exist $c \in A \cap f(R)$ and $d \in F^{\perp}$ with $bab \leq ac \vee ca + d$. (Use Theorem 2.5.9. If $0 \neq x, y \in A \cap F$, where A is a proper ℓ-ideal, then $\mathbb{Z}|x||y| < |x| \wedge |y|$.)

31. Let R be an ℓ-ring and let S be a subset of $f_\ell(R)^*$. Show that $S \neq \emptyset$, $R^2 \neq 0$ and R_R is an ℓ-simple right R-ℓ-module iff $r(a;R) = 0$ for every $a \in f_\ell(R)^*$, $S^{\perp}$ contains no nonzero right ℓ-ideals of R, and R has a minimal right ℓ-ideal.

Notes. Theorem 3.3.2 is due to Hion [HIO], Theorems 3.3.4 and 3.3.9 come from Birkhoff and Pierce [BP], and the rest is mostly due to Johnson [JOH1]. Exercise 4 comes from Henriksen and Isbell [HI], Exercise 6 is due to Hayes [HAY], Exercise 16 is due to Keimel [KE2], and Exercises 11 through 13 and 19 come from Steinberg

[ST23]. Exercise 27 comes from Ma and Wojciechowski [MW2] and Exercises 26 and 28 through 31 come from Ma [M4].

3.4 Embedding in a Unital f-Algebra

According to Exercise 3.1.2 each ℓ-ring can be embedded in a po-unital ℓ-ring. An f-ring is called *unitable* if it can be embedded in a unital f-ring. More generally, an f-ring that is a po-algebra over the po-ring C is called *C-unitable* if it can be embedded in a unital f-ring that is a po-algebra over C. Since a C-unitable f-ring is clearly an ℓ-algebra over C we will assume in this section that each po-algebra-f-ring is an ℓ-algebra. Also, in addition to $1 \in C^+$, we will assume, for convenience, that C is directed. The C-unitable f-algebras form a variety of f-algebras which, for certain base rings C, is determined by a single identity. In this case each f-algebra R in this variety has a unique unital cover, and the conditions for R to be convex in its cover will be found. Order idempotency will play a prominent role in the ensuing development.

As a consequence of Theorem 2.4.2(a) (for representable Ω-f-groups) we know that a totally ordered ℓ-algebra is C-unitable if and only if it can be embedded in a totally ordered unital ℓ-algebra. There exist f-rings that are not unitable. To see this, first recall that two idempotents e and f in a ring are *orthogonal* if $ef = fe = 0$. Clearly, a totally ordered ring does not contain a pair of nonzero orthogonal idempotents. So, in fact, in any f-ring, two idempotents are orthogonal if and only if they are disjoint. Since $e = e^2$ and $1-e$ are orthogonal the only idempotents in a totally ordered unital ring are 0 and 1. Thus, the idempotents in a direct product of unital totally ordered rings are central, and so the idempotents in any unitable f-ring are central. If R is any row or column of an $n \times n$ matrix ring ($n \geq 2$) over a totally ordered field, ordered lexicographically with the diagonal term dominating, then R is a totally ordered algebra which has noncentral idempotents. An example of a commutative totally ordered ring that is not unitable and has no nonzero idempotents is given in Exercise 22(d). However, from Exercise 3.3.16 we get that each commutative f-algebra over C whose annihilator vanishes is C-unitable.

The class of C-unitable f-algebras is clearly hereditary and productive. Before we see that it is homomorphically closed, and hence a variety, we will prove the following fundamental fact which will be useful later and which is itself interesting.

Theorem 3.4.1. *In any f-ring the sublattice that is generated by a subring is an ℓ-subring.*

Proof. We first show that the following identities hold in any f-ring R; we may assume that R is totally ordered.

$$[x^2y^+ - xy^+ + y^+]^- = 0. \tag{3.4.1}$$

$$xy^+ = [xy \wedge (x^2y + y)] \vee [0 \wedge (-x^2y - y)]. \tag{3.4.2}$$

For (3.4.1), if $y > 0$, then since $x^2y - xy + y = (x^2 - x)y + y$, either of the cases $y \geq xy$ or $x^2 \geq x$ is obvious. The remaining case has $xy > y$ and $x > x^2$. But then $x^2y \geq xy \geq x^2y$, and we are done. In order to establish (3.4.2) we first take $y \geq 0$. Then $\pm xy \leq x^2y + y$ by (3.4.1), and hence (3.4.2) reduces to $xy = xy \vee (-x^2y - y)$, which is correct. Now, if $y < 0$, then the left side of (3.4.2) is 0 and the right side becomes $[xy \wedge (x^2y + y)] \vee 0 = (xy)^+ \wedge (x^2y + y)^+ = (xy)^+ \wedge 0 = 0$.

Let S be a subring of the f-ring R, and let $L(S) = L_R(S)$ be the sublattice of R generated by S. According to Theorem 2.2.4(f), $L(S)$ is an additive subgroup of R, and if u and v are in $L(S)$, then there exist elements s_{ij} and t_{pq} in S such that

$$u = \bigvee_{i=1}^{n} \bigwedge_{j=1}^{m} s_{ij} \text{ and } v = \bigvee_{p=1}^{N} \bigwedge_{q=1}^{M} t_{pq}.$$

So if $u, v \in L(S)^+$, then

$$uv = uv^+ = \bigvee_p \bigwedge_q ut_{pq}^+ \text{ and } ut_{pq}^+ = \bigvee_i \bigwedge_j s_{ij}t_{pq}^+.$$

But $s_{ij}t_{pq}^+ \in L(S)$ by (3.4.2), and hence $uv \in L(S)$. Thus $L(S)$ is a subring. □

Theorem 3.4.2. *The class of C-unitable f-algebras is a variety of ℓ-algebras.*

Proof. As we have indicated prior to Theorem 3.4.1 it suffices to show that a homomorphic image of a C-unitable f-algebra is C-unitable. Suppose that P is an algebra ℓ-ideal of the C-unitable f-algebra R, and let S be a unital f-algebra that contains R. We may assume that S is the convex ℓ-subgroup of S generated by the subalgebra $R + C \cdot 1$. If $s \in S$, then $|s| \leq \Sigma_{i,j}|r_{ij} + \alpha_{ij} \cdot 1| \leq r + \alpha \cdot 1$ where $r \in R^+$ and $\alpha \in C^+$. Let $Q = C(P)$ be the convex ℓ-subgroup of S that is generated by P. Then Q is an algebra ℓ-ideal of S. For, if $q, s \in S$ with $|s| \leq r + \alpha \cdot 1$ and $|q| \leq p \in P$, then $|sq| \leq rp + \alpha p$ and $|qs| \leq pr + \alpha p$; so sq and $qs \in Q$. Since $Q \cap R = P, R/P$ is embedded in the unital f-algebra S/Q and R/P is C-unitable. □

Let $\mathscr{V}$ be a class of ℓ-algebras over C. An ℓ-algebra is called *$\mathscr{V}$-ℓ-unitable* (respectively, *$\mathscr{V}$-unitable*) if it can be embedded into an ℓ-unital (respectively, a unital) ℓ-algebra that belongs to $\mathscr{V}$. The preceding argument shows that the class of $\mathscr{V}$-ℓ-unitable ℓ-algebras is homomorphically closed provided $\mathscr{V}$ is homomorphically closed and hereditary. In particular, it is a variety when $\mathscr{V}$ is a variety.

The unitable f-rings form a variety that is determined by a single identity. This identity actually arises from the inequalities that arose in the proof of Theorem 3.3.8. We begin, therefore, by examining these inequalities.

It will be convenient to have a name for a positive element that is comparable with its powers (see Exercise 3.3.3). The element d in the po-ring R is called *upperpotent* if $d^2 \geq d \geq 0$, and it is called *lowerpotent* if $0 \leq d^2 \leq d$. If d is a lowerpotent element in a unital f-ring, then $d \leq 1$ since this is certainly true in each totally ordered homomorphic image. If $ab \leq a, b$ for all $a, b \in R^+$, then R is called *infinitesimal*. An ℓ-ring is infinitesimal if and only if it satisfies the inequality $|xy| \leq |x| \wedge |y|$. An infinitesimal ℓ-ring is an f-ring since if $a \wedge b = 0$ and $c \geq 0$, then $0 \leq ca \wedge b, ac \wedge b \leq a \wedge b = 0$. Since any identical inequality in ℓ-rings is equivalent to some identity, the infinitesimal ℓ-rings form a variety of f-rings. This is a good time to recall that, in general, an f-ring satisfies an identity if and only if each of its totally ordered homomorphic images satisfies the identity. Each nil f-ring is infinitesimal since if $a^n = 0 = b^n$, then $ab > |a| \wedge |b|$ is impossible in a totally ordered ring. For, if $ab > a > 0$, then $0 = ab^n \geq ab > a > 0$. In a unital f-ring any subring that has 1 as an upper bound is infinitesimal, and we will soon see that, conversely, each infinitesimal f-ring is bounded by 1 in an f-ring extension.

It is convenient to replace the two variable identity that defines an infinitesimal f-ring by a single variable identity.

Theorem 3.4.3. *The following statements are equivalent for the f-ring R.*

(a) R is infinitesimal.
(b) For each $a \in R, a^2 \leq |a|$; that is, each positive element is lowerpotent.
(c) If $a, b \in R$, then $\mathbb{Z}|ab| \leq |a| \wedge |b|$.
(d) If $a, b \in R$, then the right (respectively, left, two-sided) ideal generated by $|ab|$ is bounded above by $|a| \wedge |b|$.

Proof. The implications (a) $\Rightarrow$ (b) and (d) $\Rightarrow$ (a) are obvious. If (c) holds and $a, b, x \in R$ and $n \in \mathbb{N}$, then $|abx| + n|ab| \leq (n+1)|ab| \leq |a| \wedge |b|$. So (c) $\Rightarrow$ (d), and we only need to verify that (b) implies (c). We may assume that R is totally ordered and $0 < b \leq a$. If $nab > b$, then we have the contradiction that $(2na)^2 > 2na$, by Exercise 3.3.4. □

The next result will be used repeatedly in the sequel, and it also is of interest in its own right.

Theorem 3.4.4. *Let A be the principal ℓ-ideal generated by the element a in the C-f-algebra R. If A is ℓ-idempotent, then $A = a^{\perp\perp} = C(RaR) = C(aR) = C(Ra)$, and $R = A \oplus a^{\perp}$.*

Proof. Assume that $a \neq 0$. Let R_1 be the directed po-algebra obtained by freely adjoining C to R (Exercise 3.1.2). Then R is an f-bimodule over R_1 and $A = C(R_1 a R_1)$. Since $A = A^{[3]}$ there exists $r_1 \in R_1^+$ such that $|a| \leq (r_1|a|r_1)^3 = r_1|a|r_1^2|a|r_1^2|a|r_1 \in R|a|R$; so $A = C(RaR)$. Let $\bar{R} = R/a^{\perp}$. If $C(\bar{a}R_1)$ is a proper right ℓ-ideal of $\bar{R}$, then $C(\bar{a}R_1)$ is contained in a regular algebra right ℓ-ideal $\bar{Q}$ of $\bar{R}$ (Theorem 2.4.2(c) and Section 2.5). Let $\bar{P}$ be a minimal prime subgroup of $\bar{R}$ with $\bar{P} \subseteq \bar{Q}$. Then $\bar{a} \notin \bar{P}$ since $\bar{a}^{\perp} = 0$ (Theorem 2.4.3). If $\bar{P}_1$ is the R_1-R-value of $\bar{a}$ in the f-bimodule ${}_{R_1}\bar{R}_R$ with $\bar{P} \subseteq \bar{P}_1$, then $\bar{R}/\bar{P}_1$ is a subdirectly irreducible f-algebra with ℓ-idempotent heart $C(\bar{R}\bar{a}\bar{R}) + \bar{P}_1/\bar{P}_1$. So $\bar{R}/\bar{P}_1$ has no nonzero proper right ℓ-ideals by Theorem 3.3.8. Now, $\bar{P}_1$ and $\bar{Q}$ are comparable since they both contain $\bar{P}$, and since $\bar{a} \in \bar{Q} \setminus \bar{P}_1$ we have the contradiction $0 \neq \bar{Q}/\bar{P}_1 \subset \bar{R}/\bar{P}_1$. Thus, $C(\bar{a}R_1) = \bar{R}$ and, in fact, $C(\bar{a}R_1) = C(\bar{a}R)$. For, $|\bar{a}| \leq \bar{x}|\bar{a}|\bar{x}$ with $x \in R^+$, and $\bar{x} \leq |\bar{a}|r_1$ with $r_1 \in R_1$; so $|\bar{a}| \leq |\bar{a}|r_1|\bar{a}||\bar{x}| \in |\bar{a}|\bar{R} = |\bar{a}|R$. Hence, $R = C(aR) \oplus a^{\perp}$. Similarly, $R = C(Ra) \oplus a^{\perp}$, and it follows that $A = C(aR) = C(Ra) = a^{\perp\perp}$. □

Note that each ℓ-ideal that is generated by an upperpotent element is ℓ-idempotent. See Exercise 15 for a partial converse. In the case that a is an idempotent in a unitable f-ring R it is rather easy to see that $R = Ra \oplus a^{\perp}$. Each finitely generated ℓ-idempotent ℓ-ideal in an f-algebra is a summand since it is principal. But not every ℓ-idempotent ℓ-ideal is a summand. For example, the direct sum $\mathbb{Z}^{(I)}$, which is idempotent and countably generated if I is countable, is not a summand of the product $\mathbb{Z}^I$ if I is infinite.

The first and most basic unitability result is given next. A related result about extending a total order to an overring is given in Exercise 2.

Theorem 3.4.5. *(a) If R is an infinitesimal f-algebra over the ℓ-simple f-ring C, then the lexicographically ordered algebra obtained by freely adjoining C to R, $S = R \overleftarrow{\times} C$, is a unital f-algebra that contains R.*
(b) Suppose that R is an f-algebra over C and $\ell(R) = r(R) = 0$. Let

$$S = \{f \in \mathrm{End}_R(R_R) : Rf \subseteq R\}.$$

Then S is a unital f-subalgebra of the po-endomorphism algebra $\mathrm{End}_R(R_R)$ that contains R. If R is totally ordered, then S is totally ordered. If R is a domain or is reduced, then S is a domain or is reduced, respectively (independent of any order).

Proof. (a) The ℓ-group $S = R \overleftarrow{\times} C$ is certainly a C-f-module extension of R (Exercise 2.1.10) as well as a unital ring extension, where R is embedded in S in the obvious way. Suppose that $(a+\alpha) \wedge (b+\beta) = 0$ and $c+\gamma \in S^+$. We will check that if $u = (a+\alpha)(c+\gamma) \wedge (b+\beta)$, then $u = 0$. Since $(a+\alpha) \wedge (b+\beta) = 0$, either $\alpha > 0$ and $b+\beta = 0$, or $\beta > 0$ and $a+\alpha = 0$, or $\alpha = \beta = 0$ and $a \wedge b = 0$. In the first case, if $\gamma > 0$, then $(a+\alpha)(c+\gamma) = ac+\alpha c+\gamma a+\alpha\gamma > 0$, and $u = 0$. On the other hand, if $\gamma = 0$, then $c \geq 0$. Take $\delta \in C$ with $\delta\alpha \geq 1$. Then $\delta\alpha c \geq c \geq |\delta ac| \geq -\delta ac$, $\delta(ac+\alpha c) \geq 0$, $ac+\alpha c \geq 0$, and $u = 0$. In the second case, $u = 0 \wedge (b+\beta) = 0$. In the last case, $u = (ac+\gamma a) \wedge b$ and $ac+\gamma a \in b^{\perp}$. If $\gamma = 0$, then $ac \geq 0$ and $u = 0$. If $\gamma > 0$, take $\delta \in C$ with $\delta\gamma \geq 1$. Then $\delta\gamma a \geq a \geq |\delta ac| \geq -\delta ac$, $\delta(ac+\gamma a) \geq 0$, $ac+\gamma a \geq 0$, and $u = 0$. Thus, S is a right f-ring and, similarly, S is a left f-ring.

(b) The mapping $a \mapsto \ell_a$, where $\ell_a(x) = ax$, is clearly an embedding of the po-algebra R into the po-algebra $\mathrm{End}_R(R_R)$. If $f \in \mathrm{End}_R(R_R)^+$ and $f \geq a$, then $fx \geq (ax)^+ = a^+x$ for each $x \in R^+$. So $f \geq a^+$ and R is an ℓ-subalgebra of $\mathrm{End}_R(R_R)$ (see Exercise 3.1.26(a)). Now, S is a subalgebra of $\mathrm{End}_R(R_R)$ since it is the *idealizer* of R in $\mathrm{End}_R(R_R)$; that is, S is the largest subring of $\mathrm{End}_R(R_R)$ in which R is an ideal. Note that $r(R; \mathrm{End}_R(R_R)) = 0$ since if $Rf = 0$, then $RfR = 0$, $fR = 0$, and $f = 0$. If $s \in S$ define $f : R^+ \longrightarrow R^+$ by $f(x) = (sx)^+$. Then for any $a, x, y \in R^+$,

$$af(x+y) = a[s(x+y)]^+ = (as)^+(x+y) = a[(sx)^+ + (sy)^+] = a[f(x) + f(y)].$$

So $f(x+y) = f(x) + f(y)$ and f extends uniquely to a group homomorphism of R, which is also called f, by Exercise 2.2.11. Also,

$$f(xa) = (sxa)^+ = (sx)^+a = f(x)a$$

and

$$(af)x = a(sx)^+ = (as)^+x;$$

so $f \in S^+$. If $t \in \mathrm{End}_R(R_R)^+$ and $t \geq s$, then $tx \geq (sx)^+ = fx$; so $t \geq f$, $f = s^+$ in $\mathrm{End}_R(R_R)$, and S is an ℓ-subring of $\mathrm{End}_R(R_R)$. Note that $s^+x = (sx)^+$ for each $x \in R^+$. If s and t are in S with $s \geq 0$ and $a, x \in R^+$, then

$$a(st^+)x = as(tx)^+ = (astx)^+ = a(st)^+x$$

and

$$t^+sx = (tsx)^+ = (ts)^+x.$$

So $st^+ = (st)^+$ and $t^+s = (ts)^+$. Thus, S is a unital d-ring and hence is an f-ring.

Suppose that R is totally ordered and $s \wedge t = 0$ in S. If $t > 0$, there exists $x \in R^+$ with $tx > 0$. Then, for any $y \in R^+, sy \wedge tx = 0$; so $sy = 0$ and $s = 0$.

The last statement follows from the ring analogues of Theorem 3.2.21(a) and Exercise 3.2.17(c), but a direct proof can also easily be given. □

As one consequence of this result we have

Theorem 3.4.6. *The following statements are equivalent for the totally ordered ℓ-algebra R over the ℓ-simple f-ring C.*

(a) R can be embedded in a unital totally ordered C-algebra.
(b) R is C-unitable.
(c) R is unitable.
(d) R is infinitesimal or has a superunit.
(e) Each nonzero upperpotent element of R is a superunit.

Proof. The implications (a) $\Rightarrow$ (b) $\Rightarrow$ (c) are obvious. To see that (c) $\Rightarrow$ (d), assume that R is contained in the unital f-ring S. We may assume that S is totally ordered. If R is not infinitesimal, then it contains an overpotent element x with $x^2 > x > 0$. But then $x > 1$ and x is a superunit of R. If (d) holds, then R is contained in a totally ordered unital ring by the previous result, and hence (e) holds. Similarly, (a) is a consequence of (e). □

An example of an infinitesimal totally ordered torsion-free ℓ-algebra over a totally ordered domain C that is not C-unitable is given in Theorem 3.4.9.

We present next the identity that characterizes unitability.

Theorem 3.4.7. *An f-ring is unitable if and only if it satisfies the identities*

$$[x \wedge y \wedge (x^2 - x) \wedge (y - xy)]^+ = 0, \tag{3.4.3}$$

$$[x \wedge y \wedge (x^2 - x) \wedge (y - yx)]^+ = 0. \tag{3.4.4}$$

Proof. First, assume that R is a unitable f-ring. To show that R satisfies these identities we may assume that R is unital and totally ordered, and since $(u \wedge v)^+ = u^+ \wedge v^+$ we may take $x > 0$, $y > 0$ and $x^2 - x > 0$. But then $x > 1$ and $(y - xy)^+ = (y - yx)^+ =$

0. Conversely, suppose that R satisfies these identities, and again reduce to the case that R is totally ordered. If R is not infinitesimal, then it contains an element x with $x^2 > x > 0$; so $y \le xy, yx$ for each $y \in R^+$. Thus, R is unitable by Theorem 3.4.5(b). □

Note that the two identities in Theorem 3.4.7 can be replaced by the single identity

$$[x \wedge y \wedge (x^2 - x) \wedge ((y - xy) \vee (y - yx))]^+ = 0.$$

The analogous identities for C-unitability are given in

Theorem 3.4.8. *Let C be a totally ordered domain with totally ordered field of quotients F, and let R by a torsion-free f-algebra over C. The following statements are equivalent.*

(a) R is C-unitable.
(b) $Q = R \otimes_C F$ is unitable.
(c) R satisfies the identities (here, $\alpha \in C$)

$$[\alpha x \wedge \alpha y \wedge (x^2 - \alpha x) \wedge (\alpha y - xy)]^+ = 0, \tag{3.4.5}$$

$$[\alpha x \wedge \alpha y \wedge (x^2 - \alpha x) \wedge (\alpha y - yx)]^+ = 0. \tag{3.4.6}$$

Proof. (a) ⇒ (b). Let T be a unital C-f-algebra that contains R. By factoring out the torsion ideal of T we may assume that T is torsion-free (Exercise 3.3.18). Then Q is unitable since $Q = R \otimes_C F \subseteq T \otimes_C F$.

(b) ⇔ (c). If $\alpha = 0$, then the left sides of (3.4.5) and (3.4.6) are both 0. According to the previous result Q is unitable if and only if, for all $x, y \in R$ and $0 \neq \alpha \in C$,

$$[x/\alpha \wedge y/\alpha \wedge (x^2/\alpha^2 - x/\alpha) \wedge (y/\alpha - xy/\alpha^2)]^+ = 0 \tag{3.4.7}$$

and

$$[x/\alpha \wedge y/\alpha \wedge (x^2/\alpha^2 - x/\alpha) \wedge (y/\alpha - yx/\alpha^2)]^+ = 0. \tag{3.4.8}$$

But (3.4.5) and (3.4.6) are obtained from (3.4.7) and (3.4.8) by multiplying by α^2. So (3.4.5) and (3.4.6) are identities for R exactly when (3.4.7) and (3.4.8) are identities for Q.

(b) ⇒ (a). Let $\bar{Q}$ be a totally ordered homomorphic image of the F-algebra Q. By Theorem 3.4.6 $\bar{Q}$ is infinitesimal or superunital, and by Theorem 3.4.5 $\bar{Q}$ is F-unitable. Since Q is a subdirect product of its totally ordered F-algebra homomorphic images, Q is F-unitable and R is C-unitable. □

If C is an ℓ-simple f-ring, then, as a consequence of Theorem 3.4.6, one gets that each unitable C-f-algebra is C-unitable. This is true for a larger class of base f-rings C, as we will now show.

An f-ring R is called *pseudo-regular* if it is ℓ-idempotent and each of its proper irreducible ℓ-ideals is a maximal ℓ-ideal. An equivalent condition is that each totally ordered homomorphic image of R is ℓ-simple. In terms of the radicals ℓ-$\mathscr{I}$ and ℓ-$\mathscr{A}$ that are given in Exercises 3.2.32 through 3.2.35, an f-ring is pseudo-regular if and only if each of its homomorphic images is ℓ-$\mathscr{A}$-semisimple, or each is ℓ-$\mathscr{I}$-radical, or each (principal) ℓ-ideal is ℓ-idempotent. Also, see Exercises 15(c) and 33(d) for other characterizations of a pseudo-regular f-ring.

Recall that the ℓ-algebra R over the ℓ-ring C is strong if it satisfies $\alpha^+ x^+ = (\alpha x^+)^+$ for $\alpha \in C$ and $x \in R$, or if it satisfies $|\alpha x| = |\alpha||x|$ (Theorem 3.1.1(g)). A unital right f-algebra R is strong iff $\alpha^+ \cdot 1 = (\alpha \cdot 1)^+$ for each $\alpha \in C$; that is, if the ring homomorphism $C \longrightarrow R$ is an ℓ-homomorphism. For, $1 \in R^+$ since $1^+ = 1^+(1^+ - 1^-) = 1^{++} - 1^{-+} = 1$; and if $x \in R^+$, then $\alpha^+ x = (\alpha^+ \cdot 1)x = (\alpha \cdot 1)^+ x = (\alpha x)^+$.

Theorem 3.4.9. *If C is an ℓ-semiprime f-ring and each strong unitable f-algebra over C is C-unitable, then C is pseudo-regular. Conversely, if each proper irreducible ℓ-ideal of the ℓ-ring C is a maximal ℓ-ideal, then each strong unitable f-algebra is C-unitable. In particular, an ℓ-semiprime f-ring C is pseudo-regular if and only if each of its unitable f-algebras is C-unitable.*

Proof. Assume the unitability condition and let P be a minimal prime subgroup of the ℓ-semiprime f-ring C. Then P is a minimal ℓ-prime ℓ-ideal and $\bar{C} = C/P$ is a totally ordered domain (Theorem 3.3.5). If P is not maximal, let A be an ℓ-ideal of C with $P \subset A \subset C$. Then $\bar{A}$ is a proper ℓ-ideal of $\bar{C}$, and hence $\bar{A}$ is infinitesimal. Let R be the C-algebra given by $R = \begin{pmatrix} \bar{A} & \bar{A} \\ 0 & 0 \end{pmatrix}$ and give R the lexicographic order:

$$R^+ = \left\{ \begin{pmatrix} \bar{a} & \bar{b} \\ 0 & 0 \end{pmatrix} : \bar{a} > 0 \text{ or } \bar{a} = 0 \text{ and } \bar{b} \geq 0 \right\}.$$

Then $\begin{pmatrix} \bar{a} & \bar{b} \\ 0 & 0 \end{pmatrix}^2 = \begin{pmatrix} \bar{a}^2 & \bar{a}\bar{b} \\ 0 & 0 \end{pmatrix} \leq \left| \begin{pmatrix} \bar{a} & \bar{b} \\ 0 & 0 \end{pmatrix} \right|$ and R is an infinitesimal totally ordered strong f-algebra over C. By assumption, R is contained in a unital C-f-algebra S, and we may assume that $S = R + C \cdot 1$ is totally ordered. Since $r(R) = 0$ we may also assume that S is strong, by Exercise 3. Let $B = \ell(S;C)$. If $\alpha \in B$ and $0 \neq \bar{a} \in \bar{A}$, then $0 = \alpha \begin{pmatrix} \bar{a} & 0 \\ 0 & 0 \end{pmatrix} = \begin{pmatrix} \bar{\alpha}\bar{a} & 0 \\ 0 & 0 \end{pmatrix}$ implies that $\alpha \in P$; so $B \subseteq P$. But B is an irreducible ℓ-ideal of C by Exercise 3.1.29, and hence $B = P$. Thus, $S = R + \bar{C} \cdot 1$, R is a torsion-free $\bar{C}$-unitable f-algebra, and $Q(R)$ is unitable by Theorem 3.4.8. But if F is the totally ordered field of quotients of $\bar{C}$, then $Q(R) = \begin{pmatrix} F & F \\ 0 & 0 \end{pmatrix}$ is not unitable. Consequently, P is a maximal ℓ-ideal of C.

For the converse, let R be a totally ordered unitable strong f-algebra over the ℓ-ring C. Then $P = \ell(R;C)$ is an irreducible ℓ-ideal of C and $\bar{C} = C/P$ is ℓ-simple. Thus, R is $\bar{C}$-unitable by Theorem 3.4.6, $R \subseteq R + \bar{C} \cdot 1 = R + C \cdot 1$, and R is C-unitable. It follows that each unitable strong C-f-algebra is C-unitable. The last statement follows from the fact that each ℓ-algebra over a pseudo-regular f-ring is strong (Exercise 4). □

Let R be an ℓ-subalgebra of the unital C-f-algebra T. If T is the only ℓ-subalgebra of T that contains R and has an identity element, then T is called a *C-unital cover* *of* R. If, also, $R^{\perp_T} = 0$, then T is called a *tight* C-unital cover of R. A $\mathbb{Z}$-unital cover will be called a *unital cover*. The C-unital covers that are obtained in Theorem 3.4.5 are tight. It may be that each C-unital cover is tight. In any case, we show next that some are tight. Note that $R^{\perp_T}$ is central in T since $T/R^{\perp_T \perp_T}$ is commutative when T is a strong C-unital cover of R. It is also not known whether or not a C-unital cover T of the totally ordered ℓ-algebra R is totally ordered. However, T has no nontrivial idempotents (for any indecomposable R). For, if $e \in T$ is idempotent, then Re and $R(1-e)$ are ℓ-ideals of R and $R = Re \oplus R(1-e)$. So $R = Re \subseteq Te$ or $R = R(1-e) \subseteq T(1-e)$, and hence $e = 1$ or 0.

Theorem 3.4.10. *Let R be a C-unitable f-algebra.*

(a) R has a tight C-unital cover. Moreover, if V is a unital f-algebra generated by $R + C \cdot 1$, then V is a tight C-unital cover of R iff $R^{\perp_V} = 0$.
(b) If $R = \langle e \rangle = R^{[2]}$ for some $e \in R$ or C is a direct sum of ℓ-simple f-rings, then each C-unital cover of R is tight and each unital f-algebra that contains R contains a C-unital cover of R.

Proof. Let U be an f-algebra that contains R and has the identity element 1, and let $V = L(R + C \cdot 1)$ be the sublattice of U that is generated by $R + C \cdot 1$. According to Theorem 3.4.1 V is an ℓ-subalgebra of U. Suppose that $R \subseteq W \subseteq V$ and W is an ℓ-subalgebra with the identity element e. Then $R \subseteq \ell(1-e) \cap V = (1-e)^{\perp_V}$ (Exercise 14) and $1 - e \in R^{\perp_V}$. If $R^{\perp_V} = 0$, then V is a tight C-unital cover of R. If $R^{\perp_V} \neq 0$, then R is embedded in $V_1 = V/R^{\perp_V} = L(R + C \cdot 1)$ and $R^{\perp_{V_1}} = 0$. So V_1 is a tight C-unital cover of R. This proves (a). Suppose that $R = \langle e \rangle = R^{[2]}$. Then $V = C(eV) \oplus e^{\perp_V}$ by Theorem 3.4.4, and $W = C(eV)$ is unital and contains R. But then W is a tight C-unital cover of R since $R^{\perp_W} = e^{\perp_W} = 0$ and $W = L(R + Cf)$, where f is the identity of W. Now, suppose that C is a direct sum of ℓ-simple f-rings and $R^{\perp_V} \neq 0$. Then $R^{\perp_V}$ is isomorphic to an ℓ-ideal of $V/R^{\perp_V \perp_V} \cong C \cdot 1$. Hence, $R^{\perp_V}$ has an identity e, $R^{\perp_V} = Ve$, and $R \subseteq R^{\perp_V \perp_V} = V(1-e) \subset V$. So, V is not a C-unital cover of R, and, by (a), $V(1-e) = L(R + C \cdot (1-e))$ is a tight C-unital cover of the ℓ-algebra R. □

The following two theorems give the main results on the uniqueness of C-unital covers. A C-unital cover S of R, of a particular type, is called *unique* if any other C-unital cover of R of that type is isomorphic to S via an isomorphism that is the identity on R. An example of a tight C-unital cover that is not unique is given in Exercise 18; also, see Exercise 17.

Theorem 3.4.11. *Let R be a (strong) C-unitable f-algebra. If each totally ordered homomorphic image of R has a unique tight (and strong) C-unital cover, then so does R.*

Proof. We will first construct a tight C-unital cover of R. Let $\{P_\lambda\}_{\lambda \in \Lambda}$ be the collection of all of the proper irreducible algebra ℓ-ideals of R, and let S_λ be the tight C-unital cover of $T_\lambda = R/P_\lambda$. Then $R \subseteq T = \Pi_\lambda T_\lambda \subseteq \Pi_\lambda S_\lambda = U$. Let 1 be the identity of U and let $S = L_U(R + C \cdot 1)$. Then S is a tight C-unital cover of R. For, if

$R \subseteq W \subseteq S$ and e is the identity of the ℓ-subalgebra W, then $e_\lambda = 1$ for each $\lambda \in \Lambda$ since e_λ is a nonzero idempotent in S_λ; so $e = 1$ and $W = S$. S is a tight cover of R since $R^{\perp_U} = 0$; and S is a strong ℓ-algebra over C if each S_λ is strong.

Suppose that V is a tight C-unital cover of R. Since $V = L(R + C \cdot 1)$ there is at most one R-isomorphism of the ℓ-algebra V onto another C-unital cover of R. To show the existence of one such isomorphism it suffices to show that S and V are R-isomorphic ℓ-algebras. For each $\lambda \in \Lambda$ let

$$Q_\lambda = \{u \in V : |u| \wedge |r| \in C(P_\lambda) \text{ for every } r \in R\}.$$

Then Q_λ is an ℓ-ideal of V, $P_\lambda = C(P_\lambda) \cap R = Q_\lambda \cap R$, and $(R/P_\lambda)^\perp = 0$ in V/Q_λ, by Exercise 3.3.14. We will show that $\cap_\lambda Q_\lambda = 0$, and to do so it suffices to verify that $\cap_\lambda C(P_\lambda) = 0$. For, if $0 \leq v \in \cap_\lambda Q_\lambda$, then for each $r \in R^+, v \wedge r \in \cap_\lambda C(P_\lambda) = 0$ and $v \in R^{\perp_V} = 0$. Let $0 < w \in \cap_\lambda C(P_\lambda)$; then $w \leq r$ for some $r \in R$. If Q is a minimal prime subgroup of V with $w \notin Q$, then $Q \cap R$ is a proper irreducible algebra ℓ-ideal of R since $r \notin Q \cap R$; so $Q \cap R = P_\lambda$ for some $\lambda \in \Lambda$. But then $w \in C(P_\lambda) \subseteq Q$ gives a contradiction.

We will identify V with its image in $\Pi_\lambda V/Q_\lambda = W$. Now, V/Q_λ is a tight C-unital cover of T_λ by Theorem 3.4.10(a); so there is an isomorphism $h_\lambda : S_\lambda \longrightarrow V/Q_\lambda$ which is the identity on T_λ. Consequently, there is an isomorphism $h : U = \Pi_\lambda S_\lambda \longrightarrow \Pi_\lambda V/Q_\lambda = W$ which is the identity on $T = \Pi_\lambda T_\lambda$. Since $S = L_U(R + C \cdot 1)$ and $V = L_W(R + C \cdot 1)$ the restriction of h to S gives an isomorphism of S onto V which is the identity on R. □

Theorem 3.4.12. *Each unitable f-algebra over a pseudo-regular f-ring C has a unique tight C-unital cover.*

Proof. Let R be a unitable f-algebra over C. By Theorem 3.4.9 R is C-unitable. Let V be a C-unital cover of R. We will first show that $\ell(R;C) = \ell(V;C)$. Since V is a strong f-algebra (Exercise 4) $\ell(R;C) = \ell_\ell(R;C)$, and it suffices to show that $\ell(R;C)^+ \subseteq \ell(V;C)$. Suppose that $\alpha \in \ell(R;C)^+$. Then $C = \langle \alpha \rangle \oplus \alpha^\perp = C\beta \oplus C\gamma$ (Theorem 3.4.4) where $\beta \in \langle \alpha \rangle$ and $\gamma \in \alpha^\perp$ are orthogonal idempotents of C. So $\beta R = 0$, $R \subseteq \gamma V \subseteq V$, and $V = \gamma V$. Thus, $\alpha V = \alpha \gamma V = 0$. We may assume that R is totally ordered (Theorem 3.4.11), and hence $\bar{C} = C/\ell(V;C)$ is ℓ-simple (Exercise 3.1.29). If R is infinitesimal, then, for each $\bar{\alpha} > 0$ in $\bar{C}$, $\bar{\alpha} \cdot 1$ is an upper bound of R in V, by Exercise 6. So the mapping from the f-algebra $S = R \underleftarrow{\times} \bar{C}$ of Theorem 3.4.5(a) into V, given by $(x, \bar{\alpha}) \mapsto x + \alpha \cdot 1$, is an isomorphism of S onto $R + C \cdot 1 \subseteq V$. In particular, $R + C \cdot 1$ is totally ordered and hence $V = L(R + C \cdot 1) = R + C \cdot 1$. If R has a superunit e, then $V = C(Ve)$ by Theorem 3.4.10, and e is a regular element of V. Thus $V \cong Ve \subseteq R$ as ℓ-modules over V. So, again, V is totally ordered and $V = R + C \cdot 1$. Let $S = R + C \cdot 1 \subseteq \mathrm{End}_R(R_R)$ be the totally ordered ℓ-algebra given in Theorem 3.4.5(b). Now, the mapping $S \longrightarrow V$ given by $r + \alpha \cdot 1 \mapsto r + \alpha \cdot 1$ is an isomorphism of totally ordered groups since $S \cong Se = Ve \cong V$, and hence it is an isomorphism of ℓ-algebras. □

The verification that C-unital covers are unique in the torsion-free case is left as Exercise 16. We will denote the unital cover of the unitable f-ring R by R_u, and $R_{C\text{-}u}$ will denote a C-unital cover of the C-unitable f-algebra R.

If R is an infinitesimal f-algebra over the ℓ-simple f-ring C, then, as we saw in Theorem 3.4.5, R is an ℓ-ideal of $R_{C\text{-}u} = R + C \cdot 1$; and the converse certainly holds if R is totally ordered and is not unital. We will now describe when an f-algebra can be embedded as an ℓ-ideal into a unital f-algebra. Since R is an ideal of $R_{C\text{-}u}$ it is clear that such an embedding is possible if and only if R is a convex ℓ-subalgebra of $R_{C\text{-}u}$, or of some unital f-algebra.

Theorem 3.4.13. *Suppose that $R_{C\text{-}u}$ is a tight C-unital cover of the f-algebra R over the pseudo-regular f-ring C. The following statements are equivalent.*

(a) R is a convex ℓ-subalgebra of $R_{C\text{-}u}$.
(b) Each superunital homomorphic image of the f-algebra R is unital.
(c) If $a \in R$, then $a \wedge 1 \in R$.
(d) If $a \in R$ and $\alpha \in C^+$, then $a \wedge \alpha \cdot 1 \in R$.
(e) $R_{C\text{-}u} = R + C \cdot 1$ and each superunital direct summand of R (of the form γR with $\gamma^2 = \gamma \in C$) is unital.

Proof. (a) $\Rightarrow$ (b). Suppose that $R/A = \bar{R}$ has a superunit $\bar{e}$. Then $\bar{R} = C(\bar{R}_{C\text{-}u}\bar{e})$ is a summand of $\bar{R}_{C\text{-}u} = R_{C\text{-}u}/A$, by Theorem 3.4.4, and hence $\bar{R}$ has an identity element.

(b) $\Rightarrow$ (c). Let $a \in R$ and let A be the ℓ-ideal of R generated by the set $\{(ax - x)^-, (xa - x)^- : x \in R^+\}$; so A is the smallest ℓ-ideal of R modulo which a is a superunit. Clearly, A is an algebra ideal, and, for each $x \in R^+$, $(ax - x)^+$ and $(xa - x)^+$ are both in $A^{\perp_R}$. Thus, $ax - x = (ax - x)^+ - (ax - x)^- \in A^{\perp_R} \oplus A$, and, similarly, $xa - x \in A^{\perp_R} \oplus A$, for each $x \in R$. Let e be an element in R which is the identity element modulo A. Since each of e and a is the identity modulo $A \oplus A^{\perp_R}$ we have that $e - a = b + c$ where $b \in A$ and $c \in A^{\perp_R}$. We claim that $1 \wedge a = a + c$. To see this, let $\bar{R}_{C\text{-}u} = R_{C\text{-}u}/P$ be a nonzero totally ordered homomorphic image of the ℓ-algebra $R_{C\text{-}u}$. If $\bar{a} \geq \bar{1}$, then $A \subseteq P$, $\bar{e} \neq 0$, and $\bar{1} \wedge \bar{a} = \bar{1} = \bar{e} = \bar{e} - \bar{b} = \bar{a} + \bar{c}$. On the other hand, suppose that $\bar{a} < \bar{1}$. If $A^{\perp_R} \subseteq P$, then $\bar{1} \wedge \bar{a} = \bar{a} = \bar{a} + \bar{c}$. If $A^{\perp_R} \not\subseteq P$, then $A \subseteq P$, $\bar{R}$ is a nonzero homomorphic image of R/A, and we have the contradiction that $\bar{a} \geq \bar{e} = \bar{1}$.

(c) $\Rightarrow$ (d). If $\alpha \in C^+$, then $C = \langle \alpha \rangle \oplus \alpha^\perp$ and $\langle \alpha \rangle = C\beta$ with $\beta = \beta^2$. Let $\gamma \in C^+\beta$ with $\beta \leq \alpha\gamma$. We claim that $a \wedge \alpha \cdot 1 = a \wedge \alpha(\gamma|a| \wedge 1) \in R$. To verify this equality it suffices to show that, $a \wedge \alpha \cdot 1 \leq \alpha\gamma|a|$. But, $\beta(a \wedge \alpha \cdot 1) = \beta a \wedge \alpha \cdot 1 \leq \alpha\gamma|a|$, and $(1 - \beta)(a \wedge \alpha \cdot 1) = (1 - \beta)a \wedge 0 \leq 0$; so $a \wedge \alpha \cdot 1 = \beta(a \wedge \alpha \cdot 1) + (1 - \beta)(a \wedge \alpha \cdot 1) \leq \alpha\gamma|a|$, and $a \wedge \alpha \cdot 1 \in R$.

(d) $\Rightarrow$ (e). We first show that $R_{C\text{-}u} = R + C \cdot 1$. If $a \in R$ and $\alpha \in C^+$, then $(a + \alpha \cdot 1) \wedge 0 = a + (-a \wedge \alpha \cdot 1) \in R$, while if $\alpha \leq 0$, then $(a + \alpha \cdot 1) \wedge 0 = (a \wedge -\alpha \cdot 1) + \alpha \cdot 1 \in R + C \cdot 1$. In either case, $(a + \alpha \cdot 1)^+ \in R + C \cdot 1$. Suppose that $\alpha \wedge \beta = 0$ in C. Then there exist orthogonal idempotents γ and δ in C with $\langle \alpha \rangle = C\gamma$ and $\langle \beta \rangle = C\delta$. So,

$$\gamma[(a + \alpha \cdot 1) \vee \beta \cdot 1] = (\gamma a + \alpha \cdot 1)^+ \in R + C \cdot 1$$

and

$$\delta[(a+\alpha\cdot 1)\vee\beta\cdot 1]=\delta a\vee\beta\cdot 1=\delta a+(-\delta a+\beta\cdot 1)^{+}\in R+C\cdot 1.$$

Also, if $\rho=1-(\gamma+\delta)$, then $\rho[(a+\alpha\cdot 1)\vee\beta\cdot 1]=(\rho a)^{+}\in R$. Thus, $(a+\alpha\cdot 1)\vee\beta\cdot 1\in R+C\cdot 1$. Now, for any $\alpha\in C, (a+\alpha\cdot 1)^{+}=[(a+\alpha^{+}\cdot 1)\vee\alpha^{-}\cdot 1]-\alpha^{-}\cdot 1\in R+C\cdot 1$.

Now, suppose that $R=S\oplus T$ as ℓ-algebras and S has a superunit e. Then $R_{C\text{-}u}=C(R_{C\text{-}u}e)\oplus e^{\perp R_{C\text{-}u}}$ by Theorem 3.4.4, $C(R_{C\text{-}u}e)=R_{C\text{-}u}f$ for some idempotent $f\in R_{C\text{-}u}$, and $e\geq f$ since e is a superunit of $R_{C\text{-}u}f$. Hence, $f=e\wedge f=(e\wedge 1\cdot 1)f\in Rf=S$.

(e) $\Rightarrow$ (a). It suffices to show that if $\alpha\in C^{+}, a\in R$, and $\alpha\cdot 1\leq a$, then $\alpha\cdot 1\in R$. For, $0\leq b+\beta\cdot 1\leq c$ gives that $\beta^{\pm}\cdot 1\leq|\beta|\cdot 1=|\beta\cdot 1|\leq|b|+c$. Now, $\langle\alpha\rangle=C\gamma$ for some idempotent $\gamma, \gamma\leq\delta\alpha$ with $0\leq\delta\in C\gamma$, and $\gamma\cdot 1\leq\delta\alpha\cdot 1\leq\delta a$. So δa is a superunit of γR. Since $R=\gamma R\oplus(1-\gamma)R$ there is an idempotent $e\in\gamma R$ with $\gamma R=eR=eR_{C\text{-}u}$. So $\gamma R_{C\text{-}u}=\gamma R\oplus A, R_{C\text{-}u}=\gamma R\oplus(1-\gamma)R_{C\text{-}u}\oplus A$, and $A\subseteq R^{\perp R_{C\text{-}u}}=0$. Whence, $\gamma R_{C\text{-}u}=eR, \gamma\cdot 1=e$, and $\alpha\cdot 1=\gamma\alpha\cdot 1=\alpha e\in R$. (This is the only place that the tightness of $R_{C\text{-}u}$ is used.) □

It is easy to see that a C-unitable f-algebra is a subdirect product of an infinitesimal f-algebra and an f-algebra which is a subdirect product of superunital f-algebras. The second factor is unique. This will follow from a more general decomposition theorem that holds for any f-algebra. An upperpotent element a in an f-ring R which generates R as an ℓ-ideal is called a *dominant* element. According to Theorem 3.4.4 a dominant element is an upperpotent element with $a^{\perp}=0$. Of course, each totally ordered ring is either infinitesimal or it has a dominant element. An ℓ-ideal A of the f-ring R is called *dominated* if R/A has a dominant element. The f-algebra R is said to be *locally dominated* if it is a subdirect product of f-algebras that have dominant elements.

Theorem 3.4.14. *Let R be an f-algebra and let K be the intersection of all of the dominated ring ℓ-ideals of R. Then K is the intersection of dominated polars, and R is a subdirect product of the infinitesimal f-algebra $R/K^{\perp}$ and the locally dominated f-algebra R/K. If R is the subdirect product of an infinitesimal f-algebra S and a locally dominated f-algebra T, then $R/K^{\perp}$ is a homomorphic image of S and $T\cong R/K$.*

Proof. Let A be a proper dominated ℓ-ideal of the f-ring R, let $e\in R$ be such that $\bar{e}$ is a dominant element in $\bar{R}=R/A$, and let $f=(2e^{2}-e)^{+}$. Then $\bar{f}>0$ since $2\bar{e}^{2}>\bar{e}^{2}\geq\bar{e}>0$. Suppose that P is an irreducible ℓ-ideal of R and $f\notin P$. Then $\tilde{f}>0$ in $\tilde{R}=R/P$; so $2\tilde{e}$ is a dominant element in $\tilde{R}$. Now (see the proof of Theorem 2.4.6),

$$\begin{aligned}f^{\perp}&=\cap\{P: P \text{ is a minimal prime and } f\notin P\}\\&=\cap\{P_i: P_i \text{ is an irreducible } \ell\text{-ideal and } f\notin P_i\},\end{aligned}$$

and $2e$ is upperpotent modulo $f^{\perp}$ since it is upperpotent modulo each P_i. If $x\wedge 2e\in f^{\perp}$, then $x\in P_i$, for each i; so $2e$ is a dominant element modulo $f^{\perp}$. Also, $f^{\perp}\subseteq A$. For, if P is an irreducible ℓ-ideal that contains A, then $e+P$ is dominant in R/P.

Hence, $f \notin P$ and $f^\perp \subseteq P$. Since A is the intersection of irreducible ℓ-ideals, $f^\perp \subseteq A$. This shows that each dominated ℓ-ideal contains a dominated polar. Hence, K, being an intersection of polars, is itself a polar, and K is also the intersection of all of the dominated algebra ℓ-ideals of R. If $R/K^\perp$ is not infinitesimal, then $K^\perp$ is contained in a proper dominated ℓ-ideal A. But then, as above, $f \notin A$, $f^\perp \subseteq A$, $K \subseteq f^\perp$, and, finally, $f \in K^\perp \subseteq A$.

Suppose that $B \cap D = 0$ where B and D are algebra ℓ-ideals of R such that R/B is locally dominated and R/D is infinitesimal. Then $K \subseteq B$. If $K \subset B$, then there is an irreducible dominated algebra ℓ-ideal P with $B \nsubseteq P$. Thus, $D \subseteq P$ and R/P is infinitesimal and has a dominant element. This is absurd, and hence $K = B$, $D \subseteq K^\perp$ and $R/K^\perp$ is a homomorphic image of R/D. □

We turn our attention now to those special upperpotent elements that are idempotent. Let N be an ideal of the ring R and let $\bar{a}$ be an idempotent in $\bar{R} = R/N$. We say that $\bar{a}$ can be *lifted modulo N* if there exists an idempotent e in R with $\bar{e} = \bar{a}$. The next result concerns the lifting of idempotents, and it will be used on several occasions. Recall that a set S of idempotents in R is called *orthogonal* if the product of each pair of distinct elements in S is zero.

Theorem 3.4.15. *Let N be an ideal contained in the Jacobson radical of R and let $\bar{R} = R/N$.*

(a) If N is nil and $a \in R$ with $\bar{a}$ idempotent, then there exists a polynomial $p(x) \in x\mathbb{Z}[x]$ such that $e = p(a)$ is idempotent and $\bar{e} = \bar{a}$.

(b) If idempotents can be lifted modulo N, then each countable set of orthogonal idempotents in $\bar{R}$ can be lifted to a set of orthogonal idempotents of R.

Proof. (a) We will assume that R is an ideal in a unital ring. Since N is nil and $a - a^2 \in N$ there exists $n \in \mathbb{N}$ with $(a-a^2)^n = 0$. Since $(1-a)^n = 1 - af(a)$ with $f(x) \in \mathbb{Z}[x]$, $0 = a^n(1-a)^n = a^n(1-af(a))$. So $a^n = a^n(af(a))$ and hence $a^n = a^n(af(a))^n = a^{2n}f(a)^n$. Let $e = (af(a))^n$. Then $e^2 = a^{2n}f(a)^{2n} = a^n f(a)^n = e$. Since $\overline{1-a} = \overline{(1-a)^n} = \overline{1-af(a)}$ we have that $\bar{a} = \overline{af(a)}$, and $\bar{e} = \bar{a}$.

(b) We will show first that a pair of orthogonal idempotents $\bar{e}$ and $\bar{f}$ of $\bar{R}$ can be lifted to a pair of orthogonal idempotents of R. We may assume that e and f are idempotent, and we will construct an idempotent of R which is orthogonal to e and which lifts $\bar{f}$. Since $fe \in N$, $(1-fe)^{-1}$ exists, $g = (1-fe)^{-1}f(1-fe)$ is idempotent, $ge = 0$, and $\bar{g} = \bar{f}$. Let $h = (1-e)g$. Then $he = eh = 0$, $\bar{h} = \bar{g} = \bar{f}$, and $h^2 = (1-e)g(1-e)g = (1-e)g^2 = h$. Now, suppose that I is an ideal of the poset $\mathbb{N}$, $\{\bar{a}_i : i \in I\}$ is a set of orthogonal idempotents of $\bar{R}$, and $\bar{a}_1, \ldots, \bar{a}_k$ have been lifted to the orthogonal idempotents $e_1, \ldots, e_k$ of R. Let $e = e_1 + \cdots + e_k$ and let e_{k+1} be an idempotent of R that is orthogonal to e and $\bar{e}_{k+1} = \bar{a}_{k+1}$. Then $\{e_1, \ldots, e_{k+1}\}$ is an orthogonal set of idempotents that lifts $\{\bar{a}_1, \ldots, \bar{a}_{k+1}\}$. Thus, by induction , the set $\{\bar{a}_i : i \in I\}$ can be lifted. □

Our next goal is to describe all totally ordered rings with a nonzero idempotent. Let S and T be rings and let ${}_S M_T$ be an S-T-bimodule. Then $\triangle = \triangle(S, M, T) =$

$\begin{pmatrix} S & M \\ 0 & T \end{pmatrix}$ will denote the set of all matrices $\begin{pmatrix} s & m \\ 0 & t \end{pmatrix}$ with $s \in S$, $m \in M$ and $t \in T$. If $\triangle$ is supplied with the usual operations of matrix addition and multiplication, then $\triangle$ becomes a ring, called the *formal triangular matrix ring* constructed from S, T and M. It will also be convenient to denote the analogous formal triangular matrix ring $\begin{pmatrix} T & 0 \\ M & S \end{pmatrix}$ by $\nabla(T, M, S)$. If T and S are po-rings and M is a po-bimodule, then $\triangle = \begin{pmatrix} S & M \\ 0 & T \end{pmatrix}$ will be given the lexicographic partial order $T \underleftarrow{\oplus M \oplus} S$ and $\nabla = \begin{pmatrix} T & 0 \\ M & S \end{pmatrix}$ will be given the lexicographic partial order $S \underleftarrow{\oplus M \oplus} T$. So $\begin{pmatrix} s & m \\ 0 & t \end{pmatrix} \in \triangle^+$ if $s > 0$, or $s = 0$ and $m > 0$, or $s = 0$, $m = 0$ and $t \geq 0$.

Theorem 3.4.16. *(a) Let S be a unital totally ordered ring, T a totally ordered zero ring ($T^2 = 0$), and ${}_SM_T$ a totally ordered S-T-bimodule such that ${}_SM$ is unital and $MT = 0$. Then $\triangle(S, M, T) = \begin{pmatrix} S & M \\ 0 & T \end{pmatrix}$ is a totally ordered ring with a nonzero idempotent $e = \begin{pmatrix} 1 & 0 \\ 0 & 0 \end{pmatrix}$ and with $r(\triangle) \subseteq \ell(\triangle)$ iff $\ell(s)M = 0$ for each nonzero element s in S. If instead of ${}_SM_T$ we have ${}_TM_S$ with $TM = 0$ and M_S is unital, then $\nabla = \begin{pmatrix} S & 0 \\ M & T \end{pmatrix}$ is a totally ordered ring with the idempotent e and with $\ell(\nabla) \subseteq r(\nabla)$ iff $Mr(s) = 0$ for each nonzero element s in S.*

(b) Suppose that R is a totally ordered ring with a nonzero idempotent e. Then $R \cong \begin{pmatrix} eRe & eR(1-e) \\ 0 & (1-e)R(1-e) \end{pmatrix} = \triangle(eRe,\ eR(1-e),\ (1-e)R(1-e))$ if $r(R) \subseteq \ell(R)$, and $R \cong \begin{pmatrix} eRe & 0 \\ (1-e)Re & (1-e)R(1-e) \end{pmatrix} = \nabla(eRe,\ (1-e)Re,\ (1-e)R(1-e))$ if $\ell(R) \subseteq r(R)$.

Proof. (a) Assume that $\ell(s)M = 0$ for any $0 \neq s \in S$. Since $r(\triangle) = \begin{pmatrix} 0 & 0 \\ 0 & T \end{pmatrix} \subseteq \begin{pmatrix} 0 & M \\ 0 & T \end{pmatrix} = \ell(\triangle)$ it suffices to show that the product of two positive elements is positive. If $u_i = \begin{pmatrix} s_i & m_i \\ 0 & t_i \end{pmatrix} \in \triangle^+$ for $i = 1, 2$, then $u_1u_2 = \begin{pmatrix} s_1s_2 & s_1m_2 \\ 0 & 0 \end{pmatrix}$. If $s_1s_2 > 0$, then $u_1u_2 \in \triangle^+$. If $s_1s_2 = 0$ with $s_2 = 0$, then $m_2 \geq 0$ and $u_1u_2 \in \triangle^+$; and if $s_2 > 0$, then $u_1u_2 = 0$ since $s_1M = 0$. Conversely, suppose that Δ is a totally ordered ring, and take $0 < s_1, s_2 \in S$ with $s_1s_2 = 0$. Then for any $m_2 \in M$, using the previous notation but with $t_i = 0$, we have $u_1u_2 \in \Delta^+$, and hence $s_1m_2 \geq 0$. Thus, $s_1M \subseteq M^+$ and consequently $s_1M = 0$. Clearly, the second statement is the dual of the first.

(b) Since R is totally ordered the ℓ-ideals $\ell(R)$ and $r(R)$ are comparable, and $\ell(e) = \ell(R)$ and $r(e) = r(R)$ by Theorem 3.4.4. Assume that $r(R) \subseteq \ell(R)$. Since $r(e) = (1-e)R = (1-e)Re + (1-e)R(1-e)$ and $\ell(e) = R(1-e) = eR(1-e) + (1-e)R(1-e)$, necessarily $(1-e)Re = 0$. Also, if $x \in R^+ \backslash \ell(R)$, then $x > \ell(R)$, and if

$y \in \ell(R)^+ \backslash r(R)$, then $y > r(R)$; that is, $R = lex\, \ell(R)$ and $\ell(R) = lex\, r(R)$. Thus, $R \cong \begin{pmatrix} eRe & eR(1-e) \\ 0 & (1-e)R(1-e) \end{pmatrix} = \triangle(eRe, eR(1-e), (1-e)R(1-e))$ by Exercise 2.5.20. Similarly, if $\ell(R) \subseteq r(R)$, then $R \cong \begin{pmatrix} eRe & 0 \\ (1-e)Re & (1-e)R(1-e) \end{pmatrix} = \nabla(eRe, (1-e)Re, (1-e)R(1-e))$. □

We close this section by showing that a totally ordered domain which has a one-sided ℓ-ideal that is not an ideal contains a free ring on two generators with a specific total order that has this same property. Let M_R be a po-module over the po-ring R, and let $A \subseteq R$. If $x, y \in M$, then x is called *infinitely smaller than y with respect to A*, and we write $x \ll_A y$, if $xA \leq y$. If $X \subseteq M$, then X is called *A-archimedean* if $x, y \in X$ and $x \ll_A y$ implies that $x = 0$. When $A = \mathbb{Z} = R$, then $x \ll_A y$ will be written as $x \ll y$ and, of course, to say that M is $\mathbb{Z}$-archimedean just means that it is archimedean.

Theorem 3.4.17. *Let R be a totally ordered (unital) domain that contains a right ℓ-ideal that is not an ideal. Then R contains an isomorphic copy of the totally ordered free (unital) ring that is given in Exercise 35.*

Proof. Take $y \in R^+$ such that $K = \langle y \rangle_r$ is not an ℓ-ideal of R and let $J = \langle y \rangle$. Since R is not infinitesimal it has a superunit e. Thus, $J = C(RyR)$. If $J = R$, then $e \leq ryr$ for some $r \geq 0$, $yr^2 = (yr)r$ is a superunit, and $K = R$. Thus, $J \subset R$ and J is infinitesimal. If $r \in R^+$, then $ryrryr \leq ryr^2 \wedge yr \in K$; hence $J^2 \subseteq K$. Let $x \in R^+$ with $xy \notin K$ and let I be the largest ℓ-ideal of R contained in K. Then

$$1 << x << x^2 << \cdots \quad (\text{in } R_u), \tag{3.4.9}$$

$$y << yx^k << x^k y \text{ if } k \geq 1, \tag{3.4.10}$$

$$J < x, \tag{3.4.11}$$

$$J^{[2]} \subseteq I \subset K \subset J. \tag{3.4.12}$$

Let T be the multiplicative subsemigroup of R that is generated by x and y. We will express $z \in T$ as

$$z = y^{n_1} x^{k_1} \cdots y^{n_r} x^{k_r}$$

with $n_i, k_i \in \mathbb{N}$, except that n_1 or k_r could be 0, and $n_1 + k_1 \geq 1$. Let $N(z) = n_1 + \cdots + n_r$ be the number of times that y appears as a factor in this factorization of z. We will see that $N(z)$ is uniquely determined by z. If $N(z) = 0$, then $z = x^{k_1}$ is the only word in x and y which equals z by (3.4.9) and (3.4.11). Let $z' = y^{n'_1} x^{k'_1} \cdots y^{n'_s} x^{k'_s}$ be another element of T. If $N(z) = N(z') = 0$, then $z = x^{k_1}$, $z' = x^{k'_1}$, and $z' < z$ iff $z' \ll z$, iff $k_1 < k'_1$. If $N(z) = 0$ and $N(z') \geq 1$, then $z' \ll z$ since $x^{k_1} > J$. From now on all $z \in T$ will have $N(z) \geq 1$. We proceed in a sequence of steps.

(a) If $1 \leq N(z) < N(z')$ we claim that $z' \ll z$; if so, then $N(z)$ is well defined. We use induction on $N(z)$. If $N(z) = 1$, then $z = x^k y x^\ell$; thus, $y \leq z$ and $z \notin I$. Since $z' \in J^{[2]} \subseteq I$ we have $z' \ll z$. Now assume that the assertion is true if $N(z) < t$, and suppose that $1 < N(z) = t < N(z')$. Suppose, first, that $k'_s \leq k_r$. If $n \in \mathbb{N}$, then, by induction and (3.4.9),

$$
\begin{aligned}
z - nz' &= (y^{n_1}x^{k_1}\cdots y^{n_r}x^{k_r-k'_s} - ny^{n'_1}x^{k'_1}\cdots y^{n'_s})x^{k'_s} \\
&\geq (y^{n_1}x^{k_1}\cdots y^{n_r-1} - ny^{n'_1}x^{k'_1}\cdots y^{n'_s-1})yx^{k'_s} \geq 0.
\end{aligned}
$$

If, on the other hand, $k'_s > k_r$, then, by induction and (3.4.10),

$$
\begin{aligned}
z - nz' &= (y^{n_1}x^{k_1}\cdots y^{n_r} - ny^{n'_1}x^{k'_1}\cdots y^{n'_s}x^{k'_s-k_r})x^{k_r} \\
&> (y^{n_1}x^{k_1}\cdots y^{n_r-1} - ny^{n'_1}x^{k'_1}\cdots y^{n'_s-1}x^{k'_s-k_r})yx^{k_r} \geq 0.
\end{aligned}
$$

(b) Suppose that $N(z) = N(z')$ and the first place that the spellings of z and z' differ is in a power of y. So, for some m we have $n_1 = n'_1, k_1 = k'_1, \cdots, k_{m-1} = k'_{m-1}$, and $n_m < n'_m$. We claim that $z' << z$. Again, we induct on $N(z)$. Suppose that $N(z) = 1$. Then $z = y^{n_1}x^{k_1}y^{n_2}x^{k_2}$ and $z' = y^{n'_1}x^{k'_1}y^{n'_2}x^{k'_2}$. If $m = 2$, then $n_2 = n'_2$ since $n_1 + n_2 = 1 = n'_1 + n'_2$. Hence, $m = 1$ and $z = x^{k_1}yx^{k_2} \geq xy \notin K$ since $k_1 \geq 1$, and $z' = yx^{k_1} \in K$; so $z >> z'$. Assume, now, that the claim is true if $N(z) < t$ and suppose that $1 < N(z) = t$. We will consider four cases. Let $n \in \mathbb{N}$.

(i) If $m \geq 2$, then, by induction,

$$
z - nz' = y^{n_1}x^{k_1}y(y^{n_2-1}x^{k_2}\cdots x^{k_r} - ny^{n'_2-1}x^{k'_2}\cdots x^{k'_s}) \geq 0.
$$

(ii) If $m = 1$ and $n_1 \geq 1$, then, by induction,

$$
z - nz' = y(y^{n_1-1}x^{k_1}\cdots x^{k_r} - ny^{n'_1-1}x^{k'_1}\cdots x^{k'_s}) \geq 0.
$$

(iii) If $m = 1, n_1 = 0$, and $n'_1 \geq 2$, then, by induction and (3.4.10),

$$
\begin{aligned}
z - nz' &= x^{k_1}y^{n_2}\cdots x^{k_r} - ny^{n'_1}x^{k'_1}\cdots x^{k'_s} \\
&\geq y(x^{k_1}y^{n_2-1}\cdots x^{k_r} - ny^{n'_1-1}\cdots x^{k'_s}).
\end{aligned}
$$

(iv) Suppose that $m = 1, n_1 = 0$ and $n'_1 = 1$. If $k_r \geq k'_s$ then by induction and (3.4.9),

$$
\begin{aligned}
z - nz' &= (x^{k_1}y^{n_2}\cdots y^{n-r}x^{k_r-k'_s} - nyx^{k_1}\cdots y^{n'_s})x^{k'_s} \\
&\geq (x^{k_1}y^{n_2}\cdots y^{n_r-1} - nyx^{k'_1}\cdots y^{n'_s-1})yx^{k'_s} \geq 0,
\end{aligned}
$$

while if $k_r < k'_s$, then by induction and (3.4.10),

$$
\begin{aligned}
z - nz' &= (x^{k_1}y^{n_2}\cdots y^{n_r} - nyx^{k'_1}\cdots y^{n'_s}x^{k'_s-k_r})x^{k_r} \\
&\geq (x^{k_1}y^{n_2}\cdots y^{n_r-1} - nyx^{k'_1}\cdots y^{n'_s-1}x^{k'_s-k_r})yx^{k_r} \geq 0.
\end{aligned}
$$

(c) Suppose that $N(z) = N(z')$, and the spellings of z and z' first differ in a power of x. So $n_1 = n'_1, \ldots, n_m = n'_m$ and $k_m > k'_m$. Again, we will show by induction on $N(z)$ that $z' \ll z$. If $N(z) = 1$, then $z = x^k yx^\ell$ and $z' = x^{k'}yx^{\ell'}$ with $k > k'$ or $\ell > \ell'$. Let $n \in \mathbb{N}$. If $k > k'$, then $z - nz' = x^{k'}(x^{k-k'}yx^\ell - nyx^\ell) > 0$ since $yx^\ell \in K$ and $x^{k-k}yx^\ell \geq xy \notin K$. If $\ell > \ell'$, then $k = k'$ and $z - nz' = x^k y(x^\ell - nx^{\ell'}) \geq 0$.

Assume that $z' \ll z$ if $N(z) < t$, and suppose that $N(z) = t$. Let $n \in N$. If $n_1 \geq 1$, then

$$z - nz' = y(y^{n_1-1}x^{k_1}\cdots x^{k_r} - ny^{n_1-1}x^{k_1'}\cdots x^{k_s'}) \geq 0,$$

by induction. Finally, if $n_1 = 0$, then

$$z - nz' = x^{k_1}y(y^{n_2-1}\cdots x^{k_r} - y^{n_2'-1}\cdots x^{k_s'}) \geq 0$$

if $k_1 = k_1'$, that is, when $m \geq 2$, while

$$z - nz' = x^{k_1'}(x^{k_1-k_1'}y^{n_2}\cdots x^{k_r} - ny^{n_2'}\cdots x^{k_s'}) \geq x^{k_1'}y(y^{n_2-1}\cdots x^{k_r} - ny^{n_2'-1}\cdots x^{k_s'}) \geq 0$$

if $k_1 > k_1'$, that is, when $m = 1$.

This completes the proof that $\mathbb{Z}[T] \cong \mathbb{Z}[S]$ where S is the totally ordered free semigroup of Exercise 35.

If $1 \in R$, then we leave as an exercise the verification that the totally ordered free unital ring $\mathbb{Z}[S_1]$ that is given in Exercise 35 is isomorphic to $\mathbb{Z}[1,x,y]$. □

Exercises.

1. Let $R_1 = R \times \mathbb{Z}$ be the ring obtained by freely adjoining $\mathbb{Z}$ to R (Exercise 3.1.2). If $a \in R$ and $n \in \mathbb{Z}$, then a is called an *n-fier* of R if $(a, -n) \in \ell(R) \cap r(R)$.

   (a) Show that $I = \{n \in \mathbb{Z} :$ there is an n-fier in $R\}$ is an ideal of $\mathbb{Z}$. The positive generator of I is called the *mode* of R. If it is not zero, the mode of the torsion-free po-ring R will be denoted by k, and x will be a k-fier in R.
   (b) Show that x is central in R and $\ell(x;R) = 0$.
   (c) Show that a is an n-fier iff $a = \frac{n}{k}x$.
   (d) If R is an almost f-ring, show that $x > 0$.
   (e) Let $J = \mathbb{Z}(x, -k)$ if R has nonzero mode and, otherwise, let $J = 0$. Let $P = \{(a,n) \in R_1 : (a,n)R^+ + R^+(a,n) \subseteq R^+$ and $n \neq 0\} \cup R^+$. Show that J is an ideal of $R_1, P \cap -P = J$, and $PP \subseteq P$. If R is an f-ring or, for each $a \in R, a \in R^+$ iff $aR^+ + R^+a \subseteq R^+$, show that $P + P \subseteq P$.
   (f) Let $S = R_1/J$ and let $\bar{P}$ be the image of P in S under the natural map. Show that if R is an f-ring or satisfies the other condition in (e), then R is embedded in $S, \bar{P} \cap R = R^+$, and $\bar{P}$ is the largest partial order of S that contains R^+.
   (g) If R is an infinitesimal ℓ-ring or a totally ordered superunital ring, show that $S = R_u$.

2. Let R be a totally ordered ring which is a subring of the ring S. Suppose that for each $s \in S \backslash R$ there exists $a, b \in R$ with $\ell(b;S) = r(a;S) = 0$ and $as, sb \in R$. Show that S has a unique total order which extends the order on R.

3. Let V be a unital f-algebra over the ℓ-ring C and let A be the ℓ-ideal of V generated by $\{\alpha^+ \cdot 1 - (\alpha \cdot 1)^+ : \alpha \in C\}$. If R is a strong ℓ-subalgebra of V and

either R is an ideal of V or $V = C(R + C \cdot 1)$, show that $A \subseteq \ell(R) \cap r(R)$. Hence, if $\ell(R) \cap r(R) = 0$, then R is embedded in the strong f-algebra V/A.

4. Let ${}_RM$ be an ℓ-module.
   (a) If $a, b \in R^+$ with $ab = ba = 0$ and $a + b \in d({}_RM)$, show that, for each $x \in M^+$, $(a+b)(ax \wedge bx) = 0$.
   (b) If R is a pseudo-regular f-ring and M is a d-module, show that M is strong.
5. Let R be an ℓ-semiprime ℓ-ring that is a po-algebra over the po-ring C. If $\alpha, \beta \in C^+$ with $\alpha\beta = 0$ and $x \in R^+$, show that $\alpha x \wedge \beta x = 0$. Consequently, if C is an almost f-ring, then R is a strong ℓ-algebra over C.
6. Show that the following statements are equivalent for the f-algebra R over the ℓ-simple f-ring C.
   (a) R is infinitesimal.
   (b) $Q = R \otimes_C F$ is infinitesimal.
   (c) $\beta x^2 \leq \alpha|x|$ if $\beta \in C$, $0 < \alpha \in C$ and $x \in R$.
   (d) R is unitable, and if $0 < \alpha \in C$, then $\alpha \cdot 1$ is an upper bound of R in any unital f-algebra that contains R.
7. Show that the following statements are equivalent for the f-algebra R over the pseudo-regular f-ring C.
   (a) R is infinitesimal.
   (b) R satisfies the identities $[\alpha x \wedge (x^2 - \alpha x)]^+ = 0$.
   (c) If $0 \leq \beta \leq \alpha$ in C, then $\beta x^2 \leq \alpha|x|$ for each $x \in R$.
   (d) R is unitable, and if $\alpha \in C$, then $\alpha^2 \cdot 1$ is an upper bound of $|\alpha|R$, in each unital f-algebra that contains R.
8. (a) Show that the ℓ-idempotent f-algebra $R = \langle a \rangle$ is unitable iff it is C-unitable. If R is unitable, show that R has a unique C-unital cover. (See Exercise 14(b).)
   (b) If $\ell(R) = r(R) = 0$, show that the f-algebra R has a unique C-unital cover V with $\ell(R;V) = 0$.
   (c) If $\ell(R) = r(R) = 0$ and R is a convex ℓ-subalgebra of a unital f-algebra show that R has a unique tight C-unital cover.
9. (a) Show that an f-algebra is ℓ-simple iff it is C-unitable and has a (tight) C-unital cover that is ℓ-simple. Show that an ℓ-simple f-algebra has a unique C-unital cover.
   (b) Show that an f-algebra over the pseudo-regular f-ring C is pseudo-regular iff it is C-unitable and has a pseudo-regular (tight) C-unital cover.
10. Show that each C-unitable f-algebra satisfies (3.4.5) and (3.4.6).
11. Let R be either a torsion-free f-algebra over the totally ordered domain C or an f-algebra over the pseudo-regular f-ring C. Show that R satisfies the following identities:

$$z[\alpha x \wedge \alpha y \wedge (x^2 - \alpha x) \wedge (\alpha y - xy)]^+ = 0, \tag{3.4.13}$$

$$[\alpha x \wedge \alpha y \wedge (x^2 - \alpha x) \wedge (\alpha y - yx)]^+ z = 0. \tag{3.4.14}$$

Consequently, in the first case, each torsion-free homomorphic image of $R/(\ell(R) + r(R))$ is C-unitable and, in the second case, $R/(\ell(R) + r(R))$ is C-unitable.

12. Let R be a totally ordered ring. Let $A = 0$ if R is infinitesimal and, otherwise, let $A = r(R)$. Show that $A = \{\pm[x \wedge y \wedge (x^2 - x) \wedge (y - xy)]^+ : x, y \in R\}$.

13. Show that an f-ring is unitable if and only if it satisfies either of the following identities:

$$(ax^+ - x^+)^+ \wedge (ay^+ - y^+)^- = (ax^+ - x^+)^+ \wedge (y^+a - y^+)^- = 0 \tag{3.4.15}$$

$$(x^+a - x^+)^+ \wedge (ay^+ - y^+)^- = (x^+a - x^+)^+ \wedge (y^+a - y^+)^- = 0. \tag{3.4.16}$$

14. Let a be an element of the f-ring R.
    (a) Show that $\langle a \rangle$ is ℓ-idempotent iff $\langle a \rangle_r$ is ℓ-idempotent.
    (b) If R is unitable and $\langle a \rangle$ is ℓ-idempotent, show that $a^\perp = \ell(a) = r(a)$.
    (c) If $a^\perp = \ell(a) = r(a)$, show that $R/a^\perp$ is unitable.
    (d) Let I be a minimal ℓ-ideal or a minimal one-sided ℓ-ideal of R. If I is ℓ-idempotent, show that I is a minimal ℓ-ideal and a minimal one-sided ℓ-ideal of R.

15. (a) Suppose that $A = \langle a \rangle$ is an ℓ-idempotent ℓ-ideal in the f-ring R. If a is finite valued, show that A is generated by an upperpotent element.
    (b) Let $\mathscr{C}$ be the class of f-rings with the property that each principal ℓ-idempotent ℓ-ideal is generated by an upperpotent element. Show that $\mathscr{C}$ is productive, left and right i-hereditary, is closed under direct limits (with an index set that is directed up), and contains each finite valued f-ring and each unitable f-ring.
    (c) Show that an f-ring is pseudo-regular iff each of its principal ℓ-ideals is generated by an upperpotent element.
    (d) Show that each one-sided ℓ-ideal in a pseudo-regular f-ring is an ℓ-ideal.

16. Let R be a torsion-free C-unitable f-algebra over the totally ordered domain C. Show that R has a unique tight torsion-free C-unital cover.

17. Let C be a totally ordered domain with totally ordered quotient field F, and let R be an ℓ-subalgebra of the unital torsion-free C-f-algebra T.
    (a) If $Q(R) = R \otimes_C F$ is infinitesimal, show that T contains a C-unital cover of R, any one of which is tight, and if U and V are C-unital covers of R in T, then $U \cap V = R$.
    (b) If R is as in (a) and u is a cardinal number, construct an f-algebra T that contains R and 2^u distinct C-unital covers of R.

(c) If $R \cap C \neq 0$ in its C-unital cover, show that T contains a unique C-unital cover of R.

18. Let C be a totally ordered ring in which $N_2 = \{\alpha \in C : \alpha^2 = 0\} \neq 0$ and $C \backslash N_2$ consists of regular elements. Show that $R = N_2$ has two nonisomorphic totally ordered C-unital covers. Give an example of such a totally ordered ring C.

19. Let $R_{C\text{-}u}$ be a C-unital cover of the f-algebra R over the directed po-ring C. Show that the following implications in Theorem 3.4.13 hold.
    (a) (a) $\Rightarrow$ (b) $\Leftrightarrow$ (c), (a) $\Rightarrow$ (d) and (d) $\Rightarrow$ (c).
    (b) If C is an ℓ-ring and $R_{C\text{-}u}$ is strong, then (a) $\Rightarrow$ (e), and the first part of (e) together with (d) imply (a).
    (c) If $C \cdot 1 \subseteq R_{C\text{-}u}$ is a totally ordered homomorphic image of the ℓ-ring C, then (d) $\Rightarrow$ (a) and (e).

20. Give an example of a totally ordered torsion-free C-unitable f-algebra over the totally ordered domain C for which (b) and (e) of Theorem 3.4.13 hold, but (a) does not hold.

21. (a) Show that an idempotent in a ring is central iff it commutes with every idempotent.
    (b) Let R be a torsion-free ring with only a finite number of idempotents. Show that each idempotent is central.
    (c) Show that each idempotent in a unital sp-ℓ-ring is central.

22. Let E be the set of nonzero idempotents of the totally ordered ring R.
    (a) If $r(R) \subseteq \ell(R)$ (respectively, $\ell(R) \subseteq r(R)$), show that multiplication in E satisfies the identity $xy = y$ (respectively, $xy = x$).
    (b) If $r(R) \subseteq \ell(R)$, show that R has a left identity element iff $r(R) = 0$ and $E \neq \emptyset$.
    (c) Show that the following are equivalent when $E \neq \emptyset$.
        (i) E contains a central element;
        (ii) E is a singleton;
        (iii) $\ell(R) = r(R)$;
        (iv) $R = S \underrightarrow{\times} T$ (ring direct product) where S and T are totally ordered rings with S unital and $T^2 = 0$.
    (d) Let $\mathbb{Z}_0$ be the totally ordered group $\mathbb{Z}$ with trivial multiplication and let the polynomial ring $x\mathbb{Z}[x]$ have the lexicographic order: $x \ll x^2 \ll \cdots$. If $R = x\mathbb{Z}[x] \underrightarrow{\boxplus} \mathbb{Z}_0$ (ring direct product) show that R is not unitable.

23. (a) Show that in a right f-ring two positive orthogonal idempotents are disjoint.
    (b) Let e and f be idempotents of the f-ring R. Show that the following are equivalent:
        (i) $ef = 0$;
        (ii) e and f are orthogonal;
        (iii) e and f are disjoint.

24. Let S be a finite nonempty chain in a partially ordered semigroup. If $a^2 \in S$ whenever $a \in S$, show that S contains an idempotent element.

25. Let R be an ℓ-ring such that $R/\ell\text{-Nil}(R)$ is an f-ring. Show that R has a nonzero idempotent iff there exist $a \in R$ and $n \neq m$ in $\mathbb{N}$ with $a^n = a^m \neq 0$.

26. (a) Let R be a unital ℓ-ring. Show that the following are equivalent:
    (i) R has no nonzero proper right ℓ-ideals;
    (ii) Every element in R except 1 is right ℓ-QR;
    (iii) All but a finite number of elements of R are right ℓ-QR.
    (b) Let R be an ℓ-ring that is not right ℓ-QR and suppose that $R/\mathcal{J}(R)$ is an f-ring. Show that R is an ℓ-primitive f-ring iff all but a finite number of elements of R are right ℓ-QR. (If $e = e^2 \neq 0$ and $be = 0$, then $e + b$ is not right ℓ-QR.)

27. Show that an ℓ-ring satisfies the inequality $|xy| \leq |x|$ iff it satisfies $|x||y| \leq |x|$. Show that an ℓ-ring that satisfies this inequality is a right f-ring. Give an example of an ℓ-ring that satisfies this inequality but which is not an f-ring.

28. (a) Show that an ℓ-ring is infinitesimal iff it satisfies the identity $(|x| \circ |y|)^- = 0$.
    (b) Show that an f-ring is infinitesimal if and only if it satisfies $[|x| \circ (-n|x|)]^+ = 0$ for some (or all) $2 \leq n \in \mathbb{N}$.

29. If R is an infinitesimal ℓ-ring, show that $R = \mathcal{J}(R)$; in fact, if $a \in R$, then there exists $x \leq 0$ with $a \circ x \leq 0$ (see Exercise 3.3.13).

30. Let M_R be a po-module over the po-ring R.

    (a) Show that if R is directed and $x, y \in M$ with $0 \leq x$, then $x \ll_R y$ iff $x \ll_{R^+} y$.
    (b) If M^+ is R-archimedean (respectively, R^+-archimedean), show that $\ell(R; M^+) = 0$ (respectively, $\ell(R^+; M^+) = 0$) and each bounded submodule is trivially ordered. If R is directed and $\ell(R; M^+) = 0$ and each bounded submodule is trivially ordered, show that M^+ is R-archimedean.
    (c) Show that M is R-archimedean iff $\ell(R; M) = 0$ and M has no nonzero bounded submodules.

31. Suppose that R is directed and M_R is an ℓ-module.

    (a) Show that if M is a d-module and $x, y \in M$, then $x \ll_R y$ iff $|x| \ll_R y$.
    (b) If $|x_i| \ll_R y$ for $i = 1, \ldots, n$ and $r_1, \ldots, r_n \in R$, show that $|x_1 r_1 + \cdots + x_n r_n| \ll_R y$.
    (c) Show that if M is a d-module, then M has no nonzero bounded submodules iff it has no nonzero bounded convex ℓ-submodules.

32. Let M_R be an f-module over the directed po-ring R.

    (a) Show that M has no nonzero bounded submodules if and only if, for each polar N, M/N has no nonzero bounded submodules. (If $K/N \leq y + N$, then $K \cap N^\perp \leq y$.)

(b) Show that M is R-archimedean if and only if, for each polar N, M/N is R-archimedean. (Use Exercise 30(b).)

33. (a) Show that the class $\mathscr{P}$ of infinitesimal ℓ-rings is a radical class of f-rings.
(b) Show that $\mathscr{P} = \mathscr{V}(f) \cap \ell\text{-}\mathscr{A}$ where $\mathscr{V}(f)$ is the variety of f-rings and $\ell\text{-}\mathscr{A}$ is the antisimple radical (see Exercise 3.2.32).
(c) Show that the following are equivalent for the f-ring R:
(i) $\ell\text{-}\mathscr{A}(R) = 0$;
(ii) R has no nonzero infinitesimal ℓ-ideals;
(iii) R has no nonzero infinitesimal one-sided ℓ-ideals;
(iv) R is a subdirect product of ℓ-simple f-rings;
(v) R_R is R-archimedean.
(vi) R is locally dominated and has no nonzero bounded right ℓ-ideals.
(For (v) $\Rightarrow$ (vi) and (vi) $\Rightarrow$ (i) use Theorem 3.4.14, Exercise 32 and the fact that $\ell\text{-}\mathscr{A}(R)$ is bounded by each dominant element of R.)
(d) Prove that the f-ring R is pseudo-regular iff R/A is an R-archimedean right f-module for each (right) ℓ-ideal A of R.

34. Let S be a C-unital cover of the f-algebra R.
(a) If M is a maximal ℓ-ideal of R with $R^2 \not\subseteq M$, show that there is a unique maximal ℓ-ideal K of S with $K \cap R = M$. (Use Exercises 9 and 3.3.14.)
(b) If C is ℓ-simple, show that R is contained in at most one proper ℓ-ideal of S which, if it exists, is maximal.
(c) If K is a maximal ℓ-ideal of S, show that either $K \cap R$ is a maximal ℓ-ideal of R or $R \subseteq K$.
(d) Show that R is infinitesimal if and only if it is contained in each maximal ℓ-ideal of S.
(e) Show that R is contained in a proper ℓ-ideal of S if and only if R does not have a superunit.

35. (a) Show that the semigroup ring $F[S]$ of the totally ordered cancellative semigroup S over the totally ordered domain F (Exercise 2.1.14) is a totally ordered domain if it is given the Hahn product total order: $\alpha_1 s_1 + \cdots + \alpha_n s_n > 0$ if $0 < \alpha_1 \in F$ where $s_1 > \cdots > s_n$ in S.
(b) Let $\mathbb{Z}\{x,y\}$ be the free ring on two generators, without an identity element. If S is the free semigroup on $\{x,y\}$, then $\mathbb{Z}\{x,y\} = \mathbb{Z}[S]$. Show that the following total order of S makes it into a totally ordered semigroup. The elements in S will be written as $z = y^{n_1}x^{k_1}\cdots y^{n_r}x^{k_r}$ with $n_i, k_i \in \mathbb{N}$ for $i \geq 2$ and $n_1, k_1 \in \mathbb{Z}^+$. Let $N(z) = n_1 + \cdots + n_r$. If $z' = y^{n_1'}x^{k_1'}\cdots y^{n_s'}x^{k_s'}$ define $z' < z$ if $N(z) < N(z')$, or if $N(z) = N(z')$ and, for some m, either $n_1 = n_1'$, $k_1 = k_1', \ldots, k_{m-1} = k_{m-1}'$ and $n_m < n_m'$, or $n_1 = n_1'$, $k_1 = k_1', \ldots$, $n_m = n_m'$ and $k_m' < k_m$.
(c) Show that $z' < z$ in S iff $z' \ll z$ in $\mathbb{Z}\{x,y\}$.
(d) Show that the right ℓ-ideal $\langle y \rangle_r$ of $\mathbb{Z}\{x,y\}$ is not an ℓ-ideal.

(e) Let S_1 be the free monoid on x, y. Totally order S_1 so that it is a totally ordered monoid with S as a totally ordered subsemigroup. Show that $\mathbb{Z}[S_1]$ has a right ℓ-ideal that is not an ℓ-ideal.

(f) Does $\mathbb{Z}\{x,y\}$ have a left ℓ-ideal that is not an ℓ-ideal?

36. Let R be a unital ring in which all idempotents are central. Suppose that P is an ideal of R such that R/P has no nontrivial idempotents. If I is the ideal of R generated by the idempotents in P, show that R/I contains no nontrivial idempotents. (Suppose that $u \in R$ maps to an idempotent in R/I. Assume $u \in P$ and, from Exercise 1.2.8 (d), find $f^2 = f \in I$ with $f(u^2 - u) = u^2 - u$. Then $(1-f)u$ is an idempotent in P.)

37. Suppose R is an ℓ-subalgebra of the unital C-f-algebra V and T is the ℓ-subalgebra of V generated by $R + C \cdot 1$. If $\varphi : R \longrightarrow S$ is a homomorphism onto the unital f-algebra S, show that φ can be extended to a homomorphism of T onto S. (Use Theorem 2.4.2.)

Notes. Johnson [JOH1] showed that the class of unitable f-rings is a variety, established the uniqueness of a unital cover of a totally ordered unitable f-ring, and investigated when an f-ring can be embedded into a unital f-ring as an ℓ-ideal. These results were refined and extended by Henriksen and Isbell [HI] who produced the single unitability identity of Theorem 3.5.7, showed that each unitable f-ring had a unique unital cover,and established most of Theorem 3.5.13 for a unitable f-ring. The fundamental Theorem 3.5.1 comes from this paper and superunital and infinitesimal f-rings were introduced here, though these types of f-rings appeared in Johnson's paper. Exercise 22 on idempotents in totally ordered rings also appears here, and a version of Theorem 3.5.16 is given; see Bigard, Keimel, and Wolfenstein [BKW, p. 184]. Steinberg [ST22] extended Henriksen and Isbell's unitability theorems to f-algebras over a po-ring C. Pseudo-regular f-rings were introduced and studied by Keimel in [KE1] and [KE3]. He called them quasi-regular f-rings; we have changed their name to avoid confusion with the radical class of ℓ-quasi-regular ℓ-rings. Theorem 3.5.17 as well as the accompanying Exercise 35 is due to Johnson [JOH1], as is the order part of Exercise 1. Exercise 2 is due to Fuchs [F, p. 168] and Exercises 11 and 12 come from Henriksen and Isbell [HI].

3.5 Generalized Power Series Rings

The Hahn ℓ-groups $V(\Delta, G_\gamma)$ give nice examples of ℓ-groups, and equally important is the fact that each abelian ℓ-group G can be embedded in the Hahn group $V(\Gamma(G), \mathbb{R})$. While no analogue of the latter is available for ℓ-rings in general, Hahn ℓ-groups supplied with a semigroup ring multiplication nevertheless provide a rich source of examples of ℓ-rings. These ℓ-rings will be constructed using a rooted poset with a partial addition that preserves strict inequalities. We will determine when the full Hahn product is an f-ring, and also give conditions for it to be a lattice-ordered division ring whose lattice order extends to a total order. In the exercises the same considerations are given to crossed products. A subring of the Hahn product is identified which contains the subring of elements of finite support and which is always an ℓ-subring whether or not the full Hahn product is a ring.

Let Δ be a poset with a partial binary operation which we will usually write additively. So, for some $\alpha, \beta \in \Delta, \alpha+\beta \in \Delta$. If A is a ring, we will write an element $u = (u_\alpha)_{\alpha\in\Delta}$ in the group direct product A^Δ as a power series

$$u = \sum_{\alpha\in\Delta} u_\alpha x^\alpha. \tag{3.5.1}$$

If $v = \sum v_\alpha x^\alpha$ is another element of A^Δ, then the product uv is "defined" by

$$uv = \sum_{\gamma\in\Delta} \left(\sum_{\alpha+\beta=\gamma} u_\alpha v_\beta \right) x^\gamma. \tag{3.5.2}$$

So

$$(uv)_\gamma = \sum_{\alpha+\beta=\gamma} u_\alpha v_\beta, \tag{3.5.3}$$

where this sum is over all pairs $(\alpha, \beta) \in$ supp $u \times$ supp v with $\alpha+\beta = \gamma$, and the empty sum is defined to be 0. Thus, if $uv \in A^\Delta$, then, clearly, supp $uv \subseteq$ supp u + supp v where, for subsets X and Y of Δ, $X+Y = \{\alpha+\beta : \alpha \in X, \beta \in Y\}$. The obvious problem with the definition of uv is that the sum in (3.5.3) might not be finite. In order to insure that it is finite and to produce ℓ-rings we will need to impose some conditions on Δ and A. Of course, if u and v have finite support then so does uv. Here are some of the conditions that we will want Δ to satisfy.

$$\text{If } \alpha \le \beta \text{ and } \gamma \le \delta \text{ and } \alpha+\gamma \text{ and } \beta+\delta \text{ are defined, then } \alpha+\gamma < \beta+\delta \text{ unless } \alpha = \beta \text{ and } \gamma = \delta. \tag{3.5.4}$$

$$\text{If } \alpha < \beta \text{ and } \gamma+\alpha \text{ (respectively, } \alpha+\gamma) \text{ is defined, then } \gamma+\beta \text{ (respectively, } \beta+\gamma) \text{ is defined and } \gamma+\alpha < \gamma+\beta \text{ (respectively, } \alpha+\gamma < \beta+\gamma). \tag{3.5.5}$$

$$\text{For all } \alpha, \beta, \gamma \in \Delta, (\alpha+\beta)+\gamma = \alpha+(\beta+\gamma) \text{ if either side of the equation is defined.} \tag{3.5.6}$$

Note that (3.5.4) is a consequence of (3.5.5). Δ is called *associative* if it satisfies (3.5.6), and if it also satisfies (3.5.5) it is called a *pops* (partially ordered partial semigroup).

A poset is called *narrow* if each of its trivially ordered subsets is finite, and it is called a *W-set* if it has a.c.c. and is a union of a finite number of totally ordered subsets.

Theorem 3.5.1. *Suppose that Δ satisfies (3.5.4) and let X and Y be W-sets in Δ.*

(a) For each $\gamma \in \Delta$ the set $Z = \{(\alpha, \beta) \in X \times Y : \alpha + \beta = \gamma\}$ is finite.
(b) $X + Y$ has a.c.c.
(c) If the intersection of each closed interval in Δ with $X + Y$ is a chain, then $X + Y$ is a W-set.

Proof. It clearly suffices to prove the theorem under the assumption that X and Y are totally ordered. If (α, β) and (ρ, δ) are distinct elements of Z, then $\alpha + \beta = \rho + \delta$, and hence, by (3.5.4), $\alpha \neq \rho$ and $\beta \neq \delta$; also, $\alpha < \rho$ iff $\beta > \delta$. Thus, the projections $Z \longrightarrow X$ and $Z \longrightarrow Y$ are both injective and induce dual total orders of Z, in both of which Z has a.c.c. That is, Z has a total order in which it satisfies both a.c.c. and d.c.c.; so Z is finite. To see that $X + Y$ has a.c.c. suppose that $\alpha_1 + \beta_1 < \alpha_2 + \beta_2 < \cdots$ is a strictly increasing sequence in $X + Y$. According to Exercise 1.1.9, by taking a subsequence of (α_n), we may assume that (α_n) is a decreasing sequence and that (β_{n_k}) is a decreasing subsequence of (β_n). But then we have the contradiction that $(\alpha_{n_k} + \beta_{n_k})$ is a decreasing subsequence of $(\alpha_n + \beta_n)$.

To show that $X + Y$ is a W-set if the condition on intervals that is given in (c) holds we will first assume that $X + Y$ has an upper bound in Δ. If each pair of elements in $X + Y$ has a lower bound, then $X + Y$ is totally ordered and hence is a W-set. Suppose that the elements $\alpha + \beta$ and $\gamma + \delta$, with $\alpha, \gamma \in X$ and $\beta, \delta \in Y$, do not have a lower bound in Δ. Let $X \times Y$ have the product order, and let

$$T = \{(\rho, \sigma) \in X \times Y : (\tau, \nu) \leq (\rho, \sigma) \text{ in } X \times Y \Longrightarrow \tau + \nu \text{ is not defined}\}.$$

If $\alpha_1 = \alpha \wedge \gamma$ and $\beta_1 = \beta \wedge \delta$, then $(\alpha_1, \beta_1) \in T$; for, if $(\tau, \nu) \leq (\alpha_1, \beta_1)$ and $\tau + \nu \in \Delta$, then $\tau + \nu$ is a lower bound for $\{\alpha + \beta, \gamma + \delta\}$. Let β_0 be the largest element of Y such that $(\rho, \beta_0) \in T$ for some $\rho \in X$, and let α_0 be the largest element of X with $(\alpha_0, \beta_0) \in T$. Let $X_1 = \{\alpha \in X : \alpha > \alpha_0\}, Y_1 = \{\beta \in Y : \beta \leq \beta_0\}$ and $Z_1 = \{\alpha + \beta : \alpha \in X, \beta \in Y \text{ and } \beta > \beta_0\}$. We claim that $X + Y = Z_1 \cup (X_1 + Y_1)$ and Z_1 is totally ordered. For the latter, suppose that $\alpha_1 + \beta_1, \alpha_2 + \beta_2 \in Z_1$ and let $\alpha = \alpha_1 \wedge \alpha_2$ and $\beta = \beta_1 \wedge \beta_2$. Since $\beta > \beta_0, (\alpha, \beta) \notin T$; so for some $(\tau, \nu) \leq (\alpha, \beta), \tau + \nu \in X + Y$. But then $\tau \leq \alpha_i, \nu \leq \beta_i, \tau + \nu$ is a lower bound for $\{\alpha_1 + \beta_1, \alpha_2 + \beta_2\}$, and hence Z_1 is totally ordered. To see that $X + Y$ is the given union suppose that $\alpha + \beta \in X + Y$ with $\beta \leq \beta_0$. If also $\alpha \leq \alpha_0$ we have a contradiction since $(\alpha, \beta) \leq (\alpha_0, \beta_0) \in T$. Thus, $\alpha > \alpha_0$ and $\alpha + \beta \in X_1 + Y_1$. If $X_1 + Y_1$ is totally ordered we are done. Otherwise, we can repeat this argument to get $(\alpha_1, \beta_1) \in X_1 \times Y_1$ and $X_1 + Y_1 = Z_2 \cup (X_2 + Y_2)$ where Z_2 is totally ordered and $X_2 = \{\alpha \in X_1 : \alpha > \alpha_1\}$. Since $\alpha_1 > \alpha_0$ and X has a.c.c., after n steps, say, we obtain $X + Y = Z_1 \cup Z_2 \cup \cdots \cup Z_n$ where each Z_i is totally ordered.

Now assume that $X + Y$ does not have an upper bound in Δ. Let β_0 be the largest element of Y such that $\alpha + \beta_0 \in X + Y$ for some $\alpha \in X$, and let α_0 be the largest element of X with $\alpha_0 + \beta_0 \in X + Y$. Let $X_1 = \{\alpha \in X : \alpha > \alpha_0\}, Y_1 = \{\beta \in Y : \beta <$

$\beta_0\}, X' = X \setminus X_1$ and $Y' = Y_1 \cup \{\beta_0\}$. Then $X+Y = (X'+Y') \cup (X_1+Y_1)$ and $X'+Y'$ is totally ordered by the previous case since $\alpha_0+\beta_0$ is an upper bound for $X'+Y'$. If X_1+Y_1 does not have an upper bound we may repeat this construction to get an element $\alpha_1 \in X_1$ and a decomposition $X_1+Y_1 = (X_1'+Y_1') \cup (X_2+Y_2)$ with $X_1'+Y_1'$ totally ordered. Again, the process terminates after a finite number of steps since X has a.c.c. and $\alpha_0 < \alpha_1 < \cdots$. □

Let A be a ring. We can now show that certain additive subgroups of A^Δ are rings with respect to the power series multiplication given by (3.5.2). If $\alpha \in \Delta$, let $S(\alpha)$ be the set of summands of α. If Δ is associative, then $S(\alpha) = \{\beta \in \Delta : \alpha \in (\beta+\Delta) \cup (\Delta+\beta) \cup (\Delta+\beta+\Delta)\}$. Also, let

$$\Sigma = \Sigma(\Delta, A) = \{u \in A^\Delta : \operatorname{supp} u \text{ is finite}\},$$

$$W = W(\Delta, A) = \{u \in A^\Delta : \operatorname{supp} u \text{ is a W-set}\},$$

$$S = S(\Delta, A) = \{u \in A^\Delta : \forall \alpha \in \Delta, \operatorname{supp} u \cap S(\alpha) \text{ is a W-set}\}.$$

Theorem 3.5.2. *Suppose that Δ is associative and satisfies (3.5.4), and each closed interval in Δ is a chain. Then $\Sigma \subseteq W \subseteq S$ are rings.*

Proof. We will merely show that these subsets are closed under multiplication and leave the rest of the proof as an exercise. For Σ this is obvious. Suppose that $u, v \in S$ and $\gamma \in \Delta$. Then the sum in (3.5.3) is finite by (a) of Theorem 3.5.1, and hence $uv \in A^\Delta$. For, if $Z = \{(\alpha, \beta) \in \operatorname{supp} u \times \operatorname{supp} v : \alpha+\beta = \gamma\}$, then Z is finite since

$$Z \subseteq [\operatorname{supp} u \cap S(\gamma)] \times [\operatorname{supp} v \cap S(\gamma)]$$

and $\operatorname{supp} u \cap S(\gamma)$ and $\operatorname{supp} v \cap S(\gamma)$ are W-sets. Also, if $\sigma \in \Delta$, then

$$\operatorname{supp} uv \cap S(\sigma) \subseteq \operatorname{supp} u \cap S(\sigma) + \operatorname{supp} v \cap S(\sigma)$$

and hence $uv \in S$ by Theorem 3.5.1(c). For, if $\gamma \in \operatorname{supp} uv \cap S(\sigma)$, then $\gamma = \alpha+\beta \in \operatorname{supp} u + \operatorname{supp} v$, and for some $\tau, \delta \in \Delta, \sigma = \tau+\gamma$ or $\sigma = \gamma+\tau$ or $\sigma = \tau+\gamma+\delta$. So, $\sigma = \tau+\alpha+\beta, \sigma = \alpha+\beta+\tau$, or $\sigma = \tau+\alpha+\beta+\delta$, and $\alpha, \beta \in S(\sigma)$. Thus, S is closed under multiplication. But W is a subring of S since if $u, v \in W$, then $\operatorname{supp} uv \subseteq \operatorname{supp} u + \operatorname{supp} v$ is a W-set. □

If A is a po-ring, then $\Sigma(\Delta, A)$ and $W(\Delta, A)$ are certainly po-subgroups of the Hahn product $V(\Delta, A)$. Sufficient conditions for these subgroups to be ℓ-rings are given in the following result.

Theorem 3.5.3. *Suppose that Δ is a pops and A is a po-domain. Then Σ is a po-ring, and for $u, v \in W^+ \cup -W^+$,*

$$maxsupp\ uv = max\,(maxsupp\ u + maxsupp\ v). \tag{3.5.7}$$

Moreover, Σ (respectively, W) is an ℓ-ring if and only if Δ is rooted and either A is totally ordered, or A is an ℓ-ring and Δ is trivially ordered.

Proof. First note that if $u, v \in W$, then supp u + supp v has a.c.c. and $uv \in V(\Delta, A)$ according to Theorem 3.5.1. Since supp $u =$ supp $-u$ it suffices to verify (3.5.7) for u, $v \in W^+$. If $0 < u, v$ and $\gamma \in$ max (maxsupp u + maxsupp v), then $\gamma = \alpha + \beta$ with $\alpha \in$ maxsupp u and $\beta \in$ maxsupp v; so $u_\alpha v_\beta > 0$. If $\gamma \notin$ supp uv, then, by (3.5.3), $\gamma = \alpha_1 + \beta_1 \in$ supp u + supp v with $0 \neq u_{\alpha_1} v_{\beta_1} \not> 0$. Thus, for some α_2 and β_2, $\alpha_1 \leq \alpha_2 \in$ maxsupp u and $\beta_1 \leq \beta_2 \in$ maxsupp v with $\alpha_1 < \alpha_2$ or $\beta_1 < \beta_2$. But then we have the contradiction that $\gamma = \alpha_1 + \beta_1 < \alpha_2 + \beta_2 \leq \gamma_1$ for some $\gamma_1 \in$ max (maxsupp u + maxsupp v). So $\gamma \in$ supp uv and

$$\text{max (maxsupp } u + \text{maxsupp } v) \subseteq \text{ supp } uv. \tag{3.5.8}$$

On the other hand, suppose that $\gamma = \alpha + \beta \in$ maxsupp uv with $\alpha \in$ supp u and $\beta \in$ supp v. Then $\alpha \leq \alpha_1 \in$ maxsupp u, $\beta \leq \beta_1 \in$ maxsupp v, and $\gamma \leq \alpha_1 + \beta_1 \leq \gamma_1 \in$ max (maxsupp u + maxsupp v). By (3.5.8), $\gamma = \gamma_1$ and hence $\alpha = \alpha_1$ and $\beta = \beta_1$; so we have

$$\text{maxsupp uv } \subseteq \text{ max (maxsupp u + maxsupp v)} \tag{3.5.9}$$

and

$$\text{if } \gamma = \alpha + \beta \in \text{ maxsupp uv with } \alpha \in \text{ supp u and } \beta \in \text{ supp v,}$$
$$\text{then } \alpha \in \text{ maxsupp u and } \beta \in \text{ maxsupp v.} \tag{3.5.10}$$

Now, (3.5.7) follows from (3.5.8) and (3.5.9). Also, if $u, v \in \Sigma^+$ and $\gamma \in$ maxsupp uv, then by (3.5.10) the sum in (3.5.3) is over some subset of maxsupp $u \times$ maxsupp v. So $(uv)_\gamma > 0$, and Σ is a po-ring.

For the last statement, if Σ (or W) is an ℓ-ring, then A and Δ satisfy the given conditions by Theorem 2.6.2 and Exercise 2.6.2. Conversely, if A and Δ satisfy these conditions, then W is a ring by Theorem 3.5.2, it is an ℓ-group by the aforementioned results from Section 2.6, and it is an ℓ-ring by (3.5.10). □

If $W(\Delta, A)$ is an ℓ-ring, then the conditions that Δ must satisfy in order for W to be a (left) d-ring or an almost f-ring are given in Exercises 15, 16, and 22. The conditions for it to be an sp-ℓ-ring are examined in Section 3.7.

The next result shows that the conditions that make Σ into an f-ring are strong enough to also force V to be an f-ring. A pops that satisfies these conditions, given in (3.5.11) and (3.5.12) below, is called an *f-pops*. More generally, let $f_\ell(\Delta)$ (respectively, $f_r(\Delta)$) be the set of all those elements γ in the pops Δ which satisfy (3.5.11) (respectively, (3.5.12)); Δ is a *left f-pops* if $\Delta = f_\ell(\Delta)$ and it is a *right f-pops* if $\Delta = f_r(\Delta)$.

Theorem 3.5.4. *Suppose that A is a totally ordered domain and Δ is a rooted pops. The following statements are equivalent.*

(a) $V(\Delta, A)$ is an f-ring.
(b) $W(\Delta, A)$ is an f-ring.
(c) $\Sigma(\Delta, A)$ is an f-ring.
(d) Δ satisfies the following conditions:

$$\text{if } \alpha || \beta \text{ and } \gamma + \alpha \text{ is defined, then } \gamma + \alpha || \beta; \tag{3.5.11}$$

$$\text{if } \alpha || \beta \text{ and } \alpha + \gamma \text{ is defined, then } \alpha + \gamma || \beta. \tag{3.5.12}$$

Proof. The implications (a) $\Rightarrow$ (b) $\Rightarrow$ (c) follow from Theorems 3.5.2 and 3.5.3.

(c) $\Rightarrow$ (d). If $0 < a \in A$, $\alpha || \beta$ and $\gamma + \alpha$ exists, then $ax^\alpha \wedge ax^\beta = 0$ by Exercise 2.6.1(c). So $a^2 x^{\gamma+\alpha} \wedge ax^\beta = 0$ and $\gamma + \alpha || \beta$.

(d) $\Rightarrow$ (a). We first need to verify that $V = V(\Delta, A)$ is a ring. If $\gamma \in \Delta$ let γ'' be the set of all those elements of Δ that are comparable with each element of Δ that is comparable to γ. Suppose that $u, v \in V$. According to Exercises 3(b) and 4(a), γ'' is totally ordered and

$$\{(\alpha,\beta) \in \text{supp } u \times \text{supp } v : \alpha + \beta = \gamma\} \subseteq \text{supp } u \cap S(\gamma) \times \\ \text{supp } v \cap S(\gamma) \subseteq \text{supp } u \cap \gamma'' \times \text{supp } v \cap \gamma''; \tag{3.5.13}$$

so the sum in (3.5.3) is finite by Theorem 3.5.1(a) and $uv \in A^\Delta$. Also, if $\gamma_1 \leq \gamma_2 \leq \cdots$ is a chain in supp u + supp $v \supseteq$ supp uv, then, since $\gamma_n'' \subseteq \gamma_1''$ for every n by Exercise 3(c), $\gamma_n = \alpha_n + \beta_n \in \text{supp } u \cap \gamma_1'' + \text{supp } v \cap \gamma_1''$. But the latter has a.c.c. by Theorem 3.5.1(b), and hence $\gamma_n = \gamma_{n+1} = \cdots$ for some n. Thus, $uv \in V$. Also, if $0 < u, v$, then the argument that gives (3.5.7) and (3.5.10) in the proof if Theorem 3.5.3 is still valid and $uv \geq 0$. So V is an ℓ-ring. Suppose that $u, v, w \in V^+$ with $u \wedge w = 0$, and $\gamma \in$ maxsupp uv and $\delta \in$ maxsupp w. Then $\gamma = \alpha + \beta \in$ supp u + supp v and $\alpha || \delta$; so $\gamma || \delta$ and $uv \wedge w = 0$. Similarly, $vu \wedge w = 0$ and V is an f-ring. □

The only way that V could be an f-ring without A being totally ordered is if $V^2 = 0$ (see Exercise 18e). It is certainly possible for V to be an ℓ-ring without it being an f-ring. For example, if Δ is finite or has only a finite number of roots then $V = W$. As a consequence of the next result we will see that $V = W$ also whenever Δ is narrow.

Theorem 3.5.5. *Let Γ be a poset in which each closed interval is a chain. Then Γ is a W-set if and only if it is noetherian and narrow.*

Proof. Suppose that Γ is noetherian and narrow but it is not a W-set. Let us call the subsets X and Y of Γ incomparable if no element of X is comparable to any element of Y. Under the assumption that Γ has a largest element we claim that there is a chain M in Γ that is maximal with respect to $\Gamma \backslash M < M$ (that is, $\alpha < \beta$ if $\alpha \in \Gamma \backslash M$ and $\beta \in M$), and $\Gamma \backslash M$ is the finite union of at least two pairwise incomparable subsets each of which has a largest element. To see this let

$$\mathscr{S} = \{X \subseteq \Gamma : X \text{ is a chain and } \Gamma \backslash X < X\}.$$

$\mathscr{S}$ is not empty since if α_0 is the largest element of Γ then $\{\alpha_0\} \in \mathscr{S}$. Also, if $\{X_i\}$ is a chain in $\mathscr{S}$ then $\cup X_i$ is in $\mathscr{S}$. So $\mathscr{S}$ has a maximal element M and $\Gamma \backslash M$ is not a W-set. Let $\{\alpha_1, \ldots, \alpha_m\}$ be the maximal elements of $\Gamma \backslash M$. If $m = 1$, then $\Gamma \backslash (M \cup \{\alpha_1\}) < M \cup \{\alpha_1\}$ and $M \cup \{\alpha_1\} \in \mathscr{S}$, which contradicts the choice of M. Let $X_i = \{\beta \in \Gamma \setminus M : \beta \le \alpha_i\}$. Then $\Gamma \setminus M$ is the union of the X_i and the latter are pairwise incomparable. For, if $i \ne j$ and $\beta_i \in X_i$ and $\beta_j \in X_j$ with $\beta_i \le \beta_j$, then $\beta_i \le \alpha_i \le \alpha_0$ and $\beta_i \le \beta_j \le \alpha_j \le \alpha_0$. But the interval $[\beta_i, \alpha_0]$ is a chain, and hence we have the contradiction that α_i and α_j are comparable. Thus, the claim has been established. Now, let $\{\alpha_1, \ldots, \alpha_n\}$ be the set of maximal elements in Γ. Then $\Gamma = \cup_i L(\alpha_i)$ where $L(\alpha_i) = \{\alpha \in \Gamma : \alpha \le \alpha_i\}$, and some $L(\alpha_i)$ is not a W-set. Replacing Γ by this $L(\alpha_i)$ and applying the claim we have a chain M in Γ such that $\Gamma \setminus M = \Gamma_1 \cup \Gamma_{11} \cup \cdots$ where Γ_1 is not a W-set. Repeatedly applying the claim we get

$$\Gamma_1 \setminus M_1 = \Gamma_2 \cup \Gamma_{21} \cup \cdots$$

and

$$\Gamma_2 \setminus M_2 = \Gamma_3 \cup \Gamma_{31} \cup \cdots$$

$$\cdots$$

where no Γ_i is a W-set. If $\beta_j \in \Gamma_{j1}$, then $\{\beta_j : 1 \le j\}$ is an infinite trivially ordered subset of Γ since $\beta_k \in \Gamma_{k1} \subseteq \Gamma_j$ if $k > j$. □

Theorem 3.5.6. *Suppose that each closed interval in the poset Γ is a chain, and let A be a nonzero ring. If Γ is narrow, then $W(\Gamma,A) = V(\Gamma,A)$. If Γ has a partial addition, then $W(\Gamma,A) = S(\Gamma,A)$ if and only if for each subset Λ of Γ that is not a W-set, $\Lambda \cap S(\gamma)$ is infinite for some $\gamma \in \Gamma$.*

Proof. If Γ is narrow and $v \in V = V(\Gamma,A)$, then supp v is a W-set by the previous result; so $v \in W = W(\Gamma,A)$. Assume that $W = S = S(\Gamma,A)$, and let Λ be a subset of Γ that is not a W-set. For $0 \ne a \in A$ let $v = \sum_{\alpha \in \Lambda} a x^{\alpha}$. Then supp $v = \Lambda$ and $v \notin S$; so $\Lambda \cap S(\gamma)$ is infinite for some $\gamma \in \Gamma$. Conversely, suppose that $v \in S \setminus W$, and assume that Γ has the stated property. Since supp v is not a W-set, by Theorem 3.5.5 it contains a sequence $(\alpha_n)_{n \in \mathbb{N}}$ of distinct elements that is either increasing or its range is trivially ordered. If Λ is the range of this sequence, then no infinite subset of Λ is a W-set. But for some $\gamma \in \Gamma$ we have that $\Lambda \cap S(\gamma) \subseteq$ supp $v \cap S(\gamma)$ is an infinite W-set. This contradiction gives that $W = S$. □

We now turn our attention to finding units in $W(\Delta,A)$. In particular, we will establish and generalize the equation $(1 - x^{-1})^{-1} = \sum_{n=0}^{-\infty} x^n$ in $V(\mathbb{Z},\mathbb{Z})$.

The pops Δ is called a *mopops* if it contains an element 0 such that $\alpha + 0 = 0 + \alpha = \alpha$ for each $\alpha \in \Delta$. If $\alpha + \beta < \alpha, \beta$ (respectively, $\alpha, \beta < \alpha + \beta$) whenever $\alpha + \beta$ is defined, then Δ is called a *negative* (respectively, *positive*) pops. In any mopops the set $L(0)^*$ of nonzero lower bounds of 0 is a negative pops, whereas the set $U(0)^*$ of nonzero upper bounds of 0 is a positive semigroup. A negative pops can be embedded in a mopops (Exercise 7). We also note that a rooted negative semigroup is totally ordered whereas a rooted positive semigroup need not be totally ordered. For example, consider the multiplicative subsemigroup $\Delta = \mathbb{N} \setminus \{1\}$ of $\mathbb{N}$ with the order given by $2 || 3$ and $2, 3 < 4 < 5 < \cdots$.

Suppose that Δ is a totally ordered negative semigroup. It will be useful to mimic the definition of the infinitely large relation for elements in the negative cone of a po-group. Let $\alpha, \beta \in \Delta$. If $n\beta \ge \alpha$ for each $n \in \mathbb{N}$, then β is called *infinitely larger* than α and we write $\beta \gg \alpha$. The elements α and β are called *a-equivalent* and we write $\alpha \sim \beta$, if

$$\{\gamma \in \Delta : \gamma \gg \alpha\} = \{\gamma \in \Delta : \gamma \gg \beta\}. \tag{3.5.14}$$

We will state the main properties of these relations but we leave their verification to Exercise 9. First, $\alpha \sim \beta$ iff for some $n, m \in \mathbb{N}, \alpha \geq n\beta$ and $\beta \geq m\alpha$. The equivalence class of α will be denoted by $v(\alpha)$ and is called the *value* of α. The *value set* $v(\Delta)$ of Δ is a totally ordered set with respect to the relation $v(\beta) > v(\alpha)$ if $\beta \gg \alpha$, and the map $v : \Delta \longrightarrow v(\Delta)$ is isotone. Moreover, $v(\alpha_1 + \cdots + \alpha_n) = v(\alpha_1) \wedge \cdots \wedge v(\alpha_n)$.

By a *subpops* of the pops Δ we mean a subset X of Δ such that $\alpha + \beta \in X$ whenever $\alpha, \beta \in X$ and $\alpha + \beta$ is defined. If Γ is a subset of Δ, then for $n \in \mathbb{N}, n\Gamma = \{\alpha_1 + \cdots + \alpha_n : \alpha_i \in \Gamma\}$ and $\omega\Gamma = \bigcup_{n=1}^{\infty} n\Gamma$ is the subpops of Δ generated by Γ.

Theorem 3.5.7. *Let Δ be a rooted negative pops. Suppose that Γ is a subset of Δ with the property that each of its roots generates a subsemigroup of Δ. If Γ is noetherian, then so is $\omega\Gamma$ and each element of $\omega\Gamma$ is in only finitely many of the $n\Gamma$. If Γ is narrow or is a W-set, then $\omega\Gamma$ is narrow or is a W-set, respectively.*

Proof. Suppose that Γ is noetherian and, by way of contradiction, let $\alpha_1 < \alpha_2 < \cdots$ be a chain in $\omega\Gamma$. Let $\alpha_i = \gamma_{i1} + \cdots + \gamma_{in_i}$ with $\gamma_{ij} \in \Gamma$. Since Δ is an f-pops by Exercise 1, the set $\{\gamma_{ij} : 1 \leq i, 1 \leq j \leq n_i\}$ is a chain in Γ by Exercise 4, and it generates a totally ordered subsemigroup of Δ which, for convenience, we will assume is Δ. Now, for each $i, v(\alpha_i) = \bigwedge_j v(\gamma_{ij}) = v(\gamma_i^*)$ where $\gamma_i^* = \bigwedge_j \gamma_{ij} \in \Gamma$. Since $v(\Delta) = v(\Gamma)$ is noetherian the increasing sequence $v(\alpha_1) \leq v(\alpha_2) \leq \cdots$ is eventually constant; say $v(\alpha_i) = w$ if $i \geq n$. We may assume that this w is the largest value of any of the terms in a strictly increasing sequence in Δ. By deleting the first few terms of such a sequence we will only consider strictly increasing sequences in $\omega\Gamma$ which are constant-valued. Let α be the largest element of $w \cap \Gamma$; note that $\gamma_i^* \in w \cap \Gamma$ and $\alpha_i \leq \gamma_i^* \leq \alpha$ for each i. Since $\alpha \sim \alpha_1$ there is an integer $p \in \mathbb{N}$ with $\alpha_1 \geq p\alpha$. We may assume that this p is minimal for all sequences considered. If $p = 1$, then for every $i \geq 1, \alpha \leq \alpha_1 \leq \alpha_i \leq \gamma_i^* \leq \alpha$; thus, $p > 1$ and $(p-1)\alpha > \alpha_1 \geq p\alpha$. Also, for every i, $(p-1)\alpha > \alpha_i$ since otherwise $\alpha_i < \alpha_{i+1} < \cdots$ is a sequence with $\alpha_i \geq (p-1)\alpha$, contrary to the choice of p. So, for each i, $(p-1)\alpha > \alpha_i \geq p\alpha$. Now, for every i, $\alpha_i = \gamma_i^*$ or $\alpha_i = \alpha_i' + \gamma_i^*$ or $\alpha_i = \gamma_i^* + \alpha_i''$ or $\alpha_i = \alpha_i' + \gamma_i^* + \alpha_i''$, and we cannot have $\alpha_i = \gamma_i^*$ for infinitely many i. Thus, one of the three other forms for α_i must occur infinitely often. We will rule out the last form from such an occurrence, and in a similar manner the other two may be ruled out. By taking a subsequence of the sequence $(\alpha_i)_i$ we may assume that $\alpha_i = \alpha_i' + \gamma_i^* + \alpha_i''$ for every i. If neither of the sequences $(\alpha_i')_i$ or $(\alpha_i'')_i$ had a strictly increasing subsequence, then their ranges would have a.c.c. (Exercise 1.1.9). Hence, the set $\{\alpha_i' : 1 \leq i\} + \{\gamma_i^* : 1 \leq i\} + \{\alpha_i'' : 1 \leq i\}$ would also have a.c.c. by Theorem 3.5.1, contrary to our assumption. By again taking a subsequence of $(\alpha_i)_i$ we may assume that $\alpha_1' < \alpha_2' < \cdots$. Since $\alpha_i < \alpha_i', w = v(\alpha_i')$ by the maximality of w; and $\alpha_i' \leq \alpha$ since $\alpha_i' = \gamma_{i1} + \cdots + \gamma_i^{**} + \cdots \leq \gamma_i^{**}$ with $v(\alpha_i') = v(\gamma_i^{**})$ and $\gamma_i^{**} \leq \alpha$. If $(p-1)\alpha > \alpha_1'$, then $p\alpha = (p-1)\alpha + \alpha > \alpha_1' + \gamma_1^* > \alpha_1' + \gamma_1^* + \alpha_1'' = \alpha_1 \geq p\alpha$. So $\alpha_1' \geq (p-1)\alpha$ and this contradicts the minimality of p. Hence, $\omega\Gamma$ is noetherian.

Let α be maximal among those elements of $\omega\Gamma$ that are in infinitely many of the $n\Gamma$. Then $\alpha = \gamma_i + \beta_i \in n_i\Gamma$ with $1 < n_1 < n_2 < \cdots, \gamma_i \in \Gamma$ and $\beta_i \in (n_i - 1)\Gamma$. Now, $(\gamma_i)_i$ has a decreasing subsequence, and the corresponding subsequence of $(\beta_i)_i$ is increasing and therefore it is eventually constant. But then $\alpha < \beta = \beta_i$ for infinitely many i and this contradicts the maximality of α.

Assume now that Γ is narrow but $\omega\Gamma$ is not. Let $\{\alpha_i : i \in \mathbb{N}\}$ be a trivially ordered subset of $\omega\Gamma$. Using the previous notation, suppose that for some $i \neq j, \gamma_i^* \leq \gamma_j^*$. Then for any $1 \leq k \leq n_i$ and $1 \leq \ell \leq n_j, \gamma_i^* \leq \gamma_{ik}$ and $\gamma_i^* \leq \gamma_j^* \leq \gamma_{j\ell}$, and hence $\{\gamma_{ik} : 1 \leq k \leq n_i\} \cup \{\gamma_{j\ell} : 1 \leq \ell \leq n_j\}$ is contained in a root of Γ. Thus, $\alpha_i + \alpha_j \in \Delta$ and α_i is comparable to α_j. This contradiction shows that $\omega\Gamma$ must be narrow. That $\omega\Gamma$ is a W-set if Γ is a W-set now follows from Theorem 3.5.5. □

We are now ready to find inverses in formal power series rings.

Theorem 3.5.8. *Let Δ be a totally ordered cancellative monoid and let A a be unital ring. Suppose that $v \in V(\Delta, A)$.*

(a) If supp $v < 0$ and $(a_n)_{n\geq 0}$ is a sequence in A, then $v^\omega = \sum_{n\geq 0} a_n v^n$ is an element of $V(\Delta, A)$ where v^ω is defined by $(v^\omega)_\alpha = \sum_{n\geq 0} a_n (v^n)_\alpha$ for each $\alpha \in \Delta$.
(b) If $w = 1 - v$ with supp $v < 0$ then $w^{-1} = \sum_{n\geq 0} v^n$.
(c) Assume that A is a domain and let $\alpha =$ maxsupp v. Then v is a unit of $V(\Delta, A)$ iff v_α is a unit of A and α is a unit of Δ.
(d) $V(\Delta, A)$ is a division ring iff Δ is a group and A is a division ring.

Proof. By Theorem 3.5.7 the sum $\sum_{n\geq 0} a_n (v^n)_\alpha$ is finite since supp $v^n \subseteq n$ supp v (here, 0 supp $v = 0$), and supp v^ω has a.c.c. since supp $v^\omega \subseteq \omega$ supp $v \cup \{0\}$. So $v^\omega \in V(\Delta, A)$. To show (b), we will first verify that $v \sum_n v^n = \sum_n v^{n+1}$. If $\gamma \in \Delta$, then

$$\begin{aligned}\left(v\sum_n v^n\right)_\gamma &= \sum_{\alpha+\beta=\gamma} v_\alpha \sum_n (v^n)_\beta \\ &= \sum_n \sum_{\alpha+\beta=\gamma} v_\alpha (v^n)_\beta \\ &= \left(\sum_n v^{n+1}\right)_\gamma .\end{aligned}$$

So $(1-v)\sum_n v^n = \sum_n v^n - \sum_n v^{n+1} = 1$. Similarly, $(\sum v^n)(1-v) = 1$. For (c), if $u \in V$ and $\beta =$ maxsupp u, then $(vu)_{\alpha+\beta} = v_\alpha u_\beta \neq 0$ and $\alpha + \beta =$ maxsupp vu since if $(\alpha_1, \beta_1) \in$ supp $v \times$ supp u with $(\alpha, \beta) \neq (\alpha_1, \beta_1)$, then $\alpha_1 + \beta_1 < \alpha + \beta$. Thus, if $vu = 1$ then $\alpha + \beta = 0$ and $v_\alpha u_\beta = 1$. Conversely, suppose that $\alpha + \beta = 0$ and $v_\alpha b = 1$. Then

$$vbx^\beta = (v_\alpha x^\alpha + \textstyle\sum_{\rho<\alpha} v_\rho x^\rho) bx^\beta = 1 + \displaystyle\sum_{\rho+\beta<0} v_\beta bx^{\rho+\beta}$$

and hence v is a unit of V by (b). Since (d) is an immediate consequence of (c) the proof is complete. □

Hamilton produced the first noncommutative division ring when he constructed the ring of real quaternions. The second division ring to appear was the totally ordered twisted power series ring $V(\mathbb{Q}(y) * \mathbb{Z})$ that was constructed by Hilbert. Here, $\mathbb{Q}(y)$ is the totally ordered field of quotients of the lexicographically ordered polynomial ring $\mathbb{Q}[y]$ and $V(\mathbb{Q}(y) * \mathbb{Z}) = V(\mathbb{Z}, \mathbb{Q}(y))$ as totally ordered groups. Let σ be the ℓ-automorphism of $\mathbb{Q}(y)$ that is induced by $y \mapsto 2y$. If the elements of $V(\mathbb{Q}(y) * \mathbb{Z})$ are represented as formal power series $u = \sum x^n \alpha_n$ with the coefficients $\alpha_n \in \mathbb{Q}(y)$ on the right, then multiplication in $V(\mathbb{Q}(y) * \mathbb{Z})$ is induced by the equations $x^n \alpha x^m \beta = x^{n+m} \sigma^m(\alpha)\beta$. A generalization of this construction is given in Exercises 13, 14, 17, and 20.

Let R be a po-unital po-ring and let $\mathscr{U}(R^+)$ be the po-group of units of the multiplicative po-monoid R^+. If G is a subgroup of $\mathscr{U}(R^+)$ and a sublattice of R, then G is called an *ℓ-subgroup of positive units of R*. Note that if R is an f-ring, then $\mathscr{U}(R^+)$ is a sublattice of R. For, if $w \in \mathscr{U}(R^+)$, then $w \vee w^{-1} \geq 1$ since this inequality holds in each totally ordered homomorphic image of R. So, if $u, v \in \mathscr{U}(R^+)$, then $(u \vee v)^{-1} = u^{-1} \wedge v^{-1}$ since $(u \vee v)(u^{-1} \wedge v^{-1}) = 1 \wedge (uv^{-1} \vee vu^{-1}) = 1$. If $x^{-1} \in R$ whenever $x \geq 1$, then R is said to have *bounded inversion*.

Theorem 3.5.9. *The following statements are equivalent for the ℓ-group G.*

(a) G is representable.
(b) G is isomorphic to an ℓ-subgroup of positive units of a direct product of totally ordered division rings.
(c) G is isomorphic to an ℓ-subgroup of the ℓ-group of positive units of an f-ring which has bounded inversion.
(d) G is isomorphic to an ℓ-subgroup of positive units of an f-ring.

Proof. To see that (a) implies (b) suppose that G is embedded in the product of totally ordered groups $\Pi_i G_i$. If A_i is a totally ordered division ring, then clearly G_i is embedded in the group of positive units of the totally ordered division ring $V(G_i, A_i)$. So G is embedded in $\Pi_i \mathscr{U}(V(G_i, A_i)^+) = \mathscr{U}((\Pi_i V(G_i, A_i))^+)$.

Since the implications (b) $\Rightarrow$ (c) $\Rightarrow$ (d) $\Rightarrow$ (a) are obvious the proof is complete. □

Another application of Theorem 3.5.8 concerns free groups; see Exercise 2.4.8.

Theorem 3.5.10. *Each free group can be totally ordered.*

Proof. Let S be the free semigroup on the set Y. We can extend a given total order of Y to a total order of S in the following way. If $u = z_1 \cdots z_n$ and $v = y_1 \cdots y_m$ are two words of S in the alphabet Y, then $u < v$ iff $n > m$, or $n = m$ and for some k with $1 \leq k \leq n, z_1 = y_1, \ldots, z_{k-1} = y_{k-1}$ and $z_k < y_k$. With this order S is a totally ordered negative semigroup, and $T = S \cup \{e\}$, with $S < e$, is the (totally ordered) free monoid on Y. Let A be a totally ordered division ring. By Theorem 3.5.8, for each $y \in Y, 1 + y$ is a positive unit of the totally ordered domain $V(T, A)$. We claim that the multiplicative subgroup F of $V(T, A)$ that is generated by the set $\{1 + y : y \in Y\}$ is a free group on this set. Note that if $0 \neq g \in \mathbb{Z}$, then $(1 + y)^g = 1 + gy + w$ where

$w = 0$ or supp $w \subseteq \{y^k : k \geq 2\}$. This is clear if $g > 0$ and it follows from Theorem 3.5.8 if $f = -g > 0$ since

$$(1+y)^g = (1+fy+\cdots+y^f)^{-1} = 1-(fy+\cdots+y^f)+(fy+\cdots y^f)^2-\cdots.$$

Now, suppose that $u = (1+y_1)^{g_1}\cdots(1+y_n)^{g_n}$ where $y_i \in Y, y_i \neq y_{i+1}$ and $g_i \neq 0$ for each i. Then $u = (1+g_1y_1+w_1)\cdots(1+g_ny_n+w_n) = 1+g_1\cdots g_ny_1\cdots y_n+\sum_j v_j$ and clearly supp $v_j < y_1\cdots y_n$ for each j. Thus, $u \neq 1$ and F is free on $\{1+y : y \in Y\}$. □

Another proof of this result using polynomials and 2×2 matrices is given in Exercises 24, 25, and 26.

We conclude this section with an example of a lattice-ordered power series division ring that is not totally ordered. Of course, one such example comes from the division ring $V(\Delta,A)$ in Theorem 3.5.8 by taking A to be totally ordered and weakening the total order of $V(\Delta,A)$ to the lattice order $P = \{v \in V(\Delta,A) : v_\alpha \geq 0$ for each $\alpha \in \Delta\}$. In a loose sense the partial order of Δ has been replaced by the trivial order—loose, because with Δ trivially ordered $V(\Delta,A)$ is the full product A^Δ. This process can sometimes be reversed. Specifically, if the partial order of Δ can be extended to a total order, then the lattice order of $V(\Delta,A)$ may also be extendable to a total order. We first need some preparation.

Theorem 3.5.11. *Let H be a subgroup of finite index in the group G.*

(a) If G is finitely generated, then H is finitely generated.
(b) If H is contained in the center of G, then G', the commutator subgroup of G, is finite.

Proof. (a) Let $x_1,\ldots,x_m$ be generators for G and let $1 = y_1, y_2,\ldots,y_n$ be representatives of the $n = [G:H]$ distinct left cosets of H in G. Since left addition by x_j permutes these left cosets, for $1 \leq j \leq m$ and $1 \leq i \leq n$ there is an $i_j \in \{1,\ldots,n\}$ with $x_j+y_i+H = y_{i_j}+H$ and $i_j \neq k_j$ if $i \neq k$. Let $h_{ji} \in H$ be defined by $x_j+y_i = y_{i_j}+h_{ji}$, and let A be the subgroup of H generated by $\{h_{ji} : 1 \leq j \leq m, 1 \leq i \leq n\}$. We will show $A = H$. Let $B = \bigcup_i y_i+A$. Then for fixed j,

$$x_j+B = \bigcup_{i=1}^{n} x_j+y_i+A = \bigcup_{i=1}^{n} y_{i_j}+h_{ji}+A = \bigcup_{i=1}^{n} y_{i_j}+A = B,$$

and hence $G+B = B$ and $G = B$. But then $H = H\cap B = H\cap(y_1+A) = A$.

(b) Using the y_i again, if $u,v \in G$, then $u = h+y_i$, $v = k+y_j$ with $h,k \in H$, and hence the commutator $[u,v] = [h+y_i,k+y_j] = [y_i,y_j] = a_{ij}$. So there are at most n^2 commutators. Since $n[u,v] \in H$,

$$\begin{aligned}(n+1)[u,v] &= -u-v+u+v+n[u,v]\\ &= -u-v+u+n[u,v]+v\\ &= -u-v+u+(-u-v+u+v)+(n-1)[u,v]+v\\ &= -u-2v+u+2v-v+(n-1)[u,v]+v\\ &= [u,2v]+(n-1)[-v+u+v,v].\end{aligned}$$

We will use this to show that each element of G' is a sum of at most n^3 commutators. Suppose $w \in G'$ and $w = c_1 + \cdots + c_p$ with $p > n^3$ and each c_k is a commutator. Then some a_{ij} is equal to at least $n+1$ of these c_k's; say $c = [x,y]$ is this a_{ij}. Since

$$\begin{aligned}[u,v]+[x,y] &= [x,y]-c+[u,v]+c\\ &= [x,y]+[-c+u+c,-c+v+c]\end{aligned}$$

we can shift $n+1$ of the c's to the left and $w = (n+1)[x,y] + d_{n+2} + \cdots + d_p$ where each d_j is a commutator. Since $(n+1)[x,y]$ is a sum of n commutators w is a sum of $p-1$ commutators and an easy induction reduces the number of commutators to n^3. Thus, $|G'| \le n^{2n^3}$. □

We can now lift a total order of a subgroup to the whole group provided we impose a finiteness condition. A group is *locally finite* if each of its finitely generated subgroups is finite. If G is a rooted po-group, then according to Exercise 5 its normal subgroup $H = G^+ - G^+$ is totally ordered and the roots of G are just the cosets of H in G.

Theorem 3.5.12. *Let H be the maximal totally ordered subgroup of the torsion-free rooted po-group G. If G/H is locally finite, then there is a unique total order of G containing H^+.*

Proof. Let
$$P = \{g \in G : kg \in G^+ \text{ for some } k \in \mathbb{N}\}.$$

If T is any total order of G which contains G^+ and $g \in T$, then $kg \in H$ for some $k \in \mathbb{N}$ and hence $kg \in G^+$; otherwise, $kg \in -G^+ \cap T \subseteq -T \cap T = 0$ and $g = 0$. Thus, if P is a total order, then $T = P$. To show that P is a total order we need only verify that $P + P \subseteq P$ since P clearly has the other required properties of a total order. Let $a, b \in P$, let A be the subgroup of G generated by $\{a,b\}$ and let $B = A \cap H$. Then $A/B \cong A + H/H \subseteq G/H$, A is rooted and B is its maximal totally ordered subgroup. So we may assume that G is generated by the two elements a and b from P and G/H is finite of order $n \ge 2$. Let $C = \{h \in H : h \ll x$ for some $x \in H\}$. By Exercise 2.4.24 C is a convex normal subgroup of H which contains the derived group H' of H, and C is a proper subgroup of H since H is finitely generated by Theorem 3.5.11. We claim that $G' \cap H \subseteq C$. First we will verify that the subgroup $[G,H]$ of G generated by the commutators $[g,h]$, with $g \in G$ and $h \in H$, is contained in C. Suppose $-g+h+g < h$. Then $-ng+h+ng < -g+h+g < h$ and $0 < -g-h+g+h < -ng-h+ng+h$; that is, $0 < [g,h] < [ng,h]$. Since $ng \in H$ we have $[g,h] \in C$. On the other hand, if $h < -g+h+g$, then $-g-h+g < -h$ and the preceding gives $[g,-h] \in C$ and, again, $[g,h] = -h - [g,-h] + h \in C$. Since $H/[G,H]$ is central and of finite index in $G/[G,H]$ the derived group $(G/[G,H])' = G'/[G,H]$ is finite by Theorem 3.5.11. Let $g \in G' \cap H$. Then for some $m \in \mathbb{N}$, $mg \in [G,H] \subseteq C$ and hence $g \in C$. In order to show that $n(a+b) > 0$ and hence $a+b \in P$ consider the element $c = -na - nb + n(a+b)$ which lies in $G' \cap H$ and hence also in C. If $n(a+b) < 0$, then $c < -na, -nb < 0$ and both na and nb belong to C. But if $h \in H \backslash C$, then $h = pa + qb + x$ and $nh = pna + qnb + y$ with $x, y \in G'$ and $y \in H$; so $nh \in C$ and we have the contradiction $h \in C$. Thus, $n(a+b) > 0$ and $P + P \subseteq P$. □

The preceding result can be used to lift the lattice order of a power series ℓ-ring to a total order.

Theorem 3.5.13. *Suppose A is a totally ordered domain and Δ is a torsion-free rooted po-group which is locally finite modulo its maximal totally ordered subgroup Γ. Then $W(\Delta,A)$ is a domain and a finite valued ℓ-ring whose lattice order can be extended to a total order. If A is a division ring, then the set of basic elements in $W(\Delta,A)$ is the multiplicative group of units in $W(\Delta,A)^+$. If, in addition, Γ has finite index in Δ, then $W(\Delta,A) = V(\Delta,A)$ is a division ring.*

Proof. By Theorem 3.5.12 the partial order of Δ can be extended to a total order. Let Δ_1 be this totally ordered group. Then $W(\Delta,A)$ is an ℓ-ring by Theorem 3.5.3 and $W(\Delta,A) \subseteq W(\Delta_1,A) = V(\Delta_1,A)$ as rings. For, if $u \in W(\Delta,A)$, then supp u is contained in the union of finitely many cosets of Γ and hence $u \in V(\Delta_1,A)$. Also, $W(\Delta,A)^+ \subseteq V(\Delta_1,A)^+$ by Exercise 27. If Γ has finite index in Δ, then clearly $W(\Delta,A) = V(\Delta,A) = V(\Delta_1,A)$ and hence $W(\Delta,A)$ is a division ring by Theorem 3.5.8, provided A is a division ring. Let $\{\Gamma_j : j \in J\}$ be the set of roots of Δ. Each $\Gamma_j = \Gamma + \gamma_j$ is a coset of Δ, and $V(\Delta,A) = \Pi_j V(\Gamma_j,A)$ and $W(\Delta,A) = \oplus_j V(\Gamma_j,A)$ as A-f-bimodules (see Exercise 2.6.6). The basic elements of $W(\Delta,A)$ are the strictly positive elements that lie in one of the summands. Suppose $u \in V(\Gamma_j,A)$ and $v \in V(\Gamma_k,A)$ are basic elements. Then supp $uv \subseteq \Gamma + \gamma_j + \gamma_k$ and uv is basic. Also, if A is a division ring, then $ux^{-\gamma_j} \in V(\Gamma,A)$ has its inverse $v \in V(\Gamma,A)$, and $x^{-\gamma_j}v \in V(\Gamma - \gamma_j, A) \subseteq W(\Delta,A)$ is the inverse of u. Thus, the basic elements of $W(\Delta,A)$ constitute a multiplicative subgroup of $W(\Delta,A)^+$. But each unit of $W(\Delta,A)^+$ is basic by Theorem 3.1.3. □

Exercises.

1. Let Δ be a rooted poset with a partial addition. If $\alpha + \beta \leq \alpha, \beta$ whenever $\alpha + \beta$ exists show that Δ satisfies (3.5.11) and (3.5.12).
2. Let A be a po-domain with $A^+ \neq 0$ and let Δ be a poset with a partial addition.
   (a) Show that $\sum(\Delta,A)$ is a po-ring iff Δ is a pops.
   (b) Show that $\sum(\Delta,A)$ is a po-domain iff Δ is a pops and a semigroup.
   (c) Show that $W(\Delta,A)$ is an ℓ-domain iff Δ is a rooted pops and a semigroup, A is an ℓ-domain, and A is totally ordered or Δ is trivially ordered.

3. For each subset X of the poset Δ let X' be the set of all those elements in Δ that are comparable to each element of X. Verify the following.

   (a) $'$ is a Galois connection on the power set of Δ.
   (b) If X is a chain, then so is X''.
   (c) If the set $U(X)$ of upper bounds of X is a chain, then $U(X) \subseteq X''$.

4. Suppose that the poset Δ has a partial addition and it satisfies (3.5.11) and (3.5.12). For $\beta \in \Delta$ let $S(\beta)$ be the set of summands of β.

   (a) If $\beta \in X''$ show that $S(\beta) \subseteq X''$ (see Exercise 3).
   (b) If $\alpha \leq \beta$ show that $S(\alpha) \cup S(\beta) \cup \{\alpha, \beta\}$ is a chain.

5. Let Γ be the subgroup generated by the positive cone Δ^+ of the po-group Δ. Show that the following statements are equivalent.

   (a) Δ is rooted.
   (b) Γ is rooted.
   (c) Δ^+ is a chain.
   (d) Γ is totally ordered.
   (e) Some coset of Γ is a chain.
   (f) Each coset of Γ is a chain.
   (g) Each maximal chain in Δ is a coset of Γ.
   (h) Some maximal chain in Λ is a coset of Γ.

6. Let Δ be a rooted po-group with Γ as its largest totally ordered subgroup, and let Λ be a subgroup of Δ.

   (a) Show that Δ/Λ is a rooted quasi-ordered set (see (2.2.1)).
   (b) Show that Δ/Λ is a trivially ordered poset iff $\Gamma \subseteq \Lambda$.
   (c) Show that $\Gamma = f_\ell(\Delta) = f_r(\Delta) = f(\Delta)$.

7. Show that a pops is negative iff it is the set of nonzero lower bounds of the identity of some mopops.

8. Show that an f-pops is negative iff $2\alpha < \alpha$ for each $\alpha \in \Delta$ for which 2α is defined.

9. Let Δ be a totally ordered negative pops. For $\alpha, \beta \in \Delta$ define $\beta \gg \alpha$ iff $\forall n \in \mathbb{N}$, if $n\beta \in \Delta$, then $n\beta \geq \alpha$. Also, define the relation $\sim$ by $\alpha \sim \beta$ iff $\exists n \in \mathbb{N}$ such that $n\alpha \notin \Delta$ and $n\beta \notin \Delta$, or $\{\gamma : \gamma \gg \alpha\} = \{\gamma : \gamma \gg \beta\}$. Verify the following.

   (a) $\sim$ is an equivalence relation on Δ.
   (b) $\alpha \sim \beta$ iff $\exists m, n \in \mathbb{N}$ such that either $n\alpha \notin \Delta$ and $m\beta \notin \Delta$, or $\alpha \geq n\beta$ and $\beta \geq m\alpha$.
   (c) $\beta \gg \alpha$ and $\beta \sim \alpha$ iff $\exists n \in \mathbb{N}$ with $n\beta \in \Delta, (n+1)\beta \notin \Delta$ and $n\beta \geq \alpha$.
   (d) The relation defined on the set of equivalence classes $\nu(\Delta)$ by $\nu(\beta) \geq \nu(\alpha)$ iff $\beta \gg \alpha$ or $\beta \sim \alpha$ is a total order of $\nu(\Delta)$.
   (e) The natural map $\nu : \Delta \longrightarrow \nu(\Delta)$ is isotone.

(f) If $\alpha_1,\ldots,\alpha_n \in \Delta$, then $v(\alpha_1+\cdots+\alpha_n)=v(\alpha_1)\wedge\cdots\wedge v(\alpha_n)$.

10. A *groupoid* is a set with a binary operation.
   (a) Suppose that Δ is a rooted poset with a partial addition that satisfies (3.5.11) and (3.5.12). Let Γ be a subset of Δ with the property that each of its roots generates a subgroupoid of Δ. Show that the mapping $X \longmapsto \omega X$ induces a bijection between the roots of Γ and the roots of $\omega\Gamma$. Also, show that $\omega(\text{trunk}\,\Gamma) = \text{trunk}\,(\omega\Gamma)$. (Use Exercise 3.)
   (b) Suppose that Δ is a poset with a partial addition that satisfies (3.5.5) in the weak sense (replace "$<$" by "$\leq$" in (3.5.5)), and assume that, for each $\gamma\in\Delta$, γ satisfies (3.5.11) and $2\gamma\in\Delta$. Show that each maximal chain of Δ is a totally ordered groupoid.

11. (a) Let Λ be a totally ordered negative semigroup and let $\gamma_0\in\Lambda$. Suppose that $\{\Lambda_i : i\in I\}$ is a disjoint family of pops such that each Λ_i is disjoint from Λ and $\{\gamma\in\Lambda:\gamma<\gamma_0\}$is isomorphic to Λ_i via the map $\gamma\longmapsto\gamma^{(i)}$. Let $\Delta=\Lambda\cup\bigcup_{i\in I}\Lambda_i$ have the cardinal order supplemented by $\alpha^{(i)}<\gamma$ for each $\alpha^{(i)}\in\Lambda_i$ and each $\gamma\in\Lambda$ with $\gamma_0\leq\gamma$. Also, let Δ have the partial addition obtained by supplementing each component addition by $\alpha^{(i)}+\gamma=(\alpha+\gamma)^{(i)}$ and $\gamma+\alpha^{(i)}=(\gamma+\alpha)^{(i)}$ if $\gamma_0\leq\gamma\in\Lambda$. Show that Δ is a rooted negative pops and each root generates a totally ordered subsemigroup.
   (b) Give an example of a subset Γ of a rooted negative pops Δ such that $n\Gamma$ is a totally ordered subsemigroup of Δ for each $2\leq n$, but $\omega\Gamma$ is neither totally ordered nor a subsemigroup of Δ.

12. Let $v\in V(\Delta,A)$ or $v\in W(\Delta,A)$ where Δ is a rooted mopops and A is a unital ring. Suppose that supp $v<0$ and each root of supp v generates a subsemigroup of Δ. Show that (a) and (b) of Theorem 3.5.8 hold.

13. Suppose that A is a unital ring, $\mathscr{U}(A)$ is its group of units, and $\text{End}_1(A)$ is its semigroup of 1-preserving ring endomorphisms, acting on the right of A. Let Δ be a partial semigroup and let $\Delta_+=\{(\alpha,\beta)\in\Delta\times\Delta:\alpha+\beta\in\Delta\}$ be the domain of +. Assume that $\sigma:\Delta\longrightarrow\text{End}_1(A)$ and $\tau:\Delta_+\longrightarrow A$ are functions and that $\tau(\beta,\gamma)\in\mathscr{U}(A)$ if there is an $\alpha\in\Delta$ with $\alpha+\beta+\gamma\in\Delta$. Let $\sum(A*\Delta)$ be a free right A-module with the basis $\{x^\alpha:\alpha\in\Delta\}$. For $\alpha,\beta\in\Delta$ and $a,b\in A$ define
$$(x^\alpha a)(x^\beta b)=\begin{cases}x^{\alpha+\beta}\tau(\alpha,\beta)a^{\sigma(\beta)}b & \text{if } \alpha+\beta\in\Delta\\ 0 & \text{if } \alpha+\beta\notin\Delta\end{cases}$$
and extend this multiplication to all of $\sum(A*\Delta)$ using distributivity. So if $u=\sum x^\alpha u_\alpha$ and $v=\sum x^\alpha v_\alpha$, then
$$uv=\sum_{\gamma\in\Delta}x^\gamma\left(\sum_{\alpha+\beta=\gamma}\tau(\alpha,\beta)u_\alpha^{\sigma(\beta)}v_\beta\right). \tag{3.5.15}$$
When $\sum(A*\Delta)$ is a ring it is called a *crossed product* of Δ over A. If $a\in\mathscr{U}(A)$ let $\eta(a)=a^{-1}(\;)a$ denote the automorphism of A given by conjugation by a;

and let $\eta(\alpha,\beta) = \eta(\tau(\alpha,\beta))$ if $\tau(\alpha,\beta) \in \mathscr{U}(A)$. If A is a po-ring we will assume that for all $(\alpha,\beta) \in \Delta_+, \tau(\alpha,\beta) > 0, \tau(\alpha,\beta)^{-1} > 0$ if $\tau(\alpha,\beta) \in \mathscr{U}(A)$ and $0 < a^{\sigma(\alpha)}$ if $0 < a \in A$ and $\alpha \in \Delta$.

(a) Show that $\Sigma(A * \Delta)$ is a ring if and only if for all $\alpha,\beta,\gamma \in \Delta$ with $\alpha+\beta+\gamma \in \Delta$

$$\tau(\alpha+\beta,\gamma)\tau(\alpha,\beta)^{\sigma(\gamma)} = \tau(\alpha,\beta+\gamma)\tau(\beta,\gamma), \quad (3.5.16)$$

$$\sigma(\beta)\sigma(\gamma) = \sigma(\beta+\gamma)n(\beta,\gamma). \quad (3.5.17)$$

For the remainder of these exercises, unless stated otherwise, $\Sigma(A * \Delta)$ is a ring.

(b) Suppose that $d_\alpha \in \mathscr{U}(A)$ for each $\alpha \in \Delta$. Show that $\Sigma(A * \Delta)$ is a crossed product with respect to the basis $\{y^\alpha = x^\alpha d_\alpha : \alpha \in \Delta\}$, and determine the new twist τ_1 and the new action σ_1 in terms of τ and σ. This change of basis is called a *diagonal change of basis*.

(c) Suppose that Δ has an identity element 0. Show that the following are equivalent.
   (i) $\Sigma(A * \Delta)$ has an identity element of the form $x^\beta a$.
   (ii) $\sigma(0)$ is an automorphism.
   (iii) $\sigma(0) = n(b)$ for some $b \in \mathscr{U}(A)$.

(d) If $1 = x^0$ show that $\tau(\alpha,0) = \tau(0,\alpha) = 1$ for each $\alpha \in \Delta$, and $\sigma(0) = 1$. Also, show that if Δ is a partial monoid, then it can be assumed that $1 = x^0$.

(e) Show that Theorem 3.5.2 holds for $W(A * \Delta)$ and $S(A * \Delta)$ where, as additive groups, $W(A * \Delta) = W(\Delta,A)$ and $S(A * \Delta) = S(\Delta,A)$ and multiplication is defined by (3.5.15).

For convenience, the rings $\Sigma(\Delta,A), W(\Delta,A), V(\Delta,A)$ and $S(\Delta,A)$, when they exist, will be considered power series crossed products with trivial action and twisting even though A is not assumed to be unital.

14. Suppose that Δ is a pops and A is a po-domain. Show that Theorem 3.5.3 holds for $\Sigma(A * \Delta)$ and $W(A * \Delta)$. Moreover, if $u_1,\dots,u_n \in W(A * \Delta)^+ \cup -W(A * \Delta)^+$ show that

$$\text{maxsupp } u_1 \cdots u_n = \max(\text{maxsupp } u_1 + \cdots + \text{maxsupp } u_n).$$

15. Let Δ be a pops, let $d_\ell(\Delta) = \{\gamma \in \Delta : \forall \alpha,\beta \in \Delta$, if $\alpha \| \beta$ and $\gamma+\alpha$ and $\gamma+\beta$ are defined, then $\gamma+\alpha \| \gamma+\beta\}$ and let $d_r(\Delta)$ denote the right-sided version of $d_\ell(\Delta)$. A subset Λ of $d_\ell(\Delta)$ (respectively, $d_r(\Delta)$) is said to have *property* (d_ℓ) (respectively, *property* (d_r)) if whenever $\gamma,\delta \in \Lambda$ and $\alpha,\beta \in \Delta$ with $\alpha \| \beta$ and $\gamma \| \delta$, then $\gamma+\alpha \| \delta+\beta$ provided these elements are defined. Let A be a totally ordered domain. Verify the following.

(a) $d_\ell(\Delta)$ is a subpops of Δ.
(b) If Δ is rooted, then $d_\ell(\Delta)$ is an ideal of the poset Δ.
(c) Suppose that Δ is rooted and let $w \in W(A * \Delta)^+$ (respectively, $\Sigma(A * \Delta)^+$). Then $w \in d_\ell(W(A * \Delta))$ (respectively, $d_\ell(\Sigma(A * \Delta))$ if and only if supp w has property (d_ℓ).

(d) The crossed product power series po-ring $W(A*\Delta)$ (respectively, $\Sigma(A*\Delta)$) is a left d-ring iff Δ is rooted, $\Delta = d_\ell(\Delta)$, and Δ has property (d_ℓ).
(e) State the analogous characterization of the right d-elements of $W(A*\Delta)$ and $\Sigma(A*\Delta)$.
(f) $W(A*\Delta)$ (respectively, $\Sigma(A*\Delta)$) is a d-ring iff $\Delta = d_\ell(\Delta) = d_r(\Delta)$, Δ is rooted, and Δ has property (d_ℓ).
(g) If Δ is a group show that $\Delta = d_\ell(\Delta) = d_r(\Delta)$ and a subset Λ has property (d_ℓ) iff Λ is totally ordered.

16. Suppose that Δ is a trivially ordered pops and A is an ℓ-domain. Let w be an element of $\Sigma(A*\Delta)^+$.

(a) Show that $w \in d_\ell(\Sigma(A*\Delta))$ iff for all $\gamma, \delta \in$ supp w and for all $\alpha, \beta \in \Delta$, if $\gamma+\alpha = \delta+\beta$, then
  (i) $\alpha \neq \beta \Longrightarrow \tau(\gamma,\alpha)w_\gamma^{\sigma(\alpha)}A \subseteq (\tau(\delta,\beta)w_\delta^{\sigma(\beta)}A)^\perp$, and
  (ii) $\alpha = \beta \Longrightarrow \tau(\gamma,\alpha)w_\gamma^{\sigma(\alpha)} + \tau(\delta,\alpha)w_\delta^{\sigma(\alpha)} \in d_\ell(A)$.
(a) If A is totally ordered, show that $w \in d_\ell(\Sigma(A*\Delta))$ iff for all $\gamma, \delta \in$ supp w and for all $\alpha, \beta \in \Delta, \gamma+\alpha = \delta+\beta \Longrightarrow \alpha = \beta$.
(b) If A is totally ordered, show that $\Sigma(A*\Delta)$ is a d-ring iff $\gamma+\alpha = \delta+\beta \Longrightarrow \gamma = \delta$ and $\alpha = \beta$, for all $\alpha, \beta, \gamma, \delta \in \Delta$.
(c) Describe the right d-elements of $\Sigma(A*\Delta)$ and determine when $\Sigma(A*\Delta)$ is a right d-ring.

17. Let Δ be a pops.

(a) Show that $f_\ell(\Delta)$ is a subpops of Δ.
(b) If Δ is rooted, show that $f_\ell(\Delta)$ is an ideal of the poset Δ.
(c) If Δ is rooted and A is totally ordered show that $f_\ell(W(A*\Delta)) = W(A*f_\ell(\Delta))$ and $f_\ell(\Sigma(A*\Delta)) = \Sigma(A*f_\ell(\Delta))$ and that similar equations hold for the right f-elements.
(d) With the conditions in (c), let R be an ℓ-subgroup of $V(A*\Delta)$ which contains $\Sigma(A*\Delta)$ and which is an ℓ-ring with multiplication given by (3.5.15). Show that $F_\ell(R) = R \cap V(A*f_\ell(\Delta))$, $F_r(R) = R \cap V(A*f_r(\Delta))$ and $F(R) = R \cap V(A*f(\Delta))$.
(e) Show that Theorem 3.5.4 holds for crossed products.

18. Suppose that Δ is a trivially ordered pops and A is an ℓ-domain. Let $0 \leq w \in \Sigma(A*\Delta)$.

(a) Show that $w \in f_\ell(\Sigma(A*\Delta))$ iff $\forall(\gamma,\alpha) \in$ supp $w \times \Delta$, $\gamma+\alpha \in \Delta \Longrightarrow \gamma+\alpha = \alpha$ and $\tau(\gamma,\alpha)w_\gamma^{\sigma(\alpha)} \in f_\ell(A)$.
(b) Show that $w \in f_r(\Sigma(A*\Delta))$ iff $\forall(\gamma,\alpha) \in$ supp $w \times \Delta$, $\alpha+\gamma \in \Delta \Longrightarrow \alpha+\gamma = \alpha$, and $\forall b \in A$, $\tau(\alpha,\gamma)(b^\perp)^{\sigma(\gamma)}w_\gamma \subseteq b^\perp$.
(c) Show that $\Sigma(A*\Delta)$ is a left f-ring iff $\Delta+\Delta = \emptyset$, or, if $\gamma+\alpha \in \Delta$, then $\gamma+\alpha = \alpha$ and $\tau(\gamma,\alpha)A^{\sigma(\alpha)} \subseteq F_\ell(A)$.

(d) Show that $\Sigma(A*\Delta)$ is a right f-ring iff $\Delta+\Delta=\emptyset$, or, if $\alpha+\gamma\in\Delta$, then $\alpha+\gamma=\alpha$ and $\tau(\alpha,\gamma)C^{\sigma(\gamma)}A\subseteq C$ for each polar C of A.

(e) Show that each of the statements in Theorem 3.5.4 (see Exercise 14) is equivalent to the following: $\Delta+\Delta=\emptyset$, or if $\alpha+\beta\in\Delta$, then $\alpha+\beta=\alpha=\beta$ and A is totally ordered.

(f) Let $e(\Delta)=\{\gamma\in\Delta:\forall\alpha\in\Delta,\gamma+\alpha$ (respectively, $\alpha+\gamma)\in\Delta\Rightarrow\gamma+\alpha=\alpha$ (respectively,$\alpha+\gamma=\alpha)\}$. Show that $e(\Delta)$ is a subpops of Δ and $F(\Sigma(\Delta,A))=\Sigma(e(\Delta),F(A))$.

19. Let Δ be a pops and let M_A be a strict po-module over the po-domain A. For $u\in M^\Delta$ and $v\in A^\Delta$ "define" $uv\in M^\Delta$ by means of the equations (3.5.2) and (3.5.3).

(a) State and prove the analogue of Theorem 3.5.3 and the equations (3.5.7) through (3.5.10) for the appropriate modules.

(b) Suppose that Δ is rooted and M and A are totally ordered. Show that if $N=W(\Delta,M)$ and $R=W(\Delta,A)$, then $f(N_R)=W(f_r(\Delta),A),d(N_R)=\{w\in R:$ supp w has property $(d_r)\}$ (see Exercise 15), and $d_N(R)=\{u\in N:$ supp u has property $(d_\ell)\}$ (see Exercise 3.1.7).

(c) If Δ is also an f-pops in (b), show that $V(\Delta,M)$ is an f-module over the f-ring $V(\Delta,A)$.

20. Let Δ be a totally ordered monoid. Assuming that $1=x^0$ (see Exercise 13 (c) and (d)) show that Theorem 3.5.8 holds for $V(A*\Delta)$.

21. Let Δ be a rooted f-pops and let A be a totally ordered domain. Show that the ℓ-group isomorphism $V(A*\Delta)\longrightarrow\Pi_i V(A*\Delta_i)$ that is given in Exercise 2.6.6 is an isomorphism of f-rings.

22. Let Δ be a rooted pops and let A be an ℓ-domain. Show that $W(A*\Delta)$ is an almost f-ring iff $W(A*\Delta)^2=0$, or A is totally ordered and Δ satisfies the condition: $\alpha+\beta\in\Delta\Longrightarrow\alpha$ and β are comparable.

23. Find an example of an ℓ-unital ℓ-ring R which is generated by its d-elements and which contains an element $u\in\mathscr{U}(R^+)$ such that $u\vee 1$ is not a unit of R and $u,u^{-1}\in 1^\perp$. So $\mathscr{U}(R^+)$ is not a sublattice of R and $u\vee u^{-1}\not\geq 1$.

24. In this and the following two exercises the group operation will be written multiplicatively. Let (G,P) be a po-group. If $\{a_1,\dots,a_n\}\subseteq G$ let $S(a_1,\dots,a_n)$ denote the normal subsemigroup of G generated by $\{a_1,\dots,a_n\}$ and let $M(a_1,\dots,a_n)=S(a_1,\dots,a_n)\cup\{1\}$. Verify each of the following.

(a) If $a\in P$, then $M(a)\subseteq P$; if $1\neq a\in P$, then $P\cap S(a^{-1})=\emptyset$; $M(a_1,\dots,a_n)=M(a_1)\cdots M(a_n)$; $S(a_1,\dots,a_n)^{-1}=S(a_1^{-1},\dots,a_n^{-1})$.

(b) P can be extended to a total order of G iff the following condition is satisfied:

$$\forall a_1,\dots,a_n\in G\backslash\{1\},\exists\text{ signs }\varepsilon_1,\dots,\varepsilon_n\in\{\pm1\}$$

such that

$$P \cap S(a_1^{\varepsilon_1}, \ldots, a_n^{\varepsilon_n}) = \emptyset. \tag{3.5.18}$$

(Show that if $a \in G$, then for some $\varepsilon \in \{\pm 1\}$ the normal submonoid $PM(a^{\varepsilon})$ satisfies (3.5.18).)

(c) Show that a group can be totally ordered iff each of its finitely generated subgroups can be totally ordered.

25. (a) Let F be the free group on the set $\{x, y\}$ and let $X = \{y^{-n}xy^n : n \in \mathbb{Z}^+\}$. Show that the subgroup of F generated by X is free on X.
(b) Show that the free product of the family of groups $\{\mathbb{Z}_i : i \in I\}$ is the free group of rank card (I) (see Exercise 1.4.21).

26. Let R be a unital ring and let $\mathscr{U}_2(R)$ donote the group of invertible 2×2 matrices over R. We will identify the matrix ring $R[t]_2$ with the polynomial ring $R_2[t]$. Let F and G be subsets of $\mathscr{U}_2(R[t])$ with the following property. If $\begin{pmatrix} f_1 & f_2 \\ f_3 & f_4 \end{pmatrix} \in F \backslash G$, then deg f_2, the degree of f_2, is larger than the degree of the other three elements, and the highest coefficient of f_2 is not a zero divisor of R; and if $\begin{pmatrix} g_1 & g_2 \\ g_3 & g_4 \end{pmatrix} \in G \backslash F$, then deg g_3 is larger than the degree of the other three elements, and the highest coefficient of g_3 is not a zero divisor of R. Suppose that $h_1 \cdots h_n$ is a nonempty product whose factors alternately come from the sets $F \backslash G$ and $G \backslash F$, and let

$$\begin{pmatrix} a \\ b \end{pmatrix} = h_1 \cdots h_n \begin{pmatrix} 1 \\ 1 \end{pmatrix}.$$

(a) If $h_1 \in F$ show that deg $a >$ deg b and if $h_1 \in G$ show that deg $b >$ deg a.
(b) If F and G are subgroups of $\mathscr{U}_2(R[t])$ such that $F \cap G = 1$ show that the subgroup generated by F and G is their free product.
(c) Suppose that R has characteristic 0 (that is, 1 has infinite order). Show that the subgroup of $\mathscr{U}_2(R[t])$ generated by $\begin{pmatrix} 1 & t \\ 0 & 1 \end{pmatrix}$ and $\begin{pmatrix} 1 & 0 \\ t & 1 \end{pmatrix}$ is free on these generators.
(d) Suppose that R is a totally ordered domain and $R[[t]] = V(\mathbb{Z}^+, R)$ is the formal power series ring constructed by using d.c.c. instead of a.c.c.; that is, the supports of the power series have d.c.c.. Let U be the subgroup of $\mathscr{U}_2(R[[t]])$ consisting of those formal power series of the form

$$\begin{pmatrix} \alpha & 0 \\ 0 & \beta \end{pmatrix} + A_1 t + A_2 t^2 + \cdots$$

with $\alpha > 0$ and $\beta > 0$ units in R (again, $R_2[[t]]$ and $R[[t]]_2$ have been identified). Make $R^{(4)}$ into a totally ordered R-module by giving it a lexicographic order and transfer this order to R_2 via an R-module-isomorphism $R^{(4)} \cong R_2$. For $a, b \in U$ define $a > b$ if

$$a - b = A_n t^n + A_{n+1} t^{n+1} + \cdots$$

and $A_n > 0$. Show that $(U, \geq)$ is a totally ordered group.

(e) Show that a free group can be totally ordered.

(f) Show that the free product of a set of orderable groups is orderable. (Let F and G be orderable groups which are subgroups of the group of positive units of a totally ordered domain R (see Exercise 2.1.14) and consider their representations in $\mathscr{U}_2(R[t])$ given by $f \longmapsto \begin{pmatrix} f & 0 \\ 0 & 1 \end{pmatrix}$ followed by conjugation by $\begin{pmatrix} 1 & t \\ 0 & 1 \end{pmatrix}$ if $f \in F$, and $g \longmapsto \begin{pmatrix} 1 & 0 \\ 0 & g \end{pmatrix}$ followed by conjugation by $\begin{pmatrix} 1 & 0 \\ t & 1 \end{pmatrix}$ if $g \in G$.)

27. Suppose $(\Delta, +, \leq)$ is a pops and $\leq_1$ is a partial order of Δ which contains $\leq$ such that $(\Delta, +, \leq_1)$ is also a pops; call it Δ_1. If A is a po-ring show that:

(a) $\Sigma(\Delta,A) = \Sigma(\Delta_1,A)$, $W(\Delta,A) \subseteq W(\Delta_1,A)$, $S(\Delta,A) \subseteq S(\Delta_1,A)$, and $V(\Delta_1,A) \subseteq V(\Delta,A)$;

(b) $\Sigma(\Delta,A)^+ \subseteq \Sigma(\Delta_1,A)^+$, $W(\Delta,A)^+ \subseteq W(\Delta_1^+,A)$, $V(\Delta,A)^+ \cap V(\Delta_1,A) \subseteq V(\Delta_1,A)^+$.

(c) If Δ is the union of a finite number of chains, show that the inclusions in (a) may be replaced by equalities.

(d) Give an example to show that the inclusions in (a) can be proper.

28. Let G be the group of basic elements of the division ring $V(\Delta,A)$ given in Theorem 3.5.13. Show that G is a finitely rooted po-group and there are po-group homomorphisms $\pi : G \longrightarrow \Delta$ and $\varphi : \Delta \longrightarrow G$ such that $\pi\varphi = 1$.

Notes. Generalized power series ℓ-rings over a pops first appeared in Conrad and McCarthy [CMC] and Theorems 3.5.3 and 3.5.4 come from this paper; generalized power series rings are studied by Conrad in [CON2] and this is the source of Theorems 3.5.1, 3.5.2, 3.5.5, and 3.5.6. Theorem 3.5.7 for a totally ordered negative semigroup and Theorem 3.5.8 come from Neumann [N]. Theorem 3.5.11 is taken from Passman [P1], Theorem 3.5.12 comes from Neumann and Sheppard [NS], and Theorem 3.5.13 is due to Conrad and Dauns [CD]. Exercise 24 gives Fuch's criteria for extending a partial order of a group to a total order [F, p. 56] and Exercise 26 gives Bergman's method of ordering a free group and the free product of orderable groups. [BER3].

3.6 Archimedean f-Rings

The ℓ-ring analogue of Theorem 2.3.9 is the false statement that each archimedean ℓ-ring is commutative. The canonically ordered $n \times n$ matrix ring over the reals is an example of a noncommutative complete ℓ-ring. However, the story is different for f-rings, as Theorem 3.3.2 suggests. Moreover, a not necessarily associative archimedean f-ring must be associative. To clarify this we need to point out that the definitions of a po-ring, ℓ-ring and f-ring apply to any not necessarily associative ring and Theorem 3.1.2 is still valid in this generalized setting. In this section we present these commutative and associative results, and we also consider the completion of an archimedean f-ring as well as the representation theorem for an archimdedean semiprime f-ring that is the analogue of the ℓ-group representation theorem given in Section 2.3. In addition, we will show that the ring generated by the semiring of f-endomorphisms of an archimedean ℓ-group is an f-ring and in the exercises we will determine when this f-ring has the same Stone space as the ℓ-group.

Theorem 3.6.1. *Let R be a not necessarily associative f-ring. Then for all $x,y,z \in R$ we have*

(a) $|xy - yx| \ll x^2 + y^2$ *and*
(b) $|(xy)z - x(yz)| \ll (x^2+y^2+z^2) + ((x^2)^2 + (y^2)^2 + (z^2)^2 + 2(|x^2x| \vee |xx^2| + |y^2y| \vee |yy^2| + |z^2z| \vee |zz^2|)$.

In particular, a not necessarily associative archimedean f-ring is a commutative f-ring.

Proof. We may assume that R is totally ordered, and we also suppose that $0 < z \leq x \leq y$. Let $n \in \mathbb{N}$. If $k \in \mathbb{Z}$ and $ky \leq nx$, then $k \leq n$ since $n < k$ gives that $nx \leq ny < ky \leq nx$. If m is the largest integer with $my \leq nx$, then $my \leq nx < (m+1)y$ gives that $nx = my + y_1$ with $0 \leq y_1 < y$. Since the commutator $[x,y] = xy - yx$ is additive in each variable we have

$$n|[x,y]| = |[my+y_1,y]| = |[y_1,y]| \leq y_1y \vee yy_1 \leq y^2 \leq x^2+y^2.$$

This completes the proof of (a) since $[y,x] = -[x,y]$. Now, with m as above, $mz = ky + y_2$ with $k \in \mathbb{Z}^+$ and $0 \leq y_2 < y$. So

$$\begin{aligned} n|(xy)z - x(yz)| &= |((my+y_1)y)z - (my+y_1)(yz)| \\ &= |y^2(ky+y_2) + (y_1y)z - y(y(ky+y_2)) - y_1(yz)| \\ &= |k(y^2y - yy^2) + (y^2y_2 - y(yy_2)) + ((y_1y)z - y_1(yz))| \\ &\leq k|[y^2,y]| + |y^2y_2 - y(yy_2)| + |(y_1y)z - y_1(yz)| \\ &\leq (y^2)^2 + y^2 + y^2y_2 \vee y(yy_2) + (y_1y)z \vee y_1(yz) \\ &\leq (y^2)^2 + y^2 + 2(y^2y \vee yy^2) \\ &\leq (x^2+y^2+z^2) + ((x^2)^2 + (y^2)^2 + (z^2)^2) + 2(x^2x \vee xx^2 + y^2y \vee yy^2 + z^2z \vee zz^2). \end{aligned}$$

The five other cases that arise from a different order of $\{x,y,z\}$ are handled in the same way, and the cases that arise if one or more of x,y, or z are negative clearly follow from the positive case. □

Since the nil radical of an f-ring is infinitesimal we expect the nilpotent elements in an archimedean f-ring to be small and to have a small index of nilpotency. In fact, this is the case in any sp-ℓ-ring. It is, of course, not true in an arbitrary archimedean ℓ-ring as one can easily verify by considering an infinite row-finite matrix ring over the reals.

Recall (from Section 3.4) that if M is a po-module over the po-ring C, then M^+ is C^+-archimedean if $C^+x \le y$ implies that $x = 0$, for any $x, y \in M^+$ (see Exercises 3.4.30 and 3.4.31).

Theorem 3.6.2. *Let R be an sp-ℓ-ring that is a po-algebra over the po-ring C, and suppose that R^+ is C^+-archimedean. Then the following hold.*

(a) $\ell_\ell(R) = r_\ell(R) = \{x \in R : |x|^2 = 0\}$, and $(R/\ell_\ell(R))^+$ is C^+-archimedean.
(b) $\ell_\ell(R^2) = r_\ell(R^2) = \{x \in R : |x|^3 = 0\} = \ell\text{-}\beta(R) = \ell\text{-}N_g(R)$, and $R/\ell\text{-}\beta(R)$ is ℓ-reduced and its positive cone is C^+-archimedean.
(c) If R is an f-ring then $\ell\text{-}\beta(R) = \ell(R)$. If R is an f-algebra over C, then $\ell\text{-}\beta(R) \cap R^{[2]} = 0$; consequently, R is a subdirect product of a semiprime f-algebra whose positive cone is C^+-archimedean and a zero f-ring on which C^+ acts as f-endomorphisms.

Proof. Let $x, y, z \in R$ with $|x|$ nilpotent of index m. Then, for each $\alpha \in C^+$, the inequality $\alpha(|x||y| + |y||x|) \le \alpha^2|x|^2 + |y|^2$ is a consequence of the inequality $0 \le (\alpha|x| - |y|)^2$. If $m \le 2$, then $|x||y| = |y||x| = 0$ and the first part of (a) follows. Now, C^+ operates on $R/\ell_\ell(R)$ since $C^+\ell_\ell(R) \subseteq \ell_\ell(R)$ and the fact that $(R/\ell_\ell(R))^+$ is C^+-archimedean has already been noted in Exercise 3.2.19, at least for $C = \mathbb{Z}$; but the same straighforward proof that works for $\mathbb{Z}$ also works for a general po-ring C. If $m \ge 4$, then from $(|x|^{m-2})^2 = 0$ and (a) we get the contradiction that $|x|^{m-1} = 0$. Thus, $m \le 3$ and $|x|^2R = R|x|^2 = 0$. If the first inequality in this paragraph is multiplied on the right by $|z|$ we get $\alpha|x||y||z| \le \alpha(|x||y| + |y||x|)|z| \le |y|^2z$ and hence $|x| \in \ell_\ell(R^2)$; similarly, $|x| \in r_\ell(R^2)$. This proves all of (b) except for $(R/\ell\text{-}\beta(R))^+$ being C^+-archimedean. But if $z + \ell\text{-}\beta(R) \ll_{C^+} y + \ell\text{-}\beta(R)$ in $(R/\ell\text{-}\beta(R))^+$, then for each $\alpha \in C^+$ there is an element $z_\alpha \in \ell\text{-}\beta(R)^+$ such that $\alpha|z| \le |y| + z_\alpha$. Hence $\alpha|z||t|^2 \le |y||t|^2$ for each $t \in R$; so $z \in \ell\text{-}\beta(R)$ and $(R/\ell\text{-}\beta(R))^+$ is C^+-archimedean. Suppose that R is an f-ring. If $\alpha \in C^+$, then $\alpha|x|^2 \le |x|$. For, if $\overline{\alpha|x|^2} > \overline{|x|}$ in a totally ordered (ring) homomomorphic image $\overline{R}$ of R, then $0 = \overline{\alpha^2|x|^3} \ge \overline{\alpha|x|^2} > 0$. So $|x|^2 = 0$ and $\ell\text{-}\beta(R) = \ell(R)$. Now assume that R is a C-f-algebra. If $x \in \ell\text{-}\beta(R) \cap R^{[2]}$, then $|x| \le y^2$ for some $y \in R^+$. Suppose that $\alpha|\bar{x}| > \bar{y}$ for some $\alpha \in C^+$ and some totally ordered homomorphic image $\bar{R}$ of R. Then $\bar{y}^2 = 0$ and hence we have the contradiction that $\bar{x} = 0$. So, $C^+|x| \le y$ and $x = 0$. □

An example of a nilpotent archimedean sp-ℓ-ring (even an almost f-ring that is a d-ring) of index of nilpotency 3 is given in Exercise 1. Also, see Exercise 3.7.25. An example of an archimedean f-ring for which the subdirect product representation in (c) above cannot be replaced by a direct product representation is given in Exercise 25.

The representation theorem for an archimedean ℓ-group that is given in Theorem 2.3.23 certainly holds for the additive ℓ-group of an archimedean ℓ-ring R, but it cannot be a ring representation unless R is a semiprime f-ring. These are the only requirements on R as we will now show. Recall that $D(X)$ denotes the lattice of extended real valued continuous functions on a topological space X which are real valued on a dense subset of X and that $D(X)$ is a complete f-ring when X is extremely disconnected (Theorem 2.3.22).

Theorem 3.6.3. *Let R be a semiprime archimedean f-ring and let X be the Stone space of the Boolean algebra of polars of R.*

(a) There is a $c\ell$-essential ℓ-ring monomorphism from R into $D(X)$.
(b) Suppose that $Y_i, i = 1,2$, are Stone spaces and $\varphi_i : R \longrightarrow D(Y_i)$ are $c\ell$-essential ℓ-ring monomorphisms. Then there is a homeomorphism $\tau : Y_1 \longrightarrow Y_2$ such that $\varphi_1(a)(y_1) = \varphi_2(a)(\tau(y_1))$ for each $a \in R$ and each $y_1 \in Y_1$.

Proof. As in the proof of Theorem 2.3.23 we may assume that R is divisible. Let S be the unital $\mathbb{Q}$-f-algebra extension of R that is given in (b) of Theorem 3.4.5. S is archimedean since if $0 \le s \ll t$ in S and $a \in R^+$, then $0 \le sa \ll ta$ in R; so $sR = 0$ and $s = 0$. If $T_a : S \longrightarrow S$ denotes right multiplication by $a \in S$ and $\Omega = \{T_a : a \in R^+\}$, then the triple (R,S,Ω) satisfies the conditions in Exercise 2.2.14(g). Thus, the Boolean algebras of polars of R and S, $\mathscr{B}(R)$ and $\mathscr{B}(S)$, respectively, are canonically isomorphic, and X is also the Stone space of $\mathscr{B}(S)$. Moreover, according to Exercise 2.3.23, S is a $c\ell$-essential extension of R. Thus, for the proof of (a) we may assume that R is unital.

Let $\varphi : R \longrightarrow D(X)$ be the ℓ-group representation of Theorem 2.3.23 that is associated with the maximal disjoint set $\{1\}$ of R. We will use the notation from the proof of this theorem; so $\varphi(a) = f_a$. According to step (vi) in the proof of Theorem 2.3.23 $\varphi(1)$ is the characteristic function of $V(1) = X$; that is, $\varphi(p) = p$ for each $p \in \mathbb{Q}$. To show that φ is a ring homomorphism it suffices to show that $f_a f_b = f_{ab}$ if $a,b \in R^+$. Suppose that $f_a(m) < p$ and $f_b(m) < q$ where $p,q \in \mathbb{Q}$ and $m \in X$ with $f_a(m)$, $f_b(m)$ and $f_{ab}(m) \in \mathbb{R}$. Since $((p-a)(q+b))^+ = (p-a)^+(q+b) \ge q(p-a)$ and

$$f_{(p-a)^+}(m) = (p - f_a(m)) \vee 0 = p - f_a(m) > 0$$

we have that $f_{((p-a)(q+b))^-}(m) = 0$ and

$$f_{(p-a)(q+b)}(m) = f_{((p-a)(q+b))^+}(m) \ge q(p - f_a(m)).$$

Since $pq - ab = (p-a)(q+b) + qa - pb$ we have that

$$pq - f_{ab}(m) \ge pq - qf_a(m) + qf_a(m) - pf_b(m) = p(q - f_b(m)) > 0.$$

But then

$$f_a(m)f_b(m) = \left(\bigwedge_{f_a(m)<p} p\right)\left(\bigwedge_{f_b(m)<q} q\right) = \bigwedge_{p,q} pq \ge f_{ab}(m).$$

The reverse inequality will follow from a dual computation. If $0 \le p < f_a(m)$ and $0 \le q < f_b(m)$ with $p,q \in \mathbb{Q}$, then from the equation $ab - pq = (a-p)(q+b) +$

$pb - qa$ we get that

$$f_{ab}(m) - pq \geq q(f_a(m) - p) + pf_b(m) - qf_a(m) = p(f_b(m) - q) > 0.$$

So

$$f_a(m)f_b(m) = \left(\bigvee_{0\leq p<f_a(m)} p\right)\left(\bigvee_{0\leq q<f_b(m)} q\right) = \bigvee_{p,q} pq \leq f_{ab}(m).$$

Thus, $f_a f_b$ and f_{ab} agree on a dense subset of X and hence $f_a f_b = f_{ab}$.

Now, (b) also follows from the analogous result for archimedean ℓ-groups. By Exercise 2.3.20 there is a homeomorphism $Y_1 \xrightarrow{\tau} Y_2$ and a positive unit f in $D(Y_2)$ such that $\varphi_1(a)(y_1) = f(\tau(y_1))\varphi_2(a)(\tau(y_1))$ for each $a \in R$ and for each $y_1 \in Y_1$. Let

$$W = f^{-1}((0,\infty)) \cap \bigcup_{a\in R} \varphi_2(a)^{-1}((0,\infty)).$$

Then W is a dense open subset of Y_2. For, if χ_U is the characteristic function of the nonempty clopen subset U of Y_2, then $0 < \varphi_2(a) \leq n\chi_U$ for some $a \in R^+$ and some $n \in \mathbb{N}$, and hence $U \cap W \neq \emptyset$. If $y_1 \in \tau^{-1}(W)$, then for some $a \in R$, $0 < f(\tau(y_1))$, $\varphi_2(a)(\tau(y_1)) < \infty$, and since

$$\begin{aligned} f(\tau(y_1))^2\varphi_2(a^2)(\tau(y_1)) &= f(\tau(y_1))^2(\varphi_2(a)(\tau(y_1)))^2 = \varphi_1(a^2)(y_1) \\ &= f(\tau(y_1))\varphi_2(a^2)(\tau(y_1)), \end{aligned}$$

we have that $f(\tau(y_1)) = 1$ and hence $f = 1$. □

One consequence of Theorem 3.6.3 is that the Dedekind completion of a semiprime archimedean $\mathbb{Q}$-f-algebra R is an f-algebra extension of R since it is just the convex ℓ-subgroup of $D(X)$ generated by the image of R. In fact, as we will soon see, this result holds for any archimedean f-ring. An ℓ-ring will be called an *infinite d-ring* if it satisfies the following condition:

$$\text{if } \bigwedge_{i\in I} x_i = 0 \text{ and } a \geq 0, \text{ then } \bigwedge_{i\in I} ax_i = \bigwedge_{i\in I} x_i a = 0. \tag{3.6.1}$$

In other words, an infinite d-ring is an ℓ-ring in which the multiplication maps induced by positive elements are all complete. As a prelude to the aforementioned result about the completion of an archimedean f-ring we present a result about the completeness of certain homomorphisms.

Theorem 3.6.4. *Suppose that $\bigwedge_{\alpha\in A} x_\alpha = 0$ in the ℓ-group G. Fix $\beta \in A$ and let $u_\alpha = (-x_\beta + \varphi(x_\alpha))^+$ and $T = \cap\{[0, \psi(x_\alpha)] : \alpha \in A\}$ where $\varphi, \psi : G^+ \longrightarrow G^+$ are functions which satisfy the following conditions:*

(i) φ and ψ are isotone;
(ii) $\wedge\varphi(x_\alpha) = 0$;
(iii) for each $\alpha \in A$, $\psi(x_\beta + u_\alpha) \leq \psi(x_\beta) + \psi(u_\alpha)$;
(iv) for each $\alpha \in A$, $\varphi\psi(x_\alpha) \leq \psi\varphi(x_\alpha)$;

(v) if $x_\alpha \in z^\perp$ for some α, then $\varphi(x_\alpha) \in z^\perp$;
(vi) if $u_\alpha \in z^\perp$ for some α, then $\psi(u_\alpha) \in z^\perp$.

Then $\varphi(T) \subseteq [0, \psi(x_\beta)]$. In particular, each f-endomorphism of an archimedean ℓ-group is complete and hence an archimedean f-ring is an infinite d-ring.

Proof. We will first show that $\cap u_\alpha^{\perp\perp} = 0$. Suppose that $0 \leq z \in \cap u_\alpha^{\perp\perp}$. Then for each $\alpha \in A$, $z \wedge x_\beta \leq x_\beta \vee \varphi(x_\alpha) = \varphi(x_\alpha) + (-x_\beta + \varphi(x_\alpha))^-$ yields that $z \wedge x_\beta \leq \varphi(x_\alpha)$ since $(-x_\beta + \varphi(x_\alpha))^- \in u_\alpha^\perp$, and so $z \wedge (-x_\beta + \varphi(x_\alpha))^- = 0$. But then $z \wedge x_\beta = 0$ by (ii) and $z \in u_\beta^\perp$ by (v); so $z = 0$. Now, suppose that $0 \leq t \leq \psi(x_\alpha)$ for each $\alpha \in A$. Then $\varphi(t) \leq \psi\varphi(x_\alpha) \leq \psi(x_\beta + u_\alpha) \leq \psi(x_\beta) + \psi(u_\alpha)$ by (iv) and (iii). But $\wedge\psi(u_\alpha) = 0$ and hence $\varphi(t) \leq \psi(x_\beta)$. For, if $0 \leq v \leq \psi(u_\alpha)$ for each $\alpha \in A$, then $v \in (\psi(u_\alpha))^{\perp\perp} \subseteq u_\alpha^{\perp\perp}$ by (vi) and hence $v = 0$. If G is archimedean and φ is multiplication by $n \in \mathbb{N}$ and ψ is an f-endomorphism of G we get that $nt \leq \psi(x_\beta)$. So $t = 0$ and $\wedge\psi(x_\alpha) = 0$. □

A complete infinite d-ring need not be an f-ring as is illustrated by the top row of the cannonically ordered matrix ring $\mathbb{R}_2$. Also, an infinite d-ring that is an f-ring need not be archimedean as is illustrated by any non-archimedean totally ordered division ring, or by the antilexicographically ordered polynomial ring $\mathbb{R}[x]$. On the other hand, the lexicographically ordered polynomial ring $\mathbb{R}[x]$ is not an infinite d-ring.

We turn next to the completion of an ℓ-ring.

Theorem 3.6.5. *The Dedekind completion $D(R)$ of an archimedean infinite d-ring R is an ℓ-ring extension of R and is itself an infinite d-ring. If R is a one-sided f-ring or an almost f-ring, then so is $D(R)$.*

Proof. By Theorem 2.3.3 $D(R)$ is an ℓ-group extension of the additive ℓ-group of R. For $x \in D(R)$ let $L(x) = \{a \in R : a \leq x\}$. If $x, y \in D(R)^+$, then we define xy by $xy = \text{lub}_{D(R)} L(x)^+ L(y)^+$. Suppose that $x, y, z \in D(R)^+$ and let a, b, and c be variables for the elements in $L(x)^+$, $L(y)^+$ and $L(z)^+$, respectively. We first show that if $\{a_\alpha\} \subseteq L(x)^+$ and $\{b_\beta\} \subseteq L(y)^+$, then

$$x = \bigvee_\alpha a_\alpha \text{ and } y = \bigvee_\beta b_\beta \Rightarrow xy = \bigvee_{\alpha,\beta} a_\alpha b_\beta.$$

Clearly, $xy \geq \bigvee a_\alpha b_\beta$. On the other hand, by Theorem 2.1.3,

$$ab = a\left(b \wedge \bigvee_\beta b_\beta\right) = a\left(\bigvee_\beta b \wedge b_\beta\right) = \bigvee_\beta (ab \wedge ab_\beta) = ab \wedge \bigvee_\beta ab_\beta$$

and

$$ab_\beta = \left(a \wedge \bigvee_\alpha a_\alpha\right) b_\beta = \left(\bigvee_\alpha a \wedge a_\alpha\right) b_\beta = ab_\beta \wedge \bigvee_\alpha a_\alpha b_\beta.$$

So

$$ab \leq \bigvee_{\beta} ab_\beta \leq \bigvee_{\alpha,\beta} a_\alpha b_\beta$$

and $xy \leq \bigvee a_\alpha b_\beta$. Now we can easily check that multiplication in $D(R)^+$ is associative and distributive. For, $(xy)z = \bigvee (ab)c = \bigvee a(bc) = x(yz)$. Also, $x(y+z) = \bigvee a(b+c)$ and $xy + xz = \bigvee (ab + a'c)$ where a' is also a variable for the elements in $L(x)^+$. But $ab + a'c \leq (a \vee a')(b+c)$ and $a \vee a' \in L(x)$; so $x(y+z) = xy + xz$ and, similarly, $(y+z)x = yx + zx$.

In exactly the same way we have that

$$x(y \vee z) = \bigvee a(b \vee c) = \bigvee (ab \vee ac) = \bigvee (ab \vee a'c) = xy \vee xz,$$

and, similarly, $(x \vee y)z = xz \vee yz$. Let $\ell_x, r_y : D(R)^+ \longrightarrow D(R)^+$ be the maps defined by $\ell_x(y) = xy = r_y(x)$. By Exercise 2.2.11 (b) ℓ_x and r_y extend uniquely to ℓ-group endomorphisms of $D(R)$ which we will also denote by ℓ_x and r_y. Now, for $u, v \in D(R)$ let ℓ_u and r_v be the group endomorphisms of $D(R)$ defined by $\ell_u = \ell_{u^+} - \ell_{u^-}$ and $r_v = r_{v^+} - r_{v^-}$. Then $\ell_u(v) = (u^+v^+ + u^-v^-) - (u^+v^- + u^-v^+) = r_v(u)$ and hence $D(R)$ is a d-ring with multiplication given by $uv = \ell_u(v) = r_v(u)$. We will now check that $D(R)$ is an infinite d-ring. Suppose that $\bigwedge x_\alpha = 0$ in $D(R)$ and $y \in D(R)^+$. Since R is dense in $D(R)$ there exists an element $b \in R$ with $y \leq b$. Now, $x_\alpha = \bigwedge_j x_{\alpha j}$ with $x_{\alpha j} \in R$. So

$$0 \leq \bigwedge_{\alpha} yx_\alpha \leq \bigwedge_{\alpha} bx_\alpha \leq \bigwedge_{\alpha} \bigwedge_{j} bx_{\alpha j} = 0$$

since $\bigwedge_{\alpha,j} x_{\alpha j} = 0$. Similarly, right multiplication by y is complete.

Now, suppose that $x, y, z \in D(R)^+$ with $y \wedge z = 0$. If R is a left f-ring, then, using the previous notation, we have that

$$xy \wedge z = \left(\bigvee_{a,b} ab \right) \wedge \bigvee_{c} c = \bigvee_{a,b,c} (ab \wedge c) = 0$$

since $ab \wedge c = 0$ for every a, b and c. If R is an almost f-ring, then $yz = \bigvee bc = 0$ since $bc = 0$ for every pair $(b,c) \in L(y)^+ \times L(z)^+$. □

It is not hard to show directly that if an archimedean infinite d-ring R is an sp-ring, then so is its completion $D(R)$ (see Exercise 4). However, an archimedean d-ring that has all its squares positive must be an almost f-ring (Exercise 3.8.22).

Each left (right) annihilator A in an infinite d-ring R is completely closed in R. Hence, if R is archimedean, then A is a polar by Exercise 2.3.22 (b). In particular, if R is an archimedean f-ring, then $D(R) = T \oplus S$ where $T^2 = 0$ and S is semiprime, and the lower ℓ-radical ℓ-$\beta(R)$ is a polar of R. Note that $S^{[2]}$ could be a proper ℓ-ideal of S but $(S^{[2]})^{\perp_S \perp_S} = S$ (in any semiprime f-ring S).

We will now use the representation theorem for an archimedean ℓ-group to give a useful representation of the ring generated by the f-endomorphisms of the ℓ-group. Since an abelian ℓ-group M is a left ℓ-module over its po-endomorphism ring $E = \mathrm{End}_{\mathbb{Z}}(M)$ we will, as usual, let $F(M) = F({}_E M)$ denote the subring of E generated by the set $f(M)$ of f-endomorphisms of M (see Theorem 3.1.1).

Theorem 3.6.6. *Let X be the Stone space of the Boolean algebra of polars of the archimedean ℓ-group M, and let $\varphi : M \longrightarrow D(X)$ be the representation given in Theorem 2.3.23. Then there is a po-ring isomorphism $\sigma : F(M) \longrightarrow \{h \in D(X) : h\varphi(M) \subseteq \varphi(M)\}$, with $\sigma(1) = 1$, such that for every $g \in M$ and every $\alpha \in F(M)$, $\varphi(\alpha g) = \sigma(\alpha)\varphi(g)$. In particular, $F(M)$ is an archimedean f-ring and M is a strong f-module over $F(M)$.*

Proof. Let $\{e_\lambda : \lambda \in \Lambda\}$ be the maximal disjoint subset of M that is used to construct φ. To simplify the notation we will identify M with $\varphi(M)$. So $M \subseteq D(X)$, each e_λ is the characteristic function of the clopen subset X_λ of X (see step (vi) in the proof of Theorem 2.3.23), $\cup_\lambda X_\lambda$ is dense in X, and $X_\lambda \cap X_\mu = \emptyset$ if $\mu \neq \lambda$. For each $x \in X$ let $M_x = \{g \in M : g(x) = 0\}$ and $M^x = \{g \in M : g(x) \in \mathbb{R}\}$. Then $M_x \subseteq M^x$ are prime subgroups of M, and if $x \in X_\lambda$, then the evaluation map $\rho : M^x/M_x \longrightarrow \mathbb{R}$ given by $\rho(g + M_x) = g(x)$ is an embedding of the nonzero totally ordered group M^x/M_x into $\mathbb{R}$. Now, let $\alpha \in f(M) = F(M)^+$. Then $Y_\lambda = X_\lambda \cap (\alpha e_\lambda)^{-1}(\mathbb{R})$ is a dense open subset of X_λ, and if $x \in Y_\lambda$, then $\alpha e_\lambda \in M^x$. Since $e_\lambda \in M^x \backslash M_x$, we have, by Exercise 2.5.2, that $\alpha M^x \subseteq M^x$ and $\alpha M_x \subseteq M_x$. Consequently, for each $x \in Y_\lambda, \alpha$ induces an endomorphism $\overline{\overline{\alpha}} : M^x/M_x \longrightarrow M^x/M_x$ defined by $\overline{\overline{\alpha}}(g + M_x) = \alpha g + M_x$; but then we have the (isotone) endomorphism $\rho\overline{\overline{\alpha}}\rho^{-1} : \rho(M^x/M_x) \longrightarrow \rho(M^x/M_x)$, and by Theorem 3.3.1 there is a unique real number $\alpha_x \in \mathbb{R}^+$ such that $(\rho\overline{\overline{\alpha}}\rho^{-1})(\rho(g + M_x)) = \alpha_x \rho(g + M_x)$ for each $g \in M^x$. Thus,

$$(\alpha g)(x) = \alpha_x g(x) \tag{3.6.2}$$

for each $\lambda \in \Lambda$ and each $x \in Y_\lambda$ and each $g \in M^x$. Let $Y = \cup_\lambda Y_\lambda$ and define $\bar{\alpha} : Y \longrightarrow \mathbb{R}$ by $\bar{\alpha}(x) = \alpha_x = \alpha_x e_\lambda(x) = (\alpha e_\lambda)(x)$ if $x \in Y_\lambda$; that is, $\bar{\alpha}$ agrees with αe_λ on Y_λ and hence $\bar{\alpha}$ is continuous. But then $\bar{\alpha}$ is the restriction to Y of a unique element of $D(X)$ which we will also call $\bar{\alpha}$, by Theorem 2.3.20. Now define $\sigma : f(M) \longrightarrow D(X)^+$ by $\sigma(\alpha) = \bar{\alpha}$. If $g \in M$ and $x \in g^{-1}(\mathbb{R}) \cap Y$, then $(\alpha g)(x) = \alpha_x g(x) = \bar{\alpha}(x)g(x)$ by (3.6.2). Thus, $\alpha g = \sigma(\alpha)g$ since $g^{-1}(\mathbb{R}) \cap Y$ is dense in X. If $\beta \in f(M)$, then $\sigma(\alpha + \beta)g = (\alpha + \beta)g = (\sigma(\alpha) + \sigma(\beta))g$. Thus, $\sigma(\alpha + \beta) - (\sigma(\alpha) + \sigma(\beta)) \in \ell(M; D(X)) = M^{\perp_{D(X)}} = 0$, and hence $\sigma(\alpha + \beta) = \sigma(\alpha) + \sigma(\beta)$; similarly, $\sigma(\alpha\beta) = \sigma(\alpha)\sigma(\beta)$. Clearly, the unique po-group homomorphism $\sigma : F(M) \longrightarrow \{h \in D(X) : hM \subseteq M\}$ that extends σ, and whose existence is guaranteed by Exercise 2.2.11, is a po-ring homomorphism, and $\gamma g = \sigma(\gamma)g$ for every $g \in M$ and every $\gamma = \alpha - \beta \in F(M)$. Moreover, σ is an isomorphism. For if $h \in D(X)$ and $hM \subseteq M$, then, denoting multiplication by h by $h\cdot$, we have that $h\cdot = h^+\cdot - h^-\cdot \in F(M)$ and $\sigma(h\cdot) = h$ since $\sigma(h\cdot)g = hg$ for every $g \in M$; also, since $h \in D(X)^+$ iff $hM^+ \subseteq M^+$ we have that $\gamma \in f(M) = F(M)^+$ iff $\sigma(\gamma) \in D(X)^+$. Since $F(M)$ is isomorphic to the archimedean f-ring $\sigma(F(M))$ it is an archimedean f-ring itself. Note that the lattice operations in $F(M)$ are pointwise and hence M is a strong $F(M)$-f-module. That is, if $\gamma \in F(M)$ and $g \in M^+$, then

$$\gamma^+(g) = \sigma(\gamma^+)g = \sigma(\gamma)^+ g = (\sigma(\gamma)g)^+ = (\gamma(g))^+.$$

□

Several interesting properties of $F(M)$ and of its elements follow quickly from this representation theorem. For example, if M has a weak order unit e, then there is a representation φ with $\varphi(e) = 1$. Consequently $\sigma(F(M)) \subseteq \varphi(M)$ and any property of ℓ-groups that is inherited by ℓ-subgroups is passed from M to $F(M)$. In particular, if M is hyper-archimedean, then so is $F(M)$ (see Exercise 2.4.15). Also, if M is finite valued, then so is $F(M)$. In this case, by Exercises 2.5.22 and 13, $M = \oplus_{i \in I} M_i$ and $F(M) = \oplus F(M_i)$ for some finite set I and some family $\{M_i\}_{i \in I}$ of subgroups of $\mathbb{R}$. If M does not have a weak order unit, then neither of these two properties is inherited by $F(M)$ as the previous example with I infinite illustrates. Also, see Exercises 8, 9, 12, and 16. Another proof that $F(M)$ is an f-ring which does not make use of the representation of M in $D(X)$ is outlined in Exercise 22. We show next that $F(M)$ is the largest almost f-subring of $\mathrm{End}_{\mathbb{Z}}(M)$ which contains 1.

Theorem 3.6.7. *Let M be an archimedean ℓ-group, and suppose that R is a po-subring of $\mathrm{End}_{\mathbb{Z}}(M)$ which contains a monomorphism of M that belongs to $F(M)$. If R is an ℓ-ring and either R is an almost f-ring or M is a strong ℓ-module over R, then R is an ℓ-subring of $F(M)$.*

Proof. We will assume that $M \subseteq D(X)$. It suffices to show that if $\alpha \in R^+$ and $a \in M^+$, then $\alpha a \in a^{\perp\perp}$. Let $0 < \gamma \in R \cap F(M)$ be a monomorphism. From Theorem 3.6.6 we have that $\gamma a = \sigma(\gamma)a$ and $\sigma(\gamma)$ is an invertible element of $D(X)$ (see Exercises 9 and 12). Since $a^{\perp\perp} = \{b \in D(X) : S(b) \subseteq S(a)\}$ (Exercise 2.3.12), where $S(a)$ denotes the support of a, we need to verify that $S(\alpha a) \subseteq S(a)$. If $(\alpha a)^{-1}((0,\infty]) \cap (X \backslash S(a)) = \emptyset$, then $S(\alpha a) \cap (X \backslash S(a)) = ((\alpha a)^{-1}((0,\infty]))^{-} \cap (X \backslash S(a)) = \emptyset$ since $S(a)$ is closed, and we're done. Otherwise, there is some element $x \in X$ such that $a(x) = 0$, $0 < (\alpha a)(x) < \infty$, $0 \leq (\alpha^2 a)(x) < \infty$, and $0 < \sigma(\gamma)(x) < \infty$ since $(\alpha a)^{-1}(\mathbb{R}) \cap (\alpha^2 a)^{-1}(\mathbb{R}) \cap \sigma(\gamma)^{-1}((0,\infty))$ is dense in X. Take $n \in \mathbb{N}$ such that $n(\sigma(\gamma)(x))(\alpha a)(x) > (\alpha^2 a)(x)$. By replacing γ by $n\gamma$ we may assume that $(\sigma(\gamma)(\alpha a))(x) = (\gamma\alpha a)(x) > (\alpha^2 a)(x)$. Note that $(\gamma a)(x) = \sigma(\gamma)(x)a(x) = 0$. Let $\beta = \alpha \wedge \gamma \in R$. Then $(\gamma - \beta) \wedge (\alpha - \beta) = 0$. We will now show that $(\gamma - \beta)(\alpha - \beta)a > 0$. Let $b = (\alpha - \beta)a$. Then $b(x) = (\alpha a)(x)$ since $0 \leq (\beta a)(x) \leq (\gamma a)(x) = 0$. Now $\beta \leq \alpha$ and $b \leq \alpha a$ give that $(\beta b)(x) \leq (\alpha b)(x) \leq (\alpha^2 a)(x) < \sigma(\gamma)(x)(\alpha a)(x) = \sigma(\gamma)(x)b(x) = \gamma(b)(x)$. So $((\gamma - \beta)(\alpha - \beta)a)(x) = ((\gamma - \beta)b)(x) > 0$. If R is an almost f-ring, then we have a contradiction to the equation $(\gamma - \beta)(\alpha - \beta) = 0$. Suppose that ${}_RM$ is a strong ℓ-module. Then $((\gamma - \beta)c)(x) \wedge ((\alpha - \beta)c)(x) = 0$ where $c = \alpha a + a$. However, we will now check that $((\gamma - \beta)c)(x) > 0$ and $((\alpha - \beta)c)(x) > 0$. For, $(\beta c)(x) = (\beta\alpha a)(x) = (\alpha^2 a \wedge \gamma\alpha a)(x) = (\alpha^2 a)(x) < (\gamma\alpha a)(x) = (\gamma c)(x)$, and hence $((\gamma - \beta)c)(x) > 0$; also $(\alpha c)(x) = (\alpha^2 a)(x) + (\alpha a)(x) > (\alpha^2 a)(x) = (\beta c)(x)$, and $((\alpha - \beta)(c))(x) > 0$. □

The condition that R is an almost f-ring in Theorem 3.6.7 can be weakened to just requiring that it have squares positive if we also assume that $R \cap \mathbb{Z} \neq 0$; see Exercise 24. However, some condition on $R \cap F(M)$ is necessary. For example, if M is divisible and $0 < \alpha \in \mathrm{End}_{\mathbb{Q}}(M)$ is an idempotent not in $F(M)$, then M is a strong ℓ-module over the totally ordered subfield $\mathbb{Q}\alpha$ of $\mathrm{End}_{\mathbb{Q}}(M)$, yet $\mathbb{Q}\alpha \cap F(M) = 0$.

Let $\{g_\lambda : \lambda \in \Lambda\}$ be a maximal disjoint subset of M^{+*} in the archimedean ℓ-group M, and let C be an ℓ-subring of $F(M)$ which contains 1. A *C-unital cover* of the pair $({}_CM, \{g_\lambda\}_{\lambda\in\Lambda})$ is a pair (R, ψ) where R is a unital f-algebra over C, $\psi : M \longrightarrow R$ is a $c\ell$-essential C-monomorphism, $1 = \vee_\lambda \psi(g_\lambda)$, and R is generated as an ℓ-algebra by $\varphi(M) \cup \{1\}$. According to Theorem 3.6.6, if $D(X)$ is considered a C-algebra via the homomorphism $\sigma : C \longrightarrow D(X)$, then the ℓ-subalgebra of $D(X)$ generated by $\varphi(M) \cup \{1\}$ is a C-unital cover of $(M, \{g_\lambda\})$. This definition could be phrased more generally by just assuming that C is an ℓ-ring and ${}_CM$ is a strong f-module over C. In view of Exercise 6, however, this more general cover reduces to the previous one. The interesting fact about these covers is that they are unique.

Theorem 3.6.8. *Suppose that (R_1, ψ_1) and (R_2, ψ_2) are both C-unital covers of the archimedean f-module $({}_CM, \{g_\lambda\}_{\lambda\in\Lambda})$. Then there is a unique C-ℓ-algebra isomorphism $\rho : R_1 \longrightarrow R_2$ with $\psi_2 = \rho\psi_1$.*

Proof. Let $R = R_1$ and, for the moment, identify M with $\psi_1(M)$. We first note that R is semiprime. For, if there is an element $0 < a \in R$ with $a^2 = 0$, then we may assume that $a \in M$ and $a \leq g_\lambda$ for some $\lambda \in \Lambda$. But this is impossible since g_λ is idempotent and hence $\mathbb{N}a \leq g_\lambda$. We show next that if T is the convex ℓ-C-subalgebra of R generated by M, then T is archimedean. Toward this end suppose that $\mathbb{N}s \leq t$ with $0 < s, t \in T$. Again, we may assume that $s \in M$ and $s \leq g_\lambda \leq 1$ for some $\lambda \in \Lambda$. Now choose $a \in M^+$ with $t \leq a^{n_1} + \cdots + a^{n_k}$. Since $\mathbb{N}s \nleq a$, for some $m \in \mathbb{N}$, $b = (ms - a)^+ > 0$, and hence $\bar{b} > 0$ in $\bar{R} = R/b^\perp$. But then $\bar{s} > 0$, $\bar{a} \leq m\bar{s}$, and $\bar{a}^{\,k} \leq (m\bar{s})^k \leq m^k \bar{s}$ for each $k \in \mathbb{N}$; so $\mathbb{N}\bar{s} \leq p\bar{s}$ for some $p \in \mathbb{N}$ and we have the contradiction that $\bar{s} = 0$. By Exercise 3.4.5 (or otherwise) R is a strong ℓ-algebra over C, and hence $R = T + C \cdot 1$ since $T + C \cdot 1$ is an ℓ-subalgebra of R. Thus, T is an ideal of R and if $\mathbb{N}u \leq v$ with $u, v \in R^+$, then $u \in T^{\perp_R} = 0$; so R is archimedean. Let X be the Stone space of the Boolean algebra of polars of M. By Exercise 2.3.23 the following diagram is commutative

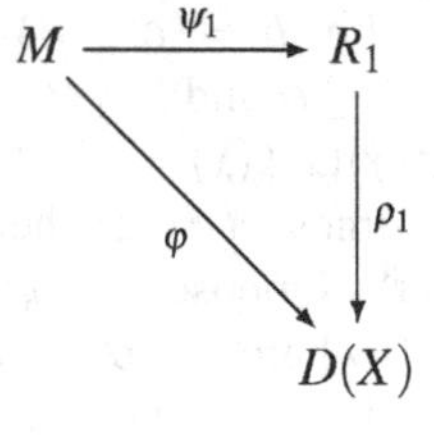

where φ and ρ_1 are the $c\ell$-essential monomorphisms attached to the disjoint set $\{g_\lambda\}$. If $\alpha \in C^+$, then $\rho_1(\alpha \cdot 1) = \rho_1(\alpha \cdot \bigvee \psi_1(g_\lambda)) = \bigvee \rho_1\psi_1(\alpha g_\lambda) = \bigvee \varphi(\alpha g_\lambda) = \alpha \cdot \bigvee \varphi(g_\lambda) = \alpha \cdot 1 = \sigma(\alpha)$ by Theorems 3.6.4 and 3.6.6. In particular, $\rho_1(1) = 1$ and ρ_1 is a ring homomorphism by Exercise 2. Also, ρ_1 is a C-homomorphism on both $C \cdot 1$ and on the subring S of R_1 generated by $\psi_1(M)$ (which is a C-subalgebra). But then ρ_1 is a C-homomorphism since, by Theorem 3.4.1, R_1 is the sublattice generated by $S + C \cdot 1$. Note that $\rho_1(R_1)$ is the C-ℓ-subalgebra of $D(X)$ generated by

$\varphi(M) \cup \{1\}$. Similarly, there is a C-ℓ-algebra embedding $\rho_2 : R_2 \longrightarrow D(X)$ with $\varphi = \rho_2 \psi_2$ and $\rho_2(R_2) = \rho_1(R_1)$. Thus $\rho = \rho_2^{-1}\rho_1 : R_1 \longrightarrow R_2$ is the desired isomorphism. If $n : R_1 \longrightarrow R_2$ is another, then $n^{-1}\rho$ fixes $\psi_1(M)$ and $C \cdot 1$ and hence $\rho = n$. □

Exercises.

1. Let T be an ℓ-ring and let R be the ℓ-subring of the canonically ordered matrix ring T_3 given by
$$R = \left\{ \begin{pmatrix} 0 & a & b \\ 0 & 0 & a \\ 0 & 0 & 0 \end{pmatrix} : a, b \in T \right\}$$
(see Exercise 3.1.24(a)). Show that R is nilpotent and is an sp-ℓ-ring, an almost f-ring, a left d-ring, or a right d-ring iff T has the corresponding property. If $T^3 \neq 0$ show that R is not an f-ring.

2. Let R be a unital semiprime archimedean f-ring. Suppose that Y is a Stone space and $\psi : R \longrightarrow D(Y)$ is a $c\ell$-essential monomorphism of ℓ-groups. Show that ψ is a ring homomorphism if and only if $\psi(1) = 1$. (Use Theorem 2.3.24.)

3. Suppose that ${}_RM_S$ is an f-bimodule and that M is T-archimedean for the subset T of R^+. Show that if multiplication by t, $t\cdot : M \longrightarrow M$, is complete for each $t \in T$, then multiplication by s, $\cdot s : M \longrightarrow M$, is complete for each $s \in S^+$.

4. If the archimedean infinite d-ring R is an sp-ring, show directly that its completion $D(R)$ is also an sp-ring.

5. If $M = \mathbb{Z} \overleftarrow{\times} \mathbb{Q}$ show that $F(M)$ is an archimedean f-subring of $\mathrm{End}_{\mathbb{Z}}(M)$ and M is not a strong F-module. (If A and B are R-modules,
$$\mathrm{End}_R(A \boxplus B) \cong \begin{pmatrix} \mathrm{End}_R(A) & \mathrm{Hom}_R(B,A) \\ \mathrm{Hom}_R(A,B) & \mathrm{End}_R(B) \end{pmatrix}.)$$

6. Suppose that M is an archimedean ℓ-group and ${}_RM$ is a strong f-module over the ℓ-ring R. Show that $R/\ell(M)$ is a semiprime archimedean f-ring.

7. (a) Show that an archimedean f-ring is unitable.
   (b) Suppose that R is a strong f-algebra over the ℓ-ring C. If R is archimedean and has a basis, show that R is C-unitable. (Use Exercise 2.5.27 and Theorems 3.3.1 and 3.4.5.)

8. Suppose that Y is a Stone space and $\varphi_1 : M \longrightarrow D(Y)$ is a $c\ell$-essential monomorphism of ℓ-groups. Show that there is a unique ℓ-ring monomorphism $\sigma_1 : F(M) \longrightarrow D(Y)$ such that $\varphi_1(\alpha a) = \sigma_1(\alpha)\varphi_1(a)$ for $\alpha \in F(M)$ and $a \in M$. Show that $\sigma_1(F(M)) = \{h \in D(Y) : h\varphi_1(M) \subseteq \varphi_1(M)\}$.

9. If $\alpha, \beta \in F(M)$, where M is an archimedean ℓ-group, show that:

   (a) $\ker \alpha = (\alpha M)^{\perp} = (\alpha M^+)^{\perp} = \ker|\alpha| = \varphi^{-1}(\sigma(\alpha)^{\perp_{\varphi(M)}})$ (see Theorem 3.6.6);

(b) $\ker(|\alpha| \vee |\beta|) = \ker\alpha \cap \ker\beta \subseteq \ker(\alpha \vee \beta)$;
(c) $\ker\alpha\beta = \ker(|\alpha| \wedge |\beta|) = \ker\alpha \vee \ker\beta$ (the sup is in $\mathscr{B}(M)$).
((b) and (c) can be verified directly; but the last equality in (a) shows, with the help of Exercises 2.2.14 (h), (e) and (f), that they are instances of Exercise 2.2.13 (e).)

10. Let R be an archimedean f-ring. Verify each of the following.
   (a) $F(R) \subseteq \mathrm{End}_R(R_R)$ (that is, R is an f-algebra over $F(R)$) and if $n : R \longrightarrow F(R)$ is the regular representation of R, then n is an $F(R)$-homomorphism and $\ker n = \ell\text{-}\beta(R)$.
   (b) If $\rho : F(R) \longrightarrow F(R/\ell\text{-}\beta(R))$ is the natural homomorphism, then $\ker\rho = \{\alpha \in F(R) : R^2 \subseteq \ker\alpha\} = n(R)^{\perp}$.
   (c) If $A_i \subseteq R$ for $1 \leq i \leq n$ and A_1 is $F(R)$-invariant, then $C(A_1 \cdots A_n)$ is an $F(R)$-submodule of R; in particular, $R^{[2]}$ is an $F(R)$-submodule and the subdirect product representation in Theorem 3.6.2 is an $F(R)$-representation.
   (d) If R is semiprime, then $F(R) = \mathrm{End}_R(R_R)$.
   (e) If R is semiprime and $\varphi_1 : R \longrightarrow D(Y)$ is a $c\ell$-essential ℓ-ring monomorphism, then the map σ_1 in Exercise 8 is the unique *ring* homomorphism that extends φ_1.
11. Show that two unital archimedean f-rings are isomorphic as ℓ-groups iff they are isomorphic f-rings.
12. Let M^e be the $c\ell$-essential closure of the archimedean ℓ-group M.
   (a) If $\alpha \in F(M)$ show that there is a unique extension of α to $\bar{\alpha} \in F(M^e)$, and that the mapping $\alpha \mapsto \bar{\alpha}$ is an embedding of f-rings.
   (b) Let N be an $F(M)$-ℓ-submodule of M^e which contains M, and for $\alpha \in F(M)$ let $\bar{\bar{\alpha}}$ be the restriction of $\bar{\alpha}$ to N; so $\alpha \longmapsto \bar{\bar{\alpha}}$ is an embedding of $F(M)$ into $F(N)$. Show that α is monic iff $\bar{\alpha}$ is epic, iff $\bar{\bar{\alpha}}$ is monic, and that if α is epic then $\bar{\bar{\alpha}}$ is an automorphism.
   (c) Show that the divisible closure $d(M)$ and the completion $D(M)$ are both $F(M)$-ℓ-submodules of M^e. In each case give an example of an $\alpha \in F(M)$ which is not epic but $\bar{\bar{\alpha}}$ is an automorphism.
13. If $\{M_\lambda\}_{\lambda\in\Lambda}$ is a family of archimedean ℓ-groups show that $F(\oplus M_\lambda) \cong \Pi F(M_\lambda) \cong F(\Pi M_\lambda)$ (see Exercise 2.4.19).
14. (a) If the archimedean ℓ-group M is complete or laterally complete show that $F(M)$ is complete or laterally complete, respectively.
   (b) In general, $F(D(M))$ is not the completion of $F(M)$. For an example, let M be the subgroup of $\mathbb{Q}$ consisting of all those rationals with square free denominator.
15. Let $\{M_\lambda : \lambda \in \Lambda\}$ be a family of subgroups of $\mathbb{R}$. If M is a subdirect product of $\{M_\lambda\}$ show that $F(M)$ is a subdirect product of $\{T_\lambda : \lambda \in \Lambda\}$ where T_λ is a subring of $F(M_\lambda)$. (Assume that $M_\lambda = M/A_\lambda$; by Exercise 2.5.2 (c) each A_λ is a convex ℓ-$F(M)$- submodule of M.)

16. Show that the following statements are equivalent for the archimedean ℓ-group M.

   (a) The embedding $\sigma : F(M) \longrightarrow D(X)$ given in Theorem 3.6.6 (or Exercise 8) is a $c\ell$-essential monomorphism.
   (b) For each $0 \neq a \in M$ there is an element $0 \neq \alpha \in F(M)$ with $\alpha M \subseteq a^{\perp\perp}$.
   (c) If A is a nonzero polar of M, then $\alpha M \subseteq A$ for some nonzero $\alpha \in F(M)$.

17. Recall from Section 2.3 that an ℓ-group is projectable if each principal polar is a summand. Let M be a projectable archimedean ℓ-group.

   (a) Show that $\sigma : F(M) \longrightarrow D(X)$ is a $c\ell$-essential monomorphism (see Exercise 16).
   (b) Show that $F(M)$ is projectable. (If $\alpha \in F(M)$ and $a \in M$ first show that $a \in (\alpha M)^{\perp\perp} \oplus (\alpha M)^{\perp}$ by considering the projection of M onto $(\alpha a)^{\perp\perp}$, and then show that $\alpha^{\perp} = f^{\perp}$ where f is the projection of M onto $(\alpha M)^{\perp\perp}$.)

18. If M is an archimedean ℓ-group with a basis show that $\sigma : F(M) \longrightarrow D(X)$ is a $c\ell$-essential monomorphism and hence $F(M)$ has a basis.

19. Let G be the group of ℓ-automorphisms of the ℓ-group M and let K be the subgroup of G consisting of the f-automorphisms of M.

   (a) Show that G is a po-group with the partial order that it inherits from $\mathrm{End}_{\mathbb{Z}}(M)$ (namely, $\alpha \geq \beta$ if $\alpha a \geq \beta a$ for each $a \in M^{+}$) and K is a convex normal subgroup of G.
   (b) If M is archimedean show that K is an ℓ-subgroup of G.
   (c) If $M = R$ is a unital archimedean f-ring and H is the group of ℓ-ring automorphisms of R show that G is the semidirect product of K by H. (Assume that $R \subseteq D(X)$ and use Exercise 2.3.20. Each $\alpha \in G$ has a unique factorization $\alpha = \ell_f \circ \tau^* \circ i$, and a unique extension to $D(X)$, where f is a positive unit of $D(X)$, τ is a homeomorphism of X, and i is the inclusion map.)
   (d) Let $M = R$ be a semiprime archimedean f-ring and let $\alpha \in G$. Show that α extends to an ℓ-automorphism of its unital closure R_u iff $\alpha(1)$ and $\alpha^{-1}(1)$ are both in R_u.

20. Show that the archimedean ℓ-group M is the additive ℓ-group of a unital f-ring if and only if it is a cyclic $F(M)$-module.

21. Let R be a unital archimedean f-ring.

   (a) Show that $(R, +, *)$ is an f-ring, where $*$ is a binary operation on R, iff there is an element $p \in R^{+}$ such that $a * b = abp$ for every $a, b \in R$.
   (b) Let $\alpha \in G$ (see Exercise 19). Show that α is a ring homomorphism for each (unital) f-ring multiplication of R iff $\alpha = 1$ (α fixes the positive units of R).
   (c) Suppose that $(R, +, *)$ and $(R, +, **)$ are f-rings with associated elements p and q, respectively. Show that the following statements are equivalent.
      (i) The f-rings $(R, +, *)$ and $(R, +, **)$ are isomorphic.

(ii) There exists an ℓ-ring automorphism α of R and a positive unit u of R such that $\alpha(p) = qu$.
(iii) There exists an ℓ-group automorphism β of R such that $\beta(p) = q$.

22. Let M be an archimedean ℓ-group. Here is a proof of the fact that $F(M)$ is an f-ring and M is a strong $F(M)$-f-module that uses Exercise 3.1.30 instead of the representation of M in $D(X)$. Let $\alpha, \beta \in f(M)$ and $a, b \in M^+$.

(a) Suppose that Q is a value of a such that $\alpha a \vee \beta a \vee b$ is an element of the cover Q^* of Q. Use Exercise 2.5.2 to show that the identity in Exercise 3.1.30 (b) holds modulo Q.
(b) Show that this identity holds modulo $a^{\perp\perp}$.
(c) Use Exercise 2.5.32 to show that the identity holds in M.

23. Show that the totally ordered free ring $\mathbb{Z}\{x,y\}$ that is given in Exercise 3.4.35 is an infinite d-ring.

24. Let R be an sp-po-subring of $\mathrm{End}_{\mathbb{Z}}(M)$ where M is an archimedean ℓ-group.

(a) Suppose R is an sp-ℓ-ring and assume there is an element $\gamma \in R \cap F(M)$ which is a right (or left) superunit in R and is an f-element of R. If $r_\ell(R;M) = 0$, show that $R \subseteq F(M)$. (Use Theorem 3.8.12.)
(b) If R contains a monomorphism of M show R is reduced. (Use Exercise 3.7.25.)
(c) Suppose $\gamma \in R \cap F(M)$ is a monomorphism, γ is central in R, and $R = R\gamma$ (or $R^+ \subseteq R^+\gamma$). Show that R is directed.

25. Let A be the ℓ-subring of the ring of real-valued continuous functions on $[0,1]$, $C([0,1])$, generated by the ring of polynomials with zero constant term, and let A_0 be the underlying ℓ-group of A with $A_0^2 = 0$. Let $R = \{(f,g) \in A \oplus A_0 : f'(0) = g'(0)\}$.

(a) Show that $A = \{f \in C[0,1] : f \text{ is piecewise polynomial and } f(0) = 0\}$.
(b) Show that R is an archimedean f-ring that is not the direct product of a nilpotent f-ring and a semiprime f-ring. (Consider $(x,x) \in R$.)

26. Suppose R is an ℓ-reduced ℓ-algebra over the directed po-ring C and R is C^+-archimedean and C^+-strict.

(a) If $0 < a \in F(R)$ show that there exist $0 < \alpha \in C$ and a minimal ℓ-prime ℓ-ideal P of R such that $\alpha a + P$ is a superunit of R/P.
(b) If $F(R) \neq 0$ and it is a strong ℓ-algebra over the ℓ-ring C show that C is an ℓ-simple f-ring.

Assume now that $F(R)^\perp$ does not contain any nonzero ℓ-ideals of R.

(c) Show that R is a subdirect product of a family of C^+-strict ℓ-algebras each of which is an ℓ-domain with an f-superunit.
(d) Show that R is a subdirect product of a family of ℓ-simple ℓ-algebras each of which has an f-superunit. (Use Exercise 3.3.27(b).)

(e) If R has a.c.c. on ℓ-annihilator ℓ-ideals show that R has an f-superunit and is isomorphic to a direct product of ℓ-simple ℓ-algebras. (See Exercise 3.3.22.)

Notes. Archimedean f-rings have been studied by many authors. Birkhoff and Pierce [BP] established Theorem 3.6.1, Henriksen and Isbell [HI] gave the f-ring version of Theorem 3.6.2, Theorem 3.6.3 is Bernaus' representation theorem [BERN1] and Theorem 3.6.4 is an altered version of a result from this paper. Theorem 3.6.5 comes from Johnson [JOH2], and Theorem 3.6.6 is due independently to Conrad and Diem [CDI], where a weaker form of Theorem 3.6.7 appears, and to Bigard and Keimel [BK] (see Exercise 22). Theorem 3.6.8 comes from Conrad [CON13]—also, see Hager and Robertson [HR]—as do most of Exercises 19–21. Most of Exercises 5, 8, 9, and 11 - 15 come from [CDI] and [BK], and Exercises 17 and 18 come from[BK]. Exercise 25 comes from Henriksen and Isbell [HI] and Exercise 26 for ℓ-rings comes from Ma [M4].

3.7 Squares Positive

We have already seen some instances in which an sp-ℓ-ring R exhibits behaviour similar to that of an f-ring. For instance, if R is a local ring, then it is necessarily an f-ring (Exercise 3.1.12 (a)), and if R is archimedean, then its generalized ℓ-nil radical coincides with its lower ℓ-nil radical and this radical is nilpotent of index at most three (Theorem 3.6.2). In this section we will examine the structure of an sp-ℓ-ring in more detail. It is shown that if R is ℓ-prime and has a nonzero f-element, then it is a domain, and if it does not have any nonzero right ℓ-quasi-regular right ℓ-ideals, then it is an ℓ-domain. As a consequence of this, in an sp-ℓ-ring the two radicals mentioned above frequently coincide with a union of nilpotent ℓ-ideals. It is also shown that an ℓ-prime sp-ℓ-algebra with an f-superunit can be embedded in a unital sp-ℓ-algebra. The conditions that a pops must satisfy in order for the generalized power series ring over it to be an sp-ℓ-ring are determined; and this sp-ℓ-ring turns out to always be an ℓ-domain when it is ℓ-prime and to be embeddable in a unital sp-ℓ-domain if it also has a nonzero f-element.

Our first two results examine elements that are nilpotent modulo certain subrings.

Theorem 3.7.1. *Let S be a convex ℓ-subring of the subring F of f-elements of the sp-ℓ-ring R.*

(a) Suppose that $k, m, n, p \in \mathbb{Z}^+$ with $1 \le p \le m+k+2$. If $a \in R$ with $S^{[k]}a^{2^n}S^{[m]} \subseteq S^{[p]}$, then $S^{[k]}aS^{[n+m]} + S^{[k+n]}aS^{[m]} \subseteq S^{[p]}$.

(b) If $a \in R$ and $a^{2^n} \in S$ for some $n \in \mathbb{Z}^+$, then for each $x \in R$, $|a|x^+ \wedge x^- \in r(S^n;R)$ and $x^+|a| \wedge x^- \in \ell(S^n;R)$.

Proof. (a) We proceed by induction on n. If $n = 0$ the implication is trivial. Suppose that it holds for some integer n and $S^{[k]}a^{2^{n+1}}S^{[m]} \subseteq S^{[p]}$. Then $S^{[k]}a^2S^{[n+m]} +$

$S^{[k+n]}a^2S^{[m]} \subseteq S^{[p]}$. If $t \in S^+$, then $0 \leq (a+t)^2$, $(a-t)^2$ yields that - $(t^2+a^2) \leq ta+at \leq t^2+a^2$; so $|ta+at| \leq t^2+a^2$. But, by Exercise 2.4.21,

$$|a|t+t|a| = |at+ta| \leq a^2+t^2 \text{ if } t \in S^+ \subseteq F \text{ and } a \in R. \tag{3.7.1}$$

Thus,

$$|t^k at^{n+m+1}| = t^k|at|t^{n+m} \leq t^{k+n+m+2} + t^k a^2 t^{n+m} \in S^{[p]}$$

and $S^{[k]}\,a\,S^{[n+m+1]} \subseteq S^{[p]}$; similarly, $S^{[k+n+1]}\,a\,S^{[m]} \subseteq S^{[p]}$.

(b) By (a), with $k = m = 0$ and $p = 1$ (and $S^o = \mathbb{Z}$) we have that $aS^n + S^n a \subseteq S$. Since the result is trivial if $n = 0$ we assume that $n \geq 1$. If $0 \leq s \in S^{[n]}$, then $s \leq t^n$ for some $t \in S^+$. So $s(|a|x^+ \wedge x^-) \leq |t^n a|x^+ \wedge t^n x^- = 0$. Thus, $|a|x^+ \wedge x^- \in r(S^{[n]}) = r(S^n)$ and similarly $x^+|a| \wedge x^- \in \ell(S^n)$. □

Theorem 3.7.2. *Let I be a maximal right ℓ-ideal of the sp-ℓ-ring R and suppose that $R^2 \nsubseteq I$. If $a \in R^+$ and $a^n \in I$ for some $n \in \mathbb{N}$, then $a \in I$.*

Proof. First assume that $n = 2$. If $x \in I^+$, then $ax \leq x^2 + a^2$ gives that $ax \in I$, and hence $aI \subseteq I$. If $a \notin I$, then $R = I + C(aR + \mathbb{Z}a)$ and so $aR \subseteq I$. Now, $J = \{x \in R : |x|R \subseteq I\}$ is a proper right ℓ-ideal of R that contains I, and hence $J = I$. But then we have the contradiction that $a \in I$. So in this case a is an element of I. Now assume that $1 \leq n$ is minimal with $a^n \in I$. If $n \geq 2$, then $(a^{n-1})^2 = a^n a^{n-2} \in I$; so $a^{n-1} \in I$. Thus $n = 1$ and $a \in I$. □

It will be convenient, as in (3.3.2), to denote the set of nilpotent elements of R of index at most n by N_n (or $N_n(R)$) and to denote the set of all nilpotent elements of R by $N = N(R)$. Also, $M_n = M_n(R) = \{x \in R : |x| \in N_n\}$ and $M = M(R) = \{x \in R : |x| \in N\}$.

Theorem 3.7.3. *Let R be an sp-ℓ-ring.*

(a) If $\mathscr{R}(R) \cap \mathscr{R}_{left}(R) = 0$ and R is ℓ-semiprime (ℓ-prime), then R is ℓ-reduced (an ℓ-domain).

(b) If R is ℓ-semiprime and $F(R)$ is not contained in any minimal ℓ-prime ℓ-ideal of R, then R is reduced. In particular, R is a domain provided it is ℓ-prime and has a nonzero f-element.

Proof. (a) According to Theorem 3.7.2 we have that $N^+ \subseteq \mathscr{R}(R) \cap \mathscr{R}_{left}(R)$, and so R is ℓ-reduced. In the remarks preceding Theorem 3.2.21 we have noted that an ℓ-reduced ℓ-prime ℓ-ring is an ℓ-domain.

(b) We will first show that $FM_2 = M_2F = 0$ in any sp-ℓ-ring in which F is semiprime. For, $(N_2^+)^2 = 0$ in any sp-ℓ-ring and $N_2F + FN_2 \subseteq F$ by (a) of Theorem 3.7.1 (with $n = p = 1$ and $m = k = 0$). Thus, $N_2^+F^+N_2^+ \subseteq N_2^+ \cap F^+ = 0$, and, again, $N_2^+F^+ + F^+N_2^+ \subseteq N_2^+ \cap F^+$; so $N_2^+ \subseteq \ell(F) \cap r(F)$ and $M_2F = FM_2 = 0$ since $N_2^+ = M_2^+$ and $M_2 = M_2^+ - M_2^+$ (see Exercise 1). Since $F^+R^+\ell(F)^+ \subseteq N_2^+$ we get that $F^2R\ell(F) = 0$. Now, if R is ℓ-prime, then F is semiprime by (e) of Theorem 3.2.13; and if $F \neq 0$, then $\ell(F) = 0$. Similarly, $r(F) = 0$. Thus, R is reduced since $N_2 = N_2 \cap F = 0$ by (b) of Theorem 3.7.1; but then R is a domain since it is an

ℓ-domain and from $ab = 0$ we get that $a^2b^2 = 0$ and hence $a = 0$ or $b = 0$. Now assume that R is ℓ-semiprime and P is a minimal ℓ-prime ℓ-ideal of R with $F \nsubseteq P$. Then $F(R/P) \neq 0$ and hence R/P is a domain. So R is reduced by Exercise 3.2.15. □

If R is an ℓ-prime sp-ℓ-algebra, then $F(R)$ is a totally ordered domain and hence it is infinitesimal or has a superunit. If $F(R)$ has a superunit which is also a superunit of R, then R is *sp-unitable* in the sense that it can be embedded in a unital (ℓ-prime) sp-ℓ-ring. This is the content of the next theorem which is given for an ℓ-algebra over the directed po-ring C.

Theorem 3.7.4. *Suppose that the C-ℓ-algebra R has a superunit which is a basic f-element. Then $S = R + C \cdot 1 \subseteq End_R(R_R)$ is an ℓ-subalgebra of $End_R(R_R)$, R is an ℓ-subalgebra of S, and $F(S) = F(R) + C \cdot 1$ is totally ordered. Moreover, R is an sp-ℓ-algebra, is semiprime, ℓ-semiprime, reduced, ℓ-reduced, prime, ℓ-prime, an ℓ-domain, or a domain iff S has the corresponding property.*

Proof. Let e be a basic f-superunit of R. From Exercise 4 we obtain that $F = F(R)$ is totally ordered, and from Theorem 2.5.9 we have that $R = U \cup (F \oplus F^{\perp})$ where $U = U(F;R) = \{a \in R : |a| \geq F\}$. If $s \in S$, then according to Exercise 7, $s \in S^+$ iff $se \in R^+$, iff $es \in R^+$; also, R is an ℓ-subalgebra of S and $F + C \cdot 1$ is totally ordered. Note that if $a \in U$, then $|a| \geq C \cdot 1$ since $|a|e \geq |a| \geq Ce$. To show that S is an ℓ-algebra we will check that if $a + \alpha \cdot 1 \in S$, then

$$(a+\alpha\cdot 1)^+ = \begin{cases} a^+ + \alpha\cdot 1 & \text{if } a^+ \in U \\ a^+ & \text{if } a^- \in U \\ (a_1+\alpha\cdot 1)^+ + a_2^+ & \text{if } a = a_1 + a_2 \in F \oplus F^{\perp}. \end{cases} \tag{3.7.2}$$

The first case will follow from the equation

$$(ae + \alpha e)^+ = a^+e + \alpha e \text{ if } a^+ \in U. \tag{3.7.3}$$

For, if (3.7.3) holds and if $b + \beta \cdot 1 \geq 0,\ a + \alpha \cdot 1$, then $(b + \beta \cdot 1)e \geq (ae + \alpha e)^+ = a^+e + \alpha e$, and $b + \beta \cdot 1 \geq a^+ + \alpha \cdot 1 \geq 0,\ a + \alpha \cdot 1$. Thus, $(a + \alpha \cdot 1)^+ = a^+ + \alpha \cdot 1$. To establish (3.7.3) note that $a^+e + \alpha e \geq ae + \alpha e$, 0, and if $r \in R$ with $r \geq ae + \alpha e$, 0, then $r - \alpha e \geq ae \vee -\alpha e$. But $ae \vee -\alpha e \geq 0$ since if $\overline{ae} \vee -\overline{\alpha e} < 0$ in some totally ordered homomorphic image $\bar{R}$ of the f-bi-module ${}_C R_F$, then $\bar{a}\bar{e} = \overline{ae} < 0$ and $-\alpha\bar{e} = \overline{-\alpha e} < 0$. Therefore, $\bar{a} < 0$ and we have the contradiction $0 < \bar{e} \leq \overline{a^+} = 0$. Consequently, $r - \alpha e \geq (ae)^+ = a^+e$ and (3.7.3) has been established.

The second case follows from the first since $(a + \alpha \cdot 1)^+ = (a + \alpha \cdot 1) + (-a - \alpha \cdot 1)^+ = a + \alpha \cdot 1 + a^- - \alpha \cdot 1 = a^+$.

For the last case, clearly $(a_1 + \alpha \cdot 1)^+ + a_2^+ \geq a + \alpha \cdot 1$, 0 if $a_1 \in F$ and $a_2 \in F^{\perp}$ with $a = a_1 + a_2$. Also,

$$[(a + \alpha \cdot 1)e]^+ = (a_1 + \alpha \cdot 1)^+e + a_2^+e. \tag{3.7.4}$$

So if $b + \beta \cdot 1 \geq a + \alpha \cdot 1$, 0, then $b + \beta \cdot 1 \geq (a_1 + \alpha \cdot 1)^+ + a_2^+$ since $(b + \beta \cdot 1)e \geq [(a + \alpha \cdot 1)e]^+ = [(a_1 + \alpha \cdot 1)^+ + a_2^+]e$.

If u and v are elements of any ℓ-ring, then $u^+v=(uv)^+$ iff $u^-v=(uv)^-$. It follows from (3.7.2), (3.7.3), and (3.7.4) that e is a right d-element of S. If $u \wedge v = 0$ in S, then $0 \leq (ue \wedge v)e = ue^2 \wedge ve \leq ue^2 \wedge ve^2 \leq (u \wedge v)e^2 = 0$ shows that e is a right f-element of S. Similarly, e is a left f-element of S and thus $F \subseteq F(S)$. Also, S is an ℓ-algebra over C since if $u \wedge v = 0$, then $(\alpha u \wedge v)e = \alpha ue \wedge ve = 0$. Thus, $F + C \cdot 1 \subseteq F(S)$ and if $a + \alpha \cdot 1 \in F(S)$, then $a \in F(S) \cap R \subseteq F(R)$; so $F(S) = F + C \cdot 1$.

Now suppose that R is an sp-ℓ-algebra. If $a \in U$ and $t \in F^+$, then from (3.7.1) we have that $|a|t + t|a| \leq a^2 + t^2 \leq a^2 + t|a|,\ a^2 + |a|t$. So $|a|t,\ t|a| \leq a^2$. Also, if $\alpha = \alpha_1 - \alpha_2 \in C$ with $\alpha_i \in C^+$ and $\beta = \alpha_1 + \alpha_2$, then $\alpha a \leq |\alpha a| \leq \beta|a| \leq \beta e|a| \leq a^2$. Thus, $(a+\alpha \cdot 1)^2 = a^2 + 2\alpha a + \alpha^2 \cdot 1 \geq 0$ since $(\alpha^2 \cdot 1)e^2 = (\alpha e)^2 \geq 0$. If $a \in F^\perp$, then we again obtain from (3.7.1) that $|a|t,\ t|a|,\ \alpha a \leq a^2$ since $|a|t,\ t|a| \in F^\perp$ and $t^2 \in F$; so the previous calculation again gives $(a + \alpha \cdot 1)^2 \geq 0$. If $a = a_1 + a_2 \in F \oplus F^\perp$, then $(a+\alpha \cdot 1)^2 = (a_1 + \alpha \cdot 1)^2 + a_2^2 + a_1a_2 + a_2a_1 + 2\alpha a_2 \geq 0$ since $a_2^2 \geq Fa_2 + a_2F + Ca_2$ by Exercise 3.4.31 (b). The verification of the other statements is straightforward and is left to the reader. □

Unlike the situation for f-rings an ℓ-prime sp-ℓ-ring with an f-superunit can be embedded in nonisomorphic minimal unital sp-ℓ-rings; see Exercise 9. However, there is a unique sp-unital cover among those sp-unital covers for which an f-superunit is preserved; see Exercise 10.

We will now consider the question of when an ℓ-prime sp-ℓ-algebra with $F \neq 0$ can be embedded into a unital sp-ℓ-algebra as an ℓ-ideal, or, equivalently, as a convex ℓ-subalgebra since then its image is an ℓ-ideal of the unital ℓ-subalgebra that it generates (see the proof of Theorem 3.2.18). Recall that when this question is restricted to f-algebras the requirement is that the f-algebra be infinitesimal or unital. Those ℓ-algebras that are dual to infinitesimal ℓ-algebras will come into play for sp-ℓ-algebras. An ℓ-algebra R is called *C-supertesimal* if $C|a| \leq a^2$ for each $a \in R$.

Theorem 3.7.5. *Let R be an ℓ-prime sp-ℓ-algebra which has a nonzero f-element and which is not unital. If R is a convex ℓ-subalgebra of a unital sp-ℓ-algebra, then either (i) R is C-supertesimal, or (ii) $R = F \oplus F^\perp$ as C-f-modules, F is a convex ℓ-subalgebra of its tight C-unital cover and $C|a| \leq a^2$ for each $a \in F^\perp$. Conversely, if R is a C-supertesimal strong ℓ-algebra or it satisfies the conditions in (ii), then R can be embedded in a unital sp-ℓ-algebra as an ℓ-ideal.*

Proof. We first note again the following easy consequence of (3.7.1) for an sp-ℓ-ring R:

$$\text{if } t \in F(R)^+ \text{ and } a \in F^\perp, \text{ then } t|a| \leq a^2 \text{ and } |a|t \leq a^2. \quad (3.7.5)$$

Suppose that R is an ℓ-ideal of the unital sp-ℓ-algebra T. By taking an ℓ-ideal P of T that is maximal with respect to being disjoint from $R^+\backslash\{0\}$ and by replacing T with T/P we may, by (d) of Theorem 3.2.13 and Theorem 3.7.3, assume that T is a domain. Now $R = U \cup (F \oplus F^{\perp_R})$ and $T = V \cup (E \oplus E^{\perp_T})$ where $E = F(T)$ and $V = \{t \in T : |t| \geq E\}$. Since $1 \notin R$ we have that $V \cap R = \emptyset$ and $R = (R \cap E) \oplus (R \cap E^{\perp_T}) = (F \cap E) \oplus E^{\perp_R}$ as C-f-modules. Also, $F^{\perp_R} \subseteq E^{\perp_R}$ since if $0 \leq a \in F^{\perp_R}$ and $0 \leq x \in E$, then $a \wedge x \in E \cap R \cap F^{\perp_R} \subseteq F \cap F^{\perp_R} = 0$. If $F \cap E = 0$, then $R \subseteq E^{\perp_T}$

and R is C-supertesimal by (3.7.5). Otherwise, according to Exercise 2.4.10, F and E are comparable and hence $F \subset E$ and $R = F \oplus F^{\perp_R}$. In this case F is a convex ℓ-subalgebra of its unique tight C-unital cover $F + C \cdot 1 \subseteq E$ (see Exercise 3.4.8), and $C|a| \leq a^2$ for each $a \in F^{\perp_R} \subseteq E^{\perp_T}$ by (3.7.5).

For the converse, first note that if R is strong, then it is a torsion-free ℓ-algebra over the totally ordered domain $C_1 = C/A$, by Exercise 3.1.29, where $A = \ell(R; C) = \ell(x; C)$ for each $0 \neq x \in R$. If R is supertesimal, then the ℓ-algebra $T = R \times C_1$ obtained by freely adjoining C_1 to R, with the coordinatewise order, is easily seen to be an sp-ℓ-algebra over C which contains R as an ℓ-ideal. In the other case let $T = R + C \cdot 1 \subseteq \mathrm{End}_R(R_R)$ be the po-subalgebra of $\mathrm{End}_R(R_R)$ generated by R and $C \cdot 1$. Then, according to Exercise 7, $T = (F + C \cdot 1) \oplus F^{\perp}$ as C-f-modules and $F + C \cdot 1$ is totally ordered. Since T is a domain $F + C \cdot 1$ is the tight C-unital cover of F and F is a convex ℓ-subalgebra of $F + C \cdot 1$; hence $R = F \oplus F^{\perp}$ is a convex ℓ-subalgebra of T. If $a \in F, b \in F^{\perp}$, and $\alpha \in C$, then $[(a + \alpha \cdot 1) + b]^2 = (a + \alpha \cdot 1)^2 + (b^2 + 2\alpha b + ab + ba) \geq 0$. For, $b^2 \geq Fb$ and $b^2 \geq bF$ by (3.7.5), and $b^2 \geq Cb$ by assumption; so $b^2 \geq Cb + Fb + bF$ by (b) of Exercise 3.4.31. Thus, T is an sp-ℓ-algebra. □

If A is a totally ordered domain and Δ is a pops the conditions that Δ must satisfy in order for the formal power series ring $W(\Delta, A)$ to be an f-ring were given in Theorem 3.5.4. Also, the conditions on Δ for $W(\Delta, A)$ to be an almost f-ring or a (left) d-ring were given in Exercises 3.5.22 and 3.5.15. We will now determine when $W(\Delta, A)$ is an sp-ℓ-ring. To avoid some trivialities we will tacitly assume that $\alpha + \beta \in \Delta$ for at least one pair of elements $\alpha, \beta \in \Delta$. The results will be stated for $W(\Delta, A)$ but they could just as well be stated for the generalized semigroup ℓ-ring $\Sigma(\Delta, A)$. So, for example, $W(\Delta, A)$ may be replaced by $\Sigma(\Delta, A)$ in Theorem 3.7.6. Moreover, these results generally also hold for $V(\Delta, A)$ when it is an ℓ-ring.

Consider the following conditions on Δ and Σ.

$$\text{If } \alpha + \beta \text{ is defined, then } \alpha + \beta < 2\alpha \text{ or } \alpha + \beta < 2\beta \text{ or } \alpha + \beta = 2\alpha = 2\beta. \tag{3.7.6}$$

$$\text{If } 2\gamma = \alpha + \beta \text{ with } \alpha \neq \beta, \text{ then } 2\gamma < 2\alpha \text{ or } 2\gamma < 2\beta. \tag{3.7.7}$$

$$\text{If } \alpha + \beta \in \Delta \text{ with } \alpha \neq \beta, \text{ then } \alpha + \beta < 2\alpha \text{ or } \alpha + \beta < 2\beta. \tag{3.7.8}$$

$$\text{If } \alpha \neq \beta \text{ with } \alpha + \beta = 2\alpha = 2\beta, \text{ then either } 2\alpha + f(\Delta) \neq \emptyset \text{ or } f(\Delta) + 2\alpha \neq \emptyset. \tag{3.7.9}$$

$$(a_1 x^{\alpha_1} + \cdots + a_n x^{\alpha_n})^2 \geq 0 \text{ for all } a_1, \ldots, a_n \in A \text{ and } \alpha_1, \ldots, \alpha_n \in \Delta. \tag{3.7.10}$$

Clearly, (3.7.6) and (3.7.7) together are equivalent to (3.7.8), and, as we will soon see, the latter is the essential condition that Δ must satisfy in order for W to be an sp-ring. A pops is called an *sp-pops* if it satisfies (3.7.8) and an *almost sp-pops* if it satisfies (3.7.6).

Theorem 3.7.6. *Suppose that A is a domain and the generalized power series ring $W(\Delta, A)$ is an ℓ-ring.*

(a) $W(\Delta, A)$ satisfies (3.7.10) for $n = 2$ iff Δ satisfies (3.7.6), A is an sp-ℓ-ring, and either Δ has the property that $\alpha + \beta \in \Delta$ iff $\beta + \alpha \in \Delta$ or A satisfies the identity $ab \leq a^2 + b^2$.

(b) The following statements are equivalent when A is an sp-ℓ-ring.

(i) Δ satisfies (3.7.8).

(ii) $W(\Delta, A)$ is an sp-ℓ-ring and Δ satisfies (3.7.9).

(iii) $W(\Delta, A)$ satisfies (3.7.10) for $n = 3$ and Δ satisfies (3.7.9).

(iv) $W(\Delta, A)$ satisfies (3.7.10) for $n = 2$, $N_2(W(\Delta,A))$ is a sublattice of $W(\Delta,A)$ and Δ satisfies (3.7.9).

Proof. (a) Suppose that $0 \leq u^2 = a^2x^\alpha x^\alpha + b^2x^\beta x^\beta + abx^\alpha x^\beta + bax^\beta x^\alpha$ for every $u = ax^\alpha + bx^\beta$ in $W = W(\Delta,A)$. Take α and β in Δ such that $\alpha + \beta$ is defined. If $\alpha + \beta \not\leq 2\alpha$ and $\alpha + \beta \not\leq 2\beta$, then the choice of coefficients $b = -a < 0$ gives the contradiction $0 \not\leq a^2x^\alpha x^\alpha + a^2x^\beta x^\beta - a^2x^{\alpha+\beta} - a^2x^\beta x^\alpha$ irrespective of the ordering in $\{\alpha+\beta, \beta+\alpha, 2\alpha, 2\beta\}$. Now, if (3.7.6) doesn't hold for the pair α, β, then $\alpha + \beta = 2\alpha \not\leq 2\beta$ or $\alpha + \beta = 2\beta \not\leq 2\alpha$. In the first case we get the contradiction $0 \not\leq -a^2x^{2\alpha} + 4a^2x^\beta x^\beta - 2a^2x^\beta x^\alpha$ by choosing $b = -2a < 0$, and in the second case we obtain a similar contradiction by choosing $a = -2b < 0$. Thus, Δ satisfies (3.7.6). Since $\Delta + \Delta \neq \emptyset$ there is an $\alpha \in \Delta$ such that 2α is defined. So if $a \in A$, then $a^2 \geq 0$ since $0 \leq a^2x^{2\alpha}$ in W. Suppose that $a, b \in A$ with $ab \not\leq a^2 + b^2$. Then A is not totally ordered and Δ is trivially ordered according to Theorem 3.5.3. Thus, if $\alpha \neq \beta$ and $\alpha + \beta \in \Delta$, then $\alpha + \beta = 2\alpha = 2\beta$ and $0 \leq (ax^\alpha - bx^\beta)^2 = (a^2 + b^2 - ab)x^{2\alpha} - bax^\beta x^\alpha$ forces $\beta + \alpha$ to be defined (and to be equal to $\alpha + \beta$). For the converse let $u = ax^\alpha + bx^\beta \in W$ with $\alpha \neq \beta$ and $u^2 \neq 0$. Then maxsupp $u^2 \subseteq \{2\alpha, 2\beta\}$ by (3.7.6). For, assume that $\alpha + \beta \in$ maxsupp $u^2 \backslash \{2\alpha, 2\beta\}$ and $\alpha + \beta < 2\alpha$ (the case $\alpha + \beta < 2\beta$ is similar). Since $2\alpha \notin$ supp u^2 either $2\alpha = 2\beta \neq \beta + \alpha$ or $2\alpha = \beta + \alpha < 2\beta$ or $2\alpha = 2\beta = \beta + \alpha$. In the first case we have the contradiction $a^2 + b^2 = 0$. In the second case we have the contradiction that $2\beta \notin$ supp u^2 yet $2\beta \notin \{2\alpha, \beta + \alpha, \alpha + \beta\}$. In the last case we have that $a^2 + b^2 + ab = 0$ which is impossible in the totally ordered domain A. If $2\alpha \in$ maxsupp u^2, then the coefficient of $x^{2\alpha}$ in u^2 is one of the elements: $a^2, a^2 + b^2, a^2 + b^2 + ab, a^2 + b^2 + ba, (a+b)^2$. Thus $u^2 \geq 0$ and W satisfies (3.7.10) for $n = 2$.

(b) In order to show that (i) implies (ii) take $u = \sum u_\alpha x^\alpha \in W$ with $u^2 \neq 0$. We claim that

$$\text{maxsupp } u^2 = \max (2 \text{ maxsupp } u). \tag{3.7.11}$$

To see this let $Y = \max (2 \text{ maxsupp } u)$ and note first that if $\alpha, \beta \in$ supp u with $\alpha + \beta$ defined, then $\alpha + \beta \leq \gamma$ for some $\gamma \in Y$. For, $\alpha \leq \alpha_1$ and $\beta \leq \beta_1$ with $\alpha_1, \beta_1 \in$ maxsupp u; so $\alpha + \beta \leq \alpha_1 + \beta_1$ and $\alpha_1 + \beta_1 \leq 2\alpha_1$ or $\alpha_1 + \beta_1 \leq 2\beta_1$ by (3.7.8). In either case we get the desired $\gamma \in Y$. But this implies that if $\gamma \in Y$, then the coefficient of x^γ in u^2 is $\sum u_\beta^2 > 0$ where this sum is over all $\beta \in$ maxsupp u with $2\beta = \gamma$. For, if $\gamma = \alpha + \beta$ with $\alpha, \beta \in$ supp u and $\alpha \neq \beta$, then, since $\gamma < 2\alpha$ or $\gamma < 2\beta$, we have the contradiction that $\gamma < \gamma_1$ for some $\gamma_1 \in Y$. So $\alpha = \beta \leq \alpha_1 \in$ maxsupp u, and $\gamma = 2\alpha \leq 2\alpha_1 \leq \gamma_1$ with $\gamma_1 \in Y$ shows that $\alpha = \alpha_1$; thus, $\gamma = 2\beta$ with $\beta \in$ maxsupp u and Σu_β^2 is the coefficient of x^γ. Now, if $\gamma_1 \in$ maxsupp u^2, then $\gamma_1 \leq \gamma \in Y$ and $(u^2)_\gamma > 0$; so $\gamma_1 = \gamma$. But if $\gamma \in Y$, then $\gamma \leq \gamma_1$ for some γ_1 in maxsupp u^2 and $\gamma_1 \leq \gamma_2$ for some $\gamma_2 \in Y$; so again $\gamma = \gamma_1$. Thus, the claim, as well as the assertion that (i) implies (ii), has been established.

Certainly (iii) is a consequence of (ii), and we now show that (iii) implies (i). Assume that W satisfies (3.7.10) for $n = 3$ and Δ satisfies (3.7.9). From (a) we get that Δ satisfies (3.7.6) and so we only have to establish that (3.7.7) also holds in Δ. If (3.7.7) does not hold, then we can find $\alpha, \beta \in \Delta$ with $\alpha || \beta$ and $\alpha + \beta = 2\alpha = 2\beta$. Since Δ satisfies (3.7.9) we may assume that $\rho + 2\alpha \in \Delta$ for some $\rho \in f(\Delta)$. Now $\rho + \alpha$ is comparable to α since $\rho + \alpha || \alpha$ would give that $\rho + \alpha || \rho + \alpha$. Also, $\rho + \alpha < 2\alpha$ since otherwise we have $\rho + \alpha \leq 2\rho$ which gives the following contradictions: if $\rho || \alpha$ then $2\rho || \rho + \alpha$; if $\rho < \alpha$ then $2\rho < \rho + \alpha$; if $\alpha \leq \rho$ then $\alpha \in f(\Delta)$. So α is comparable to 2α since α is comparable to $\rho + \alpha$ and Δ is rooted. Similarly, β is comparable to 2β. If $2\alpha = 2\beta \leq \alpha$, then we have the contradiction that α is comparable to β. Thus, $\alpha < 2\alpha < \cdots$ and also $\beta < 2\beta < \cdots$. In particular, since $3\alpha = \alpha + \beta + \alpha$ we have that $\beta + \alpha$ is defined and, in fact, $\beta + \alpha = \alpha + \beta$ since, otherwise, $\beta + \alpha < 2\alpha$ and hence $3\alpha < 3\alpha$. Now take $0 < a \in A$ and put $u = ax^\alpha - ax^\beta \in W$. Then $u^2 = 0$ and $0 \leq (u - ax^\rho)^2 = a^2 x^\rho x^\rho - (a^2 x^{\rho+\alpha} + a^2 x^\alpha x^\rho) + (a^2 x^{\rho+\beta} + a^2 x^\beta x^\rho)$. Since $\rho + \alpha || \rho + \beta$ and $\rho + \alpha || \beta + \rho$ if $\beta + \rho$ is defined we must have that $\rho + \alpha \leq 2\rho$. Thus, $\rho + \alpha \in f(\Delta)$ and hence $\rho + 2\alpha || \rho + \alpha + \beta$, which is absurd. Thus, no such α and β exist and Δ does satisfy (3.7.7).

The proof that (iv) implies (i) is similar to the proof that (iii) implies (i). We only have to establish that Δ satisfies (3.7.7), and, again, if it does not, we can produce $0 \neq u = ax^\alpha - ax^\beta \in W$ with $\alpha || \beta$ and $\alpha + \beta = \beta + \alpha = 2\alpha = 2\beta$. But then $u^+ = ax^\alpha \in N_2(W)$, yet $(u^+)^2 \neq 0$.

To complete the proof we will show that (i) implies (iv), and to do this we only have to verify that $N_2(W)$ is a sublattice of W. Let

$$N_2(\Delta) = \{\alpha \in \Delta : 2\alpha \text{ is not defined}\}.$$

If Δ is trivially ordered and $\alpha + \beta$ is defined, then $\alpha = \beta$. So if $u \in W$, then the coefficient of $\gamma = 2\alpha$ in u^2 is $\sum u_\beta^2$ where this sum is over all $\beta \in$ supp u with $2\beta = 2\alpha$. In particular, if $u^2 = 0$, then supp $u \subseteq N_2(\Delta)$ and hence $(u^+)^2 = 0$ also. Now assume that A is totally ordered, and suppose, to the contrary, that $u \in N_2(W)$ but $(u^+)^2 + (u^-)^2 = u^+u^- + u^-u^+ > 0$. Take $\gamma \in$ maxsupp $((u^+)^2 + (u^-)^2)$. Then $\gamma \in$ maxsupp $(u^+)^2 \cup$ maxsupp $(u^-)^2)$ by (2.6.1). Assume that $\gamma \in$ maxsupp $(u^+)^2 =$ max (2 maxsupp u^+) (from (3.7.10)). Then $\gamma = 2\rho$ where $\rho \in$ maxsupp u^+. Also, assuming that $\gamma \in$ maxsupp $u^+u^- =$ max (maxsupp $u^+ +$ maxsupp u^-) (from (3.5.7)), we have that $\gamma = \alpha + \beta$ where $\alpha \in$ maxsupp u^+ and $\beta \in$ maxsupp u^-. Since $2\rho = \gamma \in$ max (2 maxsupp u^+) the inequality $2\rho < 2\alpha$ is impossible, and hence

$2\rho < 2\beta$. Thus, $(u^-)^2 > 0$ (see Exercise 14) and $\gamma < 2\beta \leq 2\beta_1 \in \max$ (2 maxsupp $u^-) = \text{maxsupp } (u^-)^2$ for some $\beta_1 \in \text{maxsupp } u^-$. But then we have the contradiction $0 = (u^+)^2(2\beta_1) = -(u^-)^2(2\beta_1) < 0$. Thus, $N_2(W)$ is a sublattice. □

The ℓ-ring that appears in Exercise 3.2.30 is an example of an sp-ℓ-domain $W(\Delta, A)$ that is not sp-unitable and the almost sp-semigroup Δ is not an sp-semigroup. An example of an almost sp-monoid that is not an sp-monoid is given in Exercise 15. Nevertheless, Theorem 3.7.4 can be improved for power series rings by relaxing the f-superunit assumption to merely requiring the existence of nonzero f-elements. Recall that the element u in a po-ring is lowerpotent if $0 \leq u^2 \leq u$. According to Exercise 3.3.3 a positive f-element is lowerpotent provided it is an upper bound for one of its higher powers.

Theorem 3.7.7. *Let Δ be a rooted pops with $f(\Delta) \neq \emptyset$. For each totally ordered domain A the ℓ-ring $W(\Delta, A)$ can be embedded in a unital ℓ-prime sp-ℓ-ring in which each lowerpotent element is an f-element if and only if Δ is a semigroup and an sp-pops.*

Proof. If $W(\Delta, A)$ is contained in a unital ℓ-prime sp-ℓ-ring, then $W(\Delta, A)$ is a domain by Theorem 3.7.3. Thus, Δ is a semigroup and Δ is an sp-pops by Theorem 3.7.6. Conversely, suppose that Δ is a semigroup and an sp-pops. Then $W(\Delta, A)$ is a domain and by replacing A by its unital cover we may assume that A is unital. Also, $f(\Delta)$ is a totally ordered ideal of the poset Δ. If Δ is a monoid, then $W = W(\Delta, A)$ is itself unital. Under the assumption that Δ is not a monoid we will show that the monoid $\Delta_1 = \Delta \cup \{0\}$ can be made into a rooted sp-pops that is an extension of Δ. Then $W(\Delta_1, A)$ is the desired extension of W.

We will proceed in a sequence of steps. The order relation on Δ_1 that will be shown to extend that of Δ is defined by: for $\beta \in \Delta$, $\beta < 0$ iff $2\beta < \beta$, and $\beta > 0$ iff $2\beta > \beta > \alpha > 2\alpha$ for some $\alpha \in \Delta$. Let

$$\Gamma = \{\alpha \in \Delta : \forall \beta \in f(\Delta),\ \alpha || \beta\}$$

and

$$U = \{\alpha \in \Delta : \alpha > f(\Delta)\}.$$

Note that these subsets of Δ arise naturally from the totally ordered convex subring $F(W)$ of W by means of Theorem 2.5.9.

(i) If $\alpha \in \Delta$ and $\alpha || \gamma$ for some $\gamma \in f(\Delta)$, then $\alpha \in \Gamma$. For, if $\beta \in f(\Delta)$ and $\alpha \leq \beta$, then $\alpha \in f(\Delta)$ and hence α is comparable to γ; and if $\beta < \alpha$, then since γ is comparable to β and Δ is rooted we again get the contradiction that α is comparable to γ. Thus, $\alpha || \beta$ and $\alpha \in \Gamma$.

(ii) $\Delta = f(\Delta) \cup \Gamma \cup U$ is a disjoint union of sets. For, if $\alpha \notin f(\Delta) \cup \Gamma$, then since α is comparable to each element of $f(\Delta)$, we necessarily have that $\alpha \in U$.

(iii) If $\alpha \in \Gamma$, then $\alpha < 2\alpha$. For any $\gamma \in f(\Delta)$ we have that $\alpha + \gamma || 2\gamma$ since $\alpha || \gamma$. So $\alpha + \gamma \in \Gamma$ and $\alpha + \gamma < 2\alpha$ by (3.7.6). Now, α and $\alpha + \gamma$ are comparable, and if $\alpha \leq \alpha + \gamma$, then $\alpha < 2\alpha$ as desired. If $\alpha + \gamma < \alpha$, however, then α and 2α are still comparable. If we replace α by $\alpha + \gamma$ in the second inequality of this paragraph we obtain that $(\alpha+\gamma)+\gamma < 2(\alpha+\gamma)$, and hence $\alpha+\gamma < \alpha+\gamma+\alpha$. Thus, $\alpha+\gamma+\alpha < \alpha+\gamma+2\alpha$ and $\alpha < 2\alpha$.

(iv) If $2\beta < \beta$, then $\beta \in f(\Delta)$. From (ii) and (iii) we know that $\beta \in f(\Delta)\cup U$. Assume that $\beta \in U$ and let $\gamma \in f(\Delta)$. Then $\beta+\gamma > 2\gamma$ shows that $\beta+\gamma \notin \Gamma$. Also, $\beta+\gamma \notin U$ since otherwise $\beta+\gamma > \gamma$ and $2\beta+\gamma > \beta+\gamma > 2\beta+\gamma$. Thus, $\beta+\gamma \in f(\Delta)$, and either $\gamma \leq \beta+\gamma$ and hence $\beta+\gamma \leq 2\beta+\gamma < \beta+\gamma$, or $\beta+\gamma < \gamma$ and hence $\gamma+\beta+\gamma < 2\gamma$ and $\gamma+\beta < \gamma$. But in this latter case $\beta+\gamma$ and $\gamma+\beta$ are both elements of $f(\Delta)$, and hence $\beta \in f(\Delta)$. For, $\sigma || \tau$ yields that $\gamma+\beta+\sigma || \gamma+\tau$, which implies that $\beta+\sigma || \tau$. Similarly, $\sigma+\beta || \tau$. Since this is impossible we have that $\beta \in f(\Delta)$.

(v) If $\alpha \in \Delta$, then $\alpha < 2\alpha$ or $\alpha > 2\alpha$. Since Δ has no idempotent elements it suffices to show that α and 2α are comparable, and we already know this if $\alpha \in \Gamma \cup f(\Delta)$. But if $\alpha \in U$ and $\gamma \in f(\Delta)$, then $\alpha || 2\alpha$ is impossible since $\alpha+\gamma < 2\alpha$.

(vi) If $2\alpha < \alpha$, then, for each $\gamma \in \Delta$, $\alpha+\gamma < \gamma$ and $\gamma+\alpha < \gamma$. From (iv) we know that $\alpha+\gamma$ is comparable to γ, and hence $\alpha+\gamma < \gamma$ since $\gamma \leq \alpha+\gamma$ gives that $2\alpha+\gamma < \alpha+\gamma \leq 2\alpha+\gamma$. Similarly, $\gamma+\alpha < \gamma$.

(vii) If $2\beta > \beta > \alpha > 2\alpha$, then $\beta+\gamma > \gamma$ and $\gamma+\beta > \gamma$ for each $\gamma \in \Delta$. For, $\alpha+\gamma < \beta+\gamma$ and $\alpha+\gamma < \gamma$ yield that γ and $\beta+\gamma$ are comparable and hence $\gamma < \beta+\gamma$. Similarly, $\gamma < \gamma+\beta$.

(viii) If $2\beta_1 > \beta_1 > \alpha_1 > 2\alpha_1$ and $2\beta_2 > \beta_2 > \alpha_2 > 2\alpha_2$, then β_1 and β_2 are comparable. This follows from (iv), the fact that Δ is rooted and $f(\Delta)$ is totally ordered, and the inequalities $\beta_1, \beta_2 \geq \alpha_1 \wedge \alpha_2$.

(ix) Δ_1 is a poset. Let $\alpha, \beta, \gamma \in \Delta_1$. Clearly, $\preceq$ is an antisymmetric relation on Δ_1. Suppose that $\alpha < \beta$ and $\beta < \gamma$. If $\alpha = 0$, then $\beta, \gamma \in \Delta$ and $2\alpha_1 < \alpha_1 < \beta < 2\beta$ for some $\alpha_1 \in \Delta$. Now, $\gamma < 2\gamma$ and hence $2\alpha_1 < \alpha_1 < \beta < \gamma < 2\gamma$ shows that $\alpha < \gamma$, since otherwise, $2\gamma < \gamma$ and we have the contradiction $2\gamma < \beta+2\gamma < \beta+\gamma < 2\gamma$ by (vii). If $\beta = 0$, then $\alpha, \gamma \in \Delta$ and $\alpha < \gamma$ since $\alpha < \alpha+\gamma < \gamma$ by (vii) and (vi). If $\gamma = 0$, then $\alpha < \beta$ in Δ and $2\beta < \beta$. But then $2\alpha < \alpha+\beta < \alpha$ by (vi) and hence $\alpha < \gamma$. Thus $<$ is transitive and Δ_1 is a poset.

(x) Δ_1 is rooted. Let $\delta \in \Delta_1$ and put $X = \{\beta \in \Delta_1 : \beta \geq \delta\}$. If $\delta = 0$ then X is a chain by (viii), and if $\delta \neq 0$ and $0 \notin X$, then clearly X is a chain. Suppose that $\delta \neq 0$ and $0 \in X$, and let $\beta \in X\backslash\{0, \delta\}$. So $\beta > \delta$ and $0 > \delta$, and thus $\beta > \delta > 2\delta$. By (v), either $2\beta > \beta$ and $\beta > 0$, or $2\beta < \beta$ and $\beta < 0$. Thus, X is totally ordered.

(xi) Δ_1 is an sp-pops. Let $\alpha, \beta \in \Delta_1$ with $\alpha < \beta$ and let $\gamma \in \Delta$. If $\alpha = 0$, then $\gamma+\alpha < \gamma+\beta$ and $\alpha+\gamma < \beta+\gamma$ by (vii), and if $\beta = 0$, then $\gamma+\alpha < \gamma+\beta$ and $\alpha+\gamma < \beta+\gamma$ by (vi). Thus, Δ_1 is a pops. That Δ_1 satisfies (3.7.8) is an immediate consequence of (v).

(xii) If $u \in W(\Delta_1, A)$ and $u^2 \leq u$, then supp $u \leq 0$. If so, then $u \in f(W)$ by (iv) and Exercise 3.5.17. Suppose that $\gamma \in$ supp u and $\gamma \nleq 0$. Then $\gamma < 2\gamma$ since Δ_1 satisfies (3.7.6), and there is an element $\alpha \in$ maxsupp u with $2\gamma \leq 2\alpha \in$ max (2 maxsupp u). But then $2\alpha \in$ maxsupp u^2 by (3.7.10). Since $0 \leq u^2 \leq u$ there is an element $\beta \in$ maxsupp u with $2\alpha \leq \beta$. If $\alpha < 2\alpha$, then we have the contradiction that $\alpha < \beta$; thus, $\alpha \leq 0$ and we have the contradiction that $\gamma < 0$. □

Note that as a consequence of Theorem 3.7.3 or, more basically, of Exercise 14 (c), the assumption in Theorem 3.7.7 that Δ is a semigroup can be replaced by the assumption that $W(\Delta, A)$ is an ℓ-prime ℓ-ring.

A (rooted) pops is called *sp-unitable* or *almost sp-unitable* if it is contained in an (rooted) sp-mopops or an almost sp-mopops, respectively. When Δ is a semigroup the sp-unitability of the ℓ-ring $W(\Delta, A)$ in Theorem 3.7.7 was shown to be equivalent to the sp-unitability of the rooted sp-pops Δ. The two are not equivalent in general, however, even for an f-pops. An example of a rooted f-pops with two roots that is not sp-unitable but for which $W(\Delta, A)$ is a semiprime f-ring is given in Exercise 21. The two are equivalent for a totally ordered pops. We will obtain this result by establishing general unitability criteria for a rooted almost sp-pops.

For the pops Δ let

$$L_2(\Delta) = \{\beta \in \Delta : 2\beta < \beta\},$$

$$U_2(\Delta) = \{\beta \in \Delta : \beta < 2\beta\},$$

$$N_n(\Delta) = \{\beta \in \Delta : n\beta \notin \Delta\}$$

and

$$V_2(\Delta) = \{\beta \in U_2(\Delta) : \exists \alpha \in L_2(\Delta) \cup N_2(\Delta) \text{ with } \beta > \alpha\}.$$

These subsets of a rooted sp-pops are useful in determining whether or not it is sp-unital. The element $\alpha \in \Delta$ is a *left superunit of* Δ if, for each $\beta \in \Delta$, $\alpha + \beta \geq \beta$; if equality always holds, then α is a *left identity element of* Δ. A *right superunit* or *a right identity element* is defined analogously and a *superunit of* Δ is an element that is both a right and left superunit. A subset Γ of Δ is a *right ideal of* $(\Delta, +)$ if for all $\alpha \in \Gamma$ and $\beta \in \Delta$, if $\alpha + \beta \in \Delta$, then $\alpha + \beta \in \Gamma$. It is a *right ideal of the pops* Δ if it is both an ideal of the poset $(\Delta, \leq)$ and a right ideal of $(\Delta, +)$. *Left ideals* and *ideals* of Δ are defined analogously. Note that $N_n(\Delta)$ is an ideal of the poset Δ.

Theorem 3.7.8. *Let Δ be a rooted pops.*

(a) $N_2(\Delta) \cup L_2(\Delta) \subseteq f(\Delta)$ iff for each $\beta \in N_2(\Delta) \cup L_2(\Delta)$ and each $\gamma \in \Delta$, $\beta + \gamma < \gamma$ (respectively, $\gamma + \beta < \gamma$) whenever $\beta + \gamma$ (respectively, $\gamma + \beta$) is defined.

(b) Suppose that Δ is an (almost) sp-pops. Then Δ can be embedded in a rooted (almost) sp-mopops iff $\Delta \backslash \{0\} = N_2(\Delta) \cup L_2(\Delta) \cup U_2(\Delta)$, $N_2(\Delta) \cup L_2(\Delta) \subseteq f(\Delta)$, and $V_2(\Delta)$ is a totally ordered set of superunits of Δ.

(c) Suppose that Δ is an (almost) sp-pops and, for each $\beta \in \Delta$, $\beta + f(\Delta) \neq \emptyset$ and $f(\Delta) + \beta \neq \emptyset$. Then Δ can be embedded in a rooted (almost) sp-mopops iff $\Delta \backslash \{0\}$ has no idempotents, $L_2(\Delta) \subseteq f(\Delta)$, and $V_2(\Delta)$ has the description given in (b).

Proof. (a) If $N_2 \cup L_2 \subseteq f(\Delta)$, then, as in step (vi) in the proof of Theorem 3.7.7, the addition of $\beta \in N_2 \cup L_2$ to $\gamma \in \Delta$ decreases γ. Conversely, if $\gamma \| \delta$ and $\beta + \gamma$ and δ are comparable, then $\beta + \gamma \leq \delta$ yields that γ and δ are comparable since Δ is rooted, and $\delta < \beta + \gamma < \gamma$ is also impossible; so $\beta + \gamma \| \delta$ and hence $\beta \in f(\Delta)$.

(b) Suppose that Δ is contained in the rooted almost sp-mopops Δ_1. If $\beta \in \Delta_1 \backslash \{0\}$, then since $\beta < 2\beta$ or $\beta < 0$, and since $f(\Delta_1)$ is an order ideal of Δ_1, we have that $\beta \in N_2(\Delta_1) \cup L_2(\Delta_1) \cup U_2(\Delta_1)$ and $N_2(\Delta_1) \cup L_2(\Delta_1) \subseteq f(\Delta_1)$. Also, if $\beta_1, \beta_2 \in V_2(\Delta_1)$, then $\beta_i > \alpha_i$ with $\alpha_i < 0$; so $\beta_i > 0$, β_1 and β_2 are comparable, and β_i is a superunit. Conversely, suppose that Δ has these three properties and $0 \notin \Delta$. Note that if $\beta \in V_2(\Delta)$ and $\alpha \in N_2(\Delta) \cup L_2(\Delta)$, then $\alpha \leq \alpha + \beta < \beta$ by (a). Let $\Delta_1 = \Delta \cup \{0\}$ and extend the order relation of Δ to Δ_1 by defining $N_2(\Delta) \cup L_2(\Delta) < 0 < V_2(\Delta)$. To verify that this relation is a partial order of Δ_1 it suffices to show that it is transitive. Suppose that $\alpha < \beta$ and $\beta < \gamma$ with $\alpha, \beta, \gamma \in \Delta_1$. If $\alpha = 0$ or $\beta = 0$, then $\gamma \in V_2(\Delta)$ and $\alpha < \gamma$. If $\gamma = 0$, then $\beta \in N_2(\Delta) \cup L_2(\Delta)$; hence, if $2\alpha \in \Delta$, then $2\alpha < \alpha + \beta < \alpha$ by (a). That Δ_1 is a rooted pops which satisfies (3.7.6) (or (3.7.8)) if Δ satisfies (3.7.6) (or (3.7.8)) follows from the arguments in steps (x) and (xi) in the proof of Theorem 3.7.7.

(c) One implication is an immediate consequence of (b). Suppose that the almost sp-pops Δ has the three given properties. We will show that Δ has the first two properties of (b). Suppose that $\beta \in \Delta$ and $\gamma \in f(\Delta)$ with $\beta + \gamma \in \Delta$. If $2\beta \notin \Delta$, then $\beta + \gamma < 2\gamma$ and hence $\beta < \gamma$ since β and γ must be comparable. Thus, $N_2(\Delta) \subseteq f(\Delta)$. Suppose tht $2\beta \in \Delta$ and assume that $\beta || 2\beta$. Since $\beta + \gamma || 2\beta$ we again have that $\beta + \gamma < 2\gamma$ and $\beta < \gamma$; but then $\beta \in f(\Delta)$ and $2\beta || 2\beta$. Thus β is comparable to 2β and $\Delta \backslash \{0\} = N_2 \cup L_2 \cup U_2$. □

The sp-unitability of a totally ordered pops is equivalent to it being embeddable in a totally ordered mopops, and, in fact, there is a pops analogue of the unitability theorem for a totally ordered ring (Theorem 3.4.6). The pops Δ is called *infinitesimal* if $\Delta = N_2(\Delta) \cup L_2(\Delta)$. We will start with the infinitesimal case, and here the totally ordered assumption can be relaxed.

Theorem 3.7.9. *The following statements are equivalent for the rooted infinitesimal pops Δ.*

(a) Δ can be embedded in an almost sp-mopops.
(b) Δ can be embedded in a rooted almost sp-mopops.
(c) Δ is an f-pops.
(d) Δ can be embedded in a rooted f-mopops.
(e) For each totally ordered domain A, $V(\Delta, A)$ is an infinitesimal f-ring.
(f) For each totally ordered domain A, $\Sigma(\Delta, A)$ is an infinitesimal f-ring.

Proof. If Δ can be embedded in an almost sp-mopops, then $\Delta < 0$ and Δ is an f-pops by (a) of Theorem 3.7.8. Thus, (a) $\Rightarrow$ (c), and clearly (c) $\Rightarrow$ (b) $\Rightarrow$ (a) follows from (b) of Theorem 3.7.8. If Δ is an f-pops, then $\Delta_1 = \Delta \cup \{0\}$ with $\Delta < 0$ is certainly a rooted f-mopops. Let A be a totally ordered domain, which we may assume is unital. Then $V(\Delta_1, A)$ is a unital f-ring (Theorem 3.5.4), and if $u \in V(\Delta, A)^+$, then supp $u < 0$. So $u < x^0 = 1$ and $V(\Delta, A)$ is infinitesimal. Thus, (d) $\Leftrightarrow$ (c) $\Rightarrow$ (e), and also, (e) $\Rightarrow$ (f) $\Rightarrow$ (c). □

We are now ready for the unitability theorem for a totally ordered pops.

Theorem 3.7.10. *The following statements are equivalent for the totally ordered pops Δ.*

(a) Δ can be embedded in a totally ordered mopops.
(b) Δ is sp-unitable.
(c) Δ is almost sp-unitable.
(d) Δ is infinitesimal or has a superunit.
(e) For each totally ordered domain A, $V(\Delta, A)$ is a unitable f-ring.
(f) For each totally ordered domain A, $\sum(\Delta, A)$ is a unitable f-ring.

Proof. The implications (a) $\Rightarrow$ (b) $\Rightarrow$ (c) are trivial.

(c) $\Rightarrow$ (d). Assume that $0 \notin \Delta$. By (b) of Theorem 3.7.8 we have that $\Delta = N_2 \cup L_2 \cup U_2$ and V_2 is a set of superunits. Suppose Δ is not infinitesimal and does not possess a superunit. Then there is an element $\beta \in \Delta$ with $\beta < 2\beta$. Now, for each $\alpha \in \Delta$, $\alpha + \beta$, $\beta + \alpha$ and 2α are all defined. If $\alpha \geq \beta$ this is clear and if $\alpha < \beta$ it follows from the assumption that $V_2 = \emptyset$. If $\beta + \alpha \leq \alpha$ we have the contradiction $2\beta + \alpha \leq \beta + \alpha < 2\beta + \alpha$. Thus, $\alpha < \alpha + \beta$ and similarly, $\alpha < \beta + \alpha$, and we have the contradiction that β is a superunit.

(d) $\Rightarrow$ (a) and (e). If Δ is infinitesimal this follows from Theorem 3.7.9. Suppose that β is a superunit of Δ and Δ is not a mopops. Let A be a unital totally ordered domain, and let $0 < u \in V(\Delta, A)$ with $\gamma =$ maxsupp u. Then $\gamma < \beta + \gamma$, $\gamma + \beta$ by Exercise 19(b). But then $u < x^{\beta}u$, ux^{β} since maxsupp $x^{\beta}u = \beta + \gamma$ and maxsupp $ux^{\beta} = \gamma + \beta$. Thus, $V(\Delta, A)$ has the superunit x^{β} and $V(\Delta, A)$ is unitable. If $\alpha \in U_2(\Delta)$, then $(x^{\alpha})^2 > x^{\alpha}$, and hence x^{α} is a superunit of $V(\Delta, A)$ and α is a superunit of Δ. Moreover, if $\rho \in N_2(\Delta) \cup L_2(\Delta)$ and $\gamma \in U_2(\Delta)$, then $(x^{\rho})^2 < x^{\rho}$ and $x^{\gamma} < (x^{\gamma})^2$; so $x^{\rho} < x^{\gamma}$ and $\rho < \gamma$. Thus, if $\Delta_1 = \Delta \cup \{0\}$ with the relation $N_2(\Delta) \cup L_2(\Delta) < 0 < U_2(\Delta)$, then Δ_1 is totally ordered. It also is a pops. For, suppose that $\alpha, \rho \in \Delta_1$ and $\gamma \in \Delta$ with $\alpha < \rho$ and $\gamma + \alpha \in \Delta_1$. If $\alpha = 0$, then $\rho \in U_2(\Delta)$ and $\gamma + \alpha < \gamma + \rho$ since ρ is a superunit. If $\rho = 0$, then $\gamma + \alpha < \gamma + \rho$ by (a) of Theorem 3.7.8. Similarly, $\alpha + \gamma < \rho + \gamma$ if $\alpha + \gamma$ is defined.

(f) $\Rightarrow$ (d). Let A be a unital totally ordered domain. If $\sum(\Delta, A)$ is infinitesimal, then certainly Δ is infinitesimal. Suppose that $\sum(\Delta, A)$ has a superunit. By Exercise 17 Δ is a union of a finite number of ideals (actually, only one), each of which has a superunit, and the largest of these local superunits is a superunit of Δ.

Since (e) $\Rightarrow$ (f) is trivial the proof is complete □

We will now give an ℓ-semiprime version of Theorem 3.7.7.

Theorem 3.7.11. *Suppose that A is a totally ordered domain and $W(\Delta, A)$ is an sp-ℓ-ring such that $F(W(\Delta, A))$ is not contained in any minimal ℓ-prime ℓ-ideal of $W(\Delta, A)$. Then $W(\Delta, A)/\ell\text{-}\beta(W(\Delta, A))$ is sp-unitable.*

Proof. By Exercise 14(d) we have that $B = \ell\text{-}\beta(W(\Delta, A)) = \ell\text{-}N_g(W(\Delta, A))$ is an A-A-subbimodule of $W(\Delta, A) = W$ since B is the union of the ℓ-ideals $N_{2^n}(W) = W(N_{2^n}(\Delta), A)$. Let J be a minimal ℓ-prime ℓ-ideal of W. Then $\bar{J} = J/B$ is a minimal ℓ-prime ℓ-ideal of the reduced ℓ-ring $\overline{W} = W/B$. If $0 \leq \bar{u} \in \bar{J}$, then by Theorem 3.2.22 there is an element $0 \leq \bar{v} \in \overline{W} \backslash \bar{J}$ with $\bar{u}\bar{v} = \bar{v}\bar{u} = 0$. Then $\overline{au}\,\bar{v} = \bar{v}\,\overline{ua} = 0$ for each $a \in A^+$, and hence $a\bar{u}$, $\bar{u}a \in \bar{J}$ since $\bar{J}$ is a completely ℓ-prime ℓ-ideal of $\overline{W}$ (see Theorems 3.2.21 and 3.2.23 and Exercise 3.2.15). Thus, J is an A-A subbimodule of

$W(\Delta, A)$. From Exercise 18(c) we have that $\Delta\backslash\Delta(J)$ is a subsemigroup of Δ, where $\Delta(J)$ is the union of the supports of the elements in J. We assume temporarily that A is an ℓ-primitive f-ring. Then $J = W(\Delta(J), A)$ and $W(\Delta, A)/J \cong W(\Delta\backslash\Delta(J), A)$ by Exercises 11, 12, and 18. According to Theorem 3.7.7 W/J is sp-unitable, and hence so is W/B since B is the intersection of the minimal ℓ-prime ℓ-ideals of W. In general, let A_1 be an ℓ-primitive f-ring that contains A; the existence of A_1 is guaranteed by Exercise 3.3.22. If K is a minimal ℓ-prime ℓ-ideal of $W(\Delta, A_1) = W_1$, then $K = W(\Delta(K), A_1)$, and we will use Theorem 3.2.22 to show that $J = K \cap W$ is a minimal ℓ-prime ℓ-ideal of W. It is certainly a completely ℓ-prime ℓ-ideal of W. Let $B_1 = \ell\text{-}\beta(W_1)$, $\overline{W}_1 = W_1/B_1$ and $\overline{K} = K/B_1$. If $0 \le \bar{u} \in \bar{J}$ let $0 \le v \in W_1\backslash K$ with $\bar{u}\bar{v} = 0$, and let $\alpha \in \operatorname{maxsupp} v$ with $\alpha \notin \Delta(K)$. Take $0 < a \in A_1$ with $1 < v_\alpha a$ where v_α is the coefficient of α in v. Then $x^\alpha < va$ and if $0 < c \in A$ we have that $0 \le \bar{u}\overline{cx^\alpha} \le \bar{u}\bar{v}ac = 0$. Since $\overline{cx^\alpha} \in \overline{W}\backslash\bar{J}$ we have established the minimality of $\bar{J}$ and also of J. Since $W(f(\Delta), A) = F(W) \not\subseteq J = W(\Delta(K), A_1) \cap W(\Delta, A)$, necessarily $f(\Delta) \not\subseteq \Delta(K)$ and $F(W_1) = W(f(\Delta), A_1) \not\subseteq K$. So $W/B \subseteq W_1/B_1$ are both sp-unitable. □

Exercises.

1. Show that the following statements are equivalent in the ℓ-ring R, and if M_n is replaced by M, then the first four statements are equivalent. (Use Exercise 2.2.17. The definitions of M_n and M are given after Theorem 4.7.2)
   (a) M_n is an additive subgroup of R.
   (b) M_n is a convex ℓ-subring of R.
   (c) $M_n^+ + M_n^+ \subseteq M_n^+$.
   (d) $M_n^+ - M_n^+ \subseteq M_n^+$.
   (e) $(M_n^+)^n = 0$.

2. The C-ℓ-algebra R is called *square-archimedean* if for each pair of elements $x, y \in R^+$ there is an element $\alpha \in C^+$ such that $xy + yx \le \alpha(x^2 + y^2)$. Show that the following hold in a square-archimedean ℓ-algebra. Recall that $\ell\text{-}N(R)$ is the sum of the nilpotent ℓ-ideals of the ℓ-ring R and $\ell\text{-}N_g(R)$ is the generalized ℓ-nil radical of R.
   (a) For each $x, y \in R^+$ and $n \in \mathbb{N}$, there exists $\beta \in C^+$ such that $(x+y)^{2^n} \le \beta(x^{2^n} + y^{2^n})$ and $(xy)^{2^n} \le \beta(x^{2^{n+1}} + y^{2^{n+1}})$.
   (b) M_{2^n} is a nilpotent convex ℓ-subalgebra of R for each $n \in \mathbb{N}$.
   (c) If R is nil, then M_{2^n} is an ℓ-ideal of R. (Show that $xax = 0$ if $x \in M_2^+$ and $a \in R^+$.)
   (d) $\ell\text{-}N(R) = \ell\text{-}\beta(R) = \ell\text{-Nil}(R)$. (Apply (c) and Theorem 3.2.7 to $\ell\text{-Nil}(R)$.)
   (e) If disjoint elements commute, then $\ell\text{-}\beta(R) = \ell\text{-}N_g(R)$.
   (f) The archimedean ℓ-algebra in Exercise 3.1.24 (c) is square-archimedean but is not an sp-ℓ-algebra.

3. Let R be an ℓ-ring.

(a) Show that $\ell\text{-}\beta(R) = M(R)$ iff for $a \in M(R)^+$ and $x \in R$, $ax^+ \wedge x^- \in \ell\text{-}\beta(R)$ and $x^+ a \wedge x^- \in \ell\text{-}\beta(R)$. (See Theorem 3.2.13 (e) and Exercise 3.2.20.)
(b) If $M(R) \subseteq F(R)$ show that $\ell\text{-}N(R) = \ell\text{-}N_g(R) = \ell\text{-}\beta(F(R))$ and $M_{2^n}(R)$ is a nilpotent ℓ-ideal of R for each $n \in \mathbb{N}$.
(c) Show that $\ell\text{-}\beta(R) = N(R)$ iff for $a \in N(R)$ and $x \in R$, $|a|x^+ \wedge x^- \in \ell\text{-}\beta(R)$ and $x^+|a| \wedge x^- \in \ell\text{-}\beta(R)$.
(d) If $N(R) \subseteq F(R)$ show that $\ell\text{-}N(R) = N_g(R)$ and $N_{2^n}(R)$ is a nilpotent ℓ-ideal of R for each $n \in \mathbb{N}$.

4. Let e be a left d-element in the ℓ-ring R.

(a) Show that e is a left superunit of R iff $e^2 \geq e$ and $r(e; R) = 0$.
(b) If e is a left superunit of R, show that $e \in F_\ell(R)$.
(c) Suppose that e is a left superunit of R and $e \in F(R)$. Show that e is basic iff F is totally ordered.

5. Let S be a convex ℓ-subring of the sp-ℓ-ring R and suppose that S contains the left superunit e of R. Let x be an element of R with $x^{2^n} \in S$ for some $n \in \mathbb{N}$.

(a) Show that $x^2 \in S$, and if $x \geq 0$, then $x \in S$.
(b) If $S \subseteq F(R)$ show that $x \in S$.
(c) If e is a left identity element of R and R_S is a d-module show that $x \in S$.
(d) If e is a superunit of R show that S contains each lowerpotent element of S.
(e) If $e^\perp = (e^2)^\perp$ and $y \in e^\perp$, then $y \ll y^2$.
(f) If R_S is a d-module and $e^\perp = (e^2)^\perp$, then e is a weak order unit of R iff e^2 is a weak order unit of Re.

6. Let R be an sp-ℓ-ring.

(a) If $\ell(F(R); R) = r(F(R); R) = 0$ or $F(R)$ contains a left superunit of R, show that $N(R) \subseteq F(R)$.
(b) If $F(R)$ is not contained in any minimal ℓ-prime ℓ-ideal of R, show that the set of minimal ℓ-prime ℓ-ideals of R coincides with the set of minimal prime ideals of R. (See Theorem 3.2.23.)

7. Let R be an ℓ-algebra over C and suppose that $\ell(F(R); R) = r(F(R); R) = 0$. Let $T = \{t \in \mathrm{End}_R(R_R) : Rt \subseteq R\}$. Show each of the following.

(a) If $t \in T$, then $t \in T^+$ iff $tF^+ \subseteq R^+$, iff $F^+ t \subseteq R^+$, where $F = F(R)$.
(b) R is an ℓ-subalgebra of T.
(c) If there are elements $a, b \in F$ with $\ell(b; R) = r(a; R) = 0$, then for any $t \in T$, $t \in T^+$ iff $t|b| \in R^+$, iff $|a|t \in R^+$. Moreover, F is totally ordered iff $F + C \cdot 1$, the unital subalgebra of T generated by F, is totally ordered.

8. Show that the ℓ-ring R is totally ordered with $\ell(R) = r(R) = 0$ iff it is an almost f-ring and $F(R)$ has a convex totally ordered subring E with $\ell(E; R) = r(E; R) = 0$.

9. Let $T = \mathbb{Z}[x]$ and let $R = Tx$ be totally ordered lexicographically: $x \ll x^2 \ll \cdots$. Let P_0, P_1, and P_2 be the partial orders of the ℓ-groups $\mathbb{Z} \oplus R$, $(\mathbb{Z} \oplus \mathbb{Z}x) \underset{\longleftarrow}{\boxplus} Rx$,

and $\mathbb{Z} \overleftarrow{\boxplus} R$, respectively. Show that each (T, P_i) is an sp-unital cover of R, and each satisfies the identity $ab \leq a^2 + b^2$. (See Theorem 3.7.6.)

10. Let $\varphi : R \longrightarrow V$ be an embedding of ℓ-algebras where V is unital, and assume that $\varphi(e)$ is a basic f-superunit of V. Let $S = R + C \cdot 1$ be the unital ℓ-algebra given in Theorem 3.7.4 (also see Exercise 7). Show that φ has a unique extension $\psi : S \longrightarrow V$ and ψ is also an embedding.

11. Suppose that Γ is an ideal of the pops Δ. If $\alpha, \beta \in \Delta \backslash \Gamma$ define $\alpha +_\Gamma \beta = \alpha + \beta$ iff $\alpha + \beta \in \Delta \backslash \Gamma$. Let $\Delta / \Gamma = (\Delta \backslash \Gamma, \leq, +_\Gamma)$.

    (a) Show that Δ / Γ is a pops that inherits each of the following properties from Δ : almost sp-pops, sp-pops, f-pops, $\Delta = d_\ell(\Delta)$, property (d_ℓ) (see Exercise 3.5.15), $\alpha || \beta \Rightarrow \alpha + \beta$ is not defined (see Exercise 3.5.22).
    (b) Show that $N_n(\Delta / \Gamma) = N_n(\Delta) \cap (\Delta \backslash \Gamma) \cup \{\beta \in \Delta \backslash \Gamma : n\beta \in \Gamma\}$.

12. Suppose that A is a domain and $W(\Delta, A)$ is an ℓ-ring.

    (a) Let Γ be a subset of Δ. Show that $W(\Gamma, A)$ is a right (left) ℓ-ideal of $W(\Delta, A)$ iff Γ is a right (left) ideal of the pops Δ.
    (b) If Γ is an ideal of the pops Δ show that $W(\Delta, A)/W(\Gamma, A) \cong W(\Delta / \Gamma, A)$ (see Exercise 11); in particular, $W(\Gamma, A)$ is a completely ℓ-prime ℓ-ideal iff $\Delta \backslash \Gamma$ is a subsemigroup of Δ.
    (c) If Γ is a right ideal of Δ define an action of Δ on $\Delta \backslash \Gamma$ as in Exercise 11, and show that (b) still holds with the isomorphism now being one of right $W(\Delta, A)$-ℓ-modules.
    (d) These results hold for $\Sigma(\Delta, A)$ and also for $V(\Delta, A)$ if it is an ℓ-ring.

13. Suppose that Δ is an almost sp-pops.

    (a) If $X \subseteq \Delta$ show that $X + X = \emptyset$ iff $X \subseteq N_2(\Delta)$.
    (b) If Δ is rooted show that $N_2(\Delta)$ is an ideal of the pops Δ. (If $\alpha \in \Delta$, $\beta \in N_2(\Delta)$ and $\alpha + \beta \notin N_2(\Delta)$, show that $\beta + \alpha + \beta < \beta + 2\alpha$, $2\alpha + \beta$.)

14. Let A be a po-domain and let Δ be an almost sp-pops. Suppose that $u \in V(\Delta, A)$ and $u^2 \in V(\Delta, A)$.

    (a) If $u \in V^+$, show that $u^2 > 0$ iff supp $u \nsubseteq N_2(\Delta)$.
    (b) If A is a domain and an sp-po-ring and Δ is an sp-pops show that $u^2 > 0$ if and only if supp $u \nsubseteq N_2(\Delta)$.
    (c) Assume that A is a domain and $W(\Delta, A)$ is an ℓ-ring. Show that $M_{2^n}(W(\Delta,A)) = W(N_{2^n}(\Delta), A)$ is a nilpotent ℓ-ideal of $W(\Delta, A)$ for each $n \in \mathbb{N}$. Consequently, $W(\Delta, A)$ is ℓ-semiprime iff it is ℓ-reduced, iff $2\beta \in \Delta$ for each $\beta \in \Delta$, and $W(\Delta, A)$ is ℓ-prime iff it is an ℓ-domain, iff Δ is a semigroup. (Use induction on n and apply Exercise 12(b) to $\Gamma = N_{2^n}(\Delta)$ in order to show that $M_{2^{n+1}}(W(\Delta, A))$ is an ℓ-ideal of $W(\Delta, A)$. Use Exercise 11(b) to show that $N_{2^{n+1}}(\Delta) = N_{2^n}(\Delta) \cup N_2(\Delta / \Gamma)$ (or show this directly) and conclude that $M_{2^{n+1}}(W(\Delta, A)) = W(N_{2^{n+1}}(\Delta), A)$.)

(d) If Δ is an sp-pops in (c) show that $M_{2^n}(W(\Delta,A)) = N_{2^n}(W(\Delta,A))$ for each $n \in \mathbb{N}$.

15. (a) Let $\Delta = \{0, \beta, n\gamma (n \in \mathbb{N})\}$ with $\beta+\gamma = \gamma+\beta = 2\beta$ and $n\beta = n\gamma$ if $n \geq 2$. The partial order of Δ is given by: $0 < \beta < 2\gamma < \cdots < n\gamma < \cdots$ and $\gamma < 2\gamma$. Show that Δ is a rooted almost sp-monoid that is not an sp-monoid.
(b) Show that a po-group is an almost sp-pops iff it is totally ordered.
(c) Show that a rooted po-group Δ satisfies (3.7.7) iff it contains a totally ordered normal po-subgroup Γ such that every element of the factor group Δ/Γ has order at most two (see Exercise 3.5.5).

16. Suppose that A is a po-domain and Δ is a pops.

(a) If Δ is an almost sp-pops and $\Sigma(\Delta,A)$ is reduced show that Δ is an sp-pops.
(b) Let A be a domain and suppose that $W(\Delta,A)$ is an ℓ-ring and $(\beta + f(\Delta)) \cup (f(\Delta)+\beta) \neq \emptyset$ for each $\beta \in \Delta$. Show that $W(\Delta,A)$ is a reduced sp-ℓ-ring iff Δ is an sp-pops and $2\beta \in \Delta$ for each $\beta \in \Delta$.

17. Let A be a unital totally ordered domain and let Δ be a rooted pops.

(a) Show that $W(f(\Delta),A)$ contains a left superunit (left identity element) of $W(\Delta,A)$ iff $\Delta = \Delta_1 \cup \cdots \cup \Delta_n$ is the cardinal sum of a finite number of right ideals $\Delta_1, \ldots, \Delta_n$ of Δ such that, for $1 \leq i \leq n$, $\Delta_i \cap f(\Delta)$ has a left superunit (left identity element) of Δ_i. (Take a left f-superunit in W whose support is the union of a minimal number of disjoint totally ordered subsets.)
(b) If Δ is an almost sp-pops, or $2\beta \in \Delta$ for each $\beta \in \Delta$, show that $W(f(\Delta),A)$ contains a superunit of $W(\Delta,A)$ iff Δ is the cardinal sum of a finite number of ideals of Δ each of which contains a superunit which is an element of $f(\Delta)$.
(c) Show that $1 \in W(\Delta,A)^+$ iff $\Delta = \Delta_1 \cup \cdots \cup \Delta_n = \Gamma_1 \cup \cdots \cup \Gamma_n$ has two cardinal sum decompositions where Δ_i (respectively, Γ_i) is a right (respectively, left) ideal of Δ and $\Delta_i \cap \Gamma_i \cap f(\Delta)$ contains an element that is a left identity of Δ_i and a right identity of Γ_i.
(d) Show that (c) holds for $\Sigma(\Delta,A)$, and that if $1 \in \Sigma(\Delta,A)^+$ and T is any ring between $\Sigma(\Delta,A)$ and $V(\Delta,A)$, then $1 \in T$.
(e) Show that (a) and (b) hold for $\Sigma(\Delta,A)$ when Δ is any pops (not necessarily rooted).
(f) Let $\Delta = \{\alpha, \beta, \gamma\}$ be trivially ordered with $2\alpha = \alpha$, $2\beta = \beta$ and $\gamma = \gamma + \alpha = \beta + \gamma = \beta + \alpha$. Show that $\Sigma(\Delta,A)$ is unital but not ℓ-unital.
(g) Give an example of a unital f-ring $V(\Delta,A)$ for which $\Sigma(\Delta,A)$ is not unital.

18. Let A be a totally ordered domain and let Δ be a rooted pops. For the subset J of $W = W(\Delta,A)$ let $\Delta(J)$ be the union of the supports of the elements of J.

(a) If J is a convex ℓ-subgroup of W show that $\Delta(J) = \{\alpha \in \Delta : \exists\, 0 < a \in A$ with $ax^\alpha \in J\}$, and $J = W(\Delta(J),A)$ iff $\Delta(J) = \{\alpha \in \Delta : Ax^\alpha \subseteq J\}$.

(b) Show that the mapping $J \mapsto \Delta(J)$ is a lattice homomorphism from $\mathscr{C}(W)$ onto the lattice of ideals of the poset Δ, and if J is a right (left) ℓ-ideal of W, then $\Delta(J)$ is a right (left) ideal of the pops Δ.
(c) If J is a completely ℓ-prime ℓ-ideal of W, show that $\Delta \backslash \Delta(J)$ is a subsemigroup of $(\Delta, +)$ and $W(\Delta(J), A)$ is a completely ℓ-prime ℓ-ideal of W. Give an example to show that $W(\Delta(J), A)$ may properly contain J, and give an example to show that $\Delta \backslash \Delta(J)$ could be a subsemigroup even though J is not an ℓ-prime ℓ-ideal.
(d) If A is ℓ-simple show that $\Delta(J) = \{\alpha \in \Delta : Ax^{\alpha} \subseteq J\}$ whenever J is a convex ℓ-submodule of W_A or of ${}_AW$. Consequently, $\mathscr{C}(W_A) = \mathscr{C}({}_AW)$ and the mapping in (b) is a lattice isomorphism when restricted to $\mathscr{C}(W_A)$.
(e) Suppose that for each $\alpha \in \Delta$ there is some $\gamma \in \Delta$ with $\alpha < \alpha + \gamma$. Show that $\Delta(J) = \{\alpha \in \Delta : Ax^{\alpha} \subseteq J\}$ for each (right) ℓ-ideal J of W and $J \mapsto \Delta(J)$ is a lattice isomorphism between the lattice of (right) ℓ-ideals of W and the lattice of (right) ideals of the pops Δ.
(f) Let J be an ℓ-ideal of W. Assume that either A is ℓ-simple and J is an A-submodule of W, or Δ satisfies the condition in (e). Show that J is a completely ℓ-prime ℓ-ideal of W iff $\Delta \backslash \Delta(J)$ is a subsemigroup of Δ.

19. Let Δ be a pops and let $\gamma \in f(\Delta)$.
(a) If $\gamma = 2\gamma$ show that, for any $\beta \in \Delta$, $\gamma + \beta = \beta$ ($\beta + \gamma = \beta$) if the sum exists.
(b) If γ is a superunit of Δ show that either $\gamma = 0$, or $\beta < \gamma + \beta$, $\beta + \gamma$ for each $\beta \in \Delta$.
(c) Suppose that $0 \notin \Delta$ and Δ has a superunit. Show that $f(\Delta)$ does not have an idempotent element.

20. Let Δ be an (almost) sp-pops. Show that Δ is contained in an (almost) sp-mopops iff $\Delta \backslash \{0\} = N_2(\Delta) \cup L_2(\Delta) \cup U_2(\Delta)$, and, for each $\beta \in N_2(\Delta) \cup L_2(\Delta)$ and $\gamma \in \Delta$, if $\beta + \gamma \in \Delta$ (respectively, $\gamma + \beta \in \Delta$), then $\beta + \gamma < \gamma$ (respectively, $\gamma + \beta < \gamma$).

21. This is an example of a commutative rooted f-pops Δ with two roots, which has a superunit, with $N_2(\Delta) = \emptyset$, and which is sp-unitable as a pops but not as a rooted pops. Let Δ_1 be the free commutative semigroup on the set $\{s, \beta_1, \alpha_1\}$. Totally order Δ_1 by (here, $i, j, k, m, n, p \in \mathbb{Z}^+$):

$$s^i \alpha_1^j \beta_1^k < s^p \alpha_1^m \beta_1^n \quad \text{if} \quad \begin{cases} j > m, \text{ or} \\ j = m \text{ and } s^i \beta_1^k < s^p \beta_1^n \end{cases}$$

and

$$s^i \beta_1^k < s^p \beta_1^n \quad \text{if} \quad \begin{cases} k < n \text{ and } i \leq p, \text{ or} \\ k > n \text{ and } p \geq 1, \text{ or} \\ k = n \text{ and } i < p. \end{cases}$$

Similarly, let Δ_2 be the free commutative semigroup on the set $\{s, \beta_2, \alpha_2\}$, ordered in the same way as Δ_1. Let Δ be the pops generated by $\{s, \alpha_1, \beta_1, \alpha_2, \beta_2\}$

such that Δ_1 and Δ_2 are subpops of Δ, and if $\rho, \delta \in \Delta$, then $\rho \leq \delta$ and $\rho\delta$ is defined if and only if $\rho, \delta \in \Delta_1$ or $\rho, \delta \in \Delta_2$. Show that Δ has the desired properties. (A picture of Δ is useful—s is a root point; use Exercise 20 and Theorem 3.7.8.)

22. Suppose that A is a unital totally ordered domain and Δ is a rooted f-pops. Show that $\Sigma(\Delta, A)$ is unitable iff $V(\Delta, A)$ is unitable. (Use Exercise 12.)

23. Assume that A is a totally ordered domain and Δ is a rooted almost sp-pops. State and prove the corresponding version of Theorem 3.7.11 for $W(\Delta, A)$.

24. Suppose that A is a po-domain with $A^+ \neq 0$ and Δ is a pops. If $W(\Delta, A)$ is reduced show that Δ is an sp-pops iff it is an almost sp-pops.

25. Let R be an archimedean sp-po-ring, let $t(R) = \{a \in R : na = 0 \text{ for some } n \in \mathbb{N}\}$, $t_2(R) = \{a \in R : 2a = 0\}$ and let $a, b, c, d \in R$. Verify each of the following.

    (a) $t(R) \subseteq N_2(R)$.
    (b) $a \in N_2(R)$ iff $ab + ba = 0$.
    (c) If $a \in N_2(R)$, then, $abc = bac = bca \in t_2(R)$.
    (d) $N(R) = N_4(R) = \{a \in R : a^3 \in t_2(R)\}$. (If $a^{2m} = 0$ consider $a + na^k$ with $k = 2m - 3$ and $k = 2$.)
    (e) If $a \in N(R)$, then $ab, ba \in N_2(R)$. Consequently, $\beta(R) = N(R)$ where $\beta(R)$ is the lower nil radical of R. (First show that $(aba)^2 = 0$.)
    (f) $N_2(R)$ is a convex ideal of R.
    (g) If $a \in N(R)$, then $abc = -cab = bca$.
    (h) If $a \in N_2(R)$, then all six products obtained by permuting the factors in abc are equal.
    (i) If $a \in N(R)$, then $abcd = bacd = bcad \in t_2(R)$ and all twenty-four products obtained by permuting the factors in $abcd$ are equal.
    (j) $\ell(R) = r(R)$ and hence $\ell(R^m) = r(R^m)$ for each $m \in \mathbb{N}$.
    (k)
$$\begin{aligned} N(R) &= \{a \in R : abc + cab = 0\} \\ &= \{a \in R : bca + cab = 0\} \\ &= \{a \in R : 2a \in \ell(R^3)\} \end{aligned}$$
    is a convex ideal of R, and $R/N(R)$ is a torsion-free archimedean sp-po-ring.
    (l) $N(R)^2R = N(R)RN(R) = RN(R)^2 \subseteq t_2(R)$.
    (m) If R is directed, then it is torsion-free.
    (n) R is semiprime iff R is reduced iff $\ell(R) = t_2(R) = 0$.

26. Show that a unital sp-ℓ-ring in which $a \in Ra^2$ whenever $0 \leq a \ll a^2$ is an almost f-ring.

Notes. Birkhoff and Pierce [BP] showed that both a d-ring in which squares are positive and a unital ℓ-ring in which 1 is a weak order unit are almost f-rings. They asked whether or not a unital ℓ-semiprime ℓ-ring in which 1 is a weak order unit must be an f-ring. Diem [DI] answered this question by showing that an ℓ-

semiprime almost f-ring is an f-ring. He also showed that an ℓ-prime sp-ℓ-ring which is archimedean or in which disjoint elements commute is an ℓ-domain, and asked whether or not an ℓ-prime sp-ℓ-ring is an ℓ-domain. Steinberg [ST12] gave a partial answer by showing it is a domain provided the subring of f-elements has no nonzero left or right annihilator, and Ma [M2] strengthened this result by merely requiring the existence of a nonzero f-element. Earlier, Steinberg [ST7] showed that an ℓ-semiprime sp-ℓ-ring with no nonzero right ℓ-QR right ℓ-ideal is an ℓ-domain. Theorems 3.7.4 and 3.7.5 on the sp-unitability of an ℓ-prime ℓ-ring are due to Ma [M3], as is Exercise 10. Most of the results connected with $W(\Delta,A)$ being an sp-ℓ-ring come from Steinberg [ST11] and [ST15]. Exercise 2 for an ℓ-ring is due to Shyr and Viswanathan [SV] and Exercise 25 is due to Lavric [LAV]. Other papers dealing with sp-ℓ-rings are Steinberg [ST6] and [ST8] and Ma and Steinberg [MS].

3.8 Polynomial Constraints

We have seen that the constraint $x^2 \geq 0$ imposes restrictions on the nilpotent and idempotent elements of an ℓ-ring and forces some ℓ-prime ℓ-rings to be ℓ-domains. The question arises as to whether or not these results also hold for ℓ-rings that satisfy an identity of the form $(x^6)^- = 0$, or $(3x+2x^2+5x^3)^- = 0$ or $(-2xy^2 - yxy + y^3 - x^2y^2)^- = 0$. The answer is that they generally do hold for unital ℓ-rings. In fact, the polynomials can frequently be allowed to vary from element to element, or from pairs of elements to pairs of elements; so an ℓ-algebra with these constraints is analogous to an algebraic algebra as well as to an algebra with a polynomial identity. However, ℓ-algebras that are algebraic or satisfy a polynomial identity may not satisfy the constraints which generalize $x^2 \geq 0$ since the coefficients of our polynomials will have to be suitably conditioned. A simple counterexample is provided by the canonically ordered matrix algebras over a totally ordered field. Other commutative properties of these constrained ℓ-algebras are that the products of two positive disjoint elements generate the same ℓ-ideal, and the ℓ-algebra is an f-algebra if it is archimedean over the base ring. The identity $x^+x^- = 0$ that determines the variety of almost f-rings is a polynomial constraint. Surprisingly, an ℓ-ring with an f-superunit is an almost f-ring if and only if the f-superunit is a weak order unit, and an archimedean almost f-ring is commutative. An almost f-ring, as well as a d-ring, possesses a nilpotent ℓ-ideal modulo which it is an f-ring. An example is given in the exercises to show that a unital one-sided f-ring, though an almost f-ring, need not be an f-ring.

Throughout this section R will be a torsion-free ℓ-algebra over the commutative unital totally ordered domain C. Let $C[x_1,\dots,x_n]$ be the free unital C-algebra in the indeterminates $x_1,\dots,x_n$. When $n=2$ we will usually let $x = x_1$ and $y = x_2$. A polynomial $f(x,y) \in C[x,y]$ is *nice* if it has at least one monomial of degree 1 in x and each such monomial has a negative coefficient. Thus, a nice polynomial $f(x,y)$ has the form

$$f(x,y) = -g(x,y) + p(y) + h(x,y)$$

where $g(x, y)$ is nonzero and each of its monomials is of degree 1 in x and has a positive coefficient, and $h(x, y) = 0$ or each of its monomials is of degree at least 2 in x. If $g(x, y)$ has a monomial ending (beginning) in x, then $f(x, y)$ is called *right (left) nice*, and if $h(x, y) \in C[x^k, y]$, then $f(x, y)$ is *k-nice*. For example, $f(x, y) = -(xy + yx) + \alpha(x^2 + y^2)$ is left and right 2-nice for each $\alpha \in C$; also, $(y - x)^n$ and modifications of it obtained by inserting appropriate coefficients in front of its monomials are nice polynomials. If $f(x, y)$ is nice, then $f(x, 1)'(0) < 0$ where $p'(x)$ denotes the derivative of $p(x) \in C[x]$.

If $(a_1, \ldots, a_n)$ is an n-tuple of elements from R let $P((a_1, \ldots, a_n)) = \{f(x_1, \ldots, x_n) \in C[x_1, \ldots, x_n] \colon f(a_1, \ldots, a_n) \in R^+\}$ and $P^*((a_1, \ldots, a_n)) = P((a_1, \ldots, a_n)) \backslash C$. The set $P((a_1, \ldots, a_n))$ is a subsemiring of $C[x_1, \ldots, x_n]$ and $P((a_1, \ldots, a_n)) \cap -P((a_1, \ldots, a_n))$ is the ideal of polynomials which are 0 at $(a_1, \ldots, a_n)$. If R is not known to be unital, then $P((a_1, \ldots, a_n))$ will necessarily consist of polynomials with zero constant term. The n-tuple $(a_1, \ldots, a_n)$ is called *p-algebraic* if $f(a_1, \ldots, a_n) \geq 0$ for some $f(x_1, \ldots, x_n) \in P^*((a_1, \ldots, a_n))$. The pair $(a, b) \in R^2$ will be called a *(right, left, k-)nice p-algebraic pair* if $P((a, b))$ contains a (right, left, k-) nice polynomial, and if each element of $X \subseteq R^2$ has this property, then X is called a *(right, left, k-) nice p-algebraic set*. Note that if (a, b) is a nice p-algebraic pair, then in a general sense "$a \leq$ sums of higher powers of a." Similarly, a subset X of R will be called *p-algebraic* if each of its elements is p-algebraic; also, X is *weakly p-positive* if for each $a \in X$ there exists $p(x) \in P^*(a)$ with $p'(1) > 0$ in C, and X is *p-positive* if $P^*(a) \cap C^+[x]$ is not empty for each a in X. If the polynomials that make X p-algebraic of a particular type can be chosen so that their degrees are no larger than a fixed integer, then X is said to be of *bounded degree*.

We will first examine those nilpotent elements which produce p-algebraic units when added to ± 1. This, of course, gives a local generalization to the squares positive inequality.

Theorem 3.8.1. *Let R be a unital torsion-free C-ℓ-algebra and let a be a nilpotent element in R. The following statements are equivalent.*

(a) $|a| < 1$.
(b) *There is a polynomial $p(x)$ in $C[x^2]$ with $p(a^n \pm 1) \geq 0$ for each $n \in \mathbb{N}$, and $0 \neq p'(1) \cdot 1 \in R^+$.*
(c) *For each $n \in \mathbb{N}$ there are polynomials $p_1(x)$ and $q_1(x)$ in $C[x]$ with $p_1(a^n + 1) \geq 0$, $q_1((a^n - 1)^2) \geq 0$ and $p_1'(1)\, q_1'(1) \cdot 1 > 0$ in R.*
(d) *For each $n \in \mathbb{N}$ there are polynomials $p_2(x)$ and $q_2(x)$ in $C[x]$ with $p_2(a^n + 1) \geq 0$, $q_2(a^n - 1) \geq 0$ and $p_2'(1)\, q_2'(-1) \cdot 1 < 0$ in R.*
(e) *$1 \in R^+$ and for each b in $\{\pm a^n : n \in \mathbb{N}\}$ there is a polynomial $f(x, y) \in C[x, y]$ such that $f(b, 1) \geq 0$ and $f(x, 1)'(0) < 0$.*
(f) *$1 \in R^+$, $|a|$ is nilpotent and if $u \wedge v = 0$ with $u \leq |a|^m$ for some $m \in \mathbb{Z}^+$ and some unit $v \leq 1$, then there is a nice polynomial $f(x, y) \in C[x, y]$ with $f(u, v) \geq 0$.*
(g) *For each $n \in \mathbb{N}$ there are polynomials $p_3(x)$ and $q_3(x)$ in $C[x]$, with only odd terms, such that $p_3(b)^+ p_3(b)^- = 0$ if $b = \pm(a^n + 1)$, and $q_3(c)^+ q_3(c)^- = 0$ if $c = \pm(a^n - 1)$, and $p_3(1) p_3'(1) q_3(1) q_3'(1) \cdot 1 > 0$ in R^+.*

Proof. We assume at the outset that $1 \in R^+$ since this is implied by each of these statements. Note that each statement is a consequence of (a).
(b) $\Rightarrow$ (c). If $p(x) = h(x^2)$ is the polynomial given in (b) let $p_1(x) = p(x)$ and $q_1(x) = h(x)$.
(c) $\Rightarrow$ (d). Let $q_2(x) = q_1(x^2)$ and $p_2(x) = p_1(x)$.
(d) $\Rightarrow$ (e). Let $b = a^n$ and take $p_2(x), q_2(x) \in C[x]$ with $p_2(a^n+1) \geq 0, q_2(a^n-1) \geq 0$ and $p_2'(1)q_2'(-1) \cdot 1 < 0$. Now, if $q_2(x) = \alpha_0 + \cdots + \alpha_m x^m$, then

$$\begin{aligned} 0 \leq q_2(b-1) &= \alpha_0 + \alpha_1(b-1) + \alpha_2(b-1)^2 + \cdots + \alpha_m(b-1)^m \\ &= (\alpha_1 - 2\alpha_2 + \cdots + (-1)^{m-1} m\alpha_m)b + \beta_0 + h(b) \\ &= q_2'(-1)b + \beta_0 + h(b) \end{aligned}$$

where $h(x) \in x^2F[x]$ and $\beta_0 = q_2(-1)$. Similarly, there exists $h_1(x) \in x^2F[x]$ with $\gamma_0 = p_2(1)$ and

$$0 \leq p_2(b+1) = p_2'(1)b + \gamma_0 + h_1(b).$$

If $q_2'(-1) < 0$, then $f_+(x, y) = q_2'(-1)x + \beta_0 + h(x)$ is a nice polynomial with $f_+(b, 1) \geq 0$. Also, $p_2'(1) > 0$ since $p_2'(1)q_2'(-1) < 0$, and $f_-(x, y) = -p_2'(1)x + \gamma_0 + h_2(x)$ is a nice polynomial with $f_-(-b, 1) \geq 0$; here, if $h_1(x) = \sum \gamma_i x^i$, then $h_2(x) = \sum (-1)^i \gamma_i x^i$.

If $q_2'(-1) > 0$, then again we get two nice polynomials $f_\pm(x, y)$ with $f_+(b, 1) \geq 0$ and $f_-(-b, 1) \geq 0$.

(e) $\Rightarrow$ (a). By induction on the index of nilpotency of a we may assume that $a^k \in F = F(R)$ if $k \geq 2$. Let $f(x, y) = g(x, y) + p(y) + h(x, y)$ be a polynomial with $f(x, 1)'(0) < 0$ and $f(a, 1) = g(a, 1) + p(1) + h(a, 1) \geq 0$, where the monomials of $g(x, y)$ (respectively, $h(x, y)$) are of degree 1 (respectively, at least 2) in x. Then, since $g(a, 1) = -\beta a$ where $\beta = -f(x, 1)'(0) > 0$ and $h(a, 1) \in a^2C[a] \subseteq F$, we have $\beta a \leq s$ for some $s \in F$. By using a similar polynomial for $-a$, we get $-\gamma a \leq t$ for some $t \in F$ and $0 < \gamma \in C$. So $-\beta t \leq \gamma\beta a \leq \gamma s$ and $a \in F$.

(f) $\Rightarrow$ (a). By induction on the index of nilpotency of $b = |a|$, we may assume that $b^n = 0$, $n \geq 2$, and $b^k \in F$ if $k \geq 2$. Let $c = b \wedge 1$, and let $u = b - c$ and $v = 1 - c$. Then $c, v \in F$, v is a unit since c is nilpotent, and $u \wedge v = 0$. Let $f(x, y) = -g(x, y) + p(y) + h(x, y)$ be a nice polynomial with $f(u, v) \geq 0$. Then $0 \leq g(u, v) \leq p(v) + h(u, v)$. Each term of $h(u, v)$ is of the form $\alpha w = \alpha u^{n_1} v^{m_1} u^{n_2} v^{m_2} \cdots u^{n_t} v^{m_t}$ with $N = \sum n_i \geq 2$. Since $v \leq 1$, $0 \leq w \leq u^N \leq b^N \in F$; so $\alpha w \in F$, $h(u, v) \in F$, and $g(u, v) \in F$ since $p(v) \in F$. Now $g(u, v)$ contains a term of the form αu, $\alpha u v^m$, $\alpha v^m u$ or $\alpha v^m u v^k$, where $\alpha > 0$ and $m, k \geq 0$. Since $g(x, y)$ has positive coefficients, if d is this term, then $0 \leq d \leq g(u, v)$ and hence u, uv^m, $v^m u$ or $v^m u v^k \in F$. But then $u \in F$ and so $b = u + c \in F$.

(g) $\Rightarrow$ (d). Since $p_3(x)$ has only odd terms $p_3(-b) = -p_3(b)$; and hence $p_3(-b)^+ = p_3(b)^-$ and $p_3(-b)^- = p_3(b)^+$. So if $b = a^n + 1$, then $p_3(b)^+ p_3(b)^- = 0$ and $p_3(b)^- p_3(b)^+ = 0$, and hence

$$p_3(b)^2 = [p_3(b)^+ - p_3(b)^-]^2 = [p_3(b)^+]^2 + [p_3(b)^-]^2 \geq 0.$$

Similarly, $q_3(b)^2 \geq 0$ if $b = a^n - 1$. Let $p_2(x) = p_3(x)^2$ and $q_2(x) = q_3(x)^2$. Then $p_2(a^n+1) \geq 0$, $q_2(a^n-1) \geq 0$ and $p_2'(1)q_2'(-1) \cdot 1 < 0$ in R.

The proof is now complete. □

We also have the following two-variable generalization of Theorem 3.7.1 and Exercises 3.7.2 and 3.7.5.

Theorem 3.8.2. *Suppose that $a \in R$, $k \in \mathbb{N}$ and $a^{k^n} \in S$ for some $n \in \mathbb{Z}^+$ where S is a convex ℓ-subalgebra of $F(R)$. Assume that for each $m \in \mathbb{Z}^+$ and each $t \in S^+$ there are two k-nice polynomials $f_i(x,y) = -g_i(x,y) + p_i(y) + h_i(x,y), i = 1,2$, with $f_1(x,y) \in P((a^{k^m},t))$ and $f_2(x,y) \in P((-a^{k^m},t))$, at least one of which is right nice, such that $g_2(a^{k^m},t) \leq g_1(a^{k^m},t)$. Then for each $s \in S \cup \{1\}$ and for each $t \in S$ there is an integer $N \geq 0$ with $t^N sa \in S$.*

Proof. Let $t \in S$ and $s \in S \cup \{1\}$. We may assume that $s \geq 0$ and $t \geq 0$. For if $|t|^N|s|a \in S$, then

$$|t^N sa| \leq |t|^N|s||a| = ||t|^N|s|a| \in S$$

and so $t^N sa \in S$. Let $t_1 = t \vee s$ if $s \neq 1$ and let $t_1 = t$ if $s = 1$. We argue by induction on n. If $n = 0$, then $a \in S$ and we can let $N = 0$. Assume the result is true for the integer n and suppose that $a^{k^{n+1}} \in S$. Let $b = a^k$. Then $b^{k^n} \in S$ and hence for each $s_1 \in S \cup \{1\}$ there is an integer N_1 with $t_1^N s_1 b \in S$. Now for each integer $r \geq 1$ there is an integer N_r with $t_1^{N_r} s_1 b^r \in S$. For, if $t_1^{N_r} s_1 b^r \in S$, then there exists an integer M with $t_1^M(t_1^{N_r} s_1 b^r)b \in S$, and hence we may take $N_{r+1} = M + N_r$.

Let $f_1(x,y) \in P((a,t_1))$ and $f_2(x,y) \in P((-a,t_1))$ be the two k-nice polynomials which satisfy the given conditions. If u is a term of $h_1(a,t_1) = q_1(a^k,t_1) = q_1(b,t_1)$, then

$$u = \alpha t_1^{i_1} b^{j_1} t_1^{i_2} b^{j_2} \cdots t_1^{i_l} b^{j_l}$$

with $0 \neq \alpha \in C$, $l \geq 1$, $i_1 \geq 0$, $j_l \geq 0$ and $j_1 \geq 1$. We claim that $t_1^L u \in S$ for some L. If $l = 1$ this follows from the previous paragraph. Assume that $l \geq 2$ and $t_1^{L_1}(\alpha t_1^{i_1} b^{j_1} \cdots t_1^{i_{l-1}} b^{j_{l-1}}) = s_2 \in S$. Then, again, there is an integer L_2 with

$$t_1^{L_1+L_2} u = t_1^{L_2}(s_2 t_1^{i_l}) b^{j_l} \in S,$$

and so $L = L_1 + L_2$ works. Thus, there exists an integer L_3 with $t_1^{L_3} h_1(a,t_1) \in S$; similarly, there is an integer, which we may assume to be L_3, with $t_1^{L_3} h_2(-a,t_1) \in S$. Now, $t_1^{L_3} g_i(a,t_1) \in S$. For, $g_1(a,t_1) \leq p_1(t_1) + h_1(a,t_1)$ and $g_2(-a,t_1) \leq p_2(t_1) + h_2(-a,t_1)$. But $g_2(-a,t_1) = -g_2(a,t_1)$; so

$$-(p_2(t_1) + h_2(-a,t_1)) \leq g_2(a,t_1) \leq g_1(a,t_1) \leq p_1(t_1) + h_1(a,t_1).$$

Thus,

$$\begin{aligned} -t_1^{L_3}(p_2(t_1) + h_2(-a,t_1)) &\leq t_1^{L_3} g_2(a,t_1) \leq t_1^{L_3} g_1(a,t_1) \\ &\leq t_1^{L_3}(p_1(t_1) + h_1(a,t_1)) \end{aligned}$$

and $t_1^{L_3} g_i(a, t_1) \in S$.

Now suppose that $g_1(a, t_1)$ has a term of the form $\beta t_1^{L_4} a$. But $t_1 \geq 0$ and all the coefficients of $g_1(x, y)$ are in C^+; so $|\beta t_1^{L_4} a| \leq |g_1(a, t_1)|$, since this inequality holds in any totally ordered C-$F(R)$-$F(R)$ bimodule which is a homomorphic image of R, and R is a subdirect product of these modules. Thus, $\beta |t_1^{L_3} t_1^{L_4} a| \leq t_1^{L_3} |g_1(a, t_1)| = |t_1^{L_3} g_1(a, t_1)|$, and if $N = L_3 + L_4$ then $t_1^N a \in S$. We are done if $s = 1$.

If $N = 0$, then $a \in S$ and $t^N sa \in S$. If $N \geq 1$ and $s \neq 1$, then $0 \leq t^{N-1} s \leq t_1^N$ and hence $|t^{N-1} sa| = t^{N-1} s |a| \leq t_1^N |a| = |t_1^N a|$; so $t^{N-1} sa \in S$. □

We can use the previous two results to show that in an ℓ-ring which satisfies these polynomial constraints each (positive) nilpotent element is usually an f-element, and hence the ℓ-ring is reduced (ℓ-reduced) modulo its lower ℓ-nil radical. So, in fact, in many cases these general constraints reduce to the specific constraint $(x - y)^2 \geq 0$ with x nilpotent and y an f-element.

Theorem 3.8.3. *Each of the following conditions implies that ℓ-$\beta(R) = M(R) \subseteq F(R)$.*

(a) $F(R)$ contains a left superunit e of R, and if S is the convex ℓ-subalgebra generated by e, then for some $k \geq 2$ the set $M(R)^+ \times S^+$ is a right k-nice p-algebraic set.

(b) $F(R)$ contains a convex ℓ-subring S with $\ell(S; R) = r(S; R) = 0$, and, for some integers $k, k_1 \geq 2$, $M(R)^+ \times S^+$ is a left k-nice p-algebraic set of bounded degree and a right k_1-nice p-algebraic set of bounded degree.

(c) R is ℓ-unital and the set $\{(u, v) \in R^2 : u \wedge v = 0,\ u \in M(R),\ v \leq 1$ and v is a unit$\}$ is a nice p-algebraic set.

Proof. For (a), if $(a, t) \in M(R)^+ \times S^+$ and $f_1(x, y)$ is a right k-nice polynomial with $f_1(a, t) \geq 0$, let $f_2(x, y) = -g_1(x, y)$. Then $f_2(-a, t) = g_1(a, t) \geq 0$, and, by Theorem 3.8.2, $a \leq e^N a \in S$ for some integer N. Thus, ℓ-$\beta(R) = M(R)$ by Exercise 3.7.3. For (b), according to Exercise 4, there is an integer N with $t^N a,\ at^N \in S$ for every $a \in M(R)^+$ and every $t \in S^+$. But then, if $u \wedge v = 0$ in R we get that $t^N(au \wedge v) = t^N au \wedge t^N v = 0$; so $au \wedge v \in r(F^n) = 0$. Similarly, $ua \wedge v = 0$ and $a \in F(R)$. The equivalence of (a) and (f) in Theorem 3.8.1 gives (c). □

We also have the following result whose proof is similar to that of Theorem 3.7.3.

Theorem 3.8.4. *Suppose that the ℓ-algebra R either contains a convex ℓ-subalgebra S of f-elements with vanishing left and right annihilator ideals in R such that $N(R) \times S^+$ is a left k-nice and a right k_1-nice p-algebraic set of bounded degree for some integers $k, k_1 \geq 2$, or it is ℓ-unital and its group of units is weakly p-positive. Then ℓ-$\beta(R) = N(R) \subseteq F(R)$.* □

In any ℓ-group or, trivially, in any almost f-ring two disjoint elements commute. This property does not hold in an arbitrary unital sp-ℓ-ring; see Exercise 6. However, in such an ℓ-ring the ℓ-ideals generated by the products of two positive disjoint elements are identical, and this is still the case in more generally constrained

ℓ-algebras. Note that Theorem 3.2.24 indicates the importance of knowing when disjoint elements do commute.

Theorem 3.8.5. *Suppose that R is a torsion-free C-ℓ-algebra with $N_2(R)^+ \subseteq F(R)$. Assume that if $u \wedge v = 0$, where u is a zero divisor in R^+ and $v \in F(R)$, then either $v-u$ is p-positive or (u, v) is a right nice p-algebraic pair. Then for any $a, b \in R$ with $a \wedge b = ab = 0$ and any $e \in F(R)^+$, there exists an integer $N \in \mathbb{Z}^+$ with $ebe^N ae = 0$. Moreover, if $ea \geq a$ and $eb \geq b$ and $\ell(F(R); F(R)) = 0$, then $ba = 0$.*

Proof. Let $a, b \in R$ with $a \wedge b = ab = 0$ and take $e \in F^+ = F(R)^+$. Set $a_1 = a \wedge e$ and $b_1 = b \wedge e$. We first show that $be^m a_1 e = 0$ for each $m \in \mathbb{Z}^+$. Let $b_2 = b - b_1$ and $e_2 = e - b_1$; and let $a_2 = a - a_1$ and $f_2 = e - a_1$. Then

$$b_2 \wedge e_2 = 0 \tag{3.8.1}$$

and

$$a_2 \wedge f_2 = 0. \tag{3.8.2}$$

Let $b_0 = b$ and $a_0 = a$; then since $a_1 b_i = 0$ we have

$$f_2 b_i = eb_i \quad \text{for } 0 \leq i \leq 2. \tag{3.8.3}$$

Also, since $a_i b_1 = 0$ we get

$$a_i e_2 = a_i e \quad \text{for } 0 \leq i \leq 2. \tag{3.8.4}$$

Now $a_1 \wedge b_1 e^m = 0$ in F; so $b_1 e^m a_1 = 0$. Also, (3.8.1) implies $b_2 e^m a_1^l \wedge e_2 = 0$, for any $l, m \in \mathbb{Z}^+$. But $e_2 \in F$, and $(b_2 e^m a_1^l)^2 = 0$ (if $l \geq 1$) implies $b_2 e^m a_1^l \in N_2^+ \subseteq F$; so

$$b_2 e^m a_1^l e = 0 \quad \text{for all } m \in \mathbb{Z}^+ \text{ and } l \geq 1 \tag{3.8.5}$$

since $b_2 e^m a_1^l e = b_2 e^m a_1^l e_2 = 0$ by (3.8.4). But then,

$$be^m a_1 e = (b_2 + b_1) e^m a_1 e = b_2 e^m a_1 e + b_1 e^m a_1 e = 0.$$

By (3.8.2) $b_1 e^m a_2 \wedge f_2 = 0$, and therefore by (3.8.3) $eb_1 e^m a_2 = f_2 b_1 e^m a_2 = 0$. So

$$ebe^m ae = eb_2 e^m a_2 e \quad \text{for all } m \in \mathbb{Z}^+ \tag{3.8.6}$$

since $eb_1 e^m a_2 = be^m a_1 e = 0$, and

$$eb_2 e^m a_2 e = e(b - b_1) e^m (a - a_1) e = ebe^m ae - ebe^m a_1 e - eb_1 e^m a_2 e.$$

Since $(b_2 (f_2 e)^m a_2)(f_2 e)^s \in F^+$ we get from (3.8.2) that

$$b_2 (f_2 e)^m a_2 (f_2 e)^s a_2 \wedge f_2 = 0,$$

and hence (3.8.3) implies

$$eb_2 (f_2 e)^m a_2 (f_2 e)^s a_2 = 0 \quad \text{for all } m,\ s \in \mathbb{Z}^+. \tag{3.8.7}$$

Let $p(x)$ be a polynomial in $C[x]$ of degree ≥ 1 and with positive coefficients such that $p(f_2e-a_2)\geq 0$. Then ($\alpha_0=0$ if $1\notin R$)

$$0\leq \alpha_0+\alpha_1(f_2e-a_2)+\cdots+\alpha_n(f_2e-a_2)^n=p(f_2e-a_2), \tag{3.8.8}$$

and so

$$0\leq g(a_2,f_2e)\leq \alpha_0+\sum_{k\geq 1}\alpha_k(f_2e)^k+h(a_2,f_2e) \tag{3.8.9}$$

where $-g(a_2 f_2e)$ is the sum of all those monomials in a_2 and f_2e in (3.8.8) which contain exactly one a_2, and $h(a_2,f_2e)$ is the sum of all those monomials which contain more than one a_2. A typical term in $h(a_2,f_2e)$ is of the form $\alpha w=\alpha(f_2e)^{m_1}a_2(f_2e)^{m_2}a_2\cdots(f_2e)^{m_t}$ with $m_i\in\mathbb{Z}^+$, $t\geq 3$ and $\alpha\in C$. By (3.8.7) $eb_2w=0$ and hence $eb_2h(a_2,f_2e)=0$. From (3.8.9) we get that

$$0\leq eb_2g(a_2,f_2e)\leq \sum \alpha_k eb_2(f_2e)^k. \tag{3.8.10}$$

A typical term in $g(a_2,f_2e)$ is $\alpha(f_2e)^m a_2(f_2e)^s$. But

$$b_2(f_2e)^m a_2(f_2e)^s e\wedge b_2=0 \quad \text{for all } m,\ s\in\mathbb{Z}^+ \tag{3.8.11}$$

since $f_2\leq e$, and

$$\begin{aligned}0\leq b_2(f_2e)^m a_2(f_2e)^s e\wedge b_2 &\leq b_2(f_2e)^m a_2(e^2)^s e\wedge b_2\\ &= b_2(f_2e)^m a_2e_2e^{2s}\wedge b_2=0,\end{aligned}$$

by (3.8.4) and (3.8.1); and (3.8.11) implies

$$eb_2(f_2e)^m a_2(f_2e)^s e\wedge eb_2(f_2e)^k e=0 \quad \text{for all } m,\ s,\ k\in\mathbb{Z}^+. \tag{3.8.12}$$

Now, (3.8.10) and (3.8.12) imply that

$$0\leq eb_2g(a_2,f_2e)e=eb_2g(a_2,f_2e)e\wedge\sum\alpha_k eb_2(f_2e)^k e=0,$$

and hence

$$eb_2g(a_2,f_2e)e=0. \tag{3.8.13}$$

However, one term in $g(a_2,f_2e)$ is $\alpha(f_2e)^m a_2$ with $0<\alpha\in C$ and $m\geq 0$ (m is minimal with $\alpha_{m+1}\neq 0$); since $g(x,y)\in C^+[x,y]$, (3.8.13) implies

$$eb_2(f_2e)^m a_2e=0. \tag{3.8.14}$$

Now, for any $k\in\mathbb{Z}^+$

$$b_2(f_2e)^k a_2=b_2(e-a_1)e(e-a_1)e\cdots(e-a_1)ea_2=b_2e^{2k}a_2 \tag{3.8.15}$$

since all other terms contain a factor $b_2e^r a_1^l e$ with $l\geq 1$, and $b_2e^r a_1^l e=0$ by (3.8.5). Thus,

$$ebe^{2m}ae=eb_2e^{2m}a_2e=eb_2(f_2e)^m a_2e=0 \tag{3.8.16}$$

by (3.8.6), (3.8.15), and (3.8.14).

If there is a nice polynomial $f(x, y) = -g(x, y) + p(y) + h(x, y)$ with $f(a_2, f_2 e) \geq 0$, then we again get (3.8.9) (some α_k may be negative); and if $g(x, y)$ has a monomial which ends in x, the calculation from (3.8.9) through (3.8.16) is still valid.

Now suppose that $ea \geq a$ and $eb \geq b$; then $0 \leq bae \leq ebe^{2m}ae = 0$. If $t \in F^+$, then $e + t$ may replace e; so $ba(e + t) = 0$ and hence $bat = 0$. Consequently $ba = 0$ when $\ell(F(R); F(R)) = 0$. □

Recall from Exercise 3.3.17 that the submodule A of the C-module B is closed if B/A is torsion-free; and the closure of A in B is the submodule $\hat{A} = \{b \in B : \alpha b \in A$ for some $0 \neq \alpha \in C\}$. When C is ℓ-simple each convex ℓ-submodule of a C-f-module is closed.

Theorem 3.8.6. *Suppose that the f-subalgebra $F(R)$ contains a superunit of R. In each of the following cases the closed ℓ-ideals of R generated by u^+u^- and u^-u^+ are identical, for each u in R.*

(a) R is an sp-ℓ-algebra.
(b) For some $k \geq 2$, $R^+ \times R^+$ is a right k-nice p-algebraic set.
(c) R is unital and p-positive.
(d) R is unital and $\{(u, v) \in R^2 : u \wedge v = 0\}$ is a right nice p-algebraic set.

Proof. First note that in each case the hypotheses are preserved in every homomorphic image of R. Also, each ℓ-ideal I of R is a C-submodule since if $a \in I^+$ and $e \in R$ with $a \leq ea$, then $0 \leq \alpha a \leq (\alpha e)a \in I$ for every $\alpha \in C^+$. In (d) note that $1 \in R^+$ since $0 \leq 1^- \leq e1^- = e^- = 0$ if e is a left f-superunit of R. Suppose that $u \in R$ and let I be the closed ℓ-ideal of R generated by u^+u^-. By Theorems 3.8.3 and 3.8.4 we have that $N_2(R/I)^+ \subseteq F(R/I)$, and by Theorem 3.8.5 we obtain that $u^-u^+ \in I$. Similarly, u^+u^- is in the closed ℓ-ideal generated by u^-u^+. □

We turn our attention next to the subvariety of sp-ℓ-rings consisting of the almost f-rings. An immediate consequence of Theorem 3.2.24 is that an ℓ-prime almost f-ring is a totally ordered domain, and hence an almost f-ring modulo its lower ℓ-nil radical is an f-ring. This fact will be strengthened later when it is shown that an almost f-ring modulo a nilpotent ℓ-ideal is an f-ring. In a unital almost f-ring, the identity element is a weak order unit; that is, $1^\perp = 0$. It is surprising that the converse holds in an ℓ-unital ℓ-ring R, and, in fact, we can relax the assumption that R is unital. Before we present this result in Theorems 3.8.9 and 3.8.10, we will again examine nilpotent elements and investigate their role in showing that the product of disjoint elements is zero.

Theorem 3.8.7. *Suppose that the convex ℓ-subring S of the ℓ-ring R contains a positive weak order unit e of R. Assume that either $e \in F_r(S)$ and e is a left superunit in S, or $ex \geq 0$ implies that $x \geq 0$ for $x \in S$, and for every $x, y \in S^+$, $xy \leq x^2 + y^2$. Then $a \leq e$ if $a \in R^+$ and $a \wedge e$ is nilpotent.*

Proof. For the first case suppose that $a \wedge 2e \not\leq e$. Then, since S_G is a right f-module over $G = F_r(S)$, there is a nonzero totally ordered homomorphic image $\overline{S}_G$ of S_G in which $\overline{a \wedge 2e} > \overline{e}$. So $\overline{a \wedge e} = \overline{a \wedge 2e \wedge e} = \overline{a \wedge 2e} \wedge \overline{e} = \overline{e} \neq 0$. Since $a \wedge e$ is nilpotent, there is a smallest integer n with $\overline{(a \wedge e)^n} = 0$ and $n \geq 2$. But then $0 = \overline{(a \wedge e)^n} = \overline{a \wedge e}\overline{(a \wedge e)^{n-1}} = \overline{e}\overline{(a \wedge e)^{n-1}} = \overline{e(a \wedge e)^{n-1}} \geq \overline{(a \wedge e)^{n-1}} \geq 0$ contradicts the minimality of n. Thus $a \wedge 2e \leq e$, and so $(a-e)^+ \wedge e = 0$; whence $a \leq e$.

In the second case suppose that $(a \wedge e)^n = 0$ and $2 \leq n \leq 2^m$. Let $b = ma \wedge (m+1)e$. Since $0 \leq b \leq (m+1)(a \wedge e)$ we have that $e^{2^{m-1}} b^{2^{m-1}} \leq e^{2^m}$ and therefore $b^{2^{m-1}} \leq e^{2^{m-1}}$. Again, $e^{2^{m-2}} b^{2^{m-2}} \leq e^{2^{m-1}} + b^{2^{m-1}} \leq 2e^{2^{m-1}}$ and therefore $b^{2^{m-2}} \leq 2e^{2^{m-2}}$. Continuing we eventually get that $ma \wedge (m+1)e \leq me$ and hence $m(a-e)^+ \wedge e = 0$. Thus, $a \leq e$. □

Theorem 3.8.8. *Suppose that the convex ℓ-subring S of R is an almost f-ring, $e \in S^+$ is a right f-element of R, and for any $c \in R^+$, if $(c \wedge e)^2 = 0$, then $c \in S$. Assume that $a \in R$ and $b \in S$ with $a \wedge b = 0$.*

(a) *If e is a left superunit in S, then $ab = 0$.*
(b) *If $S \subseteq F_\ell(R)$ and $r(e; S) = 0$, then $ba = 0$.*

Proof. Let $a_1 = a - a \wedge e$ and $e_1 = e - a \wedge e$. Since $(a \wedge e) \wedge b = 0$ in S and S satisfies $x^+x^- = 0$, $(a \wedge e)b = b(a \wedge e) = 0$. Since $a = a_1 + a \wedge e$, to show that $ab = 0$ or $ba = 0$ it suffices to show that $a_1 b = 0$ or $ba_1 = 0$, respectively. Since $a_1 \wedge e_1 = 0$ and e_1 is a right f-element we have $a_1e_1 \wedge e_1 = 0 = (a_1e_1 \wedge e) \wedge e_1$. But e_1, $a_1e_1 \wedge e \in S$ and hence $e_1(a_1e_1 \wedge e) = 0$. So $0 \leq (a_1e_1 \wedge e)^2 \leq a_1e_1(a_1e_1 \wedge e) = 0$ and hence $a_1e_1 \in S$. Now, $0 = a_1e_1b = a_1(e - a \wedge e)b = a_1eb$ since a_1e_1 and b are disjoint elements of S. But $0 \leq a_1b \leq a_1(eb) = 0$ if e is a left superunit of S. This completes the proof of (a). To complete the proof of (b) we first note that $eba_1 \in S$ since $(eba_1)^2 = 0$; hence eba_1 and e_1 are disjoint elements of S since eb is a left f-element of R. Thus, $e^2ba_1 = (e_1 + a \wedge e)eba_1 = e_1eba_1 = 0$ since $a \wedge e$ and eb are disjoint elements of S, and hence $ba_1 \in r(e^2; R) = r(e^2; S) = 0$. For, $e^2x = 0$ with x in R gives that $(e|x| \wedge e)^2 = 0$, $e|x| \in S$, $e|x| = 0$, $(e \wedge |x|)^2 = 0$, and $x \in S$. □

We can now show that the cyclic right ideal that is generated by a "good" weak order unit in an ℓ-ring is an almost f-ring, and if the weak order unit is "better", then the ℓ-ring itself is an almost f-ring.

Theorem 3.8.9. *Suppose that T is an almost f-ring that is a convex ℓ-subring of $F_\ell(R)$ and e is an f-element of R belonging to T with $r(e; R) = 0$. The following statements are equivalent.*

(a) *eR is an almost f-ring.*
(b) *e is a weak order unit of R.*
(c) *If $a \in R^+$ and $a \wedge e$ is nilpotent, then $a \in T$.*
(d) *If $a \in R^+$ and $(a \wedge e)^2 = 0$, then $a \in T$.*
(e) *eR is an sp-ℓ-ring, and if $a \in R^+$ with $(a \wedge e)^2 = 0$, then $(ea)^2 = 0$.*

Proof. (a) $\Rightarrow$ (b). From $e \wedge a = 0$ it follows that $e^2 \wedge ea = 0$, $e^3 a = 0$ and $a = 0$.
(b) $\Rightarrow$ (c). This is a consequence of Theorem 3.8.7.
(c) $\Rightarrow$ (d). This is trivial.
(d) $\Rightarrow$ (a). Suppose that $a, b \in R$ with $a \wedge b = 0$. Let $a_1 = a - a \wedge e$ and $e_1 = e - a \wedge e$. From (b) of Theorem 3.8.8 we have $e_1 a_1 = 0$, $(a_1 e_1)^2 = 0$, $a_1 e_1 \in T$ and $a_1 e_1 b = 0$. Also,

$$aeb = (a_1 + a \wedge e)eb = a_1 eb = a_1(e_1 + a \wedge e)b = a_1 e_1 b = 0.$$

In particular, if $ea \wedge eb = 0$, then $a \wedge b = 0$ and $eaeb = 0$.
(a) $\Rightarrow$ (e). Suppose that $a \in R^+$ with $(a \wedge e)^2 = 0$. Then $a \leq e$ by Theorem 3.8.7 and $a^2 = 0 = (ea)^2$ by Theorem 3.3.4.
(e) $\Rightarrow$ (d). If $(a \wedge e)^2 = 0$, then $aea = 0$ and $0 \leq (a-e)e(a-e) = -ae^2 - e^2 a + e^3$; so $e^2 a \leq e^3$ and $a \leq e$. □

Theorem 3.8.10. *Let T be an almost f-ring that is a convex ℓ-subring of the ℓ-ring R, and suppose that e is an f-element of R belonging to T. Assume that either T is a right f-ring and e is a left superunit of R, or $T \subseteq F_\ell(R)$ and $r(e;R) = \ell(e;R) = 0$. Then the following statements are equivalent.*

(a) R is an almost f-ring.
(b) e is a weak order unit of R.
(c) If $a \in R^+$ and $a \wedge e$ is nilpotent, then $a \in T$.
(d) If $a \in R^+$ and $(a \wedge e)^2 = 0$, then $a \in T$.
(e) $N_2(R)^+ \subseteq T$, and R satisfies

$$\text{if } a \in R^+ \text{ with } a \wedge e \in N_2(R), \text{ then } a \in N_2(R). \tag{3.8.17}$$

(f) R is an sp-ℓ-ring that satisfies (3.8.17).

If $e = 1$ and T is a C-subalgebra of R, then each of these statements is equivalent to the following statement.

(g) R satisfies (3.8.17) and for each $a \in N_2(R)^+$ there are polynomials $p(x)$, $q(x) \in C[x]$ with $p(a+1) \geq 0$, $q(a-1) \geq 0$, and $p'(1)q'(-1) < 0$.

Proof. (a) $\Rightarrow$ (b). If $a \wedge e = 0$, then $ea = 0$ implies that $a = 0$ since $r(e;R) = 0$.
(b) $\Rightarrow$ (c). This is Theorem 3.8.7.
(c) $\Rightarrow$ (d). This is trivial.
(d) $\Rightarrow$ (a). Suppose that $a \wedge b = 0$. Let

$$a_1 = a - a \wedge e\ ,\ b_1 = b - b \wedge e$$
$$e_1 = e - a \wedge e\ ,\ e_2 = e - b \wedge e.$$

Under the assumption that e is a left superunit we have, by (a) of Theorem 3.8.8,

$$0 \leq ba = b(a_1 + a \wedge e) = ba_1 \leq bea_1 = be_1 a_1 = 0.$$

The last equality arises from the fact that $e_1a_1 \in b^{\perp_T}$ since $a_1e_1 = 0$ and consequently $(e_1a_1)^2 = 0$.

Now assume that $T \subseteq F_\ell(R)$ and $r(e;R) = \ell(e;R) = 0$. Note that since e is a weak order unit of T it follows from Theorem 3.8.7 that $a \leq e$ whenever $a \in R^+$ with $(a \wedge e)^2 = 0$. Consequently, by letting the ℓ-ring S in Theorm 3.8.8 (b) be the convex ℓ-subring generated by e we can conclude that the product of two positive disjoint elements in R is 0 provided that one of them belongs to S. In particular, we have that $e_1a_1 = b_1e_2 = 0$ and $ab = a_1b = ab_1$. From Theorem 3.8.9 we know that $bea = aeb = 0$. Thus, $(eab)^2 = 0$, $eab \leq e$, $e^2eab \leq e^4 + (eab)^2 = e^4$ and $ab \leq e$. So,

$$0 \leq eab^2 = (e_1 + a \wedge e)a_1b^2 = (a^2b \wedge eab)b \leq a^2b^2 \leq aeb = 0$$

and $ab^2 = 0$. Hence,

$$abe = ab_1(e_2 + b \wedge e) = ab^2 \wedge abe = 0$$

and $ab = 0$.

(a) $\Rightarrow$ (e) and (f). This follows from (b) and an application of Theorem 3.8.7 with S being the convex ℓ-subring generated by e.

(f) $\Rightarrow$ (d). If $a \in R^+$ and $(a \wedge e)^2 = 0$, then $ea \leq e^2 + a^2 = e^2$; so $a \leq e$.

Since (e) $\Rightarrow$ (d) and (f) $\Rightarrow$ (g) are trivial and (g) $\Rightarrow$ (e) is a consequence of Theorem 3.8.1, the proof is complete. □

Even though the previous two results can be stated more simply by taking the ℓ-subring T to be $F(R)$, or even the smaller convex ℓ-subring generated by e, the apparently larger T that is used could make it easier to verify that (c), (d), or the second (e) is satisfied in a particular ℓ-ring. Other polynomial constraints that yield that $N_2(R)^+ \subseteq T$ can, of course, also be substituted for the identity $(x^2)^- = 0$ in (f). Some examples of almost f-rings with a good weak order unit are given in Exercises 13, 14, 15, and 16. In particular, there is a right f-ring with an identity element that is not an f-ring. Also, in contrast to Theorem 3.8.10, it is possible for $F(R)$ to contain a weak order unit of R without the identity $x^+x^- = 0$ lifting from $F(R)$ to R.

Theorem 3.8.10 can be used to show that an archimedean ℓ-algebra which satisfies a polynomial identity and which has a good superunit must be an f-algebra. The essential computations are given in the next result. Note that the free algebra $C[x,y]$ is an ℓ-algebra with positive cone $C^+[x,y]$, and the positive part, negative part and absolute value of $f(x,y)$ in this ℓ-algebra will be denoted by $f^+(x,y)$ $f^-(x,y)$ and $|f|(x,y)$, respectively.

Theorem 3.8.11. *Let R be a torsion-free ℓ-algebra over C, let G be a cofinal subset of C^+, and let $a, e \in R^+$.*

(a) Suppose that $p(x) = \alpha_0 + \alpha_1x + \cdots + \alpha_nx^n \in C[x]$ has degree $n \geq 2$, $\alpha_n > 0$ and $p(\alpha e - a) \geq 0$ for each $\alpha \in G$. Assume that $a \wedge e^n = 0$, $a \wedge e^{n-1} = 0$ if $\alpha_{n-1} > 0$, and for some $0 < \delta_1, \delta_2 \in C$,

$$\delta_1 a \leq \delta_2 \sum_{i+j=n-1} e^i a e^j. \tag{3.8.18}$$

Then there are $0 < \rho \in C$ and $q(x, y) \in C^+[x, y]$ with $\rho a \ll_C q(a, e)$.

(b) Suppose that $f(x, y) = -\rho m(x, y) + u(x, y) \in C[x, y]$ where $0 < \rho \in C$ and $m(x, y)$ is a monomial of degree $n \geq 1$ in x whose y-degree exceeds the y-degree of $u^+(x, y)$. Assume that $a \leq ea$ and $f(a, \alpha e) \geq 0$ for each $\alpha \in G$. Then there are $q(x, y) \in C^+[x, y]$ and $t \in \mathbb{Z}^+$ with $\rho a^n e^t \ll_C q(a, e)$.

Proof. (a) For each $k \geq 1$ the coefficient of α^k in $p(\alpha e - a)$ comes from $\alpha_k(\alpha e - a)^k + \alpha_{k+1}(\alpha e - a)^{k+1} + \cdots + \alpha_n(\alpha e - a)^n$ and is

$$\alpha_k e^k + \sum_{m \geq k+1} (-1)^{m-k} \alpha_m \sum_{\substack{i_1+\cdots+i_t=k \\ j_1+\cdots+j_t=m-k}} e^{i_1} a^{j_1} \cdots e^{i_t} a^{j_t} = \alpha_k e^k + \mathcal{O}_k(a, e).$$

So

$$0 \leq p(\alpha e - a) = p(-a) + \sum_{k=1}^{n} [\alpha_k e^k + \mathcal{O}_k(a, e)]\alpha^k$$

gives that

$$0 \leq p(-a) + [p(\alpha e) - p(0)] + \left[\sum_{k=1}^{n-2} \mathcal{O}_k(a, e)\alpha^k\right] - \alpha_n \alpha^{n-1} \sum_{i+j=n-1} e^i a e^j. \tag{3.8.19}$$

Thus, by bringing the last sum in (3.8.19) to the left side, and by multiplying (3.8.18) by $\alpha_n \alpha^{n-1}$, and by dropping the negative terms that remain on the right side of (3.8.19) we get

$$\begin{aligned} 0 &\leq \alpha_n \delta_1 \alpha^{n-1} a \\ &\leq \delta_2 \left[u^+(a) + \sum_{k=1}^{n-2} (\alpha_k^+ e^k + \mathcal{O}_k^+(a, e))\alpha^k + \alpha_{n-1}^+ \alpha^{n-1} e^{n-1} + \alpha_n^+ \alpha^n e^n\right], \end{aligned}$$

where $u(x) = p(-x)$. So, if $\alpha \geq 1$, then

$$\begin{aligned} 0 &\leq \alpha(\delta_1 \alpha_n a) \\ &\leq \delta_2 \left[u^+(a) + \sum_{k=1}^{n-2} (\alpha_k^+ e^k + \mathcal{O}_k^+(a, e)) + \alpha_{n-1}^+ \alpha e^{n-1} + \alpha_n \alpha^2 e^n\right], \end{aligned} \tag{3.8.20}$$

and since a is disjoint from the last two terms in (3.8.20) we obtain $0 \leq \alpha \rho a \leq q(a, e)$ with $\rho = \alpha_n \delta_1$ and

$$q(x, y) = \delta_2 \left[u^+(x) + \sum_{k=1}^{n-2} \alpha_k^+ y^k + \mathcal{O}_k^+(x, y)\right].$$

(b) Let $m(x, y) = x^{i_1} y^{j_1} x^{i_2} y^{j_2} \cdots x^{i_r} y^t$. Since $a^i \leq e^k a^i$ if $i \geq 1$ and $k \geq 0$, $a^n e^t \leq m(a, e)$. If $m(x, y)$ has degree s in y, then $f(a, \alpha e) \geq 0$ implies

$$0 \leq \rho\alpha^s a^n e^t \leq \rho m(a, \alpha e) \leq u(a, \alpha e) \leq u^+(a, \alpha e).$$

If $\alpha \geq 1$, then

$$\rho\alpha a^n e^t \leq \alpha^{1-s} u^+(a, \alpha e) \leq u^+(a, e) = q(a, e),$$

since $s >$ degree of y in $u^+(x, y)$; so $q(x, y) = u^+(x, y)$. □

Let us call a nice polynomial $f(x, y) = -g(x, y) + p(y) + h(x, y)$ *y-special* if the y-degree of $g(x, y)$ exceeds the y-degree of $h^+(x, y)$.

Theorem 3.8.12. *Let R be a torsion-free C-archimedean ℓ-algebra, and suppose that R has a left superunit which is an f-element. The following statements are equivalent.*

(a) R is an f-algebra.
(b) R satisfies the identity $p(x)^- = 0$ for some $p(x) \in C[x] \backslash C$.
(c) R satisfies the identity $f(x, y)^- = 0$ where $f(x, y) \in C[x, y]$ is a y-special right k-nice polynomial with $k \geq 2$.

Proof. Since (a) implies (b) and (c) it suffices to show that the converse implications are valid. Let e be an f-element of R which is a left superunit of R.

(b) ⇒ (a). As a consequence of Theorem 3.8.11 (a) we have that e is a weak order unit of R, and hence R is an almost f-algebra by Theorem 3.8.10. By Exercise 3.7.5 each nilpotent element of R is contained in $F(R)$, and by Theorem 3.8.7 (or Exercise 3.7.5), if a is nilpotent, then $a \ll_C e$. Thus, R is reduced and hence it is an f-algebra.

(c) ⇒ (a). Again, we only have to show that e is a weak order unit of R. If $a \wedge e = 0$ and $\alpha \in C^+$, then

$$0 \leq g(a, \alpha e) \leq p(\alpha e) + h(a, \alpha e) \leq |p|(\alpha e) + h^+(a, \alpha e).$$

Since $g(a, \alpha e) \wedge |p|(\alpha e) = 0$ we have that $g(a, \alpha e) \leq h^+(a, \alpha e)$ and so $ae^t = 0$ for some $t \in \mathbb{Z}^+$ by Theorem 3.8.11(b). But then $0 \leq a^2 \leq ae^t a = 0$. By Theorem 3.8.2, $e^N a \in F(R)$ for some integer N and hence $a \in F(R)$. But $a \in F(R)^\perp$ since if $b \in F(R)^+$, then $a \wedge b \leq a \wedge eb = 0$. So $a = 0$ and e is a weak order unit. □

We will now show, as previously promised, that both a d-ring and an almost f-ring share the property of being an f-ring modulo a nilpotent ℓ-ideal. This is, of course, a stronger property than that of being an extended f-ring; see Theorems 3.3.4 and 3.3.5 and Exercises 3.3.9 through 3.3.13.

Theorem 3.8.13. *Let R be an ℓ-ring that is an almost f-ring or satisfies one of the identities: $y^+|x|y^+ = |y^+xy^+|$, $y^+x^+y^+x^-y^+ = 0$. Then R has a nilpotent ℓ-ideal A such that R/A is an f-ring, and $N(R)$ is a sum of nilpotent ℓ-ideals of R.*

Proof. The last statement is a consequence of the equations $\ell\text{-}N(R)/A = \ell\text{-}N(R/A) = N(R/A) = N(R)/A$ where, as usual, $\ell\text{-}N(R)$ denotes the sum of the nilpotent ℓ-ideals of R. In any ℓ-ring the middle annihilator $A = \{x \in R\colon\ RxR = 0\}$ is an ideal whose cube is 0 and the middle ℓ-annihilator $B = \{x \in R\colon\ R|x|R = 0\}$ is the largest ℓ-ideal of R contained in A. Now, R satisfies the identity $y|x|y = |yxy|$ for $y \in R^+$ and $x \in R$ iff the maps $x \mapsto yxy$ are ℓ-homomorphisms of the additive ℓ-group of R; and in this case $A = B$. So if $a \wedge b = 0$ in such an ℓ-ring and $c, d \in R^+$, then

$$0 \le d(ca \wedge b)d = dcad \wedge dbd \le (dc \vee d)(a \wedge b)(dc \vee d) = 0.$$

Thus, $ca \wedge b \in A$ and, similarly, $ac \wedge b \in A$; so R/A is an f-ring.

Suppose that R satisfies the identity $y^+x^+y^+x^-y^+ = 0$. Then $Rx^+Rx^-R = 0$ for each $x \in R$, and

$$(ba)^2b = b(a^+ - a^-)b(a^+ - a^-)b = (ba^+)^2b + (ba^-)^2b \ge 0$$

if $a \in R$ and $b \in R^+$. Let

$$A = \{a \in R\colon\ (b|a|)^2b = 0,\ \forall b \in R^+\}. \tag{3.8.21}$$

If $a_1, a_2 \in A$ and $b \in R^+$, then

$$0 \le (b|a_1 + a_2|)^2b \le (b|a_1| + b|a_2|)^2b = b|a_1|b|a_2|b + b|a_2|b|a_1|b.$$

But

$$0 \le (b(|a_1| - |a_2|))^2b = -(b|a_1|b|a_2|b + b|a_2|b|a_1|b) \le 0,$$

and hence A is a convex ℓ-subgroup of R. Also, if $r \in R$, then $(b|a_1r|)^2b \le b|a_1||r|b|a_1||r|b = 0$ since $R|a_1|R|a_1|R = 0$; similarly, $ra_1 \in A$. It follows that A is a nilpotent ℓ-ideal of R whose index of nilpotency is at most five. If $c \wedge d = 0$ and $u \in R^+$, then $(b(uc \wedge d))^2b \le bucbdb = 0$ and $(b(cu \wedge d))^2b \le bcubdb = 0$; consequently, R/A is an f-ring.

Now assume that R is an almost f-ring. Then $N(R)$ is an ℓ-ideal of R and, as in the previous paragraph, the set A that is defined in (3.8.21) is a nilpotent ℓ-ideal of R. Let B/A be the ℓ-ideal of R/A defined by (3.8.21). Note that $a \in B^+$ iff $y(xaxax)y(xaxax)y = 0$ for every $x, y \in R^+$. If $a, x, y \in R^+$ with $a^2 = 0$, then $axaya = 0$ since $c = xay$ is nilpotent and $aca \le (ac)^2 + a^2 = (aca)c \le (aca)c^2 \le \cdots$; so $a \in B$. In particular, if $c \wedge d = 0$ and $u \in R^+$, then $uc \wedge d$ and $cu \wedge d \in B$. Thus, B is the desired nilpotent ℓ-ideal of R. □

When R is an archimedean almost f-ring a stronger version of Theorem 3.8.13 is a consequence of Theorems 3.2.24 and 3.6.2 since $N(R)$ is an ℓ-ideal whose cube is 0 and $R/N(R)$ is an archimedean f-ring. Thus, R is close to being commutative. In fact, it is commutative as we will now show. Just as in Theorem 3.6.1 associativity is not used in the proof and this allows for a more general statement of this commutative result.

Let A and D be abelian groups and let $n \in N$. Recall that the mapping $\varphi : A^n \longrightarrow D$ is n-multilinear if it is additive in each variable, and it is symmetric if $\varphi(a_1, \ldots, a_n) = \varphi(a_{\sigma(1)}, \ldots, a_{\sigma(n)})$ for each permutation σ of $\{1, \ldots, n\}$. If A and

D are ℓ-groups φ is called *positive* if $\varphi((A^+)^n) \subseteq D^+$, and φ is an *almost f-map*, abbreviated as *af-map*, when it satisfies the following condition:

$$\text{if}\, a_i \wedge a_j = 0 \text{ for some } i \text{ and } j, \text{ then } \varphi(a_1, \ldots, a_n) = 0. \tag{3.8.22}$$

Theorem 3.8.14. *Let A and D be abelian ℓ-groups with D archimedean. If $\varphi : A^2 \longrightarrow D$ is a bilinear positive af-map, then φ is symmetric. In particular, a not necessarily associative almost f-ring is commutative.*

Proof. By replacing A and D by their divisible closures we may assume that A and D are vector lattices over $\mathbb{Q}$. For $a, b \in A$ we will write $ab = \varphi(a, b)$, and, as usual, the commutator of a and b is written as $ab - ba = [a, b]$. Since the elements $a - a \wedge b$ and $b - a \wedge b$ are disjoint we have that $(a - a \wedge b)(b - a \wedge b) = (b - a \wedge b)(a - a \wedge b) = 0$. Consequently,

$$\begin{aligned} ab &= (a - a \wedge b + a \wedge b)(b - a \wedge b + a \wedge b) \\ &= (a - a \wedge b)(a \wedge b) + (a \wedge b)(b - a \wedge b) + (a \wedge b)^2 \end{aligned}$$

and

$$ba = (b - a \wedge b)(a \wedge b) + (a \wedge b)(a - a \wedge b) + (a \wedge b)^2.$$

Thus,

$$\begin{aligned} [a, b] = [a - a \wedge b, a \wedge b] + [a \wedge b, b - a \wedge b] = \\ [a \wedge b, b - a \wedge b] - [a \wedge b, a - a \wedge b], \end{aligned} \tag{3.8.23}$$

and we must show that commutators of the latter type vanish for positive a and b. Since the argument is rather complicated we will proceed by a sequence of steps. Let $a, b \in A^+$ and $0 < p \in \mathbb{Q}$. We first show that

$$a \wedge b - a \wedge b \wedge p^{-1}(a - a \wedge b) \in (a - a \wedge (p+1)b)^{\perp}. \tag{3.8.24}$$

For,

$$\begin{aligned} 0 \le a \wedge b - (a \wedge b \wedge p^{-1}(a - a \wedge b)) &= p^{-1}[p(a \wedge b) - (p(a \wedge b) \wedge (a - a \wedge b))] \\ &= p^{-1}[p(a \wedge b) - (((p+1)(a \wedge b) \wedge a) - a \wedge b)] \\ &= p^{-1}((p+1)(a \wedge b) - (p+1)(a \wedge b) \wedge a) \\ &= p^{-1}((p+1)(a \wedge b) - a \wedge (p+1)b) \\ &\le p^{-1}((p+1)b - a \wedge (p+1)b) \end{aligned}$$

and $(p+1)b - a \wedge (p+1)b \in (a - a \wedge (p+1)b)^{\perp}$.

Now, the inequality

$$a - a \wedge b - (a - a \wedge (p+1)b) = a \wedge (p+1)b - a \wedge b$$

$$\leq a\wedge pb + a\wedge b - a\wedge b \leq pb$$

gives that

$$0 \leq ((a - a\wedge b) - (a - a\wedge(p+1)b))(a\wedge b) \leq pb(a\wedge b) \leq pb^2. \tag{3.8.25}$$

From (3.8.24) we get that

$$(a - a\wedge(p+1)b)(a\wedge b) = (a - a\wedge(p+1)b)(a\wedge b\wedge p^{-1}(a - a\wedge b)),$$

and substituting this in (3.8.25) gives

$$0 \leq (a - a\wedge b)(a\wedge b) - ((a - a\wedge(p+1)b)(a\wedge b\wedge p^{-1}(a - a\wedge b))) \leq pb^2. \tag{3.8.26}$$

The inequality

$$0 \leq (a\wedge b)(a - a\wedge b) - ((a\wedge b\wedge p^{-1}(a - a\wedge b))(a - a\wedge(p+1)b)) \leq pb^2 \tag{3.8.27}$$

is obtained in the same way.

We will use (3.8.23), (3.8.26) and (3.8.27) to approximate the commutator $[a, b]$, and it is desirable to single out certain expressions that appear in (3.8.26) and (3.8.27). We define the functions f_0 and f_1 with domain $A^+ \times A^+ \times (\mathbb{Q}^+ \backslash \{0\})$ and values in A^+, by

$$f_0(a, b, p) = a\wedge b\wedge p^{-1}(a - a\wedge b),\ f_1(a, b, p) = a - a\wedge(p+1)b.$$

Note that $0 \leq f_0(a, b, p) \leq a\wedge b$, $0 \leq f_1(a, b, p) \leq a - a\wedge b$ and

$$0 \leq f_0(a, b, p) + f_1(a, b, p) \leq a. \tag{3.8.28}$$

Using these functions the inequalities (3.8.26) and (3.8.27) become

$$0 \leq (a - a\wedge b)(a\wedge b) - f_1(a, b, p)f_0(a, b, p) \leq pb^2 \tag{3.8.29}$$

and

$$0 \leq (a\wedge b)(a - a\wedge b) - f_0(a, b, p)f_1(a, b, p) \leq pb^2, \tag{3.8.30}$$

and we will use these inequalities to show

$$\begin{aligned}|[a, b] + [f_0(a, b, p), f_1(a, b, p)] - [f_0(b, a, p), f_1(b, a, p)]| \\ \leq 2p(a^2 + b^2).\end{aligned} \tag{3.8.31}$$

For simplicity let $x = a - a\wedge b$, $y = a\wedge b$, $f_0 = f_0(a, b, p), f_1 = f_1(a, b, p)$, $g_0 = f_0(b, a, p)$, $g_1 = f_1(b, a, p)$ and $z = b - a\wedge b$. Then (3.8.29) and (3.8.30) become $0 \leq xy - f_1f_0,\ yx - f_0f_1 \leq pb^2$, and $[a, b] = [y, z] - [y, x]$ by (3.8.23). Now,

$$|[y, x] - [f_0, f_1]| = |(yx - f_0f_1) - (xy - f_1f_0)| \leq 2pb^2,$$

and, similarly, $|[y, z] - [g_0, g_1]| \leq 2pa^2$. So

$$|[a, b] + [f_0, f_1] - [g_0, g_1]| = |[y, z] - [y, x] + [f_0, f_1] - [g_0, g_1]|$$

$$= |[y, z] - [g_0, g_1] - ([y, x] - [f_0, f_1])| \leq 2p(a^2 + b^2),$$

and (3.8.31) has been established.

We will now recursively define new elements in A^+, two at a time, for the purpose of approximating the commutator $[a, b]$ with a sum of commutators of these elements, as in (3.8.31). Let $D = \{0, 1\} \subseteq \mathbb{Q}$. If $\varepsilon = (\varepsilon_1, \ldots, \varepsilon_k) \in D^k$ and $i \in D$, then by (ε, i) we mean the $k+1$ - tuple $(\varepsilon, i) = (\varepsilon_1, \ldots, \varepsilon_k, i) \in D^{k+1}$. Let $(p_n)_n$ be a sequence of strictly positive rational numbers. For $k \in \mathbb{N}$ and $\varepsilon \in D^k$ we define elements $a(\varepsilon) = a((p_n), a, b, \varepsilon)$ in A^+ by $a(0) = a, a(1) = b$ and for $\varepsilon \in D^k$ and $i \in D$,

$$a((\varepsilon, i)) = f_i(a(\varepsilon), a((\varepsilon_1, \ldots, \varepsilon_{k-1}, 1 - \varepsilon_k)), p_k). \tag{3.8.32}$$

The first property of these elements that we need is

$$a((\varepsilon_1, \ldots, \varepsilon_k)) \leq a((\varepsilon_1, \ldots, \varepsilon_r)) \text{ if } 1 \leq r \leq k. \tag{3.8.33}$$

For, if $2 \leq k$, then by (3.8.32) and (3.8.28) $a((\varepsilon_1, \ldots, \varepsilon_k)) \leq a((\varepsilon_1, \ldots, \varepsilon_{k-1}))$. The next property that we need to establish is that any two of the elments which are defined by (3.8.32) at the same level using different preceding sequences are disjoint. More specifically,

$$a((\varepsilon, i)) \wedge a((\varepsilon', j)) = 0 \text{ if } \varepsilon, \varepsilon' \in D^k \text{ with } \varepsilon \neq \varepsilon'. \tag{3.8.34}$$

To see this suppose that $\varepsilon = (\varepsilon_1, \ldots, \varepsilon_{r-1}, \varepsilon_r, \ldots, \varepsilon_k)$ and $\varepsilon' = (\varepsilon_1, \ldots, \varepsilon_{r-1}, \varepsilon'_r, \ldots, \varepsilon'_k)$ with $\varepsilon_r \neq \varepsilon'_r$. If $\overline{\varepsilon} = (\varepsilon_1, \ldots, \varepsilon_{r-1})$, then, by (3.8.33),

$$\begin{aligned} a((\varepsilon, i)) \wedge a((\varepsilon', j)) &= a((\varepsilon_1, \ldots, \varepsilon_{r-1}, \varepsilon_r, \ldots, \varepsilon_k, i)) \\ &\quad \wedge a((\varepsilon_1, \ldots, \varepsilon_{r-1}, \varepsilon'_r, \ldots, \varepsilon'_k, j)) \end{aligned} \tag{3.8.35}$$

$$\leq a((\varepsilon_1, \ldots, \varepsilon_{r-1}, \varepsilon_r, \varepsilon_{r+1})) \wedge a((\varepsilon_1, \ldots, \varepsilon_{r-1}, \varepsilon'_r, \varepsilon'_{r+1}))$$
$$= a((\overline{\varepsilon}, 0, \overline{i})) \wedge a((\overline{\varepsilon}, 1, \overline{j}))$$

for some $\overline{i}, \overline{j} \in D$. But if $c = a((\overline{\varepsilon}, 0))$ and $d = a((\overline{\varepsilon}, 1))$, then from the definitions of f_i we obtain

$$a((\overline{\varepsilon}, 0, 0)) = f_0(c, d, p_r) \leq p_r^{-1}(c - c \wedge d) = p_r^{-1}(c - d)^+,$$

$$a((\overline{\varepsilon}, 0, 1)) = f_1(c, d, p_r) \leq (c - d)^+,$$

$$a((\overline{\varepsilon}, 1, 0)) = f_0(d, c, p_r) \leq p_r^{-1}(c - d)^-,$$

and

$$a((\overline{\varepsilon}, 1, 1)) = f_1(d, c, p_r) \leq (c - d)^-.$$

So (3.8.34) is a consequence of (3.8.35) and the preceding inequalities.

Note that as a consequence of (3.8.33) we have that

$$a((\varepsilon, 1)) \leq a(\varepsilon_1) \leq a+b \quad \text{for any } \varepsilon \in D^k. \tag{3.8.36}$$

Our next goal is to establish the similar inequality

$$a((\varepsilon, 0)) \leq \frac{1}{k+1}(a+b) \quad \text{if } \varepsilon \in D^k \tag{3.8.37}$$

using induction on k. If $k = 1$, then the definitions of f_0 and $a((\varepsilon, i))$ give

$$a((0,0)), a((1,0)) \leq a \wedge b \leq \tfrac{1}{2}(a+b).$$

Assume that (3.8.37) holds for some $k \geq 1$ and all a, b and (p_n). Before we show that it also holds for $k+1$ we need to do some shifting. For $\varepsilon = (\varepsilon_1, \ldots, \varepsilon_{m+1}) \in D^{m+1}$ let $a^* = a((p_n),\ a, b, (\varepsilon_1, \varepsilon_2))$, $b^* = a((p_n),\ a, b, (\varepsilon_1, 1-\varepsilon_2))$ and $\varepsilon^* = (0, \varepsilon_3, \ldots, \varepsilon_m, \varepsilon_{m+1}) \in D^m$. We claim that

$$a((p_n), a, b, (\varepsilon, i)) = a^*((p_{n+1}), a^*, b^*, (\varepsilon^*, i)). \tag{3.8.38}$$

By $a^*((\varepsilon_1, \ldots, \varepsilon_m))$ we mean $a^*((p_{n+1}), a^*, b^*, (\varepsilon_1, \ldots, \varepsilon_m))$ for any $m \in \mathbb{N}$ and any $\varepsilon_i \in D$. For $m = 1$, we have that $\varepsilon = (\varepsilon_1, \varepsilon_2)$, $\varepsilon^* = (0)$ and

$$a((\varepsilon, i)) = f_i(a((\varepsilon_1, \varepsilon_2)), a((\varepsilon_1, 1-\varepsilon_2)), p_2) = f_i(a^*(0), a^*(1), p_2)$$

$$= a^*((\varepsilon^*, i)).$$

Suppose that (3.8.38) holds for some m and let $\varepsilon = (\varepsilon_1, \ldots, \varepsilon_{m+2}) \in D^{m+2}$ and set $\overline{\varepsilon} = (\varepsilon_1, \ldots, \varepsilon_{m+1})$. Then

$$a((\varepsilon, i)) = f_i(a((\overline{\varepsilon}, \varepsilon_{m+2})), a((\overline{\varepsilon}, 1-\varepsilon_{m+2})), p_{m+2})$$

$$= f_i(a^*((\overline{\varepsilon}^*, \varepsilon_{m+2})), a^*((\overline{\varepsilon}^*, 1-\varepsilon_{m+2})), p_{m+2})$$

$$= a^*((\varepsilon^*, i))$$

and (3.8.38) has been established. Now, if $\varepsilon \in D^{k+1}$, then $\varepsilon^* \in D^k$ and by (3.8.38) and (3.8.28) we have that

$$a((\varepsilon, 0)) = a^*((\varepsilon^*, 0)) \leq \frac{1}{k+1}(a^*(0) + a^*(1)) \tag{3.8.39}$$

$$= \tfrac{1}{k+1}(a((\varepsilon_1, \varepsilon_2)) + a((\varepsilon_1, 1-\varepsilon_2)) \leq \tfrac{1}{k+1} a(\varepsilon_1).$$

Also,

$$a((\varepsilon, 0)) = f_0(a(\varepsilon), a((\varepsilon_1, \ldots, \varepsilon_k, 1-\varepsilon_{k+1})), p_{k+1})$$

$$\leq a((\varepsilon_1, \ldots, \varepsilon_k, \varepsilon_{k+1})) \wedge a((\varepsilon_1, \ldots, \varepsilon_k, 1-\varepsilon_{k+1})) \leq a((\varepsilon_1, \ldots \varepsilon_k,), 0),$$

and continuing we get that $a((\varepsilon, 0)) \leq a((\varepsilon_1, 0)) \leq a(1-\varepsilon_1)$. But then $(k+2)a((\varepsilon, 0)) \leq a(\varepsilon_1) + a(1-\varepsilon_1) = a+b$ by (3.8.39), and the proof of (3.8.37) is complete.

For $\varepsilon = (\varepsilon_1, \ldots, \varepsilon_k) \in D^k$ let $|\varepsilon| = \varepsilon_1 + \cdots + \varepsilon_k$. For each $k \in \mathbb{N}$ let

$$C_k = \sum_{\varepsilon \in D^k} (-1)^{|\varepsilon|+k} [a((\varepsilon, 0), a((\varepsilon, 1))],$$

and let $C_0 = [a, b]$. We claim that for each $k \in \mathbb{Z}^+$

$$|C_k - C_{k+1}| \leq 2p_{k+1}(a^2 + b^2). \tag{3.8.40}$$

Since

$$\begin{aligned} C_1 &= -[a((0,0)), a((0,1))] + [a((1,0)), a((1,1))] \\ &= -[f_0(a, b, p_1), f_1(a, b, p_1)] + [f_0(b, a, p_1), f_1(b, a, p_1)] \end{aligned}$$

we have that $|C_0 - C_1| \leq 2p_1(a^2 + b^2)$ by (3.8.31); and this is (3.8.40) for $k = 0$.

For $k \geq 1$ we have that

$$\begin{aligned} C_{k+1} = &\sum_{\varepsilon \in D^k} (-1)^{|\varepsilon|+k+1} [a((\varepsilon, 0, 0)), a((\varepsilon, 0, 1))] \\ &+ \sum_{\varepsilon \in D^k} (-1)^{|\varepsilon|+k} [a((\varepsilon, 1, 0), a((\varepsilon, 1, 1))] \end{aligned}$$

and hence

$$\begin{aligned} C_k - C_{k+1} = &\sum_{\varepsilon \in D^k} (-1)^{|\varepsilon|+k} ([a((\varepsilon, 0)), a((\varepsilon, 1))] + [a((\varepsilon, 0, 0)), a((\varepsilon, 0, 1))] \\ &- [a((\varepsilon, 1, 0), a(\varepsilon, 1, 1)]). \end{aligned}$$

This gives that

$$\begin{aligned} |C_k - C_{k+1}| &\leq \sum_{\varepsilon \in D^k} 2p_{k+1}(a((\varepsilon, 0))^2 + a((\varepsilon, 1))^2) \\ &\leq 2p_{k+1} \sum_{\varepsilon \in D^k} (a((\varepsilon, 0)) + a((\varepsilon, 1)))^2 \\ &\leq 2p_{k+1} \sum_{\varepsilon \in D^k} a(\varepsilon)^2 = 2p_{k+1} \sum_{\varepsilon \in D^{k-1}} (a((\varepsilon, 0))^2 + a((\varepsilon, 1))^2), \end{aligned}$$

where the first inequality comes from (3.8.31) and the last inequality comes from (3.8.28). Repeated uses of (3.8.28), starting with this last sum, eventually gives (3.8.40).

Now let $0 < q \in \mathbb{Q}$ and choose the sequence (p_n) such that $\sum_{k=1}^{\infty} p_k < q/2$. It follows from (3.8.40) that

$$|[a, b] - C_n| = \left| \sum_{k=1}^{n} C_{k-1} - C_k \right| \leq \sum_{k=1}^{n} 2p_k(a^2 + b^2) < q(a^2 + b^2).$$

According to (3.8.34) and (3.8.36)

$$\sum_{\varepsilon \in D^k} a((\varepsilon, 1)) = \bigvee_{\varepsilon \in D^k} a((\varepsilon, 1)) \leq a+b,$$

and hence using (3.8.37) we obtain that

$$|C_k| \leq \sum_{\varepsilon \in D^k} a((\varepsilon, 0))a((\varepsilon, 1)) + a((\varepsilon, 1))a((\varepsilon, 0))$$

$$\leq \tfrac{1}{k+1} \sum_{\varepsilon \in D^k} (a+b)a((\varepsilon, 1)) + \tfrac{1}{k+1} \sum_{\varepsilon \in D^k} a((\varepsilon, 1))(a+b)$$

$$\leq \tfrac{2}{k+1}(a+b)^2.$$

Thus, for each $0 < q \in \mathbb{Q}$ and each $k \in \mathbb{N}$,

$$|[a, b]| \leq q(a^2 + b^2) + \tfrac{2}{k+1}(a+b)^2.$$

Letting $q = \frac{1}{k+1}$ we get that $(k+1)|[a, b]| \leq 3(a+b)^2$ and hence $[a, b] = 0$ since D is archimedean. □

Exercises.

1. Show that the following statements are equivalent for the unital torsion-free C-ℓ-algebra R.

   (a) The idempotents of R are contained in the interval $[0, 1]$ and are central.
   (b) Each idempotent e is p-algebraic and $[p(1) - p(0)] \cdot 1 > 0$ in R for some $p(x) \in P^*(e)$.
   (c) Each idempotent e is p-algebraic and $(p(1) - p(0))(q(1) - q(0)) \cdot 1 > 0$ in R for some $p(x) \in P^*(e)$ and some $q(x) \in P^*(1-e)$.
   (d) For each idempotent e in R there are polynomials $p(x)$, $q(x) \in C[x]x$ such that $p(e)^+p(e)^- = q(e)^-q(e)^+ = 0$ and $p(1)q(1) > 0$.

2. Suppose that the ℓ-algebra R in Exercise 1 is also a reduced ℓ-domain in which each zero divisor a is p-algebraic with $P^*(a) \cap C[x]x$ not empty. Let Q be the totally ordered field of quotients of C. Show that the following statements are equivalent.

   (a) R is an ℓ-unital domain.
   (b) If $a \in R$ and $0 < \alpha \in C$ with $a^2 = \alpha a$, then $[p(\alpha) - p(0)] \cdot 1 > 0$ in R for some $p(x) \in P^*(a)$.
   (c) Each idempotent in $R \otimes_C Q$, the C-divisible closure of R, is positive.
   (d) R is ℓ-unital and if a is a zero divisor of $R \otimes_C Q$, then there exists $p(x) \in P^*(a) \cap Q[x]x$ with $p(1) \neq 0$.

   (For (c) $\Rightarrow$ (a) show that if a is an algebraic element of $R \otimes_C Q$, then $Q[a]$ is a direct sum of fields and $a \in Q[a]a^2$.)

3. Let R be an ℓ-prime p-algebraic ℓ-unital torsion-free C-ℓ-algebra.

(a) Suppose that for each zero divisor a of R there exists $p(x) \in P^*(a)$ with zero constant term and $p(1) \neq 0$, and the group of units of R is weakly p-positive. Show that R is a domain provided that C is a field.

(b) Suppose that R is p-positive and $P^*(a) \cap C[x]x \neq \emptyset$ for each zero divisor a in R. Show that R is a domain.

(c) Let the ring $T = \mathbb{Q} \times \mathbb{Q}$ be given the positive cone $T^+ = \{(u, v) : 0 \leq v \leq u\}$. (see Exercise 3.1.24 (b)). Show that T is an ℓ-domain and $R = (2\mathbb{Z} \times 2\mathbb{Z}) + \mathbb{Z}(1, 1)$ is a weakly p-positive ℓ-subring of T that satisifies the conditions in (a) with $C = \mathbb{Z}$. (If $a = (u, v) \in T$, then $p(a) \geq 0$ or $p(a) \leq 0$ where $p(x) = vx - x^2$. If $u, v \in \mathbb{Z}$, then, for example, $p(x) = vx^2 - x^3 \in P(a)$ if $u < 0$ and $v \geq 2$, and $p(x) = v^3x - x^4 \in P(a)$ if $u > 0$ and $v > u$; find polynomials for the other cases.)

4. Assume that the k-nice polynomials $f(x, y) = -g(x, y) + p(y) + q(x^k, y)$ in Theorem 3.8.2 satisfy the conditions: the y-degree of each monomial of $g(x, y)$ which ends in x is bounded by M_1 and the x-degree of $q(x, y)$ is bounded by M_2. Show that the integer N in Theorem 3.8.2 can be chosen to satisfy $N \leq M_1(M_2^n + M_2^{n-1} + \cdots + 1)$.

5. (a) Suppose that $2 \leq n \in \mathbb{N}$ and the pops Δ satisfies the following condition.

$$\begin{aligned} &\text{If } \alpha_1, \ldots, \alpha_n \in \Delta \text{ are not all equal} \\ &\text{and } \alpha_1 + \cdots + \alpha_n \in \Delta, \text{ then } \alpha_1 + \cdots + \alpha_n < n\alpha_i \\ &\text{for some } i,\ 1 \leq i \leq n. \end{aligned} \tag{3.8.41}$$

Let A be a domain and a po-ring which satisfies the inequality $a^n \geq 0$. If $u \in W(\Delta, A)$ and $u^n \in W(\Delta, A)$ show that $u^n \geq 0$. (Imitate part of the proof of Theorem 3.7.6.)

(b) Show that each sp-pops satisfies (3.8.41).

6. Let Δ be the free monoid generated by the set Y. If $y_1, \ldots, y_p \in Y$, then the element $s = y_1 \cdots y_p \in \Delta$ has length p : $\ell(s) = p$. Given a total order of Y define the relation $<$ in Δ by: $s < t$ if either (i) $1 \leq \ell(s) < \ell(t)$, or (ii) for some $m \geq 0$, $s = y_1 \cdots y_{m+1} \cdots y_p$, $t = y_1 \cdots y_m z_{m+1} \cdots z_p$, $y_i, z_i \in Y$, $p \geq 2$ and $y_{m+1} < z_{m+1}$.

(a) Show that Δ is a rooted sp-pops.

(b) If Y has at least two elements and A is a unital totally ordered domain show that $\Sigma(\Delta, A)$ is an sp-ℓ-ring in which disjoint elements do not commute.

(c) Let e denote the identity element of Δ. Suppose that the partial order of Δ is strengthened by adding: (iii) $e < t$ if $t \in \Delta$ and $\ell(t) \geq 2$. Show that (a) and (b) still hold.

(d) Suppose that the partial order of Δ is changed in the following way. Let $n \in \mathbb{N}$. In (i) and (iii) we require that $\ell(t) \geq 2n$ while in (ii) we require that $p \geq 2n$. Let Δ_n denote the monoid Δ with this relation. Show that Δ_n is a rooted pops that satisfies the condition (3.8.41) given in Exercise 5; hence

$W(\Delta_n, A)$ is an ℓ-ring that satisfies $x^{2n} \geq 0$ but not $x^m \geq 0$ if $m < 2n$, and disjoint elements in $W(\Delta_n, A)$ do not commute.

7. (a) Let σ be an order preserving automorphism of the unital totally ordered domain A, and let $A[y; \sigma]$ be the twisted polynomial ring determined by σ (see Exercises 3.1.14 and 3.5.13). Define $p(y) = a_0 + a_1 y + \cdots + a_n y^n \geq 0$ if $n \geq 2$ and $a_n > 0$ or $p(x) = a_0 + a_1 y$ and $a_0 \geq 0$ and $a_1 \geq 0$. Show that this defines a positive cone for $A[y; \sigma]$ which makes it into an sp-ℓ-ring, and if $\sigma \neq 1$, then there are disjoint elements in $A[y; \sigma]$ which do not commute.
   (b) Let $n \in \mathbb{N}$ and modify the positive cone given in (a) so that $A[y; \sigma]$ satisfies $x^{2n} \geq 0$ but not $x^m \geq 0$ if $m < 2n$ (see Exercise 6). Show that $A[y; \sigma]$ still has disjoint elements that do not commute.

8. Suppose that R is an ℓ-unital torsion-free C-ℓ-algebra. Assume that for each $u \in R$ there is a polynomial $p(x) \in C^+[x] \backslash C$ with $p(u)^+ p(u)^- = 0$. Show that $p(u)^- p(u)^+ = 0$ for each $u \in R$ iff $p(u)^2 \geq 0$ for each $u \in R$.

9. Let G be a (multiplicative) po-group and set $K = \{g \in G : g$ has infinite order$\} \cup \{1\}$.
   (a) Show that $K = G^+ \cup (G^+)^{-1}$ iff for each $a \in G$ with $a \| 1$ there is a polynomial $f(x, y) \in C[x, y]$ with $f(a, 1) \geq 0$ in the group algebra $\Sigma(G, C)$ and $f(x, 1)'(0) < 0$.
   (b) Suppose G has an element of infinite order and either G is rooted or K is a subgroup of G. Show that the following statements are equivalent.
      (i) G is totally ordered.
      (ii) There is a polynomial $p(x) \in C^+[x] \backslash C$ such that $p(u) \geq 0$ for each u in $\Sigma(G, C)$.
      (iii) $\Sigma(G, C)$ is weakly p-positive.
      (iv) If u and v are two incomparable elements of $\Sigma(G, C)$, then there exists $f(x, y) \in P((u, v))$ with $f(x, 1)'(0) < 0$.
      (v) If g and h are incomparable elements of G, then there exists $f(x, y) \in P((g, h))$ with $f(x, 1)'(0) < 0$.

10. Give an example of an ℓ-field which is not totally ordered but in which $1^\perp = 0$.

11. Let S be an ℓ-subring of the ℓ-ring R. Suppose that S is a right f-ring and $F(S)$ has an element e with $r(e; S) = 0$. Let $n \in \mathbb{N}$; show that the following statements are equivalent.
   (a) $N_n(R)^+ \subseteq S$ and if $a \in R^+$ with $a \wedge |e| \in N_n(R)$, then $a \in N_n(R)$.
   (b) If $a \in R^+$ and $a \wedge |e| \in N_n(R)$, then $a \in S$.
   (c) $[0, |e|] \cap N_n(R) \subseteq S$ and if $a \in R^+$ with $a \wedge |e| \in N_n(R)$, then $a \leq |e|$.

   (For (b) $\Rightarrow$ (c) use Exercise 2.5.31 with $G = \Lambda = S$ and $\Omega = \{|e|\}$.)

12. Show that each of the following statements is equivalent to each statement in Theorem 3.8.9 (or Theorem 3.8.10).
   (c$'$) If $a \in R^+$ and $ea \wedge e$ is nilpotent, then $a \in T$.

(d′) If $a \in R^+$ and $(ea \wedge e)^2 = 0$, then $a \in T$.
(c″) If $a \in R^+$ and $ea \wedge e$ is nilpotent, then $ea \in T$.
(d″) If $a \in R^+$ and $(ea \wedge e)^2 = 0$, then $ea \in T$.

13. Here is an example of a generalized semigroup ℓ-ring $\Sigma(\Delta, A)$ that is a unital right f-ring but it is not a left f-ring. Let Δ be the multiplicative partial monoid that is generated by α, β, and γ and which has the relations

$$\begin{aligned}&\gamma\alpha = \gamma\beta,\\ &\gamma^0 = e,\\ &\rho\sigma \text{ is defined if and only if } \rho \text{ or } \sigma \in \{\gamma^n : n \geq 0\}.\end{aligned}$$

So

$$\Delta = \{\gamma^n : n \in \mathbb{Z}^+\} \cup \{\gamma^n\alpha\gamma^m : n, m \in \mathbb{Z}^+\} \cup \{\beta\gamma^m : m \in \mathbb{Z}^+\}.$$

The partial order on Δ is given by (see the diagram below): for all $n, m \in \mathbb{Z}^+$,

$$\begin{aligned}&e < \gamma < \gamma^2 < \cdots\\ &\alpha\gamma^n \| \beta\gamma^m\\ &\beta\gamma^n < \gamma\alpha\\ &\beta\gamma^n < \beta\gamma^m && \text{if } n < m\\ &\gamma^n\alpha\gamma^m < e\\ &\gamma^n\alpha\gamma^m < \gamma^p\alpha\gamma^q && \text{if } n < p, \text{ or } n = p \text{ and } m < q.\end{aligned}$$

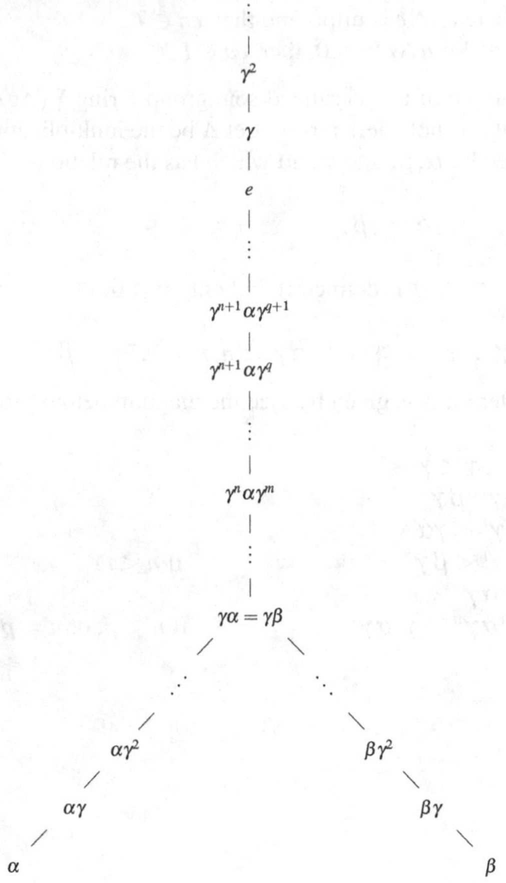

Let A be a unital totally ordered domain.

(a) Show that Δ is a right f-pops and that $\Sigma(\Delta, A) = W(\Delta, A) = V(\Delta, A)$ is a unital right f-ring that is not a left f-ring (see condition (3.5.12) and Exercise 3.5.17).

(b) Show that $F(\Sigma(\Delta, A)) = C_A(1)$.

(c) Let ${}_R J_R$ be a unital ℓ-bimodule over the unital totally ordered domain R, and give the ring

$$T = \left\{ \begin{pmatrix} r & j \\ 0 & r \end{pmatrix} : r \in R \text{ and } j \in J \right\}$$

the positive cone

$$T^+ = \left\{ \begin{pmatrix} r & j \\ 0 & r \end{pmatrix} : r > 0, \text{ or } r = 0 \text{ and } j \geq 0 \right\}.$$

Show that T is an almost f-ring, and T is a right (left) f-ring if and only if J is a right (left) f-module over R.

(d) Let $R = A[\gamma]$ be the lexicographically ordered polynomial ring in the indeterminate γ over A : $a_0 + a_1\gamma + \cdots + a_n\gamma^n > 0$ if $a_n > 0$. Let J be the R-R-bimodule described as follows: J_R is free with the basis $\{b\} \cup \{c_n : n \in \mathbb{Z}^+\}$; $ab = ba$ and $ac_n = c_n a$ for each $a \in A$ and $n \in \mathbb{Z}^+$; $\gamma b = c_1$ and $\gamma c_n = c_{n+1}$ for each n. Suppose that the positive cone of J is defined by

$$J^+ = \{bp(\gamma) + \sum_{k=0}^{n} c_k p_k(\gamma) : n \geq 1 \text{ and } p_n(\gamma) > 0,$$
$$\text{or } n = 0 \text{ and } p(\gamma), p_0(\gamma) \in R^+\}.$$

Show that the ℓ-ring T given in (c) using this R and J is isomorphic to $\Sigma(\Delta, A)$.

(e) Assume now and in (f) and (g) that A is ℓ-simple. Show that J is the only proper nonzero left ℓ-ideal of $\Sigma(\Delta, A)$.

(f) Determine the right ℓ-ideals of $\Sigma(\Delta, A)$ and show that, except for J, each is a finitely generated right ideal.

(g) Show that $N_2(\Sigma) = J = \mathscr{I}(\Sigma)$ and if A is a division ring, then J is the Jacobson radical of $\Sigma = \Sigma(\Delta, A)$.

14. Let $R = \Sigma(\Delta, A)$ where A is a unital totally ordered domain and $\Delta = \{\alpha, \beta, \gamma, \rho\}$ is the rooted pops with partial addition defined by $2\gamma = \gamma$, $\alpha + \gamma = \rho$, $\rho + \gamma = \rho$, and partial order given by $\rho, \beta < \gamma$. Let $e = x^\alpha + x^\beta + x^\rho$. Show that R is an sp-ℓ-ring that is a left f-ring but not an almost f-ring, $e \in F(R) = N(R)$, and R satisfies (a), (b) and (e) of Theorem 3.8.9 but not (d).

15. Let A be a unital totally ordered domain and let $T = A[x, y]$ be the free A-ring in two variables, totally ordered so that $x > 1$ and $y > 1$. Let R be the ideal of polynomials in T with zero constant term and with the positive cone $xT^+ + yT^+$. Show that R is a right f-domain which has a superunit which is a weak order unit, $F(R) = 0$, and R is not an almost f-ring.

16. Let A and P be totally ordered domains with P an ℓ-A-ring (see the end of Section 3.1), and let Q be a left and a right ℓ-module over each of the rings A and P. Assume that multiplication in $M = A \cup P \cup Q$ is associative whenever it is defined, and that $pp' \geq \alpha p' \vee p\alpha > 0$ and $pq \geq \alpha q > 0$ for all $0 < p, p' \in P$, $0 < \alpha \in A$ and $0 < q \in Q$. Let $2 \leq n \in \mathbb{N}$ and let $R = P \underset{\rightarrow}{\times} [A \underset{\rightarrow}{\times} (Q \oplus \cdots \oplus Q)]$ as an ℓ-group where the direct sum is of n copies of Q.

(a) Suppose that $qp \geq q\alpha > 0$ for $0 < p \in P$, $0 < q \in Q$ and $0 < \alpha \in A$. Define a product in R by

$$(p, \alpha, q_1, \ldots, q_n)(p', \alpha', q'_1, \ldots, q'_n) = (npp' + \alpha p' + p\alpha', \alpha\alpha', ps'_n +$$
$$s_n p' + \alpha q'_1 + q_1\alpha', ps'_n + s_n p' + \alpha q'_2 + q_2\alpha', \ldots, ps'_n + s_n p' + \alpha q'_n + q_n\alpha')$$

where $s_n = q_1 + \cdots + q_n$ and $s'_n = q'_1 + \cdots + q'_n$.

Show that R is an almost f-ring that is not an f-ring, and if A is superunital or is unital, then R is superunital or is unital, respectively. [Determine $F_\ell(R)$ and $F_r(R)$]. Show that R is commutative if the multiplication in M is commutative.

(b) Define another multiplication in R by (assuming $qp \geq q\alpha > 0$)

$$(p, \alpha, q_1, \ldots, q_n)(p', \alpha', q'_1, \ldots, q'_n)$$
$$= (npp' + \alpha p' + p\alpha', \alpha\alpha', ps'_n + \alpha q'_1, \ldots, ps'_n + \alpha q'_n).$$

Show that R is an almost f-ring and a right f-ring that is not an f-ring. If A has a superunit or an identity element show that R has a left superunit or a left identity element, respectively. Show that R does not have a right superunit. Determine $F_\ell(R)$.

(c) Find an example of A, P and Q that satisfies all the conditions.

17. Suppose that the sp-ℓ-ring R has a left superunit e with $r(e;R) = 0$. Let S be the convex ℓ-subring of R generated by e. Show that R is unital iff S is right unital and $\ell(R;R)$ is directed, iff S is right unital and e is a superunit of R. (Use Exercise 3.7.5.)

18. Let R be an ℓ-unital ℓ-algebra over the totally ordered field Q. Show that R is isomorphic to the canonically ordered ℓ-algebra of upper triangular 2×2 matrices over Q iff R is a 3-dimensional noncommutative algebra, $N(R)$ is a 1-dimensional ℓ-ideal, and R satisfies the identity $((x^2)^-)^2 = 0$. (Use Theorem 3.4.15.)

19. Show that a C-archimedean torsion-free ℓ-algebra R is an f-algebra if it has an f-superunit and it satisfies the identity $f(x^+, x^-)^- = 0$ where $f(x,y)$ is a y-special right k-nice polynomial with $k \geq 2$.

20. (a) Let R be an ℓ-ring that is a local ring. Show that R is an f-ring iff the inverse of each positive invertible element is positive.

(b) Let R be a unital ℓ-ring such that for each $a \in R$ there is an integer $n \in \mathbb{N}$ with $a^n \geq 0$. Show that R is an f-ring if it is either semiperfect, π-regular, left π-regular, or an algebraic algebra over a field. (R is semiperfect if R/J is left artinian where J is the Jacobson radical of R and idempotents can be lifted through J; use Theorem 3.4.15. R is *π-regular* if for each $a \in R$ there are $m \in \mathbb{N}$ and $x \in R$ with $a^m = a^m x a^m$; assume R is subdirectly irreducible. R is left π-regular if for each $a \in R$ the chain $Ra \supseteq Ra^2 \supseteq \cdots$ is finite.)

21. Let R be an ℓ-ring and let $n \in \mathbb{N}$ (see Theorem 3.8.13).

(a) If R satisfies the identity $y^+|x|y^+ = |y^+xy^+|$ show that $B_n = \{x \in R : Rx^nR = 0\}$ is an ℓ-ideal of R with $B_n^{n+2} = 0$. If $a, b \in B_2$ show that $ab \in N_2(R)$ and $RabR = 0$. Find an example with $B_1^2 \neq 0$.

(b) If R satisfies the identity $y^+x^+y^+x^-y^+ = 0$ show that $C_n = \{x \in R : R|x^n|R|x^n|R = 0\}$ is an ℓ-ideal of R with $C_n^{n+3} = 0$. Find an example with $C_1^3 \neq 0$.

(c) If R is an almost f-ring show that $D_n = \{x \in R : R^2|x^n|R|x^n|R^3|x^n|R|x^n|R^2 = 0\}$ is an ℓ-ideal of R with $D_n^{n+9} = 0$. Find an example with $D_1^9 \neq 0$.

22. Let R be a right d-ring.

(a) Suppose that R is a square-archimedean ℓ-algebra over C (see Exercise 3.7.2). Show that if $a \wedge b = 0$, then $ab \leq \alpha a^2$ for some $\alpha \in C^+$

(b) Show that a square-archimedean d-algebra is an almost f-algebra.

(c) Show that R is an almost f-ring iff any two positive disjoint elements in R commute.

(d) If R is a square-archimedean C-archimedean ℓ-algebra over C, show that R is an almost f-algebra.

23. Show that an archimedean right f-ring with zero right annihilator is an f-ring.

24. Show that the following statements are equivalent for the ℓ-ring R.

(a) $ab = (a \vee b)(a \wedge b)$ for all $a, b \in R$.

(b) $ab = (a \vee b)(a \wedge b)$ for all $a, b \in R^+$.

(c) R is a commutative almost f-ring.

(d) R is an almost f-ring and the interval $[0, b]$ is contained in the centralizer of b, for each b in R^+.

25. Let R be an ℓ-ring which satisfies the identity $x^+|y|x^+ = |x^+yx^+|$. Show that $Rx^+x^-R = 0$ and $(x^+x^-)^2 = 0$ for each $x \in R$.

26. Let R be an ℓ-ring which is a po-algebra over the po-ring C and which satisfies the identity $x^+|y|x^+ = |x^+yx^+|$. Assume further that R^+ is C^+-archimedean. Let $A = \{a \in R : RaR = 0\}$ be the middle annihilator of R.

(a) Show that R/A is a C-f-algebra whose positive cone is C^+-archimedean.

(b) If R is a d-ring show that R/A is semiprime and $N(R) = N_3(R) = A = \{a \in R : a^2R = 0\} = \{a \in R : Ra^2 = 0\}$. (If $a, b \in R^+$ with $R^2aR = 0$ use the elements $b(ab-b)^+$ and $ba(ab-b)^-$ to show that $bab \leq b^2$.)

27. Let R be a po-algebra over the po-unital po-ring C and let M be a left algebra ℓ-module over R which is a d-module over C such that M^+ is C^+-archimedean. Suppose that $a, b \in R^+$ such that $b + ba$ is a d-element on M and $ba^kM = 0$ for some $k \in \mathbb{N}$. Apply the hint in Exercise 26 (b) to the positive and negative parts of $a^{k-1}x - a^{k-2}x$, where $x \in M^+$, to show that $baM = 0$.

28. Let R be the canonically ordered $\mathbb{R}$-ℓ-algebra

$$R = \begin{pmatrix} \mathbb{R} & \mathbb{R} & 0 \\ 0 & 0 & 0 \\ \mathbb{R} & \mathbb{R} & 0 \end{pmatrix}.$$

Show that R is a complete infinite d-algebra in which $N_2(R)$ is neither convex nor an additive subgroup of R. Determine $F_\ell(R)$, $F_r(R)$ and $F(R)$, and find an element $a \in N(R)^+$ with $aR^2 \neq 0$.

29. Let M and B be abelian ℓ-groups with B archimedean, and let $\theta : M^n \longrightarrow B$ be a positive multilinear almost f-map, with $n \geq 2$.

    (a) Show that θ is symmetric.
    (b) If $\omega \in F(M)$ show that for all $i < j$, $\theta(a_1, \ldots, \omega a_i, \ldots, a_j, \ldots, a_n) = \theta\,(a_1, \ldots, a_i, \ldots, \omega a_j, \ldots, a_n)$. (Recall that $F(M)$ is the subring of $\mathrm{End}_{\mathbb{Z}}(M)$ generated by the f-endomorphisms of M).
    (c) Suppose that $\omega_1, \ldots, \omega_n$ are commuting elements of $F(M)$. Show that $\theta(\omega_1 a_1, \ldots, \omega_n a_n) = \theta(\omega_{\sigma(1)} a_{\tau(1)}, \ldots, \omega_{\sigma(n)} a_{\tau(n)}$ for all $a_1, \ldots, a_n$ in M and any two permutations σ and τ of $\{1, \ldots, n\}$.

30. Let R be a complete almost f-algebra over $\mathbb{R}$ and let $a_1, \ldots, a_n, a, b \in R^+$. Show that there are elements $u, v \in R^+$ with $u^n = a_1 \cdots a_n$ and $v^n = a^n + b^n$. (Use Theorems 2.3.25 and 3.6.3 to first show that if $1 \in R$, then each element of R^+ has a unique n^{th} root; the identity in Exercise 24 reduces this verification to elements that are comparable to 1. Next, let M be the convex ℓ-subgroup of R generated by $e = a_1 + \cdots + a_n$, and apply Theorems 2.3.25 and 3.6.6 and the previous exercise to M.)

31. Let $f(x_1, \ldots, x_n) \in \mathbb{R}^+[x_1, \ldots, x_n]$ be a homogeneous polynomial of degree m. Use the previous exercise to show that if R is a complete almost f-algebra over $\mathbb{R}$ and $a_1, \ldots, a_n \in R^+$, then $f(a_1, \ldots, a_n) = a^m$ for some $a \in R^+$.

32. Let R be a complete almost f-algebra over $\mathbb{R}$ and for $n \in \mathbb{N}$ let $P_n = \{a^n : a \in R^+\}$.

    (a) Show that (R^n, P_n) is a directed integrally closed po-$\mathbb{R}$-algebra and each element of R^n is of the form $a_1 \cdots a_n$ with $a_i \in R$.
    (b) If $n \geq 3$ show that (R^n, P_n) is an ℓ-algebra with $a^n \vee_n b^n = (a \vee b)^n$ and $a^n \wedge_n b^n = (a \wedge b)^n$ for all $a, b \in R^+$, where $a^n \vee_n b^n = \sup_{R^n}\{a^n, b^n\}$ and $a^n \wedge_n b^n = \inf_{R^n}\{a^n, b^n\}$. (If $c \in R^+$ with $c^n \geq a^n, b^n$ pass to the semiprime archimedean f-ring $R/N(R)$ (see the remarks after Theorem 3.8.13) to show that $(c - (a \vee b))^- \in N(R)$, and use Exercise 30 to see that $c^n \geq_n (a \vee b)^n$.)
    (c) If $n \geq 3$ show that (R^n, P_n) is a semiprime f-algebra; and show that it is an ℓ-subalgebra of R provided R is a d-ring.
    (d) If $n \geq 2$ and $a, b, u, v \in R^+$ with $u \wedge v = 0$ show that $a^n u \wedge_{n+1} b^n v = 0$. (Pass to R/N.)
    (e) If $n \geq 3$ and $w = a_1 \cdots a_n \in R^n$ show that $|w|_n = |a_1| \cdots |a_n|$. (Use (d) and Exercise 30.)

33. Let A be an ℓ-ring and Δ a set with a distinguished element α. Define a partial addition in Δ by: for every $\beta \in \Delta \backslash \{\alpha\}$, $2\beta = \alpha$.

    (a) Show that Δ, with the trivial order, is a pops.

(b) Show that $W(\Delta, A)$ is an ℓ-ring that is an almost f-ring or an sp-ℓ-ring iff A is an almost f-ring or an sp-ℓ-ring, respectively (see Exercise 3.5.22).
(c) Show that if $A^2 \neq 0$, then $V(\Delta, A)$ is an ℓ-ring iff Δ is finite.

34. Let A be a totally ordered ℓ-simple ℓ-ring and let Δ be a rooted pops.

(a) Show that $W(\Delta, A)$ is an A-archimedean almost f-ring iff Δ is trivially ordered and is the disjoint union of a family of subpops $\{\Delta_\lambda : \lambda \in \Lambda\}$ such that $\Delta_\lambda + \Delta_\mu = \emptyset$ if $\lambda \neq \mu$ and for each $\lambda \in \Lambda$, either $\Delta_\lambda + \Delta_\lambda = \emptyset$; or for $\alpha, \beta \in \Delta_\lambda$, $\alpha + \beta \in \Delta_\lambda$ and $\alpha + \beta = 2\alpha$ iff $\alpha = \beta$; or Δ_λ has the form described in the previous exercise.
(b) Show that $V(\Delta, A)$ is an A-archimedean almost f-ring iff Δ has the properties given in (a) and for each $\alpha \in \Delta$ the set $\{\beta \in \Delta : 2\beta = \alpha\}$ is finite.

35. Suppose R is an almost f-ring and $a \in F = F(R)$ is not a zero divisor in F. Show that $\ell(a;R) = r(a;R) = a^\perp = F^\perp = \ell(F;R) = r(F;R)$.

36. Suppose I is a right ideal of the ℓ-ring R with $I \subseteq F(R)$ and $r_\ell(I;R) = 0$, and let K be the right ℓ-ideal of R generated by I^+.

(a) Show that $r(K) = r(I) = 0$. (Use Theorem 4.3.4.)
(b) Show that R is a left f-ring.
(c) If I is semiprime and $F(R)$ contains a left superunit of R, show that R is an f-ring.

Notes. Perhaps the first appearance of a polynomial constraint which generalizes the sp-ℓ-ring identity $(x^2)^- = 0$ occurs in Shyr and Viswanathan [SV] where the square-archimedean constraints of Exercise 3.7.2 appear; Theorem 3.8.13 also comes from this paper. Birkhoff and Pierce [BP] showed that a unital ℓ-ring is an almost f-ring precisely when the identity element is a weak order unit and gave an example of a unital almost f-ring which is not an f-ring; the examples in Exercise 16 are based on their example. This result together with Diem's [DI] structural results for an sp-ℓ-ring motivated the study of the connection between constraints of the form $f(x_1,\ldots,x_n)^- = 0$ and the identity $x^+x^- = 0$ and the influence of these constraints on the structure of an ℓ-ring in Steinberg [ST7], [ST8], [ST12], [ST13], [ST14], and Ma [M1]. Theorem 3.8.14 comes from Bernau and Huijsmans [BH] who showed that an archimedean almost f-ring is commutative; Boulabiar [BO] observed that their proof did not require the ring to be associative. Exercises 29 through 32 also come from Boulabiar's paper. Exercises 33 and 34 are based on Wojciechowski [WO].

Chapter 4
The Category of f-Modules

In order to gain information about the category of f-modules it is useful to understand the free f-modules as well as the injective f-modules. Because there are generally no injectives in this category our efforts will be spent on studying those relative injectives that arise by bounding the cardinality of the f-module to which a given morphism is to be extended. Sophisticated techniques will be required to characterize these f-modules. One of the characterizing properties they have, not surprisingly, is that of being an injective module; the other properties are all order theoretic. These order properties can also be used to characterize the relative injectives in other categories of ordered structures.

We will first construct the injective hull of a module and the analogous maximal right quotient ring of a ring. With an eye toward our applications we investigate the maximal right quotient ring of a semiprime ring whose Boolean algebra of annihilator ideals is atomic and certain torsion-free modules over this ring. One fundamental question that arises is to determine when the injective hull of an f-module is an f-module extension and when the maximal right quotient ring of an f-ring is an f-ring extension. The answer is given in the more general context of rings and modules of quotients with respect to a hereditary torsion theory. Large classes of po-rings are identified over which all torsion-free f-modules have this property.

Free R-f-modules are constructed and represented in the more general context of a free f-module over a partially ordered module. We will investigate their embeddability into a product of copies of R_R, their indecomposability and the size of their disjoint subsets. Three related ordered tensor products will be constructed using free abelian ℓ-groups.

4.1 Rings of Quotients and Essential Extensions

In this section the maximal (right) quotient ring of a ring is constructed and those aspects of this construction which are pertinent to f-rings and f-modules will be considered. The maximal quotient ring is a generalization of the field of quotients of

S.A. Steinberg, *Lattice-ordered Rings and Modules*,
DOI 10.1007/978-1-4419-1721-8_4, © Springer Science + Business Media, LLC 2010

a commutative domain and, more generally, the division ring of quotients of an Öre domain. In the place of fractions one has module homomorphisms defined on large submodules of the ring. For example, $1/2$ is the homomorphism whose domain is $2\mathbb{Z}$ and which halves each even integer. Analogously, the injective hull of a module is the generalization of the divisible closure of an abelian group. It will also be considered here and dealt with for f-modules later. In many instances the maximal right quotient ring is identical to the injective hull of the ring considered as a right module over itself.

The submodule N of the right R-module M is *essential in* M, and M is *an essential extension of* N, provided $N \cap K \neq 0$ for each nonzero submodule K of M. Equivalently, N is essential in M if for each $0 \neq x \in M$, $0 \neq xr + nx \in N$ for some $r \in R$ and some $n \in \mathbb{Z}$. The set $\mathscr{E}(M_R)$ of essential submodules of M is easily seen to be a dual ideal of the lattice $\mathscr{L}(M_R)$ of submodules of M. If $X, Y \subseteq M$ let $(Y : X) = \{r \in R : Xr \subseteq Y\}$; when Y is a submodule of M it is clear that $(Y : X) = r(\{x + Y : x \in X\}; R)$ is the annihilator of the image of X in M/Y. Moreover, if $N \in \mathscr{E}(M)$, then for each $x \in M$, $(N : x) \in \mathscr{E}(R)$ since $xI \cap N \neq 0$ if I is a right ideal of R with $xI \neq 0$. More generally, a quite similar argument shows that if $\alpha : M \longrightarrow N$ is any R-homomorphism, then $\alpha^{-1}(\mathscr{E}(N)) \subseteq \mathscr{E}(M)$. One way to manufacture essential submodules is as follows. The submodule B of M is called a *complement in M of the submodule A* if B is maximal among those submodules K of M with $K \cap A = 0$. If B is a complement of A, then $A \boxplus B \in \mathscr{E}(M)$ since $(C + B) \cap A = 0$ whenever C is a submodule of M with $C \cap (A + B) = 0$. Recall that $A \boxplus B$ denotes the group (or module) direct sum. This implies that if D is a complement of B in M that contains A, where B is a complement of A, then A is essential in D. Consequently, a submodule B of M is a complement in M (of some submodule) iff it is *essentially closed in M* in the sense that the only essential extension of B in M is B itself.

We mention two other useful facts that are easily verified. Suppose that B, C and D are submodules of M with $B \subseteq C$. If B is essential in C, then $B \cap D$ is essential in $C \cap D$. If B is essentially closed in M and C is essential in M, then C/B is essential in M/B. For, if $0 \neq K/B \subseteq M/B$, then $L \cap B = 0$ for some nonzero submodule L of K. Hence $0 \neq (L \cap C) + B/B \subseteq K/B \cap C/B$.

The module E is *injective* if each R-homomorphism $f : A \longrightarrow E$ from a submodule A of a module B can be extended to an R-homomorphism $g : B \longrightarrow E$; that is, there is some g which makes the following diagram commutative.

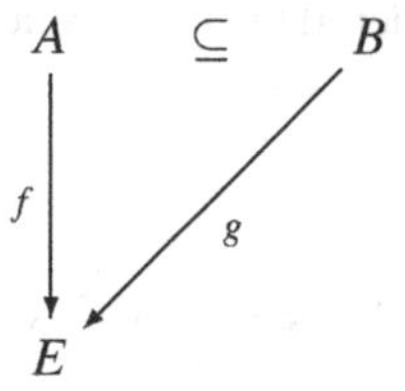

Note that the inclusion $A \subseteq B$ in the previous diagram can be replaced by any monomorphism $A \longrightarrow B$ of any modules. If R is unital, E is unital, and f can be extended to B whenever B is also unital, then E is called *u-injective*. With the aid of Exercise 3.1.20 it is easy to see that a unital module is u-injective iff it is injective. It is also clear that if R_1 is the ring obtained by freely adjoining $\mathbb{Z}$ to R (see Exercise 3.1.2), then the R-module E is injective iff the R_1-module E is u-injective. This fact affords us the opportunity to obtain injective modules from u-injective modules and this is advantageous since u-injective modules are a little bit nicer. For example, a u-injective module E is characterized by the following condition which is known as *Baer's Criterion*: Each homomorphism from a right ideal I of R into E is of the form $r \mapsto er$ for some $e \in E$; that is, it is extendible to R. For, given the homomorphism $f : A \longrightarrow E$ with $A \subseteq B$, and given some $x \in B$, then since $x(A : x) \subseteq A$ we have a homomorphism $(A : x) \longrightarrow E$ given by $r \mapsto f(xr) = er$ for some $e \in E$. But then the map $A + xR \longrightarrow E$ given by $a + xr \mapsto f(a) + er$ is a well-defined extension of f. In particular, if f has no proper extension, which can always be arranged by an application of Zorn's Lemma, then $A = B$ and hence E is injective.

A module M_R over any ring is called a *Baer* module if each R-homomorphism from a right ideal of R into M is induced by left multiplication by some element of M. For some other connetions between Baer modules and injectivity see Exercises 45 and 46.

The module M_R is called *divisible* if $M = Ma$ for each $a \in R$ which is right regular (that is, $r(a; R) = 0$). If M is a u-injective module over the unital ring R it is easy to see that M is divisible. In particular, if R is a unital domain in which each right ideal is principal, then by Baer's Criterion M is u-injective iff M is divisible.

The essential extension B of the module A is called a *maximal essential extension of* A provided that $B = C$ whenever B is an essential submodule of C. An injective module E that contains A is called a *minimal injective extension of* A provided $C = E$ whenever $A \subseteq C \subseteq E$ and C is injective. We will now see that these two kinds of extensions are identical.

Theorem 4.1.1. *The following statements are equivalent for the extension $M \subseteq E$ of R-modules.*

(a) *E is a maximal essential extension of M.*
(b) *E is a minimal injective extension of M.*
(c) *E is injective and M is an essential submodule of E.*

Moreover, each module M has an extension E with these properties, and any two such extensions of M are isomorphic via an isomorphism that is the identity on M.

Proof. We may assume that R is unital and that all modules considered are unital. For if $R_1 = R + \mathbb{Z}$ is the unital ring obtained by freely adjoining $\mathbb{Z}$ to R, then each of these three conditions for the pair of R-modules $M \subseteq E$ in the category of R-modules is equivalent to the corresponding condition for the same pair in the category of unital R_1-modules.

We first give a brief proof, leaving many details to the reader, of the fact that each module M can be embedded in an injective (actually, u-injective) module. If D is an abelian group, then $H = \mathrm{Hom}_{\mathbb{Z}}(R, D)$ is a right R-module due to the fact that R is a left R-module: $(\alpha r)(s) = \alpha(rs)$ for $\alpha \in H$ and $r, s \in R$. Moreover, if $M \subseteq D$, then

M_R is embedded in H_R via left multiplication by the elements of M. Since M is a homomorphic image of a free abelian group and the latter is a subgroup of a $\mathbb{Q}$-vector space, M is embeddable in a divisible abelian group. Now, if D is a divisible abelian group (hence $\mathbb{Z}$-injective), then H_R is injective. For an R-diagram on the left can be completed to the R-diagram on the far right

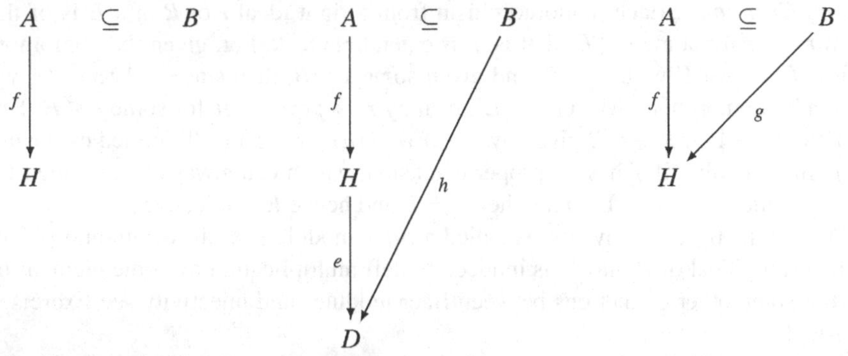

where e is the evaluation map at 1 defined by $e(\alpha) = \alpha(1)$, h is a group homomorphism extending ef and g is the R-homomorphism given by $g(b)(r) = h(br)$.

Since the direct product of modules is injective iff each factor is injective, one consequence of the previous embedding is that a module is injective iff it is a direct summand of each module which contains it.

To show that (a) implies (c) it suffices to show that E is injective. Let F be an injective module that contains E and let B be a complement of E in F. Then B is essentially closed in F and $B \boxplus E$ is essential in F; so $E \cong B \boxplus E/B$ is essential in F/B. Thus, E is an essential submodule of $G = (F/B \setminus (B + E/B)) \cup E$, and hence $F = B \boxplus E$ and E is injective. Note that G has the unique module operations induced by the obvious bijection between F/B and G.

That (c) implies (b) is a consequence of the fact that an injective module F with $M \subseteq F \subseteq E$ is a summand of E. To see that (b) implies (a) note that if A is an essentially closed submodule of an injective module L, then A is a maximal essential extension of itself since any essential extension of A is isomorphic to an essential extension of A within L; hence A is injective by the previous paragraph. Thus, if $M \subseteq A \subseteq E$ with $M \in \mathscr{E}(A)$ and A is essentially closed in E, then $A = E$, and E is a maximal essential extension of M since E has no proper essential extensions. This completes the proof of the equivalence of these three kinds of extensions of M and also shows the existence of such an extension of M: take an essential extension of M which is essentially closed in an injective module that contains M. Uniqueness follows from the fact that an essential extension of M can be M-embedded in any injective module that contains M. □

A maximal essential extension of the module M is called an (the) *injective hull of* M and will be denoted by $E(M)$. Note that any complement which contains M of a complement of M within an injective module is equal to $E(M)$. Note also that a submodule N of M is essentially closed in M if and only if $N = F \cap M$ for some injective submodule of $E(M)$. For if N is essentially closed and $F \subseteq E(M)$ is an injective hull of N, then N is essential in $F \cap M$. Conversely, suppose that $N = F \cap M$ where $E(M) = F \boxplus G$. Let $N \subseteq K \subseteq M$ with K an essential extension of N. If $k \in K \backslash N$ and $k = f + g \in F \boxplus G$, then there is some $a_1 \in R_1$ with $0 \neq ga_1 \in M$. But then $ga_1 = ka_1 - fa_1 \in G \cap K = 0$; so $N = K$ and N is essentially closed in M.

We turn now to the Utumi maximal right quotient ring of a ring. For this purpose we need a relation between rings that is stronger than the essential relation. We will first consider it for rings and later for modules in the broader context of torsion theories.

The overring S of the ring R is called a *right quotient ring of R*, or *a ring of right quotients of R*, if for every pair of elements $x, y \in S$ with $x \neq 0$ there is an $a \in R$ with $xa \neq 0$ and $ya \in R$. A right ideal D of R is called *dense* if R is a right quotient ring of D. The collection of all the dense right ideals of R will be denoted by $\mathbb{D}(R)$ or by just $\mathbb{D}$ if R is understood. Clearly, an ideal of R is a dense right ideal exactly when its left annihilator in R is zero. The following two results collect some elementary facts about rings of quotients.

Theorem 4.1.2. *Let R be a subring of S.*

(a) R has a right quotient ring iff $\ell(R,R) = 0$, iff R is a right quotient ring of R.
(b) If S is a right quotient ring of R, then S_R is an essential extension of R_R.
(c) S is a right quotient ring of R iff for each $n \in \mathbb{N}$ and all $x_1, \ldots, x_n$ in S with $x_1 \neq 0$ there is an element $a \in R$ with $x_1 a \neq 0$ and $x_i a \in R$ for $1 \leq i \leq n$.
(d) If S is a subring of T, then T is a right quotient ring of R iff S is a right quotient ring of R and T is a right quotient ring of S.
(e) The right ideal D of R is dense iff for $x, y \in R$ with $x \neq 0$ there is some $a \in R$ with $xa \neq 0$ and $ya \in D$.

Proof. We will only prove (c) and (e) and leave the verification of the other parts to the reader. For (c), suppose that S is a right quotient ring of R. Assume, by induction, that the condition holds for some n and let $x_1, \ldots, x_n, x_{n+1}$ be elements of S with $x_1 \neq 0$. Then for some $a \in R$, $x_1 a, x_2 a, \ldots, x_n a$ are elements in R and $x_1 a \neq 0$. Now take $b \in R$ with $x_1 ab \neq 0$ and $x_{n+1} ab \in R$. Then $x_i ab \in R$ for $1 \leq i \leq n+1$ and the proof of (c) is complete. Assume the condition in (e) holds and take $b \in R$ with $xab \neq 0$ and $ab \in D$. Then $yab \in D$ and hence R is a right quotient ring of D. It is obvious that conversely, each dense right ideal satisfies this condition. □

Theorem 4.1.3. *Suppose that S is a right quotient ring of each of its subrings R and T.*

(a) If U is the subring of S generated by TR, then S is a right quotient ring of U.
(b) If T is an R-submodule of S_R and $\alpha : R \longrightarrow S$ is an R-homomorphism, then $\alpha^{-1}(T)$ is a dense right ideal of R.

(c) If T is an R-submodule of S_R, then S is a right quotient ring of $T \cap R$.
(d) If $s \in S$ and $D_s = (R : s)$, then $D_s \in \mathbb{D}(R)$.
(e) $\mathbb{D}(R)$ is a dual ideal of the lattice of right ideals of R.
(f) Suppose that $J \in \mathbb{D}(R)$ and for each $a \in J$, $I_a \in \mathbb{D}(R)$. Then $D = \sum_{a \in J} aI_a \in \mathbb{D}(R)$.

Proof. (a) Let $x, y \in S$ with $x \neq 0$, and take $t \in T$ with $0 \neq xt$ and $yt \in T$. Now take $r \in R$ with $xtr \neq 0$ and $tr \in R$. Since $tr \in U$ and $ytr \in U$, S is a right quotient ring of U.

(b) Let $x, y \in R$ with $x \neq 0$. Then for some $t \in T$ we have $xt \neq 0$ and $(\alpha y)t \in T$. Also, $xta \neq 0$ and $ta \in R$ for some $a \in R$. Thus, $\alpha(yta) = (\alpha y)ta \in T$ and $\alpha^{-1}(T)$ is dense by (e) of Theorem 4.1.2.

(c) This follows from (b) since $T \cap R = \alpha^{-1}(T)$ where $\alpha : R \longrightarrow S$ is inclusion.

(d) Let $\mu_s : R \longrightarrow S$ be left multiplication by $s \in S$. Then $D_s = \mu_s^{-1}(R)$ is dense by (b).

(e) Suppose that I, J, K are right ideals of R with I and J dense and $I \subseteq K$. Then $I \cap J$ is dense by (c) and K is dense by (d) of Theorem 4.1.2.

(f) Let $0 \neq x$, $y \in R$. Then $xr \neq 0$ and $a = yr \in J$ for some $r \in J$. Let $s \in I_a$ with $xrs \neq 0$. Then $yrs = as \in aI_a \subseteq D$, and D is dense by (e) of Theorem 4.1.2. □

In the next result we present several characterizations of a dense right ideal in addition to that given in (e) of Theorem 4.1.2.

Theorem 4.1.4. *The following statements are equivalent for the right ideal D of the ring R.*

(a) D is a dense right ideal of R.
(b) If $x_1, \ldots, x_n \in R$ with $x_1 \neq 0$, then for some $a \in R$, $x_1 a \neq 0$ and $x_i a \in D$ for $1 \leq i \leq n$.
(c) Each $\alpha \in Hom_R(D, R)$ has a unique extension to an element of $Hom_R(R, E(R))$; and $\ell(R; R) = 0$.
(d) If $\alpha \in End_R(E(R))$ and $\alpha(D) = 0$, then $\alpha(R) = 0$; and $\ell(R; R) = 0$.
(e) $Hom_R(R/D, E(R)) = 0$ and $\ell(R; R) = 0$.
(f) $\ell(D; E(R)) = 0$.
(g) For each $a \in R$, $(D : a)$ is a dense right ideal of R.
(h) There is a dense right ideal J such that $(D : a)$ is dense for each $a \in J$.
(i) $\ell((D : a); R) = 0$ for each $a \in R$.

Proof. (a) ⇔ (b). This is a consequence of (c) (and its proof) and (e) of Theorem 4.1.2.

(a) ⇒ (c). Suppose that $\beta, \gamma \in \mathrm{Hom}_R(R, E(R))$ both extend α. If $\beta x \neq \gamma x$ for some $x \in R$, then $\beta x a_1 \neq \gamma x a_1 \in R$ for some $a_1 \in R_1$. Now take $b \in R$ with $\beta x a_1 b \neq \gamma x a_1 b$ and $x a_1 b \in D$ to get a contradiction.

(c) ⇒ (d). If $\alpha \in \mathrm{End}_R(E(R))$ with $\alpha(D) = 0$, then $\alpha(R) = 0$ since the restriction of α to R is an extension of $0 \in \mathrm{Hom}_R(D, R)$.

(d) ⇒ (f). Suppose that $x \in E(R)$ and $xD = 0$. Let $\alpha \in \mathrm{End}_R(E(R))$ be an extension of $\mu_x \in \mathrm{Hom}_R(R, E(R))$ where $\mu_x(r) = xr$. Then $0 = \alpha(R) = xR$. But $\ell(R; E(R)) = 0$ since otherwise $\ell(R; R) = \ell(R; E(R)) \cap R \neq 0$. So $x = 0$.

(f) ⇒ (a). If D is not dense, then there exist elements $x, y \in R$ with $x \neq 0$ such that $x(D:y) = 0$. Then $\alpha : (y+D)R \longrightarrow E(R)$ defined by $\alpha((y+D)a) = xa$ for each $a \in R$ is a well-defined R-homomorphism and has an extension $\beta \in \mathrm{Hom}_R(R_1/D, E(R))$. Since $\beta(1+D)D = 0$ we have that $\beta(1+D) = 0$ and hence $\beta = 0$. Thus, $xR = 0$ and we have the contradiction that $x = 0$.

Since the equivalence of (d) and (e) is apparent, statements (a) through (f) have been shown to be equivalent.

(d) ⇒ (g). Suppose that $\alpha \in \mathrm{End}_R(E(R))$ with $\alpha((D:a)) = 0$. As above, define $\beta : (a+D)R \longrightarrow E(R)$ by $\beta((a+D)r) = \alpha(r)$ and extend β to $\gamma \in \mathrm{Hom}_R(R/D, E(R))$. Then $\gamma = 0$ (by (e)) and hence $\alpha(R) = \gamma(R) = 0$. Thus (d) holds for $(D:a)$ and hence $(D:a)$ is dense.

(g) ⇒ (h). This is trivial.

(h) ⇒ (a). This is a consequence of (e) and (f) of Theorem 4.1.3 since $\sum a(D:a) \subseteq D$.

Since (i) is just a reformulation of (e) in Theorem 4.1.2 the proof is complete. □

We will now construct the maximal right quotient ring of R. We assume that $\ell(R;R) = 0$. If M and N are R-modules it will be convenient to denote the abelian group $\mathrm{Hom}_R(M,N)$ by $[M,N]$ or by ${}_R[M,N]$ if necessary. The partial order on $\mathbb{D}$ which is dual to inclusion makes $\mathbb{D}$ into a poset that is directed up, and the set of abelian groups (actually, left R-modules) $\{[D,R] : D \in \mathbb{D}\}$ is a direct system where, for $D_1 \subseteq D_2$, the homomorphism $[D_2,R] \longrightarrow [D_1,R]$ is the restriction map. Let $Q(R) = \varinjlim [D,R]$ be the direct limit of this system (see Exercises 1.1.7 and 1.4.23). $Q(R)$ is called the (*Utumi*) *maximal right quotient ring of* R. More concretely,

$$Q(R) = \bigcup_{D \in \mathbb{D}} [D,R] / \sim$$

where $\sim$ is the equivalence relation defined on the union by: if $\alpha : D_1 \longrightarrow R$ and $\beta : D_2 \longrightarrow R$, then $\alpha \sim \beta$ iff there is a dense right ideal $D \subseteq D_1 \cap D_2$ such that $\alpha(d) = \beta(d)$ for each $d \in D$. $Q(R)$ is a left R-module and the map $R \longrightarrow Q(R)$ defined by $x \mapsto [\mu_x]$ is a left R-monomorphism where $\mu_x : R \longrightarrow R$ is left multiplication by x and $[\mu_x]$ denotes the equivalence class of μ_x. In fact, this map is an embedding of rings. To see this take $[\alpha], [\beta] \in Q(R)$ with $\alpha : D_1 \longrightarrow R$ and $\beta : D_2 \longrightarrow R$ and let $D = \beta^{-1}(D_1)$. Then $D \in \mathbb{D}$ by (b) of Theorem 4.1.3, and hence $\alpha\beta : D \longrightarrow \mathbb{R}$ given by $(\alpha\beta)(d) = \alpha(\beta d)$ gives rise to an element of $Q(R)$. Thus $[\alpha][\beta] = [\alpha\beta]$. This is a well-defined product since if α and α_1 agree on $D_3 \in \mathbb{D}$ and β and β_1 agree on $D_4 \in \mathbb{D}$, then $\alpha\beta$ and $\alpha_1\beta_1$ agree on $\beta^{-1}(D_3) \cap D_4$ and the latter is a dense right ideal of R. It is now easily seen that $Q(R)$ is a unital ring and R is embedded in $Q(R)$ as a subring. We will identify R with its image in $Q(R)$. Moveover, if $[\alpha] \in Q(R)$, $\alpha : D \longrightarrow R$, and $x \in D$, then $\alpha\mu_x = \mu_{\alpha x}$. In particular, if $[\alpha] \neq 0$ and $[\beta] \in Q(R)$, $\beta : D_1 \longrightarrow R$, and $x \in D \cap D_1$ with $\alpha x \neq 0$, then $[\alpha][\mu_x] = [\mu_{\alpha x}] \neq 0$ and $[\beta][\mu_x] = [\mu_{\beta x}] \in R$; so $Q(R)$ is a right quotient ring of R.

If S and T are ring extensions of R, then a ring monomorphism from S into T which is the identity on R is called an *R-embedding* of S into T.

Theorem 4.1.5. *The following statements are equivalent for the right quotient ring T of R.*

(a) *T is R-isomorphic to $Q(R)$.*
(b) *Each right quotient ring of R can be R-embedded in T.*
(c) *If $D \in \mathbb{D}(R)$ and $\alpha : D \longrightarrow R$ is an R-homomorphism, then there is an element $t \in T$ such that $\alpha d = td$ for each $d \in D$.*
(d) *If U is a right quotient ring of T, then $U = T$.*

Proof. (a) $\Rightarrow$ (b). It suffices to show that if S is a right quotient ring of R, then S can be R-embedded in $Q(R)$. If $s \in S$ and $\mu_s : R \longrightarrow S$ is left multiplication by s, then $D_s = (R : s) = \mu_s^{-1}(R)$ is a dense right ideal of R according to (d) of Theorem 4.1.3. Thus, s determines the element $[\mu_s]$ of $Q(R)$ where μ_s now has D_s as its domain. Since $\mu_s + \mu_t = \mu_{s+t}$ on $D_s \cap D_t$ and $\mu_s \mu_t = \mu_{st}$ on $(D_s : t) \cap D_t = \mu_t^{-1}(D_s)$, the mapping $s \mapsto [\mu_s]$ is a ring homomorphism. It is monic since if D is dense and $D \subseteq D_s$ with $s \neq 0$, then $sD \neq 0$ since S is a right quotient ring of D. Clearly, this map is an R-embedding.

(b) $\Rightarrow$ (c). Let $\varphi : Q(R) \longrightarrow T$ be an R-embedding and put $t = \varphi([\alpha])$. Then for each $d \in D$, $td = \varphi([\alpha][\mu_d]) = \varphi([\mu_{\alpha d}]) = \alpha d$.

(c) $\Rightarrow$ (a). From the fact that (a) implies (b) (for $Q(R)$) we obtain an R-embedding $\varphi : T \longrightarrow Q(R)$. If $[\alpha] \in Q(R)$ is represented by $\alpha : D \longrightarrow R$, let $t \in T$ be such that $\alpha d = td$ for each $d \in D$. Hence $\varphi(t)[\mu_d] = \varphi(td) = \varphi(\alpha d) = [\mu_{\alpha d}] = [\alpha][\mu_d]$ and $(\varphi(t) - [\alpha])D = 0$; consequently, $\varphi(t) = [\alpha]$ and φ is an isomorphism.

(a) $\Rightarrow$ (d). It suffices to show that (d) holds for $Q(R)$. Using (d) of Theorem 4.1.2 and (b) we obtain an R-embedding $\varphi : U \longrightarrow Q(R)$. If $s \in U$ and $d \in (R : s)$, then $sd = \varphi(sd) = \varphi(s)d$; thus, $s = \varphi(s)$ and $U = Q(R)$ since $(R : s)$ is a dense right ideal of R.

(d) $\Rightarrow$ (a). Since (b) holds for $Q(R)$ there is an R-embedding $\varphi : T \longrightarrow Q(R)$. By replacing $\varphi(T)$ by T (that is, replace $Q(R)$ by $(Q(R) \backslash \varphi(T)) \cup T$) we get a right quotient ring of T. Hence, φ is an isomorphism. □

The *singular submodule* of the module M is defined by $Z(M) = \{x \in M : r(x) \in \mathscr{E}(R)\}$. It is a submodule since $r(x - y) \supseteq r(x) \cap r(y)$ and $r(xa) = (r(x) : a)$ if x, $y \in M$ and $a \in R$. If $\alpha \in [M, N]$, then clearly $\alpha(Z(M)) \subseteq Z(N)$. In particular, the *right singular ideal* $Z(R)$ of R is an ideal of R which we will sometimes denote by $Z_r(R)$ to distinguish it from its left counterpart $Z_\ell(R) = Z({}_RR)$. M is called *singular* if $M = Z(M)$ and it is *nonsingular* if $Z(M) = 0$. The class of nonsingular modules is hereditary and contains the injective hull of each of its members, and the class of singular modules is also hereditary. However, the injective hull of a singular module need not be singular. For an example let $R = F[x]/x^2F[x]$ where F is a field and let y be the image of x in R. Then R is an injective R-module, $Z(R) = yR$ and $E(yR) = R$ is not singular. However, if we enlarge the class of singular modules slightly, then this new class will be closed under injective hulls.

For a module M let $Z_1(M) = Z(M)$ and for $n \in \mathbb{N}$ define $Z_{n+1}(M)$ by $Z_{n+1}(M)/Z_n(M) = Z(M/Z_n(M))$.

Theorem 4.1.6. *For each module $M, Z_2(M) = Z_n(M)$ if $n \geq 2$. Consequently, if $Z_2(M) = M$, then $Z_2(E(M)) = E(M)$.*

Proof. It suffices to show that $Z(M)$ is essential in $Z_3(M)$. For then $(Z(M) : x) \in \mathscr{E}(R)$ for each $x \in Z_3(M)$, and so $Z_3(M) \subseteq Z_2(M)$. Suppose that $L \subseteq Z_3(M)$ with $L \cap Z(M) = 0$. If $x \in L$, then $xD \subseteq Z_2(M)$ for some essential right ideal D of R. For each $d \in D$ there is an essential right ideal D_d of R with $xdD_d \subseteq Z(M) \cap L = 0$. Thus, $xd \in Z(M) \cap L = 0$, $x \in Z(M) \cap L = 0$ and $L = 0$. Now, if $Z_2(M) = M$ and $x \in E(M)$, then $xD \subseteq M = Z_2(M) \subseteq Z_2(E(M))$ for some essential right ideal D; hence $x \in Z_3(E(M)) = Z_2(E(M))$. □

If N is a submodule of M, then the *closure of N in M*, $c\ell N$, is defined by $c\ell N/N = Z(M/N)$. So $c\ell N = \{x \in M : xD \subseteq N \text{ for some } D \in \mathscr{E}(R)\}$. The submodule N is *closed in M* if $c\ell N = N$. As a consequence of the preceding theorem we have $Z(M/N) = c\ell N/N \subseteq c\ell c\ell N/N = Z_2(M/N) = Z_3(M/N) = c\ell c\ell c\ell N/N$, and hence $c\ell c\ell N$ is the smallest closed submodule of M that contains N. For, if $Z_2(M/N) = K/N$, then

$$c\ell c\ell N / c\ell N = Z(M/c\ell N) \cong Z(M/N \,/\, c\ell N/N) = Z_2(M/N) \,/\, c\ell N/N$$

$$= K/N \,/\, c\ell N/N \cong K/c\ell N$$

and $K = c\ell c\ell N$. Similarly, $Z_3(M/N) = c\ell c\ell c\ell N/N$. Also, $c\ell N$ is essential in $c\ell c\ell N$ since $c\ell N/N$ is essential in $c\ell c\ell N/N$. If N is closed in M, then N is essentially closed in M, and the converse holds if M is nonsingular since N is then an essential submodule of $c\ell N$.

If R is a right nonsingular ring, then, as we will soon see, $Q(R)$ is all of the injective hull of R_R. It also has the following property. The element a in the ring R is *von Neumann regular* or just *regular* if $a = axa$ for some $x \in R$, and R is a *regular ring* if each of its elements is regular.

Theorem 4.1.7. *The following statements are equivalent for the ring R.*

(a) R is regular.
(b) Each right ideal of R with one generator is generated by an idempotent.
(c) Each finitely generated right ideal of R is generated by an idempotent.

Moreover, the class of regular rings is an i-hereditary radical.

Proof. (a) ⇒ (b). If $a, x \in R$ with $a = axa$, then $e = ax$ is idempotent with $a = ea$; so $aR = eR = aR + \mathbb{Z}a$.

(b) ⇒ (c). By induction on the number of generators of the right ideal it suffices to show that if e and f are idempotent, then $eR + fR = hR$ for some idempotent h. Now, $eR + fR = eR + (1-e)fR$ since $f = ef + (1-e)f$ and $(1-e)f = f - ef$, and $(1-e)fR = f_1R$ with $f_1^2 = f_1$ and $ef_1 = 0$. But then $g = f_1(1-e)$ is an idempotent orthogonal to e with $f_1g = g$ and $gf_1 = f_1$. So $eR + fR = eR + f_1R = eR + gR = (e+g)R$.

(c) ⇒ (a). If $a \in R$, then $aR + \mathbb{Z}a = eR$ with $e = e^2 \in (aR + \mathbb{Z}a)e \subseteq aR$; so $aR = eR$, $e = ax$, and $a = ea = axa$.

Let $\mathscr{P}$ be the class of regular rings. By the ring analogue of Theorem 3.2.1, in order to show that $\mathscr{P}$ is an i-hereditary radical it suffices to verify that $\mathscr{P}$ is extensionally closed. Toward this end let I be an ideal of R such that I and R/I are both regular. If $a \in R$ there exists some elment $x \in R$ with $a - axa \in I$, and also $a - axa \in (a-axa)I(a-axa)$. Thus, $a \in aRa$ and a is regular. □

A ring R is called *right self-injective* if the right R-module R_R is injective. Some examples of such rings are given in Exercises 2, 12, and 25. Here are some more examples.

Theorem 4.1.8. *The following statements are equivalent for the ring R.*

(a) *R is right nonsingular.*
(b) *Each essential right ideal of R is dense.*
(c) *R has a regular right quotient ring.*
(d) *$Q(R)$ is a regular right self-injective ring and $Q(R)$ is an injective right R-module.*

Proof. (a) ⇔ (b) If $Z_r(R) = 0$ and $D \in \mathscr{E}(R_R)$, then $\ell(D;E(R)) = 0$ since $E(R_R)$ is nonsingular. So $D \in \mathbb{D}(R)$ by Theorem 4.1.4. Conversely, if $\mathscr{E}(R) \subseteq \mathbb{D}(R)$, then trivially, $Z_r(R) = 0$.

(a) ⇒ (d). $Q = Q(R)$ is right nonsingular by Exercise 3. Let I be a right ideal of Q and let $\alpha : I \longrightarrow Q$ be a Q-homomorphism. If J is a complement of I in Q_Q, then $I \boxplus J \in \mathbb{D}(Q)$ and α can be extended to $I \boxplus J$; so we may assume that I is a dense right ideal of Q. According to Theorem 4.1.5 there exists $q \in Q$ with $\alpha a = qa$ for each $a \in I$ and hence Q_Q is injective by Baer's Criterion. If $x \in E(R_R)$ take $D \in \mathscr{E}(R_R) = \mathbb{D}(R)$ with $xD \subseteq R$. Again, by Theorem 4.1.5, there is some element $q \in Q \subseteq E(R)$ such that $xd = qd$ for each $d \in D$. But then $x - q \in \ell(D;E(R)) = 0$ by Theorem 4.1.4 and $x \in Q$. To see that Q is regular let $p \in Q$ and let L_Q be a complement of $r(p;Q)$ in Q_Q. The previous argument shows that the map $pL \longrightarrow L$ given by $px \mapsto x$ is given by left multiplication by some $q \in Q$; that is $x = qpx$ for each $x \in L$. So if $x + a \in L + r(p;Q) = D$, then $p(x+a) = px = pqp(x+a)$ and hence $p = pqp$ since D is dense.

Since (d) ⇒ (c) is trivial and (c) ⇒ (a) is a consequence of Exercise 3 the proof is complete. □

The kind of information displayed in Theorem 4.1.8 can also be obtained from certain "complete" modules. A module M_R is *quasi-injective* if each R-homomorphism $\alpha : N \longrightarrow M$ whose domain N is a submodule of M can be extended to an R-endomorphism of M. Every semisimple module (Exercise 2.6.9) as well as every injective module is quasi-injective. Other examples, as well as some useful properties of quasi-injective modules, are given in Exercises 6, 7, and 8. Recall that $J(R)$ denotes the Jacobson radical of the ring R.

Theorem 4.1.9. *Let M_R be quasi-injective, let $S = \mathrm{End}_R(M)$, and let $J = \{\alpha \in S : \ker\alpha \in \mathscr{E}(M)\}$. Then the following hold.*

(a) *$J = J(S)$ and if M is nonsingular, then $J = 0$.*

(b) S/J is a regular right self-injective ring.
(c) Idempotents can be lifted modulo J.
(d) If $T = End_R(E(M))$, then $T/J(T) \cong S/J(S)$.

Proof. If $\alpha, \beta \in J$ and $\gamma \in S$, then $\ker(\alpha - \beta) \supseteq \ker\alpha \cap \ker\beta$ and $\ker\gamma\alpha \supseteq \ker\alpha$ imply that J is a left ideal of S; also, $\gamma^{-1}\ker\alpha \subseteq \ker\alpha\gamma$ and $\gamma^{-1}\ker\alpha \in \mathscr{E}(M)$ imply that J is a right ideal. Since $\ker\alpha \cap \ker(1-\alpha) = 0$, necessarily $\ker(1-\alpha) = 0$, and the map $(1-\alpha)M \longrightarrow M$ defined by $(1-\alpha)x \mapsto x$ is the restriction of some $\gamma \in S$. Thus $\gamma(1-\alpha) = 1$ and J is a left quasi-regular ideal of S; so J is contained in $J(S)$.

The argument used in the previous result to show that $Q(R)$ is regular can be adapted to show that $\bar{S} = S/J$ is regular. If $\alpha \in S$ and L is a complement of $\ker\alpha$ in M, then there exists $\gamma \in S$ such that $\gamma\alpha x = x$ for each $x \in L$. Since $L + \ker\alpha \in \mathscr{E}(M)$ and $L + \ker\alpha \subseteq \ker(\alpha - \alpha\gamma\alpha)$ we have $\bar{\alpha} = \bar{\alpha}\bar{\gamma}\bar{\alpha}$. Thus, $\bar{S}$ is regular and $J(S) = J$. If $\alpha \in J$ and $x \in M$, then $D = (\ker\alpha : x) \in \mathscr{E}(R)$ and $\alpha(x)D = 0$; so $\alpha M \subseteq Z(M)$ and $\alpha = 0$ if M is nonsingular.

To prove (c) note that for any $\alpha \in S$, $\ker(\alpha^2 - \alpha) \subseteq \ker\alpha \boxplus \ker(1-\alpha)$. So if $\alpha^2 - \alpha \in J$, then $\ker\alpha \boxplus \ker(1-\alpha)$ is essential in M_R and $E(M) = E(\ker\alpha) \boxplus E(\ker(1-\alpha))$ by Exercise 5. Let $\beta \in \mathrm{End}_R(E(M))$ be the projection of $E(M)$ onto $E(\ker(1-\alpha))$. Then $\beta M \subseteq M$ by Exercise 7 and β and α agree on $\ker\alpha \boxplus \ker(1-\alpha)$. If γ is the restriction of β to M, then $\gamma^2 = \gamma$ and γ lifts the idempotent $\bar{\alpha} \in \bar{S}$.

Let $\varphi : T \longrightarrow S/J(S)$ be the ring epimorphism defined by $\varphi(\alpha) = \alpha_* + J(S)$ where α_* denotes the restriction of α to M. Then $\varphi(\alpha) = 0$ iff $\ker\alpha \cap M \in \mathscr{E}(M)$, iff $\ker\alpha \in \mathscr{E}(E(M))$; so $\ker\varphi = J(T)$.

The only thing left to prove is that $\bar{S}$ is right self-injective. Let $f : \bar{I} \longrightarrow \bar{S}$ be an $\bar{S}$-homomorphism where $\bar{I}$ is a right ideal of $\bar{S}$, and let $\{\bar{\alpha}_\lambda \bar{S} : \lambda \in \Lambda\}$ be a maximal independent family of principal right ideals of $\bar{S}$ contained in $\bar{I}$. Since $\bar{S}$ is regular we may assume, using (c), that each α_λ is idempotent. For each $\lambda \in \Lambda$ choose $\beta_\lambda \in S$ with $f(\bar{\alpha}_\lambda) = \bar{\beta}_\lambda$. Assume, for the moment, that the sum $N = \sum_\lambda \alpha_\lambda M$ is a direct sum in M. Then the restrictions of β_λ to $\alpha_\lambda M$ define an element in $\mathrm{Hom}_R(N, M)$, and hence there is an element $\gamma \in S$ such that $\gamma\alpha_\lambda x = \beta_\lambda \alpha_\lambda x$ for each $\lambda \in \Lambda$ and each $x \in M$; that is, $\gamma\alpha_\lambda = \beta_\lambda \alpha_\lambda$ for each λ. But then

$$f(\bar{\alpha}_\lambda) = f(\bar{\alpha}_\lambda)\bar{\alpha}_\lambda = \bar{\beta}_\lambda \bar{\alpha}_\lambda = \bar{\gamma}\bar{\alpha}_\lambda$$

for each λ. Since $\boxplus \bar{\alpha}_\lambda \bar{S}$ is essential in $\bar{I}$ and $\bar{S}$ is nonsingular, $f(\bar{\rho}) = \bar{\alpha}\bar{\rho}$ for each $\bar{\rho} \in \bar{I}$. Thus, $\bar{S}$ is right self-injective by Baer's Criterion. It remains to prove that the sum $\sum \alpha_\lambda M$ is direct, and for this purpose we may assume that Λ is finite and $\{\alpha_\lambda\} = \{\alpha_1, \ldots, \alpha_n\}$. Since $\bar{S}$ is regular $\bar{\alpha}_1 \bar{S} \boxplus \cdots \boxplus \bar{\alpha}_n \bar{S}$ is a direct summand of $\bar{S}_{\bar{S}}$ and hence there exist orthogonal idempotents $\bar{\beta}_1, \ldots, \bar{\beta}_n$ in $\bar{S}$ with $\bar{\beta}_i \bar{S} = \bar{\alpha}_i \bar{S}$ for each i. By Theorem 3.4.15 we may assume that $\{\beta_1, \ldots, \beta_n\}$ is a set of orthogonal idempotents in S. Since $\alpha_i - \beta_i\alpha_i \in J$, $K_i = \ker(\alpha_i - \beta_i\alpha_i) \in \mathscr{E}(M)$ and $K_i \cap \alpha_i M$ is essential in $\alpha_i M$. But $x \in K_i \cap \alpha_i M$ iff $x = \alpha_i x$ and $\alpha_i x = \beta_i \alpha_i x$, iff $x = \alpha_i x = \beta_i x$. Thus, $\alpha_i M \cap \beta_i M = K_i \cap \alpha_i M$ is essential in $\alpha_i M$ and, similarly, $\alpha_i M \cap \beta_i M$ is essential in $\beta_i M$. Since the sum $\sum \beta_i M$ is direct the sum $\sum \alpha_i M$ is also direct (Exercise 5). □

There are regular right self-injective rings that are not left self-injective (see Exercise 12). So the maximal right quotient ring of R need not be equal to its maximal left quotient ring. But there is a maximal two-sided quotient ring of R. If S is a right quotient ring of R as well as a left quotient ring of R, then S is called a *two-sided quotient ring of* R. The collection of dense left ideals of R will be denoted by $\mathbb{D}_{\text{left}}(R)$. Suppose that $\ell(R) = r(R) = 0$. Let

$$Q_2(R) = \{p \in Q(R) : Dp \subseteq R \text{ for some } D \in \mathbb{D}_{\text{left}}(R)\}.$$

Then $Q_2(R)$ is the unique (up to an R-isomorphism) maximal two-sided quotient ring of R. We will leave the verification of this to Exercise 13. If R is reduced, then so is $Q_2(R)$ (Exercise 14) but $Q(R)$ need not be reduced. In fact, when R is a domain $Q(R)$ is reduced precisely when it is a division ring (Exercise 26). In order to determine when $Q(R)$ is reduced we will first define a stronger version of regularity. The element a in R is *strongly regular* if $a \in a^2R$, and R is a *strongly regular ring* if each of its elements is strongly regular. A strongly regular ring is just a reduced regular ring; see Exercise 22. Since the class of regular rings is a radical class and since the class of reduced rings is extensionally closed, the class of strongly regular rings is also a radical class (Theorem 3.2.1).

In a reduced ring there is no distinction between left and right annihilators, and we will denote the annihilator of the subset X in the reduced ring R by X^*. Recall also that in a semiprime ring the left and right annihilators of an ideal coincide. Since $x^* \cap xR = 0 = x^* \cap Rx$ for each $x \in R$ both singular ideals vanish in a reduced ring; for a generalization see Exercise 20.

Theorem 4.1.10. *The following statements are equivalent for the right nonsingular ring R.*

(a) If $a, b \in R$ with $aR \cap bR = 0$, then $ab = 0$.
(b) $Q(R)$ is reduced.
(c) R has a strongly regular right quotient ring.
(d) Each essentially closed right ideal of R is the right annihilator of an ideal of R.
(e) Each essentially closed right ideal of R is an ideal.

If R is reduced, then each of the following is equivalent to (a).

(f) If I is a right ideal of R, then I is essential in I^{**} (and I^{**} is the maximal essential extension of I in R).
(g) If I is a right ideal of R, then I^* is a right complement of I (and is the unique right complement of I).

Proof. (a) $\Rightarrow$ (b). Let I be a right ideal of R. Since $J \subseteq r(I)$ if J is a right ideal with $I \cap J = 0$, the right ideal $I + r(I)$ is essential in R. So if $I^2 = 0$, then $I = 0$, and hence R is semiprime. Let $a \in R$. From $(r(aR) \cap aR)^2 = 0$ we obtain $r(aR) \cap aR = 0$ and hence $r(aR)aR = 0$, $(ar(aR))^2 = 0$, and $ar(aR) = 0$. If $a^2 = 0$, then $a(aR + r(aR)) = 0$ yields that $a = 0$; so R is reduced. Suppose that $q \in Q = Q(R)$ and D is a dense right ideal of R with $qD \subseteq R$. Then $J = qD \boxplus ((qD)^* \cap D)$ is essential in $qD \boxplus (qD)^*$ and hence is also essential in R. If $x \in (qD)^* \cap D$, then $xq = 0$, $(qx)^2 = 0$, and $qx = 0$; so if $q^2 = 0$, then $qJ = 0$ and $q = 0$.

(c) $\Rightarrow$ (a). Since the idempotents in a reduced ring are central, if T is a strongly regular right quotient ring of R, then T satisfies the condition in (a) and hence so does R. For if $a, b \in R$ with $aR \cap bR = 0$ and $0 \neq as = bt \in aT \cap bT$, then for some $D \in \mathbb{D}(R)$, $sD, tD \subseteq R$. So $0 \neq asD \subseteq aR \cap bR$. Thus, $aT \cap bT = 0$ and $ab = 0$.

(b) $\Rightarrow$ (d). Let I be an essentially closed right ideal of R and let J be a complement of I in R. Then $I = E(I) \cap R$ and $J = E(J) \cap R$ and $Q = E(R) = E(I) \boxplus E(J)$. Now $E(I)$ and $E(J)$ are ideals of Q (by Exercise 22); so I and J are ideals of R. Also, $IJ \subseteq E(I)E(J) = 0$ gives that $I \subseteq J^*$ and hence $I = J^*$ since I is a complement of J.

The implications (b) $\Rightarrow$ (c) and (d) $\Rightarrow$ (e) are trivial, and as for (e) $\Rightarrow$ (a), if $I \cap J = 0$, then $c\ell I \cap c\ell J = 0$ and $IJ \subseteq c\ell I c\ell J = 0$.

If R is now reduced, then it is easy to establish the implications (a) $\Rightarrow$ (f) $\Rightarrow$ (g) $\Rightarrow$ (d). We will check the second implication in this chain and leave the rest to the reader. If $I \cap J = 0$, then $I^{**} \cap J = 0$ and $J \subseteq I^{***} = I^*$; so I^* is the right complement of I in R. □

One consequence of the previous result is that the property of $Q(R)$ being strongly regular is inherited by right ideals of R (Exercise 34). Another consequence is that the left and right maximal quotient rings of R coincide provided R has strongly regular quotient rings on each side.

Theorem 4.1.11. *The following statements are equivalent for the ring R.*

(a) R has a strongly regular right quotient ring and a strongly regular left quotient ring.
(b) $Q_2(R)$ is strongly regular.
(c) $Q(R) = Q_2(R)$ is strongly regular.

Proof. Since the implications (c) $\Rightarrow$ (b) $\Rightarrow$ (a) are trivial we only need to show that (c) is a consequence of of (a). Since $Q = Q(R)$ is a strongly regular right self-injective ring it is also left self-injective. To see this let T be the maximal left quotient ring of Q and note that T is left self-injective. If $0 \neq t \in T$ let D be a dense left ideal of Q with $Dt \subseteq Q$. Take $d \in D$ with $dt \neq 0$. Then $dQ = eQ$ with $e = e^2$ and $0 \neq et = te \in tQ \cap Q$, that is, Q_Q is essential in T_Q and hence $Q = T$. To show that $Q = Q_2(R)$ it suffices to verify that ${}_RR$ is essential in ${}_RQ$. Let $0 \neq q \in Q$ and take $d \in R$ with $0 \neq qd \in R$. Since $qd^2 \neq 0$, $Rqd \cap Rd \neq 0$ by the left-sided version of Theorem 4.1.10. Take $a, b \in R$ with $aqd = bd \neq 0$ and let $y \in Q$ with $dyd = d$. Then $dy = yd$ is idempotent and

$$daq = dydaq = daqdy = dbdy = dydb = db \neq 0;$$

hence $Rq \cap R \neq 0$. □

In a reduced ring one can weaken the hypothesis that each finitely generated right ideal is generated by an idempotent and still obtain the conclusion that the maximal right quotient ring is strongly regular.

Theorem 4.1.12. *If R is a reduced ring in which each finitely generated right ideal is principal, then $Q(R)$ is reduced.*

Proof. Let $S = R + \mathbb{Z}1 \subseteq Q_2(R)$, and suppose $q \in Q(R)$ with $q^2 = 0$. If $d \in R$ with $qd \in R$, then $qdS + dS = cS$ for some $c \in R$. Take x, y, a, $b \in S$ with

$$qdx + dy = c,$$
$$d = ca,$$
$$qd = cb.$$

Then
$$cbyb = qdyb = q(c - qdx)b = qcb = q^2d = 0.$$

Since S is reduced $0 = cby = qdy$, and hence $qc = 0$ and $qd = 0$. Thus, $q(R : q) = 0$ and hence $q = 0$. □

If R is a semiprime ring then, in general, there is no relationship between $Q(R)$ and the maximal right quotient rings of the prime homomorphic images of R. However, if R is a large subdirect product of prime rings, then the expected satisfactory relationship does hold.

Suppose that $f : R \longrightarrow \Pi_{\lambda \in \Lambda} R_\lambda$ is a subdirect product of the family of rings $\{R_\lambda : \lambda \in \Lambda\}$. This subdirect product representation of R is called *irredundant* if for each $\mu \in \Lambda$ the homomorphism

$$R \xrightarrow{f} \prod_{\lambda \in \Lambda} R_\lambda \longrightarrow \prod_{\lambda \neq \mu} R_\lambda$$

has a nonzero kernel; here, the second map is the projection onto the indicated factor of the full product. Clearly, an equivalent condition is that $f(R) \cap R_\mu \neq 0$ for each $\mu \in \Lambda$. Note that R is an irredundant subdirect product of the family $\{R_\lambda : \lambda \in \Lambda\}$ iff R has a family of ideals $\{P_\lambda : \lambda \in \Lambda\}$ with zero intersection and, for each $\mu \in \Lambda$, $R_\mu \cong R/P_\mu$ and $\bigcap_{\lambda \neq \mu} P_\lambda \neq 0$.

The concept of an irredundant subdirect product is, of course, meaningful for groups and ℓ-groups with operators as well as other algebraic structures. In fact it has already come up in Exercises 2.5.25, 2.5.26, and 2.5.27.

We first present a ring theory analogue of Exercise 2.2.14(h). If R is a subring of S and A and B are ideals of R and S, respectively, then we will write $A'^R = \ell(A;R)$ and $B' = \ell(B;S)$ for the (left) annihilators of A and of B. Recall from Exercise 1.2.7 that when R is semiprime the set $\mathrm{Ann}(R)$ of annihilator ideals of R is a complete Boolean algebra with respect to inclusion and the operator $'^R$ and $\ell(A;R) = r(A;R)$ for any ideal A of R.

Theorem 4.1.13. *Let S be a two-sided quotient ring of the semiprime ring R. Then the mappings*

$$\varphi : Ann(R) \longrightarrow Ann(S),\ \varphi(A) = (SA'^R S)'$$

and

$$\psi : Ann(S) \longrightarrow Ann(R),\ \psi(B) = B \cap R$$

are inverse Boolean algebra isomorphisms.

Proof. We will first show that if $B = B'' \in \mathrm{Ann}(S)$, then $B \cap R = (B' \cap R)'^{R}$; so $\psi(B)$ is an element of $\mathrm{Ann}(R)$. Clearly $B \cap R \subseteq (B' \cap R)'^{R}$. For the reverse inclusion suppose $a \in R$ and $a(B' \cap R) = 0$. If $s \in B'$, then since $(R : s)$ is a dense right ideal of R and $as(R : s) \subseteq a(B' \cap R) = 0$, necessarily $as = 0$; so $aB' = 0$ and $a \in B'' = B$. Now,

$$\varphi\psi(B) = (S(B' \cap R)'^{R\prime R} S)' = (S(B' \cap R)S)'$$
$$= B'' = B$$

because $S(B' \cap R)S$ is essential in B' by Exercise 30 and hence the next to last equality is a consequence of Exercise 29.

To see that A is contained in $\psi\varphi(A) = (SA'^{R}S)' \cap R$ take $s = \sum_i s_i x_i t_i \in SA'^{R}S$ with $s_i, t_i \in S$ and $x_i \in A'^{R}$. Let D be a dense right ideal of R and F a dense left ideal of R with $t_i D + sD \subseteq R$ and $Fs_i + Fs \subseteq R$. Then $FsD \subseteq A'^{R}$; so $AFsD = 0$, $AFs = 0$, $FsA = 0$, $sA = 0$, and $A \subseteq (SA'^{R}S)'$. To see the reverse inclusion note that

$$\psi\varphi(A) = (SA'^{R}S)' \cap R = ((SA'^{R}S)'' \cap R)'^{R}.$$

By Exercise 30

$$(SA'^{R}S)' = (SA'^{R} + A'^{R}S + SA'^{R}S + A'^{R})',$$

and hence $A'^{R}(SA'^{R}S)' = 0$ and $A'^{R} \subseteq (SA'^{R}S)'' \cap R$. Thus,

$$((SA'^{R}S)'' \cap R)'^{R} \subseteq A'^{R\prime R} = A$$

and $\psi\varphi(A) = A$. □

A (semiprime) ring is called *irredundant* if it is an irredundant subdirect product of a family of prime rings.

Theorem 4.1.14. *(a) A semiprime ring R is irredundant if and only if its Boolean algebra of annihilator ideals, Ann(R), is atomic.*

(b) If R is irredundant and $\{P_\lambda : \lambda \in \Lambda\}$ is the set of maximal elements in Ann(R), then each P_λ is a minimal prime ideal of R, and R is an irredundant subdirect product of the prime rings $\{R/P_\lambda : \lambda \in \Lambda\}$.

(c) If R is an irredundant subdirect product of the prime rings $\{R_{\lambda'} : \lambda' \in \Lambda'\}$, then the set of kernels of the projections $R \longrightarrow R_{\lambda'}$ coincides with $\{P_\lambda : \lambda \in \Lambda\}$. In particular, R has a unique irredundant decomposition.

(d) If R is irredundant, then $Q_2(R) = \Pi_\lambda Q_2(R/P_\lambda)$ and $Q(R) = \Pi_\lambda Q(R/P_\lambda)$.

Proof. Suppose that R is an irredundant subdirect product of the prime rings $\{R_\lambda : \lambda \in \Lambda\}$ and identify R with its image in $S = \Pi_\lambda R_\lambda$. Since $R \cap R_\lambda$ is a nonzero ideal of R_λ the latter is a two-sided quotient ring of the former (Exercise 15). Thus, since $\Sigma_\lambda(R \cap R_\lambda) \subseteq R \subseteq S$, $Q(R) = \Pi_\lambda Q(R_\lambda)$ and $Q_2(R) = \Pi_\lambda Q_2(R_\lambda)$ by Exercise 17.

According to Theorem 4.1.13 and Exercise 32, $\mathrm{Ann}(R) \cong \mathrm{Ann}(S) \cong \mathscr{P}(\Lambda)$ via the isomorphisms $A \mapsto \Pi\{R_\lambda : A \cap R_\lambda \neq 0\} \mapsto \{\lambda : A \cap R_\lambda \neq 0\}$, and hence Ann($R$) is atomic. Thus, A is a maximal annihilator of R iff there is some $\mu \in \Lambda$ such that

$A = P_\mu = (\Pi_{\lambda \neq \mu} R_\lambda) \cap R = (R_\mu \cap R)'^R$; that is, iff A is the kernel of the projection $R \longrightarrow R_\mu$. By Exercise 33 each P_μ is a minimal prime of R. Conversely, if $\mathrm{Ann}(R)$ is atomic, then R is irredundant by Exercise 33 since in a complete atomic Boolean algebra the inf of the set of maximal elements is 0 but the inf of all the maximal elements but one is nonzero. □

We will show next that "torsion-free" modules over an irredundant ring also have unique irredundant subdirect product representations. The notation given in Theorem 4.1.14 will be fixed for the irredundant ring R, and we will consider R to be a subring of the direct product $\Pi_\lambda R_\lambda$. The rings R_λ will be called the *components of* R. If $X \subseteq R$, then X_λ denotes the projection of X into R_λ. Each R_λ-module will be considered an R-module via the projection $R \longrightarrow R_\lambda$.

The R-module M_R is called *I-torsion-free* if for each ideal J of R, $\ell(J;R) = 0$ implies $\ell(J;M) = 0$, or, equivalently, if $NJ = 0$ for some nonzero submodule N of M, then $KJ = 0$ for some nonzero ideal K of R. Each nonsingular module is I-torsion-free and the converse holds when R is commutative. However, this is not the case generally; see Exercise 38.

Theorem 4.1.15. *Let R be an irredundant ring.*

(a) If J is an ideal of R, then $J' = 0$ iff $J_\lambda \neq 0$ for each $\lambda \in \Lambda$.
(b) The R_λ-module M is I-torsion-free iff it is an I-torsion-free R-module.
(c) If M is an I-torsion-free R_λ-module, then its R-injective hull coincides with its R_λ-injective hull.

Proof. (a) If $J' = 0$, then $J_\lambda(R \cap R_\lambda) = J(R \cap R_\lambda) \neq 0$. Conversely, if each $J_\lambda \neq 0$ and $a \in R$ with $aJ = 0$, then $a = 0$ since $a_\lambda J_\lambda = 0$ for each λ.

(b) Suppose that M is I-torsion-free as an R-module, and J is an ideal of R which properly contains P_λ. If N is a submodule of M with $NJ = 0$, then $N(J(R \cap R_\lambda) + P_\lambda) = 0$. But $(J(R \cap R_\lambda) + P_\lambda)' = 0$ by (a); so $N = 0$ and M is R_λ-I-torsion-free. Conversely, suppose M is R_λ-I-torsion-free, and let J be an ideal of R with $J' = 0$. If N is a submodule of M with $NJ = 0$, then $NJ_\lambda = 0$ and hence $N = 0$.

(c) Let $E = E(M_R)$. Since $(EP_\lambda \cap M)(R \cap R_\lambda) = 0$, necessarily $EP_\lambda = 0$ and E is an R_λ-module. Clearly, E is an R_λ-essential extension of M. If E is a submodule of the R_λ-module L, then E is an R-summand and hence an R_λ-summand of L. Thus, E is the R_λ-injective hull of M. □

Before we give the irredundant decomposition of an I-torsion-free module we present an analogue of Theorem 4.1.13 which will be useful in showing the uniqueness of the decomposition.

Theorem 4.1.16. *Suppose R is an irredundant subdirect product of the prime rings $\{R_\lambda : \lambda \in \Lambda\}$, and, for each $\lambda \in \Lambda$, let M_λ be an I-torsion-free R_λ-module. Assume M_R is a subdirect product of $\{M_\lambda : \lambda \in \Lambda\}$ and let $\Gamma = \{\lambda \in \Lambda : M_\lambda \neq 0\}$.*

(a) M is I-torsion-free and is an irredundant subdirect product of $\{M_\lambda : \lambda \in \Gamma\}$.
(b) If J is an ideal of R, then $\ell(J;M) = \{x \in M : x_\lambda = 0$ whenever $J_\lambda \neq 0\}$.

(c) If N is an R-submodule of M, then $r(N;R) = \{a \in R : a_\lambda = 0$ whenever $N_\lambda \neq 0\}$.

(d) Let $\mathscr{I}(R)$ be the lattice of ideals of R and let $Ann(\mathscr{I}(R);M)$ be the lattice of annihilator submodules in M of the ideals of R. Then the mapping $\rho : \mathscr{P}(\Gamma) \longrightarrow Ann(\mathscr{I}(R);M)$ defined by

$$\rho(\Omega) = \{x \in M : x_\lambda = 0 \text{ for each } \lambda \in \Gamma\backslash\Omega\}$$

is a lattice isomorphism.

Proof. The subdirect product is irredundant since if $\lambda \in \Gamma$, then $0 \neq M(R \cap R_\lambda) \subseteq M \cap M_\lambda$. Moreover, M is I-torsion-free by (b) of Theorem 4.1.15 and Exercise 35. For (b), $xJ = 0 (x \in M)$ iff $x_\lambda J_\lambda = 0$ for each $\lambda \in \Gamma$, iff either $x_\lambda = 0$ or $J_\lambda = 0$ for each $\lambda \in \Gamma$. Similarly, if $a \in R$, then $Na = 0$ iff $a_\lambda = 0$ or $N_\lambda = 0$ for each $\lambda \in \Gamma$, and this proves (c). By (b), $\rho(\Omega) = \ell(\Sigma_{\lambda \in \Gamma\backslash\Omega} R_\lambda \cap R;M)$ is an annhilator submodule of M. Also, if J is an ideal of R and $\Omega = \{\lambda \in \Gamma : J_\lambda = 0\}$, then $\rho(\Omega) = \ell(J;M)$; so ρ is onto. To see that ρ is 1-1 suppose $\mu \in \Omega\backslash\Delta$ where Ω and Δ are subsets of Γ. Then $M \cap M_\mu \subseteq \rho(\Omega)\backslash\rho(\Delta)$ since $(M \cap M_\mu)(\Sigma_{\lambda \notin \Delta} R \cap R_\lambda) \neq 0$ but $(M \cap M_\mu)(\Sigma_{\lambda \notin \Omega} R \cap R_\lambda) = 0$. Since ρ is clearly isotone this also shows that it is a lattice isomorphism. □

Theorem 4.1.17. *Let M_R be a module over the irredundant ring R. Then M is I-torsion-free iff there is a subset Γ of Δ and a family of modules $\{M_\lambda : \lambda \in \Gamma\}$ such that M_λ is an R_λ-I-torsion-free module and M is an irredundant subdirect product of $\{M_\lambda : \lambda \in \Gamma\}$. This representation of M is unique (up to isomorphisms of the factors), and the set of kernels of the projections $M \longrightarrow M_\lambda$ coincides with the maximal annihilators in $Ann(\mathscr{I}(R);M)$. Moreover, $E(M_R) = \Pi_{\lambda \in \Gamma} E((M_\lambda)_{R_\lambda})$ if and only if $M \cap M_\lambda$ is essential in M_λ for each $\lambda \in \Gamma$.*

Proof. Suppose M is I-torsion-free. Let $N_\lambda = \ell(R_\lambda \cap R;M)$ for each $\lambda \in \Lambda$, and set $\Gamma = \{\lambda : N_\lambda \neq M\}$. Then $\bigcap N_\lambda = 0$ since $(\Sigma_\lambda R \cap R_\lambda)' = 0$ and $(\bigcap N_\lambda)(\Sigma R \cap R_\lambda) = 0$. Thus, M is a subdirect product of $\{M/N_\lambda : \lambda \in \Gamma\}$ and each M/N_λ is an $R_\lambda \cong R/P_\lambda$-module since $MP_\lambda = M(R_\lambda \cap R)' \subseteq N_\lambda$. To see that M/N_λ is an I-torsion-free R-module take an ideal J of R with $J' = 0$ and suppose $x \in M$ with $xJ \subseteq N_\lambda$. Then $x(R \cap R_\lambda)J \subseteq N_\lambda$ and if $\mu \neq \lambda$ we have $x(R \cap R_\lambda)J \subseteq N_\mu$. So $x(R \cap R_\lambda)J = 0$, $x(R \cap R_\lambda) = 0$ and $x \in N_\lambda$. By (b) of Theorem 4.1.15 M/N_λ is also an I-torsion-free R_λ-module. By the previous result the subdirect product is irredundant, and N is a maximal annihilator in $\text{Ann}(\mathscr{I}(R);M)$ iff there is an element $\mu \in \Gamma$ with $N = \rho(\Gamma\backslash\{\mu\}) = \ell(R_\mu \cap R;M) = N_\mu$. Suppose $M \subseteq \Pi_{\Gamma'} M_{\lambda'}$ is another such decomposition of M. Then, as we have just noted, $\Gamma = \{\lambda \in \Lambda : \ell(R_\lambda \cap R;M\} \neq M\} = \Gamma'$ and, for each $\lambda \in \Gamma$, $N_\lambda = \ell(R \cap R_\lambda;M) = \{x \in M : x_\lambda = 0\}$ is the kernel of the projection $M \longrightarrow M_\lambda$. Let E_λ be the R_λ-injective hull of $M_\lambda \cong M/N_\lambda$. Then E_λ is the R-injective hull of M_λ by Theorem 4.1.15 and $E = \Pi_\lambda E_\lambda$ is R-injective. Suppose each $M \cap M_\lambda$ is essential in M_λ. By Exercises 35 and 37 E_λ is I-torsion-free and M is essential in $\Pi_\lambda M_\lambda$. Let $0 \neq e = (e_\lambda) \in E$. Then for some $\mu \in \Gamma$, $e(R \cap R_\mu) = e_\mu(R \cap R_\mu) \neq 0$ and $e(R \cap R_\mu) \cap M_\mu \neq 0$. Thus, ΠM_λ is essential in E

and hence M is essential in E. Conversely, if $E(M) = E$, then $M \cap M_\lambda$ is essential in M_λ since M is essential in E. □

The modules M_λ will be called the *components* of the I-torsion-free R-module M. Occasionally, even those $M_\lambda = 0$ will be considered components of M. If each M_j is I-torsion-free the preceding theorem (and Exercise 35) easily give that $(\boxplus M_j)_\lambda = \boxplus (M_j)_\lambda$ and $(\Pi M_j)_\lambda = \Pi (M_j)_\lambda$ for each λ. Also, the module M will be called *essential* if $M \cap M_\lambda$ is essential in M_λ for each $\lambda \in \Lambda$. Note that each summand of an essential I-torsion-free R-module is also essential. If each independent family of nonzero submodules of M is finite, then it has only finitely many components. A module, over any ring, which lacks infinite direct sums is called *finite dimensional*, and a nonzero module which contains no direct sums at all is called *uniform*. For some properties of finite dimensional rings and modules see Exercises 26, 43, and 44.

Exercises.

1. A pair of maps

$$A \xrightarrow{\alpha} B \xrightarrow{\beta} C$$

in a module category is *exact* if the image of α is equal to the kernel of β. A sequence

$$\cdots \longrightarrow A_{n-1} \longrightarrow A_n \longrightarrow A_{n+1} \longrightarrow \cdots$$

is *exact* if each consecutive pair of maps is exact; and an exact sequence of the form

$$0 \longrightarrow A \longrightarrow B \longrightarrow C \longrightarrow 0 \qquad (4.1.1)$$

is called a *short exact sequence*. A covariant functor $F : \mathscr{C} \longrightarrow \mathscr{D}$ between two module categories is *left (right) exact* if

$$0 \longrightarrow F(A) \longrightarrow F(B) \longrightarrow F(C)$$

$$(F(A) \longrightarrow F(B) \longrightarrow F(C) \longrightarrow 0)$$

is exact for each short exact sequence (4.1.1). Here, we are assuming that F is *additive*; that is, $F(\alpha + \beta) = F(\alpha) + F(\beta)$ whenever $\alpha + \beta$ is defined. Analogous definitions can be given for contravariant functors; note that left exactness of the contravariant functor F means exactness of

$$0 \longrightarrow F(C) \longrightarrow F(B) \longrightarrow F(A).$$

The functor F is *exact* if it is both left and right exact.

(a) Show that the covariant functor $[M, \cdot]$ and the contravariant functor $[\cdot, M]$ are left exact for each module M.
(b) Show that M is injective iff the functor $[\cdot, M]$ is exact.
(c) Show that $M \otimes_R \cdot$ is right exact, and it is exact if M is projective. Here and in (d) modules are unital.

(d) Show that $[M,\cdot]$ is exact iff M is projective.
(e) Let ${}_HM_R$ be a bimodule. Show that the functors $\mathrm{Hom}_H(\cdot,\mathrm{Hom}_R({}_HM_R,N_R))$ and $\mathrm{Hom}_R(\cdot \otimes_H M,N_R)$ are naturally equivalent for each module N_R.

2. Let R be a commutative principal ideal domain. If I is a nonzero ideal of R show that R/I is self-injective. (Reduce to the case that I is a power of a prime ideal and use the composition series of R/I.)

3. Suppose that R is a subring of S and R_R is essential in S_R.

(a) If $Z_r(R)=0$ show that S is a right quotient ring of R.
(b) Show that R is right nonsingular iff S is right nonsingular.
(c) Show that $Z_r(R)$ does not contain a nonzero idempotent element.
(d) If R has a regular right quotient ring show that R is right nonsingular.

4. Let S be a right quotient ring of R.

(a) If A_R is an R-submodule of S_R, show that A_R is essential in $A+AS$.
(b) If A_R is an R-submodule of S_R, show that A_R is essential in S_R iff AS is essential in S_S.
(c) Show that $Z_r(R)=R\cap Z_r(S)$.
(d) Show that $D\in\mathbb{D}(R)$ iff $DS\in\mathbb{D}(S)$.

5. (a) Suppose that, for each λ in Λ, A_λ is an essential submodule of B_λ and B_λ is a submodule of M. If $\sum A_\lambda$ is direct show that $\sum B_\lambda$ is direct and is an essential extension of $\sum A_\lambda$.
(b) Show that $E(A_1\boxplus\cdots\boxplus A_n)=E(A_1)\boxplus\cdots\boxplus E(A_n)$.

6. Let Q be the field of quotients of the commutative principal ideal domain R, and let p be a prime element of R.

(a) If $Q_p=\{u\in Q/R : p^nu=0 \text{ for some } n\in\mathbb{N}\}$ is the pth-primary component of Q/R show that Q_p is an injective R-module.
(b) Show that, for each $n\in\mathbb{N}$, $A_n=R(\frac{1}{p^n}+R)$ is a quasi-injective R-submodule of Q/R and $E(A_n)=Q_p$.

7. Let $E=E(M_R)$ and $H=\mathrm{End}_R(E)$.

(a) Show that M_R is quasi-injective iff M is a left H-submodule of E. (If $\alpha,\gamma\in H$ and $M\cap\alpha^{-1}(M)\subseteq\ker(\gamma-\alpha)$ show that $(\gamma-\alpha)M\cap M=0$.)
(b) Show that HM is the smallest quasi-injective submodule of E_R that contains M.
(c) Show that each monomorphism of M_R into a quasi-injective module L can be extended to a monomorphism of HM into L.

8. Suppose that M is quasi-injective.

(a) If $E(M)=\boxplus_\lambda E_\lambda$, show that $M=\boxplus_\lambda(E_\lambda\cap M)$.
(b) Let N be a submodule of M. Show that N is essentially closed in M iff N is a summand of M.
(c) Show that each summand of M is quasi-injective.

9. If R is a unital right self-injective ring show that $J(R) = Z_r(R)$ and $R/J(R)$ is a regular right self-injective ring.

10. Show that a unital right artinian ring R is right noetherian. (Use the nilpotency of $J(R)$ and Theorem 4.2.13 to show that R_R has a composition series.)

11. Show that the following are equivalent for the ring R.

    (a) R is right noetherian.
    (b) The direct sum of any family of injective right R-modules is injective.
    (c) The direct sum of any countable family of injective right R-modules is injective.

    (Assume R and all modules considered are unital. For (c) $\Rightarrow$ (a), if $I_1 \subseteq I_2 \subseteq \cdots$ is a countable chain of right ideals of R let I be their union. Let E_n be an injective R-module containing I/I_n. Define $f : I \longrightarrow \boxplus_n E_n$ by $f(x) = (x + I_n)_n$.)

12. Let $H = \mathrm{End}_R(M)$ be the endomorphism ring of the module M_R.

    (a) If $M = \boxplus_{\lambda\in\Lambda} M_\lambda$ with Λ infinite and each $M_\lambda \neq 0$ show that ${}_HM$ is not injective. (Let $\varepsilon_\lambda \in H$ be the projection onto M_λ and take $0 \neq x_\lambda \in M_\lambda$ for each $\lambda \in \Lambda$. Consider the H-endomorphism $\sum_\lambda H\varepsilon_\lambda \longrightarrow M$ given by $\sum \alpha_\lambda \varepsilon_\lambda \mapsto \sum \alpha_\lambda x_\lambda$.)
    (b) If ${}_HM$ is not injective but it is a finitely generated projective module, show that H is not left self-injective.
    (c) If M_R is unital and R_R is isomorphic to a direct summand of a direct sum of copies of M_R, show that ${}_HM$ is a finitely generated projective H-module. (Show that R_R is isomorphic to a direct summand of M_R^n for some $n \in \mathbb{N}$ and apply $\mathrm{Hom}_R(\cdot, M)$.)
    (d) The unital ring R is called *quasi-Frobenius* if it is right self-injective and right artinian. If R is quasi-Frobenius and M_R is free and not finitely generated show that H is right self-injective but not left self-injective. (Use Exercise 1 with $N_R = M_R$ and Exercises 10 and 11 to show $\mathrm{Hom}_H(\cdot, H_H)$ is exact.)

13. If $\ell(R) = r(R) = 0$ show that $Q_2(R)$ is the unique maximal two-sided quotient ring of R.

14. If R is reduced show that $Q_2(R)$ is reduced. If R is a domain show that $Q_2(R)$ is a domain.

15. Let L be a left ideal of R. Show that R is a right quotient ring of L iff $\ell(L;R) = 0$.

16. Give an example of a ring R which is not prime, $Z_r(R) = Z_\ell(R) = 0$ and $Q(R) = Q_{\mathrm{left}}(R)$ is prime.

17. Let $\{R_\lambda : \lambda \in \Lambda\}$ be a collection of rings. Suppose that R_λ is a subring of S_λ for each $\lambda \in \Lambda$.

    (a) Show that S_λ is a right quotient ring of R_λ, for each $\lambda \in \Lambda$, iff $\Pi_\lambda S_\lambda$ is a right quotient ring of the ring direct sum $\boxplus_\lambda R_\lambda$.

(b) Show that $Q(\boxplus_\lambda R_\lambda) = \Pi_\lambda Q(R_\lambda) = Q(\Pi_\lambda R_\lambda)$.
(c) Show that $Q_2(\boxplus_\lambda R_\lambda) = \Pi_\lambda Q_2(R_\lambda) = Q_2(\Pi_\lambda R_\lambda)$.
(d) Show that $\boxplus_\lambda R_\lambda$ is right nonsingular iff $\Pi_\lambda R_\lambda$ is right nonsingular, iff R_λ is right nonsingular for each $\lambda \in \Lambda$.
(e) Suppose $\Pi_\lambda R_\lambda$ is a right quotient ring of R and T is a subring of R. Show that R is a right quotient ring of T iff R_λ is a right quotient ring of $R_\lambda \cap T$ for each $\lambda \in \Lambda$.

18. (a) Let R be a subring of S and let $n \in \mathbb{N}$. Show that S is a right quotient ring of R iff the matrix ring S_n is a right quotient ring of R_n.
(b) Show that $Q(R_n) = Q(R)_n$ and $Q_2(R_n) = Q_2(R)_n$. (First asume that R is unital and consider the centralizer of the usual matrix units.)
(c) Let J be a dense right ideal of R_n and for $1 \le k \le n$ let

$$J(k) = \{a \in R : a \text{ is an entry of a matrix in } J \cap e_{kk}R_n\}.$$

Show that $J(k)$ is a dense right ideal of R.
(d) Let J be a right ideal of R_n. Show that J is a dense right ideal of R_n iff there is a dense right ideal D of R with $D_n \subseteq J$.
(e) Show that $Z_r(R_n) = Z_r(R)_n$.

19. Let $F[x]$ be the polynomial ring over the field F and let y be the image of x in $S = F[x]/(x^4)$. Let R be the F-subalgebra of S generated by $\{1, y^2, y^3\}$. Show that R_R is essential in S_R and each is its own maximal quotient ring.

20. Let E be the injective hull of M_R. Show that the following are equivalent statements and each one holds whenever M is nonsingular.

(a) Each submodule of E has a unique injective hull in E.
(b) Each submodule of M has a unique injective hull in E.
(c) The intersection of two injective submodules of E is injective.

21. Let R be a subring of the reduced ring S. Show that $Z({}_RS) = Z(S_R)$.

22. Show that a ring is strongly regular iff it is regular and reduced.

23. Show that each one-sided ideal is an ideal in a strongly regular ring.

24. (a) If R is a regular ring and $e = e^2 \in R$ show that eRe is regular.
(b) Show that R is regular iff R_n is regular for each $n \in \mathbb{N}$. (For $A \in R_2$ find a strictly lower triangular matrix X with $A_1 = A - AXA$ lower triangular. Use the regularity of the diagonal entries of A_1 to find X_1 such that $A_1 - A_1X_1A_1$ is strictly lower triangular and hence regular.)
(c) If M is a finitely generated projective unital module over the regular ring R, show that $\mathrm{End}_R(M)$ is regular.

25. In this exercise all modules are unital. The module ${}_HM$ is *flat* if the functor $\cdot \otimes_H M$ is exact.

(a) Suppose that ${}_HM_R$ is a bimodule with ${}_HM$ flat, and N_R is an injective module. Show that $\mathrm{Hom}_R(M,N)$ is an injective right H-module (Use Exercise 1; see Exercise 12.)
(b) If R is right self-injective show that R_n is right self-injective.

26. Show that the following are equivalent for the domain R.
(a) $Q(R)$ is reduced.
(b) $Q(R)$ is a division ring.
(c) $aR \cap bR = 0 \Rightarrow a = 0$ or $b = 0$, for every $a, b \in R$. (A domain with this property is called a *right Öre domain.*)
(d) Each nonzero element of R is invertible in $Q(R)$; and if $p \in Q(R)$, then $p = ab^{-1}$ for some $a, b \in R$.
(e) R contains no infinite direct sums of right ideals. ($aR \cap bR = 0 \Rightarrow bR + abR + a^2bR + \cdots$ is direct.)

27. (a) Let I be a minimal right ideal of R. Show that $I^2 = 0$, or $I = eR$ for some idempotent e, and then eRe is a division ring.
(b) Give an example of a (unital) ring R for which eRe is a division ring yet eR is not minimal.
(c) If R is semiprime and $0 \neq e^2 = e \in R$ show that eR (respectively, Re) is a minimal right (respectively, left) ideal of R iff eRe is a division ring.

28. Suppose R is a semiprime subring of T and R_R is essential in T_R. Let $e = e^2 \in T$ and suppose eT is a minimal right ideal of T. If $K = eTe \cap R \neq 0$ show that K is a right Öre domain with $Q(K) = eTe$. (By Exercises 26 and 27 it suffices to show K_K is essential in $(eTe)_K$. If $0 \neq d \in eTe$ take $a, b \in R$ with $0 \neq da \in R$ and $0 \neq eab \in R$; show $eabRK \neq 0$.)

29. Let $A \subseteq B$ be ideals of the semiprime ring R and let A' denote the annihilator of A. Show that the following are equivalent.

(a) $A' = B'$.
(b) A_R is essential in B_R.
(c) ${}_RA$ is essential in ${}_RB$.
(d) If C is a nonzero ideal of R contained in B, then $C \cap A \neq 0$.
(e) A_B is essential in B_B.
(f) A_A is essential in B_A.
(g) B is a two-sided quotient ring of A.

30. (a) If X is a subset of the semiprime ring R show that $(RXR)' = (RX + XR + RXR)' = (RX + XR + RXR + \mathbb{Z}X)'$.
(b) Let S be a right quotient ring of the semiprime ring R, and let B be an ideal of S. Show that $S(B \cap R)S$ is essential in B.

31. Let A and B be ideals of the semiprime ring R. Show that the following are equivalent.

(a) $A' = B'$.

(b) $A \cap B$ is essential in A and in B.
(c) AB is essential in A and in B.

32. Let $S = \Pi_{\lambda \in \Lambda} R_\lambda$ be the direct product of the prime rings R_λ. Show that the mapping $\sigma : \mathscr{P}(\Lambda) \longrightarrow \mathrm{Ann}(S)$, $\sigma(\Gamma) = \Pi_{\lambda \in \Gamma} R_\lambda$, is a Boolean isomorphism between the power set of Λ and the Boolean algebra of annihilator ideals of S.

33. Let A be an annihilator ideal of the semiprime ring R. Show that the following are equivalent (see Exercise 2.4.12).
    (a) A is a maximal annihilator.
    (b) A' is a minimal annihilator.
    (c) A is a prime ideal.
    (d) A is a minimal prime ideal.
    (e) A' is a prime ring.

34. Suppose $Q(R)$ is strongly regular.
    (a) If I is a right ideal of R show that $Q(I)$ is strongly regular and $Q(I) \subseteq Q(R)$.
    (b) If P is a prime ideal of R show that $Q(P) = Q(R)$, or P^* and R/P are right Öre domains with $Q(P^*) = Q(R/P)$.

35. Let $\mathscr{F}(I)$ be the class of I-torsion-free right R-modules over the ring R. Show that $\mathscr{F}(I)$ is hereditary, productive, extensionally closed (the middle term of a short exact sequence is in $\mathscr{F}(I)$ if the end terms are in $\mathscr{F}(I)$), and is closed under essential extensions.

36. Let M_R be an I-torsion-free module over the irredundant ring R, and let J and K be ideals of R.
    (a) Show that $\ell(J;M) = \ell(J'';M)$.
    (b) Show that the following diagram is commutative

$$\begin{array}{ccc} \mathscr{P}(\Lambda) & \xrightarrow{\quad\gamma\quad} & \mathscr{P}(\Gamma) \\ \downarrow{\scriptstyle\alpha} & & \downarrow{\scriptstyle\rho} \\ Ann(R) & \xrightarrow{\quad\beta\quad} & Ann(\mathscr{I}(R);M) \end{array}$$

    where α is the composite of the isomorphisms given in Exercise 32 and Theorem 4.1.13, ρ is the isomorphism given in Theorem 4.1.16, γ is the natural Boolean homomorphism ($\gamma(\Delta) = \Delta \cap \Gamma$), and $\beta(L) = \ell(L';M)$. Consequently, $\mathrm{Ann}(\mathscr{I}(R);M)$ is a homomorphic image of the Boolean algebra $\mathrm{Ann}(R)$ and its complements are given by $\ell(J;M)' = \ell(J';M)$.
    (c) Let A be an ideal of R that is essential in $R \cap \Pi_{\lambda \notin \Gamma} R_\lambda$. Show that $\ell(J;M) = \ell(K;M)$ iff $(J+A)' = (K+A)'$.

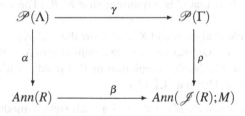

37. Let M_R be an I-torsion-free module over the irredundant ring R. Suppose that L_λ is an R_λ-module for each $\lambda \in \Lambda$ and M is a subdirect product of $\{L_\lambda : \lambda \in \Lambda\}$. Show that (a) is equivalent to (b) and it implies (c).
    (a) For each $\lambda \in \Lambda$, $L_\lambda \cap M$ is essential in L_λ.
    (b) M is essential in $\Pi_\lambda L_\lambda$.
    (c) $\sum M \cap L_\lambda$ is essential in M.

    In particular, if N is a submodule of the essential module M and $\{M_\lambda\}$ are the components of M, then $N \in \mathscr{E}(M)$ iff $N \cap M_\lambda \in \mathscr{E}(M_\lambda)$ for each $\lambda \in \Lambda$. In this case, N is also an essential module.

38. Let M be a right R-module.
    (a) If M is nonsingular show that M is I-torsion-free.
    (b) Suppose that R is irredundant and M is an R_λ-module for the component R_λ of R. Show that M is a nonsingular R-module iff it is a nonsingular R_λ-module.
    (c) Suppose R is irredundant and M is I-torsion-free. Show that M is nonsingular iff each of its components is nonsingular.
    (d) Give an example of a prime ring which has a singular I-torsion-free module (Let R be a simple right Öre domain; see Exercises 26 and 3.2.28).

39. Let Q be the maximal right quotient ring of R.
    (a) Show that the center of Q is the centralizer of R in Q.
    (b) If R is commutative show that Q is commutative.
    (c) If R is semiprime show that the centers of Q, $Q_2(R)$ and of the maximal left quotient ring of R are all equal.

40. Let B be a generalized Boolean algebra (see Exercises 1.2.2, 1.2.8, and 1.3.5).
    (a) If B is a subalgebra of the generalized Boolean algebra C and B is order dense in C show that C is a quotient ring of B. (The converse is a consequence of (c).)
    (b) If B is a Boolean algebra and $X \subseteq B$ show that $LU(X) = X^{**}$ (see Theorem 1.3.2). Here, X^{**} denotes the double annihilator of X in the reduced ring B.
    (c) Show that the MacNeille completion of B coincides with its maximal quotient ring. (Use Theorem 4.1.13.)

41. If $Z_r(R) = 0$ show that $Z_2(M) = Z(M)$ for each right R-module M. (Use Theorem 4.1.8.)

42. Let C be the center of R.
    (a) If $a \in C$ is a von Neumann regular element of R show that a is a von Neumann regular element of C. (If $a = axa$ and $b = xax$ show $br = xarx$ for each $r \in R$.)
    (b) If R is regular and right self-injective show that C is regular and self-injective and R is an injective C-module.

43. Assume $0 \neq M_R$ is finite dimensional.

(a) Show that M contains a uniform submodule. (If not, $M \supseteq A_1 \boxplus B_1$, $A_1 \supseteq A_2 \boxplus B_2, \ldots$, and $\sum B_i$ is direct.)
(b) Show that M contains an independent family of uniform submodules $\{U_1, \ldots, U_n\}$ with $U_1 \boxplus \cdots \boxplus U_n$ essential in M; and if $0 \neq V_i$ is a submodule of U_i, then $V_1 \boxplus \cdots \boxplus V_n$ is essential in M.
(c) Show that $\mathscr{E}(M_R) = \{N \in \mathscr{L}(M_R) : N \cap U_i \neq 0 \text{ for each } i = 1, \ldots, n\}$.
(d) Suppose $V_1 \boxplus \cdots \boxplus V_m \in \mathscr{E}(M_R)$ and each V_j is uniform. Show that each U_i may be replaced by some V_j. (Let $N_i = U_1 \boxplus \cdots \boxplus U_{i-1} \boxplus U_{i+1} \boxplus \cdots \boxplus U_n$; then $N_i \cap V_j = 0$ for some j, and $N_i \boxplus V_j \in \mathscr{E}(M)$.)
(e) Show that $m = n$ in (b) and (d). The number n is called the *Goldie dimension of M* and we write $\dim M = n$.
(f) Show that $N \in \mathscr{E}(M)$ iff N contains a direct sum of n uniform submodules.
(g) If N is a complement in M, show that $\dim M/N = \dim M - \dim N$.
(h) Give an example of a uniform module M and a submodule $N \subseteq M$ with M/N not finite dimensional.

44. Show that M_R is finite dimensional iff $E(M_R)$ is a direct sum of a finite number of indecomposable modules.

45. Let M_R be an R-module.

(a) Use the proof of Baer's Criterion to show that M is a Baer module iff each diagram in $\mathscr{M}_R$ on the left can be completed to a diagram in $\mathscr{M}_R$ on the right.

(b) If M is a Baer module show that $E(M)R \subseteq M$.
(c) Suppose that R has a left identity e. Show that M is injective iff M is a Baer module and $M(1-e)$ is a divisible group.
(d) Show that if M is a Baer module and $M = MR$, then M is quasi-injective (use Exercise 4.1.7).

46. (a) Show that the following are equivalent for the ring R.
(i) R is right self-injective.
(ii) R_R is quasi-injective, R has a left identity element and $\ell(R;R)$ is a divisible group.
(iii) R_R is a Baer module and $\ell(R;R)$ is divisible.

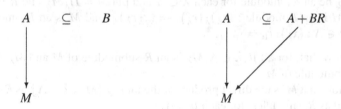

(b) Suppose the additive group of R has no nonzero elements of finite order. Show that the following are equivalent.
   (i) R is right self-injective.
   (ii) R_R is quasi-injective and has a left identity.
   (iii) R_R is a Baer module.

47. Let P be a nonzero ring and put $S = \mathrm{End}_P(P_P)$. Then we have the bimodules ${}_{P,S}P_P$ and ${}_{P,S}S_{P,S}$.
   (a) Show that the group direct sum $R = S \boxplus P$ becomes a ring with the multiplication: $(\alpha, p)(\beta, q) = (\alpha\beta + p\beta, \alpha q + pq)$.
   (b) Show that R is not unital but it has a left identity element e, and the rings P and S are embedded in R with P embedded as a left ideal and $S \cong eRe$.
   (c) If $P^2 = 0$ show that $R \cong \begin{pmatrix} S & P \\ 0 & 0 \end{pmatrix}$, $P = \ell(P;R)$, and I is a right ideal of R iff $I = \begin{pmatrix} K & J \\ 0 & 0 \end{pmatrix}$ where $K = Ie$ is a right ideal of S, J is a subgroup of P and $KP \subseteq J$.
   (d) Assume $P^2 = 0$. Show that R_R is quasi-injective iff P is a quasi-injective $\mathbb{Z}$-module. (If $f \in \mathrm{Hom}_R(I,R), I$ a right ideal of R, and $\alpha \in S$ agrees with f on J, then $(f-h)J = 0$ where h is left multiplication by $\begin{pmatrix} \alpha & 0 \\ 0 & 0 \end{pmatrix}$.)
   (e) If $P^2 = 0$ show that R is right self-injective iff P is divisible.
   (f) Show that the ring $\begin{pmatrix} \mathbb{Q} & \mathbb{Q} \\ 0 & 0 \end{pmatrix}$ is right self-injective.

48. Let M_λ be an R_λ-module for each $\lambda \in \Lambda$ and put $M = \Pi_\lambda M_\lambda$ and $R = \Pi_\Lambda R_\Lambda$. Then M is an R-module : $(x_\lambda)_\lambda (r_\lambda)_\lambda = (x_\lambda r_\lambda)_\lambda$, and M is an R_μ-module for each $\mu \in \Lambda : (x_\lambda)_\lambda r_\mu = x_\mu r_\mu$.
   (a) Show that, for all $\mu, \lambda \in \Lambda, M_\lambda$ is an R-submodule of M and M_λ is an R_μ-submodule of M.
   (b) Show that M is the direct product of the family $\{M_\lambda : \lambda \in \Lambda\}$ as R-modules and as R_μ-modules, for each $\mu \in \Lambda$.
   (c) If M is R-injective show that M is R_μ-injective, for each $\mu \in \Lambda$, and hence each M_λ is R-injective as well as R_μ-injective.

49. (a) Let S and T be rings and suppose that $R = S \times T$ is right self-injective. Show that S is right self-injective.
   (b) Suppose that the ring R is a direct product of a family of unital rings. Show that R is right self-injective iff each of its factors is right self-injective.
   (c) If S is unital and S and T are right self-injective show that $R = S \times T$ is right self-injective.
   (d) Show that the ring $R = \begin{pmatrix} \mathbb{Q} & \mathbb{Q} \\ 0 & 0 \end{pmatrix} \times \begin{pmatrix} \mathbb{Q} & \mathbb{Q} \\ 0 & 0 \end{pmatrix}$ is not right self-injective even though each factor is right self-injective (see Exercise 47).

50. Let R be a ring and let $\varphi : R \longrightarrow S = \mathrm{End}_R(R_R)$ be the canonical ring homomoprhism ($\varphi(a)$ is left multiplication of a).
    (a) Show that $\varphi(R)$ is a left ideal of S.
    (b) If R_R is quasi-injective show that $Z_r(R) = \varphi^{-1}(J(S))$ where $J(S)$ is the Jacobson radical of S.
    (c) Let $\bar{S} = S/J(S)$ and $\bar{R} = R/Z_r(R)$. Suppose R_R is quasi-injective. Show that $\bar{S} = Q(\bar{R})$ iff $Z_r(R)$ contains no nonzero direct summand of R_R. If R has a left identity, then $\bar{S} = Q(\bar{R})$.

Notes. The maximal right quotient ring of a ring was constructed by Utumi in [U1]; the nonsingular case appeared earlier in Johnson [JO1] where the singular submodule of a module is also defined. Additional references for rings of quotients include Findlay and Lambek [FL], Faith [F3], [F4], and [F5], Goodearl [GOO2], Lambek [LA], Stenström [STE2], and Lam [L1]. Irredundant semiprime rings are studied and I-torsion-free modules first appeared in Levy [LEV1]. The determination of when $Q(R)$ is reduced first appeared in Anderson [AN1] but in the context of unital f-rings. Renault [RE] gave the theorem for unital reduced rings and Steinberg [ST5] noted that Anderson's techniques work for any right nonsingular ring. Theorem 4.1.6 is due to Goldie [GO4]. The history of Theorem 4.1.9 is given in Faith [F5, p. 76]. Exercise 12 is due to Sandomierski [SA], Exercise 40 is due to Brainerd and Lambek [BL] and the injectivity part of Exercise 42 is in Armendariz and Steinberg [AS]; the fact that the center of a regular right self-injective ring is self-injective appeared earlier in Miyashita [MI, Theorem 5.12].

4.2 Torsion Theories and Rings of Quotients

In this section we will examine the connection between certain "radicals" in the category $\mathscr{M}_R$ of right R-modules over a ring R and collections of right ideals that resemble the dual ideal of dense right ideals of R. This will lead to more rings of quotients of R and of homomorphic images of R, and to modules of quotients. In particular, the classical right quotient ring of a ring will be constructed and seen to be a closer analogue of the field of quotients of a commutative domain than the maximal right quotient ring.

We will again abbreviate $\mathrm{Hom}_R(A,B)$ by $[A,B]$ or by ${}_R[A,B]$ if necessary. If $\mathscr{A}$ and $\mathscr{B}$ are classes of (right) R-modules, then $[\mathscr{A},\mathscr{B}] = 0$ means $[A,B] = 0$ for all $A \in \mathscr{A}$ and $B \in \mathscr{B}$. This relation defines a Galois connection in the power class of $\mathscr{M}_R$. For the class $\mathscr{A}$ let

$$r(\mathscr{A};\mathscr{M}_R) = r(\mathscr{A}) = \{B : [A,B] = 0 \text{ for every } A \text{ in } \mathscr{A}\}, \text{ and}$$

$$\ell(\mathscr{A};\mathscr{M}_R) = \ell(\mathscr{A}) = \{B : [B,A] = 0 \text{ for every } A \text{ in } \mathscr{A}\}$$

The pair of classes of R-modules $(\mathscr{T},\mathscr{F})$ is a *torsion theory for* $\mathscr{M}_R$ if $\mathscr{T}=\ell(\mathscr{F})$ and $\mathscr{F}=r(\mathscr{T})$. Two other equivalent formulations of this are:

$$\mathscr{T} \text{ and } \mathscr{F} \text{ are each maximal with respect to } [\mathscr{T},\mathscr{F}]=0;$$

$$\mathscr{T}=\ell r(\mathscr{T}) \text{ and } \mathscr{F}=r\ell(\mathscr{F}).$$

$\mathscr{T}$ is called a *torsion class* and $\mathscr{F}$ is called a *torsion-free class* in $\mathscr{M}_R$. The modules in $\mathscr{T}$ are *torsion modules* and those in $\mathscr{F}$ are *torsion-free* modules. A given class of modules $\mathscr{C}$ *generates* the torsion theory $(\ell r(\mathscr{C}), r(\mathscr{C}))$ and *cogenerates* the torsion theory $(\ell(\mathscr{C}), r\ell(\mathscr{C}))$. Clearly, $\ell r(\mathscr{C})$ is the smallest torsion class containing $\mathscr{C}$ and $r\ell(\mathscr{C})$ is the smallest torsion-free class containing $\mathscr{C}$. Just as for the category of rings the class $\mathscr{C}$ will be called extensionally closed if each module M is in $\mathscr{C}$ whenever it is the middle of a short exact sequence

$$0 \longrightarrow N \xrightarrow{f} M \xrightarrow{g} P \longrightarrow 0 \tag{4.2.1}$$

whose ends N and P are in $\mathscr{C}$.

Theorem 4.2.1. *A class of modules is a torsion class in $\mathscr{M}_R$ iff it is homomorphically closed, extensionally closed, and is closed under direct sums. A class of modules is a torsion-free class in $\mathscr{M}_R$ iff it hereditary, extensionally closed, and is productive.*

Proof. We will only prove the first statement and leave the proof of the second for the reader. Suppose that $(\mathscr{T},\mathscr{F})$ is a torsion theory. Clearly, $\mathscr{T}$ is homomorphically closed and if $\{M_\lambda : \lambda \in \Lambda\}$ is a subset of $\mathscr{T}$ and $F \in \mathscr{F}$, then $[\boxplus_\lambda M_\lambda, F] \cong \Pi_\lambda [M_\lambda, F]=0$. Also, given the exact sequence (4.2.1) with N and P torsion, if $\alpha \in [M,F]$ where F is torsion-free, then $\alpha=0$ on the image of N; so α factors through P and hence $\alpha=0$.

Conversely, suppose that the class $\mathscr{T}$ has these three properties. Let $M \in \mathscr{M}_R$ and let N be the sum of the $\mathscr{T}$-submodules of M. Then $N \in \mathscr{T}$ since it is a homomorphic image of a direct sum of $\mathscr{T}$-modules. If $T \in \mathscr{T}$ and $\alpha \in [T, M/N]$, then $\alpha(T)=K/N$ is a $\mathscr{T}$-module and therefore so is K. Hence, $K=N$, $\alpha=0$, and $M/N \in r(\mathscr{T})$. Now, if $M \in \ell r(\mathscr{T})$, then also $M/N \in \ell r(\mathscr{T})$; so $M=N \in \mathscr{T}$ and $\mathscr{T}=\ell r(\mathscr{T})$. □

The analogy between torsion theories in $\mathscr{M}_R$ and radical pairs in the category of rings, as indicated by Theorem 4.2.1 and its proof, is quite good. As is to be expected, however, torsion theories do have their own flavor since the category of rings and the category of R-modules are considerably different.

The class of nonsingular right R-modules is always a torsion-free class. The corresponding torsion class may be larger than the class of singular modules. For example, if F is a field and $R=F[x]/(x^2)$, then R is torsion but not singular since $Z(R)(=xR)$ and $R/Z(R)$ are both singular. In general, the associated torsion class consists of all modules M with $Z_2(M)=M$ (see Theorem 4.1.6 and Exercise 4.1.38). Another torsion-free class is the class of I-torsion-free modules (see Exercise 4.1.32).

A torsion theory is called *hereditary* if its torsion class is a hereditary class of modules. Given a torsion theory $(\mathcal{T},\mathcal{F})$ and a module M let $t_{\mathcal{T}}(M)$ denote the sum of those submodules of M which belong to $\mathcal{T}$. The submodule $t_{\mathcal{T}}(M)$ is, of course, the "$\mathcal{T}$-radical" of M. Instead of starting with a torsion theory we could start with a radical functor.

A functor $t : \mathcal{M}_R \longrightarrow \mathcal{M}_R$ is called a *radical of* $\mathcal{M}_R$ if $t(M) \subseteq M$ and $t(M/t(M)) = 0$ for each module M, and $t(\alpha)$ is the restriction of α to $t(M)$ for every $\alpha \in [M,N]$; that is $\alpha(t(M)) \subseteq t(N)$. It is easy to see that the radical t is left exact iff $t(N) = N \cap t(M)$ whenever N is a submodule of M (see Exercise 4.1.1). The class of radicals of $\mathcal{M}_R$ is partially ordered by the relation: $t_1 \leq t_2$ iff $t_1(M) \subseteq t_2(M)$ for each module M. In fact, if $\{t_\lambda : \lambda \in \Lambda\}$ is a set of radicals, then so is the functor t defined by $t(M) = \bigcap_\lambda t_\lambda(M)$; and $t = \inf\{t_\lambda : \lambda \in \Lambda\}$. If each t_λ is left exact then so is t. So both the class of radicals and the class of left exact radicals are complete lattices. In the same vein, it is clear from Theorem 4.2.1 that the classes of torsion classes and hereditary torsion classes are also complete lattices.

For the radical t let $\mathcal{T}_t = \{M \in \mathcal{M}_R : t(M) = M\}$ and $\mathcal{F}_t = \{M \in \mathcal{M}_R : t(M) = 0\}$.

Theorem 4.2.2. *The correspondence $t \mapsto (\mathcal{T}_t,\mathcal{F}_t)$ is a lattice isomorphism between the lattice of left exact radicals of $\mathcal{M}_R$ and the lattice of hereditary torsion theories for $\mathcal{M}_R$. The inverse correspondence is $(\mathcal{T},\mathcal{F}) \mapsto t_{\mathcal{T}}$.*

Proof. Let t be a left exact radical. Suppose that A is a submodule of B. If $t(B) = B$, then $t(A) = A \cap t(B) = A$ and $B/A = p(t(B)) \subseteq t(B/A)$ where $p : B \longrightarrow B/A$ is the natural map; so $\mathcal{T}_t$ is hereditary and homomorphically closed. If $\{A_\lambda\}_\lambda \subseteq \mathcal{T}_t$, then, for each $\lambda, A_\lambda = t(A_\lambda) \subseteq t(\boxplus_\lambda A_\lambda)$; so $\mathcal{T}_t$ is closed under direct sums. Clearly, for any module $B, t(B)$ is characterized as that submodule C such that $t(C) = C$ and $t(B/C) = 0$. In particular, if $A \subseteq t(B)$, then $t(B/A) = t(B)/A$. Consequently, if A and B/A are both in $\mathcal{T}_t$ so is B. Since $[\mathcal{T}_t, F] = 0$ iff $t(F) = 0$, $(\mathcal{T}_t,\mathcal{F}_t)$ is a hereditary torsion theory by Theorem 4.2.1. Conversely, suppose that $(\mathcal{T},\mathcal{F})$ is a hereditary torsion theory. The proof of Theorem 4.2.1 shows that $t_{\mathcal{T}}$ is a radical and $t_{\mathcal{T}}(N) \subseteq t_{\mathcal{T}}(M)$ for each submodule N of M. Since $\mathcal{T}$ is hereditary $N \cap t_{\mathcal{T}}(M) \subseteq t_{\mathcal{T}}(N)$; so $t_{\mathcal{T}}$ is left exact.

Given the left exact radical t, for any module M, $t(M)$ is the largest submodule N of M with $t(N) = N$, and $t_{\mathcal{T}}(M)$ is the largest submodule of M which is in $\mathcal{T}_t$; so $t = t_{\mathcal{T}_t}$. On the other hand, given the torsion theory $(\mathcal{T},\mathcal{F})$, $M \in \mathcal{T}_{t_\mathcal{T}}$ iff $M = t_{\mathcal{T}}(M)$, iff $M \in \mathcal{T}$. Since both correpondences are isotone they are lattice isomorphisms. □

From now on $(\mathcal{T},\mathcal{F})$ will be the hereditary torsion theory associated with the left exact radical t. Since M is torsion iff each of its 1-generated submodules is torsion it is not surprising that $\mathcal{T}$ is frequently determined by a collection of right ideals. The nonempty collection $\mathbb{F}$ of right ideals of R is called a (*right*) *topology of* R if the following hold. Recall that $\mathcal{L}(M_R)$ denotes the lattice of submodules of M.

T1 $D \in \mathbb{F}$ and $a \in R \Rightarrow (D : a) \in \mathbb{F}$.
T2 $J \in \mathbb{F}$ and $D \in \mathcal{L}(R_R)$ and $(D : a) \in \mathbb{F}$ for each $a \in J \Rightarrow D \in \mathbb{F}$.

According to Theorem 4.1.4 the set $\mathbb{D}(R)$ of dense right ideals of R is a topology of R. If $\mathbb{F}$ is a topology of R, then $\mathbb{F}$ determines a topology for R which makes it into a topological ring: $\mathbb{F}$ is a basis for the neighborhood system of 0 for this topology. This connection to toplogical spaces will not be used explicitly, and we leave its verification to the interested reader (Exercise 12). The set Top(R) of all toplogies of R is certainly closed under intersection and hence it is a complete lattice. We show next that each topology $\mathbb{F}$ resembles $\mathbb{D}$.

Theorem 4.2.3. *Let $\mathbb{F}$ be a topology of R.*

(a) $\mathbb{F}$ is a dual ideal of $\mathscr{L}(R_R)$.
(b) Suppose that $J \in \mathbb{F}$ and, for each $a \in J$, $D_a \in \mathbb{F}$. Then $\sum_{a \in J} aD_a \in \mathbb{F}$.
(c) If $I, J \in \mathbb{F}$, then $IJ \in \mathbb{F}$.
(d) If $M \in \mathscr{M}_R$ let $t_{\mathbb{F}}(M) = \{x \in M : r(x) \in \mathbb{F}\}$. Then $t_{\mathbb{F}}$ is a left exact radical of $\mathscr{M}_R$.
(e) The correspondence $\mathbb{F} \mapsto t_{\mathbb{F}}$ between Top(R) and the class of left exact radicals on $\mathscr{M}_R$ is one-to-one and preserves all meets.

Proof. (a) Suppose that $D \subseteq F$ are right ideals of R with $D \in \mathbb{F}$. If $a \in D$, then $R = (D : a) \in \mathbb{F}$ by T1. But then $(F : a) \in \mathbb{F}$ for each $a \in D$; so $F \in \mathbb{F}$ by T2. Now let $D, F \in \mathbb{F}$. For any $a \in D$, $(D \cap F : a) = (D : a) \cap (F : a) = (F : a) \in \mathbb{F}$; so $D \cap F \in \mathbb{F}$.

(b) Let $D = \sum aD_a$. Since $D_a \subseteq (D : a)$ for each $a \in J$, $(D : a) \in \mathbb{F}$ and necessarily $D \in \mathbb{F}$.

(c) For each $a \in I$, $J \subseteq (IJ : a)$; so $IJ \in \mathbb{F}$.

(d) Let $x, y \in t_{\mathbb{F}}(M)$ and $a \in R$. Then $r(x - y) \supseteq r(x) \cap r(y)$ and $r(xa) = (r(x) : a)$. Thus, $t_{\mathbb{F}}(M)$ is a submodule of M by (a). Suppose $x \in M$, $D \in \mathbb{F}$ and $xD \subseteq t_{\mathbb{F}}(M)$. Then, for each $a \in D$ there exists $D_a \in \mathbb{F}$ with $xaD_a = 0$. So $\sum aD_a \subseteq r(x)$, $x \in t_{\mathbb{F}}(M)$ by (b), and $t_{\mathbb{F}}(M/t_{\mathbb{F}}(M)) = 0$. Since $r(\alpha x) \supseteq r(x)$ for each $\alpha \in [M, N]$, $t_{\mathbb{F}}$ is a radical which is clearly left exact.

(e) From the definitions, $I \in \mathbb{F}$ iff $(I : a) \in \mathbb{F}$ for each $a \in R$, iff $t_{\mathbb{F}}(R/I) = R/I$. Thus, the correspondence $\mathbb{F} \mapsto t_{\mathbb{F}}$ is one-to-one. Moreover, $t_{\cap \mathbb{F}_\lambda}(M) = \bigcap t_{\mathbb{F}_\lambda}(M)$ for each module M. □

It is useful to extend a given topology to a unital overring.

Theorem 4.2.4. *Let S be a unital overring of R generated by $R \cup \{1\}$ and let $\mathbb{F}$ be a topology of R.*

(a) $\mathbb{F}(S) = \{J \in \mathscr{L}(S_S) : J \cap R \in \mathbb{F}\}$ is a topology of S and $\mathbb{F} \subseteq \mathbb{F}(S)$.
(b) The correspondence $\mathbb{F} \mapsto \mathbb{F}(S)$ is a lattice isomorphism between Top(R) and the lattice of those topologies of S which contain R.
(c) If M is an S-module, then $t_{\mathbb{F}}(M) = t_{\mathbb{F}(S)}(M)$.

Proof. (a) We first note that a right ideal J of S is in $\mathbb{F}(S)$ iff $S/J \in \mathscr{T}_{t_{\mathbb{F}}}$. This is a consequence of the fact that S/J is generated by $1 + J$ as an R-module and $r(1 + J; R) = J \cap R$. We write $(J : a)_S = \{s \in S : as \in J\}$. If $J \in \mathbb{F}(S)$ and $a \in S$, then $(J : a)_S \cap R = (J : a) \in \mathbb{F}$; so $(J : a)_S \in \mathbb{F}(S)$ and T1 holds for $\mathbb{F}(S)$. Suppose $J \in \mathbb{F}(S)$,

$I \in \mathscr{L}(S_S)$ and $(I:a)_S \in \mathbb{F}(S)$ for each $a \in J$. Then for each $a \in J \cap R$, $(I \cap R : a) = (I:a) = (I:a)_S \cap R \in \mathbb{F}$. So $I \cap R \in \mathbb{F}$, $I \in \mathbb{F}(S)$ and T2 also holds for $\mathbb{F}(S)$.

(b) Let $\mathbb{K}$ be a topology of S with $R \in \mathbb{K}$. We claim that $\mathbb{F} = \mathbb{K} \cap \mathscr{L}(R_R)$ is a topology of R. Let $I \in \mathbb{F}$ and $a \in R$. Then $(I:a) = (I:a)_S \cap R \in \mathbb{F}$. Suppose $J \in \mathbb{F}$, $D \in \mathscr{L}(R_R)$ and $(D:a) \in \mathbb{F}$ for each $a \in J$. Then $J \in \mathbb{K}$ and $(D:a)_S \in \mathbb{K}$ since $(D:a) \subseteq (D:a)_S$. So $D \in \mathbb{F}$ and $\mathbb{F}$ is a topology. Clearly, $\mathbb{F} = \mathbb{F}(S) \cap \mathscr{L}(R_R)$ for any topology $\mathbb{F}$ of R. On the other hand, if $\mathbb{K}$ is a topology of S with $R \in \mathbb{K}$, then $(\mathbb{K} \cap \mathscr{L}(R_R)(S) = \{J \in \mathscr{L}(S_S) : J \cap R \in \mathbb{K}\} = \mathbb{K}$. So the correspondence $\mathbb{F} \mapsto \mathbb{F}(S)$ is a bijection and hence a lattice isomorphism since it and its inverse are isotone.

(c) Since $r(x;S) \cap R = r(x;R)$ for each $x \in M$ it is obvious that $t_{\mathbb{F}}(M) = t_{\mathbb{F}(S)}(M)$. □

When R is a unital ring the definitions of a (hereditary) torsion theory and a (left exact) radical are meaningful for the category of unital right R-modules u-$\mathscr{M}_R$, and the previous results all hold in u-$\mathscr{M}_R$. Moreover, as shown below the mapping in (e) of Theorem 4.2.3 will now be a lattice isomorphism; so the connection between hereditary torsion theories and topologies for R is much stronger. It is also true that other parts of the theory of torsion which are presented here for $\mathscr{M}_R$ also hold for u-$\mathscr{M}_R$, with identical or simpler proofs. We will frequently use these analogous unital results without an explicit warning. However, we will occasionally point out the stronger version that holds in u-$\mathscr{M}_R$.

Theorem 4.2.5. *Let t be a left exact radical for $\mathscr{M}_R$ with associated torsion theory $(\mathscr{T}, \mathscr{F})$, and let $\mathbb{F}_t = \{I \in \mathscr{L}(R_R) : R/I \in \mathscr{T}\}$.*

(a) $\mathbb{F}_t$ is a dual ideal of $\mathscr{L}(R_R)$ which satisfies T1.
(b) If $t(M) = M$ whenever $MR = 0$, then $\mathbb{F}_t$ is a topology of R.
(c) The mapping $\mathbb{F} \mapsto t_{\mathbb{F}}$ is a lattice isomorphism between the lattice of toplogies of R and the lattice of those left exact radicals t for $\mathscr{M}_R$ with the property that $t(M) = M$ whenever $MR = 0$.
(d) If R is unital the mapping in (c) is an isomorphism between Top(R) and the lattice of left exact radicals of u-$\mathscr{M}_R$.

Proof. If I, $J \in \mathbb{F}_t$ and $K \in \mathscr{L}(R_R)$ with $I \subseteq K$, then $R/K \in \mathscr{T}$ since it is a homomomorphic image of R/I, and $R/I \cap J \in \mathscr{T}$ because of the exactness of $0 \longrightarrow R/I \cap J \longrightarrow R/I \boxplus R/J$. If $a \in R$ and $\alpha : R \longrightarrow R/I$ is defined by $\alpha(x) = ax + I$, then $\ker(\alpha) = (I:a)$; so $(I:a) \in \mathbb{F}_t$. This proves (a). Now suppose $J \in \mathbb{F}_t$, $I \in \mathscr{L}(R_R)$ and $(I:a) \in \mathbb{F}_t$ for each $a \in J$. Then $(a+I)R \subseteq t(J+I/I)$ since $(a+I)R \cong R/(I:a)$; so $JR+I/I \subseteq t(J+I/I)$. The exactness of

$$0 \longrightarrow JR+I/I \longrightarrow J+I/I \longrightarrow J+I/JR+I \longrightarrow 0$$

shows that $J+I/I \in \mathscr{T}$ provided $J+I/JR+I \in \mathscr{T}$, and the latter holds under the condition in (b). The exactness of

$$0 \longrightarrow J+I/I \longrightarrow R/I \longrightarrow R/J+I \longrightarrow 0 \qquad (4.2.2)$$

now gives that $R/I \in \mathscr{T}$ and hence $\mathbb{F}_t$ is a topology. This proves (b). As for (c), we already know that the mapping $\mathbb{F} \mapsto t_{\mathbb{F}}$ is one-to-one, and clearly $t_{\mathbb{F}}(M) = M$ if $MR = 0$. To show it is onto it suffices to verify that $t_{\mathbb{F}_t} = t$. Now, for any module M, $x \in t_{\mathbb{F}_t}(M)$ iff $xR \in \mathscr{T}$. Thus, $t(M) \subseteq t_{\mathbb{F}_t}(M)$ and equality holds since $(t_{\mathbb{F}_t}(M)/t(M))R = 0$. When R is unital (a simpler version of) the proof of (b) shows that $\mathbb{F}_t$ is a topology and (a simpler version of) the proof of (c) gives $t_{\mathbb{F}_t} = t$. □

We have already noted that the class of nonsingular modules is the torsion-free part of a hereditary torsion theory whose torsion class is the class of all modules M with $Z_2(M) = M$. This is called the *Goldie* torsion theory. Of course, Z_2 is the corresponding left exact radical. The associated topology is

$$\mathbb{G} = \{J \in \mathscr{L}(R_R) : R/J = Z_2(R/J) = c\ell c\ell J/J\}$$

$$= \{J : J \subseteq K \subset R \Rightarrow \exists a \notin K \text{ with } (K : a) \in \mathscr{E}(R)\}$$

$$= \{J : \exists I \in \mathscr{E}(R) \text{ with } J \subseteq I \text{ and } (J : a) \in \mathscr{E}(R) \text{ for each } a \in I\}.$$

The last equality is a consequence of $c\ell J$ being essential in $c\ell c\ell J$ (see the remarks after Theorem 4.1.6).

We can connect those left exact radicals for which every module with trivial action is torsion with those topologies of a unital overring which arose in (b) of Theorem 4.2.4.

Theorem 4.2.6. *Let S be a unital overring of R generated by $R \cup \{1\}$ and let t be a left exact radical of $\mathscr{M}_R$.*

(a) t induces a left exact radical t^S of u-$\mathscr{M}_S$.
(b) The following statements are equivalent.
(i) $t(S/R) = S/R$.
(ii) $t(M) = M$ if $M \in u$-$\mathscr{M}_S$ and $MR = 0$.
(iii) $\mathbb{F}_t = \mathscr{L}(R_R) \cap \mathbb{F}_{t^S}$.
(c) If $t(S/R) = S/R$, then $\mathbb{F}_t$ is a topology of R.

Proof. Since u-$\mathscr{M}_S \subseteq \mathscr{M}_R$ is invariant under t, t^S, the restriction of t to u-$\mathscr{M}_S$, is clearly a left exact radical on u-$\mathscr{M}_S$.

(i) ⇒ (iii). The inclusion $\mathscr{L}(R_R) \cap \mathbb{F}_{t^S} \subseteq \mathbb{F}_t$ always holds since $R/J \subseteq S/J \in \mathscr{T}_t$ for each $J \in \mathscr{L}(R_R) \cap \mathbb{F}_{t^S}$. But the exactness of

$$0 \longrightarrow R/J \longrightarrow S/J \longrightarrow S/R \longrightarrow 0. \tag{4.2.3}$$

shows that each $J \in \mathbb{F}_t$ is an element of $\mathscr{L}(R_R) \cap \mathbb{F}_{t^S}$.

(iii) ⇒ (ii). Let M be a unital S-module with trivial R-action. If $x \in M$, then xS is a homomorphic image of S/R and $R \in \mathbb{F}_{t^S}$. So $xS \in \mathscr{T}_t$ and hence $t(M) = M$.

(ii) ⇒ (i). This is obvious.

Suppose that S/R is torsion. Since we have already seen in Theorem 4.2.5 that $\mathbb{F}_t$ satisfies T1 we only need to verify that it satisfies T2. Suppose $J \in \mathbb{F}_t$, $I \in \mathscr{L}(R_R)$ and $(I : a) \in \mathbb{F}_t$ for each $a \in J$. From (4.2.3) we see that $S/J \in \mathscr{T}$. So $S/(I : a)_S \in \mathscr{T}$ for each $a \in J$, $J + I/I \in \mathscr{T}$, and hence $I \in \mathbb{F}_t$ is a consequence of (4.2.2). □

There are instances where $\mathbb{F}_t$ is a topology even though S/R is not torsion; see Exercise 16. It is possible to weaken the torsion definitions and still retain much of the preceding theory, with the same proofs. We will present these weaker concepts next but we note that they will not be utilized very much in the sequel. A class of modules is called a *pretorsion class* (*pretorsion-free class*) if it has the properties of a torsion class (torsion-free class) enunciated in Theorm 4.2.1 except for being extensionally closed. A functor $t : \mathscr{M}_R \longrightarrow \mathscr{M}_R$ is a *preradical* if it has the defining properties of a radical except for $t(M/t(M)) = 0$. A dual ideal of $\mathscr{L}(R_R)$ is a *pretopology* if it satisfies T1. The most prominent example of these concepts where "pre" cannot be dropped consists of the class of singular modules, the singular functor, and the dual ideal $\mathscr{E}(R)$ of essential right ideals. The left exactness of a preradical t has the same characterization as a radical, but it is also characterized by : t is idempotent $(t(t(M)) = t(M))$ and $\mathscr{T}_t$ is hereditary. The isomorphism of Theorem 4.2.2 holds for left exact preradicals and hereditary pretorsion classes. Also, (d) and (e) of Theorem 4.2.3, all of Theorem 4.2.4, and (c) and (d) of Theorem 4.2.5 all hold. Moreover, (a) and the implications (iii) $\Rightarrow$ (ii) $\Rightarrow$ (i) of Theorem 4.2.6 hold, but (i) $\Rightarrow$ (iii) need not hold for the singular preradical.

We will now imitate the construction of the maximal right quotient ring to produce a ring of quotients of R and a module of quotients of M_R. For the sequel we will fix our notation and summarize the preceding results. Starting with the topology $\mathbb{F}$ of R we have the associated left exact radical t and the hereditary torsion theory $(\mathscr{T},\mathscr{F})$ where $t(M_R) = \{x \in M : r(x) \in \mathbb{F}\}$ and $\mathscr{T} = \{M \in M_R : t(M) = M\}$. We also note:

$$\text{if } D \subseteq F \subseteq R \text{ with } D \in \mathscr{L}(R_R) \text{ and } F \in \mathbb{F},$$
$$\text{then } D \in \mathbb{F} \text{ iff } F/D \in \mathscr{T}, \tag{4.2.4}$$

$$\text{if } \alpha \in [M,N] \text{ and } K \in \mathscr{L}(N), \text{ then } M/\alpha^{-1}(K)$$
$$\text{is torsion (respectively, torsion-free)}$$
$$\text{provided } N/K \text{ is torsion (respectively, torsion-free).} \tag{4.2.5}$$

As in the previous section, left multiplication by x (on the appropriate set) will be denoted by μ_x. If $\alpha \in [D,M]$ it will be convenient to write $D = D_\alpha$.

If M is a right R-module, then the family of abelian groups $\{[D,M] : D \in \mathbb{F}\}$ is a directed direct system in u-$\mathscr{M}_{\mathbb{Z}}$. The *module of quotients of M* is defined by

$$M_{\mathbb{F}} = \varinjlim [D,M/t(M)].$$

More explicitly,

$$M_{\mathbb{F}} = \bigcup_{D \in \mathbb{F}} [D,M/t(M)]/\!\sim$$

where, as usual, $\alpha \sim \beta$ iff α and β agree on $D_\alpha \cap D_\beta$; equivalently, α and β agree on some $D \in \mathbb{F}$ with $D \subseteq D_\alpha \cap D_\beta$. If $[\alpha]$ denotes the class of α, then $[\alpha]+[\beta] = [\alpha+\beta]$ with $D_{\alpha+\beta} = D_\alpha \cap D_\beta$. Note that each map $[D,M/t(M)] \longrightarrow M_{\mathbb{F}}$ is monic. For if

$\alpha(J)=0$ with $J\subseteq D_\alpha$ and $J\in\mathbb{F}$, then $(\alpha a)(J:a)=0$ for every $a\in D_\alpha$; so $\alpha=0$. We have the map $\psi_M: M\longrightarrow M_\mathbb{F}$ given by $\psi_M(x)=[\mu_{\bar{x}}]$ which is the following composite:

$$M\longrightarrow M/t(M)\xrightarrow{\mu}[R,M/t(M)]\xrightarrow{\rho_R}M_\mathbb{F}. \tag{4.2.6}$$

Since the last two maps are monomorphisms $\ker\psi_M=t(M)$. In fact, $M_\mathbb{F}$ has a natural structure as an R-module and ψ_M is an R-homomorphism: for $[\alpha]\in M_\mathbb{F}$ and $a\in R, [\alpha]a=[\alpha\mu_a]$ where

$$(D_\alpha:a)\xrightarrow{\mu_a}D_\alpha\xrightarrow{\alpha}M/t(M).$$

Note that if $a\in D_\alpha$, then

$$[\alpha]a=[\mu_{\alpha a}]\in\psi_M(M). \tag{4.2.7}$$

This action is well-defined since if $\alpha=\beta$ on D, then $\alpha\mu_a=\beta\mu_a$ on $(D:a)$. There is a similar action of $R_\mathbb{F}$ on $M_\mathbb{F}$. If $[\alpha]\in M_\mathbb{F}$ and $[\beta]\in R_\mathbb{F}$, then $\beta: D_\beta\longrightarrow R/t(R)$ and $\alpha: D_\alpha\longrightarrow M/t(M)$. Now $[\alpha][\beta]$ is defined by $[\alpha][\beta]=[\bar{\alpha}\beta]$ where $D_{\bar{\alpha}\beta}=\beta^{-1}(D_\alpha+t(R)/t(R))$. Here, $\bar{\alpha}$ is the map $\bar{\alpha}: D_\alpha+t(R)/t(R)\cong D_\alpha/t(D_\alpha)\longrightarrow M/t(M)$ induced by α, and β is restricted to $\beta^{-1}(D_{\bar{\alpha}})\subseteq D_\beta$. From (4.2.5) and (4.2.4) we obtain that $\beta^{-1}(D_{\bar{\alpha}})\in\mathbb{F}$; so $[\bar{\alpha}\beta]\in M_\mathbb{F}$. If $\alpha\sim\alpha_1$ and $\beta\sim\beta_1$, then by restricting to domains on which each equivalent pair agree it is clear that $\bar{\alpha}\beta\sim\bar{\alpha}_1\beta_1$. So this multiplication is well-defined. Moreover, if $e: R\longrightarrow R/t(R)$ is the natural map, then $[e]$ is the identity element for this multiplication. We summarize this discussion in the following theorem. Let A be a submodule of B_R. If $\mathbb{F}$ is a pretopology of R the $\mathbb{F}$-*closure* of A in B, denoted by $c\ell_\mathbb{F}(A)$, or $c\ell_\mathbb{F}(A;B)$ if necessary, is defined by $c\ell_\mathbb{F}(A)/A=t(B/A)$. A is said to be $\mathbb{F}$-*dense* in B if $c\ell_\mathbb{F}(A)=B$ and $\mathbb{F}$-*closed* in B if $c\ell_\mathbb{F}(A)=A$.

Theorem 4.2.7. *Let $\mathbb{F}$ be a topology for the ring R. Then there are a (unital) ring of quotients $R_\mathbb{F}$ of R, a left exact functor $Q_\mathbb{F}=Q:\mathscr{M}_R\longrightarrow u\text{-}\mathscr{M}_{R_\mathbb{F}}$ and a natural transformation $\psi: 1_{\mathscr{M}_R}\longrightarrow HQ$ where $H: u\text{-}\mathscr{M}_{R_\mathbb{F}}\longrightarrow\mathscr{M}_R$ is the forgetful functor and $Q(M)=M_\mathbb{F}$. Moreover, $\ker\psi_M=t(M)$ for each M, and the action of R on $Q(M)$ is induced by the ring homomorphism $\psi_R: R\longrightarrow R_\mathbb{F}$.*

Proof. We will check that Q is left exact and leave the verification of the remaining details to the reader. We need to verify that Q transforms the short exact sequence (4.2.1) into the exact sequence

$$0\longrightarrow N_\mathbb{F}\xrightarrow{Q(f)}M_\mathbb{F}\xrightarrow{Q(g)}P_\mathbb{F}.$$

Suppose $\alpha: D\longrightarrow N/t(N)$ is an R-homomorphism and $Q(f)([\alpha])=[\bar{f}\alpha]=0$ where $\bar{f}: N/t(N)\longrightarrow M/t(M)$ is induced by f. Then $\bar{f}\alpha(D_1)=0$ for some $D_1\in\mathbb{F}$ with $D_1\subseteq D$. Since $\bar{f}$ is monic, $\alpha(D)=0$ and so $[\alpha]=0$. Now suppose $\beta: D\longrightarrow M/t(M)$ is an R-homomorphism and $[\bar{g}\beta]=0$; we may assume $\bar{g}\beta(D)=0$. But $x+t(M)\in\ker\bar{g}$ iff $g(x)\in t(P)$, iff $g(xD_1)=0$ for some $D_1\in\mathbb{F}$, iff $x\in c\ell_\mathbb{F}f(N)$. So $\beta(D)\subseteq c\ell_\mathbb{F}f(N)/t(M)$. Since $f(N)+t(M)/t(M)$ is $\mathbb{F}$-dense in $c\ell_{\mathbb{F}} f(N)/t(M)$, $D_2 = \beta^{-1}(f(N)+t(M)/t(M)) \in \mathbb{F}$ by (4.2.5) and (4.2.4). Let $\alpha : D_2 \longrightarrow N/t(N)$ be defined by $\alpha(d) = \bar{x}+t(N)$ where $\bar{x} \in N/t(N)$ is determined by $\beta(d) = f(x)+t(M)$. Then $Q(f)[\alpha] = [\bar{f}\alpha] = [\beta]$ and hence $\ker Q(g)$ is equal to the image of $Q(f)$. □

The ring $R_{\mathbb{F}}$ is, in fact, a Utumi right quotient ring of $R/t(R)$ (Exercise 11); so it is not surprising that $M_{\mathbb{F}}$ can be characterized as an essential extension of $M/t(M)$ that is relatively injective. The module M_R is called $\mathbb{F}$*-injective* if for each $D \in \mathbb{F}$ and each $\alpha \in [D,M]$ there is some $x \in M$ such that $\alpha = \mu_x$; that is, $\alpha(a) = xa$ for each $a \in D$.

Theorem 4.2.8. *Let M and E be R-modules. Then E is R-isomorphic to $M_{\mathbb{F}}$ iff E is an $\mathbb{F}$-injective essential extension of $\psi(M)_R$ with $E/\psi(M) \in \mathscr{T}$. In this case there is a unique way to make E into an $R_{\mathbb{F}}$-module so that the R-action is induced via the homomorphism $\psi_R : R \longrightarrow R_{\mathbb{F}}$. Moreover, each R-isomorphism between E and $M_{\mathbb{F}}$ is an $R_{\mathbb{F}}$-isomorphism.*

Proof. We will first check that $M_{\mathbb{F}}$ satisfies these conditions. If $0 \neq q = [\alpha] \in M_{\mathbb{F}}$ and $d \in D_\alpha$ with $\alpha d \neq 0$, then $qd = [\alpha]d = [\mu_{\alpha d}] \neq 0$ and $qd \in \psi(M)$; so $M_{\mathbb{F}}$ is an essential extension of $\psi(M)_R$. Also, $M_{\mathbb{F}}/\psi(M) \in \mathscr{T}$ since $qD \subseteq \psi(M)$. To see that $M_{\mathbb{F}}$ is $\mathbb{F}$-injective, let $\alpha \in [D, M_{\mathbb{F}}]$ where $D \in \mathbb{F}$. Then $D_1 = \alpha^{-1}(\psi(M)) \in \mathbb{F}$ by (4.2.4) and (4.2.5). Let $q = [\varphi^{-1}\alpha_1] \in M_{\mathbb{F}}$ where α_1 is the restriction of α to D_1 and $\varphi : M/t(M) \longrightarrow \psi(M)$ is the isomorphism determined by the composite of the last two maps in (4.2.6). Now, by (4.2.7),

$$qd = [\varphi^{-1}\alpha_1]d = [\mu_{(\varphi^{-1}\alpha_1)d}] = [\mu_{\varphi^{-1}(\alpha(d))}] \in \psi(M)$$

for each $d \in D_1$. Thus, $\varphi^{-1}(qd) = \varphi^{-1}(\alpha(d))$ and hence $qd = \alpha(d)$. Since $M_{\mathbb{F}}$ is torsion-free and $(\alpha(d) - qd)(D_1 : d) = 0$ for each $d \in D$, $\alpha(d) = qd$ and $M_{\mathbb{F}}$ is $\mathbb{F}$-injective.

Now let E be an $\mathbb{F}$-injective essential extension of $\psi(M)$ with $E/\psi(M)$ a torsion module. Define $\sigma : M_{\mathbb{F}} \longrightarrow E$ by $\sigma([\alpha]) = u$ where u is the unique element of E with $(\varphi\alpha)(d) = ud$ for each $d \in D_\alpha$. If $u \in E$, then $\mu_u : (\psi(M) : u) \longrightarrow \psi(M)$ determines the element $[\varphi^{-1}\mu_u]$ in $M_{\mathbb{F}}$ and $\sigma[\varphi^{-1}\mu_u] = u$. It is now easy to check that σ is an R-isomorphism. The rest follows from Exercises 8 and 9. □

A module $E_{\mathbb{F}}(M)$ is an $\mathbb{F}$*-injective hull of* M if $E_{\mathbb{F}}(M)$ is an $\mathbb{F}$-injective essential extension of M and $E_{\mathbb{F}}(M)/M$ is a torsion module. The preceding result shows that $M_{\mathbb{F}}$ is the unique $\mathbb{F}$-injective hull of $M/t(M)$; in fact, each module has an $\mathbb{F}$-injective hull (Exercise 7). Note that for the Goldie topology $\mathbb{G}$, $E_{\mathbb{G}}(M) = E(M)$ and $M_{\mathbb{G}} = E(M/Z_2(M))$.

We wish to now construct a ring of quotients by inverting certain elements. This brings the subject closer to the classical case of inverting elements of a commutative ring (see Exercise 17) or, more generally, of forming the division ring of quotients of a right Öre domain (Exercise 4.1.26).

A *basis* for the topology $\mathbb{F}$ is a subset $\mathscr{B}$ of $\mathbb{F}$ with the property that each right ideal in $\mathbb{F}$ contains some right ideal from $\mathscr{B}$. $\mathbb{F}$ is called a 1*-topology* if $\mathbb{F} \cap \{sR :$

$s \in R\}$ is a basis for $\mathbb{F}$. Let $\Sigma(\mathbb{F}) = \{s \in R : sR \in \mathbb{F}\}$. The nonempty multiplicatively closed subset Σ of R is called a *right Öre set* if R has the *common right multiple property with respect to* Σ:

$$\forall (a,s) \in R \times \Sigma, \exists (b,u) \in R \times \Sigma \text{ with } au = sb.$$

Σ is said to be *saturated* if $sR \cap \Sigma \neq \phi$ implies $s \in \Sigma$.

Theorem 4.2.9. *If $\mathbb{F}$ is a 1-topology, then $\Sigma(\mathbb{F})$ is a saturated right Öre set. If Σ is a right Öre set, then $\mathbb{F}(\Sigma) = \{J \in \mathscr{L}(R_R) : J \cap \Sigma \neq \phi\}$ is a 1-topology. The mapping $\mathbb{F} \mapsto \Sigma(\mathbb{F})$ is a bijection between the set of 1-topologies of R and the set of saturated right Öre sets in R.*

Proof. Assume $\mathbb{F}$ is a 1-topology. Suppose $s, u \in \Sigma(\mathbb{F})$. Then $su \in \Sigma(\mathbb{F})$ since $(uR : a) \subseteq (suR : sa)$ for each $sa \in sR$. If $(a,s) \in R \times \Sigma$, then $uR \subseteq (sR : a)$ for some $u \in \Sigma(\mathbb{F})$; so $\Sigma(\mathbb{F})$ is a right Öre set. It is saturated since if $sa = u \in \Sigma(\mathbb{F})$, then $uR \subseteq sR$ and hence $s \in \Sigma(\mathbb{F})$. Conversely, suppose Σ is a right Öre set. If $J \in \mathbb{F}(\Sigma)$, $a \in R$ and $s \in J \cap \Sigma$, then $sb = au$ for some $(b,u) \in R \times \Sigma$; so $u \in (J : a) \cap \Sigma$ and$(J : a) \in \mathbb{F}(\Sigma)$. Also, if $I \in \mathscr{L}(R_R)$, $J \in \mathbb{F}(\Sigma)$ and $(I : a) \in \mathbb{F}(\Sigma)$ for each $a \in J$, then there are elements $s, u \in \Sigma$ with $s \in J$ and $u \in (I : s)$; so $I \cap \Sigma \neq \phi$ and $I \in \mathbb{F}(\Sigma)$. Suppose Σ is saturated. Then $s \in \Sigma(\mathbb{F}(\Sigma))$ iff $sR \in \mathbb{F}(\Sigma)$, iff $sR \cap \Sigma \neq \phi$, iff $s \in \Sigma$; so $\Sigma = \Sigma(\mathbb{F}(\Sigma))$. On the other hand, if $\mathbb{F}$ is a 1-topology, then $J \in \mathbb{F}(\Sigma(\mathbb{F}))$ iff $J \cap \Sigma(\mathbb{F}) \neq \phi$, iff $sR \in \mathbb{F}$ for some $s \in J$, iff $J \in \mathbb{F}$; so $\mathbb{F} = \mathbb{F}(\Sigma(\mathbb{F}))$. □

A *classical right quotient ring of R with respect to the multiplicatively closed subset* Σ is a pair (T, ψ) where T is a unital ring, $\psi : R \longrightarrow T$ is a ring homomorphism, and

$$\psi(\Sigma) \text{ is contained in the units of } T; \tag{4.2.8}$$

$$T = \{\psi(a)\psi(s)^{-1} : (a,s) \in R \times \Sigma\}; \tag{4.2.9}$$

$$\ker \psi = \{a \in R : r(a) \cap \Sigma \neq \phi\}. \tag{4.2.10}$$

One condition that Σ must satisfy when T exists is:

$$\text{if } \ell(a) \cap \Sigma \neq \phi, \text{ then } r(a) \cap \Sigma \neq \phi. \tag{4.2.11}$$

A multiplicatively closed set with this property is called *right reversible*. If Σ consists of right regular elements, that is $r(s) = 0$ for each $s \in \Sigma$, then Σ is right reversible. A weaker condition for Σ to be right reversible is given in Exercise 18.

Theorem 4.2.10. *(a) A classical right quotient ring of R with respect to Σ exists iff Σ is a right reversible right Öre set.*
(b) If (T, ψ) is a classical right quotient ring of R with respect to Σ and $\varphi : R \longrightarrow S$ is a ring homomorphism such that each element of $\varphi(\Sigma)$ is a unit of S, then there is a unique ring homomorphism $\sigma : T \longrightarrow S$ such that the diagram

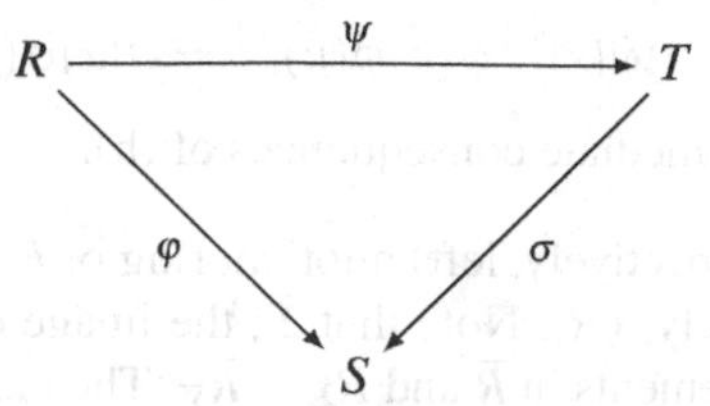

is commutative.

(c) *The classical right quotient ring of R with respect to Σ is unique up to isomorphism.*

(d) *If the classical right and left quotient rings of R with respect to Σ both exist, then they are isomorphic.*

Proof. (a) Suppose (T,ψ) is a classical right quotient ring of R with respect to Σ. If $(a,s) \in R \times \Sigma$, then for some $(b,t) \in R \times \Sigma$ we have $\psi(s)^{-1}\psi(a) = \psi(b)\psi(t)^{-1}$ and $\psi(at) = \psi(sb)$; so $atu = sbu$ for some $u \in \Sigma$ and Σ is right Öre. It is also right reversible since if $sa = 0$ with $s \in \Sigma$, then $\psi(a) = 0$ and hence $at = 0$ for some $t \in \Sigma$. Conversely, suppose Σ is a right reversible right Öre set. Then $\mathbb{F} = \mathbb{F}(\Sigma)$ is a 1-topology and we will show $(R_{\mathbb{F}}, \psi)$ has the desired properties. If $s \in \Sigma$ let $\alpha_s : sR \longrightarrow R/t(R) = \overline{R}$ be defined by $\alpha_s(sa) = \bar{a}$; α_s is an R-homomorphism since Σ is right reversible. Now, $[\mu_s][\alpha_s] = [\bar{\mu}_s\alpha_s]$ where $\bar{\mu}_s = \mu_{\bar{s}}$ and $\bar{\mu}_s\alpha_s(sa) = \overline{sa} = e(sa)$:

$$sR \xrightarrow{\alpha_s} R/t(R) \xrightarrow{\bar{\mu}_s} R/t(R);$$

so $[\mu_s][\alpha_s] = [e] = 1$. Also, $[\alpha_s][\mu_s] = [\bar{\alpha}_s\mu_s]$ where

$$\mu_s^{-1}(sR/t(sR)) \xrightarrow{\mu_s} sR/t(sR) \xrightarrow{\bar{\alpha}_s} R/t(R),$$

and $\bar{\alpha}_s\mu_s(a) = \bar{\alpha}_s(\overline{sa}) = \bar{a} = e(a)$; so $[\alpha_s][\mu_s] = 1$ and $\psi(s)$ is invertible. Moreover, if $[\beta] \in R_{\mathbb{F}}$ and $s \in D_\beta \cap \Sigma$, then $[\beta][\mu_s] = [\mu_{\beta(s)}]$ by (4.2.7), and $[\beta] = \psi(\beta(s))\psi(s)^{-1}$. Since (4.2.10) is obvious $R_{\mathbb{F}}$ is a classical right quotient ring of R.

(b) Clearly, σ must be given by $\sigma(\psi(a)\psi(s)^{-1}) = \varphi(a)\varphi(s)^{-1}$. Suppose $x = \psi(a)\psi(s)^{-1}$ and $y = \psi(b)\psi(u)^{-1} \in R_{\mathbb{F}}$. Take $(c,v) \in R \times \Sigma$ with $sc = uv$. If $\psi(a)\psi(s)^{-1} = \psi(b)\psi(u)^{-1}$, then $\psi(b) = \psi(a)\psi(s)^{-1}\psi(u) = \psi(ac)\psi(v)^{-1}$ and $acw = bvw$ for some $w \in \Sigma$. So $\varphi(a)\varphi(c) = \varphi(b)\varphi(v)$ and

$$\varphi(a)\varphi(s)^{-1} = \varphi(b)\varphi(v)\varphi(c)^{-1}\varphi(s)^{-1} = \varphi(b)\varphi(v)\varphi(v)^{-1}\varphi(u)^{-1} = \varphi(b)\varphi(u)^{-1}.$$

Thus, σ is well-defined. Also, $\psi(a)\psi(s)^{-1} = \psi(ac)\psi(uv)^{-1}$ and $\psi(b)\psi(u)^{-1} = \psi(bv)\psi(uv)^{-1}$. So

$$\sigma(x+y) = (\varphi(ac)+\varphi(bv))\varphi(uv)^{-1} = \varphi(a)\varphi(s)^{-1} + \varphi(b)\varphi(u)^{-1} = \sigma(x)+\sigma(y).$$

To see that σ preserves products take $(d,w) \in R \times \Sigma$ with $bw = sd$. Then

$$\sigma(xy) = \sigma(\psi(ad)\varphi(uw)^{-1}) = \varphi(ad)\varphi(uw)^{-1} = \varphi(a)\varphi(d)\varphi(w)^{-1}\varphi(u)^{-1}$$

$$= \varphi(a)\varphi(s)^{-1}\varphi(b)\varphi(u)^{-1} = \sigma(x)\sigma(y).$$

Both (c) and (d) are immediate consequences of (b). □

The classical right (respectively, left) quotient ring of R with respect to Σ will be denoted by R_Σ (respectively, ${}_\Sigma R$). Note that $\overline{\Sigma}$, the image of Σ in $\overline{R} = R/t(R)$, is a right Öre set of regular elements in $\overline{R}$ and $R_\Sigma = \overline{R}_{\overline{\Sigma}}$. The element $\bar{a}\bar{s}^{-1}$ of R_Σ will be denoted by as^{-1} where $a \in R$ and $s \in \Sigma$. The elements of the module of quotients M_Σ of M have a similar description.

Theorem 4.2.11. *If M is an R-module and Σ is a right reversible right Öre set of R, then $M_\Sigma \cong M \otimes_R R_\Sigma$ as right R_Σ-modules.*

Proof. Define $\tau : M_\Sigma \longrightarrow M \otimes_R R_\Sigma$ by $\tau([\alpha]) = x \otimes s^{-1}$ where $s \in D_\alpha \cap \Sigma$ and $x \in M$ with $\alpha(s) = \bar{x} \in \overline{M} = M/t(M)$. Suppose $u \in D_\alpha \cap \Sigma$ and $y \in M$ with $\alpha(u) = \bar{y}$. Take $(a,v) \in R \times \Sigma$ with $uv = sa$. Then $\overline{xa} = \overline{yv}$; so $xa = yv + z$ with $z \in t(M)$. If $zw = 0$ with $w \in \Sigma$, then $z \otimes 1 = zw \otimes w^{-1} = 0$. So

$$x \otimes s^{-1} = xa \otimes (sa)^{-1} = (yv+z) \otimes (uv)^{-1} = y \otimes u^{-1}$$

and τ is well-defined. Let $\sigma : M \otimes_R R_\Sigma \longrightarrow M_\Sigma$ be the multiplication map $\sigma(x \otimes s^{-1}) = [\mu_{\bar{x}}\alpha_s]$ where, as in Theorem 4.2.10, $\alpha_s : sR \longrightarrow \overline{R}$, $\alpha_s(sa) = \bar{a}$ and $\mu_{\bar{x}} : \overline{R} \longrightarrow \overline{M}$, $\mu_{\bar{x}}(\bar{b}) = \bar{x}\bar{b}$. Now, $\sigma\tau([\alpha]) = [\mu_{\bar{x}}\alpha_s] = [\alpha]$ since $\alpha(sa) = \bar{x}a = \bar{x}\bar{a} = \mu_{\bar{x}}\alpha_s(sa)$. Also, $\tau\sigma(x \otimes s^{-1}) = \tau([\mu_{\bar{x}}\alpha_s]) = xs \otimes s^{-2} = x \otimes s^{-1}$ since $s^2 \in D_{\mu_{\bar{x}}\alpha_s} \cap \Sigma$ and $\mu_{\bar{x}}\alpha_s(s^2) = \bar{x}\bar{s}$. Thus, σ and τ are inverse group isomorphisms. We will now show that σ is an R-homomorphism. If $a \in R$, then $x \otimes s^{-1} \cdot a = x \otimes s^{-1}a = x \otimes bu^{-1} = xb \otimes u^{-1}$ where $\overline{sb} = \overline{au}$ in $\overline{R}$; so $\sigma(x \otimes s^{-1} \cdot a) = [\mu_{\overline{xb}}\alpha_u]$. Also, $\sigma(x \otimes s^{-1})a = [\mu_{\bar{x}}\alpha_s]a = [\mu_{\bar{x}}\alpha_s\mu_a]$ where $\mu_a : (sR : a) \longrightarrow sR$. If $d \in (sR : a) \cap uR$, then $ad = sr$ and $d = ur_1$. Thus, $\mu_{\overline{xb}}\alpha_u(d) = \bar{x}\overline{br_1}$ and $\mu_{\bar{x}}\alpha_s\mu_a(d) = \bar{x}\bar{r}$. But $\bar{s}\bar{r} = \bar{a}\bar{u}\bar{r}_1 = \overline{sb}\bar{r}_1$ and hence $\bar{r} = \overline{br_1}$. By Exercise 8 σ is an R_Σ-homomorphism. □

The maps τ in Theorm 4.2.11 give a natural equivalence between the quotient functor Q_Σ and the tensor functor $\cdot \otimes_R R_\Sigma$. For, if $f \in [M,N]$ and $\bar{f} \in [\overline{M},\overline{N}]$ is the induced map, then the diagram

$$\begin{array}{ccc}
M_\Sigma & \xrightarrow{\mu_{\bar{f}}} & N_\Sigma \\
\downarrow{\scriptstyle \tau_M} & & \downarrow{\scriptstyle \tau_N} \\
M \otimes_R R_\Sigma & \xrightarrow{f \otimes 1} & N \otimes_R R_\Sigma
\end{array}$$

is commutative since $(f \otimes 1)\tau_M([\alpha]) = \bar{f}(\bar{x}) \otimes s^{-1} = \tau_N([\bar{f},\alpha]) = \tau_N\mu_{\bar{f}}([\alpha])$ where $\alpha(s) = \bar{x}$. One consequence of this is that both of these functors are exact.

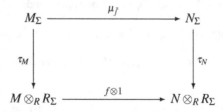

When Σ is a right reversible right Öre set in R the classical ring of quotients, and the module of quotients of M_R, can be constructed directly from ordered pairs without the use of topologies or the general theory of quotients. This is suggested by the arguments in the two previous proofs. The disadvantage of this approach is due to the calculations required to show that a ring and a module have been constructed. We will merely give the definitions and leave the verifications to the interested reader. Let $\sim$ be the relation on $R \times \Sigma$ defined by

$$(a,s) \sim (b,u) \text{ if there exist } c,d \in R \\ \text{with } ac = bd \text{ and } sc = ud \in \Sigma. \tag{4.2.12}$$

Then $\sim$ is an equivalence relation. The operations on $R \times \Sigma / \sim$ are given by

$$[(a,s)] + [(b,u)] = [(ac+bd, v)] \text{ where } sc = ud = v \in \Sigma \tag{4.2.13}$$

and

$$[(a,s)][(b,u)] = [(ac, uv)] \text{ where } v \in \Sigma \text{ and } sc = bv. \tag{4.2.14}$$

Similar definitions on $M \times \Sigma$ yield a module. On the other hand, having already constructed R_Σ and M_Σ it is less tedious to show that these definitions lead to a ring isomorphic to R_Σ and a module isomorphic to M_Σ. In fact, this works for any 1-topology; that is, for any right Öre set; see Exercise 19.

If the set of regular elements of R is a right Öre set, then R is called a *right Öre ring*. The corresponding quotient ring is denoted by $Q_c(R)$ and is called *the classical right quotient ring of R*. We wish to determine when $Q_c(R)$ is semiprime and right artinian. But first we will prove the density theorem, which represents a primitive ring as a large ring of linear transformations, and the Wedderburn-Artin theorem.

Let M be a left vector space over the division ring D. A subring R of $\mathrm{End}_D(M)$ is called *n-transitive*, where $n \in \mathbb{N}$, if for each n-dimensional D-subspace N of M and each $a \in \mathrm{End}_D(M)$ there is some $b \in R$ with $N(a-b) = 0$. This is clearly equivalent to the condition: if $X = \{x_1, \ldots, x_n\}$ is a D-independent subset of M and $y_1, \ldots, y_n$ are any elements of M, then there is some $b \in R$ with $x_i b = y_i$ for $1 \le i \le n$. Assuming R is 1-transitive, another equivalent condition is : for each n-element independent subset X, $r(X \backslash \{x_i\}; R) \not\subseteq r(x_i; R)$ for $1 \le i \le n$. For, if this condition is satisfied, then for each i, $x_i r(X \backslash \{x_i\}) = M$ by 1-transitivity; so there is an element $b_i \in R$ with $x_i b_i = y_i$ and $(X \backslash \{x_i\}) b_i = 0$. But then $x_i(b_1 + \cdots + b_n) = y_i$ for each $1 \le i \le n$. If R is n-transitive for each $n \in \mathbb{N}$, then R is called a *dense* subring of $\mathrm{End}_D(M)$. For the topological explanation of this terminology see Exercise 20.

Recall that a ring is (right) primitive if it has a modular maximal right ideal which contains no nonzero ideal; equivalently, if it has a simple faithful right module.(For the ℓ-ring analogues see the "Johnson radical" in Section 3.2 and Exercise 3.3.8.).

Theorem 4.2.12. *The ring R is primitive if and only if it is isomorphic to a dense ring of linear transformations on a left vector space.*

Proof. If R is dense in $\mathrm{End}_D({}_DM)$, then clearly M_R is simple and faithful. Conversely, suppose M_R is a faithful simple R-module. Then $D = \mathrm{End}_R(M)$ is a divsion ring since $\ker\alpha = 0$ and $\alpha M = M$ for each $0 \neq \alpha \in D$. Moreover, R embeds in $\mathrm{End}_D(M)$ and we will identify R with its image in $\mathrm{End}_D(M)$. Since $xR = M$ for each $0 \neq x \in M$, R is 1-transitive. Assume, inductively, that R is n-transitive and suppose by way of contradiction that it is not $(n+1)$-transitive. Then there is a D-independent subset $X = \{x_1, \ldots, x_{n+1}\}$ of M with $r(\{x_1, \ldots, x_n\}) \subseteq r(x_{n+1})$. Let $I = r(\{x_1, \ldots, x_{n-1}\})$. Then $x_n I = M$ and $\alpha : M \longrightarrow M$ defined by $\alpha(x_n a) = x_{n+1}a$, $a \in I$, is an element of D. But $\{x_1, \ldots, x_{n-1}, \alpha x_n - x_{n+1}\}$ is a D-independent subset of M and $I \subseteq r(\alpha x_n - x_{n+1})$. This contradiction shows R is $(n+1)$-transitive. □

For generalizations of this result see Exercises 22 and 23.

Theorem 4.2.13. *The ring R is semiprime and right artinian iff R is isomorphic to the direct product of a finite number of matrix rings over division rings. If $R \cong (D_1)_{n_1} \times \cdots \times (D_k)_{n_k}$, then the integers $k, n_1, \ldots, n_k$ are uniquely determined by R, and the division rings $D_1, \ldots, D_k$ are determined up to isomorphisms.*

Proof. If D is a division ring the ith row $e_{ii}D_n$ of D_n is a minimal right ideal of D_n since $de_{ik} = e_{ij}d_j d_j^{-1} d e_{jk}$ if $0 \neq e_{ij}d_j \in e_{ij}D_n$ and $d \in D$. So D_n is a (simple) right artinian ring and $(D_1)_{n_1} \times \cdots \times (D_k)_{n_k}$ is semiprime and right artinian. Conversely, suppose R is semiprime and right artinian. If P is a proper prime ideal of R, then R/P is right primitive since a minimal right ideal of R/P is a faithful irreducible R/P-module. So R/P is isomorphic to a dense subring of $\mathrm{End}({}_DM)$. If $\{x_n : n \in \mathbb{N}\}$ were an infinite D-independent subset of M, then $r(x_1; R/P) \supset r(\{x_1, x_2\}; R/P) \supset \cdots$ would be a strictly decreasing chain of right ideals of R/P. Thus, ${}_DM$ is finite dimensional and $R/P \cong D_m$ for some m. By Theorem 4.1.14. $R \cong \varphi(R)$ where $\varphi(R) \subseteq (D_1)_{n_1} \times \cdots \times (D_k)_{n_k}$ and $\varphi(R) \cap (D_j)_{n_j}$ is a nonzero ideal of $(D_j)_{n_j}$. Thus, $R \cong (D_1)_{n_1} \times \cdots \times (D_k)_{n_k} = S$. To see the uniqueness suppose $G_1, \ldots, G_t$ are division rings and $R \cong (G_1)_{m_1} \times \cdots \times (G_t)_{m_t}$. Since each ideal of S is a product of some of the $(D_j)_{n_j}$, $k = t$ and (after relabling) $(D_j)_{n_j} \cong (G_j)_{m_j}$. But then $n_j = m_j$ since the right $(D_j)_{n_j}$-module $(D_j)_{n_j}$ has composition series of lengths n_j and m_j (or its Goldie dimension is both n_j and m_j - see Exercise 4.1.38). Let $D = D_j$, $G = G_j$, $n = n_j$ and let $e = e_{11}$ and $f = f_{11}$ be the left corner matrix units of D_n and G_n, respectively. If M is a simple faithful right G_n-module and $0 \neq x \in M$ with $xfG_n \neq 0$, then $M = xfG_n \cong fG_n$. Let $h \in G_n$ correspond to e via the isomorphism $D_n \cong G_n$. Then

$$D \cong eD_ne \cong \mathrm{End}_{D_n}(eD_n) \cong \mathrm{End}_{G_n}(hG_n) \cong \mathrm{End}_{G_n}(fG_n) \cong G.$$

□

A semiprime right artinian ring will be called a *semisimple ring*. For some homological characterizations of a semisimple ring see Exercise 24. It is easier to determine when $Q(R)$ is semisimple than it is to determine when $Q_c(R)$ exists and is semisimple. For the module M_R let $\mathscr{C}(M) = \{N \in \mathscr{L}(M) : M/N \text{ is nonsingular}\}$. $\mathscr{C}(M)$ is a complete lattice and each submodule in $\mathscr{C}(M)$ is a complement in M. When M is nonsingular $\mathscr{C}(M)$ contains every complement in M. In the notation of Exercise 10 $\mathscr{C}(M) = \mathscr{C}_{\mathbb{G}}(M)$ where $\mathbb{G}$ is the Goldie topology of R. We will denote $\mathscr{C}(R_R)$ by $\mathscr{C}_r(R)$.

Theorem 4.2.14. *The following are equivalent for the right nonsingular ring R.*

(a) $Q(R)$ is a semisimple ring.
(b) R_R is finite dimensional.
(c) $\mathscr{C}_r(R)$ has the ascending chain condition.
(d) $\mathscr{C}_r(R)$ has the descending chain condition.

Proof. From Exercise 10 (or the discussion after Theorem 4.1.1) the lattice $\mathscr{C}_r(R)$ is isomorphic to the lattice $\mathscr{C}_r(Q(R)) = \mathscr{C}(E(R_R))$. Thus, each of these statements is a consequence of (a). Note that $\mathscr{C}_r(Q)$ consists of the principal right ideals of Q.

(b) $\Rightarrow$ (a). By Exercise 4.1.44 Q_Q is a direct sum of a finite number of indecompsable right ideals. If $0 \neq aQ \subseteq eQ$ where $e = e^2$ and eQ is indecomposable, then aQ is a summand of Q_Q since Q is regular. Hence, aQ is a summand of eQ, $aQ = eQ$, and eQ is a minimal right ideal of Q. So Q is semisimple.

(c) $\Rightarrow$ (a). Let I be a right ideal of Q and let eQ be a maximal principal right ideal of I. If $a \in I$, then $aQ + eQ$ is also principal by Theorem 4.1.7. So $I = eQ$ and Q is semisimple by Exercise 24.

(d) $\Rightarrow$ (a). If $e = e^2 \in Q$, then $r(Qe) = (1-e)Q$ and $\ell(eQ) = Q(1-e)$. Thus, the lattice anti-isomorphism between the lattice of right and left annihilators induces an anti-isomorphism between the posets of principal right and left ideals (in any regular ring). So Q has the ascending chain condition on principal left ideals and, as in the previous paragraph, Q is semisimple. □

The problem with strengthening Theorem 4.2.14 to obtain that $Q_c(R)$ is semisimple is that not only must regular elements be produced but it must be shown that R is a right Öre ring. In one case no other conditions on R are required.

Theorem 4.2.15. *The following statements are equivalent for the reduced ring R.*

(a) $Q(R)$ is a semisimple ring.
(b) $Q(R)$ is a direct product of a finite number of division rings.
(c) $Q(R)$ is reduced and $\mathrm{Ann}(R)$ has the ascending chain condition.
(d) R is a right Öre ring and $\mathrm{Ann}(R)$ has the ascending chain condition.

Proof. In all cases $\mathrm{Ann}(R)$ is finite, and by Theorem 4.1.14 there are ideals $R_1, \ldots, R_n$ of R with $R_1 \boxplus \cdots \boxplus R_n \subseteq R \subseteq Q_1 \boxplus \cdots \boxplus Q_n = Q$ where each R_j is a domain, $A = R_1 + \cdots + R_n$ is essential in R, $Q_j = Q(R_j)$, and $Q = Q(R)$. If Q is semisimple, then R, and hence each R_j, is right finite dimensional by the preceding result. By Exercise 4.1.26 each R_j is a right Öre domain and $Q_j = Q_c(R_j)$ is a division ring. By Exercise 26 $Q = Q_c(R)$. Thus, (a) is equivalent to (b) and it implies (c) and (d). If Q is reduced, then $\mathscr{C}_r(R) = \mathrm{Ann}(R)$ by Theorem 4.1.10; so (c) implies (a) by Theorem 4.2.14. If R is a right Öre ring, then so is its ideal A. For, each regular element of A is a regular element of R. So if $a, s \in A$ with s a regular element of A, and $at = sb$ with $b, t \in R$ and t is regular in R, then $ats = sbs$ with ts, $bs \in A$, and ts is regular. By Exercise 26 each R_j is a right Öre domain and hence $Q_j = Q_c(R_j)$ is a division ring. We have shown that (d) implies (a). □

The ring R in Theorem 4.2.15 has the property that a right ideal is essential if and only if it contains a regular element of R. In general, this is a characterizing property of those rings which have a semisimple classical right quotient ring. R is called a *right Goldie ring* if its lattice of right annihilators has a.c.c. and R_R is finite dimensional. If R_R is finite dimensional and for each M-sequence $(a_n)_{n\in\mathbb{N}}$ in R the chain of right annihilators $r(a_1) \subseteq r(a_2) \subseteq \cdots$ is finite, then R is called a *weak right Goldie ring*.

Theorem 4.2.16. *The following statements are equivalent for the ring R.*

(a) R is a right Öre ring and $Q_c(R)$ is semisimple (simple and right artinian).
(b) R is a semiprime (prime) right Goldie ring.
(c) R is a semiprime (prime) weak right Goldie ring.
(d) A right ideal of R is essential iff it contains a regular element (and each nonzero ideal is an essential right ideal).
(e) R is a semiprime (prime) right nonsingular ring and $\mathscr{C}_r(R)$ has a.c.c..

Proof. (a) $\Rightarrow$ (b). R_R is finite dimensional by Theorem 4.2.14, and since a.c.c. on right annihilators is inherited by subrings, R is a right Goldie ring. Suppose A is a nonzero ideal of R. Then $Q_cAQ_c = Q_ce$ for some nonzero central idempotent e in Q_c. Now, $e = \sum_i p_ia_ib_is^{-1}$ with $p_i \in Q_c$, $a_i \in A$ and $b_i, s \in R$. So $es \in Q_cA$, $sA = esA \subseteq Q_cA^2$, and $A^2 \neq 0$. If Q_c is simple, then $e = 1$ and $sr(A) = 0$. So $r(A) = 0$ and R is a prime ring.

(b) $\Rightarrow$ (c). This is trivial.

(c) $\Rightarrow$ (d). We will first show that each nonzero right or left ideal contains an element whose square is not zero. For, suppose A is a right ideal and $a^2 = 0$ for each $a \in A$. Then for any $a, b, c \in A$, $0 = (a+b)^2 = ab + ba$ yields $abac = -a^2bc = 0$. So $(aA)^2 = 0$, $aA = 0$, $A^2 = 0$, and hence $A = 0$. We claim next that R has no nonzero nil one-sided ideals. For, if $0 \neq a_1 \in R$ take $0 \neq a_2 \in a_1Ra_1$ with $r(a_1) \subset r(a_2)$, if this is possible; and then take $0 \neq a_3 \in a_2Ra_2$ with $r(a_2) \subset r(a_3)$. This process must terminate and it produces nonzero elements $a_1, a_2, \ldots, a_n$ with $a_k \in a_{k-1}Ra_{k-1}$ and $r(a_n) = r(a_nxa_n)$ whenever $0 \neq a_nxa_n \in a_nRa_n$. Take $a = ua_n \in Ra_n$ with $a^2 \neq 0$. Then $r(a_n) \subseteq r(a) \subseteq r(ya)$ for each $y \in R$. But if $ya \neq 0$, then for some $z \in R$, $0 \neq yazya = yua_nzya$ and $r(ya) \subseteq r(a_nzya) = r(a_nzyua_n) = r(a_n)$. Thus, $r(a) = r(ya)$ if $ya \neq 0$. Now, if $a^m = 0$ and $a^{m-1} \neq 0$, then $m \geq 3$ and $a \in r(a^{m-1}) = r(a^{m-2}a) = r(a)$ which contradicts $a^2 \neq 0$. This shows that a is an element of Ra_1 that is not nilpotent. It also shows that $Z_r(R) = 0$. For if $a_1 \in Z_r(R)$, then $a \in Z_r(R)$ and $r(a) \cap aR = 0$. Since a_1R is nil iff Ra_1 is nil, a_1R is not a nil right ideal. For any $b \in R$ there is an $n \in \mathbb{N}$ with $r(b^n) = r(b^{n+1}) = r(b^{n+2}) = \cdots$ since (b^{3^k}) is an M-sequence. We claim that $b^nR + r(b^n)$ is an essential right ideal of R. For, suppose J is a right ideal with $J \cap (b^nR + r(b^n)) = 0$. By Exercise 27 the sum $J + b^nJ + b^{2n}J + \cdots$ is direct. Thus, $J \subseteq r(b^{kn}) = r(b^n)$ for some k and $J = 0$. In particular, if $r(b) = 0$, then bR is essential. We show next that each right ideal J contains some element a with $r(a) \cap J = 0$. Assume $J \neq 0$ and take a non-nilpotent element $b \in J$. Since $r(b^n) = r(b^{n+1}) = \cdots$ for some n, $r(b^n) = r(b^{2n})$; let $a_1 = b^n$. If $r(a_1) \cap J \neq 0$ take $0 \neq a_2 \in r(a_1) \cap J$ with $r(a_2) = r(a_2^2)$. This process produces elements $a_1, \ldots, a_k$

with $0 \neq a_j \in r(a_1) \cap \cdots \cap r(a_{j-1}) \cap J$. The sum $a_1R + \cdots + a_kR$ is direct. For if $a_jb_j + \cdots + a_kb_k = 0$ with $j \leq k$, then $a_j^2b_j = 0$ and hence $a_jb_j = 0$. Thus, for some m, we have $r(a_1) \cap \cdots \cap r(a_m) \cap J = 0$ and $a = a_1 + \cdots + a_m \in J$ with $r(a) \cap J = 0$. If J is essential, then $r(a) = 0$. We have now verified that a right ideal is essential iff it contains a right regular element. But if $r(b) = 0$, then $\ell(b)bR = 0$ shows $\ell(b) \subseteq Z_r(R) = 0$.

(d) $\Rightarrow$ (a). R is certainly a right nonsingular ring and hence $\mathbb{D}(R) = \mathscr{E}(R_R)$ by Theorem 4.1.8. By Theorem 4.2.9, R is a right Öre ring. Since the only essential right ideal of Q_c is Q_c itself and since each right ideal is a summand of an essential right ideal, Q_c must be semisimple. If each nonzero ideal of R is essential, then each nonzero ideal of Q_c contains an invertible element; so Q_c is simple.

(a) $\Rightarrow$ (e). In the first paragraph above we have seen that R is semiprime (prime). Its right singular ideal vanishes since it has a regular right quotient ring, and, by Theorem 4.2.14, $\mathscr{C}_r(R)$ has a.c.c.

(e) $\Rightarrow$ (b). This is an immediate consequence of Theorem 4.2.14. □

If F is a right Öre domain with division ring of right quotients D and R is a subring of D_n which contains F_n, then R is a prime right Goldie ring and $Q_c(R) = D_n$ is obtained from R by inverting "scalars." In fact, this is the norm.

Theorem 4.2.17. *The ring R is a semiprime right Goldie ring if and only if it is a right quotient ring of a direct product of a finite number of rings each of which is isomorphic to a matrix ring over a right Öre domain.*

Proof. One implication is clear. For the other we assume R is a prime right Goldie ring. The case where R is semiprime then follows since, by Theorem 4.1.14 and Exercise 4.1.33, R contains an essential ideal which is a direct product of a finite number of ideals of R each of which is a prime right Goldie ring. Let $\{g_{ij}\}$ be the usual matrix units of $Q = Q_c(R) = G_n$, where G is a division ring. There is a regular element s of R with $\{g_{ij}\}s \subseteq R$. Let $e_{ij} = s^{-1}g_{ij}s$. Then $X = \{e_{ij}\}$ is a set of matrix units of Q with $sX \subseteq R$. Let $D = s^{-1}Gs$ be the centralizer of X in Q. Then D is a division ring and $Q = \sum_{ij} De_{ij}$. Let $A = \{r \in R : rX \subseteq R\}$ and $B = (R : X) = \{r \in R : Xr \subseteq R\}$. Then B contains a regular element u since it is an essential right ideal of R, and $uRs \subseteq BA \subseteq R$. If $q \in Q$ there are elements c, w in R with $u^{-1}qu = cw^{-1}$. So $q = (ucs)(uws)^{-1}$ with ucs, $uws \in BA$, and Q is the classical right quotient ring of BA since each regular element of BA is right regular in Q and hence a unit of Q. We show next that $BA = \sum Fe_{ij}$ where $F = BA \cap D$. Note that $XB \subseteq B$ since $X(XB) \subseteq X^2B \subseteq XB \cup \{0\} \subseteq R$, and similarly $AX \subseteq A$; so $XBAX \subseteq BA$. If $z \in BA$, then $z = \sum z_{ij}e_{ij}$ with $z_{ij} = \sum_k e_{ki}ze_{jk} \in BA \cap D$. Since $Fe_{ij} \subseteq BAe_{ij} \subseteq BA$ we do have $BA = \sum Fe_{ij}$. Now F_F is finite dimensional since a direct sum of right ideals in F produces a direct sum of right ideals in F_n. Thus, F is a right Öre domain by Exercise 4.1.26. Since $\sum De_{ij} = Q_c(\sum Fe_{ij}) = \sum Q_c(F)e_{ij}$ we have $D = Q_c(F)$. □

If R is a prime right Goldie ring and F is a right Öre domain with $F_n \subseteq R$, then F cannot be unital unless R is unital. Even when R is unital F need not be unital. One such example is the ring $R = (2\mathbb{Z})_2 + \mathbb{Z}\cdot 1$ which has no nontrivial idempotents. For a condition which insures $R = F_n$ see Exercise 28.

We now turn our attention to some basic properties of simple algebras over a field. These results will be used in Section 6.4 to determine the algebraic division algebras over the reals and other similar maximal totally ordered fields. Recall that the centroid of the ring R is the subring $K = \mathrm{End}_R({}_RR) \cap \mathrm{End}_R(R_R)$ of $\mathrm{End}_{\mathbb{Z}}(R)$. For $x \in R$ let μ_x and ρ_x denote the left, respectively, right multiplication map on R determined by x, and let T by the subring of $\mathrm{End}_{\mathbb{Z}}(R)$ generated by $\{\mu_x, \rho_x : x \in R\}$: $T = \{\mu_x + \rho_y + \Sigma_i \mu_{x_i}\rho_{y_i}\}$. Then K is the centralizer of T in $H = \mathrm{End}_{\mathbb{Z}}(R)$. Let $\overline{T}$ be the subring of T defined by $\overline{T} = \{\Sigma_i \mu_{x_i}\rho_{y_i}\}$. If R is a simple ring, then K is the largest field over which R is an algebra and K is also the centralizer of $\overline{T}$ in H by Exercise 3.2.9. Moreover, R_T and $R_{\overline{T}}$ are both simple modules and as a consequence of Theorem 4.2.12 we know that T and $\overline{T}$ are both dense K-subalgebras of $\mathrm{End}_K({}_KR)$. An algebra R over the field K will be called *central simple* if R is a simple ring and K is its centroid.

Theorem 4.2.18. *Suppose R and S are algebras over the field K. If R is central simple and S is simple, then $R \otimes_K S$ is simple. If R and S are both unital and central simple, then $R \otimes_K S$ is central simple. Moreover, if $R \otimes_K S$ is simple or central simple, then so are R and S.*

Proof. Assume R is central simple and S is simple and let X be a K-basis of R and I a nonzero ideal of $R \otimes_K S$. Take $0 \neq \Sigma_{i=1}^n a_i \otimes b_i \in I$ with $a_i \in X$, each $a_i \otimes b_i \neq 0$ and let $a \in R$. Since $R_{\overline{T}}$ is a faithful simple $\overline{T}$-module (notation as above) with $K = \mathrm{End}_{\overline{T}}(R_{\overline{T}})$ we get from the density theorem, Theorem 4.2.12, that there exist $c_1, \ldots, c_m, d_1, \ldots, d_m$ in R with $\Sigma_j c_j a_1 d_j = a$ and $\Sigma_j c_j a_i d_j = 0$ for $i \geq 2$. Then for any e, $f \in S$

$$\begin{aligned} a \otimes eb_1 f &= \Sigma_j c_j a_1 d_j \otimes eb_1 f = \Sigma_i (\Sigma_j c_j a_i d_j) \otimes eb_i f \\ &= \Sigma_j \Sigma_i c_j a_i d_j \otimes eb_i f \\ &= \Sigma_j (c_j \otimes e)(\Sigma_i a_i \otimes b_i)(d_j \otimes f) \in I. \end{aligned}$$

If $b \in S$, then $b = \Sigma_k e_k b_1 f_k$ with e_k and $f_k \in S$ since $S = Sb_1S$; so $a \otimes b = \Sigma_k a \otimes e_k b_1 f_k \in I$ and hence $I = R \otimes_K S$. Suppose now that R and S are both unital and central simple, and let $u = \Sigma_i a_i \otimes b_i$ be in the center of $R \otimes_K S$ with $a_i \in X$. Then for any $b \in S$, $(1 \otimes b)\Sigma_i a_i \otimes b_i = (\Sigma_i a_i \otimes b_i) 1 \otimes b$ and hence $\Sigma_i a_i \otimes (bb_i - b_i b) = 0$. So $bb_i = b_i b$, each $b_i \in K$ and $u = (\Sigma_i a_i b_i) \otimes 1 = c \otimes 1$. From $(a \otimes 1)(c \otimes 1) = (c \otimes 1)(a \otimes 1)$ for each $a \in R$ we get $ac = ca$ and so $c \in K$ and also $u \in K$. Conversely suppose $R \otimes_K S$ is simple. Then R must be simple since if I is a nonzero proper ideal of R, then $I \otimes_K S$ is a nonzero proper ideal of $R \otimes_K S$. Also, if $R \otimes_K S$ is central simple over K and L is the centroid of R, then $L \subseteq K$ since $R \otimes_K S$ is an algebra over L. Thus, $L = K$. □

We wish to investigate the primitivity of some tensor products but first we note the following consequence of Theorem 4.2.18. Suppose A and B are simple K-subalgebras of a K-algebra with A central simple and suppose A centralizes B. Then either $AB = 0$ or the multiplication map $A \otimes_K B \longrightarrow AB$ is an isomorphism.

Theorem 4.2.19. *Let R be a central simple K-algebra and let R° be the opposite algebra of R.*

(a) *$R \otimes_K R^\circ$ is isomorphic to a dense K-subalgebra of $\mathrm{End}_K({}_K R)$.*
(b) *If R is a division ring and E is a maximal subfield of R, then $R \otimes_K E$ is isomorphic to a dense E-subalgebra of $\mathrm{End}_E({}_E R)$.*

Proof. Let $A = \{\rho_x : x \in R\}$ and $B = \{\mu_x : x \in R\}$ be the images of R in $\mathrm{End}_K({}_K R)$ under the right and left regular representations, respectively. Since R is a right $\mathrm{End}_K({}_K R)$-module A is isomorphic to R and B is isomorphic to R°, and hence $R \otimes_K R^\circ$ is isomorphic to AB. We have previously noted that $AB = \overline{T}$ is dense in $\mathrm{End}_K({}_K R)$.

Now let E be a maximal subfield of the division ring R and let $C = \{\mu_x : x \in E\}$. Then C is isomorphic to $E^\circ = E$, $R \otimes_K E$ is isomorphic to AC, and R_{AC} is a simple module. But C is the centralizer of AC in $\mathrm{End}_K({}_K R)$. For, suppose S is this centralizer and $f \in S$. Then $f = \mu_a$ where $a = 1f \in R$ since $f \in \mathrm{End}_R(R_R)$; so $S \subseteq B \cong R^\circ$. Since C centralizes $A \cup C$ we have $C \subseteq S \subseteq B$. If $a \in R$ and $\mu_a \in S$, then a commutes with the elements of E since μ_a commutes with the elements of C. So $a \in E$ since $E(a)$ is a subfield of R. Thus $C = S$ and AC is dense in $\mathrm{End}_E({}_E R)$ by Theorem 4.2.12. □

A finite dimensional central simple algebra is isomorphic to a matrix algebra D_m where D is a division ring which is finite dimensional over its center. This dimension is in fact a square.

Theorem 4.2.20. *Let K be the center of the division ring D and let E be a maximal subfield of D. If $[D : K] < \infty$, then $[D : K] = n^2$ for some n and $[D : E] = [E : K] = n$. Moreover, $[D : K] < \infty$ if and only if $[E : K] < \infty$.*

Proof. Suppose $[D : K] < \infty$ and let L be the algebraic closure of K. Then $D \otimes_K L$ is a simple L-algebra by Theorem 4.2.18 and $[D \otimes_K L : L] = [D : K]$. Since L is the only algebraic division algebra over L necessarily $D \otimes_K L \cong L_n$ for some n and hence $[D : K] = n^2$. Let $m = [{}_E D : E]$. Then $D \otimes_K E$ is isomorphic to $\mathrm{End}_E({}_E D) \cong E_m$ by Theorem 4.2.19. So $n^2 = [D : K] = [D \otimes_K E : E] = m^2$ and $n = m$. Since $[D : K] = [D : E][E : K]$ we also have $[E : K] = n$. Now assume that $[E : K] < \infty$. Then $D \otimes_K E$ is a finite dimensional left vector space over D and hence $D \otimes_K E$ is left artinian. By Theorems 4.2.18 and 4.2.19 $D \otimes_K E$ is a simple E-algebra which is isomorphic to a dense subalgebra of $\mathrm{End}_E({}_E D)$. By Theorem 4.2.13 and its proof ${}_E D$ is finite dimensional and hence $[D : K] < \infty$. □

According to Exercise 37 the only K-automorphisms, K a field, of the matrix algebra K_n are the inner automorphisms. Our final result on central simple algebras (until Theorem 6.4.10) is a generalization of this fact.

Theorem 4.2.21. *Let D be a division ring with center K and let V be a left vector space over D. Suppose A and B are finite dimensional simple K-subalgebras of $R = \mathrm{End}_D({}_D V)$ each of which contains the identity element of R. Then any K-algebra isomorphism of A onto B is the restriction of an inner automorphism of R.*

Proof. V is a right $\mathrm{End}_K({}_KV)$-module and D° is isomorphic to the subalgebra C of $\mathrm{End}_K({}_KV)$ consisting of the maps $v \mapsto dv$, $v \in V$ and $d \in D$. By Theorem 4.2.19 $AC \cong A \otimes_K C$ is simple. Since A is finite dimensional over K the ring $A \otimes_K C$ is a finite dimensional right vector space over C and hence AC is right artinian. Let I be a minimal right ideal of AC. By Exercise 32 each faithful simple right AC-module is isomorphic to I, and since each unital right AC-module is certainly semisimple, V_{AC} is isomorphic to $I^{(m)}$, a direct sum of m copies of I, for some cardinal m (see Exercise 2.6.9). Suppose I_C is q-dimensional. Then $[{}_DV : D] = mq$. Now let $f : A \longrightarrow B$ be a K-algebra isomorphism. Then the induced map $\Sigma_i a_i c_i \mapsto \Sigma_i f(a_i)c_i$ is a K-algebra isomorphism of AC onto BC and since V is a unital right BC-module it is also an AC-module with respect to the action $v \cdot ac = vf(a)c$. Since q is finite $V_{\cdot AC}$ is also isomorphic to $I^{(m)}$. Let $u : V_{AC} \longrightarrow V_{\cdot AC}$ be an isomorphism of these two right AC-modules. Then $u \in R$ and $vau = vu \cdot a = vuf(a)$ for $v \in V$ and $a \in A$. Thus, $au = uf(a)$ and $f(a) = u^{-1}au$. □

Exercises.

Unless otherwise specified $\mathbb{F}$ is a topology of R.

1. Let $(\mathscr{T}, \mathscr{F})$ be a torsion theory for $\mathscr{M}_R$. Show that $\mathscr{T}$ is hereditary iff $\mathscr{F}$ is closed under essential extensions.
2. Let $\mathscr{T} \subseteq \mathscr{M}_R$.
   (a) Show that $\mathscr{T}$ is a hereditary torsion class iff there is an injective module E_R such that $\mathscr{T} = \ell(E; \mathscr{M}_R)$.
   (b) Show that $\mathscr{T}$ is a hereditary torsion class arising from a topology of R iff there is an injective module E_R such that $\ell(R; E) = 0$ and $\mathscr{T} = \ell(E; \mathscr{M}_R)$.
   (c) If R is unital and $\mathscr{T} \subseteq u\text{-}\mathscr{M}_R$ show that $\mathscr{T}$ is a hereditary torsion class for $u\text{-}\mathscr{M}_R$ iff $\mathscr{T} = \ell(E; u\text{-}\mathscr{M}_R)$ for some u-injective module E.
   (d) Show that if $\ell(R; R) = 0$, then $\mathbb{D}$ is the topology corresponding to the torsion theory cogenerated by $E(R_R)$.
3. Let $(\mathscr{T}, \mathscr{F})$ be the torsion theory cogenerated by the injective module E. Show that $M \in \mathscr{F}$ iff it can be embedded in a direct product of copies of E.
4. Show that there is a bijection between the class of torsion theories for $\mathscr{M}_R$ and the class of idempotent radicals of $\mathscr{M}_R$.
5. (a) Suppose $\mathscr{C} \subseteq \mathscr{M}_R$ is homomorphically closed. Show that $M \in \ell r(\mathscr{C})$ iff each nonzero homomorphic image of M contains a nonzero $\mathscr{C}$-submodule.
   (b) If $\mathscr{C}$ is hereditary and homomorphically closed show that the torsion theory generated by $\mathscr{C}$ is hereditary.
6. (a) Suppose that $[B, M] \longrightarrow [A, M] \longrightarrow 0$ is exact whenever A is a submodule of B with $B/A \in \mathscr{T}$. Show that M is $\mathbb{F}$-injective.

(b) If M is $\mathbb{F}$-injective and $\ell(R;M) = 0$ show that M satisfies the condition in (a).

7. (a) Show that each module has an $\mathbb{F}$-injective hull.
(b) If $\ell(R;M) = 0$ show that any two $\mathbb{F}$-injective hulls of M are isomorphic over M.

8. Let M and N be $R_{\mathbb{F}}$-modules with N torsion-free.

(a) Show that $[M,N]_R = [M,N]_{R_{\mathbb{F}}}$.
(b) Show that N is an injective R-module iff it is an injective $R_{\mathbb{F}}$-module.

9. Let E be a torsion-free $\mathbb{F}$-injective R-module. Show that there is a unique way to make E into an $R_{\mathbb{F}}$-module so that the R-action is induced from the $R_{\mathbb{F}}$-action.

10. Let $\mathscr{C}_{\mathbb{F}}(M) = \{N \in \mathscr{L}(M_R) : M/N \text{ is torsion-free}\}$.

(a) Show that $\mathscr{C}_{\mathbb{F}}(M)$ is a complete modular lattice. ($A \mapsto \{x \in M : (A : x) \in \mathbb{F}$ is a closure operator on $\mathscr{L}(M_R)$.)
(b) If M is torsion-free and $\mathbb{F}$-injective, and $N \in \mathscr{L}(M_R)$, show that $N \in \mathscr{C}_{\mathbb{F}}(M)$ iff N is $\mathbb{F}$-injective.
(c) Show that the mapping $\mathscr{C}_{\mathbb{F}}(M) \longrightarrow \mathscr{C}_{\mathbb{F}}(M_{\mathbb{F}})$ given by $N \mapsto N_{\mathbb{F}}$ is a lattice isomorphism.

11. Let $e : R \longrightarrow R/t(R)$ be the natural map. Show that $R_{\mathbb{F}}$ is a right quotient ring of $R/t(R)$ and $e(\mathbb{F})$ is a topology of $R/t(R)$ contained in $\mathbb{D}(R/t(R))$. If $\mathbb{F}$ is a pretopology show that $e(\mathbb{F})$ is a pretoplogy of $R/t(R)$.

12. If R is a ring the topological space $(R, \mathscr{V})$ is called a *topological ring* provided the negation map $a \mapsto -a$ is continuous and addition and multiplication $+, \cdot : R \times R \longrightarrow R$ are both continuous when $R \times R$ has the product topology. Similarly, if M_R is an R-module and R is a topological ring, then the topological space $(M, \mathscr{W})$ is a *topological R-module* if negation, addition and scalar multiplication are all continuous maps (in the appropriate product topologies). The topological ring $(R, \mathscr{V})$ (respectively, topological module $(M, \mathscr{W})$) is called *a (right) linear topological ring* (respectively, *linear topological module)* if the neighborhood system of 0 has a base consisting of right ideals (submodules).

(a) Show that the topological ring $(R, \mathscr{V})$ is a linear topological ring iff the set $\mathbb{F}$ of all open right ideals is a pretopology of R.
(b) Suppose $(R, \mathscr{V})$ is a linear topological ring. Show that there is a unique largest topology $\mathscr{V}(M)$ of the module M_R such that $(M, \mathscr{V}(M))$ is a linear topological module. (The set of open submodules of M is $\{N \in \mathscr{L}(M_R) : M/N \in \mathscr{T}\}$).
(c) Show that $(M, \mathscr{V}(M))$ is discrete iff $M \in \mathscr{T}$.

13. Let J be a right ideal of R and let $\mathbb{K}$ be the closed interval $[J, R]$ in $\mathscr{L}(R_R)$. Show that $\mathbb{K}$ is a topology iff J is an idempotent ideal. Show that $\mathbb{F}$ is a closed interval of $\mathscr{L}(R_R)$ iff its associated torsion class is productive.

14. Let $S = R + \mathbb{Z}\cdot 1$ be a unital overring of R. Show that $t_{\mathbb{F}}(M) = 0$ iff $t_{\mathbb{F}}(M)$ is a unital S-module and $[S/I, M] = 0$ for each $I \in \mathbb{F}$.

15. If $R^2 = R$ show that $\mathscr{A}b$ is a torsion class and a torsion-free class in $\mathscr{M}_R$ ($\mathscr{A}b = \{M \in \mathscr{M}_R : MR = 0\}$).

16. Let t be the radical defined by $t(M) = 0$ for each module M. Show that $\mathbb{F}_t$ is a topology iff $R^2 = R$.

17. Let $\mathrm{Spec}(R)$ be the set of proper prime ideals of the commutative ring R. If $P \in \mathrm{Spec}(R)$ and $M \in \mathscr{M}_R$ let $t_P(M) = \{x \in M : r(x) \not\subseteq P\}$. t_P is the classical left exact radical corresponding to P and $\mathbb{F}_P = \{J \in \mathscr{L}(R) : J \not\subseteq P\}$ is the corresponding topology. If $X \subseteq \mathrm{Spec}(R)$ let $X^* = \bigcap_{P \in X} \mathbb{F}_P$, and if $\mathbb{F} \in \mathrm{Top}(R)$ let $\mathbb{F}^* = \mathrm{Spec}(R) \backslash \mathbb{F}$.

    (a) Show that the pair of mappings $^* : \mathscr{P}(\mathrm{Spec}(R)) \longrightarrow \mathrm{Top}(R)$ and $^* : \mathrm{Top}(R) \longrightarrow \mathscr{P}\,(\mathrm{spec}(R))$ forms a Galois connection, and

    $$\mathbb{F}^{**} = \{J \in \mathscr{L}(R) : J \not\subseteq P \text{ for each } P \in \mathbb{F}^*\}$$

    and

    $$X^{**} = \bigcup_{P \in X} L_{\mathrm{Spec}(R)}(P),$$

    where $L_{\mathrm{Spec}(R)}(P)$ is the set of lower bounds of P in $\mathrm{Spec}(R)$.

    (b) Show that the following are equivalent for $\mathbb{F}$.
        (i) $\mathbb{F} = X^*$ for some $X \subseteq \mathrm{Spec}(R)$.
        (ii) $\mathbb{F} = \mathbb{F}^{**}$.
        (iii) If $J \in \mathscr{L}(R) \backslash \mathbb{F}$, then there is some $P \in \mathbb{F}^*$ with $J \subseteq P$.

    (c) If P is an ideal maximal with respect to $P \notin \mathbb{F}$ show that P is a prime ideal.

    (d) Suppose $\mathbb{F}$ has a basis of finitely generated ideals. If $J \in \mathscr{L}(R) \backslash \mathbb{F}$ show that $J \subseteq P$ for some $P \in \mathbb{F}^*$. Consequently, $\mathbb{F} = \mathbb{F}^{**}$.

18. Suppose that Σ is a right Öre set in R and for each $s \in \Sigma$ the chain of right annihilators $\{r(s^n) : n \in \mathbb{N}\}$ is finite. Show that Σ is right reversible.

19. Let Σ be a right Öre set in R and let M be a right R-module. Use R_Σ and M_Σ to show that the relations defined on $R \times \Sigma$ and $M \times \Sigma$ in (4.2.12) are equivalence relations and the operations given in (4.2.13) and (4.2.14) make $R \times \Sigma / \sim$ into a ring isomorphic to R_Σ and make $M \times \Sigma / \sim$ into an R_Σ-module isomorphic to M_Σ.

20. Let ${}_DM$ be a vector space over the division ring D and let $H = \mathrm{End}_D(M) \subseteq M^M$. Give M the discrete topology and M^M the product topology - here called the *finite topology* of M^M. If X is a finite subset of M and $f \in M^M$ let $B(X; f) = \{g \in M^M : xg = xf \ \forall x \in X\}$.

    (a) Show that the sets $B(X; f)$ form a basis for the finite topology of M^M.
    (b) Show that H is closed in M^M and H is a topological ring.

(c) Let R be a subring of H. Show that R is a dense subring of H iff R is topologically dense in H.

21. Suppose M_R is a right R-module and let $D = \mathrm{End}_R(M_R)$ and $H = \mathrm{End}_D({}_DM)$.
    (a) Show that $\mathrm{End}_R(M^n) \cong D_n$. (If $f \in \mathrm{End}_R(M^n_R)$), then $f = \sum_{i,j} \kappa_j \pi_j f \kappa_i \pi_i$ where $\kappa_i : M \longrightarrow M^n$ is the ith injection and $\pi_i : M^n \longrightarrow M$ is the ith projection.)
    (b) Show that $\mathrm{End}_{D_n}({}_{D_n}M^n) \cong H$ and the action of H on M^n is coordinatewise: $(x_1,\ldots,x_n)a = (x_1a,\ldots,x_na)$. (If $a \in \mathrm{End}_{D_n}(M^n)$, then $M\kappa_j a \subseteq M\kappa_j$ and $\kappa_1 a \pi_1 = \kappa_j a \pi_j \in H$ for $1 \le j \le n$, where now κ_j and π_j act on the right of M and M^n, respectively.)
    (c) Show that M_R is faithful iff $M^n_{R_n}$ is faithful.
    (d) Show that M_R is simple iff $M^n_{R_n}$ is simple.

22. Let M_R be semisimple. Recall from Exercise 2.6.9 that each R-submodule of M is a summand of M. We contine the notation from Exercise 21.
    (a) Show that $\ell(R;M) = 0$ iff $x \in xR$ for each $x \in M$.
    (b) Show that each R-submodule of M is an H-submodule.
    (c) Suppose $\ell(R;M) = 0$. If $x_1,\ldots,x_n \in M$ and $a \in H$ show that there is some $b \in R$ with $x_ia = x_ib$ for $1 \le i \le n$. (The case $n = 1$ is a consequence of (a) and (b) and the general case is then a consequence of Exercise 21.)

23. Let M_R be quasi-injective and let $D = \mathrm{End}_R(M_R)$.
    (a) If $N \subseteq M$ is a left annihilator of a subset of R and $x \in M$ show that $N + Dx$ is also a left annihilator. (If $I = r(N;R)$ and $y \in \ell r(N + Dx)$ consider the map $xI \longrightarrow yI$.)
    (b) If $\ell(R;M) = 0$ show that each finitely generated left D-submodule of M is a left annihilator.
    (c) Suppose $\ell(R;M) = 0$ and $x_1,\ldots,x_n$ are elements of M such that $x_i \notin N_i$ for $1 \le i \le n$ where $N_i = \sum_{j \ne i} Dx_j$. If $y_i \in x_ir(N_i)$ show that there is some $a \in R$ with $x_ia = y_i$ for $1 \le i \le n$.

24. Show that the following are equivalent for the unital ring R.
    (a) R is a semisimple ring.
    (b) R_R is a semisimple module.
    (c) Each module in u-$\mathcal{M}_R$ is semisimple.
    (d) Each module in u-$\mathcal{M}_R$ is injective.
    (e) Each module in u-$\mathcal{M}_R$ is projective in u-$\mathcal{M}_R$.
    (f) Each short exact sequence in u-$\mathcal{M}_R$ splits. (The sequence (4.2.1) *splits* if there is an $h \in [M,N]$ with $hf = 1_N$; equivalently, there is a $k \in [P,M]$ with $(gk = 1_P$.)

25. Let a be an element in a unital ring R and suppose the chain of right ideals $\{a^nR\}$ is finite. Show that a is invertible iff $r(a) = 0$.

26. Let $R = R_1 \times \cdots \times R_n$. Show that R is a right Öre ring iff each R_j is a right Öre ring. If R is a right Öre ring show that $Q_c(R) = Q_c(R_1) \times \cdots \times Q_c(R_n)$.

27. Let $a \in R$ with $r(a) = r(a^2)$. If J is a right ideal with $J \cap (aR + r(a)) = 0$ show that the sum $J + aJ + a^2J + \cdots$ is direct. (This is a generalization of Exercise 4.1.26).

28. Let B be the right ideal of R defined in the proof of Theorem 4.2.17. Suppose that B is a principal right ideal of R; that is, $B = uR + \mathbb{Z}u$ for some $u \in R$. Let $T = R + \mathbb{Z} \cdot 1 \subseteq Q$. Show that T contains a set of matrix units $\{f_{ij}\}$ such that if L is the centralizer of $\{f_{ij}\}$ in R, then L is a right Öre domain, $L + \mathbb{Z} \cdot 1$ is the centralizer of $\{f_{ij}\}$ in T, and $\sum L f_{ij} \subseteq R \subseteq (\sum L f_{ij}) + \mathbb{Z} \cdot 1$. If $B = uR$ show that $R = T = \sum L f_{ij}$.

29. If R is reduced or is a domain, show that $R/t(R)$ is reduced or is a domain, respectively.

30. If $\mathbb{G}$ is the Goldie topology of R show that $I \in \mathbb{G}$ iff $Z(K)$ is essential in K for each complement K of I.

31. Let M_R be a simple module with $MR \neq 0$. If R is an algebra over the field K show that there is one and only one way to make M into a vector space over K such that $\alpha(xa) = x(\alpha a)$ for all $\alpha \in K$, $a \in R$ and $x \in M$.

32. Let R be a prime ring which has minimal right ideals (see Exercise 4.1.27). Show that any two faithful simple right (left) R-modules are isomorphic, and if R is an algebra over the field K, then any R-isomorphism between two faithful simple R-modules is also a K-isomorphism (see Exercise 31).

33. Let K be a field and let R be a subalgebra of K_n such that K^n is a simple right R-module. Show that K^n is also a simple right R^t-module where R^t is the subalgebra of K_n whose elements are the transposes of the elements in R. (Use Exercises 32 and 2.6.9 and the K-dimensions of the modules.)

34. An idempotent $e \in D_n$, D a division ring, is called *primitive* if eD_n is a minimal right ideal of D_n. Let $Y = \{e_1, \ldots, e_m\}$ be a set of primitive orthogonal idempotents in D_n. Show that:

    (a) $e = e^2$ is primitive iff $eD_ne \cong D$;
    (b) Each nonzero idempotent is a sum of primitive orthogonal idempotents;
    (c) $m \leq n$;
    (d) $1 = e_1 + \cdots + e_m$ iff $m = n$;
    (e) Y is contained in a set of n primitive orthogonal idempotents.

    (See Exercises 2.6.9 and 4.1.27.)

35. Let e and f be idempotents in the ring R.

    (a) Show that $\mathrm{Hom}_R(eR, fR) \cong fRe$ as groups, and also as rings if $e = f$.
    (b) Show that $eR \cong fR$ iff there exist elements $a \in eRf$ and $b \in fRe$ with $ab = e$ and $ba = f$.

(c) Suppose R is unital. Show that $R \cong K_n$ for some ring K and some $n \in \mathbb{N}$ iff R contains a set of orthogonal idempotents $\{e_1, \dots, e_n\}$ such that $1 = e_1 + \cdots + e_n$ and $e_1R \cong e_jR$ for $j = 1, \dots, n$.

36. Let K be a subfield of the field L and let $V = L^n$. Suppose A is a K-subalgebra of K_n and $B = LA$ is the L-subalgebra of L_n generated by A.
    (a) Show that ${}_AV_L$ is a simple bimodule iff ${}_BV$ is a simple B-module, iff ${}_AV$ is a simple A-module.
    (b) Suppose ${}_AV$ is simple, P is a generating set for the algebra A, and $0 \neq Q$ is a nonempty subset of V which is P-invariant. Show that Q spans ${}_LV$.

37. Let V be a vector space over the field K and suppose R is a dense K-subalgebra of $S = \mathrm{End}_K(V)$ which contains minimal left ideals (see Exercise 4.1.27).
    (a) If $\varphi : R \longrightarrow R$ is a K-algebra automorphism show that there exists a unit u in S such that $\varphi(r) = uru^{-1}$ for every $r \in R$. (If I is a minimal left ideal take $v \in V$ with $Iv = V$ and let $\rho_v : I \longrightarrow V$ be given by $\rho_v(a) = av$; for $r \in R$ let $\mu_r^I : I \longrightarrow I$ be given by $\mu_r^I(a) = ra$. Then $\mu_r^I = \rho_v^{-1} r \rho_v$. Similarly, if $J = \varphi(I)$, then $\mu_r^J = \rho_w^{-1} r \rho_w$ and $\mu_{\varphi(r)}^J = \varphi \mu_r^I \varphi^{-1}$. Let $u = \rho_w \varphi \rho_v^{-1}$ and use Exercise 31.)
    (b) Let K be a central subfield of the division ring D and let V be a right vector space over D. Show that each K-automorphism of S is inner and each K-automorphism of $\mathrm{End}_D(V)$ is the restriction of an inner automorphism of S.

Notes. Good references for torsion theories are Stenstrom's books [STE1] and [STE2] in which the theory is developed for the category of unital modules. However, not too much is lost when it is developed for the category of modules as is done here and in Dauns [DA6]. This is perhaps a consequence of the fact that a torsion theory can be thought of as a radical of a complete lattice in the sense of Amitsur [AM1]. Theorem 4.2.12 is the classical density theorem of Chevalley and Jacobson and Theorem 4.2.13 is the even more classical Wedderburn-Artin Theorem. Theorem 4.2.15, which was first noticed for f-rings by Anderson [AN1], is a special case of Theorem 4.2.16, and Theorem 4.2.16 is Goldie's Theorem [GO1] and [GO2]. The proof given follows Goldie [GO5]; another proof is in Goldie [GO3]. Theorem 4.2.17 is the Faith-Utumi Theorem [UF2] and the proof follows Faith [F5, p. 407]. Good references for Theorems 4.2.17, 4.2.18, 4.2.19, 4.2.20, and 4.2.21 are Herstein [HER3] and Kaplansky [K2]. The exercises mostly come from Stenstom [STE2], Kaplansky [K2], and Faith [F3] and [F5].

4.3 Lattice-ordered Rings and Modules of Quotients

Let $\mathbb{F}$ be a topology of the ℓ-ring R. We are interested in determining when $R_{\mathbb{F}}$ is an ℓ-ring extension of $\psi_R(R)$. This occurs, for example, when $R = (A_n, A_n^+)$ where A is a totally ordered right Öre domain and $\mathbb{F} = \mathbb{D}(R)$. Moreover, if M is an ℓ-module with $A^+ \subseteq d(M_R)$, that is, each element of A^+ is a d-element on M, then $M_{\mathbb{F}}$ is an $R_{\mathbb{F}}$-ℓ-module and an R-ℓ-module extension of $\psi_M(M)$. On the other hand, it does not seem possible to lattice order the field $\mathbb{Q}(x)$ so that it becomes an ℓ-ring extension of $(\mathbb{Q}[x], \mathbb{Q}^+[x])$. For the most part the extension question for ℓ-rings will be restricted to f-rings, and the analogous question for ℓ-modules will be restricted to f-modules. The topologies that will be useful for answering the extension question are those which have as a base the set of right ideals in the topology which are generated by their positive cones. This is always the case for the dense right ideals of an f-ring and is frequently the case for a topology on an sp-po-ring. The maximal right quotient ring of a right nonsingular totally ordered ring is an f-ring extension precisely when the ring is a right Öre domain. In the same spirit whether or not every nonsingular right f-module over a nonsingular f-ring R can be embedded in a product of totally ordered nonsingular f-modules reduces to the same question for the f-module R_R. Somewhat surprising is that the argument that shows each f-module over a commutative po-ring has the extension property also shows the same to be true for an f-ring in which every right ideal is convex. Another surprising result is that there is a unique totally ordered right self-injective ring which isn't unital and it is a summand of every other such f-ring. The completeness properties of a right self-injective f-ring are also investigated.

To answer the extension question for a classical right quotient ring R_Σ we can get by with the elements of Σ being d-elements instead of requiring R to be an f-ring. Recall that $d(R)$ denotes the set of d-elements of the ℓ-ring R, and $\mathscr{U}(G)$ denotes the group of units of the monoid G.

Theorem 4.3.1. *Let R_Σ be a classical right quotient ring of the ℓ-ring R. There is a (unique) lattice order of R_Σ such that R_Σ is an ℓ-ring, $\psi_R : R \longrightarrow R_\Sigma$ is an ℓ-homomorphism and $\psi_R(\Sigma) \subseteq \mathscr{U}(R_\Sigma^+)$ iff $t(R)$ is an ℓ-ideal of R and $\psi_R(\Sigma) \subseteq d(\psi_R(R))$. These latter conditions are satisfied if $\Sigma \subseteq d(R)$. Moreover, if these conditions hold and R is a d-ring or an almost f-ring, then R_Σ is an f-ring or an almost f-ring, respectively.*

Proof. Suppose $\psi = \psi_R$ is an ℓ-homorphism and $\psi(\Sigma) \subseteq \mathscr{U}(R_\Sigma^+)$. Then $t(R) = \ker \psi$ is an ℓ-ideal of R, and $\psi(\Sigma) \subseteq d(\psi(R))$ by Theorem 3.1.3. Assume $\Sigma \subseteq d(R)$ and suppose $|a| \leq |b|$ with $b \in t(R)$. Let $s \in \Sigma \cap r(b)$. Then $|as| \leq |bs| = 0$ gives $a \in t(R)$, and since $\psi(R) \cong R/t(R)$, $\psi(\Sigma) \subseteq d(\psi(R))$. Now assume $t(R)$ is an ℓ-ideal and $\psi(\Sigma) \subseteq d(\psi(R))$. By replacing R with $\psi(R)$ we may also assume that Σ consists of regular elements of R. Let $(R_\Sigma)^+ = \{as^{-1} : a \in R^+\}$. Note that if $as^{-1} = bu^{-1}$ with $a \in R^+$ and $s, u \in \Sigma$, then $b \in R^+$. For, $sc = uv$ for some $(c, v) \in R \times \Sigma$, and $ac = bu^{-1}sc = bu^{-1}uv = bv$. Since $0 = (uv)^- = (sc)^- = sc^-$, $c \geq 0$; consequently, $0 = (ac)^- = b^- v$ and $b \geq 0$. In particular, $(R_\Sigma)^+ \cap -(R_\Sigma)^+ = 0$, and to see that $(R_\Sigma)^+$ is a positive cone of R_Σ it suffices to verify that it is closed under addition and multiplication. Let $as^{-1}, bs^{-1} \in (R_\Sigma)^+$. Then $as^{-1}+bs^{-1} = (a+b)s^{-1} \in (R_\Sigma)^+$. If $s^{-1}b = cu^{-1}$, then, as previously shown, $c \in R^+$; so $as^{-1}bs^{-1} = ac(su)^{-1} \in (R_\Sigma)^+$. Now, R is an ℓ-subring of R_Σ. For, if $a \in R$ and $s \in \Sigma$, then $a \in R^+$ iff $as \in R^+$, iff $a \in (R_\Sigma)^+$. Moreover, if $as^{-1} \geq c, 0$ with $c \in R$, then $a \geq (cs)^+ = c^+s$ and $as^{-1} \geq c^+$. Since right multiplication by s^{-1} is an order automomorphism of R_Σ, for each $a \in R$, $a^+s^{-1} = (as^{-1})^+$ and therefore R_Σ is an ℓ-ring.

To see the uniqueness of the positive cone $(R_\Sigma)^+$, suppose P is a lattice order of R_Σ with the stated properties. Then $as^{-1} \in P$ iff $\psi(a) \in P\psi(s) = P$, iff $\psi(a) \in P \cap \psi(R) = \psi(R)^+$.

Since R_Σ is unital, to show it is an f-ring when R is a d-ring it suffices to verify that it is a d-ring. Let $b \in R^+$ and $as^{-1} \in R_\Sigma$. Then $sc = bu$ for some $(c,u) \in R^+ \times \Sigma$. Thus,

$$b(as^{-1})^+ = ba^+s^{-1} = (ba)^+s^{-1} = (bas^{-1})^+$$

and

$$(as^{-1})^+b = a^+cu^{-1} = (acu^{-1})^+ = (as^{-1}b)^+.$$

So $(R_\Sigma)^+ \subseteq d(R_\Sigma)$ and R_Σ is a d-ring. Now assume R is an almost f-ring. If $1 \wedge as^{-1} = 0$, then $s \wedge a = 0$ and $as = 0$; so $as^{-1} = 0$ and R_Σ is an almost f-ring by Theorem 3.8.10. □

The module analogue of Theorem 4.3.1 is given next.

Theorem 4.3.2. *Let M be an ℓ-module over the po-ring R. Suppose R_Σ is a classical right quotient ring of R and $\Sigma \subseteq R^+$. Then $t(M)$ and $t(R)$ are convex. There is a (unique) lattice order of M_Σ such that M_Σ is an R-ℓ-module, $\psi_M : M \longrightarrow M_\Sigma$ is an ℓ-homomorphism and $\psi_R(\Sigma) \subseteq d(M_\Sigma)$ if and only if $t(M)$ is an ℓ-submodule of M and whenever $(s,c) \in \Sigma \times R$ with $sc \in \Sigma$ (respectively, $sc \in R^+$), then multiplication by c is a d-map of $\psi_M(M)$ (respectively, $\psi(M)^+c \subseteq \psi(M)^+$). Moreover, when these conditions hold and M is an f-module, then M_Σ is an R-f-module if and only if multiplication by c is an f-map of $\psi(M)$ whenever $sc \in R^+$.*

Proof. Suppose (M_Σ, P) is an ℓ-module extension of $\psi(M)$ and $\psi(\Sigma) \subseteq d(M_\Sigma)$. If $y \in P$ and $s \in \Sigma$, then $(ys^{-1})^+s = (ys^{-1}s)^+ = y$; so $Ps^{-1} \subseteq P \subseteq Ps^{-1}$, $P = \{xs^{-1} : \psi(x) \in M^+,\ s \in \Sigma\}$, and $(xs^{-1})^+ = x^+s^{-1}$ for each $x \in M$. If $sc \in \Sigma$ and $x \in M$, then $\psi(x)^+c = (xs^{-1})^+sc = \psi(xc)^+$; similarly, if $sc \in R^+$ and $x \in M^+$ then $\psi(x)c \in \psi(M)^+$. For the converse, let $(M_\Sigma)^+ = \{xs^{-1} \in M_\Sigma : \psi(x) \in \psi(M)^+\}$. Then the proof of Theorem 4.3.1 shows that $(M_\Sigma, (M_\Sigma)^+)$ is an ℓ-module with the desired properties and, in fact, the elements of $\psi(\Sigma) \cup \psi(\Sigma)^{-1}$ induce d-maps on M_Σ. Assume M is an f-module. If M_Σ is an f-module, then $\psi(\Sigma) \cup \psi(\Sigma)^{-1}$ induce f-maps on M_Σ. For, if $p \wedge q = 0$ in M_Σ, then $(ps^{-1} \wedge q)s = p \wedge qs = 0$. In particular, if $sc = a \in R^+$ and $x \wedge y = 0$ in $\psi(M)$, then $xc \wedge y = xs^{-1}a \wedge y = 0$. On the other hand, suppose this latter condition holds and $xs^{-1} \wedge ys^{-1} = 0$. If $a \in R^+$, then $s^{-1}a = cu^{-1}$, $sc = au \in R^+$ and $xc \wedge y = 0$. Also, $ub = sv$ for some $v \in \Sigma$. So $(xs^{-1}a \wedge ys^{-1})uv = xcv \wedge yb = 0$ and $xs^{-1}a \wedge ys^{-1} = 0$. □

There is a natural partial order of R_Σ over which the R-ℓ-module (respectively, R-f-module) M_Σ in Theorem 4.3.2 is an R_Σ-ℓ-module (respectively, R_Σ-f-module).

We will present it for the more general quotient ring $R_{\mathbb{F}}$. A collection $\mathscr{D}$ of right ideals of the po-ring R is called *positive* if the right ideal of R generated by D^+ belongs to $\mathscr{D}$ for each $D \in \mathscr{D}$; that is, $D^+R_1 = D^+R + D^+ - D^+ \in \mathscr{D}$. If $\mathbb{F} = \mathbb{F}(\Sigma)$ and for each $s \in \Sigma$, $s^n \in R^+$ for some $n \in \mathbb{N}$, then $\mathbb{F}$ is positive since $s^{n+1} \in D^+R \cap \Sigma$ provided $s \in D \cap \Sigma$. The po-ring R is called *essentially positive* if $\mathscr{E}(R)$ is positive. The Goldie topology of an essentially positive po-ring is positive (Exercise 3). A topology $\mathbb{F}$ is positive if and only if $D^+R \in \mathbb{F}$ for each $D \in \mathbb{F}$ since $(D^+R : a) = R$ for each $a \in D^+R_1$.

Let $\mathbb{F}$ be a pretopology of the po-ring R. For any R^+-group M the subset

$$t^+(M) = \{x \in M : xD^+ = 0 \text{ for some } D \in \mathbb{F}\}$$

is easily seen to be an R^+-subgroup of M which contains $t(M)$ whenever M is an R-module. If M is an R^+-po-group, then $t^+(M)$ is convex; and if M is an R^+-d-group, then $t^+(M)$ is a convex ℓ-R^+-subgroup of M. R is called *$\mathbb{F}$-directed* if $R^+R_1 \in \mathbb{F}$.

Theorem 4.3.3. *The following statements are equivalent for the topology $\mathbb{F}$ of the po-ring R. If $\mathbb{F}$ is a pretopology, then the implications (a) $\Leftrightarrow$ (b) $\Rightarrow$ (c) hold.*

(a) $\mathbb{F}$ is positive.
(b) R is $\mathbb{F}$-directed and, for each R-module M, $t(M) = t^+(M)$.
(c) R is $\mathbb{F}$-directed and if N is a submodule of the po-module M with $M/N \in \mathscr{T}_t$, then $M^+R_1/N^+R_1 \in \mathscr{T}_t$.

Proof. (a) $\Rightarrow$ (b). If $x \in t^+(M)$, then $xD^+ = 0$ for some $D \in \mathbb{F}$; so $xD^+R_1 = 0$ and $x \in t(M)$.

(b) $\Rightarrow$ (c). First note that $\mathbb{F}$ is positive since if $D \in \mathbb{F}$, then $t(R_1/D^+R_1) = t^+(R_1/D^+R_1) = R_1/D^+R_1$. Let $x \in M^+$ and $a \in R$. Take $D_1, D_2 \in \mathbb{F}$ with $xD_1 \subseteq N$ and $aD_2 \subseteq D_1^+R_1$. Then $xD_1^+ \subseteq N^+$ and $xaD_2 \subseteq N^+R_1$; so M^+R_1/N^+R_1 is a torsion module.

(c) $\Rightarrow$ (a). If $D \in \mathbb{F}$, then $D^+R_1 \in \mathbb{F}$ since the ends of the short exact sequence

$$0 \longrightarrow R^+R_1/D^+R_1 \longrightarrow R_1/D^+R_1 \longrightarrow R_1/R^+R_1 \longrightarrow 0$$

are both torsion modules. □

If R is totally ordered or, more generally, $\mathbb{F}$ has a basis of directed right ideals, then $\mathbb{F}$ is positive. In an sp-po-ring a topology is frequently positive as the next two results show.

Theorem 4.3.4. *Let G be a subring of the ring R, and let A be the additive subgroup of R generated by the set $\{g^2 : g \in G\}$. If $a, b, c \in G$, then $2abc \in (A + Aa + Ab + Ac) \cap (A + aA + bA + cA)$. If B is the right ideal of R generated by A, then $2G^3 \subseteq B$.*

Proof. Since $ab + ba = (a+b)^2 - a^2 - b^2 \in A$, there exist elements $s, s_1, s_2, s_3 \in A$ such that

$$\begin{aligned}2abc &= abc - bca + s\\ &= abc - (-cb + s_1)a + s = abc + cba - s_1a + s\\ &= abc - (ac + ca)b + cba + (ac + ca)b - s_1a + s\\ &= abc + (bac + s_2) + (bca + s_3) + cba + (ac + ca)b - s_1a + s\\ &= (ab + ba)c + (bc + cb)a + (ac + ca)b + s_2 + s_3 - s_1a + s.\end{aligned}$$

So $2abc \in A + Aa + Ab + Ac$, and hence $2G^3 \subseteq B$. □

Theorem 4.3.5. *Let $\mathbb{F}$ be a topology of the sp-po-ring R. If $2R \in \mathbb{F}$, then $\mathbb{F}$ is positive.*

Proof. If $D \in \mathbb{F}$, then $2D^3 \subseteq \sum_{d \in D} d^2 R_1 \subseteq D^+ R_1$ by the previous result. By Exercise 1 and (c) of Theorem 4.2.3, $2D^3 \in \mathbb{F}$; so $D^+ R_1 \in \mathbb{F}$ and $\mathbb{F}$ is positive. □

The condition $2R \in \mathbb{F}$ is by no means necessary for $\mathbb{F}$ to be positive as is shown by the example $\mathbb{F} = \{R\}$ where R is unital and 2 is not a unit of R.

Let $\mathbb{F}$ be a pretopology of the po-ring R, and let $A \subseteq B$ be R^+-groups. In analogy to the $\mathbb{F}$-closure we define the $\mathbb{F}^+$-*closure* of A in B by $c\ell_{\mathbb{F}^+}(A;B)/A = t^+_{\mathbb{F}}(B/A)$. A is called $\mathbb{F}^+$-*dense* in B if $c\ell_{\mathbb{F}^+}(A;B) = B$, and A is $\mathbb{F}^+$-*closed* in B if $c\ell_{\mathbb{F}^+}(A;B) = A$. The next result investigates the order preserving properties of the closure operators; also see Exercise 5.

Theorem 4.3.6. *Let $\mathbb{F}$ be a pretopology of the po-ring R and let M be an $\mathbb{F}^+$-dense R^+-subgroup of the R^+-po-group K.*

(a) If N is a convex R^+-subgroup of M, then $c\ell_{\mathbb{F}^+}(N;K)$ is convex in K.
(b) If M is an R^+-d-group and N is an ℓ-R^+-subgroup of M, then $c\ell_{\mathbb{F}^+}(N;M)$ is an ℓ-R^+-subgroup of M.
(c) Suppose K is an R^+-d-group and M is an ℓ-R^+-subgroup of K. If N is a prime R^+-subgroup of M, then $c\ell_{\mathbb{F}^+}(N;K)$ is a prime R^+-subgroup of K.
(d) Suppose M is an ℓ-R^+-subgroup of the f-module K and $t^+(K) = 0$. Assume that for each $D \in \mathbb{F}$ there is some $J \in \mathbb{F}$ such that $J \subseteq D$ and J^+R_1 is a finitely generated right ideal. If N is a minimal prime subgroup of M, then $c\ell_{\mathbb{F}^+}(N;K)$ is a minimal prime subgroup of K and N is $\mathbb{F}^+$-closed in M.

Proof. (a) Suppose $0 \leq x \leq y$ with $y \in c\ell_{\mathbb{F}^+}(N;K)$ and $x \in K$. Take $D \in \mathbb{F}$ with $yD^+ \subseteq N$ and $xD^+ \subseteq M$. Since $0 \leq xd \leq yd$ for each $d \in D^+$ we have $xD^+ \subseteq N$, and hence $x \in c\ell_{\mathbb{F}^+}(N;K)$.

(b) If $x \in M$ and $xD^+ \subseteq N$ with $D \in \mathbb{F}$, then $x^+d = (xd)^+ \in N$ for each $d \in D^+$; so $x^+ \in c\ell_{\mathbb{F}^+}(N;M)$.

(c) By (a) and (b) $c\ell_{\mathbb{F}^+}(N;K)$ is a convex ℓ-R^+-subgroup of K. Suppose $x \wedge y = 0$ in K with $x \notin c\ell_{\mathbb{F}^+}(N;K)$. There exists $D \in \mathbb{F}$ with $xD^+ \subseteq M, xD^+ \nsubseteq N$, and $yD^+ \subseteq M$. For any d, $e \in D^+$, $xd \wedge ye = 0$; so $yD^+ \subseteq N$ and $y \in c\ell_{\mathbb{F}^+}(N;K)$.

(d) Since N is a minimal prime $N = \cup\{x^\perp : x \notin N\}$ is a directed union by Theorem 2.4.3. Let $y \in c\ell_{\mathbb{F}^+}(N;K)$. There is some $D \in \mathbb{F}$ with $D^+R_1 = d_1R_1 + \cdots + d_nR_1$, each $d_i \in D^+$, and $yD^+ \subseteq N$. So, for some $x \notin N$, $yd_i \in x^\perp$ for every i, $(|x| \wedge |y|)D^+ = 0$, and $x \in y^\perp$. Now, if $y \in M$, then $y \in N$ and hence N is $\mathbb{F}^+$-closed in M. Thus, in general, $y^\perp \nsubseteq c\ell_{\mathbb{F}^+}(N;K)$ and $c\ell_{\mathbb{F}^+}(N;K)$ is a minimal prime of K. □

If $\mathbb{F}$ is a positive topology which has a basis consisting of finitely generated right ideals and R is directed, then (d) of the preceding result shows that each torsion-free R-f-module is a subdirect product of totally ordered torsion-free modules. Another condition for this to hold is given in Exercise 6. However, not every torsion-free f-module is a subdirect product of totally ordered torsion-free f-modules, as we will see in Theorem 14 below.

Let $\mathbb{F}$ be a topology of the po-ring R and let M_R be a po-module. Then $\psi(M) \cong M/t(M)$ is a po-module over the po-ring $\psi(R) \cong R/t(R)$. Let

$$(M_{\mathbb{F}})^+ = \{x \in M_{\mathbb{F}} : \exists\, D \in \mathbb{F} \text{ with } xD^+ \subseteq \psi(M)^+\}.$$

Then $(M_{\mathbb{F}})^+$ is closed under addition, $\psi(M)^+ \subseteq (M_{\mathbb{F}})^+$, and $(M_{\mathbb{F}})^+ \cap -(M_{\mathbb{F}})^+ = t^+(M_{\mathbb{F}})$. Moreover, if $x \in (M_{\mathbb{F}})^+$ and $p \in (R_{\mathbb{F}})^+$, then $xpA \subseteq \psi(M)$ and $xpA^+ \subseteq (M_{\mathbb{F}})^+$ for some $A \in \mathbb{F}$. To see this take $D, A \in \mathbb{F}$ with $xD \subseteq \psi(M)$, $xD^+ \subseteq \psi(M)^+$, $pA \subseteq \psi(D)$ and $pA^+ \subseteq \psi(R)^+$; A exists since $\psi(D)$ is $\mathbb{F}$-dense in $R_{\mathbb{F}}$. In the special case where $p = \psi(a) \in \psi(R)^+$ we may assume $a \in R^+$ and $aA \subseteq D$, and then $xpA^+ = xaA^+ \subseteq xD^+ \subseteq \psi(M)^+$ and $xp \in (M_{\mathbb{F}})^+$. In general, $xpA^+ \subseteq x\psi(R)^+ \subseteq (M_{\mathbb{F}})^+$ and $xpA \subseteq \psi(M)$.

The subset X of the po-module M_R is said to be *$\mathbb{F}^+$-semiclosed* if $x \in M^+$ whenever $x \in X$ and $xD^+ \subseteq M^+$ for some $D \in \mathbb{F}$. If M is $\mathbb{F}^+$-semiclosed, then clearly $t^+(M) = 0$. We summarize the preceding in the next result.

Theorem 4.3.7. *Suppose M_R is a po-module and $\mathbb{F}$ is a topology of R.*

(a) *Let K be an R-submodule of $M_{\mathbb{F}}$ which contains $\psi(M)$. If $\psi(M)$ is $\mathbb{F}^+$-semiclosed and $t^+(M_{\mathbb{F}})$ is an R-submodule, then $K \cap (M_{\mathbb{F}})^+$ is the unique $\mathbb{F}^+$-semiclosed partial order of K with $K^+ \cap \psi(M) = \psi(M)^+$. Also, $(M_{\mathbb{F}})^+(R_{\mathbb{F}})^+ \subseteq (M_{\mathbb{F}})^+$.*

(b) *If M is a d-module, then M is $\mathbb{F}^+$-semiclosed iff $t^+(M) = 0$.*

(c) *If $t^+(M) = 0$ for the d-module M, then M is an ℓ-submodule of $M_{\mathbb{F}}$.*

Proof. To complete the proof that $(M_{\mathbb{F}})^+$ is a partial order of $M_{\mathbb{F}}$ it suffices to note that $t^+(M_{\mathbb{F}}) = 0$ since $\psi(M)$ is an essential submodule of $M_{\mathbb{F}}$ and $t^+(M_{\mathbb{F}}) \cap \psi(M) = t^+(\psi(M)) = 0$. Suppose $x \in M_{\mathbb{F}}$ and $D \in \mathbb{F}$ with $xD^+ \subseteq (M_{\mathbb{F}})^+$. Take $A \in \mathbb{F}$ with $xA \subseteq \psi(M)$. Then $x(A \cap D)^+ \subseteq \psi(M) \cap (M_{\mathbb{F}})^+ = \psi(M)^+$; so $x \in (M_{\mathbb{F}})^+$ and $(M_{\mathbb{F}})^+$ is an $\mathbb{F}^+$-semiclosed partial order of $M_{\mathbb{F}}$. If P is another such partial order of $M_{\mathbb{F}}$ extending $\psi(M)^+$ and $x \in M_{\mathbb{F}}$, then $x \in P$ iff $x(\psi(M) : x)^+ \subseteq P \cap \psi(M) = \psi(M)^+$; so $P = (M_{\mathbb{F}})^+$. This completes the proof of (a) for $M_{\mathbb{F}}$, and, similarly, also for K.

Suppose M is a d-module with $t^+(M) = 0$ and $x \in M$. If $xD^+ \subseteq M^+$, then $x^-D^+ = 0$; so $x^- = 0$ and M is $\mathbb{F}^+$-semiclosed. If $z \in (M_{\mathbb{F}})^+$ with $z \geq x$, then $zd \geq (xd)^+ = x^+d$ for $d \in (M : x)^+$; so $z \geq x^+$ and M is an ℓ-submodule of $M_{\mathbb{F}}$. □

If R is an f-ring, then $R_{\mathbb{F}}$ need not be an f-ring extension of $\psi(R)$ nor even an f-ring period. For example, if R is the first column of the 2×2 matrix algebra $\mathbb{Q}_2$, ordered lexicographically from the top, then $R_{\mathbb{D}} = \mathbb{Q}_2$. However, if R has the coordinatewise order, then $\mathbb{Q}_2$ is an ℓ-ring extension of R. Another pathological example is given by the polynomial ring $(R, R^+) = (\mathbb{Q}[x], \mathbb{Q}^+[x])$ with the usual topology $\mathbb{D}$. Since $((1-x)R)^+ = 0, t^+(M) = M$ for each module M and $(M_{\mathbb{F}})^+ = M_{\mathbb{F}}$ for each po-module M.

The f-module M is called a *$q_{\mathbb{F}}f$-module* if $M_{\mathbb{F}}$ is an R-f-module extension of $\psi(M)$, and the ℓ-ring R is called a *$q_{\mathbb{F}}f$-ring* if $R_{\mathbb{F}}$ is an f-ring extension of $\psi(R)$. When $\mathbb{F} = \mathbb{D}$ a $q_{\mathbb{F}}f$-ring will be called a *(right) qf-ring*; and an f-module M will be called an *if-module* if its injective hull is an f-module extension of M. The conditions for R to be a $q_{\mathbb{F}}f$-ring and for M to be a $q_{\mathbb{F}}f$-module are given next.

Theorem 4.3.8. *Let $\mathbb{F}$ be a positive topology of the po-ring R. Suppose M_R is an f-module and $\psi(M)$ can be embedded in a product of totally ordered R-modules. The following statements are equivalent.*

(a) M is a $q_{\mathbb{F}}f$-module (that is, $M_{\mathbb{F}}$ is an $(R_{\mathbb{F}})^+$-f-group).
(b) If $x \in M_{\mathbb{F}}$ and $d_1, d_2 \in (\psi(M):x)^+$, then $(xd_1)^+ \wedge (xd_2)^- = 0$.
(c) If $x \in M_{\mathbb{F}}$ there exists $D \in \mathbb{F}$ with $D \subseteq (\psi(M):x)$ such that $(xd_1)^+ \wedge (xd_2)^- = 0$ for any $d_1, d_2 \in D^+$.
(d) If $\psi(M)$ is a totally ordered homomorphic image of $\psi(M)$ and $x \in M_{\mathbb{F}}$, then $x(\psi(M):x)^+ \subseteq \psi(M)^+$ or $x(\psi(M):x)^+ \subseteq -\psi(M)^+$.

Moreover, if M is a $q_{\mathbb{F}}f$-module and R is a right f-ring, then $M_{\mathbb{F}}$ is a strong R-f-module iff for $x \in (M_{\mathbb{F}})^+$ and $d_1, d_2 \in (\psi(M):x)^+$, if $d_1 \wedge d_2 = 0$, then $xd_1 \wedge xd_2 = 0$. If $M_{\mathbb{F}}$ is a strong R-f-module and $R_{\mathbb{F}}$ is a right f-ring, then $M_{\mathbb{F}}$ is a strong $R_{\mathbb{F}}$-f-module.

Proof. The implications (a) $\Rightarrow$ (b) $\Leftrightarrow$ (c) and (b) $\Leftrightarrow$ (d) are all straightforward. Assuming (b) let $x \in M_{\mathbb{F}}$ and consider the correspondence $\alpha : (\psi(M):x)^+R \longrightarrow M_{\mathbb{F}}$ given by $\alpha(\sum d_i r_i) = \sum (xd_i)^+ r_i$. As a consequence of (d) we obtain that if $\sum d_i r_i = 0$, then $\sum (xd_i)^+ r_i = 0$. So α is an R-homomorphism and since $M_{\mathbb{F}}$ is $\mathbb{F}$-injective there exists $y \in M_{\mathbb{F}}$ such that $ydr = (xd)^+ r$ for $d \in (\psi(M):x)^+$ and $r \in R$. Thus, $yd = (xd)^+ \geq xd, 0$ and by Theorem 4.3.7, $y \geq x^+$. If $z \in M_{\mathbb{F}}$ with $z \geq x, 0$, then there exists $D \in \mathbb{F}$ with $zD \subseteq \psi(M)$ and $D \subseteq (\psi(M):x)$; so $zd \geq yd$ for each $d \in D^+$, $z \geq y$, $y = x^+$, and $x^+ d = (xd)^+$ for each $d \in (\psi(M):x)^+$. If $a \in R^+$ and $d \in (\psi(M):xa)^+$, then $(xa)^+ d = (xad)^+ = x^+ad$. So $M_{\mathbb{F}}$ is a d-module and hence it is an f-module over R.

To see that $M_{\mathbb{F}}$ is an $(R_{\mathbb{F}})^+$-f-group, suppose $x \wedge y = 0$ in $M_{\mathbb{F}}$ and $p \in (R_{\mathbb{F}})^+$. Take $D \in \mathbb{F}$ with $pD^+ \subseteq \psi(R)^+$. Then $(xp \wedge y)D^+ = 0$ and hence $xp \wedge y = 0$.

Now suppose M is a $q_{\mathbb{F}}f$-module over the right f-ring R and for each $x \in (M_{\mathbb{F}})^+$ the homomorphism $\mu_x : (\psi(M):x) \longrightarrow \psi(M)$ preserves disjoint elements. Let $x \in (M_{\mathbb{F}})^+$ and $a \in R$ and put $D = (\psi(M):xa^+) \cap (\psi(M):xa^-)$. If $d \in D^+$, then a^+d and a^-d are disjoint elements of $(\psi(M):x)^+$; so $(xa^+ \wedge xa^-)d = 0$ and $xa^+ \wedge xa^- = 0$. Thus, $(xa)^+ = (xa^+ - xa^-)^+ = xa^+$ and $M_{\mathbb{F}}$ is a strong R-f-module. Assume $M_{\mathbb{F}}$ is a strong R-f-module and $R_{\mathbb{F}}$ is a right f-ring. Take $x \in M_{\mathbb{F}}^+$, $p \in R_{\mathbb{F}}$ and $d \in (\psi(R):p)^+$. Then $pd = \psi(a)$ and

$$xp^+d = x(pd)^+ = xa^+ = (xa)^+ = (xp)^+d.$$

Consequently, $xp^+ = (xp)^+$ and $M_{\mathbb{F}}$ is a strong $R_{\mathbb{F}}$-module. □

In the special case that $\mathbb{F} = \mathbb{G}$ is the Goldie topology and $\mathbb{G}$ is positive the conditions in Theorem 4.3.8 determine when the nonsingular f-module M is an if-module since $E(M) = M_{\mathbb{G}}$.

Theorem 4.3.9. *Let $\mathbb{F}$ be a positive topology of the right f-ring R.*

(a) $R_{\mathbb{F}}$ is a right f-ring extension of $\psi(R)$ iff for $q \in R_{\mathbb{F}}$ and $d_1, d_2 \in (\psi(R) : q)^+$, $(qd_1)^+ \wedge (qd_2)^- = 0$ (iff R_R is a $q_{\mathbb{F}}f$-module).
(b) $R_{\mathbb{F}}$ is an f-ring if and only if the condition in (a) is satisfied and if $q \in R_{\mathbb{F}}$ and $d_1, d_2 \in (\psi(R) : q)$ with $d_1 \wedge d_2 = 0$, then $(qd_1)^+ \wedge \psi(d_2) = 0$.

Proof. Since (a) is an immediate consequence of Theorems 4.3.7 and 4.3.8 it suffices to check that the condition in (b) implies that the right f-ring $R_{\mathbb{F}}$ is a left f-ring. But if $s, p, q \in (R_{\mathbb{F}})^+$ with $p \wedge q = 0$ and

$$0 \leq d \in (\psi(R) : sp) \cap (\psi(R) : sq) \cap (\psi(R) : p) \cap (\psi(R) : q),$$

then $pd = \psi(a)$ and $qd = \psi(b)$ with $a \wedge b = 0$ and $a, b \in (\psi(R) : s)$. So

$$0 = sa \wedge \psi(b) = spd \wedge qd = (sp \wedge q)d,$$

and $sp \wedge q = 0$. □

We will now apply the preceding two theorems to give some examples of $q_{\mathbb{F}}f$-modules and $q_{\mathbb{F}}f$-rings.

Theorem 4.3.10. *The following statements are equivalent for the right nonsingular right f-ring R whose maximal right quotient ring is Q.*

(a) R is a qf-ring.
(b) R_R is an if-module and R is essentially positive.
(c) Q is reduced.
(d) If $aR \cap bR = 0$, then $ab = 0$.

Proof. (a) ⇒ (b). This follows from Theorem 4.1.8 and Exercise 2 since $Q_R = E(R_R)$.

(b) ⇒ (c). Q is an ℓ-ring extension of R by Theorem 4.3.7. If $p, q, s \in Q^+$ with $p \wedge q = 0$, then $(ps \wedge q)(R : s)^+ = 0$. Consequently, Q is a right f-ring and hence it is a (strongly) regular f-ring by Theorems 3.8.10 and 3.2.24.

(c) ⇒ (a). The equivalence of (c) with (d) is given in Theorem 4.1.10 where it is also shown that $I + I^*$ is an essential right ideal of R for each right ideal I of R. Thus, R is a reduced almost f-ring, and therefore it is an f-ring by Theorem 3.2.24. In particular, polars and annihilators in R are the same: $X^* = X^{\perp}$. We will show that the two conditions given in Theorem 4.3.9 are satisfied. Let $q \in Q$ and $d_1, d_2 \in (R : q)^+$. We claim that $(d_2R : d_1) + (d_2R : d_1)^* \subseteq ((qd_1)^+ \wedge (qd_2)^-)^*$. Suppose $x \in (d_2R : d_1)$. Then $d_1x = d_2y$ and $(qd_1)^+|x| = (qd_2)^+|y| \in ((qd_2)^-)^*$; so $(qd_1)^+|x| \wedge (qd_2)^-|x| = 0$ and $x \in ((qd_1)^+ \wedge (qd_2)^-)^*$. Now, suppose $0 \leq x \in (d_2R : d_1)^*$. If $0 < d_1x \wedge d_2x$, then $d_1xu = d_2xv \neq 0$ for some $u_1v \in R$. But then

$xu \in (d_2R : d_1) \cap (d_2R : d_1)^* = 0$ gives a contradiction; so $d_1x \wedge d_2x = 0$. Since $qd \in d^{**}$ for any $d \in R$ we have $(qd_1)^+x \in (d_1x)^{**}$ and $(qd_2)^-x \in (d_2x)^{**}$. Thus, $[(qd_1)^+ \wedge (qd_2)^-]x \in (d_1x)^{**} \cap (d_2x)^{**} = 0$. This shows that $(qd_1)^+ \wedge (qd_2)^- = 0$. To complete the proof we need to check that if $d_1 \wedge d_2 = 0$, then $(qd_1)^+ \wedge d_2 = 0$. But this follows from the inequalities $0 \leq ((qd_1)^+ \wedge d_2)^2 \leq (qd_1)^+d_2 = 0$. □

The essentially positive condition in (b) is, of course, automatic when R is unital or it is an f-ring. Here is the specialization to totally ordered rings.

Theorem 4.3.11. *The following are equivalent for the totally ordered right nonsingular ring R.*

(a) R is a qf-ring.
(b) R_R is an if-module.
(c) R is a right Öre domain.
(d) R is a domain and has a nonzero right if-module that is not singular.

Proof. The equivalence of (a), (b), and (c) comes from Theorem 4.3.10 and Exercise 4.1.26, and (d) is certainly a consequence of (c).

(d) ⇒ (b). Let M be a nonzero if-module and take $0 < x \in M \backslash Z(M)$. There is a nonzero right ideal J of R with $J \cap r(x) = 0$. If $0 < a \in J$, then the mapping $r \mapsto xar$ embeds the totally ordered module R_R into M_R. So R_R is an if-module by Exercise 11. □

If S is a two-sided quotient ring of the f-ring R, then it has two partial orders extending $R^+ : (R_{\mathbb{D}})^+ \cap S$ and $({}_{\mathbb{D}'}R^+ \cap S$ where $\mathbb{D}'$ is the left topology of dense left ideals of R. However, these partial orders coincide. For, if $s \in S$ with $s(R:s)^+ \subseteq R^+$ and $a \in R^+$ with $as \in R$, then $(as)^-(R:s)^+ = 0$; so $as \geq 0$ and $(R_{\mathbb{D}})^+ \cap S \subseteq ({}_{\mathbb{D}'}R)^+ \cap S$. Similarly, $({}_{\mathbb{D}'}R)^+ \cap S \subseteq (R_{\mathbb{D}})^+ \cap S$.

Theorem 4.3.12. *The maximal two-sided quotient ring $Q_2(R)$ of the f-ring R with $\ell(R) = r(R) = 0$ is an f-ring extension of R.*

Proof. The proof is similar to that of Theorem 4.3.8. Let $p \in Q_2(R)$ and let D and A be dense right and left ideals of R, respectively, with $pD \subseteq R$ and $Ap \subseteq R$. If $d_1, d_2 \in D^+$, then $(pd_1)^+ \wedge (pd_2)^- = 0$ since $A^+((pd_1)^+ \wedge (pd_2)^-) = 0$. As in the proof of Theorem 4.3.8 this yields the existence of $p^+ \in Q(R)$ with $p^+d = (pd)^+$ for $d \in D^+$. But then $ap^+d = (ap)^+d$ gives $ap^+ = (ap)^+ \in R$ for each $a \in A^+$; so $p^+ \in Q_2(R)$. The standard argument now shows that $Q_2(R)$ is a d-ring and hence an f-ring. □

One consequence of Theorem 4.3.12 is that a commutative f-ring (with zero annihilator) is a qf-ring. Using Exercise 4.2.11 it is not hard to extend this result to any ring of quotients of a commutative f-ring. The analogous result also holds for an f-module over a commutative po-ring. Since the proof of this can be applied to other rings we will present these results in a more general context.

Let R be a po-algebra over C. The element $y \in R^+$ is called

$$\textit{right po-normal} \text{ if } 0 \le x \le y \Rightarrow xy \in yR^+;$$
$$\textit{right convex} \text{ if } 0 \le x \le y \Rightarrow x \in yR^+;$$
$$\textit{right n-convex} \text{ if } 0 \le x \le y^n \Rightarrow x \in yR^+;$$
$$\textit{right p-convex} \text{ if } 0 \le x \le y \Rightarrow \exists 0 \ne p(\lambda) \in C^+[\lambda] \text{ with } xp(y) \in yR^+.$$

If each element of R^+ is right convex, then R will be called a *right convex po-ring*. Similarly, the other definitions will be transferred to R. If R is right p-convex and the degrees of the polynomials $p(\lambda)$ are bounded we will call R a *bounded right p-convex po-algebra*. Note the implications: y is right convex $\Rightarrow$ y is right po-normal $\Rightarrow$ y is right p-convex, and y is right n-convex $\Rightarrow$ y is right p-convex. Each positive element with a positive right inverse is right convex and, at the other extreme, each positive nilpotent element is right p-convex. It is not hard to see that for a topological space X the real f-algebra $C(X)$ is n-convex for every $n > 1$ (even if $n \in \mathbb{R}$), but, for example, $C([0,1])$ is not convex. (If $f(x) = 1 - 2x$, then $f \notin C([0,1])|f|$.) An interesting example of a right convex f-ring is supplied by a unital f-ring that is left self-injective (Exercise 17). The right p-convexity property allows us to treat commutative f-rings and right convex f-rings (and direct products of such f-rings) simultaneously.

The same definition of right convexity can be given for po-modules. However we will alter the definition slightly. The positive element y in the po-module M_R is *convex* if $x \in yR$ whenever $0 \le x \le y$, and M is called a *convex module* if each element of M^+ is convex. Note that when M is a strong ℓ-module over the ℓ-ring R and $y \in M^+$, then $(yR)^+ = yR^+$. In particular, an element in an f-ring R is right convex iff it is a convex element of the f-module R_R. Some properties of right convex f-modules and f-rings are given in Theorem 4.3.15 and Exercises 24 through 36.

Theorem 4.3.13. *Let R be a directed right p-convex po-algebra over the domain C and let $\mathbb{F}$ be a positive topology of R.*

(a) If R is bounded, then each C-torsion-free f-module M_R is a $q_{\mathbb{F}}f$-module.
(b) If R is a C-torsion-free f-algebra which is bounded or semiprime, then R is a $q_{\mathbb{F}}f$-ring.

Proof. (a) Let $x \in M_{\mathbb{F}}$ and take $d_1, d_2, a \in D^+$ where $xD \subseteq \psi(M)$, and set $d = d_1 + d_2 + a$. Then $d_i p_i(d) = dr_i$ for some $p_i(\lambda) \in C^+[\lambda]$ and $r_i \in R^+$. Thus,

$$0 \le ((xd_1)^+ \wedge (xd_2)^-)p_1(a)p_2(a) \le ((xd_1)^+ \wedge (xd_2)^-)p_1(d)p_2(d)$$
$$= (xd)^+ r_1 p_2(d) \wedge (xd)^- r_2 p_1(d) = 0.$$

Since $\psi(M)$ is C-torsion-free $((xd_1)^+ \wedge (xd_2)^-)a^n = 0$ for some integer n with $0 \le n \le \deg\,(p_1(\lambda)p_2(\lambda))$. If the degrees of the polynomials are bounded by m, then for all $a_1, \ldots, a_{2m} \in D^+$, $((xd_1)^+ \wedge (xd_2)^-)a_1 \cdots a_{2m} = 0$ since $(a_1 + \cdots + a_{2m})^{2m} = a_1 \cdots a_{2m} + b$ with $b \in D^+$. Thus, $((xd_1)^+ \wedge (xd_2)^-)(D^+R_1)^{2m} = 0$, $(xd_1)^+ \wedge (xd_2)^- = 0$, and hence M is a $q_{\mathbb{F}}f$-module by Theorem 4.3.8.

(b) We continue with the notation of (a) except now $x \in R_{\mathbb{F}}$. If R is bounded, then $R_{\mathbb{F}}$ is a right f-ring by (a) and Theorem 4.3.9. Suppose R is semiprime. Then $\psi(R)$

is reduced by Exercise 4.2.29, and from $((xd_1)^+ \wedge (xd_2)^-)a^n = 0$ we deduce that $((xd_1)^+ \wedge (xd_2)^-)D^+ = 0$; so, again, $R_{\mathbb{F}}$ is a right f-ring. Now suppose $d_1 \wedge d_2 = 0$ and $x \in R_{\mathbb{F}}^+$. Then

$$\begin{aligned} 0 &\leq (xd_1 \wedge \psi(d_2))p_1(a)p_2(a) \leq ((x \vee 1)d_1 \wedge (x \vee 1)d_2)p_1(d)p_2(d) \\ &\leq (x \vee 1)dr_1p_2(d) \vee (x \vee 1)dr_2p_1(d) \\ &\leq (x \vee 1)(d_1p_1(d)p_2(d) \wedge d_2p_2(d)p_1(d)) = 0. \end{aligned}$$

As we have seen previously this gives $xd_1 \wedge \psi(d_2) = 0$, and $R_{\mathbb{F}}$ is an f-ring by Theorem 4.3.9. □

As we have previously noted some torsion-free f-modules are subdirect products of totally ordered torsion-free f-modules. For example, as a consequence of Theorems 4.2.15 and Theorem 4.3.1 a semiprime right Goldie f-ring is a qf-ring, and, as a consequence of Theorem 4.2.11 and Theorem 4.3.2, each of its nonsingular f-modules is an if-module. Hence, it is a subdirect product of totally ordered nonsingular modules by Exercise 6 since the injective hull of such a module is a direct sum of vector lattices over totally ordered division rings. On the other hand, a converse of the preceding statement will show that it is just as likely for a nonsingular f-module over a commutative semiprime f-ring to not have such a representation.

Recall that a (semiprime) ring is irredundant if it is an irredundant subdirect product of prime rings, and a nonsingular module over an irredundant ring R is an irredundant subdirect product of nonsingular modules over the components of R (Exercise 4.1.38). If R is irredundant, then R is reduced iff each of its components is a domain (Theorem 4.1.14 and the ring analogue of Theorem 3.2.21); and $Q(R)$ is reduced iff each of the components of R is a right Öre domain (Exercise 4.1.26).

Theorem 4.3.14. *The following statements are equivalent for the nonsingular qf-ring R.*

(a) R is irredundant.
(b) Each nonsingular f-module is a subdirect product of totally ordered nonsingular f-modules (and is an if-module).
(c) R_R is a subdirect product of totally ordered nonsingular f-modules.

Proof. (a) ⇒ (b). If R_λ is a component of R, then $R_\lambda \cong R/P_\lambda$ where P_λ is a maximal polar; so R_λ is a totally ordered right Öre domain. Let M_R be a nonsingular f-module. If M_λ is a component of M, then $M_\lambda \cong M/N_\lambda$ where $N_\lambda = \ell(R_\lambda \cap R; M)$ (Theorem 4.1.17). Clearly, N_λ is a convex ℓ-submodule of M, and hence M_λ is a nonsingular f-module over R_λ and it is an if-module. As such, as we have previously noted, it is a subdirect product of totally ordered nonsingular R_λ-modules which are also nonsingular R-modules. Thus, M_R also has such a representation. M_R is an if-module since it is an ℓ-submodule of the if-module ΠM_λ.

(c) ⇒ (a). Let $\{A_\lambda\}$ be a collection of closed prime submodules of R_R whose intersection is zero. By Theorem 4.3.6, for each λ, $E_\lambda = E(A_\lambda) = c\ell(A_\lambda; Q(R))$ is a closed prime submodule of $Q(R)$. Since $Q(R)$ is strongly regular E_λ is an ℓ-ideal of $Q(R)$ and $Q(R) = E_\lambda \oplus F_\lambda$ is a direct sum of f-rings. Thus, F_λ is a totally ordered division ring, and $F_\lambda \subseteq F_\mu^* = E_\mu$ if $\mu \neq \lambda$; so $F_\lambda \cap \sum_{\mu\neq\lambda} F_\mu = 0$. Since the closure operator gives an isomorphism between the lattices $C_r(R)$ and $C_r(Q)$ (Exercise 4.2.10), $\bigcap_\lambda E_\lambda = 0$. Now, the projections of $Q(R)$ onto the F_λ induce an embedding of $Q(R)$ into the product $\Pi_\lambda F_\lambda$ whose image contains the direct sum $\oplus_\lambda F_\lambda$. So $Q(R) = \Pi_\lambda F_\lambda$, and since the Boolean algebras of polars of R and Q are isomorphic (Exercise 5), $\mathscr{B}(R)$ is atomic and R is irredundant. □

A right convex f-ring has some interesting localization properties in addition to those given in Theorem 4.3.13. Recall that a unital f-ring R has bounded inversion if each element that exceeds 1 is a unit of R. Clearly, each unital right convex f-ring has bounded inversion. Moreover, each ring of quotients of a right convex f-ring has bounded inversion (Exercise 25), and a unital f-ring with bounded inversion is a classical right and left quotient ring of its subring $C(1)$ of bounded elements. (Exercise 26). Note that the subring of bounded elements in any unitable f-ring R is well-defined since the unital cover R_u of R is embedded in any unital overring of R (Theorem 3.4.10). This subring will be denoted by $C^R(1)$ whenever more than one ring is nearby. Some of the elementary properties of a right convex f-ring are given next, and, in fact, these properties hold for modules. As a consequence of Theorem 4.3.15 and of Theorem 3.4.13, a right convex f-ring is an ℓ-ideal of its unital cover.

Theorem 4.3.15. *Let M_R be a convex f-module over the directed po-ring R.*

(a) M is a divisible group.
(b) $x \in xR$ for each $x \in M$.
(c) If $M = R$ is a totally ordered ring, then R is unital.
(d) If $M = R$ is an f-ring, then R is unitable, and it is unital provided R contains a right regular element.

Proof. If $n \in \mathbb{N}$, then M/nM is both a torsion-free and a torsion abelian group; so $M = nM$ and M is divisible. If $x \in M$, then xR and $C_R(x)$ are $\mathbb{Q}$-subspaces of M. So if $x \notin xR$, then $xR \cap \mathbb{Q}x = 0$ and we have the contradiction, $C_R(x) = xR + \mathbb{Q}x = xR + \mathbb{Z}x$. Now suppose that R is a right convex f-ring. If $a \in R$ with $r(a) = 0$ and $a = ae$, then e is a left identity element of R. If R is totally ordered, then $R/\ell\text{-}\beta(R)$ is a nonzero totally ordered domain since R is not nil by (b). So $R/\ell\text{-}\beta(R)$ has an identity element which can be lifted to an idempotent f of R (Theorem 3.4.15), and since R_R is indecomposable $R = fR$. But then f is the identity element of R since $\ell(R) = 0$. In general, then, a right convex f-ring is unitable and hence $e = 1$. □

We have seen that in any f-ring the maximal one-sided ℓ-ideals are indistinguishable from the maximal ℓ-ideals (Exercise 3.4.29) and the minimal ℓ-prime ℓ-ideals are indistinguishable from the minimal prime ideals (Exercise 3.2.20). An f-ring with bounded inversion has a characterization which borrows from both of these facts.

Theorem 4.3.16. *Let R be a unital f-ring, and denote the set of maximal ideals of R by $Max(R)$, the set of maximal right ideals of R by $Max(R_R)$, and the set of maximal convex right ideals of R by c-$Max(R_R)$. The following statements are equivalent.*

(a) R has bounded inversion.
(b) $Max(R) = Max(R_R) = c\text{-}Max(R_R)$.
(c) $Max(R) \subseteq Max(R_R) \subseteq c\text{-}Max(R_R)$.
(d) Each maximal right ideal of R is an ℓ-ideal.
(e) Each simple unital R-module is an f-module.
(f) Each totally ordered homomorphic image of R is local and its Jacobson radical is a convex ideal.
(g) The Jacobson radical J of R is convex and R/J has bounded inversion.

Proof. We first note that any subgroup G of the additive group of R that contains a minimal prime ideal P of R is a sublattice. For, P is an ℓ-ideal of R and hence contains x^+ or x^- for each x in R. So if $x \in G$, then $x^+ \in G$. Any maximal ideal or maximal right ideal is such a subgroup.

(a) $\Rightarrow$ (b). Suppose that R has bounded inversion and let M be a maximal ideal of R. If $0 \leq a \leq b$ with $b \in M$ and $a \notin M$, then $R = M + RaR$ yields that $1 = x + \sum r_i a s_i$ for some $x \in M$; so $1 \leq x + rar \leq x + rbr$ for some $r \in R^+$, and hence the unit $x + rbr$ is in M. This contradiction shows that M is convex, and hence it is an ℓ-ideal. A similar argument gives that each maximal right ideal is a right ℓ-ideal. Since M is a maximal right ℓ-ideal of R, hence a maximal right ideal, R/M is a totally ordered division ring and we have $\text{Max}(R) \subseteq \text{Max}(R_R) = c\text{-Max}(R_R)$. Since a maximal right ℓ-ideal contains a minimal prime ideal which is contained in a unique maximal ideal, $c\text{-Max}(R_R) \subseteq \text{Max}(R)$.

For the remainder of the proof let P be a minimal prime ideal contained in a given maximal right ideal I.

(c) $\Rightarrow$ (d). If I is a maximal right ideal and M is a maximal ideal containing P, then M and I are both maximal right ℓ-ideals containing P; so $M = I$ since M and I are comparable.

(d) $\Rightarrow$ (f). Suppose that A is an ℓ-ideal of R for which R/A is totally ordered, and let I/A and K/A be maximal right ideals of R/A. Then, again, I and K are comparable; so $I = K$, R/A is local and its Jacobson radical I/A is convex.

(f) $\Rightarrow$ (g). Let I be a maximal right ideal of R. Since R/P is local I is an ℓ-ideal of R, and hence J, being the intersection of all such I, is convex. If $a + J \geq 1 + J$ in R/J, then $a + J$ is not in any maximal right ideal of R/J, so $a + J$ is a unit.

(e) $\Rightarrow$ (d). Since $S = R/I$ is a totally ordered right R-module I is the annihilator of a positive element of S. Thus, I is a right ℓ-ideal of R. This yields that R/P is local and hence I is an ℓ-ideal.

Now, (g) $\Rightarrow$ (a) since the units of R/J are images of units of R. Since, clearly, (b) $\Rightarrow$ (c) and (d) $\Rightarrow$ (e), the proof is complete. □

According to Exercise 24 an f-ring is right convex precisely when each of its right ideals is a right ℓ-ideal. Recall that a right valuation ring is a ring whose lattice of right ideals is a chain. Clearly, each totally ordered right convex f-ring is a right valuation ring. The localization of the integers at a prime gives an example of a totally ordered commutative valuation ring that does not have bounded inversion and neither does any (valuation) subring of the reals that is not a field. An example of a commutative totally ordered domain with bounded inversion that is not convex is obtained by localizing the anti-lexicographically ordered polynomial ring $F[x]$ over the totally ordered field F at the convex ideal (x). An example of a totally ordered right convex f-ring that is not left convex is given in Exercise 31. The first column of the 2×2 matrix ring over $\mathbb{Q}$, ordered lexicographically with the left corner dominating, is an example of a totally ordered ring which is neither right nor left convex but in which each ideal is an ℓ-ideal.

Some additional localization properties which are forced by the convexity conditions and which further illustrate their commutative character are given in the next result.

Theorem 4.3.17. *Suppose the unital f-ring R is right convex, or it has bounded inversion and is a right p-convex C-torsion-free-algebra over the directed po-ring C. Then R is a right Öre ring and for each ℓ-prime ℓ-ideal P of R the multiplicatively closed subset $\Sigma = R \backslash P$ is a right Öre set in R. If P is also a minimal prime subgroup, then the classical right quotient ring R_P of R with respect to Σ exists.*

Proof. Let $C = \mathbb{Z}$ if R is right convex, and let

$$S = \{a \in R : |a| \leq \alpha \cdot 1 \text{ for some } \alpha \text{ in } C\}$$

be the subalgebra of C-bounded elements of R. Since R has bounded inversion R is a classical left and right quotient ring of S (Exercise 26). So if $\Gamma = \{a \in S^+ : a^{-1} \in R\}$, then $R = S_\Gamma = {}_\Gamma S$. Also, S inherits right convexity, or right p-convexity and the bounded inversion property from R. For if $0 \leq x \leq y$ in S, then $xp(y) = yr$ and $0 \leq p(y) \leq \alpha \cdot 1$ for some $\alpha \in C$. So $xp(y) \leq \alpha y$ and hence

$$y(r \wedge \alpha) = yr \wedge \alpha y = xp(y) \wedge \alpha y = xp(y).$$

Now, $Q = S \cap P$ is a prime ℓ-ideal of S and $P = Q_\Gamma = {}_\Gamma Q$. Let $a \in R$, $s \in \Sigma$. Then $a = u^{-1}b$ and $s = u^{-1}v$ with $u \in \Gamma$, $b \in S$, and $v \in S \backslash Q$. Since $|b| \leq \alpha \cdot 1$ for some $\alpha \in C$, $b^+v^2p(\alpha v^2) = \alpha v^2 x$ and $b^- v^2 q(\alpha v^2) = \alpha v^2 y$ for some $x, y \in S^+$ and some polynomials p and q. But then $bh(v) = vz$ with $h(v) = v^2 p(\alpha v^2) q(\alpha v^2) \in S \backslash Q$ and $z \in S$; so $ah(v) = sz$ and Σ is a right Öre set in R. A similar computation shows that R is a right Öre ring. To show that R_P exists when P is a minimal prime subgroup we need to verify that Σ is right reversible. But if $sa = 0$, then $a \in P$ and hence $aw = 0$ for some $w \in \Sigma$ by Theorem 2.4.3. □

The f-ring R is said to *have localizations* if each maximal ideal P of R is completely prime and the classical right quotient ring R_P exists. Included in this class are those f-rings in Theorem 4.3.17 which are semiprime and have bounded inversion as well as all commutative unital f-rings.

Theorem 4.3.18. *Let R be a unital f-ring that has bounded inversion and localizations. The following statements are equivalent for the strong R-f-module M.*

(a) M is convex.
(b) For each maximal ideal P of R, M_P is a convex R_P-f-module.

(c) For each maximal ideal P of R, the lattice of R_P-submodules of M_P is a chain.
(d) Each finitely generated submodule of M is cyclic.

Proof. In view of Exercises 24 and 40 it suffices to show that (d) implies (c) and (c) implies (a).

(c) $\Rightarrow$ (a). Suppose that $0 \le x \le y$ in M. If P is a maximal ideal of R and $y \in xR_P$, then $y = xp$ with $p \in R_P^+$. Since $x(p \vee 1) = xp \vee x = y \vee x = y$ and $p \vee 1$ is a unit of R_P, $x \in yR_P$. Thus $xR_P \subseteq yR_P$ for each maximal ideal P of R and hence $xR \subseteq yR$ by Exercise 39.

(d) $\Rightarrow$ (c). Suppose that $x, y \in M$. Since $xR + yR = zR$ we have $x = zu$, $y = zv$ and $z = xr + ys$ for u, v, r, $s \in R$; so $z(1 - ur - vs) = 0$. Let P be a maximal ideal of R. If $ur + vs \in P$, then $z = 0$ in M_P and $xR_P = 0 = yR_P$. If $ur + vs \notin P$, then $ur \notin P$ or $vs \notin P$; so u or v is a unit in R_P. If u is a unit, then $yR_P \subseteq zR_P = xR_P$ and if v is a unit, then $xR_P \subseteq yR_P$. □

If only some finitely generated submodules are cyclic, then some elements of M are convex (and conversely). To be more precise call a subset X of the f-module M over the f-ring R a *Prüfer set* if it is a dual ideal of the poset M^+ and if each $x \in X$ is *regular* in the sense that $r(x;R) = 0$. For some examples note that the set of regular elements of R^+ and the set of right regular elements of R^+ are both Prüfer sets, and so is the set of units in R^+. Also, if R is semiprime or M is a strong f-module, then the dual ideal of M^+ generated by a set of regular elements of M^+ is a Prüfer set. A submodule of M is called an *X-submodule*, where X is a subset of M, if its intersection with X is not empty.

Theorem 4.3.19. *Let R be a unital f-ring which has bounded inversion and localizations. Suppose that X is a Prüfer set in the unital strong right R-f-module M. Then the following statements are equivalent.*

(a) Each finitely generated X-submodule of M is cyclic.
(b) Each element of X is convex.
(c) Each X-submodule of M is a convex submodule.
(d) Each X-submodule of M is a convex ℓ-submodule.
(e) If $x, y \in M^+$ and $x + y \in X$, then $xR + yR = (x+y)R$.

Proof. (a) $\Rightarrow$ (b). Suppose that $0 \le x \le y$ in M with $y \in X$. Then the arguments in the proofs of (d) $\Rightarrow$ (c) and (c) $\Rightarrow$ (a) of Theorem 4.3.18 give that $x \in yR$.

(b) $\Rightarrow$ (c). Suppose that B is an X-submodule of M, $0 \le a \le b \in B$ and $x \in B \cap X$. Then $b + x \in X$ and $0 \le a + x \le b + x$; so $a + x \in (b+x)R$ yields that $a \in B$.

(c) $\Rightarrow$ (d). It suffices to show that if $a, x \in M$ with $x \in X$, then $a^+ \in aR + xR = B$. Since $a^+ + x$ and $a^- + x$ are both in X, $a^+ = (a^+ + x)r$ and $a^- = (a^- + x)s$ for some $r, s \in R^+$. So $a^+(1-r) = xr$ and $a^-(1-s) = xs$. Now $r \wedge s = 0$ since A is a strong f-module and x is a regular element of A; so if P is a maximal ideal of R, then r or s

is in P. If $s \in P$, then $a^- \in xR_P \subseteq B_P$, so $a^+ \in B_P$; and if $r \in P$, then, also $a^+ \in B_P$. Thus $a^+ \in B$.

(d) $\Rightarrow$ (a). Let $B = a_1R + \cdots + a_nR$ with $a_n \in X$, and set $a = |a_1| + \cdots + |a_n| \geq a_n$. Then $a \in X$, and B and aR are X-submodules of M. Hence both are convex ℓ-submodules, and $B = aR$.

Now, (e) is certainly a consequence of (d) and it implies (b). For, if $0 \leq x \leq y \in X$, then $xR + (y-x)R = yR$ shows that $x \in yR$. □

For the ideal I of the commutative unital ring R let $I^{-1} = \{p \in Q_c(R) : Ip \subseteq R\}$. I is *invertible* if $II^{-1} = R$. It is easily seen that I is invertible iff $IK = R$ for some R-submodule K of $Q_c(R)$; and then $K = I^{-1}$. If X is a subset of R, then R will be called *X-Prüfer* if each of its finitely generated X-ideals is invertible; R is a *Prüfer ring* if it is X-Prüfer when X is the set of all regular elements of R.

Theorem 4.3.20. *Let R be a commutative f-ring with bounded inversion and let X be a Prüfer subset of R. Then R is X-Prüfer iff each element of X is convex. If R_P is X_P-Prüfer for each maximal ideal P of R, then each element of X is convex.*

Proof. If each element of X is convex and J is a finitely generated X-ideal of R, then $J = aR$ by Theorem 4.3.19. Since J contains a regular element $a^{-1} \in Q_c(R)$ and $JRa^{-1} = R$. Conversely, suppose that either R is X-Prüfer or R_P is X_P-Prüfer for each maximal ideal P. If $0 \leq x \leq y$ with $x \in R$ and $y \in X$ let $J = xR + yR$. In either case J_P is invertible in R_P. For, if $JK = R$ with K an R-submodule of $Q_c(R)$, then $J_PK_P = R_P$ and $K_P \subseteq Q_c(R_P)$ since the homomorphism $R \longrightarrow R_P$ extends to a homomorphism $Q_c(R) \longrightarrow Q_c(R_P)$. Since R_P is local and $1 = xu + yv$ with u, $v \in K_P$ either xu or yv is a unit of R_P. If xu is a unit, then $w = yu(xu)^{-1} \in R_P$ and $y = xw = x(w \vee 1)$. Thus, $x \in yR_P$ since R_P has bounded inversion. On the other hand, if yv is a unit, then again, $x = yxv(yv)^{-1} \in yR_P$. So $x \in yR$ by Exercise 39. □

It is possible to characterize rings which are right convex left Öre f-rings or are commutative Prüfer f-rings with bounded inversion without assuming that the rings are even ℓ-rings! A ring is called a *quotient ring* if each of its regular elements is a unit.

Theorem 4.3.21. *Let R be a ring which has regular elements.*

(a) R is a right convex left Öre f-ring if and only if there is a right convex quotient f-ring Q with $C^Q(1) \subseteq R \subseteq Q$.
(b) Suppose R is commutative and unital. R is a Prüfer f-ring with bounded inversion if and only if there is a quotient f-ring Q with $C^Q(1) \subseteq R \subseteq Q$.

Proof. (a) One implication is given in Exercise 36 with $Q = Q_c(R)$. Conversely, suppose Q is a right convex quotient f-ring and $S = C^Q(1) \subseteq R \subseteq Q$. Then R is a right convex f-ring by Exercise 30 and R is left Öre by Exercise 26.

(b) Let $p \in Q$ with $|p| \leq n \in \mathbb{N}$ where R is a Prüfer f-ring with bounded inversion and $Q = Q_c(R)$. Then $p = as^{-1}$ with $a \in R$ and $s \in R^+$; so a^+, $a^- \leq |a| \leq ns$, $a^+, a^- \in sR$ by Theorem 4.3.20, and $p = a^+s^{-1} - a^-s^{-1} \in R$. Conversely, suppose

$C^{\mathcal{Q}}(1) \subseteq R \subseteq Q$ where Q is a quotient f-ring. Then Q has bounded inversion; so R has bounded inversion and $Q = Q_c(R)$ since Q is a classical quotient ring of $S = C^{\mathcal{Q}}(1)$ by Exercise 26. We will now check that R is an ℓ-subring of Q. If $a \in R$, then $a^+ + 1$ and $a^- + 1$ are units of Q and hence $a^+ = (a^+ + 1)r$ and $a^- = (a^- + 1)s$ with $0 \leq r, s \leq 1$. So $a^+(1-r) = r$ and $a^-(1-s) = s$. By repeating the argument in the proof of (c) $\Rightarrow$ (d) of Theorem 4.3.19 we get that $a^+ \in R_P$ for each maximal ideal P of S, and hence $a^+ \in R$ by Exercise 39. But then R is a Prüfer f-ring by Theorem 4.3.20 since each regular element of R is convex in R. □

We now return to self-injective f-rings and give the following decomposition theorem.

Theorem 4.3.22. *The f-ring R is right self-injective if and only if $R \cong S \times T$ or $R \cong S \times T \times U$ where S is a regular right self-injective f-ring, T is a unital right self-injective f-ring whose lower radical is an essential right ideal and an essential left ideal, and U is the ring $\begin{pmatrix} \mathbb{Q} & \mathbb{Q} \\ 0 & 0 \end{pmatrix}$ supplied with the lexicographic total order: $\begin{pmatrix} p & q \\ 0 & 0 \end{pmatrix} \geq 0$ iff $p > 0$, or $p = 0$ and $q \geq 0$. Moreover, $\mathscr{B}(R)$ is the lattice of right complements (summands) of R, and it is the lattice of left complements of R precisely when R is unital.*

Proof. If R has one of these decompositions, then it is right self-injective by Exercises 4.1.47 (f) and 4.1.49. Conversely, suppose R is right self-injective and assume first that it is unital. Then R is left convex (Exercise 18) and $\mathscr{B}(R)$ is the lattice of complements of the module ${}_RR$ and of the module R_R (Exercise 15). Recall that each summand of R is a right self-injective ring (Exercise 4.1.49). Let $T = Z_2(R)$. Then T_R is essentially closed in R_R (Theorem 4.1.6) and hence it is a summand of $R : R = S \oplus T$ with $S = T^{\perp}$. Since $Z_r(R) = Z_r(S) \oplus Z_r(T) \subseteq T$ we have $Z_r(S) = 0$. Thus, S is a regular ring (Theorem 4.1.8). Now, T has no nonzero regular ideals; that is, S is the regular radical of R. For, suppose A is a regular ideal of T. Since A is an f-ring it is strongly regular and each of its ideals is an ideal of $T : IT = IAT \subseteq IA \subseteq I$ for each ideal I of A. If $x \in A$, then $xD \subseteq Z_r(R) = Z_r(T)$ for some essential right ideal D of T. For each $d \in D$ there is an essential right ideal D_d of T with $xd(D_d \cap A) \subseteq xdD_d = 0$. Since $D_d \cap A$ is an essential right ideal of A, $xd = 0$; thus, $x(D \cap A) \subseteq xD = 0$, and $x \in Z_r(A) = 0$. Suppose that $\beta(T) \cap B = 0$ for some right ideal B of T. Then $T = E(B_T) \oplus F$ where $F \supseteq \beta(T)$ and hence $E(B) = 0$ since it is a regular ideal of T. If $0 \neq C$ is a left ideal of T, then ${}_TC$ is essential in ${}_TC^{\perp_T \perp_T}$ (Exercise 15) and hence $C \cap \beta(T) \neq 0$.

Now assume that R is not unital and let e be a left identity of R. Then $A = \ell(R) = \ell(e) = R(1-e)$ is a one-dimensional $\mathbb{Q}$-subalgebra of R. For, the ring homomorphism $Re \longrightarrow \mathrm{End}_{\mathbb{Z}}(A) = \mathrm{End}_{\mathbb{Q}}(A) = \mathrm{End}_R(A_R)$ given by $a \mapsto \mu_a$, where μ_a denotes left multiplication by a, is surjective, and hence $\mathrm{End}_{\mathbb{Q}}(A) \cong Re/B$ is an f-ring since $B = \ell(A; Re)$ is an ℓ-ideal of Re. Since ${}_RR = Re \boxplus A$ and $AB = BA = 0$, B is an ideal of R. The projection of $B \boxplus A$ onto B is given by left multiplication by some element $f + u \in Re + A$. Clearly, f is a left identity of B, and since Re is unital f is the identity element of B. Let $g = e - f$. Since f is central in Re and $fA = Af = 0$, f is central in R. Suppose $y \in R$ and $x \in Rf = B$ with $0 \leq y \leq x$. Then $0 \leq gy \leq gx = 0$. So, $y = ey = fy \in B$ and B is an ℓ-ideal of R. Let $C = Rg = \ell(f; Re) = \ell(B; Re)$. Then $C + A = \ell(f; R)$. For, $(C+A)f = 0$ and if $x = xe + a \in R$ with $a \in A$ and $xf = 0$, then $xef = 0$ and $xe \in C$. Hence B and $V = C + A$ are ℓ-ideals of R and $R = B \oplus (C+A)$. Since $C \cong Re/B$ is one-dimensional over $\mathbb{Q}$, $C = \mathbb{Q}g$ and $A = \mathbb{Q}a$ for $0 < a \in A$. Since $ga = (e-f)a = ea = a$, $V \cong U$. That U has the specified order is left for Exercise 43. Since B is a unital right self-injective f-ring the previous case applies to B. Suppose $R = I \boxplus J$ as right R-modules. Since R has a left identity element $I = aR$ and $J = bR$ for a pair of orthogonal idempotents a and b. So I and J are right ℓ-ideals and hence I is a polar of R; that is, each complement right ideal of R is a polar. Conversely, if I is a polar of R, then $I = J \oplus K$ where J is a polar of B and $K = 0$ or V. Since J is a summand of B_B it is clear that I is a summand of R_R. On the other hand, $B \boxplus \mathbb{Q}\begin{pmatrix} 1 & 1 \\ 0 & 0 \end{pmatrix}$ is a summand of ${}_RR$ that is not a polar. □

Recall that a unital ring is quasi-Frobenius if it is right artinian and right self-injective. In order to characterize a quasi-Frobenius f-ring in terms of convexity we will first exhibit the right-left symmetry of quasi-Frobenius rings, and to do this requires some preparation. In the next three results $J = J(R)$ denotes the Jacobson radical of the ring R.

Theorem 4.3.23. *(Nakayama's Lemma). The following statements are equivalent for the right ideal A of R.*

(a) $A \subseteq J$.
(b) If M_R is a finitely generated module with $M = MA$, then $M = 0$.
(c) If N_R is a submodule of M_R such that $M = N + MA$ and M/N is finitely generated, then $M = N$.

Proof. By replacing R with $R_1 = R + \mathbb{Z}$ we may assume that R is unital and all modules are unital since $J(R) = J(R_1)$.

(a) $\Rightarrow$ (b). Let $\{x_1, \ldots, x_n\}$ be a minimal generating set for M. Since $M = x_1R + \cdots + x_nR = x_1RA + \cdots + x_nRA$, $x_1 = x_1a_1 + \cdots + x_na_n$ with $a_j \in RA \subseteq J$. So $(1 - a_1)^{-1} \in R$, $x_1 \in x_2R + \cdots + x_nR$ and hence $M = 0$.

(b) $\Rightarrow$ (c). This is obvious since $M/N \cdot A = MA + N/N = M/N$.

(c) $\Rightarrow$ (a). If $A \nsubseteq J$, then $A \nsubseteq B$ for some maximal right ideal B of R. So $R = B + A = B + RA$ and we have the contradiction $R = B$. □

Theorem 4.3.24. *Let M and K be finitely generated unital modules over the unital ring R and assume K is projective. Let $A \subseteq J$ be a right ideal of R.*

(a) If $\alpha : M \longrightarrow K$ is an R-homomorphism and $\bar{\alpha} : M/MA \longrightarrow K/KA$ is an isomorphism, then α is an isomorphism.
(b) Assume M is projective. Then $M \cong K$ iff $M/MA \cong K/KA$.

Proof. (a) Since $\bar{\alpha}$ is onto $K/KA = \alpha(M) + KA/KA$, $K = \alpha(M) + KA$ and $K = \alpha M$ by Theorem 4.3.23. Since K is projective and α is onto, α splits; that is, there is an R-homomorphism β such that the diagram

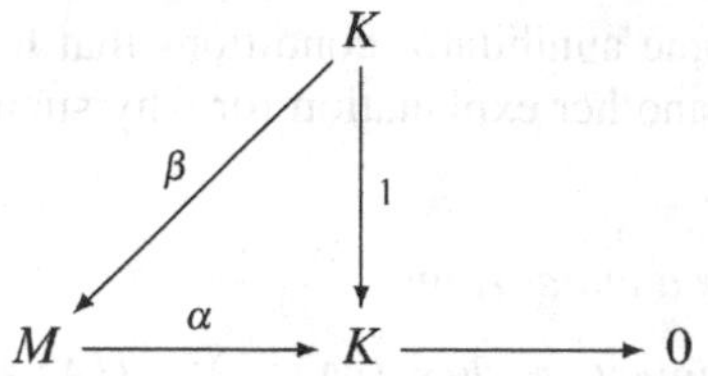

commutes. So $M = \ker\alpha \boxplus N$ with $N \cong K$. Since $\ker\alpha + MA/MA \subseteq \ker\bar{\alpha} = 0$, necessarily $\ker\alpha \subseteq MA = (\ker\alpha)A \boxplus NA$; so $\ker\alpha = (\ker\alpha)A$ and $\ker\alpha = 0$ by Theorem 4.3.23.

(b) Let $\gamma : M/MA \longrightarrow K/KA$ be an isomorphism. Since M is projective there is an R-homomorphism α such that the following diagram commutes.

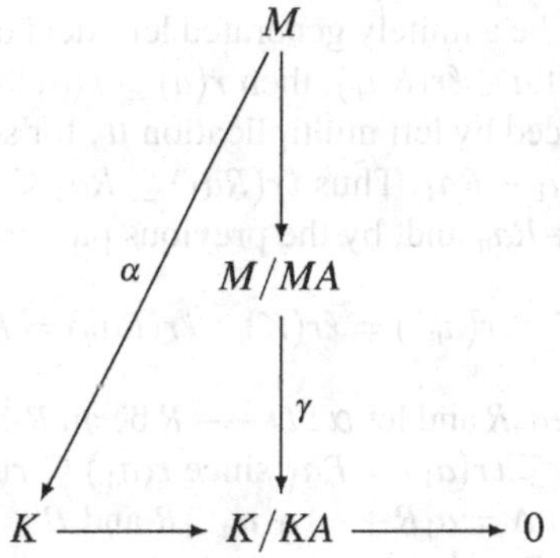

Then $\bar{\alpha} = \gamma$ and hence α is an isomorphism by (a). □

Theorem 4.3.25. *Let R be a unital right artinian ring and let $R = A_1 \boxplus \cdots \boxplus A_n$ be a decomposition of R_R into indecomposable right ideals. Then the following hold.*

(a) A_iJ is the unique largest proper submodule of $(A_i)_R$.
(b) $A_i/A_iJ \cong A_k/A_kJ$ iff $A_i \cong A_k$.
(c) If L_R is a simple unital R-module, then $L \cong A_i/A_iJ$ for some i.

Proof. (a) $A_i = e_iR$ for an idempotent e_i and $A_iJ = e_iJ = e_iR \cap J$. Suppose I is a proper submodule of A_i and assume, by way of contradiction, that $I \not\subseteq A_iJ$. Since J is the largest nilpotent ideal of R (Theorem 3.2.10) I is not nilpotent; so I contains an idempotent f with $f \neq 0$ and $f \neq e_i$. Now, $e_iR = fe_iR \boxplus (e_i - f)e_iR$ and $f = fe_if \in fe_iR$; so $(e_i - f)e_i = 0$ since e_iR is indecomposable. Thus, $e_i = fe_i \in I$ and $A_i = I$.

(b) Since each A_i is projective this follows from Theorem 4.3.24.

(c) Let $\overline{R} = R/J$ and $\overline{A}_i = A_i/A_iJ$. Then $\overline{R} = \overline{A}_1 \boxplus \cdots \boxplus \overline{A}_n$ and since $LJ = 0$, $L = L\overline{A}_1 + \cdots + L\overline{A}_n$. For some i and some $x \in L$, $x\overline{A}_i \neq 0$. So $\overline{A}_i \cong x\overline{A}_i = L$. □

We next investigate some annihilator conditions that hold in a unital right self-injective ring. This gives another explanation for why such an f-ring is left convex (see Exercise 18).

Theorem 4.3.26. *Let R be a unital ring.*

(a) If R is right self-injective, then $\ell(A \cap B) = \ell(A) + \ell(B)$ for any two right ideals A and B, and each finitely generated left ideal is a left annihilator.
(b) Conversely, if the second annihilator condition in (a) holds and the first holds for finitely generated right ideals, then each R-homomorphism $D \longrightarrow R$, where D is a finitely generated right ideal, can be extended to R.

Proof. (a) Let $u \in \ell(A \cap B)$ and define $\alpha : A + B \longrightarrow A + (1+u)B$ by $\alpha(a+b) = a + (1+u)b$. Then $\alpha = \mu_c$ for some $c \in R$; that is, $a + (1+u)b = ca + cb$ for $a \in A$ and $b \in B$. Consequently, $ca = a$, $cb = (1+u)b$ and $u = (c-1) + (u - (c-1)) \in \ell(A) + \ell(B)$. So $\ell(A \cap B) \subseteq \ell(A) + \ell(B)$ and the reverse inclusion always holds.

Let $I = Ra_1 + \cdots + Ra_n$ be a finitely generated left ideal of R. We use induction on n to show that $\ell r(I) = I$. If $a \in \ell r(Ra_1)$, then $r(a) \supseteq r(a_1)$ and the map $a_1 R \longrightarrow aR$ given by $a_1 x \mapsto ax$ is induced by left multiplication μ_b for some $b \in R$. So $ax = ba_1 x$ for each $x \in R$ and $a = ba_1 \in Ra_1$. Thus $\ell r(Ra_1) \subseteq Ra_1 \subseteq \ell r(Ra_1)$. Let $K = Ra_1 + \cdots + Ra_{n-1}$. Then $I = K + Ra_n$ and, by the previous paragraph and induction,

$$\ell r(I) = \ell(r(K) \cap r(a_n)) = \ell r(K) + \ell r(Ra_n) = K + Ra_n = I.$$

(b) Let $D = a_1 R + \cdots + a_n R$ and let $\alpha : D \longrightarrow R$ be an R-homomorphism. If $n = 1$, then $R\alpha(a_1) = \ell r(\alpha(a_1)) \subseteq \ell r(a_1) = Ra_1$ since $r(a_1) \subseteq r(\alpha(a_1))$. So $\alpha(a_1) = xa_1$ and $\alpha = \mu_x$. For $n > 1$, let $A = a_1 R + \cdots + a_{n-1} R$ and $B = a_n R$. By induction on n, there exist elements $x, y \in R$ such that $\alpha(a) = xa$ for $a \in A$ and $\alpha(b) = yb$ for $b \in B$. So $x - y \in \ell(A \cap B) = \ell(A) + \ell(B)$ and $x - y = u - v$ with $ua_n = 0$ and $vA = 0$. Let $z = x + v = y + u$. Then for $a \in A$ and $b \in B$,

$$z(a+b) = (x+v)a + (y+u)b = xa + yb = \alpha(a) + \alpha(b) = \alpha(a+b).$$

□

We now return to the symmetry of quasi-Frobenius rings.

Theorem 4.3.27. *The following statements are equivalent for the unital ring R.*

(a) R is quasi-Frobenius.
(b) R is right artinian and each right ideal is a right annihilator and each left ideal is a left annihilator.
(c) R is left artinian and left self-injective.

Proof. (a) $\Rightarrow$ (b). Since R is right artinian it has the minimum condition for right annihilators and hence it has the maximum condition for left annihilators. But then R has the maximum condition on finitely generated left ideals by Theorem 4.3.26 (a), and hence R is left noetherian. For, if $A_1 \subset A_2 \subset \cdots$ is a strictly increasing chain of left ideals and $a_1 \in A_1$ and $a_n \in A_n \backslash A_{n-1}$, then $Ra_1 \subset Ra_1 + Ra_2 \subset \cdots$ is a strictly increasing chain of finitely generated left ideals. Note that we now know that each left ideal is a left annihilator. Let A be a right ideal of R. If $A \subset r\ell(A)$, then there is a right ideal B with $A \subset B \subseteq r\ell(A)$ such that B/A is simple. As in Theorem 4.3.25 we have a decomposition of R_R into indecomposable modules: $R = A_1 \boxplus \cdots \boxplus A_n$. Since A_j is an indecomposable injective module it contains a unique minimal right ideal S_j. Moreover, if $[A_1], \ldots, [A_m]$ are the distinct isomorphism classes of $A_1, \ldots, A_n$, then $[S_1], \ldots, [S_m]$ are distinct isomorphism classes of simple modules. But by Theorem 4.3.25 there are precisely m isomorphism classes of simples; so $[S_1], \ldots, [S_m]$ are all of them. In particular, for some j, $B/A \cong S_j$, and so there is a nonzero homomorphism $\alpha : B \longrightarrow R$ with $\alpha(A) = 0$. Since $\alpha = \mu_x$ for some $x \in R$ we have $xA = 0$ and the contradiction $0 \neq \alpha(B) = xB \subseteq \ell(A)r\ell(A) = 0$. Thus, each right ideal is a right annihilator.

(b) $\Rightarrow$ (a). If A and B are right ideals of R, then

$$A \cap B = r\ell(A) \cap r\ell(B) = r(\ell(A) + \ell(B)).$$

So

$$\ell(A \cap B) = \ell r(\ell(A) + \ell(B)) = \ell(A) + \ell(B),$$

and R is right self-injective by Theorem 4.3.26 (b).

(b) $\Rightarrow$ (c). Since R is right noetherian (Exercise 4.1.10) R has the minimum condition on left annihilators; that is, R is left artinian. So the left-right dual of the preceding paragraph completes the proof.

(c) $\Rightarrow$ (b). Again, the left-right dual of the first paragraph gives that each right (left) ideal is a right (left) annihilator and R is right artinian. □

We will now give a characterization of quasi-Frobenius f-rings in which injectivity is replaced by convexity.

Theorem 4.3.28. *The following statements are equivalent for the f-ring R.*

(a) R is quasi-Frobenius.
(b) R is the direct sum of a finite number of totally ordered quasi-Frobenius rings.
(c) R is left and right convex and left artinian.
(d) R is left and right convex, left noetherian, and each prime ideal is maximal.

Proof. The equivalence of (a) with (b) is implied by Exercise 29, and that (b) implies (c) follows from Theorem 4.3.27 and Exercise 15 (or Theorem 4.3.26). Moreover, (d) is clearly a consequence of (c) since R is unital by Theorem 4.3.15 (and Exercise 29).

(d) $\Rightarrow$ (b). We may assume that R is totally ordered. Let J be the Jacobson radical of R. Since $J = \beta(R)$ is nilpotent and R is a left noetherian local ring, the series $J \supseteq J^2 \supseteq \cdots$ can be refined to a composition series for ${}_RJ$. So R is left artinian. Now, each right ideal is an ideal. For, suppose $a \in R$ and $xa \notin aR$ for some $x \in R$. Then $aR \subseteq xaR$ and $a = xay$ for some $y \in R$. If $y \notin J$, then $xa = ay^{-1} \in aR$. If

$y \in J$, then $y^n = 0$ for some $n \in \mathbb{N}$ and $a = x^n a y^n = 0$. Thus $Ra \subseteq aR$ for each $a \in R$ and $aR = Ra$ by Exercise 33. Suppose $\alpha : aR \longrightarrow R$ is an R-homomorphism. Then $\alpha(aR) \subseteq aR$ since R_R has a unique composition series and $\alpha(aR)$ does not have a larger composition series than aR. So $\alpha(a) = xa$ and $\alpha = \mu_x$ for some $x \in R$. □

Note that both left and right convexity are needed in the previous result since there are right convex right artinian f-rings that are not quasi-Frobenius (see Exercise 31).

A totally ordered unital right self-injective ring is a left valuation ring and each principal left ideal is a left annihilator. These conditions are close to characterizing right injectivity among totally ordered rings, as we will show below. A ring is called *left* (respectively, *right*) *duo* if each of its left (respectively, right) ideals is an ideal, and R is a *duo ring* if it is left and right duo. As we saw in the previous result a quasi-Frobenius f-ring is duo, and, more generally, each right convex right noetherian f-ring is right duo (Exercise 33), but it need not be left duo (Exercise 31). The next result is preparatory for the theorem that follows it.

Theorem 4.3.29. *Let R be a unital left valuation ring with maximal ideal J. Consider the following conditions on R.*

(a) J is the only prime ideal of R and R_R is uniform.
(b) $Ra = \ell r(Ra)$ for each $a \in R$.
(c) If A is a left ideal that is not a left annihilator, then there is a smallest principal left ideal Rb containing A. Moreover, $Rb = \ell r(A)$ and $A = Jb$.
(d) If $A \subset B$ are left ideals, then $\ell r(A) \subseteq B$.

Then (a) implies (b), (b) implies (c), and (c) implies (d).

Proof. (a) $\Rightarrow$ (b). If $b \in \ell r(Ra) \backslash Ra$, then $Ra \subset Rb$ and $a = xb$ with $x \in J$. Since $r(Ra) = r\ell r(Ra) \subseteq r(Rb)$ we have $r(x) \cap bR = 0$. For, if $br \in r(x)$, then $0 = xbr = ar$ and hence $br = 0$. Thus, $r(x) = 0$ yet J is a nil ideal.

(b) $\Rightarrow$ (c). Let $b \in \ell r(A) \backslash A$. Then $A \subseteq Rb \subseteq \ell r(A)$ gives $\ell r(A) \subseteq Rb \subseteq \ell r(A)$. Let $\alpha : R \longrightarrow Rb$ be right multiplication by b. Then the simple left R-module R/J maps onto Rb/Jb. So $R/J \cong Rb/Jb$, Jb is the maximal submodule of Rb, and $A \subseteq Jb$. Suppose $Jb \nsubseteq A$, and take $x \in J$ with $xb \notin A$. Since $xb \in \ell r(A) \backslash A$ we again have $Rxb = \ell r(A) = Rb$, and $b = cxb$; so $b = 0$ since $1 - cx$ is a unit of R. Thus, $Jb = A$. Now, if $A \subseteq Rd \subseteq Rb$, then $Rb = \ell r(A) \subseteq Rd \subseteq Rb$ and hence $Rb = Rd$.

(c) $\Rightarrow$ (d). If $A \subset \ell r(A)$, then $A = Jb$ and $\ell r(A) = Rb$. But then $b \in B$ since otherwise $Jb = A \subset B \subset Rb$ and Jb is not a maximal submodule of Rb. □

A ring R is called *left maximal* if whenever $\{J_\lambda : \lambda \in \Lambda\}$ is a set of left ideals of R and $\{x_\lambda : \lambda \in \Lambda\}$ is a subset of R such that the congruences $x \equiv x_\lambda (\text{mod } J_\lambda)$ are pairwise solvable, then these congruences are simultaneously solvable.

Theorem 4.3.30. *Let R be a unital left valuation ring or a totally ordered ring. If R is right self-injective, then R is left maximal and $Ra = \ell r(Ra)$ for each $a \in R$. Conversely, if R satisfies these two conditions and it is right duo, then it is right self-injective.*

Proof. We will assume R is a left valuation ring since a totally ordered ring which is either unital and right self-injective or in which each principal left ideal is a left annihilator must be a left convex left valuation ring. Suppose first that R is right self-injective. Then finitely generated left ideals are left annihilators by Theorem 4.3.26. Suppose $x \equiv x_\lambda (\text{mod } A_\lambda)$ is a pairwise solvable system of congruences in ${}_RR$. Note that the assumption that the congruences are pairwise solvable is equivalent to the condition: $x_\lambda - x_\rho \in A_\lambda + A_\rho = A_\lambda \vee A_\rho$ for all $\lambda, \rho \in \Lambda$. In particular, if $\{A_\lambda : \lambda \in \Lambda\}$ has a smallest member A_{λ_0}, then x_{λ_0} is a simultaneous solution since $x_{\lambda_0} - x_\lambda \in A_\lambda$ for every $\lambda \in \Lambda$. Now assume there is no smallest left ideal in $\{A_\lambda\}$. Let $K_\lambda = r(A_\lambda)$ and let $\mu_\lambda : K_\lambda \longrightarrow R_R$ be left multiplication by x_λ. If $b \in K_\lambda \cap K_\rho$, then $\mu_\lambda(b) = \mu_\rho(b)$ since $A_\lambda \subseteq A_\rho$, say, and therefore $x_\lambda - x_\rho \in A_\rho$ and $(x_\lambda - x_\rho)b = 0$. So these mappings are compatible and hence the homomorphism $\alpha : \cup K_\lambda \longrightarrow R$ given by $\alpha(b) = x_\lambda b$ if $b \in K_\lambda$ is induced by left multiplication by $x \in R : xb = x_\lambda b$ if $b \in K_\lambda$. So $x - x_\lambda \in \ell(K_\lambda) = \ell r(A_\lambda)$ for each $\lambda \in \Lambda$. If $A_\lambda = \ell r(A_\lambda)$, then certainly $x - x_\lambda \in A_\lambda$. Suppose $A_\lambda \subset \ell r(A_\lambda)$, and take $A_\rho \subset A_\lambda$. By (d) of Theorem 4.3.29 $x - x_\rho \in \ell r(A_\rho) \subseteq A_\lambda$, and hence $x - x_\lambda = x - x_\rho + x_\rho - x_\lambda \in A_\lambda$. So x is a solution to the system of congruences.

For the converse first note that R is a duo ring since, for each $a \in R$, $r(Ra)$ is an ideal and hence so is $Ra = \ell r(Ra)$. Let A be a right ideal of R and suppose $f : A \longrightarrow R$ is an R-homomorphism. If $a \in A$, then $f(a)r(a) = 0$; that is, $f(a) \in \ell r(Ra) = Ra$. So $f(a) = x_a a$. We claim that the system of congruences $x \equiv x_a (\text{mod } \ell(a))$ is pairwise solvable. For, suppose $a, b \in A$. Then, assuming $aR = Ra \subseteq Rb = bR$, we have $\ell(b) \subseteq \ell(a)$ and $a = br$ for some $r \in R$. Hence, $(x_b - x_a)a = x_b br - x_a a = f(b)r - f(a) = 0$, and $x_b - x_a \in \ell(a)$. Since R is left maximal we can find $x \in R$ such that $xa = x_a a = f(a)$ for each $a \in A$. Therefore, R is right self-injective. □

A ring R is called *right pre-self-injective* if R/A is right self-injective for each nonzero ideal A of R. It is called *almost left maximal* if each proper homomorphic image $R/A (A \neq 0)$ is left maximal. Since the conditions in the previous result are frequently preserved under homomorphic images they can be used to describe left convex totally ordered right pre-self-injective rings. The precise statement is given in Exercise 50. Note, however, that the totally ordered pre-self-injective ring $\mathbb{Z}$ is far from being convex. We will now give an example.

Theorem 4.3.31. *Let A be a division ring, $\Delta = -\mathbb{R}^+$, and let $R = V(A * \Delta) \subseteq V(A * \mathbb{R})$ be a crossed product formal power series ring of Δ over A with $1 = x^0$ and $\sigma : \mathbb{R} \longrightarrow Aut(A)$ (see Exercises 3.5.13 and 3.5.20). Then R is a pre-self-injective duo valuation domain and the Jacobson radical of each proper homomorphic image of R is nil. If A is totally ordered, then R is the subring of bounded elements of the totally ordered division ring $V(A * \mathbb{R})$.*

Proof. By Theorem 3.5.8 and Exercise 3.5.20 each nonzero element of R is of the form $x^\alpha u = u_1 x^\alpha$ where u and u_1 are units of R and $\alpha \leq 0$. Also, $x^\alpha x^\gamma = x^{\alpha+\gamma} w$ for some unit w. So each principal one-sided ideal is an ideal and R is a duo ring. Let I be a nonzero ideal of R and let $\beta = \text{lub } \{\gamma \in \Delta : x^\gamma \in I\}$. If $x^\beta \in I$, then $x^\alpha u = x^\beta x^{\alpha-\beta} u_1 \in x^\beta R$ for every $x^\alpha u \in I$; so $I = x^\beta R = (x^\beta)$. If $x^\beta \notin I$, then $I =$

$\{x^\gamma u : \gamma < \beta\} \cup \{0\} = (x^{<\beta})$. For, clearly, $I \subseteq (x^{<\beta})$, and if $\gamma < \beta$, then $\gamma \leq \alpha < \beta$ for some α with $x^\alpha \in I$; so $x^\gamma = x^{\gamma-\alpha}x^\alpha u \in I$.

Let $S = R/I$ where $I \neq 0$ and put $y = x + I$. We will consider S to be a formal power series ring in y. The elements of S will be denoted by $v = \sum_{\gamma \leq 0} y^\gamma a_\gamma$ with $v(\gamma) = a_\gamma$ and

$$\text{(I) } y^\gamma = 0 \; \forall \gamma \leq \beta \text{ if } I = (x^\beta),$$
$$\text{(II) } y^\gamma = 0 \; \forall \gamma < \beta \text{ if } I = (x^{<\beta});$$

so each nonzero element of S is of the form $y^\gamma u = u_1 y^\gamma$ with u and u_1 units of S, and $\gamma > \beta$ in case (I), and $\gamma \geq \beta$ in case (II). Note that $\ell(v) = r(v)$ for each $v \in S$. In case (I), $\ell(y^\gamma) = (y^{\beta-\gamma})$ and $\ell((y^{<\gamma})) = (y^{\beta-\gamma})$ if $\gamma \geq \beta$, and in case (II), $\ell(y^\gamma) = (y^{<\beta-\gamma})$ and $\ell(y^{<\gamma}) = (y^{\beta-\gamma})$. In both cases $\ell\ell(y^\gamma) = (y^\gamma)$.

We will check that S is right self-injective. Let K be an ideal of S and let $f : K_S \longrightarrow S_S$ be a homomorphism. If $K = (y^\gamma)$, then since $f(K)\ell(K) = 0$, $f(K) \subseteq \ell\ell(K) = K$; so $f(y^\gamma) = vy^\gamma$ for some $v \in S$. Therefore, f is induced by left multiplication by v. Suppose that $K = (y^{<\gamma})$. Choose a strictly increasing sequence $(\alpha_n)_{n\in\mathbb{N}}$ of real numbers with $\gamma = \text{lub}\,\{\alpha_n\}$. Then $K = \cup_n (y^{\alpha_n})$. For each n we can find $v_n \in S$ with $f(y^{\alpha_n}) = v_n y^{\alpha_n}$. If $m \geq n$, then $v_n y^{\alpha_n} = f(y^{\alpha_m})y^{\alpha_n - \alpha_m}w = v_m y^{\alpha_m} y^{\alpha_n - \alpha_m} w = v_m y^{\alpha_n}$ and $v_m - v_n \in \ell(y^{\alpha_n})$. In case (I) $v_m - v_n \in (y^{\beta-\alpha_n})$; that is $v_m(\delta) = v_n(\delta)$ if $\delta > \beta - \alpha_n$. Note that $\beta - \alpha_1 > \beta - \alpha_2 > \cdots \geq 0$ and define v by

$$v(\delta) = \begin{cases} v_1(\delta) \text{ if } & \beta - \alpha_1 < \delta \leq 0 \\ v_2(\delta) \text{ if } & \beta - \alpha_2 < \delta \leq \beta - \alpha_1 \\ & \cdots \\ v_n(\delta) \text{ if } & \beta - \alpha_n < \delta \leq \beta - \alpha_{n-1} \\ & \cdots \end{cases}$$

Then $v \in S$. For, if $\delta_1 \leq \delta_2 \leq \cdots$ is an increasing sequence in supp v, then $\delta_1 \in (\beta - \alpha_n, \beta - \alpha_{n-1}]$ for some n; consequently $\{\delta_v\} \subseteq \cup_{i=1}^n \text{supp } v_i$ and the latter has a.c.c. Now, $v - v_m \in \ell(y^{\alpha_m}) = (y^{\beta-\alpha_m})$ for each m. For, if $\delta > \beta - \alpha_m$, then $\beta - \alpha_n < \delta \leq \beta - \alpha_{n-1}$ for some $n \leq m$ ($\alpha_0 = \beta$), and $v(\delta) = v_n(\delta) = v_m(\delta)$. So if $w \in K$, then $w = y^{\alpha_m}s$ for some m and $f(w) = v_m y^{\alpha_m} s = vy^{\alpha_m}s = vw$.

A similar computation in case (II) shows that f is induced by left multiplication by some element of S. The only change is that the intervals used in the definition of v are $[\beta - \alpha_n, \beta - \alpha_{n-1})$ instead of $(\beta - \alpha_n, \beta - \alpha_{n-1}]$.

Since $J(R) = (x^{<0})$ and $J(S) = (y^{<0})$, if $y^\alpha \in J(S)$ and $n \in \mathbb{N}$ with $n\alpha < \beta$, then $(y^\alpha)^n = 0$. So $J(S)$ is nil. To see that $R = C(1)$ when A is totally ordered just note that if $0 < v \in V(A * \mathbb{R})$ and $\gamma = \text{maxsupp } v$, then $v \leq n = nx^0$ for some $n \in \mathbb{N}$ iff $\gamma \leq 0$. □

The property of being left maximal is a form of completeness that is shared by all totally ordered unital right self-injective rings. Another form of completeness that is shared by all right self-injective f-rings is lateral completeness.

Theorem 4.3.32. *A right self-injective f-ring is laterally complete. A regular f-ring is right self-injective iff it is laterally complete.*

Proof. Let R be a right self-injective f-ring. By Theorem 4.3.22 we may assume that R is unital. Also, for each $a \in R$ there is a unique idempotent e_a in R with $a^{\perp\perp} = e_aR = E(aR)$, and $a^{\perp} = (e_aR)^{\perp} = e_a^{\perp} = \ell(e_a)$ (see Exercise 3.4.14); in particular, $|a| \wedge |b| = 0$ iff $e_a \wedge e_b = 0$. Let A be a pairwise disjoint set in R^+ and put $I = AR$. Then $I^{\perp} = A^{\perp}$ since $x \in A^{\perp}$ iff $A \subseteq x^{\perp}$, iff $I = AR \subseteq x^{\perp}$, iff $x \in I^{\perp}$. Moreover, the sum $\sum_{a\in A} e_aR$ is direct. Define $\alpha : \sum_{a\in A} e_aR \longrightarrow R$ by $\alpha(\sum e_ar_a) = \sum ar_a$. α is a well-defined R-homomorphism since $\sum e_ar_a = 0$ yields $e_ar_a = 0$ and so $ar_a = ae_ar_a = 0$. Thus, there exists some $x \in R$ with $xe_a = a$ for each $a \in A$, and we may assume $0 \leq x \in I^{\perp\perp}$ since $R = I^{\perp\perp} \oplus I^{\perp}$. We will show that $x = \text{lub}_R A$. If $\overline{R}$ is a totally ordered homomorphic image of R, then $\bar{e}_a = 0$ or 1; consequently, $\bar{x} \geq \bar{a}$ for each $a \in A$, and x is an upper bound of A. Suppose y is an upper bound of A and write $y = z + u$ where $z \in I^{\perp\perp}$ and $u \in I^{\perp}$. For each $a \in A$, $ze_a = ye_a \geq ae_a = xe_a$; that is $(z-x)e_a \geq 0$. Thus, $(z-x)^- \in \ell(e_a) = a^{\perp}$ and $(z-x)^- \in I^{\perp} \cap I^{\perp\perp} = 0$; that is, $y \geq z \geq x$.

Now assume R is a regular f-ring which is laterally complete, and let $\alpha : I_R \longrightarrow R_R$ be an R-homomorphism where I is an ideal of R. Let $\{e_\lambda : \lambda \in \Lambda\}$ be a maximal set of orthogonal (equivalently, disjoint) idempotents in I. Note that $\alpha(I) \subseteq I$ since if $a \in I$, then $a = ea$ with $e^2 = e \in I$ and $\alpha(a) = \alpha(a)e \in I$. Let $\alpha(e_\lambda) = x_\lambda \in I$. Now, $\{x_\lambda^+, x_\lambda^- : \lambda \in \Lambda\}$ is a pairwise disjoint set in I since if $\rho \neq \lambda$, then $|x_\lambda| \wedge |x_\rho| = |x_\lambda|e_\lambda \wedge |x_\rho|e_\rho = 0$. Let

$$y = \bigvee_{\lambda\in\Lambda} x_\lambda^+, \qquad y_\rho = \bigvee_{\lambda\neq\rho} x_\lambda^+, \qquad z = \bigvee_{\lambda\in\Lambda} x_\lambda^-.$$

Since R is completely distributive (Theorem 2.1.3) $y \wedge z = 0$, $y = y_\rho \vee x_\rho^+ = y_\rho + x_\rho^+$, and $y_\rho \in e_\rho^{\perp} = \ell(e_\rho)$. So $ye_\rho = (y_\rho + x_\rho^+)e_\rho = x_\rho^+e_\rho = (x_\rho e_\rho)^+ = x_\rho^+$ for each $\rho \in \Lambda$. Similarly $ze_\rho = x_\rho^-$. Let $x = y - z$; then, for each $\rho \in \Lambda$, $xe_\rho = x_\rho$ and, for any idempotent $e \in I$, $(\alpha(e) - xe)e_\rho = \alpha(e_\rho)e - x_\rho e = 0$. Now, the sum $\sum_\lambda e_\lambda R$ is essential in I. So $\{e_\lambda\}^{\perp} = (\sum e_\lambda R)^{\perp} = I^{\perp}$ and $\alpha(e) - xe \in I^{\perp} \cap I = 0$. If $a \in I$, then $a = ea$ and $\alpha(a) = \alpha(e)a = xea = xa$. Thus, R_R is a Baer module; so R is unital and hence it is right self-injective. □

Exercises.

1. Let $\mathscr{C}$ be a class of R-modules which is homomorphically and extensionally closed, and let $\alpha \in \text{Hom}_R(M,N)$. Show that $N/\alpha M \in \mathscr{C}$ iff $N/\alpha K \in \mathscr{C}$ whenever $M/K \in \mathscr{C}$. In particular, if $\mathbb{F}$ is a topology of R and $\alpha \in \text{End}_R(R_R)$, then $\alpha R \in \mathbb{F}$ iff $\alpha\mathbb{F} \subseteq \mathbb{F}$.

2. Let M_R be a module over the sp-po-ring R and let $\mathbb{F}$ be a topology of R.
   (a) Show that $2t^+(M) \subseteq t(M)$.
   (b) If M has no elements of order 2 show that $t^+(M) = t(M)$.
   (c) Suppose M has no elements of order 2 and $\ell(R;M) = 0$. Show that the torsion theory cogenerated by $E(M)$ is obtained from a positive topology of R. (Use Exercise 4.2.2.)

(d) If R has no elements of order 2 show that its topology $\mathbb{D}$ of dense right ideals is positive.

3. If R is an essentially positive po-ring show that the Goldie topology $\mathbb{G}$ is positive.

4. Suppose $\mathbb{F}$ is a pretopology for the po-ring R and M_R is a d-module. Show that $t^+(M) = 0$ iff $t^+(M/A) = 0$ for every polar A which is an R^+-subgroup of M. In particular, if R is directed, $\mathbb{F}$ is a positive topology and M is an f-module, then M is torsion-free if $\mathscr{B}(M) \subseteq \mathscr{C}_{\mathbb{F}}(M)$. (See Exercise 4.2.10.)

5. Suppose $\mathbb{F}$ is a positive topology of the po-ring R and M_R is a d-module.

   (a) Show that the family of $\mathbb{F}$-closed convex ℓ-submodules of M is a complete distributive sublattice of $\mathscr{C}_{\mathbb{F}}(M)$ which satisfies the infinite distributive law $A \cap \vee_\lambda B_\lambda = \vee_\lambda A \cap B_\lambda$.
   (b) Suppose M is an $\mathbb{F}$-dense ℓ-submodule of the torsion-free f-module K. Show that $\mathscr{B}(M)$ and $\mathscr{B}(K)$ are canonically isomorphic (see Exercise 2.2.14). In particular, $\mathscr{B}(R) \cong \mathscr{B}(S)$ if the f-ring S is a right quotient ring of its ℓ-subring R.
   (c) Suppose R is directed and M and K are as in (b). If N is a convex-ℓ-submodule of M (respectively, K) show that $c\ell_{\mathbb{F}}(N^{\perp_M}; K) = (c\ell_{\mathbb{F}}(N; K))^{\perp_K}$ (respectively, $N^{\perp_K} \cap M = (N \cap M)^{\perp_M}$). In particular, if $A \in \mathscr{B}(M)$, then $c\ell_{\mathbb{F}}(A; K) = A^{\perp_M \perp_K}$.

6. Let $\mathbb{F}$ be a positive topology of the directed po-ring R and let M_R be an f-module.

   (a) If M is a torsion-free $q_{\mathbb{F}}f$-module show that M is a subdirect product of totally ordered torsion-free modules iff $M_{\mathbb{F}}$ is a subdirect product of totally ordered torsion-free $R_{\mathbb{F}}$-modules.
   (b) Suppose M_R is an irredundant f-module. Show that M is torsion-free iff it is a subdirect product of totally ordered torsion-free modules

7. Let A be a subring of the totally ordered division ring K, and suppose A is not a right Öre domain. For the infinite set Λ let $R = \{(x_\lambda) \in D^\Lambda : x_\lambda \in A$ except for finitely many $\lambda\}$. Show that R is a qf-ring that is not a right Öre ring.

8. Let R be a unital regular ℓ-ring in which some power of each element is positive. Show that R is an f-ring.

9. Show that each one-sided ideal of a regular f-ring is an ℓ-ideal.

10. Let $\mathbb{F} \subseteq \mathbb{H}$ be positive topologies of the directed po-ring R. Suppose M is an R-module with $t_{\mathbb{H}}(M) = 0$.

    (a) If M is a $q_{\mathbb{H}}f$-module show that M is a $q_{\mathbb{F}}f$-module and $M_{\mathbb{F}}$ is an ℓ-R-submodule of $M_{\mathbb{H}}$.
    (b) If R is a $q_{\mathbb{H}}f$-ring with $t_{\mathbb{H}}(R) = 0$ show that R is a $q_{\mathbb{F}}f$-ring and $R_{\mathbb{F}}$ is an ℓ-subring of $R_{\mathbb{H}}$.

11. Let $\mathbb{F}$ be a positive topology of the directed po-ring R, and let $\mathscr{C}$ be the class of torsion-free $q_{\mathbb{F}}f$-modules.

    (a) Show that $\mathscr{C}$ is hereditary and is closed under direct products, directed direct limits, and inverse limits.
    (b) Suppose M_γ is an f-module in $\mathscr{C}$ for each γ in the poset Γ. Assume $r(M_\gamma;R)^+ = 0$ for each $\gamma \in \Gamma$, and $V(\Gamma,(M_\gamma)_{\mathbb{F}})$ is an R-f-module (see Theorems 2.6.1 and 2.6.2). Show that $V(\Gamma,(M_\gamma)_{\mathbb{F}})$ is $\mathbb{F}$-injective and $V(\Gamma,M_\gamma)$ belongs to $\mathscr{C}$.
    (c) Suppose $\mathbb{F}$ is the Goldie topology and M is a module in $\mathscr{C}$. If N is a closed convex ℓ-submodule of M show that M/N belongs to $\mathscr{C}$.

12. Let $\mathbb{F}$ be a pretopology of the po-ring R. Suppose M is an $\mathbb{F}^+$-dense R^+-ℓ-subgroup of the R^+-f-group K and $t^+(K) = 0$. Let A be a convex ℓ-R^+-subgroup of M, and let $B = c\ell_{\mathbb{F}^+}(A;K)$.

    (a) If $M = \text{lex } A$ show that $K = \text{lex } B$ (see Theorem 2.5.6 and Exercise 2.5.8).
    (b) If $K = \text{lex } B$ show that $M = \text{lex } (B \cap M)$.
    (c) Suppose $\mathbb{F}$ is the Goldie topology and it is positive. Assume M_R is an f-module and $K = E(M_R)$. If $K = \text{lex } B$ and L is a complement of B in K show that L is totally ordered and $K = B \overleftarrow{\boxplus} L$.
    (d) Assume the conditions in (c) except for $K = \text{lex } B$. Let P be the partial order of L that it inherits from K/B. Show that $L \cap K^+ \subseteq P$ and use the example $R = \mathbb{Z}$, $M = R \oplus R$ to show that the containment can be proper.

13. Let $\mathbb{F}$ be a positive topology and suppose $\alpha : M \longrightarrow N$ is an R-ℓ-homomorphism between $q_{\mathbb{F}}f$-modules. Show that $Q(\alpha) : M_{\mathbb{F}} \longrightarrow N_{\mathbb{F}}$ is an ℓ-homomorphism.

14. Suppose $\mathbb{F}$ is a positive topology and M_R is a torsion-free f-module. Let $K = \{x \in M_{\mathbb{F}} : \forall d_1, d_2 \in (M:x)^+, (xd_1)^+ \wedge (xd_2)^- = 0\}$.

    (a) Show $x \in K$ iff $x^+ \in M_{\mathbb{F}}$ and $x^+ d = (xd)^+$ for each $d \in (M:x)^+$.
    (b) Show that $M \cup M_{\mathbb{F}}^+ \subseteq K \subseteq M_{\mathbb{F}}^+ - M_{\mathbb{F}}^+$, $-K \subseteq K$ and $KR_{\mathbb{F}}^+ \subseteq K$.
    (c) Show that K is a subgroup of $M_{\mathbb{F}}$ iff K is an f-module extension of M, iff $M_{\mathbb{F}}^+ - M_{\mathbb{F}}^+$ is an f-module extension of M. If K is a subgroup, then $K = M_{\mathbb{F}}^+ - M_{\mathbb{F}}^+$.
    (d) Let $R = \begin{pmatrix} \mathbb{Q} & 0 \\ \mathbb{Q} & \mathbb{Q} \end{pmatrix}$ have the coordinatewise order, let $\mathbb{F}$ be the Goldie topology, and let M be the ℓ-ideal of R given by $M = \begin{pmatrix} 0 & 0 \\ \mathbb{Q} & 0 \end{pmatrix}$. Show that $\mathbb{F}$ is positive, $E = E(M)$ is an ℓ-module extension of M and $K = M \cup E^+ \cup -E^+$ (see Exercise 4.2.13).

15. Let M_R be an f-module with $\ell(R;M) = 0$. Show that the following are equivalent.

    (a) If $x, y \in M$ with $xR \cap yR = 0$, then $|x| \wedge |y| = 0$.
    (b) Each submodule A has $A^\perp$ as its unique complement.

(c) If A is a submodule of M, then $A^{\perp\perp}$ is the unique largest essential extension of A.
(d) $\mathscr{B}(M)$ is the set of complements in M.
(e) Each complement in M is a convex ℓ-submodule.

16. Let $\mathbb{F}$ be a topology of the po-ring R. Suppose M is an $\mathbb{F}^+$-dense ℓ-submodule of the f-module K and $t^+(K) = 0$. Show that M satisfies the conditions in Exercise 15 iff K satisfies these conditions.

17. Suppose $\mathbb{F}$ is a positive topology of the directed po-ring R and M_R is an f-module. For each $D \in \mathbb{F}$ let $[D^+R_1, M/t(M)]$ have the usual partial order: $\alpha \in [D^+R_1, M/t(M)]^+$ iff $\alpha(D^+) \subseteq (M/t(M))^+$. Show that M is a $q_{\mathbb{F}}f$-module iff each $[D^+R_1, M(t(M)]$ is an ℓ-group and $\alpha^+ d = (\alpha d)^+$ for each $\alpha \in [D^+R_1, M/t(M)]$ and $d \in D^+$.

18. Let R be a left self-injective ring. Suppose R is a left d-ring and if $0 \le x \le y$, then $\ell(y) \subseteq \ell(x)$. Show that $R/r(R)$ is right convex.

19. Give an example of a po-ring R that is not an ℓ-ring but for which $Q(R)$ is an f-ring extension of R.

20. Let $\mathbb{F}$ be a positive topology of the right f-ring R, and let M_R be an f-module.
(a) Suppose $M_{\mathbb{F}}$ is a strong f-module. Show that for each $x \in M_{\mathbb{F}}$ there is an $\mathbb{F}$-dense ℓ-submodule D of R_R with $xD \subseteq \psi(M)$.
(b) Suppose R is an f-ring and D is an ℓ-submodule of R_R. Show that the poset $\{dR : d \in D\}$ is directed up iff the poset $\{dR^+ : d \in D^+\}$ is directed up.
(c) Suppose R is an f-ring and for each $q \in R_{\mathbb{F}}$ the poset $\{dR : d \in (\psi(R) : q)\}$ is directed up. Show that R is a $q_{\mathbb{F}}f$-ring iff each $(\psi(R) : q)$ contains an $\mathbb{F}$-dense ℓ-submodule of R_R.
(d) Suppose R is an f-ring in which each finitely generated right ideal is principal. Show that R is a $q_{\mathbb{F}}f$-ring iff R_R is a $q_{\mathbb{F}}f$-module, iff $(\psi(R) : q)$ contains an $\mathbb{F}$-dense ℓ-submodule of R_R for each $q \in R_{\mathbb{F}}$.

21. Let C be a commutative irredundant f-ring, and let R be an f-algebra over C which is C-nonsingular.
(a) Show that $E({}_CR)$ is an f-algebra extension of R.
(b) Show that R is C-unitable iff $E({}_CR)$ is unitable, iff R satisfies the identities (3.4.5) and (3.4.6) of Section 3.4 (see Theorem 3.4.8).
(c) Show that the components of C are all ℓ-simple iff $E({}_CR)$ is infinitesimal whenever R is infinitesimal (see Theorem 3.4.9.)

22. Suppose that the nonsingular f-ring R is an irredundant subdirect product of totally ordered rings. Show that R is a qf-ring iff R/A is a right Öre domain for each maximal polar A of R.

23. Let R be a reduced f-ring. Show that R is a qf-ring iff $aR \cap bR \neq 0$ for all $a, b \in R$ with $0 < a \le b$.

24. Let M_R be an f-module over the directed po-ring R. Show that M is convex iff each of its submodules is a convex submodule, iff each of its submodules is a convex ℓ-submodule, iff for all $x, y \in M$, if $|x| \leq |y|$, then $x \in yR$.

25. Let $\mathbb{F}$ be a topology of the right convex f-ring R. Show that $R_{\mathbb{F}}$ has bounded inversion. (For any ring R, if $p \in R_{\mathbb{F}}$, then $p^{-1} \in R_{\mathbb{F}}$ iff $r(p; \psi(R)) = 0$ and pD is $\mathbb{F}$-dense in $R_{\mathbb{F}}$ for some $D \in \mathbb{F}$.)

26. Let R be a unital f-ring with bounded inversion. Show that R is a classical left and right quotient ring of $C(1) = \{a \in R : |a| \leq n\}$. (If $r \in R^+$, then $r = r \vee 1 + (r \wedge 1 - 1)$.)

27. Show that the following statements are equivalent for the f-ring R with bounded inversion.

    (a) R is a direct sum of local f-rings.
    (b) 1 is finite valued.
    (c) $C(1)$ is a direct sum of local f-rings.

    Moreover, the number of summands is equal to the number of values of 1.

28. (a) Let T be a convex ℓ-subring of the right convex f-ring R. Show that T is right po-normal; and T is right convex iff $a \in aT$ for each $a \in T$.
    (b) Let $\mathscr{V}$ be the variety of ℓ-rings generated by the class of right convex f-rings. Show that $R \in \mathscr{V}$ iff R is an ℓ-subring of a right convex f-ring (see Exercise 1.4.16).
    (c) Let $\mathscr{V}$ be the variety of C-ℓ-algebras generated by the unital right convex C-f-algebra K. Show that $R \in \mathscr{V}$ iff R is an ℓ-subalgebra of a unital right convex member of $\mathscr{V}$ which can be taken to be a K-f-ring provided $C \cdot 1 \subseteq K$ is cofinal in K.

29. Show that the following statements are equivalent for the right convex f-ring R.

    (a) R has the maximum condition on polars.
    (b) R_R is a finite dimensional module.
    (c) R contains a finite valued regular element.
    (d) R is the direct sum of a finite number of totally ordered rings.
    (e) R is the direct sum of a finite number of local rings.

    Moreover, other statements which are equivalent to (a) are obtained by replacing R in each of these statements by $Q(R)$, $C^R(1)$ or $C^{Q(R)}(1)$. (Use Exercises 25, 26, 27, and 28.)

30. Let R be a right convex f-ring and let T be a subring of R which contains $S = C^R(1)$.

    (a) Show that R_T is a convex f-module.
    (b) Show that T is a convex ℓ-subring of R.
    (c) Suppose each element of T is bounded by a central element of T. Show that each multiplicatively closed subset of T is a right Öre set in T.

31. Let $\sigma : F \longrightarrow F$ be an order preserving monomorphism of the totally ordered field F, and let $R = F[[x;\sigma]]$ be the totally ordered twisted formal power series ring with exponents from $\mathbb{N}$ and coefficients from F. The elements of R are represented by $\sum_{n\geq 0} x^n a_n$ and $ax = xa^\sigma$ - see Exercise 3.5.13 but note that a.c.c. has been replaced by d.c.c.
    (a) Show that each nonzero element of R is of the form $x^n u$ where u is a unit of R.
    (b) Show that R is right convex.
    (c) Show that R is left convex iff σ is an automorphism. (Consider $x^2 < xa$.)
    (d) Show that $R/x^n R$ is right artinian, and it is left artinian if and only if σ is an automorphism.

32. Let R be the 2×2 upper triangular matrix ring over $\mathbb{Q}$, ordered coordinatewise. Show that R is left and right convex and R has a left (right) ideal that is not an ideal.

33. Suppose R is a right convex f-ring.
    (a) If R has a.c.c. or d.c.c. on principal right ideals show that each right ideal of R is an ideal.
    (b) If R has a.c.c. on principal right ideals show that R is right noetherian.
    (c) If R is right noetherian show that $Q_c(R) = Q(R)$ (use Exercise 29).
    (d) If $C^R(1)$ is right noetherian show that R is right noetherian, and give an example to show that the converse does not hold.

34. Let I and J be ℓ-subgroups of the right convex ℓ-ring R. If J is a right ideal show that $IJ = \{ab : a \in I, b \in J\}$.

35. Give an example of an ℓ-field that is not right convex.

36. Suppose R is a unital right convex f-ring. Show that the following conditions are equivalent, and they hold provided $a^{-1}Ra = R$ for each regular element a of R.
    (a) R is a left Öre ring.
    (b) $Q_c(R)$ is a convex right f-module over R.
    (c) R is a convex ℓ-subgroup of $Q_c(R)$.
    (d) $C^{Q_c(R)}(1) \subseteq R$.

37. Suppose Σ is a right Öre set of regular elements in the f-ring R which has bounded inversion. Assume that $|s| \in \Sigma$ whenever $s \in \Sigma$. Show that the following are equivalent.
    (a) $s^{-1}Rs = R$ for each $s \in \Sigma$.
    (b) $s^{-1}Rs = R$ for each $s \in \Sigma^+$.
    (c) $\mathscr{U}(R) \trianglelefteq \mathscr{U}(R_\Sigma)$.
    (d) $\mathscr{U}(R) \cap C^R(1) \trianglelefteq \mathscr{U}(R)_\Sigma)$.

38. Let R_Σ be a classical right quotient ring of the f-ring R and let M_R be an f-module.

    (a) If M is a strong f-module show that M_Σ is a strong R_Σ-f-module.
    (b) If M is convex show that M_Σ is a convex R_Σ-f-module.

39. Let R be a unital f-ring that has bounded inversion and localizations and let M_R be a unital R-module.

    (a) If $0 \neq x \in M$ show that $x \neq 0$ in M_P for some maximal ideal P of R.
    (b) Let B and C be submodules of M. Show that $B \subseteq C$ iff $B_P \subseteq C_P$ for each maximal ideal of P of R. (See the remarks after Theorem 4.2.11.)
    (c) Show that for each maximal ideal P, R_P is a local f-ring with bounded inversion and maximal ideal P_P.

40. Let M be a unital f-module over the f-ring R that has bounded inversion and localizations.

    (a) Show that M is a strong R-f-module iff M_P is a strong R_P-f-module for each maximal ideal P of R.
    (b) Show that M is a convex R-f-module iff each M_P is a convex R_P-f-module.
    (c) If M is convex show that, for each $P \in \text{Max}(R)$, the lattice of R_P-submodules of M_P is a chain.

41. Let R be a unital f-ring with center C and let M be a unital ℓ-module over R. Show that M is an f-module or a strong f-module or a convex f-module over R iff M_P is an f-module or a strong f-module or a convex f-module over R_P, respectively, for each maximal ideal P of C.

42. Let R be a right convex semiprime f-ring with center C. Show that if A is an ideal of C such that C/A is semiprime, then A is an ℓ-ideal of C.

43. Let $\left(\begin{pmatrix} \mathbb{Q} & \mathbb{Q} \\ 0 & 0 \end{pmatrix}, P\right)$ be an f-ring. Show that either

$$P = \left\{ \begin{pmatrix} p & q \\ 0 & 0 \end{pmatrix} : p > 0, \text{ or } p = 0 \text{ and } q \geq 0 \right\}$$

or

$$P = \left\{ \begin{pmatrix} p & q \\ 0 & 0 \end{pmatrix} : p > 0, \text{ or } p = 0 \text{ and } q \leq 0 \right\}.$$

So, up to an isomorphism, $U = \begin{pmatrix} \mathbb{Q} & \mathbb{Q} \\ 0 & 0 \end{pmatrix}$ has a unique lattice order with respect to which it is an f-ring.

44. Let J be the Jacobson radical of the ℓ-ring R. Suppose A is an ℓ-ideal of R such that $A \subseteq J$ and R/A is an f-ring. Show that J is an ℓ-ideal iff it is convex. $(0 \leq (a^+x)^2 \leq (ax)^2$ in any f-ring.)

45. Let R be a unital ℓ-ring in which some power of each element is positive, and suppose R is right self-injective. Show that the following statements are equivalent.

 (a) R is an f-ring.
 (b) R has bounded inversion.
 (c) The Jacobson radical of R is convex.
 (d) If A is a nonzero right annihilator, then $A^+ \neq 0$.

 (Show that each statement implies (b). For (c) $\Rightarrow$ (b) use Exercise 4.1.9 and Theorem 4.1.9.)

46. Let $\mathbb{F}$ be a positive topology of the right f-ring R and suppose that $R_{\mathbb{F}}$ is a right self-injective ring. Show that R is a $q_{\mathbb{F}}f$-ring iff R_R is a $q_{\mathbb{F}}f$-module and the Jacobson radical of $R_{\mathbb{F}}$ is a convex ideal.

47. Let $\mathbb{F}$ be a positive topology of the f-ring R. If R_R is a $q_{\mathbb{F}}f$-module show that $R_{\mathbb{F}}$ is a right quotient ring of its subring $F(R_{\mathbb{F}})$ of f-elements (see Exercise 4.2.11). Show that this may not be true if R is only a right f-ring.

48. Show that a right self-injective f-ring is a direct product of totally ordered right self-injective rings iff its Boolean algebra of polars is atomic.

49. (a) Let R be a unital right self-injective right duo left valuation ring. Show that R is left self-injective iff $\ell r(a) = r\ell(a)$ for each $a \in R$.
 (b) Let R be a totally ordered ring which is right duo and has only one proper prime ideal. Show that R is left and right self-injective iff R is left maximal, left convex, and $\ell r(a) = r\ell(a)$ for each $a \in R$.

50. Let R be a right duo unital left valuation ring with Jacobson radical J. Show that the following statements are equivalent.

 (a) R is right pre-self-injective.
 (b) For each $0 \neq x \in R$, R/xR is right self-injective, and either J is nil or $0 \subset J$ are all of the prime ideals of R.
 (c) R is almost left maximal, and either J is nil or R is a domain with a unique nonzero prime ideal.

51. Let M be a nonsingular f-module over the irredundant semiprime right qf-ring R, and let $\{R_\lambda : \lambda \in \Lambda\}$ and $\{M_\lambda : \lambda \in \Lambda\}$ be the components of R and M, respectively. Let $Q_\lambda = Q(R_\lambda)$, let Γ_λ be the Q_λ-value set of $E_\lambda = E(M_\lambda)$, let $V_\lambda = V(\Gamma_\lambda, (E_\lambda)^\alpha/(E_\lambda)_\alpha)$ be the Hahn product, and let Γ be the cardinal sum of the Γ_λ.

 (a) Show that the Hahn product $V(\Gamma, (E_\lambda)^\alpha/(E_\lambda)_\alpha)$ is a $Q(R)$-f-module isomorphic to $\Pi_\lambda V_\lambda$ and M_R is embedded in $V(\Gamma, (E_\lambda)^\alpha/(E_\lambda)_\alpha)$.
 (b) If M is essential and has only a finite number of nonzero components, show that Γ is the $Q(R)$-value set of $E(M)$.

52. Let R_u be the unital cover of the right convex f-ring R. Show that the following are equivalent.

(a) $R = R_u$.
(b) R_u is right convex.
(c) R_u has bounded inversion.
(d) $R_u/J(R_u)$ has bounded inversion.

53. If X is a Stone space show that $D(X)$ is a regular self-injective f-ring.

54. Let R be a semiprime archimedean f-ring and let X be the Stone space of the Boolean algebra of polars of R.

(a) Show that $Q(R) \subseteq D(X)$. (See Exercise 2.3.24.)
(b) If R is complete and divisible, show that $Q(R) = D(X)$.
(c) If R is self-injective show that the completion of R is $D(X)$. (See Theorem 2.3.31.)
(d) If R is unital and divisible show that its completion is a convex f-ring.

55. Let M_R be an f-module over the unital f-ring R.

(a) Suppose R_λ is an ℓ-primitive f-ring for each $\lambda \in \Lambda$ and $\oplus R_\lambda \subseteq R \subseteq \Pi R_\lambda$. If the multiplication maps $x\cdot : R \longrightarrow M$ are complete for each $x \in M^+$, show that M_R is unital iff it is nonsingular.
(b) Suppose M and R are archimedean and R is irredundant. Assume $R \longrightarrow F(M)$ is a complete embedding and $\sigma : F(M) \longrightarrow D(X)$ is a $c\ell$-essential monomorphism where X is the Stone space of M (see Exercises 3.6.16 through 3.6.18). Show that M_R is unital iff it is nonsingular.

56. Let X be the Stone space of the archimedean ℓ-group M.

(a) Show that $\sigma : F(M) \longrightarrow D(X)$ is a $c\ell$-essential monomorphism iff $D(X)$ is the completion of the maximal quotient ring of $F(M)$.
(b) If $\sigma(F(M))$ is $c\ell$-essential in $D(X)$ show that M is a nonsingular $F(M)$-module.

57. Let M_R be a free unital right R-module over the totally ordered domain R, and give M_R the coordinatewise lattice order. Let $F = F({}_{\mathrm{End}_R(M)}M)$ be the subring of $\mathrm{End}_R(M)$ generated by the f-endomorphisms of M_R.

(a) Show that ${}_FM$ is an injective module iff R is a division ring.
(b) Show that ${}_FM$ is an if-module iff R is a left Öre domain.

Notes. Anderson [AN2] examined the maximal right quotient ring of a unital f-ring and gave necessary and sufficient conditions for it to be an f-ring extension. He also showed that the classical right quotient ring of an f-ring, when it exists, is an f-ring extension. Steinberg [ST3] applied Anderson's technique to identify when the injective hull of a nonsingular f-module is an f-module extension. Georgoudis [G] extended this to the module of quotients of an f-module with respect to a positive topology and Bigard gave a considerable expanded version of Georgoudis' results in [BI2]. The material on convex f-rings and f-modules comes from Steinberg [ST4], [ST9], and [ST18]. Sources for quasi-Frobenius rings are Curtis and Reiner [CR] and Faith [F5]. Commutative semiprime unital Prüfer f-rings are considered in Martinez and Woodward [MAW]. Theorems 4.3.29 and 4.3.30 and Exercise 50 on self-injective valuation rings are due to Klatt and Levy [KL] in the commutative case. The example of the pre-self-injective valuation domain given in Theorem 4.3.31 is due to Levy [LEV3].

4.4 Injective f-Modules

Let $\mathscr{C}$ be a subcategory of the category Poset. Recall that the objects of Poset are posets and the morphisms are the isotone maps. An object E in $\mathscr{C}$ is *injective* if each diagram in $\mathscr{C}$ of the form

$$\begin{array}{ccc} A & \longrightarrow & B \\ \big\downarrow & & \\ E & & \end{array} \qquad (4.4.1)$$

whose row is a monomorphism in Poset can be completed to a commutative diagram in $\mathscr{C}$ by means of a morphism $B \longrightarrow E$. The usual definition of an injective requires the row in (4.4.1) to be monic. However, in the category Log of ℓ-groups and ℓ-homomorphisms the monics are precisely the monomorphisms and the same is true of the subcategories of Log with which we will mainly be concerned. Since, in fact, injectives in these categories will be scarce we will consider a restricted form of injectivity. Let $\aleph_\alpha$ be an infinite cardinal number. The object E in $\mathscr{C}$ is called $\aleph_\alpha$-*injective* if diagrams of the form (4.4.1) can be completed whenever $\text{card}(B) < \aleph_\alpha$. We will see that an $\aleph_\alpha$-injective nonsingular essential right f-module over an irredundant semiprime right qf-ring is characterized as being an injective module which has three order complete properties which, in fact, describe $\aleph_\alpha$-injective rational vector lattices. A similar but easier characterization holds for an $\aleph_\alpha$-injective po-module and also for an $\aleph_\alpha$-injective poset and these results appear in the exercises. Each nonsingular f-module can be embedded in an $\aleph_\alpha$-injective f-module. An example of a totally ordered $\aleph_\alpha$-injective abelian ℓ-group is obtained by forming a restricted Hahn product using an $\aleph_\alpha$-injective totally ordered set and with all real components. Another relative injective f-module is obtained as follows. Over a countable totally ordered right Öre domain the direct product of countably many nonsingular totally ordered right modules, modulo their direct sum, is $\aleph_1$-injective.

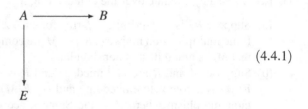

Suppose R is a po-ring and $\mathscr{C}$ is a subcategory of the category po-$\mathscr{M}_R$ of right po-modules over R. If M is a nonzero po-module in $\mathscr{C}$ and for some po-module N with $N^+ \neq 0$ the natural embedding $M \longrightarrow M \times N$ is a monomorphism in $\mathscr{C}$, then M is not injective in $\mathscr{C}$. For, if the diagram

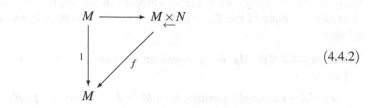

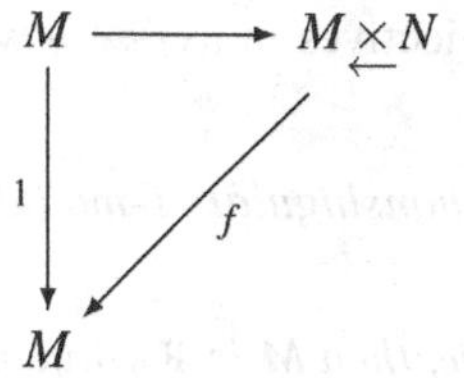

(4.4.2)

is commutative with f isotone, then $f(0,x) \geq M$ for each $0 < x \in N$. In particular, if R is a po-domain, then the category po-$\mathscr{M}_R$ has no nontrivial injectives since R, or, if necessary, the po-module obtained by freely adjoining $\mathbb{Z}$ to R, with the coordinatewise order, is such a po-module N. The same is true of the category of R-ℓ-modules (or R-f-modules) provided R has a nonzero totally ordered strict po-module.

Theorem 4.4.1. *For any po-ring R the category of right po-modules over R has no nonzero injectives.*

Proof. Let M and N be right po-modules and let $h \in \mathrm{Hom}_R(M,N)$. Define

$$P_h = \{(m,n) \in M \boxplus N : m \in M^+ \text{ and } h(m)+n \in N^+\}$$

and $P(M,N) = \sum_h P_h$. It is easily verified that $(M \boxplus N, P_h)$ and $(M \boxplus N, P(M,N))$ are po-modules over R. Note that $P_0 = (M \oplus N)^+$ and both of the injections $\kappa_1 : M \longrightarrow (M \boxplus N, P(M,N))$ and $\kappa_2 : N \longrightarrow (M \boxplus N, P(M,N))$ are monomorphisms. For, suppose $(x,0) \in P(M,N)$. Then $(x,0) = \sum_i (x_i,y_i)$ with $x_i \in M^+$; so $x = \sum_i x_i \in M^+$. Also, if $(0,y) \in P(M,N)$, then $(0,y) = \sum_i (x_i,y_i)$ yields $x_i = 0$ for each i; so $y_i \in N^+$ and $y = \sum_i y_i \in N^+$. Suppose E_R is a po-module which is injective in the category of po-modules. Then there is a po-homomorphism $h : (E \boxplus E, P(E,E)) \longrightarrow E$ with $h\kappa_2 = 1_E$; that is, $h(0,x) = x$ for each $x \in E$. Let $g = h\kappa_1 : E \longrightarrow E$; that is, $g(x) = h(x,0)$. If $x \in E^+$, then $(x,-2g(x)) \in P_{2g} \subseteq P(E,E)$; so

$$-g(x) = g(x) - 2g(x) = h(x,0) - h(0,2g(x)) = h(x,-2g(x)) \in E^+,$$

and $g(E^+) = 0$ since g is isotone. Since $(x,-x) \in P_1 \subseteq P(E,E)$ and $h(x,-x) = g(x) - h(0,x) = -x \in E^+$ we obtain that E is trivially ordered. By freely adjoining $\mathbb{Z}$ to R (with the product partial order) we may assume that $1 \in R^+$ and E_R is unital. Now, there is a po-homomorphism $f : (R \boxplus E, P(R,E)) \longrightarrow E$ with $f\kappa_2 = 1_E$. Let $\mu : R \longrightarrow E$ be left multiplication by $x \in E$. Then $(1,-x) \in P_\mu \subseteq P(R,E)$ and

$$-x = f(0,-x) = f(1,0) + f(0,-x) = f(1,-x) = 0;$$

so $E = 0$. □

Since injectives in $\mathscr{C}$ are scarce we now turn our attention to an examination of some relative injectives. We will denote the subcategory of Log consisting of the right f-modules (respectively, ℓ-modules) over the po-ring R by $f\text{-}\mathscr{M}_R$ (respectively, $\ell\text{-}\mathscr{M}_R$), and $nsf\text{-}\mathscr{M}_R$ is the subcategory of nonsingular f-modules. Our goal is to identify many of the $\aleph_\alpha$-injectives in $nsf\text{-}\mathscr{M}_R$ when R is an irredundant right qf-ring.

Theorem 4.4.2. *Let M_R be a nonsingular f-module which is $\aleph_\alpha$-injective in $nsf\text{-}\mathscr{M}_R$.*

(a) If R is essentially positive, then M is $\aleph_\alpha$-injective in $f\text{-}\mathscr{M}_R$.
(b) If M is an essential ℓ-submodule of the f-module K_R, then K is $\aleph_\alpha$-injective in $nsf\text{-}\mathscr{M}_R$.
(c) Suppose $\aleph_\alpha > card\,(R)$ and M and K are as in (b). Then $M = K$.
(d) If R has a nonzero totally ordered strict nonsingular f-module whose cardinality is exceeded by $\aleph_\alpha$, then $card\,(M) \geq \aleph_\alpha$ unless $M = 0$.

Proof. (a) Suppose M is $\aleph_\alpha$-injective in $nsf\text{-}\mathscr{M}_R$ and

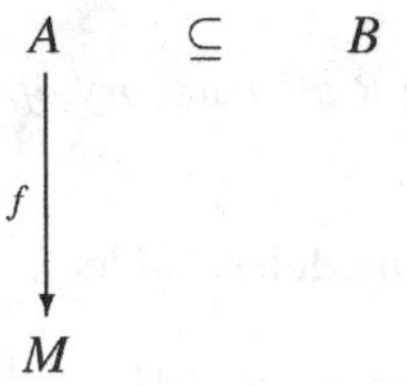

is a diagram in $f\text{-}\mathscr{M}_R$ with card$(B) < \aleph_\alpha$. Then we get the commutative diagram

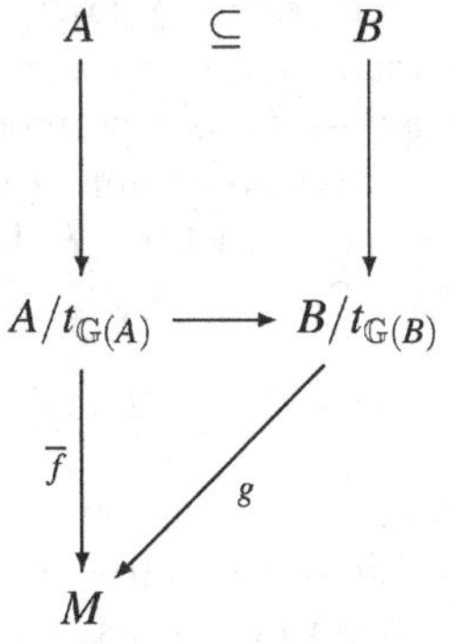

where the composite on the left is f, and hence M is $\aleph_\alpha$-injective in $f\text{-}\mathscr{M}_R$.

(b) Consider a diagram

$$\begin{array}{ccc} A & \subseteq & B \\ {\scriptstyle f}\big\downarrow & & \\ K & & \end{array}$$

in $nsf\text{-}\mathcal{M}_R$ with card $(B) < \aleph_\alpha$. We claim that $f(A) \subseteq M$. For, there is an ℓ-homomorphism $g : B \longrightarrow M$ which agrees with f on $f^{-1}(M)$. If $x \in A$, then $xD \subseteq f^{-1}(M)$ for some essential right ideal D since $f^{-1}(M)$ is essential in A. So $(f(x) - g(x))D = 0$ and $f(x) = g(x) \in M$. Thus $f^{-1}(M) = A$ and g extends f.

(c) Let $x \in K$ and let A be the ℓ-submodule of K generated by x. Since A is the union of the chain of its submodules $A_0, A_1, \ldots$ where $A_0 = xR + \mathbb{Z}x$ and A_{n+1} is the submodule generated by the sublattice generated by A_n (see Theorem 2.2.4 (f)), necessarily card $(A) < \aleph_\alpha$. From the previous paragraph, with $A = B$ and $f : A \longrightarrow K$ the inclusion map, we get $x \in M$.

(d) Suppose $0 \neq N$ is a totally ordered nonsingular strict f-module with card $(N) < \aleph_\alpha$. Then $M \underleftarrow{\times} N$ is a nonsingular f-module. If card $(M) < \aleph_\alpha$, then card $(M \underleftarrow{\times} N) < \aleph_\alpha$ and it follows from (4.4.2) that M has an upper bound in M. Thus, $M = 0$, or else card $(M) \geq \aleph_\alpha$. □

There are three order theoretic properties that are needed to characterize the $\aleph_\alpha$-injective f-modules. The first involves pairs of elements. Let u and v be elements of the ℓ-group M. The element $y \in M$ *splits* v *from* u if

$$y \geq u^+, \; y \wedge u^- = 0, \text{ and } (v - y)^+ \wedge y = 0.$$

M is called *self-splitting* if for every $u, v \in M$ there is some $y \in M$ which splits v from u. The class of self-splitting ℓ-groups is clearly productive and homomorphically closed. Note that 0 splits v from u iff $u \leq 0$, and $y = u \vee v$ splits v from u if $u \geq 0$. Hence every representable ℓ-group (f-module) is embeddable in a self-splitting ℓ-group (f-module). Here is a stronger such embedding. Let us call the pair of subsets A and B of M *pairwise disjoint* if $A \subseteq B^\perp$. M is called *pairwise almost-*$\aleph_\alpha$*-complete* if every pair of pairwise disjoint subsets of M^+ of cardinality less than $\aleph_\alpha$ have disjoint upper bounds.

Theorem 4.4.3. *Let R be a po-domain which has a nonzero strict totally ordered right R-module, and let M_R be a representable f-module over R.*

(a) M can be embedded in an f-module L with the following property:

$$\forall u \in M, \exists y(u) \in L \text{ such that } \forall v \in M, y(u) \text{ splits } v \text{ from } u.$$

(b) If $\aleph_\alpha >$ card (R) and M is $\aleph_\alpha$-injective in $f\text{-}\mathcal{M}_R$, then M is self-splitting and is pairwise almost-$\aleph_\alpha$-complete.

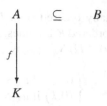

Proof. (a) Let K be a strict totally ordered right R-module and take $0 < k \in K$. Embed M in a product of totally ordered R-modules: $M \subseteq \Pi_{\lambda\in\Lambda} M_\lambda$. For each $\lambda \in \Lambda$ let $N_\lambda = M_\lambda \overleftarrow{\times} K$, and define $y(u) \in \Pi_\lambda N_\lambda = L$ by

$$y(u)_\lambda = \begin{cases} 0 & \text{if } u_\lambda \leq 0 \\ (0,k) & \text{if } u_\lambda > 0. \end{cases}$$

Then, clearly, $y(u) \geq u^+$ and $y(u) \wedge u^- = 0$. Moreover, if $y(u)_\lambda > 0$, then $v_\lambda < y(u)_\lambda$ and hence $[(v - y(u))^+ \wedge y(u)]_\lambda = 0$. Thus, $(v - y(u))^+ \wedge y(u) = 0$ and $y(u)$ splits v from u.

(b) Let $u, v \in M$ and let N be an f-module extension of M which contains an element y that splits v from u. Let A (respectively, B) be the ℓ-submodule generated by u and v (respectively, u, v, y). Then $\text{card}(B) < \aleph_\alpha$ and the injection of A into M can be extended to an ℓ-homomorphism f of B into M. But then $f(y)$ splits v from u.

Now, let A_1 and A_2 be pairwise disjoint subsets of M^+ with $\text{card}(A_1 \cup A_2) < \aleph_\alpha$. For $j = 1, 2$ let t_j be defined in L, the f-module given in (a), by:

$$\pi_\lambda(t_j) = \begin{cases} 0 & \text{if } \pi_\lambda(A_j) = 0 \\ (0,k) & \text{if } \pi_\lambda(A_j) \neq 0 \end{cases}$$

where π_λ denotes the projection of L onto N_λ. Then $t_1 \wedge t_2 = 0$ and $t_j \geq A_j$ for $j = 1, 2$. Let A be the ℓ-submodule generated by $A_1 \cup A_2$ and let B be the ℓ-submodule generated by $A \cup \{t_1, t_2\}$. Then, again, $\text{card}(B) < \aleph_\alpha$ and if $f : B \longrightarrow M$ extends the injection of A into M we have $f(t_j) \geq A_j$ and $f(t_1) \wedge f(t_2) = 0$. □

Note that an $\aleph_\alpha$-injective f-module over any po-ring whose cardinality is smaller than $\aleph_\alpha$ is self-splitting iff it can be embedded in a self-splitting f-module. In particular, a representable $\aleph_\alpha$-injective f-module over such a po-ring is self-splitting. The third order theoretic property that is needed also involves pairs of small subsets. The poset P is called an *η_α-set* (respectively, an *almost-η_α-set*) if whenever X and Y are subsets of P with $X < Y$ (X or Y could be empty) and $\text{card}(X \cup Y) < \aleph_\alpha$, then there exits an element p in P such that $X < p < Y$ (respectively, $X \leq p \leq Y$). For example, each lattice is an almost-η_0-set but it need not be an η_0-set. Also, a po-group is an almost η_0-group exactly when it is a directed Riesz group (see Exercise 2.1.18).

The infinite cardinal number $\aleph_\alpha$ is called *regular* provided $\text{card}(\cup_{\lambda\in\Lambda} X_\lambda) < \aleph_\alpha$ whenever $\{X_\lambda : \lambda \in \Lambda\}$ is a family of sets with $\text{card}(\Lambda) < \aleph_\alpha$ and $\text{card}(X_\lambda) < \aleph_\alpha$ for each $\lambda \in \Lambda$. If $\aleph_\alpha$ is not regular it is called *singular*. For example, $\aleph_0$ is regular and $\aleph_{\omega_0}$ is singular. Since $\aleph_{\alpha+1}$ is regular each cardinal is exceeded by a regular cardinal (Exercise 29). For the ordinal α, $W(\alpha) = \{\beta : 0 \leq \beta < \alpha\}$ denotes the initial segment of the ordinals determined by α and ω_α denotes the first ordinal number whose cardinality is $\aleph_\alpha$. Also, ord (T) stands for the ordinal number of the well-ordered set T.

Theorem 4.4.4. *A totally ordered module M_R can be embedded in a totally ordered module L_R with the property that whenever $X < Y$ for subsets X and Y of M, there* *is an element z in L with $X < z < Y$. Consequently, M can be embedded in a totally ordered η_α-module for any ordinal α. Moreover, if M is $\mathbb{F}$-torsion-free for some pretopology $\mathbb{F}$ of R and $\mathbb{F}$ has a basis of finitely generated right ideals, then L and the η_α-module may be chosen to be $\mathbb{F}$-torsion-free.*

Proof. By replacing M, if necessary, by its divisible closure, we may assume that if $x < y$ in M, then $x < z < y$ for some $z \in M$. Suppose $X < Y$ in M and no element of M is between X and Y. Then either X is nonempty and has no last element or Y is nonempty and has no first element. We assume the former but note that a similar construction works for the latter. For each $x \in X$ let $U_x = \{z \in X : z > x\}$. Then U_x is nonempty and $\mathscr{A} = \{U_x : x \in X\}$ is a chain of subsets of X. The dual ideal (or filter) of the power set $\mathscr{P}(X)$ of X generated by $\mathscr{A}$ is proper, and hence, by Zorn's Lemma, $\mathscr{A}$ is contained in a maximal dual ideal (or ultrafilter) $\mathscr{U}$ of $\mathscr{P}(X)$. Let $K = \{f \in M^X : f^{-1}(0) \in \mathscr{U}\}$. Then K is a prime submodule of the f-module M^X. For, suppose $h,k \in K$, $f,g \in M^X$ and $a \in R$. Then $(h-k)^{-1}(0) \supseteq h^{-1}(0) \cap k^{-1}(0)$, $(ha)^{-1}(0) \supseteq h^{-1}(0)$, and $(h^+)^{-1}(0) \supseteq h^{-1}(0)$; also, if $0 \le f \le h$, then $f^{-1}(0) \supseteq h^{-1}(0)$. So K is a convex ℓ-submodule of M^X. If $f \wedge g = 0$ and $f^{-1}(0) \notin \mathscr{U}$, then $g^{-1}(0) \supseteq X \backslash f^{-1}(0)$ and hence $g^{-1}(0) \in \mathscr{U}$ since $X \backslash f^{-1}(0) \in \mathscr{U}$ by Exercise 3. Note that $0 \le f + K$ in $N = M^X/K$ iff $\mathrm{Pos}(f) = \{x \in X : f(x) \ge 0\} \in \mathscr{U}$. For, $0 \le f + K$ iff $f^- \in K$ and $\mathrm{Pos}(f) = (f^-)^{-1}(0)$. Now, M can be diagonally embedded in M^X and in $N : m \mapsto f_m \in M^X$ where $f_m(x) = m$ for each $x \in X$. Let $\varphi : M \longrightarrow N$ be the composite of the diagonal embedding followed by the natural map: $\varphi(m) = f_m + K$. If $e : X \longrightarrow M$ is the inclusion map, then $\varphi(X) < e + K < \varphi(Y)$. For, if $m \in X$, then $\mathrm{Pos}(e - f_m) = U_m \cup \{m\} \in \mathscr{U}$ and hence $e + K \ge f_m + K$; but $(e - f_m)^{-1}(0) = \{m\} \notin \mathscr{U}$ and hence $e + K > f_m + K$. On the other hand, if $m \in Y$, then $\mathrm{Pos}(f_m - e) = X$ and $(f_m - e)^{-1}(0) = \emptyset$; so $e + K < f_m + K$. Suppose M is $\mathbb{F}$-torsion-free, $f \in M^X$, and $fD \subseteq K$ for some $D \in \mathbb{F}$ which is generated by $d_1, \ldots, d_n \in R$. Then $f^{-1}(0) = \bigcap_{i=1}^n (fd_i)^{-1}(0) \in \mathscr{U}$ and hence $f \in K$. So N is $\mathbb{F}$-torsion-free.

We can now proceed by transfinite induction. Let $\{(X_\beta, Y_\beta)\}_{\beta < \alpha_0}$ be the collection of pairs of subsets of M with $X_\beta < Y_\beta$, indexed by the initial segment of ordinals $W(\alpha_0)$. For each $\beta < \alpha_0$ we will find a totally ordered module M_β such that $M \subseteq M_\gamma \subseteq M_\beta$ if $\gamma \le \beta$ and M_β contains an element z_β with $X_\beta < z_\beta < Y_\beta$. Let $M_0 = N$ be the totally ordered module constructed above for the pair $X_0 < Y_0$. If $\beta = \gamma + 1$, let $M_\beta = (M_\gamma)_0$ be the module constructed from M_γ for the pair $X_\beta < Y_\beta$, and if β is a limit ordinal let $M_\beta = (\bigcup_{\gamma<\beta} M_\gamma)_0$ be constructed similarly from $\bigcup_{\gamma<\beta} M_\gamma$. Then clearly $L = \varinjlim M_\beta$ has the desired beteweeness property. Below we will denote L by $L(M)$.

To embed M in an η_α-module we will again use transfinite induction. Note that there is no harm in assuming that $\aleph_\alpha$ is a regular cardinal. Let $T_0 = M$ and, for $1 \le \beta < \omega_\alpha$, let $T_\beta = L(M_\gamma)$ if $\beta = \gamma + 1$, and otherwise let $T_\beta = \varinjlim_{\gamma<\beta} T_\gamma$. Now let $T = \varinjlim T_\beta$. If $Z \subseteq T$ with $\mathrm{card}\,(Z) < \aleph_\alpha$, then $Z \subseteq T_\beta$ for some $\beta < \omega_\alpha$ since each $z \in Z$ is in some T_{β_z} and $\beta = \bigvee_z \beta_z < \omega_\alpha$ by the regularity of $\aleph_\alpha$ (see Exercise 31(a)). In particular, if $X < Y$ in T with $\mathrm{card}\,(X \cup Y) < \aleph_\alpha$, then $X \cup Y \subseteq T_\beta$ for some $\beta < \omega_\alpha$. Hence, $X < z < Y$ for some $z \in T_{\beta+1} \subseteq T$. □

Following the pattern given in Theorem 4.4.3, the previous result can be used to show that an $\aleph_\alpha$-injective f-module which is representable is an almost η_α-module. The details are left for Exercise 9. The following result is useful for extending homomorphisms into totally ordered modules. It is a refinement of the easier Exercise 2.4.23.

Theorem 4.4.5. *Let G be an Ω-ℓ-subgroup of the representable Ω-f-group H. Suppose $x \in H$ and $x \wedge (g-x)^+ = 0$ for each $g \in G$. Suppose also that P is a prime Ω-subgroup of G such that $x^{\perp_G} \subseteq P$ and x is not an element of $C^H_\Omega(P)$, the convex ℓ-Ω-subgroup of H generated by P. Then there is a prime Ω-subgroup Q of H with $Q \cap G = P$ and $x \notin Q$.*

Proof. Let

$$\mathscr{S} = \{A \in \mathscr{C}_\Omega(H) : x \notin A, A \cap G = P, \text{ and } x \wedge g \in A \text{ with } g \in G^+ \text{ implies } g \in P\}.$$

We will show that $C^H_\Omega(P) \in \mathscr{S}$. Clearly $C^H_\Omega(P) \cap G = P$. Suppose $0 \leq x \wedge g \leq u$ with $u \in P$ and $g \in G$. Then $(x-u)^+ \wedge (g-u)^+ = 0$. We claim that $x \wedge (g-u)^+ = 0$. Suppose $\overline{H}$ is a totally ordered homomorphic image of H with $\overline{x} > 0$. Then $(\overline{u} - \overline{x})^+ = 0$; that is, $\overline{u} \leq \overline{x}$. If $\overline{u} = \overline{x}$, then $0 = \overline{x} \wedge (2\overline{u} - \overline{x})^+ = \overline{x}$. Thus, $\overline{u} < \overline{x}$ and $(\overline{g} - \overline{u})^+ = 0$. Now, $(g-u)^+ \in x^{\perp_G} \subseteq P$, and so $g \in P$ since $0 \leq g \leq g \vee u = (g-u)^+ + u \in P$. By Zorn's Lemma, $\mathscr{S}$ has a maximal element Q. To see that Q is prime, suppose $h_1\ h_2 \in H \backslash Q$ with $h_1 \wedge h_2 = 0$. Then $C^H_\Omega(Q, h_i)$ properly contains Q and, for each i, one of the following holds:

$$x \leq q + W_1 h_i + \cdots + q + W_n h_i$$

$$0 \leq g_i \leq q + W_1 h_i + \cdots + q + W_n h_i$$

$$0 \leq x \wedge g_i \leq q + W_1 h_i + \cdots + q + W_n h_i$$

where $g_i \in G^+ \backslash P$, $q \in Q^+$ and $W_j \in \Omega^\infty$. Since the second inequality implies the third, and because of the symmetry of these inequalities with respect to the indices 1 and 2, the nine possible cases reduce to three. Suppose the third inequality holds for both h_1 and h_2. So for some $q \in Q$, $g_i \in G^+ \backslash P$, and $W_1, \ldots, W_n, V_1, \ldots, V_m \in \Omega^\infty$,

$$x \wedge g_1 \wedge g_2 \leq (q + W_1 h_1 + \cdots + q + W_n h_1) \wedge (q + V_1 h_2 + \cdots + q + V_m h_2)$$

$$\leq q + \cdots + q \wedge V_j h_2 + \cdots + W_k h_1 \wedge q + \cdots \in Q.$$

Thus $g_1 \wedge g_2 \in P$ and we have the contradiction $g_1 \in P$ or $g_2 \in P$. The other two cases are quite similar. □

We can now give a characterization of $\aleph_\alpha$-injective f-modules.

Theorem 4.4.6. *Let R be an irredundant semiprime right qf-ring and suppose $\aleph_\alpha$ is a regular cardinal number with $\aleph_\alpha > \operatorname{card}(R)$. The following statements are equivalent for the nonsingular essential R-f-module M.*

(a) M is an $\aleph_\alpha$-injective f-module.

(b) M is an injective R-module and an $\aleph_\alpha$-injective ℓ-group.
(c) M is an R-injective, self-splitting, almost-η_α-f-module and is pairwise almost-$\aleph_\alpha$-complete.

Proof. (a) $\Rightarrow$ (c). Since M_R is an if-module by Theorem 4.3.14 we have that M is R-injective by Theorem 4.4.2, and it is an almost η_α-module by Exercise 9. By the remark after Theorem 4.4.3 we have that M is self-splitting. Let $\{R_\lambda : \lambda \in \Lambda\}$ and $\{M_\lambda : \lambda \in \Lambda\}$ be the components of R and M, respectively. By Theorem 4.3.14 each R_λ is a totally ordered right Öre domain, and, by Exercise 11, $M = \Pi_\lambda M_\lambda$ and each M_λ is an $\aleph_\alpha$-injective f-module over R_λ. By Theorem 4.4.3 any two pairwise disjoint subsets of cardinality less than $\aleph_\alpha$ in each M_λ^+ have disjoint upper bounds, and hence the same is true of M^+.

(c) $\Rightarrow$ (a). Since $M = \Pi_\lambda M_\lambda$ and each of the four conditions on the R-f-module M is equivalent to the same condition on each R_λ-f-module M_λ, we may assume M is a vector lattice over the totally ordered division ring D. Consider a diagram

$$\begin{array}{ccc} G & \subseteq & H \\ \Big\downarrow{\scriptstyle \varphi} & & \\ M & & \end{array} \qquad (4.4.3)$$

in the category f-$\mathscr{M}_D$ with card $(H) < \aleph_\alpha$. Let $x \in H$. By Theorem 4.4.3 there is a vector lattice K generated by H and some element k such that k splits each element of G from x; so

(i) $k \wedge x^- = 0$,
(ii) $k \geq x^+$, and
(iii) $(g-k)^+ \wedge k = 0$ for each $g \in G$.

Let L be the ℓ-subspace of K_D generated by G and k. We will first extend φ to L. Assume $k \notin G$. Let $X = (k^{\perp_G})^+$ and $Y = C_D^K(k) \cap G^+$ where $C_D^K(Z)$ denotes the convex ℓ-subspace of K generated by Z. Note that $Y = \{g \in G : 0 \leq g < k\}$. For, if $g \in Y$ and $\overline{g} > \overline{k}$ in some totally ordered homomorphic image of K, then $\overline{k} = 0$ by (iii) and hence also $\overline{g} = 0$. Since card $(\varphi(X) \cup \varphi(Y)) < \aleph_\alpha$ and $\varphi(X) \subseteq \varphi(Y)^\perp$ there are disjoint elements z_1 and z_2 in M^+, with $z_1 \geq \varphi(Y)$ and $z_2 \geq \varphi(X)$. Since card $(\varphi(G)) < \aleph_\alpha$ there exists $w \in M$ with $w \geq \varphi(G)$. Let $y \in M$ split w from $z = z_1 - z_2$. So

(iv) $y \geq z_1$,
(v) $y \wedge z_2 = 0$, and
(vi) $(w-y)^+ \wedge y = 0$.

The goal is to extend φ to L by mapping k to y. We will accomplish this by considering the totally ordered homomomorphic images of M. Let $\psi : M \longrightarrow \overline{M}$ be a homomorphism onto the totally ordered vector lattice $\overline{M}$. We will consider the two cases $\psi(y) = 0$ and $\psi(y) > 0$. Suppose first that $\psi(y) = 0$. Now, $C_D^L(k) \cap G = Y - Y$, and since $y \geq z_1 \geq \varphi(Y)$, $\psi\varphi(C_D^L(k) \cap G) = 0$. In fact, $\ker \psi\varphi = W \cap G$ where $W = C_D^L(k) + C_D^L(\ker \psi\varphi)$. For, if $g \in G^+ \cap W$, then $0 \leq g \leq kd + u$ with $0 \leq u \in \ker \psi\varphi$ and $0 \leq d \in D$. But then $(g-u)^+ \in Y \subseteq \ker \psi\varphi$, and hence $g \in \ker \psi\varphi$ since $0 \leq g \leq g \vee u = (g-u)^+ + u \in \ker \psi\varphi$. Since $L/W = G + W/W \cong G/\ker \psi\varphi$ it is clear that $\psi\varphi$ can be extended to L by sending k to 0.

$$\begin{array}{ccc} G & \subseteq & L \\ {\scriptstyle \psi\varphi}\downarrow & & \downarrow \\ \psi\varphi(G) & \cong & L/W \\ \downarrow & & \\ \overline{M} & & \end{array}$$

Suppose now that $\psi(y) > 0$. Then $\psi(z_2) = 0$ by (v), and $k^{\perp_G} = X - X \subseteq \ker \psi\varphi$ since $z_2 \geq \varphi(X)$. Also, $k \notin C_D^L(\ker \psi\varphi)$ since, otherwise, for some $g \in G$, $k \leq g = (2g-k) - (g-k) \in k^\perp$ by (iii). Thus, by Theorem 4.4.5 we have an epimorphism ρ onto a totally ordered vector space S such that the diagram

$$\begin{array}{ccc} G & \subseteq & L \\ {\scriptstyle \psi\varphi}\downarrow & & \downarrow{\scriptstyle \rho} \\ \psi\varphi(G) & \longrightarrow & S \\ \downarrow & & \\ \overline{M} & & \end{array}$$

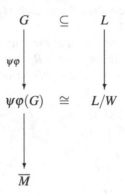

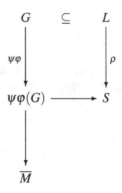

is commutative and $\rho(k) > 0$. From (iii) we see that $\rho(k) \geq \psi\varphi(G)$, and, in fact, $\rho(kd) \geq \psi\varphi(G)$ for each $0 < d \in D$; that is, $S = \psi\varphi(G) \underleftarrow{\times} \rho(k)D$. From (vi) we see that $\psi(w) \leq \psi(y)$ and hence $\psi\varphi(G) \underset{\leftarrow}{\times} \psi(y)D = \psi\varphi(G) + \psi(y)D \subseteq \overline{M}$. Thus, S embeds in $\overline{M}$ via a monomorphism σ that fixes $\psi\varphi(G)$ and sends $\rho(k)$ to $\psi(y)$:

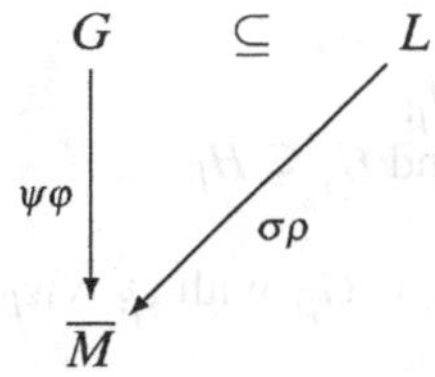

Now represent M as a subdirect product of totally ordered vector lattices $\psi : M \longrightarrow \Pi_{j\in J} M_j$. For each $j \in J$ we have a commutative diagram

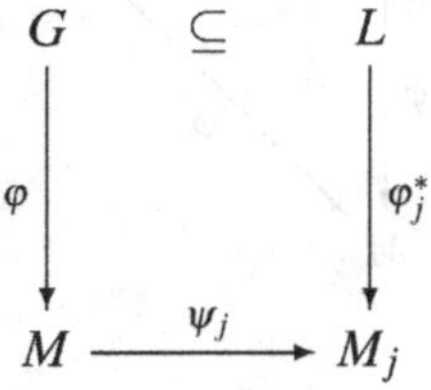

with $\varphi_j^*(k) = \psi_j(y)$. The φ_j^* induce a homomorphism $\varphi^* : L \longrightarrow \Pi M_j$ with $\varphi^*(k) = \psi(y)$ and whose restriction to G is $\psi\varphi$. Since $\varphi^*(L)$ is generated by $\varphi^*(G) = \psi\varphi(G)$ and $\varphi^*(k) = \psi(y)$ we have $\varphi^*(L) \subseteq \psi(M)$, and hence we obtain the commutative diagram

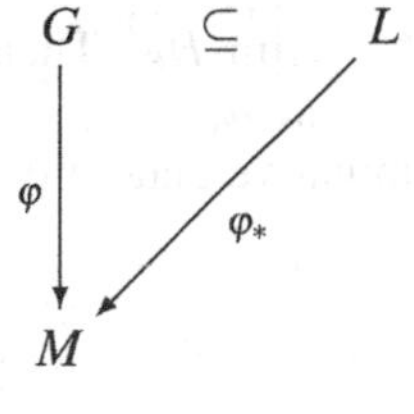

Now apply the preceding to $x = u - v$ where u and v are disjoint elements of H^+. We obtain a vector lattice K generated by H and k such that k splits each $g \in G$ from x, and we get a homomorphism φ_* from L to M which extends φ where L is the ℓ-subspace generated by G and k. Note that $k \geq u$ and $k \wedge v = 0$. Repeat, to get a vector lattice K' generated by K and k' such that k' splits each $w \in L$ from $v - k$, and to get a homomorphism $\varphi' : L' \longrightarrow M$ which extends φ_* where L' is the ℓ-subspace generated by L and k'. Note that $\text{card}(L') = \text{card}(G)$, $\text{card}(K') = \text{card}(H)$, $k \wedge k' = 0$, and $k' \geq v$.

Let $\{\{u_\beta, v_\beta\} : \beta < \alpha_0\}$ be the collection of all disjoint pairs of elements of H^+ indexed by the set of ordinals $W(\alpha_0)$. Note that card $(W(\alpha_0)) < \aleph_\alpha$. By transfinite induction we will show that for each $\beta < \alpha_0$ there are vector lattices G_β and H_β with the following properties.

(vii) $G \subseteq G_\beta \subseteq H_\beta$ and $H \subseteq H_\beta$.
(viii) If $\gamma \leq \beta$, then $G_\gamma \subseteq G_\beta$ and $H_\gamma \subseteq H_\beta$.
(ix) card $(H_\beta) < \aleph_\alpha$.
(x) There are elements $z_\beta, w_\beta \in G_\beta$ with $z_\beta \wedge w_\beta = 0$ such that $z_\beta \geq u_\beta$ and $w_\beta \geq v_\beta$.
(xi) There is a commutative diagram, and if $\gamma \leq \beta$, then φ_β is an extension of φ_γ.

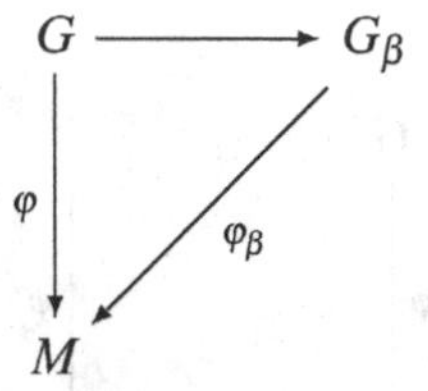

For $\beta = 0$ let $G_0 = L'$, $H_0 = K'$, $z_0 = k$, $w_0 = k'$ and $\varphi_0 = \varphi'$ in the preceding paragraph with $u = u_0$ and $v = v_0$. Suppose that $\beta < \alpha_0$ and G_γ and H_γ have been constructed with these properties for each $\gamma < \beta$. If $\beta = \gamma + 1$ apply the preceding construction with $G = G_\gamma$, $H = H_\gamma$, $u = u_\beta$, and $v = v_\beta$ to get G_β, H_β, z_β, w_β, and φ_β, while if β is a limit ordinal apply it to $G' = \varinjlim_{\gamma<\beta} G_\gamma = \bigcup_{\gamma<\beta} G_\gamma$ and $H' = \varinjlim_{\gamma<\beta} H_\gamma = \bigcup_{\gamma<\beta} H_\gamma$ with $u = u_\beta$ and $v = v_\beta$. Note that card $(H') < \aleph_\alpha$ since $\aleph_\alpha$ is regular.

Now let $G^* = \varinjlim_{\beta<\alpha_0} G_\beta$ and $H^* = \varinjlim_{\beta<\alpha_0} H_\beta$. Then $G \subseteq G^* \subseteq H^*$, $H \subseteq H^*$, card $(H^*) < \aleph_\alpha$ and we have a commutative diagram

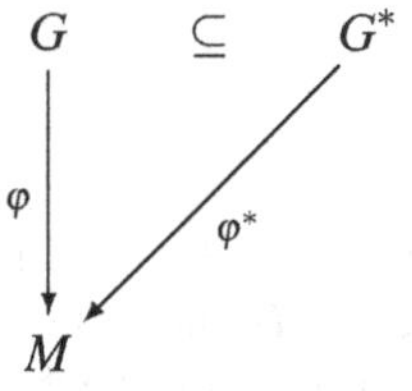

Moreover, if $u, v \in H$ with $u \wedge v = 0$, then there exist $z, w \in G^*$ with $z \geq u$, $w \geq v$ and $z \wedge w = 0$. It is desirable to extend this last statement by replacing H by H^*. For this purpose we need to enlarge G^* and H^* by iterating their construction. That is, by induction there are vector lattices G^{**} and H^{**} with $G^* \subseteq G^{**} \subseteq H^{**}$, $H^* \subseteq H^{**}$, card $(H^{**}) < \aleph_\alpha$, together with a commutative diagram

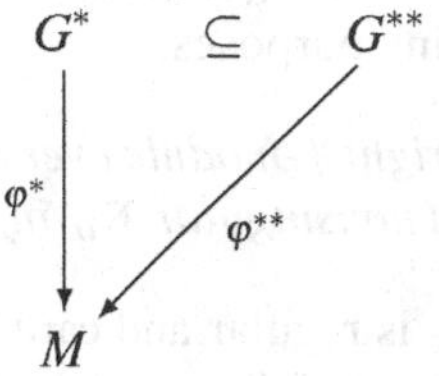

such that each pair of positive disjoint elements in H^{**} have disjoint upper bounds in G^{**}. Explicitely, let $G_0^* = G$, $H_0^* = H$, $G_{n+1}^* = (G_n^*)^*$, $H_{n+1}^* = (H_n^*)^*$, and $G^{**} = \varinjlim G_n^*$ and $H^{**} = \varinjlim H_n^*$.

By Zorn's Lemma we may assume that φ^{**} cannot be extended to any ℓ-subspace of H^{**} which properly contains G^{**}. Assuming $G^{**} \subset H^{**}$, and changing notation, we may start over with a diagram (4.4.3), but now H is generated by G and some element $x \in H^+ \backslash G$, φ cannot be extended to H and any two disjoint elements in H^+ have disjoint upper bounds in G. Note that the latter property implies that the convex ℓ-subspace of H generated by a prime subspace of G is a prime subspace of H. Let $V = \{g \in G : g < x\}$ and $U = \{g \in G : x < g\}$. Since M is an almost η_α-space there is an element $y \in M$ with $\varphi(V) \leq y \leq \varphi(U)$. Again, consider a representation $\psi : M \longrightarrow \Pi_{j\in J} M_j$ of M as a subdirect product of totally ordered vector lattices M_j with projections ψ_j, and let $P_j = \ker \psi_j\varphi$ and let Q_j be the convex ℓ-subspace of H generated by P_j. Then we have the commutative diagrams

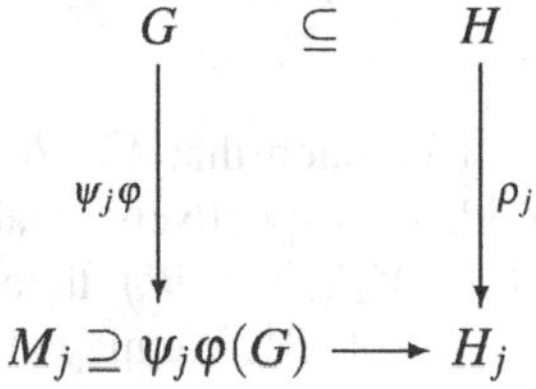

where ρ_j maps H onto the totally ordered vector lattice $H_j = H/Q_j$ which is generated by $\psi_j\varphi(G)$ and $\rho_j(x)$. We claim that for $g \in G$, $\psi_j\varphi(g) = \rho_j(g) \leq \rho_j(x)$ iff $\psi_j\varphi(g) \leq \psi_j(y)$. For, if $\rho_j(g) \leq \rho_j(x)$, then $g \leq x + p$ for some $0 \leq p \in P_j$. Thus, $g - p \in V$, $\varphi(g-p) \leq y$, $(\varphi(g) - y)^+ \leq \varphi(p)$, and $(\varphi(g) - y)^+ \in \ker \psi_j$; that is, $\psi_j\varphi(g) \leq \psi_j(y)$. Similarly, if $\rho_j(x) \leq \rho_j(g)$, then $\psi_j(y) \leq \psi_j\varphi(g)$. Thus, the mapping $\theta_j : H_j \longrightarrow M_j$ defined by $\psi_j\varphi(g) + \rho_j(x)d \mapsto \psi_j\varphi(g) + \psi_j(y)d$ is an embedding of H_j into M_j. These θ_j induce a homomorphism θ of H into $\Pi_j M_j$ which extends $\psi\varphi$ and sends x to $\psi(y)$. Since $\theta(H) \subseteq \psi(M)$ we obtain the contradiction that φ can be extended to H.

(a) $\Leftrightarrow$ (b). This is a consequence of the equivalence of (a) with (c) since (c) gives that M is a divisible group. □

Now that we have a description of $\aleph_\alpha$-injectives we can use it to show that there are enough of them for embedding purposes.

Theorem 4.4.7. *A nonsingular right f-module over an irredundant semiprime right qf-ring R can be embedded in a nonsingular $\aleph_\alpha$-injective f-module.*

Proof. We may assume that $\aleph_\alpha$ is regular and card $(R) < \aleph_\alpha$. By Theorem 4.3.14 a nonsingular f-module over R is a subdirect product of totally ordered nonsingular f-modules each of which is a nonsingular f-module over a totally ordered right Öre domain. Thus, utilizing Theorem 4.4.6 and Exercise 11, it suffices to note that according to Theorem 4.4.4 a totally ordered vector lattice can be embedded in a totally ordered η_α-vector lattice. □

When the f-module M in Theorem 4.4.6 is totally ordered the order requirements for it to be $\aleph_\alpha$-injective reduce to it being an almost-η_α-module. For other categories in which being an (almost-)η_α-set is the main order requirement for $\aleph_\alpha$-injectivity see Exercises 11–20. In particular, according to Exercise 23, the module M in Theorem 4.4.6 is $\aleph_\alpha$-injective in po-$\mathscr{M}_R$ whenever it is $\aleph_\alpha$-injective in f-$\mathscr{M}_R$, and the converse holds if M is totally ordered.

There are examples of $\aleph_\alpha$-injectives other than those constructed in Theorem 4.4.7. One such example is given in Exercise 25 for po-modules. Another construction of totally ordered η_α-groups uses Hahn products.

Theorem 4.4.8. *Suppose $\aleph_\alpha$ is a regular cardinal and Δ is a totally ordered η_α-set. For each $\delta \in \Delta$ let G_δ be equal to $\mathbb{R}$ or to $\mathbb{Z}$. Then the subgroup $V_\alpha = \{v \in V : card(supp\ v) < \aleph_\alpha\}$ of the Hahn product $V = V(\Delta, G_\delta)$ is a totally ordered η_α-group.*

Proof. Let A and B be subsets of V_α such that $A < B$ and card $(A \cup B) < \aleph_\alpha$, and let A_* (respectively, B^*) be the ideal (respectively, dual ideal) of the poset V_α generated by A (respectively, B). If $v \in V_\alpha \backslash (A_* \cup B^*)$, then $A_* < v < B^*$. Assume then, by way of contradiction, that $V_\alpha = A_* \cup B^*$. We may also assume that $0 \in A_*$ since, otherwise, $0 \in -B^* = (-B)_*$ and we may replace A and B by $-B$ and $-A$, respectively. For $0 \neq v \in V$ let $m(v) = \max \operatorname{supp} v$ and let $m(0) = -\infty < \Delta$. Clearly, $m(v+w) \leq m(v) \vee m(w)$ with equality if $m(v) \neq m(w)$. If $B = \emptyset$, then A is cofinal in V_α. In this case take $\delta \in \Delta$ with $m(A^+) < \delta$. Then we have the contradiction $A < x^\delta$ where, for $r \in G_\delta, rx^\delta$ denotes the function in V_α whose value at δ is r and whose value is 0 elsewhere. Thus, $B \neq \emptyset$. Note that $A \nleq 0$ since, otherwise, $A < x^\delta < B$ where $\delta < m(B)$.

We will establish the following by transfinite induction on $\sigma < \omega_\alpha$. There is a strictly decreasing sequence $(\delta_\rho^\sigma)_{\rho<\sigma}$ in Δ and a sequence $(r_\rho^\sigma)_{\rho<\sigma}$ in $\Pi_{\rho<\sigma} G_{\delta_\rho^\sigma}$ with each $r_\rho^\sigma > 0$ such that if $v^\sigma = \Sigma_{\rho<\sigma} r_\rho^\sigma x^{\delta_\rho^\sigma}$, then $A_\sigma \neq \emptyset$ and $B_\sigma \neq \emptyset$ where A_σ and B_σ are given by

$$A_\sigma = \{a \in A : \forall \rho < \sigma,\ m(a - v^\sigma) < \delta_\rho^\sigma\},$$
$$B_\sigma = \{b \in B : \forall \rho < \sigma,\ m(b - v^\sigma) < \delta_\rho^\sigma\}.$$

Suppose that $\tau < \omega_\alpha$ and the assertion is true for each $\sigma < \tau$. Note that an element $c \in A \cup B$ belongs to $A_\sigma \cup B_\sigma$ if and only if $c(\delta_\rho^\sigma) = r_\rho^\sigma = v^\sigma(\delta_\rho^\sigma)$ for each $\rho < \sigma$, and $c(\delta) = 0$ if $\delta \notin \{\delta_\rho^\sigma : \rho < \sigma\}$ and $\delta > \delta_\rho^\sigma$ for some $\rho < \sigma$. In particular, $m(c) = \delta_0^\sigma$ for such an element c. The sequences $(\delta_\rho^\sigma)_{\rho<\sigma}$ and $(r_\rho^\sigma)_{\rho<\sigma}$ are independent of σ. To see this take $\pi < \rho$, $\sigma < \tau$ and let $a_\sigma \in A_\sigma$, $a_\rho \in A_\rho$, $b_\sigma \in B_\sigma$ and $b_\rho \in B_\rho$. If $\delta_0^\sigma > \delta_0^\rho$, then $(a_\sigma - b_\rho)(\delta_0^\sigma) = r_0^\sigma > 0$ and $(a_\sigma - b_\rho)(\delta) = 0$ for $\delta > \delta_0^\sigma$; so we have the contradiction $a_\sigma > b_\rho$. Thus, $\delta_0^\sigma = \delta_0^\rho$. If $r_0^\sigma > r_0^\rho$, then, again, $a_\sigma > b_\rho$ since $(a_\sigma - b_\rho)(\delta_0^\sigma) = r_0^\sigma - r_0^\rho > 0$ and $(a_\sigma - b_\rho)(\delta) = 0$ for $\delta > \delta_0^\sigma$. Now, suppose $\delta_\mu^\rho = \delta_\mu^\sigma$ and $r_\mu^\rho = r_\mu^\sigma$ for all $\mu < \pi$. If $\delta_\pi^\sigma > \delta_\pi^\rho$, then $\delta_\pi^\sigma \notin \{\delta_\nu^\rho : \nu < \rho\}$ and, again, $a_\sigma - b_\rho > 0$ since $(a_\sigma - b_\rho)(\delta_\pi^\sigma) = r_\pi^\sigma > 0$ and $m(a_\sigma - b_\rho) = \delta_\pi^\sigma$. So $\delta_\pi^\sigma = \delta_\pi^\rho$ and also $r_\pi^\sigma = r_\pi^\rho$. Otherwise $r_\pi^\sigma > r_\pi^\rho$ and again $a_\sigma > b_\rho$. Thus the definitions $\delta_\rho = \delta_\rho^\sigma$ and $r_\rho = r_\rho^\sigma$, for $\rho < \sigma$, are well-defined. Note that $A_\sigma \supseteq A_\pi$ and $B_\sigma \supseteq B_\pi$ if $\pi \geq \sigma$.

First, suppose τ is a successor ordinal; so $\tau = \sigma + 1$ and we need to construct $\delta_\sigma \in \Delta$ and $0 < r_\sigma \in G_{\delta_\sigma}$ so that $A_\tau \neq \emptyset$ and $B_\tau \neq \emptyset$. Now, A_σ is cofinal in A. If not, $A_\sigma < a$ for some $a \in A$ and with $\delta = m(a - v^\sigma)$ we have $\delta \geq \delta_\rho$ for some $\rho < \sigma$. But then for $a_\sigma \in A_\sigma$ and $b_\sigma \in B_\sigma$ we have $a_\sigma(\gamma) = b_\sigma(\gamma) = v^\sigma(\gamma)$ for each $\gamma \geq \delta$, and hence $\delta = m(a - a_\sigma) = m(a - b_\sigma)$ and $0 < (a - a_\sigma)(\delta) = (a - b_\sigma)(\delta) < 0$. Similarly, B_σ is coinitial in B. Let $A'_\sigma = A_\sigma - v^\sigma$ and $B'_\sigma = B_\sigma - v^\sigma$. Since $A'_\sigma \not< 0$ or $0 \not< B'_\sigma$, either $A'^{+}_\sigma \neq \emptyset$ or $B'_\sigma \cap -V_\alpha^+ \neq \emptyset$, and by symmetry we may assume the former; so $A'^{+}_\sigma < B'_\sigma$ and $0 < B'_\sigma$. Let $\Gamma_1 = m(A'^{+}_\sigma)$ and $\Gamma_2 = m(B'_\sigma)$ and note that $\Gamma_1 \leq \Gamma_2$. If $\Gamma_1 < \Gamma_2$, then there exists $\delta \in \Delta$ with $\Gamma_1 < \delta < \Gamma_2$. But then $A'^{+}_\sigma < x^\delta < B'_\sigma$, $A'_\sigma < x^\delta < B'_\sigma$, and hence $A < v^\sigma + x^\delta < B$. Thus, $\Gamma_1 \cap \Gamma_2 = \{\delta\}$ with $\delta \in \Delta$ and $\delta < \{\delta_\rho : \rho < \sigma\}$ since $\delta = m(b_\sigma - v^\sigma)$ for some $b_\sigma \in B_\sigma$. So the decreasing sequence $(\delta_\rho)_{\rho<\sigma}$ is extended by letting $\delta_\sigma = \delta$. Now let

$$D_1 = \{r \in G_{\delta_\sigma} : \exists a' \in A'^{+}_\sigma \text{ with } m(a) = \delta_\sigma \text{ and } r \leq a'(\delta_\sigma)\}$$

and

$$D_2 = \{r \in G_{\delta_\sigma} : \exists b' \in B'_\sigma \text{ with } m(b') = \delta_\sigma \text{ and } b'(\delta_\sigma) \leq r\}.$$

Since $0 \leq A'^{+}_\sigma < B'_\sigma$ we have $D_1 \leq D_2$, and clearly D_1 is an ideal and D_2 is a dual ideal of the poset G_{δ_σ}. If $D_1 < r < D_2$ for some $r \in G_{\delta_\sigma}$, then $A < v^\sigma + rx^{\delta_\sigma} < B$. For, let $a - v^\sigma \in A'^{+}_\sigma$ and $b - v^\sigma \in B'_\sigma$ with $m(a - v^\sigma) = m(b - v^\sigma) = \delta_\sigma$. Then $a - v^\sigma < rx^{\delta_\sigma} < b - v^\sigma$ since $(a - v^\sigma)(\delta_\sigma) \in D_1$ and $(b - v^\sigma)(\delta_\sigma) \in D_2$. Now if $a_1 - v^\sigma \in A'^{+}_\sigma$ and $b_1 - v^\sigma \in B'_\sigma$, then $m(a_1 - v^\sigma) \leq \delta_\sigma \leq m(b_1 - v^\sigma)$, and if we have the strict inequalities in both cases, then $a_1 - v^\sigma < a - v^\sigma < rx^{\delta_\sigma} < b - v^\sigma < b_1 - v^\sigma$. So we do have $A < v^\sigma + rx^{\delta_\sigma} < B$ if such an r exists; hence, $G_{\delta_\sigma} = D_1 \cup D_2$. Let $r = \text{lub} D_1$. Then $r > 0$ since $0 < (a - v^\sigma)(\delta_\sigma) \in D_1$ where $a - v^\sigma$ is, as above, an element in A'^{+}_σ with $m(a - v^\sigma) = \delta_\sigma$. Suppose $D_1 \cap D_2 = \emptyset$. If $r \in D_1$, then $A''_\sigma \neq \emptyset$ where

$$A''_\sigma = \{a' \in A'^{+}_\sigma : m(a') = \delta_\sigma \text{ and } a'(\delta_\sigma) = r\},$$

and there exists some $\gamma \in \Delta$ with

$$\{m(a'' - rx^{\delta_\sigma}) : a'' \in A''_\sigma\} < \gamma < \delta_\sigma.$$

But now $A'^{+}_{\sigma} < rx^{\delta_\sigma} + x^\gamma < B'_\sigma$ contrary to our original assumption. For, if $a' \in A'^{+}_{\sigma}$ then either $m(a') < \delta_\sigma$, in which case $a' < rx^{\delta_\sigma}$, or $m(a') = \delta_\sigma$ and $a'(\delta_\sigma) < r$, in which case $a' < rx^{\delta_\sigma}$, or $a' \in A''_\sigma$ in which case $a' - rx^{\delta_\sigma} < x^\gamma$ and $a' < rx^{\delta_\sigma} + x^\gamma$. Also, if $b' \in B'_\sigma$ then either $m(b') > \delta_\sigma$, in which case $b' > rx^{\delta_\sigma} + x^\gamma$, or $m(b') = \delta_\sigma$ and then $b'(\delta_\sigma) > r$ and $b' > rx^{\delta_\sigma} + x^\gamma$. If $r \in D_2$ we get an element between A'^{+}_{σ} and B'_σ in a similar manner. Thus, $D_1 \cap D_2 \neq \emptyset$ and, of course, $D_1 \cap D_2 = \{r\}$. So the sequence $(r_\rho)_{\rho<\sigma}$ is extended by letting $r_\sigma = r$. Since r_σ is the largest element in D_1 and the smallest in D_2 there exist elements $a - v^\sigma \in A'^{+}_{\sigma}$ and $b - v^\sigma \in B'_\sigma$ with $m(a - v^\sigma) = m(b - v^\sigma) = \delta_\sigma$ and $(a - v^\sigma)(\delta_\sigma) = (b - v^\sigma)(\delta_\sigma) = r_\sigma$. So $m(a - v^\tau) = m(a - v^\sigma - r_\sigma x^{\delta_\sigma}) < \delta_\sigma$, and $m(b - v^\tau) = m(b - v^\sigma - r_\sigma x^{\delta_\sigma}) < \delta_\sigma$, and $a \in A_\tau$ and $b \in B_\tau$.

Now suppose τ is a limit ordinal. For the sequences $(\delta_\rho)_{\rho<\tau}$ and $(r_\rho)_{\rho<\tau}$ we need to check that $A_\tau \neq \emptyset$ and $B_\tau \neq \emptyset$. We first note that $A < v^\tau$ or A_τ is cofinal in A depending upon whether $A_\tau = \emptyset$ or $A_\tau \neq \emptyset$, respectively. If $A_\tau = \emptyset$ and $A \not< v^\tau$, then $a > v^\tau$ for some $a \in A$, and $m(a - v^\tau) \geq \delta_\sigma$ for some $\sigma < \tau$. As we saw in the previous paragraph $A_{\sigma+1}$ is cofinal in A; so we can find $a_{\sigma+1} \in A_{\sigma+1}$ with $v^{\sigma+1} < v^\tau < a \leq a_{\sigma+1}$, and hence $\delta_\sigma \leq m(a - v^\tau) \leq m(a_{\sigma+1} - v^{\sigma+1}) < \delta_\sigma$. On the other hand, if $A_\tau \neq \emptyset$ and $A_\tau < a$ for some $a \in A$, then $\delta = m(a - v^\tau) \geq \delta_\rho > \delta_{\rho+1}$ for some $\rho < \tau$. Take $a_\tau \in A_\tau$ and $b_{\rho+1} \in B_{\rho+1}$. Then $a_\tau(\gamma) = v^\tau(\gamma) = v^{\rho+1}(\gamma) = b_{\rho+1}(\gamma)$ for $\gamma \geq \delta$. So $\delta = m(a - a_\tau) = m(a - b_{\rho+1})$ and we have the contradiction $0 < (a - a_\tau)(\delta) = (a - b_{\rho+1})(\delta) < 0$. Similarly, $v^\tau < B$ or B_τ is coinitial in B depending upon whether $B_\tau = \emptyset$ or $B_\tau \neq \emptyset$. In particular, at least one of A_τ and B_τ is nonempty. Suppose $A_\tau = \emptyset$ and $B_\tau \neq \emptyset$. Take $\delta \in \Delta$ with $m(B_\tau - v^\tau) < \delta < \{\delta_\rho : \rho < \tau\}$. Then $A < v^\tau - x^\delta < B$. For, if $a \in A$, then $m(a - v^\tau) \geq \delta_\rho > \delta$ for some $\rho < \tau$ and $v^\tau - a > x^\delta$. Also, for $b_\tau \in B_\tau$ we have $v^\tau - b_\tau < x^\delta$ and $v^\tau - x^\delta < b_\tau$; so $v^\tau - x^\delta < B$.

The induction is now complete and we have the sequences $(\delta_\tau)_{\tau<\omega_\alpha}$ and $(r_\tau)_{\tau<\omega_\alpha}$ with $A_\tau \neq \emptyset$ and $B_\tau \neq \emptyset$ for each τ. If a belonged to every A_τ, then card (supp a) $\geq \aleph_\alpha$ since $\{\delta_\tau : \tau < \omega_\alpha\} \subseteq$ supp a. So the intersection of the A_τ is empty and for each $a \in A$ there is a minimal ordinal $\tau(a) < \omega_\alpha$ with $a \notin A_{\tau(a)}$. If $a \in A_\tau$, then $\tau(a) > \tau$; consequently, $\tau(A)$ is cofinal in $[0, \omega_\alpha)$. But now $\aleph_\alpha = \text{card}\,(\tau(A)) < \aleph_\alpha$ by the regularity of $\aleph_\alpha$; see Exercise 31. □

A construction similar to that given in Theorem 4.4.4 can be used to construct totally ordered (almost-)η_α-sets, or we can just quote Theorem 4.4.4 since a totally ordered set can be embedded in a totally ordered $\mathbb{Z}$-module; also see Exercises 13–18. Another construction uses lexicographic products. Recall that **2** denotes the 2-element Boolean algebra. The product $\mathbf{2}^{W(\beta)}$ supplied with the lexicographic order is totally ordered: $(a_\gamma)_{\gamma<\beta} < (b_\gamma)_{\gamma<\beta}$ if $a_\gamma < b_\gamma$ where γ is the first ordinal with $a_\gamma \neq b_\gamma$. Let

$$T_\alpha = \{(a_\gamma)_{\gamma<\omega_\alpha} \in \mathbf{2}^{W(\omega_\alpha)} : \exists \beta < \omega_\alpha \text{ with } a_\beta = 1 \text{ and } a_\gamma = 0 \text{ if } \beta < \gamma\}.$$

The cardinal $\aleph_\alpha$ is *admissible* if $2^{\aleph_\beta} \leq \aleph_\alpha$ for each $\beta < \alpha$. Note that $\aleph_{\beta+1}$ is admissible if and only if $2^{\aleph_\beta} = \aleph_{\beta+1}$.

Theorem 4.4.9. *(a) If $\aleph_\alpha$ is regular, then T_α is a totally ordered η_α-set.*
(b) For any ordinal α, $T_{\alpha+1}$ is a totally ordered $\eta_{\alpha+1}$-set of cardinality $2^{\aleph_\alpha}$.
(c) If $\aleph_\alpha$ is regular and admissible, then T_α has cardinality $\aleph_\alpha$.

Proof. (a) Let A and B be subsets of T_α with $A < B$ and $\mathrm{card}(A \cup B) < \aleph_\alpha$. For $a = (a_\gamma)_{\gamma<\omega_\alpha} \in T_\alpha$ let $f(a)$ be the ordinal with $a_{f(a)} = 1$ and $a_\beta = 0$ if $\beta > f(a)$. The regularity of $\aleph_\alpha$ implies the existence of the ordinal $\gamma_A < \omega_\alpha$ with $\cup_{a\in A} W(f(a)+1) = W(\gamma_A)$. So γ_A is the least ordinal such that $a_\beta = 0$ whenever $\beta \geq \gamma_A$ and $(a_\beta) \in A$. We now inductively define the terms of a sequence $x = (x_\gamma)_{\gamma<\omega_\alpha} \in T_\alpha$ with $A < x < B$. If $a_0 = 1$ for some $a \in A$, then $x_0 = 1$, and otherwise $x_0 = 0$. Suppose $\rho < \omega_\alpha$ and x_γ has been defined for all $\gamma < \rho$. Define

$$x_\rho = \begin{cases} 1 \text{ if } \exists a \in A \text{ with } a_\rho = 1 \text{ and } a_\beta = x_\beta \text{ for each } \beta < \rho, \\ 1 \text{ if } \rho = \gamma_A \vee \gamma_B, \\ 0 \text{ otherwise.} \end{cases}$$

Clearly $x = (x_\gamma)_{\gamma<\omega_\alpha} \in T_\alpha$ and $A < x$. Suppose there is some $b = (b_\gamma)_{\gamma<\omega_\alpha}$ in B with $b \leq x$. Then $b < x$ since $b_{\gamma_A \vee \gamma_B} = 0$ while $x_{\gamma_A \vee \gamma_B} = 1$. Let γ_0 be the first ordinal with $b_{\gamma_0} \neq x_{\gamma_0}$; so $b_{\gamma_0} = 0$ and $x_{\gamma_0} = 1$. If $\gamma_0 < \gamma_A$, then A contains an element $a = (a_\gamma)_{\gamma<\omega_\alpha}$ such that $a_{\gamma_0} = 1$ and $a_\gamma = x_\gamma$ if $\gamma \leq \gamma_0$. Then $a_\gamma = b_\gamma$ if $\gamma < \gamma_0$ and $b_{\gamma_0} < a_{\gamma_0}$; so $b < a$, which is impossible. Thus, $\gamma_0 \geq \gamma_A$ and $b_\gamma = x_\gamma$ if $\gamma < \gamma_A$. Suppose $f(b) < \gamma_A$. Then $x_{f(b)} = 1$ and A contains an element $a = (a_\gamma)_{\gamma<\omega_\alpha}$ such that $a_\gamma = x_\gamma = b_\gamma$ for all $\gamma \leq f(b)$. Thus, either $a = b$, or if β is the first ordinal where they differ, then $\beta > f(b)$, $b_\beta = 0$, and $b < a$. So we must have $\gamma_A \leq f(b) < \gamma_B$. Since $\gamma_0 \geq \gamma_A$ and $x_{\gamma_0} = 1$ the definition of x yields that $\gamma_0 = \gamma_A \vee \gamma_B$ and also $x_{f(b)} = 0$ since $\gamma_A \leq f(b) < \gamma_B = \gamma_A \vee \gamma_B$; but this contradicts the minimality of γ_0 since $b_{f(b)} = 1$ and $f(b) < \gamma_0$.

(b) and (c). It is easy to see that $\mathrm{card}(T_0) = \aleph_0$ and, in fact, T_0 is isomorphic to $\mathbb{Q}$ by Exercise 1.1.2. We will check that the cardinality of T_α is $\sum_{\beta<\alpha} 2^{\aleph_\beta}$ if $0 < \alpha$. To see this consider the partition $\{f^{-1}(\gamma) : \gamma < \omega_\alpha\}$ of T_α determined by the function $f : T_\alpha \longrightarrow W(\omega_\alpha)$ given in (a). Note that $f^{-1}(\gamma) = \{a = (a_\beta)_{\beta<\omega_\alpha} : a_\gamma = 1$ and $a_\beta = 0$ if $\beta > \gamma\}$ and $\mathrm{card}(f^{-1}(\gamma)) = \mathrm{card}(2^{W(\gamma)}) = 2^{\mathrm{card}(\gamma)}$. Thus, if $\mathrm{card}(\gamma) = \aleph_\beta$, then $\beta < \alpha$ and $\mathrm{card}(f^{-1}(\gamma)) = 2^{\aleph_\beta}$. Since

$$\mathrm{card}\{\gamma : \mathrm{card}(\gamma) = \aleph_\beta\} = \mathrm{card}(W(\omega_{\beta+1}) - W(\omega_\beta)) = \aleph_{\beta+1}, \tag{4.4.4}$$

we have

$$\begin{aligned} \mathrm{card}(T_\alpha) &= \sum_{\gamma<\omega_\alpha} \mathrm{card}(f^{-1}(\gamma)) = \aleph_0 + \sum_{\omega_0 \leq \gamma < \omega_\alpha} \mathrm{card}(f^{-1}(\gamma)) \\ &= \sum_{\beta<\alpha} \sum_{\mathrm{card}(\gamma)=\aleph_\beta} 2^{\aleph_\beta} = \sum_{\beta<\alpha} 2^{\aleph_\beta} \aleph_{\beta+1} = \sum_{\beta<\alpha} 2^{\aleph_\beta}. \end{aligned} \tag{4.4.5}$$

In particular, since $\aleph_{\alpha+1}$ is regular, $T_{\alpha+1}$ is a totally ordered $\eta_{\alpha+1}$-set of cardinality $2^{\aleph_\alpha}$ because $\alpha \leq \omega_\alpha$ (Exercise 33) and

$$2^{\aleph_\alpha} \leq \sum_{\beta\leq\alpha} 2^{\aleph_\beta} \leq 2^{\aleph_\alpha}\aleph_\alpha = 2^{\aleph_\alpha}.$$

Also, if $\aleph_\alpha$ is regular and admissible, then

$$\sum_{\beta<\alpha} 2^{\aleph_\beta} \leq \sum_{\beta<\alpha} \aleph_\alpha \leq \operatorname{card}(\alpha)\aleph_\alpha = \aleph_\alpha. \tag{4.4.6}$$

But we always have $\aleph_\alpha \leq \sum_{\beta<\alpha} 2^{\aleph_\beta}$. For if $\alpha = \gamma + 1$, then $\aleph_\alpha \leq 2^{\aleph_\gamma} \leq \sum_{\beta\leq\gamma} 2^{\aleph_\beta}$, and if α is a limit ordinal, then

$$\aleph_\alpha = \bigvee_{\beta<\alpha} \aleph_\beta \leq \sum_{\beta<\alpha} 2^{\aleph_\beta}.$$

□

According to Exercise 32 a totally ordered set of cardinality at most $\aleph_\alpha$ can be embedded in any totally ordered η_α-set, and hence a totally ordered η_α-set of cardinality $\aleph_\alpha$ is minimal among totally ordered η_α-sets; moreover, any two are isomorphic. If the restricted Hahn product V_α of Theorem 4.4.8 is constructed using a totally ordered η_α-set of cardinality $\aleph_\alpha$, then it too will have cardinality $\aleph_\alpha$. Before verifying this we will first show the necessity of the regularity and admissibility of $\aleph_\alpha$ for the existence of a totally ordered η_α-set of cardinality $\aleph_\alpha$.

A partially ordered set is called *$\aleph_\alpha$-free* if each of its subsets contains a coterminal subset whose cardinality is smaller than $\aleph_\alpha$. This concept will be used in the next result. It is connected to $\aleph_\alpha$-injectivity in the category Poset; see Exercise 14.

Theorem 4.4.10. *There exists a totally ordered η_α-set whose cardinality is $\aleph_\alpha$ if and only if $\aleph_\alpha$ is regular and admissible.*

Proof. One direction is given by Theorem 4.4.9. Suppose that Δ is a totally ordered η_α-set with $\operatorname{card}(\Delta) = \aleph_\alpha$. That $\aleph_\alpha$ must be regular follows from Exercise 31. As an aid in showing that $\aleph_\alpha$ must be admissible we will first establish that $\mathbf{2}^{W(\omega_\alpha)}$ (for any α) does not contain any subsets isomorphic to $W(\omega_{\alpha+1})$ or to its dual $W(\omega_{\alpha+1})^0$. Suppose, to the contrary, that $f : W(\omega_{\alpha+1}) \longrightarrow \mathbf{2}^{W(\omega_\alpha)}$ is an embedding with image $\{f_\beta : \beta < \omega_{\alpha+1}\}$. For each $\beta < \omega_{\alpha+1}$ let $\varphi(\beta) \in W(\omega_\alpha)$ be minimal with respect to $f_\beta(\varphi(\beta)) \neq f_{\beta+1}(\varphi(\beta))$; so $f_\beta(\varphi(\beta)) = 0$ and $f_{\beta+1}(\varphi(\beta)) = 1$. Since $W(\omega_{\alpha+1}) = \bigcup_{\lambda<w_\alpha} \varphi^{-1}(\lambda)$ there is some $\lambda < \omega_\alpha$ with $\operatorname{card}(\varphi^{-1}(\lambda)) = \aleph_{\alpha+1}$, and we may assume that this λ is minimal for all such embeddings f. Clearly, $W(\omega_{\alpha+1})$ is isomorphic to its well-ordered subset $\varphi^{-1}(\lambda)$; let $s : W(\omega_{\alpha+1}) \longrightarrow \varphi^{-1}(\lambda)$ be an isomorphism. Then $f \circ s$ is another embedding of $W(\omega_{\alpha+1})$ into $\mathbf{2}^{W(\omega_\alpha)}$ with image $\{f_{s(\beta)}\}$. For each β, since $f_{s(\beta)+1} \leq f_{s(\beta+1)}$ while $f_{s(\beta)+1}(\lambda) = 1$ and $f_{s(\beta+1)}(\lambda) = 0$, we have $f_{s(\beta)+1} < f_{s(\beta+1)}$, and if $\psi(\beta)$ is the first ordinal where $f_{s(\beta)+1}$ and $f_{s(\beta+1)}$ differ, then $\psi(\beta) < \lambda$. Thus, as above, there is some $\mu < \lambda$ such that $\text{card}(\psi^{-1}(\mu)) = \aleph_{\alpha+1}$. Consider the sequence obtained from $(f_{s(\beta)})_{\beta<\omega_{\alpha+1}}$ by placing $f_{s(\beta)+1}$ after $f_{s(\beta)}$; that is, define $h : W(\omega_{\alpha+1}) \longrightarrow \mathbf{2}^{W(\omega_\alpha)}$ by (see Exercise 30) $h(\beta) = f_{s(\beta)}$ if β is even and $h(\beta) = f_{s(\beta-1)+1}$ if β is odd. Then for this embedding $\text{card}(\varphi^{-1}(\mu)) = \aleph_{\alpha+1}$ and this contradicts the minimality of λ. Similarly, $W(\omega_{\alpha+1})^0$ cannot be embedded in $\mathbf{2}^{W(\omega_\alpha)}$. As a consequence we see that $\mathbf{2}^{W(\omega_\alpha)}$ is $\aleph_{\alpha+1}$-free since each subset X of $\mathbf{2}^{W(\omega_\alpha)}$ has a well-ordered cofinal subset Y by Exercise 1.1.8, and the ordinal number of Y is smaller than $\omega_{\alpha+1}$. Hence $\text{card}(Y)$ is no bigger than $\aleph_\alpha$. Dually, X has a coinitial subset whose cardinality is not bigger than $\aleph_\alpha$. Now, if $\alpha = \delta + 1$, then $\mathbf{2}^{W(\omega_\delta)}$ can be embedded in Δ by Exercise 15, and hence $2^{\aleph_\delta} \leq \text{card}(\Delta) = \aleph_{\delta+1} \leq 2^{\aleph_\delta}$. If α is a limit ordinal and $\beta < \alpha$, then $\beta + 1 < \alpha$, Δ is an $\eta_{\beta+1}$-set, $\mathbf{2}^{W(\omega_\beta)}$ can be embedded in Δ and $2^{\aleph_\beta} \leq \text{card}(\Delta) = \aleph_\alpha$. So in both cases $\aleph_\alpha$ is admissible. □

The preceding argument implies that any totally ordered $\eta_{\alpha+1}$-set T has cardinality at least $2^{\aleph_\alpha}$. For, $\mathbf{2}^{W(\omega_\alpha)}$ is $\aleph_{\alpha+1}$-free and hence it can be embedded in T (Exercise 15). In particular, this applies to a totally ordered η_α-set when $\aleph_\alpha$ is singular since it then must be an $\eta_{\alpha+1}$-set (Exercise 31(d)).

Returning to groups we have

Theorem 4.4.11. *Let $\alpha > 0$ and let Δ be a totally ordered η_α-set of cardinality $\aleph_\alpha$. Then the Hahn group V_α of Theorem 4.4.8 is a totally ordered η_α-group of cardinality $\aleph_\alpha$.*

Proof. Since $V_\alpha \subseteq V_\alpha(\Delta, \mathbb{R})$ and $\aleph_\alpha \leq \text{card}(V_\alpha)$ because V_α contains the characteristic functions of points, we only need to verify that $\text{card}(V_\alpha) \leq \aleph_\alpha$ when $V_\alpha = V_\alpha(\Delta, \mathbb{R})$. Let $\varphi : W(\omega_\alpha) \longrightarrow \Delta$ be a bijection of sets. If $v \in V_\alpha$, then by the regularity of $\aleph_\alpha$ (see Exercise 31(a)) there is some ordinal $\gamma < \omega_\alpha$ with $\varphi^{-1}(\text{supp}\, v) < \gamma$. Thus V_α is the union of the W_γ where $W_\gamma = \{v \in V_\alpha : \varphi^{-1}(\text{supp}\, v) < \gamma\}$. Since $\text{card}(W_\gamma) = \text{card}(V_\alpha(\varphi(W(\gamma)), \mathbb{R}))$ we have $\text{card}(W_\gamma) \leq 2^{\aleph_0 \text{card}(\gamma)}$, and, using (4.4.4) and (4.4.6) we have, as in (4.4.5),

$$\text{card}(V_\alpha) \leq \sum_{\gamma<\omega_\alpha} 2^{\aleph_0 \text{card}(\gamma)} = \sum_{\beta<\alpha} \aleph_{\beta+1} 2^{\aleph_\beta} \leq \aleph_\alpha. \tag{4.4.7}$$

□

In addition to large products of ultraproducts, and of Hahn products, there are other examples of $\aleph_\alpha$-injective f-modules. The next result is crucial to our presentation of these examples.

Theorem 4.4.12. *Let $\{M_n : n \in \mathbb{N}\}$ be a sequence of nonzero Riesz groups and let $\overline{M} = \Pi M_n / \oplus M_n$. Then $\overline{M}$ is an almost η_1-group. If each M_n is an ℓ-group, then $\overline{M}$ is pairwise almost $\aleph_1$-complete. Both statements also hold for homomorphic images of $\overline{M}$.*

Proof. Let $\overline{A}$ and $\overline{B}$ be countable subsets of $\overline{M}$ with $\overline{A} < \overline{B}$. By Exercise 2.2.19 there are subsets A and B of ΠM_n such that $A < B$ and $A \longrightarrow \overline{A}$ and $B \longrightarrow \overline{B}$ are order isomorphisms. Suppose $\overline{A} = \{\overline{a}_n : n \in \mathbb{N}\}, \overline{B} = \{\overline{b}_n : n \in \mathbb{N}\}$, $A = \{a_n : n \in \mathbb{N}\}$, $B = \{b_n : n \in \mathbb{N}\}$, $a_n \mapsto \overline{a}_n$ and $b_n \mapsto \overline{b}_n$. For each $n \in \mathbb{N}$ take $g_n \in M_n$ with $\{a_1(n), \ldots, a_n(n)\} \leq g_n \leq \{b_1(n), \ldots, b_n(n)\}$ and define $g \in \Pi M_n$ by $g(n) = g_n$. We claim that $\overline{A} \leq \overline{g} \leq \overline{B}$. To see this note that if $a_k(n) \nleq g(n) = g_n$ for some k and n, then $k > n$; that is, $\{n : a_k(n) \nleq g(n)\} \subseteq \{1, \ldots, k-1\}$. So if $h_k \in \Pi M_n$ is defined by $h_k(n) = 0$ if $n \geq k$ and $h_k(n) = -g(n) + a_k(n)$ if $n < k$, then $h_k \in \oplus M_n$ and $a_k \leq g + h_k$. Thus, $\overline{A} \leq \overline{g}$ and, similarly, $\overline{g} \leq \overline{B}$. This same argument works even if $\overline{B}$ (or $\overline{A}$) is the empty set; so $\overline{M}$ is an almost η_1-group.

Suppose each M_n is an ℓ-group and let $\overline{A} = \{\overline{a}_n\}$ and $\overline{B} = \{\overline{b}_n\}$ be two countable subsets of $\overline{M}$ with $\overline{a}_n \wedge \overline{b}_m = 0$ for each $n, m \in \mathbb{N}$. By replacing $\overline{a}_n$ by $\overline{a}_1 \vee \cdots \vee \overline{a}_n$ and $\overline{b}_n$ by $\overline{b}_1 \vee \cdots \vee \overline{b}_n$ we may assume that $\overline{A}$ and $\overline{B}$ are increasing sequences. By lifting $\overline{A} \cup \overline{B} \cup \{0\}$ to an isomorphic copy in ΠM_n we obtain increasing sequence $A = \{a_n\}$ and $B = \{b_n\}$ in ΠM_n and an element $z \in \oplus M_n$ with $a_n \wedge b_m = z$ for every n and m. By replacing A by $A - z$ and B by $B - z$ we may assume $z = 0$. Define $c \in \Pi M_n$ by $c(n) = a_n(n)$. Now, for each n, $\{m \in \mathbb{N} : c(m) \ngeq a_n(m)\} \subseteq \{1, \ldots, n-1\}$ since if $m \geq n$ then $c(m) = a_m(m) \geq a_n(m)$. Also, $c \in B^{\perp}$ since $(c \wedge b_m)(n) = (a_n \wedge b_m)(n) = 0$ for every n and m. Thus, $\overline{c} \in \overline{B}^{\perp}$ is an upper bound of $\overline{A}$. Now replace $\overline{A}$ by $\overline{B}$ and $\overline{B}$ by $\{\overline{c}\}$ to get an upper bound $\overline{d}$ of $\overline{B}$ with $\overline{d} \wedge \overline{c} = 0$.

Homomorphic images of $\overline{M}$ also have these properties since countable subsets can be lifted isomorphically to $\overline{M}$. □

We now introduce a class of po-rings for which Theorem 4.4.12 can be used to produce $\aleph_1$-injective f-modules. A function $\theta : R \longrightarrow T$ from the domain R into the well-ordered set T is called a *left division algorithm for R* if for all $a, b \in R$ with $b \neq 0$, there exist $q, r \in R$ such that

$$a = qb + r \text{ and } \theta(r) < \theta(b). \tag{4.4.8}$$

The domain R is called a *left Euclidean domain* if it has a left division algorithm. If R is a po-ring and θ is isotone on R^+, then θ is called a *left po-division algorithm* and R is called a *left po-Euclidean domain*. A *right (po)-Euclidean domain* is defined analogously. Just as in the commutative case a left Euclidean domain is a principal left ideal domain and is unital. Other properties are given in Exercises 36–39. Examples of left po-Euclidean domains include $\mathbb{Z}$, any po-division ring D with trivial $\theta : \theta(0) = 0$ and $\theta(R \backslash \{0\}) = 1$, as well as the polynomial ring $D[x]$ supplied with either the coordinatewise order or the lexicographic order and with θ being the degree function. This last example can be generalized to the left skew polynomial ring $D[x; \sigma, \delta]$ with coefficients written on the left of the powers of x. Here, σ is an injective po-endomorphism of D and δ is a *left σ-derivation* of $D : \delta \in \mathrm{End}_{\mathbb{Z}}(D)$ and for all $a, b \in D$, $(ab)^{\delta} = a^{\delta}b + a^{\sigma}b^{\delta}$ (δ is a *right σ-derivation* if $(ab)^{\delta} = a^{\delta}b^{\sigma} + ab^{\delta}$). The multiplication in $D[x; \sigma, \delta]$ is induced by the rule: $xa = a^{\sigma}x + a^{\delta}$ for $a \in D$. With this data $D[x; \sigma, \delta]$ is a left po-Euclidean domain in the lexicographic order, and it is left po-Euclidean in the coordinatewise order provided δ is also isotone. In the antilexicographic order $D[x; \sigma, \delta]$ is a totally ordered left Euclidean domain when D is totally ordered and σ and δ are isotone, but it is not po-Euclidean. When δ is a right σ-derivation the right skew polynomial ring with coefficients on the right will be denoted by $D[\delta,\sigma;x]$. If σ is an automorphism of D, then $D[x;\sigma,\delta] = D[-\sigma^{-1}\delta,\sigma^{-1};x]$ is also right po-Euclidean in the lexicographic order (see Exercises 41–45).

Theorem 4.4.13. *Let R be a totally ordered countable right Öre domain with totally ordered division ring of right quotients D.*

(a) If $\{M_n : n \in \mathbb{N}\}$ is a sequence of self-splitting vector lattices over D, then $\Pi M_n/\oplus M_n$ and all of its homomomorphic images are $\aleph_1$-injective R-f-modules.

(b) Suppose, additionally, that R is an ℓ-simple left po-Euclidean domain and has a left division algorithm θ with $\theta(a) = \theta(-a)$ for each a in R. Then $R^{\mathbb{N}}/C_R(1)$ and all of its homomorphic images are $\aleph_1$-injective right R-f-modules.

Proof. That (a) holds is an immediate consequence of Theorems 4.4.6 and 4.4.12. As for (b), we have by Theorem 4.4.12 that $R^{\mathbb{N}}/C_R(1)$ is a pairwise almost $\aleph_1$-complete almost η_1-ℓ-group since $R^{(\mathbb{N})} \subseteq C_R(1) = \{f \in R^{\mathbb{N}} : |f| \leq a \text{ for some } a \in R\}$. Suppose $f \in R^{\mathbb{N}}$ and $0 < a,\, b \in R$ with $|fa| \leq b$. Then $b \leq ca$ for some $c \in R$; so $|f| \leq c$ and $R^{\mathbb{N}}/C_R(1)$ is a nonsingular right R-module. To see that it is R-divisible take $f \in R^{\mathbb{N}}$ and $0 < a \in R$. Then $f(n) = q(n)a + r(n)$ with $\theta(r(n)) < \theta(a)$ for each n. Thus, $|r| < a$, $f + C_R(1) = (q + C_R(1))a$, and $R^{\mathbb{N}}/C_R(1)$ is an injective right R-module. □

Exercises.

1. Let y,u,v be elements of the ℓ-group M.

   (a) Show that y splits v from u iff y splits v^+ from u.

   (b) If $v^+ \geq u^+$ and y splits v from u show that $y \wedge v^+$ splits v from u.

   (c) Show that y splits v from u iff $y \in M^+$ and ωy splits v from u for every f-map ω of M with $\omega y \geq y$.

   (d) Suppose M is a subdirect product of the family of totally ordered groups $\{M_j : j \in J\}$. Show that the following are equivalent.

   (i) y splits v from u.

   (ii) $y \in M^+$ and for each $j \in J$,
   $$y_j > 0 \Rightarrow y_j \geq u_j \geq 0 \text{ and } y_j \geq v_j,$$
   $$y_j = 0 \Rightarrow u_j \leq 0.$$

   (iii) $y \in M^+$ and for each $j \in J$,
   $$u_j < 0 \Rightarrow y_j = 0,$$
   $$u_j > 0 \Rightarrow y_j \geq u_j \vee v_j,$$
   $$u_j = 0 \Rightarrow y_j = 0 \text{ or } y_j \geq v_j^+.$$

   (e) If M is representable and y and z split v from u show that $y+z$, $y \vee z$, and $y \wedge z$ all split v from u.

2. Let u be an element of the ℓ-group M and suppose $(u^-)^\perp$ is a summand of M. Show that if $v \in M$, then M contains an element that splits v from u. In

particular, a projectable ℓ-group is self-splitting. (If $v^+ = x+z$ with $x \in (u^-)^\perp$ and $z \in (u^-)^{\perp\perp}$ let $y = x \vee u^+$ and use Exercise 1(a).)

3. Let u be an element of the ℓ-group M. Show that the following statements are equivalent.

 (a) For every $v \in M$ there is an element $y \in M$ such that y splits v from u.
 (b) If $v \geq u^+$, then there are elements $v_1, v_2 \in M$ with $v = v_1 + v_2$, $v_1 \wedge v_2 = 0$, $u^- \wedge v_2 = 0$ and $v_2 \geq u^+$.
 (c) $M = \bigcup_{w \in M^+} [w^{\perp\perp} \cap (u^+)^\perp \oplus w^\perp \cap (u^-)^\perp]$.

 ((a) $\Rightarrow$ (b). Using Exercise 1(b) assume y splits v from u and $y \leq v$ and let $v_2 = y$. (b) $\Rightarrow$ (c). Let $z \in M$, let $v = |z| \vee u$ and let $w = v_1 \wedge |z|$. (c) $\Rightarrow$ (a). For $v \in M$ write $v \vee u^+ = v_1 + v_2$ and let $y = v_2$.)

4. Let M_R be a nonsingular f-module over the irredundant semiprime right qf-ring R.

 (a) If M is self-splitting show that its injective hull is self-splitting.
 (b) Show that the converse of (a) holds when R is pseudo-regular. (Use Exercise 1(d) and the fact that each component of R is ℓ-simple.)

5. Let Δ be the rooted po-set of proper prime subgroups of the self-splitting ℓ-group M. Show that distinct roots of Δ are disjoint. (Let Δ_1, Δ_2 be distinct roots, $P \in \Delta_1 \cap \Delta_2$, $P_1 \in \Delta_1 \backslash \Delta_2$, $P_2 \in \Delta_2 \backslash \Delta_1$, $x_1 \in P_2^+ \backslash P_1$, $x_2 \in P_1^+ \backslash P_2$, $x \in G^+ \backslash P$, $v = x \vee x_1 \vee x_2$, $u = x_1 - x_2$ and apply Exercise 3(b). Let Q be a value of v_2 containing P_1. Show that $P_2 \not\subseteq Q$, $Q \subset P$, and $v_1, v_2 \in P$.)

6. (a) Show that an ℓ-group which has a proper trunk is self-splitting iff it is totally ordered.
 (b) Show that the ℓ-group $M \underleftarrow{\times} T$ is not self-splitting if $T \neq 0$ and M is not totally ordered.
 (c) Suppose the Hahn group $V(\Gamma, G_\gamma)$ is an ℓ-group. Show that V is self-splitting iff each G_γ is self-splitting, Γ is a cardinal sum of chains and each minimal element of Γ is a root.

7. Let $\mathscr{A}$ be a proper dual ideal of the power set $\mathscr{P}(X)$ of the set X. Show that $\mathscr{A}$ is a maximal dual ideal if and only if, for each $A \in \mathscr{P}(X)$, either $A \in \mathscr{A}$ or $X \backslash A \in \mathscr{A}$.

8. Let M be a totally ordered nonsingular module over the essentially positive po-ring R. If $X < Y$ in M show that M can be embedded in a totally ordered nonsingular module N which contains an element u with $X \leq u \leq Y$.

9. Show that an $\aleph_\alpha$-injective representable f-module over a po-ring whose cardinality is exceeded by $\aleph_\alpha$ is an almost-η_α-module. (Use Theorem 4.4.4.)

10. Show that the direct product $M = \Pi_\lambda M_\lambda$ of f-modules is $\aleph_\alpha$-injective iff each M_λ is $\aleph_\alpha$-injective.

11. Let $\{R_\lambda : \lambda \in \Lambda\}$ be the components of the irredundant semiprime right qf-ring R and let $\{M_\lambda : \lambda \in \Lambda\}$ be the components (including 0) of the nonsingular right f-module M_R. Assume $\aleph_\alpha > \text{card}\,(R)$. Prove the equivalence of the following statements.

    (a) M_R is $\aleph_\alpha$-injective and essential.
    (b) $M = \Pi_\lambda M_\lambda$ and M_λ is $\aleph_\alpha$-injective in f-$\mathcal{M}_R$ for each $\lambda \in \Lambda$.
    (c) $M = \Pi_\lambda M_\lambda$ and M_λ is $\aleph_\alpha$-injective in f-$\mathcal{M}_{R_\lambda}$ for each $\lambda \in \Lambda$.

12. Let $\{E_j : j \in J\}$ be a family of nonsingular $\aleph_\alpha$-injective right f-modules over the irredundant semiprime right qf-ring R with $\text{card}\,(R) < \aleph_\alpha$. Show that $E = \{u \in \Pi_j E_j : \text{card}\,(\text{supp}\, u) < \aleph_\alpha\}$ is an $\aleph_\alpha$-injective f-module over R.

13. The triple (A,B,C) of subsets of the poset P is called a *formation in P* if $A < B$ and for all $(a,b,c) \in A \times B \times C$, $c \nleq a$ and $b \nleq c$. It is an *$\aleph_\alpha$-formation* if $\text{card}\,(A \cup B \cup C) < \aleph_\alpha$. The element $p \in P$ *splits* the formation (A,B,C) if $A < p < B$ and p is not comparable to any element in C. Show that each poset P can be embedded in a poset P_1 in which each formation in P is split. (Let $P_1 = P \cup \{(A,B) : A$ (respectively, B) is an ideal (respectively, dual ideal) of P and $A < B\}$, partially ordered by adding the following relations to the relations on $P : p \leq (A,B)$ if $p \in A, (A,B) \leq p$ if $p \in B$ and $(A,B) \leq (C,D)$ if $A \subseteq C$ and $D \subseteq B$. The formation (A,B,C) in P is split by (A_*,B^*) in P_1 where A_* (respectively, B^*) is the ideal (respectively, dual ideal) of P generated by A (respectively, B).)

14. An object E in a subcategory of Poset is called *coterminally $\aleph_\alpha$-injective* if each diagram of the form (4.4.1) in the category can be completed to a commutative diagram whenever B is $\aleph_\alpha$-free. Show that the following are equivalent for the poset E.

    (a) E is $\aleph_\alpha$-injective in the category Poset.
    (b) E is coterminally $\aleph_\alpha$-injective in Poset.
    (c) E is an almost η_α-set.

    (For (a) $\Rightarrow$ (c) use Exercise 13. For (c) $\Rightarrow$ (b) assume $B = A \cup \{x\}$ with $x \notin A$ and consider the sets $L_A(x)$ and $U_A(x)$ consisting of the lower and upper bounds, respectively, of x in A.)

15. Let $\mathscr{C}$ be the subcategory of Poset whose morphisms are the 1-1 isotone maps. Show that Exercise 14 holds for $\mathscr{C}$ provided "almost" is deleted from (c). ($U_A(x)$—or, alternatively, $L_A(x)$—needs to be slightly enlarged. Replace $U_A(x)$ by $\varphi^{-1}(V_E(\varphi(L_A(x))))$ where φ is the vertical map in (4.4.1) and $V_E(\varphi(L_A(x)))$ denotes the set of strict upper bounds of $\varphi(L_A(x))$.

16. Let $\mathscr{D}$ be the subcategory of Poset whose morphisms are the monomorphisms of Poset. Show that Exercise 14 holds for $\mathscr{D}$ provided (c) is changed to : Every $\aleph_\alpha$-formation in E is split in E (see Exercise 13).

17. Show that the poset E is injective in the category Poset iff E is complete (compare Exercise 4.1.40).

18. Show that each poset can be embedded in a poset that is $\aleph_\alpha$-injective in the category given in Exercise 15.

19. Let E be a totally ordered set and let $\mathscr{A}$ be one of the categories: Poset, $\mathscr{C}$ (in Exercise 15), $\mathscr{D}$ (in Exercise 16). Show that E is $\aleph_\alpha$-injective in $\mathscr{A}$ iff E is $\aleph_\alpha$-injective in the subcategory of $\mathscr{A}$ whose objects are the totally ordered sets.

20. Let R be a po-ring and let M_R be $\aleph_\alpha$-injective in the category po-$\mathscr{M}_R$ of po-modules. Suppose $\aleph_\alpha > \text{card}(R)$, $\aleph_0$.

    (a) Show that M is an injective R-module.
    (b) Let $x \in M$. Show that x is in the largest directed submodule of M iff the po-submodule generated by x can be embedded in a po-module which contains an upper bound for $\{-x, 0\}$. In particular, if R is a po-domain show that M is directed.
    (c) Suppose $x \in M$, D is a right ideal of R with $\ell(D;M) = 0$ and $xD^+ \subseteq M^+$. Show that $x \in M^+$. In particular, if $\mathbb{F}$ is a pretopology of R, then M is $\mathbb{F}$-torsion-free iff M is $\mathbb{F}^+$-semiclosed.

21. Suppose $\mathbb{F}$ is a positive pretopology of the po-ring R and M_R is an $\mathbb{F}$-torsion-free po-module. Show that M is $\aleph_\alpha$-injective in the category of po-R-modules iff it is $\aleph_\alpha$-injective in the category of $\mathbb{F}$-torsion-free po-R-modules.

22. Let R be a po-domain with $R^+ \neq 0$ and assume that $aR^+ \cap bR^+ \neq 0$ if $0 < a, b \in R$. Let M_R be a po-module.

    (a) If $xa \neq 0$ whenever $0 \neq x \in M$ and $0 < a \in R$ show that M^+ is contained in a (strict) total order of M_R.
    (b) Show that M^+ is the intersection of strict total orders of M_R iff M is R^+-*semiclosed*; that is, $x \in M^+$ whenever there is some $0 < a \in R$ with $xa \in M^+$ (see Exercises 2.1.7 and 2.1.8).
    (c) Suppose M is R^+-semiclosed. Show that Theorem 4.4.4 holds for M. More explicitly, show that M can be embedded in an R^+-semiclosed f-module in which each formation $(X, Y, \emptyset)$ of M splits (see Exercise 13), and M can be embedded in an R^+-semiclosed η_α-f-module; and these modules may be chosen to be $\mathbb{F}$-torsion-free if M is $\mathbb{F}$-torsion-free and $\mathbb{F}$ has a basis consisting of finitely generated right ideals.
    (d) Show that $R^+ \backslash \{0\}$ is a right Öre set in $S = R^+ - R^+$.

23. Let R be an irredundant semiprime right qf-ring and let M_R be an essential nonsingular po-module. Assume $\aleph_\alpha$ is regular and card $(R) < \aleph_\alpha$. Show that the following are equivalent.

    (a) M is $\aleph_\alpha$-injective in po-$\mathscr{M}_R$.
    (b) M is coterminally $\aleph_\alpha$-injective in po-$\mathscr{M}_R$.

(c) M_R is injective, $\mathbb{D}(R)^+$-semiclosed, and is an almost-η_α-module.
(d) M_R is injective and M is $\aleph_\alpha$-injective in po-$\mathscr{M}_{\mathbb{Z}}$. (Use Exercises 20–22 to mimic the proof of Theorem 4.4.6; it is easier here.)

24. Show that the previous exercise holds when the category is changed by requiring all morphisms to be 1-1 and "almost" is deleted from (c) (see Exercise 15).

25. (a) Let R be a directed po-unital po-ring. If G is an $\aleph_\alpha$-injective abelian po-group show that $\mathrm{Hom}_{\mathbb{Z}}(R,G)$ is an $\aleph_\alpha$-injective right R-po-module.
(b) Suppose R is directed and M_R is a po-module whose underlying po-group is semiclosed. Show that M can be embedded in an $\aleph_\alpha$-injective po-module. (Use Exercises 21 and 22.)

26. Let G be a totally ordered abelian (almost-)η_α-group. Show that its value set $\Gamma(G)$ is an (almost-)η_α-set.

27. Let G be an abelian almost-η_α-ℓ-group where $\alpha > 0$. If γ is a value of G show that G^γ/G_γ is isomorphic to $\mathbb{Z}$ or to $\mathbb{R}$. (Use Exercise 2.2.19 or 2.2.20.)

28. Let R be a semiprime right qf-ring. Verify that the following are equivalent. The object E in the category $\mathscr{C}$ is called *quasi-injective* if the diagram in (4.4.1) can be completed when $B = E$.

(a) $R = Q(R)$.
(b) R_R is quasi-injective in $\mathscr{M}_R$.
(c) R_R is quasi-injective in po-$\mathscr{M}_R$.
(d) R_R is quasi-injective in f-$\mathscr{M}_R$.

29. Show that $\aleph_{\alpha+1}$ is a regular cardinal for any ordinal α.

30. Show that each ordinal number is uniquely of the form $\tau + n$ where τ is a limit ordinal and n is a finite ordinal. $\tau + n$ is called *even* (*odd*) if n is even (odd).

31. (a) Show that $\aleph_\alpha$ is regular iff each subset of $W(\omega_\alpha)$ with fewer than $\aleph_\alpha$ elements has an upper bound in $W(\omega_\alpha)$, iff $W(\omega_\alpha)$ is an almost-η_α-set.
(b) Let T be a totally ordered set with card $(T) \leq \aleph_\alpha$. Show that T has a cofinal well-ordered subset S with ord $(S) \leq \omega_\alpha$. (If $T = \{t_\beta : \beta < \omega_\alpha\}$ let $S = \{t_\beta : \gamma < \beta \Rightarrow t_\gamma < t_\beta\}$.)
(c) If $\aleph_\alpha$ is singular show that the set T in (b) has a coterminal subset whose cardinality is less than $\aleph_\alpha$.
(d) If $\aleph_\alpha$ is singular show that each totally ordered η_α-set is an $\eta_{\alpha+1}$-set.

32. Let Q be an η_α-set.

(a) If P is a totally ordered set with card $(P) \leq \aleph_\alpha$ show that P can be embedded in Q. (If $P = \{b_\beta : \beta < \omega_\alpha\}$ use Exercise 15 and transfinite induction to construct an embedding.)
(b) If P and Q are totally ordered η_α-sets of cardinality $\aleph_\alpha$ show that P and Q are isomorphic. (Use Exercise 30 and the technique of Exercise 1.1.2.)

33. Show that $\alpha \leq \omega_\alpha$ for each ordinal α.

34. (a) Let $x, y \in \mathbf{2}^{W(\omega_\alpha)}$. Show that y is the successor of x ($y = x_+$) iff there is an ordinal $\tau < \omega_\alpha$ such that $x_\beta = y_\beta$ for $\beta < \tau$, $x_\tau = 0$ and $y_\tau = 1$, and, for $\sigma > \tau$, $x_\sigma = 1$ and $y_\sigma = 0$.
    (b) Show that $x \in T_\alpha$ iff x is the successor of some element of $\mathbf{2}^{W(\omega_\alpha)}$
    (c) Show that $\mathbf{2}^{W(\omega_\alpha)}$ is complete. (To construct the sup of a subset A, assume the components of $x = (x_\beta)_{\beta<\omega_\alpha}$ have been defined for $\beta < \gamma$, and let $x_\gamma = 1$ if, for some $a \in A$, $a_\beta = x_\beta$ for $\beta < \gamma$ and $a_\gamma = 1$; otherwise, let $x_\gamma = 0$.)
    (d) If $x < y$ in $\mathbf{2}^{W(\omega_\alpha)}$ and $y \neq x_+$ show that $x < u < y$ for some $u \in T_\alpha$.
    (e) Let Y be the subset of $\mathbf{2}^{W(\omega_\alpha)}$ consisting of all those elements that do not have a successor in $\mathbf{2}^{W(\omega_\alpha)}$. Show that Y is the completion of T_α.

35. If T is a totally ordered set with card $(T) \leq \aleph_\alpha$ show that T can be embedded in T_α. (If $T = \{t_\beta : \beta < \omega_\alpha\}$ define $\varphi : T \longrightarrow T_\alpha$ by $(\varphi(t_\beta))(\gamma) = 1$ iff $\gamma \leq \beta$ and $t_\gamma \leq t_\beta$.)

36. Let θ be a left division algorithm for the domain R and let $R^* = R \backslash \{0\}$. Show that $\theta(0) \leq \theta(R)$, and if $u \neq 0$ with $\theta(u) \leq \theta(R^*)$, then u is a unit of R.

37. Let $\theta : R \longrightarrow T$ be a function from the domain R into the well-ordered set T. Show that θ is a left division algorithm iff for every $a, b \in R$ with $b \neq 0$ and $\theta(b) \leq \theta(a)$, there exists some $q \in R$ with $\theta(a - qb) < \theta(a)$.

38. Let θ be a left division algorithm for R. Show that the remainder r in (4.4.7) is unique (and hence so is q) iff for every $a, b \in R$, $\theta(a - b) \leq \theta(a) \vee \theta(b)$ and for every $a, b \in R^*$, $\theta(b) \leq \theta(ab)$. (For only if: $a = 0(a - b) + a = 1(a - b) + b$ and $0 = -ab + ab = 0 \cdot b + 0$.)

39. For the subset X of the domain R let $X' = \{b \in R : Rb + X = R\}$. Let $R_0 = \{0\}$, and for the ordinal $\alpha > 0$ let $R_\alpha = R_\beta \cup R'_\beta$ if $\alpha = \beta + 1$ and let $R_\alpha = \bigcup_{\beta<\alpha} R_\beta$ if α is a limit ordinal. If $R_\tau = R$ for some ordinal τ, then R is called *transfinitely left Euclidean*; and if $\tau = \omega_0$, then R is called *left Euclidean*. For $b \in R$ let $\psi(b)$ denote the least ordinal α with $b \in R_\alpha$, provided such an ordinal exists.
    (a) Show that $\psi(b)$ is not a limit ordinal if $b \neq 0$.
    (b) If $R_\beta = R_{\beta+1}$ for some β show that $R_\beta = R_\gamma$ for every $\gamma \geq \beta$.
    (c) If R is transfinitely left Euclidean show that ψ is a left division algorithm for R and $\psi(b) \leq \psi(ab)$ for all $a, b \in R^*$.
    (d) If θ is an ordinal-valued or a $\mathbb{Z}^+$-valued left division algorithm for R and $\theta(b)$ is not a limit ordinal for $b \neq 0$, show that R is transfinitely left Euclidean or left Euclidean, respectively, and $\psi(b) \leq \theta(b)$ for every $b \in R$. (Show that $\theta^{-1}([0, \alpha]) \subseteq R_\alpha$.)

40. Let $R = D[x; \sigma, \delta]$ be a skew polynomial ring with coefficients in the division ring D. Use Exercise 37 to show that the degree function is a left division algorithm.

41. A *degree function* for the domain R is a function $d : R \longrightarrow \mathbb{Z}^+ \cup \{-\infty\}$ such that for all $a, b \in R$

$$d(0) = -\infty \text{ and } d(R^*) \subseteq \mathbb{Z}^+,$$

$$d(a-b) \leq d(a) \vee d(b),$$
$$d(ab) = d(a) + d(b).$$

If d is a degree function show that $d(1) = 0$ (if $1 \in R$), $d(a) = d(-a)$, and $d(a+b) = d(a) \vee d(b)$ if $d(a) \neq d(b)$.

42. Suppose that the domain R is generated by its unital subring A and an element x such that $R = \boxplus_{i=0}^{\infty} Ax^i$, and the function d given by $d(\sum a_i x^i) =$ largest i with $a_i \neq 0$ is a degree function. Show that $R = A[x; \sigma, \delta]$ where σ is a monic endomorphism of A and δ is a left σ-derivation of A.

43. Let d be a degree function for the domain R. Show that d is a left division algorithm iff there is a division ring D such that $R = D$, or $R = D[x; \sigma, \delta]$ with $d(x) > 0$. In the first case $d(R^*) = \{0\}$ and in the second case $d(f) = d(x) \deg f$. (Let $D = \{a \in R : d(a) \leq 0\}$; if $0 < d(x)$ is minimal show that $R = \boxplus_i Dx^i$ and use Exercises 41 and 42).

44. Show that the following are equivalent for the skew polynomial ring $R = D[x; \sigma, \delta]$ over the division ring D.

    (a) σ is an automorphism.
    (b) $R = D[-\sigma^{-1}\delta, \sigma^{-1}; x]$.
    (c) Every right ideal of R is principal.
    (d) R is a right Öre ring.

    (For (c) $\Rightarrow$ (d) use Exercise 4.1.26 and for (d) $\Rightarrow$ (a) consider $xR \cap axR$ where $a \in D$.)

45. Let R be a totally ordered domain. Show that $R = D[x; \sigma, \delta]$ is a lexicographically ordered skew polynomial ring over a totally ordered division ring D with σ isotone iff R has a degree function which is a left po-division algorithm and R has elements of positive degree.

46. Show that if E is injective in a subcategory $\mathscr{A}$ of the category of ℓ-groups, then E is $c\ell$-essentially closed in $\mathscr{A}$; that is, each $c\ell$-essential monomorphism $E \longrightarrow A$ in $\mathscr{A}$ is an isomorphism.

47. Let Z be a subring of the unital f-ring R with the same identity as R. Suppose Z contains an increasing sequence $T = \{a_1 < a_2 < \cdots\}$ which is cofinal in R, each a_n is a unit of R and $1 \leq a_1$. Let A be the convex ℓ-submodule of $(R^{\mathbb{N}})_R$ generated by $(a_n^{-1})_n = s$ and let B be the convex ℓ-submodule generated by 1.

    (a) Show that each unital R-ℓ-module is an f-module.
    (b) If M_R is an ℓ-module show that $\mathscr{C}_Z(M) = \mathscr{C}_R(M)$.
    (c) Show that $R^{(\mathbb{N})} \subset A \subset B \subset R^{\mathbb{N}}$.
    (d) Suppose M_Z is a unital ℓ-module and there is a nonzero Z-ℓ-homomorphism $\alpha : A \longrightarrow M$ with $R^{(\mathbb{N})} \subseteq \ker \alpha$. If there is a Z-ℓ-homomorphism $\beta : B \longrightarrow M$ extending α show that $\alpha(s)Z \leq \beta(1)$.

(e) Suppose R is archimedean and has a finite basis. Show that there are no nonzero injectives in each of the following full subcategories of u-f-$\mathscr{M}_Z$: all archimedean Z-f-modules, all archimedean Z-f-modules which are finitely generated as a convex ℓ-submodule, all archimedean Z-f-modules with a basis, all Z-f-modules which are complete ℓ-groups. (Reduce to the case that R is totally ordered and use the previous exercise to show that a nonzero injective in one of these categories contains $\mathbb{R}$ and hence it contains each ℓ-simple Z-f-module.)

(f) Suppose R is archimedean and has a basis. (Z may not exist.) Show that there are no nonzero essential injectives in each of the subcategories of the category of archimedean nonsingular R-f-modules corresponding to those categories (of Z-f-modules) listed in (e).

(g) Give examples of pairs (R,Z) for which R is neither archimedean, commutative, nor has a basis.

(h) Suppose R is an ℓ-primitive f-ring. Show that there are no nonzero injectives in any of the first three categories of unital *Z-archimedean* Z-f-modules listed in (e).

48. Let R be a countable irredundant semiprime right qf-ring, and suppose M_R is a nonsingular essential $\aleph_1$-injective f-module. If N is a closed convex ℓ-submodule of M show that M/N is $\aleph_1$-injective.

Notes. Ribenboim introduced the concept of an $\aleph_\alpha$-injective po-module in [RI2] and showed that po-$\mathscr{M}_R$ has no injectives when R is directed. Shatalova [SH3] removed this restriction by means of Theorem 4.4.1. The definition of a self-splitting ℓ-group, the characterization of an $\aleph_\alpha$-injective abelian ℓ-group given in Theorems 4.4.2–4.4.6, the embedding given in Theorem 4.4.7, the examples given in Theorems 4.4.12 and 4.4.13 for $\mathbb{Z}$-modules, and Exercises 2, 6, 13–16, and 18 and 19 come from Weinberg [WE7]; also see [WE8]. The generalization of these results to f-modules appears in Steinberg [ST3]. The definition of a totally ordered η_α-set comes from Hausdorff who also established several properties of these sets [HAU]. Theorem 4.4.9 is due to Sierpinski [SI] and the converse to (c) of Theorem 4.4.9, which is given in Theorem 4.4.10, is due to Gillman [GI]. Our presentation of this material is dependent on Rosenstein [RO, p. 163] where additional material and references may be found. Theorem 4.4.8 is due to Alling [AL2] and the proof given is due to Schwartz [SCH1]. Another proof is given in Ribenboim [RI1]. Exercises 1, 2, and 5, for the most part, come from Powell [PO2], Exercise 25 comes from Ribenboim [RI2] and Exercise 34 is in Gillman and Jerison [GJ, p. 189]. Exercises 36–44 are in Cohn [C2, p. 87] and Exercises 46 and 47 for the pair $\mathbb{Z}$ and $\mathbb{R}$ come from Conrad [CON8].

4.5 Free f-Modules

In contrast to the lack of injectives the category f-$\mathcal{M}_R$ (or u-f-$\mathcal{M}_R$ of unital f-modules) has an ample supply of projectives since it is a variety of Ω-algebras and hence has free objects; see Exercise 1.4.14. We wish to give a useful representation of a free f-module. More generally, we will construct free f-modules over po-modules. The free f-module will occur when the po-module is a trivially ordered free module. For modules over a totally ordered domain the situation is similar in quality to that of the preceding section in the sense that there always exists a free t-torsion-free f-module over a t-torsion-free po-module precisely when the ring is a right Öre domain. Also of interest is the fact that when the domain is archimedean each disjoint subset of a free f-module is countable, whereas an example using group rings or power series will show that this is not the case in general. For many f-rings R it will be seen that the free nonsingular f-modules are ℓ-submodules of the product of copies of R. By using free abelian ℓ-groups two lattice-ordered tensor products will be constructed.

Let M_R be a po-module over the po-ring R, and let $\varphi : M \longrightarrow F$ be a one-to-one po-homomorphism into the f-module F_R. The pair (F, φ) is called a *free f-module over M* if $\varphi(M)$ generates F as an f-module, and for each po-homomorphism $\psi : M \longrightarrow K$ into an f-module K there is an ℓ-homomorphism $\sigma : F \longrightarrow K$ with $\sigma\varphi = \psi$.

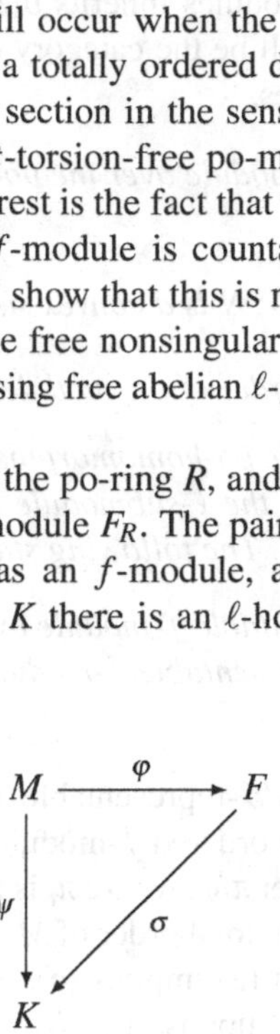

(4.5.1)

Since $\varphi(M)$ generates F the homomorphism σ is uniquely determined by φ and ψ. Moreover, if (F_1, φ_1) is another free f-module over M, then it is easily seen that there is a unique isomorphism $\sigma : F \longrightarrow F_1$ with $\sigma\varphi = \varphi_1$. If φ is an embedding, then F is called the free *f-module extension of M*. If $\mathcal{C}$ is a subcategory of f-$\mathcal{M}_R$, then the definitions of a *$\mathcal{C}$-free f-module over M* and of a $\mathcal{C}$-free *f-module extension of M* are obtained by stipulating that the morphism σ in (4.5.1) lies in $\mathcal{C}$. These free f-modules will also be called free $\mathcal{C}$-f-modules over M or extending M, respectively. In particular, suppose $\mathbb{F}$ is a positive topology of R and M is $\mathbb{F}$-torsion-free. If F is the free f-module over M or the free f-module extension of M, then $F/t_{\mathbb{F}}(F)$ is the $\mathbb{F}$-free f-module over M or the $\mathbb{F}$-free f-module extension of M, respectively, in the category of $\mathbb{F}$-torsion-free f-modules.

Let M_R be a submodule of the f-module V_R. Recall that the ℓ-submodule U of V generated by M is the union of a chain of submodules and a chain of R^+-ℓ-subgroups. Specifically, let $M_0 = M$ and let $L_0 = L(M)$ be the sublattice of V generated by M. Then $U = \bigcup_n M_n = \bigcup_n L_n$ where, inductively, M_{n+1} is the R-submodule generated by L_n and $L_{n+1} = L(M_{n+1})$.

Throughout, we will assume that $\mathscr{C}$ is a full subcategory of f-$\mathscr{M}_R$ and that the class of objects in $\mathscr{C}$ is productive, hereditary and contains each f-module that is isomorphic to a member of $\mathscr{C}$. An f-module in $\mathscr{C}$ is *$\mathscr{C}$-representable* if it is a subdirect product of a family of totally ordered modules each of which belongs to $\mathscr{C}$. The category of $\mathscr{C}$-representable f-modules inherits the previously mentioned properties of $\mathscr{C}$. For the most part $\mathscr{C}$ will be the category of all (unital) f-modules or the category of nonsingular f-modules.

Theorem 4.5.1. *Let M_R be a po-module over the po-ring R and let $\mathscr{C}$ be a subcategory of f-$\mathscr{M}_R$. Let*

$$\mathscr{S} = \{(M/N, P) \in \mathscr{C} : N \textit{ is a convex submodule of } M \textit{ and}$$

$$P \textit{ is a total order of the module } M/N \textit{ with } (M/N)^+ \subseteq P\}.$$

Let $\varphi : M \longrightarrow \Pi_{\mathscr{S}}(M/N, P)$ be the po-homomorphism induced by the natural maps $M \longrightarrow (M/N, P)$, and let F_M be the ℓ-submodule of the f-module $\Pi_{\mathscr{S}}(M/N, P)$ generated by its submodule $\varphi(M)$. The following statements are equivalent.

(a) There is a free $\mathscr{C}$-representable f-module over M.
(b) (F_M, φ) is the free $\mathscr{C}$-representable f-module over M.
(c) $\bigcap_{\mathscr{S}} N = 0$.

Proof. Suppose $(\overline{F}, \overline{\varphi})$ is the free $\mathscr{C}$-representable f-module over M and represent $\overline{F}$ as a subdirect product of totally ordered f-modules: $\overline{F} \longrightarrow \Pi_i F_i$; we will identify $\overline{F}$ with its image in $\Pi_i F_i$. If $N_i = \ker \pi_i \overline{\varphi}$, where π_i is the ith projection, then $\bigcap_i N_i = 0$ and $(M/N_i, P_i) \in \mathscr{S}$ where P_i is the total order of M/N_i induced by the isomorphism $M/N_i \cong \pi_i \overline{\varphi}(M)$. This shows that (a) implies (c), and we will complete the proof by showing that (c) implies (b). Suppose $\bigcap_{\mathscr{S}} N = 0$; then φ is a one-to-one po-homomorphism. Let $\psi : M \longrightarrow K$ be a po-homomorphism into the $\mathscr{C}$-representable f-module K. We may assume that K is totally ordered. For, K is a subdirect product of totally ordered R-modules: $K \subseteq \Pi K_i$. Let p_i denote the projection of ΠK_i onto K_i, and suppose, for each $i, \sigma_i : F_M \longrightarrow K_i$ is an ℓ-homomorphism with $\sigma_i \varphi = p_i \psi$. Then the induced ℓ-homomorphism $\sigma : F_M \longrightarrow \Pi_i K_i$ has its image in K and $\sigma\varphi = \psi$ since, first, $\sigma(L_0) = \sigma(L(\varphi(M))) = L(\sigma\varphi(M)) = L(\psi(M)) \subseteq K$, and, second, if $\sigma(L_n) \subseteq K$, then $\sigma(L_{n+1}) = \sigma(L(L_n + L_n R)) = L(\sigma(L_n + L_n R)) \subseteq K$. Now, assuming that K is totally ordered, let P_ψ be the total order of $M/\ker\psi$ induced by the module isomorphism $\psi_* : M/\ker\psi \cong \psi(M)$. Then $(M/\ker\psi, P_\psi) \in \mathscr{S}$ and ψ_* is an isomorphism of totally ordered modules. Define $\sigma : F_M \longrightarrow K$ to be the composite of the projection of F_M onto $(M/\ker\psi, P_\psi)$ followed by ψ_*; so $\sigma(u) = \psi_*(u_{P_\psi})$. Clearly, σ is an ℓ-homomorphism with $\sigma\varphi = \psi$. □

Note that for any po-module M, the f-module F_M given in Theorem 4.5.1 is the free $\mathscr{C}$-representable f-module over $M/\ker\varphi$, and if M is trivially ordered it is the free $\mathscr{C}$-representable f-module extension of $M/\ker\varphi$. We will now utilize Theorem

4.5.1 to construct some free f-modules. If F is the free $\mathscr{C}$-f-module on the set X and M is the R-submodule generated by X it is easy to see that F is the free $\mathscr{C}$-f-module extending (M, M^+), and it is also the free $\mathscr{C}$-f-module over the trivially ordered module M.

Theorem 4.5.2. *Let M_R be the trivially ordered free (unital) R-module with basis X over the (unital) po-ring R. Then the $\mathscr{C}$-free (unital) f-module over M, if it exists, is a $\mathscr{C}$-free (unital) f-module on X. In particular, if R is unital and it has a partial order for which it is a $\mathscr{C}$-representable f-module over R, and M is unital, then the f-module F_M that is constructed in Theorem 4.5.1 is the free unital $\mathscr{C}$-representable f-module on X.*

Proof. Given a function $f : X \longrightarrow K$ into the (unital) $\mathscr{C}$-f-module K we obtain the commutative diagram

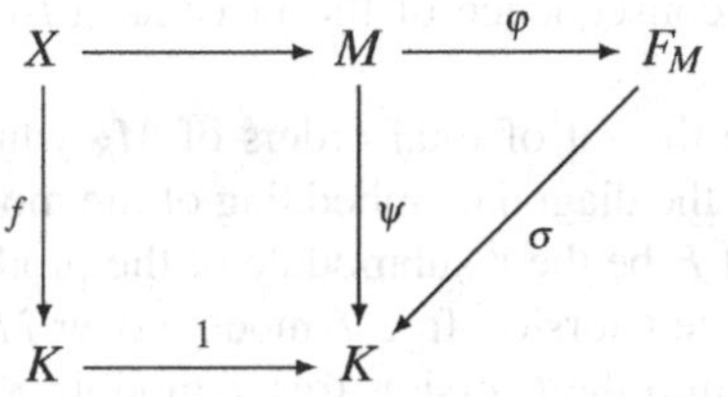

Here, ψ is the R-homomorphism extending f, which certainly is isotone, and σ is the ℓ-homomorphism extending ψ. Now suppose R is unital and (R,T) is a $\mathscr{C}$-representable R-f-module. Then the free unital module $M \cong \boxplus_X xR$ can be made into a $\mathscr{C}$-representable f-module, and hence F_M is the free unital $\mathscr{C}$-representable f-module with basis X since the condition (c) in Theorem 4.5.1 is satisfied. □

We note that if the right R-f-module $S = (R,T)$, which is assumed to exist in Theorem 4.5.2, is a right f-ring, then the free unital S-f-module on X is an R-f-module homomorphic image of the free unital representable R-f-module on X.

In many cases the free nonsingular f-module over a nonsingular po-module M can be constructed from the total orders of M instead of from the total orders of the homomorphic images of M. The module M_R over the domain R is called *t-torsion-free* if $xa = 0$ implies $x = 0$ or $a = 0$ for $x \in M$ and $a \in R$. Because a torsion-free module is now associated with a topology $\mathbb{F}$ we have introduced this new terminology for a module which previously was just called torsion-free; see Exercises 3.3.17 and 3.3.20. Recall that M is R^+-semiclosed if $x \in M^+$ whenever $xa \in M^+$ for some $0 < a \in R$.

Theorem 4.5.3. *Let R be a domain which is a po-ring with $R^+ \neq 0$. The following statements are equivalent.*

(a) If $0 < a, b$, then $aR^+ \cap bR^+ \neq 0$.

(b) *The positive cone of each t-torsion-free po-module can be extended to a total order of the module.*
(c) *The positive cone of each t-torsion-free po-module is contained in an R^+-semiclosed partial order of the module.*
(d) *If M is a t-torsion-free po-module, then there is a free t-torsion-free f-module over M.*

Proof. The implication (a) $\Rightarrow$ (b) is given in Exercise 4.4.22 (a) and the implication (b) $\Rightarrow$ (c) is trivial.

(c) $\Rightarrow$ (a). Suppose a and b are nonzero elements of R^+ with $aR^+ \cap bR^+ = 0$. Let $S = R + \mathbb{Z}1 \subseteq \mathrm{End}_R(R_R)$. Then S is a domain and $S^+ = \{s \in S : sR^+ \subseteq R^+\}$ is a partial order for the ring S with $R^+ \subseteq S^+$ (see Theorem 3.4.5). Since $aS^+ \cap bS^+ = 0$ it is clear that $aS^+ - bS^+$ is a partial order for S_R. Hence $aS^+ - bS^+ \subseteq P$ where P is an R^+-semiclosed partial of S. But then we have the contradiction $1 \in P \cap -P$ since $1a \in P$ and $(-1)b \in P$.

(d) $\Rightarrow$ (c). This is a consequence of the fact that a t-torsion-free f-module is R^+-semiclosed.

(a) $\Rightarrow$ (d). Let $\mathscr{U}$ be the set of total orders of M_R which contain M^+, and let $\varphi : M \longrightarrow \Pi_{\mathscr{U}}(M,P)$ be the diagonal embedding of the module M into the product. Clearly φ is isotone. Let F be the ℓ-submodule of the product generated by $\varphi(M)$. We will show F is the free t-torsion-free f-module over M. Suppose $\psi : M \longrightarrow K$ is a po-homomorphism into the t-torsion-free f-module K. By Exercise 4.4.22(b) K is a subdirect product of totally ordered t-torsion-free modules. As in the proof of Theorem 4.5.1 we may assume K itself is totally ordered. By (b) there is a total order T of $\ker\psi$ which contains $(\ker\psi)^+$. Let $P = \{x \in M : \psi(x) > 0\} \cup T$. Then $P \in \mathscr{U}$ and we have the commutative diagram

$$\begin{array}{ccc} M & \xrightarrow{\ \varphi\ } & F \\ {\scriptstyle\psi}\downarrow & & \downarrow{\scriptstyle\pi} \\ K & \xleftarrow{\ \psi\ } & (M,P) \end{array}$$

where the projection π and the bottom map ψ are ℓ-homomorphisms. □

Note that the free t-torsion-free f-module over M constructed in Theorem 4.5.3 is the free t-torsion-free f-module extension of M if and only if M is R^+-semiclosed since, according to Exercise 4.4.22(b), the latter holds exactly when M^+ is the intersection of total orders of M_R.

The previous theorem can be extended to po-modules over an irredundant semiprime right qf-ring. An I-torsion-free module over a reduced irredundant semiprime ring is called *reduced* if each of its components is a t-torsion-free module over the corresponding component ring. When the maximal right quotient ring is reduced each nonsingular module is a reduced module since its components are nonsingular modules over right Öre domains.

Theorem 4.5.4. *Let R be an irredundant semiprime f-ring. Then R is a right qf-ring if and only if, for each reduced po-module M_R, there is a free reduced f-module over M.*

Proof. Let $\{R_\lambda\}$ be the components of R and let $\{M_\lambda\}$ be the components of the nonsingular po-module M_R. Recall from Theorem 4.1.17 and Exercise 4.1.38 that $M_\lambda = M/N_\lambda$ is a nonsingular R_λ-module where $N_\lambda = \ell(R \cap R_\lambda : M)$ is a convex submodule of M. Suppose R is a right qf-ring. As noted prior to Theorem 4.3.14 each R_λ is a totally ordered right Öre domain. Let

$$\mathscr{V} = \{(M/N, P) : N \text{ is a closed convex submodule of } M \text{ and } P \text{ is a}$$

$$\text{total order of the module } M/N \text{ with } (M/N)^+ \subseteq P\}.$$

Let $\varphi : M \longrightarrow \Pi_{\mathscr{V}}(M/N,P)$ be defined by $\varphi(x) = (x+N)_P$ and let G_M be the ℓ-submodule of the product generated by $\varphi(M)$. By Exercise 4.4.22(a) each M/N_λ has a total order P_λ with $(M/N_\lambda, P_\lambda) \in \mathscr{V}$. Thus, $\bigcap_{\mathscr{V}} N \subseteq \bigcap_\lambda N_\lambda = 0$ and φ is one-to-one. Since each nonsingular f-module is a subdirect product of totally ordered nonsingular f-modules (Theorem 4.3.14) Theorem 4.5.1 gives that (G_M, φ) is the free nonsingular f-module over M.

For the converse, let M be a t-torsion-free po-module over the component R_λ of R, and let (F, φ) be the free reduced R-f-module over M. Since F is generated by $\varphi(M), F(R \cap R_\mu) = 0$ and $F_\mu = 0$ for each $\mu \neq \lambda$. Thus, F is a free t-torsion-free R_λ-f-module over M, R_λ is a right Öre domain by Theorem 4.5.3, and R is a right qf-ring by the remarks preceeding Theorem 4.3.14. □

From Theorem 4.5.2 we know that the free nonsingular f-module of a given rank over an irredundant semiprime right qf-ring is the free nonsingular f-module over the trivially ordered free module of the same rank. It is also the free f-module modulo its singular submodule.

Let R be a directed po-ring which is a unital right Öre domain with right quotient division ring Q, and suppose $aR^+ \cap bR^+ \neq 0$ if $0 < a, b \in R$. For the cardinal number n let F_n be the free nonsingular R-f-module of rank n. Note that F_n is the free unital R-f-module of rank n. If $n < m$, then F_m cannot be embedded in F_n and hence each free generating set for F_n must have cardinality n. To see this let P be a total order of $R^{(m)}$ and let A be the kernel of an ℓ-epimorphism $F_m \longrightarrow (R^{(m)}, P)$. Suppose $F_m \subseteq F_n$. Let B be a convex ℓ-submodule of F_n maximal with respect to $B \cap F_m = A$. Then B is a prime submodule of F_n (see Exercise 2.4.23). Now, $R^{(m)} \cong F_m/A \subseteq F_n/B = V$ as R-modules, and also $R^{(m)}$ can be embedded in the nonsingular R-module $U = V/Z(V)$. This is impossible since $Q^{(m)} = Q(R^{(m)})$ has Q-dimension m whereas, $Q(U)$ has Q-dimension at most n. The same thing is true if R is an irredundant semiprime right qf-ring since, if R_λ is a component of R, then $(R^{(n)})_\lambda = R_\lambda^{(n)}$ and $(F_n)_\lambda = (F_\lambda)_n$ by Exercise 1, where $(F_\lambda)_n$ denotes the free nonsingular R_λ-f-module of rank n. We record this in the following result.

Theorem 4.5.5. *Let F_n be the free unital f-module of rank n over a unital irredundant semiprime right qf-ring. If m and n are cardinals with $n < m$, then F_m cannot be embedded in F_n.* □

Let R be a unital directed po-ring and assume that the free unital f-module F of rank one is given as in Theorem 4.5.2 (or Theorem 4.5.3) and is not trivial. So F is the free unital f-module over the trivially ordered module $M = R_R$. Suppose J is a right ideal of R and $(R/J,P)$ is a totally ordered right R-module. Then $1+J \in P$ or $-1+J \in P$ and both $(R/J,P)$ and $(R/J,-P)$ are in $\mathscr{S}$. Let $\mathscr{S}_1 = \{(R/J,P) : 1+J \in P\}$ and $\mathscr{S}_2 = \{(R/J,P) : -1+J \in P\}$. Then the map $(R/J,P) \mapsto (R/J,-P)$ is a bijection from $\mathscr{S}_1$ onto $\mathscr{S}_2$ and we have the decomposition

$$\varphi : M \longrightarrow \Pi_{\mathscr{S}}(R/J,P) = \Pi_{\mathscr{S}_1}(R/J,P) \oplus \Pi_{\mathscr{S}_2}(R/J,P).$$

Now, for each $a \in R^+$ we have $\varphi(a)^+ = ((a+J)_{\mathscr{S}_1}, (a+J)_{\mathscr{S}_2})^+ = ((a+J)_{\mathscr{S}_1}, 0)$ and $\varphi(a)^- = (0, (a+J)_{\mathscr{S}_2})$; so $\varphi(a)^+ \wedge \varphi(b)^- = 0$ for $a, b \in R^+$. Let F_1 be the convex ℓ-submodule of F generated by $\{\varphi(a)^+ : a \in R^+\}$ and let F_2 be the convex ℓ-submodule generated by $\{\varphi(a)^- : a \in R^+\}$. Then $F = F_1 \oplus F_2$ since $\varphi(M) \subseteq F_1 \oplus F_2$, and $F_i \neq 0$ for $i = 1,2$ since $R^+ \not\subseteq J$. Now suppose R is an f-ring. Then $G = R \oplus R$ is the free strong unital f-module of rank one. To see this take $G = (R, R^+) \oplus (R, -R^+)$ and let $\varphi : M \longrightarrow G$ be the diagonal map. Note that $\varphi(M)$ generates G as a lattice since $(0,1) = (2,2) - (2,1) = (2,2) - (2,2) \vee (1,1)$. Suppose $\psi : M \longrightarrow K$ is an R-homomorphism into a unital totally ordered strong f-module. If $\psi(1) > 0$ define $\sigma : G \longrightarrow K$ by $\sigma((a,b)) = \psi(a)$. Then σ is an R-ℓ-homomorphism with $\sigma\varphi = \psi$ since

$$\sigma((a,b)^+) = \psi(a^+) = \psi(1)a^+ = \psi(a)^+ = \sigma((a,b))^+.$$

Similarly, if $\psi(1) < 0$, then the map σ defined by $\sigma((a,b)) = \psi(b)$ is an ℓ-homomorphism with $\sigma\varphi = \psi$.

Since the free f-module of any rank is the free product of rank one free f-modules (see Exercise 1.4.21) the previous paragraph suggests that free f-modules have nontrivial summands. We will see later that this is not always the case.

An R-module is *torsionless* if it can be embedded in a direct product of copies of the module R_R, and an ℓ-module over the ℓ-ring R is called *ℓ-torsionless* if it can be embedded in a direct product of copies of the ℓ-module R_R. We have just seen that the free strong unital f-module of rank 1 over an f-ring is ℓ-torsionless. Other free f-modules are also ℓ-torsionless.

Theorem 4.5.6. *Let R be a countable ℓ-simple totally ordered domain that is left po-Euclidean and right Öre. Then every free unital right f-module over R is ℓ-torsionless.*

Proof. Each unital f-module M over R is t-torsion-free since if $xa = 0$ with $x \in M$ and $0 < a \in R$, then $1 \leq ab$ for some $b \in R$ and $|x| \leq |xa|b = 0$. Let F be a free unital right f-module over R. To show that F is ℓ-torsionless we may assume that it has finite rank since F is a subdirect product of free f-modules of finite rank (see Exercise 1.4.11). Since F is the lattice generated by a finitely generated free module it is countable. If E is any nonzero unital $\aleph_1$-injective right f-module and M_R is a countable totally ordered unital module, then M can be embedded in E since R can be embedded in E and hence the morphism g in the following diagram must be injective.

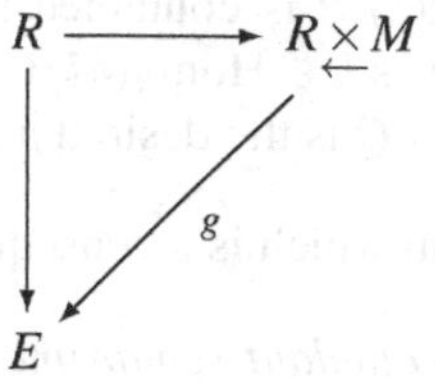

Take $E = R^{\mathbb{N}}/C_R(1)$, which is $\aleph_1$-injective by Theorem 4.4.13. Represent F as a subdirect product of totally ordered modules $T_i : F \subseteq \Pi T_i$. Then T_i can be embedded in E and thus there is an ℓ-submodule A_i of $R^{\mathbb{N}}$ together with a homomorphism of A_i onto T_i. But then ΠA_i maps onto ΠT_i and hence F is a homomorphic image of some ℓ-submodule A of $\Pi_i A_i \subseteq \Pi_i R^{\mathbb{N}} = R^J$. Since F is a projective f-module it is embeddable in A and hence in R^J:

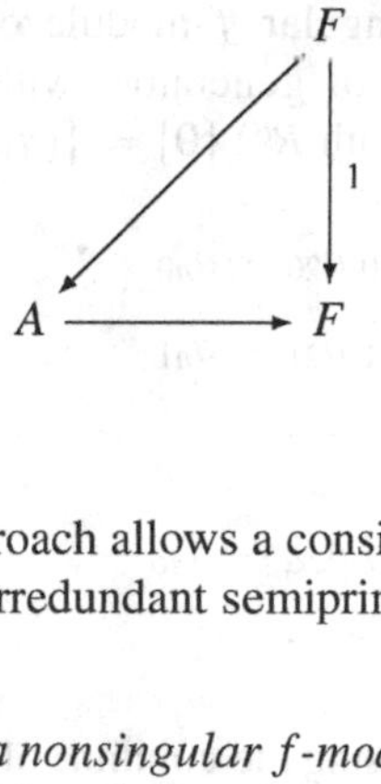

. □

Another more direct approach allows a considerable improvement of the preceding result. It holds for any irredundant semiprime two-sided qf-ring. We start with a technical result.

Theorem 4.5.7. *Let M_R be a nonsingular f-module over the irredundant semiprime right qf-ring R. Suppose X is a finite subset of M and* $\inf X \not\leq 0$. *Then there is an element $g \in Hom_R(M_R, Q(R)_R)$ with* $\inf g(X) > 0$.

Proof. As usual, we may assume that R is a totally ordered right Öre domain and M is totally ordered. Let $Q = Q(R_R)$. Since $E(M_R) = XQ \boxplus G$ as Q-modules we may also assume that $M = XQ$. We argue by induction on $\dim M_Q$. If $\dim M = 1$, then $M = xQ$ for a fixed $x \in X$ and for each $y \in X$ there is an element $0 < p_y \in Q$ with $y = xp_y$. In this case define $g : M \longrightarrow Q$ by $g(xp) = p$. Suppose $\dim M \geq 2$ and let $k \in \mathbb{N}$ be minimal such that there is a subset $\{x_1, \ldots, x_k\}$ of X with $X \subseteq x_1Q^+ + \cdots + x_kQ^+$. Note that $k \geq 2$, and we suppose $x_1 > x_2$. We claim that

$$x_1Q^+ + \cdots + x_kQ^+ \cap (x_1 - x_2)Q = 0. \tag{4.5.2}$$

If not, take $0 \neq (x_1 - x_2)p = x_1p_1 + \cdots + x_kp_k$ with each $p_i \in Q^+$. Then $p > 0$ and $x_1 - x_2 = x_1q_1 + \cdots + x_kq_k$ with each $q_i \in Q^+$. So, $x_1(1 - q_1) = x_2(1 + q_2) + \cdots + x_kq_k > 0$ and this gives the contradiction $X \subseteq x_2Q^+ + \cdots + x_kQ^+$. Let $C = (x_1 - x_2)Q$, and let $P = \sum(x_i + C)Q^+ \subseteq M/C$. As a consequence of (4.5.2) we have that $P \cap -P = 0$, and hence P is a positive cone of the Q-module M/C with $x + C \in P\backslash\{0\}$ for each $x \in X$. By Exercise 4.4.22(a) P is contained in a total order of M/C. Since $\dim M/C = \dim M - 1$, there exists $h \in \mathrm{Hom}_R(M/C, Q)$ with $\bigwedge_x h(x + C) > 0$; thus the composition $M \longrightarrow M/C \longrightarrow Q$ is the desired map. □

Here is an embedding theorem which is a consequence of the previous result.

Theorem 4.5.8. *Let R be an irredundant semiprime right qf-ring and let F be the free nonsingular f-module extension of the trivially ordered module M. Then F can be embedded in a product of copies of $Q(R)_R$.*

Proof. If $0 < x \in F$, then $x = \bigvee_i \bigwedge_j x_{ij}$ with $x_{ij} \in M$ and for some i, $\bigwedge_j x_{ij} \nleq 0$. By Theorem 4.5.7 there is an R-homomorphism $\rho : F \longrightarrow Q$ with $\bigwedge_j \rho(x_{ij}) > 0$, and ρ must be an ℓ-homomorphism since the restriction of ρ to M extends to an ℓ-homomorphism of F. So $\rho(x) > 0$ and hence F can be embedded in Q^I where $I = F^+\backslash\{0\}$. □

We will show that a free nonsingular f-module over R of finite rank is ℓ-torsionless by producing a free set of generators within some power of R. Let $2 \leq n \in \mathbb{N}$ and let τ be an ordinal with $R^n\backslash\{0\} = \{(a_{1\alpha}, \ldots, a_{n\alpha}) : \alpha < \tau\}$. Consider the array:

$$a_{10}\ a_{20} \cdots a_{n0}$$

$$a_{11}\ a_{21} \cdots a_{n1}$$

$$\vdots$$

$$a_{1\alpha}\ a_{2\alpha} \cdots a_{n\alpha}$$

$$\vdots$$

Let $z_1, \ldots, z_n \in R^{W(\tau)}$ be defined by using the columns of this array. So, $(z_j)_\alpha = a_{j\alpha}$:

$$z_1 = (a_{10}, a_{11}, \ldots, a_{1\alpha}, \ldots)$$

$$z_2 = (a_{20}, a_{21}, \ldots, a_{2\alpha}, \ldots)$$

$$\vdots$$

$$z_n = (a_{n0}, a_{n1}, \ldots, a_{n\alpha}, \ldots).$$

For the po-ring R and $(r_1, \ldots, r_n) \in R^n$ define the *right open half-plane*

$$H(r_1, \ldots, r_n) = \{(s_1, \ldots, s_n) \in R^n : s_1r_1 + \cdots + s_nr_n > 0\}.$$

Theorem 4.5.9. *Let R be a reduced sp-po-ring. With the notation above let $A = z_1R + \cdots + z_nR \subseteq R^{W(\tau)}$.*

(a) $\{z_1, \ldots, z_n\}$ is an R-independent subset of $R^{W(\tau)}$.

(b) Suppose R is a totally ordered domain. Let $x_1, \ldots, x_k$ be elements of $A : x_i = \sum_{j=1}^n z_j r_{ij}$. The following are equivalent.

(i) $\bigwedge x_i \leq 0$.

(ii) $\bigvee x_i \geq 0$.

(iii) $\bigcap_{i=1}^k H(r_{i1}, \ldots, r_{in}) = \emptyset$.

(iv) $\bigcap_{i=1}^k H(-r_{i1}, \ldots, -r_{in}) = \emptyset$.

(c) Let R be a unital totally ordered two-sided Öre domain. If $g : A \longrightarrow N$ is an R-homomorphism into the nonsingular f-module N, then $\bigwedge x_i \leq 0$ implies $\bigwedge g(x_i) \leq 0$ and $\bigvee x_i \geq 0$ implies $\bigvee g(x_i) \geq 0$.

Proof. (a) If $\sum_{j=1}^n z_j a_{j\alpha} = 0$ for some $\alpha < \tau$, then $0 = (\sum_j z_j a_{j\alpha})_\alpha = \sum_j a_{j\alpha}^2$ and we have the contradiction $a_{1\alpha} = \cdots = a_{n\alpha} = 0$.

(b) (i) $\Rightarrow$ (iii). If $(a_{1\alpha}, \ldots, a_{n\alpha}) \in \bigcap_{i=1}^k H(r_{i1}, \cdots, r_{in})$ for some $\alpha < \tau$, then

$$0 \geq \bigwedge_i (x_i)_\alpha = \bigwedge_i \sum_j a_{j\alpha} r_{ij} > 0.$$

(iii) $\Rightarrow$ (i). For each $\alpha < \tau$ there is some m, $1 \leq m \leq k$, with $\sum_{j=1}^n a_{j\alpha} r_{mj} \leq 0$. So

$$\left(\bigwedge_i x_i\right)_\alpha \leq (x_m)_\alpha = \sum_{j=1}^n a_{j\alpha} r_{mj} \leq 0.$$

Similarly, (ii) is equivalent to (iv), and clearly (iii) and (iv) are equivalent.

(c) Suppose $\bigwedge x_i \leq 0$ but $\bigwedge g(x_i) \nleq 0$. By Theorem 4.5.7 we may assume that N is the totally ordered division ring of quotients of R and $\bigwedge g(x_i) > 0$. Take $0 < a \in R$ with $ag(z_j) \in R$ for $j = 1, \ldots, n$. Then $0 < ag(x_i) = \sum_{j=1}^n ag(z_j) r_{ij}$ for each $i = 1, \ldots, k$; but this is impossible by (b). If $\bigvee x_i \geq 0$, then $\bigwedge -x_i \leq 0$ and hence $\bigvee g(x_i) \geq 0$. □

Theorem 4.5.10. *Let R be a unital irredundant semiprime two-sided qf-ring. Each free nonsingular R-f-module is ℓ-torsionless.*

Proof. As noted previously, it suffices to prove the theorem for the case in which R is totally ordered and the free f-module has finite rank. Let B be the ℓ-submodule of $R^{W(\tau)}$ generated by the R-module A of Theorem 4.5.9. We claim that B is a free nonsingular f-module on the set $\{z_1, \ldots, z_n\}$. For, by (a) of Theorem 4.5.9 a function $\{z_1, \ldots, z_n\} \longrightarrow N$ into the nonsingular f-module N extends to an R-homomorphism $g : A \longrightarrow N$. Define $h : B \longrightarrow N$ by

$$h\left(\bigvee_{i \in I} \bigwedge_{j \in J} x_{ij}\right) = \bigvee_i \bigwedge_j g(x_{ij}), \quad x_{ij} \in A.$$

To see that h is well-defined, suppose

$$0 = \bigvee_{i \in I} \bigwedge_{j \in J} x_{ij} = \bigwedge_{f \in J^I} \bigvee_{i \in I} x_{if(i)}.$$

Then for each $i \in I$, $\bigwedge_j x_{ij} \leq 0$, and for each $f \in J^I$, $\bigvee_i x_{if(i)} \geq 0$. By (c) of Theorem 4.5.9 we have $\bigwedge_j g(x_{ij}) \leq 0$ and $\bigvee_i g(x_{if(i)}) \geq 0$. So

$$0 \leq \bigwedge_{f \in J^I} \bigvee_{i \in I} g(x_{if(i)}) = \bigvee_{i \in I} \bigwedge_{j \in J} g(x_{ij}) \leq 0.$$

Thus, h is an R-ℓ-homomorphism by Exercise 2.2.19 and B is free on $\{z_1, \ldots, z_n\}$. □

If R is the direct product of two copies of $\mathbb{R}$, then the free unital R-f-module of rank 2 is the direct sum of two copies of the free unital $\mathbb{R}$-f-module of rank 2 (see Exercise 2). Consequently, to show that free f-modules are indecomposable it is reasonable to assume that R is an indecomposable po-ring. We will further restrict R to a subring of $\mathbb{R}$. The next result is clearly related to indecomposability and so is the one that follows.

Theorem 4.5.11. *Let R be a subring of $\mathbb{R}$ and let F be the free nonsingular f-module extension of the trivially ordered R-module M. Suppose that $u, v \in F$ and $x_1, \ldots, x_n$ are elements of M such that $u \wedge v = 0$ and $u + v = x_1^+ \wedge \cdots \wedge x_n^+$. Then $u = 0$ or $v = 0$.*

Proof. By Theorem 4.5.8 we may assume that F is contained in a product of copies of $\mathbb{R}_R$. Thus, it suffices to show that $Tu = 0$ or $Tv = 0$ where T is the set of R-ℓ-homomorphisms from F into $\mathbb{R}$. If $\psi \in T$ and $\bigwedge_i \psi(x_i) \leq 0$ then $\psi(u) = \psi(v) = 0$; so we only need to consider those ψ with $\bigwedge_i \psi(x_i) > 0$. Now suppose that ψ and θ are two such homomomorphisms and $\psi(u) = 0$. For $t \in [0,1]$ let φ_t be that element of T that lifts the R-homomomorphism $t\psi + (1-t)\theta \in \mathrm{Hom}_R(M, \mathbb{R})$. Let $s = \bigwedge_i \psi(x_i) \wedge \bigwedge_i \theta(x_i)$. Then $s > 0$ and since $z \wedge w \leq tz + (1-t)w$ for $t \in [0,1]$ and $w, z \in \mathbb{R}^+$, we have

$$\bigwedge_i \varphi_t(x_i) = \bigwedge_i (t\psi(x_i) + (1-t)\,\theta(x_i)) \geq \bigwedge_i \psi(x_i) \wedge \bigwedge_i \theta(x_i) = s.$$

Note that $\varphi_1(u) = \psi(u) = 0$, and if $\varphi_t(u) > 0$ then $\varphi_t(u) = \bigwedge_i \varphi_t(x_i) \geq s$. Now, $u = \bigvee_i \bigwedge_j m_{ij}$ where $m_{ij} \in M$. Define the functions $f_{ij} : [0,1] \longrightarrow \mathbb{R}$ by $f_{ij}(t) = t\psi(m_{ij}) + (1-t)\,\theta(m_{ij})$. Then each f_{ij} is continuous and hence so is $f = \bigvee_i \bigwedge_j f_{ij}$. Also,

$$f(t) = \bigvee_i \bigwedge_j \varphi_t(m_{ij}) = \varphi_t(u)$$

for each $t \in [0,1]$. Since $f(1) = 0$ and $f(t) \geq s$ whenever $f(t) > 0$, we must have $f = 0$ by the continuity of f. Thus, $\theta(u) = \varphi_0(u) = f(0) = 0$. □

Theorem 4.5.12. *Let M be a nonsingular po-module over the totally ordered right Öre domain R, and let (F_0, φ_0) be the free nonsingular f-module extension of the trivially ordered module $M_0 = M$. If C is the convex ℓ-subgroup of F_0 generated by the set $\{\varphi_0(x)^- : x \in M^+\}$, then $F_0/c\ell C$ is the free nonsingular f-module over M.*

Proof. Clearly, C is a submodule of F_0 and the composite $\varphi : M \longrightarrow F_0/c\ell C$ given by $\varphi(x) = \varphi_0(x) + c\ell C$ is isotone. We will show that it is injective. Suppose $\varphi_0(x)a \in C$ with $0 < a \in R$. Then there exist nonzero elements $x_1, \ldots, x_n$ in M^+ with

$$|\varphi_0(xa)| \leq \varphi_0(x_1)^- + \cdots + \varphi_0(x_n)^-$$

and hence

$$\varphi_0(xa)^+ \wedge \bigwedge_j \varphi_0(x_j)^+ \leq \sum_j \varphi_0(x_j)^- \wedge \bigwedge_j \varphi_0(x_j)^+ = 0.$$

By Exercise 4.4.22(a) there is a total order P of M_R which contains M^+. By replacing x by $-x$ if necessary, we can assume that $x \in P$. But from the construction of F_0 in Theorem 4.5.3 we see that if $xa \neq 0$, then $0 < \varphi_0(xa)^+ \wedge \bigwedge_j \varphi_0(x_j)^+$. Thus, $xa = 0$ and hence $x = 0$. Now, suppose $\psi : M \longrightarrow K$ is a po-homomorphism into the nonsingular f-module K. Since $\psi : M_0 \longrightarrow K$ is also a po-homomorphism it has a lifting to $\sigma : F_0 \longrightarrow K$, a homomorphism of f-modules. If $x \in M^+$, then $\sigma\varphi_0(x)^- = \psi(x)^- = 0$ since $\psi(x) \in K^+$. Thus $c\ell C \subseteq \ker \sigma$ since K is nonsingular, and σ induces an f-module homomorphism on $F_0/c\ell C$ which extends ψ. This shows that $(F_0/c\ell C, \varphi)$ is the free nonsingular f-module over M. □

We will now use the previous two results to show the indecomposability of some free f-modules.

Theorem 4.5.13. *Let R be a subring of $\mathbb{R}$ and let F be the free nonsingular f-module over the nonsingular po-module M_R. Then F is decomposable if and only if M_R is uniform and trivially ordered.*

Proof. If M_R is uniform and trivially ordered, then F is decomposable by Exercise 3. Conversely, suppose that F is decomposable and assume first that M is trivially ordered. Let $F = F_1 \oplus F_2$ be a decomposition of F with each $F_i \neq 0$. We will use the construction of F that is given in Theorem 4.5.3. If $0 \neq x \in M$, then $\varphi(x)$ is a weak order unit of F since all of its components in $\Pi_{\mathscr{S}}(M,P)$ are nonzero. Hence, $\varphi(x) \notin F_1 \cup F_2$, or else $\varphi(x) \in F_2$, say, and $F_1 \subseteq F_2^\perp \subseteq \varphi(x)^\perp = 0$. Now, $\varphi(x) = u + v$ with $u \in F_1$ and $v \in F_2$. By Theorem 4.5.11 $\varphi(x)^+ \in \{u^+, v^+\}$ and $\varphi(x)^- \in \{u^-, v^-\}$. So, either $\varphi(x)^+ \in F_1$ and $\varphi(x)^- \in F_2$, or $\varphi(x)^- \in F_1$ and $\varphi(x)^+ \in F_2$. Suppose x and y are nonzero elements of M with $X \cap Y = 0$ where X (respectively, Y) is the submodule of M generated by x (respectively, y). We may assume $\varphi(x)^+ \in F_1$ and $\varphi(y)^- \in F_2$; the other three cases can be dealt with in an analogous way. Since $(xR^+ + \mathbb{Z}^+x) - (yR^+ + \mathbb{Z}^+y)$ is a partial order of M_R it is contained in some total order P by Exercise 4.4.22(a). But now we have the contradiction $\varphi(x)^+ \wedge \varphi(y)^- = 0$, yet

$$(\varphi(x)^+)_P \wedge (\varphi(y)^-)_P = x \wedge -y > 0$$

in (M,P). So M_R is uniform.

Now assume that $M^+ \neq 0$. Using the notation of Theorem 4.5.12 we have $F = F_0/c\ell C = F_1/c\ell C \oplus F_2/c\ell C$ where each F_j is a convex ℓ-submodule of F_0 which properly contains $c\ell C$ and $F_1 \cap F_2 = c\ell C$. If $0 \neq x \in M^+$, then $\varphi_0(x) = u_1 + u_2$ with $u_j \in F_j$. Take $0 < a \in R$ and $x_1, \ldots, x_n \in M^+ \backslash \{0\}$ with

$$(|u_1| \wedge |u_2|)a \leq \varphi_0(x_1)^- + \cdots + \varphi_0(x_n)^-.$$

Since M^+ is contained in a total order of M_R, if w is defined by

$$w = \varphi_0(x)^+ \wedge \varphi_0(x_1)^+ \wedge \cdots \wedge \varphi_0(x_n)^+,$$

then $0 < w \leq u_1^+ + u_2^+$. So $w = v_1 + v_2$ where $0 \leq v_j \leq u_j^+$ and hence

$$(v_1 \wedge v_2)a \leq \left[\bigwedge \varphi_0(x_i)^+ \wedge \sum \varphi_0(x_i)^-\right] a = 0.$$

Thus, $v_1 \wedge v_2 = 0$ and either $v_1 = 0$ or $v_2 = 0$ by Theorem 4.5.11. Assume that $v_2 = 0$; so $w = v_1 \in F_1$ and $w \wedge |v| \in c\ell C$ for each $v \in F_2$. From the construction of F in Theorem 4.5.3 we know that the image of each $0 \neq y \in M^+$ is a weak order unit of F. Thus, the image of w is a weak order unit in F and we have that $F_2 \subseteq c\ell C$. This contradiction gives that F is indecomposable when M is not trivially ordered. □

We wish to examine the size of disjoint sets in free f-modules, and for our first result we will utilize some topological considerations. Recall that the topological space X is separable if it has a countable dense subset. Recall also that if $\{X_j : j \in J\}$ is a family of topological spaces, then the product topology on $X = \Pi_{j\in J} X_j$ has as a base the family of all subsets of X of the form

$$U(V_j : j \in K) = \{f \in X : f(j) \in V_j \text{ for every } j \in K\},$$

where K is a finite subset of J and each V_j is an open subset of X_j. So the open sets in X are unions of these basic open subsets. Now a separable space clearly has the property that every family of pairwise disjoint open sets is countable. More generally, this latter property holds in any space in which every family $\mathscr{U}$ of open sets with card $(\mathscr{U}) = \aleph_1$ has a subfamily $\mathscr{V}$ with card $(\mathscr{V}) = \aleph_1$ and $\bigcap_{V\in\mathscr{V}} V \neq \emptyset$; a space with this property is said to have *caliber* $\aleph_1$. Of course, a separable space has caliber $\aleph_1$.

Theorem 4.5.14. *Let $X = \Pi_{j\in J} X_j$ be a product space.*

(a) If each X_j is separable and card $(J) \leq$ card $(\mathbb{R})$, then X is separable.
(b) If each X_j is separable, then X has caliber $\aleph_1$, and hence each family of pairwise disjoint open sets is countable.

Proof. (a) Since this is obvious if J is finite we will assume J is infinite and is a subset of $\mathbb{R}$. For each $j \in J$, let $D_j = \{x_{jm} : m \in \mathbb{N}\}$ be a dense subset of X_j. Define the countable set D by

$$D = \{s = (p_1, \ldots, p_{n-1}; m_1, \ldots, m_n) : 2 \leq n \in \mathbb{N},\ p_i \in \mathbb{Q},\ m_i \in \mathbb{N} \text{ and } p_1 < \cdots < p_{n-1}\}.$$

For $s \in D$ let $f_s \in X$ be defined by

$$f_s(j) = \begin{cases} x_{jm_1} & \text{if } j \le p_1 \\ x_{jm_i} & \text{if } p_{i-1} < j \le p_i \\ x_{jm_n} & \text{if } p_{n-1} < j. \end{cases}$$

Now, $\{f_s : s \in D\}$ is dense in X. For, if $U = U(V_j : j \in K)$ is a nonempty basic open set with $K = \{j_1 < \cdots < j_n\}$ and $n \ge 2$ take $\{p_1, \ldots, p_{n-1}\} \subseteq \mathbb{Q}$ with $j_1 < p_1 < j_2 < p_2 < \cdots < j_{n-1} < p_{n-1} < j_n$, and take $m_i \in \mathbb{N}$ such that $x_{j_i m_i} \in V_{j_i}$. Then $s = (p_1, \ldots, p_{n-1}; m_1, \ldots, m_n) \in D$ and $f_s \in U$.

(b) Let $\mathscr{U} = \{U_\gamma : \gamma \in \Gamma\}$ be a collection of nonempty open subsets of X with card $(\mathscr{U}) = \aleph_1$. For each $\gamma \in \Gamma$ let $W_\gamma = U(V_j : j \in K_\gamma)$ be a nonempty basic open set with $W_\gamma \subseteq U_\gamma$. If $\mathscr{W} = \{W_\gamma : \gamma \in \Gamma\}$ is countable, then for some $\gamma \in \Gamma$ the set $\Delta = \{\delta \in \Gamma : W_\gamma \subseteq U_\delta\}$ has cardinality $\aleph_1$ and $\bigcap_{\delta \in \Delta} U_\delta \neq \emptyset$. Now, suppose card $(\mathscr{W}) = \aleph_1$ and let $A = \bigcup_{\gamma \in \Gamma} K_\gamma$, $Y = \Pi_{j \in A} X_j$ and $Z = \Pi_{j \notin A} X_j$. Then $X = Y \times Z$ and Y is separable by (a). For each $\gamma \in \Gamma$ we have $W_\gamma = W_\gamma(Y) \times Z$ where $W_\gamma(Y) = \{f \in Y : f(j) \in V_j \text{ for each } j \in K_\gamma\}$ is a basic open set in Y. But then $\{W_\gamma(Y) : \gamma \in \Gamma\}$ has a subfamily $\{W_\delta(Y) : \delta \in \Delta\}$ whose cardinality is $\aleph_1$ and such that $\bigcap_{\delta \in \Delta} W_\delta(Y) \neq \emptyset$. Let $\mathscr{V} = \{U \in \mathscr{U} : W_\delta(Y) \times Z \subseteq U \text{ for some } \delta \in \Delta\}$. Then card $(\mathscr{V}) = \aleph_1$ and $\bigcap_{U \in \mathscr{V}} U \neq \emptyset$. □

The last result will now be applied to show that disjoint subsets of some free f-modules over some archimedean f-rings must be countable.

Theorem 4.5.15. *Let R be a unital archimedean f-ring.*

(a) If R has a finite basis, then each disjoint subset of a free unital f-module is countable.
(b) If R has a countable basis, then each disjoint subset of a free nonsingular f-module is countable.

Proof. (a) By Theorems 3.6.2 and 4.1.14 and Exercises 2.4.13 and 2.5.27 $R = R_1 \oplus \cdots \oplus R_n$ where each R_i is a subring of $\mathbb{R}$. The ℓ-simplicity of R_i gives that each unital R_i-f-module is nonsingular and hence the same is true for unital R-f-modules. By Theorem 4.5.10 the variety of unital R-f-modules is generated by R_R. Let F be a free unital f-module of rank m and let J be a set with card$(J) = m$. By Exercise 4 we may take F to be the R-ℓ-submodule of $R^{(R^J)}$ generated by the projections $\{\pi_j : j \in J\}$. Now, any product S of copies of $\mathbb{R}$ is a topological f-ring in the product topology. This is most easily seen as a consequence of the commutativity of the diagrams

$$
\begin{array}{ccc}
S\times S & \xrightarrow{\ \rho\ } & S \\
{\scriptstyle p\times p}\downarrow & & \downarrow{\scriptstyle p} \\
\mathbb{R}\times\mathbb{R} & \xrightarrow{\ \rho\ } & \mathbb{R}
\end{array}
\qquad\qquad
\begin{array}{ccc}
S & \xrightarrow{\ \sigma\ } & S \\
{\scriptstyle p}\downarrow & & \downarrow{\scriptstyle p} \\
\mathbb{R} & \xrightarrow{\ \sigma\ } & \mathbb{R}
\end{array}
$$

Here, p denotes one of the projections, σ is inversion and ρ is any one of the operations $+, \cdot, \wedge, \vee$. Since R is a subring of a finite product of copies of $\mathbb{R}$ it is a separable topological f-ring. Because each π_j is continuous when R^J is given the product topology, F is an f-submodule of the R-f-module of continuous functions from R^J into R. Suppose $u, v \in F$ with $u \wedge v = 0$. For $i = 1, \ldots, n$ let $p_i : R \longrightarrow R_i$ be the projections. Then for each i, $(p_i u)^{-1}(R_i \backslash \{0\})$ and $(p_i v)^{-1}(R_i \backslash \{0\})$ are disjoint open subsets of R^J. Suppose that $\{u_\alpha : \alpha < \omega_1\}$ is a set of disjoint nonzero elements in F^+. For some i, the set $\{\alpha : p_i u_\alpha \neq 0\}$ is uncountable, and hence we may assume $p_1 u_\alpha \neq 0$ for every $\alpha < \omega_1$. But then $\{(p_1 u_\alpha)^{-1}(R_1 \backslash \{0\}) : \alpha < \omega_1\}$ is a collection of nonempty pairwise disjoint open subsets of R^J and this is impossible by Theorem 4.5.14. Thus, each disjoint subset of F must be countable.

(b) There is a countable family of subrings R_n of $\mathbb{R}$ with $\oplus R_n \subseteq R \subseteq \Pi R_n$. As in (a) R is a separable topological f-ring. According to Exercise 4 a free nonsingular f-module F may be described as in (a) and hence the previous argument completes the proof. □

The preceding result depends mainly on the fact that the ring is archimedean. It is surprising that it holds without any restriction on the rank of the free f-module. Equally surprising is that when it fails the rank of the free f-module is again irrelevant. We will present an example to illustrate this. The ring will be a totally ordered group ring or power series ring. In order to prepare for this example we will consider partial orders on the free module of rank 2.

Theorem 4.5.16. *Let $V = V(G,A)$ be the totally ordered formal power series ring where A is a totally ordered domain and G is a totally ordered cancellative semigroup. For $k \in G$ define $P_k \subseteq V \boxplus V$ by $P_k = (V^+ \times V^+) \cup Q_k$ where $(u,v) \in Q_k$ provided: $u \neq 0$, $v \neq 0$,*

$$
\begin{aligned}
u &= u_g x^g + \cdots \\
v &= v_h x^h + \cdots
\end{aligned}
\tag{4.5.3}
$$

$g = maxsupp\ u$, $h = maxsupp\ v$, and

$$
\begin{array}{ll}
u_g > 0 & \textit{if } g > h + k, \\
v_h > 0 & \textit{if } g < h + k, \\
u_g + v_h > 0 & \textit{if } g = h + k.
\end{array}
$$

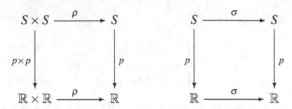

Then P_k is a partial order for the module $(V \boxplus V)_{V(C(k),A)}$ where $C(k)$ is the centralizer of k in G.

Proof. Since V is a totally ordered domain and clearly $P_k V(C(k),A)^+ \subseteq P_k$ and $P_k \cap -P_k = 0$ we only need to show that $P_k + P_k \subseteq P_k$. There are, of course, numerous cases to check. If $(u,v),(u',v') \in P_k$ we will write (u,v) as in (4.5.3) and $(u,v)+(u',v') = (u'',v'')$ where

$$u' = u'_{g'}x^{g'} + \cdots$$

$$v' = v'_{h'}x^{h'} + \cdots$$

$$u'' = u''_{g''}x^{g''} + \cdots$$

$$v'' = v''_{h''}x^{h''} + \cdots$$

If any one of these elements is zero, then the corresponding sum is, of course, only formal. We proceed with the various cases, starting with $(u,v),(u',v') \in Q_k$.

(i) $g > h+k$ and $g' > h'+k$. Then $u_g > 0, u'_{g'} > 0$ and $g'' = g \vee g' > (h+k) \vee (h'+k) \geq h''+k$; so $u''_{g''} > 0$.

(ii) $g > h+k$ and $g' < h'+k$. Then $u_g > 0$ and $v'_{h'} > 0$. There are three cases to consider. If $g > h'+k$, then $g'' = g > h''+k$ and $u''_{g''} = u_g > 0$. If $g < h'+k$, then $g'' < h'+k = h''+k$ and $v''_{h''} = v'_{h'} > 0$. If $g = h'+k$, then $g'' = g = h''+k$ since $h'+k > h+k$ and therefore $h'' = h'$. But $u''_{g''} + v''_{h''} = u_g + v'_{h'} > 0$.

(iii) $g > h+k$ and $g' = h'+k$. Again, we consider three cases. If $g > g'$, then $g = g''$, $u''_{g''} = u_g > 0$ and $g > h''+k$. If $g < g'$, then $g'' = g'$, $u''_{g''} = u'_{g'}$, $h'' = h'$ and $v''_{h''} = v'_{h'}$; so $g'' = h''+k$ and $u''_{g''} + v''_{h''} = u'_{g'} + v'_{h'} > 0$. If $g = g'$, then $u_g > 0$, $h'' = h'$ and $v''_{h''} = v'_{h'}$. If $u+u' = 0$, then since $u'_{g'} + v'_{h'} > 0$, $u_g + u'_{g'} = 0$ and $u_g > 0$, we have $u'_{g'} < 0$ and $v'_{h'} > 0$; so $(u'',v'') = (0,v'') \in V^+ \times V^+$. On the other hand, if $u'' \neq 0$, then either $u''_{g''} = u_g + u'_g \neq 0$, $g'' = g = h'+k = h''+k$, and $u''_{g''} + v''_{h''} = u_g + u'_{g'} + v'_{h'} > 0$, or $u_g + u'_g = 0$, $g'' < g < h''+k$, and $v''_{h''} = v'_{h'} > 0$ since $u'_g + v'_{h'} > 0$ and $u'_g = -u_g < 0$ forces $v'_{h'} > 0$.

(iv) $g < h+k$ and $g' > h'+k$. This is case (ii).

(v) $g < h+k$ and $g' < h'+k$. Here, the leading term of v'' is $v_h x^h$ or $v'_{h'}x^{h'}$ or $(v_h + v'_h)x^h$ and in all cases $v''_{h''} > 0$, and $g'' < h''+k$ or $u'' = 0$.

(vi) $g < h+k$ and $g' = h'+k$. If $h < h'$, then $g < g'$, $g'' = h''+k$ and $u''_{g''} + v''_{h''} = u'_{g'} + v'_{h'} > 0$. If $h > h'$, then $v''_{h''} = v_h > 0$ and $u+u' = 0$ or $g'' \leq g \vee g' < h+k = h''+k$. Suppose $h = h'$. Then $g < g' = g''$ and $u'' = u'_{g'}x^{g'} + \cdots$. If $v_h + v'_h = 0$, then $u'_{g'} > 0$ since $u'_{g'} + v'_h > 0$ and $v'_h = -v_h < 0$. Thus, whether $v'' = 0$ or $v'' \neq 0$, in which case $h''+k < h+k = g''$, we have $(u'',v'') \in P_k$.

(vii) If $g = h+k$ and $g' > h'+k$ or $g' < h'+k$ we have case (iii) or (vi), respectively.

(viii) $g = h+k$ and $g' = h'+k$. Two cases arise depending upon whether g and g' are equal or unequal. If $g < g'$, then $h < h'$ and $g'' = g'$, $h'' = h'$ and $u''_{g''} + v''_{h''}$, $u'_{g'} + v'_{h'} > 0$. Suppose $g = g'$ and $h = h'$. If $u_g + u'_g = 0$, then $v_h + v'_h = v_h + v'_h + u_g +$

$u'_g > 0$, and whether $u'' = 0$ or $u'' \neq 0$ and hence $g'' < g < h'' + k$, we have $(u'', v'') \in P_k$. Similarly, if $v_h + v'_h = 0$, then $u_g + u'_g > 0$ and either $g'' = g = h + k > h'' + k$ or $v'' = 0$; hence $(u'', v'') \in P_k$. If $u_g + u'_g \neq 0$ and $v_h + v'_h \neq 0$, then $g'' = g$, $h'' = h$ and $u''_g + v''_h = u_g + v_h + u'_g + v'_h > 0$.

This completes the proof of $Q_k + Q_k \subseteq Q_k$. Now, suppose $(u, v) \in Q_k$ and $0 \neq (u', v') \in V^+ \times V^+$.

(ix) Suppose $u' = 0$. Then $u'' = u_g x^g + \cdots$ and $v'_{h'} > 0$. If $h' < h$, then $v'' = v_h x^h + \cdots$ and clearly $(u'', v'') \in P_k$ irrespective of the relation between g and $h + k$. Suppose $h' = h$. Then $g > h + k$ implies $u_g > 0$ and $(u'', v'') \in P_k$ since $h'' \leq h$. If $g = h + k$, then either $v_h + v'_h \neq 0$, $g'' = h'' + k$ and $u''_{g''} + v''_{h''} = u_g + v_h + v'_h > 0$, or $v_h + v'_h = 0$ and hence $u_g > 0$ since $u_g + v_h > 0$. If $g < h + k$, then $v_h > 0$ and $g'' < h'' + k$ with $v''_{h''} = v_h + v'_h > 0$. Suppose, finally, that $h' > h$. Then either $g > h' + k$ and $u_g > 0$, or $g < h' + k$ and $v'_{h'} > 0$, or $g = h' + k$ and $u_g + v'_{h'} > 0$ since $u_g > 0$.

(x) Suppose $v' = 0$. If $g > g'$, then $u''_{g''} = u_g$ and $v''_{h''} = v_h$ and $(u'', v'') \in P_k$ since $(u, v) \in Q_k$. If $g < g'$, then $g'' = g'$ and $u''_{g''} = u'_{g'}$. Now, either $g' > h + k$ and $(u'', v'') \in P_k$ since $u'_{g'} > 0$, or $g' = h + k > g$ and $u''_{g''} + v''_{h''} = u'_{g'} + v_h > 0$, or $g' < h + k$ and $v''_{h''} = v_h > 0$ since $h + k > g$. If $g = g'$, then either $g > h + k$, $u''_{g''} = u_g + u'_g > 0$ and $g'' = g$, or $g < h + k$, $g'' \leq g < h + k$ and $v''_{h''} = v_h > 0$, or $g = h + k$, $u_g + v_h > 0$. In the latter case, either $u_g + u'_g \neq 0$, $g'' = g = h'' + k$ and $u''_{g''} + v''_{h''} = u_g + u'_g + v_h > 0$, or $u_g + u'_g = 0$, $v_h > -u_g = u'_g > 0$ and $(u'', v) \in P_k$ since either $u'' = 0$ or $g'' < g = h + k$.

In the remaining cases we have $u' > 0$ and $v' > 0$; so $u'_{g'} > 0$ and $v'_{h'} > 0$. We assume $u'' < 0$ or $v'' < 0$ since otherwise $(u'', v'') \in P_k$.

(xi) Suppose $u'' < 0$. Then $u < -u' < 0$ and $g \leq h + k$ since $(u, v) \in Q_k$. If $g < h + k$ then $v_h > 0$, and if $g = h + k$ then $u_g + v_h > 0$ and $v_h > -u_g > 0$; so in either case $v > 0$. Note that $g', g'' \leq g$ since, otherwise, $g'' = g' > g$, $u''_{g''} = u'_{g'} > 0$, and $u'' > 0$. If $g'' < h'' + k$, then $(u'', v'') \in Q_k$ since $v'' > 0$. On the other hand, if $g'' \geq h'' + k$, then $g \leq h + k \leq h'' + k \leq g'' \leq g$; so $g = h + k$, $g'' = h'' + k$, and $h = h'' \geq h'$. Thus, we have, since $g > g'$ or $g = g' = g''$

$$u''_{g''} = \begin{cases} u_g & \text{if } g > g' \\ u_g + u'_{g'} & \text{if } g = g' \text{ and } u_g + u'_g \neq 0, \end{cases}$$

$$v''_{h''} = \begin{cases} v_h & \text{if } h' < h \\ v_h + v'_{h'} & \text{if } h = h', \end{cases}$$

and in all cases $u''_{g''} + v''_{h''} > 0$. So $(u'', v'') \in Q_k$.

(xii) Suppose $v'' < 0$. Then $v < 0$ and $h + k \leq g$. By (xi) we may assume $u'' \geq 0$. If $u'' = 0$, then as in (xi) we would have $v > 0$; so $u'' > 0$. If $h + k = g$, then $u_g > -v_h > 0$, and if $h + k < g$, then $u_g > 0$; so $u > 0$, and therefore $g'' = g \vee g'$. Now, $h'' \vee h' \leq h$ since otherwise $h'' = h' > h$ and $v'' > 0$. If $h'' + k < g''$, and this occurs when $h'' < h$, then $(u'' v'') \in Q_k$. Suppose, then, that $h = h''$ and $g'' \leq h'' + k$. So, $g'' \leq h'' + k = h + k \leq g \leq g''$, and $g = g''$, $h = h''$, $g = h + k$ and $g'' = h'' + k$. Then $v''_{h''}$ is given as in (xi) and

$$u''_{g''} = \begin{cases} u_g & \text{if } g' < g \\ u_g + u'_{g'} & \text{if } g = g'. \end{cases}$$

Consequently, $u''_{g''} + v''_{h''} > 0$ and $(u'', v'') \in Q_k$. □

We will now apply this construction to produce large disjoint sets in free f-modules.

Theorem 4.5.17. *Let A be a unital totally ordered domain and let G be an abelian totally ordered group. Let R be a subring of the power series ring $V(G,A)$ which contains the group ring $\Sigma(G,A)$ and assume R is a right Öre domain. Then each free unital R-f-module whose rank is at least 2 has a disjoint set of cardinality equal to that of G.*

Proof. Let F be a free unital R-f-module of rank 2. Since a free unital f-module of larger rank contains an isomorphic copy of F it suffices to construct the desired disjoint set in F. Since R is ℓ-simple each unital R-f-module is nonsingular and we may use the representation of F given in Theorem 4.5.3. So $\varphi : M \longrightarrow F \subseteq \Pi_{\mathscr{U}}(M,P)$ where $M = R \boxplus R$ is trivially ordered and $\mathscr{U}$ is the set of total orders of M. For $g \in G$ let ${}_g y \in F$ be defined by

$${}_g y = [\varphi((-x^g, 2x^{-g})) \wedge \varphi((2x^g, -x^{-g}))]^+.$$

To see that ${}_g y > 0$ let $k = 2g$ and let P_k be the partial order of M_R given in Theorem 4.5.16 (technically, this P_k is $P_k \cap M$). Clearly, $(-x^g, 2x^{-g})$, $(2x^g, -x^{-g}) \in P_k$. By Exercise 4.4.22(a) there is a total order P of M_R which contains P_k. Since $({}_g y)_P > 0$, we have that ${}_g y > 0$. To show that $\{{}_g y : g \in G\}$ is a disjoint subset of F we need to verify that for any two elements $g < h$ in G and any total order P of M_R the set

$$\{(-x^g, 2x^{-g}), (2x^g, -x^{-g}), (-x^h, 2x^{-h}), (2x^h, -x^{-h})\}$$

is not a subset of P. Suppose that it is contained in P. Now, $(1,0), (0,1) \in P$, and hence $R^+ \times R^+ \subseteq P$ since

$$(3,0) = [(-x^g, 2x^{-g}) + 2(2x^g, -x^{-g})]x^{-g} \in P$$

and

$$(0,3) = [2(-x^g, 2x^{-g}) + (2x^g, -x^{-g})]x^g \in P.$$

Also, $a = x^h - 2x^g \in R^+$ and $b = x^{-g} - 2x^{-h} \in R^+$, but

$$-(a,b) = (2x^g - x^h, 2x^{-h} - x^{-g}) = (2x^g, -x^{-g}) + (-x^h, 2x^{-h}) \in P.$$

This contradiction gives that $\{{}_g y : g \in G\}$ is a disjoint set. □

If the ring A in Theorem 4.5.17 is a right Öre domain, then so is the group ring $\Sigma(G,A)$; see Exercise 36. To complete the picture we show next that the cardinality of disjoint sets in free f-modules is bounded by the cardinality of the ring. The proof is quite general and works for categories of lattice-ordered Ω-algebras.

Theorem 4.5.18. *Let $\mathscr{C}$ be a subcategory of the category of f-modules over the po-ring R, and let F be a $\mathscr{C}$-free f-module. If $\aleph_\alpha$ is a regular cardinal number with card$(R) < \aleph_\alpha$, then the cardinality of a disjoint subset of F is less than $\aleph_\alpha$ unless $\alpha = 0$ in which case it is bounded by $\aleph_0$.*

Proof. We assume that R is infinite and leave the other case to Exercise 5. Let $F = F(X)$ be $\mathscr{C}$-free on X. Since $\mathscr{C}$ is hereditary the ℓ-submodule $F(Y)$ of F generated by $Y \subseteq X$ is $\mathscr{C}$-free on Y. Suppose $U = \{u_\lambda\}_{\lambda\in\Lambda}$ is a disjoint subset of $F^+\backslash\{0\}$ whose cardinality is $\aleph_\alpha$. For each $\lambda \in \Lambda$ let T_λ be a finite subset of X with $u_\lambda \in F(T_\lambda)$. From the remarks preceeding Theorem 4.5.1 an easy induction gives such a subset T_λ and also that $\text{card}(F(T_\lambda)) \leq \text{card}(R)$. For each $n \in \mathbb{N}$ let $\Lambda_n = \{\lambda \in \Lambda : \text{card}(T_\lambda) = n\}$. Since Λ is the union of the Λ_n the regularity of $\aleph_\alpha$ forces some Λ_n to have $\text{card}(\Lambda_n) = \aleph_\alpha$. We will assume that $\Lambda = \Lambda_n$. Now the set $\{T_\lambda : \lambda \in \Lambda\}$ has cardinality $\aleph_\alpha$. For, if not, then $\{T_\lambda\} = \{T_\rho : \rho < \tau\}$ for some ordinal $\tau < \omega_\alpha$. Let $U_\rho = \{u_\lambda : T_\lambda = T_\rho\}$. Then $U = \bigcup_{\rho<\tau} U_\rho$ and $\text{card}(U_\rho) = \aleph_\alpha$ for some $\rho < \tau$, yet $\text{card}(U_\rho) \leq \text{card}(F(T_\rho)) \leq \text{card}(R) < \aleph_\alpha$.

Let $k \in \mathbb{Z}^+$ be maximal such that there is a subset Y of X with k elements and Y is contained in $\aleph_\alpha$ many T_λ. Note that $0 \leq k < n$; otherwise, there exist $T_\lambda \neq T_\mu$ yet $T_\lambda = Y = T_\mu$. For each T_λ with $Y \subseteq T_\lambda$ let $Y_\lambda = T_\lambda - Y$, and note that $\text{card}(\mathscr{Y}) = \aleph_\alpha$ where $\mathscr{Y} = \{Y_\lambda\}$ is the collection of these Y_λ. By Zorn's Lemma we can find a subset Γ of Λ such that $\overline{\mathscr{Y}} = \{Y_\lambda : \lambda \in \Gamma\}$ is maximal with respect to $Y_\lambda \cap Y_\mu = \emptyset$ for each $\lambda \neq \mu$ in Γ. Suppose that $\text{card}(\Gamma) < \aleph_\alpha$. Then $W = \bigcup_{\lambda\in\Gamma} Y_\lambda$ has cardinality $< \aleph_\alpha$ and by the maximality of $\overline{\mathscr{Y}}$, $Y_\lambda \cap W \neq \emptyset$ for each $Y_\lambda \in \mathscr{Y}$. For $w \in W$ let $\mathscr{Y}_w = \{Y_\lambda \in \mathscr{Y} : w \in Y_\lambda\}$. Then $\mathscr{Y} = \bigcup_{w\in W} \mathscr{Y}_w$ and hence $\text{card}(\mathscr{Y}_w) = \aleph_\alpha$ for some $w \in W$. But this contradicts the maximality of k since $w \notin Y$ and $Y \cup \{w\}$ is contained in $\aleph_\alpha$ many T_λ. Thus, $\text{card}(\Gamma) = \aleph_\alpha$.

Let Z be a set with $n-k$ elements which is disjoint from Y, and for each $\lambda \in \Gamma$ let

$$g_\lambda : T_\lambda = Y \cup Y_\lambda \longrightarrow Y \cup Z$$

be a bijection which is the identity on Y. Then g_λ extends to an isomorphism

$$h_\lambda : F(T_\lambda) \longrightarrow F(Y \cup Z).$$

For $\lambda, \mu \in \Gamma$ there is a morphism

$$f : F(T_\lambda \cup T_\mu) \longrightarrow F(Y \cup Z)$$

in $\mathscr{C}$ determined by

$$f(y) = \begin{cases} y & \text{if } y \in Y \\ g_\lambda(y) & \text{if } y \in Y_\lambda \\ g_\mu(y) & \text{if } y \in Y_\mu \end{cases}.$$

If $\lambda \neq \mu$, then since f extends both h_λ and h_μ we have

$$h_\lambda(u_\lambda) \wedge h_\mu(u_\mu) = f(u_\lambda) \wedge f(u_\mu) = 0.$$

Thus $\{f(u_\lambda) : \lambda \in \Gamma\}$ is a disjoint subset of $F(Y \cup Z)^+ \backslash \{0\}$ whose cardinality is $\aleph_\alpha$ but $\operatorname{card}(F(Y \cup Z)) < \aleph_\alpha$. □

We can imitate the construction of the tensor product of two modules to construct two different ordered tensor products of po-modules, a po-group tensor product and an ℓ-group tensor product. For the latter we will utilize the existence of free abelian ℓ-groups over po-groups.

Let M_R be a right po-module and ${}_RN$ a left po-module over the po-ring R. Recall that a mapping $f : M \times N \longrightarrow G$ into the abelian group G is *balanced* if it is additive in each variable and $f(xa, y) = f(x, ay)$ for all $x \in M$, $y \in N$ and $a \in R$. If G is a po-group and $f(M^+ \times N^+) \subseteq G^+$, then f will be called *po-balanced.* A *po-tensor product* of M and N is a pair (T, t) where T is an abelian po-group, $t : M \times N \longrightarrow T$ is po-balanced, and for any po-balanced map $f : M \times N \longrightarrow G$ there is a unique po-homomorphism g which makes the following diagram commutative

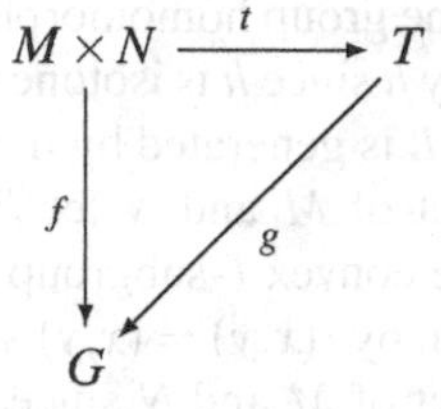

An *ℓ-tensor product* of M and N is defined analogously by stipulating that T and G are ℓ-groups and requiring g to be an ℓ-homomorphism. Clearly, if (T, t) and (T', t') are two po-tensor products (ℓ-tensor products) of M and N, then there is a unique isomorphism $\alpha : T \longrightarrow T'$ with $\alpha t = t'$. The po-tensor product and ℓ-tensor product will be denoted by $M \otimes_R^{po} N$ or $M \otimes^{po} N$ and $M \otimes_R^{\ell} N$ or $M \otimes^{\ell} N$, respectively.

In the following, by the free $\mathbb{Z}$-f-module over the abelian po-group M we mean, as usual, the free unital $\mathbb{Z}$-f-module over M (see Exercises 9–11). We will also refer to it as the free (abelian) ℓ-group over M.

Theorem 4.5.19. *If M_R and ${}_RN$ are po-modules, then $M \otimes_R^{po} N$ and $M \otimes_R^{\ell} N$ exist.*

Proof. Let A be the free abelian group on the set $M \times N$ and let A^+ be the free abelian submonoid of A generated by $M^+ \times N^+$:

$$A = \boxplus_{(x,y) \in M \times N} \mathbb{Z}(x,y), \qquad A^+ = \sum_{(x,y) \in M^+ \times N^+} \mathbb{Z}^+(x,y)\,.$$

Then (A, A^+) is a semiclosed po-group. Let K be the subgroup of A generated by the elements of the form

$$\begin{aligned} &(x_1 + x_2, y_1) - (x_1, y_1) - (x_2, y_1) \\ &(x_1, y_1 + y_2) - (x_1, y_1) - (x_1, y_2) \\ &(x_1 a, y_1) - (x_1, a y_1) \end{aligned} \tag{4.5.4}$$

where $x_1, x_2 \in M$, $y_1, y_2 \in N$ and $a \in R$. The tensor product $M \otimes_R N$ is, of course, the group A/K. Let L be the convex subgroup of A generated by K. Then A/L is a po-group, $t : M \times N \longrightarrow A/L$ given by $t(x,y) = (x,y) + L$ is po-balanced, and $(A/L, t)$ is a po-tensor product of M and N since a po-balanced map f produces the commutative diagram

$$\begin{array}{ccccc}
M \times N & \longrightarrow & A & \longrightarrow & A/L \\
\downarrow f & & \downarrow h & & \downarrow g \\
G & \xrightarrow{1} & G & \xrightarrow{1} & G
\end{array} .$$

Here, the top composite is t, h is the group homomorphism which extends f, and g is the po-homomorphism induced by h since h is isotone and $h(L) = 0$. The uniqueness of g follows from the fact that A/L is generated by the image of t.

To obtain the ℓ-tensor product of M and N let $F(A)$ be the free $\mathbb{Z}$-f-module extension of (A, A^+), let C be the convex ℓ-subgroup of $F(A)$ generated by K, and let $t : M \times N \longrightarrow F(A)/C$ be given by $t(x,y) = (x,y) + C$. Then t is po-balanced and $(F(A)/C, t)$ is an ℓ-tensor product of M and N since a po-balanced map f into an abelian ℓ-group produces the commutative diagram

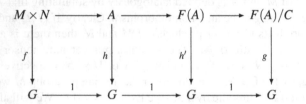

$$\begin{array}{ccccccc}
M \times N & \longrightarrow & A & \longrightarrow & F(A) & \longrightarrow & F(A)/C \\
\downarrow f & & \downarrow h & & \downarrow h' & & \downarrow g \\
G & \xrightarrow{1} & G & \xrightarrow{1} & G & \xrightarrow{1} & G
\end{array} .$$

Here, the top composite is t, h' is the ℓ-homomorphism induced by h and g is the ℓ-homomorphism which is induced by h' since $h'(C) = 0$. Again, this g is unique since $F(A)/C$ is generated by $t(M \times N)$. □

If we denote the canonical images of (x,y) in $M \otimes^{po} N$ and $M \otimes^{\ell} N$ by $x \otimes^{po} y$ and $x \otimes^{\ell} y$, respectively, then

$$M \otimes^{po} N = \left\{ \sum_i x_i \otimes^{po} y_i : x_i \in M, y_i \in N \right\},$$

$$(M \otimes^{po} N)^+ = \left\{ \sum_i x_i \otimes^{po} y_i : x_i \in M^+, y_i \in N^+ \right\},$$

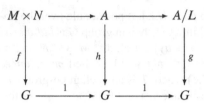

$$(M \otimes^{\ell} N) = \left\{ \bigvee_i \bigwedge_j \sum_k x_{ijk} \otimes^{\ell} y_{ijk} : x_{ijk} \in M, y_{ijk} \in N \right\},$$

$$M^+ \otimes^{\ell} N^+ \subseteq (M \otimes^{\ell} N)^+.$$

We have canonical group homomorphisms

$$M \otimes_R N \longrightarrow M \otimes_R^{po} N \longrightarrow M \otimes_R^{\ell} N \tag{4.5.5}$$

with the first being surjective and the second being isotone.

A third tensor product of M_R and ${}_RN$ can be given when they are both ℓ-modules. A balanced map $f : M \times N \longrightarrow G$ into the abelian ℓ-group G is called *ℓ-balanced* if $f(x, \ )$ and $f(\ ,y)$ are ℓ-homomorphisms whenever $x \in M^+$ and $y \in N^+$. Let $F_0(M,N)$ be the free abelian ℓ-group on the set $M \times N$ and let U be the convex ℓ-subgroup of $F_0(M,N)$ generated by the elements in (4.5.4) together with the elements

$$\begin{aligned} &(x_1 \vee x_2, y) - (x_1, y) \vee (x_2, y) \\ &(x, y_1 \vee y_2) - (x, y_1) \vee (x, y_2) \end{aligned} \tag{4.5.6}$$

where $x_1, x_2 \in M$, $y_1, y_2 \in N$, $y \in N^+$ and $x \in M^+$. Then the abelian ℓ-group $F_0(M,N)/U$, which will be denoted by $M \otimes_R^{\ell\ell} N$, is universal with respect to ℓ-balanced maps in the sense that each ℓ-balanced map f determines a unique ℓ-homomorphism g which makes the diagram

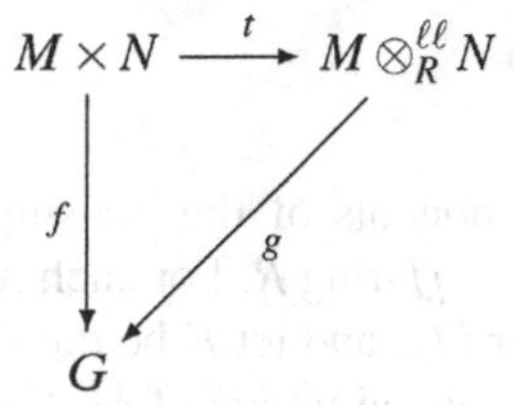

commutative. The top map is, of course, $t(x,y) = (x,y) + U$. As usual, this universal property of $M \otimes_R^{\ell\ell} N$ determines it up to isomorphism. Note that in $M \otimes_R^{\ell\ell} N$ we have for all $x \geq 0$ and $y \geq 0$

$$\begin{aligned} x \otimes^{\ell\ell} (y_1 \vee y_2) &= (x \otimes^{\ell\ell} y_1) \vee (x \otimes^{\ell\ell} y_2), \\ x \otimes^{\ell\ell} (y_1 \wedge y_2) &= (x \otimes^{\ell\ell} y_1) \wedge (x \otimes^{\ell\ell} y_2), \\ (x_1 \vee x_2) \otimes^{\ell\ell} y &= (x_1 \otimes^{\ell\ell} y) \vee (x_2 \otimes^{\ell\ell} y), \\ (x_1 \wedge x_2) \otimes^{\ell\ell} y &= (x_1 \otimes^{\ell\ell} y) \wedge (x_2 \otimes^{\ell\ell} y). \end{aligned} \tag{4.5.7}$$

Since t is ℓ-balanced it is po-balanced and hence the sequence (4.5.5) can be extended for ℓ-modules to

$$M \otimes_R N \longrightarrow M \otimes_R^{po} N \longrightarrow M \otimes_R^{\ell} N \longrightarrow M \otimes_R^{\ell\ell} N \tag{4.5.8}$$

where the last map is a surjective ℓ-homomorphism.

These ordered tensor products have some of the properties of the ordinary tensor product. For example, $M\dot{\otimes}_R$ is a functor which takes surjections to surjections. On the other hand, $M\otimes_R^\ell$ does not preserve direct sums whereas the other two do. The verification of these and other properties are left for the exercises. Here, we will be content with providing sufficient conditions for the first map in (4.5.8) to be a group isomorphism.

Theorem 4.5.20. *Let M_R and $_RN$ be nonsingular po-modules over the totally ordered Öre domain R. Then $M\otimes_R N = M\otimes_R^{po} N$ as groups.*

Proof. Using the notation from Theorem 4.5.19 it suffices to show that the subgroup K is convex in (A,A^+). If $0 < u \leq v = \sum_i a_i(x_i,y_i)$ in (A,A^+) with $a_i \in \mathbb{N}$ and $(x_i,y_i) \in M^+ \times N^+$, then $u = \sum_i b_i(x_i,y_i)$ with $0 \leq b_i \leq a_i$ for each i. Thus, it suffices to show that if $0 < v \in K$, then each $(x_i,y_i) \in K$; in fact, we claim that for each i, $x_i = 0$ or $y_i = 0$. If not, then since $(\{0\}\times N^+)\cup(M^+\times\{0\}) \subseteq K$ we may assume $x_i > 0$ and $y_i > 0$ for each i. By Exercise 4.3.22(a) there are total orders P_M and P_N of M_R and $_RN$ with $M^+ \subseteq P_M$ and $N^+ \subseteq P_N$, and by Theorem 4.5.7 there are R-homomorphisms $\alpha : M_R \longrightarrow Q(R)_R$ and $\beta : {}_RN \longrightarrow {}_RQ(R)$ with $\alpha(x_i) > 0$ and $\beta(y_i) > 0$ for every i. Then the function $f : M\times N \longrightarrow Q(R)$ defined by $f(x,y) = \alpha(x)\beta(y)$ is balanced and induces the homomorphism $g : M\otimes_R N \longrightarrow Q(R)$. But now we have the contradiction

$$0 = \textstyle\sum_i a_i g(x_i \otimes y_i) = \sum_i a_i \alpha(x_i)\beta(y_i) > 0. \qquad \square$$

Exercises.

1. Let $\{M_\lambda\}_{\lambda\in\Lambda}$ be the components of the nonsingular po-module M over the semiprime irredundant right qf-ring R. For each $\lambda \in \Lambda$, let F_λ be the free nonsingular R_λ-f-module over M_λ, and let F be the R-ℓ-submodule of the product $\Pi_\lambda F_\lambda$ generated by the canonical image of M. Show that F is the free nonsingular R-f-module over M and $\{F_\lambda\}_{\lambda\in\Lambda}$ is the set of components of F.
2. If R is the direct product of two copies of $\mathbb{R}$ show that the free unital R-f-module of rank two is the direct sum of two copies of the free unital $\mathbb{R}$-f-module of rank two.
3. Let R and M be as in Exercise 1 and suppose M is trivially ordered and uniform. Show that if F is the free nonsingular f-module over M, then $F = A\oplus A$ where $A_R \cong M_R$ as modules and A_R can be embedded in a component of $Q(R)$ as f-modules.
4. Let R be a unital right nonsingular right f-ring and let F be a free nonsingular f-module of rank u. If F is R-ℓ-torsionless and J is a set of cardinality u show that F is isomorphic to the ℓ-submodule of $R^{(R^J)}$ generated by the projections.
5. If F is a $\mathscr{C}$-free R-f-module and $FR = 0$ show that each disjoint subset of F is countable.

6. If R is a unital f-ring show that R_R is projective in u-f-$\mathcal{M}_R$ but is not free.

7. Let $\mathcal{D}$ be a nontrivial subcategory of po-$\mathcal{M}_R$ where R is a po-ring. Suppose $M \in \mathcal{D}$ and $(M,M^+) \oplus (M,-M^+) \in \mathcal{D}$. If M is generated by X and $X \cap (M^+ \cup -M^+) \neq \emptyset$ show that M is not $\mathcal{D}$-free on X. In particular, if M is a totally ordered nonzero module, then M is not free in $\mathcal{D}$.

8. (a) Let M_R be a nonsingular f-module over the totally ordered right Öre domain R and let (F,φ) be the free f-module over M. Show that the following are equivalent.
   (i) $\varphi(M) = F$.
   (ii) M is totally ordered.
   (iii) po-$\mathcal{M}_R[M,K] = f$-$\mathcal{M}_R[M,K]$ for every nonsingular f-module K.
   (iv) po-$\mathcal{M}_R[M,K] = f$-$\mathcal{M}_R[M,K]$ for every f-module K.

   (b) Let R be a semiprime irredundant right qf-ring. Show that R is totally ordered iff each nonsingular f-module which is free over itself is totally ordered.

9. (a) Show that each free R-module is a direct sum of free R-modules of rank one.
   (b) If R is unital show that the R-module direct sum $R \boxplus \mathbb{Z}$, with $\mathbb{Z}R = 0$, is the free R-module of rank one.
   (c) If R is not unital show that as a right R-module the ring $R_1 = R + \mathbb{Z}$ obtained by freely adjoining $\mathbb{Z}$ to R is the free R-module of rank one.

10. Let M_R be the trivially ordered free R-module of rank one over the unital po-ring R.
   (a) Show that N is a submodule of M for which, for some P, $(M/N,P)$ is an element of the set $\mathcal{S}$ given in Theorem 4.5.1 iff $N = J$ or $N = J \boxplus \mathbb{Z}$ where J is a convex right ideal of R with the property that the partial order of R/J can be extended to a total order of $(R/J)_R$ (see Exercise 3.1.20(f)).
   (b) Suppose R is an ℓ-ring. Show that $(M/N,P) \in \mathcal{S}$ and $1+J$ is a d-element on R iff J is a prime submodule of R_R.
   (c) Suppose R is an ℓ-primitive f-ring. Determine the free R-f-module of rank one.

11. Let R be an f-ring with a left identity element. Show that $R \oplus R$ is the free strong right f-module over the trivially ordered module R_R.

12. Let R be a directed po-ring, $\mathcal{C}$ a subcategory of f-$\mathcal{M}_R$ and $\mathcal{A}$ the subcategory of po-$\mathcal{M}_R$ consisting of those po-modules M for which the free $\mathcal{C}$-representable f-module over M, F_M, exists.
   (a) Show that $F : \mathcal{A} \longrightarrow \mathcal{C}$ is a functor and the function $\varphi : 1 \longrightarrow F$ that is given in Theorem 4.5.1 is a natural transformation.
   (b) If $f : M \longrightarrow N$ is a surjection in $\mathcal{A}$ show that $F(f)$ is a surjection.
   (c) Suppose that either R is a domain satisfying the conditions in Theorem 4.5.3 and $\mathcal{C}$ is the category of t-torsion-free f-modules, or R is an irredundant

semiprime right qf-ring and $\mathscr{C} = nsf\text{-}\mathscr{M}_R$. If $f : M \longrightarrow N$ is an injection in $\mathscr{A}$ show that $F(f)$ is an embedding. (In the first case show that each total order of $f(M)$ which contains $f(M^+)$ can be extended to a total order of N which contains N^+.

13. Let $R = \mathbb{Z}[x]/(2x^2)$ be the homomorphic image of the polynomial ring $\mathbb{Z}[x]$ supplied with the partial order $R^+ = \mathbb{Z}^+$. Let $M = N = R$ as R-modules with $M^+ = N^+ = \mathbb{Z}^+ + \mathbb{Z}^+x$. Show that $M \otimes_R N \neq M \otimes_R^{po} N$.

14. Let M_R and ${}_RN$ be po-modules and suppose (M, P_M) and (N, P_N) are also po-modules with $P_M \subseteq M^+$ and $P_N \subseteq N^+$. If $M \otimes_R^{po} N = M \otimes_R N$ show that $(M, P_M) \otimes_R^{po} (N, P_N) = M \otimes_R N$.

15. (a) If M_R is trivially ordered show that for any po-module ${}_RN$, $M \otimes_R^{po} N \cong M \otimes_R^{po} (N, 0)$ is trivially ordered.
    (b) If $M \otimes_R N$ is also a torsion-free group, show that $M \otimes_R^{po} N = M \otimes_R N$ and the free abelian ℓ-group extension of the trivially ordered po-group $M \otimes_R N$ is $(M \otimes_R^{\ell} N, \varphi')$ where $\varphi' : M \otimes_R N \longrightarrow M \otimes_R^{\ell} N$ is the map in (4.5.5).

16. Suppose ${}_RN_S$ is a po-bimodule and S is directed.
    (a) Show that for every po-module M_R, $M \otimes_R^{po} N$ is a right po-S-module and $M \otimes_R^{\ell} N$ is a right ℓ-module over S.
    (b) If M_R and ${}_RN_S$ are ℓ-modules and N_S is a d-module show that $M \otimes_R^{\ell\ell} N$ is a right d-module over S.
    (c) If R, M_R and ${}_SU$ are directed show that $(M \otimes_R^{po} N) \otimes_S^{po} U \cong M \otimes_R^{po} (N \otimes_S^{po} U)$ and $(M \otimes_R^{\ell} N) \otimes_S^{\ell} U \cong M \otimes_R^{\ell} (N \otimes_S^{\ell} U)$.

17. (a) Let M_R and ${}_RN$ be nonsingular po-modules over the totally ordered Öre domain R. If $M \otimes_R N$ is a torsion-free group show that the free abelian ℓ-group over the po-group $M \otimes_R N = M \otimes_R^{po} N$ coincides with $M \otimes_R^{\ell} N$.
    (b) If Q is the totally ordered division ring of quotients of R show that $M \otimes_R Q = M \otimes_R^{\ell} Q$ iff M is totally ordered (see Exercise 8).

18. Let $\alpha : M'_R \longrightarrow M_R$ and $\beta : {}_RN' \longrightarrow {}_RN$ be po-R-homomorphisms. Show that there is a unique po-homomorphism $\alpha \otimes^{po} \beta : M' \otimes_R^{po} N' \longrightarrow M \otimes_R^{po} N$ and a unique ℓ-homomorphism $\alpha \otimes^{\ell} \beta : M' \otimes_R^{\ell} N' \longrightarrow M \otimes_R^{\ell} N$ given by $x' \dot{\otimes} y' \mapsto \alpha(x') \dot{\otimes} \beta(y')$. If the modules are ℓ-modules and α and β are ℓ-homomorphisms show that $\alpha \otimes^{\ell\ell} \beta : M' \otimes_R^{\ell\ell} N' \longrightarrow M \otimes_R^{\ell\ell} N$ is an ℓ-homomorphism.

19. Show that $M \otimes_R^{po} \cdot$, $M \otimes_R^{\ell} \cdot$, and $M \otimes_R^{\ell\ell} \cdot$ are functors, the first and third of which preserve direct sums. Give an example to show that $M \otimes_R^{\ell} \cdot$ need not preserve direct sums (see Exercise 15).

20. (a) Let M_R be a po-module and let $G = M^+R + M^+ - M^+$ be the submodule of M generated by M^+. For $N \in$ po-$\mathscr{M}_R$, let $[M,N] = \mathrm{Hom}_R(M,N)$ and $[M,N]^+ = $ po-$\mathscr{M}_R[M,N]$. Show that the following are equivalent.
        (i) For every $N \in$ po-$\mathscr{M}_R$, $[M,N]^+$ is a partial order of $[M,N]$.
        (ii) $[M, M/G]^+$ is a partial order of $[M, M/G]$.

(iii) $M = G$.

(b) Let R be a unital po-ring. Show that $([R,R], [R,R]^+) \cong R_R$ as po-modules iff $1 \in R^+$, iff $([R,N], [R,N]^+) \cong N_R$ for every unital po-module N_R.

21. Let R and S be po-rings with S directed, let ${}_SM_R$ be a po-bimodule, and let ${}_RN$ and ${}_SL$ be po-modules. Assume $M = M^+R + M^+ - M^+$ and $N = RN^+ + N^+ - N^+$ (see Exercise 20).

(a) Show that $\mathrm{Hom}_S(M \otimes_R^{po} N, L)^+ \cong \mathrm{Hom}_R(N, \mathrm{Hom}_S(M,L))^+$ as po-monoids.

(b) Show that there is a natural bijection between $\ell\text{-}_S\mathcal{M}[M \otimes_R^{\ell} N, L]$ and $\mathrm{Hom}_R(N, \mathrm{Hom}_S(M,L))^+$ whenever L is an ℓ-module.

22. Let M_R be a unital po-module over the po-unital po-ring R. Show that $M \otimes_R^{po} R \cong M$ and if R is an ℓ-ring and M is a strong d-module, then $M \otimes_R^{\ell\ell} R \cong M$.

23. Let R and S be po-rings with S po-unital and directed, and suppose M_R and ${}_RS$ are po-modules. Assume that $\ell\text{-}\mathcal{M}_S = f\text{-}\mathcal{M}_S$ (see Theorem 3.1.3 and Exercise 3.1.11).

(a) Show that the map $\varphi : M \longrightarrow M \otimes_R^{\ell} S$ given by $\varphi(x) = x \otimes 1$ is an isotone group homomorphism which is an R-homomorphism if R is directed, ${}_RS_R$ is a po-bimodule and $1 \cdot r = r \cdot 1$ for each $r \in R$.

(b) Suppose R is directed and $R \longrightarrow S$ is a po-ring homomorphism which induces the actions of R on S. Show that if $\psi : M \longrightarrow K_S$ is an R-po-homomorphism into the unital f-module K_S, then there is a unique S-ℓ-homomorphism $\sigma : M \otimes_R^{\ell} S \longrightarrow K$ with $\sigma\varphi = \psi$.

(c) Suppose $R = S$ and the homomorphism $R \longrightarrow S$ in (b) is the identity map. Show that the free unital f-module over M exists iff φ is one-to-one, iff $(M \otimes_R^{\ell} R, \varphi)$ is the free unital f-module over M.

24. Let M_R be a unital po-module over the po-unital po-ring R and assume M is a torsion-free group.

(a) If R is directed and the functor $M \otimes_R^{\ell} \cdot \; : u\text{-po-}_R\mathcal{M} \longrightarrow \ell\text{-}\mathcal{M}_{\mathbb{Z}}$ preserves monics show that M is trivially ordered, and if M is also a divisible group show that M_R is a flat module.

(b) If M_R is flat and M is divisible and trivially ordered show that $M \otimes_R^{\ell} \cdot$ preserves monics.

(Use Exercises 15, 23, and 12.)

25. If R is a commutative po-ring show that $M \dot{\otimes}_R N \cong N \dot{\otimes}_R M$ for each of the three ordered tensor products.

26. Suppose R and S are directed po-algebras over the directed po-ring C. Show that $R \otimes_C^{po} S$ is a directed po-algebra and $R \otimes_C^{\ell} S$ is an ℓ-algebra.

27. Let M_R and ${}_RN$ be ℓ-modules and let $x \in M$, $y \in N$. Verify the following:

(a) $(x \otimes^{\ell\ell} y)^+ = x^+ \otimes^{\ell\ell} y^+ + x^- \otimes^{\ell\ell} y^-$.

(b) $|x \otimes^{\ell\ell} y| = |x| \otimes^{\ell\ell} |y|$.

(c) If $u \in (M \otimes^{\ell\ell} N)^+$, then $u \leq x \otimes^{\ell\ell} y$ for some $x \in M^+$ and some $y \in N^+$.

28. Give the definitions of a free representable ℓ-group over or extending the po-group M and prove the analogues of Theorems 4.5.1–4.5.3 in this context (see Exercise 2.4.3).

29. Suppose $\bigvee_{i\in I} \bigwedge_{j\in J} x_{ij} \neq 0$ in the ℓ-group G where I and J are finite. Show that there is a left order P of G extending G^+ such that $\bigvee_i \bigwedge_j x_{ij} \neq 0$ in (G,P). (Use Exercises 2.4.1 and 2.4.2 and consider the two cases that arise depending on whether or not $\bigwedge_j x_{ij} \not\leq 0$ for some $i \in I$.)

30. Assume that $\bigvee_{i\in I} \bigwedge_{j\in J} x_{ij} \neq 0$ (I and J finite) in the left O-group G. If $\rho : G \longrightarrow \mathrm{Aut}(G)$ is the left regular representation of G ($(\rho(g))(x) = g + x$) show that $\bigvee_i \bigwedge_j \rho(x_{ij}) \neq 1$.

31. Let $\psi : M \longrightarrow G$ be an isotone homomorphism between left po-groups. Suppose T is a left order of G containing G^+ and S is a left order of $K = \ker \psi$ containing K^+. Show that $P = \{x \in M : 0 \neq \psi(x) \in T\} \cup S$ is a left order of M containing M^+ and $\psi : (M,P) \longrightarrow (G,T)$ is isotone.

32. Suppose M is a po-group, $\psi : M \longrightarrow G$ is a po-homomorphism into the ℓ-group G, and $\ker \psi$ has a left order containing $(\ker \psi)^+$. If $\bigvee_i \bigwedge_j \psi(x_{ij}) \neq 0$ where $\{x_{ij} : i \in I, j \in J\} \subseteq M$ and I and J are finite, show there is a left order P of M such that $\bigvee_i \bigwedge_j \rho(x_{ij}) \neq 1$ in $\mathrm{Aut}(G,P)$ where $\rho : (M,P) \longrightarrow \mathrm{Aut}(M,P)$ is the left regular representation of (M,P). (Use Exercises 28–30.)

33. Let M be a po-group such that $M^+ = \bigcap_\lambda P_\lambda$ where $\{P_\lambda : \lambda \in \Lambda\}$ is the set of left orders of M containing M^+. Let $\varphi : M \longrightarrow \Pi_\lambda \mathrm{Aut}(M,P_\lambda) = H$ be the embedding induced by the left regular representations $\rho_\lambda : M \longrightarrow \mathrm{Aut}(M,P_\lambda)$ of M on the totally ordered sets (M,P_λ). If F is the ℓ-subgroup of H generated by $\varphi(M)$ show that (F,φ) is the free ℓ-group extension of M. (Use Exercises 31 and 2.4.22.)

34. Show that there is a free ℓ-group extension of the po-group M iff M can be embedded in an ℓ-group, iff M^+ is the intersection of left orders. If M is trivially ordered these conditions reduce to M being left orderable.

35. Show that Exercise 8(a) holds for the category of ℓ-groups.

36. Let X be a free generating set for the free group M.

    (a) Show that the free ℓ-group F over the trivially ordered po-group M is the free ℓ-group on X.
    (b) If K is the convex ℓ-subgroup of F generated by the derived subgroup F' of F show that F/K is the free abelian ℓ-group on the set $\{x + K : x \in X\}$.
    (c) If Y is a set of free generators for the ℓ-group F show that $\mathrm{card}(Y) = \mathrm{card}(X)$. Thus, the *rank* of a free ℓ-group is well-defined.
    (d) Show that a free ℓ-group of rank 2 contains a free ℓ-group of countable rank. (Consider the commutator subgroup M'.)

37. Let A be a right Öre domain.

    (a) Show that the polynomial ring $A[x_1,\ldots,x_n]$ in the commuting indeterminates $x_1,\ldots,x_n$ is a right Öre domain.
    (b) If G is a torsion-free abelian group show that the group ring $A[G]$ is a right Öre domain. (Reduce to the case G is finitely generated and use (a).)

38. Show that there are exactly four varieties of right f-modules over a unital ℓ-simple totally ordered left and right Öre domain. If R is the direct product of n such totally ordered rings determine all the varieties of right R-f-modules.

Notes. In [WE2] and [WE3] Weinberg gave the representations of and initiated the study of free abelian ℓ-group extensions of abelian po-groups and of free abelian ℓ-groups. He showed that the variety of abelian ℓ-groups is generated by $\mathbb{Z}$. Bigard [BI3] showed that Weinberg's construction holds for all t-torsion-free po-modules over a totally ordered domain if and only if the domain is a right Öre domain. Powell [PO1] modified the construction by making use of total orders of homomorphic images to obtain free unital f-modules and free f-module extensions of unital po-modules over right f-rings. The other results on free f-modules are mainly generalizations to f-modules of results about abelian ℓ-groups (or real vector lattices) from Baker [B], Bernau [BERN2], Bleier [BL], and Conrad [CON11]. Theorem 4.5.14 comes from Ross and Stone [RS] and Theorems 4.5.16 and 4.5.17 are due to Powell and Tsinakis [PT]. The po-tensor product comes from Viswanathan [V3] and the ℓ-tensor product (for ℓ-groups) comes from Martinez [MART]. The representation of free ℓ-groups given in Exercises 29–36 is due to Conrad [CON9].

Chapter 5
Lattice-ordered Fields

In this chapter we will concentrate on lattice-ordered fields. Since more is known about totally ordered fields than about ℓ-fields in general most of this chapter will be concerned with totally ordered fields. Examples of ℓ-fields come from power series ℓ-rings and from constructing lattice orders on the reals and other similar totally ordered fields. We will develop the algebraic properties of totally ordered fields, including the existence and uniqueness of its real closure, which culminates in a description of those fields whose algebraic closure is a finite extension. In order to show that some commutative ℓ-domains can be embedded in power series ℓ-fields with real coefficients and with exponents in the associated value po-group—the Hahn Embedding Theorem—we will need to develop enough valuation theory to first carry out the embedding for totally ordered fields. We will also see that a totally ordered division ring can be enlarged to one whose center contains the reals.

5.1 Totally Ordered Extensions of Totally Ordered Fields

It is not surprising that the totally ordered fields are the most studied and best understood ℓ-rings and the easiest in which to compute. In this section we will present the algebraic condition which determines whether or not a field can be totally ordered and we will see that each totally ordered field has a unique largest totally ordered algebraic extension. These are the fields of codimension 2 in their algebraic closures. Another property they share with the real numbers is that each positive element is a square. Any such field which is also an η_α-set for the regular cardinal $\aleph_\alpha$ is quite large in the sense that it contains each totally ordered field whose cardinality is bounded by $\aleph_\alpha$.

A field can be totally ordered exactly when it has a partial order with respect to which it is an sp-po-ring. More generally, we can determine when a partial order of a field can be extended to a total order of an extension field. If P is a nonzero partial order of a commutative domain R, then P_{P^*} is a partial order of the ring of quotients R_{P^*} which contains P and which will be denoted by $Q(P)$.

S.A. Steinberg, *Lattice-ordered Rings and Modules*,
DOI 10.1007/978-1-4419-1721-8_5,

Theorem 5.1.1. *Let F be a field which contains the partially ordered field K with $K^+ \neq 0$. The following statements are equivalent.*

(a) *K^+ is contained in a total order of F.*
(b) *-1 is not a linear combination of squares in F with coefficients from $Q(K^+)$.*
(c) *If $k_1 f_1^2 + \cdots + k_n f_n^2 = 0$ with $k_i \in Q(K^+)$ and $f_i \in F$, then $k_i f_i = 0$ for $i = 1, \ldots, n$.*

Proof. Since the implications (a) $\Rightarrow$ (b) $\Rightarrow$ (c) are clear we need only show that (a) is a consequence of (c). Let

$$P = \{k_1 f_1^2 + \cdots + k_n f_n^2 : k_i \in Q(K^+), f_i \in F\}.$$

Then P is a subsemiring of F which contains K^+. If $u = \sum k_i f_i^2 \in P \cap -P$, then $0 = u + (-u) = \sum k_i f_i^2 + \sum k_j f_j^2$ and, as a consequence of (c), each $k_i f_i = 0$ and hence $u = 0$. Thus P is an sp-partial order of F. By Zorn's lemma there is a maximal partial order F^+ of F which contains P. If $x \in F \backslash F^+$, then $F^+ - F^+ x$ is easily seen to be a partial order of F (or use Exercise 1); hence $F^+ - F^+ x = F^+$, $-x \in F^+$ and F^+ is a total order of F. □

We record an easy consequence of Theorem 5.1.1.

Theorem 5.1.2. *The following are equivalent for the field F.*

(a) *F has a total order.*
(b) *-1 is not a sum of squares in F.*
(c) *If $f_1^2 + \cdots + f_n^2 = 0$, then $f_1 = \cdots = f_n = 0$.*

Proof. Under the assumption of (c) the characteristic of F is 0 and hence the previous result can be applied with $K = \mathbb{Q}$. Specifically, if $m, m_i \in \mathbb{Z}^+$, $f_i \in F$ and

$$\frac{m_1}{m} f_1^2 + \cdots + \frac{m_n}{m} f_n^2 = 0,$$

then each $f_i = 0$. Consequently, F has a total order. Since the other implications are trivial the proof is complete. □

With some more work this theorem can be given for a division ring or even for a domain; see Exercises 2, 3, and 4. We will now apply Theorem 5.1.1 to the situation where F is a finite extension of K. The polynomial $f(x) \in K[x]$, K a totally ordered field, is said to *change signs in K* if there are elements $a, b \in K$ with $f(a)f(b) < 0$.

Theorem 5.1.3. *Let K be a totally ordered field and suppose $f(x) \in K[x]$ is irreducible and changes signs in K. Then the field $K[x]/(f(x))$ has a total order containing K^+.*

Proof. We induct on $n = \deg f(x)$, the case $n = 1$ being trivial. Suppose $n > 1$ and the conclusion is true for all irreducible polynomials of degree $< n$. If it is not true for $f(x)$, then there are nonzero elements $a_1, \ldots, a_m \in K^+$ and nonzero polynomials $f_1(x), \ldots, f_m(x)$, $g(x) \in K[x]$ with $1 + a_1 f_1(x)^2 + \cdots + a_m f_m(x)^2 = f(x)g(x)$

and $\deg f_i(x) \leq n-1$; hence $\deg g(x) \leq n-2$. If $a,b \in K$ with $f(a)f(b) < 0$, then, since $f(c)g(c) > 0$ for every $c \in K$, necessarily, $g(a)g(b) < 0$. But then $g(x)$ has an irreducible factor $h(x)$ with $h(a)h(b) < 0$. This gives the contradiction that $\overline{K[x]} = K[x]/(h(x))$ has a total order containing K^+ yet $-1 = a_1\overline{f}_1^2 + \cdots + a_m\overline{f}_m^2$ in $\overline{K[x]}$. □

Theorem 5.1.4. *Suppose $F = K(a)$ is a simple extension of the totally ordered field K. If $a^2 \in K^+$ or $[F:K]$ is odd, then F has a total order containing K^+.*

Proof. Assume $F \neq K$ and let $f(x)$ be the irreducible polynomial of a over K. Then $f(x)$ changes signs in K. For if $f(x) = x^2 - a^2$, then $f(0) = -a^2 < 0 < a^4 + a^2 + 1 = f(a^2+1)$. On the other hand, suppose $f(x) = a_0 + a_1x + \cdots + x^n$ with n odd. Take $a \in K$ with $1 \vee (|a_0| + \cdots + |a_{n-1}|) < |a|$. Then $|a^{-k}| < 1$ if $k \geq 1$,

$$|a_0a^{1-n} + a_1a^{2-n} + \cdots + a_{n-1}| \leq |a_0| + \cdots + |a_{n-1}| < |a|,$$

and $|a_0a^{-n} + a_1a^{1-n} + \cdots + a_{n-1}a^{-1}| < 1$. So the sign of $f(a) = a^n(a_0a^{-n} + \cdots + a_{n-1}a^{-1} + 1)$ is the same as the sign of a. The conclusion is now a consequence of Theorem 5.1.3. □

A field is called *real closed* if it has a total order for which no proper algebraic extension is a totally ordered extension. We will see that any real closed field, like $\mathbb{R}$, has a unique total order. A *real closure* of a totally ordered field F is a real closed algebraic extension $\overline{F}$ of F whose total order extends that of F. It is easy to see that each totally ordered field F has a real closure. Just take, by Zorn's lemma, a maximal totally ordered extension of F within its algebraic closure. The uniqueness of the real closure will require more of an effort. For now, we will give a characterization of a real closed field in terms of the familiar relationship between $\mathbb{R}$ and the field of complex numbers.

We will generically denote a root of the polynomial x^2+1 by i, and $\sqrt{a}$ denotes a positive element whose square is a.

Theorem 5.1.5. *Let F be a totally ordered field.*

(a) *If F is real closed, then $F^+ = \{a^2 : a \in F\}$.*

(b) *If every element of F^+ is a square, then every element of $F(i)$ is a square.*

Proof. If F is real closed, $b \in F^+$, and a is root of $x^2 - b$, then $F = F(a)$ by Theorem 5.1.4 and $b = a^2$. For (b), suppose $a + bi \in F(i)$. We need to find $x, y \in F$ with $(x+yi)^2 = a + bi$; equivalently, $x^2 - y^2 = a$ and $2xy = b$. Now, $a^2 + b^2 = (x^2+y^2)^2$ and $x^2 + y^2 = \sqrt{a^2+b^2}$. Since $|a| \leq \sqrt{a^2+b^2}$ the right sides of the equations

$$x^2 = \frac{a + \sqrt{a^2+b^2}}{2}, \quad y^2 = \frac{-a + \sqrt{a^2+b^2}}{2}$$

are in F^+; so $|x|$ and $|y|$ are determined and we need only choose x and y so that the equation $2xy = b$ is satisfied. □

Theorem 5.1.6. *The following statements are equivalent for the totally ordered field F.*

(a) $F(i)$ is algebraically closed.
(b) F is real closed.
(c) Every positive element of F is a square in F and each polynomial in $F[x]$ of odd degree has a root in F.

Proof. That (a) implies (b) follows from the fact that $F(i)$ is the only proper algebraic extension of F, and that (b) implies (c) follows from Theorems 5.1.4 and 5.1.5 since a polynomial of odd degree has an irreducible factor of odd degree.

(c) implies (a). Let L be a finite extension of $F(i)$. We may assume that L is a Galois extension of F. Let G be the Galois group of L over F and let H be a Sylow 2-subgroup of G. If H' denotes the fixed field of H, then $[H' : F] = [G : H]$ is odd. So, if $a \in H'$ and if $f(x)$ is the irreducible polynomial of a over F, then $\deg f(x) = 1$. Thus, $H' = F$ and $G = H$ is a 2-group. If $L \neq F(i)$, then $F(i)'$, the Galois group of L over $F(i)$, has a subgroup H of index 2. But then H' is an extension of $F(i)$ of degree 2. Consequently $H' = F(i)(a)$ with $a^2 \in F(i)$, and this is impossible according to Theorem 5.1.5. Thus $L = F(i)$ and $F(i)$ is algebraically closed. □

Interestingly, Theorem 5.1.6 can be considerably improved and put in a completely algebraic setting by dropping the order hypothesis and replacing $F(i)$ by a finite proper extension of F.

Theorem 5.1.7. *The field K is real closed if and only if it is not algebraically closed but it has an extension of finite degree that is algebraically closed.*

Proof. Let L be an algebraically closed extension of K of finite degree. We claim L is a Galois extension of K. If not, since L is a splitting extension of K it cannot be separable over K. But then there is an element a in K which is not a pth power in K where p is the characteristic of K. Let $u \in L$ with $u^p = a$. The polynomial $x^p - a$ is irreducible over K. For, the roots of $x^p - a$ are all of the form tu with $t^p = 1$, and if $f(x)$ is an irreducible factor of $x^p - a$ of degree $m \leq p-1$, then $\pm f(0) = t_1 u \cdots t_m u = su^m$ with $s^p = 1$. There exist integers q and k with $1 = qm + kp$; so $u = u^{mq}u^{pk} = (\pm f(0)s^{-1})^q a^k$, $us^q = (\pm f(0)^q a^k \in K$ and $(us^q)^p = a$. This contradiction shows that $x^p - a$ is irreducible. Now, u is not a pth power in $K(u)$ since the pth powers of elements of $K(u)$ are in $K : (\sum a_j u^j)^p = \sum a_j{}^p a^j \in K$. If $u_1 \in L$ is a root of $x^p - u$, then we have $K \subset K(u) \subset K(u, u_1)$ and we could continue this process to get an extension of K inside L of degree $p^n > [L : K]$. Thus, L is a Galois extension of K.

Let q be a prime divisor of the Galois group G of L over K, let H be a subgroup of G of order q and let E be the fixed field of H. So $[L : E] = q$. We claim that q is not the characteristic of K. Otherwise, by Exercise 5(d), $L = E(u)$ where u is a root of the irreducible polynomial $x^q - x - a$ in $E[x]$. In this case let $\psi : L \longrightarrow L$ be the function given by $\psi(b) = b^q - b$ and let $\varphi : E \longrightarrow E$ be the restriction of ψ. Clearly, ψ is onto and φ is not since $a \notin \varphi(E)$. Let $T : L \longrightarrow E$ be the trace function (see Exercise 5). If σ is a generator for H, then for $b \in L$

$$(T \circ \psi)(b) = T(b^q - b) = \sum_{j=0}^{q-1} (\sigma^j(b)^q - \sigma^j(b))$$
$$= T(b)^q - T(b) = (\varphi \circ T)(b)$$

and $\varphi \circ T = T \circ \psi$ is onto. This contradiction shows that q is not the characteristic of K. Let $t \in L$ be a primitive qth root of unity. Since t is a root of $1 + x + \cdots + x^{q-1}$ and $[L:E] = q$ we have $t \in E$, and hence $L = E(b)$ where b is a root of the polynomial $x^q - a \in E[x]$ (Exercise 5(e)). Now,

$$x^q - a = (x - b_1)\cdots(x - b_q) = (-1)^q b_1 \cdots b_q + \cdots + x^q$$

and $b_1 \cdots b_q = (-1)^{q+1} a$. Take $c \in L$ with $c^q = b$ and let $N: L \longrightarrow E$ be the norm function (Exercise 5). Then $N(c)^q = N(b) = b_1 \cdots b_q = (-1)^{q+1} a$. Since a is not a qth power in E we get that $q = 2$. If i were in E, then a would be a square in E since $a = (iN(c))^2$. So $i \notin K$. If $K(i) \neq L$, then by replacing K by $K(i)$ we would have $i \notin K(i)$. So $L = K(i)$, and hence by the computation in the proof of Theorem 5.1.5 the sum of two squares in K is also a square in K. An application of Theorems 5.1.2 and 5.1.6 completes the proof. □

We now turn our attention to showing the uniqueness of the real closure. An essential ingredient for this purpose is the fact that if a polynomial has a root in one real closure, then it has a root in every real closure. In order to establish this we will develop a procedure that allows us to count the number of roots of the polynomial in a closed interval.

A finite sequence of polynomials $f_0(x), f_1(x), \ldots, f_m(x)$ in $F[x]$, F a totally ordered field, is called a *Sturm sequence* for $f(x) = f_0(x)$ on the interval $[a,b]$ if it satisfies the following four conditions.

(i) $f(a)f(b) \neq 0$.
(ii) $f_m(x)$ has no roots in $[a,b]$.
(iii) If $c \in [a,b]$ is a root of $f_j(x)$, $1 \leq j \leq m-1$, then $f_{j-1}(c)f_{j+1}(c) < 0$.
(iv) If $c \in [a,b]$ is a root of $f(x)$, then there exist $c_1, c_2 \in [a,b]$ with $c_1 < c < c_2$ such that $f(t)f_1(t) < 0$ if $t \in (c_1, c)$ and $f(t)f_1(t) > 0$ if $t \in (c, c_2)$.

We hasten to construct such a sequence using the usual algorithm for finding the greatest common divisor of $f(x)$ and $f'(x)$ except that the remainders, but not the divisors, are replaced by their negatives. The *standard sequence* for $f(x) \in F[x] \backslash F$ is the sequence $f_0(x), f_1(x), \ldots, f_m(x)$ defined by

$$
\begin{aligned}
f_0(x) &= f(x) \text{ and } f_1(x) = f'(x) \\
f_0(x) &= q_1(x)f_1(x) - f_2(x),\ \deg f_2(x) < \deg f_1(x) \\
&\vdots \\
f_{j-1}(x) &= q_j(x)f_j(x) - f_{j+1}(x),\ \deg f_{j+1}(x) < \deg f_j(x) \\
&\vdots \\
f_{m-2}(x) &= q_{m-1}(x)f_{m-1}(x) - f_m(x),\ \deg f_m(x) < \deg f_{m-1}(x) \\
f_{m-1}(x) &= q_m(x)f_m(x).
\end{aligned}
\tag{5.1.1}
$$

Clearly, $f_m(x)$ is the greatest common divisor of $f_{j-1}(x)$ and $f_j(x)$ for $j = 1,\ldots,m$. In particular, $0 \neq f_m(x) \in F$ if $f(x)$ has no multiple roots. Let $g_j(x) = f_j(x)f_m(x)^{-1} \in F[x]$ for $j = 0,\ldots,m$. Then $g_m(x) = 1$, $f(x)$ and $g_0(x)$ have the same distinct roots, and $g_0(x)$ has no multiple roots. The sequence $g_0(x),\ldots,g_m(x)$ is called the *modified standard sequence* for $f(x)$.

Theorem 5.1.8. *Let F be a real closed field, and let $f(x)$ be a polynomial of positive degree in $F[x]$. Suppose $f(a)f(b) \neq 0$ where $a < b$ in F.*

(a) *If $f(x)$ has no multiple roots, then its standard sequence is a Sturm sequence for $f(x)$ on $[a,b]$.*
(b) *The modified standard sequence for $f(x)$ is a Sturm sequence on $[a,b]$.*

Proof. Since a Sturm sequence remains a Sturm sequence if all of its elements are multiplied by the same nonzero element of F we only need to establish (b). As above, let $g_0(x),\ldots,g_m(x)$ be the modified standard sequence for $f(x)$. We obtain the equations

$$g_{j-1}(x) = q_j(x)g_j(x) - g_{j+1}(x), \qquad 1 \leq j \leq m-1 \tag{5.1.2}$$

by dividing the equations in (5.1.1) by $f_m(x)$. Clearly, the conditions (i) and (ii) in the definition of a Sturm sequence hold for this sequence. Suppose $c \in [a,b]$ and $g_k(c) = 0$ for some $k \geq 1$. Then from (5.1.2) we have $g_{k-1}(c) = -g_{k+1}(c)$ and if $g_{k-1}(c) = 0$, then $0 = g_{k+1}(c) = \cdots = g_m(c) = 1$; so $g_{k-1}(c)g_{k+1}(c) = -g_{k+1}(c)^2 < 0$. Thus (iii) also holds. As for (iv), suppose $g_0(c) = 0$ with $c \in [a,b]$. Then $f(x) = (x-c)^n p(x)$ with $n \geq 1$ and $p(c) \neq 0$. Also,

$$f_m(x) = (x-c)^{n-1}h(x) \text{ with } p(x) = h(x)r(x)$$

and

$$f'(x) = (x-c)^n p'(x) + n(x-c)^{n-1}p(x).$$

Now, $p'(x) = h(x)s(x)$ since $h(x)$ is a divisor of both $p(x)$ and $f'(x)$ and $h(c) \neq 0$. Thus, we have

$$g_0(x) = (x-c)r(x)$$

and

$$g_1(x) = (x-c)s(x) + nr(x).$$

Since $g_1(c)r(c) = nr(c)^2 > 0$ there exists an interval $[c_1, c_2] \subseteq [a,b]$ with $c \in (c_1, c_2)$ and $g_1(t)r(t) > 0$ for $t \in [c_1, c_2]$: just exclude the roots of $g_1(x)r(x)$ from $[c_1, c_2]$ and then $g_1(x)r(x)$ has constant sign in $[c_1, c_2]$ (Exercise 6(a)). Then $g_0(t)g_1(t) = (t-c)g_1(t)r(t) < 0$ if $t \in (c_1, c)$ and $g_0(t)g_1(t) > 0$ if $t \in (c, c_2)$. □

The *number of variations in sign* of the sequence of nonzero elements $a_1, a_2, \dots, a_m$ in the totally ordered field F is the number of j with $a_j a_{j+1} < 0$, and the *number of variations in sign* of an arbitrary sequence of elements in F is the number of variations in sign of the sequence obtained by omitting all zero terms.

Theorem 5.1.9. *Let $f_0(x), f_1(x), \dots, f_m(x)$ be a sequence of polynomials in $F[x]$ where F is a real closed field, and suppose $V(c) = V_{f_0}(c)$ denotes the number of variations in sign of the sequence $f_0(c), f_1(c), \dots, f_m(c)$. If the sequence is a Sturm sequence for $f(x)$ on $[a,b]$ or is the standard sequence for $f(x)$ and $f(a)f(b) \neq 0$, then the number of distinct roots of $f(x)$ in the interval $[a,b]$ is $V(a) - V(b)$.*

Proof. The fact that a polynomial which has no roots in an interval must have constant sign in that interval will be used repeatedly (Exercise 6). Suppose that $f_0(x), \dots, f_m(x)$ is a Sturm sequence for $f(x)$ on $[a,b]$ and let $h(x) = f_0(x) \cdots f_m(x)$ be the product of these polynomials. We will initially assume that there is an element $c \in [a,b]$ which is the only possible root of $h(x)$ in $[a,b]$. If $h(c) \neq 0$ then $V(a) - V(b) = 0$. Suppose $f_j(c) = 0$ with $1 \leq j \leq m-1$. Then $f_{j-1}(c)f_{j+1}(c) < 0$ by (iii), and hence both $f_{j-1}(x)$ and $f_{j+1}(x)$ have constant sign in $[a,b]$. Hence, each of the subsequences

$$f_{j-1}(a), f_j(a), f_{j+1}(a)$$

$$f_{j-1}(c), f_j(c) = 0, f_{j+1}(c)$$

$$f_{j-1}(b), f_j(b), f_{j+1}(b)$$

has one variation in sign. Now suppose $f(c) = 0$. Then $a < c < b$ and, by (iv), $f_0(t)f_1(t) < 0$ if $t \in [a,c)$ and $f_0(t)f_1(t) > 0$ if $t \in (c,b]$. Since $f_0(x)$ and $f_1(x)$ have constant signs in each of the intervals $[a,c)$ and $(c,b]$ the subsequence $f_0(a), f_1(a)$ has a variation in sign whereas the subsequence $f_0(b), f_1(b)$ does not. So $V(a) - V(b) = 0$ if $f(c) \neq 0$, $V(a) - V(b) = 1$ if $f(c) = 0$, and the assertion is proven in this special case.

In general, let $c_1 < c_2 < \cdots < c_k$ be all the roots of $h(x)$ in $[a,b]$ and put $c_0 = a$. For each $j = 1, \dots, k$ choose $a_j \in F$ with $c_{j-1} < a_j < c_j$ except choose $a_1 = a$ if $c_1 = a$. Then

$$c_0 = a \leq a_1 \leq c_1 < a_2 < c_2 < \cdots < a_j < c_j < a_{j+1} < \cdots < a_k < c_k \leq b = a_{k+1}$$

and

$$V(a) - V(b) = V(a) - V(a_1) + \cdots + V(a_j) - V(a_{j+1}) + \cdots + V(a_k) - V(b).$$

The previous paragraph gives that $V(a) - V(a_1) = 0$ and, for $j = 1, \dots, k$, $V(a_j) - V(a_{j+1}) = 1$ or 0 depending on whether c_j is or is not a root of $f(x)$. Thus, $V(a) - V(b)$ is the number of different roots of $f(x)$ in $[a,b]$.

Now, let $f_0(x),\ldots,f_m(x)$ be the standard sequence for $f(x)$ and let $g_0(x),\ldots,g_m(x)$ be the modified standard sequence for $f(x)$. According to Theorem 5.1.8 the latter is a Sturm sequence for $g_0(x)$ on $[a,b]$ and hence the number of roots of $g_0(x)$ in $[a,b]$ is equal to $\overline{V}(a)-\overline{V}(b)$ where $\overline{V}(c)$ denotes the number of variations in sign of the sequence $g_0(c),\ldots,g_m(c)$. Since $f(x)$ has the same roots as $g_0(x)$, $f(a)f(b)\neq 0$, and

$$(f_0(c),f_1(c),\ldots,f_m(c)) = f_m(c)(g_0(c),g_1(c),\ldots,g_m(c)),$$

it is clear that $V(a)=\overline{V}(a), V(b)=\overline{V}(b)$ and $V(a)-V(b)$ is the number of distinct roots of $f(x)$ in $[a,b]$. □

We can now verify the uniqueness of real closures. We do a little more.

Theorem 5.1.10. *Let F be a real closure of the totally ordered field K and suppose $\sigma: K \longrightarrow L$ is an embedding of K into the real closed field L. Then σ can be extended to an embedding of F into L. If L is a real closure of $\sigma(K)$, then this extension is an isomorphism and is the unique field monomorphism which extends σ. In particular, two real closures of K are K-isomorphic.*

Proof. We will first show that if E is a finite dimensional extension of K within F, $K\subseteq E\subseteq F$, then σ can be extended to an embedding of the totally ordered field E into L. The extension of σ to $K[x]$ will also be denoted by σ and we will write $\sigma(f(x)) = f^{\sigma}(x)$. For some $c\in E$ we have $E=K(c)$. Let $f(x)=a_0+a_1x+\cdots+x^n$ be the irreducible polynomial of c over K. If $a\in K$ with $a>1\vee(|a_0|+\cdots+|a_{n-1}|)$ then the computation in Theorem 5.1.4 shows that $f(a)>0$ and all of the roots of $f(x)$ in F lie in the F-interval $(-a,a)$. Let $f_0(x),f_1(x),\ldots,f_m(x)$ be the standard sequence for $f(x)$. Since σ commutes with differentiation it is clear from (5.1.1) that $f_0^{\sigma}(x),\ldots,f_m^{\sigma}(x)$ is the standard sequence for $f^{\sigma}(x)$. Also, σ takes each of the two sequences (all +'s or all –'s) $f_0(\pm a),f_1(\pm a),\ldots,f_m(\pm a)$ into the corresponding sequence of the pair $f_0^{\sigma}(\pm\sigma(a)),f_1^{\sigma}(\pm\sigma(a)),\ldots,f_m^{\sigma}(\pm\sigma(a))$. So by the previous theorem $1\le V_f(-a)-V_f(a)=V_{f^{\sigma}}(-\sigma(a))-V_{f^{\sigma}}(\sigma(a))=k$ and $f^{\sigma}(x)$ has $k\geq 1$ roots in L. Corresponding to these k roots are k field embeddings $\sigma_1,\ldots,\sigma_k$ of E into L, each of which extends σ. If none of them were isotone we could find elements $b_1,\ldots,b_k\in E^+$ with $\sigma_j(b_j)$ not a square in L. Take $d_j\in F^+$ with $b_j=d_j^2$. Then, as we have just shown, σ extends to a field embedding $\tau: E(d_1,\ldots,d_k)\longrightarrow L$. Since τ is an extension of some σ_j we have the contradiction $\sigma_j(b_j)\notin L^+$ yet $\sigma_j(b_j)=\tau(d_j)^2\in L^+$.

By Zorn's lemma there is an intermediate field $K\subseteq M\subseteq F$ and an embedding $\rho: M\longrightarrow L$ of totally ordered fields which extends σ and which cannot be extended to any subfield of F which properly contains M. But then $M=F$ by the preceding paragraph and ρ is the desired extension of σ. Now assume that L is a real closure of $\sigma(K)$; then $\rho(F)=L$ since $\sigma(K)\subseteq\rho(F)\subseteq L$ and $\rho(F)$ is real closed by Theorem 5.1.6. Suppose $\tau: F\longrightarrow L$ is a field monomorphism that extends σ. Since each positive element in F is a square τ is isotone and, again, $\tau(F)=L$. Suppose $a\in F$ and $f(x)$ is its irreducible polynomial over K. If $a_1<\cdots<a_r$ are all of the roots of $f(x)$ in F, then $\tau(a_1)<\cdots<\tau(a_r)$ are all of the roots of its mate $f^{\sigma}(x)$ in L. Thus,

$\rho(a_j) = \tau(a_j)$ for each j and $\rho(a) = \tau(a)$. The last statement follows from taking σ to be inclusion. □

If G is a totally ordered η_α-group, then the formal power series ring $V_\alpha(G,\mathbb{R})$ is a totally ordered η_α-division ring according to Theorems 3.5.3, 3.5.8, and 4.4.8. In view of the Hahn embedding theorem for totally ordered abelian groups and the $\aleph_\alpha$-injectivity of totally ordered divisible abelian η_α-groups it seems reasonable to expect that a totally ordered field can be embedded in a formal power series field (η_α-)field with real coefficients. This is almost a consequence of the next result.

Theorem 5.1.11. *Suppose $\aleph_\alpha$ is a regular cardinal with $\alpha \geq 1$ and F is a real closed η_α-field. Then each totally ordered field whose cardinality is bounded by $\aleph_\alpha$ can be embedded in F.*

Proof. Let L be a totally ordered field with $\text{card}(L) \leq \aleph_\alpha$. Suppose, first, that $L = K(x)$ where K is real closed, x is transcendental over K, $\text{card}(K) < \aleph_\alpha$, and $\sigma : K \longrightarrow F$ is an embedding. We claim that σ can be extended to L. Let A (respectively, B) be the set of lower (respectively, upper) bounds of x in K. Then there exists an element $y \in F$ with $\sigma(A) < y < \sigma(B)$. Since $\sigma(K) = \sigma(A) \cup \sigma(B)$ is real closed y must be transcendental over $\sigma(K)$. Let $\tau : L \longrightarrow \sigma(K)(y)$ be the unique field isomorphism which extends σ and sends x to y. If $c \in K$, then $x > c$ iff $c \in A$, iff $y > \sigma(c)$; that is, $x - c > 0$ iff $\tau(x-c) > 0$. This shows that τ is isotone on $K[x]$ and hence on L since if $0 \neq f(x) \in K[x]$, then $f(x)$ has the factorization

$$f(x) = a(x-a_1)\cdots(x-a_n)[(x+b_1)^2 + c_1^2]\cdots[(x+b_m)^2 + c_m^2]$$

with $c_1 \cdots c_m \neq 0$ (Exercise 8).

In the general case we may assume by Theorem 5.1.10 that L is real closed since the real closure of L has the same cardinality as L. For convenience we also will assume that $\text{card}(L) = \aleph_\alpha$; this can be accomplished by adjoining indeterminates to L. Let $X = \{x_\beta : \beta < \omega_\alpha\}$ be a transcendence base of L over $\mathbb{Q}$ and set $X_\beta = \{x_\gamma : \gamma < \beta\}$. Let L_β be the real closure of $\mathbb{Q}(X_\beta)$ in L; that is, L_β is the set of those elements in L that are algebraic over $\mathbb{Q}(X_\beta)$ (Exercise 9). Then $\{L_\beta\}$ is a chain of subfields of L and for each $\beta < \omega_\alpha$ we have: (i) x_β is transcendental over L_β; (ii) $L_{\beta+1}$ is the real closure of $L_\beta(x_\beta)$ in L; (iii) $\bigcup_{\gamma<\beta} L_\gamma = L_\beta$ if β is a limit ordinal. That (i) holds is a consequence of the fact that x_β is transcendental over $\mathbb{Q}(X_\beta)$ and hence is transcendental over the algebraic extension L_β of $\mathbb{Q}(X_\beta)$; as for (ii), $\mathbb{Q}(X_\beta) \subseteq \mathbb{Q}(X_{\beta+1}) \subseteq L_\beta(x_\beta) \subseteq L_{\beta+1}$ and $L_{\beta+1}$ is algebraic over $\mathbb{Q}(X_{\beta+1})$, and for (iii), by Theorem 5.1.6 the union is a real closed subfield of L_β which contains $\mathbb{Q}(X_\beta)$. Note that by (ii) and (iii) and the regularity of $\aleph_\alpha$ we have, by induction, that $\text{card}\,(L_\beta) < \aleph_\alpha$ for each $\beta < \omega_\alpha$. Using transfinite induction we will construct monomorphisms $\sigma_\beta : L_\beta \longrightarrow F$, for each $\beta < \omega_\alpha$, such that if $\gamma < \beta$, then σ_β extends σ_γ. Let σ_0 be the isomorphism between the real closures of $\mathbb{Q}$ in L and F. Suppose σ_γ has been constructed for all $\gamma < \beta$. If $\beta = \delta + 1$, then by the previous paragraph σ_δ can be extended to $L_\delta(x_\delta)$ and then to the real closure L_β of $L_\delta(x_\delta)$ by the previous theorem. If β is a limit ordinal, then by (iii) there is a unique σ_β which extends each σ_γ with $\gamma < \beta$. Since $\mathbb{Q}(X) \subseteq \bigcup_{\beta<\omega_\alpha} L_\beta \subseteq L$ and the latter two fields are both real closed and algebraic over $\mathbb{Q}(X)$, necessarily $L = \bigcup_{\beta<\omega_\alpha} L_\beta$. So there is a unique embedding $\sigma : L \longrightarrow F$ which extends each σ_β. □

Let K be a totally ordered field and suppose $\aleph_\alpha$ is a regular cardinal with card $(K) \leq \aleph_\alpha$ and $\alpha \geq 1$. Returning to the discussion prior to Theorem 5.1.11, if we knew that $V_\alpha(G,\mathbb{R})$ were real closed (where G is a divisible totally ordered abelian η_α-group), then we would have the desired embedding of K into $V_\alpha(G,\mathbb{R})$. This is indeed the case, but to show it and to also show that K can be embedded in $V(\Gamma(K),\mathbb{R})$ we need to develop some valuation theory. We will do this in the next section.

Exercises.

1. Let x be an element of the po-ring R.
   (a) If $x^2 \in R^+$, $pxq = pqx$ for all $p,q \in R^+$, $\ell(x)^+ = 0$ and $R^+ \cap R^+x = 0$, show that $R^+ - R^+x$ is a partial order of R.
   (b) If $1 \in R$, x is in the center of $R, x^2 \in R^+$ and each nonzero positive element has a positive inverse, show that $R^+ - R^+x$ is a partial order of R iff $R^+ \cap R^+x = 0$.
   (c) Suppose R is an sp-po-division ring and x is a central element. Show that $R^+ - R^+x$ is a partial order of R iff $R^+ \cap R^+x = 0$.
   (d) Show that an sp-po-field R is totally ordered iff R^+ is a maximal partial order of R.

2. If X is a subset of a ring R, then a product of elements from X is called a monomial on X. If $a_1,\ldots,a_n \in R$ let $S(X;a_1,\ldots,a_n)$ denote the set of sums of monomials on $X \cup \{a_1,\ldots,a_n,0\}$; this set is the subsemiring of R generated by $X \cup \{a_1,\ldots,a_n,0\}$. Show that the following statements are equivalent for the po-ring R.
   (a) R^+ is contained in a total order of R.
   (b) For any finite number of elements $a_1,\ldots,a_n$ in R there exist $\varepsilon_1,\ldots,\varepsilon_n$ in $\{1,-1\}$ such that $S(R^+;\varepsilon_1 a_1,\ldots,\varepsilon_n a_n)$ is a partial order of R.
   (c) For each $x \in R$ one can choose $\varepsilon \in \{1,-1\}$ such that $S(R^+;\varepsilon x)$ is a partial order of R which satisfies (b).

   (For (c) $\Rightarrow$ (a) use Zorn's lemma on the set of partial orders of R which contain R^+ and satisfy (b).)

3. Let P be a partial order of the domain R. If $a_1,\ldots,a_n \in R^*$, a monomial μ on $P^* \cup \{a_1,\ldots,a_n\}$ is *even with respect to* $a_1,\ldots,a_n$ if each a_i not in P* occurs an even number of times in μ. Let $[P^*;a_1,\ldots,a_n]$ denote the set of sums of the even monomials on $P^* \cup \{a_1,\ldots,a_n\}$; $[P^*;a_1,\ldots,a_n]$ is a subsemiring of R. The partial order P is *even* if $[P^*;a_1,\ldots,a_n] \subseteq P^*$ whenever $a_1,\ldots,a_n \in R^*$. Show the equivalence of the following statements.

   (a) P is contained in a total order of R.
   (b) P is contained in an even partial order of R.

(c) If $a_1,\dots,a_n \in R^*$, then $0 \notin [P^*;a_1,\dots,a_n]$.

(For (c) $\Rightarrow$ (a) show: (i) the union of $[P^*;a_1,\dots,a_n]$ over all $a_1,\dots,a_n$ in R^* together with 0 is an even partial order of R; (ii) if T is even, $a \in R^*$ and Q is the subsemiring generated by $T^* \cup \{a\}$, then Q is contained in an even partial order iff $0 \notin T^* + aT^*$ (each element of Q is of the form $q_0 + q_1$ where $q_0, aq_1 \in T^*$, and $0 \notin [Q;a_1,\dots,a_n]$); (iii) a maximal even partial order containing P is a total order.)

4. (a) Let P be a partial order of the division ring R. Show that P is contained in a total order of R iff 0 is not a sum of terms of the form $pa_1^2\cdots a_n^2$ where $p \in P^* \cup \{1\}$ and $a_i \in R^*$. (Use Exercise 3 and the equations $aba = b(b^{-1})^2(ba)^2$, $ab = ba[a,b]$ where $[a,b] = a^{-1}b^{-1}ab$, and $[a,b] = a^{-2}(ab^{-1})^2b^2$.)
   (b) Show that R can be totally ordered iff -1 is not a sum of products of the form $a_1^2\cdots a_n^2$.

5. Let F/K be a finite Galois extension of fields with Galois group $G = \{1 = \sigma_1, \sigma_2, \dots, \sigma_n\}$. The trace and norm functions T and N, respectively, are defined from F to K by: $T(a) = \Sigma_i \sigma_i(a)$ and $N(a) = \Pi_i \sigma_i(a)$.
   (a) Show that T is a K-linear transformation that is onto.
   (b) Suppose G is cyclic with generator σ. Show that $T(a) = 0$ iff $a = b - \sigma(b)$ for some $b \in F$. (If $T(c) = 1$ and $T(a) = 0$ let $d_j = \sigma^j(c)(a + \sigma(a) + \cdots + \sigma^j(a))$ for $j = 0,\dots,n-2$, and $b = d_0 + \cdots + d_{n-2}$; compute $d_{j+1} - \sigma(d_j)$ for $0 \le j \le n-3$ and $\sigma(d_{n-2})$.)
   (c) With the same hypothesis as (b) show that $N(a) = 1$ iff $a = b\sigma(b)^{-1}$ for some $b \in F$. (If $N(a) = 1$, and $a_j = a\sigma(a)\cdots\sigma^j(a)$ the independence of G gives $c \in F$ with $b = a_0c + a_1\sigma(c) + \cdots + a_{n-1}\sigma^{n-1}(c) \neq 0$.)
   (d) If $n = p$ is the characteristic of K show that $F = K(u)$ where u is a root of an irreducible polynomial over K of the form $x^p - x - a$. (Use (b) with $1 = -u - \sigma(-u)$).
   (e) With the same hypothesis as (b) assume that the characteristic of K is 0 or is relatively prime to n and that $x^n - 1$ splits in $K[x]$. Show that $F = K(u)$ where u is a root of a polynomial $x^n - a \in K[x]$. (If t is a primitive nth root of unity, use (c) with $t = \sigma(u)u^{-1}$.)

6. Let a and b be elements in the real closed field F and let $f(x) \in F[x]$.
   (a) If $f(a) < 0 < f(b)$ show that $f(c) = 0$ for some c between a and b.
   (b) If $f(a) < y < f(b)$ show that $f(c) = y$ for some c between a and b.
   (c) If $a < b$ show that there exists an element $c \in (a,b)$ with $f'(c)(b-a) = f(b) - f(a)$. (Reduce to the case a and b are the only roots of $f(x)$ in $[a,b]$.)
   (d) If $f(x)$ has only simple roots in F and $c \in F$ is a root of $f(x)$ show that $f(x)$ is strictly increasing or strictly decreasing in some interval about c.

7. Show that a totally ordered field which satisfies the condition in Exercise 6(a) must be real closed.

8. If F is a real closed field show that $f(x) \in F[x]$ is monic and irreducible iff it is of the form $x-a$ or $(x-a)^2+b^2$ with $b \neq 0$. Conversely, show that if the irreducible polynomials over the field F have this characterization, then F is real closed.
9. Let K be a subfield of the real closed field L and let F be the algebraic closure of K in L; so $a \in F$ iff $a \in L$ and it is algebraic over K. Show that F is the unique real closure of K in L.
10. Let K be a subfield of the algebraically closed field C and suppose K is totally ordered. Show that there is a real closed field F with $K \subseteq F \subseteq C$ such that $K^+ \subseteq F^+$ and $C = F(i)$.
11. Let R be a unital convex subring of the totally ordered field F and let J be the Jacobson radical of R.
    (a) If F is real closed show that R/J is real closed.
    (b) Give an example to show that R/J could be real closed without F being real closed.
12. Show that a totally ordered division ring is an η_α-set iff it is an almost η_α-set. (If $0 < X$ has no least element, then X^{-1} has no largest element.)
13. (a) Show that an ultraproduct of real closed fields is a real closed field. (Recall that if $\{F_\lambda : \lambda \in \Lambda\}$ is a set of fields and $\mathscr{F}$ is an ultrafilter of Λ, then $\Pi_\lambda F_\lambda/\mathscr{F} = \Pi_\lambda F_\lambda/I$ is an ultraproduct where $I = \{f \in \Pi_\lambda F_\lambda : Z(f) = \{\lambda : f(\lambda) = 0\} \in \mathscr{F}\}$.)
    (b) Let $\mathscr{F}$ be an ultrafilter on $\mathbb{N}$ which contains the complement of each finite subset of $\mathbb{N}$. If K is a totally ordered field show that $K^{\mathbb{N}}/\mathscr{F}$ is an η_1-field which contains K and which is real closed provided K is real closed. (Use Theorem 4.4.12 and Exercise 12.)
14. Let $F(x)$ be the rational function field in one variable over the real closed field F.
    (a) If $(F(x),P)$ is a totally ordered field show it is an ℓ-algebra over F.
    (b) Suppose $(F(x),\leq_1)$ and $F(x),\leq_2)$ are totally ordered fields and the restrictions of these total orders to $F \cup \{x\}$ are identical. Show that $\leq_1 = \leq_2$.
    (c) Show that each total order of $\mathbb{R}(x)$ arises from the lexicographic or antilexicographic order of $\mathbb{R}[y]$ where $y = x-c$ or $y = -(x-c)$ with $c \in \mathbb{R}$. (Consider $\{a \in \mathbb{R} : a < x\}$ and $\{b \in \mathbb{R} : x < b\}$.)
15. Suppose F is a totally ordered field and there is a polynomial of odd degree over F with no root in F. Let L be the real closure of F and let M be the subfield of L consisting of all elements a for which there is a tower $F = F_0 \subseteq F_1 \subseteq \cdots \subseteq F_n \subseteq L$ with $a \in F_n$ and $F_i = F_{i-1}(a_i)$ with $a_i^2 \in F_{i-1}$ for $i = 1,\ldots,n$. Show that M has a unique total order and M is not real closed.
16. Suppose K is a totally ordered field, L is its real closure and F is a finite extension of K. Show that F has exactly m distinct total orders which extend K^+

where m is the number of field K-monomorphisms of F into L. (F is a simple extension of K.)

17. Let F be a field of characteristic $\neq 2$, and let $S(F)$ be the set of sums of squares in F. Suppose the po-field K is a subfield of F and put $Q = Q(K^+)$ if $K^+ \neq 0$ and $Q = \mathbb{Z}^+$ if $K^+ = 0$.
    (a) Show that F has a total order which contains K^+ iff $QS(F) \subset F$. (Use Theorems 5.1.1 and 5.1.2 and $4a = (1+a)^2 - (1-a)^2$.)
    (b) Suppose $a \in F$ and L is an algebraic extension of F maximal with respect to $a \notin QS(L)$. Show that L has a total order which contains K^+ and $-a$ is a square in L. (If not, $L \subset L(ci)$ with $c^2 = a$.)
    (c) Show that $QS(F)$ is the intersection of the family of total orders of F which contain K^+.
    (d) In particular, $S(F)$ is the intersection of all the total orders of F.
18. Find a subfield of $\mathbb{C}$ distinct from $\mathbb{R}$ and isomorphic to $\mathbb{R}$.
19. Suppose $\aleph_\alpha$ is regular with $\alpha > 0$, F is a real closed η_α-field, and K is a subfield of the totally ordered field L with $|K| < \aleph_\alpha$ and $|L| \leq \aleph_\alpha$. Show that each embedding of K into F can be extended to an embedding of L into F.

Notes. The concept and properties of the real closure of a totally ordered field as well as most of the theory of totally ordered fields presented here is due to Artin and Schreier [AS]; also see Lang and Tate [LT]. Theorems 5.1.1 and 5.1.3 are due to Serre [SE], and our proof of Sturm's Theorem, Theorems 5.1.8 and 5.1.9, follows Jacobson [J4]. The embedding of a totally ordered field into a real closed η_α-field that is given in Theorem 5.1.11 comes from Alling [AL2] and Erdos, Gillman, and Henriksen [EGH]; also, see Gillman and Jerison [GJ, p. 193] and Prestel and Delzell [PD, p. 95]. The proof of Theorem 5.1.7 that is given is due to Leicht [LEI]; also see Kaplansky [K2, p. 66] and Prieß-Crampe [PC2, p. 27]. Exercises 2 and 3 are due to Fuchs and Exercise 4 is due independently to Johnson [JO2] and Podderjugin – see Fuchs [F, Chapter VII]. The proof outlined in Exercise 13 of the fact that the ultraproduct is a totally ordered η_1-field is due to Weinberg [W6, p. III-41].

5.2 Valuations and the Hahn Embedding Theorem for Totally Ordered Fields

In this section enough of the valuation theory of fields will be developed so that a totally ordered field can be embedded into the field of formal power series with real coefficients and exponents from the value group of the field. The key to this embedding will be to show the existence of a unique totally ordered field extension which is maximal with respect to having the same value group and residue class field as the original field, and the characterization of such fields as those which are complete in the sense that certain transfinite sequences have a limit. As a consequence of this, or of even a simpler embedding, each totally ordered field is a subfield of a totally ordered field which contains the reals. A similar statement can be made for a totally ordered division ring and the intricate and complicated verification of this involves the adjunction of those missing real numbers in the correct order.

Suppose D is a division ring and Γ is a totally ordered group. Let $-\infty$ be an element with $-\infty < \Gamma$. A function $v : D \longrightarrow \Gamma \cup \{-\infty\}$ is called a *valuation on D* if for all $a, b \in D$

$$v(D) = \Gamma \cup \{-\infty\}, \tag{5.2.1}$$

$$v(a) = -\infty \text{ iff } a = 0, \tag{5.2.2}$$

$$v(ab) = v(a) + v(b), \tag{5.2.3}$$

$$v(a+b) \leq v(a) \vee v(b). \tag{5.2.4}$$

The pair (D, v) is called a *valued division ring*. The valuation is *trivial* or *non-trivial* as Γ is zero or nonzero, respectively. Recall that a unital domain is a valuation domain if the lattices of left ideals and of right ideals are both chains. Such a ring is characterized as a subring R of a division ring D with the property that $a \in R$ or $a^{-1} \in R$ for each a in D. R is a *normal valuation ring* if $a^{-1}Ra = R$ for each nonzero element a in D. We collect some simple consequences of the definition of a valuation. R^*, as usual, denotes the set of nonzero elements in the monoid R.

Theorem 5.2.1. *Let (D, v) be a valued division ring.*

$$v(1) = v(-1) = 0; \tag{5.2.5}$$

$$v(-a) = v(a); \tag{5.2.6}$$

$$v(a_1 + \cdots + a_n) \leq v(a_1) \vee \cdots \vee v(a_n); \tag{5.2.7}$$

$$v(a_1 + \cdots + a_n) = v(a_1) \vee \cdots \vee v(a_n) \textit{ if } v(a_i) \neq v(a_j) \textit{ for } i \neq j; \tag{5.2.8}$$

$$v^{-1}(0) \textit{ is a normal subgroup of the multiplicative group of } D; \tag{5.2.9}$$

$$R = \{a \in D : v(a) \leq 0\} \textit{ is a normal valuation ring in } D \text{ with unique maximal ideal } J = \{a \in D : v(a) < 0\}; \tag{5.2.10}$$

each one-sided R-submodule of D is an R-R-bimodule; (5.2.11)

the lattice of R-submodules of D is totally ordered. (5.2.12)

Proof. Since $2v(-1) = v(1) = 2v(1)$ we obtain $v(\pm 1) = 0$ and $v(-a) = v(-1) + v(a) = v(a)$. An easy induction gives (5.2.7). Suppose $v(a_1) < v(a_2)$. If also, $v(a_1 + a_2) < v(a_2)$, then $v(a_2) = v(a_1 + a_2 - a_1) \leq v(a_1 + a_2) \vee v(a_1) < v(a_2)$; so $v(a_1 + a_2) = v(a_1) \vee v(a_2)$ and another easy induction gives (5.2.8). It is clear that R is a valuation ring whose group of units is $v^{-1}(0)$ and consequently J is the maximal ideal of R. If $x \in R$, then $v(a^{-1}xa) = -v(a) + v(x) + v(a) \leq 0$ with equality holding if $v(x) = 0$. This verifies (5.2.9) and (5.2.10) and (5.2.11) follows easily from the normality of R. For any nonzero elements $a, b \in D$, $ab^{-1} \in R$ or $ba^{-1} \in R$. So $aR \subseteq bR$ or $bR \subseteq aR$ and (5.2.12) follows. □

The totally ordered group Γ is called the *value group* of the valuation, the division ring R/J is called the *residue class division ring* of the valuation, and $v^{-1}(0)$ is called the *group of units* of the valuation. Both the value group and the valuation can be recaptured from R. For suppose R is a normal valuation ring with group of units U and maximal ideal J. Then U is a normal subgroup of D^* since $x, x^{-1} \in R$ implies $a^{-1}xa$, $a^{-1}x^{-1}a \in R$ for any $a \in R^*$; also, $a^{-1}Ja = J$ since $J = R\backslash U$. We define a relation on D^*/U by

$$aU \leq bU \text{ iff } b^{-1}a \in R \text{ (iff } aR \subseteq bR). \tag{5.2.13}$$

Since $aU = bU$ iff $aR = bR$ this relation is a total order of D^*/U which makes it into a totally ordered group. Also, the extended natural homomorphism $w : D \longrightarrow D^*/U \cup \{-\infty\}$ is a valuation which has R as its valuation ring. If R were originally the valuation ring of a valuation $v : D \longrightarrow \Gamma \cup \{-\infty\}$, then the mapping $h : \Gamma \longrightarrow D^*/U$ given by $h(v(a)) = aU$ is an isomorphism of totally ordered groups and $w = hv$.

The data associated to the valued division ring (D, v) will frequently be denoted by Γ_D, R_D, J_D and U_D. For our purposes the most important example of a valuation ring arises from a totally ordered division ring D. Here, we take $R = C(1)$, the convex subgroup of D generated by 1; R is, of course, a left and right convex totally ordered domain and the residue class field R/J is isomorphic to a totally ordered subfield of the reals. The associated valuation will be called the *natural valuation* of D. This valuation $w : D \longrightarrow D^*/U \cup \{-\infty\}$ is isotone when it is restricted to D^+. Any valuation on a totally ordered division ring which is isotone on the positive cone will be called an *isotone valuation*. The valuation ring of an isotone valuation always contains $C(1)$. Note also that $C(a) = aC(1)$, and $w(a) = w(b)$ precisely when $C(a) = C(b)$, where w is the natural valuation. A related example arises from the formal power series ring $V = V(\Gamma, D)$. Here, Γ is a totally ordered group and D is a division ring. The valuation $\pi : V \longrightarrow \Gamma \cup \{-\infty\}$ is given by $\pi(f) = \max\operatorname{supp} f$. The value group of π is Γ and the residue class division ring is D (by Theorem 3.5.8). If D is totally ordered, then the valuation ring of π is $C_D(1) = \{f \in V : |f| \leq d \text{ for some } d \in D\}$. If D is not archimedean, then $C(1) \subset C_D(1)$ and the isotone valuation π is not the natural valuation of $V(\Gamma, D)$. The same remarks apply to the crossed product formal power series division ring $V(D * \Gamma)$; see Exercise 3.5.20.

The valued division ring (E, w) is an *extension* of the valued division ring (D, v) if there is a commutative diagram

$$\begin{array}{ccc} D & \longrightarrow & E \\ \downarrow v & & \downarrow w \\ \Gamma_D \cup \{-\infty\} & \longrightarrow & \Gamma_E \cup \{-\infty\} \end{array} \tag{5.2.14}$$

whose top row is a ring monomorphism and whose bottom row is an extended monomorphism of totally ordered groups. If (E,w) is an extension of (D,v) and the top row of (5.2.14) is the inclusion map, then $R_D = D \cap R_E$, $J_D = D \cap J_E$ and $U_D = D \cap U_E$. If the rows are isomorphisms, then the top row is a *value preserving isomorphism.*

We now turn our attention back to fields and we show first that a valuation on a field can always be extended to a larger field.

Theorem 5.2.2. *Let K be a subfield of the field F. Then each valuation on K can be extended to F.*

Proof. Suppose S is a unital subring of F and I is a proper ideal of S. We will find a valuation ring R in F with $I \subseteq J(R)$. To do this, apply Zorn's lemma to the poset

$$\mathscr{S} = \{(T,A) : T \text{ is a subring of } F \text{ containing } S \text{ and } A \text{ is a proper ideal of } T \text{ containing } I\}.$$

The partial order in $\mathscr{S}$ is given by $(T,A) \leq (T_1,A_1)$ if $T \subseteq T_1$ and $A \subseteq A_1$. Let (R,J) be a maximal element in $\mathscr{S}$. If R is not a valuation ring of F, then there is an element u in F with neither u nor u^{-1} in R. We claim that $J[u] \subset R[u]$ or $J[u^{-1}] \subset R[u^{-1}]$ and this will contradict the maximality of (R,J). Suppose to the contrary that $R[u] = J[u]$ and $R[u^{-1}] = J[u^{-1}]$. Then there are elements a_i, $b_j \in J$ with

$$a_0 + a_1 u + \cdots + a_n u^n = 1, \tag{5.2.15}$$
$$b_0 + b_1 u^{-1} + \cdots + b_m u^{-m} = 1. \tag{5.2.16}$$

We may assume that $n \geq m$ and n is minimal among all equations expressing 1 as a polynomial in $J[u]$. If we multiply (5.2.16) by u^n and (5.2.15) by $1 - b_0$ we get

$$(1-b_0)u^n = b_m u^{n-m} + \cdots + b_1 u^{n-1}, \tag{5.2.17}$$
$$\begin{aligned}(1-b_0) &= (1-b_0)a_0 + \cdots + a_n(1-b_0)u^n \\ &= (1-b_0)a_0 + \cdots + a_{n-1}(1-b_0)u^{n-1} \\ &\quad + a_n(b_m u^{n-m} + \cdots + b_1 u^{n-1}).\end{aligned} \tag{5.2.18}$$

Since (5.2.18) has the form of (5.2.15) we have a contradiction to the minimality of n. So R is a valuation ring of F and $J = J(R)$.

Now suppose $v : K \longrightarrow \Gamma \cup \{-\infty\}$ is a valuation on K with valuation ring R_K, maximal ideal J_K and group of units U_K. We assume that $\Gamma = K^*/U_K$. Let R_F be a valuation ring in F with data J_F and U_F and with $R_K \subseteq R_F$ and $J_K \subseteq J_F$. Since $R_K = J_K \cup U_K$ and $R_F = J_F \cup U_F$ are disjoint unions we have $J_K = J_F \cap K$ and $U_K = U_F \cap K$. The latter equation implies that the map $K^*/U_K \longrightarrow F^*/U_F$ which sends aU_K to aU_F is an embedding of groups. Since $R_K \subseteq R_F$ this map is isotone. Thus, $w : F \longrightarrow F^*/U_F \cup \{-\infty\}$ is a valuation which extends v. □

Let (F,v) be a valued field. For an infinite limit ordinal α a sequence of elements $(a_\rho)_{\rho<\alpha}$ in F is called *pseudo-convergent* if

$$v(a_\sigma - a_\rho) > v(a_\tau - a_\sigma) \text{ for all } \rho < \sigma < \tau. \tag{5.2.19}$$

If $(a_p)_{p<\alpha}$ is pseudo-convergent, then so is $(c(a_\rho - d))_{\rho<\alpha}$ for $c, d \in F$ with $c \neq 0$. Also, for a given sequence $(a_\rho)_{\rho<\alpha}$, if there is some μ such that (5.2.19) holds for all $\mu < \rho < \sigma < \alpha$, then we will call this sequence *eventually pseudo-convergent*.

Note that if $(a_\rho)_{\rho<\alpha}$ is a sequence in F and $(v(a_\rho))_{\rho<\alpha}$ is a strictly decreasing sequence in Γ, then $(a_\rho)_{\rho<\alpha}$ is pseudo-convergent. Otherwise, we could find $\rho < \sigma < \tau$ with $v(a_\sigma - a_\rho) \leq v(a_\tau - a_\sigma)$ and hence, by (5.2.8),

$$v(a_\rho) = v(a_\sigma - a_\rho) \leq v(a_\tau - a_\sigma) = v(a_\sigma) < v(a_\rho).$$

Here is a partial converse.

Theorem 5.2.3. *Let $(a_\rho)_{\rho<\alpha}$ be a pseudo-convergent sequence in the valued field (F, v). Then the sequence of values $(v(a_\rho))_{\rho<\alpha}$ in Γ is either strictly decreasing or is constant from some point on. Moreover, if $\gamma_\rho = v(a_{\rho+1} - a_\rho)$, then $v(a_\sigma - a_\rho) = \gamma_\rho$ for all $\rho < \sigma$, and $(\gamma_\rho)_{\rho<\alpha}$ is a strictly decreasing sequence.*

Proof. Suppose $(v(a_\rho))_\rho$ is not strictly decreasing; so there exists $\rho < \sigma$ with $v(a_\rho) \leq v(a_\sigma)$. Then $v(a_\tau) = v(a_\sigma)$ for every $\tau > \sigma$. Otherwise,

$$v(a_\tau - a_\sigma) = v(a_\tau) \vee v(a_\sigma) \geq v(a_\sigma) = v(a_\rho) \vee v(a_\sigma)$$

$$\geq v(a_\sigma - a_\rho) > v(a_\tau - a_\sigma).$$

For the second part take $\rho < \rho + 1 < \sigma$. Then

$$v(a_\sigma - a_\rho) = v((a_\sigma - a_{\rho+1}) + (a_{\rho+1} - a_\rho)) = v(a_{\rho+1} - a_\rho) = \gamma_\rho$$

since $v(a_{\rho+1} - a_\rho) > v(a_\sigma - a_{\rho+1})$. Also, if $\rho < \sigma$, then $\gamma_\rho = v(a_\sigma - a_\rho) > v(a_{\sigma+1} - a_\sigma) = \gamma_\sigma$. □

The next result is quite technical but it will be used several times.

Theorem 5.2.4. *Suppose $\alpha > 0$ is a limit ordinal and $(\gamma_\rho)_{\rho<\alpha}$ is a strictly decreasing sequence in the totally ordered abelian group Γ. Let $t_1, \ldots, t_n$ be distinct integers in $\mathbb{N}$ and let $\beta_1, \ldots, \beta_n$ be elements in Γ. Then there exists an ordinal $\mu < \alpha$ and an integer k such that $\beta_i + t_i\gamma_\rho < \beta_k + t_k\gamma_\rho$ whenever $\rho > \mu$ and $i \neq k$.*

Proof. In the divisible hull $d(\Gamma)$ of Γ the desired inequalities amount to

$$\gamma_\rho > \frac{\beta_i - \beta_k}{t_k - t_i} \qquad \text{if } t_k - t_i > 0 \tag{5.2.20}$$

and

$$\gamma_\rho < \frac{\beta_i - \beta_k}{t_k - t_i} \qquad \text{if } t_k - t_i < 0. \tag{5.2.21}$$

Let $X = \left\{ \frac{\beta_i - \beta_j}{t_j - t_i} : i \neq j \right\} = \{x_1 < x_2 < \cdots < x_m\}$. We will consider three cases.

Suppose first that $\gamma_\mu \leq x_1$ for some μ. Let

$$\beta_k + t_k\gamma_{\mu+1} = \bigvee_{i=1}^{n} \beta_i + t_i\gamma_{\mu+1}. \tag{5.2.22}$$

Then

$$\beta_k + t_k\gamma_{\mu+1} > \beta_i + t_i\gamma_{\mu+1} \qquad \text{for } i \neq k, \tag{5.2.23}$$

$\beta_k - \beta_i > (t_i - t_k)\gamma_{\mu+1}$ and $t_k - t_i < 0$ since $\gamma_{\mu+1} < X$. So (5.2.21) holds for all $\rho > \mu$ and all $i \neq k$.

Suppose next that $\gamma_\rho > x_m$ for all $\rho < \alpha$. Then from (5.2.22) we again get (5.2.23). This time $t_k - t_i > 0$ since $X < \gamma_\rho$, and (5.2.20) holds for all ρ and all $i \neq k$.

The conditions for the third case are that $x_1 < \gamma_\rho$ for all ρ and $\gamma_\rho \leq x_m$ for some ρ. Let j be maximal with $x_j < \gamma_\rho$ for all ρ; then $1 \leq j < m$ and for some μ, $x_j < \gamma_\rho < \gamma_\mu \leq x_{j+1}$ for every $\rho > \mu$. Again, (5.2.22) leads to (5.2.23). Now, if $t_k - t_i > 0$, then $\gamma_{\mu+1} > \frac{\beta_i - \beta_k}{t_k - t_i}$ and $\frac{\beta_i - \beta_k}{t_k - t_i} \leq x_j < \gamma_\rho$ for $\rho > \mu$; so (5.2.20) holds. But if $t_k - t_i < 0$, then $\gamma_\rho \leq \gamma_{\mu+1} < \frac{\beta_i - \beta_k}{t_k - t_i}$ for $\rho > \mu$; so (5.2.21) holds. □

For the remainder of this section we will assume that all residue class fields are of characteristic 0. For much of the theory this is not necessary but it is certainly true for an isotone valuation on a totally ordered field. Any unnamed valuation on a totally ordered field will always be the natural valuation.

Let F be a field of characteristic 0, let $f(x)$ be a polynomial over F of degree $n \geq 1$ and let $f^{(k)}(x)$ denote its kth derivative. We will develop the Taylor expansion of $f(x)$ for later use.

$$f(x) = \sum_{i=0}^{n} a_i x^i \tag{5.2.24}$$

$$f^{(k)}(x) = \sum_{i=k}^{n} i(i-1)\cdots[i-(k-1)]a_i x^{i-k} = \sum_{i=k}^{n} \binom{i}{k} k! a_i x^{i-k}$$

Let

$$f_k(x) = \sum_{i=k}^{n} \binom{i}{k} a_i x^{i-k} = \frac{f^{(k)}(x)}{k!}, \qquad 0 \leq k \leq n. \tag{5.2.25}$$

Then $\deg f_k(x) = n - k$. Now, for $a \in F$

$$x^m = (a + (x-a))^m = \sum_{k=0}^{m} \binom{m}{k} a^{m-k}(x-a)^k$$

$$= \sum_{k=0}^{m} \frac{(x^m)^{(k)}}{k!}|_{x=a}(x-a)^k.$$

So,

$$f(x) = \sum_{i=0}^{n} a_i \sum_{k=0}^{i} \frac{(x^i)^{(k)}}{k!}|_{x=a}(x-a)^k$$

$$= \sum_{k=0}^{n}\sum_{i=k}^{n} a_i \frac{(x^i)^{(k)}}{k!}|_{x=a}(x-a)^k$$

$$= \sum_{k=0}^{n} \frac{f^{(k)}(a)}{k!}(x-a)^k.$$

Thus,

$$f(x) = \sum_{k=0}^{n} f_k(a)(x-a)^k \tag{5.2.26}$$

and we have

$$f(x) - f(a) = \sum_{k=1}^{n} f_k(a)(x-a)^k, \tag{5.2.27}$$

$$f(x+a) = \sum_{k=0}^{n} f_k(a)x^k. \tag{5.2.28}$$

Theorem 5.2.5. *Suppose $(a_\rho)_{\rho<\alpha}$ is a pseudo-convergent sequence in the valued field (F,v) and $f(x)$ is a polynomial in $F[x]$ of positive degree. Then $(f(a_\rho))_{\rho<\alpha}$ is eventually pseudo-convergent.*

Proof. Let $n = \deg f(x)$. Since the case $n = 1$ has already been noted we assume $n \geq 2$. Suppose first that (F,v) has an algebraic extension (L,w) in which there is some element a with $w(a_\sigma - a) < w(a_\rho - a)$ for all $\rho < \sigma$. Let $b_\rho = a_\rho - a$ and let $h(x) = f(x+a) - f(a)$. Clearly, $h(b_\sigma) - h(b_\rho) = f(a_\sigma) - f(a_\rho)$ and it suffices to show that $(w(h(b_\rho)))_{\rho<\alpha}$ is eventually strictly decreasing. For then $(h(b_\rho))_\rho$ is eventually pseudo-convergent and, for some μ, $v(f(a_\sigma) - f(a_\rho)) = w(h(b_\sigma) - h(b_\rho)) > w(h(a_\tau) - h(a_\sigma)) = v(f(a_\tau) - f(a_\sigma))$ for all $\mu < \rho < \sigma < \tau$. Since $h(b_\rho) = f(a_\rho) - f(a)$ we have, by (5.2.27),

$$h(b_\rho) = \sum_{j=1}^{n} f_j(a)b_\rho^j, \tag{5.2.29}$$

and by Theorem 5.2.4 we can find an integer k and an ordinal μ such that for all $\rho > \mu$ and for $j \neq k$,

$$w(f_k(a)) + kw(b_\rho) > w(f_j(a)) + jw(b_\rho).$$

From (5.2.29) we then have that $w(h(b_\rho)) = w(f_k(a)b_\rho^k)$, and for $\mu < \rho < \sigma$,

$$\begin{aligned} w(h(b_\rho)) &= w(f_k(a)) + kw(b_\rho) \\ &> w(f_k(a)) + kw(b_\sigma) \\ &= w(h(b_\sigma)). \end{aligned}$$

Now assume there is no such extension of (F,v). Let L be a splitting field of $g(x) = f_0(x)f_1(x)\cdots f_n(x)$ over F. By Theorem 5.2.2 there is a valuation w on L that extends v. For each $a \in L$ the sequence $(a_\rho - a)_{\rho<\alpha}$ is pseudo-convergent, and by Theorem 5.2.3 and our assumption there is an ordinal $\lambda = \lambda(a)$ with

$$w(a_\rho - a) = w(a_\sigma - a) \text{ for all } \lambda \leq \rho, \sigma. \tag{5.2.30}$$

By taking λ large enough we may assume that (5.2.30) holds for each root a of $g(x)$ and also that $g(a_\rho) \neq 0$ (for $\rho \geq \lambda$). Since $f_j(x) = c_j(x-d_1)\cdots(x-d_r)$ we then have that $w(f_j(a_\rho)) = w(f_j(a_\sigma))$ for $\lambda \leq \rho$, σ and $0 \leq j \leq n$. From (5.2.27) with $a = a_\rho$ and $x = a_\sigma$ we obtain

$$f(a_\sigma) - f(a_\rho) = \sum_{j=1}^{n} f_j(a_\rho)(a_\sigma - a_\rho)^j. \tag{5.2.31}$$

From Theorem 5.2.4 we get an ordinal μ with $\mu \geq \lambda$ and an integer k with $0 \leq k \leq n$ such that

$$v(f_k(a_\rho)) + kv(a_\sigma - a_\rho) > v(f_j(a_\rho)) + jv(a_\sigma - a_\rho)$$

for all $j \neq k$ and for all $\mu < \rho < \sigma$. From (5.2.31)

$$\begin{aligned} v(f(a_\sigma) - f(a_\rho)) &= v(f_k(a_\rho)(a_\sigma - a_\rho)^k) > v(f_k(a_\rho)(a_\tau - a_\sigma)^k) \\ &= v(f(a_\tau) - f(a_\sigma)) \end{aligned}$$

if $\mu < \rho < \sigma < \tau$. □

Given a pseudo-convergent sequence $(a_\rho)_{\rho<\alpha}$ in (F,v), according to the previous result and Theorem 5.2.3, either there is a polynomial $f(x) \in F[x]$ such that $(v(f(a_\rho)))_{\rho<\alpha}$ is eventually strictly decreasing, in which case the sequence $(a_\rho)_{\rho<\alpha}$ is called *algebraic*, or for every polynomial $f(x)$ the sequence $(v(f(a_\rho)))_{\rho<\alpha}$ is eventually constant, in which case $(a_\rho)_{\rho<\alpha}$ is called *transcendental*. We wish to find limits for pseudo-convergent sequences and to do so we will consider the transcendental and algebraic sequences separately. First, we need some definitions.

The element a in F is called a *pseudo-limit* of the pseudo-convergent sequence $(a_\rho)_{\rho<\alpha}$ if $v(a - a_\rho) = \gamma_\rho (= v(a_{\rho+1} - a_\rho))$ for every ρ, and the *breadth* of $(a_\rho)_{\rho<\alpha}$ is defined to be

$$B = B((a_\rho)_\rho) = \{b \in F : v(b) < \gamma_\rho \text{ for every } \rho < \alpha\}.$$

If a is a pseudo-limit of $(a_\rho)_\rho$, then $a + B$ consists of all the pseudo-limits of $(a_\rho)_\rho$ in F; the easy verification is left for Exercise 1.

The extension (L,w) of (F,v) is called an *immediate extension* of (F,v) if L and F have the same value group and the same residue class field. The valued field (F,v) is *maximally complete* if its only immediate extension is itself. Our goal is to relate maximal completeness to the existence of pseudo-limits.

Theorem 5.2.6. *Let (L,v) be an immediate extension of (F,v). If $t \in L\backslash F$, then t is a pseudo-limit of a pseudo-convergent sequence in F that has no pseudo-limit in F.*

Proof. If $a \in F$, then $v(t-a) = v(b)$ for some $b \in F$. Then $(t-a)b^{-1} \in R_L$ and $(t-a)b^{-1}+J_L = c+J_L$ for some $c \in F$. So $t-a = cb+db$ with $d \in J_L$ and $v(t-(a+cb)) < v(b)$. This shows that $v(t-F)$ has no smallest element. Consequently, if α is an ordinal for which there is a coinitial strictly decreasing sequence $(v(t-a_\rho))_{\rho<\alpha}$ in $v(t-F)$ (see Exercise 1.1.8), then α is a limit ordinal. Now, $(a_\rho)_{\rho<\alpha}$ is a pseudo-convergent sequence in F which has t as a pseudo-limit. For if $\rho < \sigma < \tau$, then

$$\begin{aligned} v(a_\sigma - a_\rho) &= v((t-a_\rho)-(t-a_\sigma)) = v(t-a_\rho) \vee v(t-a_\sigma) = v(t-a_\rho) \\ &> v(t-a_\sigma) = v(a_\tau - a_\sigma). \end{aligned}$$

Suppose $y \in F$ is also a pseudo-limit of $(a_\rho)_{\rho<\alpha}$. Then $t-y$ is an element of the breadth of $(a_\rho)_{\rho<\alpha}$ and by Exercise 1, for all $\rho < \sigma$, we have $v(t-y) < v(a_\sigma - a_\rho) = v(t-a_\rho)$. This is impossible since $\{v(t-a_\rho) : \rho < \alpha\}$ is coinitial in $v(t-F)$. □

We will now construct pseudo-limits of pseudo-convergent sequences.

Theorem 5.2.7. *Let $(a_\rho)_{\rho<\alpha}$ be a transcendental pseudo-convergent sequence in the valued field (F,v) and suppose $(a_\rho)_{\rho<\alpha}$ has no pseudo-limit in F. There is a unique valuation w on the rational function field $F(x)$ which extends v and for which x is a pseudo-limit of $(a_\rho)_{\rho<\alpha}$. The valued field $(F(x),w)$ is an immediate extension of (F,v). Moreover, any pseudo-limit of $(a_\rho)_{\rho<\alpha}$ in an extension of (F,v) is transcendental over F.*

Proof. By Exercise 3 it suffices to define w on $F[x]$. For $0 \neq f(x) \in F[x]$ there is some $\lambda < \alpha$ such that $v(f(a_\rho)) = v(f(a_\sigma))$ for all $\lambda < \rho, \sigma$; define $w(f(x))$ by $w(f(x)) = v(f(a_\rho))$. Clearly, $w(f(x)) \neq -\infty$. For $0 \neq g(x) \in F[x]$ and a sufficiently large λ, $v(g(a_\rho))$ is also constant for $\rho > \lambda$ and $w(f(x)g(x)) = v(f(a_\rho)g(a_\rho)) = v(f(a_\rho)) + v(g(a_\rho)) = w(f(x)) + w(g(x))$. Also, $w(f(x)+g(x)) = v(f(a_\rho)+g(a_\rho)) \leq v(f(a_\rho)) \vee v(g(a_\rho)) = w(f(x)) \vee w(g(x))$. Thus, w is a valuation which extends v and the two value groups are the same. To see that the residue class fields are the same take $r(x) = h(x)g(x)^{-1} \in F(x)$ with $w(h(x)) = w(g(x))$. By multiplying the numerator and denominator of $r(x)$ by $c \in F$ with $v(c) = -w(h(x))$ we may assume $w(h(x)) = 0$. Since $(h(a_\rho))_\rho$ and $(g(a_\rho))_\rho$ are eventually pseudo-convergent there exists a $\lambda < \alpha$ such that if $\lambda < \rho < \sigma < \tau$, then $w(h(x)) = v(h(a_\rho))$, $w(g(x)) = v(g(a_\rho))$, $v(h(a_\sigma) - h(a_\rho)) > v(h(a_\tau) - h(a_\sigma))$ and $v(g(a_\sigma) - g(a_\rho)) > v(g(a_\tau) - g(a_\sigma))$. Fix $\sigma > \rho$ (with $\rho > \lambda$). For some $\lambda_1 > \sigma$ we have $v(h(a_\tau) - h(a_\sigma)) = v(h(a_\mu) - h(a_\sigma))$ for all $\tau > \mu > \lambda_1$. Consequently,

$$w(h(x) - h(a_\sigma)) = v(h(a_\tau) - h(a_\sigma)) < v(h(a_\sigma) - h(a_\rho)) \leq 0$$

and $h(x)+J_{F(x)} = h(a_\rho)+J_{F(x)}$. Similarly, $w(g(x)-g(a_\rho)) < 0$, $g(x)+J_{F(x)} = g(a_\rho)+J_{F(x)}$ and

$$h(x)g(x)^{-1}+J_{F(x)} = h(a_\rho)g(a_\rho)^{-1}+J_{F(x)}.$$

To see that x is a pseudo-limit of $(a_\rho)_{\rho<\alpha}$, for a fixed ρ take $\lambda > \rho$ such that $v(a_\tau - a_\rho) = v(a_\sigma - a_\rho)$ for all $\tau, \sigma > \lambda$. Then $w(x-a_\rho) = v(a_\tau - a_\rho) = \gamma_\rho$.

Now suppose that w' is a valuation on $F(x)$ which extends v and x is a pseudo-limit of $(a_\rho)_{\rho<\alpha}$ in $(F(x), w')$. So $w'(x-a_\rho) = \gamma_\rho$ for every $\rho < \alpha$. Let $f(x) \in F[x]$ be of positive degree n. Choose λ such that for all $\lambda \leq \rho$, σ we have $v(f_j(a_\rho)) = v(f_j(a_\sigma))$ for $0 \leq j \leq n$ (see (5.2.25)). By (5.2.27)

$$f(x)-f(a_\rho) = \sum_{j=1}^{n} f_j(a_\rho)(x-a_\rho)^j \tag{5.2.32}$$

and by Theorem 5.2.4 there is some $\mu \geq \lambda$ and some $k \geq 1$ with $v(f_j(a_\rho))+j\gamma_\rho < v(f_k(a_\rho))+k\gamma_\rho$ for each $1 \leq j \neq k$ and each $\rho > \mu$. So for $\sigma > \rho > \mu$, we have $\gamma_\rho > \gamma_\sigma$ and

$$\begin{aligned} w'(f(x)-f(a_\rho)) &= v(f_k(a_\rho))+k\gamma_\rho = w(f(x)-f(a_\rho)) > \\ v(f_k(a_\sigma))+k\gamma_\sigma &= w(f(x)-f(a_\sigma)) = w'(f(x)-f(a_\sigma)). \end{aligned}$$

If $w'(f(x)) > w'(f(a_\rho))$, then

$$w'(f(x)) = w'(f(x)-f(a_\rho)) > w'(f(x)-f(a_\sigma)) = w'(f(x)),$$

and if $w'(f(x)) < w'(f(a_\rho))$, then

$$w'(f(a_\rho)) = w'(f(x)-f(a_\rho)) > w'(f(x)-f(a_\sigma)) = w'(f(a_\rho)).$$

So $w'(f(x)) = v(f(a_\rho)) = w(f(x))$.

Suppose a is a pseudo-limit of $(a_\rho)_{\rho<\alpha}$ in some extension of F. If a is algebraic over F let $f(x)$ be its irreducible polynomial over F. Then from (5.2.27) we get, for large ρ,

$$-f(a_\rho) = \sum_{j=1}^{n} f_j(a_\rho)(a-a_\rho)^j,$$

and, by Theorem 5.2.4, for some $k \geq 1$, $v(f(a_\rho)) = v(f_k(a_\rho))+k\gamma_\rho$. This is impossible since, for large ρ, $v(f(a_\rho))$ and $v(f_k(a_\rho))$ are independent of ρ but $(\gamma_\rho)_{\rho<\alpha}$ is strictly decreasing. □

We need a similar result for algebraic pseudo-convergent sequences.

Theorem 5.2.8. *Let $(a_\rho)_{\rho<\alpha}$ be an algebraic pseudo-convergent sequence in the valued field (F, v). Suppose $f(x)$ is a polynomial of least degree in $F[x]$ such that the sequence $(v(f(a_\rho)))_{\rho<\alpha}$ is eventually strictly decreasing. Then*

(a) $f(x)$ is irreducible.

(b) *If a is a root of $f(x)$, then there is a unique valuation w on $F(a)$ which extends v and such that a is a pseudo-limit of $(a_\rho)_\rho$.*

(c) *$(F(a), w)$ is an immediate extension of (F, v).*

Proof. If $f(x) = c(x-a)$, then, since $(a_\rho - a)_{\rho<\alpha}$ is pseudo-convergent and $(v(a_\rho - a))_\rho$ is eventually decreasing, $v(a_\rho - a) > v(a_\sigma - a)$ for all $\rho < \sigma$ by Theorem 5.2.3. But then $v(a_\rho - a) = v((a_\sigma - a) - (a_\rho - a)) = v(a_\sigma - a_\rho) = \gamma_\rho$ for all $\rho < \sigma$, and a is a pseudo-limit of $(a_\rho)_\rho$. Thus, we will assume that deg $f(x) = n \geq 2$. If $f(x) = g_1(x)g_2(x)$ with $g_i(x) \in F[x]$ and $1 \leq$ deg $g_i(x) < n$, then for some λ and for all $\lambda < \rho < \sigma$, $v(g_i(a_\rho)) = v(g_i(a_\sigma))$. This gives a contradiction since

$$v(f(a_\rho)) = v(g_1(a_\rho)) + v(g_2(a_\rho)) = v(f(a_\sigma)) < v(f(a_\rho))$$

for large ρ and σ. Now, for $g(a) \in F(a)$ with deg $g(x) < n$, the sequence $v(g(a_\rho))_{\rho<\alpha}$ is eventually constant and we define $w(g(a))$ to be this constant. Suppose $g_1(x)$, $g_2(x) \in F[x]$ and both have degree smaller than n. Then for large ρ,

$$\begin{aligned} w(g_1(a) + g_2(a)) &= v(g_1(a_\rho)) + g_2(a_\rho)) \leq v(g_1(a_\rho)) \vee v(g_2(a_\rho)) \\ &= w(g_1(a)) \vee w(g_2(a)). \end{aligned}$$

Also, $g_1(x)g_2(x) = f(x)q(x) + r(x)$ with $r(x) \neq 0$ and deg $r(x) < n$. Now, for large ρ, $v(g_1(a_\rho))$, $v(g_2(a_\rho))$, $v(q(a_\rho))$ and $v(r(a_\rho))$ are each independent of ρ and $v(f(a_\rho)q(a_\rho))$ is strictly decreasing. By Exercise 4 we then have $v(g_1(a_\rho)g_2(a_\rho)) = v(r(a_\rho))$ and

$$w(g_1(a)g_2(a)) = v(r(a_\rho)) = v(g_1(a_\rho)) + v(g_2(a_\rho)) = w(g_1(a)) + w(g_2(a)).$$

Thus, w is a valuation on $F(a)$ that extends v and these valuations have the same value group. Since each element $g(a) \in F(a)$ comes from a polynomial $g(x) \in F[x]$ for which the sequence $(g(a_\rho))$ is eventually constant valued the remaining verifications are the same as for the transcendental case given in the previous result. □

We can now give the tight connection between maximal completeness and pseudo-limits.

Theorem 5.2.9. *A valued field (F, v) is maximally complete if and only if every pseudo-convergent sequence in F has a pseudo-limit in F.*

Proof. Let $(a_\rho)_{\rho<\alpha}$ be a pseudo-convergent sequence in F. By Theorems 5.2.7 and 5.2.8, $(a_\rho)_\rho$ has a pseudo-limit t in an immediate extension of F. If F is maximally complete, then $t \in F$. The converse is a consequence of Theorem 5.2.6. □

Criteria will now be given for a maximally complete field to be algebraically closed and then it will be applied to fields of formal power series.

Theorem 5.2.10. *Suppose (F, v) is a valued field. If F is algebraically closed, then its value group is divisible and its residue class field is algebraically closed. Conversely, these two conditions imply that each algebraic extension of F is an immediate extension.*

Proof. Suppose F is algebraically closed. For $n \in \mathbb{N}$ and $a \in F$ take $b \in F$ with $b^n = a$. Then $v(a) = nv(b)$ and $v(F^*)$ is divisible. Let $\overline{f}(x) \in (R/J)[x]$ be a monic polynomial of positive degree and let $f(x) \in R[x]$ be a monic lifting of $\overline{f}(x)$ such that each of its nonzero coefficients is a unit of R. If $f(x) = \sum a_i x^i$ and $0 \neq a \in F$ is a root of $f(x)$, then from $\sum a_i a^i = 0$ we have that for some $i < j$, $-\infty \neq v(a_i a^i) = v(a_j a^j)$; that is, $(j-i)v(a) = v(a_i a_j^{-1}) = 0$. So $a \in R$ and $f(x) = (x-a)g(x)$ with $g(x) \in R[x]$ and $g(x)$ monic. Therefore, $\overline{f}(x)$ has the root $\overline{a}$ in R/J. Conversely, suppose the two conditions hold and L is an algebraic extension of F. Let w be a valuation on L that extends v. If $a \in L$, then there exists $f(x) \in R_F[x] \backslash J_F[x]$ with $f(a) = 0$. Just as above, $(j-i)w(a) \in v(F)$ and hence $w(a) \in v(F)$ by the divisibility of $v(F^*)$. Suppose $a \in R_L \backslash J_L$. Then $\overline{f}(\overline{a}) = 0$ shows that $\overline{a} = a + J_L$ is algebraic over R_F/J_F. Thus, $\overline{a} \in R_F/J_F$ and L is an immediate extension of F. □

If G is a group and Γ is a totally ordered set, then a mapping $v : G \longrightarrow \Gamma \cup \{-\infty\}$ is a valuation provided (5.2.1), (5.2.2) and (5.2.4) hold. Then (5.2.6), (5.2.7), and (5.2.8) also hold and so does Theorem 5.2.3 since the concepts of pseudo-convergent sequences and of pseudo-limits still makes sense in this more general setting.

Theorem 5.2.11. *Suppose G_γ is a group for each γ in the totally ordered set Γ. Then every pseudo-convergent sequence in the Hahn product $V(\Gamma, G_\gamma)$ has a pseudo-limit.*

Proof. Recall that the valuation here is given by $\pi(f) = \text{maxsupp } f$. Suppose $(f_\rho)_{\rho<\alpha}$ is pseudo-convergent in V. Let $\gamma_\rho = \pi(f_{\rho+1} - f_\rho)$. By Theorem 5.2.3 we have $\gamma_\rho = \pi(f_\sigma - f_\rho)$ and $\gamma_\sigma < \gamma_\rho$ if $\rho < \sigma$; so $f_\rho(\gamma) = f_\sigma(\gamma)$ if $\gamma_\rho < \gamma$ and the definition

$$f(\gamma) = \begin{cases} f_\rho(\gamma) & \text{if } \gamma_\rho < \gamma \text{ for some } \rho < \alpha \\ 0 & \text{if } \gamma \leq \gamma_\rho \text{ for all } \rho < \alpha \end{cases} \tag{5.2.33}$$

gives an element of the product $\Pi_\gamma G_\gamma$. If $\delta_1 \leq \delta_2 \leq \cdots$ in supp f, then $\gamma_\rho < \delta_1$, for some ρ and $\{\delta_n\} \subseteq \text{supp } f_\rho$. So $\delta_n = \delta_{n+1} = \cdots$ for some n and $f \in V(\Gamma, G_\gamma)$. Now, $\pi(f - f_\rho) \leq \gamma_\rho$ since f and f_ρ agree on $\{\gamma \in \Gamma : \gamma_\rho < \gamma\}$; but $f(\gamma_\rho) = f_{\rho+1}(\gamma_\rho) \neq f_\rho(\gamma_\rho)$ by (5.2.33) since $\gamma_{\rho+1} < \gamma_\rho$. Thus, $\pi(f - f_\rho) = \gamma_\rho$ for each $\rho < \alpha$ and f is a psuedo-limit of $(f_\rho)_{\rho<\alpha}$. □

We can now apply the previous results to get an embedding theorem for totally ordered fields.

If $V(K * \Gamma)$ is the formal power series crossed product division ring given in Exercise 3.5.20, then $V(K * \Gamma)$ will be a field exactly when Γ is abelian, K is a field, the action is trivial and the twist satisfies $\tau(\gamma, \delta) = \tau(\delta, \gamma)$ for all γ, $\delta \in \Gamma$. It also follows from this exercise and the proof of Theorem 3.5.8 that for an ordinal number $\alpha > 0$ the subring $V_\alpha(K * \Gamma) = \{v \in V : |\text{supp } v| < \aleph_\alpha\}$ of $V(K * \Gamma)$ is a subfield.

Theorem 5.2.12. *Let $V(K * \Gamma)$ be a formal power series crossed product field and for the ordinal $\alpha > 0$ let $V_\alpha(K * \Gamma) = \{v \in V(K * \Gamma) : |\text{supp } v| < \aleph_\alpha\}$. The following statements are equivalent.*

(a) $V(K*\Gamma)$ is algebraically (real) closed.
(b) K is algebraically (real) closed and Γ is divisible.
(c) $V_\alpha(K*\Gamma)$ is algebraically (real) closed.

Proof. By Theorems 5.2.9 and 5.2.11 $(V(K*\Gamma),\pi)$ is maximally complete. Since K is the residue class field and Γ is the value group of both fields the equivalence of (a) with (b) for the algebraically closed case and the fact that (c) implies (b) for this case is a consequence of Theorem 5.2.10. The same equivalence and implication for the real closed case now follows from Theorem 5.1.6 and the isomorphisms $V(K*\Gamma)(i) \cong V(K(i)*\Gamma)$ and $V_\alpha(K*\Gamma)(i) \cong V_\alpha(K(i)*\Gamma)$ given in Exercise 5. To finish the proof it suffices to show that (a) implies (c). Suppose $f(y) = v_0 + v_1 y + \cdots + v_n y^n \in V_\alpha(K*\Gamma)[y]$. Let Δ be the $\mathbb{Q}$-subspace of Γ spanned by $X = \bigcup_{j=0}^{n} \operatorname{supp} v_j$. Then $|\Delta| < \aleph_\alpha$ since $|X| < \aleph_\alpha$. From the equivalence of (a) with (b) we know that $V(K*\Delta)$ is algebraically closed or real closed. Suppose $f(y)$ has positive degree in the first case and has odd degree or $f(y) = y^2 - v_0$ with $v_0 > 0$ in the second case. By identifying $f(y)$ with the polynomial in $V(K*\Delta)[y]$ obtained by restricting each v_j to Δ we see that $f(y)$ has a root in the subfield $V(K*\Delta)$ of $V_\alpha(K*\Gamma)$. Thus, by Theorem 5.1.6, $V_\alpha(K*\Gamma)$ has the same closure property as $V(K*\Gamma)$. □

Theorem 5.2.13. *Let K be a totally ordered field and let Γ be a totally ordered divisible abelian η_α-group where $\aleph_\alpha$ is a regular cardinal with card $(K) \leq \aleph_\alpha$ and $\alpha > 0$. Then K can be embedded in $V_\alpha(\Gamma,\mathbb{R})$.*

Proof. By Theorem 4.4.8 $V_\alpha(\Gamma,\mathbb{R})$ is a totally ordered η_α-field which is real closed by the previous result. Thus, K can be embedded in $V_\alpha(\Gamma,\mathbb{R})$ by Theorem 5.1.11. □

In order to replace Γ in Theorem 5.2.13 by the smaller value group of K with respect to its natural valuation we require two more properties of maximally complete fields. A valued field (F,v) is *henselian* if whenever $f(x) \in R_F[x]$ is a monic polynomial and a is an element of R_F with the property that $\overline{a} \in \overline{R}_F = R_F/J_F$ is a simple root of $\overline{f}(x) \in \overline{R}_F[x]$, then $f(b) = 0$ for some $b \in R_F$ with $\overline{b} = \overline{a}$.

Theorem 5.2.14. *A maximally complete valued field (F,v) is henselian.*

Proof. Suppose $f(x)$ and a are as above and $f(a) \neq 0$. Let $0 < \alpha$ be a limit ordinal. By transfinite induction we will construct a sequence $(a_\rho)_{\rho<\alpha}$ in R_F with the following properties for all $\rho < \sigma < \alpha$:

(i) $f'(a_\rho) \in U_F$;
(ii) $v(a_\rho - a_\sigma) = v(f(a_\rho))$;
(iii) if $f(a_\rho) \neq 0$, then $v(f(a_\sigma)) < v(f(a_\rho))$.

Let $a_0 = a$. Suppose $\mu < \alpha$ and the sequence $(a_\rho)_{\rho<\mu}$ has been constructed with these three properties. If $f(a_\rho) = 0$ for some ρ let $a_\mu = a_\rho$. Suppose each $f(a_\rho) \neq 0$. Note that by (ii) and (iii) both $a_\rho - a_0$ and $f(a_\rho)$ are elements of J_F since $f(a_0) \in J_F$. Assume first that μ is a limit ordinal. Then $(a_\rho)_{\rho<\mu}$ is pseudo-convergent since for

$\rho < \sigma < \tau < \mu$ we have $v(a_\sigma - a_\rho) = v(f(a_\rho)) > v(f(a_\sigma)) = v(a_\tau - a_\sigma)$. According to Theorem 5.2.9 the sequence $(a_\rho)_{\rho<\mu}$ has a pseudo-limit $a_\mu \in F$. Now, (ii) is satisfied by a_μ, and $a_\mu \in R_F$ since $v(a_\rho - a_\mu) = v(a_{\rho+1} - a_\rho) = v(f(a_\rho)) \leq 0$ and $a_\mu = (a_\mu - a_\rho) + a_\rho$. To verify (iii) for $\rho < \mu$ we use (5.2.26) with base point $a_{\rho+1}$. Then, since $f_0(x) = f(x)$,

$$v(f(a_\mu)) = v\left(\sum_{k=0}^{n} f_k(a_{\rho+1})(a_\mu - a_{\rho+1})^k\right)$$

$$\leq \bigvee_{k=0}^{n} v(f_k(a_{\rho+1})f(a_{\rho+1})^k)$$

$$= v(f(a_{\rho+1})) < v(f(a_\rho)).$$

In order to show that $f'(a_\mu) \in U_F$ we again use (5.2.26) but now take the base point to be a_μ. Since $f_1(x) = f'(x)$ we have

$$f(x) - f(a_\mu) = f'(a_\mu)(x - a_\mu) + \sum_{k=2}^{n} f_k(a_\mu)(x - a_\mu)^k$$

and

$$f'(a_\mu) = \frac{f(a) - f(a_\mu)}{a - a_\mu} - \sum_{k=2}^{n} f_k(a_\mu)(a - a_\mu)^{k-1}. \tag{5.2.34}$$

From (ii) and (iii) for a_0 and a_μ we have

$$v\left(\frac{f(a) - f(a_\mu)}{a - a_\mu}\right) = v(f(a)) - v(f(a)) = 0,$$

and $f_k(a)(a - a_\mu)^{k-1} \in J_F$ for $k \geq 2$. So $v(f'(a_\mu)) = 0$.

Now suppose $\mu = \lambda + 1$, and use (5.2.26) with base point a_λ. Then

$$f(x) = f(a_\lambda) + f'(a_\lambda)(x - a_\lambda) + \sum_{k=2}^{n} f_k(a_\lambda)(x - a_\lambda)^k. \tag{5.2.35}$$

Let $c = f(a_\lambda)$, $d = -f'(a_\lambda)^{-1}$ and $a_\mu = a_\lambda + cd$; $c \in J_F$, $d \in U_F$, and $a_\mu \in R_F$. From (5.2.35) with $x = a_\mu$,

$$f(a_\mu) = f(a_\lambda) - f'(a_\lambda)f(a_\lambda)f'(a_\lambda)^{-1} + \sum_{k=2}^{n} f_k(a_\lambda)c^k d^k$$

$$= \sum_{k=2}^{n} f_k(a_\lambda)c^k d^k;$$

so $f(a_\mu)c^{-1} \in J_F$ and $v(f(a_\mu)) < v(f(a_\lambda))$. Thus (iii) is satisfied. For (ii), $v(a_\lambda - a_\mu) = v(cd) = v(c) = v(f(a_\lambda))$, and if $\rho < \lambda$, then $v(a_\rho - a_\mu) = v((a_\rho - a_\lambda) +$

$(a_\lambda - a_\mu)) = v(f(a_\rho)) \vee v(f(a_\lambda)) = v(f(a_\rho))$ by the induction hypothesis. From (5.2.26) we have, as in (5.2.34) but using a_λ in place of a,

$$f'(a_\mu) = \frac{f(a_\lambda) - f(a_\mu)}{a_\lambda - a_\mu} - \sum_{k=2}^{\infty} f_k(a_\mu)(a_\lambda - a_\mu)^{k-1}.$$

But the sum on the right is in J_F and we have just shown that $v(f(a_\lambda) - f(a_\mu)) = v(f(a_\lambda)) = v(a_\lambda - a_\mu)$; so $f'(a_\mu) \in U_F$ and the new sequence has the desired properties.

Now, if $\aleph_\alpha > \operatorname{card}(F)$, then the terms of the sequence $(a_\rho)_{\rho<\omega_\alpha}$ cannot be distinct. So $f(a_\rho) = 0$ for some $\rho < \omega_\alpha$, and $\overline{a}_\rho = \overline{a}$ by (ii) since $a_0 = a$. □

An interesting property of henselian fields that we will make use of is that they have intermediate fields with specified data.

Theorem 5.2.15. *Let K be a subfield of the henselian field (M, v) and suppose that the residue class field of M has characteristic* 0. *Then there is an intermediate field F, $K \subseteq F \subseteq M$, such that K and F have the same value group and F and M have the same residue class field.*

Proof. Apply Zorn's lemma to the set of all intermediate fields E of the extension $K \subseteq M$ with the property that $\Gamma_K = \Gamma_E$ to get a maximal such intermediate field F. Suppose $a + J_M \in (R_M/J_M) \backslash (R_F/J_F)$. Then $v(a) = 0$. If $a + J_M$ is transcendental over R_F/J_F, then a is transcendental over F. Otherwise, it would satisfy a polynomial in $R_F[x] \backslash J_F[x]$ and hence $a + J_M$ would be algebraic over R_F/J_F. If $a + J_M$ is algebraic over R_F/J_F, then by lifting its irreducible polynomial to a monic polynomial $g(x) \in R_F[x]$ we get an element $b \in R_M$ with $b - a \in J_M$ and $g(b) = 0$. Let $b = a$ if $a + J_M$ is transcendental over R_F/J_F. We claim that $F(b)$ and F have the same value group. Suppose $0 \neq f(b) \in F[b]$ and assume that $f(x) = f_0 + f_1 x + \cdots + f_m x^m$ has smaller degree than $g(x)$ if b is algebraic over F. Take $c \in F$ with $v(c) = \bigvee_{j=0}^{m} v(f_j)$. Then $c^{-1} f(x) \in R_F[x] \backslash J_F[x]$. If $v(f(b)) < v(c)$, then $c^{-1} f(b) \in J_M$ and we have the contradiction that $a + J_M$ satisfies a nonzero polynomial over R_F/J_F, of smaller degree than that of $g(x)$ if $a + J_M$ is algebraic over R_F/J_F. Thus, $v(f(b)) = v(c) \in v(F)$ and $v(F(b)) = v(F)$. By the maximality of F we must have $R_M/J_M = R_F/J_F$. □

The next two theorems will be used to show that a totally ordered field has a unique largest immediate extension.

Theorem 5.2.16. *Let $(a_\rho)_{\rho<\alpha}$ be an algebraic pseudo-convergent sequence in the valued field (F, v). Suppose $f(x)$ is a polynomial in $F[x]$ of least degree n such that $v(f(a_\rho))$ is eventually strictly decreasing and let $f_j(x)$ be the polynomial given by (5.2.25). Then, with $\gamma_\rho = v(a_{\rho+1} - a_\rho)$, there is an ordinal $\tau_0 < \alpha$ such that for all σ, $\rho > \tau_0$ and for $1 \leq j \leq n-1$*

(a) *$v(f_j(a_\rho)) = v(f_j(a_\sigma)) = \beta_j$ (this defines β_j),*
(b) *$\beta_j + j\gamma_\rho < \beta_1 + \gamma_\rho$ if $2 \leq j$,*
(c) *$v(f(a_\rho)) = \beta_1 + \gamma_\rho$.*

Moreover, if a is a pseudo-limit of $(a_\rho)_{\rho<\alpha}$, *then*

(d) $v(f(a)) < \beta_1 + \gamma_\rho$ *for all* ρ,

(e) $v(f_j(a)) = \beta_j$ *for* $j = 1,2,\ldots,n$.

Proof. We assume $n \geq 2$ since for $n = 1$ this is trivial and almost vacuous. As we have previously noted, the existence of an ordinal τ_1 for which (a) holds for $\rho > \tau_1$ is just a consequence of the minimality of n. By differentiating equation (5.2.27) with $a = a_\rho$ we obtain

$$f'(x) - f'(a_\rho) = \sum_{j=2}^{n} f_j(a_\rho) j (x - a_\rho)^{j-1}$$

and

$$f_1(a_\sigma) - f_1(a_\rho) = \sum_{j=2}^{n} f_j(a_\rho) j (a_\sigma - a_\rho)^{j-1}.$$

By Theorem 5.2.4 there is some k, $2 \leq k \leq n$, and some $\mu \geq \tau_1$ such that if $2 \leq j \neq k \leq n$, then $\beta_j + (j-1)\gamma_\rho < \beta_k + (k-1)\gamma_\rho$ for $\rho > \mu$. So, for $\sigma > \rho > \mu$,

$$\beta_1 = v(f_1(a_\sigma)) \vee v(f_1(a_\rho)) \geq v(f_1(a_\sigma) - f_1(a_\rho)) = \beta_k + (k-1)\gamma_\rho \geq \beta_j + (j-1)\gamma_\rho,$$

and the last inequality is strict for $j \neq k$. If $\beta_1 > \beta_k + (k-1)\gamma_\rho$ for all $\rho > \mu$, then (a) and (b) hold with $\tau_0 = \mu$. If $\beta_1 = \beta_k + (k-1)\gamma_\tau$ for some $\tau > \mu$, then $\beta_1 > \beta_k + (k-1)\gamma_\rho$ if $\rho > \tau$ and hence (a) and (b) hold for $\tau_0 = \tau + 1$. We may assume that $(v(f(a_\rho)))_{\tau_0 < \rho}$ is strictly decreasing. Since for $\sigma > \rho > \tau_0$ we have, from (5.2.27),

$$f(a_\sigma) = f(a_\rho) + \sum_{j=1}^{n} f_j(a_\rho)(a_\sigma - a_\rho)^j$$

and since the value of the large sum on the right, which we will call u_σ, is $\beta_1 + \gamma_\rho$ (by (b)), we must have $v(f(a_\rho)) = \beta_1 + \gamma_\rho$, also. If not, for $\tau > \sigma$

$$v(f(a_\sigma)) = v(f(a_\rho)) \vee v(u_\sigma) = v(f(a_\rho)) \vee v(f(a_\tau)) = v(f(a_\tau)) < v(f(a_\sigma)).$$

This proves (c).

For (d), from (5.2.26),

$$f(a) = f(a_\rho) + f_1(a_\rho)(a - a_\rho) + \sum_{j=2}^{n} f_j(a_\rho)(a - a_\rho)^j$$

and for $\rho > \tau_0$ each of the first two terms have the largest value, $\beta_1 + \gamma_\rho$, of all the terms on the right. So $v(f(a)) \leq \beta_1 + \gamma_\rho$. If the inequality is not strict, then $v(f(a)) = \beta_1 + \gamma_\rho > \beta_1 + \gamma_\sigma \geq v(f(a))$ for $\rho < \sigma$. Since $(\gamma_\rho)_\rho$ is strictly decreasing (d) holds for every ρ.

By differentiating

$$f(x) = \sum_{j=0}^{n} f_j(a_\rho)(x - a_\rho)^j$$

i times and using (5.2.25) we get

$$f_i(a) - f_i(a_\rho) = \sum_{j=i+1}^{n} \binom{j}{i} f_j(a_\rho)(a - a_\rho)^{j-i}. \tag{5.2.36}$$

By Theorem 5.2.4 there is some k, $i+1 \le k \le n$, and some $\lambda \ge \tau_0$ such that if $i+1 \le j \le n$ and $j \ne k$, then $(j-i)\gamma_\rho + \beta_j < (k-i)\gamma_\rho + \beta_k$ for all $\rho > \lambda$. If $v(f_i(a)) \ne v(f_i(a_\rho))$ for some $\rho > \lambda$, then, for all $\sigma > \rho$, either $v(f_i(a)) = (k-i)\gamma_\sigma + \beta_k$ or $v(f_i(a_\rho)) = v(f_i(a_\sigma)) = (k-i)\gamma_\sigma + \beta_k$. Since (γ_ρ) is a strictly decreasing sequence this is impossible and hence $v(f_i(a)) = v(f_i(a_\rho)) = \beta_i$ for all $\rho > \lambda$. □

We have seen in Theorem 5.2.8 that $(a_\rho)_{\rho<\alpha}$ has a pseudo-limit which is a root of $f(x)$ in some immediate extension of F. We show next that this is always the case in any immediate extension that is maximally complete.

Theorem 5.2.17. *With the same setup as in Theorem 5.2.16, suppose a is a pseudo-limit of $(a_\rho)_{\rho<\alpha}$ in some immediate extension M of F and $f(a) \ne 0$.*

(a) M contains a pseudo-limit b of $(a_\rho)_{\rho<\alpha}$ such that $v(f(b)) < v(f(a))$ and $v(b-a) = v(f(a)) - \beta_1$.

(b) If M is maximally complete, then $f(x)$ has a root in M that is a pseudo-limit of $(a_\rho)_{\rho<\alpha}$.

Proof. (a) Let $v(f(a)) = \gamma$ and put $\delta = \gamma - \beta_1$. By (d) of Theorem 5.2.16, $\gamma < \beta_1 + \gamma_\rho$ for each $\rho < \alpha$; i.e. $\delta < \gamma_\rho$. Take $d \in M$ with $v(d) = \delta$. From (5.2.26), for any $c \in M$ we have

$$\begin{aligned} f(a+dc)f(a)^{-1} &= \sum_{j=0}^{n} f_j(a)f(a)^{-1}d^j c^j \\ &= 1 + (f_1(a)f(a)^{-1}d)c + \sum_{j=2}^{n} (f_j(a)f(a)^{-1}d^j)c^j. \end{aligned} \tag{5.2.37}$$

If $j \ge 2$ the coefficient of c^j in (5.2.37) is in J_M. For, by (e), (d), and (b) of the previous theorem $v(f_j(a)f(a)^{-1}d^j) = \beta_j - \gamma + j\delta$ and

$$\beta_j + (j-1)(\gamma - \beta_1) < \beta_j + (j-1)\gamma_\rho < \beta_1.$$

So

$$\beta_j - \gamma + j\delta = \beta_j - \gamma + j(\gamma - \beta_1) = \beta_j + (j-1)\gamma - j\beta_1 < 0.$$

The coefficient of c in (5.2.37) is a unit since its value is 0. Now let $c = -f_1(a)^{-1}d^{-1}f(a)$ and $b = a + dc$. Then (5.2.37) gives that $f(b)f(a)^{-1} \in J_M$ and hence $v(f(b)) < v(f(a))$. Also, $v(b-a) = v(cd) = v(f_1(a)^{-1}f(a)) = \gamma - \beta_1 < \gamma_\rho$ for all $\rho < \alpha$. So $b - a$ is in the breadth of the sequence, and, by Exercise 1, b is a pseudo-limit with the desired properties.

(b) Suppose M does not contain a pseudo-limit of $(a_\rho)_{\rho<\alpha}$ which is a root of $f(x)$. Let α_0 be a limit ordinal. We will construct a sequence $(x_\mu)_{\mu<\alpha_0}$ of pseudo-limits of $(a_\rho)_\rho$ in M such that $(v(f(x_\mu)))_{\mu<\alpha_0}$ is strictly decreasing and $v(x_\nu -$

$x_\mu) = v(f(x_\mu)) - \beta_1$ for all $\mu < \nu$. Of course this is impossible if card $(\alpha_0) >$ card (M); so some pseudo-limit must be a root of $f(x)$. Let $x_0 \in M$ be a pseudo-limit of $(a_\rho)_{\rho<\alpha}$. The existence of x_0 is given by Theorem 5.2.9. Suppose for some ordinal $\lambda < \alpha_0$ the sequence $(x_\mu)_{\mu<\lambda}$ has been constructed with the desired properties. We will first consider the case where λ is a limit ordinal. Let $\delta_\mu = v(f(x_\mu))$. If $\mu < \nu < \sigma < \lambda$, then $v(x_\nu - x_\mu) = \delta_\mu - \beta_1 > \delta_\nu - \beta_1 = v(x_\sigma - x_\nu)$ and hence $(x_\mu)_{\mu<\lambda}$ is pseudo-convergent. Let $x_\lambda \in M$ be a pseudo-limit of $(x_\mu)_{\mu<\lambda}$. Then $v(x_\lambda - x_\mu) = v(x_{\mu+1} - x_\mu) = \delta_\mu - \beta_1 < \gamma_\rho$ for all $\lambda < \mu$ and all $\rho < \alpha$, by (d) of Theorem 5.2.16. So $x_\lambda - x_\mu$ is in the breadth of $(a_\rho)_{\rho<\alpha}$ and x_λ is a pseudo-limit of this sequence. We still need to check that $v(f(x_\lambda)) < v(f(x_\mu))$ for $\mu < \lambda$. From (5.2.26)

$$f(x_\lambda) = \sum_{j=0}^{n} f_j(x_\mu)(x_\lambda - x_\mu)^j.$$

Now, by (e) of Theorem 5.2.16, $v(f_1(x_\mu)(x_\lambda - x_\mu)) = \beta_1 + (\delta_\mu - \beta_1) = \delta_\mu$ and, for $j \geq 2$,

$$v(f_j(x_\mu)(x_\lambda - x_\mu)^j) = \beta_j + j(\delta_\mu - \beta_1) < \beta_j + j\gamma_\rho < \beta_1 + \gamma_\rho = \delta_\mu$$

by (d), (b), and (c) of Theorem 5.2.16. So, $v(f(x_\lambda)) \leq \delta_\mu$ and $v(f(x_\lambda)) \leq \delta_{\mu+1} < \delta_\mu$.

Suppose now that $\lambda = \sigma + 1$. By (a) there is a pseudo-limit x_λ of $(a_\rho)_{\rho<\alpha}$ such that $v(f(x_\lambda)) < \delta_\sigma$ and $v(x_\lambda - x_\sigma) = \delta_\sigma - \beta_1$. If $\mu < \sigma$, then $v(x_\lambda - x_\mu) = v(x_\lambda - x_\sigma) \vee v(x_\sigma - x_\mu) = \delta_\mu - \beta_1$ since $\delta_\sigma < \delta_\mu$. This completes the construction and the proof. □

We can now show that a totally ordered field has a unique maximally complete immediate extension. This is true for any valued field whose residue class field has characteristic 0. The only missing ingredient for the proof of this more general result is the set theoretic fact that the cardinality of a valued field (F, v) is bounded by a function of $|\Gamma|$ and $|R/J|$.

We first extend the total order to an immediate extension.

Theorem 5.2.18. *Let v be the natural valuation of the totally ordered field F. If (L, v) is an immediate extension of (F, v), then there is a unique total order of L extending that of F and for which v is the natural valuation of L.*

Proof. Let

$$L^+ = \{0 \neq a \in L : \exists b \in F^+ \text{ with } v(a) = v(b) \text{ and } ab^{-1} + J_L > 0\} \cup \{0\}.$$

Clearly, $L^+ \cup -L^+ = L$. Suppose $0 \neq a \in L$ and $b, c \in F^+$ with $v(a) = v(b) = v(c)$. Then $bc^{-1} \in U_F^+$ and $(ab^{-1} + J_L)bc^{-1} = ac^{-1} + J_L$; so $ab^{-1} + J_L > 0$ if and only if $ac^{-1} + J_L > 0$, and $L^+ \cap -L^+ = 0$. Now take $a, a_1 \in L^+$ and assume $-\infty < v(a_1) \leq v(a)$. Then $(a + a_1)b^{-1} \in R_L$ and $(a + a_1)b^{-1} + J_L = ab^{-1} + J_L + a_1 b^{-1} + J_L > 0$; so $a + a_1 \in L^+$. If $b_1 \in F^+$ with $a_1 b_1^{-1} + J_L > 0$, then $aa_1(bb_1)^{-1} + J_L = (ab^{-1} + J_L)(a_1 b_1^{-1} + J_L) > 0$, and $aa_1 \in L^+$. Thus, L^+ is a total order of L.

Now, the homomorphism $R_L \longrightarrow R_L/J_L$ is isotone since if $a \in R_L^+ \backslash J_L$, then $a + J_L = (ab^{-1} + J_L)b > 0$. If $a \in R_L^+$; then $a + J_L < n + J_L$ for some $n \in \mathbb{N}$ since R_L/J_L is archimedean; so $0 \le a < n$ and $a \in C(1)$. On the other hand, if $0 < a < n$ and $a \notin R_L$, then $a^{-1} \in J_L$, and we have the contradiction $1 \in J_L$ since $0 < 1 < na^{-1}$ and J_L is convex in R_L. Thus $R_L = C(1)$. Moreover, for $a, b \in L$, $v(a) = v(b)$ is equivalent to $C(a) = R_L|a| = R_L|b| = C(b)$. This shows that v is the natural valuation of the totally ordered field L. If P is another total order of L extending F^+ and for which v is the natural valuation, then $(R_L, P \cap R_L) \longrightarrow R_L/J_L$ is isotone. So if $0 \ne a \in P$ and $b \in F^+$ with $v(a) = v(b)$, then $ab^{-1} + J_L > 0$, $a \in L^+$ and $P = L^+$. □

Different total orders of a field can give the same natural valuation. In fact, all archimedean total orders of a field produce the same valuation. The construction of a non-archimedean example is left for Exercise 2.

Theorem 5.2.19. *Each totally ordered field has a maximally complete immediate extension. Suppose $\sigma : F \longrightarrow F_1$ is an isomorphism of totally ordered fields and M (respectively, M_1) is a maximally complete immediate extension of F (respectively, F_1). Then σ can be extended to an isomorphism between the totally ordered fields M and M_1.*

Proof. If Γ is the value group of the totally ordered field F, then the additive group of F can be embedded in $V(\Gamma, \mathbb{R})$ by Theorem 2.6.5. So the cardinality of F, and also of any immediate extension of F, is bounded by $2^{|\Gamma|\aleph_0}$ by the previous theorem. We can now apply Zorn's Lemma in the usual manner. Let S be a set with $\operatorname{card}(S) > 2^{|\Gamma|\aleph_0}$ and which contains F, and let $\mathscr{S}$ be the set of all immediate extensions (L, w) of (F, v) with $L \subseteq S$. The partial order of $\mathscr{S}$ is given by: $(L_1, w_1) \le (L_2, w_2)$ if (L_2, w_2) is an immediate extension of (L_1, w_1). Since the union of a chain of immediate extensions of F is easily seen to be an immediate extension of F, $\mathscr{S}$ has a maximal element (M, w). Suppose (K, w_1) is an immediate extension of (M, w). Since $|K \backslash M| \le |K| < |S| = |S \backslash M|$ there is a bijection f from $K = M \cup K \backslash M$ to $M \cup S \backslash M$ which is the identity on M. If the data for K is transferred to $f(K)$ via f, then $f(K) \in \mathscr{S}$ and $M \le f(K)$. So $f(M) = M = f(K)$, $M = K$, and M is a maximal immediate extension of F.

In order to extend σ to M we apply Zorn's Lemma to the set $\mathscr{T}$ which consists of all triples (N, N_1, φ) where $F \subseteq N \subseteq M$, $F_1 \subseteq N_1 \subseteq M_1$ and $\varphi : N \longrightarrow N_1$ is a value preserving F-isomorphism of fields which extends σ. The partial order in $\mathscr{T}$ is: $(N, N_1, \varphi) \le (K, K_1, \psi)$ if $N \subseteq K$, $N_1 \subseteq K_1$ and φ is the restriction of ψ to N. If $\{(N_j, N_{j1}, \varphi_j) : j \in \Delta\}$ is a chain in $\mathscr{T}$, then $(\cup N_j, \cup N_{j1}, \cup \varphi_j)$ is clearly the least upper bound of the chain in $\mathscr{T}$. Let (N, N_1, φ) be a maximal element in $\mathscr{T}$. We of course need to verify that $N = M$ and $N_1 = M_1$. Suppose $a \in M \backslash N$. By Theorem 5.2.6 a is a pseudo-limit of a pseudo-convergent sequence $(a_\rho)_{\rho<\alpha}$ in N which has no pseudo-limit in N. If $(a_\rho)_{\rho<\alpha}$ is transcendental, then a is transcendental over N by Theorem 5.2.7, $(\varphi(a_\rho))_{\rho<\alpha}$ is a transcendental pseudo-convergent sequence in N_1 and it has a pseudo-limit a_1 in M_1 by Theorem 5.2.9. But according to the uniqueness part of Theorem 5.2.7, φ can be extended to $\psi : N(a) \longrightarrow N_1(a)$ and $(N, N_1, \varphi) < (N(a), N_1(a), \psi)$. Suppose now that $(a_\rho)_{\rho<\alpha}$ is algebraic and let $f(x) \in N[x]$ be a monic polynomial of least degree such that $(v(f(a_\rho)))_\rho$ is eventually strictly decreasing. From Theorem 5.2.17 we get that

$(a_\rho)_{\rho<\alpha}$ has a pseudo-limit a in M which is a root of $f(x)$, and $(\varphi(a_\rho))_{\rho<\alpha}$ has a pseudo-limit a_1 in M_1 which is a root of $f_1(x)$ where $f_1(x) = \varphi(f(x)) \in N_1[x]$. Since $f(x)$ is irreducible the uniqueness part of Theorem 5.2.8 shows that the isomorphism $\psi : N(a) \longrightarrow N_1(a_1)$ which extends φ and takes a to a_1 is value preserving: $v(g(a)) = v(g(a_\rho)) = v_1(g_1(\varphi(a_\rho))) = v_1(g_1(a_1)) = v_1(\psi(g(a)))$ for large ρ where $\deg g(x) < \deg f(x)$ and $g_1(x) = \varphi(g(x))$. Thus, $M = N$ and also $\varphi(M) = M_1$ since $\varphi(M)$ is maximally complete. That φ is isotone follows from Exercise 22. □

We now have all the tools needed to prove the Hahn embedding theorem for totally ordered fields.

Theorem 5.2.20. *Let A be a maximal archimedean subfield of the totally ordered field F and let B be the real closure of the residue class field R_F/J_F. Suppose Δ is a $\mathbb{Z}$-independent subset of the value group Γ_F and, for each δ in Δ, let t_δ be a strictly positive element of F with value δ. Then there is an embedding $\varphi : F \longrightarrow V(\Gamma_F, B)$ of totally ordered fields such that $\varphi(at_\delta) = ax^\delta$ for $a \in A$ and $\delta \in \Delta$.*

Proof. Let L be a maximally complete immediate extension of the real closure K of F. By Theorems 5.2.14 and 5.2.15 there is a subfield E of L which contains F such that $\Gamma_E = \Gamma_F$ and $R_E/J_E = R_L/J_L$. Let M be a maximally complete immediate extension of E. According to Theorem 5.2.18 L and M are totally ordered field extensions of K and E, respectively, and by Exercise 5.1.11, R_M/J_M is the real closure of R_F/J_F since $R_M/J_M = R_K/J_K$ is algebraic over R_F/J_F; that is, $R_M/J_M = B$ since R_F/J_F is algebraic over $A + J_F/J_F$ by Exercise 9(a). The subgroup $G = U_M \cap M^+$ of the multiplicative group M^{+*} of positive elements of M is divisible. For, suppose $d \in G$ and $n \in \mathbb{N}$. Then the polynomial $x^n - (d + J_M)$ has a root in R_M/J_M and hence the polynomial $x^n - d$ has a root in R_M since M is henselian. Let H be the subgroup of M^{+*} generated by $\{t_\delta : \delta \in \Delta\}$. If $\delta_1, \ldots, \delta_n$ are distinct elements of Δ and $k_1, \ldots, k_n$ are nonzero integers, then $v(t_{\delta_1}^{k_1} \cdots t_{\delta_n}^{k_n}) = k_1\delta_1 + \cdots + k_n\delta_n \neq 0$; so $G \cap H = 1$. Let T be a subgroup of M^{+*} which contains H and is maximal with respect to $G \cap T = 1$; that is, T is a complement of G in M^{+*}. Then $M^{+*} = GT$ is a direct product since G is $\mathbb{Z}$-injective. Now, let C be a maximal subfield of R_M which contains A. By Exercise 9 the map $C \longrightarrow R_M/J_M$ is an isomorphism and hence C is naturally isomorphic to B. We claim that T is a C-independent set; that is, the subring $C[T]$ of R_M is the lexicographically ordered group algebra of T over C. For, if t_1 and t_2 are distinct elements of T, then $v(t_1) \neq v(t_2)$ since, otherwise, $t_1t_2^{-1} \in T \cap G = 1$. But then if $c_1, \ldots, c_n$ are nonzero elements of C and $t_1, \ldots, t_n$ are distinct elements of T we have $v(\sum_{i=1}^n c_it_i) = \bigvee_{i=1}^n v(t_i)$. Also, if $t_1 < t_2$, then $Ct_1 < t_2$ since $t_2 \leq ct_1$ gives $v(t_1) < v(t_2) \leq v(ct_1) = v(t_1)$. Thus, if $t_1 < \cdots < t_n$ then $\sum_{i=1}^n c_it_i > 0$ exactly when $c_n > 0$.

Define $\varphi : C[T] \longrightarrow V(\Gamma_F, C)$ by

$$\varphi\left(\sum c_it_i\right) = \sum c_ix^{v(t_i)}.$$

Then φ is an isomorphism between the totally ordered rings $C[T]$ and $\Sigma(\Gamma_F, C) \subseteq V(\Gamma_F, C)$, and $\varphi(at_\delta) = ax^\delta$ for $a \in A$ and $\delta \in \Delta$. Of course, φ can be extended to the field of quotients $C(T)$ of $C[T]$ in M. Now, M is an immediate extension of $C(T)$ since $\Gamma_F = v(M) = v(GT) = v(T) \subseteq v(C(T)) \subseteq v(M)$ and $C \longrightarrow R_{C(T)}/J_{C(T)} \longrightarrow R_M/J_M$ is an isomorphism. By Theorems 5.2.9, 5.2.11 and 5.2.19 φ extends to an isomorphism of M onto $V(\Gamma_F, C)$. □

According to Theorem 5.2.20, or Theorem 5.2.13, a totally ordered field can be embedded in a totally ordered field which contains the real numbers. We will establish the analogue of this for a totally ordered division ring D by embedding it into a division ring whose center contains a field isomorphic to $\mathbb{R}$. A perhaps shorter proof of this compatibility of the reals with a totally ordered field can be given with the aid of Theorem 6.3.13; see Exercise 4.3.28(c). By contrast, there is no known short proof of its compatibility with a totally ordered division ring. We first construct the topological completion of a totally ordered field in a more general setting.

Let D be a totally ordered division ring. Recall that the interval topology for D has as a subbase the sets of the form $\{x : x < a\}$ and $\{x : x > a\}$ for $a \in D$, and the open intervals (a,b) form a base for this topology. For $0 < \varepsilon \in D$ and $x \in D$ the open interval $(x-\varepsilon, x+\varepsilon)$ will be denoted by $N_\varepsilon(x)$.

Theorem 5.2.21. *The totally ordered division ring D is a Hausdorff topological ring in its interval topology.*

Proof. Let $g,h : D \times D \longrightarrow D$ be the addition and multiplication functions, respectively. Suppose $(a,b) \in g^{-1}(N_\varepsilon(u))$; that is, $|a+b-u| < \varepsilon$. Take $0 < 2\delta < \varepsilon - |a+b-u|$. Then if $(x,y) \in N_\delta(a) \times N_\delta(b)$ we have

$$\begin{aligned}|x+y-u| &= |x-a+y-b+a+b-u| \\ &\leq |x-a| + |y-b| + |a+b-u| \\ &< 2\delta + |a+b-u| < \varepsilon,\end{aligned}$$

and $N_\delta(a) \times N_\delta(b) \subseteq g^{-1}(N_\varepsilon(u))$. Now suppose $(a,b) \in h^{-1}(N_\varepsilon(u))$; that is, $|ab - u| < \varepsilon$. Take

$$0 < \delta < 1 \wedge (2(|a|+|b|+1))^{-1}(\varepsilon - |ab-u|) \wedge (\varepsilon - |ab-u|)(2(|a|+|b|+1))^{-1}.$$

Then if $(x,y) \in N_\delta(a) \times N_\delta(b)$ we have

$$\begin{aligned}|xy-u| &= |(x-a+a)(y-b+b)-u| \\ &\leq |x-a||b| + |a||y-b| + |x-a||y-b| + |ab-u| \\ &\leq \delta|b| + |a|\delta + \delta^2 + |ab-u| \\ &\leq \delta(|a|+|b|+1) + (|a|+|b|+1)\delta + |ab-u| \\ &< \frac{\varepsilon-|ab-u|}{2} + \frac{\varepsilon-|ab-u|}{2} + |ab-u| \\ &= \varepsilon,\end{aligned}$$

and $N_\delta(a) \times N_\delta(b) \subseteq h^{-1}(N_\varepsilon(u))$. □

Let I be a poset which is directed up. A function from I to the set X is called a *net in* X and it will be denoted by $(x_n)_{n\in I}$. The net $(x_n)_{n\in I}$ in the totally ordered division ring D *converges* to the element x in D if for each $0<\varepsilon\in D$ there is some $n\in I$ such that $|x_j-x|<\varepsilon$ for all $j\geq n$. Since D is Hausdorff the net $(x_n)_{n\in I}$ converges to at most one point and we will write $\lim x_n = x$ if $(x_n)_{n\in I}$ converges to x. The net $(x_n)_{n\in I}$ is a *Cauchy net* if for every $0<\varepsilon\in D$ there exists an $n\in I$ such that $|x_j-x_k|<\varepsilon$ for all $j, k\geq n$. Each convergent net is certainly Cauchy. If each Cauchy net in D converges, then D is called *Cauchy complete*. Suppose X is a coinitial subset of D^{*+} which has a.c.c. Then there is an ordinal $\xi>0$ such that X is the range of the strictly decreasing net $(\varepsilon_\rho)_{\rho<\xi}$ in D which converges to 0. The ordinal ξ is a limit ordinal since if $\xi=\mu+1$, then ε_μ would be the smallest strictly positive element of D. For any limit ordinal ξ a net $(x_\rho)_{\rho<\xi}$ in D is called an ξ*-net* and D is called ξ*-Cauchy complete* if each Cauchy ξ-net in D converges.

Theorem 5.2.22. *Let D be a totally ordered division ring and let $(\varepsilon_\rho)_{\rho<\xi}$ be a strictly decreasing ξ-net in D^{*+} which converges to* 0. *Then D is Cauchy complete if and only if it is ξ-Cauchy complete.*

Proof. Suppose D is ξ-Cauchy complete and let $(x_n)_{n\in I}$ be a Cauchy net in D. For each $\rho<\xi$ there exists $n_\rho\in I$ such that $|x_n-x_m|<\frac{\varepsilon_\rho}{2}$ if $n,m\geq n_\rho$. The ξ-net $(x_{n_\rho})_{\rho<\xi}$ is Cauchy. For, if $0<\varepsilon\in D$ take $\varepsilon_\lambda<\frac{\varepsilon}{2}$; then for $\rho\geq\sigma\geq\lambda$ and $n\geq n_\rho$, n_σ we have

$$|x_{n_\rho}-x_{n_\sigma}|\leq|x_{n_\rho}-x_n|+|x_n-x_{n_\sigma}|<\varepsilon_\rho+\varepsilon_\sigma\leq 2\varepsilon_\lambda<\varepsilon.$$

Let $\lim x_{n_\rho}=x$. Then $\lim x_n = x$. For, if $0<\varepsilon\in D$, again take $\varepsilon_\lambda<\frac{\varepsilon}{2}$. There exists a $\rho\geq\lambda$ such that for any $\sigma\geq\rho$ we have $|x_{n_\sigma}-x|<\varepsilon_\lambda$. So if $n\geq n_\rho$, then

$$|x_n-x|\leq|x_n-x_{n_\rho}|+|x_{n_\rho}-x|<\frac{\varepsilon_\rho}{2}+\varepsilon_\lambda<2\varepsilon_\lambda<\varepsilon.$$

□

Suppose $F\subseteq E$ are totally ordered division rings. According to Exercise 2.3.1 F is order dense in E if and only if it is topologically dense in E; so the assertion that F is dense in E is unambiguous. E is a *Cauchy completion* of D if E is Cauchy complete and D is isomorphic to a dense subring of E.

Theorem 5.2.23. *Suppose $f:D\longrightarrow E$ and $g:D\longrightarrow K$ are embeddings of totally ordered division rings and $f(D)$ and $g(D)$ are dense in E and K, respectively. Suppose also that K is Cauchy complete. Then there is a unique embedding $h:E\longrightarrow K$ such that $hf=g$:*

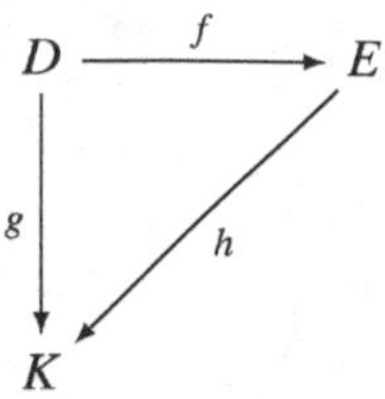

Moreover, if E is Cauchy complete, then h is an isomorphism.

Proof. Let $(I, \leq')$ be the dual of the totally ordered set D^{+*} and let $x \in E$. For $\varepsilon \in I$ choose $x_\varepsilon \in D$ with $f(x_\varepsilon) \in f(D) \cap N_{f(\varepsilon)}(x)$. Then $(f(x_\varepsilon))_{\varepsilon \in I}$ is a net in $f(D)$ which converges to x. For, if $0 < a \in E$, $0 < f(\varepsilon) < a$, and $\eta \geq' \varepsilon$, then $|f(x_\eta) - x| < f(\eta) \leq f(\varepsilon) < a$. Thus, each element of E is the limit of some Cauchy net in $f(D)$. Now, suppose $(x_n)_{n \in J}$ and $(t_p)_{p \in P}$ are Cauchy nets in D with $\lim f(x_n) = \lim f(t_p) = x$. Then $(g(x_n))_{n \in J}$ and $(g(t_p))_{p \in P}$ are Cauchy nets in K since $g(D)$ is dense in K, and hence they have limits x' and x'', respectively. Let $0 < \varepsilon \in D$. There exist $n \in J$ and $p \in P$ such that $4|g(x_m) - x'| < g(\varepsilon)$ and $4|f(x_m) - x| < f(\varepsilon)$ if $m \geq n$, and $4|g(t_q) - x''| < g(\varepsilon)$ and $4|f(t_q) - x| < f(\varepsilon)$ if $q \geq p$. Now,

$$|f(x_n) - f(t_p)| \leq |f(x_n) - x| + |x - f(t_p)| < \frac{f(\varepsilon)}{2}$$

and hence $2|x_n - t_p| < \varepsilon$. So,

$$\begin{aligned} |x' - x''| &\leq |x' - g(x_n)| + |g(x_n) - g(t_p)| + |g(t_p) - x''| \\ &< \frac{g(\varepsilon)}{4} + \frac{g(\varepsilon)}{2} + \frac{g(\varepsilon)}{4} = g(\varepsilon) \end{aligned}$$

and hence $x' = x''$ since $g(D)$ is dense in K. Thus, the function h defined by $h(x) = x'$ whenever $(x_n)_{n \in J}$ is a Cauchy net in D with $\lim f(x_n) = x$ and $\lim g(x_n) = x'$ is well defined. Note that $h(\lim f(x_n)) = \lim g(x_n)$ for any Cauchy net $(x_n)_{n \in J}$ in D. For $x, y \in E$ let $(x_\varepsilon)_{\varepsilon \in I}$ and $(y_\varepsilon)_{\varepsilon \in I}$ be the Cauchy nets in D given above with $\lim f(x_\varepsilon) = x$ and $\lim f(y_\varepsilon) = y$. An easy consequence of Theorem 5.2.21 is that h is a ring homomorphism: $\lim f(x_\varepsilon + y_\varepsilon) = x + y$, $\lim f(x_\varepsilon y_\varepsilon) = xy$, $(x_\varepsilon + y_\varepsilon)_{\varepsilon \in I}$ and $(x_\varepsilon y_\varepsilon)_{\varepsilon \in I}$ are Cauchy, and $\lim g(x_\varepsilon + y_\varepsilon) = \lim g(x_\varepsilon) + \lim g(y_\varepsilon)$, and $\lim g(x_\varepsilon y_\varepsilon) = \lim g(x_\varepsilon) \lim g(y_\varepsilon)$. Suppose $h' : E \longrightarrow K$ is another embedding with $h'f = g$. Then $h'(E)$ is dense in K since $g(D) \subseteq h'(E)$ and hence h' preserves limits (that is, h' is continuous). For, if $\lim x_n = x$ in E and $0 < a \in K$ take $\varepsilon \in E$ with $0 < h'(\varepsilon) < a$. Then for some n and all $m \geq n$, $|x_m - x| < \varepsilon$ and $|h'(x_m) - h'(x)| < h'(\varepsilon) < a$. In particular, with $x = \lim f(x_n)$ we have $h'(x) = \lim h'f(x_n) = \lim g(x_n) = h(x)$. If E is also Cauchy complete, then there is a unique embedding h_1 of K into E with $h_1 g = f$. Then $hh_1 g = hf = 1_K g$ and $h_1 hf = h_1 g = 1_E f$ and hence $hh_1 = 1_K$ and $h_1 h = 1_E$ by the uniqueness of the embeddings 1_K and 1_E in the diagrams

□

Let F be a central subfield of the totally ordered division ring D and let ξ be the smallest ordinal for which there is a strictly decreasing net $(\varepsilon_\rho)_{\rho<\xi}$ in F^{*+} converging to 0. An ξ-net $(x_\rho)_{\rho<\xi}$ in D is called *F-Cauchy* if for every $0<\varepsilon\in F$ there is some $\lambda<\xi$ such that $|x_\rho - x_\sigma|<\varepsilon$ for all $\rho,\sigma\geq\lambda$. Let $R=C_F(1)$ be the convex F-subspace of D generated by 1. R is a valuation ring of D which contains F and we will construct a Cauchy completion of F by considering the F-Cauchy ξ-nets in R. Let

$$\mathscr{R}=\{(x_\rho)_{\rho<\xi}\in R^{W(\xi)}\ :\ (x_\rho)_{\rho<\xi}\text{ is }F\text{-Cauchy}\}.$$

Let us check that $\mathscr{R}$ is an F-subalgebra of $R^{W(\xi)}$. Suppose $(x_\rho)_{\rho<\xi}$ and $(y_\rho)_{\rho<\xi}$ are elements of $\mathscr{R}$. For any $0<\varepsilon\in F$ there is some $\lambda<\xi$ such that if $\rho,\sigma\geq\lambda$, then $2|x_\sigma-x_\rho|<\varepsilon$ and $2|y_\sigma-y_\rho|<\varepsilon$. So $|(x_\sigma-y_\sigma)-(x_\rho-y_\rho)|\leq|x_\sigma-x_\rho|+|y_\sigma-y_\rho|<\varepsilon$ for $\rho,\sigma\geq\lambda$; and also $|x_\rho|-|x_{\lambda+1}|\leq|x_\rho-x_{\lambda+1}|<\varepsilon$ or $|x_\rho|<|x_{\lambda+1}|+\varepsilon\leq a\in F$ for all $\rho\geq\lambda$. We may also assume that $|y_\rho|<a$ for all $\rho\geq\lambda$. Now choose $\mu\geq\lambda$ such that $|x_\sigma-x_\rho|<(2a)^{-1}\varepsilon$ and $|y_\sigma-y_\rho|<(2a)^{-1}\varepsilon$ for all $\sigma,\rho\geq\mu$. Then

$$\begin{aligned}|x_\sigma y_\sigma - x_\rho y_\rho| &\leq |x_\sigma-x_\rho||y_\sigma|+|x_\rho||y_\sigma-y_\rho|\\ &<(2a)^{-1}\varepsilon a+a(2a)^{-1}\varepsilon=\varepsilon\end{aligned}$$

whenever $\sigma,\rho\geq\mu$. Note that $\mathscr{R}$ contains each F-Cauchy ξ-net in F. The ξ-net $(x_\rho)_{\rho<\xi}$ in D is *F-convergent* to x in D if for every $0<\varepsilon\in F$ there exists $\lambda<\xi$ such that $|x_\rho-x|<\varepsilon$ for every $\rho\geq\lambda$; we write $F\text{-}\lim x_\rho=x$. Note that an F-convergent ξ-net is F-Cauchy. Let

$$N=\{(x_\rho)_{\rho<\xi}\in\mathscr{R}\ :\ (x_\rho)_{\rho<\xi}\ F\text{-converges to }0\}.$$

Then N is an ideal of $\mathscr{R}$. For suppose $(x_\rho)_{\rho<\xi}\in N$ and $(y_\rho)_{\rho<\xi}\in\mathscr{R}$. For any $0<\varepsilon\in F$ we can find $0<a\in F$ and $\lambda<\xi$ such that if $\rho\geq\lambda$, then $|y_\rho|<a$ and $|x_\rho|<a^{-1}\varepsilon$; so $|x_\rho y_\rho|<\varepsilon$ and $|y_\rho x_\rho|<\varepsilon$. This information is summarized in the next theorem. Recall that $Z(U)$ denotes the center of the ring U.

Theorem 5.2.24. *Let F be a central subfield of the totally ordered division ring D and let ξ be the smallest ordinal for which there is a strictly decreasing net $(\varepsilon_\rho)_{\rho<\xi}$ in F^{*+} with F-$\lim\varepsilon_\rho=0$. Let $R=C_F(1)\subseteq D$, $J=J(R)$ the maximal ideal of R,*

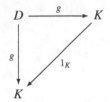

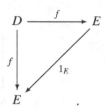

$$\mathscr{R} = \{(x_\rho)_{\rho<\xi} \in R^{W(\xi)} \ : \ (x_\rho)_{\rho<\xi} \textit{ is F-Cauchy}\},$$
$$N = \{(x_\rho)_{\rho<\xi} \in \mathscr{R} \ : \ F\text{-}\lim x_\rho = 0\}$$

and

$$P = \{(x_\rho)_{\rho<\xi} \in \mathscr{R} \ : \ \textit{for some } 0 < \varepsilon \in F \textit{ and some } \lambda < \xi,\ x_\rho > \varepsilon \textit{ for all } \rho \geq \lambda\} \cup \{0\}. \tag{5.2.38}$$

Then $(\mathscr{R}, P)$ *is a po-unital po-algebra over* F, N *is a convex ideal of* $\mathscr{R}$ *and* $\overline{\mathscr{R}} = \mathscr{R}/N$ *is a totally ordered division algebra over* F. *Moreover, if* $\psi : \mathscr{R} \longrightarrow \overline{\mathscr{R}}$ *is the natural map,* $i : R \longrightarrow \mathscr{R}$ *is the map that sends* $a \in R$ *to the constant* ξ*-net* $(a_\rho)_{\rho<\xi}$ *with* $a_\rho = a$ *for each* ρ, *and* $\varphi : R \longrightarrow \overline{\mathscr{R}}$ *is the composite* $\varphi = \psi i$, *then* φ *is isotone,*

$$F \cong \overline{F} \subseteq \hat{F} = \psi(F^{W(\xi)} \cap \mathscr{R}) \subseteq Z(\overline{\mathscr{R}}),$$

$Z(\overline{R}) \subseteq Z(\overline{\mathscr{R}})$ *and* $\hat{F}$ *is a Cauchy completion of* F. *Also,* $\overline{F}^{*+}$ *is coinitial in* $\overline{\mathscr{R}}^{*+}$.

Proof. We will first check that $\mathscr{R}/N$ is a division ring. Suppose $(x_\rho)_{\rho<\xi} \in \mathscr{R}\backslash N$. Then for some $0 < a \in F$ there exists $\lambda < \xi$ with $|x_\rho| > a$ for all $\rho \geq \lambda$. If not, for $0 < \varepsilon \in F$ take $\lambda < \xi$ with $|x_\rho - x_\sigma| < \frac{\varepsilon}{2}$ for $\rho, \sigma \geq \lambda$. Then, for some $\sigma \geq \lambda$, $|x_\sigma| \leq \frac{\varepsilon}{2}$ and hence $|x_\rho| \leq |x_\rho - x_\sigma| + |x_\sigma| < \varepsilon$ for any $\rho \geq \lambda$; that is $(x_\rho)_{\rho<\xi}$ is in N contrary to our assumption. Let $y_\rho = 1$ if $\rho < \lambda$ and let $y_\rho = x_\rho^{-1}$ if $\rho \geq \lambda$ where $|x_\rho| > a > 0$ for all $\rho \geq \lambda$. We claim that $(y_\rho)_{\rho<\xi} \in \mathscr{R}$. Since $|x_\rho^{-1}| < a^{-1}$ for $\rho \geq \lambda$ we certainly have $y_\rho \in R$ for each $\rho < \xi$. For $0 < \varepsilon \in F$ choose $\mu \geq \lambda$ such that $|x_\rho - x_\sigma| < \varepsilon a^2$ for all $\rho, \sigma \geq \mu$. Then

$$|x_\rho^{-1} - x_\sigma^{-1}| = |x_\rho|^{-1}|x_\rho - x_\sigma||x_\sigma|^{-1} < a^{-2}\varepsilon a^2 = \varepsilon,$$

and $(y_\rho)_{\rho<\xi} \in \mathscr{R}$. Moreover, for $\rho \geq \lambda$ we have $|y_\rho x_\rho - 1| = |x_\rho y_\rho - 1| = 0 < \varepsilon$; so $(y_\rho)_{\rho<\xi} + N = ((x_\rho)_{\rho<\xi} + N)^{-1}$ and $\mathscr{R}/N$ is a division ring.

Clearly, $P + P \subseteq P$, $PP \subseteq P$ and $P \cap -P = 0$. Suppose $(x_\rho)_{\rho<\xi}, (y_\rho)_{\rho<\xi} \in \mathscr{R}$ and $0 < (y_\rho)_{\rho<\xi} < (x_\rho)_{\rho<\xi} \in N$. There exist $\lambda < \xi$ and $0 < \varepsilon_1, \varepsilon_2 \in F$ such that $x_\rho - y_\rho > \varepsilon_1$ and $y_\rho > \varepsilon_2$ for all $\rho \geq \lambda$. Given $0 < \varepsilon \in F$ there exists $\mu \geq \lambda$ such that $|x_\rho| < \varepsilon$ for all $\rho \geq \mu$. Thus, for $\rho \geq \mu$ we have $\varepsilon > x_\rho > y_\rho + \varepsilon_1 > y_\rho = |y_\rho|$ and $(y_\rho)_{\rho<\xi} \in N$. To see that $\mathscr{R}/N$ is totally ordered take $(x_\rho)_{\rho<\xi} \in \mathscr{R}\backslash P \cup -P$. Given $0 < \varepsilon \in F$ take $\lambda < \xi$ such that $2|x_\rho - x_\sigma| < \varepsilon$ for all $\rho, \sigma \geq \lambda$. There exist $\sigma_1, \sigma_2 \geq \lambda$ such that $2x_{\sigma_1} \leq \varepsilon$ and $-2x_{\sigma_2} \leq \varepsilon$. So, for $\rho \geq \lambda$ we have

$$x_\rho = x_\rho - x_{\sigma_1} + x_{\sigma_1} < \frac{\varepsilon}{2} + \frac{\varepsilon}{2} = \varepsilon$$

and

$$-x_\rho = x_{\sigma_2} - x_\rho - x_{\sigma_2} < \frac{\varepsilon}{2} + \frac{\varepsilon}{2} = \varepsilon.$$

So $|x_\rho| < \varepsilon$ and $(x_\rho)_{\rho<\xi} \in N$.

Now, $\varphi : R \longrightarrow \mathscr{R}/N$ is an isotone F-algebra homomorphism whose kernel is the maximal ideal J of R since

$$J = \{x \in D : Fx \leq 1\} = \{x \in D : |x| \leq F^{*+}\}$$

and $i(x) \in N$ if and only if $|x| < \varepsilon$ for all $0 < \varepsilon \in F$. Let $\mathscr{R}(F) = \mathscr{R} \cap F^{W(\xi)}$ and $N(F) = N \cap F^{W(\xi)}$. Then $\varphi(F) \subseteq \mathscr{R}(F)/N(F)$ and $\mathscr{R}(F)/N(F)$ is the field extension of $\varphi(F)$ obtained by replacing D by F in the construction of $\mathscr{R}/\mathscr{N}$. It is, of course, a central subfield of $\mathscr{R}/N$. Note that $\varphi(F^{*+})$ is coinitial in $(\mathscr{R}/N)^{*+}$ and hence also in $\mathscr{R}(F)/N(F)$. For if $(b_\rho)_{\rho<\xi} + N > 0$ in $\mathscr{R}/N$, then there exist $\lambda < \xi$ and $0 < \varepsilon \in F$ such that $b_\rho > 2\varepsilon$ for all $\rho \geq \lambda$; that is, $(b_\rho)_{\rho<\xi} + N > i(\varepsilon) + N > 0$ since $b_\rho - \varepsilon > \varepsilon$ if $\rho \geq \lambda$. Thus, in order to see that $\varphi(F)$ is dense in $\mathscr{R}(F)/N(F)$ it suffices to show that $(a_\rho)_{\rho<\xi} + N = F\text{-}\lim \varphi(a_\rho)$ for each Cauchy net $(a_\rho)_{\rho<\xi}$ in $\mathscr{R}(F)$. Now, given $0 < \varepsilon \in F$ there exists $\lambda < \xi$ such that if σ, $\mu \geq \lambda$, then $2|a_\sigma - a_\mu| < \varepsilon$; so $0 < \frac{\varepsilon}{2} < \varepsilon - |a_\sigma - a_\mu|$, and for $\sigma \geq \lambda$ we have $0 < \varphi(\varepsilon) - |i(a_\sigma) - (a_\rho)_{\rho<\xi}| + N$ and $|\varphi(a_\rho) - (a_\rho)_{\rho<\xi} + N| < \varphi(\varepsilon)$. We show next that $\mathscr{R}(F)/N(F)$ is ξ-Cauchy complete and hence Cauchy complete by Theorem 5.2.22. Suppose $(\alpha_\rho)_{\rho<\xi}$ is a Cauchy net in $\mathscr{R}(F)/N(F)$. For each $\rho < \xi$ there exists $a_\rho \in F$ with $|\varphi(a_\rho) - \alpha_\rho| < \varphi(\varepsilon_\rho)$. The net $(a_\rho)_{\rho<\xi}$ is Cauchy. For, given $0 < \varepsilon \in F$ there exists $\lambda < \xi$ such that for any $\sigma, \rho \geq \lambda$ we have $3\varepsilon_\rho < \varepsilon$ and $3|\alpha_\rho - \alpha_\sigma| < \varphi(\varepsilon)$, and hence

$$|\varphi(a_\rho) - \varphi(a_\sigma)| \leq |\varphi(a_\rho) - \alpha_\rho| + |\alpha_\rho - \alpha_\sigma| + |\alpha_\sigma - \varphi(a_\sigma)| < \varphi(\varepsilon).$$

Let $\alpha = \lim \varphi(a_\rho)$ in $R(F)/N(F)$. By choosing λ in the previous sentence so that we also have $3|\alpha - \varphi(a_\rho)| < 2\varphi(\varepsilon)$ for any $\rho \geq \lambda$ we obtain

$$|\alpha - \alpha_\rho| \leq |\alpha - \varphi(a_\rho)| + |\varphi(a_\rho) - \alpha_\rho| < \varphi(\varepsilon)$$

and hence $\lim \alpha_\rho = \alpha$.

If $a \in R$ with $\overline{a} \in Z(\overline{R})$, then $|ab - ba| < \varepsilon$ for every $b \in R$ and every $0 < \varepsilon \in F$. So $(ab_\rho - b_\rho a)_{\rho<\xi} \in N$ for every $(b_\rho)_{\rho<\xi} \in \mathscr{R}$ and $\overline{a} \in Z(\overline{\mathscr{R}})$. □

With very little changes in their proofs Theorems 5.2.21, 5.2.22, 5.2.23, and 5.2.24 give Cauchy completions of division rings, which again are division rings, and of abelian groups D in which the set of strictly positive elements has no least element. In the latter case the Cauchy completion is an abelian group. The main change is to take $F = D$.

We wish to find a totally ordered division ring which contains D and whose center contains $\mathbb{R}$. If F is archimedean and $J = 0$, or equivalently, $R = D$, then $\overline{\mathscr{R}}$ will do. So we will assume $J \neq 0$; equivalently, F^{*+} is bounded in D^{*+}. Let K be a maximal subfield of $Z(R)$. For $\overline{\theta} \in \hat{F} \backslash \overline{K}$ we will construct a totally ordered simple extension $D(\theta)$ of D. Then an application of Zorn's Lemma will produce the desired extension of D. For much of the work the requirement that F is archimedean is not necessary and will be dropped since no additional complications arise. The main problem that arises in the construction of $D(\theta)$ comes from the verification that it has an appropriate total order, especially when $\overline{\theta}$ is algebraic over $\overline{R}$. In addition to the notation given in Theorem 5.2.24 we will let $S = \varphi^{-1}(\hat{F})$ and $\mathscr{Z} = \varphi^{-1}(Z(\overline{R}))$. Note that $J \subseteq S \subseteq \mathscr{Z} \subseteq R$, S and $\mathscr{Z}$ are local rings with common maximal ideal J, and $Z(J) \subseteq Z(S) \subseteq Z(\mathscr{Z}) \subseteq Z(R) \subseteq Z(D)$ since D is the classical quotient ring of J. Also, via conjugation, the group D^* acts on the po-ring $\mathscr{R}$, and N is a D^*-invariant ideal. This action of $s \in D^*$ on some element w on which it acts will be denoted by w^s; so $a^s = s^{-1}as$ if $a \in D$ and $\overline{u}^s = \overline{u^s}$ if $u \in \mathscr{R}$. In particular, S and $\mathscr{Z}$ are also D^*-invariant. $S = R$ exactly when $\overline{R} \subseteq \hat{F}$ and this is the case if F is archimedean. For then $\overline{R}$ is archimedean since J is the maximal convex subgroup of R, and each element of $\overline{R}$ is the limit of a Cauchy sequence in $\overline{F}$ and hence is in $\hat{F} \cong \mathbb{R}$. If K is a subfield of $Z(R)$ and $\overline{f}(x) \in \overline{K}[x]$, then $f(x)$ will denote the polynomial in $K[x]$ whose image in $\overline{R}[x]$ is $\overline{f}(x)$.

We collect some useful facts below. Note that just as for fields if $D \subseteq E$ are division rings and $a \in Z(E)$ is a root of $0 \neq g(x) \in D[x]$, then $g(x)$ is irreducible in $D[x]$ (that is, if $g(x) = u(x)v(x)$ in $D[x]$, then $u(x)$ or $v(x)$ is a unit of $D[x]$) if and only if $\deg g(x) \leq \deg h(x)$ for any $0 \neq h(x) \in D[x]$ with $h(a) = 0$.

Theorem 5.2.25. *Suppose $s \in D^*$, $g(x) \in R[x]$ and $\overline{\theta} \in \hat{F}$.*

(a) $\overline{\theta}$ is fixed by s; in particular, $\overline{g}(x) = \overline{g^s}(x)$ if $g(x) \in S[x]$.
(b) $\overline{g}(\overline{\theta}) > 0$ if and only if $\overline{g^s}(\overline{\theta}) > 0$.
(c) If $\overline{g}(x)$ is monic and irreducible in $\overline{R}[x]$ and $\overline{g}(\overline{\theta}) = 0$, then $\overline{g}(x) = \overline{g^s}(x)$.
(d) If $\overline{g}(x) = \overline{g^s}(x)$ for each $s \in D^*$, then $g(x) \in \mathscr{Z}[x]$.

Proof. (a) $\overline{\theta} = (\alpha_\rho)_{\rho<\xi} + N$ where $(\alpha_\rho)_{\rho<\xi}$ is an F-Cauchy net in F. So $\overline{\theta}^s = (\alpha_\rho^s)_{\rho<\xi} + N = \overline{\theta}$.

(b) It suffices to show that $\overline{g^s}(\overline{\theta}) > 0$ provided $\overline{g}(\overline{\theta}) > 0$. Again, $\overline{\theta} = (\alpha_\rho)_{\rho<\xi} + N$, and since $\mathscr{R}$ is an R-po-bimodule and $a(\alpha_\rho)_{\rho<\xi} = (\alpha_\rho)_{\rho<\xi}a$ for each $a \in R$ we have $g((\alpha_\rho)_{\rho<\xi}) = (g(\alpha_\rho))_{\rho<\xi} \in P^*$. Thus, there exist $0 < \varepsilon \in F$ and $\lambda < \xi$ such that $g(\alpha_\rho) > \varepsilon$ for all $\rho \geq \lambda$; so $g^s(\alpha_\rho) > \varepsilon$, $g^s((\alpha_\rho)_{\rho<\xi}) \in P^*$ and $\overline{g^s}(\overline{\theta}) = g^s((\alpha_\rho)_{\rho<\xi}) + N > 0$.

(c) From (b) we deduce that $\overline{g^s}(\overline{\theta}) = 0$, and since $\overline{g}(x) - \overline{g^s}(x)$ has smaller degree than $\overline{g}(x)$ if it is not 0, necessarily $\overline{g}(x) = \overline{g^s}(x)$.

(d) If $s \in R^*$, then $\overline{s}\overline{g}(x) = \overline{s}\overline{g^s}(x) = \overline{sg^s}(x) = \overline{g(x)s} = \overline{g}(x)\overline{s}$ and $\overline{g}(x) \in Z(\overline{R})[x]$. □

For $\overline{\theta} \in \hat{F} \backslash \overline{K}$ there are three separate cases we will deal with in constructing $D(\theta)$ and supplying it with a total order. The case we deal with below arises when $\overline{\theta}$ is transcendental over $\overline{K}$. The other two cases evolve from $\overline{\theta}$ being algebraic over $\overline{K}$.

Theorem 5.2.26. *Suppose K is a subfield of $Z(R)$ and $\overline{\theta} \in \hat{F} \backslash \overline{F}$ is transcendental over $\overline{K}$. Let θ be an indeterminate over D. If $\overline{\theta}$ is algebraic over $\overline{R}$ let $\overline{g}(x) \in \overline{R}[x]$ be its monic irreducible polynomial and if $\overline{\theta}$ is transcendental over $\overline{R}$ let $\overline{g}(x) = 1$. For $0 \neq P(\theta) = \Sigma_j s_j \theta^j \in D[\theta]$ let $b = \vee_j |s_j|$ and let $p(\theta) \in R[\theta] \backslash J[\theta]$ be defined by $P(\theta) = bp(\theta)$. Factor $\overline{p}(x)$ in $\overline{R}[x]$ as $\overline{p}(x) = \overline{g}(x)^m \overline{q}(x)$ with $\overline{q}(\overline{\theta}) \neq 0$. Then $T = \{P(\theta) : \overline{q}(\overline{\theta}) > 0\} \cup \{0\}$ is a total order of $D[\theta]$ which extends that of D, and*

$$\alpha < \overline{\theta} \text{ or } \overline{\theta} < \beta \quad \Rightarrow \quad \alpha < \theta \text{ or } \theta < \beta, \text{ for all } \alpha, \beta \in F. \tag{5.2.39}$$

Proof. Since $\overline{g}(x) \in Z(\overline{R})[x]$ by Theorem 5.2.25, the factorization of $\overline{p}(x)$ given above is always possible. First note that for any $0 < s \in D$ and $P(\theta) \in D[\theta]$, $P(\theta) \in T$ iff $sP(\theta) \in T$. To show that T is a total order of $D[\theta]$ it suffices to verify that it is a subsemiring since clearly $T \cap -T = 0$ and $T \cup -T = D[\theta]$. Suppose $P_1(\theta)$, $P_2(\theta) \in T^*$ and as above write $P_i(\theta) = b_i p_i(\theta)$ and $\overline{p}_i(x) = \overline{g}_i(x)^{m_i}\overline{q}_i(x)$ for $i = 1, 2$. Without loss of generality we may assume $s = b_1^{-1}b_2 \leq 1$. Then $P_1(\theta) + P_2(\theta) = b_1(p_1(\theta) + sp_2(\theta)) = b_1 bp(\theta)$ is in T provided $bp(\theta)$ is in T. Now

$$\overline{b}\overline{p}(x) = \begin{cases} \overline{g}(x)^{m_2}(\overline{g}(x)^{m_1-m_2}\overline{q}_1(x) + \overline{s}\overline{q}_2(x)) & \text{if } m_1 > m_2 \\ \overline{g}(x)^{m_1}(\overline{q}_1(x) + \overline{s}\overline{g}(x)^{m_2-m_1}\overline{q}_2(x)) & \text{if } m_2 > m_1 \\ \overline{g}(x)^{m_1}(\overline{q}_1(x) + \overline{s}\overline{q}_2(x)) & \text{if } m_1 = m_2. \end{cases}$$

In each case the polynomial on the right, when evaluated at $\overline{\theta}$, is > 0, and hence $bp(\theta) \in T$ except in the first case when $\overline{s} = 0$. But then $\overline{b}\overline{p}(x) = \overline{p}_1(x) \neq 0$, $\overline{p}(x) = \overline{g}_1(x)^{m_1}\overline{b^{-1}q_1}(x)$ (in fact, $\overline{b} = 1$) and $\overline{b^{-1}q_1}(\overline{\theta}) > 0$; so again $bp(\theta) \in T$. Also, $P_1(\theta)P_2(\theta) \in T$ since $P_1(\theta)P_2(\theta) = b_1 b_2 p_1^{b_2}(\theta)p_2(\theta)$, $\overline{p_1^{b_2}}(x)\overline{p}_2(x) = \overline{g}(x)^{m_1+m_2}\overline{q_1^{b_2}}(x)\overline{q}_2(x)$ and $\overline{q_1^{b_2}}(\overline{\theta})\overline{q}_2(\overline{\theta}) > 0$ by Theorem 5.2.25. Clearly, $D^+ \subseteq T$ and (5.2.38) holds since $\overline{g}(x)$, if of degree 1, is not in $\overline{F}[x]$. □

Now suppose $\overline{\theta} \in \hat{F} \backslash \overline{F}$ is algebraic over $\overline{K}$ and let $f(x)$ be the monic irreducible polynomial in $K[x]$ with $\overline{f}(\overline{\theta}) = 0$. If $f(x)$ is irreducible in $D[x]$, then, just as for fields, $f(x)D[x]$ is a completely prime ideal which is a maximal one-sided ideal of $D[x]$ and $D[x]/f(x)D[x] = D(\theta)$ is a division ring with the central root $\theta = x + f(x)D[x]$. To see that $f(x)$ is irreducible under the right circumstances will require some preparation.

Let A be a subring of B which has the same identity as B. We will make use of the following and its dual obtained by interchanging $g(x)$ and $h(x)$ (usually with $r(x) = 0$). If $f(x) = g(x)h(x) + r(x)$ is an equation in $B[x]$ with $f(x)$, $g(x) \in A[x]$, the leading coefficient of $g(x)$ is a unit of A and $r(x) = 0$ or $\deg r(x) < \deg g(x)$, then $h(x)$ and $r(x)$ are in $A[x]$. This is, of course, an immediate consequence of the division algorithms in $A[x]$ and $B[x]$.

As usual, the nonzero polynomial $f(x) \in R[x]$ is irreducible if $g(x)$ or $h(x)$ is a unit of R whenever $g(x), h(x) \in R[x]$ with $f(x) = g(x)h(x)$. The factorization $f(x) = f_1(x) \cdots f_m(x)$ in $D[x]$ (or $R[x]$) is called a monic factorization if each $f_j(x)$ is monic, and $u(x)$ is a monic factor of $f(x)$ if it is one of the factors in a monic factorization of $f(x)$. Note that a monic polynomial $f(x) \in R[x]$ (or $D[x]$) is irreducible in $R[x]$ (or $D[x]$) precisely when $g(x) = 1$ or $h(x) = 1$ whenever $f(x) = g(x)h(x)$ is a monic factorization in $R[x]$ (or $D[x]$).

Theorem 5.2.27. *Suppose $f(x)$ is a monic polynomial in $R[x]$.*

(a) Each monic factor of $f(x)$ in $D[x]$ belongs to $R[x]$.
(b) $f(x)$ is irreducible in $R[x]$ if and only if it is irreducible in $D[x]$.

(c) If $a \in D$ is a root of $f(x)$, then $a \in R$. Moreover, if $f(x) \in Z(R)[x]$, then $a \in Z(R)$.

(d) Suppose $f(x) \in \mathscr{L}[x]$, $\overline{f}(x)$ has only simple roots (in its splitting field over $Z(\overline{R})[x]$) and $f(x) = g(x)h(x) = g_1(x)h_1(x)$ are monic factorizations in $D[x]$ with $g(x) - g_1(x) \in J[x]$ and $\overline{f}(x) = \overline{f^s}(x)$ and $\overline{g}(x) = \overline{g^s}(x)$ for each $s \in D^*$. Then $g(x) = g_1(x)$ and $h(x) = h_1(x)$.

(e) Assume $f(x) \in Z(R)[x]$ and $\overline{f}(x)$ has only simple roots. If $f(x) = g_1(x) \cdots g_t(x)$ is a monic factorization in $D[x]$ and $\overline{g}_j(x) = \overline{g_j^s}(x)$ for each j and each $s \in D^*$, then $g_j(x) \in Z(R)[x]$ for $j = 1, \ldots, t$.

Proof. (a) It suffices to show that if $f(x) = g(x)h(x)$ is a monic factorization in $D[x]$, then $g(x)$, $h(x) \in R[x]$. Let

$$f(x) = a_0 + a_1x + \cdots + a_{n-1}x^{n-1} + x^n,$$
$$g(x) = b_0 + b_1x + \cdots + b_{m-1}x^{m-1} + x^m,$$
$$h(x) = c_0 + c_1x + \cdots + c_{r-1}x^{r-1} + x^r.$$

Assume $g(x) \notin R[x]$ and let k be maximal with $b_k \notin R$. Then $a_{k+r} = b_k + b_{k+1}c_{r-1} + \cdots$ and since a_{k+r} and $b_{k+j} (j \geq 1) \in R$, some $c_{r-j} \notin R$. Let $b = \vee_{j=0}^{m-1} |b_j|$, $c = \vee_{j=0}^{r-1} |c_j|$, and put $g_1(x) = c^{-1}b^{-1}g(x)c$ and $h_1(x) = c^{-1}h(x)$. Then $b, c \in D \backslash R$, $g_1(x)$, $h_1(x) \in C(1)[x] \backslash J[x]$ and $f(x) = bcg_1(x)h_1(x)$. But $bc \notin R$ since $bc \geq FF = F$, and $g_1(x)h_1(x)$ has a coefficient d in $R \backslash J$ since $J[x]$ is a completely prime ideal of $R[x]$. This gives the contradiction $a_j = (bc)d \in D \backslash R$ for some j. So $g(x) \in R[x]$ and also $h(x) \in R[x]$.

(b) This is a consequence of (a).

(c) Since $f(x) = (x-a)h(x)$ in $D[x]$ we get $a \in R$ from (a). Also, $a \in Z(D)$ by Theorem 6.2.2 provided $f(x) \in Z(R)[x]$. Note that if $a \in S$ and $f(x) \in K[x]$ where K is a subfield of $Z(R)$, then the centrality of a can also be obtained from (a) of Theorem 5.2.25 and (d).

(d) By (a) we have $g(x)$, $g_1(x)$, $h(x)$, $h_1(x) \in R[x]$ and since $\overline{g}(x)\overline{h}(x) = \overline{g^s}(x)\overline{h^s}(x) = \overline{g}(x)\overline{h^s}(x)$, $\overline{h}(x) = \overline{h^s}(x)$ for each $s \in D^*$; and $\overline{g}(x)$, $\overline{h}(x) \in Z(\overline{R})[x]$ by (d) of Theorem 5.2.25. Let $g(x)$ be as above and let $g_1(x) = d_0 + d_1x + \cdots + x^m$. If $g(x) \neq g_1(x)$, then $b = \vee_{j=0}^{m-1} |b_j - d_j| > 0$. Let $g_2(x) = (g(x) - g_1(x))b^{-1} \in C(1)[x]$, $h_2(x) = h^{b^{-1}}(x) \in R[x]$ and $h_3(x) = (h(x) - h_1(x))b^{-1}$. Note that $\overline{g}_2(x) \neq 0$. By multiplying the equation

$$(g(x) - g_1(x))h(x) + g_1(x)(h(x) - h_1(x)) = 0$$

on the right by b^{-1} we get $g_2(x)h_2(x) + g_1(x)h_3(x) = 0$ and hence $h_3(x) \in R[x[$. Now, $\overline{g}_2(x)\overline{h}(x) = \overline{g}_2(x)\overline{h}_2(x) = -\overline{g}(x)\overline{h}_3(x)$ and if $\overline{g}(x)$ and $\overline{h}(x)$ were relatively prime in $Z(\overline{R})[x]$, then $\overline{g}_2(x) \in \overline{R}[x]\overline{g}(x)$. Thus, $\overline{g}(x)$ and $\overline{h}(x)$ have a common divisor of positive degree in $Z(\overline{R})[x]$ and this contradicts the assumption that $\overline{f}(x) = \overline{g}(x)\overline{h}(x)$ has simple roots. So, $g(x) = g_1(x)$ and $h(x) = h_1(x)$.

(e) Let $g(x) = g_1(x)$ and $h(x) = g_2(x) \cdots g_t(x)$. For $s \in D^*$, $f(x) = g(x)h(x) = g^s(x)h^s(x)$, and therefore from (d) we have $g(x) = g^s(x) \in Z(R)[x]$ and $h(x) \in$

$Z(R)[x]$. Since $\overline{h}(x) = \overline{g}_2(x)\cdots\overline{g}_t(x)$ has simple roots we get $g_j(x) \in Z(R)[x]$ for $j = 2,\ldots,t$, by induction on t. □

We can now give conditions which will guarantee that a central element associated to an element $\overline{\theta} \in \hat{F}\backslash F$ which is algebraic over $Z(\overline{R})$ can be adjoined to D.

Theorem 5.2.28. *Suppose K is a maximal subfield of $Z(R)$, $\overline{\theta} \in \hat{F}\backslash\overline{F}$ is algebraic over $\overline{K}$ and $\overline{f}(x) = \mathrm{Irr}(\overline{\theta},\overline{K})$. The following statements are equivalent.*

(a) $f(x)$ is irreducible in $D[x]$.
(b) If $g(x)$ is a monic divisor of $f(x)$ in $D[x]$, then $\overline{g^s}(x) = \overline{g}(x)$ for each $s \in D^*$.
(c) If $g(x)$ is a monic irreducible divisor of $f(x)$ in $D[x]$, then $\overline{g^s}(x) = \overline{g}(x)$ for each $s \in D^*$.

Proof. The implications (a) ⇒ (b) ⇒ (c) are trivial. We use the previous result to show that (c) ⇒ (a). Suppose $f(x) = g(x)h(x)$ is a monic factorization of $f(x)$ in $R[x]$ with $\overline{g}(\overline{\theta}) = 0$. Then $g(x)$ and $h(x)$ are in $Z(R)[x]$ since each monic irreducible factor of $g(x)$ is fixed by D^* and hence so is $g(x)$. Each coefficient a of $g(x)$ is an elementary symmetric function of the roots of $f(x)$ and hence is algebraic over K. So $K \subseteq K(a) \subseteq Z(R)$ and hence $K = K(a)$ and $g(x) = f(x)$. □

We now deal with the easier of the two algebraic cases.

Theorem 5.2.29. *Suppose K is a subfield of $Z(R)$ with $\overline{K} = \overline{R}$ and $\overline{\theta} \in \hat{F}\backslash\overline{R}$ is algebraic over $\overline{K}$ of degree n. Let $\overline{f}(x) = Irr(\overline{\theta},\overline{K})$. Then $f(x)$ is irreducible in $D[x]$, $D(\theta) = D[x]/D[x]f(x)$ is a division ring, and*

$$T = \{P(\theta) \in D(\theta) : \deg P(x) < n,\ P(x) = bp(x) \text{ with } 0 < b \in D,\ p(x) \in R[x]\backslash J[x] \text{ and } \overline{p}(\overline{\theta}) > 0\} \cup \{0\}$$

is a total order of $D(\theta)$ which contains D^+ and for which (5.2.38) holds.

Proof. K is certainly a maximal subfield of $Z(R)$ and if $f(x) = g(x)h(x)$ is a monic factorization in $D[x]$, then $g(x), h(x) \in R[x]$ by Theorem 5.2.27, and we may assume $\overline{h}(x) = 1$. So $\overline{g^s}(x) = \overline{f^s}(x) = \overline{f}(x) = \overline{g}(x)$ and $f(x)$ is irreducible in $D[x]$ by Theorem 5.2.28. Suppose $0 \neq P(\theta) \in D(\theta)$ with $\deg P(x) < n$ and write $P(x) = bp(x)$ with $b \in D^+$ and $p(x) \in R[x]\backslash J[x]$. If $\overline{p}(\overline{\theta}) = 0$, then $\overline{p}(x) = \overline{f}(x)\overline{g}(x)$ for some $\overline{g}(x)$, and, necessarily, $\overline{p}(x) = 0$. So $\overline{p}(\overline{\theta}) > 0$ or $\overline{p}(\overline{\theta}) < 0$ and $P(\theta) \in T \cup -T$. We claim that $T \cap -T = 0$. If not, take $P(\theta) \in T^* \cap -T^*$ with $\deg P(x) < n$. Then $P(x) = bp(x) = b_1p_1(x)$ with $0 < b, b_1$ and $\overline{p}_1(\overline{\theta}) < 0 < \overline{p}(\overline{\theta})$. But $p_1(x) = b_1^{-1}bp(x)$ with $b_1^{-1}b \in R^+\backslash J$ and hence $0 < \overline{b_1^{-1}b}\overline{p}(\overline{\theta}) = \overline{p}_1(\overline{\theta}) < 0$. To see that $T + T \subseteq T$ and $TT \subseteq T$ take $P_i(\theta) \in T^*$ with canonical expressions $P_i(x) = b_ip_i(x)$ for $i = 1,2$ and assume $b_1 \geq b_2$. Then $P_1(x) + P_2(x) = b_1(p_1(x) + bp_2(x))$ with $b = b_1^{-1}b_2 \in R$ and $P_1(\theta) + P_2(\theta) \in T^*$ since $\overline{p}_1(\overline{\theta}) + \overline{b}\overline{p}_2(\overline{\theta}) \geq \overline{p}_1(\overline{\theta}) > 0$. Also, $P_1(x)P_2(x) = b_1b_2p_1^{b_2}(x)p_2(x)$ and $p_1^{b_2}(\theta) \in T^*$ by (b) of Theorem 5.2.25. By the division algorithm $p_1^{b_2}(x)p_2(x) = q(x)f(x) + r(x)$ and $\overline{r}(\overline{\theta}) = \overline{p_1^{b_2}}(\overline{\theta})\overline{p}_2(\overline{\theta}) > 0$. So

$$P_1(\theta)P_2(\theta) = b_1 b_2 p_1^{b_2}(\theta)p_2(\theta) = b_1 b_2 r(\theta) \in T^*$$

since clearly $D^+T \subseteq T$. As for (5.2.38), if $\alpha < \overline{\theta}$, then $p(\theta) \in T^*$ with $p(x) = x - \alpha$. □

In order to totally order $D(\theta)$ when $\overline{\theta}$ is algebraic over $\overline{K}$ but without assuming $\overline{K} = \overline{R}$ it is necessary to examine commutators in S and to give a detailed analysis of the evaluation of polynomials at elements of S.

Theorem 5.2.30. *Let $a,\ b \in S$ and $m \in \mathbb{N}$.*

(a) $[a,b] \in bJ$ (here b need only be in R).
(b) $a^m - b^m = (\Sigma_{k=0}^{m-1} a^{m-(k+1)} b^k (a-b)) + c(a-b)$ for some $c \in J$.
(c) If $b - a \in Ja$, then $a^m - b^m = ma^{m-1}(a-b) + c(a-b)$ for some $c \in J$.
(d) If $ab = ba$, then

$$a^m - b^m = ma^{m-1}(a-b) + c(a-b)^2 \text{ for some } c \in S.$$

Proof. (a) Assuming $b \neq 0$ we have $b^{-1}ab - a = c \in J$; so $[a,b] = ab - ba = bc \in bJ$.

(b)
$$\begin{aligned} a^m - b^m &= \sum_{k=0}^{m-1} (a^{m-k}b^k - a^{m-(k+1)}b^{k+1}) \\ &= \sum_{k=0}^{m-1} a^{m-(k+1)}(a-b)b^k \\ &= \sum_{k=0}^{m-1} a^{m-(k+1)}(b^k(a-b) + [a-b, b^k]) \\ &= (\sum_{k=0}^{m-1} a^{m-(k+1)}b^k(a-b)) + c(a-b) \end{aligned}$$

where $c \in J$, by (a).

(c) Since $b = a(1+d)$ with $d \in J$, for each $k \geq 0$,

$$\begin{aligned} b^k &= a^k(a^{-(k-1)}(1+d)a^{k-1})(a^{-(k-2)}(1+d)a^{k-2})\cdots(a^{-1}(1+d)a)(1+d) \\ &= a^k(1+d_k) \end{aligned}$$

with $d_k \in J$. From (b) we have

$$\begin{aligned} a^m - b^m &= \sum_{k=0}^{m-1} a^{m-(k+1)}a^k(1+d_k)(a-b) + c(a-b) \\ &= \sum_{k=0}^{m-1} a^{m-1}(a-b) + \sum_{k=0}^{m-1} a^{m-1}d_k(a-b) + c(a-b) \\ &= ma^{m-1}(a-b) + d(a-b) \end{aligned}$$

with $d \in J$.

(d) If $0 \leq k \leq m-2$, then $a^k b^{m-(k+1)} - a^{m-1} = a^k(b^{m-(k+1)} - a^{m-(k+1)}) =$

$a^k c_k(b-a)$ with $c_k \in \mathbb{Z}[a,b] \subseteq S$. So

$$\begin{aligned} a^m - b^m &= \sum_{k=0}^{m-1} a^k b^{m-(k+1)}(a-b) \\ &= \left(\sum_{k=0}^{m-2} a^{m-1} + a^k c_k(b-a) + a^{m-1} \right)(a-b) \\ &= ma^{m-1}(a-b) + c(a-b)^2 \end{aligned}$$

with $c \in S$. □

Applying these approximations to polynomials gives the following facts.

Theorem 5.2.31. *Let a, $b \in S$ and $p(x) \in R[x]$.*

(a) $p(a) - p(b) \in R(a-b)$.
(b) If $b - a \in Ja$, then $p(a) - p(b) = p'(a)(a-b) + c(a-b)$ for some $c \in J$. Also, if $p'(a) \notin J$, then $R(a-b) = R(p(a) - p(b))$ and $p(a) - p(b) = p'(a)(a-b)(1+d)$ with $d \in J$.
(c) If $ab = ba$, then $p(a) - p(b) = p'(a)(a-b) + c(a-b)^2$ for some $c \in S$.

Proof. Let $p(x) = \Sigma_i d_i x^i$. Then $p(a) - p(b) = \Sigma_i d_i(a^i - b^i) \in R(a-b)$ by (b) of Theorem 5.2.30, and

$$\begin{aligned} p(a) - p(b) &= \sum_i d_i[ia^{i-1}(a-b) + c_i(a-b)] \\ &= p'(a)(a-b) + c(a-b) \end{aligned}$$

with $c \in J$ by (c) of Theorem 5.2.30. If $p'(a)$ is a unit of R, then so is $p'(a) + c$ and $p(a) - p(b) = (p'(a) + c)(a-b)$. Moreover,

$$\begin{aligned} p(a) - p(b) &= p'(a)((a-b) + p'(a)^{-1}c(a-b)) \\ &= p'(a)((a-b) + (a-b)d) \\ &= p'(a)(a-b)(1+d) \end{aligned}$$

with $d \in J$. This proves (a) and (b), and (c) follows from (d) of Theorem 5.2.30. □

Theorem 5.2.32. *Suppose $p(x) \in R[x]$, $a \in S \backslash J$, $p(a) \in J$ and $p'(a) \notin J$.*

(a) There exists $b \in a + Ja$ such that $p(b) \in Jp(a)$ and $p'(b) \notin J$.
(b) If $p(x) \in Z(R)[x]$ then b can be chosen so that $p(b) \in Rp(a)^2$.

Proof. Let $b = a - p'(a)^{-1}p(a)$. Then $b - a = -p'(a)^{-1}p(a) \in J = Ja$ and by (b) of Theorem 5.2.31, for some $c \in J$,

$$\begin{aligned} p(a) - p(b) &= p'(a)(a-b) + c(a-b) \\ &= p(a) + cp'(a)^{-1}p(a). \end{aligned}$$

So $p(b) = -cp'(a)^{-1}p(a) \in Jp(a)$, and, from (a) of Theorem 5.2.31, $p'(b) - p'(a) \in R(b-a) \subseteq J$ and hence $p'(b) \notin J$. If $p(x) \in Z(R)[x]$, then a commutes with b since it commutes with $p(a)$ and $p'(a)$. By (c) of Theorem 5.2.31, $p(a) - p(b) = p'(a)(a - b) + d(a-b)^2$ with $d \in R$, and hence $p(b) = -d(a-b)^2 = -dp'(a)^{-2}p(a)^2$. □

We can now give a first approximation to how far the evaluation map $R[x] \longrightarrow R$ at $a \in S$ is from being a homomorphism. For $t \in D$ δ_t is the left inner derivation determined by $t : \delta_t(s) = [t,s] = ts - st$.

Theorem 5.2.33. *Suppose $p(x)$, $q(x) \in R[x]$ and $a \in S$. Then $(pq)(a) - p(a)q(a) \in Jq(a)$, and it is also in Ja provided $q(x) \in S[x]$. If $p(a) \notin J$, then $(pq)(a) = p(a)q(a)(1+c)$ for some $c \in J$.*

Proof. Let $p(x) = \Sigma_i a_i x^i$ and $q(x) = \Sigma_j b_j x^j$. Then

$$\begin{aligned} p(a)q(a) - (pq)(a) &= \sum_{i,j} a_i a^i b_j a^j - \sum_{i,j} a_i b_j a^i a^j \\ &= \sum_i a_i \left(\sum_j (a^i b_j a^j - b_j a^j a^i) \right) \\ &= \sum_i a_i \left(\sum_j \delta_{a^i}(b_j a^j) \right) \\ &- \sum_i a_i \delta_{a^i}(q(a)). \end{aligned} \tag{5.2.40}$$

However, this last sum is in $Jq(a)$ by (a) of Theorem 5.2.30 which also gives that it is in Ja whenever $q(a) \in S$. If $p(a) \notin J$, then $J = Jp(a)$ and $Jq(a) = Jp(a)q(a) = p(a)q(a)J$. So $(pq)(a) - p(a)q(a) = p(a)q(a)c$ for some $c \in J$ and $(pq)(a) = p(a)q(a)(1+c)$. □

We require a better estimate than that given in Theorem 5.2.33.

Theorem 5.2.34. *Suppose $p(x)$, $q(x) \in R[x]$, $f(x) \in Z(S)[x]$ and a, $b \in S$.*

(a) If the degree of $p(x)$ is n, then

$$(pq)(a) = p(a)q(a) + \sum_{k=1}^{n} (-1)^k \frac{1}{k!} p^{(k)}(a) \delta_a^k(q(a))$$

(b) If $q(a)$, $p'(a) \in S$, then $(pq)(a) = p(a)q(a) + c\delta_a(q(a))$ with $c \in S$.
(c) $\delta_{f(a)}(b) = f'(a)\delta_a(b) + d\delta_a^2(b)$ for some $d \in S$.
(d) If $f'(a) \notin J$, then $\delta_a(b) = e\delta_{f(a)}(b)$ for some $e \in S \backslash J$.
(e) If $f'(a) \notin J$, $q(a)$, $p'(a) \in S$ and $f(a) \in R(pq)(a)$, then $(pq)(a) = p(a)q(a)(1+g)$ for some $g \in J$.

Proof. (a) Let μ_a and $\rho_a \in \mathrm{End}_{\mathbb{Z}}(R)$ denote left and right multiplication by a, respectively. If a and c commute, then δ_a commutes with μ_c and ρ_c. Since $\rho_a = \mu_a - \delta_a$ and $\rho_a^i = \rho_{a^i}$ and $\mu_a^i = \mu_{a^i}$, we have

$$\rho_{a^i} = (\mu_a - \delta_a)^i = \mu_{a^i} - \sum_{k=1}^{i} (-1)^{k-1} \binom{i}{k} \mu_{a^{i-k}} \delta_a^k$$

and

$$\delta_{a^i} = \mu_{a^i} - \rho_{a^i} = \sum_{k=1}^{i} (-1)^{k-1} \binom{i}{k} \mu_{a^{i-k}} \delta_a^k.$$

Since $\binom{i}{k} x^{i-k} = \frac{1}{k!}(x^i)^{(k)}$ we get from (5.2.39)

$$\begin{aligned}(pq)(a) - p(a)q(a) &= \sum_{i=1}^{n} a_i \sum_{k=1}^{i} (-1)^k \binom{i}{k} a^{i-k} \delta_a^k(q(a)) \\ &= \sum_{i=1}^{n} \sum_{k=1}^{i} (-1)^k \binom{i}{k} a_i a^{i-k} \delta_a^k(q(a)) \\ &= \sum_{k=1}^{n} (-1)^k \left(\sum_{i=k}^{n} \frac{1}{k!} a_i a^{i-k} \right) \delta_a^k(q(a)) \\ &= \sum_{k=1}^{n} (-1)^k \frac{1}{k!} p^{(k)}(a) \delta_a^k(q(a)).\end{aligned}$$

(b) By (a) of Theorem 5.2.30 $\delta_a(b) \in Jb$, and using induction it is easily seen that $\delta_a^k(b) \in J^{k-i}\delta_a^i(b)$ if $0 \leq i \leq k$. So if $q(a) \in S$, then

$$\begin{aligned}(pq)(a) &= p(a)q(a) - p'(a)\delta_a(q(a)) + \frac{p^{(2)}(a)}{2!}\delta_a^2(q(a)) - \frac{p^{(3)}(a)}{3!}\delta_a^3(q(a)) + \cdots \\ &= p(a)q(a) - p'(a)\delta_a(q(a)) + s\delta_a^2(q(a)) \qquad (5.2.41)\end{aligned}$$

with $s \in R$; and if $p^{(2)}(a) \in S$, then $s \in S$. Also, if $p'(a) \in S$, then $(pq)(a) = p(a)q(a) + c\delta_a(q(a))$ with $c \in S$.

(c) If $q(a) \in S$ then from (5.2.40) we have

$$(fq)(a) - f(a)q(a) = -f'(a)\delta_a(q(a)) + s\delta_a^2(q(a))$$

with $s \in S$, and from (5.2.39) we see that $q(a)f(a) = (qf)(a) = (fq)(a)$. So

$$q(a)f(a) - f(a)q(a) = -f'(a)\delta_a(q(a)) + s\delta_a^2(q(a))$$

and $\delta_{f(a)}(q(a)) = f'(a)\delta_a(q(a)) + d\delta_a^2(q(a))$ with $d = -s \in S$. The result follows by letting $q(x) = b$.

(d) Since $\delta_a^2(b) = z\delta_a(b)$ with $z \in J$, from (c) we obtain $\delta_{f(a)}(b) = (f'(a) + dz)\delta_a(b)$ and hence $\delta_a(b) = (f'(a) + dz)^{-1}\delta_{f(a)}(b)$ with $(f'(a) + dz)^{-1} \in S$.

(e) From (d), and from (a) of Theorem 5.2.30, we have $\delta_a(q(a)) = e\delta_{f(a)}(q(a)) = ezf(a) = ezr(pq)(a) = (pq)(a)s$ with $e \in S$, z, $s \in J$ and $r \in R$. From (b) we then have, since $c\delta_a(q(a)) = \delta_a(q(a))c_1$ with $c_1 \in R$, $(pq)(a) = p(a)q(a) + (pq)(a)sc_1$

and $(pq)(a)(1-sc_1) = p(a)q(a)$. Since $sc_1 \in J$, $(pq)(a) = p(a)q(a)(1+g)$ for some $g \in J$. $\square$

To each element $\overline{\theta}$ in $\hat{F} \cap \overline{R}$ which is algebraic over $\overline{K}$ we can attach a prime ideal of R which will allow us to specialize to certain $\overline{\theta}$. Each prime ideal of R is, of course, completely prime since it is a convex ideal.

Theorem 5.2.35. *Suppose K is a subfield of $Z(R)$ and $\overline{\theta} \in \hat{F} \cap \overline{R}$ is algebraic over $\overline{K}$. Let $\overline{f}(x) = Irr(\overline{\theta}, \overline{K})$ and let*

$$\mathscr{P}(\overline{\theta}) = \bigcap \{Rg(a) : a \in \varphi^{-1}(\overline{\theta}) \text{ and } g(x) \in K[x]^*\}.$$

Then $\mathscr{P}(\overline{\theta})$ is a completely prime D^*-invariant ideal of R which is properly contained in J. Moreover,

$$\begin{aligned}\mathscr{P}(\overline{\theta}) &= \bigcap_{a \in \varphi^{-1}(\overline{\theta})} Rf(a)\\ &= \bigcap \{Jg(a) : a \in \varphi^{-1}(\overline{\theta}) \text{ and } g(x) \in K[x]^*\}\\ &= \bigcap_{a \in \varphi^{-1}(\overline{\theta})} Jf(a).\end{aligned}$$

Proof. If $\overline{\theta} \in \overline{K}$, then $\mathscr{P}(\overline{\theta}) = 0$ and the rest is trivial. So we will assume $\overline{\theta} \notin \overline{K}$. Let $A = \cap_{a \in \varphi^{-1}(\overline{\theta})} Rf(a)$. If $d \in R$ and $d \notin A$, then $Rf(a) \subset Rd$ for some $a \in \varphi^{-1}(\overline{\theta})$ and $f(a) = cd$ with $c \in J$. From Theorem 5.2.32 we get $b = a + ea$ with $e \in J$ and $f(b) = rf(a)^2$ with $r \in R$. So $f(b) = rcdcd = sd^2$ with $s \in J$ and hence $d^2 \notin Rf(b)$. Thus, $d^2 \notin A$ and A is (completely) prime since R/A is reduced and totally ordered. Clearly, $\mathscr{P}(\overline{\theta}) \subseteq A$. For the other inclusion take $d \in R \backslash \mathscr{P}(\overline{\theta})$. Then $d \notin Rg(a)$ for some $a \in \varphi^{-1}(\overline{\theta})$ and some $0 \neq g(x) \in K[x]$. So $g(a) = cd$ with $c \in J$. Write $g(x) = f(x)^m h(x)$ in $K[x]$ with $h(x)$ relatively prime to $f(x)$. Then $f(a)^m = g(a)h(a)^{-1} = cdh(a)^{-1} \in Jd$. If d were in A then $f(a) \in A$ and, since A is a proper ideal, (a) of Theorem 5.2.32 guarantees the existence of an element $b = a + ea$ with $e \in J$ and such that $f(b) = tf(a)$ with $t \in J$. But $f(a) \in Rf(b)$ implies $t \notin J$. So $d \notin A$ and $A \subseteq \mathscr{P}(\overline{\theta})$.

Since $\cap_{a,g} Jg(a) \subseteq \cap_a Jf(a) \subseteq \mathscr{P}(\overline{\theta})$ it suffices to verify that $\mathscr{P}(\overline{\theta}) \subseteq \cap_{a,g} Jg(a)$ in order to establish the equalities. If $0 \neq b \in \mathscr{P}(\overline{\theta})$, $a \in \varphi^{-1}(\overline{\theta})$ and $0 \neq g(x) \in K[x]$, then $b = rg(a)$ with $r \in R$. Assuming $r \notin J$ we get $g(a) = r^{-1}b = r^{-1}sf(a)g(a)$ with $s \in R$, and $1 = r^{-1}sf(a)$. This is a contradiction since $f(a) \in J$.

$\mathscr{P}(\overline{\theta})$ is properly contained in J since $\mathscr{P}(\overline{\theta}) \subseteq Jf(a) \subset J$ if $a \in \varphi^{-1}(\overline{\theta})$. Also, $\overline{a^s} = \overline{\theta}^s = \overline{\theta}$ for any $s \in D^*$ by (a) of Theorem 5.2.25. Thus, $\varphi^{-1}(\overline{\theta})^s = \varphi^{-1}(\overline{\theta})$ and

$$\mathscr{P}(\overline{\theta})^s = \bigcap_{a \in \varphi^{-1}(\overline{\theta})} Rf(a)^s = \bigcap_{a \in \varphi^{-1}(\overline{\theta})} Rf(a^s) = \mathscr{P}(\overline{\theta}).$$

$\square$

The definition of $\mathscr{P}(\overline{\theta})$ makes sense for any $\overline{\theta} \in \hat{F}$. The two other cases arise when either $\overline{\theta} \in \overline{R}$ and is transcendental over $\overline{K}$ or when $\overline{\theta} \notin \overline{R}$, and in both these cases $\mathscr{P}(\overline{\theta}) = R$.

If $\mathscr{P}$ is a proper prime ideal of R, then $\overline{\overline{R}} = R/\mathscr{P} = C_{\overline{\overline{F}}}(\overline{\overline{1}})$ in the totally ordered classical quotient division ring $\overline{\overline{D}}$ of $\overline{\overline{R}}$. The mappings $R \longrightarrow \overline{\overline{R}}$ and $\overline{\overline{R}} \longrightarrow \overline{R}$ will be denoted by $a \mapsto \overline{\overline{a}}$ and $\overline{\overline{a}} \mapsto \overline{a}$, respectively. The next result determines when $f(x)$ picks up a root in $\overline{\overline{R}}$.

Theorem 5.2.36. *Suppose K is a subfield of $Z(R)$, $\overline{\theta} \in \hat{F} \cap \overline{R}$ is algebraic over $\overline{K}$, $0 \neq \mathscr{P}$ is a prime ideal of R, and $\overline{\overline{R}} = R/\mathscr{P}$. Then there exists $a \in R$ with $\overline{a} = \overline{\theta}$ and $\overline{\overline{K}}(\overline{\overline{a}}) \subseteq Z(\overline{\overline{R}})$ if and only if $\mathscr{P}(\overline{\theta}) \subset \mathscr{P}$.*

Proof. Let $\overline{f}(x) = \mathrm{Irr}(\overline{\theta}, \overline{K})$. $\mathscr{P}(\overline{\theta}) \subset \mathscr{P}$ iff there exist $b \in \mathscr{P}$ and $a \in \varphi^{-1}(\overline{\theta})$ with $f(a) = cb$ for some $c \in J$. So if $\mathscr{P}(\overline{\theta}) \subset \mathscr{P}$, then $\overline{\overline{f}}(\overline{\overline{a}}) = 0$ and $\overline{\overline{K}}(\overline{\overline{a}}) \subseteq Z(\overline{\overline{R}})$ by (c) of Theorem 5.2.27. On the other hand, given $a \in \varphi^{-1}(\overline{\theta})$ with $\overline{\overline{K}}(\overline{\overline{a}}) \subseteq Z(\overline{\overline{R}})$ we have $\overline{\overline{f(a)}} = \overline{\overline{f}}(\overline{\overline{a}}) = 0$ since $\overline{\overline{K}}(\overline{\overline{a}}) \cong \overline{K}(\overline{\theta})$ and hence $f(a) \in \mathscr{P}$. If $f(a) = 0$, then $\mathscr{P}(\overline{\theta}) = 0$ and we're done; otherwise $f(a) \notin \mathscr{P}(\overline{\theta})$ is a consequence of Theorem 5.2.35 since $f(a) \notin Jf(a)$. □

The previous result will now be used to compare the ideals $\mathscr{P}(\overline{\theta})$ for different values of $\overline{\theta}$.

Theorem 5.2.37. *Suppose K is a subfield of $Z(R)$ and $\overline{\theta}$, $\overline{\theta}_1$, $\overline{\theta}_2 \in \hat{F}$.*

(a) If $\overline{\theta}_1 \in \overline{K}(\overline{\theta})$, then $\mathscr{P}(\overline{\theta}_1) \subseteq \mathscr{P}(\overline{\theta})$.
(b) If $\overline{K}(\overline{\theta}) = \overline{K}(\overline{\theta}_1, \overline{\theta}_2)$, then $\mathscr{P}(\overline{\theta}) = \mathscr{P}(\overline{\theta}_1) + \mathscr{P}(\overline{\theta}_2)$.

Proof. If $\overline{\theta} \notin \overline{R}$ or $\overline{\theta}$ is transcendental over $\overline{K}$, then $\overline{\mathscr{P}(\theta)} = \overline{R}$ and both (a) and (b) are obvious. We therefore assume $\overline{\theta} \in \overline{R}$ is algebraic over $\overline{K}$. If there is a subfield L of $Z(R)$ with $\overline{K}(\overline{\theta}) \subseteq \overline{L}$, then there exists an $a \in \varphi^{-1}(\overline{\theta})$ with $K(a) \cong \overline{K}(\overline{\theta})$ and $f(a) = 0$, where $\overline{f}(x) = \mathrm{Irr}(\overline{\theta}, \overline{K})$. In this case $\mathscr{P}(\overline{\theta}_1) = \mathscr{P}(\overline{\theta}) = 0$ and we are done. In any case let $\overline{\overline{R}} = R/\mathscr{P}(\overline{\theta}_1)$. Suppose first that there is a subfield $\overline{\overline{L}}$ of $Z(\overline{\overline{R}})$ with $\overline{K}(\overline{\theta}_1) \subseteq \overline{L}$ and let $\overline{h}(x) = \mathrm{Irr}(\overline{\theta}_1, \overline{K})$. Then there exists some $a_1 \in \varphi^{-1}(\overline{\theta}_1)$ with $\overline{\overline{K}}(\overline{\overline{a}}_1) \cong \overline{K}(\overline{\theta}_1)$ and $\overline{\overline{h}}(\overline{\overline{a}}_1) = 0$. So $h(a_1) \in \mathscr{P}(\overline{\theta}_1) \subseteq Jh(a_1)$, $h(a_1) = 0$ and $\mathscr{P}(\overline{\theta}_1) = 0$. Now assume there is no such subfield of $Z(\overline{\overline{R}})$ and $\mathscr{P}(\overline{\theta}_1) \neq 0$. Then there is no subfield $\overline{\overline{L}}$ of $Z(\overline{\overline{R}})$ with $\overline{K}(\overline{\theta}) \subseteq \overline{L}$ and hence $\mathscr{P}(\overline{\theta}_1) \subseteq \mathscr{P}(\overline{\theta})$ by Theorem 5.2.36. This concludes the proof of (a) and it gives the inclusion $\mathscr{P}(\overline{\theta}_1) + \mathscr{P}(\overline{\theta}_2) \subseteq \mathscr{P}(\overline{\theta})$ of (b). If this inclusion were proper, then again by Theorem 5.2.36 there exist $a_1 \in \varphi^{-1}(\overline{\theta}_1)$ and $a_2 \in \varphi^{-1}(\overline{\theta}_2)$ such that in $\overline{\overline{R}} = R/\mathscr{P}(\overline{\theta})$ we have $\overline{\overline{K}}(\overline{\overline{a}}_1), \overline{\overline{K}}(\overline{\overline{a}}_2) \subseteq Z(\overline{\overline{R}})$. But then $\overline{K}(\overline{\theta}) = \overline{K}(\overline{a}_1, \overline{a}_2)$ and for some $a \in \varphi^{-1}(\overline{\theta})$, $\overline{\overline{K}}(\overline{\overline{a}}) = \overline{\overline{K}}(\overline{\overline{a}}_1, \overline{\overline{a}}_2) \subseteq Z(\overline{\overline{R}})$. This produces the contradiction $\mathscr{P}(\overline{\theta}) \subset \mathscr{P}(\overline{\theta})$. □

Let K be a subfield of $Z(R)$. An element $\overline{\theta} \in \overline{S} \backslash \overline{K}$ is called *solid* if it is algebraic over $\overline{K}$ and $\overline{\overline{f}}(x)$ is irreducible in $\overline{\overline{R}}[x] \cong R[x]/\mathscr{P}(\overline{\theta})[x]$ where $\overline{f}(x) = \mathrm{Irr}(\overline{\theta}, \overline{K})$. This property depends on K but no confusion will result from not using "K" in its definition.

Theorem 5.2.38. *Suppose $R = S$ and K is a maximal subfield of $Z(R)$. If there is an algebraic element in $\overline{R}\backslash\overline{K}$, then there is a solid element in $\overline{R}\backslash\overline{K}$.*

Proof. Let $\overline{\theta'} \in \overline{R}\backslash\overline{K}$ be algebraic over $\overline{K}$ with $\overline{p}(x) = \mathrm{Irr}(\overline{\theta'},\overline{K})$, let L be the splitting field of $\overline{p}(x)$ over $\overline{K}(\overline{\theta'})$ and let $\overline{E} = L \cap \overline{R}$. Then $\overline{K} \subset \overline{K}(\overline{\theta'}) \subseteq \overline{E}$. Let $\{\overline{K}(\overline{\theta}), \overline{K}(\overline{\theta}_1), \ldots, \overline{K}(\overline{\theta}_t)\}$ be all the subfields of $\overline{E}$ which properly contain $\overline{K}$ indexed so that $\mathscr{P}(\overline{\theta}) \subseteq \mathscr{P}(\overline{\theta}_1) \subseteq \cdots \subseteq \mathscr{P}(\overline{\theta}_t)$. We will show that $\overline{\theta}$ is solid. Let $\overline{f}(x) = \mathrm{Irr}(\overline{\theta},\overline{K})$. If $\mathscr{P}(\overline{\theta}) = 0$, then $\overline{\theta}$ is solid since $f(x)$ is irreducible in $R[x]$ by Theorem 5.2.28 and (a) of Theorems 5.2.27 and 5.2.25. Suppose $\mathscr{P}(\overline{\theta}) \neq 0$. Let $g(x) = b_0 + b_1x + \cdots + b_{m-1}x^{m-1} + x^m$ and $h(x)$ be monic in $R[x]$ such that $\overline{\overline{f}}(x) = \overline{\overline{g}}(x)\overline{\overline{h}}(x)$ in $\overline{\overline{R}}[x]$ where $\overline{\overline{R}} = R/\mathscr{P}(\overline{\theta})$. By (e) of Theorem 5.2.27 the monic factors of $\overline{\overline{f}}(x)$ in $\overline{\overline{R}}[x]$ are central and we may assume $\overline{\overline{g}}(x)$ is irreducible in $\overline{\overline{R}}[x]$ and $\overline{g}(\overline{\theta}) = 0$. Each $\overline{b}_j$ is in $\overline{E}$ since it is an elementary symmetric function of some of the roots of $\overline{f}(x)$, all of which are in L. So $\overline{\overline{K}}(\overline{\overline{b}}_j) \subseteq Z(\overline{\overline{R}})$ and by Theorem 5.2.36 $\mathscr{P}(\overline{b}_j) \subset \mathscr{P}(\overline{\theta})$ for each j. But then from Theorem 5.2.37 we see that $\overline{K}(\overline{b}_j) \notin \{\overline{K}(\overline{\theta}), \ldots, \overline{K}(\overline{\theta}_t)\}$ and hence $\overline{K}(\overline{b}_j) = \overline{K}$. So $\overline{g}(x), \overline{\overline{h}}(x) \in \overline{\overline{K}}[x]$ and hence $\overline{\overline{f}}(x) = \overline{\overline{g}}(x)$ is irreducible. □

Solid elements will be used to obtain a total order for a simple algebraic extension of D. To do this we need to connect them to a useful property of polynomials in $R[x]$.

Theorem 5.2.39. *Suppose $R = S$. Let K be a subfield of $Z(R)$, $\overline{\theta} \in \overline{R}\backslash\overline{K}$, $\overline{f}(x) = Irr(\overline{\theta},\overline{K})$, and*

$$I(\overline{\theta}) = \{q(x) \in R[x] : q(a) \in Rf(a) \text{ for each } a \in \varphi^{-1}(\overline{\theta})\}.$$

Suppose $g(x)$ and $h(x)$ are monic polynomials in $R[x]$ such that $\overline{\overline{f}}(x) = \overline{\overline{h}}(x)\overline{\overline{g}}(x)$ in $\overline{\overline{R}}[x]$, where $\overline{\overline{R}} = R/\mathscr{P}(\overline{\theta})$ and $\overline{\overline{g}}(x)$ is irreducible in $\overline{\overline{R}}[x]$ with $\overline{g}(\overline{\theta}) = 0$. Then:

(a) $I(\overline{\theta})$ is a completely prime D^*-invariant ideal of $R[x]$;
(b) $I(\overline{\theta}) = \mathscr{P}(\overline{\theta})[x] + R[x]g(x)$;
(c) $\overline{\theta}$ is solid if and only if $I(\overline{\theta}) = \mathscr{P}(\overline{\theta})[x] + R[x]f(x)$.

Proof. $I(\overline{\theta})$ is D^*-invariant since $\varphi^{-1}(\overline{\theta})$ is D^*-invariant. From Theorem 5.2.33 we have $(pq)(a) = p(a)q(a) + cq(a)$ with $c \in J$, for any $p(x)$, $q(x) \in R[x]$ and $a \in R$; and also $(pq)(a) = p(a)q(a)(1+d)$ with $d \in J$ provided $p(a) \notin J$. The first equation shows that $I(\overline{\theta})$ is a left ideal since it is clearly an additive subgroup of $R[x]$. The existence of $g(x)$ and $h(x)$, as well as of the uniqueness of $\overline{\overline{g}}(x)$ and $\overline{\overline{h}}(x)$ in $Z(\overline{\overline{R}})[x]$ and the fact that $\overline{h}(\overline{\theta}) \neq 0$, is guaranteed by (b) and (e) of Theorem 5.2.27 applied to $\overline{\overline{R}}$. To see that $\mathscr{P}(\overline{\theta})[x] + R[x]g(x) \subseteq I(\overline{\theta})$ first note that for any $b \in \mathscr{P}(\overline{\theta})$, $m \in \mathbb{Z}^+$ and $a \in \varphi^{-1}(\overline{\theta})$, $ba^m \in Jf(a)a^m \subseteq Jf(a)$ and $\mathscr{P}(\overline{\theta})[x] \subseteq I(\overline{\theta})$. Now $f(x) = h(x)g(x) + r(x)$ with $r(x) \in \mathscr{P}(\overline{\theta})[x]$. So, for $a \in \varphi^{-1}(\overline{\theta})$,

$$f(a) = (hg)(a) + r(a) = h(a)g(a) + cg(a) + df(a)$$

with $c, d \in J$, and hence $g(a) = (c + h(a))^{-1}(1-d)(f(a) \in Rf(a)$. Thus, $g(x) \in I(\overline{\theta})$. For the other inclusion suppose $p(x) \in R[x]$ and $p(x) \notin \mathscr{P}(\overline{\theta})[x] + R[x]g(x)$.

Then $\overline{\overline{p}}(x) \notin \overline{\overline{\overline{D}}}[x]\overline{\overline{g}}(x)$, $\overline{\overline{\overline{D}}}[x]\overline{\overline{g}}(x) + \overline{\overline{\overline{D}}}[x]\overline{\overline{p}}(x) = \overline{\overline{\overline{D}}}[x]$, and $1 = \overline{\overline{\overline{u}}}_1(x)\overline{\overline{\overline{p}}}(x) + \overline{\overline{\overline{v}}}_1(x)\overline{\overline{\overline{g}}}(x)$ for some $\overline{\overline{u}}_1(x)$, $\overline{\overline{v}}_1(x) \in \overline{\overline{\overline{D}}}[x]$. If $\overline{\overline{b}}$ is the maximum of the absolute values of the coefficients of $\overline{\overline{\overline{u}}}_1(x)$ and $\overline{\overline{\overline{v}}}_1(x)$, then $\overline{\overline{u}}_1(x) = \overline{\overline{b}}\overline{\overline{u}}(x)$ and $\overline{\overline{v}}_1(x) = \overline{\overline{b}}\overline{\overline{v}}(x)$ with $\overline{\overline{u}}(x)$, $\overline{\overline{v}}(x) \in \overline{\overline{R}}[x]$, $\overline{\overline{u}}(x)\overline{\overline{p}}(x) + \overline{\overline{v}}(x)\overline{\overline{g}}(x) = \overline{\overline{b}}^{-1} = \overline{\overline{d}} \in \overline{\overline{R}}$, and

$$u(x)p(x) + v(x)g(x) = d + s(x)$$

with $s(x) \in \mathscr{P}(\overline{\theta})[x]$. For each $a \in \varphi^{-1}(\overline{\theta})$ both $s(a)$ and $(vg)(a)$ belong to $Rf(a)$. Since $d \notin \mathscr{P}(\overline{\theta})$ there exists some $a_1 \in \varphi^{-1}(\overline{\theta})$ with $d \notin Rf(a_1)$, and hence $(up)(a_1) \notin Rf(a_1)$. Since $(up)(a_1) = u(a_1)p(a_1) + ep(a_1)$ with $e \in J$, necessarily $p(a_1) \notin Rf(a_1)$. So $p(x) \notin I(\overline{\theta})$ and (b) has been established. $I(\overline{\theta})$ is an ideal since $I(\overline{\theta})/\mathscr{P}(\overline{\theta})[x] = \overline{\overline{R}}[x]\overline{\overline{g}}(x)$ is an ideal of $\overline{\overline{R}}[x]$, and $I(\overline{\theta})$ is completely prime since $R[x]/I(\overline{\theta}) \cong \overline{\overline{R}}[x]/\overline{\overline{R}}[x]\overline{\overline{g}}(x)$ and the latter is embedded in the division ring $\overline{\overline{D}}[x]/\overline{\overline{D}}[x]\overline{\overline{g}}(x)$. Finally, (c) is just a restatement of the uniqueness of $\overline{\overline{g}}(x)$ since $I(\overline{\theta}) = \mathscr{P}(\overline{\theta})[x] + R[x]f(x)$ iff $\overline{\overline{R}}[x]\overline{\overline{f}}(x) = \overline{\overline{R}}[x]\overline{\overline{g}}(x)$, iff $\overline{\overline{f}}(x) = \overline{\overline{g}}(x)$. □

The following result, in a sense, will allow us to free $\overline{\theta}$ (actually, θ) from its dependence on a particular $a \in \varphi^{-1}(\overline{\theta})$.

Theorem 5.2.40. *Suppose K is a subfield of $Z(R)$, $\overline{\theta} \in \overline{S}\backslash\overline{K}$ is algebraic over K, $f(x) = Irr(\overline{\theta}, \overline{K})$, $p(x)$, $p_1(x)$, $p_2(x), \ldots, p_m(x) \in R[x]$ and $a_1, a_2, \ldots, a_m \in \varphi^{-1}(\overline{\theta})$.*

(a) If $\overline{\theta}$ is solid and $p(x) \notin \mathscr{P}(\overline{\theta})[x]$ and has smaller degree than $f(x)$, then $f(a) \in Jp(a)$ for some $a \in \varphi^{-1}(\overline{\theta})$.
(b) If $f(a_1) \in Jp(a_1)$ and $f(a_2) \in Rf(a_1)$, then $f(a_2) \in Jp(a_2)$.
(c) If $f(a_1) \in Jp(a_1)$ and $f(a_2) \in Jp(a_2)$, then $p(a_2) = p(a_1)(1+d)$ for some $d \in J$.
(d) If $f(a_i) \in Jp_i(a_i)$ for $i = 1, \ldots, m$, and $f(a_k) \in Rf(a_i)$ for every i, then $f(a_k) \in \cap_{i=1}^{m} Jp_i(a_k)$.

Proof. (a) The proof of the previous theorem shows that when $\overline{\theta}$ is solid $I(\overline{\theta}) = \mathscr{P}(\overline{\theta})[x] + R[x]f(x)$ is a completely prime ideal of $R[x]$ (without $R = S$). Since $0 \neq \overline{\overline{p}}(x) \notin \overline{\overline{R}}[x]\overline{\overline{f}}(x)$, where $\overline{\overline{R}} = R/\mathscr{P}(\overline{\theta})$, $p(x) \notin I(\overline{\theta})$ and hence $p(a) \notin Rf(a)$ for some $a \in \varphi^{-1}(\overline{\theta})$. So $f(a) \in Jp(a)$.

(b) $Ra_1 = Ra_2$ since if $a_1 = ra_2$, then $r \notin J$ as $\overline{\theta} \neq 0$; also, $a_2 - a_1 \in Ja_1$ since $a_2 - a_1 = sa_1$ gives $0 = \overline{s}\overline{\theta}$. From (a) and (b) of Theorem 5.2.31 we have

$$p(a_2) - p(a_1) \in R(a_2 - a_1) = R(f(a_2) - f(a_1)) \subseteq Rf(a_1) \subseteq Jp(a_1);$$

so $p(a_2) = (1+c)p(a_1)$ with $c \in J$. Thus, $f(a_2) \in Rf(a_1) \subseteq Jp(a_1) = Jp(a_2)$.

(c) By symmetry we may assume $f(a_2) \in Rf(a_1)$ and as we have just seen $p(a_2) = (1+c)p(a_1) = p(a_1)(1+d)$ with $d \in J$ since $bJ = Jb$ for any b.

(d) This is an immediate consequence of (b). □

Now suppose $R = S$, K is a maximal subfield of $Z(R)$, $\overline{\theta} \in \overline{R}\backslash\overline{K}$ is a solid element of degree n over K and $\overline{f}(x) = \mathrm{Irr}(\overline{\theta}, \overline{K})$. By Theorems 5.2.25 (a) and 5.2.28 $D(\theta) =$

$D[x]/D[x]f(x)$ is a division ring extension of D which has $1, \theta, \theta^2, \ldots, \theta^{n-1}$ as a D-basis, where $\theta = x + D[x]f(x)$, $f(\theta) = 0$ and $Z(D)(\theta) \subseteq Z(D(\theta))$. We proceed to define a total order of $D(\theta)$. Let $0 \neq P(x) \in D[x]$ with $\deg P(x) < n$. A triple $(b, p(x), a) \in D \times (R[x]\backslash\mathscr{P}(\overline{\theta})[x]) \times \varphi^{-1}(\overline{\theta})$ is called an *O-determiner* for $P(x)$ if $P(x) = bp(x)$ and $f(a) \in Jp(a)$. In fact, $p(x) \notin I(\overline{\theta})$ since $p(a) \notin Rf(a)$. An O-determiner for $P(x)$ exists. For if $P(x) = b_0 + b_1x + \cdots + b_mx^m$ and $b = \vee_j |b_j|$, then $p(x) = b^{-1}P(x) \in R[x]\backslash J[x] \subseteq R[x]\backslash\mathscr{P}(\overline{\theta})[x]$, and by (a) of Theorem 5.2.40, $f(a) \in Jp(a)$ for some $a \in \varphi^{-1}(\overline{\theta})$. Note that $P(a) \neq 0$ since otherwise $f(a) = 0$ and $a \in K$ by (c) of Theorem 5.2.27. Let

$$T = \{P(\theta) \in D(\theta) : \deg P(x) < n \text{ and } P(a) \in D^{+*} \text{ for some O-determiner } (b, p(x), a) \text{ for } P(x)\} \cup \{0\}.$$

Theorem 5.2.41. *Assume the conditions: $R = S$, K is a maximal subfield of $Z(R)$, and $\overline{\theta} \in \overline{R}\backslash\overline{K}$ is solid. Then T is a total order of $D(\theta)$ which contains D^+ and for which (5.2.38) is valid.*

Proof. We have just seen that $T \cup -T = D(\theta)$. We proceed in steps to show that T is a total order.

(i) If $(b_1, p_1(x), a_1)$ and $(b_2, p_2(x), a_2)$ are O-determiners for $P(x)$, then $P(a_1)P(a_2) > 0$ and hence $T \cap -T = 0$. Assume first that $p_2(x) = p_1(x)$ and $b_2 = b_1$. Then by (c) of Theorem 5.2.40 $p_1(a_2) = p_1(a_1)(1+d)$ with $d \in J$. Since $1 + d > 0$, $P_1(a_1)P_2(a_2) = b_1p_1(a_1)b_1p_1(a_1)(1+d) > 0$. In general, according to (d) of Theorem 5.2.40 we may assume $f(a_2) \in Jp_1(a_2)$; that is, $(b_1, p_1(x), a_2)$ is an O-determiner for $P(x)$. Thus, we are reduced to the previous case and $P(a_1)P(a_2) > 0$. Since $(b, p(x), a)$ is an O-determiner for $P(x)$ if and only if it is an O-determiner for $-P(x)$ we do have $T \cap -T = 0$.

(ii) $T + T \subseteq T$. Suppose $P_1(\theta)$, $P_2(\theta) \in T^*$ and let $(b_i, p_i(x), a_i)$ be an O-determiner for $P_i(x)$ for $i = 1, 2$. By (d) of Theorem 5.2.40 we may assume $a_1 = a_2$ and certainly $(P_1 + P_2)(a_1) = P_1(a_1) + P_2(a_1) > 0$. Since $(-b_i, -p_i(x), a_i)$ is an O-determiner for $P_i(x)$ we may also assume that $b_1 \geq b_2 > 0$. Now,

$$P_1(x) + P_2(x) = b_1p_1(x) + b_2p_2(x) = b_1(p_1(x) + b_1^{-1}b_2p_2(x)) = b_1p(x),$$

$p(x) \in R[x]$, and $p(a_1) = p_1(a_1) + b_1^{-1}b_2p_2(a_1) > p_1(a_1) > 0$. Thus, $f(a_1) \in Jp_1(a_1) \subseteq Jp(a_1)$, $p(a_1) \notin Rf(a_1)$ and hence $p(x) \notin \mathscr{P}(\overline{\theta})[x]$. So $(b_1, p(x), a_1)$ is an O-determiner for $P_1(x) + P_2(x)$ and we have $P_1(\theta) + P_2(\theta) \in T^*$.

(iii) If $P(x)$ in $D[x]$ is divided by $f(x)$ to give a nonzero remainder $H(x)$, $P(x) = Q(x)f(x) + H(x)$, and there exist $b \in D$, $p(x) \in R[x]\backslash I(\overline{\theta})$ and $a \in \varphi^{-1}(\overline{\theta})$ with $P(x) = bp(x)$ and $f(a) \in Jp(a)$, then we claim that $H(\theta) \in T^*$ if and only if $P(a) > 0$. We have $p(x) = b^{-1}P(x) = q(x)f(x) + r(x)$ with $q(x) = b^{-1}Q(x)$ and $r(x) = b^{-1}H(x)$ both in $R[x]$. We will show that $(b, r(x), a)$ is an O-determiner for $H(x)$. First, $r(x) \notin \mathscr{P}(\overline{\theta})[x]$ since $f(x) \in I(\overline{\theta})$ and $\mathscr{P}(\overline{\theta})[x] \subseteq I(\overline{\theta})$ by (c) of Theorem 5.2.39, and $r(x) + I(\overline{\theta}) = p(x) + I(\overline{\theta}) \neq 0$ in $R[x]/I(\overline{\theta})$. Also, by Theorem 5.2.33, for some $c \in J$,

$$\begin{aligned} r(a) &= p(a) - q(a)f(a) + cf(a) \\ &= p(a) + dp(a) \\ &= (1+d)p(a) = p(a)(1+e) \end{aligned}$$

with $d, e \in J$, and hence $f(a) \in Jr(a)$. Thus, $(b, r(x), a)$ is an O-determiner for $H(x)$ and $H(\theta) \in T^*$ iff $H(a) > 0$. But

$$H(a) = br(a) = bp(a)(1+e) = P(a)(1+e) > 0$$

iff $P(a) > 0$.

(iv) T is D^*-invariant. Suppose $P(\theta) \in T^*$, $(b, p(x), a)$ is an O-determiner for $P(x)$ and $s \in D^*$. Then $(b^s, p^s(x), a^s)$ is an O-determiner for $P^s(x)$ since J and $\mathscr{P}(\overline{\theta})$ are D^*-invariant. So $P^s(\theta) \in T^*$ since $P^s(a^s) = P(a)^s > 0$.

(v) $TT \subseteq T$. Suppose $P_1(\theta)$, $P_2(\theta) \in T^*$ and let $P(x) = P_1(x)P_2(x)$. Let $(b_i, p_i(x), a_i)$ be an O-determinator for $P_i(x)$, $i = 1, 2$ with $b_i > 0$. Then $P(x) = b_1p_1(x)b_2p_2(x) = b_1b_2p_1^{b_2}(x)p_2(x) = bp(x)$ with $b = b_1b_2$ and $p(x) = p_1^{b_2}(x)p_2(x) \in R[x]\backslash I(\overline{\theta})$ by (a) of Theorem 5.2.39. Divide $P(x)$ by $f(x)$ to get $P(x) = Q(x)f(x) + H(x)$. Since $0 \neq P_1(\theta)P_2(\theta) = P(\theta) = H(\theta), H(x) \neq 0$. Since $p(x)$, $p_1^{b_2}(x) \notin I(\overline{\theta})$, $f(a_3) \in Jp(a_3)$ and $f(a_4) \in Jp_1^{b_2}(a_4)$ for some a_3, $a_4 \in \varphi^{-1}(\overline{\theta})$. By (d) of Theorem 5.2.40 we can find $a \in \varphi^{-1}(\overline{\theta})$ with $f(a) \in Jp(a) \cap Jp_1^{b_2}(a) \cap Jp_2(a)$. From (iii) we have that $P_1(\theta)P_2(\theta) = H(\theta) \in T^*$ iff $bp(a) = P(a) > 0$. But $p(a) = (p_1^{b_2}p_2)(a) = p_1^{b_2}(a)p_2(a)(1+c)$ for some $c \in J$ by (e) of Theorem 5.2.34, and $p_1^{b_2}(\theta) \in T^*$ by (iv). Since $(1, p_1^{b_2}(x), a)$ and $(1, p_2(x), a)$ are O-determiners for $p_1^{b_2}(x)$ and $p_2(x)$, respectively, we have $p_1^{b_2}(a) > 0$ and $p_2(a) > 0$ by (i). Thus, $p(a) > 0$ and $P_1(\theta)P_2(\theta) \in T^*$.

(vi) Finally, $D^+ \subseteq T$, and $\overline{\alpha} < \overline{\theta}$ or $\overline{\theta} < \overline{\beta}$ with $\alpha, \beta \in F$ implies $\theta - \alpha \in T^*$ or $\beta - \theta \in T^*$, respectively. To see this take $a \in \varphi^{-1}(\overline{\theta})$. If $0 < s \in D$, then $(s, 1, a)$ is an O-determiner for s. Therefore $s \in T$. Suppose $\overline{\alpha} < \overline{\theta}$. Then $\alpha < a + c$ with $c \in J$, and $(1, x - \alpha, a + c)$ is an O-determiner for $x - \alpha$ since $a + c - \alpha \notin Rf(a+c)$; so $\theta - \alpha \in T^*$. Similarly, $\overline{\theta} < \overline{\beta}$ gives $\beta - \theta \in T^*$. □

All of the cases of a simple extension have now been considered and the totally ordered real division algebra extension of D will now be given.

Theorem 5.2.42. *Each totally ordered division ring can be embedded into a totally ordered division algebra over* $\mathbb{R}$.

Proof. Let $D \subseteq E$ be totally ordered division rings with $Z(D) \subseteq Z(E)$ and let F be a subfield of $Z(D)$. Assume the notation given in Theorem 5.2.24 for both D and E. The various sets determined by E will be distinguished by those determined by D by using E as a subscript. So $R_E = C_F^E(1)$, etc. Then $R = R_E \cap D$, $J = J_E \cap R$, $\mathscr{R} = \mathscr{R}_E \cap R^{W(\xi)}$, $N = N_E \cap \mathscr{R}$ and $P = P_E \cap \mathscr{R}$. So the totally ordered division algebra $\overline{\mathscr{R}}$ (over F) is naturally embedded in the totally ordered division algebra $\overline{\mathscr{R}}_E$ and similarly $\overline{R}$ is embedded in $\overline{R}_E$. Of course, $\hat{F}$ is isomorphic with $\hat{F}_E$ via the embedding $\overline{R} \longrightarrow \overline{R}_E$.

Now let K be a maximal subfield of $Z(D)$ with $F \subseteq K$ and let $E = D(\theta)$ be one of the totally ordered division rings given in Theorem 5.2.26, 5.2.29, or 5.2.41. Then $\varphi_E(\theta) = \overline{\theta}$; that is, $(\theta_\rho)_{\rho<\xi} + N_E = \overline{\theta}$ where $\theta_\rho = \theta$ for each $\rho < \xi$. For, $\overline{\theta} = (\alpha_\rho)_{\rho<\xi} + N_E = F\text{-}\lim \overline{\alpha}_\rho$ where $(\alpha_\rho)_{\rho<\xi}$ is an F-Cauchy net in F. So, for each $0 < \varepsilon \in F$ there exists $\lambda < \xi$ such that $|\overline{\theta} - \overline{\alpha}_\rho| < \overline{\varepsilon}$ and hence $|\theta - \alpha_\rho| < \varepsilon$ (by (5.2.38)) for all $\rho \geq \lambda$. But then $|\varphi_E(\theta) - \overline{\alpha}_\rho| < \overline{\varepsilon}$ and $\varphi_E(\theta) = F\text{-}\lim \overline{\alpha}_\rho = \overline{\theta}$ since F^{+*} is coinitial in $\overline{\mathscr{R}}_E^{*+}$. Note that $\overline{\theta}$ is the image of θ in $\overline{R}_E$ since $\theta \in R_E$; for $|\theta| < \alpha$ for some $\alpha \in F$ because $|\overline{\theta}| < \overline{\alpha}$ for some $\alpha \in F$. In fact, $K(\theta) \subseteq Z(R_E)$ and $K(\theta) \longrightarrow \overline{K(\theta)}$ is an isomorphism. For, $K[\theta] \subseteq Z(R_E)$ and if $P(\theta) \in K[\theta]$ with $P(\theta) > 0$, then $\overline{P(\theta)} > 0$. So $\overline{P(\theta)} > \overline{\varepsilon} > 0$ for some $\varepsilon \in F$, $P(\theta) > \varepsilon$, $P(\theta)^{-1} \in R_E$ and $K(\theta) \subseteq Z(R_E)$.

Now let D be a totally ordered division ring and suppose F is archimedean. Then $\hat{F} = \overline{\mathbb{R}} \cong \mathbb{R}$, $R = C\ (1)$ and $\overline{R} \subseteq \hat{F}$. Let U be a set which contains D and whose cardinality exceeds $|\mathbb{R}||D|$. Let $\mathscr{S}$ be the family of all those subsets E of U such that E is a totally ordered division ring extension of D, $E = D(X)$ for a set X with $F(X) \subseteq Z(R_E)$ and $\overline{X} \subseteq \overline{\mathbb{R}} \backslash \overline{F}$. Partially order $\mathscr{S}$ by: $E_1 \leq E_2$ iff $E_1 = D(X_1)$, $E_2 = D(X_2)$, $X_1 \subseteq X_2$ and E_1 is a totally ordered division subring of E_2. Note that if $E_1 = D(X_1) \leq D(X_2) = E_2$, then $Z(E_1) \subseteq Z(E_2)$. For, E_1 is the division ring of quotients of $D[X_1]$ and hence $u \in Z(E_1)$ iff D is contained in the centralizer of u. In particular, $Z(R_{E_1}) = Z(E_1) \cap R_{E_1} \subseteq Z(E_2) \cap R_{E_2} = Z_1(R_{E_2})$. Let $\{E_\lambda = D(X_\lambda) : \lambda \in \Lambda\}$ be a chain in $\mathscr{S}$ and let $X = \bigcup_\lambda X_\lambda$. Then $E = \bigcup_\lambda E_\lambda = D(X)$ is a totally ordered division ring extension of D, $X \subseteq \bigcup_\lambda Z(E_\lambda) = Z(E)$, $R_E = \bigcup_\lambda R_{E_\lambda}$ and

$$\begin{aligned} F(X) = \bigcup_\lambda F(X_\lambda) \subseteq \bigcup_\lambda Z(R_{E_\lambda}) = \bigcup_\lambda (Z(E) \cap R_\lambda) \\ = Z(E) \cap R_E = Z(R_E). \end{aligned}$$

Thus, $E \in \mathscr{S}$ since $\overline{X} = \bigcup_\lambda \overline{X}_\lambda \subseteq \overline{\mathbb{R}} \backslash \overline{F}$, and clearly $E_\lambda \leq E$ for each $\lambda \in \Lambda$. By Zorn's Lemma $\mathscr{S}$ has a maximal element $E = D(X)$. Let K be a maximal subfield of $Z(R_E)$ with $F(X) \subseteq K$. If $\overline{K} = \mathbb{R}$, then E is a division algebra over $\mathbb{R}$. Suppose $\overline{K} \subset \mathbb{R}$ and take $\overline{\theta} \in \mathbb{R} \backslash \overline{K}$. If $\overline{\theta}$ is transcendental over $\overline{K}$, then, by Theorem 5.2.26 and the preceding paragraph, $E \subseteq E(\theta)$ with $K(\theta) \subseteq Z(R_{E(\theta)})$ and $\overline{K(\theta)} = \overline{K}(\overline{\theta}) \subseteq \mathbb{R}$. Since

$$|E(\theta) - E| = |E| \leq |D||X| < |U| = |U - E|$$

there is a bijection $E(\theta) = E \cup (E(\theta) \backslash E) \longrightarrow E \,\dot{\cup}\, A \subseteq E \cup (U - E) = U$ which is the identity on E and which becomes an isomorphism of totally ordered division rings $E(\theta) \cong E(\theta_1)$ when the structure of $E(\theta)$ is transferred to its image. Since $E < E(\theta_1)$ we have a contradiction to the maximality of E. Thus, $\overline{\theta}$ is algebraic over $\overline{K}$. If $\overline{\theta} \in \overline{R}_E$, then by Theorem 5.2.38 we may assume $\overline{\theta}$ is solid and by Theorem 5.2.41 we get an extension $E(\theta)$ of E which again contradicts the maximality of E. So $\overline{K} = \overline{R}_E$ and now Theorem 5.2.29 produces an extension $E(\theta_1) > E$. This contradiction shows $\overline{K} = \mathbb{R}$. □

Exercises.

1. Let B be the breadth of the pseudo-convergent sequence $(a_\rho)_{\rho<\alpha}$ in the valued field (F,v) and let $a \in F$ be a pseudo-limit of $(a_\rho)_{\rho<\alpha}$.
   (a) Show that B is an R-submodule of F and $c \in F$ is also a pseudo-limit of $(a_\rho)_{\rho<\alpha}$ iff $c - a \in B$.
   (b) Show that 0 is a pseudo-limit of $(a_\rho)_{\rho<\alpha}$ iff $(v(a_\rho))_{\rho<\alpha}$ is strictly decreasing, iff B is the set of pseudo-limits of $(a_\rho)_{\rho<\alpha}$ in F.

2. Let $E = \mathbb{Q}(x)$ have either the lexicographic total order $(1 \ll x \ll x^2 \ll \cdots)$ or the antilexicographic total order $(1 \gg x \gg \cdots)$ and let F be the real closure of E. If v is the natural valuation show that $\Gamma_E \cong \mathbb{Z}$, $v(x) = 1$ and $\Gamma_F \cong \mathbb{Q}$.

3. Show that a valuation on a right Öre domain has a unique extension to its division ring of right quotients.

4. Let $0 < \alpha$ be a limit ordinal and let $(a_\rho)_{\rho<\alpha}$, $(b_\rho)_{\rho<\alpha}$ and $(c_\rho)_{\rho<\alpha}$ be sequences in the valued field (F,v) such that for some $\tau_0 < \alpha$ and for all τ and σ with $\tau_0 \le \tau < \sigma < \alpha$ we have $a_\tau = b_\tau + c_\tau$, $v(a_\sigma) = v(a_\tau)$, $v(c_\sigma) = v(c_\tau)$ and $v(b_\tau) > v(b_\sigma)$. Show that for some $\tau_1 \ge \tau_0$ and for all $\sigma \ge \tau_1$, $v(c_\sigma) > v(b_\sigma)$ and $v(a_\sigma) = v(c_\sigma)$.

5. If $V(K * \Gamma)$ is a formal power series crossed product field show that $V(K * \Gamma)(i) \cong V(K(i) * \Gamma)$ and $V_\alpha(K * \Gamma)(i) \cong V_\alpha(K(i) * \Gamma)$ for any ordinal α.

6. Show that each field with the trivial valuation is maximally complete.

7. For $j = 1,2$ let F_j be a subfield of the totally ordered field L_j and suppose that L_j is an immediate extension of F_j, where all valuations are natural. Show that a value preserving isomorphism $\psi : L_1 \longrightarrow L_2$ which extends an isotone isomorphism $\varphi : F_1 \longrightarrow F_2$ is itself isotone.

8. Let $E \subseteq D$ be division rings and suppose v is a valuation on D. An element $a \in D$ is transcendental over E if the elements $1, a, a^2, \ldots, a^n, \cdots$ are independent in the vector spaces ${}_D E$ and E_D. Show that v is trivial iff its restriction to E is trivial and each transcendental element over E is a unit.

9. Let C be a maximal subfield of the valuation ring $R = C(1)$ of the totally ordered field F and let J be the maximal ideal of R. Show each of the following:
   (a) $\overline{R} = R/J$ is algebraic over $\overline{C}$.
   (b) If F is algebraic over its subfield K, then F is K-archimedean.
   (c) C is algebraically closed in F; that is, each element of F that is algebraic over C is in C.
   (d) If F is real closed, then $\overline{R}$ is real closed and $\overline{C} = \overline{R}$.
   (e) If F is henselian and $\overline{R}$ is real closed, then $\overline{C} = \overline{R}$.

10. Let $(a_\rho)_{\rho<\alpha}$ be a pseudo-convergent sequence in the valued field (F,v). Show the equivalence of the following statements.
    (a) $(a_\rho)_{\rho<\alpha}$ is transcendental.

(b) If (E,w) is an extension of (F,v) and E is algebraic over F, then, for each $a \in E$, the sequence $(w(a_\rho - a))_\rho$ is eventually constant.
(c) For every $f(x) \in F[x]\backslash F$ and each $\lambda < \alpha$, if $(f(a_\rho))_{\lambda \leq \rho < \alpha}$ is pseudo-convergent, then it has no pseudo-limit in F.
(d) If a is a pseudo-limit of $(a_\rho)_{\rho<\alpha}$ in an extension (E,w), then a is transcendental over F.

11. (a) Show that the pseudo-convergent sequence $(a_\rho)_{\rho<\alpha}$ in (F,v) is algebraic iff there is a polynomial $f(x) \in F[x]$ of positive degree and an ordinal $\lambda < \alpha$ such that $(f(a_\rho))_{\lambda \leq \rho}$ has a pseudo-limit in F.
(b) Suppose $(a_\rho)_{\rho<\alpha}$ is algebraic and $f(x) \in F[x]$ is a polynomial of least degree such that $v(f(a_\rho))_{\rho<\alpha}$ is eventually strictly decreasing. Show that $n = \deg f(x)$ iff for each $\lambda < \alpha$ and each $g(x) \in F[x]\backslash F$, if $\deg g(x) < n$, then $(g(a_\rho))_{\lambda \leq \rho < \alpha}$ has no pseudo-limit in F.

12. Here is an example of a totally ordered field F for which there is no field embedding $F \longrightarrow V(\Gamma_F, R_F/J_F)$ (see Theorem 5.2.20). The torsion divisible group $\mathbb{Q}/\mathbb{Z}$ has the decomposition $\mathbb{Q}/\mathbb{Z} = \boxplus_p \mathbb{Z}(p^\infty)$ where the direct sum is over all primes p and $\mathbb{Z}(p^\infty)$, the subgroup of $\mathbb{Q}/\mathbb{Z}$ generated by $\{\frac{1}{p^n} + \mathbb{Z} : n \in \mathbb{N}\}$, is the set of all elements in $\mathbb{Q}/\mathbb{Z}$ of pth power order. Each element of $\mathbb{Z}(p^\infty)$ is a coset $\frac{s}{p^n} + \mathbb{Z}$ with $0 \leq s < p^n$ and uniquely so if p doesn't divide s. Fix the prime p and for $a \in \mathbb{Q}$ let $a^* = \frac{s}{p^n} \in \mathbb{Q} \cap [0,1)$ where $\frac{s}{p^n} + \mathbb{Z}$ is the pth component of $a + \mathbb{Z}$ in $\mathbb{Q}/\mathbb{Z}$, and let $f_a = 2^{a^*} x^a \in V((\mathbb{Q},+),\mathbb{R})$. Let R be the $\mathbb{Q}$-subalgebra of $V(\mathbb{Q},\mathbb{R})$ generated by $\{f_a : a \in \mathbb{Q}\}$ and let F be its field of quotients within $V(\mathbb{Q},\mathbb{R})$. Verify each of the following.

(a) If $a_1, a_2 \in \mathbb{Q}$, then $(a_1+a_2)^* = a_1^* + a_2^*$ or $(a_1+a_2)^* = a_1^* + a_2^* - 1$ depending on whether $a_1^* + a_2^* < 1$ or $a_1^* + a_2^* \geq 1$.
(b) $f_{a_1} f_{a_2} = f_{a_1+a_2}$ or $f_{a_1} f_{a_2} = 2 f_{a_1+a_2}$ depending on whether $a_1^* + a_2^* < 1$ or $a_1^* + a_2^* \geq 1$.
(c) The value group of F with respect to the natural valuation is $\mathbb{Q}$ and the valuation ring of F is the field $\mathbb{Q}$.
(d) $(f_{\frac{1}{p}})^p = 2 f_1$.
(e) If q is a prime different from p and $n \in \mathbb{N}$ then $(f_{\frac{1}{q^n}})^{q^n} = f_1$.
(f) Show that $V(\mathbb{Q},\mathbb{Q})$ does not contain a nonzero element g which satisfies the equations in (e) and (d) (that is, an element g which has a q^nth root for each $n \in \mathbb{N}$ and $2g$ has a pth root), and hence the field F cannot be embedded in $V(\mathbb{Q},\mathbb{Q})$. (Consider the leading coefficient of g and of the pth root of $2g$.)

13. Here is an example of a commutative totally ordered domain R with value semigroup Δ for which there is no value preserving ring monomorphism of R into $V(\Delta,\mathbb{R})$ (see Theorem 5.2.20.) Let $S = \mathbb{R}[x]$ have the antilexicographic order $1 \gg x \gg \cdots \gg x^n \gg \cdots$ and let R be the $\mathbb{R}$-subalgebra of S generated by x^4 and $x^6 + x^7$. The valuation v is the natural valuation.

(a) Show that the value group of $\mathbb{R}(x)$ is $(\mathbb{Z}, -\mathbb{Z}^+)$.

(b) Show that if $k,n \in \mathbb{Z}^+$ with $0 \le k \le n$ and $(n,k) \notin \{(0,0),(1,1)\}$, then there are integers $p,q,h \in \mathbb{Z}^+$ with $n6+k = p4+q6+h13$. (Use induction on n.)
(c) Show that the value semigroup $\Delta = v(R)$ of R is generated by 4, 6, and 13 : $\Delta = (\mathbb{Z}^+4+\mathbb{Z}^+6+\mathbb{Z}^+13)\backslash\{0\}$.
(d) Show that $v((x^4)^3-(x^6+x^7)^2) = v(x^{13})$ but there are no elements $f,g \in V(\Delta,\mathbb{R}) = \mathbb{R}[[x^4,x^6,x^{13}]]$ with $v(f)=4$, $v(g)=6$ and $v(f^3-g^2)=13$. (A nonzero element of $V(\Delta,\mathbb{R})$ has only even degree terms up to x^{13}.)

14. Suppose K is a subdivision ring of the valued division ring (D,v).

(a) Let $\{a_1+J_D,\ldots,a_r+J_D\}$ be a left R_K/J_K-independent subset of R_D/J_D. Show that $v(\Sigma_i\alpha_ia_i) = \bigvee_i v(\alpha_i)$ for all $\alpha_1,\ldots,\alpha_r \in K$ and $\{a_1,\ldots,a_r\}$ is an independent set in ${}_KD$. (If $v(\alpha_1) = \bigvee_i v(\alpha_i) \neq -\infty$, then $\Sigma_i\alpha_1^{-1}\alpha_ia_i$ is a unit of R_D.)
(b) If $\Gamma_K + v(b_1),\ldots,\Gamma_K+v(b_s)$ are distinct cosets show that $v(\Sigma_j\alpha_jb_j) = \bigvee_j v(\alpha_jb_j)$ provided $\{v(\alpha_1),\ldots,v(\alpha_s)\} \subseteq \Gamma_K \cup \{-\infty\}$ and $\{b_1,\ldots,b_s\}$ is an independent set in ${}_KD$.
(c) If a_i and b_j are as in (a) and (b), respectively, show that $v(\Sigma_{i,j}\alpha_{ij}a_ib_j) = \bigvee_{i,j} v(\alpha_{ij}b_j)$ for all $\alpha_{ij} \in K$ and $[R_D/J_D : R_K/J_K][\Gamma_D : \Gamma_K] \le [D:K]$ where the vector spaces are left vector spaces.

15. Let R be a subring of the division ring D.

(a) Show that R is a right Öre domain if and only if $\{ab^{-1} : a,b \in R, b \neq 0\}$ is a subring of D.
(b) Suppose R is right Öre. If $X \subseteq Z(D)$ show that $R[X]$ is right Öre and $Q(R[X]) = Q(R)(X)$.

16. Let M be a nonzero totally ordered vector lattice over the totally ordered division ring D and let $\hat{D}$ be the Cauchy completion of D. Show that the following statements are equivalent.

(a) M can be embedded in $\hat{D}$.
(b) For each $0 \neq x_0 \in M$, if $x<y$ in M, then $x < \alpha x_0 < y$ for some $\alpha \in D$.
(c) There exists some $0 \neq x_0 \in M$ with the property in (b).
(For (c) $\Rightarrow$ (a) take ξ and $(\varepsilon_\rho)_{\rho<\xi}$ as in Theorem 5.2.24. Show that for $x \in M$ there exists an ξ-net $(\alpha_\rho)_{\rho<\xi}$ in D with $|\alpha_\rho x_0 - x| < \varepsilon_\rho x_0$ for every ρ, and if, for some $\alpha \in D$ and some ξ-net $(\beta_\rho)_{\rho<\xi}$ in D, $|\beta_\rho x_0 - x| < \alpha\varepsilon_\rho x_0$ for every ρ, then $(\beta_\rho)_{\rho<\xi}$ is Cauchy.)

17. Let ${}_RM$ be a totally ordered left f-module over the totally ordered left Öre domain R. Suppose $\hat{D}$ is the Cauchy completion of the division ring D of left quotients of R. Show that ${}_RM$ can be embedded in ${}_R\hat{D}$ if and only if for any $0 \neq x_0$ in M and any $x<y$ in M, $ax < bx_0 < ay$ for some $a,b \in R$ with $a>0$.

18. Let $M = \mathbb{Z}[x,y]$ be the polynomial ring over $\mathbb{Z}$ in the commuting indeterminates x and y and let $S = \{x^iy^j\}$ be the free abelian monoid generated by x and y. S is a totally ordered monoid with the partial order: $x^iy^j < x^py^q$ if $i+j<p+q$ or $i+j=p+q$ and $i<p$, and the semigroup ring $M = \mathbb{Z}[S]$ is a totally ordered

ring when it is given the lexicographic Hahn order: $\alpha_1 s_1 + \cdots + \alpha_n s_n > 0$ if $\alpha_n > 0$ where $s_1 < s_2 \cdots < s_n$ in S and $\alpha_j \in \mathbb{Z}$. Let $\hat{F}$ be the Cauchy completion of the totally ordered quotient field F of the totally ordered subring $R = \mathbb{Z}[x]$ of M. Show that ${}_RM$ cannot be embedded in ${}_R\hat{F}$. ($y^2 < xy$ and no element of Fx is between y^2 and xy; use Exercise 16.)

19. Let $\hat{D}$ be the Cauchy completion of the totally ordered division ring D with center F. Show that $a \in Z(\hat{D})$ if and only if a centralizes D. If F^{+*} is coinitial in D^{+*} show that $Z(\hat{D})$ is Cauchy complete and $\hat{F} \subseteq Z(\hat{D})$.

Notes. Most of the general theory of valuations and pseudo-convergence presented here comes from Kaplansky [K1] as well as from Schilling [SC]; Prieß-Crampe [PC2] is a very good reference for this section. Original references are Krull [KR] and Ostrowski [O]. Other sources are MacLane [MAC2] for Theorem 5.2.12, Alling [AL2] for Theorem 5.2.13, Hartmann [HAR] for the proof of Theorem 5.2.14 and MacLane [MAC1] for Theorem 5.2.15. The Hahn embedding theorem given in Theorem 5.2.20 originated in Hahn [H]. The first complete proof, with the coefficients of the power series coming from $\mathbb{R}$, appeared in Conrad and Dauns [CD] but with a weak point. The first fairly elementary proof for a real closed field is due to Prieß-Crampe [PC1] and the proof given for an arbitrary totally ordered field comes from Prieß-Crampe [PC2, p. 124]; in both cases no reference is made to a $\mathbb{Z}$-independent set in the value group. Exercise 12 comes from Prieß-Crampe [PC2, p. 124] where it is attributed to Panek [P] and Greither [GR] and Exercise 13 is due to Bergman [BER2]. The embedding of a totally ordered division ring into a totally ordered division algebra over the reals is due to Neumann [N]; we have used some of the refinements given by Jaeger [JAE]. The extension of Hölder's theorem to modules over a commutative totally ordered domain given in Exercises 16, 17, and 18 is due to Viswanathan [V2]. We have simplified his presentation and extended it to left Öre domains by using Cauchy nets instead of Cauchy filters to construct the Cauchy completion.

5.3 Lattice-ordered Fields

Examples of lattice-ordered fields that are not totally ordered have previously appeared in several places in the text. Additional examples will be provided here with an ingenious construction of archimedean lattice orders on each subfield of $\mathbb{R}$ with the exception of $\mathbb{Q}$. There is an interesting and surprising converse in that each archimedean ℓ-field has a largest subfield which can be totally ordered and over which it is an ℓ-algebra, and if it is algebraic over this subfield then it is a subfield of $\mathbb{R}$. Some topological algebra will be used to verify this converse.

Having successfully embedded a totally ordered field into a closely connected totally ordered power series field we will first turn our attention to a similar embedding of an ℓ-field into a power series ℓ-field of the type given in Theorem 3.5.13. To do this requires that the basic elements form a multiplicatively closed subset of the field.

Let $S(R)$, or just S, denote the set of positive special elements of the ℓ-ring R (see Section 2.5) and let $B(R)$, or B, denote its subset of basic elements (see Exercises 2.4.11 – 2.4.13). B is a subset of S by Exercise 2.4.11 and Theorem 2.5.3. Recall that $d(R)$ denotes the set of d-elements in R and $\mathscr{U}(X)$ denotes the group of units of the monoid X. The ℓ-ring R is called a *D-ring* if $S \neq \emptyset$ and $S \subseteq d(R)$. If R is a unital ℓ-ring, then $d(R) \cap \mathscr{U}(R) = \mathscr{U}(R^+)$ by Theorem 3.1.3. Since multiplication by $a \in \mathscr{U}(R^+)$ is an ℓ-automorphism of the additive ℓ-group of R, $\mathscr{U}(R^+) \subseteq S$ or $\mathscr{U}(R^+) \subseteq B$ if and only if 1 is special or basic, respectively, or $1 \not\geq 0$. So, if $\emptyset \neq S \subseteq \mathscr{U}(R^+)$, then R is a D-ring and necessarily $S = \mathscr{U}(R^+)$. The generalized semigroup ring $A[G] = \Sigma(G,A)$ of the trivially ordered pops G over the totally ordered ring A is a D-ring and so is the power series division ring given in Theorem 3.5.13. These examples illustrate the fact, and we will see more of this in the next theorem, that frequently S is multiplicatively closed in a D-ring. The converse doesn't hold, however, as the example in Exercise 2 shows.

Theorem 5.3.1. *Let R be an ℓ-domain.*

(a) If $d(R)^*$ is a right Öre set in R, then R is a D-ring if and only if $d(R)^* = S (= B)$. In this case the right quotient ring $Q = R_{d(R)^*}$ is also a D-ℓ-domain and $S(Q) = \{as^{-1} : a, s \in S(R)\} = d(Q)^*$.

(b) Suppose R is unital, $d(R)^* \subseteq \mathscr{U}(R)$ and R has a special element. Then the following statements are equivalent.

(i) R is a D-ring.

(ii) $S = d(R)^* = \mathscr{U}(R^+)$.

(iii) S is a multiplicative subgroup of R.

(iv) $S \subseteq \mathscr{U}(R^+)$.

Proof. (a) Since R is an ℓ-domain and $\ell(a) = \ell_\ell(a)$ for $a \in d(R)$, $d(R)^*$ consists of regular elements of R. By Theorem 4.3.1 Q is an ℓ-ring extension of R and $d(R)^* \subseteq \mathscr{U}(Q^+) \subseteq d(Q)$. Also, Q is an ℓ-domain since if $as^{-1}bt^{-1} = 0$ with $a, b \in R^+$, then $s^{-1}b = cu^{-1}$, $acu^{-1}t^{-1} = 0$, $a = 0$ or $c = 0$, and $as^{-1} = 0$ or $bt^{-1} = 0$. Since $1 \in Q^+$, $C(1)$ is a totally ordered domain. Thus, if R is a D-ring, then

$$d(R)^* \subseteq \mathscr{U}(Q^+) \cap R \subseteq B(Q) \cap R \subseteq B(R) \subseteq S(R) \subseteq d(R)^* \qquad (5.3.1)$$

and $d(R)^* = S(R) = B(R)$. The converse is trivial. If R is a D-ring and $as^{-1} \in S(Q)$, then $a = as^{-1}s \in S(Q)$ and $a \in S(R)$ by Exercise 2.5.4; so $a \in d(R)$, $as^{-1} \in \mathscr{U}(Q^+)$, and $as^{-1} \in d(Q)$. Thus, Q is a D-ring. Let $G = \{as^{-1} : a, s \in d(R)^*\}$. We will verify the nontrivial inclusions in the sequence

$$S(Q) \subseteq G \subseteq \mathscr{U}(Q^+) \subseteq d(Q)^* \subseteq G \subseteq S(Q). \qquad (5.3.2)$$

Suppose $as^{-1} \in S(Q)$. Then $a \in S(Q)$ and $C^Q(a) = \operatorname{lex} N$ with $N \subset C^Q(a)$ by Theorem 2.5..8, where $C^Q(a)$ denotes the convex ℓ-subgroup of Q generated by a. But then $C^R(a) = \text{lex}\,(N \cap R)$, $a \in S(R)$, $as^{-1} \in G$, and we have the first inclusion. Suppose $as^{-1} \in d(Q)^*$. Then $sas^{-1} \in d(Q)^*$, $a \in d(Q)^*$, $a \in d(R)^*$ and $d(Q)^* \subseteq G$. The last inclusion comes from $G \subseteq \mathscr{U}(Q^+) \subseteq B(Q) \subseteq S(Q)$.

(b) That (i) implies (ii) follows from (5.3.1) (or (5.3.2)) since now $Q = R$, and the remaining circular implications are obvious. □

Suppose R is a D-ℓ-domain in which $d(R)^*$ is a right Öre set. The additive subgroup of R generated by $S = B$ is the largest convex ℓ-subring of R which is a finite valued D-domain (Exercise 1). We are interested in the case where this subring is R. Assume R is finite valued. Then its additive ℓ-group is a direct sum of totally ordered groups by Exercise 2.4.11 or Exercise 2.5.22. Moreover, if $\Gamma = \Gamma(R)$ is the value set of R, then the mapping $v = v_R : S \longrightarrow \Gamma$ which sends a special element to its value is onto. This follows from Theorem 2.5.11, but it is easier here because of the direct sum decomposition of R, and it follows from Exercise 2.5.13. The function v has the following properties.

$$v(s) \leq v(t) \text{ if and only if } s \in C(t). \tag{5.3.3}$$

$$v(s) = v(t) \text{ if and only if } C(s) = C(t). \tag{5.3.4}$$

$$v(s) < v(t) \text{ if and only if } s \ll t. \tag{5.3.5}$$

$$\text{If } v(s) = v(t) \text{ and } a \in S, \text{ then } v(as) = v(at) \text{ and } v(sa) = v(ta). \tag{5.3.6}$$

Clearly, (5.3.3) is an easy consequence of Exercise 2.5.13 and it implies (5.3.4) through (5.3.6). These properties allow us to make Γ into a strict rooted partially ordered semigroup with addition given by $v(s) + v(t) = v(st)$. From Theorem 2.5.11 we may deduce that the right quotient ring $Q = R_{d(R)^*}$ of R is also finite valued. For, if $0 \neq as^{-1} \in Q$, then $a = a_1 + \cdots + a_n$ where each $|a_i| = \pm a_i$ is basic and $|a_i| \wedge |a_j| = 0$ for $i \neq j$. So $as^{-1} = a_1 s^{-1} + \cdots + a_n s^{-1}$, each $|a_i s^{-1}|$ is a basic element of Q by the previous theorem, and $|a_i s^{-1}| \wedge (a_j s^{-1}| = (|a_i| \wedge |a_j|)s^{-1} = 0$ if $i \neq j$. Since $S(R) \subseteq S(Q)$ the mapping $\Gamma(R) \longrightarrow \Gamma(Q)$ given by $v_R(s) \mapsto v_Q(s)$ is an embedding by (5.3.4); in fact, since $S(Q)$ is the right quotient group of $S(R)$ by (5.3.1) and (5.3.2), $\Gamma(Q)$ is the right quotient po-group of the po-semigroup $\Gamma(R)$. Recall that R is finite valued whenever it is finitely rooted; that is, the number of roots in $\Gamma(R)$ is finite or, equivalently, the Boolean algebra of polars of R is finite (see Exercise 2.5.11)

Theorem 5.3.2. *Let R be a finite valued D-ℓ-domain whose set of nonzero d-elements is a right Öre set in R. If the value group $\Gamma(Q)$ of the right quotient ring $Q = R_{d(R)^*}$ of R is torsion-free and is locally finite modulo its maximal totally ordered subgroup, then R is a domain and the lattice order of R can be extended to a total order of R.*

Proof. Since Q inherits these properties from R we will assume $R = Q$. By Theorem 3.5.12 the partial order $\leq$ of $\Gamma = \Gamma(R)$ can be extended to a total order $\leq_1$ of Γ. As an ℓ-group, or even as an $F(R)$-f-bimodule, $R = \oplus_j G_j$ where each G_j is a totally ordered submodule, and if $0 \neq a \in G_j$, then its value $R_{v(|a|)}$ in R is the convex ℓ-subgroup $\oplus_{k \neq j} G_k \oplus M$ where M is the maximal convex subgroup of $C(a)$.

If $0 \neq a \in R$, then $a = a_1 + \cdots + a_n$ where the a_i's come from different components and $v(|a_1|) >_1 \cdots >_1 v(|a_n|)$. Let $T = \{a \in R : a_1 > 0\}$. Clearly, $R^+ \subseteq T \cup \{0\}$ and we claim that $T \cup \{0\}$ is a total order of R. Take $a, b \in T$ with a as previously given and write $b = b_1 + \cdots + b_m \in T$ with $v(|b_1|) >_1 \cdots >_1 v(|b_m|)$, and assume, by way of symmetry, that $v(|a_1|) \geq_1 v(|b_1|)$. If a_λ and b_μ come from the same G_j, then $a_\lambda + b_\mu = 0$ or $v(|a_\lambda + b_\mu|) \leq v(|a_\lambda|) \vee v(|b_\mu|) <_1 v(a_1)$ provided $\lambda > 1$ and $\mu > 1$. These inequalities also hold when $\lambda > 1$ and $\mu = 1$. For, if $v(|a_\lambda|) \vee v(b_1) = v(a_1)$, then $v(b_1) = v(a_1)$ and we have the contradiction that a_1 and a_λ are in the same G_j. So if $a + b = c_1 + \cdots + c_t$ with $v(|c_1|) >_1 \cdots >_1 v(|c_t|)$ and the c's are in different components, then $c_1 = a_1$ or $c_1 = a_1 + b_1$ or $c_1 = a_1 + b_\mu$ with $\mu \geq 2$. In this last case $v(a_1) \geq_1 v(b_1) >_1 v(|b_\mu|)$ gives $v(a_1) > v(|b_\mu|)$ and hence $c_1 = a_1 + b_\mu > 0$. So $a + b \in T$ in all cases. Also, $v(a_1 b_1) = v(a_1) + v(b_1) >_1 v(|a_\lambda|) + v(|b_\mu|) = v(|a_\lambda b_\mu|)$ if $(\lambda, \mu) \neq (1,1)$. Consequently, if d_1 is the sum of those $a_\lambda b_\mu$ in the same G_j as $a_1 b_1$, then $d_1 = a_1 b_1 + \sum a_\lambda b_\mu$ with $a_1 b_1 \gg a_\lambda b_\mu$, $a_1 b_1 \gg \sum a_\lambda b_\mu$, and $d_1 > 0$; and if d_k is a nonzero sum of those $a_\lambda b_\mu$ in another summand, then $v(|d_k|) \leq \bigvee v(|a_\lambda b_\mu|) <_1 v(a_1 b_1) = v(d_1)$. Since $ab = d_1 + d_2 + \cdots + d_p$ is the canonical decomposition of ab, we have $ab \in T$. Since $T \cap -T = \emptyset$ and $T \cup -T = R^*$ are obvious $T \cup \{0\}$ is a total order of R. This also shows that if $a \neq 0$ and $b \neq 0$, then $ab \neq 0$. □

We will now extend the Hahn embedding theorem for totally ordered fields (Theorem 5.2.20) to a class of ℓ-fields.

Theorem 5.3.3. *Let R be a commutative finitely rooted D-ℓ-domain with n roots, Γ the value group of its quotient ℓ-ring $Q = R_{d(R)^*}$, and C the real closure of a maximal archimedean subfield A of the subfield $F(Q)$ of f-elements of Q. If Γ is torsion-free, then Q is an n-dimensional field extension of $F(Q)$ and there is a value preserving ℓ-monomorphism of R into the power series ℓ-field $V(\Gamma, C)$.*

Proof. The case $n = 1$ is Theorem 5.2.20; so assume $n \geq 2$. Since Q also has n roots (see Exercise 2.5.11) and the inclusion $R \longrightarrow Q$ is value preserving we may assume $R = Q$. Let $F = F(Q)$ be the totally ordered subring of f-elements of Q. If $0 < a \in F$, then $a^{-1} \in Q^+$ since a is special (Theorem 5.3.1) and $a^{-1} \in F$ since multiplication by a^{-1} is the inverse of the polar preserving map $y \mapsto ay$. As we noted after Theorem 5.3.1, Q has the decomposition

$$Q = G_1 \oplus \cdots \oplus G_n \tag{5.3.7}$$

into totally ordered subgroups and, in fact, each G_j is a vector lattice over F. Let $1 \in G_1$. Then $F = G_1$. For, if $0 < a \in F$, then $a \in G_1$ since it is a special element comparable to $1 \in G_1$, and if $0 < a \in G_1$, then $a \in F$ since $1 + a$ is a d-element and hence a is an f-element. Let $H = v(F^{+*})$. Then $H = \Gamma^+ \cup -\Gamma^+$ is the largest totally ordered subgroup of the rooted po-group $\Gamma = v(S)$. For, by (5.3.3), $v(a)$ is comparable to $v(1) = 0$ if and only if a is comparable to 1. So the roots of Γ are the cosets of H:

$$\Gamma = \gamma_1 + H \cup \cdots \cup \gamma_n + H$$

and we will take $\gamma_1 = 0$. Note that $G_j^{+*} = v^{-1}(\gamma_j + H)$. Since Γ / H is a finite abelian group it is the direct sum of nonzero cyclic groups

$$\Gamma/H = \mathbb{Z}(\delta_1 + H) \boxplus \cdots \boxplus \mathbb{Z}(\delta_p + H)$$

where the order of $\delta_i + H$ is d_i and d_{i+1} divides d_i for $i = 1, \ldots, p-1$. The sum $\sum_i \mathbb{Z}\delta_i$ is a direct sum in Γ. For if $\sum_i m_i\delta_i = 0$ with $m_1, \ldots, m_p$ not all zero and relatively prime, then $\sum_i m_i(\delta_i + H) = 0$ and hence we have the contradiction that d_p is a divisor of each m_i. Take $z_i \in S$ with $v(z_i) = \delta_i$. For $1 \le j \le n$ write

$$\gamma_j + H = \sum_{i=1}^{p} n_{ji}(\delta_i + H), \text{ with } 0 \le n_{ji} < d_i. \tag{5.3.8}$$

Then $G_j = F z_1^{n_{j1}} \cdots z_p^{n_{jp}}$. For, $z_1^{n_{j1}} \cdots z_p^{n_{jp}} \in G_j$ since its value is in $\gamma_j + H$, and if $0 < b \in G_j$, then $v(b) = v(z_1^{n_{j1}} \cdots z_p^{n_{jp}}) + v(a) = v(az_1^{n_{j1}} \cdots z_p^{n_{jp}})$ for some $0 < a \in F$; so $b = caz_1^{n_{j1}} \cdots z_p^{n_{jp}}$ for some $0 < c \in F$. This shows that Q is n-dimensional over F and hence is a field since it is a domain by Theorem 5.3.2. Since $d_1\delta_1, \ldots, d_p\delta_p$ are independent elements of $H = v(F)$ there is, according to Theorem 5.2.20, a value preserving isomorphism of totally ordered fields

$$\varphi : F \longrightarrow V(H,C) \subseteq V(\Gamma,C)$$

with $\varphi(az_i^{d_i}) = ax^{d_i\delta_i}$ for $a \in A$. By Theorem 3.5.13 $V(\Gamma,C)$ is an ℓ-field and, clearly, $V(H,C)$ is the maximal totally ordered subfield of the ℓ-field $V(\Gamma,C)$. Let $\rho_j = \sum_i n_{ji}\delta_i$ for $j = 1, \ldots, n$. Then $\rho_i, \ldots, \rho_n$ are distinct elements of Γ, and the $V(H,C)$-vector lattice decomposition of $V(\Gamma,C)$ corresponding to (5.3.7) is

$$V(\Gamma,C) = V(H,C)x^{\rho_1} \oplus \cdots \oplus V(H,C)x^{\rho_n}$$

since $\gamma_j - \rho_j \in H$ and hence $V(H,C)x^{\gamma_j} = V(H,C)x^{\rho_j}$ for each j. The function $\psi : Q \longrightarrow V(\Gamma,C)$ induced on the direct sum (5.3.7) by

$$\psi(az_1^{n_{j1}} \cdots z_p^{n_{jp}}) = \varphi(a)x^{\rho_j}, \qquad a \in F$$

is a value preserving ℓ-monomorphism of the F-vector lattice Q into the $V(H,C)$-vector lattice $V(\Gamma,C)$ which extends φ and satisfies $\psi(au) = \varphi(a)\psi(u)$ for $a \in F$ and $u \in Q$. To verify that ψ preserves multiplication it suffices to show that it preserves products of special elements since the special elements span the group Q.

For each $j = 1, \ldots, p$ we will write $u_j = z_1^{n_{j1}} \cdots z_p^{n_{jp}}$. Take $0 < s \in G_j$ and $0 < t \in G_k$. Then $s = au_j$ and $t = bu_k$ where $0 < a, b \in F$. Suppose that $st \in G_\ell$. Then $\gamma_j + \gamma_k + H = v(st) + H = \gamma_\ell + H$. So

$$\sum_{i=1}^{p}(n_{ji} + n_{ki})(\delta_i + H) = \sum_{i=1}^{p} n_{\ell i}(\delta_i + H)$$

and there are integers q_i with $n_{ji} + n_{ki} = d_iq_i + n_{\ell i}$ for $1 \le i \le p$. Thus,

$$u_ju_k = z_1^{d_1q_1} \cdots z_p^{d_pq_p} u_\ell \text{ with } z_i^{d_iq_i} \in F$$

and

$$\rho_j + \rho_k = \sum_i d_i q_i \delta_i + \rho_\ell.$$

So,

$$\begin{aligned} \psi(st) &= \psi(abu_j u_k) = \varphi(abz_1^{d_1 q_1} \cdots z_p^{d_p \delta_p}) x^{\rho_\ell} \\ &= \varphi(a)\varphi(b) x^{d_1 q_1 \delta_1} \cdots x^{d_p q_p \delta_p} x^{\rho_\ell} \\ &= \varphi(a)\varphi(b) x^{\rho_j + \rho_k} = \psi(s)\psi(t). \end{aligned}$$ □

In any finite valued commutative D-ℓ-domain R, $d(R) \subseteq f(R) \cup f(R)^{\perp +}$ and equality holds when R has at most two roots. For ℓ-fields in general we will soon see that this equality actually forces R to be a D-domain.

By relaxing the restrictions on Γ and C in Theorem 5.3.3 one can embed a D-field with a suitable basis into a commutative crossed product power series ℓ-algebra (see Exercise 3.5.13). Let Δ be a rooted pops. A lattice ideal $\mathscr{S}$ of the power set $\mathscr{P}(\Delta)$ of Δ is called a *supporting subset* of $\mathscr{P}(\Delta)$ if for all $\Gamma, \Gamma_1, \Gamma_2$ in $\mathscr{S}$ and $\gamma \in \Delta$:

(P_1) Γ is noetherian;
(P_2) $\{\gamma\} \in \mathscr{S}$;
(P_3) $\Gamma_1 + \Gamma_2 \in \mathscr{S}$;
(P_4) $\{(\alpha, \beta) \in \Gamma_1 \times \Gamma_2 : \alpha + \beta = \gamma\}$ is finite.

When Δ is a group (P_4) holds for any γ provided it holds for $\gamma = 0$. If $\mathscr{S}$ is a supporting subset of $\mathscr{P}(\Delta)$ and A is a totally ordered unital domain, then the restricted formal power series crossed product

$$V_{\mathscr{S}}(A * \Delta) = \{v \in V(A * \Delta) : \text{ supp } v \in \mathscr{S}\}$$

is an ℓ-subring of the Hahn product $V(A * \Delta)$ which contains $\Sigma(A * \Delta)$ in the sense that it is an A-A-f-subbimodule of $V(A * \Delta)$ and it is an ℓ-ring with the crossed product multiplication. Moreover, Exercises 3.5.13–3.5.21 hold for $V_{\mathscr{S}}(A * \Delta)$. One example of a supporting subset arises from a rooted f-pops $(\Delta, \leq)$ and a partial order $\leq_1$ of Δ such that $\Delta_1 = (\Delta, \leq_1)$ is a rooted pops and $\leq_1 \subseteq \leq$. Then $\mathscr{S} = \{\Gamma \subseteq \Delta : (\Gamma, \leq) \text{ is noetherian}\}$ is a supporting subset of $\mathscr{P}(\Delta_1)$ and $V_{\mathscr{S}}(A * \Delta_1) = V(A * \Delta)$ as rings.

Let K be an ℓ-unital ℓ-domain. K is *f-embeddable* if there is an $F(K)$-embedding $\varphi : K \longrightarrow V_{\mathscr{S}}(F(K) * \Delta)$ of K into a formal power series crossed product ℓ-ring with trivial action for some rooted po-group Δ and some supporting subset $\mathscr{S}$ of $\mathscr{P}(\Delta)$ such that $\Sigma(F(K) * \Delta) \subseteq \varphi(K)$. For a subset B of K^{+*} and $u \in K$ let $B(u) = \{b \in B : b \wedge |u| > 0\}$. Since $(|b| \wedge |u|)^{\perp\perp} = b^{\perp\perp} \cap u^{\perp\perp}$ a basic element b in B is in $B(u)$ iff $b^{\perp\perp} \subseteq u^{\perp\perp}$. B is called *enabling* if $B(uv) \subseteq (B(u)B(v))^{\perp\perp}$ and $\{(a, c) \in B(u) \times B(v) : ac \wedge 1 > 0\}$ is finite for all $u, v \in K^{+*}$.

Theorem 5.3.4. *Let $B(K)$ be the set of basic elements in the ℓ-unital division ℓ-ring K and let $F = F(K)$. Then (a) implies (b) and (b) implies (c). If K has a central basis,* *then (c) implies (a) and K is f-embeddable into* $V_{\mathscr{S}}(F * B(K)/F^{+*})$*; moreover, if* F^{+*} *is a direct factor of* $B(K)$*, then K can be f-embedded into* $V_{\mathscr{S}}(B(K)/F^{+*}, F)$.

(a) *K is f-embeddable.*

(b) $B(K) \subseteq d(K)$, $K = b^{\perp} \oplus b^{\perp\perp}$ *for each* $b \in B(K)$ *and K has an enabling basis.*

(c) *K is a D-ring, the set of special elements in* $\Gamma(K)$ *is plenary and K has a maximal disjoint subset of special elements which is enabling.*

Proof. (a) $\Rightarrow$ (b). Suppose $\varphi : K \longrightarrow V_{\mathscr{S}}(F * \Delta)$ is an F-embedding of ℓ-rings with $\Sigma(F(K) * \Delta) \subseteq \varphi(K)$. Since $F = F(\varphi(K)) = \varphi(K) \cap V_{\mathscr{S}}(F(K) * f(\Delta))$ by Exercise 3.5.17, we must have $f(\Delta) = \{0\}$ and hence Δ is trivially ordered since $f(\Delta)$ is its maximal totally ordered subgroup by Exercise 3.5.6. So the order in $V(F * \Delta)$ is coordinatewise. We identify K with $\varphi(K)$. By Exercise 2.6.12 the basic elements of K are of the form ax^{α} with $0 < a \in F$, and $K = (ax^{\alpha})^{\perp_K} \oplus (ax^{\alpha})^{\perp_K \perp_K}$ since $(ax^{\alpha})^{\perp_K} = \{u \in K : \alpha \notin \text{supp}\, u\}$ and $(ax^{\alpha})^{\perp_K \perp_K} = Fx^{\alpha}$. Since $(ax^{\alpha})^{-1} = \tau(\alpha, -\alpha)^{-1} a^{-1} x^{-\alpha} > 0$ we have $B(K) \subseteq d(K)$. The subset $B = \{x^{\alpha} : \alpha \in \Delta\}$ is clearly a basis of K. If $x^{\alpha} \in B(uv)$, then $x^{\alpha} \wedge uv > 0$, $\alpha \in \text{supp}\, uv$ and $\alpha = \beta + \gamma$ with $\beta \in \text{supp}\, u$ and $\gamma \in \text{supp}\, v$; so $x^{\alpha} = x^{\beta} x^{\gamma} \in B(u)B(v) \subseteq (B(u)B(v))^{\perp\perp}$. Also, $(x^{\alpha}, x^{\beta}) \in B(u) \times B(v)$ with $\tau(\alpha, \beta) x^{\alpha} x^{\beta} \wedge 1 > 0$ iff $(\alpha, \beta) \in \text{supp}\, u \times \text{supp}\, v$ and $\alpha + \beta = 0$. Thus, B is an enabling basis.

(b) $\Rightarrow$ (c). Since K has a basis each positive element of K is the sup of a disjoint set of basic elements by Exercise 2.5.26 and hence the set of special elements in $\Gamma(K)$ is plenary by Theorem 2.5.14. By Theorem 2.5.8 and Exercise 2.5.35 each special element is basic and hence K is a D-ring.

(c) $\Rightarrow$ (a). $B(K) = S(K)$ by Theorem 5.3.1, and by Theorem 2.5.14 and Exercise 2.5.26 K has a basis B and there is an F-isomorphism of bi-vector lattices ${}_F K_F \cong G$ where $\oplus_{b \in B} b^{\perp\perp} \subseteq G \subseteq \Pi_{b \in B} b^{\perp\perp}$. In particular, if $0 \leq u \in K$, then $u = \bigvee_b u_b$ where $0 \leq u_b \in b^{\perp\perp}$ and for $u \in K$, $u_b = (u^+)_b - (u^-)_b$. By Exercise 8 $b^{\perp\perp} = Fb = bF$ for each $b \in B(K)$. Note that $u_b \neq 0$ iff $b \in B(u)$; also, $u^{\perp} = B(u)^{\perp}$. We show next that each basis B_1 of K is enabling. For if B is an enabling basis, then each element in B_1 is uniquely of the form $t_b b = b s_b$ for some $b \in B$ and some $0 < s_b, t_b \in F$. If $0 < s, t \in F$ and $x, y \in K$, then $x \wedge y = 0$ iff $txs \wedge y = 0$. Thus, for $u, v \in K^+$, $B_1(u) = \{t_b b : b \in B(u)\}$ and $(B_1(u)B_1(v))^{\perp} = (B(u)B(v))^{\perp}$ since $x \wedge t_b b d s_d = 0$ for all $t_b b \in B_1(u)$ and $d s_d \in B_1(v)$ iff $x \wedge bd = 0$ for all $b \in B(u)$ and $d \in B(v)$. So $B_1(uv) \subseteq (B_1(u)B_1(v))^{\perp\perp}$ since if $t_b b \in B_1(uv)$, then $b \in B(uv) \subseteq (B(u)B(v))^{\perp\perp} = (B_1(u)B_1(v))^{\perp\perp}$ and $t_b b \in (B_1(u)B_1(v))^{\perp\perp}$. Clearly, $\{(t_b b, d s_d) \in B_1(u) \times B_1(v) : t_b b d s_d \wedge 1 > 0\}$ is finite, and hence B_1 is an enabling basis.

We claim that for $u, v \in K$ and $b \in B$

$$(uv)_b = \sum_{ac \wedge b > 0} u_a v_c. \tag{5.3.9}$$

To see that this sum is finite consider the function

$$\psi : \{(a,c) \in B(u) \times B(v) : ac \wedge b > 0\} \longrightarrow \{(d,e) \in B(b^{-1}u) \times B(v) : de \wedge 1 > 0\}$$

given by $\psi(a,c) = (d,c)$ where $B \cap (b^{-1}a)^{\perp\perp} = \{d\}$. ψ is a bijection since if $\psi(a,c) = \psi(a_1,c_1)$, then $c = c_1$, $b^{-1}a^{\perp\perp} = d^{\perp\perp} = b^{-1}a_1^{\perp\perp}$, and $a = a_1$; and if $(d,c) \in B(b^{-1}u) \times B(v)$ with $dc \wedge 1 > 0$, then $(bd)c \wedge b > 0$, $bd \wedge |u| > 0$, $a \wedge |u| > 0$ where $a \in B \cap (bd)^{\perp\perp}$, and $\psi(a,c) = (d,c)$. Let $\{(a_1,c_1),\ldots,(a_n,c_n)\}$ be the domain of ψ and note that $a_i \neq a_j$ and $c_i \neq c_j$ if $i \neq j$. Now,

$$u = \sum_i u_{a_i} + \bigvee_a (u^+)_a - \bigvee_a (u^-)_a = x + \sum_i u_{a_i}$$

where $a \notin \{a_1,\ldots,a_n\}$ and, similarly, $v = y + \Sigma_i v_{c_i}$; so $|x| \wedge a_i = |y| \wedge c_i = 0$ for $i = 1,\ldots,n$, and $b \in (B(x)B(y))^{\perp}$. Thus,

$$uv = \sum_i u_{a_i} v_{c_i} + \sum_{i \neq j} u_{a_i} v_{c_j} + \sum_i x v_{c_i} + \sum_i u_{a_i} y + xy.$$

However, $u_{a_i} v_{c_j} \in F a_i c_j F = b^{\perp\perp}$ if $i = j$, and $u_{a_i} v_{c_j} \in b^{\perp}$ if $i \neq j$. Moreover, $x v_{c_i} \in (a_i v_{c_i})^{\perp} = (a_i c_i)^{\perp} = b^{\perp}$, $u_{a_i} y \in b^{\perp}$, and $xy \in B(xy)^{\perp\perp} \subseteq (B(x)B(y))^{\perp\perp} \subseteq b^{\perp}$ if $x, y \in K^+$. Thus, (5.3.9) holds for $u, v \in K^+$. The general case follows from the fact that the domain of ψ is partitioned into four subsets, one for each term in the expression $uv = u^+v^+ + u^-v^- - u^+v^- - u^-v^+$, and $(uv)_b = (u^+v^+)_b + (u^-v^-)_b - (u^+v^-)_b - (u^-v^+)_b$.

We will now assume $1 \in B$. Let $H = F^{+*} = f(B(K))$ and let Δ be the trivially ordered po-group $\Delta = B(K)/H$ (Exercise 3.5.6). For $0 < u \in K$ let $\Delta_u = \{Hb \in \Delta : b \wedge u > 0\}$. We claim that the set

$$\mathscr{S} = \{\Lambda \subseteq \Delta : \Lambda \subseteq \Delta_u \text{ for some } 0 < u \in K\}$$

is a supporting subset of $\mathscr{P}(\Delta)$. That $\mathscr{S}$ is an ideal is a consequence of the equation $\Delta_u \cup \Delta_v = \Delta_{u+v}$. ($P_1$) is trivial and ($P_2$) is a consequence of the fact $\Delta_b = \{Hb\}$ for each $b \in B(K)$. If $Ha \in \Delta_u$ and $Hb \in \Delta_v$, then $Hab \in \Delta_{uv}$ since $ab \wedge uv \geq (a \wedge u)(b \wedge v) > 0$; thus, using additive notation for $\Delta, \Delta_u + \Delta_v \subseteq \Delta_{uv}$ and (P_3) holds. As for (P_4),

$$\begin{aligned}&|\{(Ha,Hb) \in \Delta_u \times \Delta_v : Hab = H\}| = \\ &|\{(a,b) \in B(u) \times B(v) : ab \wedge 1 > 0\}|\end{aligned}$$

is finite. For each $\alpha \in \Delta$ there is a unique $b_\alpha \in B$ with $\alpha = Hb_\alpha$ since B is a transversal of H in $B(K)$. If α, $\beta \in \Delta$, then $Hb_\alpha b_\beta = \alpha + \beta = Hb_{\alpha+\beta}$ and $\tau(\alpha,\beta) = b_\alpha b_\beta b_{\alpha+\beta}^{-1} \in H$ with $\tau(\alpha,0) = \tau(0,\alpha) = 1$. Under the assumption that any two elements of B commute we have Δ is abelian and

$$\begin{aligned}\tau(\alpha+\beta,\gamma)\tau(\alpha,\beta) &= b_{\alpha+\beta} b_\gamma b_{\alpha+\beta+\gamma}^{-1} b_\alpha b_\beta b_{\alpha+\beta}^{-1} = b_\alpha b_\beta b_\gamma b_{\alpha+\beta+\gamma}^{-1} \\ &= b_\alpha b_{\beta+\gamma} b_{\alpha+\beta+\gamma}^{-1} b_\beta b_\gamma b_{\beta+\gamma}^{-1} = \tau(\alpha,\beta+\gamma)\tau(\beta,\gamma)\end{aligned}$$

for all α, β, $\gamma \in \Delta$; that is $V_{\mathscr{S}}(F * \Delta)$ is an ℓ-ring. Since right multiplication by b_α^{-1} is an isomorphism of ${}_F b_\alpha^{\perp\perp}$ onto ${}_F F$ we can redefine the original vector lattice embedding as $\varphi : K \longrightarrow V_{\mathscr{S}}(F * \Delta)$ where

$$\varphi(u) = \sum_{\alpha \in \Delta} x^{\alpha} u_{b_\alpha} b_\alpha^{-1}.$$

For $u, v \in K$ and $\gamma \in \Delta$,

$$(\varphi(uv))_\gamma = (uv)_{b_\gamma} b_\gamma^{-1} = \sum_{\beta_\alpha b_\beta \wedge b_\gamma > 0} u_{b_\alpha} v_{b_\beta} b_\gamma^{-1} = \sum_{\alpha+\beta=\gamma} u_{b_\alpha} v_{b_\beta} b_\gamma^{-1}$$

by (5.3.9), and

$$\begin{aligned}(\varphi(u)\varphi(v))_\gamma &= \sum_{\alpha+\beta=\gamma} \tau(\alpha,\beta) u_{b_\alpha} b_\alpha^{-1} v_{b_\beta} b_\beta^{-1} = \sum_{\alpha+\beta=\gamma} b_\alpha b_\beta b_{\alpha+\beta}^{-1} u_{b_\alpha} b_\alpha^{-1} v_{b_\beta} b_\beta^{-1} \\ &= \sum_{\alpha+\beta=\gamma} u_{b_\alpha} v_{b_\beta} b_\gamma^{-1}\end{aligned}$$

provided B is central. Thus, φ preserves products.

Suppose H is a direct factor of $B(K)$. Then we can take B to be a subgroup of $B(K)$ and hence $b_{\alpha+\beta} = b_\alpha b_\beta$ and $\tau(\alpha,\beta) = 1$ for all $\alpha, \beta \in \Delta$. □

An example of a division ℓ-ring which satisfies (b) and (c) of Theorem 5.3.4 but which is not f-embeddable is given in Exercise 25, and an example of an f-embeddable ℓ-field for which the twisting function τ is never trivial is given in Exercise 26.

Theorem 5.3.5. *Let R be a lattice-ordered division ring with $1 > 0$ and which is distinct from its subring F of f-elements. Then the following statements are equivalent.*

(a) $R = F \oplus Fa$ for some $a > 0$ with $a^2 \in F$.
(b) $R =$ lex $(F \oplus Fa)$ for some $a > 0$ with $a^2 \in F$.
(c) $d(R) = f(R) \cup f(R)^{\perp +}$.
(d) R is a D-ℓ-ring with exactly two roots.

Proof. (b) $\Rightarrow$ (c). We have $a \in d(R)$ since $a^{-1} = a^{-2}a > 0$, and $aF = Fa$ since conjugation by a is an ℓ-ring automorphism of R. If $b \in F^{\perp} \backslash (F \oplus Fa)$, then $|b| > F \oplus Fa$ and $1 = 1 \wedge |b| = 0$. So $F^{\perp} \subseteq F \oplus Fa$, $F^{\perp} \subseteq Fa$ and certainly $Fa \subseteq F^{\perp}$; hence $f(R)^{\perp +} = F^{+}a \subseteq d(R)$ and $f(R) \cup f(R)^{\perp +} \subseteq d(R)$. The reverse inclusion is given in Exercise 6.

(c) $\Rightarrow$ (b). If $0 < a \in f(R)^{\perp +}$, then a is basic because $a \in d(R)$ and hence $a^{-1} > 0$, and 1 is basic. If $0 < b \in f(R)^{\perp +}$, also, then $0 < a, b \leq a + b$ implies that a and b are comparable. Hence $F^{\perp}$ is totally ordered, and $ab \wedge a = a(b \wedge 1) = 0$ shows that $ab \in a^{\perp} = F^{\perp\perp}$ (Exercise 2.4.11); thus, $ab \in f(R)$ since $ab \in d(R)$. So $b = aa^{-2}ab \in aF$, $F^{\perp} = aF$ and, similarly, $F^{\perp} = Fa$. Note that $F^{\perp\perp} = (Fa)^{\perp} = F^{\perp}a = Fa^2 = F$. By Theorem 2.5.9

$$\begin{aligned}R &= \{x \in R : |x| > F\} \cup (F \oplus F^{\perp}) \\ &= \{x \in R : |x| > F^{\perp}\} \cup (F^{\perp} \oplus F).\end{aligned}$$

So if $x \in R^+ \backslash (F \oplus F^{\perp})$, then $x > F \oplus F^{\perp}$ and $R = $ lex $(F \oplus Fa)$.

(b) $\Rightarrow$ (a). From the first paragraph above, $aF = Fa$ and $(Fa)^2 = F$. So $F \oplus Fa$ is a subring of R and it is a division ring. To see this, take $0 \neq b + ca$ with $b, c \in F$. If $b = 0$, then $(ca)^{-1} = (ca)^{-2}ca \in Fa$. If $b \neq 0$, then $b + ca = b(1 + b^{-1}ca)$ and $(1+b^{-1}ca)^{-1} = (1 - b^{-1}ca)[1 - (b^{-1}ca)^2]^{-1} \in F \oplus Fa$; so $(b+ca)^{-1} \in F \oplus Fa$. Now, if $x > F \oplus Fa$, then x^{-1} cannot be in $F \oplus Fa$ and so $x^{-1} = |x^{-1}| > F \oplus Fa$. This gives the contradiction $1 = xx^{-1} > 1$. So $R = F \oplus Fa$.

Since (a) trivially implies (b) the first three statements are equivalent, and (a) and (c) together with the fact that Fa is totally ordered clearly imply (d). On the other hand, if R is a D-ℓ-ring with two roots, then, as in (5.3.7), $R = F \oplus F^{\perp}$ and $d(R) = S(R) \cup \{0\} = f(R) \cup f(R)^{\perp +}$ by Theorem 5.3.1. □

A version of the preceding theorem holds for ℓ-domains. However, now (b) doesn't imply (a); see Exercise 11.

The ℓ-fields in the previous results all have $1 > 0$ and consequently they have a good totally ordered subfield, namely, the subfield of f-elements or, more precisely, the additive subgroup generated by the f-elements which, according to Exercise 3.1.12, is the largest local sp-ℓ-subring. By considering, more generally, those elements which induce an f-map under multiplication we will see that there still is a "largest totally ordered subfield" even if $1 \not> 0$, provided the ℓ-field is archimedean.

Theorem 5.3.6. *Let E be the divisible hull of the archimedean ℓ-group G. If $0 < a \in E$, then*

$$E_a = \{x \in E : \mathbb{Q}a \cup \{x\} \text{ is totally ordered}\}$$

is a $\mathbb{Q}$-subspace of E and, for $a \in G$, $E_a \cap G = G_a$ is the largest totally ordered subgroup of G which contains a. Moreover, $G_a = G_b$ if and only if $0 < b \in E_a$.

Proof. We may assume that $G = E$ since this result for E certainly implies it for G. Certainly, $\mathbb{Q}a \cup \{x\}$ is a chain exactly when $\mathbb{Z}a \cup \mathbb{Z}x$ is a chain. Let $x, y \in G_a$ and $p \in \mathbb{Q}$. Then px, $x + pa \in G_a$ since $\mathbb{Q}a \cup \{px\} = p(\mathbb{Q}a \cup \{x\})$ if $p \neq 0$ and $\mathbb{Q}a \cup \{x + pa\} = (\mathbb{Q}a \cup \{x\}) + pa$. Let $L_x = \{ra \in \mathbb{Q}a : ra < x\}$ and $U_x = \{ra \in \mathbb{Q}a : x < ra\}$ be the cut in $\mathbb{Q}a$ determined by x. If $L_x = L_y$, then $x = y$. For, if $m \in \mathbb{N}$, then since $\mathbb{Z}a \cup \{mx\}$ is a chain and G is archimedean, there is an integer $n \in \mathbb{Z}$ with $na < mx \leq (n+1)a$. But then $na < my \leq (n+1)a$ also, and $-a < m(x-y) < a$; that is, $m|x-y| < a$. Since m is arbitrary $x = y$. Suppose, to the contrary, that $x + y \notin G_a$. Then for some $p \in \mathbb{Q}$ we have $pa \| x+y$ or, $pa - x \| y$. Since $\mathbb{Q}a \cup \{pa - x\}$ is a chain and $L_y < y < U_y$, necessarily, $L_y < pa - x < U_y$. Consequently, $L_y = L_{pa-x}$ and we have the contradiction $y = pa - x$. So $x + y \in G_a$ and G_a is a subgroup which contains a. If $x \neq y$, then $L_x \subset L_y$ or $L_y \subset L_x$. Assuming the former there is an $r \in \mathbb{Q}$ with $ra < y$ and $x \leq ra$; so $x < y$ and G_a is totally ordered. If H is any totally ordered subgroup of G with $a \in H$, and $x \in H$, then $\mathbb{Z}a \cup \mathbb{Z}x$ is a chain and hence so is $\mathbb{Q}a \cup \{x\}$; so $x \in G_a$ and $H \subseteq G_a$. If $0 < b \in G_a$, then $G_a \subseteq G_b$ since G_a is totally ordered and $a \in G_b$; so $G_b \subseteq G_a$ since G_b is totally ordered. □

For the archimedean ℓ-domain R define

$$L(R) = L = \{x \in R : xR_a \subseteq R_a \text{ for each } 0 < a \in R\} \tag{5.3.10}$$

and, for $a > 0$,

$$P(a) = \{x \in L : xa \geq 0\}. \tag{5.3.11}$$

Clearly, L is a subring of R, $P(a) + P(a) \subseteq P(a)$ and $L = P(a) \cup -P(a)$. If R is commutative, then $P(a) \cap -P(a) = \ell(R)$. For, if $x \in P(a) \cap -P(a)$ and $0 < b \in R$, then xb is comparable to 0 and $xa = 0$; so $xb = 0$ or else $0 = xab = xba \neq 0$.

Recall from Exercise 3.1.8 that in any ℓ-ring R

$$\overline{f}(R) = \{x \in R : a \wedge b = 0 \Rightarrow xa \wedge b = ax \wedge b = 0\}$$

is a convex subsemiring of R and the additive subgroup $\overline{F}(R)$ it generates is a convex subring which we will call the *subring of $\overline{f}$-elements.*

Theorem 5.3.7. *Let R be a commutative archimedean ℓ-domain with $\ell(R) = 0$ and let L and $P = P(a)$ be given by (5.3.10) and (5.3.11). Then the following hold.*

(a) $P(a) = P(b)$ for all $0 < a, b \in R$.

(b) (L, P) is a totally ordered archimedean domain and if $x \in L$ is a unit of R, then $x^{-1} \in L$.

(c) $P = \overline{f}(R)$ and either L is the totally ordered subring of f-elements of R or it is the trivially ordered subring of $\overline{f}$-elements.

(d) L is the largest subring of R which has a directed partial order and over which R is an ℓ-algebra.

Proof. We saw previously that $(L, P(a))$ is a totally ordered group. Let $0 < a, b \in R$. If $x \in P(a)$, then $(xa)b \geq 0$ and $x \in P(ab)$. But if $x \in P(ab)$, then $xab \geq 0$ and hence $xa \geq 0$ since xa is comparable to 0. So $x \in P(a)$ and $P(a) = P(ab) = P(b)$. If $x, y \in P$, then $xya^2 = xaya \geq 0$ and $xy \in P$. So (L, P) is a totally ordered ring and R is a po-algebra over (L, P). If $0 \neq x \in P$ and $0 < a \in R$, then we saw above that $0 < xa$ since $\ell(R) = 0$. Consequently, L is a domain and (L, P) is archimedean since R is archimedean. Suppose $x \in P$. Since R_a is totally ordered and archimedean $0 \leq xa \leq na$ for some $n \in \mathbb{N}$. So if $a \wedge b = 0$, then $xa \wedge b = 0$. Thus, $x \in \overline{f}(R)$ and, if $x^{-1} \in R$, then multiplication by x^{-1} is an f-automorphism of the ℓ-group R; that is, $x^{-1} \in \overline{f}(R)$. Since $xR_a \subseteq R_a$, we have $xR_a = R_{xa} = R_a$ by Theorem 5.3.6; so $x^{-1}R_a = R_a$ and $x^{-1} \in L$. We have just seen that $P \subseteq \overline{f}(R)$ and now we will show that $\overline{f}(R) \subseteq P$. Let $0 \neq x \in \overline{f}(R)$ and let μ_x denote multiplication by x. Then $\mu_x \in F(R, +) = F$ where F is the subring of $\mathrm{End}_{\mathbb{Z}}(R)$ generated by the f-endomorphisms of R. By Theorem 3.6.6 F is a semiprime f-ring and R is a strong f-module over F. Let $0 < a \in R$ and put $b = (\mu_x - 1)^+ a$ and $c = (\mu_x - 1)^- a$. Then $(\mu_x - 1)b = ((\mu_x - 1)^+)^2 a \geq 0$ and $xb \geq b \geq 0$. Also, $(\mu_x - 1)c = -((\mu_x - 1)^-)^2 a \leq 0$ and $0 \leq xc \leq c$. Consequently, $xbc = bc$. Let $\alpha = \mu_x - 1$. If $\alpha \neq 0$, then $\alpha^2 d > 0$ for some $0 < d \in R$ and

$$0 \leq (\alpha^2 d)bc = (x-1)^2 dbc = (x-1)^2 bcd = 0.$$

So either $\alpha = 0$ and $x = 1$, or $b = 0$ or $c = 0$. Suppose $x \neq 1$. Then $(xa-a)^+ = 0$ or $(xa-a)^- = 0$, and hence, either $0 < xa < a$ or $0 < a < xa$. By replacing R by its divisible hull we may assume that R is a $\mathbb{Q}$-algebra. Replacing x by $r^{-1}x$ where r is a nonzero positive rational number we obtain $r^{-1}x = 1$ or $0 < r^{-1}xa < a$ or $0 < a < r^{-1}xa$; that is, $x = r$ or $xa < ra$ or $ra < xa$. Thus, $\mathbb{Q}a \cup \{xa\}$ is a chain and $xa \in R_a$. This shows $x \in P$ since $xa \geq 0$. If R has an f-element $a > 0$, then $0 = (xa)^- = x^-a$ and $x^- = 0$ whenever $x \in \overline{f}(R)$; so $\overline{f}(R) = f(R)$ and L is the subring of f-elements. On the other hand, if $f(R) = 0$, then $L = \overline{F}(R)$ is trivially ordered since $L \cap R^+ = (\overline{f}(R) \cup -\overline{f}(R)) \cap R^+ = f(R) \cup -f(R) = 0$.

Now suppose K is a subring of R, (K,T) is a po-ring and R is an ℓ-algebra over (K,T). Then $T \subseteq \overline{f}(R) = P$ and $T - T \subseteq L$. □

We wish to determine the structure of the L-submodule R_a when a is algebraic over L.

Theorem 5.3.8. *Let R be a commutative domain and an archimedean ℓ-ring whose subring L of $\overline{f}$-elements is different from 0. Suppose $0 < a \in R$ is algebraic over L. Then $\alpha R_a \subseteq La$ for some $0 \neq \alpha \in L$. If $Q(L)$ is the totally ordered field of quotients of L and $Q = R \otimes_L Q(L)$ is the $Q(L)$-ℓ-algebra of quotients of R, then $R_a = Q(L)a \cap R$. In particular, $Q_a = Q(L)a$.*

Proof. We have $R \subseteq S \subseteq Q$ where S is the $\mathbb{Z}$-divisible hull of R and Q is the L-divisible hull of R. Let $f(x) \in L[x]$ be a polynomial of minimal degree with $f(a) = 0$. So $f(a) = \alpha_0 + \alpha_1 a + \cdots + \alpha_n a^n = 0$ with $\alpha_i \in L$ and $\alpha_0\alpha_n \neq 0$. We will show that $\alpha_0 R_a \subseteq La$ by verifying that for each $0 < b \in R_a$ the element $c = \alpha_1 b + \alpha_2 ab + \cdots + \alpha_n a^{n-1}b$ is in L and hence $-\alpha_0 b = ca \in La$. Let $0 < x \in R$. In order to show that $cx \in R_x$ we first note the following. Since S_a is totally ordered so is $\mathbb{Q}a^i x \cup \{a^{i-1}bx\} \subseteq S_a a^{i-1}x$ for each $i \in \mathbb{N}$; and

$$\text{if } r \in \mathbb{Q} \text{ and } a^{i-1}bx \leq ra^i x \text{ for some } i \in \mathbb{N}, \text{ then}$$
$$a^{j-1}bx \leq ra^j x \text{ for each } j \in \mathbb{N}. \tag{5.3.12}$$

For, suppose $i < j$. Then either $a^{i-1}bx \leq ra^i x$ or $a^{i-1}bx > ra^i x$ and since multiplication by a^{j-i} is strictly isotone, we have $a^{j-1}bx \leq ra^j x$ or $a^{j-1}bx > ra^j x$, respectively. Let

$$I = \{i \in \{1,\ldots,n\} : 0 \neq \alpha_i \in P\} \text{ and } J = \{i \in \{1,\ldots,n\} : 0 \neq \alpha_i \in -P\},$$

and put

$$c_1 = \sum_{i \in I} \alpha_i a^{i-1} b\,,\ c_2 = \sum_{i \in J} \alpha_i a^{i-1} b$$
$$e_1 = \sum_{i \in I} \alpha_i a^i\,,\ e_2 = \sum_{i \in J} \alpha_i a^i.$$

Then $c = c_1 + c_2$, $\alpha_0 + e_1 + e_2 = 0$, and c_1, $e_1 > 0$ and c_2, $e_2 < 0$. Of course, we may assume $I \neq \emptyset$. If $J = \emptyset$, then $c_2 = e_2 = 0$ and in the following ignore all references to c_2 and e_2. We claim that $c_k x \in R_{|e_k x|}$ for $k = 1, 2$. For, if $r \in \mathbb{Q}$, then, by

(5.3.12), either $c_1x \leq re_1x$ and $c_2x \geq re_2x$, or $c_1x > re_1x$ and $c_2x < re_2x$. Consequently, $\mathbb{Q}|e_kx| \cup \{c_kx\}$ is totally ordered. These inequalities determine a cut (A,B) in $\mathbb{Q}$ where

$$A = \{r \in \mathbb{Q} : re_1x < c_1x\} = \{r \in \mathbb{Q} : c_2x < re_2x\},$$
$$B = \{r \in \mathbb{Q} : c_1x \leq re_1x\} = \{r \in \mathbb{Q} : re_2x \leq c_2x\}.$$

Let X be the Stone space of the Boolean algebra of polars of S. By Theorem 2.3.23 S can be embedded in $D(X)$ as a vector lattice over $\mathbb{Q}$ (and as an $L(S)$-f-module-see Exercise 2.3.19). Let α be the real number determined by the cut (A,B). So, in $\mathbb{R}$ and in $D(X)$

$$\alpha = \bigvee_{r \in A} r = \bigwedge_{r \in B} r.$$

By Theorem 3.6.4 $D(X)$ is an infinite d-ring and hence $c_1x = \alpha e_1x$. For, $e_1x \geq 0$ gives

$$\alpha e_1x = \bigvee_{r \in A} re_1x \leq c_1x \leq \bigwedge_{r \in B} re_1x = \alpha e_1x.$$

Similarly, $c_2x = \alpha e_2x$, and

$$cx = c_1x + c_2x = \alpha(e_1x + e_2x) = -\alpha_0\alpha x.$$

Thus, $\mathbb{Q}x \cup \{cx\}$ is totally ordered and $cx \in R_x$. Since $Q(L) = L(Q)$ by Exercise 13 and $Q_a = L(Q)a$, as we have just shown, $R_a = Q_a \cap R = Q(L)a \cap R$. □

If the commutative domain R in Theorem 5.3.8 is algebraic over its subring L, then Q is its field of quotients, and we will show that the lattice order of Q can be extended to a total order. In order to do this we will first develop some topological and order-theoretic properties of $\mathbb{R}^n$.

A vector space V over $\mathbb{R}$ which is also a topological space is called a *topological vector space* if addition and scalar multiplication are continuous functions:

$$V \times V \longrightarrow V\ ,\ (v,w) \mapsto v + w, \text{ is continuous} \tag{5.3.13}$$
$$\mathbb{R} \times V \longrightarrow V\ ,\ (a,v) \mapsto av, \text{ is continuous}\ . \tag{5.3.14}$$

Here, $V \times V$ and $\mathbb{R} \times V$ have the product topology. From (5.3.14) we see that the function $V \longrightarrow V$ given by $v \longrightarrow -v$ is continuous; so V is a topological group (see Exercise 4.2.12). Also, it is clear that the translations $w \mapsto v + w$ and $w \mapsto aw$ are homeomorphisms for each $v \in V$ and each $0 \neq a \in \mathbb{R}$. If $\mathscr{N}_V(x)$, or just $\mathscr{N}(x)$, denotes the collection of neighborhoods of $x \in V$, then $x + \mathscr{N}(0) = \mathscr{N}(x)$. In particular, a linear transformation between two topological vector spaces is continuous if and only if it is continuous at 0 (or at some point of its domain).

The subset E of the real vector space V is *symmetric* if $E = -E$, *balanced* if $aE \subseteq E$ whenever $a \in \mathbb{R}$ and $|a| \leq 1$, and *absorbing* if for each $v \in V$ there is an $0 < \varepsilon \in \mathbb{R}$ such that $bv \in E$ whenever $b \in \mathbb{R}$ and $|b| \leq \varepsilon$.

We collect some information about $\mathscr{N}(0)$ below. Recall that a subset of $\mathscr{N}(x)$ which is coinitial in $\mathscr{N}(x)$ with respect to inclusion is called a base for the neighborhood system at x.

Theorem 5.3.9. *Let V be a topological vector space and let U be a neighborhood of 0.*

(a) *There is a neighborhood A of 0 with $A+A \subseteq U$.*
(b) *U contains a symmetric neighborhood of 0.*
(c) *U contains a closed neighborhood of 0, and hence V is a regular topological space.*
(d) *U contains a balanced neighborhood of 0.*
(e) *U is absorbing.*

Proof. (a) By (5.3.13) there are neighborhoods U_1 and U_2 of 0 such that $U_1+U_2 \subseteq U$. Let $A = U_1 \cap U_2$.

(b) $U \cap -U$ is a symmetric neighborhood of 0 contained in U.

(c) Since the subtraction function $V \times V \longrightarrow V$, $(v,w) \mapsto v-w$, is continuous there exists $A \in \mathscr{N}(0)$ such that $A - A \subseteq U$. If $x \in A^-$, then there exists $y \in (x+A) \cap A$. So $y = x+w$ with $w \in A$ and $x = y-w \in A-A \subseteq U$. Thus, $A^- \subseteq U$.

(d) By (5.3.14) there exists $0 < a \in \mathbb{R}$ and $A \in \mathscr{N}(0)$ with $bw \in U$ if $|b| \leq |a|$ and $w \in A$. Let U_1 be the set of all these elements bw. Then $U_1 \in \mathscr{N}(0)$ since it is the union of the translates bA with $|b| \leq |a|$. If $|c| \leq 1$, then $|cb| \leq |a|$ and $cbw \in U_1$; so U_1 is balanced and $V_1 \subseteq U$.

(e) Since $0v = 0$, by (5.3.14), if $v \in V$ there exists $A \in \mathscr{N}(v)$ and $0 < a \in \mathbb{R}$ with $U_1 = \{bw : |b| \leq |a|, w \in A\} \subseteq U$; so U is absorbing. □

There is a converse to Theorem 5.3.9 in the sense that a base for the neighborhood system at 0 determines the topology of V.

Theorem 5.3.10. *Let $\mathscr{B}$ be a nonempty family of nonempty subsets of the vector space ${}_{\mathbb{R}}V$ with the following properties:*

(a) *If B_1, $B_2 \in \mathscr{B}$, then there exists $B_3 \in \mathscr{B}$ with $B_3 \subseteq B_1 \cap B_2$.*
(b) *For every $B \in \mathscr{B}$ there exists $B_1 \in \mathscr{B}$ with $B_1 + B_1 \subseteq B$.*
(c) *Each $B \in \mathscr{B}$ is balanced and absorbing.*

Then there is a unique topology $\mathscr{T}$ for V such that $\mathscr{B}$ is a base for the neighborhood system at 0 and $(V, \mathscr{T})$ is a topological vector space, or, equivalently, each additive translation is continuous.

Proof. Let $\mathscr{N}_0 = \{X \subseteq V :$ there exists $B \in \mathscr{B}$ with $B \subseteq X\}$ and, for $x \in V$, let $\mathscr{N}_x = x + \mathscr{N}_0$. Since each $B \in \mathscr{B}$ is balanced, $0 \in B$, and, clearly, $\mathscr{N}_x$ satisfies the first three conditions in Exercise 15. To see that it satisfies the fourth, take $A \in \mathscr{B}$ and $B, C \in \mathscr{B}$ with $B+B \subseteq A$ and $C+C \subseteq B$. If $x \in V$ and $y \in x+B$, then $y+C \subseteq x+B \subseteq x+A$ and $x+B \in \mathscr{N}_y$. Thus, by Exercise 15, $\mathscr{T} = \{U \subseteq V : x \in U$ implies $U \in \mathscr{N}_x\}$ is a topology for V and $\mathscr{N}(x) = \mathscr{N}_x$ for each $x \in V$. Note that $x + \mathscr{B}$ is a base for $\mathscr{N}(x)$. To see that addition is continuous suppose that $x, y \in V$ and $A \in \mathscr{B}$; so $x+y+A$

is a basic neighborhood of $x+y$. If $B \in \mathscr{B}$ with $B+B \subseteq A$, then $(x+B)+(y+B) \subseteq x+y+A$, and addition is continuous at $(x,y) \in V \times V$. We will now check that scaler multiplication is continuous at $(a,x) \in \mathbb{R} \times V$. It suffices to show that if $A \in \mathscr{B}$, then there exist $0 < \varepsilon \in \mathbb{R}$ and $B \in \mathscr{B}$ with $(a-\varepsilon, a+\varepsilon)(x+B) \subseteq ax+A$; that is, if $|b-a| < \varepsilon$, then $b(x+B) \subseteq ax+A$ or, equivalently, $(b-a)x+bB \subseteq A$. There exists $C \in \mathscr{B}$ with $C+C \subseteq A$. Since C is absorbing there exists $\varepsilon > 0$ such that $|b-a| \leq \varepsilon$ implies $(b-a)x \in C$. Using induction on $n \in \mathbb{N}$ it is easy to see that, for each n, there exists $B_n \in \mathscr{B}$ with $2^n B_n \subseteq C$. Take n with $2^n > |a| + \varepsilon$ and let $B = B_n$. Since $|b| \leq |b-a| + |a| < 2^n$ and B is balanced $bB \subseteq 2^n B \subseteq C$. Thus, $(b-a)x + bB \subseteq C+C \subseteq A$, as desired.

Since a topology is determined by the neighborhood systems of its points the topology on V is unique. □

Let W be a vector subspace of the topological vector space V and let $\rho_W = \rho : V \longrightarrow V/W$ be the natural homomorphism. Recall that the quotient topology $\mathscr{Q}$ on V/W is defined by $\mathscr{Q} = \{T \subseteq V/W : \rho^{-1}(T) \text{ is open}\}$ and is characterized by the condition: a function $f : V/W \longrightarrow Z$ into a topological space Z is continuous if and only if the composite $f\rho$ is continuous. Clearly, $\rho^{-1}\rho(U) = U+W$ for each subset U of V and ρ is a continuous and open mapping.

Theorem 5.3.11. *If W is an $\mathbb{R}$-subspace of the topological vector space V, then V/W, with the quotient topology, is a topological vector space. V/W is Hausdorff if and only if W is closed.*

Proof. By Theorem 5.3.9 the set $\mathscr{B}$ of balanced neighborhoods of 0 is a base for $\mathscr{N}(0)$ and $\mathscr{B}$ satisfies the conditions in Theorem 5.3.10. Now, $\mathscr{N}(\rho(0)) = \rho(\mathscr{N}(0))$. For, if $0 \in U \subseteq A \subseteq V$ with U open, then $\rho(0) \in \rho(U) \subseteq \rho(A)$ and $\rho(U)$ is open; and if $\rho(0) \in \rho(U) \subseteq \rho(A)$ with $\rho(U)$ open, then $W \subseteq W+U \subseteq W+A$ with $W+U$ open and $\rho(A) = \rho(W+A) \in \rho(\mathscr{N}(0))$. Consequently, we have that $\rho(\mathscr{B})$ is a base for $\mathscr{N}(\rho(0))$ and since the image of a balanced or absorbing subset is balanced or absorbing, respectively, $\rho(\mathscr{B})$ satisfies the conditions in Theorem 5.3.10. If τ_x denotes translation by x, then, clearly, $\tau_{\rho(x)}\rho = \rho\tau_x$ and hence $\tau_{\rho(x)}$ is continuous. Thus, by Theorem 5.3.10, V/W with its quotient topology is a topological vector space. If V/W is Hausdorff, then $\{\rho(0)\}$ is closed and so is $W = \rho^{-1}(\rho(0))$. On the other hand, if W is closed, then so is $\{\rho(0)\}$ and hence each point in V/W is closed. Thus, if $\rho(x) \neq \rho(y)$, then by (c) of Theorem 5.3.10 there exists $\rho(U)$ open with $\rho(x) \in \rho(U) \subseteq \rho(U)^- \subseteq \rho(V) \backslash \{\rho(y)\}$. But then $\rho(U)$ and $\rho(V) \backslash \rho(U)^-$ are disjoint neighborhoods of $\rho(x)$ and $\rho(y)$. □

Suppose that W and X are $\mathbb{R}$-subspaces of V and $V = W \boxplus X$. Then X is called a *topological complement* of W and V is called the *topological direct sum of W and X* if the topology of V is the product topology of $W \times X$ where W and X have the subspace topologies; that is, if the continuous isomorphism $q : W \times X \longrightarrow V$ given by $(w,x) \mapsto w+x$ is a homeomorphism. Note that if $p = p_W : V \longrightarrow W$ denotes the projection $p(w+x) = w$ onto W, then p is continuous precisely when it is continuous as a map of V into itself.

Theorem 5.3.12. *Suppose V is a topological vector space and $V = W \boxplus X$. The following statements are equivalent.*

(a) V is the topological direct sum of W and X.
(b) $p_W : V \longrightarrow V$ is continuous.
(c) The isomorphism $h_X : V/W \longrightarrow X$ is a homeomorphism.

Proof. (a) $\Rightarrow$ (b). This is a consequence of the equation $p_W = \pi_W q^{-1}$ where $\pi_W : W \times X \longrightarrow W$ is the projection.

(b) $\Rightarrow$ (a). The continuity of 1_V and p_W gives that $p_X = 1_V - p_W$ is continuous. But then q^{-1} is continuous since $\pi_W q^{-1} = p_W$ and $\pi_X q^{-1} = p_X$ are continuous.

(b) $\Leftrightarrow$ (c). h_X is continuous if and only if $h_X \rho_W$ is continuous. But $h_X \rho_W = p_X$ and p_X is continuous if and only if $p_W = 1_V - p_X$ is continuous. □

We show next that a finite dimensional vector space has a unique Hausdorff topology that makes it into a topological vector space. An $\mathbb{R}$-algebra is a *topological algebra* if it is a topological vector space and multiplication is continuous.

Theorem 5.3.13. *Let $(V, \mathscr{T})$ be an n-dimensional Hausdorff topological vector space. Then each $\mathbb{R}$-isomorphism $\mathbb{R}^n \longrightarrow V$ is a homeomorphism and hence $\mathscr{T}$ is unique. If V is an $\mathbb{R}$-algebra, then it is a topological algebra.*

Proof. We proceed by induction on n. For $n = 1$ let $0 \neq v_0 \in V$. The map $\mathbb{R} \longrightarrow V$, $a \mapsto av_0$, is continuous since it is the composite of the continuous map $\mathbb{R} \longrightarrow \mathbb{R} \times V$, $a \mapsto (a, v_0)$, followed by scalar multiplication. To see that its inverse φ is continuous at 0 let $0 < \varepsilon \in \mathbb{R}$ and let B and U be disjoint neighborhoods of 0 and εv_0, respectively, with B balanced. Then $\varphi(B) \subseteq (-\varepsilon, \varepsilon)$. For if $av_0 \in B$ and $|a| \geq \varepsilon$, then $|\varepsilon a^{-1}| \leq 1$ and we obtain the contradiction $\varepsilon v_0 = \varepsilon a^{-1} av_0 \in B$.

Suppose that V is n-dimensional, $n > 1$, and the first statement holds for $(n-1)$-dimensional Hausdorff spaces. Let $\psi : \mathbb{R}^n \longrightarrow V$ be an $\mathbb{R}$-isomorphism. Since $\mathbb{R}^n = \mathbb{R}^{n-1} \boxplus \mathbb{R}$ is a topological direct sum and since by induction and the case $n = 1$ the restrictions of ψ to $\mathbb{R}^{n-1}$ and to $\mathbb{R}$ are homeomorphisms onto their respective images $W = \psi(\mathbb{R}^{n-1})$ and $\psi(\mathbb{R})$, it suffices to show that W is closed. For then, V/W is Hausdorff by Theorem 5.3.10, $V/W \longrightarrow \psi(\mathbb{R})$ is a homeomorphism by the case $n = 1$, W and $\psi(\mathbb{R})$ are topological complements by Theorem 5.3.12, and $\psi : \mathbb{R}^{n-1} \boxplus R \longrightarrow W \boxplus \varphi(\mathbb{R})$ is a homeomorphism. The brief reason why W is closed is that each Cauchy net in W converges in W and limits in the Hausdorff space V are unique. We will supply more details.

We first will find a descending sequence of symmetric neighborhoods $A_1 \supseteq A_2 \supseteq \cdots$ in $\mathscr{N}_V(0)$ such that $A_{m+1} + A_{m+1} \subseteq A_m$ for each $m \in \mathbb{N}$ and $\{A_m \cap W : m \in \mathbb{N}\}$ is a base for $\mathscr{N}_W(0)$. Since W is homeomorphic to $\mathbb{R}^{n-1}$ there is a countable base $B'_1 \supseteq B'_2 \supseteq \cdots$ for $\mathscr{N}_W(0)$ and, for each m, $B'_m = A'_m \cap W$ with $A'_m \in \mathscr{N}_V(0)$. Let $A'_1 = A_1$ and proceed by induction. Given the chain $A_1 \supseteq \cdots \supseteq A_m$ with each A_j symmetric, $A_j \subseteq A'_j$, and $A_j + A_j \subseteq A_{j-1}$ for $2 \leq j \leq m$, we can find $A_{m+1} \in \mathscr{N}_V(0)$ such that A_{m+1} is symmetric, $A_{m+1} \in \mathscr{N}_V(0)$, and $A_{m+1} + A_{m+1} \subseteq A'_{m+1} \cap A_m$. Let $B_m = A_m \cap W$. Then $\{B_m : m \in \mathbb{N}\}$ is a base for $\mathscr{N}_W(0)$ since $B_m \subseteq B'_m$ for each m.

Suppose x is in the closure of W. For each $A \in \mathcal{N}_V(0)$ choose $x_A \in (x+A) \cap W$. Note that the relation $A \geq C$ if $A \subseteq C$ is an upward directed partial order of $\mathcal{N}_V(0)$ and the net $(x_A)_A$ converges to x : if $U \in \mathcal{N}_V(x)$, then $U = x + A_0$ for some $A_0 \in \mathcal{N}_V(0)$ and $A \geq A_0$ gives $x_A \in x + A \subseteq x + A_0 = U$. Let $x_m = x_{A_m}$ Then $(x_m)_m$ is a Cauchy sequence in W; that is, for each C in $\mathcal{N}_W(0)$ there exists an integer N such that if $p, m \geq N$, then $x_p - x_m \in C$. For, there exists $N \in \mathbb{N}$ with $B_N \subseteq C$. So if $p, m \geq N+1$, then $x_p = x + v_p$, $x_m = x + v_m$ with $v_p \in A_p \subseteq A_{N+1}$ and $v_m \in A_M \subseteq A_{N+1}$. Thus, $x_p - x_m = v_p - v_m \in A_{N+1} + A_{N+1} \subseteq A_N$ and $x_p - x_m \in A_N \cap W = B_N \subseteq C$. Since W is Cauchy-complete the sequence $(x_m)_m$ converges to some $y \in W$; that is, each neighborhood of y contains a tail of this sequence. But then $(x_A)_A$ converges to y. For if $C \in \mathcal{N}_W(y)$, then $B_N \subseteq -y + C$ for some integer N and there exists $N' \geq N+1$ such that if $m \geq N'$, then $x_m \in B_{N+1} + y$. Thus, if $A \geq A_{N'+1}$, then $x_A - x_{N'+1} \in (x+A) - (x + A_{N'+1}) \subseteq A_{N'+1} - A_{N'+1} \subseteq A_{N'} \subseteq A_{N+1}$ and $x_A - x_{N'+1} \in B_{N+1}$; also, $x_{N'+1} - y \in B_{N+1}$. So $x_A - y = x_A - x_{N'+1} + x_{N'+1} - y \in B_{N+1} + B_{N+1} \subseteq B_N$; that is, $x_A \in y + B_N \subseteq C$. Now, if $x \neq y$, then there exist $C_0, D_0 \in \mathcal{N}_V(0)$ with $(y + C_0) \cap (x + D_0) = \emptyset$. But there exists $A \geq C_0, D_0$ with $x_A \in (y + C_0) \cap (x + D_0)$. This contradiction shows that $x = y \in W$ and W is closed.

Now suppose V is an $\mathbb{R}$-algebra. To show that multiplication in V is continuous it suffices to verify that each bilinear map $f : \mathbb{R}^n \times \mathbb{R}^n \longrightarrow \mathbb{R}^n$ is continuous. Since f is continuous if and only if f followed by each of the n projections is continuous it suffices to verify that any bilinear map $g : \mathbb{R}^n \times \mathbb{R}^n \longrightarrow \mathbb{R}$ is continuous. Let $\{e_1, \ldots, e_n\}$ be the standard basis for $\mathbb{R}^n$ and let $G = (g(e_i, e_j))$ be the $n \times n$ matrix whose (i, j)th entry is $g(e_i, e_j)$. Then $g(x, y) = x^t G y$ where x^t is the transpose of x. Let π_j be the jth coordinate projection in $\mathbb{R}^n$. Since the sequence $(x_m)_m$ in $\mathbb{R}^n$ converges to x if and only if each of the sequences $\pi_j(x_m)$ converges to $\pi_j(x)$, and since a function from $\mathbb{R}^n$ to $\mathbb{R}^k$ is continuous if and only if it preserves limits of sequences, it is easy to see that multiplication by the matrix G is continuous and that the usual inner product on $\mathbb{R}^n$ is continuous; thus, g is continuous. □

The next goal is to show that an isotone linear operator on a finite dimensional partially ordered vector space over $\mathbb{R}$ has a positive eigenvector. The next result will be used for this purpose.

Theorem 5.3.14. *Each complex number that is not on the positive real axis is a root of a nonzero polynomial in* $\mathbb{R}^+[x]$.

Proof. Let $z \in \mathbb{C} \backslash [0, \infty)$. If $z = -a \in \mathbb{R}$, then z satisfies $a + x$. If $z = bi$, then z satisfies $b^2 + b^2 x + x^2 + x^3$, and using this a simple computation gives that if $z = -1 + bi$ with $b \neq 0$, then, since $z + 1$ satisfies the previous polynomial, z is a root of $2(b^2+1) + (b^2+5)x + 4x^2 + x^3$. We can reduce the verification to the case $z = 1 + bi$ with $b > 0$. For if $z = 1 + bi$ with $b < 0$, then z is a root of the same polynomial as its conjugate $\overline{z}$. Now, if $z = a + bi$ with $ab \neq 0$ and $\alpha_0 + \alpha_1 z a^{-1} + \cdots + \alpha_n z^n a^{-n} = 0$, then $\alpha_0 a^n + (\alpha_1 a^{n-1}) z + \cdots + \alpha_n z^n = 0$ and we have the reduction if $a > 0$. If $a < 0$, then $-za^{-1} = -1 - ba^{-1} i = -1 - ci$ and $2(c^2+1) + (c^2+5)(-za^{-1}) + 4(za^{-1})^2 - (za^{-1})^3 = 0$, as we have seen, and hence $-2a^3(c^2+1) + (c^2+5)a^2 z - 4az^2 + z^3 = 0$.

Let $z = 1 + bi$ with $b > 0$. It is straightforward to check that $1 + i$ satisfies $10 + x + x^2 + x^3 + x^4 + 1.25x^5$. Since

$$\alpha_0 + \alpha_1 z + \alpha_2 z^2 + \alpha_3 z^3 =$$

$$\alpha_0 + \alpha_1 + \alpha_2(1-b^2) + \alpha_3(1-3b^2) + [\alpha_1 + 2\alpha_2 + \alpha_3(3-b^2)]bi,$$

z is a root of the third degree polynomial with $\alpha_3 = 3(b^2-3)^{-1}$, $\alpha_1 = \alpha_2 = 1$ and $\alpha_0 = b^2 - 2 + 3(3b^2-1)(b^2-3)^{-1}$, and if $b^2 > 3$ this polynomial has positive coefficients. Similarly, when $b^2 = 3$ z is a root of the fourth degree polynomial with coefficients $\alpha_0 = 12$, $\alpha_1 = 6$, and $\alpha_2 = \alpha_3 = \alpha_4 = 1$, and if $1 < b^2 < 3$, then the coefficients $\alpha_1 = \alpha_2 = \alpha_3 = 1$, $\alpha_4 = (6-b^2)(4(b^2-1))^{-1}$, and $\alpha_0 = (b^6 + 4b^4 + 9b^2 + 6)(4(b^2-1))^{-1}$ give a polynomial which z satisfies.

We are reduced to the case $z = 1 + bi$ with $0 < b < 1$; that is, the argument θ of z is in the interval $(0, \frac{\pi}{4})$. Since $b = \tan\theta$, if we can find a polynomial for $z\cos\theta = e^{i\theta}$, then we will have one for z since $\cos\theta > 0$. We will show that $z = e^{i\theta}$ has a power z_1 in the second quadrant and another power z_2 in the third quadrant whose argument exceeds $\theta + \pi$. Then 0 is an interior point of the triangle with vertices z, z_1 and z_2 and hence $0 = rz + sz_1 + tz_2$ with $r,s,t > 0$. To find $z_2 = z^n = e^{in\theta}$ with $\pi + \theta < n\theta < \frac{3}{2}\pi$, that is, with

$$\frac{1+\frac{\theta}{\pi}}{n} < \frac{\theta}{\pi} < \frac{3}{2n} \tag{5.3.15}$$

take $n \geq 1$ minimal such that the first inequality in (5.3.15) holds. Then $n \geq 2$ and $\frac{\pi}{n-1} < \theta$; so $n \geq 6$ since $\theta < \frac{\pi}{4}$. Now,

$$\frac{\theta}{\pi} \leq \frac{1+\frac{\theta}{\pi}}{n-1} \tag{5.3.16}$$

and the inequality

$$\frac{1+\frac{\theta}{\pi}}{n-1} < \frac{3}{2n} \tag{5.3.17}$$

is equivalent to $\frac{3\pi}{\pi-2\theta} < n$. Since the inequality $\frac{3\pi}{\pi-2\theta} < 6$ is equivalent to $\theta < \frac{\pi}{4}$ we do have (5.3.17) and hence the second inequality in (5.3.15) is also valid.

To find $z_1 = z^n = e^{in\theta}$ with $\frac{\pi}{2} < n\theta < \pi$, or, equivalently, $\frac{1}{2n} < \frac{\theta}{\pi} < \frac{1}{n}$, take $n \geq 1$ minimal such that $\frac{1}{2n} < \frac{\theta}{\pi}$. Then $n \geq 3$ and $\frac{\theta}{\pi} \leq \frac{1}{2(n-1)}$. Since $\frac{1}{2(n-1)} < \frac{1}{n}$ we do have $\frac{\theta}{\pi} < \frac{1}{n}$. □

Let $A \in \mathbb{R}_n$. The *spectral radius* $\rho(A)$ of A is defined to be the maximum of the absolute values of the complex numbers that are eigenvalues of A.

Theorem 5.3.15. *Suppose that $V = \mathbb{R}^n$ is a directed partially ordered vector space such that V^+ is closed and $A \in \mathbb{R}_n$ with $AV^+ \subseteq V^+$. Then the spectral radius $\rho(A)$ of A is an eigenvalue of A and V^+ contains an eigenvector of A belonging to $\rho(A)$.*

Proof. If $\rho(A) = 0$, then A is nilpotent, and if k is its index of nilpotency, then there is a vector $v \in V^+$ with $A^{k-1}v \neq 0$; so $A^{k-1}v$ is an eigenvector of A. Suppose that $\rho = \rho(A) > 0$ and let

$$J = \begin{pmatrix} J_1 & & \\ & \ddots & \\ & & J_k \end{pmatrix}$$

be the Jordan canonical form of A. So each of the Jordan blocks J_i is an $m_i \times m_i$ complex matrix, and for some eigenvalue λ_i,

$$J_i = \begin{pmatrix} \lambda_i & 0 & \cdots & 0 \\ 1 & \lambda_i & \cdots & 0 \\ \vdots & & \lambda_i & 0 \\ 0 & \cdots & 1 & \lambda_i \end{pmatrix}.$$

Let $\{v_{ij} : 1 \le i \le k, 1 \le j \le m_i\}$ be a basis of $\mathbb{C}^n$ going with J but in reverse order; so for every pair i, j, and with $v_{i0} = 0$,

$$Av_{ij} = \lambda_i v_{ij} + v_{i,j-1}. \tag{5.3.18}$$

Note that since A is real $\overline{J} = J$ except for a rearrangement of the blocks, and there is a permutation $i \mapsto \bar{i}$ of $\{1, 2, \ldots, k\}$ of order 2 such that $\bar{i} = i$ if λ_i is real and $\lambda_{\bar{i}} = \overline{\lambda}_i$ for each i; so $J_{\bar{i}} = \overline{J}_i$. We assume that the Jordan basis is chosen so that $v_{\bar{i}j} = \overline{v}_{ij}$ for each pair i, j. Then an element v in $\mathbb{C}^n$ is, in fact, in V precisely when it can be written as

$$v = \sum_{i=1}^{k} \sum_{j=1}^{m_i} a_{ij} v_{ij}, \text{ with } a_{\bar{i}j} = \overline{a}_{ij}. \tag{5.3.19}$$

The following notation will be used:

$$\begin{aligned} \rho = \rho(A) = |\lambda_1| = \cdots = |\lambda_v| > |\lambda_{v+1}| \ge \cdots \ge |\lambda_k|, \\ m = m_1 = \cdots = m_h > m_{h+1} \ge \cdots \ge m_v, \\ \lambda_i = \rho_i z_i \text{ where } |\lambda_i| = \rho_i \text{ for } 1 \le i \le k. \end{aligned} \tag{5.3.20}$$

According to Exercise 16 there is an element $v \in V$ and an $\varepsilon > 0$ such that the real ε-ball $N_\varepsilon(v)$ is contained in V^+. Suppose v is given by (5.3.19) and $M = \sum_{i,j} ||v_{ij}||$ where $||v_{ij}||$ denotes the length of v_{ij} in $\mathbb{C}^n = \mathbb{R}^{2n}$, $0 < \delta < \frac{\varepsilon}{M}$, and $\delta \notin \{-a_{ij} : 1 \le i \le k, 1 \le j \le m_i\}$. Then $b_{ij} = a_{ij} + \delta \ne 0$ for each pair (i, j), and

$$0 \ne w = \sum_{i=1}^{k} \sum_{j=1}^{m_i} b_{ij} v_{ij} \in N_\varepsilon(v) \tag{5.3.21}$$

since $w \in V$ by (5.3.19) and

$$||w - v|| = \delta || \sum_i \sum_j v_{ij} || \le \delta M < \varepsilon.$$

We will show that $V^+ \cap \sum_{i=1}^{h} \mathbb{C} v_{i1} \neq 0$. Toward this goal we first use induction on r to verify that for each i and j,

$$A^r v_{ij} = \sum_{s=0}^{j-1} \binom{r}{s} \lambda_i^{r-s} v_{i,j-s}. \tag{5.3.22}$$

If $r = 0$ this is obvious, and if $r = 1$ (5.3.22) reduces to (5.3.18). Here, of course, the binomial coefficient $\binom{r}{s}$ is 0 when $s > r$. Assuming (5.3.22) for some r we get

$$\begin{aligned}
A^{r+1} v_{ij} &= \sum_{s=0}^{j-1} \binom{r}{s} \lambda_i^{r-s} A v_{i,j-s} \\
&= \sum_{s=0}^{j-1} \binom{r}{s} \lambda_i^{r-s} (\lambda_i v_{i,j-s} + v_{i,j-s-1}) \\
&= \sum_{s=0}^{j-1} \binom{r}{s} \lambda_i^{r+1-s} v_{i,j-s} + \sum_{s=0}^{j-1} \binom{r}{s} \lambda_i^{r-s} v_{i,j-(s+1)} \\
&= \lambda_i^{r+1} v_{ij} + \sum_{s=1}^{j-1} \binom{r}{s} \lambda_i^{r+1-s} v_{i,j-s} + \sum_{s=1}^{j-1} \binom{r}{s-1} \lambda_i^{r+1-s} v_{i,j-s} \\
&\qquad + \binom{r}{j-1} \lambda_i^{r+1-j} v_{i,0} \\
&= \sum_{s=0}^{j-1} \binom{r+1}{s} \lambda_i^{(r+1)-s} v_{i,j-s}.
\end{aligned}$$

Applying A^r to the vector w in (5.3.21) gives, for every $r \in \mathbb{N}$,

$$A^r w = \sum_{i=1}^{k} \sum_{j=1}^{m_i} \sum_{s=0}^{j-1} \binom{r}{s} b_{ij} \lambda_i^{r-s} v_{i,j-s}. \tag{5.3.23}$$

We wish to find an estimate of these A-iterates of w for large values of r. Note that $\binom{r}{s}$ is a polynomial in r of degree s with 0 constant term and s is bounded. Consider a term in (5.3.23) with $i \leq h$ and assume first that $j - s \neq 1$. Then $s \leq j - 2 \leq m - 2 < m - 1$ and for some $\gamma_1, \ldots, \gamma_s \in \mathbb{Q}$

$$\begin{aligned}
&\binom{r}{s} b_{ij} \lambda_i^{r-s} v_{i,j-s} = (\gamma_s r^s + \gamma_{s-1} r^{s-1} + \cdots + \gamma_1 r) b_{ij} \rho^{r-s} z_i^{r-s} v_{i,j-s} \\
&= r^{m-1} \rho^{r-m+1} \Big(\frac{\gamma_s}{r^{m-s-1}} + \frac{\gamma_{s-1}}{r^{m-s}} + \cdots + \frac{\gamma_1}{r^{m-2}}\Big) b_{ij} \rho^{m-1-s} z_i^{r-s} v_{i,j-s} \\
&= r^{m-1} \rho^{r-m+1} g_s(r) b_{ij} \rho^{m-1-s} z_i^{r-s} v_{i,j-s}
\end{aligned} \tag{5.3.24}$$

where

$$g_s(r) = \frac{\gamma_s}{r^{m-s-1}} + \frac{\gamma_{s-1}}{r^{m-s}} + \cdots + \frac{\gamma_1}{r^{m-2}} \tag{5.3.25}$$

and

$$\lim_{r \longrightarrow \infty} g_s(r) = 0. \tag{5.3.26}$$

For terms with $j - s = 1$ (still $i \leq h$) and $j < m$ we again have (5.3.26) since $s = j - 1 < m - 1$; but if $j = m$, then $s = m - 1$,

$$g_{m-1}(r) = \frac{1}{(m-1)!} + \frac{\gamma_{m-2}}{r} + \cdots + \frac{\gamma_1}{r^{m-2}} \tag{5.3.27}$$

and (5.3.24) becomes

$$\binom{r}{m-1} b_{im} \lambda_i^{r-m+1} v_{i1} = r^{m-1} \rho^{r-m+1} g_{m-1}(r) b_{im} z_i^{r-m+1} v_{i1}. \tag{5.3.28}$$

For terms with $h < i \leq v$ we have $s \leq j - 1 \leq m_i - 1 < m - 1$ and again (5.3.26) holds.

Now consider a term with $v < i \leq k$. Then $\rho_i < \rho$ and

$$\binom{r}{s} b_{ij} \lambda_i^{r-s} v_{i,j-s} = r^{m-1} \rho^{r-m+1} g_s(r) b_{ij} \left(\frac{\rho_i}{\rho}\right)^r \frac{\rho^{m-1}}{\rho_i^s} z_i^{r-s} v_{i,j-s}. \tag{5.3.29}$$

Since $0 \leq \rho_i \rho^{-1} < 1$ we have $\lim_{r \rightarrow \infty} r^t (\rho_i \rho^{-1})^r = 0$ for each integer t and hence, by (5.3.25)

$$\lim_{r \longrightarrow \infty} g_s(r)(\rho_i \rho^{-1})^r = 0.$$

To summarize,

$$A^r w = r^{m-1} \rho^{r-m+1} \left[\sum_{i=1}^{h} \frac{b_{im}}{(m-1)!} z_i^{r-m+1} v_{i1} + \sum_{i,j,s} f_{ijs}(r) v_{i,j-s} \right] \tag{5.3.30}$$

and $\lim_{r \rightarrow \infty} f_{ijs}(r) = 0$ for each of the finitely many functions $f_{ijs}(r)$. Note that $A^r w \neq 0$ for each r since if $A^{r_0} w = 0$, then for all $r \geq r_0$

$$\frac{b_{1m}}{(m-1)!} z_1^{r-m+1} + \sum_{j-s=1} f_{1js}(r) = 0,$$

and hence $0 \neq |b_{1m}| = \lim_{r \longrightarrow \infty} (m-1)! |\sum f_{1js}(r)| = 0$. Since V^+ is closed and the $(n-1)$-sphere S^{n-1} in $\mathbb{R}^n$ is compact the sequence $(||A^r w||^{-1} A^r w)_r$ in the compact set $V^+ \cap S^{n-1}$ has a convergent subsequence $(||A^{r_t} w||^{-1} A^{r_t} w)_t$. Let $u_t = ||A^{r_t} w||^{-1} A^{r_t} w$, $c_t = ||A^{r_t} w||^{-1} r_t^{m-1} \rho^{r_t - m+1}$, $d_{it} = ((m-1)!)^{-1} b_{im} z_i^{1-m} z_i^{r_t}$ and set w_t equal to the second sum in (5.3.30) for $A^{r_t} w$. By compactness of the unit circle in $\mathbb{C}$ and by successively taking convergent subsequences of $(z_1^{r_t})_t, \ldots, (z_h^{r_t})_t$ we may assume that

$(z_i^{r_t})_t$ converges for $i = 1, \ldots, h$, and hence each sequence $(d_{it})_t$ converges to d_i, say. Now,

$$u_t = c_t \left(\sum_{i=1}^{h} d_{it} v_{i1} + w_t \right), \tag{5.3.31}$$

and since $||u_t|| = 1$,

$$\lim_{t\to\infty} |c_t| = \lim_{t\to\infty} \frac{1}{||\sum_{i=1}^{h} d_{it} v_{i1} + w_t||}$$
$$= \frac{1}{||\sum_{i=1}^{h} d_i v_{i1}||}.$$

Thus, $(c_t)_t$ is bounded and by taking more subsequences we may assume that $(c_t)_t$ converges to c. But then

$$u = \lim_{t\to\infty} u_t = \sum_{i=1}^{h} c d_i v_{i1} \tag{5.3.32}$$

is an element of $V^+ \cap S^{n-1}$.

Now, take $0 < x = \sum_{i=1}^{h} f_i v_{i1} \in V$ with a minimal number of nonzero coefficients f_i. Suppose that $\lambda_i \neq \rho$ for some i with $f_i \neq 0$. By Theorem 5.3.14 we can find $a_0, \ldots, a_q \in \mathbb{R}^+$ with $a_0 > 0$ such that $\sum_{p=0}^{q} a_p \lambda_i^p = 0$. Let $y = \sum_{p=0}^{q} a_p A^p x$. Since $a_p A^p x \geq 0$ for each p and $a_0 x > 0$, necessarily $y > 0$. Now,

$$\begin{aligned} y &= \sum_{p=0}^{q} a_p \sum_{\ell=1}^{h} f_\ell \lambda_\ell^p v_{\ell 1} \\ &= \sum_{\substack{\ell=1 \\ \ell\neq i}}^{h} f_\ell \left(\sum_{p=0}^{q} a_p \lambda_\ell^p \right) v_{\ell 1} + f_i \left(\sum_{p=0}^{q} a_p \lambda_i^p \right) v_{i1} \\ &= \sum_{\ell \neq i} g_\ell v_{\ell 1}. \end{aligned}$$

Since y has fewer nonzero coefficients than x we have a contradiction. So $f_i \neq 0$ implies that $\lambda_i = \rho$, and $0 < x$ is an eigenvector of A belonging to the eigenvalue ρ. □

We can now show that the lattice order of an archimedean ℓ-field which is algebraic over its subfield of $\overline{f}$-elements can be extended to a total order. We will first verify this locally.

Theorem 5.3.16. *Let R be a commutative domain and an archimedean ℓ-ring. Suppose $0 < a \in R$ is algebraic over the subring L of $\overline{f}$-elements. Then the partial order of the L-subalgebra $L[a] + \mathbb{Z}[a]a$ of R generated by a and L can be extended to a total order of $L[a] + \mathbb{Z}[a]a$ which contains the total order of L given in Theorem 5.3.7.*

Proof. By Exercise 13 we may assume L is a field and by Theorems 5.3.7 and 3.3.2 we may also assume that L is a totally ordered subfield of $\mathbb{R}$ with $(L\cap\mathbb{R}^+)R^+\subseteq R^+$. Let $S=\mathbb{R}\otimes_L L(a)$. Then $S=\mathbb{R}[a]$ is a finite-dimensional $\mathbb{R}$-algebra. By Theorem 5.3.13 it has a unique Hausdorff topology which makes it into a topological $\mathbb{R}$-algebra and it is homeomorphic to some $\mathbb{R}^n$ as a topological vector space over $\mathbb{R}$. Let P be the closure of $L(a)^+=R^+\cap L(a)$ in S. We first note that P is a subsemiring of S with $\mathbb{R}^+P\subseteq P$. For, if $(x_n)_n$ and $(y_n)_n$ are sequences in $L(a)^+$ converging to the elements x and y in P, respectively, then $x+y$, $xy\in P$ since x_n+y_n converges to $x+y$ and x_ny_n converges to xy. Also, if $\alpha\in\mathbb{R}^+$, then $\alpha x\in P$ since if $(\alpha_n)_n$ is a sequence in $L\cap\mathbb{R}^+$ which converges to α, then $(\alpha_n x_n)_n$ is a sequence in $L(a)^+$ which converges to αx. Now, $(L(a), L(a)^+)$ is directed: if $f(a)\in L(a)$, then $f(a)=\sum\alpha_i a^i-\sum\beta_j a^j$ where α_i, $\beta_j\in L\cap\mathbb{R}^+$. So $P-P=S$ since $P-P$ is an $\mathbb{R}$-subalgebra of S which contains 1 and a.

Let $I=P\cap -P$. We claim that I is a proper ideal of S and $\overline{P}=P/I$ is a closed and directed partial order of the $\mathbb{R}$-algebra $\overline{S}=S/I$. Clearly, I is an additive subgroup of S, and it is an ideal since IP, $PI\subseteq I$ and $S=P-P$. $\overline{S}$ has, of course, the quotient topology, and $\overline{P}$ is closed since its inverse image $P+I=P$ is closed. Since $\overline{P}\cap-\overline{P}=0$ and $\overline{S}=\overline{P}-\overline{P}$, it is clear that $(\overline{S},\overline{P})$ is a directed po-algebra over $\mathbb{R}$. Suppose that $I=S$; then $S=P=-P$ and $L(a)^+$ is dense in S. Let $B=\{a_1,\ldots,a_n\}\subseteq L(a)^+$ be an L-basis of $L(a)$ and note that $B\subseteq -P$. Because of the linear homeomorphism of $\mathbb{R}^n$ with the product $\oplus_i\mathbb{R}a_i=S$, for each $i=1,\ldots,n$,

$$U_i=\sum_{j\neq i}\left[-\frac{1}{2n},\frac{1}{2n}\right]a_j+\left[-\frac{3}{2},-\frac{1}{2}\right]a_i$$

is a neighborhood of $-a_i$ and there exist elements $x_i\in U_i\cap L(a)^+$. So there are elements $\alpha_{ij}\in L$ with $\{\alpha_{ij}:i\neq j\}\subseteq[-\frac{1}{2n},\frac{1}{2n}]$ and $\{\alpha_{ij}:i=j\}\subseteq[-\frac{3}{2},-\frac{1}{2}]$ such that $x_i=\sum_{j=1}^n\alpha_{ij}a_j$. But then

$$\sum_{i=1}^n x_i=\sum_{j=1}^n\left(\sum_{i=1}^n\alpha_{ij}\right)a_j$$

and, for each $j=1,\ldots,n$,

$$\sum_{i=1}^n\alpha_{ij}\leq(n-1)\left(\frac{1}{2n}\right)-\frac{1}{2}=-\frac{1}{2n}<0.$$

Thus, we have the contradiction $0\neq\sum_{i=1}^n x_i\in L(a)^+\cap-L(a)^+$.

Clearly, since $a\notin I$ the field $L(a)$ is embedded in $\overline{S}$ via the natural homomorphism, and, upon identification, we have $L(a)^+\subseteq\overline{P}$ and $\overline{S}=\mathbb{R}[a]$. Applying Theorem 5.3.15 to the linear operator on $\overline{S}$ given by $x\mapsto ax$ we obtain a real number $\alpha_0\in\mathbb{R}^+$ and $0\neq x_0\in\overline{P}$ such that $ax_0=\alpha_0x_0$. Then if $b\in\overline{S}$, necessarily $bx_0=f(\alpha_0)x_0$ for any polynomial $f(x)\in\mathbb{R}[x]$ with $b=f(a)$. The mapping φ from $\overline{S}$ to $\mathbb{R}$ given by $b\mapsto f(\alpha_0)$ for $b=f(a)$ is clearly an identity preserving algebra homomorphism. In particular, it embeds the field $L(a)$ in $\mathbb{R}$. To complete the proof note that if $b \in L(a)^+$ or $b \in L \cap \mathbb{R}^+$ then $b\overline{P} \subseteq \overline{P}$; so $f(\alpha_0)x_0 \in \overline{P}$, $f(\alpha_0) \in \mathbb{R}^+$, and hence $\varphi(L(a)^+) \subseteq \mathbb{R}^+$ and $\varphi(b) = b$ if $b \in L$. □

The global extension of the previous result is given next.

Theorem 5.3.17. *Suppose R is an archimedean ℓ-ring and a commutative domain that is algebraic over its subring L of $\overline{f}$-elements. Then the lattice order of R can be extended to an archimedean total order of R which contains the total order $\overline{f}(R)$ of L.*

Proof. As in the proof of Theorem 5.3.16 we may assume L is a field, and then R is also a field. We will first show that if $a_1, \dots, a_n \in R^+$, then $L(a_1, \dots, a_n) = L(a)$ for some $a \in R^+$. Again, as in Theorem 5.3.16, $S = \mathbb{R} \otimes_L L(a_1, \dots, a_n)$ is a finite dimensional Hausdorff topological $\mathbb{R}$-algebra. Since $L(a_1, \dots, a_n)$ is a directed po-algebra over the totally ordered field L, $L(a_1, \dots, a_n)^+$ contains an L-basis $\{b_1, \dots, b_m\}$ of $L(a_1, \dots, a_n)$ by Exercise 16. Since $\sum_{i=1}^m \mathbb{R}^+ b_i$ is a homeomorphic image of $(\mathbb{R}^+)^m$ it has a nonempty interior. Let K be one of the finitely many proper intermediate fields between L and $L(a_1, \dots, a_n)$. By Exercise 17 $\mathbb{R} \otimes_L K$ is a proper topological direct summand of S and hence is a closed nowhere dense subset of S. Thus, the union X of all of these subspaces $\mathbb{R} \otimes_L K$ is nowhere dense in S, and hence there is a nonempty open subset U of $\sum_{i=1}^m \mathbb{R}^+ b_i$ which is disjoint from X. But $L(a_1, \dots, a_n) = \boxplus_i L b_i$ is dense in $S = \boxplus_i \mathbb{R} b_i$ by Theorem 5.3.13 since L is dense in $\mathbb{R}$, and hence there exists some $a \in U \cap L(a_1, \dots, a_n)$. Then $a \in L(a_1, \dots, a_n)^+$ and $L(a)$ is not a proper subfield of $L(a_1, \dots, a_n)$; so $L(a) = L(a_1, \dots, a_n)$.

Now, if K is any finitely generated intermediate extension of $L \subseteq R$, then since $K = L(b) \subseteq L(b^+, b^-) = L(a)$ for some $a \in R^+$, the partial order of K can be extended to a total order of K which contains $\overline{f}(R)$ by Theorem 5.3.16, and hence R^+ is contained in a total order of R which contains $\overline{f}(R)$ by Theorem 5.1.1. □

If $R = L$, then R is totally ordered and the total order given in the previous result is unique. There is also only one total order when R is a finitely generated L-module.

Theorem 5.3.18. *Suppose the commutative domain R is an archimedean ℓ-ring that is a finitely generated module over its subring L of $\overline{f}$-elements. Then the lattice order of R can be extended to a unique total order of R.*

Proof. Using Exercise 13 again, without loss of generality, R and L are fields and R is finite dimensional over L. Since R^+ contains a basis of ${}_L R$ the proof of Theorem 5.3.17 gives that $R = L(a)$ for some $a \in R^+$. Since R is archimedean and finitely rooted it is a direct sum of totally ordered L-subspaces (Exercise 2.4.11 or 2.5.22) each of which is one-dimensional by Theorem 5.3.8. So $R = La_1 \oplus \cdots \oplus La_n$ with $\{a_1, \dots, a_n\} \subseteq R^+$. Using the notation from the proof of Theorem 5.3.16 we have $S = \mathbb{R} \otimes_L R = \boxplus_i \mathbb{R} a_i$ and the closure P of $R^+ = \sum_i (L \cap \mathbb{R}^+) a_i$ in S is now given by $P = \sum_i \mathbb{R}^+ a_i$. So P is a lattice order of S and $S = \oplus_{i=1}^n \mathbb{R} a_i$ is an ℓ-ring extension of R. The eigenvector x_0 belonging to a is in S^+ and $ax_0 = \alpha_0 x_0$ with $\alpha_0 > 0$. Now

$$x_0 = \sum_{i=1}^n \alpha_i a_i \tag{5.3.33}$$

with $\alpha_i \geq 0$ and we claim that $\alpha_i > 0$ for $i = 1,\ldots,n$. If not we can relabel the a_i so that $x_0 = \sum_{i=1}^m \alpha_i a_i$ with $m < n$ and each $\alpha_i > 0$. Since $A = \sum_{i=1}^m La_i$ is not an ideal of R there exist $1 \leq \ell, j \leq n$ with $j \leq m$ such that $a_\ell a_j \notin A$; assume $j = 1$. So $a_\ell a_1 = \sum_{i=1}^n \beta_i a_\ell$ and $0 \neq \beta_k \in L \cap \mathbb{R}^+$ for some $k > m$. Since x_0 is also an eigenvector for $a_\ell \in \mathbb{R}[a]$,

$$a_\ell x_0 = \alpha x_0 = \sum_{i=1}^m \alpha \alpha_i a_i \tag{5.3.34}$$

for some $\alpha \in \mathbb{R}^+$. Now,

$$a_\ell x_0 = \sum_{i=1}^m \alpha_i a_\ell a_i = \sum_{i=2}^m \alpha_i a_\ell a_i + \sum_{i=1}^n \alpha_1 \beta_i a_i. \tag{5.3.35}$$

From (5.3.34) we have $(a_\ell x_0)_k = 0$ and from (5.3.35) we obtain

$$\left(\sum_{i=2}^m \alpha_i a_\ell a_i \right)_k = -\alpha_1 \beta_k < 0$$

where $(v)_i$ denotes the ith coordinate of $v \in S$ with respect to the basis $\{a_1, \ldots, a_n\}$. This is a contradiction since each $\alpha_i a_\ell a_i \in S^+$.

Let $\Sigma = R^{+*}$. Then $(R^+)_\Sigma = \{uv^{-1} : u, v \in R^+ \text{ with } v \neq 0\}$ is a partial order of R which contains R^+ (see Exercise 18). We will show that $(R^+)_\Sigma$ is a total order of R and hence it is the unique total order of R which contains R^+.

Let $0 \neq b \in R$. Then $bx_0 = \beta x_0 = \sum_{i=1}^n \beta \alpha_i a_i$ for some $\beta \in \mathbb{R}$. If $\beta = 0$, then $b = 0$ since S is a vector space over R. Suppose $\beta > 0$. For $i = 1, \ldots, n$ let U_i be an open interval about $\beta \alpha_i$ which excludes 0. Since multiplication by b is continuous on S and x_0 is in the interior of S^+, there exists an open set U contained in the interior of S^+ with $x_0 \in U$ and $bU \subseteq U_1 a_1 + \cdots + U_n a_n$. Since R^+ is dense in S^+ there exists $x \in R^+ \cap U$. So $bx \in R^+$ and $b = bxx^{-1} \in (R^+)_\Sigma$. If $\beta < 0$, then $-b \in (R^+)_\Sigma$ and thus $(R^+)_\Sigma$ is a total order of R. □

Note that the proof of Theorem 5.3.18 shows, more generally, that if $0 < a \in R$ is algebraic over L, then there is a unique total order of $L[a] + \mathbb{Z}[a]a$ which extends its partial order provided $Q(L)(a)^+$ contains a $Q(L)$-basis of $Q(L)(a)$ for which (5.3.33) holds (that is, each $\alpha_i > 0$) for some eigenvector $x_0 \in \overline{P}$ belonging to a positive eigenvalue α_0 of the map given by multiplication by a on $\overline{S} = \mathbb{R}[a]$.

If E is a subfield of $\mathbb{R}$ distinct from $\mathbb{Q}$, then E has an archimedean lattice order that is not a total order. In order to construct such an order we view E as an algebraic extension of a proper subfield and successively lattice order a chain of simple extensions from this subfield to E. The main step in this construction is to extend a lattice order of a subfield K of E to a lattice order of a finite extension $K(b)$ by finding an element a in $K(b)$ with $K(a) = K(b)$ and such that $K^+ + K^+a + K^+a^2 + \cdots$ is a positive cone. So, if a is of order n over K, then, a^n needs to be a K^+-linear combination of its lower powers. This means $\mathrm{Irr}(a, K)$, the irreducible polynomial of a over K, must have the special form $x^n - \alpha_{n-1}x^{n-1} - \cdots - \alpha_0$ with $\alpha_j \in K^+$. The construction will be carried out in a number of steps.

The n elementary symmetric polynomials in the commuting indeterminates $x_1, \ldots, x_n$ over a field K will be denoted by $\sigma_1, \ldots, \sigma_n$. Recall that

$$\sigma_k(x_1, \ldots, x_n) = \sum_{1 \le j_1 < \cdots < j_k \le n} x_{j_1} \cdots x_{j_k} \tag{5.3.36}$$

and that $(-1)^k \sigma_k(x_1, \ldots, x_n)$ is the coefficient of x^{n-k} in the polynomial $(x - x_1) \cdots (x - x_n) \in K[x_1, \ldots, x_n][x]$.

Theorem 5.3.19. *Let $f(x) = \alpha_0 + \alpha_1 x + \cdots + \alpha_{n-1}x^{n-1} + x^n \in \mathbb{R}[x]$ and suppose that all of the roots of $f(x)$ in $\mathbb{C}$ lie inside the circle of radius $\varepsilon = \frac{1}{(n+1)^2}$ about the point $-r$ where $r = \frac{1}{n+1}$. Then $0 < \alpha_0 < \cdots < \alpha_{n-1} < 1$.*

Proof. Let $a_1, a_2, \ldots, a_{2s-1}, a_{2s}, a_{2s+1}, \ldots, a_n$ be the roots of $f(x)$ in $\mathbb{C}$ where $0 \le 2s \le n$, $a_{2j} = \bar{a}_{2j-1}$ is the complex conjugate of a_{2j-1}, for $1 \le j \le s$, and $a_j \in \mathbb{R}$ for $2s+1 \le j \le n$. We will use the following inequalities in which $Re(z)$ denotes the real part of $z \in \mathbb{C}$:

$$r - \varepsilon < Re(-a_j) \quad \text{for} \quad 1 \le j \le n \tag{5.3.37}$$

$$|a_j| < r + \varepsilon \quad \text{for} \quad 1 \le j \le n \tag{5.3.38}$$

$$\binom{n}{k}(r-\varepsilon)^k > \binom{n}{k+1}(r+\varepsilon)^{k+1} \quad \text{for} \quad 0 \le k \le n-1. \tag{5.3.39}$$

The first inequality comes from the inequalities $Re(a_j) + r \le |Re(a_j) + r| \le |a_j + r| < \varepsilon$ and the second from the inequalities $|a_j| = |r - a_j - r| \le |r| + |a_j + r| < r + \varepsilon$. The third is equivalent to

$$\left(\frac{r-\varepsilon}{r+\varepsilon}\right)^k \left(\frac{k+1}{n-k}\right) > r + \varepsilon$$

which becomes, after the substitutions $r - \varepsilon = n(n+1)^{-2}$ and $r + \varepsilon = (n+2)(n+1)^{-2}$,

$$\left(\frac{n}{n+2}\right)^k \left(\frac{k+1}{n-k}\right) > \frac{n+2}{(n+1)^2}.$$

For $k = 0$, 1, 2 this is easy to check. For $k \ge 3$ consider the function $g(x) = \left(\frac{n}{n+2}\right)^x \frac{x+1}{n-x}$ of the real variable x. It suffices to verify that $g(x)$ is a strictly increasing function of x for $2 \le x \le n-1$. Note that its logarithmic derivative is $(\ln g(x))' = \ln\left(\frac{n}{n+2}\right) + \frac{1}{x+1} + \frac{1}{n-x}$. Now, $\frac{1}{x+1} + \frac{1}{n-x} \ge \frac{1}{n} + \frac{1}{n} = \frac{2}{n}$ and we claim that $\ln\left(\frac{n}{n+2}\right) > -\frac{2}{n}$ and hence $(\ln g(x))' > 0$ and $g(x)$ is increasing. To verify the claim let $h(x) = \ln \frac{x}{x+2} + \frac{2}{x}$ for $x \ge 1$. Then $h(1) > 0$, $\lim_{x \to \infty} h(x) = 0$ and $h(x)$ is a decreasing function since $h'(x) = \frac{-4}{x^2(x+2)} < 0$. Thus, $h(x) > 0$.

Now, by Exercise 19, $\alpha_{n-k} = (-1)^k \sigma_k(a_1, \ldots, a_n) \geq \sigma_k(Re(-a_1), \ldots, Re(-a_n)) > 0$ for $0 \leq k \leq n$, and hence by (5.3.37), (5.3.38) and (5.3.39) we have

$$\begin{aligned}
\alpha_{n-k} &\geq \sum_{1 \leq j_1 < \cdots < j_k \leq n} Re(-a_{j_1}) \cdots Re(-a_{j_n}) > \binom{n}{k} (r-\varepsilon)^k \\
&> \binom{n}{k+1} (r+\varepsilon)^{k+1} > \sum_{1 \leq j_1 \leq \cdots \leq j_k \leq n} |a_{j_1} \cdots a_{j_{k+1}}| \\
&\geq |\sigma_{k+1}(a_1, \ldots, a_n)| = \alpha_{n-(k+1)}
\end{aligned}$$

for $0 \leq k \leq n-1$. □

We will now apply this result to finite extensions but first we fix some notation. Let L be a subfield of $\mathbb{R}$ and suppose that L is a po-field with partial order $\leq$ and positive cone $P = P(L)$. The extended positive cone of L obtained by localizing P at P^+ will be denoted by $Q(P)$ and the associated partial order will be denoted by $\leq_q$. So $Q(P) = P_{P^*} = \{uv^{-1} : u \in P, v \in P^*\}$ (see Exercise 18). The usual total order of $\mathbb{R}$ will be denoted by $\leq_u$ and $L^+ = L \cap \mathbb{R}^+$. We note that $L^+ \subseteq Q(P)$ iff $L^+ = Q(P)$, iff $Q(P) \subseteq \mathbb{R}^+$, and $Q(P)$ is a total order of L iff $\leq_u = \leq_q$ on L.

Theorem 5.3.20. *Suppose $L \subseteq K \subseteq \mathbb{R}$ where K is a finite dimensional field extension of L, and (L, P) is a po-field with $\mathbb{Q}^+ P \subseteq P$ and such that $\leq_u = \leq_q$ on L. Then there exists an element $a \in K$ such that $K = L(a)$, $a >_u 1$, and $Irr(a, L) = x^n - \alpha_{n-1} x^{n-1} - \cdots - \alpha_0$ with $0 \neq \alpha_j \in P$.*

Proof. Since K is a finite separable extension of L, $K = L(b)$ for some $b \in K$. Let the irreducible polynomial of b over L be $f(x) = x^n - \beta_{n-1}x^{n-1} - \cdots - \beta_0$, and let $b_1 = b, b_2, \ldots, b_n$ be all the roots of $f(x)$ in $\mathbb{C}$. By considering several cases, and reducing each to the previous one, we will first find an element $c >_u 1$ with $K = L(c)$ such that the nonleading coefficients of $\mathrm{Irr}(c, L)$ lie in $-L^+$. Write $f(x) = (x-b)g(x)$ where $g(x) = (x-b_2)\cdots(x-b_n) = x^{n-1} + \gamma_{n-2}x^{n-2} + \cdots + \gamma_0$. Let $\varepsilon = \frac{1}{n^2}$ and $r = \frac{1}{n}$.

If $b >_u 1$ and, for $j \geq 2$, $|b_j + r| <_u \varepsilon$, let $c = b$. By Theorem 5.3.19 $\gamma_{n-1} = 1 >_u \gamma_{n-2} >_u \cdots >_u \gamma_0 >_u 0$. Consequently,

$$\begin{aligned}
f(x) &= (x-b) \sum_{j=1}^{n} \gamma_{n-j} x^{n-j} \\
&= x^n - \sum_{j=1}^{n-1} (b\gamma_{n-j} - \gamma_{n-(j+1)}) x^{n-j} - b\gamma_0
\end{aligned}$$

with $\beta_0 = b\gamma_0 >_u 0$ and $\beta_{n-j} = b\gamma_{n-j} - \gamma_{n-(j+1)} >_u \gamma_{n-j} - \gamma_{n-(j+1)} >_u 0$ for $j = 1, \ldots, n-1$. Suppose, next, that $b >_u 1 + r$ and, for $j \geq 2$, $|b_j| <_u \varepsilon$. Let $c = b - r$. Then $L(c) = L(b)$, $\mathrm{Irr}(c, L) = f(x+r)$ has the roots $b_1 - r, \ldots, b_n - r$, and $|(b_j - r) + r| = |b_j| <_u \varepsilon$ if $j \geq 2$. So the previous case applies to c. Next, assume $0 <_u b <_u \frac{1}{2}$ and $|b_j| >_u \frac{1}{\varepsilon}$ if $j \geq 2$. Let $c = b^{-1}$. Then Irr $(c, L) = -\beta_0^{-1} x^n f(\frac{1}{x}) = x^n + \beta_1 \beta_0^{-1} x^{n-1} + \cdots + \beta_{n-1} x - \beta_0$ and its roots $b^{-1}, b_2^{-1}, \ldots, b_n^{-1}$ satisfy the previous conditions $c = b^{-1} >_u 2 \geq_u 1+r$, and $|b_j^{-1}| <_u \varepsilon$ for $j \geq 2$. The last special case assumes that $|b_j - b| >_u \frac{1}{\varepsilon} + \frac{1}{2}$ for $j \geq 2$. Take $p \in \mathbb{Q}$ with $p <_u b <_u p + \frac{1}{2}$ and let $c_j = b_j - p$ for $j = 1, \ldots, n$. Then $L(c_1) = L(b)$, $\mathrm{Irr}(c_1, L) = f(x+p)$ and its roots $c_1, \ldots, c_n$ satisfy the previous conditions: $0 <_u c_1 <_u \frac{1}{2}$ and, for $j \geq 2$,

$$\frac{1}{\varepsilon} + \frac{1}{2} <_u |b_j - p + p - b| \leq_u |c_j| + |c_1| = |c_j| + c_1 <_u |c_j| + \frac{1}{2},$$

or $\frac{1}{\varepsilon} <_u |c_j|$. In the general case, let $0 <_u t \in \mathbb{Q}$ with

$$t\left(\frac{1}{\varepsilon} + \frac{1}{2}\right) <_u \bigwedge_{j=2}^{n} |b - b_j|,$$

and let $c = t^{-1}b$. Then $L(c) = L(b)$, $\mathrm{Irr}(c, L) = t^{-n} f(tx)$ has the roots $c_j = t^{-1}b_j$ for $j = 1, \ldots, n$ and, for $2 \leq j$, $|c_j - c| = t^{-1}|b_j - b| >_u \frac{1}{\varepsilon} + \frac{1}{2}$. Note that since each case is reduced to the first all the coefficients in $\mathrm{Irr}(c, L)$ are nonzero.

Now, given $c >_u 1$ with $K = L(c)$, $\mathrm{Irr}(c, L) = x^n - \sum_{j=1}^{n} \gamma_{n-j} x^{n-j}$, and $0 \neq \gamma_{n-j} \in L^+$, write $\gamma_{n-j} = \beta_{n-j}\delta^{-1}$ with $\beta_{n-j}, \delta \in P$ and $\delta \geq_u 1$. Then $a = \delta c >_u 1$ and $\mathrm{Irr}(a, L) = x^n - \sum_{j=1}^{n} \gamma_{n-j}\delta^j x^{n-j}$ with $0 \neq \gamma_{n-j}\delta^j = \beta_{n-j}\delta^{j-1} \in P$. □

By reading the proof of Theorem 5.3.20 backwards we get formulas for elements c and a whose irreducible polynomials have the desired form. Suppose $[L(b) : L] = n$ and $b = b_1, b_2, \ldots, b_n$ are the conjugates of b in $\mathbb{C}$. Let $p, t \in \mathbb{Q}$ with $0 <_u t <_u 2(2n^2+1)^{-1}|b - b_j|$ for each $j = 2, \ldots, n$ and $pt <_u b <_u pt + \frac{1}{2}t$. Then

$$c = \frac{(p+n)t - b}{n(b - pt)} \text{ and } a = \delta c \tag{5.3.40}$$

where $1 \leq_u \delta \in P$ is a common denominator of the nonleading coefficients of $\mathrm{Irr}(c, L)$ when they are written as elements of $Q(P)$.

One consequence of the previous theorem is that $P(K) = P(K; a) = P(L) + P(L)a + \cdots + P(L)a^{n-1}$ and $P(K)a$ are positive cones of K, and clearly, $Q(P(K)) = Q(P(K)a)$. We will denote the partial order of K corresponding to $Q(P(K))$ by $\leq_{q(a)}$. Note that the restriction of $\leq_{q(a)}$ to L is $\leq_q$; that is, $Q(P(L)) = L \cap Q(P(K))$. For, $Q(P(L)) \subseteq Q(P(K))$ since $P(L) \subseteq P(K)$, and, if $u = (\Sigma_{i=0}^{n-1} \beta_i a^i)(\Sigma_{i=0}^{n-1} \gamma_i a^i)^{-1} \in L \cap Q(P(K))$, then $\gamma_i u = \beta_i$ and $u = \beta_i \gamma_i^{-1} \in Q(P(L))$ for any $\gamma_i \neq 0$. In order to see that $\leq_{q(a)} = \leq_u$ we will first relate this equality to the denseness of $\mathbb{Q}$ in K with respect to $\leq_{q(a)}$.

Let $(L, \leq)$ be a po-field with $L \subseteq \mathbb{R}$. The element $y \in L$ is called *$\mathbb{Q}$-approximable with respect to* $\leq$ if for each $0 < \varepsilon \in \mathbb{Q}$ there exists $s \in \mathbb{Q}$ such that $y - \varepsilon < s < y$, or, equivalently, $y < t < y + \varepsilon$ for some $t \in \mathbb{Q}$, and L is called *$\mathbb{Q}$-approximable with respect to* $\leq$ if each of its elements has this property.

Theorem 5.3.21. *Let $(L, P) = (L, \geq)$ be a po-field with L a subfield of $\mathbb{R}$ and $0 \neq P \subseteq \mathbb{R}^+$.*

(a) L is $\mathbb{Q}$-approximable with respect to $\leq_q$ if and only if $\leq_u = \leq_q$ on L.
(b) The set $A = \{y \in L : y$ is $\mathbb{Q}$-approximable with respect to $\leq_q\}$ is a subring of L.
(c) Suppose L is $\mathbb{Q}$-approximable with respect to $\leq_q$ and $1 <_u a \in \mathbb{R}$ is algebraic over L with $Irr(a,L) = x^n - \alpha_{n-1}x^{n-1} - \cdots - \alpha_0$ and $\alpha_j \in P$. Then $L(a)$ is $\mathbb{Q}$-approximable with respect to $\leq_{q(a)}$.

Proof. (a) Suppose L is $\mathbb{Q}$-approximable with respect to $\leq_q$. We need to verify that $L^+ \subseteq Q(P)$. Note first that $\mathbb{Q}^+ \subseteq Q(P)$. For, if $n \in \mathbb{N}$ and $0 \neq b \in P$, then $n = (nb)b^{-1} \in Q(P)$; but $Q(Q(P)) = Q(P)$. Let $y \in L^+$ and take $p, s \in \mathbb{Q}$ with $0 <_u p <_u y$ and $y - p <_q s <_q y$. Then $s \in \mathbb{Q}^+$ since $Q(P) \subseteq \mathbb{R}^+$ gives $0 <_u y - p <_u s$, and hence $y = y - s + s >_q 0$. Conversely, suppose $L^+ \subseteq Q(P)$, $y \in L$ and $0 <_u \varepsilon \in \mathbb{Q}$. Then $y - \varepsilon <_u s <_u y$ for some $s \in \mathbb{Q}$ and hence $y - \varepsilon <_q < s <_q y$.

(b) As in (a) we have $\mathbb{Q}^+ \subseteq Q(P)$. Thus, $\leq_q = \leq_u$ on $\mathbb{Q}$ and hence $\mathbb{Q} \subseteq A$. Suppose $y_1, y_2 \in A$, $0 <_u \varepsilon \in \mathbb{Q}$, and $p_1, p_2 \in \mathbb{Q}$ with $y_1 - \frac{\varepsilon}{2} <_q p_1 <_q y_1$ and $y_2 <_q p_2 <_q y_2 + \frac{\varepsilon}{2}$. Then $(y_1 - y_2) - \varepsilon <_q p_1 - p_2 <_q y_1 - y_2$ shows that $y_1 - y_2 \in A$. The verification that $L^+ \subseteq Q(P)$ in (a) shows that $A \cap \mathbb{R}^+ \subseteq Q(P) \cap A$ and hence $\leq_q = \leq_u$ on A. To see that $y_1 y_2 \in A$, we may assume that $0 <_u y_1, y_2$. Take $t_1, t_2 \in \mathbb{Q}$ with $y_i <_q t_i <_q t_i + 1$ and let $\delta = \frac{\varepsilon}{t_1 + t_2}$. We can find $s_i \in \mathbb{Q}$ such that $y_i - \delta <_q s_i <_q y_i$. We claim that $y_1 y_2 - \varepsilon <_q s_1 s_2 <_q y_1 y_2$. For, $y_1 y_2 - s_1 s_2 = y_1(y_2 - s_2) + s_2(y_1 - s_1) \in Q(P)^*$ since each factor of each term is in $Q(P)^*$; but also

$$y_1(y_2 - s_2) + s_2(y_1 - s_1) <_q t_1 \delta + t_2 \delta = \varepsilon.$$

(c) We assume that $n \geq 2$. Let B be the set of elements in $L(a)$ which are $\mathbb{Q}$-approximable with respect to $\leq_{q(a)}$. As we noted after Theorem 5.3.20, $\leq_q = \leq_{q(a)}$ on L, and since $L \subseteq B$ by (a) and B is a subring of $L(a)$ by (b) we only need to verify that $a \in B$. Let $0 <_u \varepsilon \in \mathbb{Q}$. Note that $f(x) = \mathrm{Irr}(a,L)$ changes sign at a (with respect to $\leq_u$) since a is a simple root of $f(x)$. Since $1 <_u a$ there is some $s \in \mathbb{Q}$ with $1 <_u s$ and $0 <_u f(s) <_u \varepsilon$. So $0 <_{q(a)} f(s) <_{q(a)} \varepsilon$ and we wish to verify that $a <_{q(a)} s <_{q(a)} a + \varepsilon$, or, $0 <_{q(a)} s - a <_{q(a)} \varepsilon$. We will verify that

$$0 <_{q(a)} (s-a)f(s)^{-1} <_{q(a)} 1, \tag{5.3.41}$$

and, thus, $0 <_{q(a)} s - a <_{q(a)} f(s) <_{q(a)} \varepsilon$, as desired. Let $\delta_n = 1$ and $\delta_j = -\alpha_j$ for $0 \leq j \leq n-1$. Then

$$\begin{aligned} f(x) - f(z) &= \sum_{j=1}^{n} \delta_j (x^i - z^j) \\ &= (x-z) \sum_{j=1}^{n} \delta_j \sum_{k=0}^{j-1} z^{j-1-k} x^k \\ &= (x-z) \sum_{k=0}^{n-1} \left(\sum_{j=k+1}^{n} \delta_j z^{j-(k+1)} \right) x^k, \end{aligned}$$

and

$$f(x) = f(x) - f(a) = (x-a)\sum_{k=0}^{n-1} b_k x^k \tag{5.3.42}$$

where, for $0 \leq k = n-1$, $b_k = \sum_{j=k+1}^{n} \delta_j a^{j-(k+1)}$ and $b_k a^{k+1} = \sum_{j=k+1}^{n} \delta_j a^j$. If we let $c_k = \sum_{j=0}^{k} \alpha_j a^j$, then $c_k \in P(L(a))$, $b_k a^{k+1} - c_k = f(a) = 0$ and $b_k = c_k(a^{k+1})^{-1} \in Q(P(L(a)))$. From (5.3.42) we have $f(s)(s-a)^{-1} = \sum_{k=0}^{n-1} b_k s^k \geq_{q(a)} b_{n-1}s^{n-1} = s^{n-1} >_{q(a)} 1$ and (5.3.41) follows. □

We can now easily lattice order subfields of $\mathbb{R}$.

Theorem 5.3.22. *Let L be a proper subfield of the subfield K of $\mathbb{R}$ such that K is algebraic over L. Then K has a partial order for which it is an archimedean ℓ-field and $(L, L\cap\mathbb{R}^+)$ is its subfield of f-elements, and K also has a partial order for which it is an archimedean ℓ-field and L is its trivially ordered subfield of $\overline{f}$-elements. In either case, if $P(K)$ is its positive cone, then $Q(P(K)) = \mathbb{R}^+ \cap K$.*

Proof. Let L have its usual total order $L^+ = L\cap\mathbb{R}^+$ and let E denote an intermediate field between L and K. For a subset B of E let $P_B = \sum_{b\in B} L^+ b$. To produce a lattice order of the first kind we apply Zorn's Lemma to the set

$$\mathscr{S} = \{(E,B) : B \text{ is a basis for } {}_LE, 1 \in B \subseteq \mathbb{R}^+$$

$$\text{and } (E, P_B) \text{ is a po-field with } Q(P_B) = E\cap\mathbb{R}^+\}.$$

$\mathscr{S}$ is not empty since $(L,\{1\}) \in \mathscr{S}$. We partially order $\mathscr{S}$ by: $(E_1,B_1) \leq (E_2,B_2)$ if $E_1 \subseteq E_2$ and $B_1 \subseteq B_2$. Suppose $\{(E_\lambda, B_\lambda) : \lambda \in \Lambda\}$ is a chain in $\mathscr{S}$. Then if $E = \cup_\lambda E_\lambda$ and $B = \cup_\lambda B_\lambda$ it is easy to check that $(E,B) \in \mathscr{S}$. Suppose (E_0,B_0) is a maximal element in $\mathscr{S}$. If $b \in K\backslash E_0$, then by Theorems 5.3.20 and 5.3.21 we can find $a \in K$ such that $E_0(a) = E_0(b)$ and $(E_0(a), B_0\{1,a,\ldots,a^{n-1}\}) \in \mathscr{S}$. Thus $K = E_0$. Of course, as a vector lattice over L, $K = \oplus_{b\in B_0} Lb$ is archimedean and the subring of f-elements is clearly L.

To obtain a lattice order of the second type we consider the set $\mathscr{S}_1$ of ordered pairs (E,B) as in $\mathscr{S}$ but change the one condition $1 \in B$ to the condition $1 \notin P_B$. To see that $\mathscr{S}_1 \neq \emptyset$ we use Theorems 5.3.20 and 5.3.21 to obtain an element $a \in K\backslash L$ and the basis $B = \{a,\ldots,a^n\}$ such that $(L(a),B) \in \mathscr{S}_1$. If $1 = \sum_{j=1}^{n} \beta_j a^j$ with $\beta_j \in L^+$, then $a^n = \beta_n^{-1} - \sum_{j=1}^{n-1} \beta_n^{-1}\beta_j a^j$ and $0 <_u -\beta_n^{-1}\beta_{n-1} <_u 0$ since a is obtained from Theorem 20; so $1 \notin P_B$. The proof that a maximal element (E_0,B_0) in $\mathscr{S}_1$ must have $E_0 = K$ is the same as before except $B_0\{1,a,\ldots,a^{n-1}\}$ is replaced by $B_0\{a,\ldots,a^n\} = C_0$. Again, if 1 were in P_{C_0}, then $1 = \sum_k \gamma_k c_k$ with $\gamma_k \in L^+$ and $c_k \in C_0$; so $1 = \sum_{j=1}^{n} \beta_j a^j$ with $\beta_j \in P_{B_0}$ and we have the contradiction $0 \neq \beta_n^{-1}\beta_{n-1} \in P_{B_0} \cap -P_{B_0}$. Now, if $(K,B) \in \mathscr{S}_1$, then (K,P_B) is a vector lattice over (L,L^+) and L is contained in the subfield of $\overline{f}$-elements. On the other hand, if u is in this subfield, then $ub \in b^{\perp\perp} = Lb$ for any $b \in B$; so $u \in L$. □

Exercises.

1. Let R be a D-ℓ-domain in which $d(R)^*$ is a right Öre set. Show that the additive subgroup of R generated by $d(R)$ is the largest convex ℓ-subring of R that is a finite valued D-domain.

2. Let A be a totally ordered division ring, $R = A[x]$ and $R^+ = \{a_0 + a_1x + \cdots + a_nx^n : a_0 \geq 0 \text{ and } a_n \geq 0\}$. Show that R is an ℓ-ring with two roots, $d(R)^* = f(R)^*$ is an Öre set, and $S(R)$ is multiplicatively closed, but R is not a D-ring.

3. Let F be a totally ordered field, $R = F[x]$ and $R^+ = F^+[x]$. Show that R is a finite valued D-ℓ-domain, $\Gamma(R_{d(R)^*})$ is a torsion-free rooted po-group that is not locally finite modulo its maximal totally ordered subgroup, and $R_{d(R)^*}$ is not a field.

4. Let R be a finite valued D-ℓ-domain in which $d(R)^*$ is a right Öre set. Show that $R_{d(R)^*}$ is an ℓ-simple ℓ-ring.

5. Let Δ be a finitely rooted torsion-free rooted po-group which is locally finite modulo its maximal totally ordered subgroup and let Δ_1 be the group Δ with the total order extending Δ^+. If A is a totally ordered domain and T is the convex subring of $V(\Delta_1, A)$ consisting of those elements f with maxsupp $f \leq 0$, show that $R = T \cap W(\Delta, A)$ is an ℓ-subring of $W(\Delta, A)$ and is a finite valued D-ring (see Theorems 3.5.12 and 3.5.13).

6. Let ${}_RM$ be an ℓ-module over the ℓ-ring R. Suppose $0 \in B \subseteq f(M)$ and B is convex. If B^* consists of regular elements on M, that is, $r(b;M) = 0$ for $0 < b \in B$, show that $d(M) \subseteq f(M) \cup B^\perp$. (If $a \in d(M)$, $b \in B$, $c = a \wedge b$ and $x \wedge y = 0$, then $ax \wedge cy = 0 = c(ax \wedge y)$.)

7. Let R be an ℓ-ring and ${}_RM$ an ℓ-module.
   (a) If $d_1, d_2 \in R^+$ with $d_1d_2 \in d(M)$ and $r(d_1;M)^+ = 0$, show that $d_2 \in d(M)$.
   (b) Suppose $d \in d(M)$ and $u \in M^+$ is a d-element on R with $\ell(u;R) = 0$. If $\{d^nu : n \in \mathbb{N}\} \subseteq u^\perp$ show that $d^i \wedge d^j = 0$ for $i \neq j$.

8. Let R be an ℓ-ring with $f(R) \neq 0$. Suppose that either $f_\ell(R)^*$ consists of regular elements on ${}_RR$ and $f_r(R)^*$ consists of regular elements on R_R, or $f(R)^*$ consists of regular elements on ${}_RR$ and R_R. Show that:
   (a) $d(R) \subseteq f(R) \cup f(R)^\perp$ (use Exercise 6).
   (b) If $d \in d(R)$, then either $\{d^n : n \in \mathbb{N}\}$ is a disjoint subset of distinct elements of $f(R)^\perp$ or $d^n \in f(R)$ for some n (use Exercise 7 with $u \in f(R)$).
   (c) If $0 < d \in R$ is regular, then $d^n \in f(R)$ for some $n \in \mathbb{N}$ iff $d \in d(R)$ and there are elements $a_0, a_1, \ldots, a_m \in F(R)$ not all 0 such that $a_0 + a_1d + \cdots + a_md^m = 0$ or $a_0 + da_1 + \cdots d^ma_m = 0$.
   (d) If R is an ℓ-unital ℓ-domain then $f(R) = d(R) \cap 1^{\perp\perp}$, and hence $F(R) = 1^{\perp\perp}$ iff $(1^{\perp\perp})^+ \subseteq d(R)$.
   (e) If R is a local ℓ-unital ℓ-domain, then $F(R) = 1^{\perp\perp}$ iff $\mathscr{U}(R) \cap (1^{\perp\perp})^+ \subseteq \mathscr{U}(R^+)$, iff $Fa = a^{\perp\perp} = aF$ for each $a \in \mathscr{U}(R^+)$.

9. Let $R = A[x]$ be the polynomial ring over the totally ordered domain A and suppose R is an ℓ-ring extension of A. Show that $A \subseteq F(R)$ and $A^+x \subseteq d(R)$ iff R is totally ordered, or $R^+ = A^+[x]$, or, for some $k \geq 2$, $F(R) = A[x^k]$ and $R = F(R) \oplus F(R)x \oplus \cdots \oplus F(R)x^{k-1}$ as an $F(R)$-f-bimodule. (Use Exercise 8; either $Ax^n \subseteq f(R)^\perp$ for all n or $Ax^k \not\subseteq f(R)^\perp$ for some minimal k.)

10. Suppose $0 < u$ is a d-element of the ℓ-ring R. If the centralizer of u is an ℓ-domain and contains $[0,u]$ show that u is a basic element. (If $a, b \in [0,u]$, then $ab \leq au \wedge bu$.)

11. Let R be an ℓ-unital ℓ-domain in which $F = F(R)$ is a division ring and is a proper subring of R.
    (a) Show that the following are equivalent:
        (i) $R =$ lex $(F \oplus Fa)$ for some $a > 0$ with $a^2 \in F$.
        (ii) $d(R) = f(R) \cup f(R)^{\perp +}$ and $F^\perp$ is totally ordered.
        (iii) $d(R) = f(R) \cup f(R)^{\perp +}$ and each element of $F^{\perp +}$ is left (right) algebraic over F in the sense of Exercise 8(c).
    (b) If R satisfies the conditions in (a) show that $F \oplus Fa$ is a division ring iff R is a domain, iff $b^2 \neq 1$ for every $b \in F^\perp$.
    (c) Let F be a subfield of the totally ordered field $L = F(a)$ where $a > 0$, $a^2 \in F$ and a^2 is not a square in F. Show that $L[x]$ with the positive cone $L[x]^+ = \{c_0 + c_1 a + d_1 x + \cdots + d_n x^n : c_0, c_1 \in F, d_j \in L,\ d_n > 0 \text{ if } n \geq 1, \text{ and } c_0 \geq 0 \text{ and } c_1 \geq 0 \text{ if } n = 0\}$ is an ℓ-ring which satisfies the conditions in (a) and (b) and $R \neq F \oplus Fa$.
    (d) Give an example of an ℓ-unital ℓ-domain that satisfies the conditions in (a) and is a D-ℓ-ring but which isn't a domain. (Use diagonal matrices.)
    (e) Show that a totally ordered division ring which has a positive central element that is not a square admits a division ℓ-ring extension of the type given in Theorem 5.3.5.

12. This exercise shows that in an archimedean ℓ-field with $1 \not> 0$ the subfield L in Theorem 5.3.7 need not be a maximal trivially ordered subfield. Let $R = \mathbb{Q}(\sqrt{2}, \sqrt{3}) = \mathbb{Q}(\sqrt{2} + \sqrt{3}) \oplus \mathbb{Q}(2 + \sqrt{6}) \oplus \mathbb{Q}(3 + \sqrt{6}) \oplus \mathbb{Q}(2\sqrt{3} + 3\sqrt{2})$ as an ℓ-group. Show that R is an ℓ-field with $L = \mathbb{Q}$ and $\mathbb{Q}(\sqrt{2})^+ = \mathbb{Q}(\sqrt{3})^+ = 0$. (Note that the basis used for R is $(\sqrt{2} + \sqrt{3})\{1, \sqrt{2}, \sqrt{3}, \sqrt{6}\}$.)

13. Let R be a commutative archimedean ℓ-domain whose subring L of $\overline{f}$-elements is not 0 and let $Q(L)$ be the totally ordered field of quotients of L. Show that the L-divisible hull $Q = R_{\overline{f}(R)^*} = Q(L) \otimes_L R$ of ${}_L R$ is an archimedean ℓ-domain extension of R whose subring of $\overline{f}$-elements is $Q(L)$.

14. Let R be as in the previous exercise and assume that R is a finitely generated L-module. Show that Q is the field of quotients of R and there exist $0 < a_1, \ldots, a_n \in R$ and $0 \neq \alpha \in L$ such that $Q = Q(L)a_1 \oplus \cdots \oplus Q(L)a_n$, $R = R_{a_1} \oplus \cdots \oplus R_{a_n}$ and $\alpha R \subseteq La_1 \oplus \cdots \oplus La_n$.

15. Let V be a set and for each $x \in V$ let $\mathcal{N}_x$ be a family of subsets of V. Consider the following conditions.

(i) If $x \in V$ and $A \in \mathcal{N}_x$, then $x \in A$.
(ii) If $A, B \in \mathcal{N}_x$, then $A \cap B \in \mathcal{N}_x$.
(iii) If $A \in \mathcal{N}_x$ and $A \subseteq B \subseteq V$, then $B \in \mathcal{N}_x$.
(iv) If $A \in \mathcal{N}_x$ there exists $B \in \mathcal{N}_x$ such that $B \subseteq A$ and $B \in \mathcal{N}_y$ for every $y \in B$.

Show that if (i), (ii) and (iii) hold for each $x \in V$, then $\mathcal{T} = \{U \subseteq V : U \in \mathcal{N}_x$ for each $x \in U\}$ is a topology for V and $\mathcal{N}_x \subseteq \mathcal{N}(x)$ for each $x \in V$. If (iv) is also satisfied for every $x \in V$ show that $\mathcal{N}_x = \mathcal{N}(x)$ for each x.

16. (a) Show that a po-vector space V over a directed po-division ring is directed iff V^+ contains a basis of V.
(b) Show that if $V = {}_{\mathbb{R}}\mathbb{R}^n$ is a po-vector space, then V is directed iff V^+ has a nonempty interior. (If the ε-ball about v is contained in V^+ and $0 \neq u \in V$, then $\varepsilon(2\|u\|)^{-1}u + v \in V^+$. For the converse show that V^+ contains a homeomorphic copy of $(\mathbb{R}^+)^n$.)

17. Let V be a Hausdorff finite dimensional topological vector space over $\mathbb{R}$. Show that each algebraic complement of a vector subspace W of V is a topological complement of W.

18. Let R be a po-domain and set $\Sigma = R^+ \backslash \{0\}$. Suppose Σ consists of regular elements of R, the classical right quotient ring R_Σ of R with respect to Σ exists, and $s\Sigma \cap t\Sigma \neq \emptyset$ for all $s, t \in \Sigma$. Show that $(R^+)_\Sigma = \{as^{-1} : a, s \in R^+$ with $s \neq 0\}$ is a partial order of R_Σ which contains R^+, and $(R^+)_\Sigma$ is contained in any total order of R_Σ which contains R^+.

19. Let C be the algebraic closure of the real closed field F, and let $\sigma_k(x_1, \ldots, x_n)$ be the kth elementary symmetric function (see (5.3.36)). Suppose that $a_1, \bar{a}_1, \ldots, a_s, \bar{a}_s, a_{2s+1}, \ldots, a_n \in C$ with $a_j \in F$ for $j > 2s$ and $Re(a_j) < 0$ for every a_j. Show that for $1 \leq k \leq n$, $(-1)^k \sigma_k(a_1, \bar{a}_1, \ldots, a_n) \geq \sigma_k(Re(-a_1), Re(-a_1), \ldots, Re(-a_n)) > 0$.

20. Let $f(x) = \alpha_0 + \alpha_1 x + \cdots + \alpha_n x^n \in F[x]$ where F is a real closed field. For $r \in F$ let $f(x+r) = \beta_0 + \beta_1 x + \cdots + \beta_n x^n$. Show that there exists $0 < \varepsilon \in F$ such that for all $r \in F$ with $|r| < \varepsilon$, if $\alpha_j \neq 0$ then $\alpha_j \beta_j > 0$.

21. Let L be a subring of the po-ring $(R, R^+) = (R, \leq)$ and let (L, P) be a po-ring such that R is a left po-module over (L, P). Put $L^+ = L \cap R^+$.

(a) If $\ell(R^+) \cap L^+ = 0$ and P is a total order of L, show that $L^+ \subseteq P$.
(b) Suppose $1 \in R \cap L$, $0 < a \in R$, and $f(x) = \alpha_0 + \alpha_1 x + \cdots + \alpha_{n-1}x^{n-1} + x^n \in L[x]$ with $f(a) = \alpha_0 + \alpha_1 a + \cdots + \alpha_{n-1}a^{n-1} + a^n \leq 0$. If $\gamma \in L$ and $\gamma + \alpha_j - 1 \in P \cup L^+$ for $j = 0, \ldots, n-1$ show that $a \not\geq \gamma$. (Otherwise, $\alpha_j a^j \geq (1-a)a^j$.)

(c) Suppose $1 \in R \cap L$ and $a \in R^+$ with $Pa \le b$. If (L,P) is directed show that there do not exist an element $c \in R^+$ and a monic polynomial $f(x) \in L[x]$ with $1 \le ac$ and $f(bc) \le 0$.
(d) Suppose the positive elements in the lattice-ordered division ring R are left algebraic over its subdivision ring L in the sense that each is a root of some nonzero polynomial over L with coefficients on the left. If (L,P) is a totally ordered ring and $PR^+ \subseteq R^+$ show that ${}_LR$ is archimedean over (L,P).

22. Let $L \subseteq K \subseteq \mathbb{R}$ be fields and let trK/L be the transcendence degree of K over L.

(a) Show that the number of lattice orders of K which can be constructed using Theorem 5.3.22 is equal to the number of subfields of K over which K is algebraic.
(b) Suppose that $trK/\mathbb{Q} = \aleph_\alpha$. Show that K has as many lattice orders as it does archimedean lattice orders and this number is $2^{\aleph_\alpha}$. (If X is a transcendence base of K over $\mathbb{Q}$ find subfields of $\mathbb{Q}(X)$ over which it is algebraic.)
(c) If $trK/\mathbb{Q} = n \ge 1$ show that K has at least $\aleph_0$ lattice orders.
(d) If K is a finite proper extension of $\mathbb{Q}$ show that K has exactly $\aleph_0$ lattice orders and each is archimedean; contrast this with Exercise 5.1.16. (Use Theorem 5.3.20, Exercise 20, and Exercise 21(d) or Exercise 3.2.33.)
(e) Show that the real closure of $\mathbb{Q}$ has exactly $2^{\aleph_0}$ (archimedean) lattice orders and its algebraic closure has no lattice orders.

23. Let K be a field with a unique total order K^+ and let E be a subfield of K with K algebraic over E. Let $E^+ = E \cap K^+$. Show the equivalence of the following.

(a) K is archimedean.
(b) E is archimedean.
(c) E has a total order P such that (E,P) is archimedean and $PK^+ \subseteq K^+$.
(d) K has an archimedean lattice order P_1 with $E^+P_1 \subseteq P_1$.

24. (a) Let $b = p + q\sqrt{2} \in \mathbb{Q}(\sqrt{2})$ with $q \ne 0$. Show that $b^2 \in \mathbb{Q}^+ + \mathbb{Q}^+ b$ iff $p \ge 0$ and $2q^2 > p^2$. Generalize.
(b) Use (5.3.40) to first lattice order $\mathbb{Q}(\sqrt{2})$ and then to extend this lattice order to a lattice order of $\mathbb{Q}(\sqrt[4]{2})$.

25. Let K be an archimedean ℓ-field, L its subfield of $\bar{f}$-elements, $0 < a \in K$ algebraic over L, $S = \mathbb{R} \otimes_L L(a)$, P the closure of $L(a) \cap K^+$ in S, $I = P \cap -P$ and $\overline{S} = S/I$. (This is the set-up in the proof of Theorem 5.3.15.) Suppose $0 \ne x_0 \in \overline{S}$ and $ax_0 = \alpha_0 x_0$ with $\alpha_0 \in \mathbb{R}$. Show that x_0 is a unit of $\overline{S}$ iff $\overline{S} = \mathbb{R}$.

26. Let $\sigma : F \longrightarrow F$ be an automorphism of the totally ordered field F and let $\Delta = \mathbb{Z}$ with the trivial partial order. Show that $V(F * \Delta)$ has a commutative enabling basis, $V(F * \Delta) = b^\perp \oplus b^{\perp\perp}$ for each basic element b, and $V(F * \Delta)$ is f-embeddable iff $\sigma = 1$.

27. Suppose K is an f-embeddable ℓ-field and $B(K)/F^{+*}$ is not torsion-free. Show that there does not exist an f-embedding $\varphi : K \longrightarrow V_{\mathscr{S}}(F,\Delta)$. Give an explicit example of such an ℓ-field K.

Notes. The extension of the Hahn embedding theorem for totally ordered fields to ℓ-fields given in Theorem 5.3.3 as well as the extension of the lattice order to a total order given in Theorem 5.3.2 is due to Conrad and Dauns [CD]; the generalization to D-domains comes from Steinberg [ST2]. The two-root case given in Theorem 5.3.5 is in Ma and Steinberg [MS] and the embedding theorem of Theorem 5.3.4 is due to Redfield [R4]. The theory of archimedean ℓ-fields is due to Schwartz [SCH2] and much of the extension of his results to domains is in Ma [M5]. Birkhoff and Pierce [BP] asked if $\mathbb{R}$ had any lattice orders other than the usual one and Wilson [WI1] and [WI2] answered their question by constructing the lattice orders presented here. A good source of material on topological vector spaces is Narice and Beckenstein [NB], and Theorems 5.3.13 and 5.3.14 are due to Birkhoff [BIR4] and appear in Berman and Plemmons [BPL]. Most of the exercises come from the papers of Ma and Steinberg [MS], Schwartz, [SCH2] and Wilson [WI2].

Chapter 6
Additional Topics

Two themes that occur in this chapter concern the attempt to describe all lattice orders of a particular type on some familiar rings and the attempt to recognize some familiar lattice-ordered rings. Thus, a characterization is given of the real group ℓ-algebra of a locally finite group with the coordinatewise partial order and of the canonically ordered ℓ-group ring of a finite cyclic group with coefficients in a matrix ring over a totally ordered field. All of the lattice orders of a matrix algebra over a totally ordered subfield of the reals and of a 2×2 matrix algebra over any totally ordered field are determined. Also, for a polynomial ring in one variable over an ℓ-simple totally ordered Öre domain all nontotal lattice orders which extend the order of the coefficient ring and in which all squares are positive are described. This description is a specialization of a description of some lattice orders on a semigroup ring over a mopops.

Another theme that occurs concerns the centrality of f-elements. Along this vein it is shown that an algebraic f-element in an ℓ-domain is central and also that the subalgebra of f-elements is central provided that each commutator in which one factor is an f-element is bounded by a fixed power of the f-element. Another commutativity result which is given is that a totally ordered domain in which each nonzero left ideal contains a strictly positive lower bound of 1 is commutative provided the domain has a positive semidefinite form which has a nontrivial solution. Those totally ordered fields for which all positive semidefinite polynomials are sums of squares of rational functions will be identified, and so will the rings which have the property that each of their partial orders is contained in a total order.

6.1 Lattice-ordered Semigroup Rings

Some aspects of lattice-ordered semigroup rings will be developed in more detail in this section. At the outset we should mention that there is a definite distinction between the theory of group algebras and the theory of lattice-ordered group algebras. It is easy to verify that the lattice-ordered group algebra of a finite group determines

S.A. Steinberg, *Lattice-ordered Rings and Modules*,
DOI 10.1007/978-1-4419-1721-8_6, © Springer Science + Business Media, LLC 2010

the group (Exercise 1). However, there exist nonisomorphic finite groups whose group algebras for any field are isomorphic. An abstract description of the semigroup algebra over $\mathbb{R}$ of a locally finite left cancellative semigroup will be given. The main ingredient in this description is the familiar (in $\mathbb{R}$) expression of the product x^+y^+ as the least upper bound of the smaller products $[0,x^+]y$. Interestingly, another ingredient is the vanishing of the second cohomology group of the semigroup with coefficients in a totally ordered field in which the group of positive elements is divisible. Also, the partial order of a mopops which is a monoid will be altered to produce several new mopops. These alterations will be used to characterize a large set of lattice orders of a semigroup ring in which all squares are positive. In this way we will determine the lattice orders of a polynomial ring in one variable over a totally ordered division ring for which all squares are positive.

Let $K[S;\tau]$ be a crossed product semigroup ℓ-algebra of the trivially ordered multiplicative semigroup S over the totally ordered field K with trivial action. Recall from Exercise 3.5.13 this means $\tau : S \times S \longrightarrow K^{+*}$ is a *2-cocycle*, that is, $\tau(st,u)\tau(s,t) = \tau(s,tu)\tau(t,u)$ for all s, t, $u \in S$, $K[S;\tau] = \oplus_{s\in S} K\bar{s}$ as a vector lattice over K with $S \longrightarrow \{\bar{s} : s \in S\}$ a bijection and each $\bar{s} > 0$, and the multiplication of these basis elements is given by $\bar{s}\bar{t} = \tau(s,t)\overline{st}$. $K[S;\tau]$ will be referred to as a twisted semigroup ℓ-algebra. For the purpose of characterizing and studying these ℓ-algebras we make the following definition. An ℓ-algebra R over K is called *basic* if its underlying vector lattice has a canonical basis; that is, $R = \oplus_b Kb$. According to Exercise 2.4.11 (or 2.5.22) an ℓ-algebra is basic if and only if it is finite valued, K-archimedean, and each basic element is K-convex. When $K = \mathbb{R}$ the last condition can be omitted since a totally ordered archimedean $\mathbb{R}$-vector space is isomorphic to $\mathbb{R}$ (for example, see Exercise 3.1.21). Basic ℓ-fields have already been encountered in Theorems 5.2.3, 5.2.17, and 5.2.31.

Theorem 6.1.1. *An ℓ-algebra over the totally ordered field K is a twisted semigroup ℓ-algebra if and only if it is basic and the product of basic elements is basic.*

Proof. Suppose R is a basic ℓ-algebra in which the product of basic elements is basic. So $R = \oplus_{s\in B} Ks$. Let G be the multiplicative group K^{+*} and let $S = \{Gs : s \in B\}$. If $s,t \in B$, then $st = \tau(Gs,Gt)u$ for some $u \in B$ and some element $\tau(Gs,Gt) \in G$. So $GsGt = Gst = Gu$ and associativity of R gives that S is a semigroup and τ is a 2-cocycle. If we put $\overline{Gs} = s$ for $s \in B$, then $R = K[S;\tau]$. The converse is obvious. □

It will be convenient to express some relations of twisted semigroup ℓ-algebras in terms of the cohomology of semigroups. For each $n \in \mathbb{N}$ the set of functions $(K^{+*})^{S^n}$ is an abelian group denoted by $C^n(S,K^{+*})$ and called the *group of (positive) n-cochains of the semigroup S with coefficients in the totally ordered field K*. The set of 2-cocycles, $Z^2(S,K^{+*})$, is easily seen to be a subgroup of $C^2(S,K^{+*})$. The *coboundary map* $d_S = d : C^1(S,K^{+*}) \longrightarrow C^2(S,K^{+*})$, defined by $dc(s,t) = c(s)c(t)c(st)^{-1}$, is a homomorphism and its image, $B^2(S,K^{+*})$, the *group of coboundaries*, is a subgroup of $Z^2(S,K^{+*})$. The factor group $H^2(S,K^{+*}) = Z^2(S,K^{+*})/B^2(S,K^{+*})$ is called the *second cohomology group of S with coefficients in* K^{+*}. The assignment $S \mapsto H^2(S,K^{+*})$ is a contravariant functor from the category of semigroups to the category of abelian groups. To see this note that a homomorphism

$f : S \longrightarrow T$ induces group homomorphisms $\hat{f} : C^n(T, K^{+*}) \longrightarrow C^n(S, K^{+*})$ given by $\hat{f}(g)(s_1, \ldots, s_n) = g(f(s_1), \ldots, f(s_n))$. It is easy to check that $\hat{f}$ takes a 2-cocycle to a 2-cocycle and commutes with d: $\hat{f} d_T = d_S \hat{f}$. So $\hat{f}$ takes coboundaries to coboundaries and it induces a homomorphism $f^* = H^2(f, K^{+*}) : H^2(T, K^{+*}) \longrightarrow H^2(S, K^{+*})$. Two twisted semigroup ℓ-algebras over K are isomorphic if and only if the semigroups are isomorphic by a twist preserving isomorphism. The details are left for Exercise 4. Note that $K[S;\sigma]$ is isomorphic to the semigroup ℓ-algebra $K[S]$ if and only if σ is a coboundary.

Not unexpectedly more information is obtained when the semigroup is finite and we now turn our attention to this case. A semigroup is *locally finite* if each of its finitely generated subsemigroups is finite, and an algebra over a field is *locally finite* if each of its finitely generated subalgebras is finite dimensional. It is clear that a twisted semigroup ℓ-algebra is locally finite if and only if its base semigroup is locally finite. A 2-cocycle $\sigma \in Z^2(S, K^{+*})$ is called a *fundamental 2-cocycle* if, for all $s, t \in S$, $\sigma(s,t) = \sigma(t,s)$ and $\sigma(s,t) = 1$ whenever $st = ts$. The set of fundamental 2-cocycles is clearly a subgroup of $Z^2(S, K^{+*})$.

Theorem 6.1.2. *Let S be a locally finite semigroup and suppose K is a totally ordered field for which K^{+*} is a divisible group. Let H be the subgroup of $Z^2(S, K^{+*})$ consisting of the fundamental 2-cocycles.*

(a) $Z^2(S, K^{+*})$ is the direct product of $B^2(S, K^{+*})$ and H.

(b) A 2-cocycle σ is in H if and only if $\sigma(s^n, s^m) = 1$ for every $s \in S$ and all m, $n \subset \mathbb{N}$.

(c) If S is commutative, then $H^2(S, K^{+*}) = 1$.

Proof. Let $G = K^{+*}$ and fix $\sigma \in Z^2(S, G)$; put $R = K[S;\sigma]$. Note that $S \cong \{G\bar{s} : s \in S\}$. We proceed in steps to verify (a).

(i) Suppose $u, v \in G\bar{s}$ with $u^{j+p} = u^j$ and $v^{k+q} = av^k$ for some j, k, p, $q \in \mathbb{N}$ and $a \in G$. Then $u^{k+q} = u^k$. Since $(Gu)^{k+q} = (Gv)^{k+q} = (Gv)^k = Gu^k$, $u^{k+q} = bu^k$ for some $b \in G$. Then

$$\begin{aligned} b^{j+p} u^{jk} &= b^{j+p}(u^{j+p})^k = (bu^k)^{j+p} \\ &= u^{(k+q)(j+p)} = u^{j(k+q)} = b^j u^{kj}. \end{aligned}$$

So $b^{j+p} = b^j$ and hence $b = 1$.

(ii) For each $s \in S$ there is a unique $u \in G\bar{s}$ with $u^{j+p} = u^j$ for some j, $p \in \mathbb{N}$. Since S is locally finite there are integers $j, p \in \mathbb{N}$ with $s^{j+p} = s^j$. Then $G\bar{s}^j = G\bar{s}^{j+p}$ and $\bar{s}^{j+p} = a\bar{s}^j$ with $a \in G$. Now,

$$\left(a^{-\frac{1}{p}}\bar{s}\right)^{j+p} = a^{-\left(\frac{j+p}{p}\right)} a\bar{s}^j = a^{-\frac{j}{p}}\bar{s}^j = \left(a^{-\frac{1}{p}}\bar{s}\right)^j;$$

so $u = a^{-\frac{1}{p}}\bar{s}$ is a desired element. If $v \in G\bar{s}$ with $v^{k+q} = v^k$, then, by (i), $u^{k+q} = u^k$, and since $u = bv$, $b^k v^k = u^k = u^{k+q} = b^{k+q} v^{k+q} = b^{k+q} v^k$ and $b = 1$.

(iii) If $u, v \in G\bar{s}$ with $u^{j+p} = u^j$ and $v^{k+q} = av^k$, then $u = a^{-\frac{1}{q}} v$. For, $(a^{-\frac{1}{q}} v)^{k+q} = a^{-\frac{(k+q)}{q}} av^k = (a^{-\frac{1}{q}} v)^k$, and hence $u = a^{-\frac{1}{q}} v$ by the uniqueness part of (ii).

(iv) Suppose $u \in G\bar{s}$, $v \in G\bar{t}$ and $u^{j+p} = u^j$, $v^{k+q} = v^k$, and $uv = vu$. Then $st = ts$ and $(uv)^{jk+pq} = (uv)^{jk}$. Since $G\overline{st} = Guv = Gvu = G\overline{ts}$, necessarily $st = ts$. Using the fact that $x^m e^n = x^m$ whenever $xe = x$ we have

$$(uv)^{jk+pq} = u^{jk}u^{pq}v^{jk}v^{pq} = u^{jk}v^{jk} = (uv)^{jk}.$$

Now, for each $s \in S$ let $u_s \in G\bar{s}$ be that element with $u_s^{j+p} = u_s^j$ for some $j, p \in \mathbb{N}$. Then $u_s u_t = \tau(s,t)u_{st}$ and τ is a 2-cocyle.

(v) τ is a fundamental 2-cocyle. Take s, $t \in S$. Let $j \in \mathbb{N}$ be minimal such that $(Gu_s u_t)^j$ is equal to a larger power of $Gu_s u_t$ and let $p \in \mathbb{N}$ be minimal with $(Gu_s u_t)^{j+p} = (Gu_s u_t)^j$. Similarly, let k and q be minimal with $(Gu_t u_s)^{k+q} = (Gu_t u_s)^k$. By Exercise 5(b), $p = q$ and $k-1 \leq j \leq k+1$. Since $(u_s u_t)^{j+p} = c(u_s u_t)^j$ and $(u_t u_s)^{k+p} = e(u_t u_s)^k$ for some $c, e \in G$, $u_{st} = c^{-\frac{1}{p}} u_s u_t$ and $u_{ts} = e^{-\frac{1}{p}} u_t u_s$ by (iii). Thus, $\tau(s,t) = c^{\frac{1}{p}}$ and $\tau(t,s) = e^{\frac{1}{p}}$ and we claim that $c = e$. If $j = k+1$, then

$$\begin{aligned} c(u_s u_t)^j &= (u_s u_t)^{k+1+p} = u_s(u_t u_s)^{k+p}u_t = e u_s (u_t u_s)^k u_t \\ &= e(u_s u_t)^j; \end{aligned}$$

so $c = e$. By symmetry, $c = e$ if $j = k-1$. The third possibility is that $j = k$, and then, as above, $c(u_s u_t)^{j+1} = e(u_s u_t)^{j+1}$ and, again, $c = e$. In fact, when $j = k$ we have $c = e = 1$ since $e(u_s u_t)^{j+1}$ is easily seen to be equal to $(u_s u_t)^{j+1}$. So $\tau(s,t) = \tau(t,s)$. Suppose, now, that $st = ts$. Then $u_s u_t = \tau(s,t)u_{st} = \tau(t,s)u_{ts} = u_t u_s$ and, by (iv) and (ii), $u_s u_t = u_{st}$ since $u_s u_t \in G\overline{st}$. So $\tau(s,t) = 1$ and τ is a fundamental 2-cocycle.

Now we can give the proof of (a). Since $\bar{s} = c(s)u_s$ with $c(s) \in G$ for each $s \in S$, we have

$$\sigma(s,t)c(st)u_{st} = \sigma(s,t)\overline{st} = \bar{s}\bar{t} = c(s)c(t)u_s u_t = c(s)c(t)\tau(s,t)u_{st},$$

and hence $\sigma = (dc)\tau \in B^2 H$. If $\sigma \in B^2 \cap H$, then $\sigma = dc$ for some 1-cochain c, and, for $s \in S$ and $m, n \in \mathbb{N}$, $1 = \sigma(s^n, s^m) = c(s^n)c(s^m)c(s^{n+m})^{-1}$; so $c(s^n s^m) = c(s^n)c(s^m)$ and c restricted to $\{s^n : n \in \mathbb{N}\}$ is a homomorphism. Since $s^m = s^n$ for some $m > n$ we have $c(s)^m = c(s)^n$, and hence $c(s) = 1$ and $\sigma = 1$.

A similar argument proves (b). For, $\sigma = (dc)\tau$ for some c and some fundamental 2-cocyle τ, and again $c = 1$ and hence $\sigma = \tau$.

Clearly, (c) follows from (a) since if S is commutative, then $1 = H \cong H^2$. □

We next wish to give a condition which will simultaneously force the semigroup in Theorem 6.1.1 to be a group and the 2-cocycle to be a coboundary. The positive element x in the ℓ-ring R is called a *super left d-element* if, for all $y \in R$,

$$xy^+ = \bigvee\{x_1 y : 0 \leq x_1 \leq x\}. \tag{6.1.1}$$

This condition is easily seen to be equivalent to the condition, for all $y, z \in R$,

$$x(y \vee z) = \bigvee\{x_1 y + x_2 z : 0 \leq x_1, x_2 \leq x, \quad x_1 + x_2 = x\}. \tag{6.1.2}$$

The details are left to Exercise 6. A *super right d-element* is defined analogously, and a *super d-element* is an element which satisfies both (6.1.1) and its right-sided version. The same terminology will be used for the ℓ-ring if each element of R^+ satisfies (6.1.1) or its right counterpart, or both. If x is a super left d-element in an ℓ-algebra over K and x is K-convex, then x is a left d-element. For,

$$xy^+ = \bigvee\{x_1y : 0 \le x_1 \le x\} = \bigvee\{axy : 0 \le a \le 1, a \in K\} \le (xy)^+ \le xy^+$$

since $axy \le a(xy)^+ \le (xy)^+$ for each $0 \le a \le 1$. In order to insure that the required sups exist in the ℓ-algebras under consideration we next let $K = \mathbb{R}$.

Theorem 6.1.3. *The following statements are equivalent for the twisted semigroup ℓ-algebra $R = \mathbb{R}[S;\sigma]$.*

(a) *S is left cancellative.*
(b) *R is a super left d-algebra.*
(c) *For each $s \in S$, $\overline{s}$ is a left d-element of R.*

Proof. (a) $\Rightarrow$ (b). Let $\alpha, \beta \in R$ with $\alpha \ge 0$. It suffices to verify that

$$\alpha\beta^+ \le \bigvee\{\alpha_1\beta : 0 \le \alpha_1 \le \alpha, \alpha_1 \in R\} \tag{6.1.3}$$

because this sup exists since R is complete and $\alpha_1\beta \le \alpha\beta^+$ for each $0 \le \alpha_1 \le \alpha$. Write $\alpha = \Sigma_s a_s\overline{s}$, $\beta = \Sigma_s b_s\overline{s}$. Then

$$\alpha\beta = \sum_s \left(\sum_{xy=s} a_x b_y \sigma(x,y) \right) \overline{s}.$$

The coefficient of $\overline{t}$ in $\alpha\beta^+$ is $\Sigma a_x b_y \sigma(x,y)$ where the sum is over all $x, y \in S$ with $xy = t$ and $b_y \ge 0$. Define $\gamma^t = \Sigma_s c_s\overline{s}$ where

$$c_s = \begin{cases} a_s & \text{if, for some } r \in S,\ t = sr \text{ and } b_r \ge 0 \\ 0 & \text{otherwise} \end{cases}.$$

Since S is left cancellative c_s and γ^t are well-defined. Now $0 \le \gamma^t \le \alpha$ and

$$\gamma^t\beta = \sum_s \left(\sum_{xy=s} c_x b_y \sigma(x,y) \right) \overline{s}$$
$$= \sum_s \left(\sum_{\substack{xy=s,\ t=xr \\ b_r \ge 0}} a_x b_y \sigma(x,y) \right) \overline{s}.$$

So

$$(\gamma^t\beta)_t = \sum_{xy=t,\ b_y \ge 0} a_x b_y \sigma(x,y) = (\alpha\beta^+)_t,$$

and, for each $t \in S$,

$$\left(\bigvee\{\alpha_1\beta : 0 \le \alpha_1 \le \alpha\}\right)_t \ge (\gamma^t\beta)_t = (\alpha\beta^+)_t.$$

This establishes (6.1.3).

We have previously noted that (b) implies (c).

(c) $\Rightarrow$ (a). Suppose s, t, $w \in S$ with $t \neq w$ and $st = sw$. Then

$$\begin{aligned}\sigma(s,t)\overline{st} = \overline{s}\overline{t} = \overline{s}(\overline{t}-\overline{w})^+ = (\overline{s}\overline{t}-\overline{s}\overline{w})^+ = ((\sigma(s,t)-\sigma(s,w))\overline{st})^+\\ = (\sigma(s,t)-\sigma(s,w))^+\overline{st},\end{aligned}$$

and hence we have the contradiction $\sigma(s,t) = (\sigma(s,t)-\sigma(s,w))^+$. □

In a finite dimensional super left d-algebra the basic elements from a multiplicatively closed subset.

Theorem 6.1.4. *Let R be an ℓ-domain which is a locally finite basic ℓ-algebra over K in which each basic element is a left d-element. Then the product of basic elements is basic.*

Proof. Suppose first that R is finite dimensional. So $R = Ku_1 \oplus \cdots \oplus Ku_n$. For each j, $\{u_ju_1,\ldots,u_ju_n\}$ is a set of n nonzero disjoint elements. So $0 \neq u_ju_i \in Ku_k$ for some k. In general, $R = \oplus_\lambda Ku_\lambda$. Let u and v be basic elements and let T be the K-subalgebra of R generated by u and v. Since T is finite dimensional there exists a finite subset $\{u_1,\ldots,u_n\}$ of $\{u_\lambda\}$ with $T \subseteq Ku_1 \oplus \cdots \oplus Ku_n = A$ and we assume that n is minimal. For each i, $1 \le i \le n$, there is a monomial $m_i(u,v)$ in u and v with

$$m_i(u,v) = a_1u_1 + \cdots + a_nu_n, \quad a_i > 0 \text{ and each } a_k \ge 0.$$

So if $m_j(u,v) = b_1u_1 + \cdots + b_nu_n$, then $0 < a_ib_ju_iu_j \le m_i(u,v)m_j(u,v) \in T \subseteq A$, and hence $u_iu_j \in A$. Thus, A is an ℓ-subalgebra and uv is basic in A and also in R. □

To get the desired characterization of a semigroup ℓ-algebra we need to see that the second cohomology group vanishes.

Theorem 6.1.5. *Suppose S is a locally finite left cancellative semigroup and K is a totally ordered field whose group of strictly positive elements is divisible. Then $H^2(S,K^{+*}) = 1$.*

Proof. Suppose first that S is finite of order n, and let σ be a 2-cocyle. Define $c \in C^1(S,K^{+*})$ by $c(s)^n = \Pi_{u\in S}\sigma(s,u)$. Since $tS = S$,

$$\begin{aligned} dc^n(s,t) &= \Pi_{u\in S}\,\sigma(s,u)\,\sigma(t,u)\,\sigma(st,u)^{-1}\\ &= \sigma(s,t)^n\Pi_{u\in S}\,\sigma(s,u)\,\sigma(t,u)\,\sigma(st,u)^{-1}\sigma(s,t)^{-1}\\ &= \sigma(s,t)^n\Pi_{u\in S}\,\sigma(s,u)\,\sigma(t,u)\,\sigma(s,tu)^{-1}\sigma(t,u)^{-1}\\ &= \sigma(s,t)^n.\end{aligned}$$

So $\sigma(s,t) = dc(s,t)$, $\sigma = dc \in B^2(S,K^{+*})$ and $H^2(S,K^{+*}) = 1$. If now S is locally finite and σ is a fundamental 2-cocyle, then σ is the identity on each finite subsemigroup of S. Thus, $\sigma = 1$ and $H^2(S,K^{+*}) = 1$ by Theorem 6.1.2. □

We now have assembled all the necessary ingredients to get a semigroup ℓ-algebra.

Theorem 6.1.6. *Let R be an ℓ-algebra over $\mathbb{R}$. R is isomorphic to a semigroup (group) ℓ-algebra over a left cancellative locally finite semigroup (group) if and only if R is an ℓ-domain which is basic and is a locally finite left (and right) super d-algebra.*

Proof. If $R \cong \mathbb{R}[S]$ where S is locally finite and left cancellative (a group), then R is a left super d-algebra (a super d-algebra) by Theorem 6.1.3. Conversely, suppose R has the listed left properties. Then, by Theorems 6.1.4, 6.1.1, 6.1.5, and 6.1.3, R is isomorphic to a semigroup ℓ-algebra over a semigroup S with the desired properties. If R is also a right super d-algebra, then S is also right cancellative and is a group since it is locally finite. □

The polynomial ring $A[x]$ over the totally ordered division ring A has numerous partial orders in which all squares are positive. We are going to determine all those sp-partial orders for which $A[x]$ is a nontotally ordered sp-ℓ-ring extension of A and x is comparable to 0. A description of these lattice orders for $x > 0$ is given by

$$A[x] = [A \oplus \{(Ay \underset{\leftarrow}{\boxplus} \cdots \underset{\leftarrow}{\boxplus} Ay^{n-1}) \underset{\leftarrow}{\boxplus} A(y^n - a)\}] \underset{\leftarrow}{\boxplus} [Ay^{n+1} \underset{\leftarrow}{\boxplus} \cdots],$$

where $n \in \mathbb{N}$, $y = x - c$, and a and c are in the center of A^+. These orders may be viewed as arising from perturbations of the total order of the additive monoid $\mathbb{Z}^+$. Thus, we proceed, more generally, to modify the partial order of a given mopops that is a monoid so that in the new partial order we still have a pops which retains some of the good properties of the original. We will return to additive notation for the pops, which now will not be trivially ordered, and to the formal power series notation for the elements of the semigroup ring.

As usual, $X^* = X \backslash \{0\}$ for any subset X of a larger set which may contain 0. The element n of the pops Δ is *positive* if $n < 2n$. It is *strongly right (left) positive* if $n < n + k\,(n < k + n)$ for each $k \in \Delta^*$ and it is *strongly positive* if it is both strongly right and strongly left positive. The pops Δ is *weakly positive* if each of its nonzero elements is positive. Recall that Δ is positive if $n, m < n + m$ whenever $n, m \in \Delta^*$ and $n + m$ is defined. A positive pops is weakly positive provided $2n$ exists for every $n \in \Delta$, but the converse is not true in general; see Exercises 8 and 9. The following information will be useful.

Theorem 6.1.7. *Suppose Δ is a pops with a strongly right positive element.*

(a) Δ is a semigroup.
(b) Δ^* is a subpops of Δ.
(c) If Δ is a mopops, then 0 is a minimal element of Δ.
(d) Each element of Δ that is comparable to every element of Δ^* is strongly left positive.

If Δ is a weakly positive pops, then (b) and (c) hold and so does (d) with "left" deleted.

Proof. Let n be a strongly right positive element of Δ. Then $n < n+k$, $n+p$ for $k,p \in \Delta^*$; so $n < n+p < n+k+p$ and Δ^* is a subsemigroup of the semigroup Δ. Also, 0 must be minimal since if $k < 0$, then $n < n+k < n$. Suppose m is comparable to every element of Δ^*. If $m = 0$ then m is strongly positive. If $m \neq 0$, the inequality $k+m \leq m$ produces the contradiction $n+m < n+k+m \leq n+m$.

Suppose now that Δ is weakly positive. If $p, n \in \Delta^*$ with $p+n \in \Delta$ the inequality $p+n < 2p+2n$ shows $p+n \in \Delta^*$, and clearly 0 is minimal. Suppose $m \in \Delta^*$ is comparable to every element of Δ^* and let $k \in \Delta^*$. Since $k \leq m$ or $m \leq k$, $m+k$ and $k+m$ are defined since $2m$ and $2k \in \Delta$. If $m+k \leq m$, then $m+2k \leq m+k < m+2k$; so $m < m+k$ and, similarly, $m < k+m$. □

A monoid that is a rooted mopops can be weakly positive without having all of its elements strongly positive; see Exercise 8.

In all of the perturbations of the mopops that will be given it is only the partial order that will be changed, not addition. The first modification of the partial order of a mopops that we need is that obtained by isolating 0. Let Δ be a mopops and let Δ_0 be the poset which is the cardinal sum $\Delta^* \dot\cup \{0\}$ of the posets Δ^* and $\{0\}$.

Theorem 6.1.8. *The following statements are equivalent for the mopops* Δ.

(a) Δ *is an (almost) sp-pops and* 0 *is a minimal element.*
(b) Δ^* *is a weakly positive (almost) sp-subpops of* Δ.
(c) Δ_0 *is an (almost) sp-pops.*

Proof. We will only treat the almost sp-pops case; the sp-pops case is quite similar.

(a) ⇒ (b). Suppose m and n are distinct elements of Δ^* with $m+n \in \Delta$. Since $p = p+0 < 2p$ for each $p \in \Delta^*$, $m+n < 2m+2n$ and hence $m+n \in \Delta^*$. So Δ^* is a weakly positive almost sp-subpops of Δ.

(b) ⇒ (a). 0 is minimal since if $n < 0$, then $2n < n < 2n$. Also, Δ is clearly an almost sp-pops.

(b) ⇒ (c). Since Δ^* is a subpops of Δ, Δ_0 is clearly a pops and Δ^* is a subpops of Δ_0. So Δ_0 is an almost sp-pops by the previous implication.

(c) ⇒ (b). This is a consequence of the fact that (a) implies (b). □

The second modification of the partial order applies to a rooted mopops and unlike the first may not quite be a weakening of the original order. Suppose Δ is a rooted mopops and let $\Delta^* = \Gamma_1 \cup \Gamma_2$ with $\Gamma_1 < \Gamma_2$ where Γ_2 is contained in the trunk of Δ^*, and assume that $m+n$ and $n+m$ exist for all $(m,n) \in \Delta \times \Gamma_2$. Let $\Delta_{\Gamma_1,\Gamma_2}$ be the poset which is the ordinal sum of $\Gamma_1 \dot\cup \{0\}$ and Γ_2 with Γ_2 on top; so $m \leq_{\Gamma_1,\Gamma_2} n$ in $\Delta_{\Gamma_1,\Gamma_2}$ if and only if $m \leq n$ in $\Gamma_1 \dot\cup \{0\}$ or in Γ_2, or $m \in \Gamma_1 \dot\cup \{0\}$ and $n \in \Gamma_2$. Note that $\leq$ and $\leq_{\Gamma_1,\Gamma_2}$ agree in Δ^*.

Theorem 6.1.9. *Suppose* Δ *is a rooted mopops that is either weakly positive or has a strongly positive element. Then* $\Delta_{\Gamma_1,\Gamma_2}$ *is a rooted mopops that is weakly positive or has a strongly positive element, respectively. Moreover,* Δ *is an (almost) sp-pops if and only if* $\Delta_{\Gamma_1,\Gamma_2}$ *is an (almost) sp-pops.*

Proof. Certainly $\Delta_{\Gamma_1,\Gamma_2}$ is a rooted poset, and Δ^* is a subpops of Δ by Theorem 6.1.7. Suppose that $m <_{\Gamma_1,\Gamma_2} n$ and $p \in \Delta^*$ with $p+m \in \Delta$. If $m \neq 0$, then $p+m <_{\Gamma_1,\Gamma_2} p+n$ since this inequality is the inequality $p+m < p+n$. If $m = 0$, then $n \in \Gamma_2$ and n is strongly positive in Δ by Theorem 6.1.7. So, again, $p <_{\Gamma_1,\Gamma_2} p+n$. If Δ is weakly positive, then clearly so is $\Delta_{\Gamma_1,\Gamma_2}$, and if $n \in \Delta^*$ is a strongly positive element of Δ, then it is strongly positive in $\Delta_{\Gamma_1,\Gamma_2}$. On the other hand, if 0 is strongly positive in Δ, then $0 < \Delta^*$ and each element of Δ^* is strongly positive in $\Delta_{\Gamma_1,\Gamma_2}$. The last statement is a consequence of Theorem 6.1.8. □

We now rephrase these facts using po-rings. Let A be a po-domain and let Δ be a mopops. The partial orders, as an A-A-bimodule, of the ring $A[\Delta] = \Sigma(\Delta_0, A) = \Sigma(\Delta_{\Gamma_1,\Gamma_2}, A)$ induced by Δ_0 and $\Delta_{\Gamma_1,\Gamma_2}$ will be denoted by $P_0(A[\Delta]) = P_0$ and $P_{\Gamma_1,\Gamma_2}(A[\Delta]) = P_{\Gamma_1,\Gamma_2}$, respectively. These partial orders are also described as follows:

$$P_0 = A[\Delta]^+ \text{ where } A[\Delta] = A \oplus \Sigma(\Delta^*, A), \tag{6.1.4}$$

$$P_{\Gamma_1,\Gamma_2} = A[\Delta]^+ \text{ where } A[\Delta] = [A \oplus \Sigma(\Gamma_1, A)] \overleftarrow{\boxplus} \Sigma(\Gamma_2, A). \tag{6.1.5}$$

Theorem 6.1.10. *Suppose that A is a po-domain and Δ^* is a subpops of the mopops Δ.*

(a) P_0 is a partial order of the generalized semigroup ring $A[\Delta]$.
(b) If A is a domain and 0 is a minimal element of Δ, then $(A[\Delta], P_0)$ is an sp ring iff A is an sp-ring and Δ is an sp-pops.
(c) If Δ is rooted and is either weakly positive or has a strongly positive element, then P_{Γ_1,Γ_2} is a partial order of the ring $A[\Delta]$.
(d) If A is a totally ordered domain and Δ is a weakly positive rooted po-monoid, then A is the subring of left (right) f-elements of $(A[\Delta], P)$ for $P = P_0$ or $P = P_{\Gamma_1,\Gamma_2}$, unless $\Gamma_1 = \emptyset$, in which case $\Sigma(\Delta_{\Gamma_1,\Gamma_2}, A)$ is totally ordered. Moreover, if Δ is an sp-pops then $A[\Delta]$ is a domain and $(A[\Delta], P)$ is an sp-ℓ-ring.

Proof. Since Δ^* is a subpops of Δ Δ_0 is a pops and (a) follows from Theorem 3.5.3. If A is totally ordered and Δ is rooted, then (b) is a consequence of Theorems 6.1.8, 3.5.3, and 3.7.6. In general, we need to use Exercise 3.7.14(b) in place of Theorem 3.7.6. Clearly, (c) is a consequence of Theorems 6.1.9 and 3.5.3. For (d), each of the po-rings is an ℓ-ring and the first part follows from Exercise 3.7.17 since $f_\ell(\Delta_0) = f_r(\Delta_0) = f_\ell(\Delta_{\Gamma_1,\Gamma_2}) = f_r(\Delta_{\Gamma_1,\Gamma_2}) = 0$. The second part is an immediate consequence of Theorem 3.7.6 since Δ_0 and $\Delta_{\Gamma_1,\Gamma_2}$ are rooted sp-pops. □

We will now construct other partial orders of $A[\Delta]$ which make it into an sp-ℓ-ring. The initial construction will only be concerned with the additive structure of $A[\Delta]$ and will be carried out for any poset. Recall that $U(X)$ (respectively, $L(X)$) denotes the set of upper (respectively, lower) bounds of the subset X of a poset. The set of strict upper (respectively, lower) bounds of X will be denoted by $U_s(X)$ (respectively, $L_s(X)$) : $U_s(X) = U(X) \backslash X$. Let Γ be a poset with a distinguished element 0, and let $n \in \Gamma^* = \Gamma \backslash \{0\}$ and $V(n) = \Gamma^* \backslash U(n)$. For the subset X of Γ we will use the previous notation $X^* = X \backslash \{0\}$ to exclude 0. Suppose that A is a po-ring with

$A^+ \neq 0$ and $u(A)$ is a po-unital po-ring which contains A. We assume that the pair $(\alpha, \beta) \in A^+ \times u(A)^+$ satisfies the following conditions:

$$A\beta \subseteq A \quad \text{and} \quad \text{if } \gamma \in A \text{ with } \gamma\beta \geq 0 \text{ then } \gamma \geq 0; \tag{6.1.6}$$

$$A^+\alpha \subseteq A^+\beta \cup U(A^+\beta). \tag{6.1.7}$$

One example of a pair that satisfies (6.1.7) is obtained by taking A totally ordered with $A\beta$ convex and $\alpha \in A^+$. Another example is given by taking $u(A)$ totally ordered, I a left ℓ-ideal of $u(A)$, J a left ideal of A with $I \cap J = 0$, and $\alpha \in J^+$ and $\beta \in I^+$. The canonical totally ordered domain with a left ℓ-ideal that is not an ideal given in Exercise 3.4.35 and Theorem 3.4.17 is one such example.

Each element $f \in A[\Gamma]$, the direct sum of Γ copies of the bimodule ${}_AA_A$, can be written uniquely as

$$f = \alpha_0 + f_1 + \alpha_n x^n + f_2$$
$$\text{with} \quad \operatorname{supp} f_1 \subseteq V(n) \quad \text{and} \quad \operatorname{supp} f_2 \subseteq U_s(n)^*. \tag{6.1.8}$$

Here, α_0 denotes $\alpha_0 x^0$. Let $P_{n,\alpha,\beta}(A[\Gamma]) = P_{n,\alpha,\beta}$ be defined by

$$P_{n,\alpha,\beta} = \{f \in A[\Gamma] : 0 < f_2 \in \Sigma(U_s(n)^*, A),$$
$$\text{or } f_2 = 0, \alpha_n > 0, \text{ and } \alpha_0\beta + \alpha_n\alpha \geq 0 \text{ whenever } \alpha_n\alpha \in A^+\beta,$$
$$\text{or } f_2 = \alpha_n = 0, f_1 \in (\Sigma(V(n), A))^+, \text{ and } 0 \leq \alpha_0\}. \tag{6.1.9}$$

We note that for each $0 < \gamma \in A$, $\gamma\alpha > A^+\beta$ iff $\alpha_0 + f_1 + \gamma x^n \in P_{n,\alpha,\beta}$ for any $f_1 \in \Sigma(V,n), A)$ and any $\alpha_0 \in A$. For, if all such elements are in $P_{n,\alpha,\beta}$, then $\gamma\alpha \geq A^+\beta$ by the definition of $P_{n,\alpha,\beta}$. But $A^+\beta \neq 0$ by (6.1.6) and hence $\gamma\alpha \notin A^+\beta$; so $\gamma\alpha > A^+\beta$. If $\beta = 1$, then we will denote $P_{n,\alpha,\beta}$ by $P_{n,\alpha}$. Since $\alpha_0 + \alpha_n x^n = (\alpha_0 + \alpha_n\alpha) + \alpha_n(x^n - \alpha)$, $P_{n,\alpha}$ is the positive cone of the po-group

$$(A[\Gamma], P_{n,\alpha}) = [A \oplus (\Sigma(V(n), A) \underleftarrow{\boxplus} A(x^n - \alpha))] \underleftarrow{\boxplus} \Sigma(U_s(n)^*, A). \tag{6.1.10}$$

If $\gamma\alpha > A^+\beta$ for each $0 < \gamma \in A$, then $P_{n,\alpha,\beta}$ is the positive cone of

$$[A \oplus \Sigma(V(n), A)] \underleftarrow{\boxplus} \Sigma(U(n)^*, A). \tag{6.1.11}$$

Note, also, that $P_{n,\alpha,\beta} = P_{n,0}$ for any β, if $A^+\alpha = 0$.

These positive cones are related to the previous ones. Suppose that $\Gamma = \Delta$ is a rooted mopops. If Γ_1 is empty, and 0 is the minimal element of Δ, then $\Delta_{\Gamma_1,\Gamma_2} = \Delta$, and if Γ_2 is empty, then $\Delta_{\Gamma_1,\Gamma_2} = \Delta_0$. On the other hand, if Γ_1 has a largest element n, then $P_{\Gamma_1,\Gamma_2}(A[\Delta]) = P_{n,0}(A[\Delta])$. These are the only possibilities, of course, when $\Delta = \mathbb{Z}^+$.

Theorem 6.1.11. *Let Γ be a poset, $0, n \in \Gamma$ with $n \neq 0$, and let A be a po-subring of the po-unital po-ring $u(A)$. Suppose $\alpha \in A^+$, $\beta \in u(A)^+$, and α and β satisfy*

(6.1.6) and (6.1.7). Let $R = (A[\Gamma], P_{n,\alpha,\beta})$ *where* $P_{n,\alpha,\beta}$ *is the subset of* $A[\Gamma]$ *defined by (6.1.9).*

(a) *R is a po-group.*
(b) *If*
  (i) *A is totally ordered,*
  (ii) $U_s(n)^*$ *is totally ordered, and*
  (iii) $V(n)$ *is rooted,*

then R is an ℓ*-group. Moreover, if* $f \in R$ *and* $f \| 0$*, then* $f = \alpha_0 + f_1 + \alpha_n x^n$*,* $\alpha_n \alpha = \delta\beta$ *for some* $\delta \in A$*, and*

$$f^+ = \begin{cases} -\delta + f_1 + \alpha_n x^n, & \text{if } \alpha_n > 0, \\ \alpha_0 + \delta, & \text{if } \alpha_n < 0, \\ \alpha_0^+ + f_1^+, & \text{if } \alpha_n = 0, \end{cases} \tag{6.1.12}$$

where $f_1^+ = f_1 \vee 0$ *in* $\Sigma(V(n), A)$*. Conversely, if* Γ *has at least three elements and R is an* ℓ*-group, then (i), (ii) and (iii) hold.*
(c) *R is totally ordered iff A is totally ordered,* Γ^* *is totally ordered, n is the least element of* Γ^**, and* $\gamma\alpha > A\beta$ *for each* $0 < \gamma \in A$*.*
(d) *Suppose the three conditions in (b) are satisfied and* $0 < f, g$*.* *Then* $f \wedge g = 0$ *if and only if (i) or (ii) or the analogue of (i) obtained by interchanging f and g is satisfied:*

  (i) $f = \alpha_0$, $g = \beta_0 + g_1 + \beta_n x^n$ *with* $\beta_n > 0$ *and* $\beta_n \alpha = -\beta_0 \beta$
  (ii) $f = \alpha_0 + f_1$, $g = \beta_0 + g_1$ *and* $\alpha_0 \wedge \beta_0 = 0$ *and* $f_1 \wedge g_1 = 0$*.*

Proof. Let $f = \alpha_0 + f_1 + \alpha_n x^n + f_2$ and $g = \beta_0 + g_1 + \beta_n x^n + g_2$ be two elements of R that are decomposed as in (6.1.8).

(a) Suppose that $f, g \in P_{n,\alpha,\beta}$. If $f_2 \neq 0$ or $g_2 \neq 0$, then $0 < f_2 + g_2 \in \Sigma(U_s(n)^*, A)$; or if $f_2 = g_2 = 0$ and, either $\alpha_n \alpha > A^+\beta$ or $\beta_n \alpha > A^+\beta$, or $\alpha_n = \beta_n = 0$, then, either $(\alpha_n + \beta_n)\alpha > A^+\beta$ since $\alpha_n \alpha \geq 0$ and $\beta_n \alpha \geq 0$, or $f_1 + g_1 \in (\Sigma(V(n), A))^+$ and $\alpha_0 + \beta_0 \geq 0$. In these three cases $f + g \in P_{n,\alpha,\beta}$. The remaining case has $f_2 = g_2 = 0$, $0 < \alpha_n + \beta_n$, and $\alpha_n \alpha$, $\beta_n \alpha \in A^+\beta$. Then

$$(\alpha_0 + \beta_0)\beta + (\alpha_n + \beta_n)\alpha = (\alpha_0 \beta + \alpha_n \alpha) + (\beta_0 \beta + \beta_n \alpha) \geq 0;$$

so $f + g \in P_{n,\alpha,\beta}$. Since $P_{n,\alpha,\beta} \cap -P_{n,\alpha,\beta} = 0$, $P_{n,\alpha,\beta}$ is a partial order of $A[\Gamma]$.

(b) Assume these three conditions hold and suppose $f \| 0$ and $g \geq 0, f$. Then $f = \alpha_0 + f_1 + \alpha_n x^n$ and $|\alpha_n|\alpha \not> A^+\beta$; so $\alpha_n \alpha = \delta\beta$ for some $\delta \in A$. Let h be the element of R defined by (6.1.12). If $g_2 > 0$, or $g_2 = 0$ and $\beta_n \alpha > A^+\beta$, then $g > h$ since, in the latter case, when $\alpha_n > 0$ we have $\delta \geq 0$ and so $(\beta_n - \alpha_n)\alpha = \beta_n \alpha - \delta\beta > (A^+ + \delta)\beta - \delta\beta = A^+\beta$. Suppose that $g_2 = 0$ and $\beta_n \alpha = \rho\beta$. We need to check that $g \geq h$ in this case and that $h \geq 0, f$.

If $\alpha_n > 0$, then $h = -\delta + f_1 + \alpha_n x^n \geq 0$ since $-\delta\beta + \alpha_n \alpha = 0$; and $h \geq f$ since $(\alpha_0 + \delta)\beta = \alpha_0 \beta + \alpha_n \alpha < 0$ and $h - f = -(\delta + \alpha_0) > 0$ by (6.1.9). Now,

$$0 \le g - f = (\beta_0 - \alpha_0) + (g_1 - f_1) + (\beta_n - \alpha_n)x^n$$

forces $\beta_n \ge \alpha_n > 0$. Since

$$g - h = (\beta_0 + \delta) + (g_1 - f_1) + (\beta_n - \alpha_n)x^n$$

and

$$(\beta_0 + \delta)\beta + (\beta_n - \alpha_n)\alpha = \beta_0\beta + \beta_n\alpha \ge 0,$$

if $\beta_n > \alpha_n$ then $g > h$; and if $\beta_n = \alpha_n$, then $g_1 \ge f_1$ and $\beta_0 + \delta = \beta_0 + \rho \ge 0$, and so $g \ge h$. Thus, $h = f^+$ when $\alpha_n > 0$.

Assume now that $\alpha_n < 0$. Then

$$f^+ = f + (-f)^+ = \alpha_0 + f_1 + \alpha_n x^n + (\delta - f_1 - \alpha_n x^n) = \alpha_0 + \delta$$

by the previous case.

Finally, suppose that $f = \alpha_0 + f_1$ and $h = \alpha_0^+ + f_1^+$. Then $h \ge 0$, f, and

$$0 \le g - f = (\beta_0 - \alpha_0) + (g_1 - f_1) + \beta_n x^n$$

yields that $(\beta_0 - \alpha_0)\beta + \beta_n\alpha \ge 0$ and $\beta_0\beta + \beta_n\alpha \ge (\alpha_0\beta)^+ = \alpha_0^+\beta$ since $g \ge 0$; so

$$0 \le g - h = (\beta_0 - \alpha_0^+) + (g_1 - f_1^+) + \beta_n x^n,$$

whether $\beta_n > 0$ or $\beta_n = 0$, since

$$(\beta_0 - \alpha_0^+)\beta + \beta_n\alpha = (\beta_0\beta + \beta_n\alpha) - \alpha_0^+\beta \ge 0.$$

Thus, $h = f^+$.

Conversely, suppose that R is an ℓ-group. Then $\Sigma(V(n), A)$ is an ℓ-subgroup of R. For, if $V(n) \ne \emptyset$ and $f \in \Sigma(V(n), A)$ is not comparable to 0, then, since

$$R = [A + \Sigma(V(n), A) + Ax^n] \underset{\leftarrow}{\boxplus} \Sigma(U_s(n)^*, A), \tag{6.1.13}$$

necessarily $f^+ = \gamma_0 + h_1 + \gamma_n x^n$. If $0 < \gamma_n$ and $m \in V(n)$, then $0, f < k = \gamma_0 + (f_1 - \gamma_n x^m) + \gamma_n x^n < f^+$. For, $k > 0$ and $k - f > 0$ for the same reasons that make $f^+ > 0$, and $f^+ - k = \gamma_n x^m > 0$. So $\gamma_n = 0$ and hence $\gamma_0 = 0$, also. Thus, $V(n)$ is rooted. If $U_s(n)^* \ne \emptyset$, then $U_s(n)^*$ and A are both totally ordered by (6.1.13). If $U_s(n)^* = \emptyset$, then $V(n) \ne \emptyset$ and A is totally ordered since $\Sigma(V(n), A) + Ax^n = \Sigma(V(n), A) \underset{\leftarrow}{\boxplus} Ax^n$.

(c) Suppose R is totally ordered. If $0 < \gamma \in A$ with $\gamma\alpha = \delta\beta$ take $\alpha_0 \in A$ with $\alpha_0 < -\delta$. Then we have the contradiction that $f = \alpha_0 + \gamma x^n \in R$ is not comparable to 0 since $\gamma > 0$ and $\alpha_0\beta + \gamma\alpha < 0$. Thus, $\gamma\alpha > A^+\beta$ for each $0 < \gamma \in A$ and, by (6.1.11), $\Gamma^* = U(n)^*$ is totally ordered and so is A. Conversely, these conditions together with (6.1.11) imply that R is totally ordered.

(d) If f and g have the form in (i), then $f \| g$ and by (6.1.12), $(g - f)^+ = \beta_0 + g_1 + \beta_n x^n = g$ and $f \wedge g = 0$. If they have the form in (ii), then $f \wedge g = 0$ since $A \oplus \Sigma(V(n), A)$ is an ℓ-subgroup of R by (6.1.12). Conversely, suppose that $f \wedge g = 0$. If $g_2 > 0$ and $g_2 \ge f_2$, then $2g_2 > f_2$ and $2g > f$. So $g_2 = f_2 = 0$. If $\beta_n \ge \alpha_n > 0$, then $2\beta_n > \alpha_n$ and again by (b), $(2\beta_n - \alpha_n)\alpha = \delta\beta$ and

$$(2g-f)^+ = -\delta + (2g_1 - f_1) + (2\beta_n - \alpha_n)x^n \neq 2g,$$

which is nonsense. If $\beta_n = \alpha_n = 0$, then f and g have the form of (ii). Finally, if $\beta_n > 0 = \alpha_n$, then $(g-f)^+ = g$ implies that $-\delta + (g_1 - f_1) + \beta_n x^n = \beta_0 + g_1 + \beta_n x^n$ with $\beta_n\alpha = \delta\beta$; so $f_1 = 0$ and f and g have the form of (i). □

We will now determine the relations between these partial orders. If $X \subseteq A$ and $\rho \in u(A)$, then $(\rho : X) = \{\gamma \in A : \gamma\rho \in X\}$.

Theorem 6.1.12. *Suppose (n,α,β) and (n',α',β') satisfy the conditions in Theorem 6.1.11. Then $P_{n,\alpha,\beta} \subseteq P_{n',\alpha',\beta'}$ if and only if n and n' are comparable and one of the following conditions is satisfied.*

(i) $n' < n$ and $U_s(n')^* = (n',n)^* \cup U(n)^*$.

(ii) $n < n'$, $A^+\alpha = 0$, $\Gamma^* = L(n)^* \cup U(n')^*$, and for every $0 < \gamma \in A$ we have that $\gamma\alpha' > A^+\beta'$.

(iii) $n = n'$, $(\alpha' : A\beta')^+ \subseteq (\alpha : A\beta)^+$, and, if $\gamma \in (\alpha' : A\beta')^+$ with $\gamma\alpha = \delta\beta$ and $\gamma\alpha' = \delta'\beta'$, then $\delta' \geq \delta$.

In case (ii) $P_{n,\alpha,\beta} = P_{n,0} = P_{n',\alpha',\beta'}$ and $(n,n')^* = \emptyset$. In general, $P_{n,\alpha,\beta} = P_{n,\alpha'}$ precisely when $A^+(\alpha - \alpha'\beta) = 0$.

Proof. Let $0 < \gamma \in A$. If $n \| n'$, then $n \in V(n')$, $n' \in V(n)$, and $-\gamma x^{n'} + \gamma x^n \in P_{n,\alpha,\beta} \backslash P_{n',\alpha',\beta'}$. Assume n and n' are comparable.

Suppose first that $n' < n$. If $P_{n,\alpha,\beta} \subseteq P_{n',\alpha',\beta'}$, then each upper bound of n' in Γ^* is comparable to n. For, if $m > n'$ and $m \| n$, then $-\gamma x^m + \gamma x^n \in P_{n,\alpha,\beta} \backslash P_{n',\alpha',\beta'}$. Conversely, suppose that $U_s(n')^* = (n',n)^* \cup U(n)^*$, and take $f \in P_{n,\alpha,\beta}$ as in (6.1.8). Then we can decompose f_1 as $f_1 = f_3 + \alpha_{n'}x^{n'} + f_4$ where supp $f_3 \subseteq V(n')$ and supp $f_4 \subseteq (n',n)$. Now, clearly,

$$f = \alpha_0 + f_3 + \alpha_{n'}x^{n'} + (f_4 + \alpha_n x^n + f_2)$$

is the decomposition of f that is given in (6.1.8) relative to n'. Let $m \in \operatorname{maxsupp}(f_4 + \alpha_n x^n + f_2)$ with coefficient α_m. Then $m \in \operatorname{maxsupp} f_2$, or $f_2 = 0$ and $m = n$, or $f_2 = 0$, $\alpha_n = 0$, and $m \in \operatorname{maxsupp} f_4$. In the latter case $m \in \operatorname{maxsupp} f$, and hence in all these cases $\alpha_m > 0$ and $f \in P_{n',\alpha',\beta'}$. If, however, $f_4 + \alpha_n x^n + f_2 = 0$, then $\alpha_n = 0$ and $f_2 = f_4 = 0$, and $f = \alpha_0 + f_3 + \alpha_{n'}x^{n'}$ with $\alpha_0 \geq 0$. So, if $\alpha_{n'} \neq 0$, then $n' \in \operatorname{maxsupp} f_1$, $\alpha_{n'} > 0$ and $f \in P_{n',\alpha',\beta'}$. On the other hand, if $\alpha_{n'} = 0$, then $f_1 = f_3$ and $f = \alpha_0 + f_1 \in P_{n',\alpha',\beta'}$.

Suppose next that $n < n'$. Assume first that $P_{n,\alpha,\beta} \subseteq P_{n',\alpha',\beta'}$. If $\gamma\alpha > A^+\beta$, then $-\gamma + \gamma x^n \in P_{n,\alpha,\beta} \backslash P_{n',\alpha',\beta'}$ since $-\gamma < 0$; and if $\gamma\alpha = \delta\beta$ with $\delta > 0$, then $-\delta + \gamma x^n \in P_{n,\alpha,\beta} \backslash P_{n',\alpha',\beta'}$ since $-\delta < 0$. So $A^+\alpha = 0$ and $P_{n,\alpha,\beta} = P_{n,0}$ by the remark after (6.1.11). Suppose, by way of contradiction, that $m \in \Gamma^* \backslash (L(n)^* \cup U(n')^*)$. If $m \| n$, then $m \in V(n)$, $m, n \in V(n')$ and $-\gamma x^m + \gamma x^n \in P_{n,0} \backslash P_{n',\alpha',\beta'}$; so $m > n$. Now $m \| n'$ or $m < n'$, and in either case $-\gamma + \gamma x^m \in P_{n,\alpha,\beta} \backslash P_{n',\alpha',\beta'}$. Thus, $\Gamma^* = L(n)^* \cup U(n')^*$, and $P_{n,\alpha,\beta} = P_{n',\alpha',\beta'}$ by (i) with the roles of n and n' reversed since each upper bound of n in Γ^* is comparable to n'. To see that $\gamma\alpha' > A^+\beta$ we assume to the contrary that $\gamma\alpha' = \delta'\beta'$ and take $\alpha_0 < -\delta'$. Then $\alpha_0 + \gamma x^{n'} \in P_{n,\alpha,\beta} \backslash P_{n',\alpha',\beta'}$ since $\alpha_0\beta' + \gamma\alpha' = (\alpha_0 + \delta')\beta' < 0$. Thus, $\gamma\alpha' > A^+\beta'$. Conversely, if these three conditions hold and $f = \alpha_0 + f_1 + \alpha_n x^n + f_2 \in P_{n,0}$ as in (6.1.8), then the analogous decomposition of f relative to n' is given below and shows that $P_{n,\alpha,\beta} \subseteq P_{n',\alpha',\beta'}$.

$$f = \alpha_0 + (f_1 + \alpha_n x^n) + \alpha_{n'} x^{n'} + (f_2 - \alpha_{n'} x^{n'})$$

For (iii), first assume that $P_{n,\alpha,\beta} \subseteq P_{n,\alpha',\beta'}$. If $0 < \gamma \in (\alpha' : A\beta') \backslash (\alpha : A\beta)$, then $\gamma\alpha' = \delta'\beta'$ with $\delta' \geq 0$ and $\gamma\alpha > A^+\beta$. So, if $\alpha_0 < -\delta'$, then $\alpha_0 + \gamma x^n \in P_{n,\alpha,\beta} \backslash P_{n,\alpha',\beta'}$ since $\alpha_0\beta' + \gamma\alpha' = (\alpha_0 + \delta')\beta' < 0$. Thus, $(\alpha' : A\beta')^+ \subseteq (\alpha : A\beta)^+$. If $0 < \gamma \in (\alpha' : A\beta')$ and $\gamma\alpha = \delta\beta$ and $\gamma\alpha' = \delta'\beta'$, then $-\delta + \gamma x^n \in P_{n,\alpha,\beta}$ and hence $-\delta + \gamma x^n \in P_{n,\alpha',\beta'}$. Thus, $0 \leq -\delta\beta' + \gamma\alpha' = (-\delta + \delta')\beta'$ and $\delta \leq \delta'$. Conversely, assume these two conditions hold, and let $f = \alpha_0 + f_1 + \gamma x^n \in P_{n,\alpha,\beta}$ with $\gamma > 0$. If $\gamma \in (\alpha' : A\beta')$, then, using the previous notation, $0 \leq \alpha_0\beta + \gamma\alpha = (\alpha_0 + \delta)\beta$ gives that $\alpha_0 + \delta' \geq \alpha_0 + \delta \geq 0$ and hence $\alpha_0\beta' + \gamma\alpha' \geq 0$. Thus, $f \in P_{n,\alpha',\beta'}$ and $P_{n,\alpha,\beta} \subseteq P_{n,\alpha',\beta'}$ since the other types of elements in $P_{n,\alpha,\beta}$ are clearly all in $P_{n,\alpha',\beta'}$.

That $P_{n,\alpha,\beta} = P_{n,\alpha'}$ if and only if $A^+(\alpha - \alpha'\beta) = 0$ follows from (iii). For, $P_{n,\alpha,\beta} = P_{n,\alpha'} = P_{n,\alpha',1}$ iff $(\alpha : A\beta)^+ = (\alpha' : A)^+ = A^+$ and, for any $\gamma \in A^+$, $\gamma\alpha = \delta\beta$ and $\gamma\alpha' = \delta'$ implies that $\delta = \delta'$. So $P_{n,\alpha,\beta} = P_{n,\alpha'}$ precisely when $A^+(\alpha - \alpha'\beta) = 0$. □

We will now identify some ℓ-subgroups of $(A[\Gamma], P_{n,\alpha,\beta})$. Recall that $C(X)$ denotes the convex ℓ-subgroup generated by X.

Theorem 6.1.13. *Suppose A is totally ordered and B is a subgroup of the additive group of A.*

(a) If $R = (A[\Gamma], P_{n,\alpha,\beta})$ is an ℓ-group and $|\Gamma| \geq 3$, then $B[\Gamma]$ is an ℓ-subgroup of R if and only if $(\beta : B\alpha) \cap C(B) \subseteq B$.

(b) If Γ is rooted and B is a convex subgroup of A, then $B[\Gamma]$ is an ℓ-subgroup of the ℓ-group $(A[\Gamma], P_{n,\alpha,\beta})$, for any n, α, β.

(c) If Γ is rooted and B is a right ideal of A, then $B[\Gamma]$ is an ℓ-subgroup of each ℓ-group $(A[\Gamma], P_{n,\alpha})$.

(d) Assume Γ is rooted and B is a subring of A with the property that $\alpha \in B$ whenever $B\alpha \subseteq B$. If $B[\Gamma]$ is an ℓ-subgroup of each ℓ-group $(A[\Gamma], P_{n,\alpha})$, then B is a convex subring of A.

Proof. For (a), assume $B[\Gamma]$ is an ℓ-subgroup, and let $0 < \delta \in (\beta : B\alpha) \cap C(B)$. Then $0 < \delta\beta = \gamma\alpha$ with $\gamma \in B$ and $\delta < \alpha_0 \in B$. Now, $f = -\alpha_0 + \gamma x^n \in B[\Gamma]$ and $f \| 0$ since $\gamma > 0$ and $-\alpha_0\beta + \gamma\alpha = (-\alpha_0 + \delta)\beta < 0$ by (6.1.12); so $f^+ = -\delta + \gamma x^n \in B[\Gamma]$ and $\delta \in B$. For the converse, suppose $f = \alpha_0 + f_1 + \alpha_n x^n \in B[\Gamma]$ with $\alpha_n \neq 0$, $f \| 0$, and $\alpha_n\alpha = \delta\beta$ with $\delta \in A$. If $\alpha_n > 0$, then $(\alpha_0 + \delta)\beta = \alpha_0\beta + \alpha_n\alpha < 0$; so $0 \leq \delta < -\alpha_0$ and $\delta \in (\beta : B\alpha) \cap C(B) \subseteq B$. If $\alpha_n < 0$, then $-\delta \in B$. In both cases $f^+ \in B[\Gamma]$ is a consequence of (6.1.12). Both (b) and (c) are obvious consequences of (a), and, as for (d), suppose that $0 \leq \alpha \leq \gamma$ with $\gamma \in B$. Then, for each $\delta \in B^+$, $\delta\alpha \in B\alpha \cap C(B) \subseteq B$ by (a). So, $\alpha \in B$ and B is convex. □

We now return to the ring $A[\Delta]$ and determine when the partial order $P_{n,\alpha,\beta}$ makes $A[\Delta]$ into an ℓ-ring with squares positive. For $\gamma \in A$ let D_γ denote the right inner derivation determined by γ; so if $\delta \in A$, then $D_\gamma(\delta) = [\delta, \gamma] = \delta\gamma - \gamma\delta$.

Theorem 6.1.14. *Suppose that A is a directed po-domain, Δ is a rooted mopops, and n is a strongly positive element in the trunk of Δ^*. Let $R = (A[\Delta], P_{n,\alpha,\beta})$.*

(a) R is a po-ring (equivalently, a po-domain) if and only if,

$$\text{for all } (\gamma_0, \gamma_1, \gamma) \in A \times A^{+*} \times A^+, \text{ if } \gamma_0\beta + \gamma_1\alpha \geq 0, \quad \text{then } \gamma_0\gamma\beta + \gamma_1\gamma\alpha \geq 0. \tag{6.1.14}$$

In particular, suppose β centralizes A. If $A^+D_\alpha(A^+) \subseteq A^+$, then R is a po-ring, and the converse holds provided $A = \beta A$.

(b) If A is a domain and R is a po-ring, then R is an sp-ring iff Δ is an sp-pops and A is an sp-ring.

(c) If R is an ℓ-ring that is not totally ordered, then $F_\ell(R) = A$ and

$$F_r(R) = \{\gamma \in A : \forall \rho, \delta \in A, \rho\alpha = \delta\beta \Rightarrow \rho\gamma\alpha = \delta\gamma\beta\}. \tag{6.1.15}$$

Moreover, if β centralizes A, then $F_r(R)$ is the centralizer of α in A.

Proof. (a) We first note that the condition in (6.1.14) is precisely what is needed for R_A to be a (strict) po-module. For, suppose R_A is a po-module and $\gamma_0, \gamma_1, \gamma \in A$ with $0 < \gamma, \gamma_1$ and $\gamma_0\beta + \gamma_1\alpha \geq 0$. Then $f = \gamma_0 + \gamma_1 x^n > 0$ and hence $f\gamma = \gamma_0\gamma + \gamma_1\gamma x^n > 0$. If $\gamma_1\gamma\alpha > A^+\beta$ then $\gamma_1\gamma\alpha > A\beta$ since A is directed; so $\gamma_1\gamma\alpha > -\gamma_0\gamma\beta$. Otherwise, $\gamma_1\gamma\alpha \in A^+\beta$ and $\gamma_0\gamma\beta + \gamma_1\gamma\alpha \geq 0$. Conversely, assume that (6.1.14) holds and let $f = \alpha_0 + f_1 + \alpha_n x^n + f_2 > 0$, and take $0 < \gamma \in A$. If $f_2 > 0$, then $f_2\gamma > 0$ and $f\gamma > 0$, and if $f_2 = \alpha_n = 0$, then $f\gamma = \alpha_0\gamma + f_1\gamma > 0$. The remaining case to be considered has $f_2 = 0$ and $\alpha_n > 0$. Now, whether $\alpha_n\alpha > A\beta$ or $\alpha_n\alpha \in A^+\beta$ we have $\alpha_0\beta + \alpha_n\alpha \geq 0$. So $\alpha_0\gamma\beta + \alpha_n\gamma\alpha \geq 0$ and $f\gamma \geq 0$. Thus, it suffices to show that if R_A is a po-module, then R is a po-ring. Note that $\Sigma(\Delta^*, A)^+ = P_{n,\alpha,\beta} \cap A[\Delta^*]$, and $\Sigma(\Delta^*, A)$ is an ℓ-domain by Theorem 3.5.3 and a strict A-A-po-bimodule by Theorem 2.6.1. Let $0 < f, g \in R$, and write $f = \alpha_0 + f^*$ and $g = \beta_0 + g^*$ where $f^*, g^* \in A[\Delta^*]$. Then $f^*, g^* \in \Sigma(\Delta^*, A)^+$, and hence $f^*g^* \in \Sigma(\Delta^*, A)^+$. If $f^* = 0$ then $\alpha_0 > 0$ and $fg = \alpha_0 g > 0$, and if $g^* = 0$ then $\beta_0 > 0$ and $fg = f\beta_0 > 0$. Suppose that $f^* > 0$ and $g^* > 0$; then $f^*g^* > 0$ by Exercise 3.5.2. If there is an element $m \in \text{maxsupp } f^*g^*$ with $m \geq n$, then $\text{maxsupp } f^*g^* = \{m\}$ since m is comparable to each element of $\text{supp } f^*g^*$. Now, $\text{supp } f^* \cup \text{supp } g^* < m$. For if $k \in \text{supp } f^*$ with $m \leq k$, then k is strongly positive by Theorem 6.1.7, and for any $\ell \in \text{supp } g^*$ we have the contradiction $k < k + \ell \leq m \leq k$ by Theorem 3.5.3. If $m > n$, then $fg = \alpha_0\beta_0 + \alpha_0 g^* + f^*\beta_0 + f^*g^* > 0$ since $(fg)_2 = (f^*g^*)_2 > 0$. If $m = n$, then $f = \alpha_0 + f_1$ and $g = \beta_0 + g_1$ with $\alpha_0, \beta_0 \geq 0$; so, $fg > 0$. If there is no such m, then $\max(\text{maxsupp } f^* + \text{maxsupp } g^*) = \text{maxsupp } f^*g^* < n$ and $\text{supp } f^* \cup \text{supp } g^* < n$, as above, since any upper bound of n is strongly positive. Then $0 \leq \alpha_0, \beta_0$ and hence $0 < fg$.

Since $\gamma_0\gamma\beta + \gamma_1\gamma\alpha = (\gamma_0\beta + \gamma_1\alpha)\gamma + \gamma_1[\gamma,\alpha] + \gamma_0[\gamma,\beta]$, if β centralizes A^+ and $A^+[\gamma,\alpha] \subseteq A^+$, then A satisfies (6.1.14). Conversely, suppose R is a po-ring and $A = \beta A$. Take $0 < \gamma$, $\gamma_1 \in A$. Then $\gamma_0\beta + \gamma_1\alpha = 0$ for some $\gamma_0 \in A$ and therefore $\gamma_1[\gamma,\alpha] = \gamma_1\gamma\alpha - \gamma_1\alpha\gamma = \gamma_1\gamma\alpha + \gamma_0\gamma\beta \geq 0$.

(b) Let $f = \alpha_0 + f^*$ where supp $f^* \subseteq \Delta^*$. If $\alpha_0 \geq 0$, then $f \in P_{n,\alpha,\beta}$ iff $f^* \in \Sigma(\Delta^*,A)^+$. Thus $f^2 \in P_{n,\alpha,\beta}$ iff $f^*\alpha_0 + \alpha_0 f^* + f^{*2} \in \Sigma(\Delta^*,A)^+$; so R has squares positive iff $\Sigma(\Delta_0,A)$ has squares positive. But according to Theorem 6.1.10(b), $\Sigma(\Delta_0,A)$ has squares positive iff Δ is an sp-pops and A is an sp-domain.

(c) Assume R is an ℓ-ring that is not totally ordered. Take $\gamma \in A^+$ and $0 < f, g \in R$ with $f \wedge g = 0$. By (d) of Theorem 6.1.11, either $f = \alpha_0$ and $g = \beta_0 + g_1 + \beta_n x^n$ with $\beta_n > 0$ and $\beta_0\beta + \beta_n\alpha = 0$, or $f = \alpha_0 + f_1$ and $g = \beta_0 + g_1$ with $\alpha_0 \wedge \beta_0 = 0$ and $f_1 \wedge g_1 = 0$. In either case $\gamma f \wedge g = f \wedge \gamma g = 0$, and in the second case $f\gamma \wedge g = 0$. In the first case $f\gamma \wedge g = 0$, but $f \wedge g\gamma = 0$ iff $\beta_n\gamma\alpha = -\beta_0\gamma\beta$. Thus, $A \subseteq F_\ell(R)$ and the right side of (6.1.15) is $F_r(R) \cap A$. Suppose that $h \in R^+ \backslash A$. If n is not minimal in Δ^*, then there exists $m < n$ and $0 < \gamma \in A$ with $0 < \gamma x^m < 2h$. Then $\gamma \wedge \gamma x^m = 0$ but $(\gamma x^m)\gamma \wedge \gamma x^m > 0$ and $\gamma(\gamma x^m) \wedge \gamma x^m > 0$; so $h \notin f_r(R) \cup f_\ell(R)$. If n is minimal, then by (c) of Theorem 6.1.11, there are elements $0 < \gamma \in A$ and $\delta \in A^+$ with $\gamma\alpha = \delta\beta$. If $g = -\delta + \gamma x^n$, then $\gamma \wedge g = 0$, but $\gamma h \wedge g > 0$ and $h\gamma \wedge g > 0$; so $h \notin f_r(R) \cup f_\ell(R)$, and hence $F_\ell(R) = A$ and $F_r(R)$ has the description given in (6.1.15). Now suppose β centralizes A. If $\gamma \in F_r(R)$, then $\beta\gamma\alpha = \alpha\gamma\beta$ gives that γ centralizes α. Conversely, $\gamma\alpha = \alpha\gamma$ and $\rho\alpha = \delta\beta$ yield $\rho\gamma\alpha = \rho\alpha\gamma = \delta\gamma\beta$. □

When $\beta = 1$ we have just seen that there are connections between the po-ring properties of $(A[\Delta], P_{n,\alpha})$ and the commutative properties of α. More instances of this connection appear below and in Exercises 11 and 12, but we do note that D_α could be isotone without α being central; see Exercise 13.

The *extended centroid* of the domain A is the center C of its maximal right quotient ring $Q_r(A)$. Recall from Exercise 4.1.39 that C is also the center of its maximal left quotient ring, and hence the *central closure* $T = C + AC$ of A is a subring of the maximal two sided quotient ring $Q_2(A)$ of A. So T is a domain by Exercise 4.1.14 and each total order of A can be extended to a unique total order of T by Theorem 4.3.12. Since $Q_r(A)$ is a regular prime ring it is easy to check that C is a field.

Theorem 6.1.15. *Let C be the extended centroid of the domain A, let $\alpha, \beta \in A$, and let F be the subring of A defined in (6.1.15).*

(a) If $\alpha\gamma\beta = \beta\gamma\alpha$ for all $\gamma \in A$, then $\alpha \in C\beta$.
(b) Suppose $A\alpha \cap A\beta \neq 0$. Then $\beta \in F$ if and only if α and β commute, and $A = F$ if and only if $\alpha \in C\beta$.

Proof. (a) Assume $\beta \neq 0$ and define $\psi : A\beta A \longrightarrow A$ by $\psi(\Sigma_i\gamma_i\beta\delta_i) = \Sigma_i\gamma_i\alpha\delta_i$. If $\Sigma_i\gamma_i\beta\delta_i = 0$, then $\Sigma_i\alpha\gamma_i\beta\delta_i = 0$ yields $\Sigma_i\beta\gamma_i\alpha\delta_i = 0$ and hence $\Sigma_i\gamma_i\alpha\delta_i = 0$. So ψ is a left and right A-homomorphism. Since $A\beta A$ is a dense right ideal of A there exists $\sigma \in Q_r(R)$ such that $\sigma\gamma\beta\delta = \gamma\alpha\delta$ for all $\gamma, \delta \in A$. For any $\rho \in A$,

$$(\sigma\rho - \rho\sigma)(\gamma\beta\delta) = \rho\gamma\alpha\delta - \rho\gamma\alpha\delta = 0.$$

So $\sigma\rho = \rho\sigma$ and hence $\sigma \in C$. Since $\gamma(\sigma\beta - \alpha)\delta = 0$ we have $\alpha = \sigma\beta$.

(b) Suppose that $0 \neq \rho\alpha = \delta\beta$. Then $\rho\beta\alpha = \delta\beta^2 = \rho\alpha\beta$ iff $\beta\alpha = \alpha\beta$ and we have the first part. If $\rho\gamma\alpha = \delta\gamma\beta$ for any $\gamma \in A$, then $\rho\gamma\beta\gamma\alpha = \delta\gamma\beta\gamma\beta = \rho\gamma\alpha\gamma\beta$; so $\beta\gamma\alpha = \alpha\gamma\beta$ for each $\gamma \in A$ and hence $\alpha = \sigma\beta$ with $\sigma \in C$ by (a). Conversely, if $\alpha = \sigma\beta$ and $\rho\alpha = \delta\beta$, then $\rho\sigma = \delta$ and $\rho\gamma\alpha = \rho\gamma\sigma\beta = \delta\gamma\beta$ for any $\gamma \in A$; so $A = F$. □

We will now show that the lattice orders which have just been constructed can be identified abstractly. In the next three theorems A is a totally ordered domain and Δ is a monoid and a rooted weakly positive pops such that Δ^* has a nonempty trunk. The elements in trunk (Δ^*) are strongly positive by Theorem 6.1.7.

Theorem 6.1.16. *Suppose $R = A[\Delta]$ is an ℓ-ring such that $A[\Delta^*] = \Sigma(\Delta^*, A)$ is an ℓ-subring of R, $A \subseteq F_\ell(R)$ and $Ax^m \subseteq A^\perp$ for each $m \in \Delta^* \backslash \text{trunk}(\Delta^*)$. Then $A = F_\ell(R)$, or $R = \Sigma(\Delta_{\emptyset,\Delta^*}, A)$ and Δ^* and R are totally ordered.*

Proof. Assume that $\gamma x^n \in F_\ell(R)$ for some $n \in \Delta^*$ and some $\gamma > 0$. If $n \notin \text{trunk}(\Delta^*)$, then $\gamma^2 x^n \wedge \gamma = 0$ gives that $0 = \gamma^2 x^n \wedge \gamma^2 x^n$. Thus, $n \in \text{trunk}(\Delta^*)$, n is strongly positive by Theorem 6.1.7, and $n + k$ and $n + \ell$ are comparable for any two elements k and ℓ in Δ^* since $n < n + k, n + \ell$. Now, $\gamma x^n \in F_\ell(\Sigma(\Delta^*, A)) = \Sigma(f_\ell(\Delta^*), A)$ by Exercise 3.5.17, and k and ℓ must be comparable. So Δ^* is totally ordered and hence R is totally ordered since $R \cong \gamma x^n R \subseteq \Sigma(\Delta^*, A)$ as right R-ℓ-modules. In fact, $R = \Sigma(\Delta_{\emptyset,\Delta^*}, A)$ since if $0 < \rho x^m < \delta$ for some $m \in \Delta$, then $\rho^2 x^{2m} < \rho\delta x^m < \rho^2 x^{2m}$ unless $m = 0$. So A is convex in R and R has the Hahn order determined by the ordinal sum $\{0\} \overleftarrow{\cup} \Delta^*$. Now, if there exists some element $0 < f = \alpha_0 + f^* \in F_\ell(R)$ with $0 \neq f^* \in A[\Delta^*]$, then since $f^*, (f^*)^+, (f^*)^- \in F_\ell(R) \cap \Sigma(\Delta^*, A)$ we may assume $\alpha_0 = 0$. This gives that $0 < \gamma x^n \in F_\ell(R)$ for some $n \in \Delta^*$ and hence R is totally ordered. □

Theorem 6.1.17. *Let $R = A[\Delta]$ be a po-ring. The following two statements are equivalent.*

(I) (a) There is a partition $\{\Gamma_1, \Gamma_2\}$ of Δ^* with $\Gamma_2 \subseteq$ trunk (Δ^*) and $\Gamma_1 < \Gamma_2$ such that $R = \Sigma(\Delta_{\Gamma_1,\Gamma_2}, A)$; or
(b) there is an element $0 < \sigma$ in the extended centroid of A with $A^+\sigma \subseteq A \cup U(A)$ and an element n in the trunk of Δ^* such that R is a po-subring of $(T[\Delta], P_{n,\sigma})$, where T is the central closure of A.

(II) R is an ℓ-ring with the following properties.
(i) $A[\Delta^*] = \Sigma(\Delta^*, A)$ is an ℓ-subring of R;
(ii) $A \subseteq F(R)$;
(iii) $Ax^m \subseteq A^\perp$ for each $m \in \Delta^* \backslash \text{trunk}(\Delta^*)$.

Moreover, if the conditions in (II) are satisfied, then $A = F(R) = F_\ell(R) = F_r(R)$; and R is an sp-ℓ-ring if and only if Δ is an sp-pops.

Proof. If $R = \Sigma(\Delta_{\Gamma_1,\Gamma_2},A)$, then R is an ℓ-ring and the three conditions in (II) hold by (6.1.5) and Theorem 6.1.10. If $R^+ = R \cap P_{n,\sigma}$, then $(T[\Delta],P_{n,\sigma})$ is an ℓ-ring by Theorems 6.1.11(b) and 6.1.14, and R is an ℓ-subring by Theorem 6.1.13 since $(1 : A\sigma) \cap C(A) = A\sigma \cap C(A) \subseteq A$. Also, (i) and (iii) hold for T and $(T[\Delta],P_{n,\sigma})$ by (6.1.10) and hence also for R, and, similarly, (ii) holds by Theorem 6.1.14.

For the converse, suppose R is not totally ordered, $A \subseteq F_\ell(R)$, and both (i) and (iii) hold. Then $A = F_\ell(R)$ by Thoerem 6.1.16 and $R^+ = U(A) \cup (A \oplus A^\perp)^+$ by Theorem 2.5.9. Let

$$M = \{n \in \text{trunk}(\Delta^*) : \gamma x^n \not> A \text{ for some } 0 < \gamma \in A\}$$

and set $N = \text{trunk}(\Delta^*) \backslash M$ and $K = \Delta^* \backslash \text{trunk}(\Delta^*)$; then $K < M < N$. We first note that if $m < n$ in Δ^* and $\gamma x^n \not> A$ for some $\gamma > 0$, then $Ax^m \subseteq A^\perp$. For, if $0 < \rho \in A$, then $\rho x^m < \gamma x^n$; so $\rho x^m \not> A$ and $\rho x^m = \tau + b$ with $\tau \in A^+$ and $b \in (A^\perp)^+$. If $\tau > 0$, then for any $\delta > 0$ we have $\tau\delta \leq \rho\delta x^m < \tau\gamma x^n$ and $\delta < \gamma x^n$. Thus, $\tau = 0$ and $\rho x^m \in A^\perp$. So if M is either the empty set or is not empty but does not have a largest element, then $A[M \cup K] \subseteq A^\perp$. But also $A^\perp \subseteq A[M \cup K]$ since if $0 < f \in A^\perp$ and $n \in N \cap \text{maxsupp} f$ with coefficient α_n, then $A < \alpha_n x^n < 2f \in A^\perp$. Thus,

$$R = [A \oplus \Sigma(M \cup K, A)] \overleftarrow{\boxplus} \Sigma(N,A);$$

that is, $R^+ = P_{M \cup K,N}$. Suppose that n is the largest element of M. Then there exists $0 < \beta \in A$ with $\beta x^n = \alpha + b \in A \oplus A^\perp$. Now, for each $\gamma > 0$, if $\gamma x^n > A$, then $\gamma\alpha + \gamma b = \gamma\beta x^n > A\beta$ and $\gamma\alpha > A\beta$. Thus, if $\gamma\alpha \not> A\beta$, then $\gamma x^n = \alpha_1 + b_1 \in A \oplus A^\perp$, $\gamma\beta x^n = \alpha_1\beta + b_1\beta = \gamma\alpha + \alpha b$, and $\gamma\alpha = \alpha_1\beta$ if we now assume that $\beta \in F_r(R)$. So $P_{n,\alpha,\beta}$ is defined and we will check that $R^+ = P_{n,\alpha,\beta}$. Let $f = \alpha_0 + f_1 + \alpha_n x^n + f_2$ with supp $f_1 \subseteq L_s(n)^*$ and supp $f_2 \subseteq U_s(n) = N$. If $f_2 \neq 0$ let $p = \text{maxsupp} f_2$. Then $f = u + \alpha_p x^p$ and because of (i) and the inequality $A < |\alpha_p| x^p$ we have $\pm u < |\alpha_p| x^p$ by Exercise 3.4.31. So $f > 0$ iff $\alpha_p > 0$, iff $f \in P_{n,\alpha,\beta}$. If $f_2 = \alpha_n = 0$, then since we have just seen that $f_1 \in A^\perp$, clearly, $0 < f = \alpha_0 + f_1$ iff $f \in P_{n,\alpha,\beta}$. Suppose, then, that $f_2 = 0$ and $\alpha_n \neq 0$. Now, $f\beta = (\alpha_0\beta + \alpha_n\alpha) + (f_1\beta + \alpha_n b) \in A \oplus A^\perp$. If $f > 0$, then $f|\alpha_n|x^n = \alpha_0|\alpha_n|x^n + f_1|\alpha_n|x^n + |\alpha_n|\alpha_n x^{2n} > 0$ and hence $\alpha_n > 0$. Since $f\beta > 0$ we have $\alpha_0\beta + \alpha_n\alpha \geq 0$ and hence $f \in P_{n,\alpha,\beta}$. On the other hand, if $f \in P_{n,\alpha,\beta}$, then $\alpha_n > 0$ and $\alpha_0\beta + \alpha_n\alpha \geq 0$. But also $\alpha_n b \geq -f_1\beta$ since $\alpha_n b + \alpha_n\alpha = \alpha_n\beta x^n > \Sigma(L_s(n)^*,A)$ and $\Sigma(L_s(n)^*,A) \subseteq A^\perp$. Thus, $f > 0$ and we have shown $R^+ = P_{n,\alpha,\beta}$. If $A\alpha \cap A\beta = 0$, then $\alpha = 0$ or $\gamma\alpha > A\beta$ for each $\gamma > 0$; so $P_{n,\alpha,\beta} = P_{n,0} = P_{L(n)^*,U(n)}$ or $P_{n,\alpha,\beta} = P_{L_s(n)^*,U(n)}$. Now, suppose $A \subseteq F(R)$. Then $A = F_\ell(R) = F(R) \subseteq F_r(R) \subseteq A$ by (6.1.15). So, assuming that $A\alpha \cap A\beta \neq 0$, we have that $\alpha = \sigma\beta$ for some $0 < \sigma$ in the extended centroid C of A, by Theorem 6.1.15. Moreover, the condition $A^+\alpha \subseteq A\beta \cup U(A\beta)$ in A now becomes $A^+\sigma \subseteq A \cup U(A)$ in the central closure T of A. Also, since $T\alpha = T\beta$, $P_{n,\alpha,\beta}(T[\Delta])$ is defined and $P_{n,\alpha,\beta}(T[\Delta]) = P_{n,\sigma}(T[\Delta])$ by Theorem 6.1.12. Since the pair (α,β) satisfies (6.1.6) and (6.1.7) in both A and T, it is clear that $P_{n,\alpha,\beta}(T[\Delta]) \cap A[\Delta] = P_{n,\alpha,\beta}(A[\Delta])$. The last statement is a consequence of Theorems 6.1.9, 6.1.14, and 3.7.6. □

When A is ℓ-simple the condition $A \subseteq F(R)$ in (II) can be relaxed to $A \subseteq F_\ell(R)$ - see Exercise 15. The lattice orders of $A[x]$ in which squares are positive can also be identified.

Theorem 6.1.18. *Let A be a totally ordered domain which has an ℓ-simple classical one-sided quotient ring L. Suppose the polynomial ring $R = A[x]$ is an sp-ℓ-ring which is not totally ordered, $A \subseteq F(R)$ and $A^+x \subseteq R^+$. Then $A = F(R)$, and there are central elements $\sigma, \gamma \in L^+$ with $y = x - \gamma > 0$ and an integer $n \in \mathbb{N}$ such that*

$$R^+ = P_0(L[y]) \cap R \text{ or } R^+ = P_{n,\sigma}(L([y]) \cap R.$$

Proof. Let $F = F(R)$. We will first assume that $A = L$. Suppose $x \ll_A x^2$. Since A is ℓ-simple $u \ll_A v$ implies $u \ll_A \rho v$ for each $0 < \rho \in A$ and hence $A[x]x = \Sigma(\mathbb{N}, A)$. Thus, by Exercise 15 and Theorem 6.1.17 and its proof, $A = F$, and $R^+ = P_0(A[x])$ or $R^+ = P_{n,\alpha,\beta}(A[x]) = P_{n,\sigma}(A[x])$ where $\alpha = \sigma\beta$ and σ is in the center of A. In this case $\gamma = 0$. Since $R^+ = U(F) \cup (F \oplus F^\perp)^+$ by Theorem 2.5.9 the other possibility is that $x = \gamma + y \in F \oplus F^\perp$. Note that y and γ are central elements of R since, for any $\delta \in A$, $\delta\gamma + \delta y = \delta x = x\delta = \gamma\delta + y\delta$ and $\delta\gamma = \gamma\delta$ and $\delta y = y\delta$ since F and $F^\perp$ are A-A-bimodules. Also, $y \neq 0$ since otherwise $A[x] = F$ is totally ordered. From Exercise 14 applied to $F[y] = A[x]$ we have that $y \ll_F \delta y^2$ for each $0 < \delta \in F$. Now, $F = A$. For, suppose $f(x) \in F^+ \backslash A$; so $f(x) = \alpha_0 + \alpha_1 x + \cdots + \alpha_n x^n$ with $\alpha_j \in A$, $n \geq 1$, and $\alpha_n \neq 0$. Then $f(x) = f(\gamma + y) = f(\gamma) + \beta_1 y + \cdots + \alpha_n y^n$ with $\beta_j \in F$. If $n = 1$ we have the contradiction $0 \neq \alpha_1 y = f(x) - f(\gamma) \in F \cap F^\perp$. If $n \geq 2$, then from $0 < f(x)y = f(\gamma)y + \beta_1 y^2 + \cdots + \alpha_n y^{n+1}$ we get $\alpha_n > 0$ and the contradiction $y \in F$ since $0 < y < \beta_1 y + \cdots + \alpha_n y^n = f(x) - f(\gamma) \in F$. So $R = A[y]$ and R^+ has the desired form by the previous case. Now suppose that $L = A_S$ where S is a right Öre subset of A^+. Then S is a right Öre subset of R and $L[x] = R_S$ is an ℓ-ring extension of $A[x]$ by Exercise 16. Let $E = F(L[x])$. According to Exercise 16 L is a subring of E. Now, either $x \ll_L x^2$ or $x = \gamma + y \in E \oplus E^\perp$. In the first case we have, as above, that $L = E = F_\ell(L[x]) = F_r(L[x])$ and $L[x]^+ = P_0(L[x])$ or $L[x]^+ = P_{n,\sigma}(L[x])$ with σ in the center of L. Also, $F \subseteq F_\ell(L[x])$ since if $0 < \tau \in F$ and $f\rho^{-1} \wedge g\rho^{-1} = 0$ with $f, g \in A[x]$ and $\rho \in S$, then $f \wedge g = 0$, $\tau f \wedge g = 0$ and $\tau f\rho^{-1} \wedge g\rho^{-1} = 0$. So $F \subseteq L \cap A[x] = A$ and $F = A$. For the second case, again, as above, it suffices to show that $\rho y \leq \delta y^2$ for any $0 < \rho, \delta \in E$. Take $\tau \in S$ such that $\rho\tau$, $\delta\tau$ and $y\tau$ are all in $A[x]$. Then $\rho\tau, \delta\tau \in E \cap A[x] \subseteq F$ and $y\tau \in E^{\perp_R}$. But F (respectively, E) is a totally ordered convex subring of R (respectively, $L[x]$) and $A \subseteq F \cap E$; so if $0 \neq \gamma \in A$, then $E^{\perp_R} = \gamma^{\perp_R} = F^{\perp_R} = \mu^{\perp_R}$ for any $0 \neq \mu \in F$. Now, by Exercise 14, $(\rho\tau^2)(\delta\tau)(y\tau) \leq (\delta\tau y\tau)(\delta\tau y\tau)$; that is $\rho y \leq \delta y^2$. □

Other perturbations of the partial order of the mopops Δ which preserve good properties of Δ are given in Exercises 22, 23, and 24. In particular, when G is a totally ordered group and $\Delta = (G^+)_0$ these constructions can be used to make the semigroup ring $A[\Delta]$ into an sp-ℓ-ring in various ways. By taking G to be a free abelian group or a free group of rank n, with positive generators, sp-lattice orders of polynomial rings and free rings in n variables are obtained.

Exercise 5.3.9 shows that it is not always the case that $A = F(A[x])$ whenever $A[x]$ is an ℓ-ring extension of the totally ordered domain A with $A \subseteq F(A[x])$. However, we do have equality of these subrings if $A[x]$ is an sp-ℓ-ring even when x is not comparable to 0.

Theorem 6.1.19. *Let A be a totally ordered domain and suppose $A[x]$ is an sp-ℓ-ring extension of A with $A \subseteq F(A[x]) \subset A[x]$. Then $A = F(A[x])$. If A is ℓ-simple, then $A[\rho x^2]$ is an ℓ-subring of $A[x]$ for each positive element ρ in the center of A. If A is unital there are sp-lattice orders of $A[x]$ with $x^+ = \rho$, for any central element ρ, but none with x^+ of degree one or two.*

Proof. We will repeatedly use the following fact and its analogue on the right: if ${}_RM$ is an ℓ-module and $\alpha, \alpha\beta \in f({}_RM)$ with $\beta \in R^+$ and $r(\alpha; M)^+ = 0$, then $\beta \in f({}_RM)$ - see the proof of Theorem 2.4.8. We will also use Theorem 2.5.9 for $F = F(A[x])$: for $u \in A[x]$ either $|u| > F$ or $u \in F \oplus F^\perp$. Suppose $\beta x^2 \in F$ for some $0 < \beta \in F$. Then for any $0 < \gamma \in F$, $\beta\gamma x^2$ and $\gamma x^2 \beta$ are in F; so $Fx^2 \subseteq F$. From $0 \le (\gamma \pm \gamma x)^2$ we obtain

$$\gamma^2 |x| \le \gamma^2 + \gamma^2 x^2. \tag{6.1.16}$$

So $\gamma^2|x| = \gamma|x|\gamma \in F$, $\gamma x \in F$, $\gamma^{m-1}\gamma x^m = \gamma x^m \gamma^{m-1} \in F$ for $m \in \mathbb{N}$, $Ax^m \subseteq F$, and we have the contradiction $F = A[x]$. Now, suppose $f = \alpha_0 + \alpha_1 x + \cdots + \alpha_n x^n = \alpha_0 + gx \in F$ with $n \ge 1$, $\alpha_i \in A$ and $\alpha_n \ne 0$. Then $gx = f - \alpha_0 \in F$. If $\beta x^2 \ge \alpha > 0$ for some $\alpha, \beta \in F$, then $0 < g^2\alpha \le g^2x^2\beta$, $0 < \alpha g^2 \le \beta x^2 g^2$, and hence $g^2 \in F$ and $\beta x^2 \in F$. Thus, $\beta x^2 \notin U(\alpha)$ for any $0 < \alpha \in A$, and $Ax^2 \subseteq F^\perp$. By an application of (6.1.16), for each $\beta \in F^+$, $\beta^2 x^2 \le \beta^2 x^4$ and $\beta x^2 \le \beta x^4$. Since $g^2x^2\beta \le g^2x^4\beta$ and $g^2x^4\beta = (g^2x^2)(x^2\beta) \in F^\perp$, $g^2x^2\beta \in F \cap F^\perp = 0$. So, $A = F$.

Suppose A is ℓ-simple. If $\gamma x^2 > A$ for some $0 < \gamma \in A$, then $\rho x^2 > A$ for any $0 < \rho \in A$. In this case $A[x^2]$ has the lexicographic total order. The alternative is that $Ax^2 \subseteq A \oplus A^\perp$. If $0 < \rho$ is in the center of A, then $\rho x^2 = \alpha + y \in A \oplus A^\perp$ with α and y central. So $A[\rho x^2] = A[y]$ is an ℓ-subring of $A[x]$ by Exercise 18, and by Theorem 6.1.18 $A[\rho x^2]^+$ is either $P_0(A[y])$ or $P_{n,\sigma}(L[y]) \cap A[y]$ for some $n \in \mathbb{N}$ and some σ in the center of L, the classical totally ordered quotient ring of A obtained by inverting the powers of ρ. Of course, if A is unital, then σ is in the center of A.

Now assume A is unital and x is not comparable to 0. Since $1 \wedge x^+ \wedge x^- = 0$ either $x^+ \in A^\perp$ or $x^- \in A^\perp$ and we will assume the latter. Suppose $n \in \mathbb{Z}^+$, $\alpha_n \ne 0$, and

$$x^+ = \alpha_0 + \alpha_1 x + \cdots + \alpha_n x^n$$

and

$$x^- = \alpha_0 + (\alpha_1 - 1)x + \cdots + \alpha_n x^n.$$

Since x^+ centralizes A each α_j is in the center of A. If S is the set of strictly positive central elements of A, then according to Exercise 16, $A_S[x]$ is an sp-ℓ-ring extension of $A[x]$ and $A_S = F(A_S[x])$. If $n = 0$, then $x^- = \alpha_0 - x$, $A[x] = A[x^-]$ and again, by Exercise 18, $A[x]^+ = P_0(A[\alpha_0 - x])$ or $A[x]^+ = P_{m,\sigma}(A[\alpha_0 - x])$ with $\sigma = 0$ when $m = 1$. Since $x = \alpha_0 - (\alpha_0 - x) \in A \boxplus A(\alpha_0 - x)$ each strictly positive central element α_0 of A does produce these lattice orders of $A[x]$ with $x^+ = \alpha_0$. Suppose $n = 1$. Then

$\alpha_1 \neq 1$ since $x^- \notin A$. Because $A_S[x] = A_S[\alpha_0 + (\alpha_1 - 1)x] = A_S[x^-]$ and

$$x = -\alpha_0(\alpha_1 - 1)^{-1} + (\alpha_1 - 1)^{-1}(\alpha_0 + (\alpha_1 - 1)x) \in A_S \oplus A_S(\alpha_0 + (\alpha_1 - 1)x) \subseteq A_S[x]$$

we have

$$x^+ = (-\alpha_0(\alpha_1 - 1)^{-1})^+ + ((\alpha_1 - 1)^{-1})^+ \alpha_0 + ((\alpha_1 - 1)^{-1})^+(\alpha_1 - 1)x.$$

So $((\alpha_1 - 1)^{-1})^+(\alpha_1 - 1) = \alpha_1$, and $\alpha_1 = 0$ or $\alpha_1 = 1$, neither of which is possible.

Suppose $n = 2$. We will first show that $\alpha_1 = 0$. Note that x is not comparable to any element of A_S. For, suppose $\gamma \in A_S$. If $x \geq \gamma$, then $x \geq 0$ or $x^- \in A_S$ depending on whether $\gamma \geq 0$ or $\gamma < 0$, and if $x \leq \gamma$, then, similarly, $x \leq 0$ or $x^+ \in A_S$ depending on whether $\gamma \leq 0$ or $\gamma > 0$. Now, if $\alpha_2 < 0$, then $\alpha_0 + \alpha_1 x = x^+ - \alpha_2 x^2 \geq 0$; and if $\alpha_2 > 0$, then from $0 \leq (1 - \alpha_2 x)^2$ we have $\alpha_2\alpha_0 + \alpha_2\alpha_1 x + \alpha_2^2 x^2 = \alpha_2 x^+ \leq 1 + \alpha_2^2 x^2$ and $\alpha_2\alpha_1 x \leq 1 - \alpha_2\alpha_0$. In either case x is comparable to some element of A unless $\alpha_1 = 0$. If $x^2 \geq A_S$, then, for each $\alpha \in A_S^+$, $2\alpha|x| \leq \alpha^2 + x^2 \leq 2x^2$ and $x^2 \geq A_S + A_S x$. But from $\alpha_2 > 0$ we obtain the contradiction $x^- = \alpha_2 x^2 + (\alpha_0 - x) > A_S$, and from $\alpha_2 < 0$ we obtain the contradiction $\alpha_0 - x > -\alpha_2 x^2 > A_S + A_S x$. Thus, $x^2 = \alpha + y \in A_S \oplus A_S^\perp$, and, for any $\gamma \in A_S^+$,

$$\gamma(\alpha_0 + \alpha_2\alpha) + \gamma\alpha_2 y = \gamma x^+ \leq \gamma|x| \leq \gamma^2 + x^2 = \gamma^2 + \alpha + y.$$

This implies $A_S = A_S\alpha_2 \leq 1$. □

Exercise.

1. Let D and D_1 be totally ordered division rings and suppose G and H are groups with G periodic. If $(D[G], D^+[G]) \cong (D_1[H], D_1^+[H])$ show that $D \cong D_1$ and $G \cong H$. (Proceed directly or use Exercises 3.5.16(b) and 3.5.18(a).)

2. Suppose R is a finite valued K-archimedean ℓ-algebra over the totally ordered field K in which the product of special elements is special. Write $R = \oplus_i R_i$ where each R_i is a totally ordered vector lattice over K (Exercise 2.5.22). Verify the following.

   (a) R is an ℓ-domain.
   (b) For every i, j there exists some k with $R_i R_j \subseteq R_k$.
   (c) If $1 \in R$ is special and $1 \in R_0$, then R_0 is a K-subalgebra and $R_0 R_i = R_i = R_i R_0$ for every i.
   (d) If the special elements form a group, then equality holds in (b).
   (e) The special elements form a group and 1 is a convex element of ${}_K R$ iff R is a twisted group ℓ-algebra over K.

3. If A is a right (respectively, left) algebra ℓ-ideal of the twisted semigroup ℓ-algebra $K[S;\tau]$ show that $A = K[S_1;\tau]$ where S_1 is a right (respectively, left) ideal of S; that is, $S_1 S \subseteq S_1$ (respectively, $SS_1 \subseteq S_1$).

4. Let $f : S \longrightarrow T$ be an injective homomorphism of semigroups and suppose $\sigma \in Z^2(S, K^{+*})$, $\tau \in Z^2(T, K^{+*})$ and $f^*(\overline{\tau}) = \overline{\sigma}$; so $\hat{f}(\tau) = d_S(c)\sigma$ for some c in $C^1(S, K^{+*})$.

   (a) If $a \in C^1(S, K^{+*})$ show that the K-linear transformation $\varphi : K[S;\sigma] \longrightarrow K[T;\tau]$ induced by $\varphi(\overline{s}) = a(s)\overline{f(s)}$ is a monomorphism of ℓ-algebras if and only if ac is in the kernel of the coboundary map d_S.
   (b) If f is an isomorphism and $d(ac) = 1$ show that φ is an isomorphism.
   (c) Conversely, if $\varphi : K[S;\sigma] \longrightarrow K[T;\tau]$ is an isomorphism of ℓ-algebras show that there is a semigroup isomorphism $f : S \longrightarrow T$ with $f^*(\overline{\tau}) = \overline{\sigma}$ and which induces φ as in (a).
   (d) Show that the group of automorphisms of the ℓ-algebra $K[S;\sigma]$ is isomorphic to the group of automorphisms of the semigroup S.

5. Let S be a finite semigroup.

   (a) For $s \in S$ let $j \in \mathbb{N}$ be minimal such that $s^j = s^k$ for some $k > j$, and let $p \in \mathbb{N}$ be minimal such that $s^j = s^{j+p}$. Show that if $k > i \geq 1$, then $s^k = s^i$ iff $i \geq j$ and $k = i + np$ for some $n \in \mathbb{N}$.
   (b) Suppose $s, t \in S$ and j and p are minimal with $(st)^j = (st)^{j+p}$ and k and q are minimal with $(ts)^k = (ts)^{k+q}$. Show that $p = q$ and $k - 1 \leq j \leq k + 1$. ($(st)^{k+q+1} = s(ts)^{k+q}t$.)

6. Let M and N be ℓ-groups with N abelian and, as usual, let the homomorphism group $E = \mathrm{Hom}_Z(M, N)$ be the abelian po-group whose positive cone consists of those $f \in E$ with $f(M^+) \subseteq N^+$. Assume all the necessary sups exist; for example, N is complete. Let $x, y \in M$, $f \in E^+$, and $A = [0, f]$.

   (a) Show that $f(x \vee y) \geq \{g(x) + h(y) : g, h \in A, g + h = f\}$.
   (b) Let B be a subset of A such that $g \in B$ implies $f - g \in B$. Show that for all $x, y \in M$,
   $$f(x \vee y) = \bigvee \{g(x) + h(y) : g, h \in B, g + h = f\}$$
   if and only if, for all $x \in M$,
   $$f(x^+) = \bigvee \{g(x) : g \in B\}.$$

7. If X is a nonempty set, then $S = X \times X$ is a semigroup with product given by $(a,b)(c,d) = (a,d)$. Let $X = \{0,1\} \subseteq \mathbb{R}$. For each $b \in \mathbb{R}$ define the function σ_b on S^2 by $\sigma_b((x_1,y_1),(x_2,y_2)) = \exp(b(x_1 - x_2)(y_1 - y_2))$. Show that σ_b is a fundamental 2-cocyle, that every fundamental 2-cocyle is some σ_b, and $H^2(S, \mathbb{R}^{+*}) \cong \mathbb{R}$.

8. (a) Let Δ be a positive pops which contains $2n$ for each $n \in \Delta$ and which has a partition $\{\Delta_i : i \in I\}$ with $|I| \geq 2$, each Δ_i is totally ordered, and, for all i, $j \in I$, there is a $k \in I$ with $\Delta_i + \Delta_j \subseteq \Delta_k$. Reorder Δ so that its new partial order is that of the cardinal sum of the Δ_i. Show that $|I| \geq 3$, each Δ_i is a subsemigroup of Δ, and Δ, with its new order, is weakly positive but not positive.

(b) As an instance of (a) let $\Delta = \Gamma_1 \times \Gamma_2$ be the direct product of two totally ordered positive monoids.

9. Show that a weakly positive f-pops is positive.

10. Suppose A is a directed po-domain, Γ is a poset and R is the po-group $R = (A[\Gamma], P_{n,\alpha,\beta})$.

(a) Show that ${}_AR$ is a strict po-module.
(b) If the conditions in (b) of Theorem 6.1.11 are satisfied show that $F({}_AR) = A$.

11. Let A be a po-unital po-ring and suppose the inner derivation $D_\alpha = [\ \ ,\alpha]$ is isotone. If u, $u^{-1} \in A^+$ show that $u\alpha = \alpha u$.

12. Let $R = (A[\Delta], P_{n,\alpha,\beta})$ be an ℓ-ring where Δ is a rooted mopops and n is a strongly positive element in the trunk of Δ^*.

(a) If $\alpha = \alpha'\beta$ show that the inner derivation $D_{\alpha'}$ is isotone.
(b) If β, $\beta^{-1} \in A^+$ show that $D_{\alpha\beta^{-1}}$ is isotone.
(c) If A is a division ring show that $\alpha\beta^{-1}$ is central.
(d) If β centralizes A and $A = C(Z(A))$ show that α is central.

13. Let R be the totally ordered free unital ring with free generators x and y given in Exercise 3.4.35. Show that the derivations $D_y = [\ \ ,y]$ and ${}_xD = [x,\ \ ]$ are both isotone.

14. Suppose D is a multiplicatively closed subset of the positive cone of a Riesz ring R and $a \in R$. If, for every $\delta, \rho \in D$, $\delta a \wedge \rho = 0 \leq (\rho - \delta a)^2$, show that, for each $\delta \in D, D\delta a \leq (\delta a)^2$. (See (3.7.5).)

15. Suppose that A, Δ and $R = A[\Delta]$ satisfy all the conditions given in (II) of Theorem 6.1.17 except that now $A \subseteq F_\ell(R)$ and $F(R) \neq 0$, and additionally A is ℓ-simple. Show that R is totally ordered or $R^+ = P_{\Gamma_1,\Gamma_2}$ or $R^+ = P_{n,\alpha,\beta}$ for some element n in the trunk of Δ^* and some elements α, $\beta \in A$ with $A\alpha \subseteq A\beta$ and $\alpha\beta = \beta\alpha$. (In the proof of Theorem 6.1.17 choose $\beta \in F(R)$.)

16. Let S be a right Öre set of positive regular elements in the f-ring A. Suppose that the polynomial ring $R = A[x]$ is an ℓ-ring extension of A and $A^+ \subseteq d(R)$.

(a) Show that S is a right Öre set in R and $R_S = A_S[x]$ is an ℓ-ring extension of R.
(b) If $A \subseteq F_\ell(R)$ (respectively, $F_r(R)$), show that $A_S \subseteq F_\ell(R_S)$ (respectively, $F_r(R_S)$).
(c) If $A = F_\ell(R)$, $F_r(R)$, or $F(R)$, show that $A_S = F_\ell(R_S)$, $F_r(R_S)$, or $F(R_S)$, respectively.

17. Suppose R is an ℓ-ring and the domain A is a totally ordered ℓ-subring of R with $AA^\perp \subseteq A^\perp$. Let $0 < a \in A^\perp$ have the following properties: a centralizes A, $A[a]$ has squares positive, and $A[a]$ is an A-semiclosed left or right po-module. Show that a is transcendental over A and $P_0(A[a]) \subseteq A[a]^+ \subseteq P_{1,0}(A[a])$.

18. Let R be an ℓ-ring, A a totally ordered domain that is a convex subring of $F_\ell(R)$, and $0 < a$ an element of R that centralizes A. Show that (a) and (c) are equivalent, (b) implies (c), and if $A \subseteq F(A[a])$, then (c) implies (b).
    (a) $A[a]$ is an sp-subring of R, ${}_AA[a]$ is semiclosed and $a \in A^\perp$.
    (b) a is transcendental over A and $A[a]^+$ is P_0, $P_{1,0}$, or $P_{n,\alpha,\beta}$ for some $2 \leq n \in \mathbb{N}$ and some $\alpha, \beta \in A^+$ such that $\alpha = \sigma\beta$ where σ is in the extended centroid of A and $A^+\sigma \subseteq A \cup U(A)$.
    (c) $A[a]$ is an sp-ℓ-subring of R, ${}_AA[a]$ is semiclosed and $a \in A^\perp$.

19. Let R be a commutative sp-ℓ-ring and suppose that A is a convex subring of $F(R)$ such that ${}_AR$ is torsion-free. Show that

$$R_0 = \{a \in R : A[|a|]^+ = P_0\} \cup \{0\}$$

is a convex ℓ-A-subalgebra of R, and it is the largest subring of R contained in $A^\perp = F(R)^\perp$.

20. Suppose the commutative unital domain R is an sp-ℓ-algebra over the totally ordered field B. Let $0 < a \in 1^\perp$ and let A be the convex subgroup of R generated by B. If $B[a]$ is not an ℓ-subalgebra of R show that there are elements $0 < \rho$, $\delta < \alpha$ in R with $\alpha \in B$ such that $B\rho \cap A \not\subseteq \delta B$.

21. Suppose $R = (A[\Delta], P_{n,\alpha,\beta})$ is an ℓ-ring where Δ is a rooted weakly positive mopops and n is an element in the trunk of Δ^*. If $r \in \mathbb{N}$ and $n < (2r)\Delta^*$ show that R satisfies the identity $y^{2r} \geq 0$.

22. Let Δ be a monoid and a rooted pops and let m be a strongly positive element in the trunk of Δ^*. Let Δ_m be the reordering of Δ given by the ordinal sum $\Delta_m = (\{0\} \cup L(m)_t) \underleftarrow{\cup} U_s(m)$ where $L(m)_t$ is the set $L(m)$ with the trivial partial order. Verify each of the following:
    (a) Δ_m is a rooted po-monoid whose trunk is $U_s(m)$, and m is a strongly positive element of Δ_m.
    (b) Δ_m is an sp-pops iff Δ is an sp-pops and $m < 2k$ for each $k \in L_s(m)$.
    (c) If Γ is a subset of the totally ordered group G with $0 < \Gamma < m < 2\Gamma$, then $\Delta = \{0\} \cup \Gamma \cup U(m)$ is an example of an sp-pops which satisfies (b).
    (d) For the totally ordered domain A and $\alpha, \beta \in A$ satisfying (6.1.6), (6.1.7) and (6.1.14) and $m < n$, let $P_{m,n,\alpha,\beta}$ be the lattice order $P_{n,\alpha,\beta}(A[\Delta_m])$ of $A[\Delta]$. If $2 \leq r \in \mathbb{N}$, then $(A[x], P_{2r-2,2r-1,\alpha,\beta})$ is an ℓ-ring satisfying $y^{2r} \geq 0$ but not $y^{2s} \geq 0$ if $1 \leq s < r$.

23. Let m and k be elements of the mopops Δ such that:
    (i) $m < k$;
    (ii) m is *strongly irreducible*: $m = a + b \Rightarrow a = 0$ or $b = 0$;
    (iii) if $p < 0$, then $m + p$ and $p + m$ are not defined;
    (iv) if $0 < q \neq m$ and $q + m$ (respectively, $m + q$) is defined, then $k \leq q + m$ (respectively, $k \leq m + q$).

Define the relation $\leq_{m,k}$ on Δ by $p <_{m,k} q$ iff $m \notin \{p,q\}$ and $p < q$, or $p = m$ and $k \leq q$. Let $\Delta_{m,k}$ be the generalized monoid Δ together with the relation $\leq_{m,k}$. Verify each of the following:

(a) $\leq_{m,k}$ is a weaker partial order than $\leq$.
(b) $\Delta_{m,k} = \Delta$ iff m is a minimal element of Δ and $U_s(m) = U(k)$; $\Delta_{0,k} = \Delta$ provided $\Delta_{0,k}$ exists.
(c) $\Delta_{m,k}$ is a mopops.
(d) Δ is rooted iff $\Delta_{m,k}$ is rooted and $U_s(m)$ and $L_s(k)$ are rooted subsets of Δ.
(e) If $m \neq 0$, then $\Delta_{m,k}$ is weakly positive iff Δ is weakly positive and $k \leq 2m$.
(f) Suppose 0 is a minimal element of Δ and $m \neq 0$. Then Δ is an (almost) sp-pops and $k \leq 2m$ iff $\Delta_{m,k}$ is an (almost) sp-pops.
(g) $(\Delta_0)_{m,k}$ exists if m is strongly irreducible.

24. Let $\Delta = (\mathbb{Z}^+ \underleftarrow{\boxplus} \mathbb{Z}^+)_{\Gamma_1,\Gamma_2}$ be the rooted mopops obtained from the totally ordered mopops $\mathbb{Z}^+ \underleftarrow{\boxplus} \mathbb{Z}^+$ with $\Gamma_2 = U((1,1))$ and $\Gamma_1 = L_s((1,1))^*$, and take $m = (0,1)$ and $k = (0,2)$ in the previous exercise.

(a) Show that $\Delta_{m,k}$ is a rooted sp-po-monoid.
(b) Determine the partial order of $\Delta_{m,k}$ explicitly.
(c) Let $x = x^{(1,0)}$, $y = x^{(0,1)}$, and $A[x,y] = \Sigma(\Delta_{m,k}, A)$ where A is a totally ordered domain. Give an explicit description of when the polynomial

$$f = (\alpha_0 + \alpha_1 x + \cdots + \alpha_r x^r) + (\beta_0 + \beta_1 x + \cdots + \beta_s x^s)y + \cdots + f_n(x)y^n$$

is in $A[x,y]^+$.
(d) Show that $A[x]$ and $A[y]$ are ℓ-subrings of $A[x,y]$ and identify their lattice orders.

Notes. The characterization of real group ℓ-algebras and twisted group ℓ-algebras over finite groups comes from Rieffel [RIE], and the extension of Rieffel's results to locally finite ℓ-algebras appears in Steinberg [ST18]. The lattice orders constructed on semigroup rings over positive rooted monoids comes from Ma and Steinberg [MS]. The bounds on the partial order of a polynomial subring of an sp-ℓ-ring that are given in Exercise 17 appear in Steinberg [ST6]. Additional material on semigroup ℓ-algebras may be found in Ma [M8] and [M10].

6.2 Algebraic f-Elements Are Central

Several aspects of commutativity have arisen in previous sections. For instance, the smallness of a commutator relative to the squares of its factors in an f-ring (Theorem 3.6.1), the commutativity of an archimedean almost f-ring (Theorem 3.8.14), and the cohesion of sets of nilpotent elements into an ℓ-ideal in an ℓ-ring which is polynomial constrained (Theorems 3.8.3 and 3.8.4). The centrality of algebraic elements will be discussed here. It will be seen that algebraic f-elements in ℓ-reduced ℓ-rings are central, and a property of totally ordered domains will be identified which is dual to ℓ-simplicity and which forces the domain to be commutative provided it is constrained by a suitable polynomial. Rings with the property that each of its partial orders is contained in a total order are algebraic over the integers and will be completely classified. Interestingly, it turns out that the seemingly weaker property that each partial order can be extended to a lattice order over which the ring is an almost f-ring is not weaker at all. Initially, the Galois character of a division ring relative to its center will be established, and this will be used to show that the center of a totally ordered division ring is relatively algebraically closed.

If G is a set of endomorphisms of the ring R the fixed ring of G is the subring of R consisting of all a in R such that $\sigma(a) = a$ for each $\sigma \in G$. Note that if E is the centralizer in R of a subset X of units of R, then E is the fixed ring of the group of inner automorphisms of R associated to X. That is, if G is the group of inner automorphisms generated by $\{x(\ )x^{-1} : x \in X\}$, then $a \in E$ if and only if $xax^{-1} = a$ for each $x \in X$. This remark establishes that Theorem 6.2.1 can be applied to centralizers and, in particular, to the center of a division ring.

For the ring R, $b \in R$ and $f(x) = \Sigma_i a_i x^i \in R[x]$, let $f(b) = \Sigma_i a_i b^i$.

Theorem 6.2.1. *Let G be a group of automorphisms of the division ring D and let E be its fixed ring. Suppose $b \in D$ and $f(x) \in E[x]$ is monic of minimal degree $n \geq 1$ with $f(b) = 0$. Then there exist $\sigma_1, \ldots, \sigma_n \in G$, $d_1, \ldots, d_n \in D^*$, and $b_j = d_j\sigma_j(b)d_j^{-1}$, for $j = 1, \ldots, n$, with $b_1 = b$, such that*

(a) $f(x) = (x-b_n)(x-b_{n-1})\cdots(x-b_1)$.
(b) If $a \in D$ with $f(a) = 0$, then $a = db_jd^{-1}$ for some $d \in D^*$ and some b_j.
(c) If E is central, then for each $j = 1, \ldots, n, f(x) = (x-b_j)(x-b_{j-1})\cdots(x-b_1)(x-b_n)\cdots(x-b_{j+1})$.

Proof. If $n = 1$, then $f(x) = x - b$ and the statements hold trivially; so assume $n \geq 2$. By Exercise 1 $f(x) = g(x)(x-b)$ with $g(x) \in D[x]$ and, for each $\sigma \in G$,

$$f(x) = \sigma(g(x))(x - \sigma(b)). \tag{6.2.1}$$

Since $b \notin E$ there exists $\sigma \in G$ with $c = \sigma(b) \neq b$ and $f(x) = \sigma(g(x))(x-c)$. So, $x-c$ is a right factor of $f(x)$ but not of $(x-b)$; by Exercise 1 some conjugate of $x-c$ is a right factor of $g(x)$: $f(x) = g_1(x)(x-b_2)(x-b_1)$ with b_2 a conjugate of c and $b_1 = b$. This process can be continued until we get $m \in \mathbb{N}$ with $m \leq n$ and $p(x) \in D[x]$ and elements $b = b_1, \ldots, b_m$ of the specified form with

$$f(x) = p(x)(x-b_m)\cdots(x-b_1) \tag{6.2.2}$$

and, for each $\sigma \in G$, $x - \sigma(b)$ is a right factor of $k(x) = (x-b_m)\cdots(x-b_1)$; that is, $k(\sigma(b)) = 0$. In particular, for ρ, $\tau \in G$ we have $k(\tau^{-1}\rho(b)) = 0$ and hence $\tau(k(x))(\rho(b)) = 0$. Now, suppose $k(x) \notin E[x]$ and take $\sigma \in G$ with $\sigma(k(x)) \neq k(x)$. Then $k(x) - \sigma(k(x))$ is a nonzero polynomial in $D[x]$ of degree $< m$ which has every $\rho(b)$ as a root. If $k(x) - \sigma(k(x)) \notin E[x]$ then we continue until we get a nonzero polynomial in $E[x]$ of positive degree $< m$ which has b as a root. This contradicts the minimality of n. So $k(x) \in E[x]$, $p(x) = 1$ and (a) has been established.

For (b), let $f_0(x) = 1$ and $f_i(x) = (x - b_i)(x - b_{i-1}) \cdots (x - b_1)$ for $1 \le i \le n$. Then for some i, $0 \le i \le n-1$, we have $f_i(a) \ne 0$ and $f_{i+1}(a) = 0$. Since $f_{i+1}(x) = (x - b_{i+1}) f_i(x)$, by Exercise 1, $x - f_i(a) a f_i(a)^{-1}$ is a right factor of $x - b_{i+1}$; that is, $b_{i+1} = f_i(a) a f_i(a)^{-1}$.

If E is central, then since $x - b$ is a divisor of $f(x)$ in $E(b)[x]$ we get from (a) that $(x - b_n) \cdots (x - b_2) \in E(b)[x]$, and hence $f(x) = (x - b)(x - b_n) \cdots (x - b_2)$. Iteration of this gives (c). □

Theorem 6.2.2. *Let D be a totally ordered division ring with center K. Each element of D that is algebraic over K must be in K.*

Proof. Suppose $b \in D$ is algebraic over K of degree $n \ge 2$ and let $f(x) = a_0 + a_1 x + \cdots + a_{n-1} x^{n-1} + x^n$ be its irreducible polynomial over K. Identify b with the K-linear transformation on $K(b)$ given by $u \mapsto bu$. From the matrix of b with respect to the basis $1, b, \ldots, b^{n-1}$ we see that $\operatorname{tr} b = -a_{n-1}$ where $\operatorname{tr} b$ denotes the trace of b. So, if $c = |b + \frac{a_{n-1}}{n}|$, then $\operatorname{tr} c = 0$ and $K(b) = K(c)$; that is, we may assume $\operatorname{tr} b = 0$ and $b > 0$. Since K is its own double centralizer we may apply Theorem 6.2.1 to get $f(x) = (x - b_n) \cdots (x - b_1)$ with $b_1 = b$ and $b_i = d_i b d_i^{-1} > 0$ for each i. So, $0 = \operatorname{tr} b = a_{n-1} = -(b_1 + \cdots + b_n) < 0$. Thus, $n = 1$ and $b \in K$. □

In order to generalize the previous result to totally ordered domains and other ℓ-rings we will use a technique which in its overview is analogous to the one just employed but which is technically much more complicated. Specifically, we will obtain a result on higher commutators that holds in any ring and which will force centrality when applied to ordered rings. The proof of this commutator equation will be highly combinatorial.

In order to get information about the signs of terms in a commutator sum that will arise later we need to consider partitions of the sets $\mathbb{N}_n = \{1, 2, \ldots, n\}$. Suppose $b_1, \ldots, b_n \in \mathbb{N}$ and $b_1 + \cdots + b_n = a_n$. The partition $\{B_1, \ldots, B_n\}$ of $\mathbb{N}_{a_n}$ is called $(b_1, \ldots, b_n)$-*admissible* if $|B_i| = b_i$ for $i = 1, \ldots, n$ and $B_1 \not> B_2 \not> \cdots \not> B_n$ if $n \ge 2$. Let $g_n(b_1, \ldots, b_n)$ denote the number of $(b_1, \ldots, b_n)$-admissible partitions of $\mathbb{N}_{a_n}$.

Theorem 6.2.3. *(a) $g_1(b_1) = 1$.*

(b) $g_2(b_1, b_2) = \binom{b_1 + b_2}{b_1} - 1$.

(c) $g_n(b_1, \ldots, b_n) \ge g_{n-1}(b_1, \ldots, b_{n-1}) \ge 1$ for $n \ge 2$.

(d) Let $a_k = b_1 + \cdots + b_k$. Then, for each $n \in \mathbb{N}$,

$$g_n(b_1, \ldots, b_n) = (-1)^{n-1} \left(1 + \sum_{j=1}^{n-1} (-1)^j \binom{a_n}{a_j} g_j(b_1, \ldots, b_j) \right).$$

Proof. (a) is trivial, and as for (b), if $\{B_1, B_2\}$ is a partition of $\mathbb{N}_{b_1 + b_2}$ with $|B_1| = b_1$ and $|B_2| = b_2$ and $B_1 > B_2$, then, clearly, $B_2 = \mathbb{N}_{b_2}$. Suppose $\{B_1, \ldots, B_{n-1}\}$ is a $(b_1, \ldots, b_{n-1})$-admissible partition of $\mathbb{N}_{a_{n-1}}$. Then $\{B_1, \ldots, B_{n-1}, B_n\}$ is a $(b_1, \ldots,$

b_n)-admissible partition of $\mathbb{N}_{a_n}$ where $B_n = \{a_{n-1}+1,\ldots,a_n\}$, and (c) follows. It remains to verify (d) for $n \geq 2$; note that for $n = 2$ (d) reduces to (b). For each partition $\{B_1,\ldots,B_n\}$ of $\mathbb{N}_{a_n}$ with $|B_i| = b_i$ there is a unique order isomorphism between $B_1 \cup \cdots \cup B_{n-1}$ and $\mathbb{N}_{a_{n-1}}$, and hence there is a unique partition of $\mathbb{N}_{a_{n-1}}$ that is "order-isomorphic" to $\{B_1,\ldots,B_{n-1}\}$. Consequently, $\binom{a_n}{b_n} g_{n-1}(b_1,\ldots,b_{n-1})$ is the number of partitions $\{B_1,\ldots,B_{n-1},B_n\}$ with $|B_i| = b_i$ and $B_1 \not> B_2 \not> \cdots \not> B_{n-1}$. Now, $g_{n-1}(b_1,\ldots,b_{n-2},b_{n-1}+b_n)$ is the number of partitions $\{B_1,\ldots,B_n\}$ with $|B_i| = b_i$ and $B_1 \not> \cdots \not> B_{n-1} > B_n$. For, $B_{n-2} \not> B_{n-1}$ gives $B_{n-2} \not> B_{n-1} \cup B_n$ and hence $\{B_1,\ldots,B_{n-2},B_{n-1} \cup B_n\}$ is $(b_1,\ldots,b_{n-2},b_{n-1}+b_n)$-admissible; and if $(B_1,\ldots,B_{n-2},B)$ is $(b_1,\ldots,b_{n-2},b_{n-1}+b_n)$-admissible, then there is a unique partition $\{B_{n-1},B_n\}$ of B with $B_{n-1} > B_n$ and $|B_{n-1}| = b_{n-1}$. We now have

$$\binom{a_n}{a_{n-1}} g_{n-1}(b_1,\ldots,b_{n-1}) = g_{n-1}(b_1,\ldots,b_{n-2},b_{n-1}+b_n) + g_n(b_1,\ldots,b_n)$$

or

$$g_n(b_1,\ldots,b_n) = \binom{a_n}{a_{n-1}} g_{n-1}(b_1,\ldots,b_{n-1}) - g_{n-1}(b_1,\ldots,b_{n-2},b_{n-1}+b_n).$$

Now, suppose $n > 2$ and, by induction, (d) holds for $n-1$. Then, using the previous equation and a reversal of terms in (d),

$$\begin{aligned} g_n(b_1,\ldots,b_n) &= \binom{a_n}{a_{n-1}} g_{n-1}(b_1,\ldots,b_{n-1}) \\ &\quad - \left(\sum_{j=1}^{n-2}(-1)^{j-1}\binom{a_n}{a_{n-1-j}} g_{n-1-j}(b_1,\ldots,b_{n-1-j}) + (-1)^{n-2}\right) \\ &= \binom{a_n}{a_{n-1}} g_{n-1}(b_1,\ldots,b_{n-1}) \\ &\quad + \sum_{j=2}^{n-1}(-1)^{j-1}\binom{a_n}{a_{n-j}} g_{n-j}(b_1,\ldots,b_{n-j}) + (-1)^{n-1}, \end{aligned}$$

and this is (d). □

We next use this last equation to investigate signs.

Theorem 6.2.4. *Let $(s_n)_n$ be a sequence in $\{1,-1\} \subseteq \mathbb{Z}$ with $s_1 = 1$. Define the sequence $(d_n)_n$ in $\mathbb{Z}$ recursively by $d_1 = 1$ and for $n \geq 2$*

$$d_n = 1 - \sum_{\substack{1 \leq j \leq n-1 \\ s_j s_{j+1} = -1}} \binom{n}{j} d_j.$$

Then $s_n d_n > 0$ for each n.

Proof. We assume $n \geq 2$. If $1 = s_1 = \cdots = s_n$, then $d_n = 1$ and we are done. Otherwise, let

$$\{1 \leq a_1 < a_2 < \cdots < a_k \leq n-1\} = \{j : 1 \leq j \leq n-1 \text{ and } s_j s_{j+1} = -1\};$$

also, put $a_{k+1} = n$. Then

$$d_{a_j} = 1 - \sum_{i=1}^{j-1} \binom{a_j}{a_i} d_{a_i} \text{ for } j = 1,2,\ldots,k+1.$$

Let $b_1 = a_1$ and put $b_j = a_j - a_{j-1}$ for $2 \leq j \leq k+1$; so $a_j = \Sigma_{i=1}^{j} b_i$ for $j = 2,\ldots,k+1$. Now, define $h_1(b_1) = d_{a_1} = 1$ and, for $2 \leq j \leq k+1$, $h_j(b_1,\ldots,b_j) = (-1)^{j-1} d_{a_j}$. Then, for $j = 1,\ldots,k+1$,

$$(-1)^{j-1} h_j(b_1,\ldots,b_j) = d_{a_j} = 1 - \sum_{i=1}^{j-1} (-1)^{i-1} \binom{a_j}{a_i} h_i(b_1,\ldots,b_i),$$

or

$$h_j(b_1,\ldots,b_j) = (-1)^{j-1} \left(1 + \sum_{i=1}^{j-1} (-1)^i \binom{a_j}{a_i} h_i(b_1,\ldots,b_i) \right).$$

Thus, by Theorem 6.2.3, $h_j(b_1,\ldots,b_j) = g_j(b_1,\ldots,b_j) \geq 1$, and, in particular, $s_n d_n = s_{a_k} d_{a_{k+1}} = (-1)^k (-1)^k g_{k+1}(b_1,\ldots,b_{k+1}) \geq 1$. □

We will now determine a relation which holds in any ring and which expresses a commutator in which one of the factors is a polynomial expression in terms of higher commutators and derivatives. We will formulate it for a module.

For integers $n,k \in \mathbb{Z}$ let $\binom{n}{k}$, as usual, denote the binomial coefficient when $0 \leq k \leq n$ and let it be 0 otherwise; so $\binom{n}{k} = 0$ if $k < 0$ or $k > n$. We will use the following relations which hold for all $n,k \in \mathbb{Z}$.

$$\binom{n}{k} = \binom{n-1}{k-1} + \binom{n-1}{k} \text{ unless } n = k = 0 \tag{6.2.3}$$

$$\binom{n}{k}\binom{k}{r} = \binom{n}{r}\binom{n-r}{k-r} \tag{6.2.4}$$

$$\binom{n}{k} = \binom{n}{n-k} \tag{6.2.5}$$

For sequences $(p_j)_j$ and $(q_j)_j$ in an abelian group and $n \in \mathbb{N}$ define, for $i = 0,1,\ldots,n$,

$$L_i^{(n)}((p_j),(q_j)) = L_i^{(n)} = \sum_{j=1}^{n} \left[\binom{n-i}{n-j} p_j + (-1)^{n-j} \binom{i}{n-j} q_j \right]. \tag{6.2.6}$$

Theorem 6.2.5. *For $i = 0, 1, \ldots, n-1$,*

$$L_{i+1}^{(n)} = L_i^{(n)} - L_i^{(n-1)} \tag{6.2.7}$$

and

$$L_{i+1}^{(n)} = L_0^{(n)} - L_0^{(n-1)} - \sum_{k=1}^{i} L_k^{(n-1)}. \tag{6.2.8}$$

Proof. Using (6.2.6) and (6.2.3) we have

$$\begin{aligned}
L_i^{(n)} &= \sum_{j=1}^{n-1} \left[\left(\binom{n-i-1}{n-j-1} + \binom{n-i-1}{n-j} \right) p_j + (-1)^{n-j} \left(\binom{i+1}{n-j} \right.\right. \\
&\qquad \left.\left. - \binom{i}{n-j-1} \right) q_j \right] + p_n + q_n \\
&= \sum_{j=1}^{n-1} \left[\binom{n-1-i}{n-1-j} p_j + (-1)^{n+1-j} \binom{i}{n-1-j} q_j \right] + \\
&\sum_{j=1}^{n-1} \left[\binom{n-(i+1)}{n-j} p_j + (-1)^{n-j} \binom{i+1}{n-j} q_j \right] + p_n + q_n = L_i^{(n-1)} + L_{i+1}^{(n)}.
\end{aligned}$$

To verify (6.2.8) we use induction on i. For $i = 0$ this is (6.2.7). Assuming (6.2.8) holds for $i-1$ we have

$$L_{i+1}^{(n)} = L_i^{(n)} - L_i^{(n-1)} = L_0^{(n)} - L_0^{(n-1)} - \sum_{k=1}^{i-1} L_k^{(n-1)} - L_i^{(n-1)}.$$

□

Let R be a unital ring with no 2-torsion. If $(r_j)_j$ is a sequence in R define the new sequence $(c_j)_j$ recursively by $c_1 = 1$ and, for $n \geq 2$,

$$c_n = 1 + \tfrac{1}{2} \sum_{j=1}^{n-1} (r_j - 1) \binom{n}{j} c_j. \tag{6.2.9}$$

Let $M_i^{(n)} = L_i^{(n)}((p_j), (q_j))$ where $p_j = \frac{1}{2}(1 - r_j)c_j$ and $q_j = \frac{1}{2}(1 + r_j)c_j$. So

$$M_i^{(n)} = \tfrac{1}{2} \sum_{j=1}^{n} \left[\binom{n-i}{n-j} (1 - r_j) + (-1)^{n-j} \binom{i}{n-j} (1 + r_j) \right] c_j. \tag{6.2.10}$$

Now, we have

$$M_0^{(n)} = 1, \tag{6.2.11}$$

$$M_n^{(n)} = (-1)^{n-1}, \tag{6.2.12}$$

and

$$M_i^{(n)} = 0 \text{ for } i = 1,2,\ldots,n-1. \tag{6.2.13}$$

For,

$$\begin{aligned} M_0^{(n)} &= \sum_{j=1}^{n-1} \binom{n}{j} \tfrac{1}{2}(1-r_j)c_j + \tfrac{1}{2}(1+r_n)c_n + \tfrac{1}{2}(1+r_n)c_n \\ &= 1 - c_n + c_n = 1 \end{aligned}$$

and

$$M_1^{(1)} = \tfrac{1}{2}(1-r_1)c_1 + \tfrac{1}{2}(1+r_1)c_1 = 1.$$

From (6.2.8), by induction on $n \geq 2$ and for $1 \leq j \leq n$,

$$M_j^{(n)} = M_0^{(n)} - M_0^{(n-1)} - \sum_{k=1}^{j-1} M_k^{(n-1)} = 1 - 1 - 0 = 0$$

if $j \leq n-1$, and $M_n^{(n)} = 1 - 1 - (-1)^{n-2} = (-1)^{n-1}$.

Suppose R is an algebra over the commutative ring C, and let $R_1 = R$ if R has an identity element and, otherwise, let $R_1 = R \times C$ be the C-algebra obtained by freely adjoining C to R. Let t, $a \in R$ and let δ be the left inner derivation determined by a: $\delta(x) = ax - xa$. Let $T = C[t,a]$ be the subalgebra of R generated by t and a, and let $S = C[t_0, t_1, \ldots, t_m]$ be the subalgebra of T generated by $t_0, \ldots, t_m$ where $t_j = \delta^j(t)$.

Let T_1 denote the subalgebra of R_1 generated by T and C. Suppose that M is a unital C-module which is a left and a right S-module, and let $f_0, \ldots, f_m : T_1 \longrightarrow M$ be left and right S-homomorphisms. Put

$$f(a) = \sum_{i=0}^{m} f_i(a^i) \tag{6.2.14}$$

and for $j = 0, 1, \ldots, m$ define

$$w_j = \sum_{i=j}^{m} \binom{i}{j} f_i(a^{i-j}), \tag{6.2.15}$$

$$f^{(j)}(a) = j!\,w_j, \quad u_j = t_j w_j, \quad v_j = w_j t_j. \tag{6.2.16}$$

Note that on S each f_i is left multiplication by $x_i = f_i(1) \in M$ and, for $j = 0, 1, \ldots, m$, $x_i t_j = t_j x_i$; also, if $x \in M$ "centralizes" S in this way, then left multiplication by x is a left and a right T-homomorphism from T_1 to M provided M is a left S-right T-bimodule.

Theorem 6.2.6. *Suppose R is a C-algebra, a, $t \in R$, and let t_j, S, T, M and f_i be as given above, and let f, w_j, u_j and v_j be as defined in (6.2.14), (6.2.15), and (6.2.16). Let $(r_n)_n$ be a sequence of odd integers and let $(c_n)_n$ be the sequence in $\mathbb{Z}$ defined by (6.2.9). Then*

$$f(a)t - tf(a) = \sum_{j=1}^{m} \frac{(1+r_j)}{2} c_j u_j + \sum_{j=1}^{m} \frac{(1-r_j)}{2} c_j v_j$$

Proof. From (6.2.10), (6.2.4), and (6.2.5) we have

$$\begin{aligned}
\binom{n}{i} M_i^{(n)} &= \frac{1}{2} \sum_{j=1}^{n} \Bigg[\binom{n}{n-i} \binom{n-i}{n-j} (1-r_j) \\
&\qquad + (-1)^{n-j} \binom{n}{i} \binom{i}{n-j} (1+r_j) \Bigg] c_j \\
&= \frac{1}{2} \sum_{j=1}^{n} \Bigg[\binom{n}{n-j} \binom{j}{j-i} (1-r_j) \\
&\qquad + (-1)^{n-j} \binom{n}{n-j} \binom{j}{j-n+i} (1+r_j) \Bigg] c_j \\
&= \frac{1}{2} \sum_{j=1}^{n} \binom{n}{j} \Bigg[\binom{j}{i} (1-r_j) + (-1)^{n-j} \binom{j}{n-i} (1+r_j) \Bigg] c_j,
\end{aligned}$$

and from (6.2.11), (6.2.12), and (6.2.13) we have

$$(-1)^i \binom{n}{i} M_i^{(n)} = \begin{cases} 1 & \text{if } i = 0 \\ 0 & \text{if } 1 \le i \le n-1 \\ -1 & \text{if } i = n. \end{cases}$$

So,

$$\begin{aligned}
a^n t - t a^n &= \sum_{i=0}^{n}(-1)^i \binom{n}{i} M_i^{(n)} a^{n-i} t a^i \\
&= \sum_{i=0}^{n}(-1)^i (\tfrac{1}{2}\sum_{j=1}^{n}\binom{n}{j}\Big[\binom{j}{i}(1-r_j) \\
&\qquad\qquad +(-1)^{n-j}\binom{j}{n-i}(1+r_j)\Big] c_j) a^{n-i} t a^i \\
&= \tfrac{1}{2}\sum_{j=1}^{n}\binom{n}{j} c_j \sum_{i=0}^{n}\Big[(-1)^i\binom{j}{i}(1-r_j)a^{n-i} t a_i \\
&\qquad\qquad +(-1)^{n-j-i}\binom{j}{n-i}(1+r_j)a^{n-i} t a^i\Big] \\
&= \tfrac{1}{2}\sum_{j=1}^{n}\binom{n}{j} c_j \Big[(1-r_j)a^{n-j}\sum_{i=0}^{j}(-1)^i\binom{j}{i} a^{j-i} t a^i \\
&\qquad\qquad + (1+r_j)\Big(\sum_{i=n-j}^{n}(-1)^{n-j-i}\binom{j}{n-i} a^{n-i} t a^{i+j-n}\Big) a^{n-j}\Big] \\
&= \tfrac{1}{2}\sum_{j=1}^{n}\binom{n}{j} c_j \left[(1-r_j)a^{n-j} t_j + (1+r_j) t_j a^{n-j}\right]
\end{aligned}$$

by Exercise 2. Applying f_n we obtain

$$f_n(a^n t - t a^n) = \tfrac{1}{2}\sum_{j=1}^{n}\binom{n}{j} c_j \left[(1-r_j) f_n(a^{n-j}) t_j + (1+r_j) t_j f_n(a^{n-j})\right].$$

Now,

$$f(a)t - t f(a) = \sum_{i=1}^{m} f_i(a^i)t - t f_i(a^i) = \sum_{i=1}^{m} f_i(a^i t - t a^i)$$

$$\begin{aligned}
&= \tfrac{1}{2}\sum_{i=1}^{m}\sum_{j=1}^{i}\binom{i}{j} c_j \left[(1-r_j) f_i(a^{i-j}) t_j + (1+r_j) t_j f_i(a^{i-j})\right] \\
&= \tfrac{1}{2}\sum_{j=1}^{m}(1-r_j)c_j\sum_{i=j}^{m}\binom{i}{j} f_i(a^{i-j}) t_j + \tfrac{1}{2}\sum_{j=1}^{m}(1+r_j)c_j t_j \sum_{i=j}^{m}\binom{i}{j} f_i(a^{i-j}) \\
&= \sum_{j=1}^{m}\frac{(1+r_j)}{2} c_j u_j + \sum_{j=1}^{m}\frac{(1-r_j)}{2} c_j v_j.
\end{aligned}$$

□

By using Theorem 6.2.4 we get the following sharpened form of the previous result.

Theorem 6.2.7. *Assume all of the data in the first sentence of Theorem 6.2.6. Then for any sequence $(s_n)_n$ in $\{1,-1\}$ with $s_1 = 1$, there is a sequence $(d_n)_n$ in $\mathbb{Z}$ with*

$d_1 = 1$ such that $s_n d_n > 0$ for each n and

$$f(a)t - tf(a) = \sum_{j=1}^{m} d_j z_j,$$

where $z_j = u_j$ or v_j depending on whether $s_j s_{j+1} = 1$ or -1.

Proof. Let $r_j = s_j s_{j+1}$ for each $j \in \mathbb{N}$ and let $c_n \in \mathbb{Z}$ be defined as in (6.2.9). Then, in fact, $c_n = d_n$ for each n where $(d_n)_n$ is the sequence defined in Theorem 6.2.4. So $s_n d_n > 0$ and, by Theorem 6.2.6,

$$f(a)t - tf(a) = \sum_{\substack{1 \leq j \leq m \\ s_j s_{j+1} = 1}} d_j u_j + \sum_{\substack{1 \leq j \leq m \\ s_j s_{j+1} = -1}} d_j v_j. \qquad \square$$

By applying this result to ordered rings and modules, as we will now do, we get that algebraic elements are central.

Theorem 6.2.8. *Suppose M in Theorem 6.2.7 is also a po-group and, for each $j, \{u_j, v_j\} \subseteq M^+$ or $\{u_j, v_j\} \subseteq -M^+$. If $u_1 > 0$ and $v_1 > 0$, then $f(a)t - tf(a) > 0$.*

Proof. Define a sequence $(s_j)_j$ by $s_1 = 1$ and for $j \geq 2$,

$$s_j = \begin{cases} 1 & \text{if } u_j > 0 \text{ or } v_j > 0 \\ -1 & \text{if } u_j < 0 \text{ or } v_j < 0 \\ s_{j-1} & \text{if } u_j = v_j = 0. \end{cases}$$

Then by Theorem 6.2.7 there is a sequence of integers $(d_j)_j$ with $d_1 = 1$ and $d_j s_j > 0$. Suppose $z_j = u_j$. If $u_j > 0$, then $d_j > 0$ and if $u_j < 0$, then $d_j < 0$; in either case $d_j z_j > 0$. Similarly, $d_j z_j \geq 0$ if $z_j = v_j$. So

$$f(a)t - tf(a) = \sum_{j=1}^{m} d_j z_j \geq z_1 > 0. \qquad \square$$

Theorem 6.2.9. *Let R be an ℓ-ring and let a, t, $x_0, \ldots, x_m \in R$ with $a \in F = F(R)$ and with $t_j x_i = x_i t_j$ for $0 \leq i,\ j \leq m$, where $t_j = \delta^j(t)$ and δ is the left inner derivation on R determined by a. Let*

$$f(a) = \sum_{i=0}^{m} x_i a^i.$$

Assume that each of the derivatives $f^{(j)}(a)$ lies in F and is comparable to 0 and $f'(a)$ is not a zero divisor in R. Then $[f(a), t] \neq 0$ provided $t_1 = [a, t] \neq 0$.

Proof. Let P be a minimal prime subgroup of R such that $[a,t] \notin P$. We may assume $\overline{[a,t]} > 0$ in $\overline{R} = R/P$, and also that $f'(a) > 0$. Since P is an F-F-subbimodule of R, and since

$$\overline{u}_j = \frac{1}{j!}\overline{t}_j f^{(j)}(a) \quad \text{and} \quad \overline{v}_j = \frac{1}{j!} f^{(j)}(a)\overline{t}_j,$$

$\overline{u}_j$ and $\overline{v}_j$ have the same sign, if both are not zero. Also, $\overline{u}_1 > 0$ and $\overline{v}_1 > 0$. For,

$$\overline{u}_1 = \overline{t}_1 f'(a) = \overline{[a,t]} f'(a) \geq 0;$$

and if $\overline{u}_1 = 0$, then $[a,t]f'(a) \in P$. So by Theorem 2.4.3, for some $y \notin P$, $|[a,t]|f'(a) \wedge y = 0$, and hence,

$$0 = |[a,t]|f'(a) \wedge yf'(a) = (|[a,t]| \wedge y)f'(a).$$

Thus $|[a,t]| \wedge y = 0$ and $[a,t] \in P$, which contradicts our assumption that $[a,t] \notin P$. Similarly, $\overline{v}_1 > 0$.

Now define the sequence $(s_j)_j$ as in Theorem 6.2.8, but using $\overline{u}_j$ in place of u_j and $\overline{v}_j$ in place of v_j. Then by Theorem 6.2.12

$$\overline{[f(a),t]} = \sum_{j=1}^{m} d_j \overline{z}_j \geq \overline{z}_1 > 0$$

since $d_j \overline{z}_j \geq 0$. □

This, of course, yields that some algebraic elements in F are central in R.

Theorem 6.2.10. *Let R be a torsion-free ℓ-algebra over the totally ordered domain C. If R is also t-torsion-free as both a left and a right F-module, then each element in F which is algebraic over C is central in R. In particular, if R is an ℓ-reduced ℓ-algebra, then each algebraic f-element is central.*

Proof. Let $f(x) \in C[x]$ be a polynomial of minimal degree that $0 \neq a \in F$ satisfies. If F_u is the C-unital cover of F (Theorem 3.4.5 and Exercise 3.4.16), then each $f^{(j)}(a) \in F_u$ and $f'(a) \neq 0$. Since R is easily seen to be a t-torsion-free F_u-F_u-f-bimodule the proof of Theorem 6.2.9 is still valid. Hence $[a,t] = 0$ for each t in R.

Suppose R is ℓ-reduced and let P be a minimal ℓ-prime ℓ-ideal of R. Then $\overline{R} = R/P$ is an ℓ-domain by the proof of Theorem 3.2.21, and the nonzero elements of the totally ordered domain $F(\overline{R})$ are regular elements of $\overline{R}$. Moreover, using the characterization of P given in Theorem 3.2.22 it is easy to see that P is an algebra ideal and $\overline{R}$ is torsion-free over C. Since $F(R)$ maps into $F(\overline{R})$ an algebraic element a in $F(R)$ maps into a central element $\overline{a}$ of $\overline{R}$. But according to Theorem 3.2.21 R is a subdirect product of the set $\{R/P : P \text{ a minimal } \ell\text{-prime } \ell\text{-ideal of } R\}$; so a is central in R. □

Note that the two-sided condition in Theorem 6.2.10 is needed. For if R is the top row of the 2×2 matrix algebra over a totally ordered field, then ${}_F R$ but not R_F, is t-torsion-free, and R is centerless. Here the lattice-order of R can be either the canonical order or the lexicographic total order with the left corner dominating.

By emphasizing the factor t_j in u_j and v_j instead of w_j we get module analogues of Theorems 6.2.9 and 6.2.10. Let M be both a left and a right R-module. The element $a \in R$ is called *algebraic on M* if there exist $x_0, x_1, \ldots, x_m$ in M, not all 0, with $x_j t = t x_j$ for $j = 0, \ldots, m$, and every $t \in R$ and $f(a) = x_0 a + \cdots + x_m a^m = 0$.

Theorem 6.2.11. *Let R be a po-algebra over the po-ring C, and let M be a left and right ℓ-module over R. Let $F = F(M_R) \cap F(_RM)$ be the subring of f-elements on M, and suppose that M_F and $_FM$ are t-torsion-free modules. Assume that $a \in F$ and that all the commutators $[a,t], t \in R$, are comparable to 0 and lie in F. If $a \in R$ is algebraic on M and satisfies $f(a) = 0$, while $f'(a) \neq 0$, then a is central in R.*

Proof. If a is not central, take $t \in R$ with $t_1 = [a,t] \neq 0$. Let P be a minimal prime subgroup of M with $f'(a) \notin P$. In $\overline{M} = M/P$ we may assume that $\overline{f'(a)} > 0$. Now, $\overline{u}_1 = \overline{t_1 f'(a)} = t_1 \overline{f'(a)}$. If $\overline{u}_1 = 0$, then $t_1 f'(a) \in P$ and hence $|t_1||f'(a)| \wedge y = 0$ for some $y \notin P$ by Theorem 2.4.3. But then $t_1(|f'(a)| \wedge y) = 0$ implies $|f'(a)| \wedge y = 0$. Thus $f'(a) \in P$, which is impossible. Thus $\overline{u}_1 \neq 0$ and we may assume $\overline{u}_1 > 0$. Similarly, $\overline{v}_1 > 0$. Also, $\overline{u}_j = t_j \overline{w}_j$ and $\overline{v}_j = \overline{w}_j t_j$ are both positive or negative. Thus, as in Theorem 6.2.8 or 6.2.9

$$0 = \overline{f(a)t - tf(a)} = \sum_{j=1}^{m} d_j \overline{z}_1 \geq \overline{z}_1 > 0.$$

This contradiction yields that a is central. □

The following module-theoretic version of Theorem 6.2.2 is an immediate consequence of Theorem 6.2.11.

Theorem 6.2.12. *Let R be a totally ordered domain and suppose a is a nonzero element of R. The following are equivalent.*

(a) a is central.
(b) a is algebraic over the center of R.
(c) There is an abelian ℓ-group M such that $_RM$ and M_R are both t-torsion-free f-modules and a is algebraic on M.

□

If each partial order of the ring R is contained in a total order of R, then R is called an O^*-ring. Similarly, if $\mathscr{A}$ is a class of ℓ-rings, then an *$\mathscr{A}^*$-ring* is a ring with the property that each of its partial orders is contained in a lattice order for which the resulting ℓ-ring is a member of $\mathscr{A}$. In particular, we will discuss f^*-ring, almost f^*-ring and sp^*-ℓ^*-ring. We will show that an almost f^*-ring is an O^*-ring and we will determine all the O^*-rings. As will be seen below, an O^*-ring is algebraic over $\mathbb{Z}$ and is either commutative or is very close to being commutative. Clearly each subring of an O^*-ring is an O^*-ring. If the po-ring (R,P) is a subring of the torsion-free ring S, then $d(P;S) = \{x \in S : nx \in P \text{ for some } n \in \mathbb{N}\}$ is a partial order of S which contains P. In particular, if P is a partial order of $\mathbb{Q}$ and $\frac{m}{n} \in P$ with $n, -m \in \mathbb{N}$, then $m = (-m)(-1) \in P$ and $-1 \in d(P;\mathbb{Q})$. This shows that $P \subseteq \mathbb{Q}^+$ and $\mathbb{Q}$ is an O^*-ring.

Theorem 6.2.13. *A ring R is an O^*-ring if and only if there is a rational vector space N and an embedding of R*

(i) *into the direct product $E \times N$ where E is an O^*-subfield of the reals that is algebraic over $\mathbb{Q}$ and $N^2 = 0$, or*

(ii) *into the algebra $\begin{pmatrix} \mathbb{Q} & N \\ 0 & 0 \end{pmatrix}$ or its dual $\begin{pmatrix} \mathbb{Q} & 0 \\ N & 0 \end{pmatrix}$, the $\mathbb{Q}$-algebras obtained by freely adjoining a one-sided identity to N, or*

(iii) *into $\left\{ \begin{pmatrix} a & b \\ 0 & a \end{pmatrix} : a \in \mathbb{Q},\ b \in N \right\}$, the $\mathbb{Q}$-algebra obtained by freely adjoining an identity to N.*

Proof. Suppose R is an O^*-ring and $a \in R$. If a is not algebraic over $\mathbb{Z}$, then $\mathbb{Z}^+[-a^2]$ is a partial order of R which is not contained in any total order of R. Thus, R is algebraic over $\mathbb{Z}$. Suppose a is nilpotent of index $n > 2$. Let $b = -a^{n-2}$ if n is even and let $b = -a^{n-1}$ if n is odd. Then $b^2 = 0$ and $\mathbb{Z}^+b$ is a partial order which is not contained in any total order. So $n \leq 2$ and $\beta(R) = N_2(R) = N$. If $S = d(R)$ is the divisible hull of R, then S is also an O^*-ring. For, if P is a partial order of S and Q is a total order of R which contains $P \cap R$, then clearly $d(Q;S)$ is a total order of S and $P \subseteq d(Q;S)$. So, without loss of generality, we will assume that R is a $\mathbb{Q}$-algebra. If $R = N$, then R can be embedded in a ring of the type in (i). Assume $R^2 \neq 0$. Then $E = R/N$ is a domain which is algebraic over $\mathbb{Q}$ and hence E is a division ring by Exercise 5. Since E can be totally ordered it is a field by Theorem 6.2.2, and we may assume it is a subfield of $\mathbb{R}$ since, by Theorem 5.1.10, it can be embedded in the real closure of $\mathbb{Q}$. So if $N = 0$, then, again, R can be embedded in a ring of the type in (i). Assume $N \neq 0$, and note that N is a vector space over E since $N^2 = 0$.

Suppose R is unital. If E is a proper extension of $\mathbb{Q}$ let $\{a_i : i \geq 1\}$ be a basis of ${}_{\mathbb{Q}}E$ with $a_1 = 1$. Then $E = \overrightarrow{\boxplus}_{i \geq 1} \mathbb{Q}a_i$ is a totally ordered group with positive cone E^+. Let $0 \neq x \in N$. Then E^+x is a partial order of R and hence is contained in a total order T of R. This total order induces a total order T_E of the field E. Since (E, T_E) is archimedean, $T_E \not\subseteq E^+$. Let $a \in T$ with $a + N \notin E^+$. Then $(a+N)x = ax \in T \cap Ex = E^+x$ yields the contradiction that $a + N \in E^+$. Thus $E = \mathbb{Q}$ and $R = \mathbb{Q}1 + N$ is isomorphic to a ring of the type given in (iii).

Suppose now that R is not unital. By Theorem 3.4.15 R has a nonzero idempotent e. Since the left and right annihilator ideals of R are convex ideals in any total order one of them is contained in the other. Suppose the right annihilator $r(R)$ is contained in the left annihilator $\ell(R)$. According to Theorem 3.4.16 the Pierce decomposition of R is $R = B \boxplus C \boxplus D$ where $B = eRe$, $D = r(R) = (1-e)R$, $C = eR(1-e)$, and $C \boxplus D = \ell(R) = R(1-e)$. Also, any total order of R is of the form $(B \overrightarrow{\boxplus} C \overrightarrow{\boxplus} D)^+$. If $C \neq 0$ and $D \neq 0$, then a total order $(C \overleftarrow{\boxplus} D)^+$ of $C \boxplus D$ could be extended to a total order of R. Thus, one of C or D is zero but the other is nonzero. If $C = 0$, then B and D are ideals of R. If B is not a field and $0 \neq b \in B$ with $b^2 = 0$, and $0 \neq d \in D$, then the partial order $(\mathbb{Z}b \overleftarrow{\boxplus} \mathbb{Z}d)^+$ could be extended to a total order of R. Thus B is a field and R is of the type given (i). Suppose, then, that $D = 0$. If $0 \neq b \in B$ with $b^2 = 0$, then, since in any total order of R either $b \geq C$ or $-b \geq C$, we must have $bC = 0$. But then for $0 \neq c \in C$ there is a total order of R containing $(\mathbb{Z}b \overleftarrow{\boxplus} \mathbb{Z}c)^+$. So

B is a field. By an argument similar to the one given in the previous paragraph we have that $B = \mathbb{Q}$. Thus R is of the type given in (ii).

We show next that each of these rings is an O^*-ring. If $R = E \boxplus N$ is of type (i) and P is a partial order of R, then $P_E = \{\alpha \in E : \alpha + x \in P \text{ for some } x \in N\}$ is a partial order of E. For, P_E is closed under addition and multiplication, and if $\alpha + x$ and $-\alpha + y$ are in P then $-\alpha^2 \in P \cap E$. Thus $\alpha = 0$ since E is an O^*-field. Now, if T_E is a total order of the field E with $T_E \supseteq P_E$ and T_N is a total order of the group N with $T_N \supseteq P \cap N$, then $R^+ = [(E, T_E) \underset{\rightarrow}{\boxplus} (N, T_N)]^+$ is a total order of R which contains P. Suppose R is of the type (iii) and P is a partial order of R. If $x = \begin{pmatrix} a & b \\ 0 & a \end{pmatrix} \in P$ with $a < 0$, then, by passing to $d(P;R)$, we may assume $a = -1$. But now $-1 = x^2 + 2x \in P$ and this is impossible. So if T_N is a total order of the group $\begin{pmatrix} 0 & N \\ 0 & 0 \end{pmatrix}$ which contains $\begin{pmatrix} 0 & N \\ 0 & 0 \end{pmatrix} \cap P$, then $\mathbb{Q}\begin{pmatrix} 1 & 0 \\ 0 & 1 \end{pmatrix} \underset{\rightarrow}{\boxplus} (\begin{pmatrix} 0 & N \\ 0 & 0 \end{pmatrix}, T_N)$ gives a total order of R which contains P. Similarly, each ring of the type given in (ii) is an O^*-ring. □

If E is a subfield of $\mathbb{R}$ that is algebraic over $\mathbb{Q}$, then E need not be an O^*-field. To decide if E is an O^*-field it suffices to assume E is a finite extension of $\mathbb{Q}$. For, suppose each subfield of E that is finite dimensional over $\mathbb{Q}$ is an O^*-field and let P be a partial order of E. If $\Sigma_j \frac{p_j}{q_j} a_j^2 = 0$ with $a_j \in E$ and $p_j, q_j \in P$ let $K = \mathbb{Q}(a_j, p_j, q_j)$. Since K is an O^*-field and $p_j, q_j \in K \cap P$ we have $p_j a_j = 0$ for all j, and hence E is an O^*-field by Theorem 5.1.1. The next theorem produces O^*-fields as well as non-O^*-fields.

Theorem 6.2.14. *Let $e \in K$ and let $E = K(\sqrt{e}) \subseteq \mathbb{R}$ be a quadratic extension of the O^*-field K. The following statements are equivalent.*

(a) E is an O^*-field.

(b) e is totally positive in K and for each partial order P of E there is a total order T of K such that PT is a partial order of E.

(c) Each maximal partial order of E contains e and a total order of K.

Proof. (a) ⇒ (b). Each total order T of K is contained in a total order T' of E which must contain e since e is a square in E. So $e \in T' \cap K = T$ and e is totally positive in K. If T' is a total order of E which contains P, then $T = T' \cap K$ is a total order of K and $PT \subseteq T'$ is a partial order of E.

(b) ⇒ (c). Let P be a maximal partial order of E and let T be a total order of K such that PT is a partial order of E. Then $P = PT$ and $e \in T \subseteq P$.

(c) ⇒ (a). Since each total order of K is contained in a maximal partial order of E, e is totally positive. Let P be a maximal partial order of E. By hypothesis $T = P \cap K$ is a total order of K. By Theorem 5.1.4 there is a total order T_1 of E extending T and we may assume $\sqrt{e} \in T_1$. Let

$$T_2 = \{a + b\sqrt{e} \in E : a, b \in K \text{ and } a - b\sqrt{e} \in T_1\}.$$

Then T_2 is a total order of E and we claim that $P \subseteq T_1$ or $P \subseteq T_2$. All of the inequalities that appear below are taken with respect to T_1. If $x = a + b\sqrt{e} \in E$ let

$\bar{x} = a - b\sqrt{e}$ and note that $\overline{P} \cap K = T$, $\overline{T}_1 = T_2$ and $\overline{T}_2 = T_1$. We show first that if $x = a + b\sqrt{e} \in P$ with $a < 0$, then

$$x\bar{x} = a^2 - b^2 e < 0, \tag{6.2.17}$$

$$b > 0 \Leftrightarrow \sqrt{e} \in P \Leftrightarrow \bar{x} < 0, \tag{6.2.18}$$

$$b < 0 \Leftrightarrow -\sqrt{e} \in P \Leftrightarrow x < 0. \tag{6.2.19}$$

For, $b\sqrt{e} = x - a \in P$; so $b > 0$ (respectively, $b < 0$) $\Leftrightarrow \sqrt{e}$ (respectively, $-\sqrt{e}$) $\in P$. Also, $x^2 - ax = eb^2 + ab\sqrt{e} \in P$; so $1 + \frac{a}{eb}\sqrt{e}$ and $-(1 + \frac{b}{a}\sqrt{e})$ are in P, and consequently their sum $(\frac{a^2 - b^2 e}{abe})\sqrt{e} = (\frac{a}{eb} - \frac{b}{a})\sqrt{e}$ is also in P. Thus, $a^2 - b^2 e < 0$ in both cases. If $x < 0$ and also $b > 0$, then $0 < b\sqrt{e} < -a$ and $b^2 e < a^2$; so, in fact, $b < 0$. Trivially, $b < 0$ gives $x < 0$ and (6.2.19) has been verified. The proof of (6.2.18) is similar.

Assume to the contrary that $P \not\subseteq T_1, T_2$. Then there are $x \in P \backslash T_1$ and $y \in P \backslash T_2$. So $x = a + b\sqrt{e} < 0$ and $y = c + d\sqrt{e}$ with $\bar{y} < 0$; hence $a < 0$ or $b < 0$, and $c < 0$ or $d > 0$. We consider each of the four cases separately.

(I) $a < 0$ and $c < 0$. This case is impossible since $\pm\sqrt{e} \in P$ by (6.2.18) and (6.2.19).

(II) $a < 0$ and $d > 0$. By (6.2.19) $-\sqrt{e} \in P$ and hence $c > 0$. But then $y_1 = -\sqrt{e}y = -de - c\sqrt{e} \in P$ and $\bar{y}_1 = -de + c\sqrt{e} = \sqrt{e}\bar{y} < 0$. This is case I for x and y_1.

(III) $b < 0$ and $c < 0$. After passing to $\overline{P}$ this is case II for $\bar{y}$ and $\bar{x}$.

(IV) $b < 0$ and $d > 0$. To avoid the other cases $a \geq 0$ and $c \geq 0$. If $a = 0$, then $-\sqrt{e} \in P$, and hence $\sqrt{e}y = -de - c\sqrt{e} \in P$; so $c^2 e > d^2 e^2$ and $c^2 > d^2 e$ by (6.2.17). But $c < d\sqrt{e}$ since $\bar{y} < 0$. Thus $a > 0$. If $c = 0$, then $\sqrt{e} \in P$ and $\sqrt{e}x = be + a\sqrt{e} \in P$. This is case II for $\sqrt{e}x$ and y. Thus, $c > 0$ and $xy = (ac + bde) + (ad + bd)\sqrt{e} \in P$ with $ac + bde < 0$, since $a < -b\sqrt{e}$ and $c < d\sqrt{e}$. By (6.2.18) and (6.2.19) $\sqrt{e} \in P$ or $-\sqrt{e} \in P$. If the former holds, then $\sqrt{e}x = be + a\sqrt{e} \in P$ and, again, this is case II for $\sqrt{e}x$ and y. If the latter holds, $y_1 = -\sqrt{e}y = -de - c\sqrt{e} \in P$, $-de < 0$, and $\bar{y}_1 = \sqrt{e}\bar{y} < 0$. This contradicts (6.2.18). □

An immediate consequence of Theorem 6.2.14 is that each real quadratic extension of $\mathbb{Q}$ is an O^*-field but $\mathbb{Q}(\sqrt[4]{2})$ is not.

Theorem 6.2.15. *Each almost f^*-ring is an O^*-ring.*

Proof. Let R be an almost f^*-ring. Just as for an O^*-ring R is algebraic over $\mathbb{Z}$ and $N_2(R) = N$. In fact, these two properties hold in any sp^*-ℓ^*-ring. By Theorems 3.2.14 and 3.2.24 $\beta(R) = N$ and it is an ℓ-ideal for any almost f-order of R. Also, it is easily seen that the divisible hull of R is also an almost f^*-ring; so we can and will assume R is a $\mathbb{Q}$-algebra. We claim that N is a completely prime ideal of R. To see this suppose (R, R^+) is an almost f-ring and take $a, b \in R$ with $a \wedge b = 0$ and $a \notin N$. Then the positive cone P of $((\mathbb{Z}[a], \mathbb{Z}^+[a]) \underrightarrow{\boxplus} (\mathbb{Z}b^2, \{0\}))$ is a partial order of R and therefore is contained in an almost f-order Q of R. But then $a >_Q b^2$ in (R, Q) and $0 = ab^2 \geq_Q b^4 \geq_Q 0$. So $b \in N$ and the homomorphic image R/N of (R, R^+) is totally ordered, and consequently it is a domain. As in the proof of Theorem 6.2.13 R/N is a field. In fact, it is an O^*-field since, for each partial order $\overline{P}$ of R/N, the lifted partial order $P = \{x \in R\backslash N : x+N \in \overline{P}\} \cup \{0\}$ of R is contained in an almost f-order P_1 of R, and, clearly, $\overline{P}$ is contained in the total order $P_1 + N/N$ of R/N.

We can now show that R is an O^*-ring; of course, we assume $0 \neq N \subset R$. If $N \subset \ell(N;R) \cap r(N;R)$, then $\ell(N;R) = r(N;R) = R$ by the maximality of N, and $NR = RN = 0$. Let e be a nonzero idempotent in R, the existence of which is guaranteed by Theorem 3.4.15. Then R is the direct sum of its ideals N and $Re = eR$. For, $R = N + Re = N + eR$ since e maps to the identity of R/N, and $x \in N \cap Re$ yields $x = xe = 0$. Thus, $eR = eRe = Re$ is an ideal and $R = N + Re$ is a direct product. Since $Re \cong R/N$, R is an O^*-ring by Theorem 6.2.13. Now, assume $N = \ell(N;R) \cap r(N;R)$. Suppose (R,P) is an almost f-ring and let P_N be a total order of the group N with $P \cap N \subseteq P_N$. There exists an almost f-order Q of R with $P_N \subseteq Q$ and hence $P_N = Q \cap N$. Now, $N^\perp = 0$ in (R,Q) since $N^\perp \subseteq \ell(N;R) \cap r(N;R) = N$. So $(R,Q) = N^{\perp\perp}$ is totally ordered since N is totally ordered in (R,Q). We claim $P \subseteq Q$. If $x \in P\backslash Q$, then since $P \cap N \subseteq P_N \subseteq Q$, necessarily $x \notin N$. So $xN \neq 0$ or $Nx \neq 0$ and the two cases are similar. Suppose $y \in N$ with $yx \neq 0$. We may take $y \in P$, since $y^+x \neq 0$ or $y^-x \neq 0$ in (R,P). So y, $yx \in P \cap N \subseteq Q$ and $-x \in Q$ gives the contradiction $0 \neq -yx \in Q$. □

A module M_R over the po-ring R is called an *O^*-module* (respectively, a *strong O^*-module*) if each of its partial orders is contained in a total order (respectively, a strong total order). Each torsion-free abelian group is an O^*-module over $\mathbb{Z}$ (Exercise 2.1.7). A generalization of this fact for po-domains is given in Exercise 4.4.22 and Theorem 4.5.3; for totally ordered domains it is equivalent to the domain having a division ring of quotients; see Exercise 12.

Theorem 6.2.16. *Suppose R is an ℓ-ring and M is a right R-module with $\ell(R;M) = 0$. If M is a strong O^*-module, then for each $x \in M$:*

(a) $r(x;R)$ is a right ℓ-ideal;
(b) $xaR^+ \cap xbR^+ \neq 0$ whenever $a, b \in R^+\backslash r(x;R)$.

Conversely, if R is a right f-ring, then these two conditions imply M is a strong O^*-module.

Proof. Assume M is a strong O^*-module. By Exercise 9 $r(x;R)$ is a prime subgroup of R. Take $a, b \in R^+$ with $xa \neq 0$ and $xb \neq 0$. Then (M, xaR^+) is a po-module over R since if $xac = -xad$ with $c, d \in R^+$, then $0 \leq ac \leq ac + ad \in r(x;R)$ and hence $xac = 0$. Suppose $xaR^+ \cap xbR^+ = 0$. Then $xaR^+ - xbR^+$ is a partial order of M and there exists a total order T of M such that (M,T) is a strong f-module with $xaR^+ - xbR^+ \subseteq T$. If $0 \leq r \leq s$ in R, $y \in M$ and $ys \in T$, then $yr \in T$ since

$$0 \leq_T (yr)^- = y^- r \leq_T y^- s = (ys)^- = 0.$$

Thus, for c, $d \in R^+$, $x(ac \wedge bd) \in T$, $-x(ac \wedge bd) \in T$, $ac \wedge bd \in r(x;R)$, and hence $ac \in r(x;R)$ or $bd \in r(x;R)$. This is impossible since there exist elements c, $d \in R^+$ with $xac \neq 0$ and $xbd \neq 0$. So $xaR^+ \cap xbR^+ \neq 0$.

For the converse, suppose T is a partial order of M maximal with respect to the property (M,T) is an R-po-module and suppose $x \in M\backslash(T \cup -T)$. As in the previous paragraph the convexity of $r(x;R)$ gives that xR^+ is a partial order of M. Since the set $P = \{y \in M : yR^+ \subseteq T\}$ is a partial order of M which contains T, and hence is T, if $-xR^+ \cap T = 0$, then $xR^+ + T$ is a partial order of M and we have the contradiction $x \in T$. Thus, $-xR^+ \cap T \neq 0$ and similarly $xR^+ \cap T \neq 0$. Therefore, there are elements a, $b \in R^+$ with $0 \neq xa \in T$ and $0 \neq -xb \in T$ and we now have the contradiction $0 \neq xaR^+ \cap -xbR^+ \subseteq T \cap -T = 0$; so (M,T) is totally ordered. To see that (M,T) is a strong f-module it suffices, by Exercise 9, to show that $r(x;R)$ is prime for each x in M. Suppose $a \wedge b = 0$ in R but neither a nor b annihilates x. Then $0 \neq xac = xbd$ for some $c, d \in R^+$. Then $ac - bd \in r(x;R)$ and $ac \wedge bd = 0$; so $ac = (ac-bd)^+ \in r(x;R)$. □

Exercises.

1. Let R be a unital ring, $f(x), g(x) \in R[x]$, and suppose the leading coefficient of $g(x)$ is a unit. Verify the following.
   (a) There exist unique polynomials $q(x)$ and $r(x)$ with $r(x) = 0$ or deg $r(x) <$ deg $g(x)$ such that $f(x) = q(x)g(x) + r(x)$.
   (b) If $b \in R$, then $f(x) = q(x)(x-b) + f(b)$ for some $q(x)$.
   (c) If $f(x) = g(x)h(x)$, $x-b$ is a right factor of $f(x)$ but not of $h(x)$ and $h(b) = s$ is a unit of R, then $x - sbs^{-1}$ is a right factor of $g(x)$.

2. Let $a \in R$ and let δ be the left inner derivation determined by a; $\delta(t) = at - ta$. Show that, for $0 \le j \le n$,

$$\delta^j(t) = \sum_{i=0}^{j}(-1)^i\binom{j}{i}a^{j-i}ta^i = \sum_{i=n-j}^{n}(-1)^{n-j-i}\binom{j}{n-i}a^{n-i}ta^{i+j-n}.$$

3. Let R be a po-unital ℓ-algebra over the totally ordered field C. If K is a totally ordered subdivision ring of R with the same identity as R show that each element of K which is algebraic over C is central in R.

4. Let K' be the centralizer of the subring K of the totally ordered ring R, and let $a_1, a_2 \in K'$ and $f_1(x), f_2(x) \in K[x]$.
   (a) Suppose $f_1'(a_1) > 0$ and $r(f_1'(a_1);K') = \ell(f_1'(a_1);K') = 0$. Show that $a_1a_2 > a_2a_1$ iff $f_1(a_1)a_2 > a_2f_1(a_1)$.
   (b) Suppose K is commutative, $f_i'(a_i) > 0$ and $r(f_i'(a_i);K') = \ell(f_i'(a_i);K') = 0$ for $i = 1,2$. Show that $a_1a_2 > a_2a_1$ iff $f_1(a_1)f_2(a_2) > f_2(a_2)f_1(a_1)$.
   (c) Suppose $f_1(a_1) = 0$ and $r(f_1'(a_1);K') = \ell(f_1'(a_1);K') = 0$. Show that $a_1 \in K''$.

5. Show that a reduced ring (respectively, a domain) which is algebraic over a field is regular (respectively, a division ring).

6. Show that a unital ring is an O^*-ring if and only if it is an sp^*-ℓ^*-ring and its lower nil radical is a completely prime ideal.

7. Show that the following are equivalent for the O^*-ring R.
   (a) R has a one-sided identity element.
   (b) If (R,R^+) is an sp-ℓ-ring, then (R,R^+) is an almost f-ring.

8. (a) Let B be a nonzero domain that can be totally ordered and let A be a nonzero ring with $\ell(A)\cap r(A)=0$. Show that $A\times B$ is not an ℓ^*-ring. (Use Exercise 3.1.1.)
   (b) Show that for $n\geq 2$ the $n\times n$ matrix ring R_n over the ℓ-simple totally ordered domain R is not an ℓ^*-ring. (Let $\{e_{ij}: 1\leq i,j\leq n\}$ be the usual matrix units, let $e_{nn}R_ne_{nn}$ have the total order of R and let P be the positive cone of $(e_{11}R_n \boxplus\cdots\boxplus e_{n-1,n-1}R_n) \underleftarrow{\boxplus} e_{nn}R_ne_{nn}$ where $e_{11}R_n\boxplus\cdots\boxplus e_{n-1,n-1}R_n$ is trivially ordered. By multiplying on the right by $\alpha e_{in}+\beta e_{nn}$ where $i=1,\ldots,n-1$ and $\alpha,\beta\in R$, show that P is a maximal partial order of R_n.)
   (c) If R is an sp-po-domain show that $R[x]$ is not an ℓ^*-ring. (Let P consist of 0 and all polynomials $f=\alpha_0+\cdots+\alpha_nx^n$ of degree n with $\alpha_n>0$ if $n\equiv 0 \pmod 4$ and $\alpha_n<0$ if $n\equiv 2 \pmod 4$. Show that P is a maximal partial order that is not directed.)

9. (a) Let G and H be ℓ-groups and let $\varphi: G\longrightarrow H$ be a homomorphism of po-groups. Show that φ is an ℓ-homomorphism and $\varphi(G)$ is totally ordered iff $\ker\varphi$ is a prime subgroup of G.
   (b) Let M_R be an ℓ-module over the ℓ-ring R and suppose $x\in M$ with $xR^+\subseteq M^+$. Show that left multiplication by x, $\mu_x: R\longrightarrow M$, is an ℓ-homomorphism and xR is a totally ordered submodule of M iff $r(x;R)$ is a prime subgroup of R.

10. Let M_R be a module with $\ell(R;M)=0$ and let $x\in M$ and $I=r(x;R)$. Show the equivalence of the following.
   (a) I is meet-irreducible in the lattice of right ideals of R: if $I=J\cap K$ where J and K are right ideals of R, then $I=J$ or $I=K$.
   (b) R/I is uniform.
   (c) If $a,b\in R\backslash I$, then $xaR\cap xbR\neq 0$.

11. Suppose R is a totally ordered ring and M_R is a module with $\ell(R;M)=0$.
   (a) If M is an O^*-module show that $r(x;R)$ is meet-irreducible and convex for each $x\in M$.
   (b) If R is a right valuation ring and each right annihilator $r(x;R)$, $x\in M$, is convex show that M is an O^*-module.

12. Show the equivalence of the following statements for the totally ordered domain R; see Theorem 4.5.3.
   (a) R has a nonzero t-torsion-free right O^*-module.

(b) Each t-torsion-free right R-module is an O^*-module.
(c) R is a right Öre domain.

13. Suppose M_R is an O^*-module over the po-ring R with $\ell(R^+;M)=0$. If (M,P) is a po-module show that P is the intersection of total orders iff for all $a \in R^+$ and $x \in M$, $xa \in P$ implies $x \in P$ or $xa = 0$ (xR^+ is a partial order of M.)

14. Consider the following conditions for the po-module M_R with $\ell(R;M)=0$ over the ℓ-simple f-ring R.
(a) M^+ is the intersection of total orders.
(b) M_R can be embedded in a product of totally ordered t-torsion-free po-modules.
(c) M is R^+-semiclosed.
Show that (a) $\Rightarrow$ (b) $\Rightarrow$ (c) and if M is an O^*-module, then (c) $\Rightarrow$ (a).

Notes. Albert [A1] showed that an algebraic element in a totally ordered division ring is central (Theorem 6.2.2) and we have given his proof which uses Wedderburn's Theorem 6.2.1 [W]. Actually, Theorem 6.2.1 with $E = Z(D)$ is due to Wedderburn and the generalization given is due to Cohn [C3, Prop. 3.3.7]. Tamhankar [T] generalized Albert's theorem by showing that an algebraic element a in a totally ordered ring is central provided $f'(a)$ is a regular element where $f(x)$ is a central polynomial with $f(a)=0$, and Exercise 4 also comes from Tamhankar's paper. Steinberg [ST17] showed that Tamhankar's method (Theorems 6.2.3–6.2.7) can be applied to ℓ-algebras and ℓ-modules to show that certain algebraic f-elements are central. Leung [LE2] is also concerned with Tamhankar's theorem. The characterization of O^*-rings, which answers a question posed by Fuchs [F, p. 288], is due to Steinberg [ST19] and the fact that an f^*-ring is an O^*-ring comes from Ma and Wojciechowski [MW1]. Kreinovich [KR] noted that an O^*-ring is algebraic over $\mathbb{Z}$ and its nilpotent elements have index bounded by 2. The discussion of O^*-modules comes from Bigard [BI1], and Exercise 8 is based on Wojciechowski and Kreinovich [WK].

6.3 More Polynomial Constraints on Totally Ordered Domains

Let $R[x_1,\ldots,x_n]$ be the polynomial ring in the noncommuting indeterminates $x_1,\ldots,$ x_n over the ring R. A nonzero homogeneous polynomial $f(x_1,\ldots,x_n) \in R[x_1,\ldots,x_n]$ will be called a *form of degree k* if

$$f(x_1,\ldots,x_n) = \sum_{i_1+\cdots+i_n=k} a_{(i)}x_1^{i_1}\cdots x_n^{i_n}.$$

If R is a po-ring the form $f(x_1,\ldots,x_n)$ is called *positive semidefinite* (P.S.D.) if for any elements $y_1,\ldots,y_n$ in R, $f(y_1,\ldots,y_n) \geq 0$. Using the centrality of algebraic elements it will be shown that a totally ordered domain that satisfies a condition dual to ℓ-simplicity and that has a P.S.D. form which vanishes for some elements $y_1, \ldots, y_n$ from R^* must be commutative. Of course, a commutativity result of this form requires restrictive polynomials since $x_1^2 + x_2^2 \pm (x_1x_2 + x_2x_1) \geq 0$ are constraints with nontrivial solutions in any nonzero sp-po-ring. It is also shown that a semiprime f-ring must be commutative if the polynomial $x^{2n} + y^{2n} - 2x^ny^n$ is P.S.D., and also that the ℓ-subalgebra of f-elements in an ℓ-algebra is central provided it consists of regular elements and its commutators are bounded. The variety of ℓ-algebras generated by a real closed field is shown to be independent of the field and this result is used to verify that a P.S.D. polynomial for the field in commuting variables is a sum of squares of rational functions. Moreover, the totally ordered fields which have this property are determined. In addition, the free f-rings in the variety of ℓ-rings generated by $\mathbb{R}$ are represented in terms of the total orders of the free commutative rings.

We start by giving a specific construction of the coproduct of two algebras even though a general construction is given in Exercise 1.4.21. First, we will review the construction of the tensor algebra obtained from two modules since the coproduct will be obtained as a homomorphic image of the tensor algebra. In the following discussion C is a commutative unital ring, all C-modules are unital, $\alpha x = x\alpha$ for each module element x and each $\alpha \in C$, and tensor products are over C. We will freely use the identifications $C \otimes X = X \otimes C = X$ but not $X \otimes Y = Y \otimes X$, in general. Let V and W be modules over C. For $n \in \mathbb{Z}^+$, the *n-fold tensor product of* V *and* W is the C-module direct sum T_n of the 2^n C-modules $X_1 \otimes \cdots \otimes X_n$ where each X_i is either V or W; note that $T_0 = C$. The tensor algebra T of V and W over C is the C-module direct sum $T = \boxplus_n T_n$. Before we give the multiplication in T we recall that there is a bijection between the C-algebra multiplications of the C-module A and the C-module maps $\mu : A \otimes A \longrightarrow A$ for which the following diagram commutes:

$$
\begin{array}{ccc}
(A \otimes A) \otimes A & \xrightarrow{\mu \otimes 1} & A \otimes A \\
\Big\downarrow & & \Big\downarrow \mu \\
A \otimes (A \otimes A) & \xrightarrow{1 \otimes \mu} & A
\end{array}
\qquad (6.3.1)
$$

Moreover, (A, μ) is a unital C-algebra if and only if there is a C-homomorphism $\varepsilon : C \longrightarrow A$ such that the diagrams

$$\begin{array}{ccc} A\otimes C & \xrightarrow{1\otimes\varepsilon} & A\otimes A \\ & \searrow & \downarrow \mu \\ & & A \end{array} \qquad \begin{array}{ccc} C\otimes A & \xrightarrow{\varepsilon\otimes 1} & A\otimes A \\ & \searrow & \downarrow \mu \\ & & A \end{array} \qquad (6.3.2)$$

both commute.

Since $T\otimes T$ is isomorphic to the direct sum of C and all $(X_1\otimes\cdots\otimes X_i)\otimes(Y_1\otimes\cdots\otimes Y_j)$ with $X_r,\ Y_s\in\{V,W\}$, the mappings $(x_1\otimes\cdots\otimes x_i)\otimes(y_1\otimes\cdots\otimes y_j)\mapsto x_1\otimes\cdots\otimes x_i\otimes y_1\otimes\cdots\otimes y_j$ induce a C-homomorphism $\mu: T\otimes T\longrightarrow T$ for which (6.3.1) holds. We identify C, V and W with their images in T. T is a unital C-algebra since the inclusion map $C\longrightarrow T$ satisfies (6.3.2). It is easily seen that the C-algebra T has the following universal mapping property: For any unital C-algebra B and any C-homomorphisms $\varphi: V\longrightarrow B$ and $\psi: W\longrightarrow B$ there is a unique algebra homomorphism $\rho: T\longrightarrow B$ which extends both φ and ψ. ρ is given by $\rho(x_1\otimes\cdots\otimes x_j)=\tau(x_1)\cdots\tau(x_j)$ where $\tau=\varphi$ or ψ depending on whether $x_k\in V$ or $x_k\in W$.

We now recall the definition of the coproduct in the category of unital C-algebras. Let A_1 and A_2 be unital C-algebras. A *coproduct of* A_1 *and* A_2 is a triple (A,σ_1,σ_2) where A is a unital algebra and $\sigma_i: A_i\longrightarrow A$ are unital algebra homomorphisms, such that the following hold:

(i) A is generated as a C-algebra by $\sigma_1(A_1)\cup\sigma_2(A_2)$.
(ii) For any unital C-algebra B and any unital algebra homomorphisms $\varphi_i: A_i\longrightarrow B$ for $i=1,2$ there is a unital algebra homomorphism $\varphi: A\longrightarrow B$ for which the following diagram is commutative

$$\begin{array}{ccccc} A_1 & \xrightarrow{\sigma_1} & A & \xleftarrow{\sigma_2} & A_2 \\ & {\scriptstyle\varphi_1}\searrow & \downarrow{\scriptstyle\varphi} & \swarrow{\scriptstyle\varphi_2} & \\ & & B & & \end{array} \qquad (6.3.3)$$

.

Theorem 6.3.1. *The coproduct of two unital C-algebras A_1 and A_2 exists and is unique up to isomorphism. Moreover, if C is a module direct summand of A_1 and of A_2, then $\sigma_1(A_1)\cap\sigma(A_2)=C$ and σ_1 and σ_2 are embeddings.*

Proof. Uniqueness of the coproduct follows quickly from its definition. For existence, let I be the ideal of the tensor algebra T over the C-modules A_1 and A_2 that is generated by all elements of the form

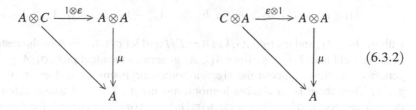

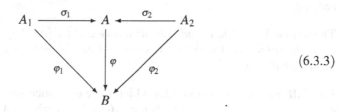

$$a_1 \otimes b_1 - a_1 b_1, a_2 \otimes b_2 - a_2 b_2, \qquad 1_{A_1} - 1_{A_2}, 1_{A_1} - 1_C, 1_{A_2} - 1_C,$$

with $a_1, b_1 \in A_1$, and $a_2, b_2 \in A_2$. Let $A = T/I$ and let $\sigma_i : A_i \longrightarrow A$ be the restrictions of the natural map $T \longrightarrow A$. Since $A_1 \cup A_2$ generates the algebra T, $\sigma_1(A_1) \cup \sigma_2(A_2)$ generates A. Now, suppose the algebra homomorphisms φ_1 and φ_2 in (6.3.3) are given. Then there is an algebra homomorphism $\rho : T \longrightarrow B$ which extends both φ_1 and φ_2. Since $\rho(a_i \otimes b_i) = \varphi_i(a_i)\varphi_i(b_i) = \varphi_i(a_i b_i) = \rho(a_i b_i)$ for $i = 1,2$ and $\rho(1_{A_i}) = \varphi_i(1_{A_i}) = 1_B$, and $\rho(1_C) = \rho(1_T) = 1_B$, we have $\rho(I) = 0$ and hence ρ induces an algebra homomorphism $\varphi : A \longrightarrow B$ which makes (6.3.3) commutative. So A is the coproduct of A_1 and A_2.

Let $\varphi_1 : A_1 \longrightarrow A_1 \otimes A_2$ and $\varphi_2 : A_2 \longrightarrow A_1 \otimes A_2$ be the algebra homomorphisms given by $\varphi_1(a_1) = a_1 \otimes 1$ and $\varphi_2(a_2) = 1 \otimes a_2$, and let φ be the homomorphism given in (6.3.3) with $B = A_1 \otimes A_2$. Now suppose C is a summand of both A_1 and A_2. This means, of course, $C \cong C \cdot 1_{A_i}$ and $C \cdot 1_{A_i}$ is a summand of A_i. Let $\pi : A_1 \longrightarrow C$ be the projection. The balanced map $A_1 \times A_2 \longrightarrow A_2$ given by $(a_1, a_2) \mapsto \pi(a_1)a_2$ induces a C-homomorphism $\tau : A_1 \otimes A_2 \longrightarrow A_2$. If $\sigma_2(a_2) = 0$, then $1 \otimes a_2 = \varphi_2(a_2) = \varphi\sigma_2(a_2) = 0$, and hence $a_2 = \tau(1 \otimes a_2) = 0$. So σ_2 is an injection and by symmetry, so is σ_1. In particular, $C \subseteq \sigma_1(A_1) \cap \sigma_2(A_2)$. Suppose $\sigma_1(a_1) = \sigma_2(a_2)$ for some $a_i \in A_i$. Then, from (6.3.3), $a_1 \otimes 1 = 1 \otimes a_2$,

$$a_2 = \pi(1)a_2 = \tau(1 \otimes a_2) = \tau(a_1 \otimes 1) = \pi(a_1)1,$$

and $\sigma_2(a_2) = a_2 \in C$. □

Let R be a domain, C its extended centroid and $A = RC + C$ its central closure. We will denote the C-algebra coproduct of A and the polynomial ring $C[t]$ by $A\langle t\rangle$. Since C is a field the previous theorem tells us that A and $C[t]$ are subalgebras of $A\langle t\rangle$. Note that, for each $z \in A$ (or $z \in A\langle t\rangle$), there is a unique C-algebra homomorphism from $A\langle t\rangle$ to A (or to $A\langle t\rangle$) which sends t to z and is the identity on A. We will now proceed to collect some useful information about $A\langle t\rangle$, mainly when R, and also A, is totally ordered.

Theorem 6.3.2. *Let R, C and A be as above and let $\{a_i : 1 \le i \le n\}$ and $\{d_i : 1 \le i \le n\}$ be subsets of A with $\{a_i\}$ C-independent and suppose $\Sigma_{i=1}^n a_i t d_i = 0$ in $A\langle t\rangle$. Then each $d_i = 0$.*

Proof. If not, let n be minimal for which there exist such sets with $\Sigma a_i t d_i = 0$ and $d_i \neq 0$ for $i = 1, \ldots, n$. If $n = 1$, then $a_1 x d_1 = 0$ for each $x \in A$ so $d_1 = 0$. If $n > 1$, then $\Sigma a_i t d_i x d_1 = 0$ and $\Sigma a_i t (d_1 x) d_i = 0$ for each $x \in A$. This yields $\Sigma_{i=2}^n a_i t(d_i x d_1 - d_1 x d_i) = 0$. Thus, for each $i \ge 2$, $d_i x d_1 = d_1 x d_i$ and by Theorem 6.1.15 there exist $c_i \in C$ with $d_i = c_i d_1$. But then, with $c_1 = 1$, $\Sigma_{i=1}^n a_i c_i t d_1 = 0$, and hence each $c_i = 0$ because $\{a_i\}$ is an independent set. We again have a contradiction. □

A totally ordered domain R is called *co-ℓ-simple* if for all c, $d \in R^*$ there is an element $x \in R^*$ with $|cx| < |d|$. This one-sided condition is left-right symmetric and is dual to the condition for ℓ-simplicity of an f-ring. In particular, there is a dual of Theorem 3.3.8 which is given in Exercise 1. The duality is not perfect, however. For example, the antilexicographically ordered polynomial ring $R[x]$ over any totally ordered domain R is co-ℓ-simple whereas the ℓ-simplicity of $R[x]$ with the lexicographic order requires R to be ℓ-simple. If R is unital and the units of R^+ are cofinal in R, then R is co-ℓ-simple since $|c| < u$ with $u^{-1} \in R$ gives $|cu^{-1}d| < |d|$. So the group ring $R[G]$, where $G \neq e$ is a totally ordered group, is co-ℓ-simple with either the Hahn order or its dual; it is also ℓ-simple.

Let

$$A[t] = \left\{ \sum_{i=0}^{n} d_i t^i : d_i \in A \right\} \subseteq A\langle t \rangle,$$

and let

$$\mathscr{P} = \{a_1 t a_2 t \cdots a_n t a_{n+1} : n \geq 2, a_i \in A^*\}.$$

Theorem 6.3.3. *Let R be a co-ℓ-simple totally ordered domain and let A be its central closure and C its extended centroid.*

(a) Suppose $f(t) = a_r t^r + a_{r+1} t^{r+1} + \cdots + a_s t^s$ is an element of $A[t]$ with $a_r a_s \neq 0$ and $f(x) \geq 0$ for each x in R. Then $a_r > 0$ and r is even.

(b) If $p_1, \ldots, p_m \in \mathscr{P}$ and $a, c \in R^*$, then $|axc| > \Sigma_{j=1}^m |p_j(x)|$ for some $x \in R$.

(c) Suppose $a_1, \ldots, a_n$, $b_1, \ldots, b_n \in A^*$, $p_1(t), \ldots, p_m(t) \in \mathscr{P}$, and, for each x in R,

$$\left| \sum_{i=1}^{n} a_i x b_i \right| \leq \sum_{j=1}^{m} |p_j(x)|. \tag{6.3.4}$$

Then $n \geq 2$, $\Sigma_{i=1}^n a_i t b_i = 0$ in $A\langle t \rangle$, and $\{a_1, \ldots, a_n\}$ and $\{b_1, \ldots, b_n\}$ are C-dependent sets.

Proof. (a) We may assume each $a_i \in R$ since there exists $0 < d \in R$ with $da_i \in R$ for $i = r, r+1, \ldots, s$. If $r = s$ the assertion is obvious. Suppose $r < s$ and let $b = (s-r) \bigvee_{i=r+1}^{s} |a_i|$. Since R is co-ℓ-simple there exists $x \in R^*$ with $|x| < 1$ and $|bx| < |a_r|$. Then, for $i = r+1, \ldots, s$, $(s-r)|a_i x| \leq |bx| < |a_r|$ and

$$(s-r)|a_i x^i| \leq |bx^i| < |a_r x^{i-1}| \leq |a_r x^r|.$$

So

$$(s-r) \left| \sum_{i=r+1}^{s} a_i x^i \right| < \sum_{i=r+1}^{s} |a_r x^r| = (s-r)|a_r x^r|,$$

$$\left| \sum_{i=r+1}^{s} a_i x^i \right| < |a_r x^r|,$$

and

$$-|a_r x^r| < -a_{r+1} x^{r+1} - \cdots - a_s x^s \leq a_r x^r.$$

Thus, $a_r x^r > 0$, and since x can be replaced by $-x$, necessarily r is even and $a_r > 0$.

(b) Suppose $p_j(t) = d_1 t d_2 t \cdots d_n t d_{n+1}$ and $x \in A$ with $|x| \leq 1$. Let $b_j = |d_1 \cdots d_{n-1}| \vee |d_n| \vee |d_{n+1}|$. Then $|p_j(x)| \leq |b_j x b_j x b_j|$ since $n \geq 2$, and, for $d = \vee_j b_j$ and $b = md$ we have

$$\sum_{j=1}^{m} |p_j(x)| \le \sum_{j=1}^{m} |dxdxd| = m|dxdxd| \le |bxbxb|.$$

Take $y \in R^*$ with $|y| < 1$ and $|by| \vee |yb| < |a| \wedge |c| \wedge 1$. Then

$$|by^3by^3b| = |byy^2byyyb| < |ay^3c|,$$

and $x = y^3$ is the desired element.

(c) By (b) we have $n \ge 2$. Suppose $\Sigma_{i=1}^{n} a_i t b_i \ne 0$ and assume n is minimal for which there exist $a_1, \ldots, a_n$, $b_1, \ldots, b_n$, $p_1(t), \ldots, p_m(t)$ such that (6.3.4) holds but $\Sigma a_i t b_i \ne 0$. If $a_1 = \Sigma_{i=2}^{n} c_i a_i$ with $c_i \in C$, then

$$\sum_{i=1}^{n} a_i x b_i = \sum_{i=2}^{n} c_i a_i x b_1 + \sum_{i=2}^{n} a_i x b_i = \sum_{i=2}^{n} a_i x (c_i b_1 + b_i).$$

So, the minimality of n requires $\{a_1, \ldots, a_n\}$ and $\{b_1, \ldots, b_n\}$ to be C-independent sets. Now, by (6.3.4) we get for each x, $d \in R$, $|\Sigma_{i=1}^{n} a_i x b_i d b_1| \le \Sigma_{j=1}^{m} |p_j(x) d b_1|$ and $|\Sigma_{i=1}^{n} a_i x b_1 d b_i| \le \Sigma_{j=1}^{m} |p_j(x b_1 d)|$. So

$$\left|\sum_{i=2}^{n} a_i x (b_1 d b_i - b_i d b_1)\right| = \left|\sum_{i=1}^{n} a_i x b_1 d b_i - \sum_{i=1}^{n} a_i x b_i d b_1\right| \le \sum_{j=1}^{m} [|p_j(x b_1 d)| + |p_j(x) d b_1|].$$

Thus, by the minimality of n, in $A\langle t \rangle$,

$$\sum_{i=2}^{n} a_i t (b_1 d b_i - b_i d b_1) = 0.$$

So $b_1 d b_i = b_i d b_1$ for $i = 2, \ldots, n$ (or $a_i = 0$ for $i \ge 2$) by Theorem 6.3.2 and hence $\{b_1, \ldots, b_n\}$ is a C-dependent set by Theorem 6.1.15. This contradiction gives that $\Sigma a_i t b_i = 0$ and hence $\{a_1, \ldots, a_n\}$ and $\{b_1, \ldots, b_n\}$ are both C-dependent sets. □

This last result will be used to produce a central element and then commutativity will follow.

Theorem 6.3.4. *Let R, C and A be as in Theorem 6.3.3 and let $0 \ne g(t) \in A[t]$. Suppose $y \in A^*$ and $g(x+y) \ge 0$ for each $x \in R$. If $g(y) = 0$, then $y \in C$.*

Proof. Let $g(t) = a_0 + \cdots + a_n t^n$ with $a_n \ne 0$. Then $n \ge 2$. Otherwise, $g(t) = a_0 + a_1 t$ and $a_1 x = g(x+y) \ge 0$ for every $x \in R$. For each $x \in R$,

$$g(y+x) = \sum_{i=0}^{n} a_i (y+x)^i =$$

$$g(y) + a_1 x + \sum_{i=2}^{n} a_i (y^{i-1} x + y^{i-2} xy + \cdots + xy^{i-1}) + \sum_{j=1}^{m} p_j(x) =$$

$$\sum_{i=0}^{n-1}\left(\sum_{j=i+1}^{n} a_j y^{j-(i+1)}\right) xy^i + \sum_{j=1}^{m} p_j(x) \geq 0,$$

where $p_j(t) \in \mathscr{P}$ and $m \geq 1$ since $n \geq 2$. For $0 \leq i \leq n-1$ let

$$b_i = \sum_{j=i+1}^{n} a_j y^{j-(i+1)}.$$

Then, for each x in R,

$$\sum_{j=1}^{m} |p_j(x)| \geq \sum_{j=1}^{m} p_j(x) \geq -\sum_{i=0}^{n-1} b_i x y^i.$$

Since x may be replaced by $-x$ we get that for each x in R

$$\left|\sum_{i=0}^{n-1} b_i x y^i\right| \leq \sum_{j=1}^{m} |p_j(x)|.$$

But $b_{n-1} = a_n \neq 0$, and hence by Theorem 6.3.3 y is algebraic over C. But then $y \in C$ by Theorem 6.2.10. □

Theorem 6.3.5. *Let R be a co-ℓ-simple totally ordered domain and let $A = RC + C$ be its central closure. If there is a form over A which is P.S.D. in R and which has a solution in R^*, then R is commutative.*

Proof. Let

$$f(x_1, \ldots, x_n) = \sum_{i_1+\cdots+i_n=k} a_{(i)} x_1^{i_1} \cdots x_n^{i_n}$$

be a form with $a_{(i)} \in A$ such that $f(z_1, \ldots, z_n) \geq 0$ for all $z_1, \ldots, z_n \in R$, and let $y_1, \ldots, y_n \in R^*$ with $f(y_1, \ldots, y_n) = 0$. We first verify the following:

$$\begin{aligned}&\text{If } C \cap R \neq 0, \text{ then there exist } \overline{y}_1, \ldots, \overline{y}_n \in (C \cap R)^* \\ &\text{with } f(\overline{y}_1, \ldots, \overline{y}_n) = 0, \text{ whereas if } C \cap R = 0, \\ &\text{then } f(1, 1, \ldots, 1) = 0.\end{aligned} \tag{6.3.5}$$

To see this, suppose that $1 \leq r \leq n$ and $y_{r+1}, \ldots, y_n \in C$. Let $g_r(t) = f(y_1, \ldots, y_{r-1}, t, y_{r+1}, \ldots, y_n) \in A[t]$. Then

$$g_r(t) = \sum_{(i)} a_{(i)} y_1^{i_1} \cdots y_{r-1}^{i_{r-1}} t^{i_r} y_{r+1}^{i_{r+1}} \cdots y_n^{i_n}$$

and $g_r(y_r) = 0$. If $g_r(t) \neq 0$, then $y_r \in C$ by Theorem 6.3.4. If $g_r(t) = 0$, then $g_r(z) = 0$ for each $z \in A$. So if $C \cap R \neq 0$, then $y_1, \ldots, y_{r-1}, \overline{y}_r, y_{r+1}, \ldots, y_n$ is a nontrivial solution of $f(x_1, \ldots, x_n)$ with $0 \neq \overline{y}_r \in C \cap R$. Continuing in this way gives a solution $\overline{y}_1, \ldots, \overline{y}_r, y_{r+1}, \ldots, y_n$ with $y_i, \overline{y}_j \in (C \cap R)^*$. On the other hand, if $C \cap R = 0$, then

$r = n$, $g_n(t) = 0$, and hence $y_1,\dots,y_{n-1},1$ is a solution of $f(x_1,\dots,x_n)$; continuing gives $f(1,1,\dots,1) = 0$.

We now proceed by induction on n, with $n \geq 2$, assuming $f(x_1,\dots,x_n)$ has a central solution $y_1,\dots,y_n$ of the type given in (6.3.5). The inductive step and the case $n = 2$ will be treated simultaneously. Let $g_n(t) = f(y_1,\dots,y_{n-1},t) \in A[t]$. First suppose that $g_n(t) \neq 0$. Since f is homogeneous and $y_1,\dots,y_n$ are central, $f(y_1x,\dots,y_nx) = x^k f(y_1,\dots,y_n) = 0$ for each $x \in R^*$. Fix x and let $h_n(t) = f(y_1x,\dots,y_{n-1}x,t)$. Thus, if $g_n(t) = \Sigma b_it^i$, then $h_n(t) = \Sigma b_ix^{k-i}t^i \neq 0$. Now $h_n(y_nx) = 0$ and $h_n(y_nx+u) \geq 0$ for each u in R. Thus $y_nx \in C$ by Theorem 6.3.4 and hence $x \in C$. Note that when $n = 2$, $f(x_1,x_2) = \Sigma_{i=r}^s a_ix_1^{k-i}x_2^i$ with $a_ra_s \neq 0$, and $g_2(t) = \Sigma_{i=r}^s a_iy_1^{k-i}t^i \neq 0$.

On the other hand, suppose $g_n(t) = 0$. Now,

$$f(x_1,\dots,x_n) = \sum a_{(i)}x_1^{i_1}\cdots x_n^{i_n} = \sum_{i=r}^{s} f_i(x_1,\dots,x_{n-1})x_n^i$$

where $r \leq s$, $f_i(x_1,\dots,x_{n-1})$ is a form of degree $k-i$, and $f_r(x_1,\dots,x_{n-1})$ and $f_s(x_1,\dots,x_{n-1})$ are nonzero. If $z_1,\dots,z_{n-1} \in R$, then $g(t) = f(z_1,\dots,z_{n-1},t) = \Sigma_{i=r}^s f_i(z_1,\dots,z_{n-1})t^i \in A[t]$, and $g(z) \geq 0$ for each $z \in R$. So, by Theorem 6.3.3, $f_r(z_1,\dots,z_{n-1}) \geq 0$. Since $f_r(y_1,\dots,y_{n-1}) = 0$, $f_r(x_1,\dots,x_{n-1})$ is a P.S.D. form on R with a solution in R^* and hence R is commutative by induction. □

An example of a noncommutative totally ordered domain which has a P.S.D. form with a nontrivial solution is given in Exercise 2. Nevertheless, specific P.S.D. forms on totally ordered domains and other ℓ-rings do force commutativity. Since small commutators play a roll we will first develop some more properties of commutators in an arbitrary ring.

If X and Y are subsets of the ring R, then $[X,Y] = \{[x,y] : x \in X,\ y \in Y\}$. Here, $[x,y] = xy - yx = \delta_xy$. The additive subgroup U of R is a *Lie ideal* if $[U,R] \subseteq U$.

Theorem 6.3.6. *Suppose U is both a subring and a Lie ideal of R. If $x,y \in U$, then $R[x,y]R \subseteq U$. If U is commutative, 2-torsion-free, and either R or U is semiprime, then U is central.*

Proof. If $z \in R$, then $\delta_x(yz) = (\delta_xy)z + y(\delta_xz) \in U$ and $y(\delta_xz) \in U$; so $(\delta_xy)R \subseteq U$. If $v \in R$, then $((xy-yx)z)v - v((xy-yx)z) \in U$ since U is a Lie ideal and $(xy-yx)zv \in U$; so $v(xy-yx)z \in U$ and hence $R[x,y]R \subseteq U$. Now suppose U is commutative and 2-torsion-free. If $u \in U$ and $\delta = \delta_u$, then $\delta^2 = 0$ since $\delta R \subseteq U$ and $\delta = 0$ on U. For any $r, s \in R$ we have $\delta(rs) = (\delta r)s + r(\delta s)$ and

$$0 = \delta^2(rs) = (\delta^2 r)s + (\delta r)(\delta s) + (\delta r)(\delta s) + r(\delta^2 s).$$

So $2(\delta r)(\delta s) = 0$ and $(\delta r)(\delta s) = 0$. Let $s = vr$ where $v \in R$. Then $0 = (\delta r)((\delta v)r + v(\delta r))$ yields $(\delta r)v(\delta r) = 0$; that is, $(\delta r)R(\delta r) = 0$. If R is semiprime, then $\delta r = 0$. If U is semiprime, then $\delta r = 0$ since $(\delta r)^2 = 0$. In either case, $\delta = 0$, u is central and U is contained in the center of R. □

Applying this result to an ℓ-prime ℓ-ring gives a dichotomy for the subring of f-elements.

Theorem 6.3.7. *Let R be an ℓ-prime ℓ-ring. Then $F(R)$ is a Lie ideal of R if and only if R is an f-ring or $F(R)$ is central.*

Proof. Suppose $F = F(R)$ is a Lie ideal of R and assume F is not central. By Theorem 3.2.13 F is semiprime, and by Theorem 6.3.6 F contains a nonzero ideal I of R. Let

$$K = C(I^+) = \{u \in R : |u| \leq a_1 + \cdots + a_n,\ a_j \in I^+\}$$

be the convex ℓ-subgroup of R generated by I^+. Then $K \subseteq F$ is a nonzero ℓ-ideal of R. If $u, v \in F^+$ with $uv = 0$, then $C(uK)C(vK) = 0$, $uK = 0$ or $vK = 0$, and $u = 0$ or $v = 0$ since R is ℓ-prime. So F is a totally ordered domain. If $x \in R$ with $x^+ \neq 0$, then $ux^+ \neq 0$ for some $u \in K^+$. But then, for any $v \in K^+$, $ux^+ \wedge vx^- = 0$, $vx^- = 0$, and $Kx^- = 0$. Thus, $x^- = 0$ and R is totally ordered. □

The previous result fails for an ℓ-semiprime ℓ-ring. As an example take the direct product of a noncommutative semiprime f-ring with a nonzero ℓ-semiprime ℓ-ring which does not have any nonzero f-elements.

Suppose R is a torsion-free ℓ-algebra over the totally ordered domain C and let $F = F(R)$. Assume F is C-unitable with C-unital cover $F_{C\text{-}u}$. Then the following subsets of R, actually of F, are well-defined:

$$R(1) = \{x \in R : |x| \leq \alpha \cdot 1 \text{ for some } \alpha \in C\}$$

$$M(1) = \{x \in R : C|x| \leq 1\}.$$

$R(1)$ will be called the convex-ℓ-subalgebra of R generated by 1 and $M(1)$ is an algebra ℓ-ideal of $R(1)$. In particular, if R_F is a t-torsion-free f-module, then F is a totally ordered domain and it has a unique totally ordered C-torsion-free C-unital cover $F_{C\text{-}u}$ by Exercise 3.4.16 and Theorem 3.4.5. Moreover, $R_{F_{C\text{-}u}}$ is a t-torsion-free f-module, and $M(1) = R(1)$ or $M(1)$ is the unique maximal one-sided ℓ-ideal of $R(1)$. For the last part, if $x \in R(1)^+ \backslash M(1)$ and $y \in R(1)^+$, then $1 < \beta x$ for some $\beta \in C^+$; so, $y \leq \beta xy \in C(xR(1))$ and $C(xR(1)) = R(1)$. Thus, if I is a right ℓ-ideal of $R(1)$, then $I \subseteq M(1)$ or $I = R(1)$.

If X is a subset of R we say the *commutators of X are bounded* if, for each x in X and each y in R there exists $\alpha \in C$ and $2 \leq k \in \mathbb{N}$ such that $|[x,y]| \leq \alpha|x|^k$. We say the *commutators of X are of bounded degree n* if they are bounded and $k \leq n$. By considering the totally ordered homomorphic images of F it is easy to see that when $X \subseteq F$ this is equivalent to $|[x,y]| \leq \alpha(x^2 \vee |x|^n)$. If the commutators of $M(1)$ are bounded, then they are of bounded degree 2 since $|x| \leq \beta$ implies $|x|^k \leq \beta^{k-2}|x|^2$ for each $k \geq 2$.

Theorem 6.3.8. *Suppose R is a torsion-free ℓ-algebra over the totally ordered domain C and R_F is a t-torsion-free f-module over $F = F(R)$.*

(a) *If the commutators of $R(1)$ are bounded and $M(1) \neq 0$, then $R(1)$ is central.*

(b) If the commutators of F are of bounded degree, then $F\backslash R(1)$ is central.

Proof. (a) We first note that ${}_FR$ is also t-torsion-free. For if $sr = 0$ with $s \in F$ and $r \in R$, then $(rs)^2 = 0$ and $rs = [r,s] \in C_C(s^2) \subseteq F$ gives that $rs = 0$.

Let $x \in R(1)^+$, $y \in R^+$, and take $0 < a \in M(1)$. Since xa and ax are comparable we may assume $xa \leq ax$. Then $(ax)^n \leq a^n x^n$ for each $n \geq 1$. Now,

$$a[x,y] = [ax,y] + [y,a]x \tag{6.3.6}$$

implies that for some $\alpha, \beta, \gamma \in F$,

$$a|[x,y]| = |a[x,y]| \leq \alpha(ax)^2 + \beta a^2 x \leq \alpha a^2 x^2 + \beta a^2 x \leq \gamma a^2$$

since $x \in R(1)$. Since ${}_FR$ is t-torsion-free we get $|[x,y]| \leq \gamma a$ and $[x,y] \in M(1)$. In particular, if $a = |[x,y]| \neq 0$, then

$$a = |[x,y]| \leq \gamma a^2 < a;$$

so $a = 0$. If $ax < xa$, a similar computation using $[x,y]a$ gives $a = 0$. Thus, x is in the center of R.

(b) Let $x \in F^+\backslash R(1)$, $y \in R^+$ and let n be the bound. Then for some $\alpha \in C^+$

$$x^{n+1}|[x,y]| = |[x,x^{n+1}y]| \leq \alpha x^n$$

and

$$|[x,y]| \leq x|[x,y]| \leq \alpha.$$

But now $a = |[x,y]| \in M(1)$. For if $\beta \in C^+$, then $\alpha\beta a \leq xa \leq \alpha$, implies $\beta a \leq 1$. So if $a \neq 0$, then $R(1)$ is central by (a). Thus, xa and a are both central and hence so is x. Consequently, $a = 0$. □

We can now draw some conclusions about when $F(R)$ is a central subring.

Theorem 6.3.9. *Let R be a torsion-free C-ℓ-algebra which is also a t-torsion-free right module over $F = F(R)$. Assume that the commutators of F are of bounded degree $n \geq 2$. So, for each $x \in F$ and $y \in R$ there exists $\alpha = \alpha(x,y) \in C$ such that*

$$|[x,y]| \leq \alpha(x^2 \vee |x^n|). \tag{6.3.7}$$

Then each of the following conditions implies F is central in R.

(a) $R(1)$ is a proper subring of F.
(b) F is not archimedean over C.
(c) F is archimedean.
(d) α is independent of x and y.

Proof. If (a) holds and $x \in F\backslash R(1)$ and $y \in R(1)$, then x and $x+y$ are central by Theorem 6.3.8; so y is central also. If (b) holds we may assume $F = R(1)$ by (a), and then $M(1) \neq 0$; so $F \subseteq Z(R)$ by Theorem 6.3.8. If F is archimedean then $F \subseteq Z(R)$ by Theorem 6.3.6.

(d) By (a) and (b) we may assume $F = R(1)$ is archimedean over C. From (6.3.6), if $0 \neq a, x \in F^+$ with $xa \leq ax$, then for any $y \in R$

$$\begin{aligned} a|[x,y]| &\leq |[ax,y]| + |[y,a]|x \\ &\leq \alpha a^m x^m + \alpha a^k x, \end{aligned}$$

with $2 \leq k, m \leq n$. Now, if $1 \leq \gamma \in C$ with $x, a \leq \gamma$, then

$$\begin{aligned} |[x,y]| &\leq \alpha a \gamma^{m-2} \gamma^m + \alpha a \gamma^{k-2} \gamma \\ &\leq 2\alpha \gamma^{2n-2} a. \end{aligned}$$

Replacing y by $4\alpha\gamma^{2n-2}y$ we get that if $[x,y] \neq 0$, then

$$|[x,y]| < 2|[x,y]| \leq a.$$

Thus $[x,y] = 0$ and $x \in Z(R)$ since otherwise we can let $a = |[x,y]|$. A similar argument using $|[x,y]|a$ gives $F \subseteq Z(R)$ when $ax < xa$. □

Theorem 6.3.10. *Let R be a torsion-free ℓ-algebra over C and suppose U is a Lie ideal and a semiprime f-subalgebra of R. If the commutators of U are of bounded degree $n \geq 2$ and either C is archimedean, or, in (6.3.7) α is independent of $x \in U$ and $y \in R$, or C is bounded by an element of U, then $U \subseteq Z(R)$.*

Proof. Let P be a minimal prime ideal of U. Since P is a minimal prime subgroup of U, as a consequence of Theorem 2.4.3 U/P is C-torsion-free. By Theorem 6.3.9 U/P is commutative. Therefore, U is commutative and U is central by Theorem 6.3.6. □

We now show that particular P.S.D. forms force a semiprime f-ring to be commutative.

Theorem 6.3.11. *The following statements are equivalent for the semiprime f-ring R.*

(a) For every $x, y \in R$ there is an integer $n = n(x,y) \in \mathbb{N}$ such that

$$2x^n y^n \leq x^{2n} + y^{2n}.$$

(b) For every $x, y \in R$ there is an integer $m = m(x,y) \in \mathbb{N}$ such that

$$|xy - yx| \leq m(x^2 \wedge y^2).$$

(c) R is commutative.

Proof. (a) implies (b). We will assume R is a totally ordered domain and, without loss of generality, $x > y > 0$ and $yx < xy$. Then

$$y^\ell x^k \leq x^k y^\ell \qquad \text{for all } \ell, k \geq 0. \tag{6.3.8}$$

Take $n \in \mathbb{N}$ with $x^{2n} + (x+y)^{2n} \geq 2x^n(x+y)^n$. Expansion of these binomial terms gives

$$x^{2n} + x^{2n} + \sum_{k=1}^{n} x^{2n-k}yx^{k-1} + \sum_{k=1}^{n} x^{n-k}yx^{n+k-1} + u \geq$$
$$2x^n \left(x^n + \sum_{k=1}^{n} x^{n-k}yx^{k-1} + v \right)$$

where u and v are the sums of the nonlinear terms in y from $(x+y)^{2n}$ and $(x+y)^n$, respectively. So

$$u - 2x^n v \geq \sum_{k=1}^{n} (x^{2n-k}yx^{k-1} - x^{n-k}yx^{n+k-1}) \geq 0 \tag{6.3.9}$$

since $x^{2n-k}yx^{k-1} \geq x^{n-k}yx^n x^{k-1} = x^{n-k}yx^{n+k-1}$, for each k, by (6.3.8). We claim that the sum in (6.3.9) is no smaller than $x^{2n-1}y - x^{2n-2}yx = |x^{2n-1}y - x^{2n-2}yx|$. If $n = 1$ this is clear and for $n \geq 2$,

$$\begin{aligned}\sum_{k=1}^{n} (x^{2n-k}yx^{k-1} - x^{n-k}yx^{n+k-1}) &\geq x^{2n-1}y - x^{n-1}yx^n + x^{2n-2}yx - x^{n-2}yx^{n+1} \\ &\geq x^{2n-1}y - x^{2n-2}yx\end{aligned}$$

where the last inequality comes from (6.3.8) in the form $x^{2n-2}yx \geq x^{n-1}yx^n$, $x^{n-2}yx^{n+1}$. Now, each term in u is bounded above by $x^{2n-2}y^2$. To see this note that a typical term a in u is of the form

$$a = x^{k_1}y^{\ell_1}x^{k_2}y^{\ell_2}\cdots x^{k_r}y^{\ell_r} \text{ with } \sum_i \ell_i \geq 2 \text{ and } \sum_i k_i + \sum_i \ell_i = 2n.$$

But by moving each x^{k_i} to the left and by using (6.3.8) we get $x^{\Sigma k_i}y^{\Sigma \ell_i} \geq a$; and $x^{2n-2}y^2 \geq x^{\Sigma k_i}y^{\Sigma \ell_i}$ since $x > y$ and $2n-2 \geq \Sigma k_i$. Thus, for some $m \in \mathbb{N}$,

$$mx^{2n-2}y^2 \geq u \geq u - 2x^n v \geq |x^{2n-1}y - x^{2n-2}yx|,$$

and

$$|xy - yx| \leq my^2 = m(x^2 \wedge y^2).$$

That (b) implies (c) is a consequence of Theorem 6.3.10 and that (c) implies (a) follows from the fact that R is an sp-po-ring. □

An example of a noncommutative totally ordered ring for which $f(x,y) = x^2 + y^2 - 2xy$ is a P.S.D. form is given in Exercise 6. On the other hand a unital ℓ-semiprime ℓ-ring which admits $f(x,y)$ as a P.S.D. form is commutative provided it satisfies an additional semiclosed condition; see Exercise 7.

We now consider polynomial constraints in totally ordered fields and examine the variety of f-rings generated by a totally ordered field. A description of the free objects in this variety will be given.

Let C be a commutative unital totally ordered ring and let $\mathscr{V}_C(R)$ denote the variety of C-ℓ-algebras generated by the C-ℓ-algebra R. An ℓ-algebra in $\mathscr{V}_C(R)$ will be called *formally R-real*. From Section 1.4 we know that a C-ℓ-algebra is formally R-real exactly when it satisfies all of the identities satisfied by the C-ℓ-algebra R, and according to Exercise 1.4.20 this is equivalent to it being a C-ℓ-homomorphic image of an ℓ-subalgebra of some direct product R^I of copies of R. When $C = \mathbb{Z}$ references to C will be dropped and an ℓ-ring in $\mathscr{V}(R)$ is called a *formally R-real ℓ-ring*. An ℓ-ring in $\mathscr{V}(\mathbb{R})$ is called *formally real*. Suppose $\mathscr{V}$ is a variety of C-ℓ-algebras which contains a C-ℓ-algebra $(C[X]_0, P)$ where X is a set of indeterminates over C and $C[X]_0$ is the free C-algebra on X without an identity element. Let $A_{\mathscr{V}}(X)$ be the free $\mathscr{V}$-algebra on X. Then the ℓ-algebra $(C[X]_0, P)$ is a homomorphic image of $A_{\mathscr{V}}(X)$ and hence the subalgebra of $A_{\mathscr{V}}(X)$ generated by X is the free C-algebra on X. Of course, $A_{\mathscr{V}}(X)$ is generated as an ℓ-algebra by its subalgebra $C[X]_0$. When $\mathscr{V}$ is a variety of f-algebras each element of $A_{\mathscr{V}}(X)$ has the form $\vee_i \wedge_j p_{ij}(x_1, \ldots, x_n)$ for some finite subset $\{p_{ij}(x_1, \ldots, x_n)\}$ of $C[X]_0$ (Theorem 3.4.1). Similarly, if $\mathscr{V}$ is a variety of commutative ℓ-algebras which contains an ℓ-algebra whose underlying algebra is a free commutative algebra on the commuting indeterminates X, then the free $\mathscr{V}$-algebra on X is generated as an ℓ-algebra by the polynomial algebra $C[X]_0$.

Theorem 6.3.12. *Let C be a unital totally ordered subring of the real closed field K. Then each commutative C-torsion-free f-algebra which is semiprime, has square* 0*, or is C-archimedean is formally K-real.*

Proof. Let R be a commutative C-torsion-free f-algebra. Assume first that R is semiprime. Since R can be embedded in a product of totally ordered field extensions of C it suffices to show that a totally ordered field L which contains C is formally K-real. Suppose $g(x_1, \ldots, x_n)$ is a word in the free commutative C-f-algebra that K satisfies and let $\alpha_1, \ldots, \alpha_m$ be all the elements of C which occur in $g(x_1, \ldots, x_n)$. Let $a_1, \ldots, a_n \in L$ and let $\mathscr{F}$ be an ultrafilter on $\mathbb{N}$ which contains all complements of finite subsets. By Exercise 5.1.13 the ultraproduct $K^{\mathbb{N}}/\mathscr{F}$ is a real closed η_1-field and hence by Theorem 5.1.11 (actually, Exercise 5.1.19) $\mathbb{Q}(\alpha_1, \ldots, \alpha_m, a_1, \ldots, a_n)$ can be embedded in $K^{\mathbb{N}}/\mathscr{F}$:

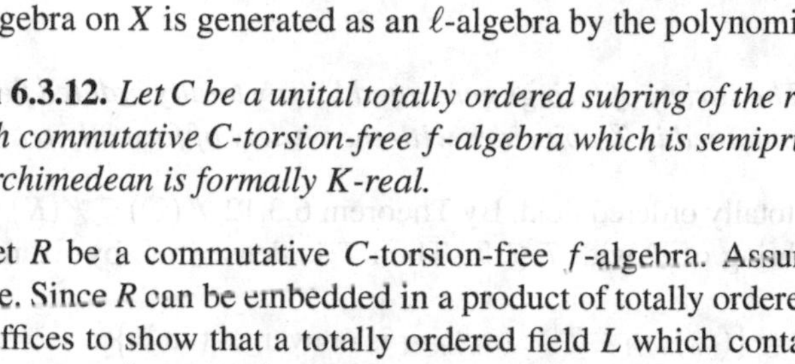

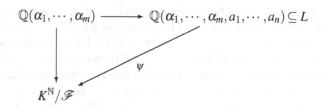

Here, the vertical map is the inclusion $\mathbb{Q}(\alpha_1, \ldots, \alpha_m) \subseteq K \subseteq K^{\mathbb{N}}/\mathscr{F}$. Since ψ $(g(a_1, \ldots, a_n) = g(\psi(a_1), \ldots, \psi(a_n)) = 0$ we have $g(a_1, \ldots, a_n) = 0$. So L satisfies $g(x_1, \ldots, x_n) = 0$ and L is formally K-real. Now suppose $R^2 = 0$. By Theorem 4.5.10 $R \in \mathscr{V}_C(C_0)$ where C_0 is the ℓ-algebra whose underlying C-f-module is

C and $C_0^2 = 0$. Thus, in order to show that R is formally K-real it suffices to verify that C_0 is formally K-real. Let τ be an ordinal and $a = (a_\rho)_{\rho<\tau}$ a sequence in $Q(C)^{W(\tau)} \subseteq K^{W(\tau)}$ such that $\{a_\rho : \rho < \tau\} \leq 1$ and is coinitial in $Q(C)^{+*}$. Let A and B be the convex ℓ-C-submodules of $K^{W(\tau)}$ generated by a and a^2, respectively. Since $0 < a < 1$, A is a C-ℓ-subalgebra of $K^{W(\tau)}$ and $B = A^{[2]}$ is an ℓ-ideal of A. Also, $c\ell B$ (in A) is a proper ℓ-ideal since if $a \in c\ell B$, then $\alpha a \leq \beta a^2$ for some $0 < \alpha, \beta \in C$, and hence $\beta^{-1}\alpha \leq a_\rho$ for each $\rho < \tau$. But for some ρ, $a_\rho < \beta^{-1}\alpha$. Thus, C_0 can be embedded in the formally K-real C-ℓ-algebra $A/c\ell B$ and hence C_0 is formally K-real. Lastly, suppose R is C-archimedean. By Theorem 3.6.2 $N \cap R^{[2]} = 0$ where $N = \ell\text{-}\beta(R) = \ell(R;R)$. Since R is C-torsion-free $N = c\ell N$ and $N \cap c\ell R^{[2]} = 0$. So R is formally K-real since it is a subdirect product of two C-torsion-free f-algebras, one which is semiprime and one whose square is 0. □

The preceding result shows that the variety of formally K-real C-ℓ-algebras where K is a real closed field which contains C is quite large and is independent of K. According to Exercise 4.3.28 each formally K-real ℓ-algebra can be embedded in a convex formally K-real ℓ-algebra. Note that a non-unitable C-f-algebra is not formally K-real.

Theorem 6.3.13. *The variety of ℓ-rings generated by any totally ordered field or any subring of $\mathbb{R}$ not contained in $\mathbb{Z}$ coincides with the variety of formally real ℓ-rings.*

Proof. Let K be a totally ordered field. By Theorem 6.3.12 $\mathscr{V}(\mathbb{Q}) \subseteq \mathscr{V}(K) \subseteq \mathscr{V}(\mathbb{R})$. Suppose T is a subring of $\mathbb{R}$ with $T \not\subseteq \mathbb{Z}$. Then T is dense in $\mathbb{R}$ by Exercise 2.3.2 and

$$R = \{(a_n)_{n\in\mathbb{N}} \in T^{\mathbb{N}} : (a_n)_{n\in\mathbb{N}} \text{ is convergent in } \mathbb{R}\}$$

is an ℓ-subring of $T^{\mathbb{N}}$. The mapping $(a_n)_n \mapsto \lim a_n$ is an ℓ-homomorphism of R onto $\mathbb{R}$. So $\mathbb{R} \in \mathscr{V}(T)$, $\mathscr{V}(\mathbb{R}) = \mathscr{V}(T)$, and $\mathscr{V}(\mathbb{Q}) = \mathscr{V}(K) = \mathscr{V}(\mathbb{R})$. □

The ring $\mathbb{Z}$ is conspicuously absent from Theorem 6.3.13 because $\mathscr{V}(\mathbb{Z})$ is a proper variety of the variety of formally real rings; see Exercise 11.

A polynomial over a totally ordered field which is a sum of squares certainly gives a polynomial constraint for that field. The converse holds for some totally ordered fields, including all real closed fields, provided the squares permitted are enlarged to include all rational functions. The rational function $r(x_1,\ldots,x_n)$ in the rational function field $K(x_1,\ldots,x_n)$, K a totally ordered field, is called *positive semidefinite* if $r(\alpha_1,\ldots,\alpha_n) \geq 0$ for all $\alpha_1,\ldots,\alpha_n$ in K for which $r(\alpha_1,\ldots,\alpha_n)$ is defined. A P.S.D. polynomial in more than one variable need not be a sum of squares of polynomials (see Exercise 13); hence rational functions are needed for a converse. One of the main tools that will be used to verify that a P.S.D. rational function is a sum of squares is given in Exercise 5.1.17. Recall from this exercise that $S(K)$ denotes the set of sums of squares of elements in the commutative ring K and for a field K of characteristic 0 $S(K)$ is the intersection of the total orders of K. Also, for the totally ordered field K, $K^+S(K(x_1,\ldots,x_n))$ is the intersection of those total orders of $K(x_1,\ldots,x_n)$ which make it into an ℓ-algebra over K.

A totally ordered field K is called a *Hilbert field* if for each $n \geq 1$ every P.S.D. rational function in $K(x_1,\ldots,x_n)$ is in $S(K(x_1,\ldots,x_n))$. Since $K \cap S(K(x_1,\ldots,x_n)) \subseteq S(K)$ for any infinite field a Hilbert field K has the unique total order $S(K)$. This is one of two conditions that characterize a Hilbert field; the other being that K is dense in its real closure, or, equivalently, its real closure is a subfield of its Cauchy completion; see Theorems 5.2.23 and 5.2.24 and Exercise 2.3.1 (b). The following result will be used to show the connection between the Hilbert property of a totally ordered field and its denseness in its real closure.

Theorem 6.3.14. *Let F be the real closure of the totally ordered field K and let L be the set of limit points of K in F. Then L is a subfield of F which contains K. Moreover, if $b \in F$ is a root of $f(x) \in L[x]$ and $0 < \varepsilon \in K$, then there is a polynomial $g(x) \in K[x]$ of the same degree as $f(x)$ such that $g(x)$ has a root in the F-interval $N_\varepsilon(b) = (b-\varepsilon, b+\varepsilon)$.*

Proof. The verification that L is an intermediate field is left for Exercise 16. By taking a monic irreducible factor of $f(x)$ which has b as a root we may assume $f(x) = a_0 + a_1x + \cdots + x^n$ is monic and irreducible. By Exercise 15 K^{+*} is coinitial in F^{+*} and by Exercise 5.1.6 there exists $0 < \delta \in K$, $\delta < \varepsilon$, such that $f(x)$ is strictly monotonic in $N_\delta(b)$; so $f(b-\delta)f(b+\delta) < 0$. Let $\rho \in K$ with $0 < \rho < |f(b-\delta)| \wedge |f(b+\delta)|$ and take $a \in K$ with $\vee_{j=0}^{n}(|b+\delta|^j \vee |b-\delta|^j) < a$. Since a_j is a limit point of K there exists $b_j \in K$ with $|a_j - b_j| < \frac{\rho}{na}$ for $j = 0,\ldots,n-1$. Let $g(x) = b_0 + b_1x + \cdots + b_{n-1}x^{n-1} + x^n$. Then

$$|f(b+\delta) - g(b+\delta)| \leq \sum_{j=0}^{n-1} |a_j - b_j||b+\delta|^j$$
$$< \sum_{j=0}^{n-1} \frac{\rho}{na} a = \rho < |f(b+\delta)|,$$

$-|f(b+\delta)| < f(b+\delta) - g(b+\delta) < |f(b+\delta)|$ and hence $f(b+\delta)g(b+\delta) > 0$. Similarly, $f(b-\delta)g(b-\delta) > 0$. Consequently, $g(b-\delta)g(b+\delta) < 0$ and $g(x)$ has a root in $N_\delta(b)$ by Exercise 5.1.6. □

Theorem 6.3.15. *The following statements are equivalent for the totally ordered field K with real closure F.*

(a) K is dense in F.
(b) For every $n \in \mathbb{N}$, if $r(x_1,\ldots,x_n) \in K(x_1,\ldots,x_n)$ is P.S.D. on K, then $r(x_1,\ldots,x_n)$ is P.S.D. on F.
(c) For every $n \in \mathbb{N}$, if $r(x_1,\ldots,x_n) \in K(x_1,\ldots,x_n)$ is P.S.D. on K, then $r(x_1,\ldots,x_n) \in K^+S(K(x_1,\ldots,x_n))$.
(d) If $r(x) \in K[x]$ is P.S.D. on K, then $r(x) \in K^+S(K(x))$.
(e) If $r(x) \in K[x]$ is P.S.D. on K, then $r(x)$ is P.S.D. on F.

Proof. (a) ⇒ (b). By considering $h = rg^2 = fg$ where $r = fg^{-1}$ it suffices to show that if $h(x_1,\ldots,x_n) \in F[x_1,\ldots,x_n]$ and $h(\beta_1,\ldots,\beta_n) \neq 0$ for some $\beta_1,\ldots,\beta_n \in F$, then $h(\alpha_1,\ldots,\alpha_n)h(\beta_1,\ldots,\beta_n) > 0$ for some $\alpha_1,\ldots,\alpha_n \in K$. We use induction on n and we will assume $h(\beta_1,\ldots,\beta_n) > 0$. For $n = 1$ take $\gamma_1 \in F$ with $\gamma_1 > \beta_1$ and such that $h(x_1)$ has no roots in $[\beta_1,\gamma_1]$. Then $h(\alpha) > 0$ for each α in $[\beta_1,\gamma_1]$ by Exercise 5.1.6 and by assumption there is an $\alpha_1 \in K \cap [\beta_1,\gamma_1]$. Now apply the case $n = 1$ to the polynomial $h(x_1,\beta_2,\ldots,\beta_n)$ to get $\alpha_1 \in K$ with $h(\alpha_1,\beta_2,\ldots,\beta_n) > 0$ and then apply the induction assumption to the polynomial $h(\alpha_1,x_2,\ldots,x_n)$ to get $\alpha_2,\ldots,\alpha_n \in K$ with $h(\alpha_1,\alpha_2,\ldots,\alpha_n) > 0$.

(b) $\Rightarrow$ (c). Suppose $r(x_1,\ldots,x_n) = f(x_1,\ldots,x_n)g(x_1,\ldots,x_n)^{-1} \in K(x_1,\ldots,x_n)$ is P.S.D. on K and let $h(x_1,\ldots,x_n) = f(x_1,\ldots,x_n)g(x_1,\ldots,x_n)$. Then $h(\alpha_1,\ldots,\alpha_n) \geq 0$ for all $\alpha_1,\ldots,\alpha_n \in F$ and hence $h(x_1,\ldots,x_n)^- = 0$ is an identity for the K-ℓ-algebra F. Let P be a total order of $K(x_1,\ldots,x_n)$ which extends K^+ and let E be the real closure of $(K(x_1,\ldots,x_n),P)$. Then $\mathscr{V}_K(F) = \mathscr{V}_K(E)$ by Theorem 6.3.12 and hence $h(x_1,\ldots,x_n)^- = 0$ is also an identity for the K-ℓ-algebra E. So $h(x_1,\ldots,x_n) \in P$ and hence $r(x_1,\ldots,x_n) \in K^+S(K(x_1,\ldots,x_n))$.

(e) $\Rightarrow$ (a). Suppose K is not dense in F. Then there exists $a \in F$ and $0 < \varepsilon \in K$ such that $N_\varepsilon(a) \cap K = \emptyset$ and we may assume that a has the least degree $n \geq 2$ over K among all such elements of F. Let $f(x)$ be the irreducible polynomial of a over K. Then $f(x)$ has no roots in the subfield L of limit points of K. For, if $b \in L$ were a root of $f(x)$, then $f(x)(x-b)^{-1} \in L[x]$ has the root a, and by Theorem 6.3.14 there would exist $g(x) \in K[x]$ of degree $n-1$ with a root c in $N_{\varepsilon/2}(a)$. This contradicts the minimality of n. Suppose $a_1 < a_2 < \cdots < a_r$ are all of the roots of $f(x)$ in F. We may assume $N_\varepsilon(a_i) \cap K = \emptyset$ for every i and also $N_\varepsilon(a_i) \cap N_\varepsilon(a_j) = \emptyset$ for $i \neq j$. So $a_1 < a_1 + \frac{\varepsilon}{2} < a_1 + \varepsilon < a_2 - \varepsilon < \cdots < a_r < a_r + \frac{\varepsilon}{2} < a_r + \varepsilon$. Let $g(x)$ be the irreducible polynomial of $a + \frac{\varepsilon}{2}$ over K. Then $g(x + \frac{\varepsilon}{2}) = f(x)$ and $a_1 + \frac{\varepsilon}{2} < \cdots < a_r + \frac{\varepsilon}{2}$ are the roots of $g(x)$ in F. We claim that $f(\alpha)g(\alpha) > 0$ for each $\alpha \in K$. Since neither $f(x)$ nor $g(x)$ has a root $< a_1$ each has constant sign in $\{b \in F : b < a_1\}$ equal to the sign of b^n. Similarly, $f(b)g(b) > 0$ if $b > a_r$. Now, $f(x)$ and $g(x)$ change their signs once in the intervals $(a_1 - \varepsilon, a_1 + \varepsilon)$ and $(a_2 - \varepsilon, a_2 + \varepsilon)$ and have constant sign in $[a_1 + \varepsilon, a_2 - \varepsilon]$. Since no elements of K lie in $N_\varepsilon(a_1) \cup N_\varepsilon(a_2)$ we have $f(\alpha)g(\alpha) > 0$ for all $\alpha \in K$ with $\alpha < a_2 + \varepsilon$. Continuing we get $f(\alpha)g(\alpha) > 0$ for all $\alpha \in K$ with $\alpha < a_r$ and the claim has been established. Thus, $f(x)g(x)$ is P.S.D. on K but not on F since $f(b)g(b) < 0$ for $b \in (a_1, a_1 + \frac{\varepsilon}{2})$.

Since the implications (c) $\Rightarrow$ (d) and (d) $\Rightarrow$ (e) are trivial the proof is complete. □

Before we give the main characterization of Hilbert fields we will show that polynomials suffice to represent a polynomial in one variable as a sum of squares.

Theorem 6.3.16. *If K is a totally ordered field, then $K[x] \cap S(K(x)) = S(K[x])$.*

Proof. Suppose $f(x) \in K[x]$ has degree $n \geq 1$ and is a sum of squares in $K(x)$. By induction we will assume that all polynomials in $S(K(x))$ of degree $< n$ are in $S(K[x])$. If there exists a polynomial $g(x)$ of positive degree with $f(x) = g(x)^2 h(x)$, then $h(x)$, and hence also $f(x)$, is a sum of squares in $K[x]$. Thus, we will assume $f(x)$ is square free. Among all equations in $K[x]$ of the form $g(x)^2 f(x) = \Sigma_j f_j(x)^2$ with $g(x) \neq 0$ take one for which $g(x)$ has minimal degree. Write $f_j(x) = q_j(x)f(x) + h_j(x)$ with $\deg h_j(x) < n$ for each j. Then

$$\begin{aligned} g(x)^2 f(x) &= \sum_j (q_j(x)^2 f(x)^2 + 2q_j(x)h_j(x)f(x) + h_j(x)^2) \\ &= f(x)q(x) + \sum_j h_j(x)^2. \end{aligned}$$

If every $h_j(x) = 0$, then $g(x)^2 = f(x)\Sigma_j q_j(x)^2$ and hence $f(x)$ is a factor of $g(x)$ since it is square free. But the equation $(g(x)f(x)^{-1})^2 f(x) = \Sigma_j q_j(x)^2$ gives a contradiction since $g(x)f(x)^{-1}$ has lower degree than $g(x)$. Thus we have a polynomial $h(x) = g(x)^2 - q(x)$ with

$$h(x)f(x) = \sum_j h_j(x)^2 \neq 0, \text{ deg } h_j(x) \leq n-1. \tag{6.3.10}$$

Among all equations of the form (6.3.10) take one in which the degree of $h(x)$ is minimal. Since $\deg h(x) \leq n-2$ and $h(x) = f(x)^{-1}\Sigma_j h_j(x)^2 \in S(K(x))$, we have $h(x) \in S(K[x])$. If $h(x) \in K$ we are done; otherwise, $\deg h(x) = m \geq 1$. Write $h_j(x) = p_j(x)h(x) + s_j(x)$ with $\deg s_j(x) \leq m-1$ for each j. If every $s_j(x) = 0$, then from (6.3.10) we have $h(x)f(x) = h(x)^2\Sigma_j p_j(x)^2$ and $f(x) = h(x)\Sigma_j p_j(x)^2 \in S(K[x])$. On the other hand, if some $s_j(x) \neq 0$, then

$$\begin{aligned} \sum_j s_j(x)^2 &= \sum_j (h_j(x) - p_j(x)h(x))^2 = \sum_j h_j(x)^2 + h(x)q(x) \\ &= h(x)(f(x) + q(x)) \end{aligned}$$

and

$$0 \neq \sum_j s_j(x)^2 = h(x)p(x) \text{ with } \deg p(x) \leq m-1. \tag{6.3.11}$$

From (6.3.10) and (6.3.11) and the equation

$$\left(\sum_{j=1}^t u_j^2\right)\left(\sum_{j=1}^t v_j^2\right) = \left(\sum_{j=1}^t u_j v_j\right)^2 + \sum_{1\leq j<k\leq t} (u_j v_k - v_k u_j)^2 \tag{6.3.12}$$

we get

$$h(x)^2 p(x) f(x) = \left(\sum_j h_j(x)s_j(x)\right)^2 + \sum_{j<k} (h_j(x)s_k(x) - h_k(x)s_j(x))^2.$$

But modulo $h(x)$ we have

$$\sum_j h_j(x)s_j(x) \equiv \sum_j h_j(x)^2 \equiv 0$$

and $h_j(x)s_k(x) - h_k(x)s_j(x) \equiv 0$; so

$$h(x)^2 p(x) f(x) = h(x)^2 u(x)^2 + \sum_i h(x)^2 v_i(x)^2$$

and $p(x)f(x) = u(x)^2 + \Sigma_i v_i(x)^2$. This equation gives a contradiction since $\deg p(x) < \deg h(x)$. □

Note that (6.3.12) and Exercise 5.1.8 show that a P.S.D. polynomial in one variable over a real closed field is a sum of two squares.

Theorem 6.3.17. *The following statements are equivalent for the totally ordered field K with real closure F.*

(a) K is a Hilbert field.
(b) K has a unique total order and K is dense in F.
(c) If $r(x) \in K[x]$ is a P.S.D. polynomial on K, then $r(x)$ is a sum of squares of polynomials in $K[x]$.
(d) If $r(x) \in K[x]$ is P.S.D. on K, then $r(x)$ is P.S.D. on F and K has a unique total order.

Proof. That (a) implies (b) is a consequence of Theorem 6.3.15 and that (b) implies (a) follows from Theorem 6.3.15 and the fact that $K^+S(K(x_1,\dots,x_n)) = S(K(x_1,\dots,x_n))$ provided K has a unique total order (Exercise 5.1.17). That (a) implies (c) is a consequence of Theorem 6.3.16, that (c) implies (d) is obvious, and that (d) implies (b) also follows from Theorem 6.3.15. □

The two conditions in (b) of the previous theorem are independent. The field $\mathbb{Q}(\sqrt{2})$ is dense in its real closure but has two total orders. An example of a field with a unique total order which is not dense in its real closure and of a P.S.D. polynomial that is not a sum of squares is given in Exercise 14.

We conclude this section with a description of the free formally real f-rings which is analogous to the description of the free f-modules given in Section 4.5. But first we note the following.

Theorem 6.3.18. *Let C be a totally ordered subring of the real closed field K. Suppose A is the free formally K-real C-f-algebra on X and B is the free commutative C-f-algebra on X. Then $A = B/c\ell(\beta(B))$.*

Proof. By Exercise 1.4.20 A is the C-ℓ-subalgebra of $K^{(K^X)}$ generated by the projections $\pi_x : K^X \longrightarrow K$, $x \in X$; hence A is semiprime and C-torsion-free and if $\varphi : B \longrightarrow A$ is the homomorphism which extends the identity on X, then $c\ell(\beta(B)) \subseteq \ker\varphi$. Now, $B/c\ell(\beta(B))$ is semiprime since $\alpha b^2 \in \beta(B)$ with $0 \neq \alpha \in C$ and $b \in B$ gives $(\alpha b)^{2n} = 0$ for some n and hence $b \in c\ell(\beta(B))$. By Theorem 6.3.12 $B/c\ell(\beta(B))$ is K-formally real and therefore there is a homomorphism ψ determined by $\psi(x) = \eta(x)$ and which makes the diagram

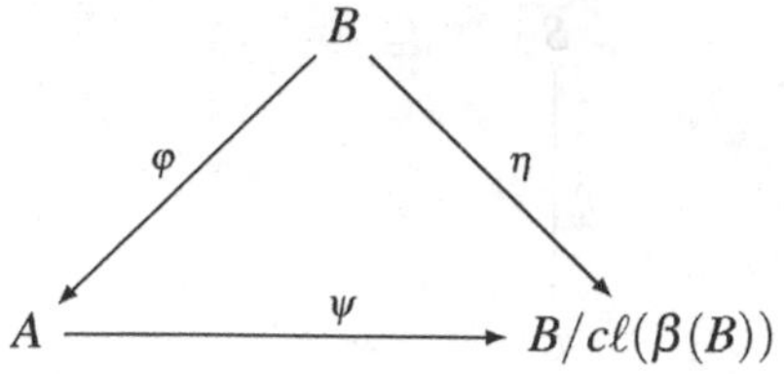

commutative, where η is the natural map; thus, $\ker\varphi \subseteq c\ell(\beta(B))$ and ψ is an isomorphism. □

Theorem 6.3.19. *Let C be a subring of $\mathbb{R}$ with its usual total order and let X be a set of commuting indeterminates over C. Let*

$$\mathscr{S} = \{P : (C[X]_0, P) \text{ is a totally ordered } C\text{-}\ell\text{-algebra}\}$$

and let $\varphi : C[X]_0 \longrightarrow \Pi_{\mathscr{S}}(C[X]_0, P)$ be the diagonal embedding. Then the C-ℓ-subalgebra A of the product generated by $\varphi(C[X]_0)$ is the free formally $\mathbb{R}$-real C-f-algebra on X.

Proof. We may assume X is a set of indeterminates over $\mathbb{R}$. Suppose $\psi : C[X]_0 \longrightarrow R$ is a homomorphism of C-algebras where R is a formally $\mathbb{R}$-real C-f-algebra. We need to lift ψ to an ℓ-algebra homomorphism of A into R, and as in the proofs of Theorems 4.5.1 and 4.5.3 we may assume that R is totally ordered. Thus, it suffices to show that there is a $P \in \mathscr{S}$ with respect to which ψ is isotone. For, given such a P we have the commutative diagram

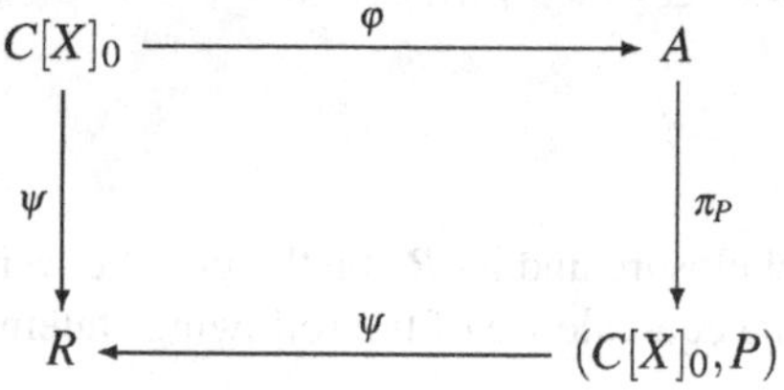

where π_P is the projection onto $(C[X]_0, P)$. Let

$$P' = \{\Sigma_i \alpha_i f_i^2 g_{i1} \cdots g_{im_i} : 0 < \alpha_i \in C, m_i \in \mathbb{Z}^+, f_i, g_{ij} \in C[X]_0^* \\ \text{and } \psi(g_{ij}) > 0\}.$$

Then P' is a subsemiring of $C[X]_0$ with $C^{+*}P' \subseteq P'$, and in order to show that $P' \cup \{0\}$ is a partial order we need to verify that $P' \cap -P' = \emptyset$, or, equivalently, that $0 \notin P'$. Suppose $f = \Sigma_i \alpha_i f_i^2 g_{i1} \cdots g_{im_i}$ is an element of P' with $f_i, g_{ij} \in C[x_1, \ldots, x_n]_0$ for all i and j. Since R is a formally $\mathbb{R}$-real C-ℓ-algebra there is an epimorphism of C-ℓ-algebras

$$\begin{array}{ccc} S & \subseteq & \mathbb{R}^{\Lambda} \\ \Big\downarrow{\scriptstyle\rho} & & \\ R & & \end{array}$$

Take $a_k \in S$ with $\rho(a_k) = \psi(x_k)$ for $k = 1, \ldots, n$. Then

$$\begin{aligned} \rho\Big(\bigwedge_{i,j} g_{ij}(a_1, \ldots, a_n)\Big) &= \bigwedge_{i,j} g_{ij}(\psi(x_1), \ldots, \psi(x_n)) \\ &= \bigwedge_{i,j} \psi(g_{ij}(x_1, \ldots, x_n)) > 0 \end{aligned}$$

and hence, for some $\lambda \in \Lambda$, $g_{ij}(a_1(\lambda), \ldots, a_n(\lambda)) > 0$ for every i and j. So there is a nonempty open subset U of $\mathbb{R}^n$ such that $g_{ij}(\gamma_1, \ldots, \gamma_n) > 0$ for $(\gamma_1, \ldots, \gamma_n) \in U$ and for every i and j. Let $Z(f_i) = \{(\gamma_1, \ldots, \gamma_n) \in \mathbb{R}^n : f_i(\gamma_1, \ldots, \gamma_n) = 0\}$. By Exercise 10 $Z(f_i)$ is a closed nowhere dense subset of $\mathbb{R}^n$ and hence $W = \mathbb{R}^n \backslash \bigcup_i Z(f_i^2)$ is an open dense subset of $\mathbb{R}^n$. So

$$f(\gamma_1, \ldots, \gamma_n) = \Sigma_i \alpha_i f_i^2(\gamma_1, \ldots, \gamma_n) g_{i1}(\gamma_1, \ldots, \gamma_n) \cdots g_{im_i}(\gamma_1, \ldots, \gamma_n) > 0$$

for all $(\gamma_1, \ldots, \gamma_n) \in U \cap W$ and $0 \notin P'$. Now, the quotient partial order $Q(P')$ is an sp-partial order for the quotient field $Q(C[X]_0)$ of $C[X]_0$ and hence P' is contained in a total order P of $C[X]_0$ by Theorem 5.1.1. If $g \in P$ and $\psi(g) < 0$, then we have the contradiction $-g^3 \in P' \subseteq P$. So $\psi(g) \geq 0$ and ψ is isotone relative to P. □

Exercises.

1. Let A be the central closure and let R_u be the unital cover of the totally ordered domain R. Verify the equivalence of the following statements.

   (a) R is co-ℓ-simple.
   (b) For each c in R^* there exists x in R^* with $|cx| < 1$.
   (c) For each c in R^* there exists x in R^* with $|xc| < 1$.
   (d) For all c and d in R^* there exists x in R^* with $|xc| < |d|$.
   (e) For each c in R^* there exist x and y in R^* with $|xcy| < 1$.
   (f) For all c and d in R^* there exist x and y in R^* with $|xcy| < |d|$.
   (g) For all $p_1, \ldots, p_m \in \mathscr{P}$ and for all $u, v \in A^*$ there exists x in R^* with $|uxv| > \Sigma_{i=1}^m |p_i(x)|$. (See Theorem 6.3.3.)
   (h) For all $u, v, b \in A^*$ there exists $x \in R^*$ with $|x| < 1$ and $|uxv| > |bxbxb|$.
   (i) For all $0 < a < b$ in R there exists $x \in R^*$ with $0 < x < 1$ and $(bx)^3 < (ax)^2$.
   (j) Each one-sided ideal of R is co-ℓ-simple.
   (k) R_u is co-ℓ-simple.

(l) For all a and b in R^*, aRb is co-ℓ-simple.
(m) The positive cone of each one-sided ideal of R has no lower bound in R^+ except 0.
(n) If S is a right (respectively, left) f-ring extension of R and S is a right (respectively, left) quotient ring of R, then S is co-ℓ-simple.

2. Show that an ultraproduct of a family of co-ℓ-simple totally ordered domains is co-ℓ-simple.

3. Let $f(x_1,\ldots,x_n)$ be a P.S.D. form over the semiprime f-ring R. Suppose $f(x_1,\ldots,x_n)$ has a regular coefficient and a solution consisting of regular elements.

   (a) If for every a, $b \in R$ there exists $d \in R$ with $|ad| \leq |b|$ show that R is commutative.
   (b) If R is infinitesimal show that R is commutative.

4. If the inequality $2xy \leq x^2 + y^2$ holds in the po-ring R show that the inequalities $\pm[x,y] \leq x^2, y^2$ also hold.

5. Let $\sigma : \mathbb{Q}[y] \longrightarrow \mathbb{Q}[y]$ be the automorphism of $K = \mathbb{Q}[y]$ determined by $\sigma(y) = \frac{1}{2}y$ and let $R = K[x;\sigma]$ be the skew polynomial ring with the scalars from K on the left; so $xy = \frac{1}{2}yx$. Give both K and R the lexicographic order; so $1 \ll y \ll y^2 \ll \cdots$ and $1 \ll_K x \ll_K \cdots$. Show that R is ℓ-simple, $f(x_1,x_2) = y^2x_1^2 - 2yx_1x_2 + x_2^2$ is a P.S.D. form over R with a solution in R^*, and R has bounded commutators.

6. Let $C[x,y]$ be the free unital algebra in two indeterminates over the totally ordered domain C, and let $R = C[x,y]/I$ where I is the ideal of $C[x,y]$ generated by all monomials of degree 3.

   (a) Show that R is a free unital C-module with basis $e_1 = 1$, $e_2 = \overline{x}$, $e_3 = \overline{y}$, $e_4 = \overline{x}^2$, $e_5 = \overline{xy}$, $e_6 = \overline{y}^2$, $e_7 = \overline{xy - yx}$.
   (b) Show that R is a totally ordered ring with the total order $Ce_1 \overrightarrow{\boxplus} \cdots \overrightarrow{\boxplus} Ce_7$.
   (c) Show that for any $f, g \in R$, $2fg \leq f^2 + g^2$ and $|[f,g]| \leq f^2 \wedge g^2$.

7. Suppose R is a unital ℓ-semiprime ℓ-ring which satisfies $2xy \leq x^2 + y^2$. Suppose also that in each ℓ-prime homomorphic image of R, $y \geq 0$ whenever $x \geq 1$ and $xy \geq 0$. Show that R is commutative. Give an example of such an ℓ-ring which is not an f-ring. (For $x \geq 1$, use $x^2[x,y] = [x, x^2y]$ to show $[x,y] \in F$, and use $[x^2,y] = 2x[x,y]$ to show $x \in Z(R)$.)

8. Let G be a rooted po-group, C a totally ordered domain and $C[G] = \Sigma(G,C)$ the group ℓ-algebra with the Hahn product order.

   (a) If G is trivially ordered and for all a, $b \in G$ there exists $p(x) \in C[x]$ with $|[a,b]| \leq |p(a)|$ show that G is abelian.
   (b) Suppose that for every r, $s \in C[G]^+$ there are polynomials $p(x)$, $q(x) \in C[x]x^2$ with $|[r,s]| \leq |p(r)| \wedge |q(s)|$. Show that G is abelian.

9. (a) In (a) of Theorem 6.3.3 assume R is ℓ-simple instead of co-ℓ-simple. Show that $a_s > 0$ and s is even.
   (b) Give examples to show that the conclusions in (a) need not hold when R is co-ℓ-simple and neither do those in Theorem 6.3.3(a) when R is ℓ-simple.

10. Suppose $\alpha_j < \beta_j$ for $j = 1,\ldots,n$ in the totally ordered field K and $f(x_1,\ldots,x_n) \in K[x_1,\ldots,x_n]$ with $f([\alpha_1,\beta_1]\times\cdots\times[\alpha_n,\beta_n]) = 0$. Show that $f(x_1,\ldots,x_n) = 0$. (Use induction on n.)

11. (a) Show that $\mathbb{Z}$ is the largest totally ordered subring of $\mathbb{R}$ which satisfies the identity $x^2 \vee x = x^2$.
   (b) Suppose $f(x_1,\ldots,x_n) \in \mathbb{Z}[x_1,\ldots,x_n]$, and for some $m \geq 1$, $f(\alpha_1,\ldots,\alpha_n) = 0$ implies $\alpha_1 = \cdots = \alpha_m = 0$ for all $(\alpha_1,\ldots,\alpha_n) \in \mathbb{Z}^n$. Show that

$$[x_1 \wedge x_2 \wedge \cdots \wedge x_m \wedge (z + f(x_1,\ldots,x_n)z) \wedge (z - f(x_1,\ldots,x_n)z)]^+ = 0$$

   is an identity which holds in $\mathbb{Z}$.
   (c) Give some examples of polynomials of this type in two variables which give an ℓ-ring identity for $\mathbb{Z}$ but not for $\mathbb{R}$.

12. Let $K = E(x)$ be the rational function field in one variable over the totally ordered field E. Show that if K has either the lexicographic or the antilexicographic total order, then $(\sqrt{x}, 2\sqrt{x}) \cap K = \emptyset$ and K is not dense in its real closure.

13. Let F be a real closed field and let $f(x,y) = 1 + x^2y^4 + x^4y^2 - 3x^2y^2 \in F[x,y]$. Show that:

$$f(x,y) = \frac{(1 - x^2y^2)^2 + x^2(1-y^2)^2 + x^2(1-x^2)^2y^2}{1+x^2}$$

   is a sum of six squares in $F(x)[y]$ but is not a sum of squares in $F[x,y]$.

14. Let $K = \mathbb{Q}(y)$ have the antilexicographic total order, let F be the real closure of K and let M be the intermediate field with a unique total order constructed in Exercise 5.1.15. If v is the natural valuation on F, then $v(F) = \mathbb{Q}$, $v(K) = \mathbb{Z}$ and $v(y) = 1$ by Exercise 5.2.2. Let $f(x) = (x^3 - y)^2 - y^3 \in F[x]$.
   (a) If $\alpha \in F$ and $f(\alpha) < 0$ show that $v(\alpha^3) = 1$. (Rule out all of the other possibilities.)
   (b) If $\alpha \in M$ show that $2^n v(\alpha) \in \mathbb{Z}$ for some $n \in \mathbb{N}$. (Use Exercise 5.2.14.)
   (c) Show that $f(x)$ is P.S.D. on M but is not a sum of squares in $F(x)$. (Verify $f(1)f(y^{\frac{1}{3}}) < 0$.)

15. Suppose $K \subseteq D$ are division rings and D is totally ordered. Show that K is cofinal in D iff K^{+*} is coinitial in D^{+*}, iff D_K is archimedean, iff ${}_KD$ is archimedean. If D is a field which is algebraic over K show that K is cofinal in D. (See the proof of Theorem 5.1.4.)

16. Suppose F is the real closure of the totally ordered field K and L is the set of limit points of K in F where F has the interval topology. Show that L is a subfield of F which contains K.

17. Suppose K is a field that can be totally ordered and, for each total order P of K, (K,P) is dense in its real closure. (For example, K is a real algebraic number field; see Exercise 5.1.16.) Suppose also that $r = fg^{-1} \in K(x_1,\ldots,x_n)$ has the property that for every total order P of K and for all $\alpha_1,\ldots,\alpha_n \in K$, $f(\alpha_1,\ldots,\alpha_n)g(\alpha_1,\ldots,\alpha_n) \in P$. Show that r is a sum of squares in $K(x_1,\ldots,x_n)$.

18. State and prove Theorem 6.3.19 for the category of unital formally $\mathbb{R}$-real C-f-algebras.

Notes. Leung [LE1] showed that a totally ordered division ring which has a P.S.D. form with a nontrivial solution is commutative and Steinberg [ST16] used Leung's methods to extend this result to a co-ℓ-simple totally ordered domain. Theorem 6.3.11 and Exercises 5 and 6 also come from Leung [LE1] and the other results on bounded commutators in an ℓ-ring as well as Theorem 6.3.7 and Exercise 7 come from Steinberg [ST17]; Exercise 8 is in Steinberg [ST14]. Theorem 6.3.2 is due to Martindale [MAR] and Theorem 6.3.6 is due to Herstein [HER1] or [HER2]. The results on formally real f-rings come from Henriksen and Isbell [HI] and our presentation is based on that given by Weinberg [WE6]. The fact that a real closed field or an archimedean totally ordered field with a unique total order is a Hilbert field constitutes Artin's [AR1] (or [LT]) solution to Hilbert's 17th problem, and the generalization to fields with a unique total order which are dense in their real closure that is given in Theorem 6.3.17 appears in Jacobson [J3]; also see p. 295 of Jacobson [J2]. The converse and the other equivalences in Theorem 6.3.17 as well as Theorems 6.3.14 and 6.3.15 are due to McKenna [MC]. The use of f-rings in the presentation of Artin's solution, which is motivated by Henriksen and Isbell [HI], appears to be new; see Jacobson [J3], Pfister [PF], and Prestel and Delzell [PD]. Theorem 6.3.16 comes from Artin [AR1]; Exercise 13 which gives a P.S.D. polynomial that is not a sum of squares of polynomials comes from Motzkin [MO]; other polynomials of this kind are given in Robinson [RO]. Exercise 14 is due to Dubois [DU].

6.4 Lattice-ordered Matrix Algebras

The usual lattice order for the matrix algebra K_n over the totally ordered field K is $(K^+)_n$. The first main conjecture for these matrix algebras is that the usual lattice order is, up to an isomorphism, the only lattice order in which 1 is positive. The verification of this conjecture when $n = 2$ or when K is archimedean is given here, and all of the lattice orders of K_n are determined for these two cases. The second main conjecture is that the lattice orders that are described for K_n when K is archimedean also give all of the lattice orders for any K_n. To establish the first main conjecture for $n \geq 3$ we will make use of some of the topological properties of Euclidean space. In fact, the conjecture will be established more generally for subalgebras of K_n which are $\mathbb{R}$-irreducible. If such an algebra can be made into an ℓ-algebra it will have to be all of K_n. As a consequence one gets a uniform verification of the fact that matrix algebras over the complex numbers and over quaternion algebras are not ℓ-algebras. The first main conjecture will also be established in several special cases: if $F(K_n)$ has a set of n disjoint elements, or if it is never one-dimensional, or if a positive element in K_n is similar to a permutation matrix determined by an n-cycle. The techniques used for this last case will be embellished so as to describe canonically ordered matrix algebras over group algebras of a finite cyclic group.

We will repeatedly use Exercises 3.2.37 and 2.5.22. Explicitly, a finite dimensional ℓ-semiprime ℓ-algebra R over a totally ordered field K is a direct sum of totally ordered ℓ-simple subspaces. Moreover, $F(R)$ is reduced by Theorem 3.2.15 (or Exercise 3.2.37) and it is the direct sum of some of the totally ordered summands of R. Each of the summands of $F(R)$ is a totally ordered field extension of K by Theorems 4.2.13 and 6.2.2. When R is the matrix algebra K_n, $F(R)$ has at most n summands since n is an upper bound for the size of an orthogonal set of idempotents in R by Exercise 4.2.34; other properties of idempotents in K_n that are given in this exercise will also be used. Each of the convex ℓ-subalgebras $F_\ell(R)$ and $F_r(R)$ is also a sum of some of the totally ordered summands of R. Since a unital one-sided f-ring is an almost f-ring by Theorem 3.8.9 and its nilpotent elements are bounded by 1 according to Theorem 3.8.1, it is easy to see that $F_\ell(R)$ (respectively, $F_r(R)$) is an f-algebra itself precisely when it is unital or reduced.

Theorem 6.4.1. *Let K be a totally ordered field.*

(a) For each $1 < \beta \in K$ there is a lattice order P_β of K_2 for which $R = (K_2, P_\beta)$ is an ℓ-algebra, and there are four idempotents f_1, f_2, f_3, f_4 in P_β which form a canonical K-basis for R and such that $1 = (1-\beta)(f_1+f_2)+\beta(f_3+f_4)$.

(b) There are idempotents g_1, g_2 and n in K_2 and a nilpotent element g_3 such that $1 = g_1+g_2-g_3$, and if P_1 is the positive cone of the vector lattice which has $\{g_1, g_2, g_3, n\}$ as a canonical K-basis, then (K_2, P_1) is an ℓ-algebra over K.

(c) If $R = (K_2, P)$ is an ℓ-algebra over K, then R is isomorphic to $(K_2, (K^+)_2)$ if $1 \in P$, and, otherwise, R is isomorphic to (K_2, P_β) for exactly one $\beta \geq 1$.

Proof. For (a), let

$$f_1 = \begin{pmatrix} 1 & 0 \\ 0 & 0 \end{pmatrix}, \qquad f_2 = \begin{pmatrix} \beta(\beta-1)^{-1} & 1 \\ -\beta(\beta-1)^{-2} & -(\beta-1)^{-1} \end{pmatrix},$$

$$f_3 = \begin{pmatrix} 1 & (\beta-1)\beta^{-1} \\ 0 & 0 \end{pmatrix}, \qquad f_4 = \begin{pmatrix} 1 & 0 \\ (1-\beta)^{-1} & 0 \end{pmatrix}, \tag{6.4.1}$$

and note that these idempotents are K-independent and satisfy the following multiplication table.

	f_1	f_2	f_3	f_4
f_1	f_1	$\beta(\beta-1)^{-1}f_3$	f_3	f_1
f_2	$\beta(\beta-1)^{-1}f_4$	f_2	f_2	f_4
f_3	f_1	f_3	f_3	$(\beta-1)\beta^{-1}f_1$
f_4	f_4	f_2	$(\beta-1)\beta^{-1}f_2$	f_4

Let P_β be the lattice order of K_2 which has these idempotents as a canonical K-basis. For (b), let

$$g_1 = \begin{pmatrix} 0 & -1 \\ 0 & 1 \end{pmatrix}, \; g_2 = \begin{pmatrix} 1 & 0 \\ 0 & 0 \end{pmatrix},$$

$$g_3 = \begin{pmatrix} 0 & -1 \\ 0 & 0 \end{pmatrix}, \; n = \begin{pmatrix} 1 & 0 \\ -1 & 0 \end{pmatrix}, \tag{6.4.2}$$

and note that these matrices are K-independent and satisfy the multiplication table given below.

	g_1	g_2	g_3	n
g_1	g_1	0	0	n
g_2	g_3	g_2	g_3	g_2
g_3	g_3	0	0	g_2
n	g_1	n	g_1	n

Let P_1 be the lattice order of K_2 which has $\{g_1, g_2, g_3, n\}$ as a canonical K-basis.

To prove (c) suppose $R = (K_2, P)$ is an ℓ-algebra over K. Then R is the vector lattice direct sum of at most four totally ordered K-archimedean subspaces and certainly there are at least two such summands. The proof will proceed by considering the 19 cases that arise depending on the number of summands, the dimensions of the summands, and the nature of the components of 1 in these summands. Only three of the cases lead to lattice orders. We first point out a few properties of K_2 that will be used in the proof. According to the Cayley–Hamilton theorem each $a \in K_2$ satisfies the equation $a^2 - (tr\,a)a + \det a = 0$ where, of course, $tr\,a$ is the trace of a and $\det a$ is its determinant; in particular, $a^2 \in Ka$ if a is not a unit. If e and f are nonzero orthogonal idempotents, then $1 = e + f$, $eK_2e \cong K$ and eK_2 (respectively, K_2e) is a minimal right (respectively, left) ideal. Each proper nonzero one-sided ideal is minimal and is a two-dimensional subspace, and a nilpotent subalgebra is at most one-dimensional. If A and B are totally ordered subspaces of R and C is a convex subspace, then $AB \subseteq C$ whenever there exist $0 \neq a \in A$ and $0 \neq b \in B$ with $ab \in C$. Also, R does not have a totally ordered subalgebra whose dimension exceeds one and which does not contain 1. For such a subalgebra would be a finite dimensional domain over K and hence a division ring with an identity element $e \neq 1$, and would be contained in the one-dimensional subalgebra eRe. All references to summands, idempotents, and the like tacitly assumes they are nonzero.

Under the assumption $1 > 0$, which we now make, three main cases arise depending upon whether R has four, three, or two totally ordered summands. In all of these cases 1 has at most two nonzero components since the components form an orthogonal set of idempotents. Let $F = F(R)$.

(I) Suppose $R = T_1 \oplus T_2 \oplus T_3 \oplus T_4$ where each T_i is totally ordered.

(Ia) Assume $1 = g_1 + g_2$ with $0 \neq g_i \in T_i$. Then $F = T_1 \oplus T_2$ as f-rings where $T_i = Fg_i = Kg_i = g_iF \cong K$ and $A = T_3 \oplus T_4$ is a unital f-bimodule over F. Now, $Ag_1 \neq 0$ or else $Rg_2 + A \subseteq \ell(g_1;R)$; similarly, Ag_2 and g_iA are nonzero and hence $\{Ag_1, Ag_2\} = \{g_1A, g_2A\}$ since $A = Ag_1 \oplus Ag_2 = g_1A \oplus g_2A$. Let $Ag_1 = Ka$ and $Ag_2 = Kb$ with $0 < a, b$. If $Ag_1 = g_1A$, then $Ag_2 = g_2A$ and $a^2 \in Ka$ since $g_2a = 0$; also, $ab = ag_1b = 0$ and hence Ka is a right ideal of R. So, $Ag_1 = g_2A$, $Ag_2 = g_1A$, and hence $(Ag_i)^2 = 0$. If $ab = 0$, then $aR = a(Kg_1 + Kg_2 + Ka + Kb) = Ka$; so $ab \neq 0$ and $ab = \alpha g_1 + \gamma g_2 + \delta a + \rho b$ with α, γ, δ, $\rho \in K^+$. Now, $0 = a^2b = \alpha a + \rho ab$ yields $\alpha = \rho = 0$, and hence $ab = abg_2 = \gamma g_2$ with $\gamma > 0$. Similarly, $ba = \alpha g_1$ with $\alpha > 0$. By replacing a with $\gamma^{-1}a$ we have $ab = g_2$ and $\alpha = 1$ since $a = g_2a = aba = \alpha a g_1 = \alpha a$. Let $g_{11} = g_1$, $g_{22} = g_2$, $g_{12} = b$ and $g_{21} = a$. Then $\{g_{ij} : 1 \leq i, j \leq 2\}$ is a set of matrix units and $R = Kg_{11} \oplus Kg_{12} \oplus Kg_{21} \oplus Kg_{22} \cong (K_2, (K^+)_2)$.

(Ib) The other possibility is that $1 \in T_4$ and $F = K = T_4$. Let $0 < x_i \in T_i$ for $i = 1,2,3$. For $1 \leq i,\ j \leq 3$, $K + T_i$ and $K + T_i + T_j$ are convex ℓ-subalgebras since $0 \leq x_ix_j \leq (x_i + x_j)^2 \in K + K(x_i + x_j)$. We will now consider three subcases which arise from the possible number of idempotents in $T_1 \cup T_2 \cup T_3$.

(Ib1) Suppose $T_1 \cup T_2 \cup T_3$ contains no idempotents. If x_1 is a unit, then $x_1^2 = \alpha + \gamma x_1$ with $\alpha > 0$, $\gamma \in K^+$ and $x_1^{-1} = \alpha^{-1}(x_1 - \gamma) \in T_1 + K$. Now, $x_1x_2 = \rho + \sigma x_1 + \tau x_2$ with ρ, σ, $\tau \in K^+$ and $\tau > 0$ since otherwise $x_2 = \rho x_1^{-1} + \sigma \in T_1 + K$. We have

$$\begin{aligned}\alpha x_2 + \gamma x_1x_2 = x_1^2x_2 = \rho x_1 + \sigma x_1^2 + \tau x_1x_2 &= \rho x_1 + \sigma(\alpha + \gamma x_1) + \tau x_1x_2 \\ &= \sigma\alpha + (\rho + \sigma\gamma)x_1 + \tau x_1x_2.\end{aligned}$$

So $(\gamma - \tau)x_1x_2 = \sigma\alpha + (\rho + \sigma\gamma)x_1 - \alpha x_2 \in K \oplus T_1 \oplus T_2$. Since $(\gamma - \tau)x_1x_2$ is comparable to 0 and $-\alpha < 0$ we have $\sigma\alpha \leq 0$ and $\rho + \sigma\gamma \leq 0$. So $\sigma = 0$, $\rho = 0$ and hence $x_1x_2 = \tau x_2$. From $x_2^2 = \eta + \lambda x_2$ we get

$$\tau\eta + \tau\lambda x_2 = \tau x_2^2 = x_1x_2^2 = \eta x_1 + \lambda x_1x_2 = \eta x_1 + \lambda\tau x_2.$$

So $\eta = 0$ and also $\lambda = 0$ since otherwise $\lambda^{-1}x_2$ is an idempotent. Thus, $x_2^2 = 0$ and, similarly, $x_3^2 = 0$. On the other hand, if none of the x_i are units, then $x_i^2 \in Kx_i = T_i$ and $x_i^2 = 0$ for $i = 1,2,3$. In both cases we have $x_2^2 = x_3^2 = 0$. From $x_2x_3 = \alpha + \gamma x_2 + \delta x_3$ with α, γ, $\delta \in K^+$ we obtain

$$0 = x_2x_3^2 = \alpha x_3 + \gamma x_2x_3 = \alpha x_3 + \gamma\alpha + \gamma^2x_2 + \gamma\delta x_3$$

and hence $\alpha = \gamma = 0$. Also,

$$0 = x_2^2x_3 = \delta x_2x_3 = \delta^2x_3$$

gives $\delta = 0$. So $x_2x_3 = 0$; by symmetry $x_3x_2 = 0$ and we have the contradiction that $T_2 \oplus T_3$ is a nilpotent subalgebra.

(Ib2) Suppose T_1 and T_2 both contain idempotents. So $T_i = Kg_i$ and $g_i = g_i^2 > 0$ for $i = 1, 2$. There exist α, γ, $\delta \in K^+$ with

$$\begin{aligned} g_1g_2 = \alpha + \gamma g_1 + \delta g_2 &= \alpha g_2 + \delta g_2 + \gamma g_1 g_2 \\ &= (\alpha + \delta)g_2 + \gamma\alpha + \gamma^2 g_1 + \gamma\delta g_2 \\ &= \gamma\alpha + \gamma^2 g_1 + (\alpha + \delta + \gamma\delta)g_2. \end{aligned}$$

So $\alpha = \gamma\alpha$, $\gamma = \gamma^2$, $\alpha + \gamma\delta = 0$, and hence $\alpha = \gamma\delta = 0$ and $\gamma = 0$ or 1. Thus, $g_1g_2 = g_1$ or $g_1g_2 = \delta g_2$, and in the latter case $g_1g_2 = \delta g_1g_2 = \delta^2 g_2$ and $\delta = 0$ or 1; the possibilities are $g_1g_2 = g_1$, or $g_1g_2 = g_2$, or $g_1g_2 = 0$. There exist ρ, σ, $\tau \in K^+$ with

$$g_1x_3 = \rho + \sigma g_1 + \tau x_3 = \rho g_1 + \sigma g_1 + \tau g_1 x_3.$$

So $\tau g_1x_3 = \rho + \tau x_3 - \rho g_1 \geq 0$ implies $\rho = 0$ and

$$g_1x_3 = \sigma g_1 + \tau g_1 x_3 = \sigma g_1 + \tau\sigma g_1 + \tau^2 x_3 = \sigma(1+\tau)g_1 + \tau^2 x_3.$$

Thus, $\sigma = \sigma(1+\tau)$ and $\tau = 0$ or 1. Consequently, $g_1x_3 = \sigma g_1$ or $g_1x_3 = x_3$, and similar equations hold for g_2x_3, x_3g_1 and x_3g_2. Now, assume $g_1g_2 = g_1$. If $g_1x_3 = \sigma g_1$ then $g_1R = T_1$; so $g_1x_3 = x_3$ and x_3 is not a unit. If $g_2x_3 = x_3$, then $Rx_3 = T_3$; so $g_2x_3 = \alpha g_2$ for some $\alpha \in K^+$ and we have the contradiction

$$\alpha g_1 = g_1(\alpha g_2) = g_1g_2x_3 = g_1x_3 = x_3.$$

So $g_1g_2 \neq g_1$. Assume $g_1g_2 = g_2$. If $x_3g_2 = \alpha g_2$, then $Rg_2 = T_2$, and if $x_3g_2 = x_3$, then either $x_3g_1 = x_3$ and $x_3R = T_3$, or $x_3g_1 = \gamma g_1$ and

$$x_3 = x_3g_2 = x_3g_1g_2 = \gamma g_1g_2 = \gamma g_2.$$

So we must have $g_1g_2 = 0$ and similarly $g_2g_1 = 0$. This gives $1 = g_1 + g_2 \in T_4 \cap (T_1 + T_2) = 0$.

(Ib3) Suppose T_1 contains the idempotent g_1 but $T_2 \cup T_3$ has no idempotents. There exist α, γ, $\delta \in K^+$ with

$$\begin{aligned} g_1x_2 = \alpha + \gamma g_1 + \delta x_2 &= \alpha g_1 + \gamma g_1 + \delta g_1 x_2 \\ &= \alpha g_1 + \gamma g_1 + \delta\alpha + \delta\gamma g_1 + \delta^2 x_2 \\ &= \delta\alpha + (\alpha + \gamma + \delta\gamma)g_1 + \delta^2 x_2. \end{aligned}$$

So $\alpha = \delta\alpha$, $\alpha + \delta\gamma = 0$ and $\delta = 0$ or 1, and hence $\alpha = 0 = \delta$, or $\alpha = 0$, $\delta = 1$ and $\gamma = 0$. Thus, $g_1x_2 = \gamma g_1$ or $g_1x_2 = x_2$. Similarly, $g_1x_3 = \rho g_1$ or $g_1x_3 = x_3$. If g_1x_2 and g_1x_3 are both in T_1, then $g_1R = T_1$, and if $g_1x_2 = x_2$ and $g_1x_3 = x_3$, then $g_1R \supseteq T_1 + T_2 + T_3$. The remaining possibilities are $g_1x_2 = \gamma g_1$ and $g_1x_3 = x_3$, or

$g_1x_2 = x_2$ and $g_1x_3 = \rho g_1$. By symmetry we will assume the latter and now there are two cases depending upon whether ρ is zero or nonzero.

If $\rho = 0$ we have $g_1x_2 = x_2$ and $g_1x_3 = 0$. Since T_3 has no idempotents $x_3^2 = 0$. From $x_2x_3 = \alpha + \sigma x_2 + \tau x_3$ with $\alpha, \sigma, \tau \in K^+$ we obtain $\alpha x_3 + \sigma x_2x_3 = 0$ and $\alpha = \sigma x_2x_3 = 0$. If $\sigma = 0$, then $x_2x_3 = \tau x_3 = g_1x_2x_3 = \tau g_1x_3 = 0$; and if $\sigma \neq 0$, then $x_2x_3 = 0$. So $Rx_3 = Kx_3$ and the case $\rho = 0$ is impossible.

If $\rho \neq 0$, then by replacing x_3 by $\rho^{-1}x_3$ we have $g_1x_2 = x_2$ and $g_1x_3 = g_1$. Calculations analogous to those in the first paragraph of this case give $x_3g_1 = x_3$ or $x_3g_1 \in K^+g_1$, and $x_2g_1 = x_2$ or $x_2g_1 \in K^+g_1$. If $x_3g_1 = x_3$, then $x_3^2 = x_3g_1x_3 = x_3g_1 = x_3$. Suppose $x_3g_1 = \tau g_1$; then $g_1 = (g_1x_3)^2 = g_1\tau g_1x_3 = \tau g_1$, $\tau = 1$ and g_1 commutes with x_3. If $x_2g_1 = x_2$, then g_1 also commutes with x_2 and we have the contradiction $g_1 \in Z(R) = T_4$. The remaining possibility is that $x_2g_1 \in K^+g_1$. But then $Rg_1 = T_1$.

(II) Suppose $R = T_1 \oplus T_2 \oplus T_3$ has three totally ordered summands. We assume T_1 is the two-dimensional summand.

(IIa) Suppose $1 = g_1 + g_2 \in T_1 + T_2$. Then $F = T_1 \oplus T_2$ is the direct product of the totally ordered fields T_1 and T_2 and T_3 is a unital F-bimodule. If $g_2T_3 = 0$, then $T_1 + T_3 \subseteq r(g_2;R)$. So $g_2T_3 = T_3$ and also $T_3g_2 = T_3$. But now we have the contradiction $T_2 \oplus T_3 \subseteq g_2Rg_2$.

(IIb) Suppose $1 = g_2 + g_3 \in T_2 \oplus T_3$. Then $F = T_2 \oplus T_3$ as f-rings and T_1 is a unital f-bimodule over F. As in the previous case $T_1 = g_2T_1 = g_3T_1$ and $T_1 = g_2g_3T_1 = 0$.

(IIc) Suppose $1 \in T_1$. Then T_2 is a vector lattice over $T_1 = F$ and hence is at least two-dimensional over K.

(IId) Suppose $1 \in T_3$. Then $T_3 = K$ and since T_1 is not a subalgebra of R its nonzero elements are units in R. Take $x_1, x_2 \in T_1^+$ and $x \in T_2^+$ with $T_1 = Kx_1 \oplus Kx_2$ and $T_2 = Kx$. We will show that $T_1x \subseteq T_2$; this is impossible since T_1x is two-dimensional. Note that $T_1 + K$ is a subalgebra. For, $0 \leq x_1x_2 \leq (x_1 + x_2)^2 \in K + K(x_1 + x_2) \subseteq T_1 + K$. Now, $x_1^2 = \alpha + \gamma x_1$ with $\alpha > 0$ and $\gamma \in K^+$ and

$$0 < x_1x = \delta + (\rho x_1 + \tau x_2) + \sigma x$$

with $\delta, \rho, \tau, \sigma \in K^+$. So,

$$\alpha x + \gamma x_1x = x_1^2x = \delta x_1 + x_1(\rho x_1 + \tau x_2) + \sigma x_1x$$

and

$$(\gamma - \sigma)x_1x = x_1(\delta + \rho x_1 + \tau x_2) - \alpha x.$$

Because $(\gamma - \sigma)x_1x$ is comparable to 0 and $-\alpha x < 0$ and $x_1(\delta + \rho x_1 + \tau x_2) \in T_1^+ + K^+$, necessarily, $\delta + \rho x_1 + \tau x_2 = 0$. So $\delta = \rho = \tau = 0$ and $x_1x = \sigma x \in T_2$, as desired.

(III) Suppose $R = T_1 \oplus T_2$ has two totally ordered summands. Then $1 \in T_1 \cup T_2$ since otherwise $R = F(R)$. Moreover, 1 cannot be in a three-dimensional totally ordered summand F because then the other summand would be a vector lattice over F and would also be at least three-dimensional. Suppose $1 \in T_1$. If T_2 is three-dimensional, then $T_1 = K$ and since T_2 is not an ideal it also is not a subalgebra. Thus, T_2^* consists of units. If $a \in T_2^*$ with $tr\,a = 0$, then $a^2 \in K$ and hence $aT_2 \subseteq K$;

this is impossible since aT_2 is three-dimensional. So $tr : T_2 \longrightarrow K$ is monic and this is also impossible. On the other hand, suppose T_2 is two-dimensional. Since T_1 is a division algebra it is central by Exercise 6.2.3; this is absurd.

We have now considered all the cases for which $1 > 0$, and the first part of (c) has been established. From now on we assume $1 \notin R^+$.

(IV) Suppose $R = T_1 \oplus T_2$ and $1 = g_1 + g_2$ with $g_i \in T_i$ and $g_1 < 0 < g_2$. If $g_i T_i \subseteq T_i$ for $i = 1$ or 2, then $(1 - g_i)T_i \subseteq T_i$ and T_i is an ideal. So neither T_1 nor T_2 is a subalgebra and the nonzero elements in $T_1 \cup T_2$ are units. Suppose $0 \neq a \in T_1 \cup T_2$ with $a^2 \in K$. Then $a^2 = a^2 g_1 + a^2 g_2 \geq 0$ gives $a^2 g_1 = 0$ and hence $a^2 = 0$. So $a^2 \notin K$ and $tr : T_i \longrightarrow K$ is monic for $i = 1, 2$. This is impossible since T_1 or T_2 is at least 2-dimensional.

(V) Suppose $R = T_1 \oplus T_2 \oplus T_3$, T_1 is 2-dimensional, and $1 = g_1 + g_2 + g_3$ with $g_i \in T_i$. At least two of g_1, g_2, g_3 are nonzero.

(Va) Now, $T_1 \cong T_1|g_2 + g_3| = T_1(g_2 + g_3) = A$ as vector lattices over K since the nonzero elements of T_1 are units. Let $p : R \longrightarrow T_2 \oplus T_3$ be the projection. If $p(A) \neq 0$, then $A \cong p(A)$ since A is totally ordered and K-archimedean. But now $p(A) = T_2 \oplus T_3$ since they are both two-dimensional. So $p(A) = 0$, $A \subseteq T_1$ and $T_1(g_2 + g_3) = T_1$. However, $g_1^2 = \alpha + \gamma g_1$ with $\alpha \neq 0$ and

$$\begin{aligned} g_1(g_2 + g_3) &= g_1 - g_1^2 = g_1 - \alpha(g_1 + g_2 + g_3) - \gamma g_1 \\ &= (1 - \alpha - \gamma)g_1 - \alpha g_2 - \alpha g_3 \notin T_1. \end{aligned}$$

(Vb) If $g_1 \neq 0$ and g_2 and g_3 have opposite signs, then

$$g_1^2 = \alpha + \gamma g_1 = (\alpha + \gamma)g_1 + \alpha g_2 + \alpha g_3 \geq 0$$

gives $\alpha = 0$. Hence T_1 is a two-dimensional subalgebra which doesn't contain 1.

(Vc) Suppose $g_1 = 0$. Then g_2 and g_3 have opposite signs and if $0 < a \in T_1$ we have $a^2 = \alpha + \gamma a = \alpha g_2 + \alpha g_3 + \gamma a > 0$; but a is invertible yet $0 = \alpha = \det a$.

(Vd) Suppose $g_3 = 0$ and $g_1 < 0 < g_2$; so $1 = g_1 + g_2$. We first note that $T_1 \oplus T_2$ is a subalgebra. For if $a_i \in T_i$, then $|a_1 a_2|, |a_2 a_1| \leq (a_1 + a_2)^2 = \alpha(g_1 + g_2) + \gamma(a_1 + a_2) \in T_1 + T_2$ and $a_i^2 = \rho_i(g_1 + g_2) + \sigma_i a_i \in T_1 + T_2$. We show next that T_3 is a non-nilpotent subalgebra. Let $0 < a \in T_3$. Then $a^2 = \alpha g_1 + \alpha g_2 + \gamma a \geq 0$ gives $\alpha = 0$ and $a^2 = \gamma a$. Suppose $\gamma = 0$. If $g_1 a = 0$, then $Ra = Ka$. So $0 > g_1 a = b + \rho g_2 + \sigma a$ with $b \in -T_1^+$ and $\rho, \sigma \leq 0$; and $0 = g_1 a^2 = ba + \rho g_2 a$ gives $ba = \rho g_2 a = 0$. Now, $g_2 a \neq 0$ since otherwise $a = g_1 a$ and T_3 contains the two-dimensional subspace $T_1 a$; also, $b = 0$ since otherwise $T_1 + T_3 \subseteq \ell(a; R)$. So $g_1 a = \sigma a \in T_3$ and, again, T_3 contains $T_1 a$. Thus, $a^2 \neq 0$ and $T_3 = Ke$ for some idempotent e. Again,

$$g_1 e = b + \rho g_2 + \sigma e \tag{6.4.3}$$

with $-b \in T_1^+$ and $\rho, \sigma \leq 0$. So

$$g_1 e = be + \rho g_2 e + \sigma e = be + \rho(1 - g_1)e + \sigma e$$

and $g_1 e + \rho g_1 e - be = (\rho + \sigma)e$; that is,

$$((1+\rho)g_1 - b)e = (\rho + \sigma)e = 0. \tag{6.4.4}$$

For if $((1+\rho)g_1 - b)e \neq 0$, then $T_1e \subseteq T_3$ and $Re = T_3$, which is nonsense. So $\rho = \sigma = 0$ and $b = g_1$; that is, $g_1e = g_1$ and $g_2e = e - g_1 \in T_1 + T_3$. Since $\ell(e;T_1) = 0$ or T_1 we have $\ell(e;T_1) = 0$ and $T_1e = T_1$. This gives the contradiction

$$Re = (T_1 + T_2 + T_3)e = T_1 + T_3 + T_2e = T_1 + T_3.$$

(Ve) Suppose $g_3 = 0$ and $1 = g_1 + g_2$ with $g_2 < 0 < g_1$. Use the same computations as in (Vd) (with an occasional change of signs) down to (6.4.4); so $\rho + \sigma = 0$. Now, however, $b \in T_1^+$, $\rho \leq 0$ and $\sigma \geq 0$. Since the nonzero elements in T_1 are units, $b = (1+\rho)g_1$ and $1 + \rho \geq 0$. Now, from (6.4.3),

$$\begin{aligned} g_2e &= (1 - g_1)e = e - g_1e = e - (1+\rho)g_1 - \rho g_2 + \rho e \\ &= -(1+\rho)g_1 - \rho g_2 + (1+\rho)e \leq 0; \end{aligned}$$

so $1 + \rho = 0$, $\rho = -1$ and $\sigma = 1$. Thus, $g_1e = -g_2 + e$ and $g_2e = g_2$. So $T_1e = T_2 + T_3 = T_2e + T_3e$, $T_2e = T_2$ and we have the contradiction $Re = T_1e$ is totally ordered. (VI) Suppose $R = T_1 \oplus T_2 \oplus T_3 \oplus T_4$ and $1 = g_1 + g_2 + g_3 + g_4$ with $g_i \in T_i$ (g_i could be 0). If there exist distinct indices i, j, k such that g_j and g_k have opposite signs, then from the Cayley–Hamilton equation for g_i we get $g_i^2 \in T_i$.

(VIa) Suppose $g_1 < 0 < g_2, g_3, g_4$. Then $g_i^2 \in T_i$ for $i = 2, 3, 4$, and, in fact, $A = T_2 + T_3 + T_4$ is a subalgebra. To verify this it suffices to show $g_jg_k \in A$ for $2 \leq j, k \leq 4$. But, for example, from the inequality,

$$\begin{aligned} 0 \leq (g_2 + g_3)^2 &= \alpha g_1 + \alpha g_2 + \alpha g_3 + \alpha g_4 + \gamma(g_2 + g_3) \\ &= \alpha g_1 + (\alpha + \gamma)g_2 + (\alpha + \gamma)g_3 + \alpha g_4 \end{aligned}$$

we get $\alpha = 0$ and $0 \leq g_2g_3, g_3g_2 \leq (g_2 + g_3)^2 \in A$. Since $Ag_1 = A(1 - g_2 - g_3 - g_4) \subseteq A$ we have the contradiction that A is a right ideal.

(VIb) Suppose $g_1, g_2 < 0 < g_3, g_4$. This case will produce the lattice orders P_β for $\beta > 1$. For each i, $g_i^2 = \alpha_i g_i$. Also, from the usual Cayley–Hamilton argument, when $i = 1$ or 2 and $j = 3$ or 4 we have $(g_j - g_i)^2 \in K^+(g_j - g_i)$. So $T_i + T_j$ is a subalgebra for these indices. Write $g_1g_3 = \gamma g_1 + \delta g_3$. Then

$$\alpha_1\gamma g_1 + \alpha_1\delta g_3 = \alpha_1 g_1g_3 = g_1^2g_3 = \gamma g_1^2 + \delta g_1g_3 = \gamma\alpha_1 g_1 + \delta g_1g_3 \tag{6.4.5}$$

and

$$\alpha_3\gamma g_1 + \alpha_3\delta g_3 = \alpha_3 g_1g_3 = g_1g_3^2 = \gamma g_1g_3 + \delta g_3^2 = \gamma g_1g_3 + \delta\alpha_3 g_3. \tag{6.4.6}$$

So $\alpha_1\delta g_3 = \delta g_1g_3$ and $\alpha_3\delta g_1 = \gamma g_1g_3$; and either $g_1g_3 = 0$, or $g_1g_3 = \alpha_1 g_3$, or $g_1g_3 = \alpha_3 g_1$. Analogous computations give similar equations for g_ig_j and g_jg_i with $1 \leq i \leq 2$ and $3 \leq j \leq 4$. We show next that these products are not 0. Suppose, for example, that $g_1g_3 = 0$. Then

$$g_1 = g_1(g_1+g_2+g_3+g_4) = \alpha_1 g_1 + g_1 g_2 + g_1 g_4 \tag{6.4.7}$$

and $g_1g_2 = (1-\alpha_1)g_1 - g_1g_4$. If $g_1g_4 \in T_1$, then $g_1R = Kg_1$; so $g_1g_4 = \alpha_1 g_4$ and $0 \le g_1g_2 = (1-\alpha_1)g_1 - \alpha_1 g_4$. This inequality gives $1 \le \alpha_1 \le 0$; so $g_1g_2 \neq 0$ and, analogously, $g_ig_j \neq 0$ and $g_jg_i \neq 0$ for $i = 1,2$ and $j = 3,4$. We now assume

$$g_1g_3 = \alpha_1 g_3, \quad \alpha_1 < 0. \tag{6.4.8}$$

As in (6.4.7) $g_1 = \alpha_1 g_1 + g_1g_2 + \alpha_1 g_3 + g_1g_4$ and $g_1g_2 = (1-\alpha_1)g_1 - \alpha_1 g_3 - g_1g_4$. If $g_1g_4 = \alpha_1 g_4$, then $g_1g_2 = (1-\alpha_1)g_1 - \alpha_1 g_3 - \alpha_1 g_4$ yields $1 \le \alpha_1 < 0$. So

$$g_1g_4 = \alpha_4 g_1, \quad \alpha_4 > 0 \tag{6.4.9}$$

and

$$g_1g_2 = (1-\alpha_1-\alpha_4)g_1 - \alpha_1 g_3. \tag{6.4.10}$$

From $g_3 = 1g_3 = \alpha_1 g_3 + g_2g_3 + \alpha_3 g_3 + g_4g_3$ we get $g_4g_3 = (1-\alpha_1-\alpha_3)g_3 - g_2g_3$. If $g_2g_3 \in Kg_3$, then $Rg_3 = T_3$; thus, $g_2g_3 \notin Kg_3$ and

$$g_2g_3 = \alpha_3 g_2, \quad \alpha_3 > 0. \tag{6.4.11}$$

Also, $g_4g_3 = (1-\alpha_1-\alpha_3)g_3 - \alpha_3 g_2$. We compute $g_1g_2g_3$ twice using (6.4.11), and (6.4.10) and (6.4.8).

$$\begin{aligned} g_1(g_2g_3) &= \alpha_3 g_1g_2 = \alpha_3(1-\alpha_1-\alpha_4)g_1 - \alpha_3\alpha_1 g_3, \\ (g_1g_2)g_3 &= (1-\alpha_1-\alpha_4)\alpha_1 g_3 - \alpha_1\alpha_3 g_3. \end{aligned}$$

So $\alpha_1 + \alpha_4 = 1$ and

$$g_1g_2 = -\alpha_1 g_3. \tag{6.4.12}$$

From $g_1g_2^2 = g_1\alpha_2 g_2 = -\alpha_1\alpha_2 g_3$ and $(g_1g_2)g_2 = -\alpha_1 g_3g_2$ we get

$$g_3g_2 = \alpha_2 g_3. \tag{6.4.13}$$

Now,

$$g_4g_2 = (1-g_1-g_2-g_3)g_2 = g_2 + \alpha_1 g_3 - \alpha_2 g_2 - \alpha_2 g_3 = (1-\alpha_2)g_2 + (\alpha_1-\alpha_2)g_3.$$

Since $g_4g_2 \in E_2 \cup E_4$ we have $\alpha_1 = \alpha_2$, and, in fact,

$$g_4g_2 = \alpha_4 g_2 \tag{6.4.14}$$

and $\alpha_4 = 1-\alpha_2$; so $\alpha_1 = \alpha_2$. Using the expression for g_4g_3 given after (6.4.10), as well as (6.4.14) and (6.4.11) we have

$$\begin{aligned} g_4g_1 &= g_4 - g_4g_2 - g_4g_3 - g_4^2 = g_4 - (1-\alpha_2)g_2 - (1-\alpha_1-\alpha_3)g_3 + \alpha_3 g_2 - \alpha_4 g_4 \\ &= (1-\alpha_4)g_4 + (\alpha_1+\alpha_3-1)g_2 - (1-\alpha_1-\alpha_3)g_3 \\ &= (1-\alpha_4)g_4 + (\alpha_3-\alpha_4)g_2 - (1-\alpha_1-\alpha_3)g_3 < 0. \end{aligned}$$

So $1 \le \alpha_4$ and since $g_4g_1 \in T_1 \cup T_4$, $\alpha_3 = \alpha_4$, $\alpha_1 + \alpha_3 = 1$ and

$$g_4g_1 = \alpha_1 g_4 \tag{6.4.15}$$

Finally, we need to compute g_3g_4 in two ways. From (6.4.13)

$$g_3g_4 = g_3(1-g_1-g_2-g_3) = g_3 - g_3g_1 - \alpha_2 g_3 - \alpha_3 g_3 = -g_3g_1 = \begin{cases} -\alpha_3 g_1 & \text{or} \\ -\alpha_1 g_3 \end{cases}$$

since $\alpha_2 + \alpha_3 = 1$, and from (6.4.9)

$$\begin{aligned} g_3g_4 &= (1-g_1-g_2-g_4)g_4 = g_4 - \alpha_4 g_1 - g_2g_4 - \alpha_4 g_4 \\ &= (1-\alpha_4)g_4 - \alpha_4 g_1 - g_2g_4 \\ &= \begin{cases} -\alpha_4 g_1 & \text{if } g_2g_4 = \alpha_2 g_4 \\ (1-\alpha_4)g_4 - \alpha_4 g_1 - \alpha_4 g_2 & \text{if } g_2g_4 = \alpha_4 g_2. \end{cases} \end{aligned}$$

So,

$$g_2g_4 = \alpha_2 g_4 \tag{6.4.16}$$

$$g_3g_4 = -\alpha_4 g_1. \tag{6.4.17}$$

All of the twelve products g_mg_n with $m \neq n$ have now been computed and appear in (6.4.8), (6.4.9), and (6.4.11)–(6.4.17) with the exception of g_2g_1, g_3g_1, and g_4g_3. These remaining products are now easily computed to be

$$g_2g_1 = -\alpha_2 g_4 \tag{6.4.18}$$

$$g_3g_1 = \alpha_3 g_1 \tag{6.4.19}$$

$$g_4g_3 = -\alpha_4 g_2. \tag{6.4.20}$$

For example,

$$\begin{aligned} g_2g_1 &= g_2(1-g_2-g_3-g_4) = g_2 - \alpha_2 g_2 - \alpha_3 g_2 - \alpha_2 g_4 \\ &= (1-\alpha_2-\alpha_3)g_2 - \alpha_2 g_4 = -\alpha_2 g_4. \end{aligned}$$

Let $\beta = \alpha_3 = \alpha_4$ and recall that $1-\beta = \alpha_1 = \alpha_2 < 0$; so $\beta > 1$. Let $\overline{g}_i = \alpha_i^{-1} g_i$ for $1 \le i \le 4$. Then each $\overline{g}_i$ is idempotent, $1 = (1-\beta)(\overline{g}_1 + \overline{g}_2) + \beta(\overline{g}_3 + \overline{g}_4)$, and $T_i = K\overline{g}_i$. Moreover, using the equations (6.4.8), (6.4.9), and (6.4.11) through (6.4.20) it is easy to see that these idempotents satisfy the multiplication table given in the proof of (a). So this is a lattice order isomorphic to P_β.

This copy of P_β arose from our assumption that (6.4.8) holds. Suppose instead of (6.4.8) we have the alternative

$$g_1g_3 = \alpha_3 g_1. \tag{6.4.21}$$

We compute the other products, as in the just completed calculation. First,

$$g_1 g_2 = g_1(1 - g_1 - g_3 - g_4) = (1 - \alpha_1 - \alpha_3)g_1 - g_1 g_4. \tag{6.4.22}$$

So

$$g_1 g_4 = \alpha_1 g_4, \tag{6.4.23}$$

or else $g_1 g_4 = \alpha_4 g_1$ and $g_1 K_2 = K g_1$. Now, from (6.4.23),

$$g_2 g_4 + g_3 g_4 = (1 - \alpha_1 - \alpha_4)g_4,$$
$$g_3 g_4 = (1 - \alpha_1 - \alpha_4)g_4 - g_2 g_4,$$

and

$$g_2 g_4 = \alpha_4 g_2, \tag{6.4.24}$$

or else $K_2 g_4 = K g_4$. From (6.4.22),

$$g_1(g_2 g_4) = \alpha_4 g_1 g_2 = \alpha_4(1 - \alpha_1 - \alpha_3)g_1 - \alpha_1 \alpha_4 g_4,$$
$$(g_1 g_2)g_4 = (1 - \alpha_1 - \alpha_3)\alpha_1 g_4 - \alpha_1 \alpha_4 g_4,$$

and hence $\alpha_1 + \alpha_3 = 1$ and

$$g_1 g_2 = -\alpha_1 g_4. \tag{6.4.25}$$

Also,

$$g_4 g_2 = \alpha_2 g_4 \tag{6.4.26}$$

since $-\alpha_1 g_4 g_2 = g_1 g_2^2 = \alpha_2 g_1 g_2 = -\alpha_1 \alpha_2 g_4$; and $\alpha_1 = \alpha_2$ and

$$g_3 g_2 = \alpha_3 g_2 \tag{6.4.27}$$

since, from (6.4.25) and (6.4.26),

$$g_3 g_2 = (1 - g_1 - g_2 - g_4)g_2 = (1 - \alpha_2)g_2 + (\alpha_1 - \alpha_2)g_4.$$

By using (6.4.26) to compute $g_4 g_3 = (1 - \alpha_2 - \alpha_4)g_4 - g_4 g_1$ and noting that if $g_4 g_1 \in T_4$, then $g_4 K_2 = K g_4$, we get

$$g_4 g_1 = \alpha_4 g_1. \tag{6.4.28}$$

From (6.4.21) we have

$$(1 - \alpha_2 - \alpha_4)g_4 - \alpha_4 g_1 = g_4 g_3 = (1 - \alpha_3)g_3 - \alpha_3 g_1 - g_2 g_3.$$

So

$$g_2 g_3 = \alpha_2 g_3 \tag{6.4.29}$$

and $\alpha_3 = \alpha_4$ and

$$g_4 g_3 = -\alpha_4 g_1. \tag{6.4.30}$$

From (6.4.24) and (6.4.29) we get $g_2 g_1 = (1 - \alpha_2 - \alpha_4)g_2 - \alpha_2 g_3$; so

$$g_2 g_1 = -\alpha_2 g_3. \tag{6.4.31}$$

From (6.4.28) and (6.4.31), $g_3g_1 = (1-\alpha_1-\alpha_4)g_1 + \alpha_2 g_3$; so

$$g_3g_1 = \alpha_2 g_3 \tag{6.4.32}$$

and hence

$$g_3g_4 = -\alpha_3 g_2 \tag{6.4.33}$$

since $g_3g_4 = (1-\alpha_2-\alpha_3)g_3 - \alpha_3 g_2$ by (6.4.27) and (6.4.33).

As before, let $\beta = \alpha_3 = 1-\alpha_1 > 0$ and $\overline{g}_i = \alpha_i^{-1} g_i$. From the equations (6.4.21) and (6.4.23)–(6.4.33) it is easily seen that the idempotents $\overline{g}_1$, $\overline{g}_2$, $\overline{g}_3$, $\overline{g}_4$ form a canonical basis for K_2 and they satisfy the multiplication table given below.

	$\overline{g}_1$	$\overline{g}_2$	$\overline{g}_3$	$\overline{g}_4$
$\overline{g}_1$	$\overline{g}_1$	$\beta(\beta-1)^{-1}\overline{g}_4$	$\overline{g}_1$	$\overline{g}_4$
$\overline{g}_2$	$\beta(\beta-1)^{-1}\overline{g}_3$	$\overline{g}_2$	$\overline{g}_3$	$\overline{g}_2$
$\overline{g}_3$	$\overline{g}_3$	$\overline{g}_2$	$\overline{g}_3$	$\beta^{-1}(\beta-1)\overline{g}_2$
$\overline{g}_4$	$\overline{g}_1$	$\overline{g}_4$	$\beta^{-1}(\beta-1)\overline{g}_1$	$\overline{g}_4$

This table is the transpose of the multiplication table for the f_i given in the definition of P_β. Thus, if Q denotes the lattice order of K_2 that has just been constructed and a^t denotes the transpose of a, then the mapping $\overline{g}_i \mapsto f_i^t$ determines an isomorphism from (K_2, Q) onto (K_2, P_β^t). However, it is easy to check that the mapping

$$f_1 \mapsto f_2^t, f_2 \mapsto f_1^t, f_3 \mapsto f_3^t, f_4 \mapsto f_4^t$$

determines an isomorphism of (K_2, P_β) with (K_2, P_β^t). So (K_2, Q) is also isomorphic with (K_2, P_β).

(VIc) Suppose g_1, g_2, $g_3 < 0 < g_4$. This won't work since just as in (VIa), $T_1 + T_2 + T_3$ is a right ideal.

(VId) Suppose $1 = g_1 + g_2 + g_3$ and g_1 and g_2 have the same sign which is opposite to that of g_3. Then $g_1^2 = \alpha_1 g_1$ and $g_2^2 = \alpha_2 g_2$. Take $0 < n \in T_4$. Since $n^2 = \gamma g_1 + \gamma g_2 + \gamma g_3 + \alpha n \geq 0$, necessarily $\gamma = 0$ and $n^2 = \alpha n$. Now, $T_1 + T_4$, $T_2 + T_4$ and $A = T_1 + T_2 + T_3$ are all subalgebras. For $i = 1, 2$ we have $|g_i|n, n|g_i| \leq (|g_i| + n)^2$, and

$$(|g_i| + n)^2 = \gamma + \delta(|g_i| + n) = (\gamma \pm \delta)g_1 + \gamma g_2 + \gamma g_3 + \delta n$$

gives $\gamma = 0$; so $T_i + T_4$ is a subalgebra. For $1 \leq i, j \leq 3$ with $i \neq j$, $(g_i + g_j)^2 \in K1 + K(g_i + g_j) \subseteq A$; so g_ig_j and g_jg_i belong to A and so does $g_3^2 = g_3(1 - g_1 - g_2)$.

We show next that $g_2n = 0$, or $g_2n = \alpha_2 n$ or $g_2n = \alpha g_2$. From $g_2n = \gamma g_2 + \delta n$ we have

$$\gamma\alpha_2 g_2 + \delta\alpha_2 n = \alpha_2 g_2 n = g_2^2 n = \gamma\alpha_2 g_2 + \delta g_2 n,$$

$$\gamma\alpha g_2 + \delta\alpha n = \alpha g_2 n = g_2 n^2 = \gamma g_2 n + \delta\alpha n.$$

So $\delta g_2 n = \delta\alpha_2 n$ and $\gamma g_2 n = \gamma\alpha g_2$. Similar equations hold for $g_1 n$, ng_1, and ng_2. Clearly, not both of g_2n and g_1n can be in Kn since then $K_2 n = Kn$. None of the products g_in or ng_i is zero for $i = 1,2$. Suppose, for example, that $g_2n = 0$. Then $g_1n = \alpha g_1 \neq 0$ and $g_3n = (1-g_1-g_2)n = n-\alpha g_1$. From $g_1g_2 = \rho g_1 + \sigma g_2 + \tau g_3$ we have

$$0 = g_1g_2n = \rho\alpha g_1 + \tau n - \tau\alpha g_1 = \alpha(\rho - \tau) + \tau n;$$

so $\rho = \tau = 0$ and $g_1g_2 = \sigma g_2$. This implies we cannot have $ng_2 \in T_2$ and hence $ng_2 = \alpha_2 n \neq 0$. But then $0 = ng_2n = \alpha_2\alpha n \neq 0$. Thus, $g_2n \neq 0$ and either $\gamma \neq 0$ and $g_2n = \alpha g_2$, or $\delta \neq 0$ and $g_2n = \alpha_2 n$.

Assume that $g_2n = \alpha g_2$. By replacing n by $\alpha^{-1}n$ we have $\alpha = 1$ and $g_2n = g_2$. Now, $g_1n \notin Kg_1$ or else $K_2n = Kg_1n + Kg_2n + K(1-g_1-g_2)n + Kn = A$. So $g_1n = \alpha_1 n$. Again, $g_1g_2 = \rho g_1 + \sigma g_2 + \tau g_3$ and

$$\begin{aligned} g_1g_2 = g_1g_2n &= \rho\alpha_1 n + \sigma g_2 + \tau(1-g_1-g_2)n \\ &= (\rho\alpha_1 + \tau - \tau\alpha_1)n + (\sigma - \tau)g_2. \end{aligned}$$

So $\rho = \tau = 0$ and $g_1g_2 = \sigma g_2$. Now $ng_2 = \alpha_2 n$, or else $ng_2 \in Kg_2$ and $K_2g_2 = Kg_2$. Also $ng_1 = g_1$, or else $ng_1 = \alpha_1 n$ and $nK_2 = Kn$. But now

$$\sigma g_2 = g_1g_2 = ng_1g_2 = \sigma ng_2 = \sigma\alpha_2 n.$$

So $\sigma = 0$ and $g_1g_2 = 0$. Now $g_2g_1 \notin Kg_1$, or else $K_2g_1 = Kg_1$. So, for some ρ, σ, $\tau \in K$ we have $0 < g_2g_1 = \rho g_1 + \sigma g_2 + \tau g_3$. Since

$$0 = g_2g_1g_2 = \sigma\alpha_2 g_2 + \tau(1-g_1-g_2)g_2 = (\sigma\alpha_2 + \tau(1-\alpha_2))g_2$$

we have $\sigma = \tau(\alpha_2 - 1)\alpha_2^{-1}$. If $g_3 > 0 > g_1$, g_2, then $\tau > 0$ and $\sigma \leq 0$; but $\alpha_2 < 0$ and hence $\sigma = \tau(\alpha_2 - 1)\alpha_2^{-1} > 0$. So we must have $g_3 < 0 < g_1, g_2$, and therefore $\tau < 0$, $\sigma \geq 0$, and $0 < \alpha_2 \leq 1$. Since

$$0 = g_1g_2g_1 = \rho\alpha_1 g_1 + \tau g_1g_3$$

and $0 \leq \rho\alpha_1 g_1$, τg_1g_3 we have $\rho = 0$ and $g_1g_3 = 0$; so $g_1^2 = g_1(1-g_2-g_3) = g_1$ and $\alpha_1 = 1$. From

$$0 \geq g_3g_2 = (1-g_1-g_2)g_2 = (1-\alpha_2)g_2 \geq 0$$

we get $\alpha_2 = 1$ and g_2 is idempotent, $g_3g_2 = 0$, $\sigma = 0$, and $g_2g_1 = \tau g_3$. Now,

$$g_1 = ng_1 = (ng_2)g_1 = \tau ng_3 = \tau n(1-g_1-g_2) = \tau(n-g_1-n) = -\tau g_1.$$

So $\tau = -1$ and $g_2g_1 = -g_3$. Furthermore, $g_2g_3 = g_2(1-g_1-g_2) = g_3$ and

$$\begin{aligned} g_3^2 &= ((1-g_1)-g_2)((1-g_1)-g_2) = (1-g_1)^2 + g_2^2 - g_2 + g_1g_2 - g_2 + g_2g_1 \\ &= 1 - g_1 - g_2 - g_3 = 0. \end{aligned}$$

So $g_3g_1 = g_3(1-g_2-g_3) = g_3$ and $g_3n = (1-g_1-g_2)n = n-n-g_2 = -g_2$. From $ng_1 = g_1$ and $ng_2 = n$ we obtain $ng_3 = -e_1$.

We have now shown that $g_1, g_2, -g_3$ and n satisfy the multiplication table given in the proof of (b). So $K^+e_1+K^+e_2+K^+(-g_3)+K^+n$ is a lattice order isomorphic to P_1.

The other possibility is that $g_2n = \alpha_2 n$ instead of $g_2n = \alpha g_2$. In this case $g_1n = \alpha g_1$ since otherwise $K_2n = Kn$. These are the same two equations that appear at the beginning of the preceding computation two paragraphs ago but with g_1 and g_2 interchanged. So, $\overline{g}_1 = g_2$, $\overline{g}_2 = g_1$, $-\overline{g}_3 = -g_3$ and n are a canonical basis for a lattice order of K_2 that is isomorphic to P_1.

(VIe) Suppose $1 = g_1 + g_2$ with $g_1 < 0 < g_2$. Take $0 < x_3 \in T_3$ and $0 < x_4 \in T_4$. Since $x_j^2 = \alpha g_1 + \alpha g_2 + \gamma g_j = \gamma g_j$ and, similarly, $(x_3+x_4)^2 \in K(x_3+x_4)$, each of the elements x_3, x_4, x_3+x_4 is nilpotent or idempotent and $A = T_3+T_4$ is a subalgebra. Also, x_3+x_4 must be idempotent since otherwise A is nilpotent. We note next that $g_1x_3 \in T_1+T_2+T_3 = B$. For, $g_1^2 = (\alpha+\gamma)g_1 + \alpha g_2 \in B$ and

$$0 \leq -g_1x_3 \leq -(g_1x_3+x_3g_1) = (-g_1+x_3)^2 - g_1 - x_3^2 \in B.$$

Suppose $x_3^2 = 0$. Since $x_4x_3 = \alpha x_3 + \gamma x_4$ we get $\gamma x_4x_3 = 0$ and $x_4x_3 \in Kx_3$. Write $g_1x_3 = \rho g_1 + \sigma g_2 + \tau x_3 \leq 0$. Then $\rho g_1x_3 + \sigma g_2x_3 = 0$ and hence $\rho g_1x_3 = \sigma g_2x_3 = 0$ since $\rho g_1x_3, \sigma g_2x_3 \leq 0$. So $\rho = 0$ and also $\sigma = 0$ since otherwise $g_2x_3 = 0$ and $0 \geq g_1x_3 = (1-g_2)x_3 = x_3 > 0$. Therefore $g_1x_3 \in T_3$ and $K_2x_3 = Kx_3$. So x_3 is idempotent and in the same way x_4 is idempotent. But then $x_3x_4 + x_4x_3 = 0$ since $(x_3+x_4)^2 = x_3+x_4$ and we have the contradiction $1 = x_3+x_4$ since x_3 and x_4 are orthogonal.

In order to complete the proof of (c) we need to check that if $\varphi : (K_2, P_\beta) \longrightarrow (K_2, P_\gamma)$ is an isomorphism, then $\beta = \gamma$. Let $\{f_1, f_2, f_3, f_4\}$ and $\{h_1, h_2, h_3, h_4\}$ be the canonical basis of idempotents for P_β and P_γ, respectively. There is a permutation σ such that $\varphi(f_j) = h_{\sigma(j)}$ for $1 \leq j \leq 4$. So

$$h_{\sigma(1)}h_{\sigma(2)} = \varphi(f_1f_2) = -\beta(1-\beta)^{-1}h_{\sigma(3)} \neq h_{\sigma(3)}$$

and therefore $-\beta(1-\beta)^{-1} = -\gamma(1-\gamma)^{-1}$ or $-\beta(1-\beta)^{-1} = -\gamma^{-1}(1-\gamma)$. In the first case we get $\beta = \gamma$ and in the second we get $2 < \beta + \gamma = 1$. □

K does not have to be a totally ordered field in order for P_β to be a partial or lattice order for K_2. If K is any unital po-ring, then (K_2, P_1) is a K-po-ring. If K is po-unital, $\beta > 1$ is central and β and $\beta - 1$ are units in K^+, then (K_2, P_β) is also a K-po-ring. Moreover, these partial orders can be obtained from a general construction of partial orders for K_n which will be given prior to Theorem 6.4.10; see Exercise 6.

Let L be a field and let $V = L^n$ be the L-space of column vectors with n entries. A subset S of L_n is called *L-irreducible* or just *irreducible* if V is a simple left module over the subalgebra B of L_n generated by S. The set S is not irreducible precisely when there is an integer $m < n$ and a unit u in L_n such that, for each $s \in S$, $u^{-1}su$ has the form $u^{-1}su = \begin{pmatrix} a & b \\ 0 & c \end{pmatrix}$ where a is an $m \times m$ matrix. The first m columns of u then form a basis for a proper B-submodule of V since if $e_1, \ldots, e_n$ denote the standard basis vectors for V and $v_j = ue_j$, then $u^{-1}sv_j = \Sigma_{i=1}^m \alpha_{ij}e_i$ and $sv_j = \Sigma_{i=1}^m \alpha_{ij}v_i$ for $j = 1, \ldots, m$.

The context in which irreducibility will be used is the following. Let K be a subfield of the real field $\mathbb{R}$ and let A be a K-subalgebra of K_n which is $\mathbb{R}$-irreducible. Then A is a simple K-algebra and $B = \mathbb{R}A$ is a simple $\mathbb{R}$-algebra by Theorems 4.2.12 and 4.2.13 since V is a simple A-module - see Exercise 4.2.36. If A can be made into an ℓ-algebra we will see that $A = K_n$; and $A^+ \cong (K^+)_n$ provided $1 \in A^+$. Moreover, all of the lattice orders of K_n will be described.

We first wish to lift a module partial order of K^n to a module partial order of $\mathbb{R}^n$. Recall that Y^- and Y° denotes the closure and the interior, respectively, of the subset Y of $\mathbb{R}^n$.

Theorem 6.4.2. *Let K be a subfield of $\mathbb{R}$ and let A be a K-subalgebra of K_n which is $\mathbb{R}$-irreducible and is generated by its subset P. Suppose Q is a subset of K^n such that $K^+Q \subseteq Q$ and $PQ \subseteq Q$. If $Q \cap -Q = 0$, then $Q^- \cap -Q^- = 0$.*

Proof. Suppose $Q^- \cap -Q^- \neq 0$. By Exercise 4.2.36 $Q^- \cap -Q^-$ contains a basis X of $\mathbb{R}^n$. By Theorem 5.3.12 $U = (\mathbb{R}^+X)^\circ \neq \emptyset$ and certainly $0 \notin U$. Since $\emptyset \neq U \cap -Q \subseteq Q^- \cap -Q$, there is a basis Y of $\mathbb{R}^n$ contained in $Q^- \cap -Q$, again by Exercise 4.2.36. Let $U_1 = (\mathbb{R}^+Y)^\circ$. Then $0 \notin U_1$ and $\emptyset \neq U_1 \cap Q \subseteq Q \cap -Q$. □

The following result gives some relations between partial orders of K^n and of $\mathbb{R}^n$ and it will be used to get minimal module partial orders for K^n.

Theorem 6.4.3. *Suppose K is a subfield of $\mathbb{R}$ and Q is a finite spanning subset of ${}_KK^n$ such that $(\mathbb{R}^n, \mathbb{R}^+Q)$ is a po-vector space over $\mathbb{R}$. Then there exists a family of bases $\{Q_\lambda : \lambda \in \Lambda\}$ of ${}_KK^n$ such that*

$$\mathbb{R}^+Q = \bigcup_\lambda \mathbb{R}^+Q_\lambda, \quad K^+Q = \bigcup_\lambda K^+Q_\lambda,$$

and $\mathbb{R}^+Q \cap K^n = K^+Q$.

Proof. The last equation is a consequence of the fact that $\mathbb{R}^+Q_\lambda \cap K^n = K^+Q_\lambda$ for each λ since Q_λ is a basis for $\mathbb{R}^n$. For $0 \neq v \in \mathbb{R}^+Q$ let $X = \{w_1, \ldots, w_m\}$ be a subset of K^+Q with m minimal such that $v \in \mathbb{R}^+X$. We claim that X is a K-independent set. If not, there exist $\beta_1, \ldots, \beta_m \in K$ with $\beta_1 w_1 + \cdots + \beta_m w_m = 0$ and $\beta_i\beta_j < 0$ for some i and j. By relabeling we have

$$w = \beta_1 w_1 + \cdots + \beta_p w_p = \beta_{p+1} w_{p+1} + \cdots + \beta_m w_m$$

with each $\beta_i \geq 0$, and there exist $1 \leq i \leq p$ and $p+1 \leq j \leq m$ such that $\beta_i > 0$ and $\beta_j > 0$. Write $v = \alpha_1 w_1 + \cdots + \alpha_m w_m$ with $\alpha_i \in \mathbb{R}^+$ and let

$$\alpha_s\beta_s^{-1} = \bigwedge\{\alpha_i\beta_i^{-1} : \beta_i > 0,\ 1 \leq i \leq p\},$$
$$\alpha_t\beta_t^{-1} = \bigwedge\{\alpha_i\beta_i^{-1} : \beta_i > 0,\ p+1 \leq i \leq m\}.$$

Then for $1 \leq i \leq p$ and $p+1 \leq j \leq m$, $\alpha_i - \alpha_s\beta_s^{-1}\beta_i \geq 0$ and $\alpha_j - \alpha_t\beta_t^{-1}\beta_j \geq 0$. Now,

$$(\alpha_s\beta_s^{-1} + \alpha_t\beta_t^{-1})w + \sum_{i=1}^{p}(\alpha_i - \alpha_s\beta_s^{-1}\beta_i)w_i + \sum_{i=p+1}^{m}(\alpha_i - \alpha_t\beta_t^{-1}\beta_i)w_i =$$

$$\sum_{i=1}^{p}(\alpha_s\beta_s^{-1}\beta_i + \alpha_i - \alpha_s\beta_s^{-1}\beta_i)w_i + \sum_{i=p+1}^{m}(\alpha_t\beta_t^{-1}w_i + \alpha_i - \alpha_t\beta_t^{-1}\beta_i)w_i =$$

$$\sum_{i=1}^{m}\alpha_i w_i = v,$$

and the coefficients of w_s and w_t in the first two sums are both 0. This contradicts the minimality of m, and hence X is independent. Since Q contains a basis of ${}_KK^n$ the subset X can be enlarged to a basis X_v of K^n with $X_v \subseteq K^+Q$. Clearly, $\{X_v : 0 \neq v \in \mathbb{R}^+Q\}$ is the desired family of bases. □

The po-module M over the po-ring R is called *minimal* if M^+ is a *minimal module partial order*; that is, $M^+ \neq 0$ and if (M,Q) is an R-po-module with $0 \neq Q \subseteq M^+$, then $Q = M^+$. The positive cone M^+ is *finitely generated over* R^+ if there exist $x_1,\ldots,x_n \in M^+$ such that $M^+ = R^+x_1 + \mathbb{Z}^+x_1 + \cdots + R^+x_n + \mathbb{Z}^+x_n$. In order to see that an $\mathbb{R}^+$-finitely generated partial order of $\mathbb{R}^n$ is closed we will make a diversion into homogeneous linear inequalities in the next three theorems. All inequalities involving matrices in these theorems will be with respect to the usual coordinatewise vector lattice partial order.

Theorem 6.4.4. *Suppose K is a totally ordered field and a is an $m \times n$ matrix over K with columns $v_1,\ldots,v_n$. Then there exist vectors $0 \leq x \in K^n$ and $z \in K^m$ such that $z^t a \geq 0$, $ax = 0$ and $z^t v_1 + \alpha_1 > 0$ where $x^t = (\alpha_1,\ldots,\alpha_n)$.*

Proof. If $a = 0$ let $x = (1,0,\ldots,0)$ and $z = 0$. We now use induction on n. If $n = 1$ and $a \neq 0$ let $x = 0$ and $z = v_1$. Assuming the result for n let $\overline{a} = (a v_{n+1})$ be an $m \times (n+1)$ matrix and let x and z be vectors satisfying the conditions relative to a. If $z^t v_{n+1} \geq 0$, let $\overline{x} = \begin{pmatrix} x \\ 0 \end{pmatrix}$ and $\overline{z} = z$. Then $\overline{z}^t\overline{a} = (z^t a\; z^t v_{n+1}) \geq 0$, $\overline{a}\overline{x} = ax = 0$ and $\overline{z}^t v_1 + \alpha_1 > 0$. Suppose, then, that $z^t v_{n+1} < 0$. For $j = 1,\ldots,n$ let $\beta_j = -(z^t v_j)(z^t v_{n+1})^{-1}$, and note that $\beta_j \geq 0$ since $z^t a = (z^t v_1 \cdots z^t v_n) \geq 0$. Let $w_j = v_j + \beta_j v_{n+1}$ for $j = 1,\ldots,n$ and let $b = (w_1 \cdots w_n)$. Then $z^t b = 0$ since

$$z^t w_j = z^t v_j - (z^t v_j)(z^t v_{n+1})^{-1}(z^t v_{n+1}) = 0$$

for each j. By induction there are vectors $v \in K^m$ and $0 \leq y \in K^n$ with $v^t b \geq 0$, $by = 0$ and $v^t w + \gamma_1 > 0$ where $y^t = (\gamma_1,\ldots,\gamma_n)$. Let $0 \leq \overline{x} \in K^{n+1}$ be defined by $\overline{x}^t = (y^t\; \Sigma_{j=1}^n \beta_j\gamma_j)$. Then

$$\begin{aligned}\overline{a}\overline{x} &= (a v_{n+1}) \begin{pmatrix} y \\ \Sigma_j \beta_j \gamma_j \end{pmatrix} = ay + \Sigma_j \beta_j \gamma_j v_{n+1} \\ &= (v_1 \cdots v_n) y + (\beta_1 v_{n+1} \cdots \beta_n v_{n+1}) y \\ &= by = 0.\end{aligned}$$

Let $\gamma = -(v^t v_{n+1})(z^t v_{n+1})^{-1}$ and $\overline{z} = v + \gamma z \in K^m$. Then $\overline{z}^t v_{n+1} = v^t v_{n+1} + \gamma z^t v_{n+1} = 0$ and $\overline{z}^t \overline{a} = (\overline{z}^t a\ 0) \geq 0$ since

$$\begin{aligned}\overline{z}^t a &= \overline{z}^t (v_1 + \beta_1 v_{n+1} \cdots v_n + \beta_n v_{n+1}) = \overline{z}^t b \\ &= (v^t + \gamma z^t) b = v^t b \geq 0.\end{aligned}$$

Also, from $\overline{z}^t a = v^t b$ we have $\overline{z}^t v_1 + \gamma_1 = v^t w_1 + \gamma_1 > 0$ and hence $\overline{x}$ and $\overline{z}$ are solutions to the linear inequalities determined by $\overline{a}$. □

For the subset A of K^n let

$$A' = \{x \in K^n : v^t x \geq 0 \text{ for each } v \in A\}.$$

Suppose $A = \{v_1, \ldots, v_p\}$ and let $\mathbb{N}_p = \{1, \ldots, p\}$. For a subset J of $\mathbb{N}_p$ the associated subset F_J of A' is defined by $F_J = O_J \cap L_J$ where

$$O_J = \{x \in K^n : v_j^t x > 0 \text{ for each } j \in J\}$$

and

$$L_J = \{x \in K^n : v_j^t x = 0 \text{ for each } j \notin J\}.$$

F_J, which could be empty, is called a *face* of A'. Clearly, the set of faces $\{F_J : J \subseteq \mathbb{N}_p\}$ is a partition of A', $F_\emptyset = L_\emptyset$, and if $I \subseteq J \subseteq \mathbb{N}_p$, then $O_J \subseteq O_I$ and $L_I \subseteq L_J$. If $F_J \neq \emptyset$ let $V_J = \Sigma_{j \notin J} K v_j$ and let $r_J = \dim_K V_J$. Then $d_J = n - r_J = \dim_K L_J$ and d_J is called the *dimension* of F_J. Let $r = r_\emptyset$ and $d = d_\emptyset$; then $d \leq d_J$ for each subset J of $\mathbb{N}_p$ since $L_\emptyset \subseteq L_J$. If $I \subset J \subseteq \mathbb{N}_p$ and F_I and F_J are both nonempty, then F_I is called *a boundary face of* F_J. In this case $d_I < d_J$. For, let $j_0 \in J \backslash I$ and take $x \in F_J$; so $v_{j_0}^t x > 0$. If $d_I = d_J$, then $V_I = V_J$ since $V_J \subseteq V_I$ and $r_J = r_I$. But now $v_{j_0} = \Sigma_{j \notin J} \alpha_j v_j$ and hence $0 < v_{j_0}^t x = \Sigma_{j \notin J} \alpha_j v_j^t x = 0$.

A subset T of K^n is *linearly convex* if whenever $v, w \in T$ and $\alpha, \beta \in K^+$ with $\alpha + \beta = 1$, then $\alpha v + \beta w \in T$. For each subset A of K^n certainly A' is linearly convex. The intersection of all the linearly convex subsets of K^n which contain the subset S is called the *convex hull of* S. It is, of course, the smallest linearly convex subset of K^n which contains S. For any subset X of K^n, clearly, $K^+ X$ is linearly convex, and it is easy to see that the convex hull of $\cup_i K^+ X_i$ is $K^+(\cup_i X_i)$ for any family of subsets $\{X_i\}$ of K^n.

Theorem 6.4.5. *Let K be a totally ordered field and let $A = \{v_1, \ldots, v_p\} \subseteq K^n$.*

(a) If F_J is a face of A' of dimension d_J with $d_j \geq d + 2$ and $x_0 \in F_J$, then there exist boundary faces F_{J_1} and F_{J_2} of F_J, each of dimension $\geq d + 1$, and vectors $x_1 \in F_{J_1}$, $x_2 \in F_{J_2}$ such that $x_0 = x_1 + x_2$.

(b) A' is the convex hull of $\cup\{F_J : d_J \le d+1\}$.

Proof. (a) Since $L_\emptyset + Kx_0$ is at most $(d+1)$-dimensional and $L_\emptyset + Kx_0 \subseteq L_J$ there exists $y \in L_J \backslash (L_\emptyset + Kx_0)$, and $L = Kx_0 + Ky$ is a two-dimensional subspace of L_J. Let $M = L \cap A'$. For any $x = tx_0 + sy \in L$ we have $v_j^t x = 0$ if $j \notin J$ and hence $x \in M$ if and only if $v^t x = tv_j^t x_0 + sv_j^t y \ge 0$ for every $j \in J$; that is, if we let $m_j = -(v_j^t y)(v_j^t x_0)^{-1}$, then $x \in M$ precisely when $t \ge m_j s$. Moreover, $t > m_j s$ if and only if $v_j^t x > 0$. Let $m' = \vee_{j\in J} m_j$ and $m'' = \wedge_{j\in J} m_j$. Then $m'' < m'$. Otherwise, $v_j^t(m'x_0 + y) = v_j^t(m_j x_0 + y) = 0$ for $j \in J$, $m'x_0 + y \in L_\emptyset$, and we have the contradiction $y \in L_\emptyset + Kx_0$. Let $m = m' - m''$, $s_1 = m^{-1}$, $t_1 = m'm^{-1}$, $s_2 = -m^{-1}$, $t_2 = -m''m^{-1}$, $x_1 = t_1 x_0 + s_1 x'$ and $x_2 = t_2 x_0 + s_2 x'$; then $x_0 = x_1 + x_2$. For $i = 1,2$ let $J_i = \{j \in J : t_i > m_j s_i\}$ and note that since $t_1 s_1^{-1} = m'$, $j \in J_1$ iff $m' > m_j$ or, equivalently, $v_j^t x_1 > 0$, and since $t_2 s_2^{-1} = m''$, $j \in J_2$ iff $m'' < m_j$, or equivalently, $v_j^t x_2 > 0$. In particular, if $m' = m_{j'}$ and $m'' = m_{j''}$, then $j'' \in J_1 \backslash J_2$, $j' \in J_2 \backslash J_1$, $x_i \in F_{J_i}$ for $i = 1,2$, and F_{J_i} is a boundary face of F_J of dimension $\ge d+1$ since $\emptyset \ne J_i \subset J$.

(b) If $A' = F_\emptyset$ we are done. Assume $F_\emptyset \subset A'$. If F_J is a face of A' with $d_J \ge d+2$, then by repeatedly using (a) we get that any $x \in F_J$ may be written as $x = x_1 + \cdots + x_q$ where each x_i is in a $(d+1)$-dimensional face of A'. Since qx_i is in the same face as x_i and $x = q^{-1}(qx_1) + \cdots + q^{-1}(qx_q)$ is in the convex hull of $\cup\{F_J : d_J \ge d+1\}$, A' is contained in this convex hull. Since A' is linearly convex it contains the convex hull of $\cup\{F_J : d_j \le d+1\}$. □

Theorem 6.4.6. *Suppose K is a totally ordered field and A is a finite subset of K^n. Then $A'' = K^+A$. Moreover, there is a finite subset B of K^n such that $A' = K^+B$ and $B' = K^+A$.*

Proof. Let $A = \{v_1, \ldots, v_p\}$ and let d be the dimension of $F_\emptyset$. We claim that for each face F_J of A' with dimension $d_J \le d+1$ there is a finite subset X_J of K^n with $F_J = K^+X_J$. Consider $F_\emptyset$ first. If $d = 0$, let $X_\emptyset = 0$, and if $d > 0$ let $X_\emptyset = Y \cup -Y$ where Y is a basis for $L_\emptyset = F_\emptyset$. Now consider a face F_J with $d_J = d+1$. If $v \in F_J$, then $L_J = L_\emptyset + Kv$. So if $y \in F_J$, then $y = x + \gamma v$ with $x \in L_\emptyset$ and $\gamma \in K$, and hence $0 < v_j^t y = \gamma v_j^t v$ for each $j \in J$. Thus, $\gamma > 0$ and we can let $X_J = X_\emptyset \cup \{v\}$. By Theorem 6.4.5 A' is the convex hull of $\cup_J\{K^+X_J : d_J \le d+1\}$ which, as previously noted is K^+B where $B = \cup_J\{X_J : d_j \le d+1\}$; so $A' = K^+B$.

To see that $A'' \subseteq K^+A$ let a be the $n \times p$ matrix $a = (v_1 \cdots v_p)$ and take $v \in A''$. By Theorem 6.4.4 we can find $0 \le \begin{pmatrix} \alpha \\ x \end{pmatrix} \in K^{p+1}$ and $z \in K^n$ such that $z^t(-v\ a) \ge 0$, $(-v\ a)\begin{pmatrix} \alpha \\ x \end{pmatrix} = 0$ and $-z^t v + \alpha > 0$. Since $v_j^t z = z^t v_j \ge 0$ for $j = 1, \ldots, p$, $z \in A'$ and hence $z^t v \ge 0$. So $\alpha > z^t v \ge 0$ and from $-\alpha v + ax = 0$ we get $v = \alpha^{-1}ax = \Sigma_{i=1}^p \alpha^{-1}\alpha_i v_i \in K^+A$ where $x^t = (\alpha_1, \ldots, \alpha_p)$. For the reverse inclusion $K^+A \subseteq A''$ take $v = \Sigma_j \alpha_j v_j$ with $\alpha_j \in K^+$ and suppose $x \in A'$. Then $v^t x = \Sigma_j \alpha_j v_j^t x \ge 0$ and hence $v \in A''$. Since $(K^+B)' = B'$ (for any B) we have $K^+A = A'' = (K^+B)' = B'$.

□

Theorem 6.4.7. *Let K be a subfield of $\mathbb{R}$ and suppose A is a K-subalgebra of K_n which is $\mathbb{R}$-irreducible and is an ℓ-algebra.*

(a) *If $v \in K^n$ and A^+v is a partial order of K^n, then A^+v is finitely generated over K^+.*
(b) *The module ${}_AK^n$ has nonzero partial orders and each nonzero partial order contains a minimal A-partial order.*

Proof. (a) Since A is a prime ℓ-algebra, as a vector lattice $A = \oplus_{i=1}^m T_i$ is the direct sum of totally ordered K-archimedean subspaces. Clearly, (T_iv, T_i^+v) is a totally ordered vector lattice over K. But $((T_iv)^+)^- \cap (-(T_iv)^+)^- \subseteq (A^+v)^- \cap (-(A^+v))^- = 0$ by Theorem 6.4.2, and therefore $T_iv = Ka_iv$ with $a_i \in T_i^+$ by Exercise 8. So $A^+v = \Sigma_{i=1}^m T_i^+v = \Sigma_{i=1}^m K^+a_iv$.

(b) The existence of a nonzero partial order of ${}_AK^n$ follows from Exercise 7. If $(K^n)^+$ is such a partial order and $0 \neq v \in (K^n)^+$ let $\{A^+v_\lambda : \lambda \in \Lambda\}$ be a maximal chain in $\{A^+w : w \in (K^n)^+$ and $0 \neq A^+w \subseteq A^+v\}$. We claim that $\cap_\lambda A^+v_\lambda \neq 0$. To see this note first that if S is the $(n-1)$-sphere in $\mathbb{R}^n$, then $S \cap (A^+v_\lambda)^- \neq \emptyset$ for each $\lambda \in \Lambda$ since $(A^+v_\lambda)^-$ is closed under scalar multiplication by elements of $\mathbb{R}^+$. Since S is compact $S \cap (\cap_\lambda A^+v_\lambda) \neq \emptyset$. From (a) we have that $A^+v_\lambda = K^+X_\lambda$ for some finite subset X_λ of K^n; consequently, $(A^+v_\lambda)^- = \mathbb{R}^+X_\lambda$ since $\mathbb{R}^+X_\lambda = X_\lambda'$ by Theorem 6.4.6, and it is easily seen that $X_\lambda' = \{v \in \mathbb{R}^n : x^t v \geq 0 \text{ for every } x \in X_\lambda\}$ is closed in $\mathbb{R}^n$. Since $A^+\mathbb{R}^+X_\lambda \subseteq \mathbb{R}^+X_\lambda$ for each λ, $\cap_\lambda (A^+v_\lambda)^-$ contains a basis X of ${}_\mathbb{R}\mathbb{R}^n$ by Exercise 4.2.35 and each X_λ also spans K^n. Since $(A^+v_\lambda)^-$ is a partial order of $\mathbb{R}^n$ by Theorem 6.4.2 we have $(A^+v_\lambda)^- \cap K^n = K^+X_\lambda = A^+v_\lambda$ by Theorem 6.4.3, and hence $\mathbb{R}^+X \cap K^n \subseteq A^+v_\lambda$ for each $\lambda \in \Lambda$. Since there is an invertible matrix a in $\mathbb{R}_n$ with $a(\mathbb{R}^+)^n = \mathbb{R}^+X$ the interior U of $\mathbb{R}^+X$ is not empty and $0 \notin U$. Since K^n is dense in $\mathbb{R}^n$, $\emptyset \neq U \cap K^n \subseteq \cap_\lambda A^+v_\lambda$ and hence $\cap_\lambda A^+v_\lambda \neq 0$ as claimed. Now, if $0 \neq u \in \cap_\lambda A^+v_\lambda$, then $0 \neq A^+u$ is a minimal A-partial order of K^n. For, suppose (K^n, P) is a po-module over A with $0 \neq x \in P \subseteq A^+u$. Then $A^+x = Av_\mu$ for some $\mu \in \Lambda$ and hence $P = A^+u$ since $A^+v_\mu \subseteq P \subseteq A^+u \subseteq A^+v_\mu$. □

We now show that minimal A-partial orders for K^n are lattice orders and then use this to identify the lattice orders of K_n.

Theorem 6.4.8. *Suppose K is a totally ordered field and A is an irreducible K-subalgebra of K_n which is an ℓ-algebra. Suppose, also, $V = K^n$ is a minimal po-module over A and V^+ is finitely generated over K^+. Then $A = K_n$ and there exists $h \in K_n$ such that $h^{-1}A^+h \subseteq (K^+)_n$ and $h^{-1}V^+ = (K^+)^n$. If A has a canonical K-basis and $1 \in A^+$, then $h^{-1}A^+h = (K^+)_n$.*

Proof. According to Exercise 4.2.36 V^+ contains a basis $\{v_1, \ldots, v_n\}$ of V. Let $v = v_1 + \cdots + v_n$. Then $V^+ = A^+v$ and with the aid of Theorem 6.4.7 we may assume $V^+ = K^+X$ with X finite and $K^+Y \subset V^+$ if $Y \subset X$. Let $x \in X$. There exists an $a \in A^+$ such that $x = av$. For each $i = 1, \ldots, n$, $av_i \leq x$ and hence by Exercise 9 there exists $\alpha_i \in K^+$ with $av_i = \alpha_i x$. So the rank of a is 1, and if $N = \ker a$, then $V^+ = \ell(N;A)^+v$ by Exercise 7. Now, if $w \in V$, then $w = bv - cv = dv$ where b, $c \in A^+$, $bN = cN = 0$, and $d = b - c$. We claim that $w^+ = d^+v$; that is $(dv)^+ = d^+v$.

Clearly, $d^+v \geq 0, w$. Suppose $u \in V$ and $u \geq 0, w$. Then $u = fv$ and $u - w = gv$ with $f, g \in A^+$ and $fN = gN = 0$. Since $dv = w = (f-g)v$, $V = N + Kv \subseteq \ker(d - f + g)$, and hence $d = f - g$. So $f = d + g \geq d^+$ and $u = fv \geq d^+v$. Now, X is a disjoint set of basic elements in ${}_KV$ and it spans ${}_KV$; so X is a canonical basis for the vector lattice ${}_KV$. Let $X = \{w_1, \ldots, w_n\}$ and let h be the matrix with columns $w_1, \ldots, w_n$. Then $h^{-1}V^+ = \Sigma_i h^{-1}K^+ w_i = (K^+)^n$ since $h^{-1}w_i = e_i$ is the ith standard basis vector. If $a \in A^+$, then $h^{-1}ahe_i \in h^{-1}V^+$ and hence $h^{-1}ah \in (K^+)_n$. Thus, we may assume $A^+ \subseteq (K^+)_n$ and $V^+ = (K^+)^n$. We will also assume that $v_i = e_i$, the ith standard basis vector of V. Since $A^+v = (K^+)^n$, for each $i = 1, \ldots, n, e_i = a_i v$ with $a_i \in A^+$; and because $a_i \in (K^+)_n$ the jth row of a_i is 0 for $j \neq i$. So $a_i^t N_i = 0$ where N_i is the K-span of $\{e_1, \ldots, e_{i-1}, e_{i+1}, \ldots, e_n\}$. Now, $(V, \ell(N_i; A^t)^+ e_i)$ is a po-module over the ℓ-algebra $(A^t, (A^+)^t)$ by Exercise 7, and A^t is K-irreducible by Exercise 4.2.33. So $\ell(N_i; A^t)^+ e_i$ contains a basis $\{b_{ij}^t e_i : 1 \leq j \leq n\}$ of V for each $i = 1, \ldots, n$, by Exercise 4.2.36, and the set $\{b_{ij} : 1 \leq i, j \leq n\}$ must be independent. For, suppose $\Sigma_{i,j} \alpha_{ij} b_{ij} = 0$ with $\alpha_{ij} \in K$. Since $b_{ij}^t N_i = 0$, for each $k = 1, \ldots, n$ we have $0 = \Sigma_{i,j} \alpha_{ij} b_{ij}^t e_k = \Sigma_j \alpha_{kj} b_{kj}^t e_k$ and hence $\alpha_{kj} = 0$ for every j. Thus, $A = K_n$. Now, suppose Z is a canonical basis for A. From the fact that $b_{ij}^t N_i = 0$ we get that all rows of b_{ij} except the ith row are 0. If $g \in Z$, then for some b_{ij},

$$b_{ij} = \alpha g + \sum_{h \neq g} \alpha_h h$$

with $\alpha > 0$ and $\alpha_h \in K^+$, and hence only the i-th row of g is nonzero since $Z \subseteq (K^+)_n$. Suppose $1 \in A^+$. Then there exist $g_1, \ldots, g_m \in Z$ and $0 < \alpha_1, \ldots, \alpha_m \in K$ with $1 = \alpha_1 g_1 + \cdots + \alpha_m g_m$ and necessarily $g_i = \alpha_i^{-1} e_{ii}$. So $m = n$ and the matrix unit $e_{ii} \in A^+$ for each i. Given the pair of indices i, j let g be an element of Z with a nonzero (i, j)-th entry β. Then $0 < ge_{jj} = \beta e_{ij} \in A$ and $e_{ij} \in A^+$. So $A^+ = (K^+)_n$. □

The previous theorem becomes more definitive when the field is archimedean.

Theorem 6.4.9. *Let A be an $\mathbb{R}$-irreducible K-subalgebra of K_n where K is a subfield of $\mathbb{R}$. If A is an ℓ-algebra over K then $A = K_n$, and if $1 \in A^+$ then (A, A^+) is isomorphic to $(K_n, (K^+)_n)$.*

Proof. By Theorem 6.4.7 the A-module $V = K^n$ has a partial order V^+ for which it is a minimal po-module and such that V^+ is K^+-finitely generated. Thus, $A = K_n$. Moreover, A is a direct sum of totally ordered K-subspaces. If T is one of these subspaces, then, since we may assume $A^+ \subseteq (K^+)_n$, we have $(T^+)^- \cap (-T^+)^- \subseteq (\mathbb{R}^+)_n \cap (-\mathbb{R}^+)_n = 0$. So T is one-dimensional by Exercise 8 and hence A has a canonical K-basis. So if $1 \in A^+$, then $A^+ = (K^+)_n$. □

One consequence of Theorem 6.4.9 is that the field $K(i)$ and the division ring $\mathbb{H}(K)$ of quaternions over K, as well as their matrix algebras, are not ℓ-algebras over K; see Exercise 16. This is a good place to verify that there are no other finite dimensional division algebras over $\mathbb{R}$.

Theorem 6.4.10. *The only noncommutative algebraic division algebra over a real closed field is the quaternion algebra over this field.*

Proof. Let K be a real closed field and let D be an algebraic division algebra over K with a proper center Z. Now $[Z:K] \leq 2$ by Theorem 5.1.6 and if $[Z:K] = 2$, then Z would be algebraically closed and hence $D = Z$. So $Z = K$. By Zorn's Lemma D has a maximal subfield E and necessarily $E = K(i)$ with $i^2 = -1$; so $[D:K] = 4$ by Theorem 4.2.20. By Theorem 4.2.21 there is an element $u \in D$ such that $u^{-1}iu = -i$; that is, the automorphism $a+bi \mapsto a-bi$ of E is the restriction of the inner automorphism of D determined by u. We must have $u^2 < 0$ in K. First of all u^2 is in E since it commutes with each of the elements in E. Since $u \notin E$ and $K \subseteq K(u) \cap E \subset E$ we have $K = K(u) \cap E$ and $u^2 \in K$. If $u^2 > 0$, then $u^2 = y^2$ with $y \in K$ and u is an element of K; so $u^2 = -x^2$ with $0 < x \in K$. Let $j = ux^{-1}$. Then $j^2 = -1$ and $j^{-1}ij = u^{-1}iu = -i$; that is, $ij = -ji$. Let $k = ij$. Then $k^2 = ijij = -i^2j^2 = -1$ and $1, i, j, k$ are K-independent. For if $a+bi+(c+di)j = 0$ with $a, b, c, d \in K$ and $c+di \neq 0$, then $-j = (c+di)^{-1}(a+bi) \in E$. So $c+di = 0$ and hence $a = b = c = d = 0$. Thus, $D = K + Ki + Kj + Kk$ is the division ring of quaternions over K. □

In order to identify all of the ℓ-algebra lattice orders of K_n, K a subfield of $\mathbb{R}$, we will first construct a plenary subset of them. Let K be a unital po-ring and let $a = (\alpha_{ij}) \in (K^+)_n$ be a matrix with the property that a^t is a unit in K_n. For $1 \leq i$, $j \leq n$ let

$$a_{ij} = e_{ij}a^t. \tag{6.4.34}$$

Then, denoting the jth column and the ith row of a matrix u by $u^{(j)}$ and $u_{(i)}$ respectively, $(a_{ij})_{(i)} = (a^{(j)})^t$ and $(a_{ij})_{(p)} = 0$ if $p \neq i$. Also, for all $1 \leq i, j, p, q \leq n$ and $\beta, \gamma \in K$

$$\beta a_{ij} \gamma a_{pq} = \beta \alpha_{pj} \gamma a_{iq} \tag{6.4.35}$$

and $\{a_{ij} : 1 \leq i, j \leq n\} = \{e_{ij}\}a^t$ is a K-basis for the left K-module K_n. Let $P(a)$ be the partial order of K_n which has the set $\{a_{ij} : 1 \leq i, j \leq n\}$ as a canonical left K-basis. Then $(K_n, P(a)) = \oplus_{i,j} K a_{ij}$, $P(a) = (K^+)_n a^t$, and as a consequence of (6.4.35) $(K_n, P(a))$ is a po-ring and each of the summands Ka_{ij} are subrings. Each a_{ij}, being part of a canonical basis for ${}_KK_n$, is a $\overline{d}$-element on K which is a d-element on K provided K is po-unital; that is, $(\gamma \vee \delta)a_{ij} = \gamma a_{ij} \vee \delta a_{ij}$ provided $\gamma, \delta \in K$ and $\gamma \vee \delta$ exists, and, dually. Moreover, $(K_n, P(a))$ is a left f-module over K precisely when K is a left f-ring, and it is a strong left ℓ-module over K precisely when K is a right f-ring. Also, $(K_n, P(a))$ is a right po-module over K if and only if $a^t K^+ (a^t)^{-1} \subseteq (K^+)_n$.

Note that if u is any unit of K_n, then the right ideal of K generated by the entries in a row of u and the left ideal generated by the entries in a column of u are both equal to K. It is possible to have u but not u^t a unit; see Exercise 20.

Theorem 6.4.11. *Let K be a totally ordered field and suppose the matrix algebra $R = K_n$ is a po-algebra over K. Then R is isomorphic to $(K_n, P(a))$ for some invertible matrix a in $(K^+)_n$ if and only if R has a canonical K-basis and K^n is a minimal po-module over R. If K is archimedean and R is an ℓ-algebra, then R is isomorphic to $(K_n, P(a))$ for some a.*

Proof. Suppose the po-algebra R is isomorphic to $(K_n, P(a))$. Then R certainly has a canonical basis. Moreover, by Exercise 4.2.37 $P(a) = b^{-1}R^+b$ for some $b \in K_n$ and $(K^n, b(K^+)^n)$ is an ℓ-module over R since $(K^n, (K^+)^n)$ is an ℓ-module over $(K_n, P(a))$:

$$R^+b(K^+)^n = bb^{-1}R^+b(K^+)^n = bP(a)(K^+)^n \subseteq b(K^+)^n.$$

To show that $(K^n, b(K^+)^n)$ is a minimal R-po-module it suffices to verify that $(K^n, (K^+)^n)$ is a minimal $(K_n, P(a))$-po-module. Let $0 \neq v \in (K^+)^n$ and suppose the qth-component of $a^t v$ is $\alpha > 0$. Then, for $j = 1, \ldots, n$, $a_{jq}v = e_{jq}a^t v = \alpha e_j$ where e_j is the jth standard basis vector of K^n. So $(K^+)^n \subseteq P(a)v$ and hence $(K^n, (K^+)^n)$ is a minimal $(K_n, P(a))$-po-module.

For the converse we may assume by Theorem 6.4.8 that $R^+ \subseteq (K^+)_n$. Also, if Z is a canonical K-basis for K_n, then from the proof of Theorem 6.4.8 we know that each element of Z has exactly one nonzero row. Let $Z_i = \{b \in Z : e_{ii}b \neq 0\}$. Then $Z_i \subseteq e_{ii}K_n$ and hence Z_i is a basis for $e_{ii}K_n$. If $b \in Z_i$, then for any j we have $0 < e_{ji}b \in K^+Z_j$. Suppose $e_{ji}b = \alpha u + \beta v + \cdots$ with u, $v \in Z_j$ and $\alpha, \beta > 0$. Then $b = e_{ij}e_{ji}b = \alpha e_{ij}u + \beta e_{ij}v + \cdots$. But $e_{ij}u > 0$ and $e_{ij}v > 0$ since left multiplication by e_{ij} is an isomorphism between the right ideals $e_{jj}K_n$ and $e_{ii}K_n$. Since b is K-convex we have the contradiction $e_{ij}u = \gamma e_{ij}v$ for some $0 < \gamma \in K$. So, taking $j = 1$, $b \in K^+e_{i1}Z_1$ and

$$R^+ = \sum_{i=1}^{n} K^+e_{i1}Z_1 = \sum_{i,p=1}^{n} K^+e_{i1}b_p$$

where $Z_1 = \{b_1, \ldots, b_n\}$. Let $a = b_1^t e_{11} + \cdots + b_n^t e_{1n}$. Then a is a unit since, for $p = 1, \ldots, n$, the pth column of a is the first column of b_p. Now, for $1 \leq i, p \leq n$,

$$a_{ip} = e_{ip}a^t = e_{ip}(e_{11}b_1 + \cdots + e_{p1}b_p + \cdots + e_{n1}b_n) = e_{i1}b_p;$$

hence, $R^+ = \Sigma_{i,p}K^+a_{ip}$ and $R^+ = P(a)$.

If K is archimedean and R is an ℓ-algebra, then K^n is R-minimal by Theorem 6.4.7 and as we have seen in Theorem 6.4.6 R has a canonical K-basis. □

Some properties of the lattice orders $P(a)$ are given in Exercises 10–14. The previous theorem leads to the second main conjecture for matrix algebras which we state again. If K is a totally ordered field, then each ℓ-algebra lattice order of K_n is isomorphic to $P(a)$ for some unit a in K_n with $a \in (K^+)_n$. Of course, the second main conjecture implies the first since if $1 \in P(a)$, then $P(a) = (K^+)_n$.

We now determine some relations among the various $P(a)$ which hold for ℓ-rings more general than totally ordered fields. Let K be a po-unital po-ring and for a fixed $n \in \mathbb{N}$ let G be the subgroup of the general linear group $GL(n, K)$ over K generated by the group of permutation matrices S and the group D of those diagonal matrices $d(\delta_1, \ldots, \delta_n)$ whose diagonal entries are in $\mathscr{U}(K)$: that is, each δ_j is a unit in K. If σ is a permutation on n letters and $(e_{\sigma(1)} \cdots e_{\sigma(n)})$ is the corresponding element of S obtained by permuting the columns of 1, then for any diagonal matrix $d(\delta_1, \ldots, \delta_n)$, $d(\delta_1, \ldots, \delta_n)(e_{\sigma(1)} \cdots e_{\sigma(n)}) = (e_{\sigma(1)} \cdots e_{\sigma(n)})\, d(\delta_{\sigma(1)}, \ldots, \delta_{\sigma(n)})$ and $G = SD$ is the semidirect product of D by S. Each element in G is obtained from a permutation matrix by replacing each occurrence of 1 by an element from $\mathscr{U}(K)$. Let H be the subgroup of G generated by S and the subgroup of D consisting of all $d(\delta_1,\ldots,\delta_n)$ with each δ_p in the center of K and $\delta_p\delta_q^{-1} > 0$ for $1 \le p, q \le n$, and let G_+ be the subgroup of G generated by S and the subgroup of D consisting of all $d(\delta_1,\ldots,d_n)$ will $\delta_i \in \mathscr{U}(K^+)$.

If Z is the center of K and $a \in Z_n$ is invertible in K_n, then it is already invertible in Z_n. For if $\det a$ is not a unit of Z, then it belongs to some maximal ideal P of K. But since the domain $\overline{Z} = Z/P \cap Z$ is embeddable in $\overline{K} = K/P$ and $\det\overline{a} = 0$ we get the contradiction that $\overline{a}\overline{c} = 0$ for some nonzero $\overline{c} \in \overline{Z}_n$.

Theorem 6.4.12. *Let K be a po-unital po-ring with center Z and suppose a, $b \in (K^+)_n$ with a^t, $b^t \in GL(n,K)$. Let G and H be the subgroups of $GL(n,K)$ defined above.*

(a) If $b = ag$ for some $g \in G_+$ and at least one of a, b or g is in Z_n, then $P(a) = P(b)$.
(b) If $P(a) = P(b)$, ${}_KK$ is an indecomposable ℓ-module, and a or $b \in Z_n$, then there exists $g \in G_+$ such that $b = ag$.
(c) If $b = gah$ for some g, $h \in G_+$ with $g \in Z_n$ and either a or b or $h \in Z_n$, then $P(b) = cP(a)c^{-1}$ where $c = (g^{-1})^t \in G_+$.
(d) Suppose ${}_KK$ is an indecomposable ℓ-module and $P(b) = cP(a)c^{-1}$ for some $c \in Z_n$. Then there is a permutation σ such that $c = \Sigma_i\delta_i e_{\sigma(i)i} \in G$, $\delta_i\delta_k^{-1}a_{(k)}$ has all its entries in K^+ for all i and k, and there exists $d \in G_+$ such that $b^t = dca^tc^{-1}$. Moreover, $b = gah$ with g, $h \in G$ if $a \in Z_n$ or $b \in Z_n$.

Proof. (a) By symmetry we may assume a or g is in Z_n. Then b is obtained from a by permuting the columns of a and multiplying each of these columns on the left by an element of $\mathscr{U}(K^+)$. So, for $i = 1,\ldots,n$,

$$\{Ka_{ij} : 1 \le j \le n\} = \{Kb_{ij} : 1 \le j \le n\} \tag{6.4.36}$$

and $P(a) = P(b)$ since $Ka_{ij} = Kb_{iq}$ implies $K^+a_{ij} = K^+b_{iq}$.

(b) Since a_{ij}, $b_{ij} \in e_{ii}K_n$, $Ka_{ij} = \Sigma_q Ka_{ij} \cap Kb_{iq}$ and the indecomposability of ${}_KK$ gives $Ka_{ij} \subseteq Kb_{iq}$ for some q. Similarly, $Kb_{iq} \subseteq Ka_{ip}$ for some p and hence $Ka_{ij} = Kb_{iq}$, (6.4.36) holds, and $a_{ij} = \gamma b_{iq} = b_{iq}\gamma$ with $\gamma \in \mathscr{U}(K^+)$. In particular, $b_{1j} = \delta_j a_{1\sigma(j)}$ for $j = 1,\ldots,n$, where $\delta_j \in \mathscr{U}(K^+)$ and σ is a permutation; that is, $b^{(j)} = a^{(\sigma(j))}\delta_j$. Let $g = \delta_1 e_{\sigma(1)1} + \cdots + \delta_n e_{\sigma(n)n}$. Then $g \in G_+$ and since $(ce_{pq})^{(k)} = c^{(p)}$ or 0 depending on whether $k = q$, or $k \ne q$, $(ag)^{(j)} = a^{(\sigma(j))}\delta_j = b^{(j)}$ for each j and hence $b = ag$.

(c) By symmetry we may assume b or h is in Z_n. Since $ga = bh^{-1}$ and $P(ga) = P(bh^{-1}) = P(b)$ by (a), we may also assume that $h = 1$ and $b = ga$. So $P(b) = (K^+)_nb^t = (g^{-1})^t(K^+)_na^tg^t = cP(a)c^{-1}$.

(d) Since conjugation by c is an isomorphism between the ℓ-rings $(K_n,P(a))$ and $(K_n,P(b))$ which fixes K elementwise, as in (b) the indecomposability of K gives that there are permutations $i \mapsto i'$ and $j \mapsto j'$ such that $Kca_{ij}c^{-1} = Kb_{i'j'}$ and $ca_{ij}c^{-1} = \gamma_{ij}b_{i'j'}$ with $\gamma_{ij} \in \mathscr{U}(K^+)$. So $ce_{ij}a^t = \gamma_{ij}e_{i'j'}b^tc$ and

$$ae_{ji}c^t = \gamma_{ij}c^t be_{j'i'}. \tag{6.4.37}$$

Write $c = (\delta_{ij})$. Then $(e_{ji}c^t)_{(j)} = (c^{(i)})^t = (\delta_{1i} \cdots \delta_{ni})$ and the other rows are 0; hence $ae_{ji}c^t = (\delta_{1i}a^{(j)} \cdots \delta_{ni}a^{(j)})$. If $k \neq i'$, then $(c^t be_{j'i'})^{(k)} = 0$ and hence (6.4.37) gives $\delta_{ki}a^{(j)} = 0$ and $\delta_{ki} = 0$. Thus, there is a permutation σ with $c = \Sigma_i \delta_i e_{\sigma(i)i} \in G$ and

$$\begin{aligned} ca_{ij}c^{-1} &= \left(\sum_p \delta_p e_{\sigma(p)p}\right)\left(\sum_k \alpha_{kj}e_{ik}\right)\left(\sum_q \delta^{-1}_{\sigma^{-1}(q)} e_{\sigma^{-1}(q)q}\right) \\ &= \sum_{p,k,q} \delta_p \delta^{-1}_{\sigma^{-1}(q)} \alpha_{kj} e_{\sigma(p)p} e_{ik} e_{\sigma^{-1}(q)q} \\ &= \sum_k \delta_i \delta_k^{-1} \alpha_{kj} e_{\sigma(i)\sigma(k)} \\ &= \gamma_{ij} b_{i'j'}. \end{aligned}$$

So, for every i, j, k, $\sigma(i) = i'$ and $\delta_i\delta_k^{-1}\alpha_{kj} = \gamma_{ij}\beta_{\sigma(i)j'} \geq 0$ where $b = (\beta_{pq})$. Let $j' = \tau(j)$. Then $a^t = \Sigma_p e_{pp} a^t = \Sigma_p a_{pp}$ and

$$ca^t c^{-1} = \sum_p \gamma_{pp} b_{\sigma(p)\tau(p)} = \left(\sum_p \gamma_{pp} e_{\sigma(p)\tau(p)}\right) b^t.$$

Let $d^{-1} = \Sigma_p \gamma_{pp} e_{\sigma(p)\tau(p)}$. Then $d \in G_+$, $ca^t c^{-1} = d^{-1}b^t$, $b^t = dca^t c^{-1}$, and $b = (c^{-1})^t ac^t d^t$ if a or b is in Z_n. □

If a and $(a^t)^{-1}$ (or b and $(b^t)^{-1}$) and c are in $F(K)_n$ or if the entries of c are comparable to zero, then in (d) above we may replace c by $u = \Sigma_i |\delta_i| e_{\sigma(i)i} \in G_+$ to get $b^t = dua^t u^{-1}$. For, the inequalities $\delta_i\delta_k^{-1}\alpha_{kj} \geq 0$ yield $\delta_i\delta_k^{-1} > 0$ since in the first case $F(K) = \Sigma_j F(K)\alpha_{kj}$, and hence in each nontrivial totally ordered homomorphic image $\overline{F(K)}$ of $F(K)$ some $\overline{\alpha}_{kj} > 0$ and consequently $\overline{\delta_i\delta_k^{-1}} > 0$. In either case $\delta_i\delta_k^{-1} = |\delta_i\delta_k^{-1}| = |\delta_i||\delta_k|^{-1}$ and $ua_{ij}u^{-1} = ca_{ij}c^{-1}$.

Note that as a consequence of (b) $P(a) = (K^+)_n$ if and only if $a \in G_+$, provided ${}_K K$ is an indecomposable ℓ-module.

We now return to the first main conjecture for an arbitrary totally ordered field.

Theorem 6.4.13. *The following statements are equivalent for the ℓ-algebra $R = K_n$ over the totally ordered field K.*

(a) R is isomorphic to K_n with its usual order.
(b) $F(R)$ has an element with n values.
(c) $F(R)$ has an element with at least n values.
(d) $1 > 0$ and 1 has at least n values.

Proof. Since $F(K_n, (K^+)_n)$ is the subalgebra consisting of all of the diagonal matrices (a) implies (b) and, trivially, (b) implies (c).

(c) implies (d). Since $F(R)$ is isomorphic to a direct product of n totally ordered division rings it contains a set of n orthogonal idempotents whose sum must be 1.

(d) implies (a). Again, the vector lattice R is the direct sum of totally ordered ℓ-simple subspaces and $F(R)$ is the sum of exactly n of them. So $1 = g_1 + \cdots + g_n$ where $\{g_1, \ldots, g_n\}$ is a complete set of orthogonal disjoint idempotents. If T is one of the totally ordered summands of R, then g_iT and Tg_i are convex subspaces of T. Thus, there exist unique i and j such that $T = g_iT = Tg_j \subseteq g_iRg_j$ and hence $T = g_iRg_j$ is one-dimensional and $T^+ = g_iR^+g_j$. So there are n^2 summands which must be $\{g_iRg_j : 1 \leq i, j \leq n\}$. For each $j \geq 2$ the minimal right ideals g_1R and g_jR are isomorphic and hence by Exercise 4.2.35 there exists a set of matrix units $\{g_{ij} : 1 \leq i, j \leq n\}$ defined as follows. Take $g_{1j} \in g_1R^+g_j$ and $g_{j1} \in g_jR^+g_1$ with $g_{1j}g_{j1} = g_1$ and $g_{j1}g_{1j} = g_j$. Let $g_{ii} = g_i$ and $g_{ij} = g_{i1}g_{1j}$. Then $\{g_{ij}\}$ is a set of positive disjoint matrix units in R and $R = \oplus Kg_{ij} \cong (K_n, (K^+)_n)$. □

The previous result verifies the first main conjecture for those lattice orders of K_n in which $F(K_n)$ has its maximum possible number of totally ordered summands. This conjecture is also true provided, for each $n \geq 2$, K_n has no ℓ-unital lattice order in which $F(K_n)$ has its minimum possible number of summands.

Theorem 6.4.14. *The following statements are equivalent for the totally ordered field K.*

(a) For every $n \geq 1$ each ℓ-unital algebra lattice order of K_n is isomorphic to the usual lattice order of K_n.

(b) For every $n \geq 2$ K_n has no ℓ-unital algebra lattice order for which $F(K_n)$ is one-dimensional.

(c) For every $n \geq 2$ K_n has no ℓ-unital algebra lattice order for which $F(K_n)$ is totally ordered.

Proof. It suffices to show that (b) implies (a) since, trivially, (a) implies (c) and (c) implies (b). Suppose K_n is an ℓ-algebra with $1 > 0$ and $F(K_n)$ is the direct sum of m totally ordered division subalgebras. If $m = 1$, then $F(K_n)$ is central by Exercise 6.2.3 and hence $F(K_n) = K1$; so $m \geq 2$. Write $1 = g_1 + \cdots + g_m$ where $g_1, \ldots, g_m$ are the identity elements of the totally ordered summands of $F(K_n)$. For each i, $g_iK_ng_i$ is a convex ℓ-subalgebra of K_n which is isomorphic to the algebra K_t for some $1 \leq t \leq n$. Suppose $t \geq 2$. Then, again, $g_i = h_1 + \cdots + h_s$ where $s \geq 2$ and $h_1, \ldots, h_s$ are disjoint idempotents in $F(K_n) \cap g_iK_ng_i$. This is impossible since the h_j are in a totally ordered summand of $F(K_n)$. Thus, each $g_iK_ng_i$ is one-dimensional, $m = n$ by Exercise 4.1.34, and K_n is isomorphic to $(K_n, (K^+)_n)$ by Theorem 6.4.13. □

Another necessary condition for an ℓ-unital matrix ℓ-algebra of size n to have the usual order that turns out to also be sufficient is the positivity of a permutation matrix corresponding to an n-cycle. In order to show this we first establish the following technical result which will be used to produce a matrix ℓ-subalgebra with the usual order in a suitably conditioned ℓ-algebra.

Theorem 6.4.15. *Let R be a unital ℓ-ring and let g, $h \in G$ where G is a finite subgroup of the group of units $\mathcal{U}(R^+)$ of R^+.*

(a) If x and y are comparable elements of R and $y = gxh$, then $y = x$.

(b) If A is a totally ordered subset of R and A and gAh are comparable subsets, then $A = gAh$.

(c) Suppose A is a totally ordered convex subgroup of the additive group of R and $(A \cap F(R))^2 \neq 0$. Then $gA = Ah$ if and only if $gA = hA = Ag = Ah$.

(d) Suppose A is a subset of R, g is of order n, and m and $r \in \mathbb{N}$ are minimal with $g^m A = A$ and $g^r A = Ag^r$. Then m and r are divisors of n. If A is a totally ordered convex subgroup of R and $(A \cap F(R))^2 \neq 0$, then r is a divisor of m.

Proof. For (a), if $x \leq y$ then $y = gxh \leq gyh \leq \cdots \leq g^p y h^p = y$ for some p; so $gxh = gyh$ and $x = y$. For (b), if $gAh \subseteq A$ and $x \in A$, then $y = gxh \in A$ and $x = y$ by (a); so $x = gxh \in gAh$ and $A = gAh$. For (c), suppose $gA = Ah$. Since gA and hA are totally ordered convex subgroups of R they are comparable or disjoint by Exercise 2.4.10, and if they are comparable they are equal by (b). Suppose $gA \cap hA = 0$ and take $0 < a \in A \cap F(R)$ with $a^2 > 0$. Then $ga = ah$ by (a) and from $ga \wedge ha = 0$ we have $0 = ga \wedge aha = ga \wedge ga^2 = g(a \wedge a^2)$. So $0 = a \wedge a^2 = a^2 \wedge a^2$ gives the contradiction $a^2 = 0$. Similarly, $Ag = Ah$. For (d), let H be the subgroup of G generated by g. H acts on the subsets of R by left multiplication and the stabilizer of A under this action, $\{g^p : g^p A = A\}$, is generated by g^m with m a divisor of n. Similarly, H acts on the subsets of R by conjugation and the stabilizer of A under this action, $\{g^p : g^p A = Ag^p\}$, is generated by g^r with r a divisor of n. If A is a totally ordered convex subgroup of R with $(A \cap F(R))^2 \neq 0$, then g^m is in this latter stabilizer by (c). So $\frac{n}{m}$ is a divisor of $\frac{n}{r}$ and hence r is a divisor of m. □

Theorem 6.4.16. *Let R be a unital ℓ-algebra over the totally ordered field K and let g be a unit of R^+ which has finite order $n \geq 2$. Suppose A is a totally ordered convex subspace of R and let $B = \Sigma_{i,j=1}^n g^i A g^j$. Let m and $r \in \mathbb{N}$ be minimal with $g^m A = A$ and $g^r A = Ag^r$. Assume that $B \cap F(R)$ contains a nonzero idempotent that commutes with g. Then B contains an ℓ-subalgebra of R which is isomorphic to the matrix ℓ-algebra $(K_r, (K^+)_r)$. If $r = 1$, then B contains an ℓ-subalgebra of R which is isomorphic to the group ℓ-algebra $(K[W], K^+[W])$ where W is the cyclic group of order m.*

Proof. Let $0 < b \in B \cap F(R)$ with $b^2 = b$ and $bg = gb$. From (b) of Theorem 6.4.15, $g^i A g^j$ and $g^p A g^q$ are equal or disjoint and hence B is the direct sum of the distinct $g^i A g^j$. So $b = a + c$ with $0 < a \in g^p A g^q \cap F(R)$ for some p and q and some $c \in B \cap F(R)$ with $a \wedge c = 0$. Then a and c are disjoint idempotents and since $B = \Sigma_{i,j=1}^n g^i (g^p A g^q) g^j$ we may assume $a \in A$. By (c) and (d) of Theorem 6.4.15, m is also minimal with $A = Ag^m$ and r is a divisor of m while m is a divisor of n. Note that r is minimal such that g^r is in the centralizer of A and m is minimal such that multiplication by g^m fixes the elements of A. If $r = 1$, then $B = A \oplus Ag \oplus \cdots \oplus Ag^{m-1}$; and $ga = ag$ as well as $(ga)^m = g^m a = a$ by (a) of Theorem 6.4.15. Let $S = Ka \oplus Kag \oplus \cdots \oplus Kag^{m-1} \subseteq B$. Then a is the identity of S and ag has order m in $\mathscr{U}(S^+)$.

Suppose $r \geq 2$. We first note that $B = \oplus_{i=1}^r \oplus_{j=1}^m g^i A g^j$. To see this, suppose $g^i A g^j = g^p A g^q$ with $1 \leq i, p \leq r$ and $1 \leq j, q \leq m$. Then $g^{i-p} A = A g^{q-j}$ and hence

$Ag^{i-p} = g^{i-p}A = Ag^{q-j} = g^{q-j}A$ by (c) of Theorem 6.4.15. Thus, $i = p$ and $j = q$ and the sum is direct by (b) of Theorem 6.4.15. For $1 \le i, j \le n$ write $i = pr + t$ and $j + pr = qm + s$ with $1 \le t \le r$ and $1 \le s \le m$. Then $g^iAg^j = g^tg^{pr}Ag^j = g^tAg^{qm}g^s = g^tAg^s$ and this sum is all of B. Now, $b = \Sigma_{i=1}^r \Sigma_{j=1}^m g^i b_{ij} g^j$ with $b_{ij} \in A^+$ and $\{g^ib_{ij}g^j : 1 \le i \le r, 1 \le j \le m\}$ is a set of disjoint idempotents in $F(R)$. From $gb = bg$ we have $\Sigma_{i=1}^r \Sigma_{j=1}^m g^{i+1} b_{ij} g^j = \Sigma_{i=1}^r \Sigma_{j=1}^m g^i b_{ij} g^{j+1}$ and by comparing components we have $g^{i+1}b_{ij}g^j = g^{i+1}b_{i+1,j-1}g^j$ for $1 \le i \le r-1$ and $1 \le j \le m$ where the second subscript j is taken modulo m using the representatives $1, 2, \ldots, m$. So $b_{ij} = b_{i+1,j-1}$ for $1 \le i \le r-1$ and $1 \le j \le m$ and hence $b_{1j} = b_{i,j-(i-1)} = b_{i,j+1-i}$ and $b_{ij} = b_{1,j+i-1}$ for $1 \le i \le r$ and $1 \le j \le m$. If $b_{1j} \le b_{1q}$ with $j \ne q$, then $b_{1j} = 0$. For,

$$0 \le gb_{1j}g^jg^q = (gb_{1j}g^j)^2g^q \le (gb_{1j}g^j)(gb_{1q}g^q)g^j = 0$$

and so $b_{1j} = 0$. Thus, since A is totally ordered there is a unique q with $1 \le q \le m$ and $b_{1q} > 0$, and

$$\begin{aligned} b &= \sum_{i=1}^{r}\sum_{j=1}^{m} g^i b_{1,j+i-1} g^j = \sum_{i=1}^{r}\sum_{k=i}^{m+i-1} g^i b_{1k} g^{k-i+1} \\ &= \sum_{k=1}^{m}\sum_{i=1}^{r} g^i b_{1k} g^{k+1-i} = \sum_{i=1}^{r} g^i b_{1q} g^{q+1-i} \in \bigoplus_{i=1}^{r} g^i A g^{q+1-i}. \end{aligned}$$

So $a \in A \cap (\oplus_{i=1}^r g^iAg^{q+1-i})$ and $A = g^iAg^{q+1-i}$ for some i. By (c) of Theorem 6.4.15 $A = g^ig^{q+1-i}A = g^{q+1}A$. So $q = m-1$ and $b = \Sigma_{i=1}^r g^ib_{1,m-1}g^{-i}$. For $1 \le i, j \le r$ let $h_{ij} = g^ib_{1,m-1}g^{-j}$. Since the terms in the preceding sum are orthogonal idempotents $b_{1,m-1}$ is an idempotent and $b_{1,m-1}g^{p-j}b_{1,m-1} = 0$ if $1 \le p \ne j \le r$. Thus, $\{h_{ij} : 1 \le i, j \le r\}$ is a set of positive matrix units and $\oplus_{i,j=1}^r Kh_{ij}$ is an ℓ-subalgebra of R which is contained in B and is isomorphic to $(K_r, (K^+)_r)$. □

Recall from Exercise 2.4.11 that the basic subgroup of an ℓ-group is the subgroup generated by its basic elements and it is a direct sum of its maximal convex totally ordered subgroups and each of these latter subgroups is a minimal polar.

Theorem 6.4.17. *Let R be a unital ℓ-algebra over the totally ordered field K and suppose 1 is in the basic subgroup B of R. Then B is an ℓ-subalgebra of R which is isomorphic to $(K_n, (K^+)_n)$ for some n if and only if there is a unit g in R^+ of order n such that the intersection of the centralizer of B with the multiplicative subgroup generated by g is $\{1\}$ and $\ell(g-1; B) \cap r(g-1; B)$ is one-dimensional.*

Proof. Suppose $B \cong (K_n, (K^+)_n)$. Since the case $n = 1$ is trivial we assume $n \ge 2$. Let g be the permutation matrix in B corresponding to an n-cycle. Specifically, let $g = \sum_{j=1}^{n-1} e_{j,j+1} + e_{n1}$. If $u \in \ell(g-1; B) \cap r(g-1; B)$, then all of the columns of u are the same since $ug^j = u$ and all of the rows of u are the same since $g^ju = u$. So $u \in K(\Sigma_{i,j}e_{ij})$ and since $\Sigma_{i,j}e_{i,j} \in \ell(g-1; B) \cap r(g-1; B)$ the intersection is one-dimensional. For the converse we again assume $n \ge 2$. Let A be a maximal convex totally ordered subgroup and put $E = \Sigma_{i,j=1}^n g^iAg^j$ and $t(a) = \Sigma_{i,j=1}^n g^iag^j$ for $a \in A$.

Then $gt(a) = t(a)g = t(a)$ and $1 + g + \cdots + g^{n-1} \in Kt(a) = \ell(g-1;B) \cap r(g-1;B)$ if $a \neq 0$. So $1 \in E$ and hence $B = E$. For, if A_1 is another maximal convex totally ordered subgroup, then $1 \in E \cap \Sigma_{i,j=1}^{n} g^i A_1 g^j$ and hence $A_1 = g^i A g^j \subseteq E$ for some i and j. Now, $g^i A \neq A g^i$ for $1 \leq i \leq n-1$ since otherwise g^i is in the centralizer of B by (a) of Theorem 6.4.15. So B is the direct sum of the $g^i A g^j$ and B contains an ℓ-subalgebra isomorphic to $(K_n, (K^+)_n)$ by Theorem 6.4.16. To show that B is this subalgebra it suffices to verify that A is one-dimensional. But if $a, b \in A$ with $b \neq 0$, then $t(a - \alpha b) = t(a) - \alpha t(b) = 0$ for some $\alpha \in K$ and hence $a = \alpha b$. □

As an immediate consequence of Theorem 6.4.17 we have the following characterization of the usual matrix ℓ-algebra.

Theorem 6.4.18. *Suppose K is a totally ordered field and K_n is an ℓ-algebra over K. Then K_n is isomorphic to $(K_n, (K^+)_n)$ if and only if $(K_n)^+$ contains a conjugate of an n-cycle.* □

Let K be a unital po-ring and let G be a group. For each $p \in \mathbb{N}$ the ring $K_p[G] \cong (K[G])_p$ is a po-ring with the canonical K-basis $\{ge_{ij} : g \in G,\ 1 \leq i, j \leq p\}$. The technique given in Theorems 6.4.16 and 6.4.17 permits a characterization of these matrix group ℓ-algebras when K is a totally ordered field and G is a finite cyclic group.

Theorem 6.4.19. *Let R be an ℓ-unital finite dimensional ℓ-algebra over the totally ordered field K. There is an integer r in $\mathbb{N}$ and a finite cyclic group G such that R is isomorphic to the ℓ-algebra $K_r[G]$ if and only if R is ℓ-semiprime and $\mathscr{U}(R^+)$ has an element g of finite order such that $\ell(g-1;R) \cap r(g-1;R)$ is one-dimensional. In this case, G is of order q, g is of order rq, and r is the least positive integer for which g^r is a central element.*

Proof. The existence of such a positive unit in $K_r[G]$ is left for Exercise 18. For the converse, as in the proof of Theorem 6.4.17, we have $R = \Sigma_{i,j=1}^{n} g^i A g^j$ where A is a maximal convex totally ordered subgroup of R and n is the order of g. Since the case $n = 1$ is trivial we assume $n \geq 2$. Let m and r be the integers that are given in Theorem 6.4.16: $g^m A = A$ and $g^r A = A g^r$. By (a) of Theorem 6.4.15 $g^m a = a$ for $a \in A$ and hence $g^m a = a$ for each a in R. So $m = n$ and from Theorem 6.4.15 we have that r is a divisor of $n = rq$ and it is the smallest positive integer such that g^r lies in the center of R. Let 1 be the idempotent b in $F(R)$ that is used in the proof of Theorem 6.4.16. From that proof we have $R = \oplus_{i=1}^{r} \oplus_{j=1}^{n} g^i A g^j$, $1 = \Sigma_{i=1}^{r} g^i a g^{-i}$ for an idempotent a in A and $a g^{i-j} a = 0$ for $1 \leq i \neq j \leq r$. From the last part of the proof of Theorem 6.4.17 we see that $A = Ka$. Let $h_{ij} = g^{1-i} a g^{j-1}$ for $1 \leq i \leq r$ and $1 \leq j \leq n$. Then

$$R = \oplus_{i=1}^{r} \oplus_{j=1}^{n} K g^i a g^j = \oplus_{i=1}^{r} \oplus_{j=1}^{n} K h_{ij}$$

since $h_{ij} = g^{r+1-i} a g^{j-1-r}$. Let x be a generator for the cyclic group G of order q and let $\psi : R \longrightarrow K_r[G]$ be the K-vector lattice isomorphism defined by $\psi(h_{ij}) =$

$e_{is}x^{\ell}$ where $j = \ell r + s$ with $1 \le s \le r$ and $0 \le \ell < q$. To see that $\psi(h_{ij}h_{i_1 j_1}) = \psi(h_{ij})\psi(h_{h_1 j_1})$ write $j_1 = \ell_1 r + s_1$ with $1 \le s_1 \le r$ and $0 \le \ell_1 < q$. Then

$$\begin{aligned} h_{ij}h_{i_1 j_1} &= g^{1-i}ag^{j-1}g^{1-i_1}ag^{j_1-1} = g^{1-i}ag^{\ell r+s-i_1}ag^{j_1-1} \\ &= g^{\ell r+1-i}ag^{s-i_1}ag^{j_1-1}. \end{aligned}$$

If $s \neq i_1$, then $h_{ij}h_{i_1 j_1} = 0$ and also $e_{is}x^{\ell}e_{i_1 s_1}x^{\ell_1} = 0$. Now, suppose that $s = i_1$. Then $h_{ij}h_{i_1 j_1} = g^{\ell r+1-i}ag^{j_1-1} = g^{1-i}ag^{\ell r+j_1-1}$, and since $\ell r + j_1 = \ell_2 n + s_2$ with $0 \le \ell_2$ and $1 \le s_2 \le n$, $h_{ij}h_{i_1 j_1} = g^{1-i}ag^{s_2-1} = h_{is_2}$. Now, $s_2 = \ell_3 r + s_3$ with $1 \le s_3 \le r$ and $0 \le \ell_3 < q$; so $\psi(h_{ij}h_{i_1 j_1}) = e_{is_3}x^{\ell_3}$. But $s_1 = s_3$ since $s_3 \equiv s_2 \equiv j_1 \equiv s_1 \pmod r$, and

$$(\ell + \ell_1)r = \ell r + j_1 - s_1 = \ell_2 n + s_2 - s_1 = \ell_2 n + \ell_3 r.$$

So $\ell + \ell_1 = \ell_2 q + \ell_3$ and hence

$$\psi(h_{ij})\psi(h_{i_1 j_1}) = e_{is}x^{\ell}e_{i_1 s_1}x^{\ell_1} = e_{is_1}x^{\ell_3}.$$

□

Exercises.

1. Let K be a totally ordered ring with $\ell(K) = r(K) = 0$. For a cardinal n let K_n be the ring of all column finite $n \times n$ matrices over K. Show that $(K^+)_n$ is maximal among those partial orders P of K_n with $K^+P + PK^+ \subseteq P$.
2. Let K be an ℓ-unital ℓ-ring with no positive idempotents except 0 and 1. Suppose $c \in GL(n,K) \cap (Z(K) \cap F(K))_n$. Show that $c(K^+)_n c^{-1} = (K^+)_n$ iff $c \in H$, the group defined prior to Theorem 6.4.12.
3. Let $\beta \ge 1$ and let $R = (K_2, P_\beta)$ be the ℓ-algebra given in Theorem 6.4.1 with K a totally ordered field.
   (a) Show that $P_\beta \neq P_\beta^t$.
   (b) Determine the automorphism group of R.
   (c) Show that $P_\beta \subseteq P_\gamma$ iff $\beta = \gamma$.
   (d) For $1 < \beta$ and $\beta(1-\beta)^{-1} \le \alpha \le (1-\beta)^{-1}$ let $a_\alpha = e_{11} + e_{12} + \alpha e_{21}$ and $Q_\alpha = a_\alpha(K^+)_2 a_\alpha^{-1}$. Show that $P_\beta \subseteq Q_\alpha$ and $Q_\alpha \neq Q_\rho$ unless $\rho = \alpha$.
   (e) Suppose $\gamma < 0$ and $0 \le \alpha \le -\gamma$. Let $a_{\alpha,\gamma} = e_{11} + \alpha e_{12} + \gamma e_{22}$ and $Q_{\alpha,\gamma} = a_{\alpha,\gamma}(K^+)_2 a_{\alpha,\gamma}^{-1}$. Show that $P_1 \subseteq Q_{\alpha,\gamma}$.
   (f) If $\alpha > 0$ and $\gamma \neq \delta$ show that $Q_{\alpha,\gamma} \neq Q_{\alpha,\delta}$. (Use Exercise 2.)
4. Let K be an ℓ-unital ℓ-ring and let $R = (K_2, P_\beta)$ where P_β is the lattice order given after Theorem 6.4.1.
   (a) Determine $d_\ell(R)$, $d_r(R)$, $f_\ell(R)$, and $f_r(R)$.
   (b) Identify the elements in the partial orders $P_\ell(P_\beta) = \{a \in R : aP_\beta \subseteq P_\beta\}$ and $P_r(P_\beta) = \{a \in R : P_\beta a \subseteq P_\beta\}$.
   (c) Identify the elements in $\overline{d}_\ell(R)$, $\overline{d}_r(R)$, $\overline{f}_\ell(R)$ and $\overline{f}_r(R)$ where $a \in \overline{d}_\ell(R)$ or $a \in \overline{f}_\ell(R)$ means left multiplication by a is a lattice homomorphism or is

an f-homomorphism, respectively, and, similarly, $a \in \overline{d}_r(R)$ or $a \in \overline{f}_r(R)$ means right multiplication by a has the corresponding property.

5. Let K be the totally ordered quotient field of the totally ordered commutative domain A. Suppose (A_2, P) is an ℓ-ring with $A^+P \subseteq P$ and let (K_2, P^e) be the ℓ-algebra of quotients of (A_2, P). Show that $(K_2, P^e) \cong (K_2, (K^+)_2)$ iff $F((A_2, P)) \neq 0$.

6. Let K be a po-unital po-ring and let P_β be the partial order of K_2 given after Theorem 6.4.1.

   (a) If $a = \begin{pmatrix} 0 & 1 \\ 1 & 1 \end{pmatrix} \in K_2$ find $c \in GL(2,K)$ such that $P(a) = cP_1c^{-1}$.

   (b) If $\beta > 1$ and $a = \begin{pmatrix} 1 & 1 \\ \beta & \beta - 1 \end{pmatrix}$ find $c \in GL(2,K)$ with $P(a) = cP_\beta c^{-1}$.

7. Suppose ${}_RM$ is a faithful R-module and R is a po-ring with $R^+ \neq 0$. Let N be maximal in $\{r(a;M) : 0 \neq a \in R^+\}$, let $x \in M \backslash N$, and let $M^+ = \ell(N;R)^+x$. Show that $M^+ \neq 0$ and (M, M^+) is a po-module over R.

8. Suppose K is a subfield of $\mathbb{R}$ and T is a nonzero subspace of K^n such that (T, T^+) is a totally ordered vector lattice over K. If $(T^+)^- \cap (-T^+)^- = 0$ show that T is 1-dimensional. (Use Exercise 3.1.21 and Theorem 5.3.12.)

9. Let V be a po-vector space over the totally ordered division ring D and let X be a subset of V^+ such that $V^+ = D^+X$ and $D^+Y \subset V^+$ if $Y \subset X$. Show that each element of X is a D-convex element of V.

10. Let K be a po-unital po-ring, let $a = (\alpha_{ij}) \in (K^+)_n$ with a^t a unit of K_n and let a_{ij} be the matrices given in (6.4.34).

    (a) If the ideal of K generated by its set of positive nilpotent elements is proper show that at most $n^2 - n$ of the a_{ij} are nilpotent.
    (b) Show that $(K_n, P(a))$ is a po-domain iff K is a po-domain and 0 is not an entry of a.
    (c) Suppose K is a totally ordered domain. Show that $P(a) = (K^+)_n$ iff $P(a)$ contains a set of $n^2 - n$ disjoint nilpotent elements.

11. Let K and a be as in Exercise 10 with K an ℓ-ring and let $R = (K_n, P(a))$, $S = (K_n, (K^+)_n)$ and $V = (K^n, (K^+)^n)$ where the elements of V are $n \times 1$ matrices. By V^t we mean the K-ℓ-module direct sum K^n whose elements are $1 \times n$ matrices. Let $\alpha \in K$. Verify each of the following.

    (a) $\Sigma_{i,j}\beta_{ij}a_{ij} \in P_\ell(P(a))$ iff $(\beta_{ij})a^t \in (K^+)_n$ (see Exercise 4).
    (b) $\Sigma_{i,j}\beta_{ij}a_{ij} \in P_r(P(a))$ iff $a^t(\beta_{ij}) \in (K^+)_n$.
    (c) $\alpha a_{ij} \in \overline{d}_\ell(R)$ iff $b \wedge c = 0$ in S implies $\alpha(a^t)_{(j)}b \wedge \alpha(a^t)_{(j)}c = 0$ in V^t.
    (d) $\alpha a_{ij} \in \overline{d}_r(R)$ iff $b \wedge c = 0$ in S implies $b(a^t)^{(i)}\alpha \wedge c(a^t)^{(i)}\alpha = 0$ in V.
    (e) $\alpha a_{ij} \in \overline{f}_\ell(R)$ iff $b \wedge c = 0$ in S implies $\alpha(a^t)_{(j)}b \wedge c_{(i)} = 0$ in V^t.
    (f) $\alpha a_{ij} \in \overline{f}_r(R)$ iff $b \wedge c = 0$ in S implies $b(a^t)^{(i)}\alpha \wedge c^{(j)} = 0$ in V.

12. Continue with the notation in Exercise 11 but now assume K is an almost f-ring. For $1 \le i, j \le n$ let $A_{ij} = B_{ij} \cap C_{ij}$ where

$$B_{ij} = \ell_\ell(\{\alpha_{pj} : p \neq i\}; K) \cap \{\alpha \in K : |\alpha|\alpha_{ij} \in f_\ell(K)\}$$

and

$$C_{ij} = r_\ell(\{\alpha_{iq} : q \neq j\}; K) \cap \{\alpha \in K : \alpha_{ij}|\alpha| \in f_r(K)\}.$$

(a) Show that B_{ij} is a convex ℓ-$F_\ell(K)$-submodule of ${}_{F_\ell(K)}K$ and C_{ij} is a convex ℓ-$F_r(K)$-submodule of $K_{F_r(K)}$.
(b) Show that $\alpha \in B_{ij}$ iff $\alpha a_{ij} = \alpha\alpha_{ij}e_{ii}$ and $\alpha\alpha_{ij} \in F_\ell(K)$; and $\alpha \in C_{ij}$ iff $(a\alpha)_{(i)} = e'_j\alpha_{ij}\alpha$ and $\alpha_{ij}\alpha \in F_r(K)$.
(c) Show that $F_\ell(R) = \Sigma_{i,j}B_{ij}a_{ij}$, $F_r(R) = \Sigma_{i,j}C_{ij}a_{ij}$ and $F(R) = \Sigma_{i,j}A_{ij}a_{ij}$.
(d) Let $B_i = \Sigma_j B_{ij}\alpha_{ij}$ and $A_i = \Sigma_j A_{ij}\alpha_{ij}$. Show that $F_\ell(R) = \Sigma_i B_i e_{ii}$ and $F(R) = \Sigma_i A_i e_{ii}$.
(e) If A_{ij} contains a regular element show that α_{ij} is a unit and $a_{ij} = \alpha_{ij}e_{ii}$.

13. Continue with the notation in Exercise 12 and suppose the nonzero entries of a are regular elements of the almost f-ring K. Let

$$I = \{i : \text{ for some } j,\ \alpha_{ij} \text{ is the only nonzero entry in } a^{(j)}\},$$
$$J = \{i : a_{(i)} \text{ has a single nonzero entry}\},$$

and $Y = I \cap J$. Assume that $\alpha_{ij} \in f(K)$ if $a^{(j)} = \alpha_{ij}e_i$ or $a_{(i)} = \alpha_{ij}e'_j$. Let $m_\ell = |I|$, $m_r = |J|$ and $m = |Y|$.

(a) Show that $F_\ell(R) = \Sigma_{i\in I}F_\ell(K)e_{ii}$, $F_r(R) = \Sigma_{i\in J}F_r(K)a_{ij}$ where j is determined by $a_{(i)} = \alpha_{ij}e'_j$, and $F(R) = \Sigma_{i\in Y}F(K)e_{ii}$.
(b) Show that each summand $F_r(K)a_{ij}$ in $F_r(R)$ is isomorphic to $F_r(K)$.
(c) Show that $0 \le m \le m_\ell,\ m_r \le n$; $m \neq n-1$; $m = n \Leftrightarrow m_\ell = n \Leftrightarrow m_r = n$; $m_\ell = n-1 \Rightarrow m_r = m+1$ and $0 \le m \le n-2$; $m_r = n-1 \Rightarrow m_\ell = m+1$ and $0 \le m \le n-2$.
(d) Let $m, m_\ell, m_r \in \mathbb{Z}^+$ satisfy the conditions in (c). Find an $a \in (\mathbb{Z}^+)_n \subseteq (K^+)_n$ whose associated triple of integers is m, m_ℓ, m_r; compare with Exercises 4 and 5.
(e) Show that R is isomorphic to $(K_n, (K^+)_n)$ if and only if $F(R)$ is isomorphic to the direct product of n copies of $F(K)$.

14. Let $a \in (K^+)_n$ with $a, a^t \in GL(n,K)$ where K is a po-unital po-ring.

(a) Find an example of a symmetric matrix $a \in K_2$ such that $P(a)$ is not isomorphic to $(K^+)_2$.
(b) Suppose $\frac{1}{2} \in K$. Find an example of a nonsymmetric matrix $a \in K_3$ such that $P(a) = P(a^t)$ and $P(a)$ is not isomorphic to $(K^+)_3$.
(c) If $a = \begin{pmatrix} 1 & 1 \\ 0 & 1 \end{pmatrix}$ show that $P(a) \nsubseteq P(a^t)$ and $P(a^t) \nsubseteq P(a)$.

(d) Let $\beta \geq 1$ and let $a = \begin{pmatrix} 1 & 1 \\ \beta & \beta - 1 \end{pmatrix}$ (see Exercise 6). Show that $P(a^t) \subseteq P(a)$ iff $\beta < 2$ and $(\beta - 1)^2 = 0$.

(e) Let a be the matrix in (d). Show that $P(a) \subseteq P(a^t)$ iff $P(a) = P(a^t)$, iff a is symmetric.

15. Let K be a po-unital po-ring, $1 \leq \beta \in K$ and $a = \begin{pmatrix} 1 & 1 \\ \beta & \beta - 1 \end{pmatrix}$. Show that $(K_2, P(a))$ is a right po-module over K iff β is in the centralizer of K^+.

16. Let L be a subfield of $\mathbb{R}$ and let $\mathbb{H}(L) = L + Li + Lj + Lk$ be the subring of the real quaternions $\mathbb{H}$ with coefficients from L. Show that for each $n \geq 1$, $L(i)_n$ and $\mathbb{H}(L)_n$ cannot be made into ℓ-algebras over L. (Embed $\mathbb{C}$ in $\mathbb{R}_2$ and $\mathbb{H}$ in $\mathbb{R}_4$ and use Theorem 6.4.9 and Exercise 4.2.21.)

17. Let R be an n^2-dimensional unital K-archimedean ℓ-algebra over the totally ordered field K.

(a) If n is a prime show that $R \cong (K_n, (K^+)_n)$ if and only if $\mathscr{U}(R^+)$ has an element of order n which is not in the centralizer of $F(R)$.

(b) Find an example of a 16-dimensional ℓ-algebra that satisfies the condition in (a) but which is not a matrix algebra.

18. Let K be a po-unital po-ring and let G be a cyclic group of order q. Show that $\mathscr{U}((K_r[G])^+)$ has an element of order rq which satisfies the conditions given in Theorem 6.4.19 when "one-dimensional" is replaced by "a free left and right K-module of rank 1."

19. Let R be an ℓ-unital finite dimensional ℓ-algebra over the totally ordered field K. Show that R is isomorphic to a direct product of matrix ℓ-algebras $(K[G])_r$ for finite cyclic groups G if and only if R is ℓ-semiprime and $\mathscr{U}(R^+)$ has an element g of finite order such that $\dim_K[\ell(g-1;R) \cap r(g-1;R)]$ is equal to the number of indecomposable factors of R.

20. Let α and β be elements of a unital ring K and suppose $\alpha\beta - \beta\alpha$ is a unit in K. Show that $u = \begin{pmatrix} 1 & \alpha \\ \beta & \alpha\beta \end{pmatrix}$ is invertible in K_2 but u^t is a left and right zero divisor of K_2 and $\ell(u^t; K_2) \cap r(u^t; K_2) \neq 0$.

21. Let R be a po-unital po-ring. Suppose $u \in R^+$ is a unit in R and ${}_RM$ is a unital R-module and a totally ordered abelian group. Show that $R^+M^+ \subseteq M^+$ iff $R^+uM^+ \subseteq M^+$, iff $M^+ = R^+M^+ = uM^+$, iff $uR^+M^+ \subseteq M^+$.

22. The po-ring T *lacks right* (respectively, *left*) *d-modules* if $MT = 0$ (respectively, $TM = 0$) for each right (respectively, left) d-module M over T.

(a) Suppose T is unital and directed. Show that T lacks right d-modules iff $MT = 0$ for each totally ordered right T-module M, iff 0 is the only unital right d-module over T.

(b) Let R be a directed po-unital po-ring and suppose $u \in R^+$ is a unit of R. Let $S = (R, R^+u)$ be the po-ring with $S^+ = R^+u$. Show that R lacks right (respectively, left) d-modules iff S lacks right (respectively, left) d-modules.

(c) Let K be a directed po-unital po-ring and let $S = (K_n, P(a))$ with $n \geq 2$. Show that S lacks left and right d-modules.

(d) Suppose $S = (K_n, P)$ is an ℓ-algebra over the totally ordered field K with $n \geq 2$ and assume K is archimedean when $n \geq 3$. Show that S lacks left and right d-modules.

Notes. The investigation of lattice orders of matrix algebras was initiated by Weinberg in [WE5] where (a) and (c) of Theorem 6.4.1 are given for the field of rational numbers for ($\beta > 1$) and where it is conjectured that the usual lattice order for the matrix ring $\mathbb{Q}_n$ is the only lattice order with $1 > 0$. The modifications of Weinberg's analysis needed to produce the lattice order P_1 that appears in (b) of Theorem 6.4.1 and to extend the theorem to all totally ordered fields comes from Steinberg [ST20], and some further modifications are given in the current proof. The affirmation of Weinberg's conjecture under either the maximum or minimum condition on $F(K_n)$ given in Theorems 6.4.13 and 6.4.14 is also in Steinberg [ST20]. The proof of Weinberg's conjecture for K_n when K is any totally ordered subfield of the real numbers, given in Theorems 6.4.2, 6.4.3, 6.4.7, 6.4.8, and 6.4.9, is due to Ma and Wojciechowski [MW3] while the extension to $\mathbb{R}$-irreducible K-subalgebras of K_n is due to Ma [M7]. Theorems 6.4.4, 6.4.5, and 6.4.6 on linear inequalities come from Tucker [TU] and Goldman and Tucker [GT], respectively. Theorem 6.4.10 is the classical theorem of Frobenius. The determination of all of the lattice orders of the algebra K_n when K is a totally ordered subfield of $\mathbb{R}$ that is given in Theorems 6.4.11 and 6.4.12 is due to Ma and Wojciechowski [MW4]. The characterization of the usual lattice order for K_n, K any totally ordered field, given in Theorems 6.4.15–6.4.18 appears in Ma [M6], and its extension to a description of the canonically lattice-ordered cyclic group algebra with coefficients in a matrix algebra comes from Ma [M8]. The exercises either appear in or are generalizations of results that do appear in the previously referenced papers: 1–5, Steinberg [ST20]; 6–15, Ma and Wojciechowski [MW3] and [MW4]; 16, Ma [M7]; 17–18, Ma [M6]; and 19, Ma [M8].

Open Problems

1. Does the lower radical construction for ℓ-rings stop at the first infinite ordinal (Theorems 3.2.5 and 3.2.8)? This is the case for rings; see [ADS].

2. If $\mathscr{P}$ is a radical class of ℓ-rings and A is an ℓ-ideal of R, is $\mathscr{P}(A)$ an ℓ-ideal of R (Theorem 3.2.16 and Exercises 3.2.24 and 3.2.25)?

3. (Diem [DI]) Is an sp-ℓ-prime ℓ-ring an ℓ-domain (Theorem 3.7.3)?

4. If R is a torsion-free ℓ-prime ℓ-algebra over the totally ordered domain C and $p(x) \in C[x] \setminus C$ with $p'(1) > 0$ in C and $p(a) \geq 0$ for every $a \in R$, is R an ℓ-domain (Theorems 3.8.3 and 3.8.4)?

5. Does each f-algebra over a commutative directed po-unital po-ring C whose left and right annihilator ideals vanish have a unique tight C-unital cover (Theorems 3.4.5, 3.4.11, and 3.4.12 and Exercises 3.4.16 and 3.4.18)?

6. Can an f-ring contain a principal ℓ-idempotent ℓ-ideal that is not generated by an upperpotent element (Theorem 3.4.4 and Exercise 3.4.15)?

7. Is each C-unital cover of a C-unitable (totally ordered) f-algebra (totally ordered) tight (Theorem 3.4.10)?

8. Is an archimedean ℓ-group a nonsingular module over its f-ring of f-endomorphisms (Exercise 4.3.56)?

9. If an archimedean ℓ-group is a finitely generated module over its ring of f-endomorphisms, is it a cyclic module (Exercise 3.6.20)?

10. Is $E({}_{F(M)}M) \subseteq {}_{F(M)}D(X)$ (Exercises 4.3.54 and 4.3.56)? Here, M is an archimedean ℓ-group, $F(M)$ is its f-ring of f-endomorphisms, X is the Stone space of its Boolean algebra of polars, and $E({}_{F(M)}M)$ denotes its injective hull as an $F(M)$-module.

11. Determine $F(M)$ when M is a free abelian ℓ-group (Theorems 4.5.6 and 4.5.10).

12. If an f-algebra over C is $\mathscr{V}$-ℓ-unitable for a suitable variety of ℓ-algebras $\mathscr{V}$, must it be a C-unitable f-algebra (Theorem 3.4.2 and remarks after it)? For example, $\mathscr{V}$ is the variety of almost f-algebras or the variety of sp-ℓ algebras.

13. Conjecture: A regular sp-ℓ-ring is an f-ring. (True if $1 \in R$ by Theorem 3.7.3).

14. Is a unital right self-injective f-ring necessarily left self-injective (Theorems 4.3.22 and 4.3.30)?

15. Is the maximal right quotient ring Q of the right nonsingular right f-ring R an ℓ-ring extension of R provided Q_R is an f-module extension of R_R?

16. If an Ω-f-group G has the property that each of its values is normal in its cover must each of its Ω-values be normal in its cover (Exercises 2.5.2, 2.5.18, and 2.5.19)? See [D] for descriptions of normal valued ℓ-groups.

17. Develop the theory of f-modules over the f-ring $C(X)$ where X is a topological space and over the f-ring $D(X)$ where X is a Stone space or just an extremally disconnected space.

18. Is a free nonsingular R-f-module ℓ-torsionless for $R = C(X)$ or $R = D(X)$ where X is as in the previous problem (Theorems 4.5.6 and 4.5.10)?

19. Determine the nonsingular $\aleph_\alpha$-injective right f-modules over a semiprime right qf-ring (Theorem 4.4.6).

20. Does the category of unital right f-modules over a unital f-ring have nonzero injectives (Theorem 4.4.1)? The answer is no when the ring is an irredundant semiprime right qf-ring.

21. Conjecture: The polynomial ring $D[x]$ over the totally ordered division ring D does not have an sp-lattice order extending that of D in which deg $x^+ \geq 3$ (Theorem 6.1.19).

22. Identify which fields, necessarily real algebraic extensions of the rationals, are O^*-fields (Theorems 6.2.13 and 6.2.14).

23. Is a free sp-ℓ-algebra or a free f-algebra or a free ℓ-algebra ℓ-semiprime (Theorems 6.3.18 and 6.3.19)? The same question for the free unital ℓ-algebras.

24. Is a totally ordered division ring which satisfies the identity $|[[x,y],z]| \leq x^2 \wedge y^2 \wedge z^2$ a field (Theorem 6.3.9 and 6.3.11)?

25. Can a finite valued ℓ-unital lattice-ordered division ring be embedded in a lattice-ordered division algebra over the reals (Theorem 5.2.42)?

26. Which unital f-rings R have the property that each lattice order of a finite matrix ring over R which extends the lattice order of R is isomorphic to the usual lattice order (Theorems 6.4.1 and 6.4.9)?

27. If K is a totally ordered division ring can each lattice order of the matrix ring K_n for which ${}_K(K_n)$ and $(K_n)_K$ are vector lattices be obtained as in Theorem 6.4.11?

References

[A1] A. A. Albert, On ordered algebras, *Bull. Am. Math. Soc.*, **46** (1940), 521–522.
[A2] A. A. Albert, A property of ordered rings, *Proc. Amer. Math. Soc.*, **8** (1957), 128–129.
[AL1] N. L. Alling, A characterization of abelian η_α-groups in terms of their natural valuation, *Proc. Nat. Acad. Sci.*, **47** (1961), 711–713.
[AL2] N. L. Alling, On the existence of real closed fields that are η_α-sets of power $\aleph_\alpha$, *Trans. Amer. Math. Soc.*, **103** (1962), 341–352.
[AL3] N. L. Alling, On exponentially closed fields, *Proc. Amer. Math. Soc.*, **13** (1962), 706–711.
[AM1] S. A. Amitsur, A general theory of radicals, I, *Amer. J. Math.*, **74** (1952), 774–786.
[AM2] S. A. Amitsur, A general theory of radicals, II, *Amer. J. Math.*, **76** (1954), 100–125.
[AM3] S. A. Amitsur, A general theory of radicals, III, *Amer. J. Math.*, **76** (1954), 126–136.
[AM4] S. A. Amitsur, Generalized polynomial identities and pivotal polynomials, *Trans. Amer. Math. Soc.*, **114** (1965), 210–226.
[AN1] F. W. Anderson, On f-rings with the ascending chain condition, *Proc. Amer. Math. Soc.*, **13** (1962), 715–721.
[AN2] F. W. Anderson, Lattice-ordered rings of quotients, *Canad. J. Math.*, **17** (1965), 434–448.
[ACK] M. Anderson, P. F. Conrad, and G. O. Kenny, Splitting properties in archimedean ℓ-groups, *J. Austral. Math. Soc.*, **23** (1977), 247–256.
[AF] M. Anderson and T. Feil, *Lattice-ordered Groups*, D. Reidel, Dordrecht, 1988.
[ADS] T. Anderson, N. Divinsky, and A. Sulinski, Hereditary radicals in associative and alternative rings, *Canad. J. Math.*, **17** (1965), 594–603.
[AR] V. Andrunakievic and Ju. M. Rjabuhin, Rings without nilpotent elements, and completely simple ideals, *Soviet Math. Dokl.*, **9** (1968), 565–567.
[AK] R. F. Arens and I. Kaplansky, Topological Representation of Algebras, *Trans. Amer. Math. Soc.*, **63** (1948), 457–481.
[AS] E. P. Armendariz and S. A. Steinberg, Regular self-injective rings with a polynomial identity, *Trans. Amer. Math. Soc.*, **190** (1974), 417–425.
[AR1] E. Artin, Über die Zerlegung definiter Funktionen in Quadrate, *Hamb. Abh.*, **5** (1927), 100–115.
[AR2] E. Artin, *Geometric Algebra*, Interscience Publishers, New York, 1957.
[AS1] E. Artin and O. Schreier, Algebraische Konstruktion reeler Körper, *Hamb. Abh.*, **5** (1926), 85–99.
[AS2] E. Artin and O. Schreier, Eine Kennzeichnung der reel abgeschlossenen Körper, *Hamb. Abh.*, **5** (1927), 225–231.
[B] K. A. Baker, Free vector lattices, *Canad. J. Math.*, **20** (1968), 58–66.
[BA] B. Banaschewski, Totalgeornete Moduln, *Arch. Math.*, **7** (1956), 430–440.
[BMM] K. I. Beidar, W. S. Martindale III and A. V. Mikhalev, *Rings with Generalized Identities*, Marcel Dekker, New York, 1996.

[BE] H. E. Bell, Duo rings: Some applications to commutativity theorems, *Canad. Math. Bull.*, **2** (1968), 375–380.

[BER1] G. M. Bergman, Conjugates and nth roots in Hahn–Laurent group rings, *Bull. Malaysian Math. Soc.*, **1** (1978), 29–41.

[BER2] G. M. Bergman, Private communication, 1981.

[BER3] G. M. Bergman, Ordering coproducts of groups and semigroups, *J. Algebra*, **133** (1990), 313–339.

[BPL] A. Berman and R. J. Plemmons, *Nonnegative Matrices in the Mathematical Sciences*, Academic Press, New York, 1979.

[BERN1] S. J. Bernau, Unique representation of archimedean lattice groups and normal lattice rings, *Proc. London Math. Soc.*, **15** (1965), 599–631.

[BERN2] S. J. Bernau, Free abelian lattice groups, *Math. Ann.*, **180** (1969), 48–59.

[BERN3] S. J. Bernau, Free non-abelian lattice groups, *Math. Ann.*, **186** (1970), 249–262.

[BERN4] S. J. Bernau, Lateral and Dedekind completions of archimedean lattice groups, *J. London Math. Soc.*, **12** (1976), 320–322.

[BH] S. J. Bernau and C. B. Huijsmans, Almost f-algebras and d-algebras, *Math. Proc. Cambridge Philos. Soc.*, **107** (1990), 287–308.

[BI1] A. Bigard, *Contribution à la théorie des groupes réticulés*, Ph.D. Thesis, University of Paris, Paris, 1969.

[BI2] A. Bigard, Theories de torsion et f-modules, Seminaire P. Dubreil, Algebre, Exp. No. 5, *Secretariat Mathematique*, 1973, 1–12.

[BI3] A. Bigard, Modules ordonnés injectifs, *Mathematica*, **15** (1973), 15–24.

[BI4] A. Bigard, Free lattice-ordered modules, *Pacific J. Math.*, **49** (1973), 1–6.

[BK] A. Bigard and K. Keimel, Sur les endomorphismes conservant les polaires d'un groupe réticulé archimédean, *Bull. Soc. Math., France*, **97** (1970), 81–96.

[BKW] A. Bigard, K. Keimel and S. Wolfenstein, *Groupes et Anneaux Réticulés*, Lecture Notes in Math. **608**, Springer, Berlin, 1977.

[BIR1] G. Birkhoff, On the structure of abstract algebras, *Proc. Cambridge Phil. Soc.*, **31** (1935), 433–454.

[BIR2] G. Birkhoff, Lattice-ordered groups, *Ann. Math.*, **43** (1942), 298–331.

[BIR3] G. Birkhoff, *Lattice Theory*, 3rd ed., Colloquium Publications No. 25, Amer. Math. Soc., Providence, 1967.

[BIR4] G. Birkhoff, Linear transformations with invariant cones, *Amer. Math. Monthly*, **72** (1967), 274–276.

[BP] G. Birkhoff and R. S. Pierce, Lattice-ordered rings, *An. Acad. Brasil. Cr.*, **28** (1956), 41–69.

[BL] R. Bleier, Free ℓ-groups and vector lattices, *J. Austral. Math. Soc.*, **19** (1975), 337–342.

[BO] K. Boulabiar, Products in almost f-algebras, *Comment. Math. Univ. Carolinae*, **41** (2000), 747–759.

[BM] W. Bradley and J. Ma, Lattice-ordered 2×2 triangular matrix algebras, *Linear Alg. and Its Applications*, **404** (2005), 262–274.

[BRL] B. Brainerd and J. Lambek, On the ring of quotients of a Boolean ring, *Canad. Math. Bull*, **2** (1959), 25–29.

[C1] P. M. Cohn, *Universal Algebra*, Revised edition, Reidel, Dordrecht, 1981.

[C2] P. M. Cohn, *Free Rings and Their Relations*, Second edition, Academic Press, London, 1985.

[C3] P. M. Cohn, *Skew Field Constructions*, Vol. 27, London Math. Soc. Lecture Notes, Cambridge Univ. Press, 1977.

[CON1] P. F. Conrad, On ordered division rings, *Proc. Amer. Math. Soc.*, **5** (1954), 323–328.

[CON2] P. F. Conrad, Generalized semigroup rings, *J. Indian Math. Soc.*, **21** (1957), 73–95.

[CON3] P. F. Conrad, Methods of ordering a vector space, *J. Indian Math. Soc.*, **22** (1958), 1–25.

[CON4] P. F. Conrad, The structure of lattice-ordered groups with a finite number of disjoint elements, *Michigan Math. J.*, **7** (1960), 171–180.

[CON5] P. F. Conrad, Some structure theorems for lattice-ordered groups, *Trans. Amer. Math. Soc.*, **99** (1961), 212–240.

[CON6] P. F. Conrad, The lattice of all convex ℓ-subgroups of a lattice-ordered group, *Czechoslovak Math. J.*, **15** (1965), 101–123.

[CON7] P. F. Conrad, A characterization of lattice-ordered groups by their convex ℓ-subgroups, *J. Austral. Math. Soc.*, **7** (1967), 145–189.

[CON8] P. F. Conrad, The essential closure of an archimedean ℓ-group, *Duke Math. J.*, **38** (1970), 151–160.

[CON9] P. F. Conrad, Free lattice-ordered groups, *J. Algebra*, **16** (1970), 191–203.

[CON10] P. F. Conrad, *Lattice-ordered groups*, Tulane Lecture Notes, Tulane University, 1970.

[CON11] P. F. Conrad, Free-abelian ℓ-groups and vector lattices, *Math. Ann.*, **190** (1971), 306–312.

[CON12] P. F. Conrad, Countable vector lattices, *Bull. Austral. Math. Soc.*, **10** (1974), 371–376.

[CON13] P. F. Conrad, The additive group of an f-ring, *Canad. J. Math.*, **26** (1974), 1157–1168.

[CD] P. F. Conrad and J. Dauns, An embedding theorem for lattice-ordered fields, *Pacific J. Math.*, **30** (1969), 385–398.

[CDI] P. F. Conrad and J. E. Diem, The ring of polar preserving endomorphisms of an abelian lattice-ordered group, *Illinois J. Math.*, **15** (1971), 222–240.

[CHH] P. F. Conrad, J. Harvey, and C. Holland, The Hahn embedding theorem for abelian lattice-ordered groups, *Trans. Amer. Math. Soc.*, **108** (1963), 143–169.

[CM] P. F. Conrad and D. McAlister, The completion of a lattice-ordered group, *J. Austral. Math. Soc.*, **9** (1969), 182–208.

[CMC] P. F. Conrad and P. McCarthy, The structure of f-algebras, *Math. Nachr.*, **58** (1973), 169–191.

[CR] C. W. Curtis and I. Reiner, *Representation Theory of Finite Groups and Associative Algebras*, Wiley-Interscience, New York, 1962.

[D] M. R. Darnel, *Theory of Lattice-ordered Groups*, Marcel Dekker, New York, 1995.

[DA1] J. Dauns, Representation of ℓ-groups and f-rings, *Pacific J. Math.*, **31** (1969), 629–654.

[DA2] J. Dauns, Integral domains that are not embeddable in division rings, *Pacific J. Math.*, **34** (1970), 27–31.

[DA3] J. Dauns, Power series semigroup rings, *Pacific J. Math.*, **34** (1970), 365–369.

[DA4] J. Dauns, Ordered domains, *Symposia Math.*, **21** (1977), 565–587.

[DA5] J. Dauns, Generalized skew polynomial rings, *Trans. Amer. Math. Soc.*, **271** (1982), 575–586.

[DA6] J. Dauns, Quotient rings and one-sided primes, *J. Reine Angew. Math.*, **278/279** (1975); Corrigendum, **280** (1976), 205; Correction, **283/284** (1976), 221.

[DA7] J. Dauns, *A Concrete Approach to Division Rings*, Heldermann-Verlag, Berlin, 1982.

[DI] J. E. Diem, A radical for lattice-ordered rings, *Pacific J. Math.*, **25** (1968), 71–82.

[DIV] N. Divinsky, *Rings and Radicals*, University of Toronto Press, Toronto, 1965.

[DU] D. W. Dubois, Note on Artin's solution of Hilbert's 17th problem, *Bull. Amer. Math. Soc.*, **73** (1967), 540–541.

[DUB] N. I. Dubrovin, Rational closures of group rings of left-ordered groups, *Russian Acad. Sci. Sb. Math.*, **79** (1994), 231–263.

[ES] B. Eckmann and A. Schopf, Uber injective Moduln, *Arch. Math.* (Basel), **4** (1953), 75–78.

[EGH] P. Erdos, L. Gillman and M. Henriksen, An isomorphism theorem for real-closed fields, *Ann. of Math.*, **61** (1955), 542–554.

[F1] C. Faith, Rings with ascending condition on annihilators, *Nagoya Math. J.*, **27** (1966), 179–191.

[F2] C. Faith, On Kothe rings, *Math. Ann.*, **164** (1966), 207–212.

[F3] C. Faith, *Lectures on Injective Modules and Quotient Rings*, Springer-Verlag, New York, 1967.

[F4] C. Faith, *Algebra: Rings, Modules and Categories I*, Springer-Verlag, New York, 1973.

[F5] C. Faith, *Algebra II: Ring Theory*, Springer-Verlag, Berlin, 1976.

[FU1] C. Faith and Y. Utumi, Baer modules, *Arch. Math.*, **15** (1964), 266–270.

[FU2] C. Faith and Y. Utumi, On noetherian prime rings, *Trans. Amer. Math. Soc.*, **114** (1965), 53–60.

[FL] G. Findlay and J. Lambek, A generalized ring of quotients, I and II, *Canad. Math. Bull.*, **1** (1958), 77–85, 155–167.

[F] L. Fuchs, *Teilweise geordnete algebraische Strukturen*, Vandenhoeck and Ruprecht in Gottingen, 1966.
[GW] G. Gardner and R. Wiegandt, *Radical Theory of Rings*, Marcel Dekker, New York, 2004.
[G] J. Georgoudis, *Torsion Theories and f-Rings*, Ph.D. Thesis, McGill University, Montreal, 1972.
[GI] L. Gillman, Some remarks on η_α-sets, *Fund. Math.*, **43** (1956), 77–82.
[GJ] L. Gillman and M. Jerison, *Rings of Continuous Functions*, Van Nostrand, Princeton, 1960.
[GIL] R. Gilmer, *Multiplicative Ideal Theory*, Marcel Dekker, New York, 1972.
[GL1] A. M. W. Glass, *Ordered Permutation Groups*, Cambridge University Press, Cambridge, 1981.
[GL2] A. M. W. Glass, *Partially Ordered Groups*, World Scientific, Singapore, 1999.
[GO1] A. W. Goldie, The structure of prime rings under ascending chain conditions, *Proc. London Math. Soc.*, **8** (1958), 589–608.
[GO2] A. W. Goldie, Semi-prime rings with maximum condition, *Proc. London Math. Soc.*, **10** (1960), 201–220.
[GO3] A. W. Goldie, *Rings with Maximum Condition*, Yale University Lecture Notes, New Haven, 1964.
[GO4] A. W. Goldie, Torsion-free modules and rings, *J. Algebra*, (1964), 268–287.
[GO5] A. W. Goldie, Some aspects of ring theory, *Bull. London Math. Soc.*, **1** (1969), 129–154.
[GT] A. J. Goldman and A. W. Tucker, Polyhedral convex cones, *Linear Inequalities and Related Systems*, Eds. H. W. Kuhn and A. W. Tucker, Annals of Mathematical Studies, **38**, Princeton University Press, Princeton (1956), 19–40.
[GOO1] K. R. Goodearl, Prime ideals in regular self-injective rings, *Canad. J. Math.*, **25** (1973), 829–839.
[GOO2] K. R. Goodearl, *Ring Theory*, Marcel Dekker, New York, 1976.
[GOO3] K. R. Goodearl, *Von Neumann Regular Rings*, 2nd ed., Krieger Publ. Co., Malabar, 1991.
[GOO4] K. R. Goodearl, *Partially Ordered Abelian Groups With Interpolation*, Mathematical Surveys and Monographs, No. 20, Amer. Math. Soc., 1986.
[GR] C. Greither, The isomorphism classes of maximal fields with valuations, preprint, referenced in [PC1, p. 124].
[H] H. Hahn, Über die nichtarchimedean Grössensysteme, *S.-B. Akad. Wiss. Wien. IIa*, **116** (1907), 601–655.
[HR] A. W. Hager and L. C. Robertson, Representing and Ringifying a Riesz Space, *Symposia Mathematica*, **21** (1977), 411–431.
[HA] P. R. Halmos, *Lectures on Boolean Algebras*, Springer-Verlag, New York, 1974.
[HAR] P. Hartmann, Ein Beweis des Henselschen Lemmas für maximal bewertete Körper, *Arch. Math.*, **37** (1981), 163–168.
[HAU] F. Hausdorff, *Grundzüge Mengenlehre*, Leipzig, 1914, Reprinted Chelsea Publishing Co., New York, 1949.
[HAY] A. Hayes, A characterization of f-rings without non-zero nilpotent elements, *J. London Math. Soc.*, **39** (1964), 706–707.
[HE] M. Henriksen, Semiprime ideals of f-rings, *Symposia Mathematica*, **21** (1977), 401–409.
[HI] M. Henricksen and J. Isbell, Lattice-ordered rings and function rings, *Pacific J. Math.*, **12** (1962), 533–565.
[HIJ] M. Henricksen, J. R. Isbell and D. G. Johnson, Residue class fields of lattice-ordered algebras, *Fund. Math.*, **50** (1961), 107–117.
[HL] M. Henriksen and S. Larson, Semiprime f-rings that are subdirect products of valuation domains, *Ordered Algebraic Structures*, J. Martinez and W. C. Holland, eds., Kluwer, Dordrecht, (1993), 159–168.
[HER1] I. N. Herstein, On the Lie and Jordan rings of a simple associative ring, *Amer. J. Math.*, **77** (1955), 279–285.
[HER2] I. N. Herstein, *Topics in Ring Theory*, University of Chicago Lecture Notes, 1965.
[HER3] I. N. Herstein, *Non-commutative Rings*, Carus Monograph 15, Math. Assoc. of America, 1968.

[HI] G. Higman, Ordering by divisibility in abstract algebras, *Proc. London Math. Soc.*, **2** (1952), 326–336.

[HIO] Ja. V. Hion, Archimedovski uporjadočennye kol'tsa, *Uspechi Mat. Nauk*, **9** (1954), 237–242.

[HO] W. C. Holland, The lattice-ordered group of automorphisms of an ordered set, *Michigan Math. J.*, **10** (1963), 399–408.

[HP] C. B. Huijsmans and B. dePagter, Ideal theory in f-algebras, *Trans. Amer. Math. Soc.*, **269** (1982), 225–245.

[I1] J. R. Isbell, Embedding two ordered rings in one ordered ring, Part I, *J. Algebra*, **2** (1966), 341–364.

[I2] J. R. Isbell, Notes on ordered rings, *Alg. Universalis*, **1** (1972), 393–399.

[J1] N. Jacobson, *Structure of Rings*, rev. ed., Colloquium Publications, No. 36, Amer. Math. Soc., Providence, 1964.

[J2] N. Jacobson, *Lectures in Abstract Algebra III*, Van Nostrand, Princeton, 1964.

[J3] N. Jacobson, *Basic Algebra II*, Freeman, San Francisco, 1980.

[J4] N. Jacobson, *Basic Algebra I*, Second Ed., Freeman, New York, 1985.

[JA] A. Jaeger, Adjunction of subfield closures to ordered division rings, *Trans. Amer. Math. Soc.*, **73** (1952), 35–39.

[JAF] P. Jaffard, Contribution a l'étude des groupes ordonnés, *J. Math. Pure Appl.*, **32** (1953), 203–280.

[JOH1] D. G. Johnson, A structure theory for a class of lattice-ordered rings, *Acta. Math.*, **104** (1960), 163–215.

[JOH2] D. G. Johnson, The completion of an archimedean f-ring, *J. London Math. Soc.*, **40** (1965), 493–496.

[JK] D. G. Johnson and J. Kist, Prime ideals in vector lattices, *Canad. J. Math.*, **14** (1962), 517–528.

[JO1] R. E. Johnson, Extended centralizer of a ring over a module, *Proc. Amer. Math. Soc.*, **2** (1951), 891–895.

[JO2] R. E. Johnson, On ordered domains of integrity, *Proc. Amer. Math. Soc.*, **3** (1952), 414–416.

[JW] R. E. Johnson and E. T. Wong, Quasi-injective modules and irreducible rings, *J. Lond. Math. Soc.*, **36** (1961), 260–268.

[K1] I. Kaplansky, Maximal fields with valuations, *Duke Math. J.*, **9** (1942), 303–321.

[K2] I. Kaplansky, *Fields and Rings*, Second ed., Chicago Lectures in Math., University of Chicago Press, Chicago, 1972.

[K3] I. Kaplansky, *Commutative Rings*, Revised ed., University of Chicago Press, Chicago, 1974.

[KE1] K. Keimel, Anneaux réticulés quasi-réguliers et hyper-archimédiens, *C. R. Acad. Sc. Paris*, **266** (1968), A524–A525.

[KE2] K. Keimel, Le centroide et le bicentroide de certains anneaux réticulés, *C. R. Acad. Sc. Paris*, **267** (1968), A589–A591.

[KE3] K. Keimel, The representation of lattice-ordered groups and rings by sections in sheaves, *Lecture Notes in Math.*, Springer-Verlag, **248** (1971), 1–96.

[KE4] K. Keimel, Radicals in lattice-ordered rings, *Proc. Coll. Math. Soc. J. Bolyai*, **6** *Rings, Modules and Radicals* (1973), 237–254.

[KEL] J. E. Kelley, *General Topology*, Van Nostrand, Princeton, 1955.

[KL] G. B. Klatt and L. S. Levy, Pre-self-injective rings, *Trans. Amer. Math. Soc.*, **137** (1969), 407–419.

[KR] V. Kreinovich, If a polynomial identity guarantees that every partial order on a ring can be extended, then this identity is true only for a zero-ring, *Alg. Universalis*, **33** (1995), 237–242.

[KRU] W. Krull, Allgemeine Bewertungstheorie, *J. Reine Angew. Math.*, **167** (1932), 160–196.

[KU] A. Kurosh, Radicals of rings and algebras, *Math. Sbor.*, **33** (1953), 13–26.

[L1] T. Y. Lam, *Lectures on Modules and Rings*, Springer-Verlag, New York, 1999.

[L2] T. Y. Lam, *A First Course in Noncommutative Rings*, Second Ed., Springer-Verlag, New York, 2001.

[LA] J. Lambek, *Lectures on Rings and Modules*, Blaisdell, Waltham, 1966.

[LT] S. Lang and J. T. Tate, Eds., *The Collected Papers of Emil Artin*, Addison-Wesley, Reading, 1965.

[LAV] B. Lavric, The nil radical of an archimedean partially ordered ring with positive squares, *Comment. Math. Univ. Carolinae*, **35** (1994), 231–238.

[LAR] S. Larson, Convexity condition on f-rings, *Canad. J. Math.*, **38** (1986), 48–64.

[LEI] J.B. Leicht, Zur Charakterisierung reell abgeschlossener Körper, *Monatsh. Math.*, **70** (1966), 452–453.

[LE1] K. H. Leung, Positive semidefinite forms over ordered skew fields, *Proc. Amer. Math. Soc.*, **106** (1989), 933–942.

[LE2] K. H. Leung, An application of the theory of order completions, *Contemp. Math.*, **155** (1994), 321–325.

[LEV1] L. S. Levy, Unique direct sums of prime rings, *Trans. Amer. Math. Soc.*, **106** (1963), 64–76.

[LEV2] L. S. Levy, Torsion-free and divisible modules over non-integral domains, *Canad. J. Math.*, **15** (1963), 132–151.

[LEV3] L. S. Levy, Commutative rings whose homomorphic images are self-injective, *Pacific J. Math.*, **18** (1966), 149–153.

[M1] J. Ma, On lattice-ordered rings with polynomial constraints, *J. Math. Research and Exposition*, **11** (1991), 325–330.

[M2] J. Ma, Lattice-ordered rings with f-elements which have zero annihilators, *J. Math. Research and Exposition*, **14** (1991), 455–460.

[M3] J. Ma, The unitability of ℓ-prime lattice-ordered rings with squares positive, *Proc. Amer. Math. Soc.*, **121** (1994), 991–997.

[M4] J. Ma, On lattice-ordered rings with chain conditions, *Comm. Algebra*, **25** (1997), 3483–3495.

[M5] J. Ma, The quotient ring of a class of lattice-ordered Öre domains, *Alg. Universalis*, **44** (2000), 299–304.

[M6] J. Ma, Lattice-ordered matrix algebras with the usual lattice order, *J. Algebra*, **228** (2000), 406–416.

[M7] J. Ma, Finite dimensional simple algebras that do not admit a lattice order, *Comm. Algebra*, **32** (2004), 1615–1617.

[M8] J. Ma, Finite dimensional lattice-ordered algebras with d-elements, *J. Algebra*, **280** (2004), 232–243.

[M9] J. Ma, Pure ℓ-ideals in lattice-ordered rings, *Comm. Algebra*, **33** (2005), 3797–3810.

[M10] J. Ma, Lattice-ordered algebras with a d-basis, *J. Algebra*, **299** (2006), 731–746.

[MR1] J. Ma and R. Redfield, Fields of quotients of lattice-ordered domains, *Alg. Universalis*, **52** (2004), 383–401.

[MR2] J. Ma and R. Redfield, Lattice-ordered matrix rings over the integers, *Comm. Algebra*, **35** (2007), 2160–2170.

[MS] J. Ma and S. A. Steinberg, Construction of lattice orders on the semigroup ring of a positive rooted monoid, *J. Algebra*, **260** (2003), 592–616.

[MW1] J. Ma and P. J. Wojciechowski, F^*-rings are O^*, Order, **17** (2000), 125–128.

[MW2] J. Ma and P. J. Wojciechowski, Structure spaces of maximal ℓ-ideals of lattice-ordered rings, Proc. Ordered Algebraic Structure Conf., J. Martinez, ed.,Kluwer, 2002, 261–274.

[MW3] J. Ma and P. J. Wojciechowski, A proof of Weinberg's conjecture on lattice-ordered matrix algebras, *Proc. Amer. Math Soc.*, **130** (2002), 2845–2851.

[MW4] J. Ma and P. J. Wojciechowski, Lattice orders on matrix algebras, *Alg. Universalis*, **47** (2002), 435–441.

[MAC1] S. MacLane, The uniqueness of the power series representation of certain fields with valuations, *Ann. of Math.*, **39** (1938), 370–382.

[MAC2] S. MacLane, The universality of formal power series fields, *Bull. Amer. Math. Soc.*, **45** (1939), 888–890.

[MAR] W. S. Martindale III, Prime rings satisfying a generalized polynomial identity, *J. Algebra*, **12** (1969), 576–584.

[MART] J. Martinez, Tensor products of partially ordered groups, *Pacific J. Math.*, **41** (1972), 771–789.
[MAW] J. Martinez and S. Woodward, Bezout and Prüfer f-rings, *Comm. Algebra*, **20** (1992), 2975–2989.
[MC] K. McKenna, New facts about Hilbert's seventeenth problem, *Model Theory and Algebra*, D. H. Saracino and V. B. Weispfenning, eds., Springer-Verlag, Berlin (1975), 220–230.
[MI] Y. Miyashita, On quasi-injective modules, *J. Fac. Sci.*, Hokkaido Univ. Ser. I, **18** (1964/1965), 158–187.
[MO] T. S. Motzkin, *The arithmetic-geometric inequality, Inequalities*, O. Shisha, ed., Academic Press, New York (1967), 205–224.
[NB] L. Narici and E. Beckenstein, *Topological Vector Spaces*, Marcel Dekker, New York, 1985.
[N] B. H. Neumann, On ordered division rings, *Trans. Amer. Math. Soc.*, **66** (1949), 202–252.
[NS] B. H. Neumann and J. A. H. Shepperd, Finite extensions of fully ordered groups, *Proc. Royal Soc.*, London, **239** (1957), 320–327.
[O] A. Ostrowski, Untersuchungen zur arithmetischen Theorie der Körper, *Math. Z.*, **39** (1935), 269–404.
[P] D. Panek, *Einbettborkeit angeordeneter Körper in Körper von formalen Potenzreihen*, Diplomarbeit München, 1979.
[PA1] D. S. Passman, *The Algebraic Structure of Group Rings*, J. Wiley-Interscience, New York, 1977.
[PA2] D. S. Passman, *Infinite Crossed Products*, Academic Press, Boston, 1989.
[PF] A. Pfister, Hilbert's seventeenth problem and related problems on definite forms, *Mathematical Developments Arising from Hilbert Problems*, Amer. Math. Soc., Providence, 1976, 483–489.
[PI1] R. S. Pierce, Radicals in function rings, *Duke Math. J.*, **23** (1956), 253–261.
[PI2] R. S. Pierce, *Associative Algebras*, Springer-Verlag, New York, 1982.
[PO1] W. B. Powell, Projectives in a class of lattice ordered modules, *Alg. Universalis*, **9** (1981), 24–40.
[PO2] W. B. Powell, Injectives in a class of lattice ordered modules, *Houston J. Math.*, **9** (1983), 275–287.
[PT] W. B. Powell and C. Tsinakis, Disjoint sets in free lattice ordered modules, *Houston J. Math.*, **15** (1989), 417–424.
[PD] A. Prestel and C. N. Delzell, *Positive Polynomials*, Springer-Verlag, Berlin, 2001.
[PC1] S. Prieß-Crampe, Zum Hahnschen Einbettungssatz für angeordnete Körper, *Arch. Math.*, **24** (1973), 607–614.
[PC2] S. Prieß-Crampe, *Angeordnete Strukturen*: Gruppen, Körper, projektive Ebenen, Springer-Verlag, Berlin, 1983.
[R1] R. H. Redfield, Embedding into power series rings, *Manuscripta Math.*, **56** (1986), 247–268.
[R2] R. H. Redfield, Dual spaces of totally ordered rings, *Czechoslovak Math. J.*, **38** (1988), 95–102.
[R3] R. H. Redfield, Nonembeddable *o*-rings, *Comm. Algebra*, **17** (1989), 59–71.
[R4] R. H. Redfield, Lattice-ordered fields as convolution algebras, *J. Algebra*, **153** (1992), 319–356.
[R5] R. H. Redfield, Lattice-ordered power series fields, *J. Austral. Math. Soc.*, **52** (1992), 299–321.
[R6] R. H. Redfield, Subfields of lattice-ordered fields that mimic maximal totally ordered subfields, *Czechoslovak Math. J.*, **51** (2001), 143–161.
[RE] G. Renault, Anneaux reduits non commutatifs, *J. Math. Pures Appl.*, **4** (1967), 203–214.
[RI1] P. Ribenboim, On the existence of totally ordered abelian groups which are η_α-sets, *Bull. Acad. Polon. Sci.*, **13** (1965), 545–548.
[RI2] P. Ribenboim, On ordered modules, *J. fur die reine und angew. Math.*, **225** (1967), 120–146.
[RI3] P. Ribenboim, Some examples of lattice-orders in real closed fields, *Arch. Math.*, **61** (1993), 59–63.
[RIE] M. A. Rieffel, A characterization of the group algebras of finite groups, *Pacific J. Math.*, **16** (1966), 347–363.

[RO] R. M. Robinson, Some definite polynomials which are not sums of squares of real polynomials, Selected Questions of Algebra and Logic, *Acad. Sci. USSR*, (1973), 264–282.

[ROS] J. G. Rosenstein, *Linear Orderings*, Academic Press, New York, 1982.

[RS] K. A. Ross and A. H. Stone, Products of separable spaces, *Amer. Math. Monthly*, **71** (1964), 398–403.

[ROT] J. J. Rotman, *An Introduction to Homological Algebra*, Academic Press, Boston, 1979.

[S] F. L. Sandomierski, Some examples of right self-injective rings that are not left self-injective, *Proc. Amer. Math. Soc.*, **26** (1970), 244–245.

[SC] O. Schilling, *The Theory of Valuations*, Amer. Math. Soc. Surveys IV, Providence, 1950.

[SCH1] N. Schwartz, η_α-Strukturen, *Math. Zeit.*, **158** (1978), 147–155.

[SCH2] N. Schwartz, Lattice-ordered fields, *Order*, **3** (1986), 179–194.

[SCH3] N. Schwartz, Gabriel filters in real closed rings, *Comment. Math. Helv.*, **72** (1997), 434–465.

[SE] J. P. Serre, Extensions de corps ordonnés, *Comptes Rendus Acad. Sci. Paris*, **229** (1949), 576–577.

[SH1] M. A. Shatalova, ℓ_A and ℓ_I-rings, *Siber. Math. J.*, **7** (1966), 1084–1094.

[SH2] M. A. Shatalova, The theory of radicals in lattice-ordered rings, *Math. Notes*, **4** (1968), 875–880.

[SH3] M. A. Shatalova, Nonexistence of *o*-injective partially ordered modules, *Math. Notes*, **7** (1970), 419–420.

[SHY] H. J. Shyr, *A Study of Partially-ordered Rings*, Ph.D. Thesis, University of Western Ontario, 1971.

[SV] H. J. Shyr and T. M. Viswanathan, On the radicals of lattice-ordered rings, *Pacific J. Math.*, **54** (1974), 257–260.

[SI] W. Sierpinski, Sur une propriété des ensembles ordonnés, *Fund. Math.*, **36** (1949), 56–67.

[ST1] S. A. Steinberg, Finitely-valued f-modules, *Pacific J. Math.*, **40** (1972), 723–737.

[ST2] S. A. Steinberg, An embedding theorem for commutative lattice-ordered domains, *Proc. Amer. Math. Soc.*, **31** (1972), 409–416.

[ST3] S. A. Steinberg, Lattice-ordered injective hulls, *Trans. Amer. Math. Soc.*, **169** (1972), 365–388.

[ST4] S. A. Steinberg, Quotient rings of a class of lattice-ordered rings, *Canad. J. Math*, **25** (1973), 627–645.

[ST5] S. A. Steinberg, Rings of quotient rings without nilpotent elements, *Pacific J. Math*, **49** (1973), 493–506.

[ST6] S. A. Steinberg, On lattice-ordered rings in which the square of every element is positive, *J. Australian Math. Soc.*, **22** (1976), 362–370.

[ST7] S. A. Steinberg, Radical theory in lattice-ordered rings, *Symposia Mathematica*, **21** (1977), 379–400.

[ST8] S. A. Steinberg, Identities and nilpotent elements in lattice-ordered rings, *Ring Theory*, S. K. Jain, ed., Lecture Notes in Mathematics, No. 25, Dekker, New York (1977), 191–212.

[ST9] S. A. Steinberg, Lattice-ordered modules of quotients, *J. Australian Math. Soc.*, **30** (1980), 243–251.

[ST10] S. A. Steinberg, Special elements in semiprime rings, *Algebra*, Ed. R. K. Amayo, Lecture Notes in Mathematics, No. 848, Springer-Verlag, New York (1981), 274–277.

[ST11] S. A. Steinberg, Examples of lattice-ordered rings, *J. Algebra*, **72** (1981), 223–236.

[ST12] S. A. Steinberg, Unital ℓ-prime lattice-ordered rings with polynomial constraints are domains, *Trans. Amer. Math. Soc.*, **276** (1983), 145–164.

[ST13] S. A. Steinberg, Archimedian, semiperfect and π-regular lattice-ordered algebras with polynomial constraints are f-algebras, *Proc. Amer. Math. Soc.*, **89** (1983), 205–210.

[ST14] S. A. Steinberg, On lattice-ordered algebras that satisfy polynomial identities, *Ordered Algebraic Structures*, W. B. Powell and C. Tsinakis, eds., Lecture Notes in Mathematics, No. 99, Marcel Dekker, New York (1985), 179–187.

[ST15] S. A. Steinberg, On the unitability of a class of partially ordered rings that have squares positive, *J. Algebra*, **100** (1986), 325–343.

[ST16] S. A. Steinberg, Positive semidefinite forms over totally ordered domains, *Comm. Algebra*, **21** (1993), 3455–3473.

[ST17] S. A. Steinberg, Central f-elements in lattice-ordered algebras, *Ordered Algebraic Structures*, J. Martinez and C. Holland, eds., Kluwer Academic Publishers, Dordrecht (1993), 203–223.

[ST18] S. A. Steinberg, Convex elements in lattice-ordered rings and modules, *Comm. Algebra*, **25** (1997), 683–708.

[ST19] S. A. Steinberg, A characterization of rings in which each partial order is contained in a total order, *Proc. Amer. Math. Soc.*, **125** (1997), 2555–2558.

[ST20] S. A. Steinberg, On the scarcity of lattice-ordered matrix algebras II, *Proc. Amer. Math. Soc.*, **128** (2000), 1605–1612.

[ST21] S. A. Steinberg, Finitely valued f-modules, an addendum, *Czechoslavak Math. J.*, **51** (2001), 387–394.

[ST22] S. A. Steinberg, f-algebras that are embeddable in unital f-algebras, *Comm. Algebra*, **30** (2002), 3991–4006.

[ST23] S. A. Steinberg, The J-radical and the Pierce radicals of a lattice-ordered ring, *Comm. Algebra*, **31** (2003), 4273–4290.

[STE1] B. Stenstrom, *Rings and Modules of Quotients*, Springer-Verlag, Berlin, 1971.

[STE2] B. Stenstrom, *Rings of Quotients*, Springer-Verlag, New York, 1975.

[SZ] F. A. Szász, *Radicals of Rings*, Wiley, New York, 1981.

[T] M. V. Tamhankar, On algebraic extensions of subrings in an ordered ring, *Alg. Universalis*, **14** (1982), 25–35.

[TU] A. W. Tucker, Dual systems of homogeneous linear relations, *Linear Inequalities and Related Systems*, H. W. Kuhn and A. W. Tucker, eds., Annals of Mathematics Studies, 38, Princeton Univ. Press, Princeton (1956), 3–18.

[U1] Y. Utumi, On quotient rings, *Osaka Math. J.*, **8** (1956), 1–18.

[U2] Y. Utumi, On rings of which any one-sided quotient rings are two-sided, *Proc. Amer. Math. Soc.*, **14** (1963), 141–147.

[V1] T. M. Viswanathan, Ph.D. Thesis, Queens University, Kingston, Canada, 1967.

[V2] T. M. Viswanathan, Generalization of Hölder's theorem to ordered modules, *Canad. J. Math*, **21** (1969), 149–157.

[V3] T. M. Viswanathan, Ordered modules of fractions, *J. Reine Angew. Math.*, **235** (1969), 78–107.

[V4] T. M. Viswanathan, Ordered fields and sign-changing polynomials, *J. Reine Angew. Math.*, **296** (1977), 1–9.

[W] J. H. M. Wedderburn, On division algebras, *Trans. Amer. Math. Soc.*, **22** (1921), 129–135.

[WE1] E. C. Weinberg, Completely distributive lattice-ordered groups, *Pacific J. Math.*, **12** (1962), 1131–1137.

[WE2] E. C. Weinberg, Free lattice-ordered abelian groups, *Math. Ann.*, **151** (1963), 187–199.

[WE3] E. C. Weinberg, Free lattice-ordered abelian groups II, *Math. Ann.*, **159** (1965), 217–222.

[WE4] E. C. Weinberg, o-projective ordered abelian groups, *J. Reine Angew. Math.*, **224** (1966), 219–220.

[WE5] E. C. Weinberg, On the scarcity of lattice-ordered matrix rings, *Pacific J. Math.*, **19** (1966), 561–571.

[WE6] E. C. Weinberg, *Lectures on Ordered Groups and Rings*, Lecture Notes, University of Illinois, Urbana, 1968.

[WE7] E. C. Weinberg, Relative injectives and universals for categories of ordered structures, preprint, 1969.

[WE8] E. C. Weinberg, Relative injectives, *Symposia Math.*, **21** (1977), 555–564.

[WI1] R. R. Wilson, *Lattice Orders on Real Fields*, Ph.D. Thesis, University of California, Los Angeles, 1974.

[WI2] R. R. Wilson, Lattice orderings on the real field, *Pacific J. Math.*, **63** (1976), 571–577.

[WO] P. J. Wojciechowski, Archimedean almost f-algebras that arise as generalized semigroup rings, *Ordered Algebraic Structures*, J. Martinez and C. Holland, eds., Kluwer Academic Publishers, Dordrecht (1993), 225–233.

[WK] P. J. Wojciechowski and V. Kreinovich, On lattice extensions of partial orders of rings, *Comm. Algebra*, **25** (1997), 935–941.

[WON] E. T. Wong, Regular rings and integral extensions of a regular ring, *Proc. Amer. Math. Soc.*, **33** (1972), 313–315.

[WJ] E. T. Wong and R. E. Johnson, Self-injective rings, *Canad. Math. Bull.*, **2** (1959), 167–173.

Index